T0315118

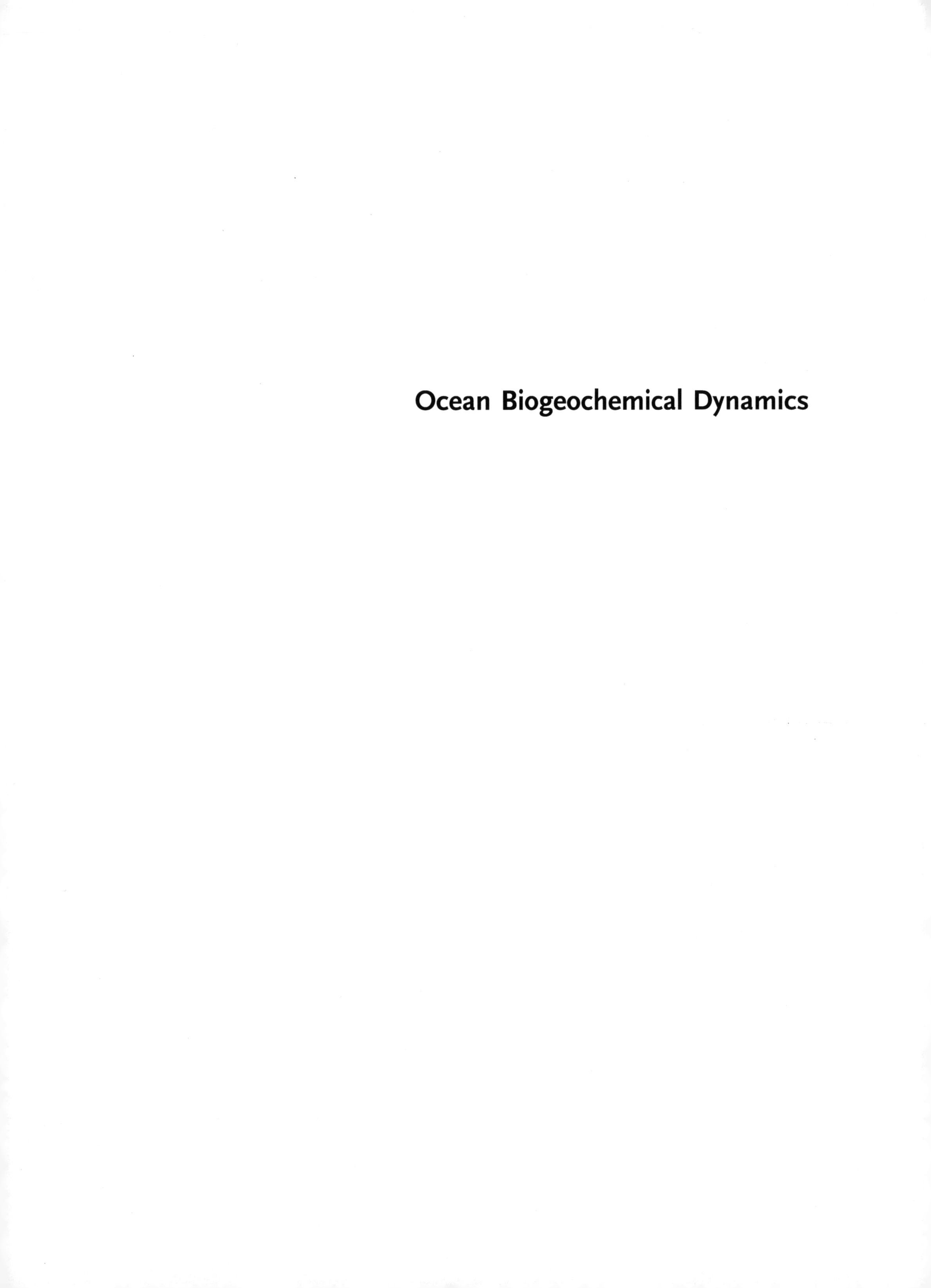

Ocean Biogeochemical Dynamics

Ocean Biogeochemical Dynamics

Jorge L. Sarmiento and
Nicolas Gruber

PRINCETON UNIVERSITY PRESS | PRINCETON AND OXFORD

Published by Princeton University Press, 41 William Street, Princeton, New Jersey 08540
In the United Kingdom: Princeton University Press, 3 Market Place, Woodstock,
Oxfordshire OX20 1SY

Library of Congress Cataloging-in-Publication Data

Sarmiento, Jorge Louis, 1946–
Ocean biogeochemical dynamics / Jorge L. Sarmiento and Nicolas Gruber.
p. cm.
Includes bibliographical references and index.
ISBN-13: 978-0-691-01707-5 (cloth : alk. paper)
ISBN-10: 0-691-01707-7 (cloth : alk. paper)
1. Chemical oceanography. 2. Biogeochemistry. 3. Chemical ocenaography—Mathematical models.
4. Biogeochemistry—Mathematical models. I. Gruber, Nicolas, 1968– II. Title.
GC116 S27 2006
551.46'6—dc22 2005050465

British Library Cataloging-in-Publication Data is available

This book has been composed in Scala and Scala Sans
Printed on acid-free paper. ∞

pup.princeton.edu

Printed in the United States of America

10

Contents

Preface

The aim of this book is to provide a broad theoretical framework for the study of the biogeochemical processes that control the chemical composition of the ocean. The book is based on lectures that the two authors have given over many years. JLS has taught primarily graduate students and postdocs at Princeton University, at the Friday Harbor Laboratory of the University of Washington, at the University of Bern, at the University of Gothenburg, and at the University of Paris VI. NG has taught both upper-level undergraduates and graduates at the University of California, Los Angeles. Most of these students had a good training in physics and mathematics. Because of this we have not been reticent to use mathematics where it makes concepts easier to explain. It would be difficult to do theory without it. On the other hand, all of the mathematics is at a level that should be understandable by any student with a year or two of college-level mathematics and/or physics. Another characteristic of the students we have taught is that many if not most of them have had only a modest amount of training in chemistry, biology, and physical oceanography. The material we cover thus tries not to assume an advanced background in these areas.

The philosophy we have tried to follow in writing this book begins with the basic premise that two of the most important lessons for students to learn in trying to understand how the ocean or any other natural system functions are (1) to put forward problems in a manner that provides guidance on how to seek solutions, and (2) to learn how to solve problems by example. The presentation of the material follows to the greatest extent possible a pattern of posing problems and seeking solutions to them. An essential step in asking good questions is to have a paradigm or hypothesis about the processes that are under scrutiny, while at the same time recognizing that such hypotheses are very likely to change over time. The solution of problems by observational and theoretical studies will either be consistent with the hypotheses, or have the more interesting outcome of requiring a modification of the hypotheses. We have endeavored to provide such paradigms or hypotheses to the greatest extent possible, along with many examples of old paradigms and hypotheses that are no longer accepted. This approach is intended to invite critical questioning by the student, the teacher, and the investigator. We have taken as a guiding principle the observation of Thomas Henry Huxley that truth is more likely to emerge from error than from vagueness.

An additional point we would make about the philosophy underlying this book is that looking at the ocean from the point of view of its chemical composition invites large-scale synthesis. If biology is the glorification of the particular, as one of our biological colleagues has described it, biogeochemistry could perhaps be characterized as the exaltation of the general. However, we believe that a real understanding of oceanic processes requires us to work at the boundaries between the particular and the general. We ocean biogeochemists must progress beyond simple box model approaches and first-order kinetics if we are to truly understand the processes that control the ocean's chemical composition.

The outline of the book is:

- Chapter 1: Introduction
- Chapter 2: Tracer Conservation and Ocean Transport
- Chapter 3: Air-Sea Interface
- Chapter 4: Organic Matter Production
- Chapter 5: Organic Matter Export and Remineralization
- Chapter 6: Remineralization and Burial in the Sediments
- Chapter 7: Silicate Cycle
- Chapter 8: Carbon Cycle
- Chapter 9: Calcium Carbonate Cycle
- Chapter 10: Oceanic Carbon Cycle, Atmospheric CO_2, and Climate

The logic of the organization is as follows: Chapter 1 sets the scene with an overview of chemical distributions and box models of what controls those distributions. Chapter 2 introduces the basic conservation equation and provides a review of ocean circulation. Chapters 3 through 6 survey oceanic biogeochemical processes progressing down the water column from the air-sea interface to the sediments, discussing the cycles of organic carbon, oxygen, nitrate, and phosphate along the way. Chapters 7, 8, and 9 apply the tools learned in earlier chapters to a system analysis of the cycles of silicon, carbon, and $CaCO_3$. The final chapter examines important issues having to do with the control of atmospheric CO_2 and climate over time, including the anthropogenic transient.

This book has been a long time in gestation. The idea for the book first took root when JLS co-taught a summer course with Peter Rhines at Friday Harbor Laboratories to a group of outstanding students from around the world. First drafts of chapters 1 through 5 were written while JLS gave a class to graduate students from a wide range of disciplines at the Physics Institute of the University of Bern in 1995. We are grateful to Thomas Stocker and Uli Siegenthaler for arranging and sponsoring this visit to Bern, which is where the coauthors first met and began their long and fruitful collaboration. The only regret of this otherwise delightful sabbatical is that Uli Siegenthaler passed away two weeks before the visit began. He was a good friend as well as an admired colleague of remarkable sagacity. The remaining five chapters were written by the coauthors in the intervening years, most of them in the last two years. During the last year, the first five chapters were thoroughly updated and expanded, a challenging task in view of the great advances that have taken place during the last decade due in large part to the Joint Global Ocean Flux Study (JGOFS).

A large number of individuals contributed to this project over the years. We mention just a few whose efforts had a major impact on the book. Geoff Evans was extremely helpful in recasting the opening chapter, and had excellent suggestions for chapter 4. Tony Michaels made extensive comments on chapters 1 and 4, wrote most of the discussion on classification of organisms and brief descriptions of major organisms in chapter 4, and was extremely helpful in preliminary discussions of the content of chapter 5. JLS thanks him and Tony Knap for arranging a very pleasant visit to Bermuda, where most of this interaction took place. Perspicacious comments on various sections of chapter 2 by Mitsuhiro Kawase, Brian Arbic, Anand Gnanadesikan, and Robbie Toggweiler helped correct important errors and improve the presentation. Rik Wanninkhof was very helpful in reviewing chapter 3 and providing data and other information for that chapter. Rob Armstrong provided invaluable comments on chapter 4 that helped to make it more coherent, and several long and delightful conversations with Dick Barber and generous help from John Dunne and many other colleagues were central to bringing it up-to-date. A final review by Robbie Toggweiler and Anand Gnanadesikan of the updated chapter was crucial. Craig Carlson reviewed portions of an early version of chapter 5 having to do with dissolved organic matter. An enjoyable visit to Rick Jahnke was crucial in getting started on chapter 6, and a review of this chapter by Jack Middelburg was extremely helpful. Chapter 7 would not have been possible without a long and fruitful interaction with Mark Brzezinski, including a wonderful visit to Santa Barbara, as well as excellent reviews of two versions of the chapter by Olivier Ragueneau. Several chapters profited from a very helpful review by Rick Jahnke. Robbie Toggweiler and Danny Sigman provided detailed comments on an earlier version of chapter 10.

Students and various colleagues at the University of Bern caught a number of errors in a preliminary version of this manuscript passed out as lecture notes, and we have received comments from numerous others who have used preliminary versions of the text over time, including students at Princeton University, UCLA, and Pennsylvania State University.

Both authors thank the wonderful group of students and postdocs they have had over the years and the colleagues they have worked with that have stimulated their thinking and made their lives as scientists such a pleasure. Dozens of these students and colleagues responded willingly and gladly to our appeals for help with figures, ideas, and just to serve as sounding boards. They are too many to list individually, but the numerous personal communications and references to their publications in our book will be some measure of acknowledgment, even if small and inadequate.

Finally, we acknowledge Wally Broecker and Tsung-Hung Peng's extraordinarily stimulating textbook *Tracers in the Sea*, which was the first book that JLS used in his teaching and which set a standard of originality and vision that has been a major influence on our field for over two decades. In both our cases, but especially JLS's, whose most exciting years in this field were spent as a graduate student with Wally Broecker, this debt is as much to him as a person as to the book that he and Tsung-Hung Peng wrote.

We, of course, bear full responsibility for any errors that remain.

We dedicate this book to our families and friends, who supported us through the many ups and downs that are invariably associated with the writing of such a textbook over many years and who had to cover for us and make many sacrifices to enable us to bring this project to completion. We would not have succeeded without their love and friendship as well as their encouragement and understanding.

Jorge L. Sarmiento
Princeton, New Jersey

Nicolas Gruber
Los Angeles, California

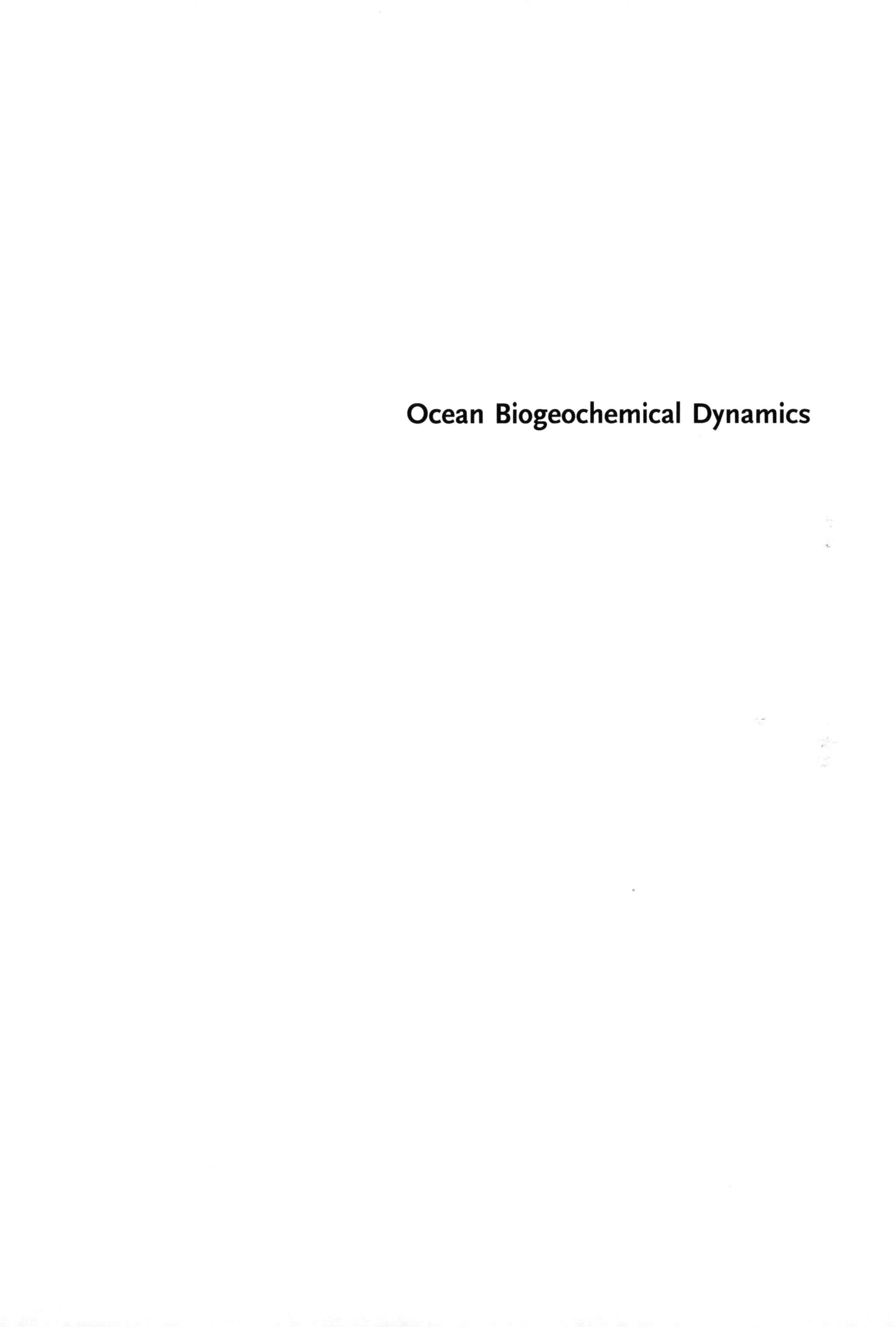

Ocean Biogeochemical Dynamics

CHAPTER 1

Introduction

This book is about the distribution of chemical elements in the sea and the processes that control it. We address the questions: What controls the mean abundance of the elements? What controls their variation in space? What controls changes in time (for example during ice ages)? The primary focus is on those elements for which "biological processes" is part of the answer, the main example naturally being carbon. The whole subject is in a state of active research, and this book will highlight important growing points. It will concentrate more on how we know than on what, thereby introducing readers to the tools with which they can add to our knowledge of ocean biogeochemistry.

The approach of this book combines two principles. First, measurements of the concentration of dissolved inorganic chemicals are among the most accurate we can make in the ocean. Therefore, we will generally start with these data and ask what needs to be explained. Possible explanations will arise from a variety of sources including many direct observations of processes. The second basic principle will be to express these qualitative explanations in terms of quantitative models of processes that make predictions that can be compared with the observations. We then return to the data to assess how accurate the predictions, and therefore the models, are.

In the following section we make our approach more concrete by taking a first look at perhaps the most important question in chemical oceanography: What controls the mean chemical composition of the ocean?

1.1 Chemical Composition of the Ocean

Table 1.1.1 and figure 1.1.1 show the mean concentrations of chemical elements and a few compounds in the ocean (ignoring the elements in water molecules, hydrogen and hydroxyl ions, including only dissolved O_2 and SO_4^{3-}, and including dissolved N_2 as well as nitrate). The mean oceanic concentration of the elements ranges over almost 12 orders of magnitude. At the low end are some of the rare earth elements with concentrations of order 1 nmol m^{-3}, for example, elements 67, 69, and 71 (holmium, thulium, and lutetium). At the high end is chlorine, with a concentration of almost 600 mol m^{-3}.

We discuss here three different hypotheses for what controls the composition of the ocean. The first two are the *accumulation hypothesis* and the *kinetic control hypothesis*. The accumulation hypothesis proposes that the oceanic concentrations represent simply the accumulated inflow from rivers since the ocean came into existence. The kinetic control hypothesis proposes that the composition of the ocean results from a balance between the input to the ocean from external sources and the rate of removal, with many of the removal processes being biologically driven. We will test each of these hypotheses by building simple models of them and using observations to see if the predictions of the models are consistent with what we know of the ocean. The third hypothesis that we discuss is the *equilibrium hypothesis*, which proposes that the oceanic composition is controlled by equilibria between seawater and chemical precipitates from seawater, the solid particles of continental origin that sink though the water column and accumulate in the sediments, and the oceanic crust that underlies the ocean.

A first obvious guess at what determines oceanic concentrations is that they represent simply the accumulated inflow from rivers since the ocean came into existence. The contrast between the high oceanic and low river concentrations of the elements (see table 1.1.1) clearly suggests that accumulation is occurring in the ocean, but can it explain the observed concentrations? Let us develop a mathematical description of how the oceanic mean concentration of an element A (which we denote alternatively by C_{oc} or [A] in units of mmol m^{-3})

TABLE 1.1.1

Concentrations of chemicals in the ocean and rivers.

Based on the compilation of Quinby-Hunt and Turekian *[1983] converted to common units of* μ*mol* m^{-3} *and then rounded off to the same number of significant figures reported by them. The final column of accumulation time and residence time was calculated using (1.1.5). Note that there are some inconsistencies between the table and Figure 1.1.1 (e.g. gold). The more recent data in Figure 1.1.1 is more reliable.*

Z	Element		Atomic Weight	Ocean Concentration (μmol m^{-3})	River Dissolved Concentration (μmol m^{-3})	τ_a and τ_r (yr)
1	Hydrogen	H	1.0079			
2	Helium	He	4.0026	1.9		
3	Lithium	Li	6.941	26,300	1,700	530,000
4	Beryllium	Be	9.01218	0.02		
5	Boron	B	10.81	420,000	1,700	8,500,000
6	Carbon	C	12.011	2,300,000		
7	Nitrogen	N	14.0067	605,000		
8	Oxygen	O	15.9994	162,000		
9	Fluorine	F	18.9984	1.3		
10	Neon	Ne	20.179	8		
11	Sodium	Na	22.9898	480,670,000	315,000	52,600,000
12	Magnesium	Mg	24.305	54,000,000	160,000	11,600,000
13	Aluminum	Al	26.9815	1,000	1,900	20,000
14	Silicon	Si	28.0855	84,000	193,000	15,000
15	Phosphorus	P	30.9738	2,000	1,300	53,000
16	Sulfur	S	32.06	28,200,000		
17	Chlorine	Cl	35.453	559,520,000	230,000	83,900,000
18	Argon	Ar	39.948	16,000		
19	Potassium	K	39.0983	10,500,000	34,500	10,500,000
20	Calcium	Ca	40.08	10,600,000	364,000	1,000,000
21	Scandium	Sc	44.9559	<0.02	0.09	8,000
22	Titanium	Ti	47.88	<0.02	200	4
23	Vanadium	V	50.9415	<20	20	40,000
24	Chromium	Cr	51.996	6.5	20	11,000
25	Manganese	Mn	54.938	0.2	150	50
26	Iron	Fe	55.847	0.7	700	40
27	Cobalt	Co	58.9332	0.03	3	400
28	Nickel	Ni	58.69	8.4	37	7,800
29	Copper	Cu	63.546	1.9	160	410
30	Zinc	Zn	65.39	6.1	500	420
31	Gallium	Ga	69.72	0.2	1.3	5,000
32	Germanium	Ge	72.59	0.07		
33	Arsenic	As	74.9216	30	23	45,000
34	Selenium	Se	78.96	2.2		
35	Bromine	Br	79.904	860,000	250	119,000,000
36	Krypton	Kr	83.8	3.8		
37	Rubidium	Rb	85.4678	1,490	18	2,900,000
38	Strontium	Sr	87.62	81,000	700	4,000,000
39	Yttrium	Y	88.9059	0.15		
40	Zirconium	Zr	91.224	<10,000		
41	Niobium	Nb	92.9064	0.01		
42	Molybdenum	Mo	95.94	120	5	800,000
43	Technetium	Tc	(98)			
44	Ruthenium	Ru	101.07	0.005		
45	Rhodium	Rh	102.906			
46	Palladium	Pd	106.42			
47	Silver	Ag	107.868	0.03	3	400
48	Cadmium	Cd	112.41	0.6		
49	Indium	In	114.82	0.002		
50	Tin	Sn	118.71	0.004		
51	Antimony	Sb	121.75	2	8	9,000

TABLE 1.1.1
Continued

Z	Element		Atomic Weight	Ocean Concentration ($\mu mol\ m^{-3}$)	River Dissolved Concentration ($\mu mol\ m^{-3}$)	τ_a and τ_r (yr)
52	Tellurium	Te	127.6			
53	Iodine	I	126.905	480		
54	Xenon	Xe	131.29	0.5		
55	Cesium	Cs	132.905	0.002	0.26	300
56	Barium	Ba	137.33	87	400	8,000
57	Lanthanum	La	138.906	0.03	0.4	3,000
58	Cerium	Ce	140.12	0.03	0.6	2,000
59	Praseodymium	Pr	140.908	0.004	0.05	3,000
60	Neodymium	Nd	144.24	0.03	0.3	4,000
61	Promethium	Pm	(145)			
62	Samarium	Sm	150.36	0.004	0.05	3,000
63	Europium	Eu	151.96	0.001	0.007	5,000
64	Gadolinium	Gd	157.25	0.005	0.05	4,000
65	Terbium	Tb	158.3925	0.001	0.006	6,000
66	Dysprosium	Dy	165.5	0.006		
67	Holmium	Ho	164.93	0.001	0.006	6,000
68	Erbium	Er	167.26	0.006	0.02	10,000
69	Thulium	Tm	168.934	0.001	0.006	6,000
70	Ytterbium	Yb	173.04	0.005	0.02	9,000
71	Lutetium	Lu	174.967	0.001	0.006	6,000
72	Hafnium	Hf	178.49	<0.05		
73	Tantalum	Ta	180.948	<0.014		
74	Tungsten	W	183.85	<0.006		
75	Rhenium	Re	186.207	0.02		
76	Osmium	Os	190.2			
77	Iridium	Ir	192.22			
78	Platinum	Pt	195.08			
79	Gold	Au	196.967	0.057	0.01	200,000
80	Mercury	Hg	200.59	0.031		
81	Thallium	Tl	204.383	0.060		
82	Lead	Pb	207.2	0.005	5	40
83	Bismuth	Bi	208.98	0.05		
84	Polonium	Po	(209)			
85	Astatine	At	(210)			
86	Radon	Rn	(222)			
87	Francium	Fr	(223)			
88	Radium	Ra	226.025			
89	Actinium	Ac	227.028			
90	Thorium	Th	232.038	<0.003	0.4	300
91	Protactinium	Pa	231.036			
92	Uranium	U	238.029	14	0.17	2,800,000

will change with time. The total number of moles of this element in the ocean is the product of its mean oceanic concentration multiplied by the oceanic volume, V_{oc} (m^3). The time rate of change of the total number of moles of this element, dM^A_{oc}/dt, is equal to the sum of all inputs and losses of this element, i.e.,

$$\frac{dM^A_{oc}}{dt} = \frac{dV_{oc}C_{oc}}{dt} = inputs - losses \qquad (1.1.1)$$

In the case of the accumulation hypothesis, we assume that there are no losses. We further assume that the inputs are controlled only by the addition of element A by rivers, which we estimate by taking the product of river flow v_{river} ($m^3\ yr^{-1}$) and river concentration of the element, C_{river}. This gives:

$$V_{oc} \cdot \frac{dC_{oc}}{dt} = v_{river} \cdot C_{river} \qquad (1.1.2)$$

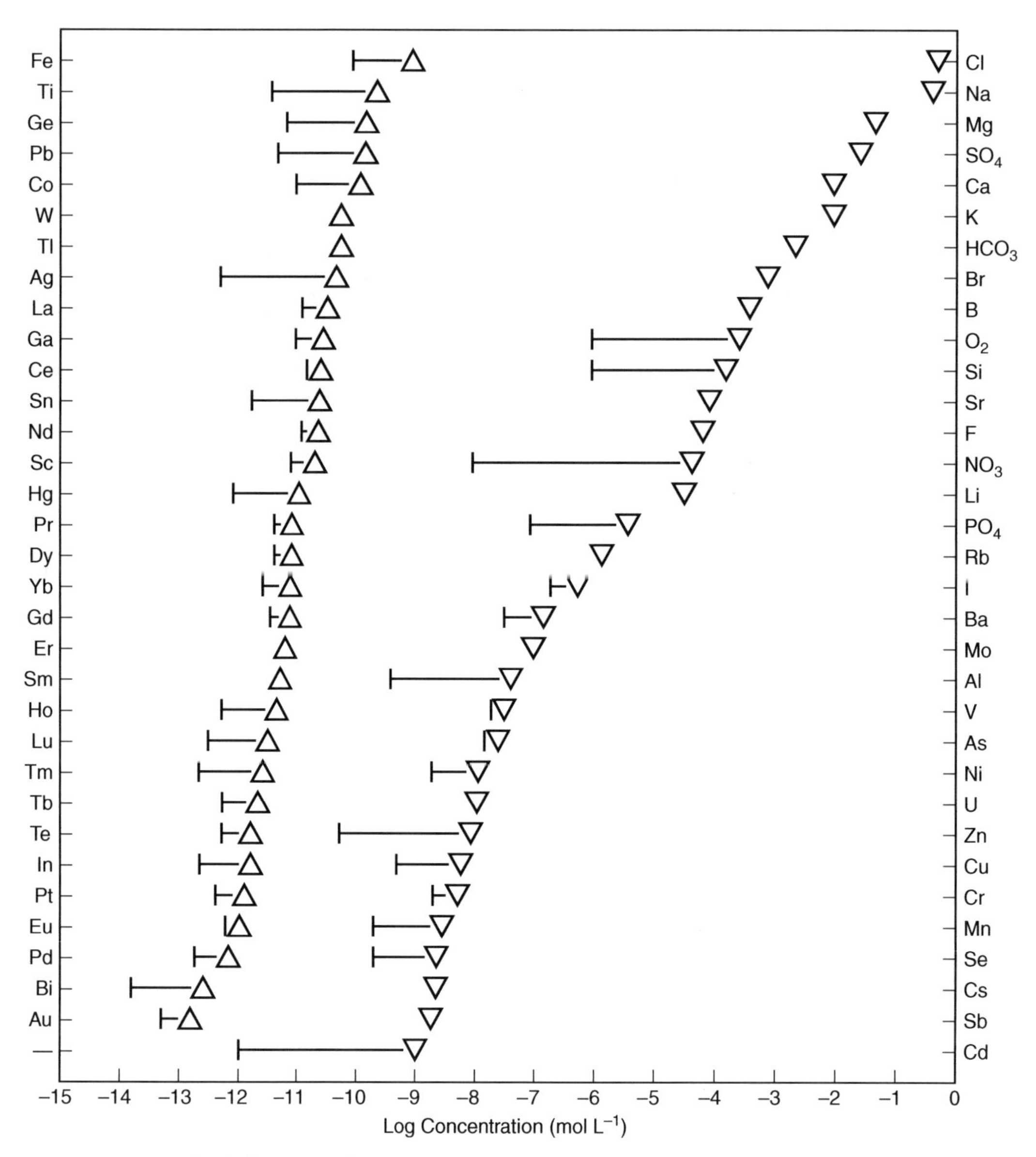

FIGURE 1.1.1: A graphical illustration of the dissolved concentrations of elements and some compounds expressed as log to the base 10 [*Johnson and Jannasch*, 1994]. The higher concentration elements are given on the right-hand side and the lower concentration elements are given on the left-hand side. The bars represent the range of concentrations in the ocean. The full range of concentrations covers almost 12 orders of magnitude.

We take V_{oc} out of the time-derivative assuming that the oceanic volume has remained constant through time. This differential equation has the solution

$$C_{oc}(t) - C_{oc}(t{=}0) = \frac{v_{river} \cdot C_{river}}{V_{oc}} \cdot \Delta t \tag{1.1.3}$$

for the change in concentration in the time interval Δt between t and $t=0$. If we assume that the ocean came into being at $t=0$ as essentially fresh water, i.e., $C_{oc}(t=0)=0$, we obtain

$$C_{oc}(t) = \frac{v_{river} \cdot C_{river}}{V_{oc}} \cdot \Delta t \tag{1.1.4}$$

This equation predicts that today's mean ocean concentration of any element A is directly proportional to its river concentration C_{river}, and hence that the ratio of various elements in the ocean should be equal to the ratio of these elements in rivers. Rather than testing for this particular prediction, we infer for each element an accumulation age, $\tau_a = \Delta t$, which represents a time in the past when, given today's river input and ocean mean concentration, the oceanic mean concentration of this element must have been zero. If the accumulation hypothesis is correct, then we expect this age to (i) reflect the age of the ocean, and (ii) be equal for all elements. We obtain τ_a, the number of years of accumulation that the present concentration and river inflow represent, by rearranging (1.1.4):

$$\tau_a = \frac{V_{oc}}{v_{river}} \cdot \frac{C_{oc}}{C_{river}} = \frac{1.29\times10^{18}\,\text{m}^3}{3.7\times10^{13}\,\text{m}^3\,\text{yr}^{-1}} \cdot \frac{C_{oc}}{C_{river}} = 34{,}500\ \text{yr} \cdot \frac{C_{oc}}{C_{river}} \tag{1.1.5}$$

where we inserted today's V_{oc} (1.29×10^{18} m^3) and v_{river} (3.7×10^{13} m^3 yr^{-1}). Both C_{oc} and C_{river} are given in table 1.1.1. The river concentration shown in the table is for the dissolved component only, not including any suspended material, which we assume is mostly removed before getting into the open ocean. Note that the above river flow includes a direct groundwater discharge of ~6% [*Berner and Berner*, 1987].

Looking at the accumulation times shown in the last column of table 1.1.1, we are struck by two things. (1) All of the accumulation times are at least a factor of 30 less than what we think the age of the ocean is (3.85 billion years). (2) The accumulation times for different elements vary by almost eight orders of magnitude. As this is in strong violation of our prediction on the basis of the accumulation hypothesis, we conclude that this hypothesis is unlikely to be correct. An alternative interpretation, that the river input of elements was smaller in the past, is the opposite of what we might expect, given that the removal of chemicals from land would gradually deplete the land of those chemicals.

We therefore turn to the kinetic control hypothesis, which considers removal as well as addition of elements to the mean ocean. We modify (1.1.2) by adding a removal term, R, in mol yr^{-1}.

$$V_{oc} \cdot \frac{dC_{oc}}{dt} = v_{river} \cdot C_{river} - R \tag{1.1.6}$$

This model is illustrated in figure 1.1.2. Given that the accumulation times we previously calculated are much shorter than the age of the ocean, it might be reasonable to suppose that the ocean concentrations have achieved a *steady state*, i.e., that $dC_{oc}/dt = 0$. We see from (1.1.6) that this implies that removal is in equilibrium with addition by rivers. Unfortunately, the steady-state solution of (1.1.6), $R = v_{river} \cdot C_{river}$, does not solve our problem of determining what controls the ocean concentration, since all it says is that the removal equals the input without specifying what the removal mechanism is. In order to solve this problem we need to specify how R is related to the concentration C_{oc}.

The easiest and most common approach that geochemists take in solving a *box model* such as the above is to assume that R is linearly proportional to the concentration in the water, i.e.,

$$R = V_{oc} \cdot k \cdot C_{oc} \tag{1.1.7}$$

where k (yr^{-1}) is the rate constant for the removal and where we included V_{oc} to account for the fact that R reflects the whole ocean removal. Chemists refer to reactions such as this as being *first order* in the concentration. Substituting (1.1.7) into (1.1.6) and dividing by V_{oc} gives:

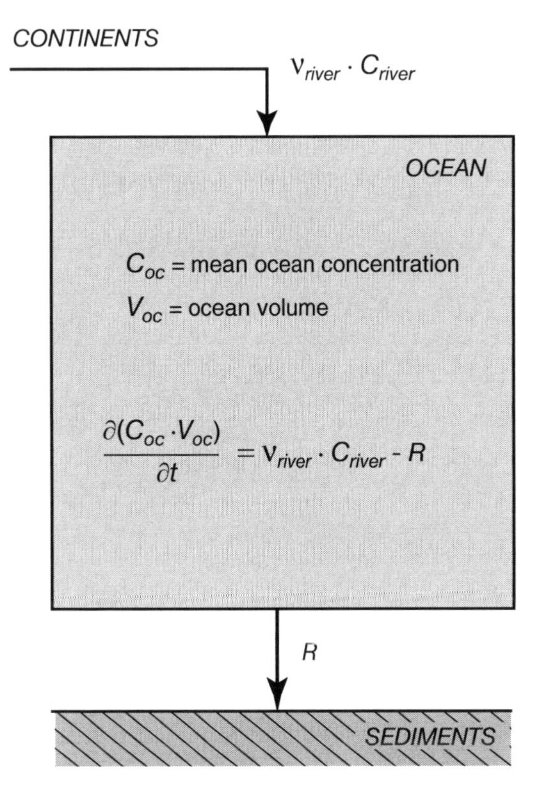

FIGURE 1.1.2: Schematic illustration of a one-box model of the ocean with river input and a removal term R

$$\frac{dC_{oc}}{dt} = \frac{v_{river} \cdot C_{river}}{V_{oc}} - k \cdot C_{oc} \tag{1.1.8}$$

In a steady state it is easy to solve (1.1.8) for the ocean concentration:

$$C_{oc} = \frac{1}{k} \cdot \frac{v_{river} \cdot C_{river}}{V_{oc}} \tag{1.1.9}$$

We show in panel 1.1.1 that this steady-state concentration is achieved with a characteristic time constant τ_r, which is determined by the inverse of the rate constant k, i.e.,

$$\tau_r = \frac{1}{k} \tag{1.1.10}$$

Elements whose removal timescale is very fast, i.e., whose k is large, will therefore tend to have a fast response time after a perturbation and quickly recover their new steady-state concentration. The opposite is the case for slowly reacting elements, i.e., those with a small k. It turns out that this characteristic response time, τ_r, is mathematically identical to the accumulation time, τ_a, (solve (1.1.9) for $1/k$ and insert it into (1.1.10)), so that the last column of table 1.1.1 actually represents both τ_r and τ_a although it is more appropriate to think of the former as representing a residence time of the chemical with respect to the addition by rivers and removal by reaction *R*.

We now have a model of what controls the ocean concentration with a steady-state solution given by (1.1.9). This equation tells us that for a constant removal rate constant k and ocean volume V_{oc}, there is a linear relationship between concentration and river input. For example, a doubling of the river input would lead to a doubling of concentration. Similarly, for constant river input, there is an inverse linear dependence on the removal rate constant. A halving of the removal rate constant (slower removal) would double the ocean concentration. Our model thus explains the wide range of concentrations in table 1.1.1 as resulting in part from differences in the river concentrations, but primarily from variations in the removal rate constant k. Our model also enables us to predict how the ocean might respond to changes in time. Panel 1.1.1 gives two examples of analytical solutions to the time-dependent equation (1.1.8) that show how the ocean would adjust to an abrupt or cyclic change in the river input.

The reader should keep in mind that (1.1.9) applies to removal of tracer by first-order processes, in other words, to removal processes that are directly proportional to the concentration of the tracer. While simple models such as this provide powerful insights into oceanic processes that we shall return to again and again, one should keep in mind that actual removal processes may have a more complex behavior than this, as we shall see in subsequent chapters.

Our analysis demonstrates the crucial role of the removal mechanisms and rate constants in controlling the chemical concentrations. The assumption that the removal of a chemical is first order in its concentration is essentially an assumption that the concentration is controlled by kinetics. An alternative approach, explored in a seminal paper published by the Swedish chemist *Sillén* [1961], is that ocean concentrations are determined instead by thermodynamic equilibria between seawater and mineral phases. In such a model, the chemical composition would remain constant through time at the equilibrium concentration $C_{equilibrium}$, i.e., $C_{oc} = C_{equilibrium}$. This is in contrast to our kinetic model (which is similar to the one developed by Broecker in his response to Sillén's proposal [*Broecker*, 1971]), which permits ocean concentrations to vary in response to changes in the river input and the removal rate constant.

Many of the inorganic chemical equilibria proposed by Sillén and others play an important role in influencing ocean chemical composition. However, observations show that seawater is usually undersaturated with respect to minerals found in deep-sea sediments. Furthermore, many of the postulated mineral phases either do not exist or are insufficient to explain the removal rate required to balance the river inflow. The evidence is thus very strong that the removal processes generally involve a more complex blend of biological, chemical, and geological processes such as the formation of evaporation basins and circulation of waters through the crust. *Berner and Berner* [1987] give detailed analyses of these processes for a variety of chemicals. An interesting alternative we discuss later in the book (chapter 6) is that some chemicals might be controlled

Panel 1.1.1: Temporal Response of Ocean Chemistry to Perturbations

We can use the simple one-box model of the ocean represented by (1.1.8) to obtain valuable information about how ocean concentrations might respond to perturbations. One such perturbation might be a change in the rate of input by rivers associated with an increase in ice cover over land. For example, if the river flux doubled due to increased dryness and erosion during periods of glaciation and all else remained the same, (1.1.9) tells us that the new steady-state concentration, C_{oc}^{final}, would be double the previous steady-state concentration, $C_{oc}^{initial}$.

It is also of interest to ask how this new equilibrium would be achieved. We can obtain an answer to this question by solving the time-dependent equation (1.1.8). An analytical solution exists if the change in river flux is assumed to occur instantaneously at an initial time $t = t_o$:

$$C_{oc}(t) = C_{oc}^{final} + (C_{oc}^{initial} - C_{oc}^{final})e^{-k(t-t_0)} \tag{1}$$

This equation shows that the approach to the final concentration is exponential, with the time scale determined by the rate constant k. Chemicals with a short residence time, i.e., large k, will approach equilibrium more rapidly than chemicals with a long residence time.

Another simple example for which an analytical solution is possible is one in which the source varies cyclically. Suppose, for example, that the glacial cycles have a large impact on the concentration of chemicals in rivers, and that this causes the river concentration to vary sinusoidally with time:

$$v_{river} \cdot C_{river} = a + b \cdot \sin \omega t \tag{2}$$

The constant a is the mean river flux, and b is the amplitude of the variation in the flux (units = mol yr^{-1}). The frequency

by a hybrid of the kinetic control and equilibrium mechanisms, where the removal rate is first order in the deviation of the concentration from the equilibrium concentration, i.e., $R = V_{oc} \cdot k \cdot (C_{oc} - C_{equilibrium})$.

The most important message of this discussion is the crucial role of the removal mechanisms and rate constants in controlling the chemical concentration. The complexity of removal processes makes the ocean in many ways a more interesting place than if only chemical equilibria controlled its composition. The geological record yields dramatic evidence of changes in oceanic composition that are consistent with the kinetic model. For example, substantial reductions in deep-sea oxygen occurred during the lower Cretaceous (125–90 million years before present), and a major redistribution of dissolved inorganic carbon in the ocean led to a reduction of atmospheric CO_2 by almost one-third during the last ice age.

1.2 Distribution of Chemicals in the Ocean

Figure 1.2.1 is our first information about the nature of variation in space: it shows vertical profiles of various elements in the Pacific Ocean. We notice that some elements vary and others do not. Let us think first about those that do not, for example, sodium, magnesium, potassium, calcium, and rubidium, on the left-hand side; and chlorine, sulfur, and bromine on the right-hand side. In combination (and with sulfur converted to sulfate ion), elements such as these make up the vast majority of the total concentration of dissolved elements in the ocean (table 1.2.1). They are also the ones with the longest residence times (table 1.1.1). We might then think that the rate at which things happen to them is slower than the rate at which the water of the ocean mixes completely. Indeed, we shall see later in this section that the mixing time of the ocean is about one thousand years, far less than the residence time of the major constituents shown in table 1.2.1. This would explain why they are relatively uniform. Nevertheless, there are small variations in their concentration driven primarily by evaporation and rainfall in combination with ocean circulation. These processes are discussed in chapter 2.

The distribution of the remaining chemicals is far from uniform. Most have lower concentrations at the surface than at depth, but some, such as cobalt and lead, have higher concentrations at the surface. Many processes affect the vertical distribution of such chemicals. Biological processes or chemical scavenging by sinking particles generally explain profiles with reduced surface concentrations. Profiles with higher concentrations at the surface are generally influenced by atmospheric deposition. The impact of biology on some of the elements is used by *Broecker and Peng* [1982] as a basis for classifying them into *biolimiting* elements (those such as nitrogen in the form of nitrate, phosphorus in the form of phosphate, and silicon in the form of silicic acid, whose surface concentrations are nearly depleted by biological processes, or processes associated with biology); *biointermediate* elements (those

ω is $2\pi/T$, where T is the period of the oscillation, for example, ~100,000 years in the case of the glacial cycles. The solution to (1.1.8) for this case is:

$$C_{oc}(t) = \frac{1}{V} \cdot \left[\frac{a}{k} + \frac{b}{\sqrt{\omega^2 + k^2}} \cdot \sin(\omega t + \phi) \right] \tag{3}$$

The first term inside the brackets shows that the mean ocean concentration is directly proportional to the mean river concentration and inversely proportional to the removal rate constant. The amplitude of the variation in ocean concentration in response to variations in the river input of the chemical is given by the second term in the brackets. This amplitude is directly related to the amplitude of the variation in the river input and inversely related to the square root of the sum of the squares of the rate constant k and the frequency of the river flux variation ω. The ϕ inside the parentheses of the sine is the phase shift in the oceanic variation relative to the river variation, the value of which can be found from $\tan \phi = \omega/k$. Analysis of (3) suggests two extremes for the possible responses of the oceanic concentration to variations in river input: (a) if k is much larger than ω, i.e., if the oceanic residence time is short relative to the period of the oscillation in the river input, then the amplitude of the oceanic response becomes proportional to b/k. This means that oceanic concentration, which has a mean of a/k, will vary in direct proportion to the variation in the river input. (b) If ω is much larger than k, i.e., if the river concentration is varying on a time scale that is much shorter than the residence time of the chemical in the ocean, then the oceanic response to variations in river concentration will be relatively small.

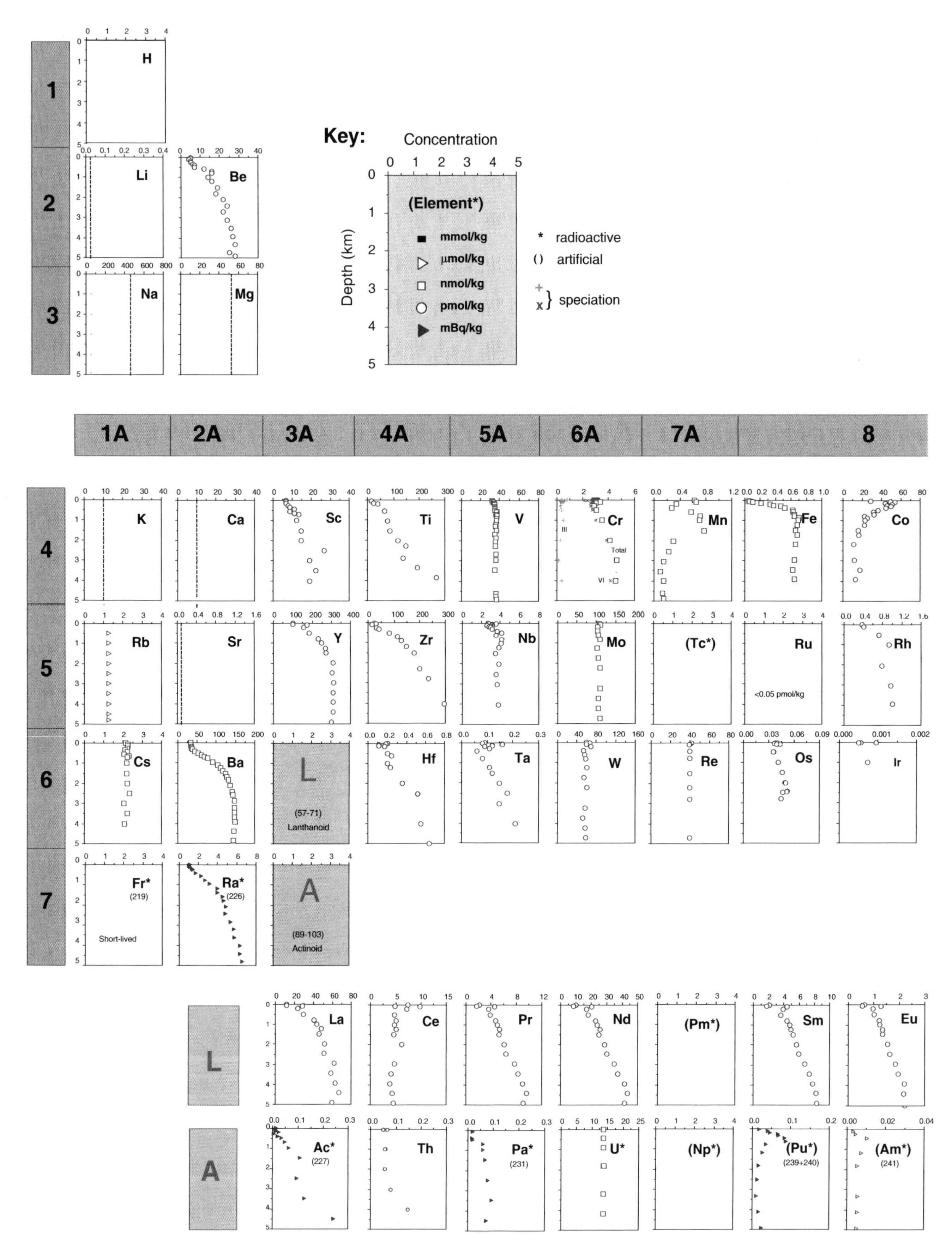

FIGURE 1.2.1: Vertical profiles of elements from the Pacific Ocean arranged as in the periodic table of elements [*Nozaki*, 1997]. The biounlimited elements have nearly uniform concentrations. Most other elements have lower concentrations at the surface than at depth due to biological removal. Biolimiting elements are nearly depleted to 0 mmol m^{-3} at the surface, whereas biointermediate elements show only partial depletion. Oxygen and the noble gases on the right side of the figure are influenced in part by their higher solubility in colder

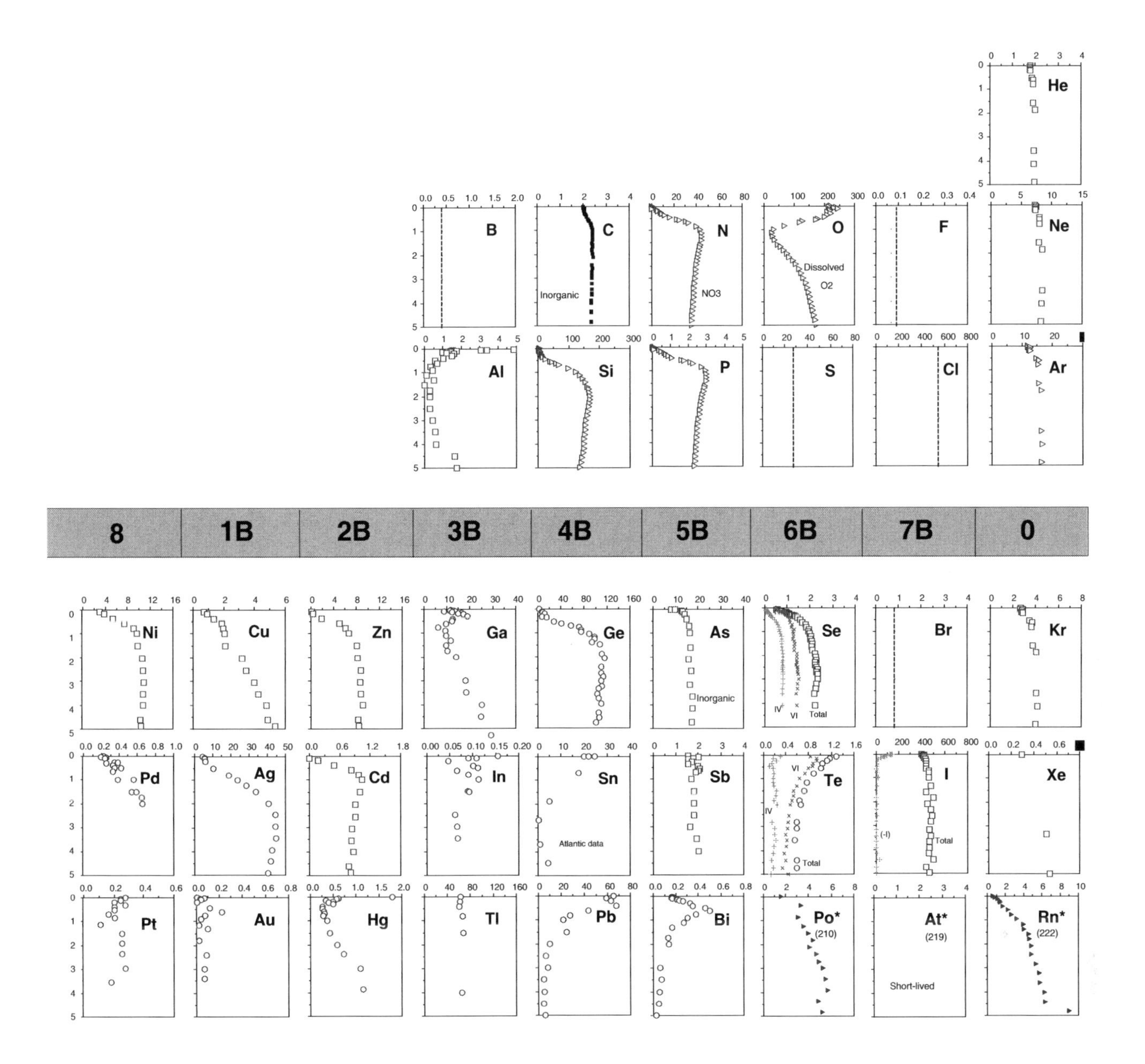

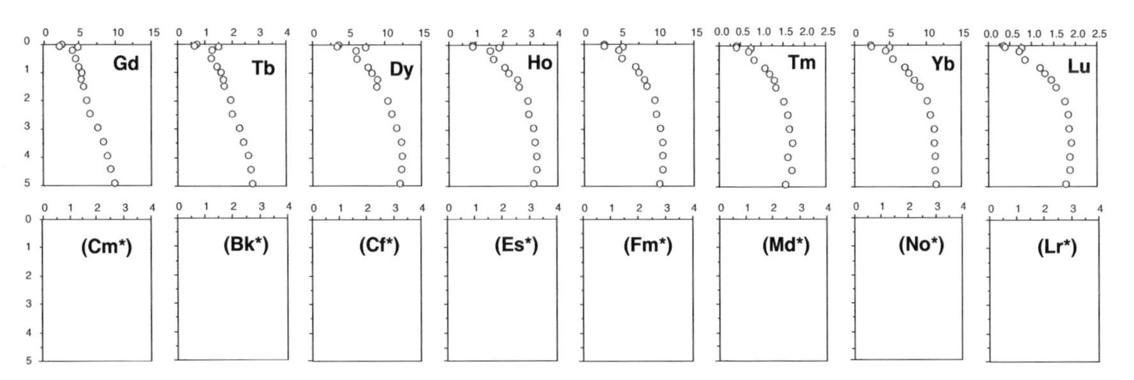

waters deep in the ocean (see chapter 3) and, in the case of oxygen, by biological production as part of photosynthesis and consumption by respiration. A few elements such as Pb have higher concentrations at the surface due to delivery by dust transport, and lower concentrations at depth due to rapid scavenging from the water column to the sediments.

TABLE 1.2.1
Concentration of major ions in seawater with a salinity of 35
Based on Table 2.3-1 of Kennish *[1989]. The percent is with respect to a total mass of major ions of 35.1589 g* kg^{-1}.

Cation	$g\,kg^{-1}$	%	*Anion*	$g\,kg^{-1}$	%
Na^+	10.76	30.60	Cl^-	19.353	55.04
Mg^{2+}	1.297	3.69	SO_4^{2-}	2.712	7.71
Ca^{2+}	0.4119	1.17			
K^+	0.399	1.13			

Note: Salinity is an approximate measure of the total concentration of chemicals in the ocean. Its original definition is the total mass in one kilogram of seawater after all organic chemicals have been completely oxidized, all carbonate converted to oxide, and all bromine and iodine replaced by chlorine. The salinity defined this way is reported in units of g salt (kg seawater)$^{-1}$, for which the symbol ‰ is used. Salinity is now reported on the Practical Salinity Scale, which refers to a conductivity ratio [*UNESCO*, 1985].

such as carbon and barium that are associated with biological processes, but are not limiting their rates, and therefore tend to be only partially depleted in surface waters); and *biounlimited* elements (those such as the major ions of table 1.2.1 that show generally uniform distributions).

For the primary example in this section, we turn to the elements that show a sharp reduction in concentration near the surface (figure 1.2.1). The obvious questions are: What causes these gradients, and why do some elements, but not all, have them? One of the main features of oceanography, and one of the main attractions of doing oceanographic research, is combining reasoning from several scientific disciplines. The following example about the control of differences between surface and subsurface concentrations shows how this happens, and in doing so serves to introduce much of the material of the book and show why it is relevant.

The key to explaining the sharp reduction in near-surface concentrations is that photosynthesis occurs in the sunlit surface ocean (figure 1.2.2). The essential effect of photosynthesis is to capture the energy from the sun and make particulate and dissolved organic matter from dissolved inorganic matter (carbon dioxide and nutrients). The organic matter that is formed can be transported downward before it is converted back into dissolved inorganic matter (*remineralized*). These remineralization processes liberate the energy contained in organic matter and provide the essential energy for the organisms that live off this organic matter. This surface production–deep remineralization loop is what causes the vertical gradients to happen. Biological production is discussed in chapter 4. One of the important messages from that discussion is that organisms utilize some chemical elements in stoichiometric (i.e., molar) ratios that can be considered nearly constant across the world. The photosynthesis and associated reactions combine inorganic carbon, nitrogen, and phosphorus together with water to form organic matter and release oxygen. This reaction is typically represented in the following form [Anderson, 1995]:

$$106\,CO_2 + 16\,HNO_3 + H_3PO_4 + 78\,H_2O \rightleftharpoons C_{106}H_{175}O_{42}N_{16}P + 150\,O_2$$

with inorganic carbon in the form of carbon dioxide, inorganic nitrogen in the form of HNO_3, and inorganic phosphorus in the form of H_3PO_4, and organic matter of composition $C_{106}H_{175}O_{42}N_{16}P$. Remineralization is the opposite of this reaction, combining oxygen with organic matter to produce the inorganic chemical forms of carbon, nitrogen, and phosphorus as well as releasing energy. We will consider the inorganic concentrations of three of these elements here: phosphorus, carbon, and O_2, in that order. From the above reaction, we see that the stoichiometric ratio of biological utilization and production of these three elements and nitrogen is $C{:}N{:}P{:}O_2 = 106{:}16{:}1{:}-150$, where C is in the form CO_2, N is in the form of HNO_3, and P is in the form H_3PO_4.

Spatial variations call for box models with more than one box. The minimum number that would be required to have surface concentrations lower than deep concentrations is two. We use the schematic in figure 1.2.3 as a basis for constructing a first model to represent the processes that give rise to sharp surface gradients. The schematic leaves out the river input and sediment burial fluxes considered in the previous section (figure 1.1.2); these external sources and sinks operate on much longer timescales than the interior ocean processes shown in figure 1.2.3 and can thus be safely ignored for purposes of this discussion. Water movement and the equations that govern it are complicated, but it often suffices to represent it very simply in terms of a volume exchange rate v ($m^3\,s^{-1}$) between two well-mixed boxes. If some substance has concentrations C_s and C_d ($mol\,m^{-3}$) in surface and deep boxes, respectively, then there is a net exchange flux $v \cdot (C_s - C_d)$ mol s^{-1} from surface to deep.

The phosphorus balance in the deep box of figure 1.2.3 is given by:

$$V_d \cdot \frac{d[PO_4^{3-}]_d}{dt} = v \cdot ([PO_4^{3-}]_s - [PO_4^{3-}]_d) + \Phi^P \qquad (1.2.1)$$

where $[PO_4^{3-}]$ is the dissolved phosphate concentration and Φ^P ($mol\,s^{-1}$) is the total surface-to-abyssal flux of organic phosphorus that is converted back into inorganic phosphorus in the deep ocean. Dissolved inorganic phosphorus exists in the ocean primarily as the phosphate ion. The organic matter flux includes sinking in the particulate form as well as volume exchange of the dissolved form, which we do not represent explicitly at this time. Note that the surface box has an equation identical to (1.2.1) except

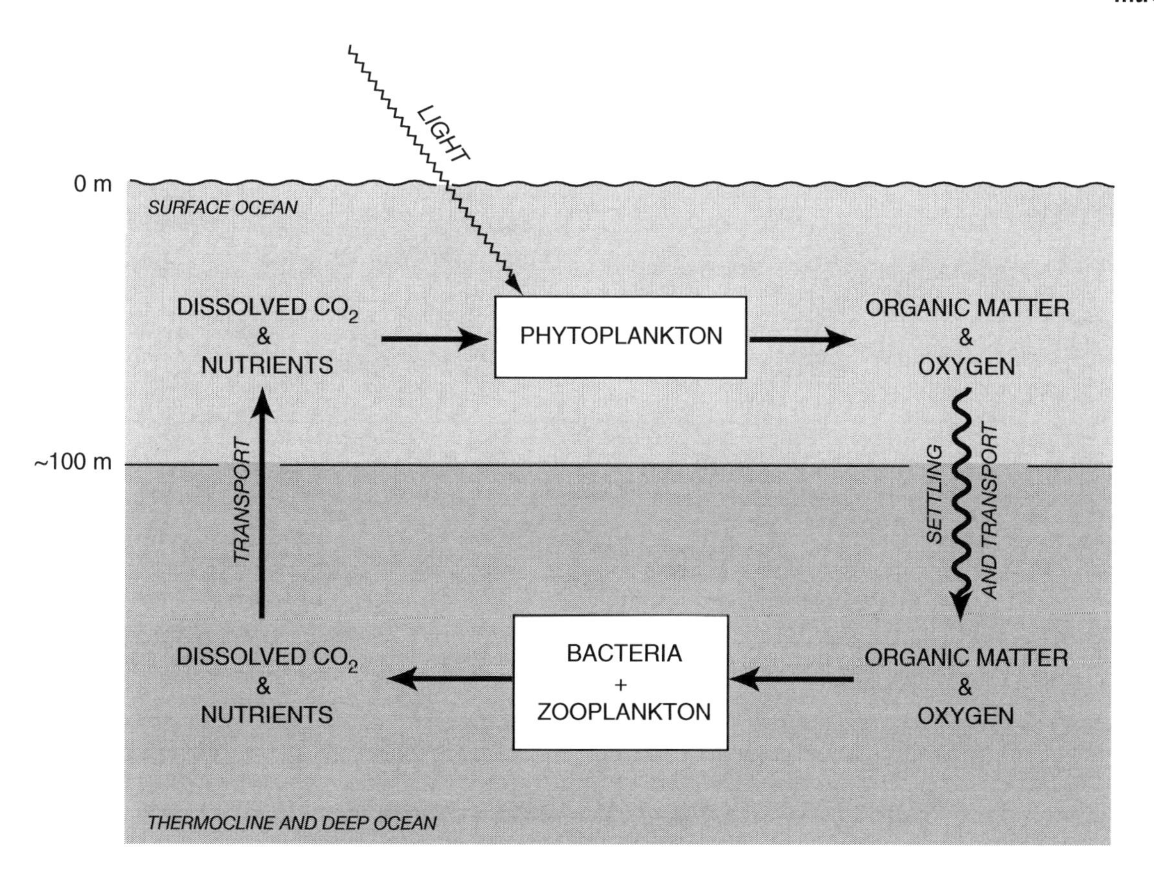

FIGURE 1.2.2: A schematic illustration of the role of biology in the oceans. Phytoplankton make organic matter from dissolved carbon dioxide and nutrients using light as a source of energy. Most of this organic matter is recycled within the surface by phytoplankton and various other organisms such as zooplankton and bacteria (see chapter 4). Eventually some of this organic carbon settles out or is transported below the surface ocean past a depth of about 100 m, below which light is insufficient for photosynthesis. Biological processes in the deep ocean consume the organic matter and oxygen, eventually converting it back into dissolved carbon dioxide and nutrients. The loop is closed by transport of these inorganic substances back into the surface ocean.

that the terms on the right-hand side have their signs reversed, i.e., $v \cdot ([PO_4^{3-}]_d - [PO_4^{3-}]_s) - \Phi^P$. The signs are reversed because the nutrient supply by volume exchange now brings in water with deep concentration and removes water with the surface concentration, and the flux of organic matter is out of the box. The steady-state solution to (1.2.1) is simple: the flux of organic phosphorus from the surface to the abyss is equal to the net transport of inorganic phosphorus from the abyss to the surface:

$$\Phi^P = v \cdot ([PO_4^{3-}]_d - [PO_4^{3-}]_s) \tag{1.2.2}$$

Note that this equation does not say anything about the mechanism by which organic phosphorus is produced by organisms. It merely states that we can determine the flux of organic phosphorus from the surface ocean if we know v, $[PO_4^{3-}]_d$, and $[PO_4^{3-}]_s$. The concentrations are measured, with deep ocean phosphate being 2.1 mmol m^{-3} and surface phosphate being near 0 over most of the ocean. However, this leaves us with one additional unknown, v. The equation therefore cannot be solved without an additional constraint. We can try using the

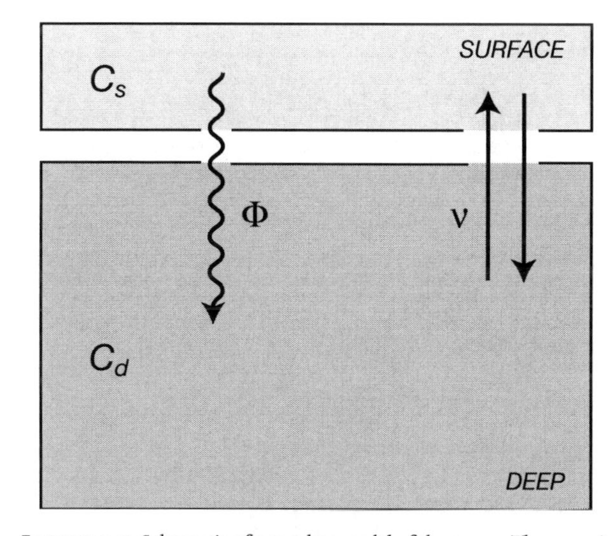

FIGURE 1.2.3: Schematic of a two-box model of the ocean. The organic matter flux Φ has units of mol s^{-1}. The volumetric exchange rate v has units of m^3 s^{-1}. We ignore river input into the surface of the ocean and loss to the sediments, both of which are negligible compared to the fluxes between the two boxes.

conservation equation for an additional biologically utilized element, such as carbon. The steady-state solution for the dissolved inorganic carbon (DIC) balance gives an equation exactly analogous to that for phosphate:

$$\Phi^C = v \cdot (DIC_d - DIC_s) \quad (1.2.3)$$

However, this new equation has also added a new unknown, the flux Φ^C of carbon to the deep ocean. Making use of the stoichiometric ratios, i.e. $\Phi^P = \Phi^c/106$, enables us to solve for the concentration of carbon with respect to that of phosphate, or vice versa, but does not give us a solution for v. The same would happen if we put down an equation for other elements that are involved in biological processes, such as oxygen.

In order to determine the magnitude of the organic matter flux, we use a *tracer* of ocean circulation that provides us with a time clock of the volume exchange rate v. Panel 1.2.1 shows how to calculate the exchange rate using radiocarbon, the radioactive isotope of carbon. The exchange rate we obtain is $v = 1.2 \times 10^{15}\ m^3\ yr^{-1}$, i.e., 38

Panel 1.2.1: Estimation of Two-Box Model Exchange Rate Using Radiocarbon

Radiocarbon follows the same biological and mixing pathways as the stable carbon isotopes ^{12}C and ^{13}C, but decays at a known rate $\partial^{14}C/\partial t = -\lambda \cdot {}^{14}C$, where $\lambda = 1/8267\ yr^{-1}$ is the decay constant of radiocarbon (the half-life is 5730 years). It is formed by cosmic ray spallation of ^{14}N in the atmosphere and enters the surface of the ocean as $^{14}CO_2$. It is the decay of this "natural" radiocarbon that gives us the time clock that we use in order to estimate the volumetric exchange rate between the boxes in the two-box model.

Radiocarbon is measured either by detecting the beta particles that a carbon sample of known mass produces as it decays, which is then reported as a ratio of ^{14}C to the total carbon C in the sample, or by using a mass spectrometer to directly measure the ratio of ^{14}C to ^{12}C. It is reported in the literature as $\Delta^{14}C$, defined as the relative deviation in per mille from a standard sample after various corrections for fractionation effects and decay of the standard [see *Stuiver and Polach*, 1977]. The standard is corrected to represent the modern atmospheric radiocarbon before the industrial revolution started adding radiocarbon-depleted fossil fuel CO_2 to the atmosphere and before nuclear bomb tests began enriching the atmospheric radiocarbon content. A simplified version of the definition with the relatively small fractionation correction removed is given by

$$\Delta^{14}C = \left[\frac{^{14}C_{sample}}{^{14}C_{standard}} - 1\right] \cdot 10^3 \quad (1)$$

For example, typical preindustrial surface water had a $\Delta^{14}C$ of −50‰, which means its radiocarbon content per unit mass or volume of total carbon was 5% lower than the preindustrial atmospheric concentration as inferred from the standard.

In order to make use of radiocarbon in our box model, we need to convert the $\Delta^{14}C$ units into units of ^{14}C concentration per unit volume. We start by converting $\Delta^{14}C$ into R^*, defined as the ratio of the radiocarbon in the measured sample to the radiocarbon in the standard:

$$R^* = \Delta^{14}C \cdot 10^{-3} + 1 = \frac{^{14}C_{sample}}{^{14}C_{standard}} \quad (2)$$

This gives $R^* = 0.95$ for the aforementioned surface water sample. In order to convert this into radiocarbon concentration in mmol m^{-3}, we multiply it by the ratio of radiocarbon atoms to total carbon atoms that the ocean would have had if it were in equilibrium with the preindustrial atmosphere, a, and by the total concentration of carbon in the water, DIC:

$$DI^{14}C = a \cdot DIC \cdot R^* \quad (3)$$

The ratio a varies by only a small amount due to temperature sensitivity of air-sea fractionation effects, which we will ignore here. We thus have for the steady-state radiocarbon balance of the deep box of the two-box model that the net input of radiocarbon by vertical exchange and organic carbon input is balanced by decay of radiocarbon:

$$v \cdot (DIC_s \cdot R_s^* - DIC_d \cdot R_d^*) + \Phi^C \cdot R_s^* = \lambda \cdot V_d \cdot DIC_d \cdot R_d^* \quad (4)$$

The constant a cancels out of the equation. Note that we specify that the radiocarbon ratio of the organic carbon is the same as the surface radiocarbon ratio, since it forms by uptake of surface inorganic carbon.

The decay term on the right-hand side in (4) provides the time clock that enables us to solve for the volume exchange rate. We eliminate the organic carbon flux Φ^C by substituting (1.2.3) into (4). This gives

$$\lambda \cdot V_d \cdot DIC_d \cdot R_d^* = v \cdot (DIC_s \cdot R_s^* - DIC_d \cdot R_d^*) + v \cdot (DIC_d - DIC_s) \cdot R_s^* \quad (5)$$

which can be rearranged to give

$$v = \frac{\lambda \cdot V_d}{R_s^*/R_d^* - 1} \quad (6)$$

The deep ocean volume is about $1.26 \times 10^{18}\ m^3$. The observed value of the preindustrial surface radiocarbon ratio is 0.95 (i.e., 5% depleted with respect to the preindustrial atmosphere). The mean radiocarbon ratio of the deep ocean is 0.84. We thus obtain the solution $v = 1.2 \times 10^{15}\ m^3\ yr^{-1}$.

Sverdrups ($1\,Sv = 10^6\,m^3\,s^{-1}$). The residence time of the deep ocean with respect to this exchange rate is $V_d/v = 1050$ years. If we divide the volume exchange rate by the area of the ocean, $3.4 \times 10^{14}\,m^2$, we find that the mean exchange rate between the surface and deep ocean in the two-box model is about $3.5\,m\,yr^{-1}$. We are now able to calculate additional details of the ocean cycle of nutrients and carbon. The net organic phosphorus flux from the surface to the deep ocean is

$$\begin{aligned}\Phi^P &= v \cdot ([PO_4^{3-}]_d - [PO_4^{3-}]_s) \\ &= 1.2 \times 10^{15}\,m^3\,yr^{-1} \cdot (2.1 - 0.0)\,mmol\,m^{-3} \\ &= 2.5 \times 10^{12}\,mol\,yr^{-1}\end{aligned} \tag{1.2.4}$$

We can estimate the export of organic carbon directly from this using the molar ratio of carbon to phosphorus in organic matter of 106. Using a molecular weight of $12\,g\,mol^{-1}$ for carbon and converting to the commonly used units of $Pg\ C\ yr^{-1}$ ($1\,Pg = 10^{15}\,g$) gives a carbon export of $3.2\,Pg\,C\,yr^{-1}$. More sophisticated ocean models tend to give a number three times this. A two-box model such as ours is too simple to capture all the important exchange processes.

Despite its limitations, which we shall examine in a moment, the two-box model we have developed provides powerful insights into the cycling of chemicals in the ocean. However, the two-box model gives erroneous results when applied to oxygen. Understanding why this is so will give us important additional insights into the box modeling approach and ocean biogeochemistry.

Photosynthesis produces oxygen and remineralization consumes it. The shape of the vertical profile of oxygen reflects these processes: it is lower at depth than at the surface (figure 1.2.1). The photosynthesis reaction above suggests that in these reactions, the ratio of moles of oxygen molecules (O_2) to moles of phosphorus atoms appears to be roughly constant at about -150. This observation permits a powerful simplifying assumption: that 150 oxygen molecules are removed for every phosphorus atom added to the water by remineralization, i.e., $\Phi^{O_2} = -150 \cdot \Phi^P$. The oxygen box model equation for the deep ocean thus becomes:

$$V_d \cdot \frac{d[O_2]_d}{dt} = v \cdot ([O_2]_s - [O_2]_d) - 150 \cdot \Phi^P \tag{1.2.5}$$

If we assume steady state and make use of Φ^P defined by (1.2.2), we obtain the following solution:

$$0 = [O_2]_s - [O_2]_d - 150 \cdot ([PO_4^{3-}]_d - [PO_4^{3-}]_s) \tag{1.2.6}$$

Note that the volume exchange rate v canceled out of the equation. We test the model by using it to predict the deep ocean oxygen concentration and comparing this against the observed average value of $162\,mmol\,m^{-3}$. Rearranging (1.2.6) and substituting in the observed mean values gives the following result:

$$\begin{aligned}[O_2]_d &= [O_2]_s - 150 \cdot ([PO_4^{3-}]_d - [PO_4^{3-}]_s) \\ &= [234 - 150 \cdot (2.1 - 0)]\,mmol\,m^{-3} \\ &= -81\,mmol\,m^{-3}\end{aligned} \tag{1.2.7}$$

The predicted concentration is negative, an impossibility. Where did we go wrong?

The answer is that we have failed to properly consider how deep water actually forms in the ocean. Most of the water sinking into the deep ocean comes from surface waters of the high latitudes. The surface waters of the high latitudes have two properties very different from the global average surface properties used in the above model. Firstly, they have much colder temperatures. Oxygen is more soluble in cold waters, therefore the oxygen concentration in the waters that sink into the deep ocean is much higher than the global surface mean, which is strongly biased by the warm waters of the low latitudes. The average surface saturation oxygen concentration in the waters that fill the abyss is about $331\,mmol\,m^{-3}$. The actual concentration is undersaturated by perhaps 10 to $20\,mmol\,m^{-3}$ [*Najjar and Keeling*, 1997], giving about $316\,mmol\,m^{-3}$ for the water that actually sinks to the deep ocean. This increases our oxygen estimate of (1.2.7) by $82\,mmol\,m^{-3}$. However, this still leaves us with a deep ocean oxygen of only $1\,mmol\,m^{-3}$.

The second and most important problem with the above model is that it assumes that biological uptake keeps the surface phosphorus at a concentration that is near zero. Again, this is a reasonable assumption for the global mean surface concentration, but not at all the case in the high-latitude regions where deep water forms (see figure 1.2.4). There the biological uptake of phosphorus is not so efficient and surface concentrations of phosphorus average about $1.3\,mmol\,m^{-3}$ [*Broecker et al.*, 1985a].

In order to deal with the issues raised by these observations, we develop a new three-box model of the ocean that splits the surface ocean into a high latitude as well as a low-latitude box (figure 1.2.5). There are two vertical exchange terms between the surface and deep ocean. The first involves vertical exchange f_{hd} between the high-latitude surface box and deep ocean, which is analogous to deep water formation processes that occur around the Antarctic. The second is an overturning term T of water flowing up into the low-latitude surface ocean, then towards the high-latitude surface ocean, before sinking. The overturning exchange T is analogous to the deep water formation processes that occur in the North Atlantic. A solution to the three-box model can be obtained by a procedure exactly analogous to that used in obtaining the two-box model solution. We solve for the organic matter fluxes Φ_h and Φ_l to the deep ocean using the phosphorus balance, and we use the ratio of oxygen to phosphorus in the biological reactions to substitute the phosphorus flux into the oxygen

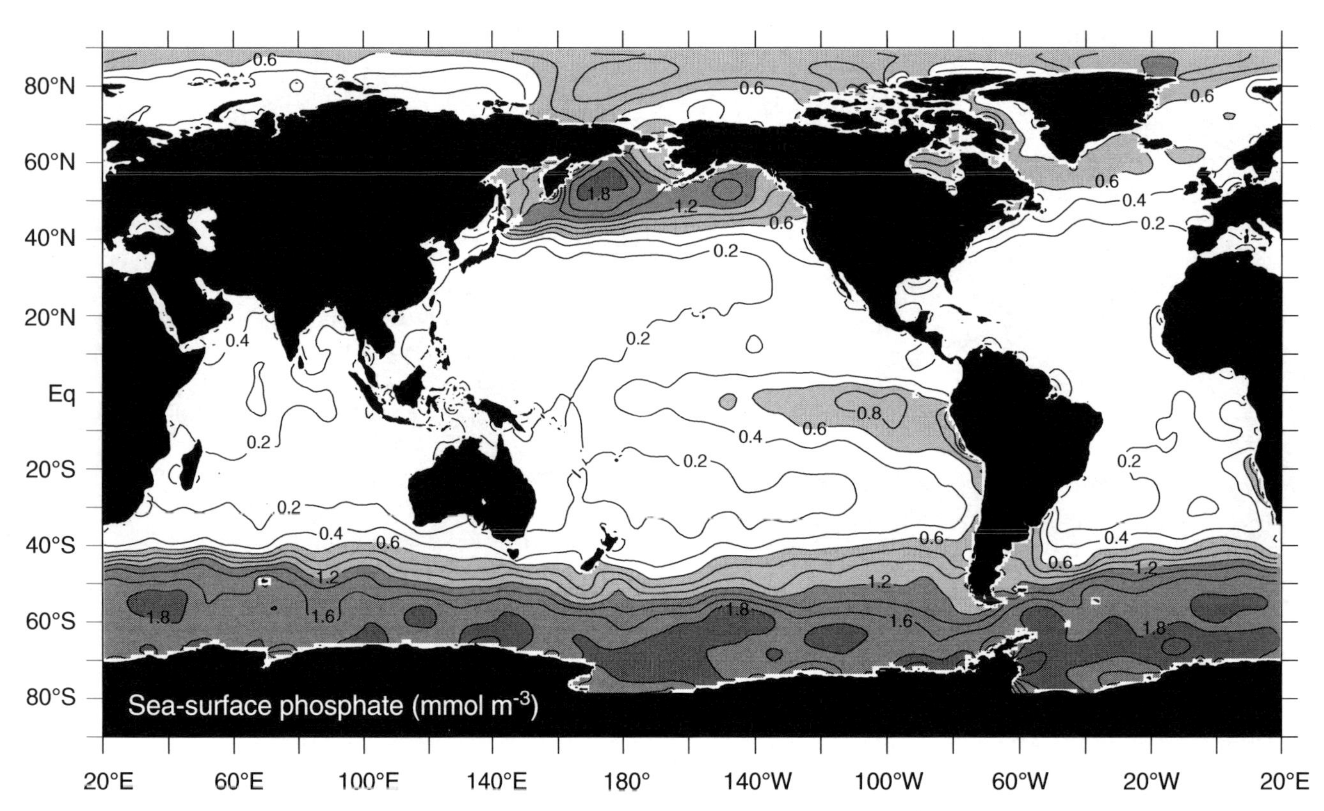

FIGURE 1.2.4: Global map of the annual mean surface phosphate concentration (see also color plate 2). Note that phosphate is nearly depleted throughout the subtropical gyres of the ocean, but that it is high in the high latitudes of the North Pacific, North Atlantic, and Southern Ocean, as well as in the equatorial Pacific and off of southwest Africa. Data are from the World Ocean Atlas 2001 [Conkright et al., 2002].

equation for the deep ocean. The final equation for the oxygen concentration in the deep ocean is:

$$\begin{aligned}[O_2]_d &= [O_2]_h - 150 \cdot ([PO_4^{3-}]_d - [PO_4^{3-}]_h) \\ &= [316 - 150 \cdot (2.1 - 1.3)]\,\text{mmol m}^{-3} \\ &= 196\,\text{mmol m}^{-3}\end{aligned} \quad (1.2.8)$$

This solution is identical to the two-box model solution of (1.2.7), except that the global mean surface oxygen and phosphorus concentrations of (1.2.7) are replaced in (1.2.8) by the high-latitude mean concentrations. The higher oxygen and phosphorus concentrations in surface waters of the high latitudes give a deep ocean oxygen prediction that is now in much better agreement with the observed concentration of 162 mmol m^{-3}. The remaining disagreement between the model prediction and observations is due in part to the fact that the organic matter that sinks into the deep ocean appears to have a composition that is rich in carbon and hydrogen. The stoichiometric ratio of oxygen consumption to inorganic phosphorus release in the deep ocean is thus closer to 170 rather than 150 [*Anderson and Sarmiento*, 1994]. This drops the oxygen prediction to 180 mmol m^{-3}.

An important theme of this book is the crucial role played by the surface nutrient content of the high latitudes in controlling the chemistry of the deep ocean. The three-box model solution (1.2.8) provides an inter-

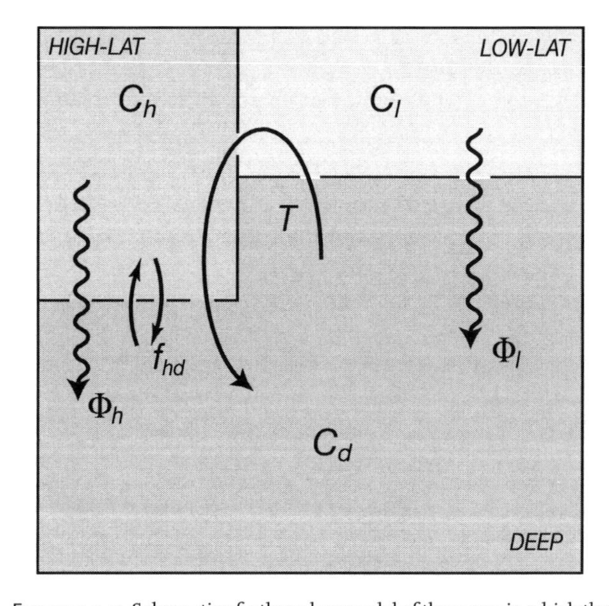

FIGURE 1.2.5: Schematic of a three-box model of the ocean in which the surface waters have been divided into high-latitude and low-latitude regions intended to represent areas where the surface nutrients tend to be high (high latitudes) and areas where the surface concentration tend to be low (low latitudes [cf. *Sarmiento and Toggweiler*, 1984]). As with the two-box model, Φ (mol s^{-1}) is the organic matter flux from the surface boxes to the deep box. The volumetric exchange rates (m^3 s^{-1}) between boxes consists of mixing f_{hd} between the high-latitude surface box and deep ocean, and an overturning circulation T, which flows from the deep ocean into the low-latitude surface and then through the high-latitude surface box back into the deep box.

esting, though oversimplified, illustration of the potential importance of this process. It suggests that the oxygen content of the deep ocean has the potential to vary between 0 and 316 mmol m^{-3} with changes in the surface phosphate concentration of the high latitudes of between 0.24 and 2.1 mmol m^{-3}. This sensitivity was exploited by *Sarmiento et al.* [1988a] to provide a hypothesis for the cause of episodes of deep ocean anoxia (total depletion of oxygen) that have been observed in the geological record of the ocean sediments. If, for example, the high-latitude phosphate were to drop, then (1.2.8) suggests that the deep ocean oxygen should also drop.

However, in subsequent work with more sophisticated three-dimensional models of ocean circulation consisting of tens of thousands of boxes, it was found that the sensitivity of the deep ocean oxygen to high-latitude nutrients was much smaller than predicted by the three-box model [*Sarmiento and Orr*, 1991]. The three-dimensional models lose only 41 mmol m^{-3} of oxygen when nutrients are reduced to zero, as contrasted with the loss of more than 180 mmol m^{-3} that our box model gives. The three-dimensional models confine the reduction of deep ocean oxygen to regions just below the high-latitude deep water formation regions so that the average deep ocean oxygen drops only modestly. This and the previous examples demonstrate clearly the importance of understanding ocean transport and of understanding how to develop box models that can represent the circulation realistically.

1.3 Chapter Conclusion and Outline of Book

The central concern of this book is to understand the processes that control the mean oceanic concentration as well as the spatial and temporal variations of chemicals that are influenced by biological processes. As illustrated by the box models, this requires understanding the transport of chemicals by processes such as advection and diffusion. It also requires us to understand sources and sinks of the chemicals, such as biological uptake and remineralization, as well as boundary conditions at the surface and bottom of the ocean such as gas exchange and sediment water fluxes.

The view taken in what follows is that kinetic processes are the primary control on the composition of the ocean, as proposed by Broecker's response to Sillén's equilibrium view [*Broecker*, 1971]. This view was explored in detail in the book *Tracers in the Sea* that Broecker wrote with Peng in 1982. The progress since then, and recent completion of several global measurement and modeling programs, suggests the need for a new look at what we have chosen to call "ocean biogeochemical dynamics." This title recognizes the important role of biological and geological, as well as chemical, processes, in controlling the oceanic composition of the chemicals we will study here, as well as the fact that these phenomena show considerable change through time.

We begin in chapter 2 with a derivation of the tracer conservation equation and a review of our understanding of ocean transport by advection and mixing. Subsequent chapters will go through each of the other processes that must be specified in order to solve the tracer conservation equation. Boundary conditions must be given at the air-sea interface and at the ocean floor. Among the boundary processes that must be considered are gas exchange at the air-sea interface, and fluxes from the sediments resulting from reactions in the sediments. We must also consider sources and sinks internal to the ocean due to chemical and biological processes. Chemical processes will not be dealt with as a separate topic, but rather will be brought into the discussion whenever relevant, primarily as part of the examples. Biological processes can remove inorganic chemicals by photosynthesis, which uses light as a source of energy, or chemosynthesis, which uses chemical reactions as a source of energy. The major region where this occurs is in the euphotic zone at the surface of the ocean, where light is sufficient to support photosynthesis. Biological processes also add inorganic chemicals by remineralization of organic matter. This occurs at all depths in the ocean as well as in the sediments.

We cover each of these topics beginning at the surface and moving down through the water column into the sediments. Chapter 3 deals with boundary conditions at the air-sea interface. Chapter 4 covers the formation of organic matter in the surface ocean. Chapter 5 covers the transport of organic matter from the surface to the abyss and its remineralization at depth. Chapter 6 covers the ocean sediments. The emphasis of these chapters is on the chemical cycles of the major nutrients (nitrate and phosphate) and oxygen.

After developing a deeper understanding of relevant oceanic processes as well as the cycling of the major nutrients and oxygen, we apply this understanding to an analysis of the chemical cycles of silicon in chapter 7 and carbon in chapter 8, and of $CaCO_3$ in chapter 9. Finally, chapter 10 applies the tools developed in chapters 1 through 9 to a discussion of three of the major outstanding problems in our understanding of the influence of the ocean on atmospheric carbon dioxide. These are the role of the ocean as a sink for anthropogenic carbon, in the interannual variability of atmospheric carbon dioxide, and in the large reductions of atmospheric carbon dioxide that occurred during ice ages.

Problems

1.1 Explain in words why we think that neither the accumulation hypothesis nor the equilibrium control hypothesis is correct for explaining the huge variations that we observe for the mean oceanic concentration of the different elements in the ocean.

   a. Describe the test that we devised for the accumulation hypothesis and what information we used to reject this hypothesis.

   b. Do the same for the equilibrium hypothesis.

1.2 If we nevertheless accept the hypothesis that the mean concentration of elements in the ocean is a result of the accumulation from river input, how large would today's ocean concentration be of magnesium, arsenic, chlorine, and gold? (Use an ocean age of 3.85 billion years and the river concentrations given in table 1.1.1).

1.3 List two elements that fall into each of the following categories: *biolimiting*, *biointermediate*, and *biounlimited*. Explain the basis of your choice.

1.4 During photosynthesis, how many free oxygen molecules are produced per nitrogen atom consumed? How many free oxygen molecules are produced per carbon atom consumed?

1.5 Equation (1.2.2) tells us that, at steady state, the upward supply of nutrients has to be balanced by the downward flux of nutrients in organic matter.

   a. Discuss how the biological export production would change if the deep and surface concentrations were suddenly halved.

   b. Discuss how biological export production would react to a sudden doubling of the vertical exchange rate v.

1.6 Describe what would happen in the one-box model if the concentration of phosphate in the ocean was suddenly doubled, but the river input and removal rate constant k remained the same. Describe first the final steady state, and then the time it would take to achieve the final steady state.

1.7 What would happen to the global mean oceanic phosphate concentration if the river concentration of phosphate in the one-box model were suddenly to double? Draw a graph of concentration versus time, showing the behavior in time as well as the final concentration. Assume that phosphate removal is first order.

a. Solve first for steady state. Then draw the solution at $t=0$ and at $t=\infty$.

b. Solve for the time-dependence. Use the solution given in panel 1.1.1. Add the time-dependence to the figure.

c. Discuss what determines the response time of the ocean.

1.8 Suppose that the input of a chemical to the ocean were to vary sinusoidally with a frequency ω of $2\pi/100{,}000\,\mathrm{yr}^{-1}$ and an amplitude equal to 50% of the mean input. What is the ratio of the amplitude in the oceanic concentration variation to the mean oceanic concentration for Mg and Si? (Assume that removal for both is first order and that the mean life for each is as given by table 1.1.1.)

1.9 Suppose that the removal rate for a substance A is second order in the concentration of A and B, i.e., $R_A = V \cdot k_1 \cdot [A] \cdot [B]$. Such might be the case if A were removed from the ocean by precipitation as the mineral AB. Suppose further that B is removed both by precipitation with A, but also by a separate first-order removal process, such that $R_B = V \cdot k_1 \cdot [A] \cdot [B] + V \cdot k_2 \cdot [B]$.

a. Derive an equation for the steady-state concentrations of A and B in terms of the river concentrations.

b. Show the approximate solution for the case where the river concentration of B is much greater than the river concentration of A. Compare this solution with the solution for first-order removal.

1.10 The water sinking into the deep box of the two-box model comes primarily from high latitudes. Suppose that the water sinking from the high latitudes has a radiocarbon ratio R_s^* of 0.92 but that everything else is as in the two-box model discussed in panel 1.2.1.

a. What is the volume exchange rate between the deep and shallow boxes?

b. The shallow box has a thickness of 100 m and the deep box a thickness of 4000 m. What is the residence time of the surface box with respect to exchange with the deep box? What is the residence time of the deep box with respect to exchange with the surface box?

c. How large is the biological export production in this case (in units of carbon, e.g., Pg C yr^{-1})?

1.11 Consider an element C that is transported to the surface in the two-box model, where g is the fraction of upwelled nutrients that is converted to organic matter and then carried into the deep ocean.

a. Derive an equation for g in terms of the surface and deep ocean concentrations.

b. Calculate the magnitude of g for the following elements using the data in table 1.1.1 for the deep ocean concentration, and the following surface concentrations.

	C_s (mmol m^{-3})
P	0.1
Ba	0.026
Ca	10,251

1.12 Add the river flux and sediment burial terms of figure 1.1.2 to the box model of figure 1.2.3. River delivery occurs into the surface box. Assume that the sediment burial flux is a fraction f of the organic matter flux Φ.

a. Derive an equation for f. Start with the conservation equation for surface and deep boxes, and then assume steady state.

b. Use the equation in (a) in combination with the information in problem 1.11 and table 1.1.1 to determine the magnitude of f for P, Ba, and Ca.

c. Discuss the solution. Is it realistic? Can you explain the differences between the three elements?

1.13 Show how to derive equation (1.2.8). Start with the conservation equation for the deep box for phosphorus and oxygen. Note that low-latitude surface phosphate is 0 and that $\Phi^{O_2} = r_{O_2:P} \cdot \Phi^P$, where $r_{O_2:P}$ is -150. Assume steady state and solve for deep O_2.

1.14 Suppose the deep phosphate concentration doubled due to an increase in the river flow.

a. How would this affect the deep ocean oxygen concentration if high-latitude phosphate remained the same?

b. How would this affect the deep ocean oxygen concentration if the high-latitude phosphate concentration also doubled?

CHAPTER 2

Tracer Conservation and Ocean Transport

The tracer conservation equation establishes the relationship between the time rate of change of tracer concentration at a given point and the processes that can change that concentration. These processes include water transport by advection and mixing, and sources and sinks due to biological and chemical transformations. This book is dedicated primarily to a study of the sources and sinks due to biological and chemical transformations. However, we begin our study in this chapter with a derivation of the conservation equation, including the transport terms, accompanied by an overview of ocean circulation. The ocean circulation overview begins with a discussion of advection driven by wind forcing. The next two subsections focus on the wind-driven circulation in the upper kilometer of the ocean in the presence of stratification, and also the deep ocean circulation below that. A final section discusses non-steady-state behavior of the oceanic circulation, with a focus on mesoscale eddies and interannual to decadal variability of the upper ocean.

Our discussion of ocean circulation focuses on only a few major features that are critical for understanding ocean biogeochemistry. Textbooks such as that of *Pond and Pickard* [1983] provide a more detailed discussion of the basic theory of ocean circulation [cf. also *Pedlosky*, 1996], and *Siedler et al.* [2001] provide an up-to-date overview of observations and modeling focused primarily on results from the World Ocean Circulation Experiment (WOCE) carried out in the 1990s [cf. also *Warren and Wunsch*, 1981].

2.1 Tracer Conservation Equation

The tracer conservation equation for a volume of water at a fixed location has the form

$$\frac{\partial C}{\partial t} = \left.\frac{\partial C}{\partial t}\right|_{advection} + \left.\frac{\partial C}{\partial t}\right|_{diffusion} + SMS(C) \qquad (2.1.1)$$

where the term $SMS(C)$ (mmol m^{-3} s^{-1}) represents internal sources minus sinks. We use C to represent tracer concentration in mmol m^{-3}, as in chapter 1. Geochemists generally prefer to report concentration in units of moles per unit mass of water (micromoles per kilogram, i.e., μmol kg^{-1}) rather than per unit volume. This is because water is compressible and the volume is thus a function of the pressure and temperature, whereas the mass is not. As a consequence, mass is conserved with changes in state, whereas volume is not. However, the use of per unit mass in the conservation equation is unwieldy because it requires including density and all of its derivatives. Furthermore, most present models of the ocean are based on approximations that imply conservation of volume. We will therefore generally use units of mmol m^{-3} for concentration. These can be converted to μmol kg^{-1} by dividing by the density in kg m$^{-3}\times 10^{-3}$.

In what follows we first derive the advection and diffusion components, and then consider the application of the conservation equation to box models such as those described in chapter 1.

Advection and Diffusion Components

We derive the advection and diffusion components of the tracer conservation equation by considering an infinitesimally small cube of water with fixed volume, such as that shown in figure 2.1.1. The total amount of tracer in the cube is equal to the concentration C times the volume of the cube V (m^3). Panel 2.1.1 shows how we derive the advective contribution to the amount of tracer contained in the volume. The final equation for the advective component of the conserva-

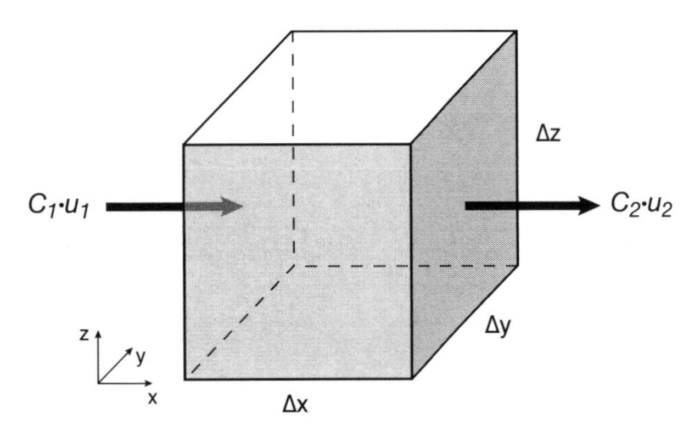

FIGURE 2.1.1: Schematic of an infinitesimally small cube, with dimensions Δx by Δy by Δz, and a fixed volume $V = \Delta x \Delta y \Delta z$. Advection u in the x-direction carries tracer of concentration C.

tion equation is

$$\left.\frac{\partial C}{\partial t}\right|_{advection} = -\frac{\partial(C\cdot u)}{\partial x} - \frac{\partial(C\cdot v)}{\partial y} - \frac{\partial(C\cdot w)}{\partial z} \qquad (2.1.2)$$

The velocities, u, v, and w, are the vector components of the advection in the x-, y-, and z- directions, respectively. By convention, x is taken as positive toward the east, y is positive toward the north, and z is positive upward. The velocities have units of m s^{-1}.

We next add the effect of diffusion across the walls of the cube. Diffusion is the flux of tracer that occurs due in part to the random motion of tracer molecules referred to as *molecular diffusion*. It also generally includes an *eddy diffusion* component that represents the net effect of small-scale motions of water parcels that do not result in net advection. Thus, if one face of the cube has equal amounts of water flowing in and out, but if the water flowing in generally has a higher tracer concentration on average than the water flowing out, there will be a net transport of tracer into the cube even though there is no net advection. One can readily see that the magnitude of this term will depend on the size of the volume under consideration, with larger volumes having larger eddy diffusion contributions. It will also depend on the amount of water flowing in and out of the face of the cube, with larger exchanges resulting in greater eddy diffusion.

The tracer flux resulting from molecular diffusion is directly proportional to the gradient in tracer concentration, with the constant of proportionality defined as the diffusivity. Fick's first law gives this relationship as:

$$\Phi_x = -\varepsilon\cdot\frac{\partial C}{\partial x}, \quad \Phi_y = -\varepsilon\cdot\frac{\partial C}{\partial y}, \quad \Phi_z = -\varepsilon\cdot\frac{\partial C}{\partial z} \qquad (2.1.3)$$

The flux Φ is in units of $\text{mol m}^{-2}\,\text{s}^{-1}$, and the molecular diffusivity ε is in $\text{m}^2\,\text{s}^{-1}$. Fick's second law relates the change in concentration to the gradient in the diffusive flux. Consider a tracer flux $\Phi_x\cdot\Delta y\cdot\Delta z$ entering the left

Panel 2.1.1: Advection Contribution to the Tracer Conservation Equation

Consider the flow of water through an infinitesimally small cube such as the one depicted in figure 2.1.1, with dimensions Δx by Δy by Δz and a volume $V = \Delta x\,\Delta y\,\Delta z$. The total amount of tracer in the cube is

$$\Delta x\,\Delta y\,\Delta z C = V\cdot C \qquad (1)$$

The response of the tracer to a velocity u in the x-direction such as that shown in figure 2.1.1 is

$$V\frac{\partial C}{\partial t} = C_1 u_1 \Delta y \Delta z - C_2 u_2 \Delta y \Delta z \qquad (2)$$

We assume that volume is conserved, which allows us to bring it out of the time-derivative. Assume that the velocity and concentration vary continuously across the cube such that

$$u_2 = u_1 + \frac{\partial u}{\partial x}\Delta x$$
$$C_2 = C_1 + \frac{\partial C}{\partial x}\Delta x \qquad (3)$$

Substituting (3) into (2) gives

$$V\frac{\partial C}{\partial t} = \Delta y \Delta z\left[C_1 u_1 - \left(C_1 u_1 + \Delta x\, C_1\frac{\partial u}{\partial x} + \Delta x\, u_1 \frac{\partial C}{\partial x} + \Delta x^2 \frac{\partial u}{\partial x}\frac{\partial C}{\partial x}\right)\right]$$
$$= -V\left(C_1\frac{\partial u}{\partial x} + u_1\frac{\partial C}{\partial x} + \Delta x\frac{\partial u}{\partial x}\frac{\partial C}{\partial x}\right) \qquad (4)$$

Because of the small size of Δx in the final term in the parentheses on the right-hand side of (4), this term is much smaller than the other two. Canceling this term out and making use of the product rule to collapse the first two terms in the parentheses into a single term gives:

$$\frac{\partial C}{\partial t} = -\frac{\partial (Cu)}{\partial x} \qquad (5)$$

The advective contributions for the y- and z-directions can be derived in the same fashion, giving the final equation

$$\frac{\partial C}{\partial t} = -\frac{\partial (Cu)}{\partial x} - \frac{\partial (Cv)}{\partial y} - \frac{\partial (Cw)}{\partial z} \qquad (6)$$

for the contribution of advection to the tracer concentration inside the cube.

face of the cube in figure 2.1.1 and a flux $[\Phi_x + (\partial \Phi_x/\partial x) \cdot \Delta x] \cdot \Delta y \cdot \Delta z$ leaving the right face of the cube. The effect of the diffusive flux on the total amount of tracer in the box $V \cdot (\partial C/\partial t)$ is obtained by subtracting the second expression from the first, i.e., $V \cdot (\partial C/\partial t) = -V \cdot (\partial \Phi_x/\partial x)$. The derivation of similar expressions for the y- and z-directions gives:

$$\left.\frac{\partial C}{\partial t}\right|_{molecular\ diffusion} = -\frac{\partial \Phi_x}{\partial x} - \frac{\partial \Phi_y}{\partial y} - \frac{\partial \Phi_z}{\partial z} \qquad (2.1.4)$$

Substituting (2.1.3) into (2.1.4) gives:

$$\left.\frac{\partial C}{\partial t}\right|_{molecular\ diffusion} = \frac{\partial}{\partial x}\left(\varepsilon \cdot \frac{\partial C}{\partial x}\right) + \frac{\partial}{\partial y}\left(\varepsilon \cdot \frac{\partial C}{\partial y}\right) + \frac{\partial}{\partial z}\left(\varepsilon \cdot \frac{\partial C}{\partial z}\right) \qquad (2.1.5)$$

The transport of tracer by eddy diffusion is generally far larger than transport by molecular diffusion. Panel 2.1.2 shows a derivation of the eddy diffusion processes that treats the water motions as consisting of a time mean component and random variations around that mean. This treatment shows that transport of tracer by eddies results from the degree of correlation between the direction of the random motions of the water parcels and the tracer concentration. For example, if water parcels moving in the positive x-direction have higher concentrations on the average than water parcels moving in the negative x-direction, the tracer will be transported in the positive x-direction. Unfortunately, we generally do not have adequate information on the flow field and concentrations to calculate the necessary correlations. The usual practice is to approximate eddy diffusion by assuming it behaves like Fickian diffusion, with one important difference: horizontal eddy diffusion in the ocean is much larger than the vertical. We therefore allow for the possibility of

Panel 2.1.2: Effect of Eddies on Tracer Transport

Eddies are the random motions of water parcels that do not result in any net transfer of water from one part of the ocean to another. However, eddies can carry tracer if the random motions in one direction have a different concentration than the random motions in the opposite direction. We can derive an equation for this process by starting with the advection equation (2.1.2):

$$\frac{\partial C}{\partial t} = -\frac{\partial (Cu)}{\partial x} - \frac{\partial (Cv)}{\partial y} - \frac{\partial (Cw)}{\partial z} \qquad (1)$$

and separating the velocity and the concentration terms into a time-average component (denoted by an overbar) defined as follows:

$$\overline{()} = \frac{1}{T}\int_0^T ()dt \qquad (2)$$

and deviations from the average (denoted by a prime) defined such that

$$u = \bar{u} + u' \qquad C = \bar{C} + C' \qquad (3)$$

Note from the definition of the deviations from the mean that the average of the deviations is equal to 0, i.e.,

$$\overline{u'} = 0 \qquad \overline{C'} = 0 \qquad \overline{\bar{C}u'} = 0 \qquad \overline{C'\bar{u}} = 0 \qquad (4)$$

Substituting (3) into (1), taking the average, and making use of (4) to simplify the resulting equation gives:

$$\frac{\partial \bar{C}}{\partial t} = -\left(\frac{\partial (\bar{C}\bar{u})}{\partial x} + \frac{\partial (\bar{C}\bar{v})}{\partial y} + \frac{\partial (\bar{C}\bar{w})}{\partial z}\right) - \left(\frac{\partial (\overline{C'u'})}{\partial x} + \frac{\partial (\overline{C'v'})}{\partial y} + \frac{\partial (\overline{C'w'})}{\partial z}\right) \qquad (5)$$

The usual convention is to drop the overbar on all the terms except those involving the products of primes. This gives:

$$\frac{\partial C}{\partial t} = -\left(\frac{\partial (Cu)}{\partial x} + \frac{\partial (Cv)}{\partial y} + \frac{\partial (Cw)}{\partial z}\right) - \left(\frac{\partial (\overline{C'u'})}{\partial x} + \frac{\partial (\overline{C'v'})}{\partial y} + \frac{\partial (\overline{C'w'})}{\partial z}\right) \qquad (6)$$

The terms in the first set of parentheses on the right-hand side represent the contribution of net advection to the tracer transport. The terms in the second pair of parentheses represent the contribution of eddies to the transport of tracer. The prime terms involve no net flow of water into or out of the box, since the average of the velocity fluctuations is zero. However, if the tracer concentration is correlated with the velocity direction, there will be a transport of tracer associated with these eddies. We assume that the eddy contribution can be represented by a Fickian-type diffusion, i.e.,

$$-\frac{\partial (\overline{C'u'})}{\partial x} - \frac{\partial (\overline{C'v'})}{\partial y} - \frac{\partial (\overline{C'w'})}{\partial z} = \frac{\partial}{\partial x}\left[D_x \frac{\partial C}{\partial x}\right] + \frac{\partial}{\partial y}\left[D_y \frac{\partial C}{\partial y}\right] + \frac{\partial}{\partial z}\left[D_z \frac{\partial C}{\partial z}\right] \qquad (7)$$

where D is an eddy diffusivity term with units of $m^2\, s^{-1}$. See text for further discussion.

different magnitudes of mixing in each direction. The eddy and molecular contributions to the diffusivity are assumed to be additive, giving:

$$\left.\frac{\partial C}{\partial t}\right|_{diffusion} = \frac{\partial}{\partial x}\left[(D_x+\varepsilon)\cdot\frac{\partial C}{\partial x}\right] + \frac{\partial}{\partial y}\left[(D_y+\varepsilon)\cdot\frac{\partial C}{\partial y}\right] + \frac{\partial}{\partial z}\left[(D_z+\varepsilon)\cdot\frac{\partial C}{\partial z}\right] \tag{2.1.6}$$

The final form of the conservation equation is therefore:

$$\frac{\partial C}{\partial t} = -\frac{\partial(C\cdot u)}{\partial x} - \frac{\partial(C\cdot v)}{\partial y} - \frac{\partial(C\cdot w)}{\partial z} + \frac{\partial}{\partial x}\left[(D_x+\varepsilon)\cdot\frac{\partial C}{\partial x}\right] + \frac{\partial}{\partial y}\left[(D_y+\varepsilon)\cdot\frac{\partial C}{\partial y}\right] + \frac{\partial}{\partial z}\left[(D_z+\varepsilon)\cdot\frac{\partial C}{\partial z}\right] + SMS(C) \tag{2.1.7}$$

We conclude the derivation by simplifying the tracer conservation equation. We expand the advection component (2.1.2) using the product rule:

$$-\frac{\partial(C\cdot u)}{\partial x} - \frac{\partial(C\cdot v)}{\partial y} - \frac{\partial(C\cdot w)}{\partial z} = -C\left(\frac{\partial u}{\partial x} + \frac{\partial v}{\partial y} + \frac{\partial w}{\partial z}\right) - \left(u\cdot\frac{\partial C}{\partial x} + v\cdot\frac{\partial C}{\partial y} + w\cdot\frac{\partial C}{\partial z}\right) \tag{2.1.8}$$

The mass conservation equation for incompressible flow derived in panel 2.1.3 enables us to cancel the first term in parentheses on the right-hand side of (2.1.8). This gives:

$$\frac{\partial C}{\partial t} = -u\cdot\frac{\partial C}{\partial x} - v\cdot\frac{\partial C}{\partial y} - w\cdot\frac{\partial C}{\partial z} + \frac{\partial}{\partial x}\left[(D_x+\varepsilon)\cdot\frac{\partial C}{\partial x}\right] + \frac{\partial}{\partial y}\left[(D_y+\varepsilon)\cdot\frac{\partial C}{\partial y}\right] + \frac{\partial}{\partial z}\left[(D_z+\varepsilon)\cdot\frac{\partial C}{\partial z}\right] + SMS(C) \tag{2.1.9}$$

Finally, we make use of the definition of the divergence operator:

$$\nabla = \mathbf{i}\frac{\partial(\,)}{\partial x} + \mathbf{j}\frac{\partial(\,)}{\partial y} + \mathbf{k}\frac{\partial(\,)}{\partial z} \tag{2.1.10}$$

to obtain:

$$\frac{\partial C}{\partial t} = -\mathbf{U}\cdot\nabla C + \nabla\cdot(\mathbf{D}\cdot\nabla C) + SMS(C) \tag{2.1.11}$$

where $\mathbf{i}$, $\mathbf{j}$, and $\mathbf{k}$ are unit vectors in the x, y, and z directions, respectively, $\mathbf{U}$ is the velocity vector, $\mathbf{i}u + \mathbf{j}v + \mathbf{k}w$, and $\mathbf{D}$ is a diffusivity tensor with diagonal terms equal to $D_x+\varepsilon$, $D_y+\varepsilon$, and $D_z+\varepsilon$.

Application to Box Models

Box models such as those described in the first chapter are the quantitative tool most frequently used by geochemists to investigate the behavior of chemicals. We define a box that encloses some region of particular interest, for example the surface ocean. The goal of developing a box model is to determine the processes that control the mean tracer concentration within that box. The usual way of proceeding is to assemble an equation that relates the change in mean concentration within the box to the fluxes across the boundaries of the box and sources and sinks internal to the box. In our example of a surface box, inputs across the boundaries of the box might be river input, air-sea fluxes, and upward flow of deep waters into the box. Losses across the boundaries of the box might include downward flow of surface waters. A sink internal to the box might be biological uptake, and an internal source might be remineralization of organic matter.

We demonstrate next how to construct a box model using the conservation equation (2.1.11). We wish to have an equation in terms of the mean concentration of the box. We accomplish this by integrating (2.1.11) over the volume of the box. The time rate of change term now represents the total change within the box. Dividing by the volume of the box will give the time rate of change of the mean concentration within the box. Note that our model does not require that the concentration be uniform within the box. A common misconception of geochemists is that box models require a uniform concentration within the box. This is useful as an assumption and is commonly employed, but it is not a necessary condition.

We next discuss the processes that can change the mean concentration within the box of our model. These include fluxes across the faces of the box. For example, a box at the surface of the ocean would have gas exchange with the atmosphere as well as advective and diffusive exchange with the layer immediately below. A box representing the deep ocean would have exchange with upper ocean boxes as well as exchange fluxes with the sediments. When we perform the volume integral of (2.1.11) as described in the previous paragraph, we also integrate the advection and diffusion terms over the entire volume. However, it is only at the boundaries of the box that these terms can provide a net flux into or out of the box that will affect the mean concentration. Advection and diffusion within the box move tracer around but do not change the mean concentration. It is therefore useful to transform the volume integral of the advection and diffusion terms into a surface integral using the divergence (or Gauss's) theorem:

$$\int[-\nabla\cdot(\mathbf{U}\cdot C) + \nabla\cdot(\mathbf{D}\cdot\nabla C)]\,dV = -\oint \mathbf{n}\cdot(\mathbf{U}\cdot C - \mathbf{D}\cdot\nabla C)\,dS \tag{2.1.12}$$

where $\mathbf{n}$ is a unit vector normal to the boundaries of the box and S is surface area in m^2. Note that we have made use of the incompressibility assumption to convert $\mathbf{U} \cdot \nabla C$ in (2.1.11) to $\nabla \mathbf{U}C$ in (2.1.12). This surface integral also includes exchanges at the air-sea and sediment-water interfaces, which usually occur through diffusive processes.

Finally, we must integrate the $SMS(C)$ term over the entire volume, since any change in this term at any location within the box will have an impact on the concentration. The final *integral form* of the tracer conservation equation is therefore:

$$\int \left(\frac{\partial C}{\partial t} - SMS(C) \right) dV = - \oint \mathbf{n} \cdot (\mathbf{U} \cdot C - \mathbf{D} \cdot \nabla C) \; dS \tag{2.1.13}$$

The problems that we encountered when we tried to simulate oxygen with the two-box model of the ocean in chapter 1 could have been foreseen by a more careful analysis of the surface transport integral in (2.1.13) and a better understanding of the ocean circulation. Consider, for example, the advection term in (2.1.13). This is the integral of the product of concentration times velocity normal to the surface of the box. One can think of this as a weighted mean of the surface concentration, where the weight is the component of the velocity that is perpendicular to the interface between the surface and deep boxes. Thus, in the two-box model of chapter 1, the surface concentration that is representative of waters sinking into the deep ocean is not the mean of the entire surface ocean, but rather a concentration that is strongly weighted toward the high-latitude regions where surface water sinks into the deep ocean. We were able to improve our box model solution in chapter 1 by separating the high-latitude surface ocean from the low-latitude surface ocean, but we could also have done this by using the velocity-weighted integral of the surface concentration as given by (2.1.13). This simple example shows how important it is to understand the major features of oceanic transport, to which the remainder of this chapter is dedicated. We begin with a discussion of the circulation that is driven by the winds, which are the main source of energy for ocean circulation.

2.2 Wind-Driven Circulation

Figure 2.2.1 summarizes the mean wind velocity at the surface of the ocean for January and July. The basic pattern of the winds, schematically illustrated in figure 2.2.2, consists of the trade winds in the tropics, the westerlies in mid-latitudes, and the polar easterlies in high latitudes. One might expect that surface water would be forced to move in the same direction as the wind. However, the water that is set directly into motion by the wind feels the effect of the Earth's rotation, and the resulting *Ekman transport*, which is confined to the top 10 to 100 m of the water column, is 90° to the right of the wind in the northern hemisphere and 90° to the left of it in the southern hemisphere (see figure 2.2.3 for the ocean flow in the top 50 m).

One can gain some insight into how the Earth's rotation affects motions in the atmosphere as well as the ocean by considering, for example, what happens to a projectile fired directly toward the north from somewhere at 30°N. The projectile carries with it not only the northward velocity imparted to it at the beginning of its trajectory, but also the eastward velocity of the Earth at 30°N. The eastward velocity of the Earth is smaller at higher latitudes because of the decreasing diameter of the circles formed by slicing the Earth at a given parallel of latitude. For example, the equator has an eastward velocity of 455 $m\,s^{-1}$, and the 30°N and 45°N latitudes have eastward velocities of 402 and 326 $m\,s^{-1}$, respectively. Thus, with respect to a fixed position on the Earth, the projectile will appear to accelerate toward the east. This simple thought experiment also shows how a projectile fired in the southern hemisphere would result in acceleration to the left of the direction in which the projectile is fired.

The Ekman transport causes water to converge in some regions of the ocean and diverge in others. This has two major impacts on ocean circulation. It causes downwelling in regions of convergent flow (*Ekman pumping*) and upwelling in regions of divergent flow (*Ekman suction*). Upwelling is particularly important for biology, because it is one of the primary mechanisms by which nutrients from the deep ocean are brought to the surface.

The other major impact of Ekman transport is that it piles up water in some regions and removes it from others, thus creating large horizontal gradients in sea surface height (figure 2.2.4). The pressure gradients that result from these sea surface height differences might be expected to lead to a downslope flow, but the Earth's rotation intervenes, and the flow that results goes at a right angle to the slope rather than down it. Figure 2.2.5 shows that the resulting upper ocean flows for depths between 0 and 500 m have a gyre-like structure that is particularly marked in the central part of the ocean basins in both hemispheres of the Atlantic and Pacific Oceans, as well as the southern hemisphere of the Indian Ocean. One particularly interesting and unexpected aspect of these gyres is that the western boundary flows are extremely narrow and intense. This westward intensification of the western boundary cur-

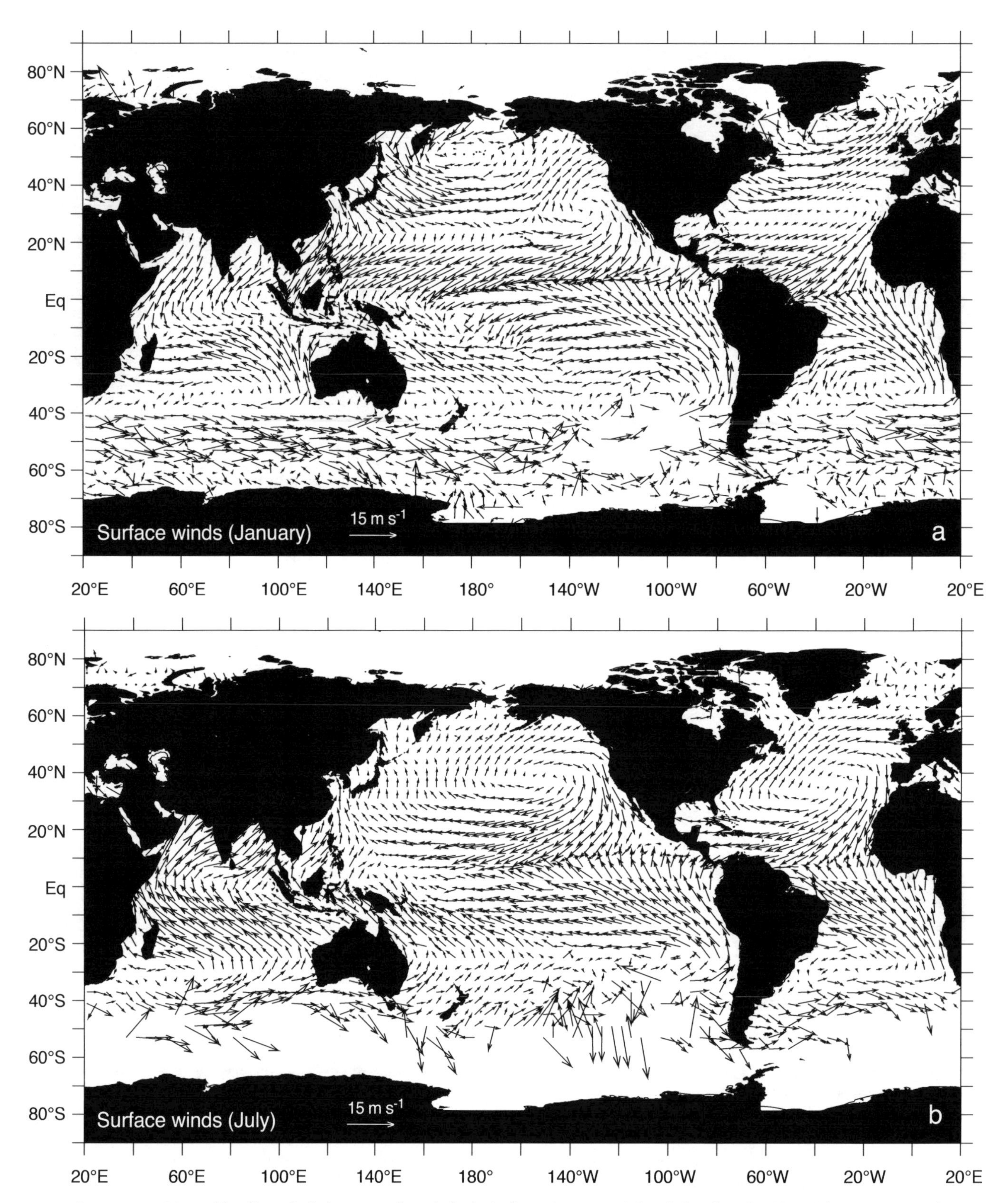

FIGURE 2.2.1: Maps of the climatological mean surface wind velocity for (a) January and (b) July based on the COADS data set [*Deser et al.*, 1996]

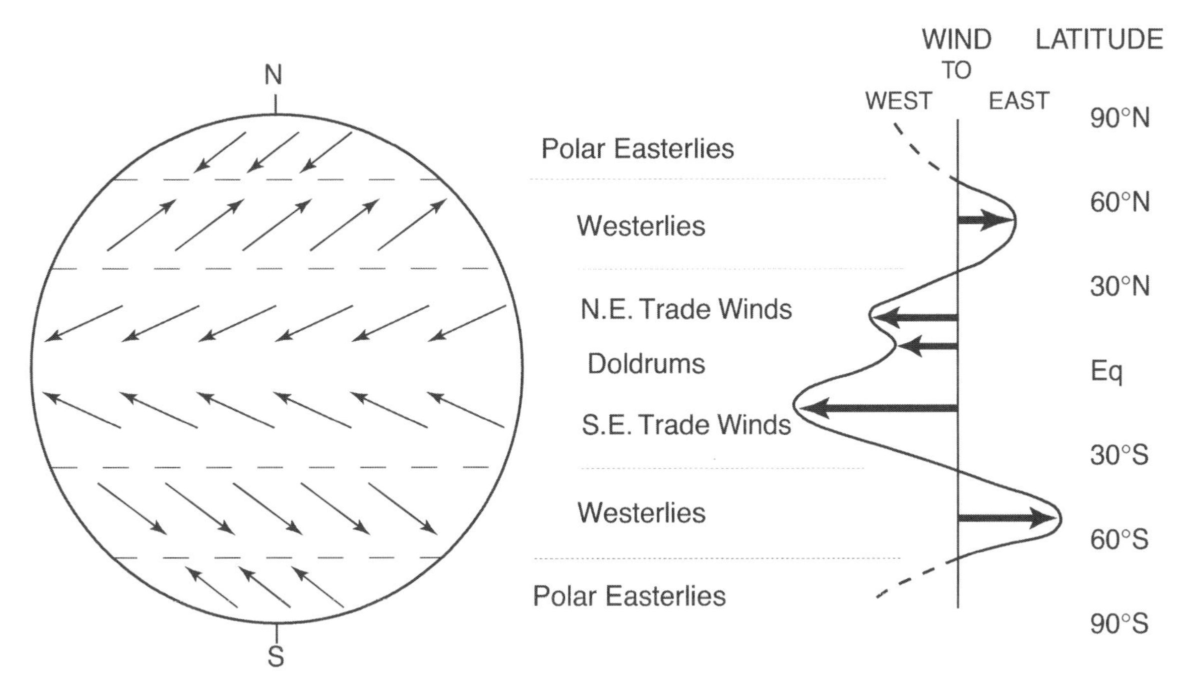

FIGURE 2.2.2: Schematic illustration of wind patterns [*Pond and Pickard*, 1983].

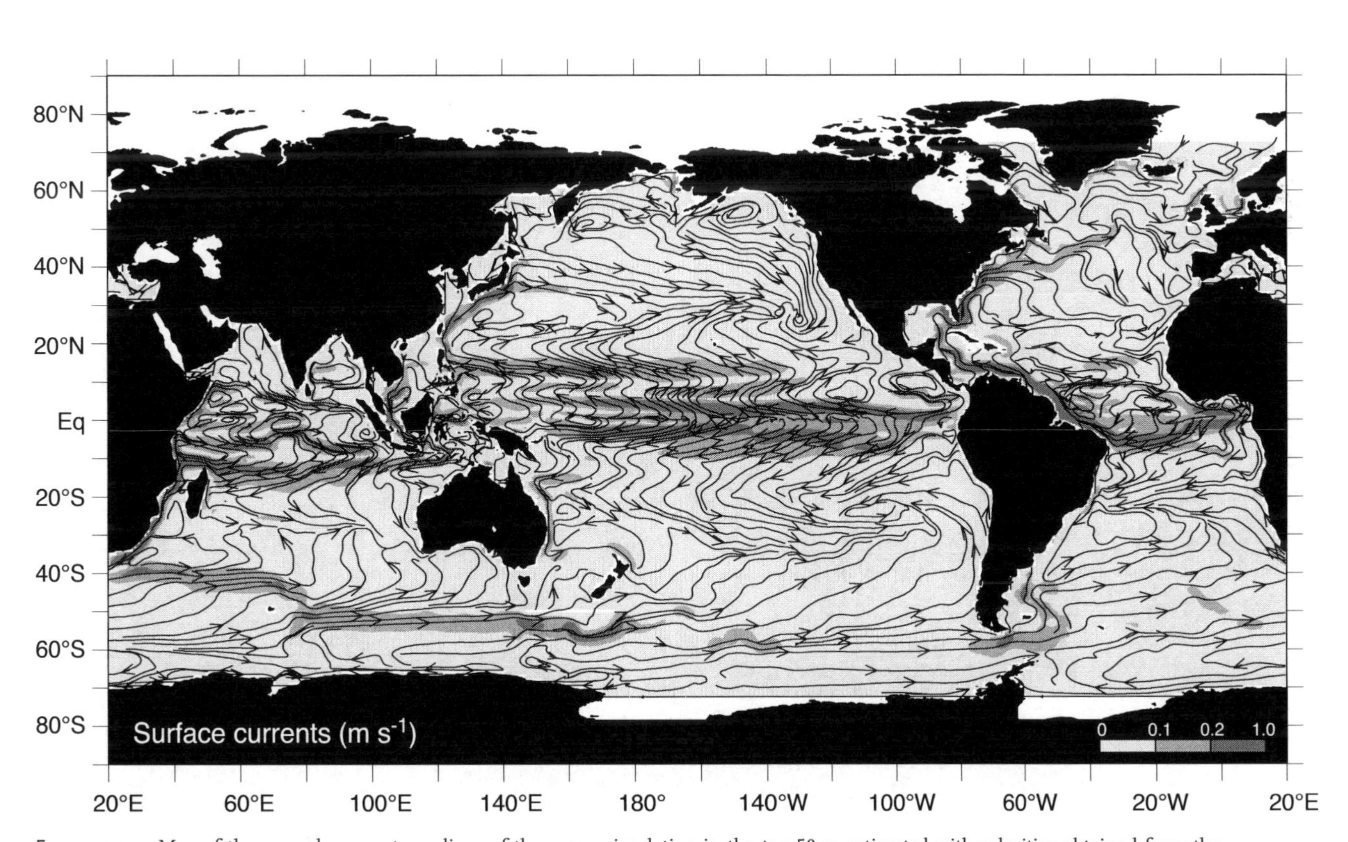

FIGURE 2.2.3: Map of the annual mean streamlines of the ocean circulation in the top 50 m estimated with velocities obtained from the Estimating the Circulation and Climate of the Ocean (ECCO) project [*Stammer et al.*, 2002]. A streamline is a line that is parallel to the velocity field. The shading indicates the mean velocity in m s^{-1} as indicated by the bar at the bottom, with relatively higher velocities shaded a darker color.

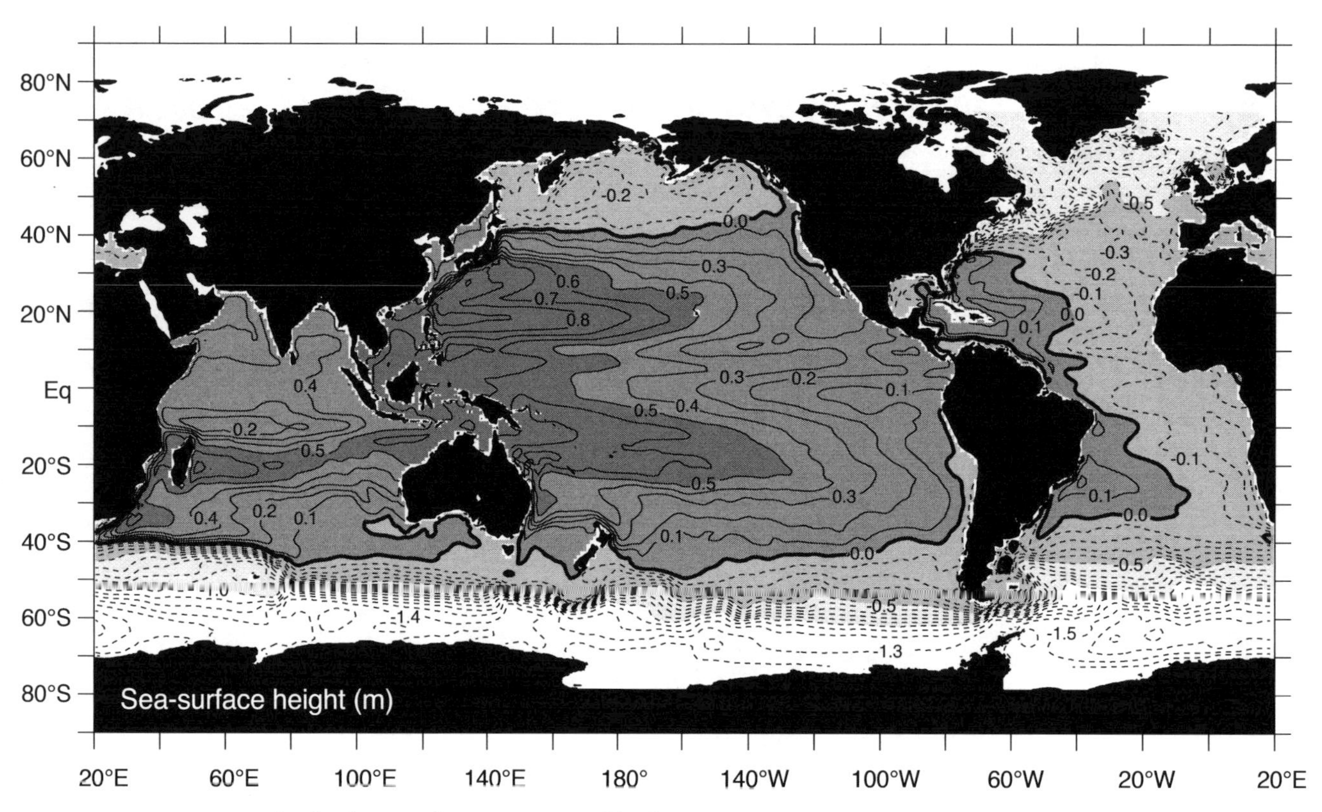

FIGURE 2.2.4: Sea-surface height anomaly in m as estimated by the ECCO project [*Stammer et al.*, 2002].

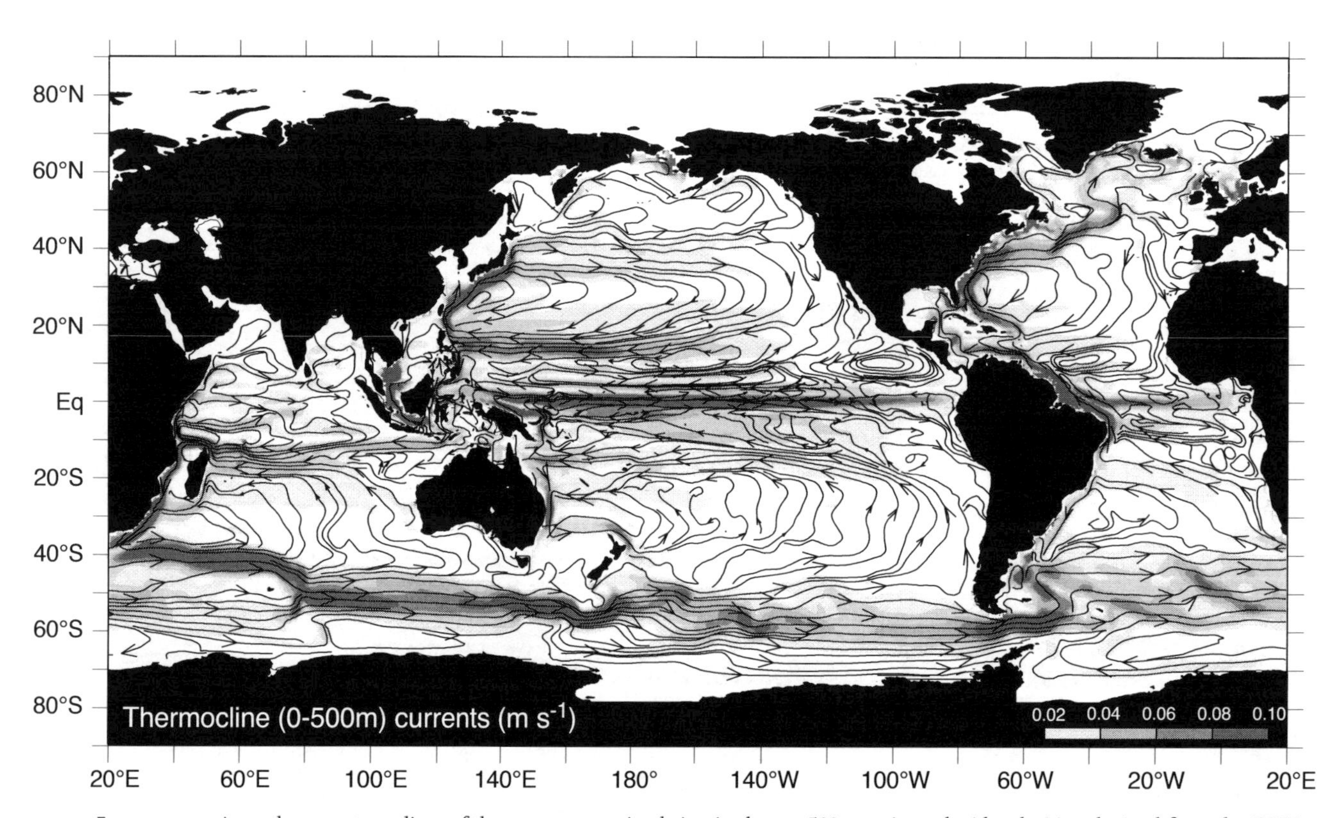

FIGURE 2.2.5: Annual mean streamlines of the mean ocean circulation in the top 500 m estimated with velocities obtained from the ECCO project [*Stammer et al.*, 2002]. The shading indicates the mean velocity in m s^{-1} as indicated by the bar at the bottom, with relatively higher velocities shaded a darker color.

rents is also due to the Earth's rotation. Note also the large sea surface height slope away from Antarctica and the resulting large circumpolar current.

In what follows, we briefly introduce the equations of motion and derive a few important analytical solutions that provide powerful insights into the major circulation patterns we have described. The emphasis is on providing a set of simple tools that are particularly useful for determining the nature of the ocean circulation that one might expect under a particular set of circumstances; and on those aspects of the circulation, such as upwelling driven by Ekman transport, that have the largest impact on biogeochemical processes.

Equations of Motion

The equations of motion are based on Newton's second law $F = ma$, where F is force (in Newtons), m is mass in kg, and a is acceleration is m s^{-2}. We use for m the volume V times the density ρ. For a full derivation of the equations of motion, the reader is referred to a standard physical oceanography textbook, such as *Pond and Pickard* [1983]. All of the solutions we consider here assume that the flow is hydrostatic (i.e., that the vertical pressure gradient is balanced by gravity) and incompressible (see panel 2.1.3). We further assume that the forces driving the ocean circulation are in balance such that there is no net acceleration a, i.e., $F/(\rho \cdot V) = 0$. However, it should be noted that the inclusion of the net acceleration term in the equations allows for a rich time-dependent array of oceanic phenomena, including waves with scales ranging from capillary to planetary, turbulent motions, formation of eddies, and convective overturning. Observations of some of these processes are discussed in section 2.5.

The forces we consider here, all expressed as accelerations $F/(\rho \cdot V)$, include the pressure gradient acceleration, the acceleration due to the earth's rotation, which is referred to as the *Coriolis acceleration*, the effect of the wind stress acting on the surface of the ocean, and the gravitational acceleration g:

$$a_x = 0 = -\overbrace{\frac{1}{\rho}\frac{\partial p}{\partial x}}^{\text{Pressure Gradient}} + \overbrace{fv}^{\text{Coriolis}} + \overbrace{\frac{1}{\rho}\frac{\partial \tau^x}{\partial z}}^{\text{Wind Stress Effect}} \tag{2.2.1a}$$

$$a_y = 0 = -\overbrace{\frac{1}{\rho}\frac{\partial p}{\partial y}}^{\text{Pressure Gradient}} - \overbrace{fu}^{\text{Coriolis}} + \overbrace{\frac{1}{\rho}\frac{\partial \tau^y}{\partial z}}^{\text{Wind Stress Effect}} \tag{2.2.1b}$$

$$a_z = 0 = -\overbrace{\frac{1}{\rho}\frac{\partial p}{\partial z}}^{\text{Pressure Gradient}} - \overbrace{g}^{\text{Gravity}} \tag{2.2.1c}$$

The first of these equations is the balance of acceleration terms in the x-direction. The second and third equations are the balances of the accelerations in the y-and z-directions, respectively. The vertical velocity w does not enter into these simplified versions of the equations. We can solve for it by using the mass conservation equation for incompressible flow (equation (12) of panel 2.1.3). And now we describe each of the terms:

1. The first term on the right-hand side of all three equations is the acceleration due to the pressure gradient force. Pressure has units of Newtons m^{-2}. If pressure increases as x, y, or z increases, the pressure gradient force will induce a flow in the opposite direction.

2. The second term on the right-hand side of (2.2.1a) and (2.2.1b) is the acceleration due to the Coriolis force. Note that the Coriolis acceleration in the x-direction (equation (2.2.1a)) depends on v, the velocity in the y direction. Conversely, the Coriolis acceleration in the y-direction (equation (2.2.1.b)) depends on the x-direction velocity u. In other words, the acceleration is always at right angles to the direction of the flow. In effect, this means that the only way the Coriolis force can balance a pressure gradient or any other force is if the velocity is at right angles to that other force. The Coriolis acceleration is derived by considering the motion in a rotating frame of reference relative to one fixed with respect to the stars. After neglecting minor terms, the Coriolis acceleration turns out to have the simple form given in the equations with a magnitude equal to the velocity times the *Coriolis parameter*:

$$f = 2 \cdot \Omega \sin\theta \tag{2.2.2}$$

where f has units of s^{-1}. Here θ is the latitude ($-\pi/2$ radians at the South Pole, 0 radians at the equator, and $\pi/2$ radians at the North Pole), and $\Omega = 7.3 \times 10^{-5}$ s^{-1} is the angular velocity of the earth in radians per second. The sign of f determines the direction of the Coriolis acceleration. In the northern hemisphere, where f is positive, the Coriolis acceleration is to the right of the flow. In the southern hemisphere, f is negative and the Coriolis acceleration is to the left of the flow. The Coriolis parameter, and thus the Coriolis acceleration, is greatest at the poles and decreases to 0 at the equator.

3. The third term on the right-hand side of (2.2.1a) and (2.2.1b) is the acceleration due to the direct influence of the winds in the top 10 to 100 m of the water column. The force of the wind on the ocean is generally represented as a horizontal stress τ_0 (Newtons m^{-2}) acting on a horizontal surface at the top of the ocean. One way of estimating the wind stress is as a function of the wind speed U as measured at a nominal height of 10 m on board a ship. In units of kg m^{-1} s^{-2}, a typical empirical relationship would be $\tau_0 = C_D \cdot \rho \cdot U^2$, where the dimensionless drag coefficient C_D would vary from 10^{-3} to 4×10^{-3} as the wind speed

increased from $2\,\mathrm{m\,s^{-1}}$ to $50\,\mathrm{m\,s^{-1}}$. Satellite measurements of microwave backscatter from wavelets are also being used to map wind stress. The surface wind stress propagates vertically into the interior of the ocean through frictional coupling of deeper layers of water by the shallower layers set directly in motion by the wind. The force resulting from this stress is equal to the vertical gradient of the stress times the volume V. Dividing this force by the mass $\rho \cdot V$ gives the acceleration expression shown in (2.2.1a) and (2.2.1b).

Ekman Transport

In the surface waters that are under the direct frictional influence of the wind, the pressure gradient forces are small relative to the Coriolis and wind stress accelerations, and can thus be neglected. The Ekman transport solution for horizontal motion in this surface layer thus results from the balance of the Coriolis acceleration

Panel 2.1.3: Mass Conservation Equation and Incompressible Flow

Consider the flow of water through a cube of fixed volume such as that illustrated in figure 2.1.1. The mass m of water in the cube is:

$$\begin{aligned} m &= \Delta x \Delta y \Delta z \rho \\ &= V\rho \end{aligned} \tag{1}$$

where ρ is density. The change of the mass due to a flow in the x-direction is:

$$\frac{\partial m}{\partial t} = V\frac{\partial \rho}{\partial t} = \rho_1 u_1 \Delta y \Delta z - \rho_2 u_2 \Delta y \Delta z \tag{2}$$

where $\rho_1 u_1 \Delta y \Delta z$ is the input from the left side of the cube, $\rho_2 u_2 \Delta y \Delta z$ is the outflow through the right side. Assume that the density and velocity change continuously across the cube such that

$$\begin{aligned} u_2 &= u_1 + \frac{\partial u}{\partial x}\Delta x \\ \rho_2 &= \rho_1 + \frac{\partial \rho}{\partial x}\Delta x \end{aligned} \tag{3}$$

Substituting (3) into (2) gives

$$\begin{aligned} V\frac{\partial \rho}{\partial t} \\ = \Delta y \Delta z\left[\rho_1 u_1 - \left(\rho_1 u_1 + \Delta x \rho_1 \frac{\partial u}{\partial x} + \Delta x u_1 \frac{\partial \rho}{\partial x} + \Delta x^2 \frac{\partial u}{\partial x}\frac{\partial \rho}{\partial x}\right)\right] \end{aligned} \tag{4}$$

which simplifies to:

$$\frac{\partial \rho}{\partial t} = -\rho_1 \frac{\partial u}{\partial x} - u_1 \frac{\partial \rho}{\partial x} - \Delta x \frac{\partial u}{\partial x}\frac{\partial \rho}{\partial x} \tag{5}$$

As in panel 2.1.1, we drop the final term on the right-hand side due to its small size relative to the other terms. We make use of the product rule to collapse the first two terms in the parentheses into a single term. This gives:

$$\frac{\partial \rho}{\partial t} = -\frac{\partial(\rho u)}{\partial x} \tag{6}$$

Similar terms can be derived for the y- and z-directions, giving:

$$\frac{\partial \rho}{\partial t} = -\frac{\partial(\rho u)}{\partial x} - \frac{\partial(\rho v)}{\partial y} - \frac{\partial(\rho w)}{\partial z} \tag{7}$$

We can expand this expression by the product rule to give:

$$\frac{\partial \rho}{\partial t} = -\rho\left(\frac{\partial u}{\partial x} + \frac{\partial v}{\partial y} + \frac{\partial w}{\partial z}\right) - \left(u\frac{\partial \rho}{\partial x} + v\frac{\partial \rho}{\partial y} + w\frac{\partial \rho}{\partial z}\right) \tag{8}$$

and make use of the definition of the total derivative:

$$\frac{d(\,)}{dt} \equiv \frac{\partial(\,)}{\partial t} + u\frac{\partial(\,)}{\partial x} + v\frac{\partial(\,)}{\partial y} + w\frac{\partial(\,)}{\partial z} \tag{9}$$

to obtain the following form of the mass conservation equation:

$$\frac{d\rho}{dt} = -\rho\left(\frac{\partial u}{\partial x} + \frac{\partial v}{\partial y} + \frac{\partial w}{\partial z}\right) \tag{10}$$

In dealing with tracers, we generally assume that seawater is incompressible. In fact, of course, it is not. Density increases by ~2% as the pressure increases from 1 atmosphere to the pressures of >400 atmospheres that are typical of the abyss. However, the effect of density variations due to compressibility is generally small compared to other tracer processes. An incompressible fluid is one in which the total derivative of density is equal to zero:

$$\frac{d\rho}{dt} = \frac{\partial \rho}{\partial t} + u\frac{\partial \rho}{\partial x} + v\frac{\partial \rho}{\partial y} + w\frac{\partial \rho}{\partial z} = 0 \tag{11}$$

From (10) we see that this condition implies that

$$\frac{\partial u}{\partial x} + \frac{\partial v}{\partial y} + \frac{\partial w}{\partial z} = 0 \tag{12}$$

Note that we do not need to consider diffusion of mass in the mass conservation equation, since, by definition, all net transfer of mass is included in the advection velocity terms.

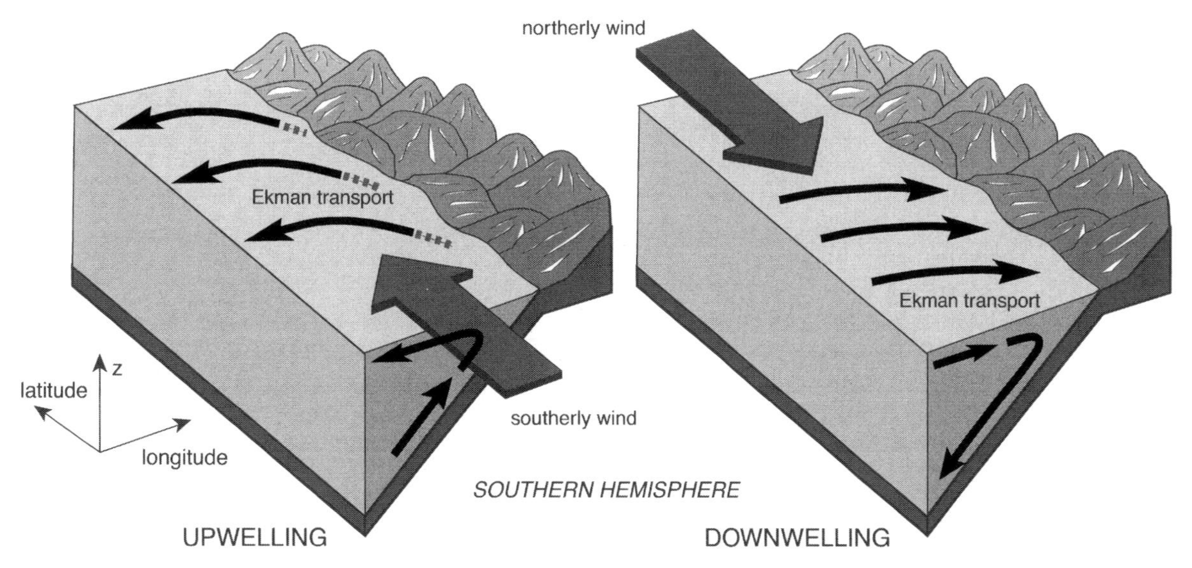

FIGURE 2.2.6: Ekman transport and associated upwelling and downwelling resulting from wind blowing parallel to shore [*Thurman*, 1990]

with the acceleration due to the wind stress. From (2.2.1a) and (2.2.1.b) we have:

$$
\begin{aligned}
0 &= fv + \frac{1}{\rho}\frac{\partial \tau^x}{\partial z} \\
0 &= -fu + \frac{1}{\rho}\frac{\partial \tau^y}{\partial z}
\end{aligned}
\qquad (2.2.3)
$$

The direct frictional influence of the wind decreases very rapidly with depth so that it is already negligible by about 10 to 100 m depth. The details of what happens within the zone of frictional influence are not important for the problems we will consider here, so we integrate (2.2.3) vertically from a depth $-h$, where the stress due to the wind is 0, to the surface, where the stress due to the wind is τ_0. We first multiply (2.2.3) by the density. We then separately integrate the wind stress acceleration term:

$$
\begin{aligned}
\int_{-h}^{0} \frac{\partial \tau^x}{\partial z} dz &= \tau_0^x \\
\int_{-h}^{0} \frac{\partial \tau^y}{\partial z} dz &= \tau_0^y
\end{aligned}
\qquad (2.2.4)
$$

where h is typically of order 10 to 100 m, and define an Ekman mass transport M^{Ek} as the vertical integral of the density times the velocity to the depth of frictional influence:

$$
\begin{aligned}
M_x^{Ek} &= \int_{-h}^{0} (\rho u) dz \\
M_y^{Ek} &= \int_{-h}^{0} (\rho v) dz
\end{aligned}
\qquad (2.2.5)
$$

Note that this mass transport has units of kg m^{-1} s^{-1}. It is the mass transport of water that occurs every second across a 1 m wide section of the water column. The final expression we obtain for the Ekman transport is

$$
\begin{aligned}
M_x^{Ek} &= \tau_0^y / f \\
M_y^{Ek} &= -\tau_0^x / f
\end{aligned}
\qquad (2.2.6)
$$

From these equations we see that the vertically integrated Ekman transport is always 90° to the right of the wind stress in the northern hemisphere, where the Coriolis parameter f is positive, and 90° to the left of the wind stress in the southern hemisphere, where f is negative.

Mass conservation requires that any water that is displaced by the Ekman transport must be replaced either by lateral transport from elsewhere or by water from below in regions of divergence (Ekman suction). In regions of convergent Ekman transport, water must be pumped downwards (Ekman pumping). There are three basic mechanisms by which the Ekman transport can lead to upwelling and downwelling. The first is by winds that have a component parallel to the coastline. If the wind blows in such a way as to drive Ekman transport away from the coast, there will be upwelling of deep waters (left panel of figure 2.2.6). Conversely, if water is driven toward the coast by Ekman transport, there will be downwelling (right panel of figure 2.2.6). Examination of the wind patterns in figure 2.2.1 show that the western margins of continents typically have conditions that favor upwelling (e.g., off California and Central America, off the coast of Perú, and off northwest and southeast Africa). This is confirmed by the circulation patterns shown in figure 2.2.3 and by the sea-surface temperature pattern of figure 2.2.7a, which shows cold waters offshore from all the upwelling regions. These are all areas of extremely high biological productivity, as we shall see in chapter 4.

The second mechanism by which Ekman transport can give rise to upwelling is the blowing of wind along the equator. Here, the trade winds have a strong eastward component both north and south of the equator (figure 2.2.1). Because of the change in sign of the Coriolis acceleration, these winds lead to northward Ekman transport north of the equator and southward Ekman transport south of the equator (figures 2.2.3 and 2.2.8a). Mass balance requires upwelling along the equator. Both the Atlantic and Pacific Oceans have a large amount of upwelling due to this phenomenon. The Indian Ocean situation is more complex because of the large seasonal changes in the winds associated with the monsoonal circulation of the Indian subcontinent.

The third mechanism by which Ekman transport can cause upwelling or downwelling is as a consequence of the large-scale pattern of the winds in the open ocean. Consider the schematic wind stress pattern of figure 2.2.2. Between about 20°N and 50°N, in the region referred to as the subtropics, the winds shift from westward moving to eastward moving. In the southern part of this zone, the westward-moving trade winds will drive an Ekman transport to the north, whereas the eastward-moving westerlies in the northern part of the subtropics will drive an Ekman transport to the south. The magnitude of the flow is largest where the wind stress is at a maximum, which occurs at the edges of the subtropics. In the middle of the gyre, the horizontal flow goes to 0. There will therefore be a convergent flow and downwelling in the central part of the subtropics (figures 2.2.3 and 2.2.8b). Applying the same considerations to other regions of the schematic wind fields of figure 2.2.2 gives upwelling in the subpolar region that lies to the north of the subtropics.

One can obtain a solution for the vertical velocity at the base of the Ekman layer at a depth $-h$ by vertically integrating the mass continuity equation (equation (12) of panel 2.1.3) to the depth $-h$, first assuming a steady state. This gives

$$0 = \frac{\partial M_x}{\partial x} + \frac{\partial M_y}{\partial y} + \rho \cdot (w_{z=0} - w_{z=-h}) \quad (2.2.7)$$

We assume that the vertical velocity is 0 at the surface of the ocean and that the mass transport terms in (2.2.7) are equal to the Ekman transport defined by (2.2.6). This gives the following solution for the velocity at the base of the Ekman layer:

$$w_{z=-h} = \frac{1}{\rho} \cdot \left[\frac{\partial}{\partial x}\left(\frac{\tau_0^y}{f}\right) - \frac{\partial}{\partial y}\left(\frac{\tau_0^x}{f}\right)\right] = \frac{1}{\rho} \cdot curl_z\left(\frac{\tau_0}{f}\right) \quad (2.2.8)$$

where $curl_z$ is defined as $[\partial(\ )/\partial x - \partial(\ )/\partial y]$. Consider again just the east-west component of the wind stress for the 20°N to 50°N subtropical region depicted in figure 2.2.2. Equation (2.2.8) gives us for the downwelling velocity $w_{z=-h} = -(1/\rho) \cdot [\partial(\tau_0^x/f)/\partial y]$. The wind stress increases towards the north at a greater rate than the Coriolis parameter. This equation therefore predicts that w will be negative, i.e., downwelling, throughout the entire subtropics except where the wind stress curl is 0, which occurs at the north and south edges. The maximum downwelling will be in the middle of the gyre, where the wind stress curl is at a maximum. Figure 2.2.8b gives a schematic illustration of this.

Applying (2.2.8) to other regions of the schematic wind fields of figure 2.2.2 gives upwelling in the subpolar region as well as the tropics. The large-scale patterns of Ekman pumping and Ekman suction have a dramatic impact on many biogeochemical processes. The lower latitudes of the subtropics, for example, are generally almost devoid of surface nutrients because there is no upwelling (and little mixing) to supply them from below. Consequently, they have substantially less biological activity than upwelling regions. As already noted, the equatorial region and many coastal upwelling zones are rich in biological activity driven by the high supply of nutrients from below. The subpolar gyres are also rich in nutrients and biological activity, but other processes such as light supply and deep winter mixing play a role in these regions; these will be discussed in chapter 4.

Gyre Circulation

As we have noted, the Ekman transport accumulates water in the subtropics and removes it from the subpolar and equatorial regions (figures 2.2.3 and 2.2.4). In addition to causing downwelling, the accumulation of water into an elevated mound within the subtropics causes a pressure gradient force directed down the slope of the mound. If the Earth did not rotate, water would accelerate in the direction of this pressure gradient. However, because of the Earth's rotation and resulting Coriolis force, the water that is set into motion by this pressure gradient force gets deflected 90° to the right in the northern hemisphere. The only way that the Coriolis force can balance the pressure gradient force is if the water flow is at right angles to the slope of the mound. The Coriolis force leads to a clockwise circulation around the center of the high-

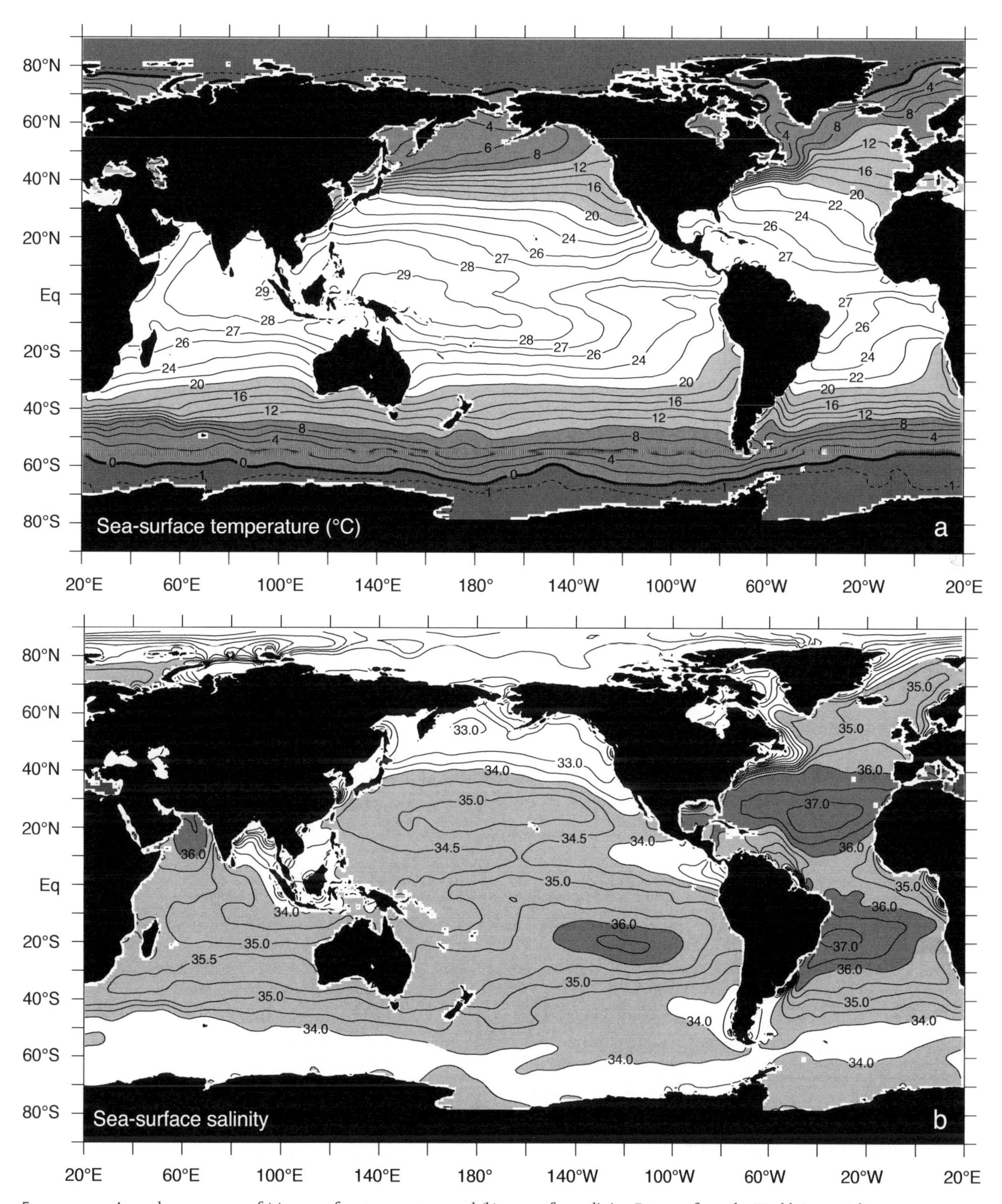

FIGURE 2.2.7: Annual mean maps of (a) sea-surface temperature, and (b) sea-surface salinity. Data are from the World Ocean Atlas 2001 [*Conkright et al.*, 2002]. See also color plate 1.

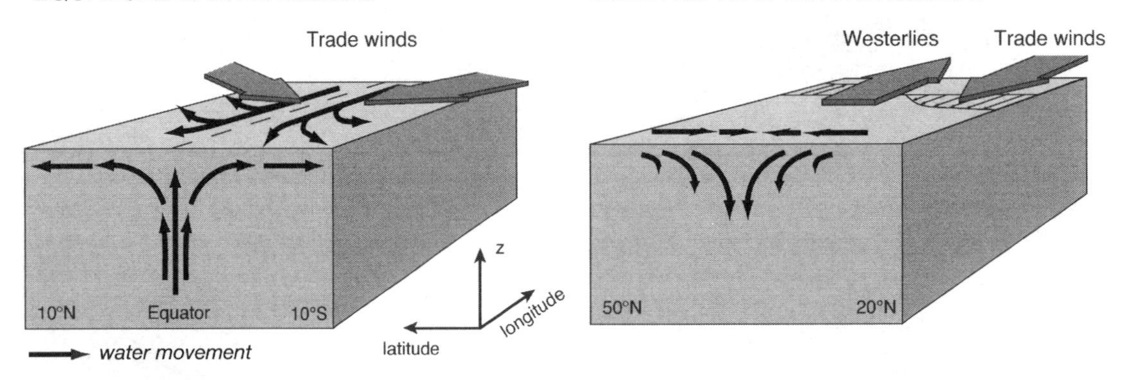

FIGURE 2.2.8: (a) Equatorial divergence and upwelling resulting from Ekman transport driven by trade winds [*Thurman*, 1990]. (b) Ekman downwelling resulting from convergent Ekman transport driven by the westerlies and trade winds.

pressure mound in the subtropical gyre regions of the northern hemisphere, and a counterclockwise rotation around the corresponding subtropical mound in the southern hemisphere (figure 2.2.5). Such a circulation around a high-pressure region is referred to as *anticyclonic*.

In the subpolar regions, the juxtaposition of the westerlies and the polar easterlies causes a divergent Ekman flow and the generation of a low-pressure region in between. The resulting pressure gradient force would lead to acceleration toward the interior of this low-pressure field, but the Coriolis force again deflects this flow 90° to the right in the northern hemisphere and to the left in the southern hemisphere. The net effect is a cyclonic circulation (opposite to anticyclonic) around the low-pressure depression of the surface.

Over most of the gyres that we have discussed, the Coriolis force roughly balances the pressure gradient force induced by the wind. The equations formed by the balance of these two terms are known as the *geostrophic equations*. We shall return to a full discussion of the geostrophic equations *per se* in section 2.3. If we add to the geostrophic equations the wind stress forcing at the top of the ocean in (2.2.1a) and (2.2.1b), and then vertically integrate the resulting equations from the ocean floor to the surface, we obtain a solution that is referred to as the *Sverdrup transport* [cf. *Pond and Pickard*, 1983]. The Sverdrup transport solution is able to predict the broad meridional (i.e., north-south) currents on the eastern sides of the gyres but not the narrow intense currents on the western side of the gyres shown in figure 2.2.5. The Gulf Stream off the coast of North America and the Kuroshio off the coast of Japan are two well-known examples of such energetic western boundary currents. The reason for the existence of these intense western boundary currents was not understood for a long time, until Stommel provided a very simple and elegant solution in 1948 [*Stommel*, 1948].

Stommel's crucial insight was to understand the central role of angular momentum or *vorticity* conservation in determining the overall pattern of the gyre circulation. The pressure gradient, Coriolis, and wind stress forces together are able to conserve vorticity with an equatorward flow of water in the subtropical gyre, but poleward movement is not possible without violating this constraint. The only way that poleward movement is possible is by removing vorticity from the flow. Stommel was able to demonstrate with a simple analytical solution of the equations of motion that frictional forces acting on narrow currents with high velocity shear provide a way of removing vorticity. A central feature of his analysis is the fundamental role played by the fact that the Coriolis parameter varies with latitude. If the Coriolis parameter is set equal to zero or to a constant, in other words, if it is not allowed to vary with latitude, Stommel's analytical model gives a symmetric flow regime, with the poleward flow being as broad as the equatorward flow. Stommel gives an excellent discussion of why this is so in his book on the Gulf Stream [*Stommel*, 1965]. It is important to note that there are alternative theories for the existence of narrow and intense western boundary currents, but all have in common the importance of meridional variations in the Coriolis force. The interested reader is referred to more advanced texts.

Figure 2.2.9 shows an idealized summary of the wind-driven circulation based on a model similar to Stommel's but with a different expression for the friction term (lateral rather than bottom friction [*Munk and Carrier*, 1950]). In the northern hemisphere, subpolar gyres are set up between the westerlies and the polar easterlies. At mid-latitudes, anticyclonic subtropical gyres are dominant, with broad interior circulation and strong western boundary currents. The equatorial circulation set up by the winds has a westward-flowing Equatorial Current in both hemispheres and an Equatorial Countercurrent at the equator. A comparison of this idealized circulation

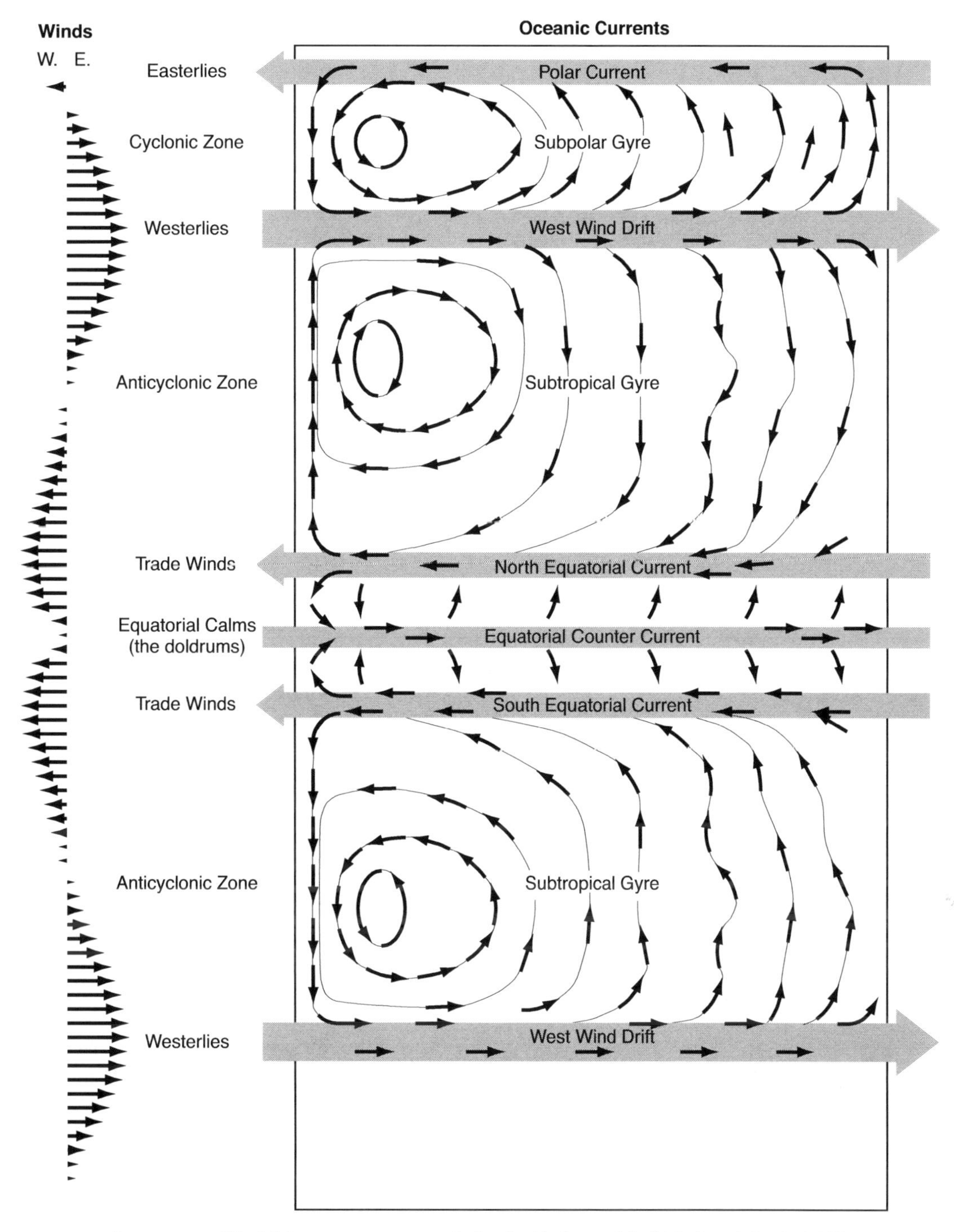

FIGURE 2.2.9: Wind-driven gyre transport calculated with the model of *Munk and Carrier* [1950].

pattern with the actual circulation patterns illustrated in figure 2.2.5 shows that this theory provides valuable insights into the basic structure of the gyre circulation of the ocean.

2.3 Wind-Driven Circulation in the Stratified Ocean

The discussion in section 2.2 ignored variations in the density of the ocean. The stratification of the water column, which is due to vertical variations in the ocean's temperature and salinity, has a dramatic impact on the wind-driven circulation. As we shall see, there is a strong tendency for stratification to trap the entire wind-driven

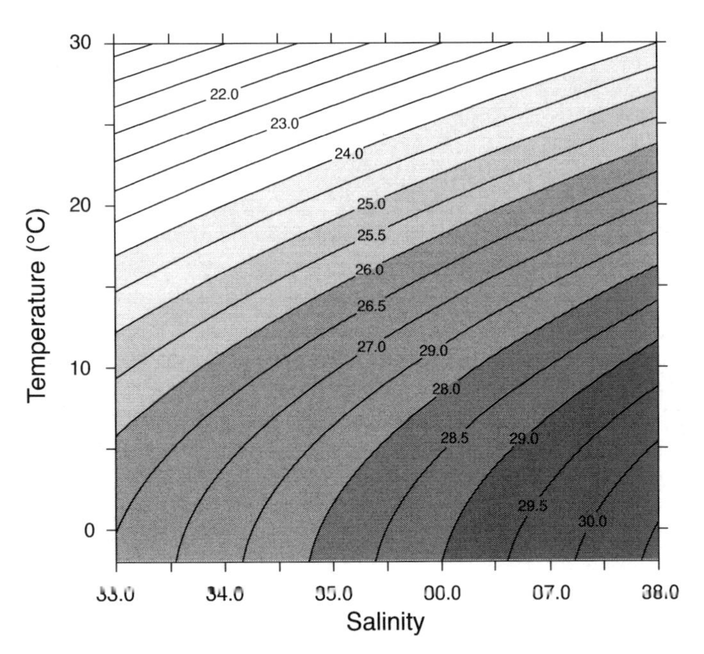

FIGURE 2.3.1: Potential density at the surface of the ocean as a function of temperature and salinity.

gyre transport in the surface ocean. There are two basic concepts that will help us understand why this is so and how the ocean is able to escape this constraint: (1) currents in the interior of the ocean are strongly constrained to occur along surfaces of constant density, and (2) the geostrophic equations need to be satisfied by the flow along the layers defined by the surfaces of constant density. In the following subsection, we discuss both of these in turn before applying them in the subsequent subsection to understand the gyre transport.

BASIC CONCEPTS

OCEAN STRATIFICATION

Density is a function of temperature T, salinity S (defined in table 1.2.1), and pressure obeying the equation of state $\rho = f(T, S, p)$. Oceanographers always report density as $\sigma = (\rho - 1000)$, so that, for example, a density of $\rho = 1027\ \mathrm{kg\ m^{-3}}$ becomes $\sigma = 27$. Figure 2.3.1 shows how density at the surface of the ocean varies as a function of temperature and salinity. The full equation of state, given by *Fofonoff* [1985], is quite complex. A simple rule of thumb is that σ increases by ~ 1 when temperature increases by 5°C and when salinity increases by 1. Because water is compressible, the density also increases as the pressure increases, at a rate of about 1 sigma unit per 200 db (about 200 m depth).

The density of a water parcel at the depth where it is located is referred to as the *in situ density*. We are commonly interested in identifying water parcels that would have the same density if they were at the same pressure. We thus find it convenient to define a *potential density*, σ_θ, as the density a water parcel would have if moved adiabatically and while conserving salinity to a reference pressure, commonly 1 atm at the surface of the ocean. Surfaces of constant potential density are referred to as *isopycnal surfaces* or *isopycnals*. In addition to considering the flow along isopycnal surfaces, we will often find it useful to consider the distribution of deep ocean properties along such surfaces.

While useful, the potential density concept suffers from the fact that two water types that have the same density at one reference pressure, e.g., at a pressure of 4000 db, but a different temperature and salinity, will most likely have a different density if referenced to another depth, e.g., the surface. This is because the compressibility of seawater varies with both temperature and salinity. It is thus generally not clear exactly how to define an isopycnal surface along which a water parcel will actually move if a significant change in depth is involved. *Reid and Lynn* [1971] addressed this problem by using a reference pressure as close as possible to the depth range of the water mass that is being examined. For example, in examining deep ocean observations, they used a reference pressure of 4000 db (about 4000 m). The potential density referenced to 4000 db is referred to as σ_4. The difficulty of spanning a broad depth range in the analysis of a particular water type has led to the introduction of the idea of a *neutral surface* [*McDougall*, 1987], which in effect continually readjusts the reference depth for calculation of the potential density of adjacent locations so as to make it as close as possible to the actual depth of the water type being considered at the two locations.

Most of the ocean stratification is due to temperature, and in what follows we will often use *potential temperature* distributions to depict the main features of the oceanic stratification. By analogy with potential density, the potential temperature, for which we use the symbol θ, is defined as the temperature a water parcel would have if it were brought adiabatically to the surface. Equations for calculating it are given by *Fofonoff* [1977, 1978]. Surfaces of constant potential temperature are referred to as *isothermal surfaces*.

Figure 2.3.2a gives a schematic illustration of the vertical profile of potential temperature at several locations. Typically, one finds that the deep ocean has relatively uniform temperatures below a depth of 1 or 2 km down to the floor at a depth of 5 or 6 km. Above this, the vertical temperature profile depends on the region of the ocean. Cold high-latitude regions have temperatures similar to the deep ocean. The mid-latitudes and tropics have a steep temperature gradient beginning at a depth of ~ 1 km and extending all the way to a *mixed layer* of uniform temperature at the surface that is typically of order 50 to 100 meters thick. The mixed layer depth is determined by the direct frictional forcing of the wind, which only penetrates to

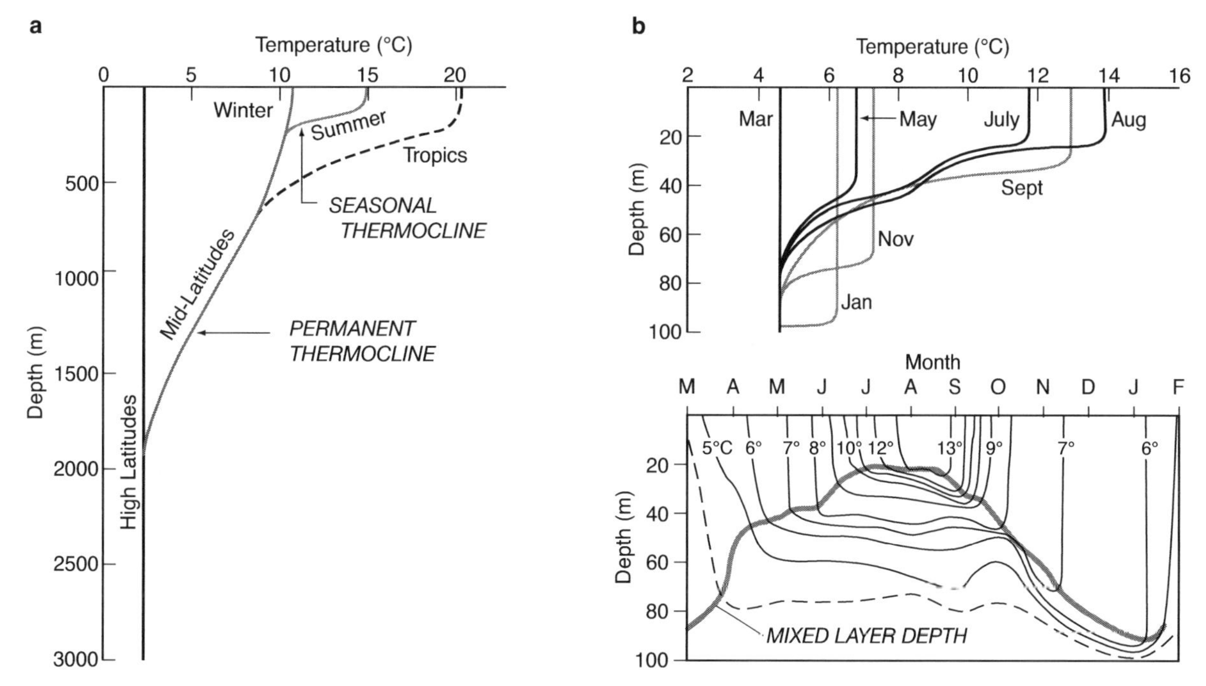

FIGURE 2.3.2: (a) Schematic illustration of terminology used to describe various features of the thermocline [*Knauss*, 1997]. (b) Seasonal behavior of the mixed layer at 50°N, 145°W in the eastern North Pacific [*Pickard and Emery*, 1990].

depths of approximately 10 to 100 m, as well as vertical overturning caused by densification of surface waters by surface cooling, an increase of salinity due to evaporation, or lateral inflow of dense waters at the surface. The mixed layer depth varies with season (figure 2.3.2b) and can achieve depths of several hundred meters in the high latitudes, where surface cooling destabilizes the water column in wintertime. As previously noted, the region of steep temperature gradients is known as the thermocline. The corresponding density gradient region is referred to as the *pycnocline*. Gradient regions in salinity, which do not always correspond to those in temperature, are referred to as *haloclines*. The thermocline is further divided into the *main* or *permanent* thermocline, which is below the reach of seasonal fluctuations at the surface of the ocean, and the *seasonal* thermocline nearer the surface, which becomes part of the surface mixed layer during the winter.

Figure 2.3.3b shows a section of observed potential temperature that begins on the left-hand side with the North Atlantic, then goes south to the Southern Ocean, the ocean that rings the Antarctic, before turning back north into the Pacific. The contours are lines of constant potential temperature. The data are from the WOCE tracks shown in figure 2.3.3a. The sections show that the main thermocline within the subtropical gyre has a basin-like convex downward structure confined to each hemisphere. The basin-like structure is particularly striking in the North Atlantic. Cold waters bound the subtropical gyre basins to both the north and south. On the poleward side, the cold waters rise all the way to the surface and the high-latitude thermocline is very weak or essentially nonexistent. In the equatorial regions, the surface waters have very high temperatures, but the thermocline is much thinner, a few hundred meters, than in the subtropical regions to the north and south, where the thermocline has a thickness in excess of a thousand meters. The vertical temperature gradient is much greater in the equatorial region than elsewhere. The places where isothermal surfaces encounter the surface of the ocean are referred to as *outcrops*. The outcropping of the isothermal surfaces (and thus also of the isopycnal surfaces) allows for penetration of water from the surface into the interior of the thermocline, referred to as *ventilation*. The ventilation of the thermocline from the outcrop regions will play a central role in the discussion of the gyre circulation that is to follow.

The ocean circulation in the interior of the ocean is strongly favored to occur along isopycnal surfaces. This is because of stratification: any water parcel that is forced away from an isopycnal to deeper waters of greater density or shallower waters of lower density will have a strong restoring force due to buoyancy that will push the water back to its original isopycnal surface. The only way to escape this constraint is to alter the density of the water parcel. One way to alter the density of water is at the surface by adding or removing heat through solar forcing and latent and sensible heat transfer, or by changing the salinity by exchanging water with the atmosphere. The surface temperature and salinity shown in figures 2.2.7a and b reflect the effect of these surface processes as well as the underlying ocean circulation and mixing. River

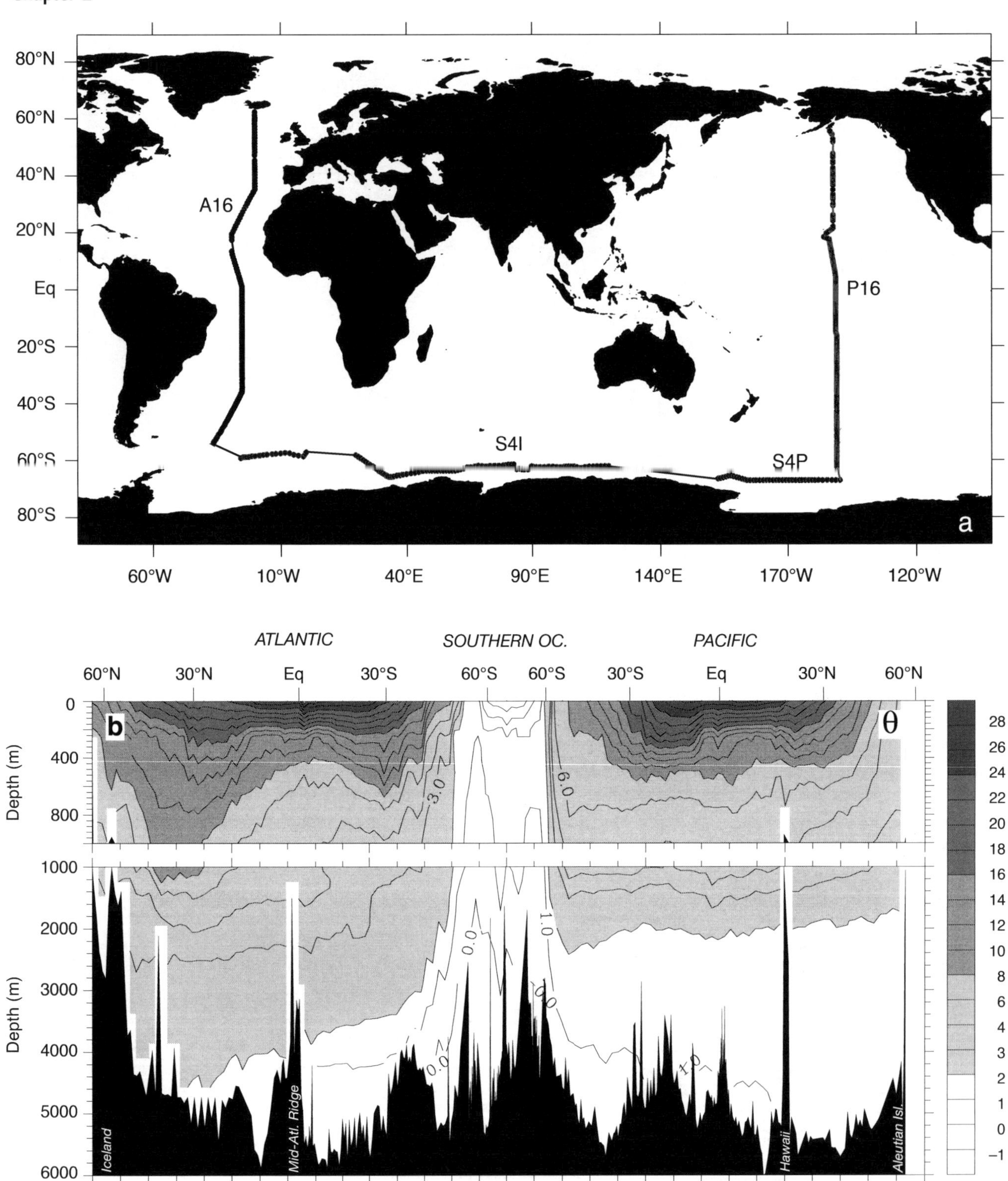

FIGURE 2.3.3: Top panel: Map showing the location of the hydrographic sections used to depict the large-scale distribution of properties in the world ocean. Bottom panel: Vertical section of potential temperature along the tracks in the top panel. Data are predominantly from the World Ocean Circulation Experiment (WOCE) using the following lines: Atlantic: WOCE A16, augmented with the Ocean Atmosphere Carbon Exchange Study (OACES), North Atlantic 1993 (NATL93), the A16N repeat occupation in 2003, and the South Atlantic Ventilation Experiment (SAVE) cruises, and WOCE S1/2 A21/A22; Southern Ocean: WOCE S41 and S4P; Pacific: WOCE P16S, P16C, and WOCE P16N. Data are from the CLIVAR and Carbon Hydrographic Data Office (http://cchdo.ucsd.edu/). The *x*-axis in the Southern Ocean (south of 57°S) was compressed by a factor of 5.

input and ice formation and melting can also change the surface salinity. Another way to alter the density is by mixing of the water parcel with water of a different density. Such exchange can be quite large over rough topography, for example [e.g., *Polzin et al.* 1997; *Ledwell et al.* 2000; *Mauritzen et al.* 2002; *Garabato et al.* 2004]. However, the amount of mixing in the interior of the ocean away from the surface and the sides and floor of the ocean is generally very small [e.g., *Ledwell et al.*, 1993, 1998] because the input of energy for mixing is low. We therefore assume in what follows that ocean transport within the interior of the ocean away from the surface, sides, and bottom occurs largely along isopycnal surfaces.

Geostrophic Equations

As noted in section 2.2, the steady-state motion in the interior of the ocean can be characterized by a balance between the pressure gradient and Coriolis forces, referred to as the geostrophic equations:

$$\frac{1}{\rho}\frac{\partial p}{\partial x} = fv \tag{2.3.1}$$

$$\frac{1}{\rho}\frac{\partial p}{\partial y} = -fu \tag{2.3.2}$$

We derive here an expression for how these equations constrain the flow along layers of thickness H defined by two isopycnal surfaces. An approximation to the above equations that simplifies our derivation is to assume that the density in the pressure gradient term can be represented by its mean, ρ_0. This is one of a set of assumptions that are collectively referred to as the *Boussinesq approximation*. Multiplying through by this density and then differentiating (2.3.1) with respect to y and subtracting the differential of (2.3.2) with respect to x to eliminate the pressure gradient term gives:

$$\rho_0\frac{\partial}{\partial x}(fu) + \rho_0\frac{\partial}{\partial y}(fv) = 0 \tag{2.3.3}$$

Here the density has been taken out of the differentials based on the assumption that its variations are negligible. In the next step, we note that f is only a function of y, so that the differential of f with respect to x is 0; and we define $\partial f/\partial y$ as β. This gives:

$$f\left(\frac{\partial u}{\partial x} + \frac{\partial v}{\partial y}\right) + \beta v = 0 \tag{2.3.4}$$

We next substitute the incompressible mass continuity equation

$$-\frac{\partial w}{\partial z} = \frac{\partial u}{\partial x} + \frac{\partial v}{\partial y} \tag{2.3.5}$$

into (2.3.4) to obtain the *geostrophic vorticity equation*:

$$\beta v - f\frac{\partial w}{\partial z} = 0 \tag{2.3.6}$$

We now proceed through a further series of manipulations to arrive at an expression of the geostrophic vorticity conservation in terms of the isopycnal thickness. We use the total derivative definition given by equation (9) of panel 2.1.3, to note that $df/dt = v\,\partial f/\partial y = \beta v$ (f is constant with time and with respect to the x and z coordinates). Substituting this into (2.3.6) gives:

$$\frac{df}{dt} - f\frac{\partial w}{\partial z} = 0 \tag{2.3.7}$$

Next we consider how to interpret the $\partial w/\partial z$ term. If we follow a parcel of water flowing in a layer of thickness H defined by two isopycnals, this term can be thought of as representing essentially the vertical compression or expansion of the water parcel as the layer defined by the two isopycnals thins or thickens. The change in thickness of the water parcel will be equal to the difference between vertical velocity at the top of the water parcel dz_{top}/dt and that at the bottom of the parcel dz_{bottom}/dt divided by the thickness of the layer H. Since $z_{top} - z_{bottom} = H$, we have that $\partial w/\partial z$ is equal to $(1/H)(dH/dt)$ (see *Pedlosky* [1987] for various ways of integrating (2.3.7) that give similar solutions). Substituting this into (2.3.7) and combining the two terms thus obtained gives:

$$H\frac{d}{dt}\left(\frac{f}{H}\right) = 0 \tag{2.3.8}$$

which requires that

$$\frac{f}{H} = \text{constant} \tag{2.3.9}$$

We refer to f/H as the *potential vorticity* or *PV*.

A simple example of the application of (2.3.9) is the behavior of a water column flowing over topography where H is the depth of the ocean. As the water column is forced to thin, the water, in order to conserve potential vorticity, must flow towards regions of smaller f, i.e., towards the equator. On the downstream side of a topographic feature the effect is reversed. This helps to explain important excursions in, for example, the Circumpolar Current around the Antarctic, which is observed to deflect towards the north as it flows over the mid-ocean ridges.

Gyre Circulation with Stratification

According to the arguments we have made thus far, the flow is strongly favored to occur on density surfaces and to follow contours of constant *PV*. Consider the idealized case of flat layers such as those depicted in figure 2.3.4. Because H is constant, a line of constant *PV* must be along a line of constant f, i.e., it must be east-west. This is depicted in the lower part of the diagram. The lines of *PV* intersect the eastern boundary (we do not draw *PV* lines in the western boundary region, where the force balance

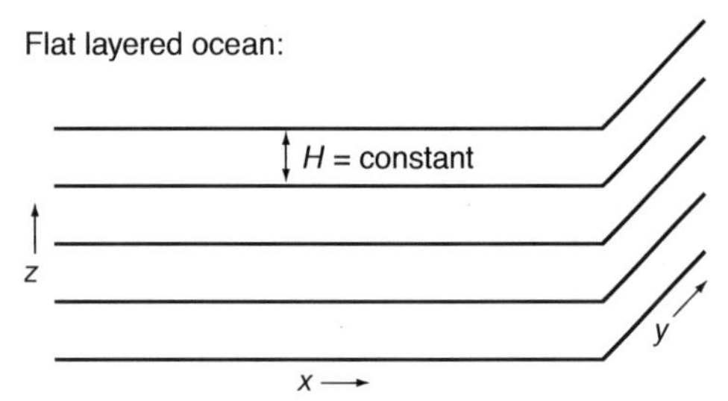

Constant $PV = f/H$ contours on a layer. Because H = constant, f must also be constant.

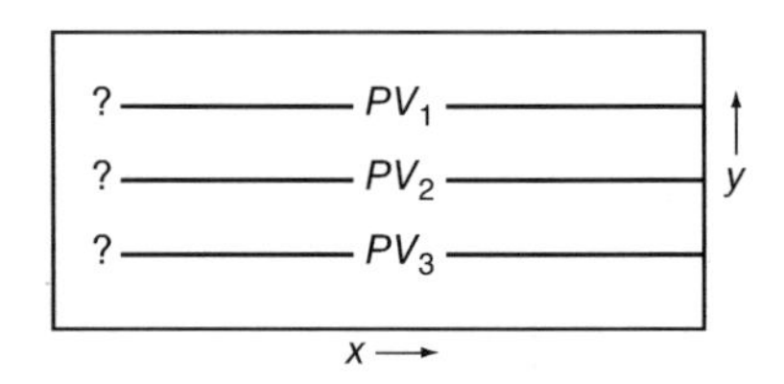

FIGURE 2.3.4: Schematic illustration of *PV* contours in an ocean with flat density layers.

includes contributions from other than the geostrophic terms). Because the flow must be along lines of constant *PV*, and because there can be no flow out of the boundary, we are forced to conclude that flat layers such as those depicted in the figure can have no transport. In such a case, the gyre transport would be trapped in the surface layer of the ocean. What does the real ocean look like?

Insights from the Potential Vorticity Distribution

In this and subsequent subsections, our discussion focuses primarily on the North Atlantic subtropical gyre, though the basic approach we use can be applied to the other subtropical gyres of the world as well. Figure 2.3.5 shows a plot of depth and *PV* contours on two isopycnal surfaces. These are based on density observations in the main thermocline of the North Atlantic. The thickness *H* is calculated as described in the figure caption. The first thing to note from the depths and outcrops of these two density surfaces is that both isopycnals outcrop at the surface of the ocean within the subtropical gyre (compare the outcrop positions with the extent of the subtropical gyre depicted in figure 2.2.5). The possibility thus exists of the gyre circulation being pumped down from the surface onto these density surfaces. We note next that there are a large number of *PV* contours in the tropics that do in fact intersect the eastern boundary as in our idealized case of figure 2.3.4. As we might infer from our previous discussion, this region defines a *shadow zone* of low circulation and slow exchange with the surface of the ocean at the outcrop of the isopycnal [*Luyten et al.*, 1983]. Figure 2.3.6 shows one-year-long flow trajectories on the 26.5 density surface as calculated by an ocean general circulation model. The dot marks the end of the trajectory. This model simulation confirms that the shadow zone is indeed a region of slow circulation and poor ventilation from the outcrop of the density surface.

Both isopycnals in figure 2.3.5a and c also show some *PV* contours that emanate from the outcrop at the surface and then curve around the subtropical gyre. Horizontal currents, such as those that form part of the subtropical gyre circulation (cf., figure 2.2.5), can flow in from the surface along these constant *PV* lines (a process referred to as *horizontal induction*), as can water that is pumped down from the surface by Ekman convergence. The particle trajectories in figure 2.3.6 illustrate this circulation very nicely. The region where inward flow from the surface is permitted is referred to as the *ventilated region* [*Luyten et al.*, 1983]. Both layers also have what are referred to as *pool regions* of homogenous vorticity on the western side of the subtropical gyre. *Rhines and Young* [1982] postulated the existence of such pool regions, arguing that they would form by frictional coupling of deep isopycnal surfaces with the gyre transport on shallower surfaces, even if the friction is weak. The closed *PV* contours of the pool region permit gyre transport that is not required to flow in or out of the outcrop of the density surface, as illustrated by the particle trajectories in figure 2.3.6.

We thus see that the gyre transport escapes being trapped at the surface by entering the main thermocline along lines of constant *PV* that emanate from the surface outcrops of the density surfaces; and by weak frictional coupling of deeper surfaces with shallower surfaces, which generates regions of homogeneous potential vorticity around which the gyre transport can occur. One of the most remarkable consequences of *PV* conservation is the shallowing of the thermocline toward the equator noted in figure 2.3.3b and which can be seen in the isopycnal depth contours of figure 2.3.6. This can readily be understood as a precondition for the equatorward transport of the subtropical gyre circulation. The equatorward flow towards regions of lower Coriolis parameter f can only occur if the thickness of the layers thins.

We end this subsection by noting that the thermocline in the subpolar gyre and equatorial regions is fundamentally different from the ventilated thermocline of the subtropical gyres. The subpolar and equatorial regions are places where the Ekman divergence leads to upwelling, and the interior gyre circulation, which carries less dense low-latitude waters towards the pole, is in the wrong direction for it to penetrate into the denser waters of the thermocline. If this were all that were going on in these regions, the upwelling would thin the thermocline, as is indeed observed to be the case. The *diffusive theories* of the thermocline assume that downward mixing of heat from above counterbalances this thinning tendency of the upwelling. The diffusive theories can be contrasted with the so-called *adiabatic theories* we have been discussing, in which it is assumed that there is no exchange of heat or salinity

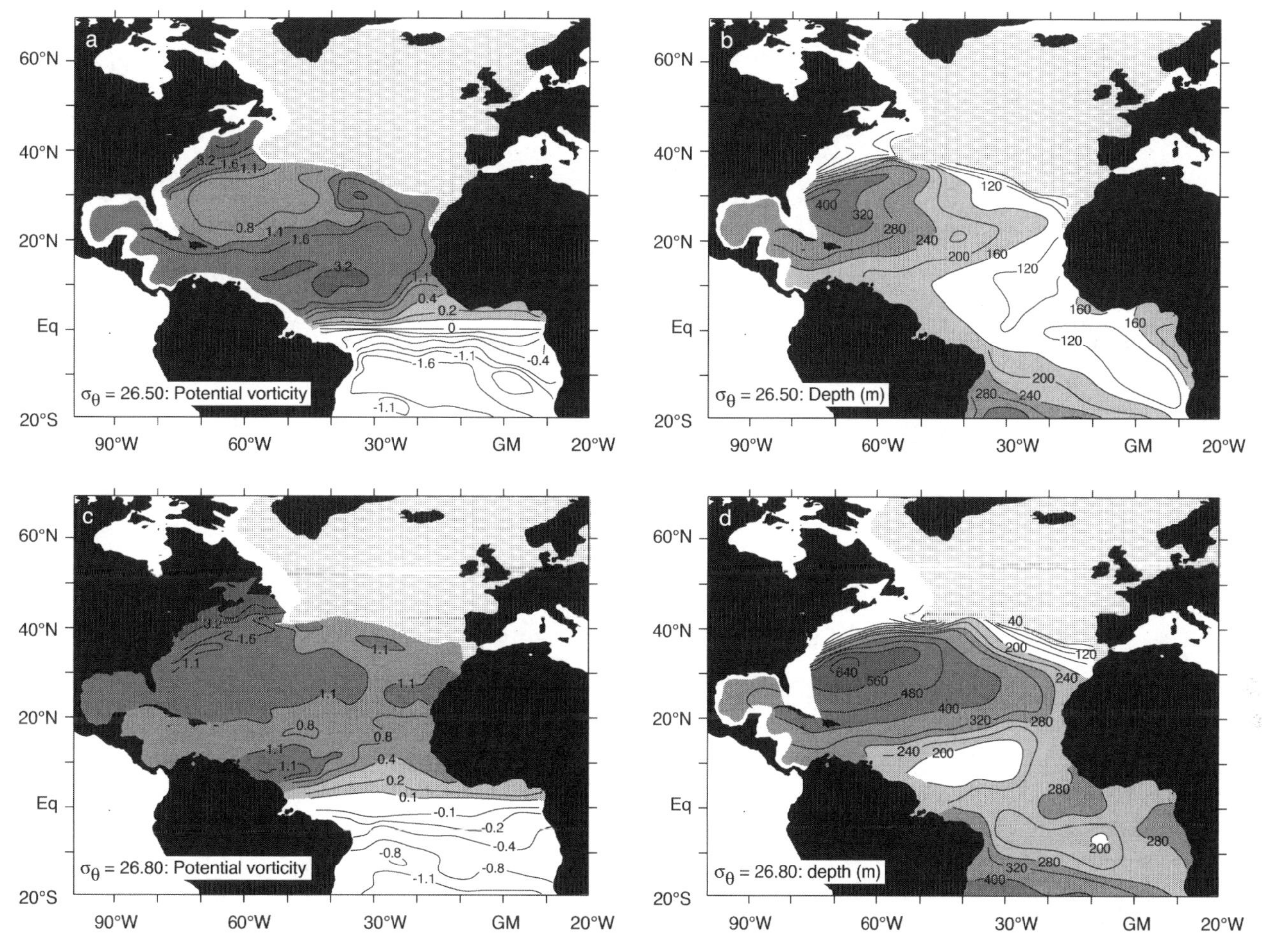

FIGURE 2.3.5: (a) and (c) Potential vorticity (units of $\times 10^{-10}$ m^{-1} s^{-1}), (b) and (d) depth (m) on two isopycnal surfaces in the North Atlantic [*Sarmiento et al.*, 1982]. The potential vorticity, defined as f/H, is calculated from observations of the potential density distribution, with H defined as $[1/(\sigma_\theta + 1000) \cdot (\partial\sigma_\theta/\partial z)]^{-1}$.

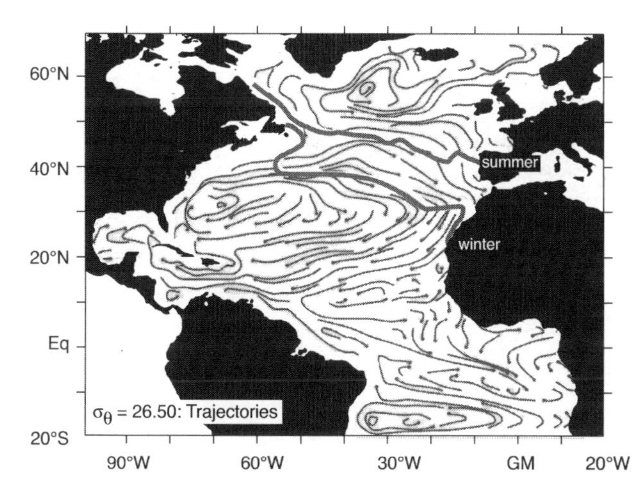

FIGURE 2.3.6: One-year-long particle trajectories on the $\sigma_\theta = 26.5$ isopycnal surface calculated with an ocean general circulation model. The grey lines denote the outcrop of this isopycnal surface in winter and summer. The dot marks the end of the trajectory [*Sarmiento et al.*, 1982].

between layers once water leaves the surface. The thermocline layer that is predicted by the diffusive theory is referred to as a boundary layer. It has been proposed that this diffusive boundary layer is a continuous blanket lying near the surface in the subpolar gyre and curving deep beneath the so-called *ventilated thermocline* of the subtropical gyre that we discussed above [*Samelson and Vallis*, 1997]. The deep boundary layer in the subtropical gyre manifests itself as a region of sharper vertical temperature gradient at a depth of several hundred meters underneath the *mode water* layer, defined in the North Atlantic by a thick layer of relatively uniform temperature at 18°C.

INSIGHTS FROM TRACERS

Among the most powerful tracers for studying thermocline ventilation are tritium, T, produced by nuclear bomb tests and its decay product helium-3, as well as chlorofluorocarbons. We will discuss tritium and helium-3 observations here, as these have been more thoroughly investigated. Tritium, the radioactive isotope of hydrogen, has a half-life of 12.32 years and decays to helium-3 by emitting a weak beta particle. It is produced in the upper atmosphere by cosmic ray spallation and fast neutron interactions with ^{14}N to form one tritium atom and ^{12}C. The global tritium

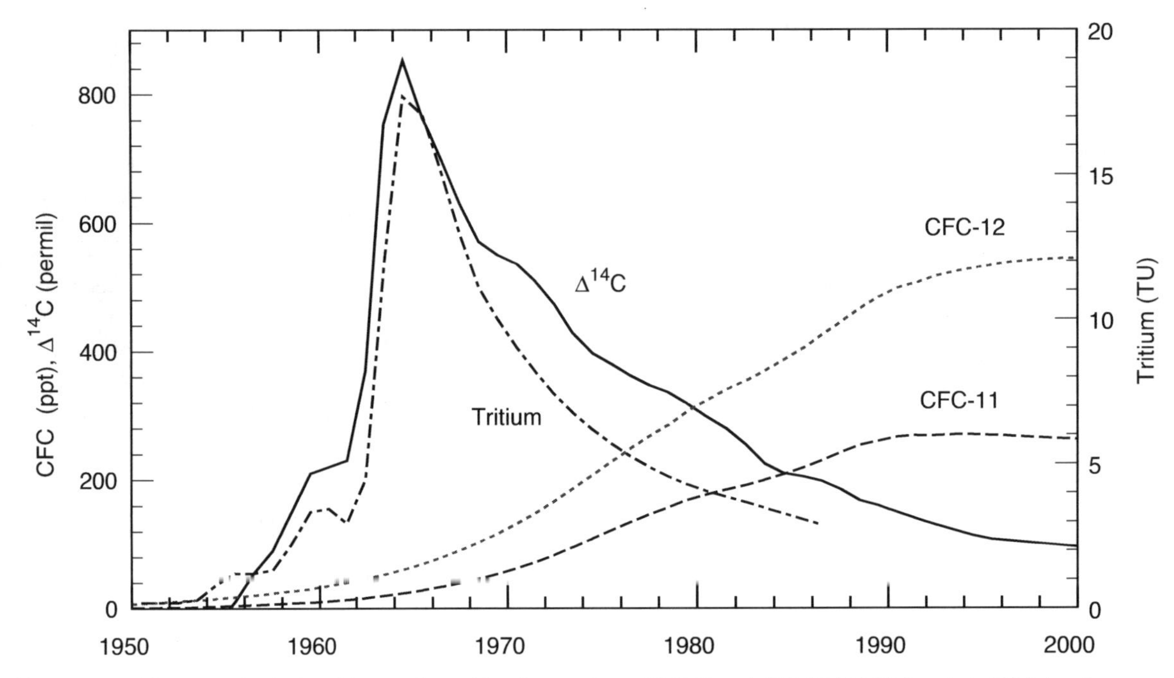

FIGURE 2.3.7: The atmospheric time history of two chlorofluorocarbons (CFC-11 and CFC-12) [cf. *Walker et al.*, 2000], bomb radiocarbon [*Rubin and Key*, 2002], and tritium in precipitation [cf. *Doney, et al.*, 1992], reported in TU units (1 tritium unit = 1 tritium atom per 10^{18} hydrogen atoms)

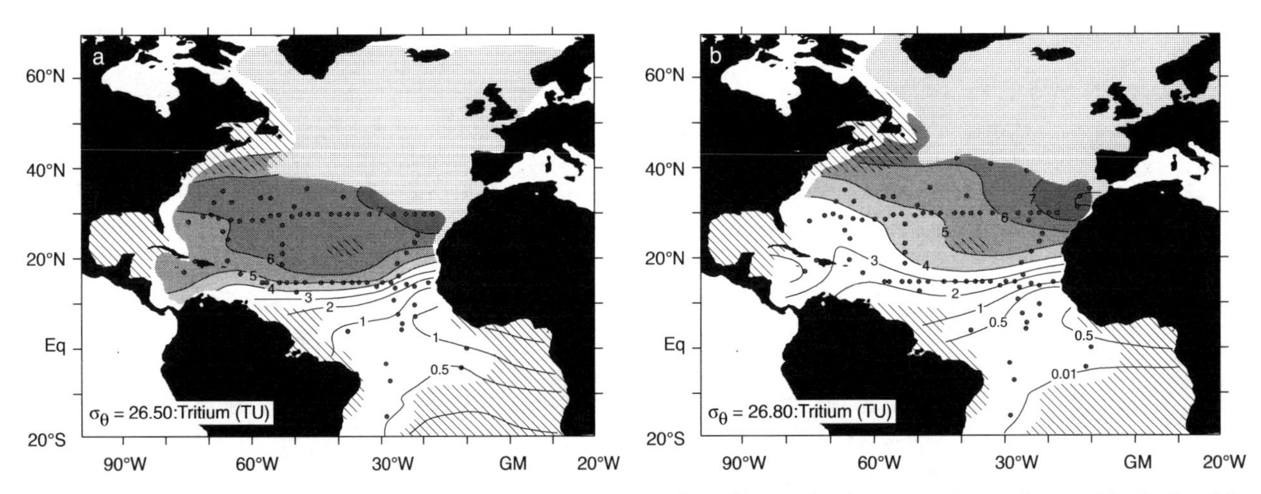

FIGURE 2.3.8: Maps of tritium on two isopycnal surfaces in the main thermocline of the North Atlantic [*Sarmiento et al.*, 1982]. The depths of the surfaces are given in figure 2.3.5.

inventory maintained by the balance between these formation processes and decay is about 3.6 kg. Nuclear bomb tests of the late 1950s and early 1960s introduced about 550 kg of tritium into the atmosphere, totally overwhelming the natural signal (figure 2.3.7). The bomb-produced tritium washed out onto the land and into the ocean as water, HTO. Its total inventory has been decreasing with time as it decays away. In what follows, we first show the distribution of tritium in the main thermocline and then follow with a discussion of age maps determined from the combined tritium/helium-3 distribution. The discussion centers on the North Atlantic, although much work has been done also in the North Pacific and will be mentioned briefly at the end of this section.

Figure 2.3.8 shows the tritium distribution in 1972 as mapped on two density surfaces within the main thermocline of the North Atlantic by *Sarmiento et al.* [1982]. The depths of these surfaces are shown in figure 2.3.5b and d. The outstanding feature is the strong front centered at about 15°N that reflects the boundary between the ventilated and pool regions of the gyre to the north, and the shadow zone of poor ventilation to the south. This provides strong confirmation

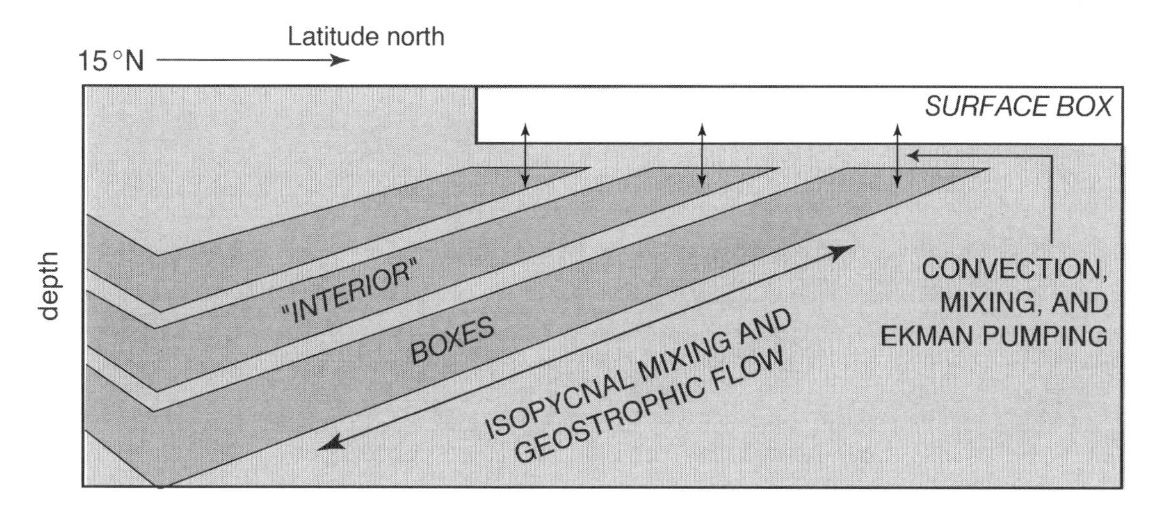

FIGURE 2.3.9: Thermocline ventilation box model of *Sarmiento* [1983].

of the thermocline theories discussed earlier, including the circulation inferred from the potential vorticity pattern of figure 2.3.5, and the particle trajectories of figure 2.3.6. Within the ventilated and pool regions, we see that the highest values are near the outcrop. On the $\sigma_\theta = 26.8$ surface, the tritium drops by quite a large amount as one traces the westward path of the ventilation region. Lower concentrations are found in the recirculation region in the northwestern corner of the gyre, which is ventilated less efficiently than the so-called ventilated region, presumably by exchange of tracer across the flow lines.

Sarmiento [1983] developed a simple box model of thermocline ventilation in order to estimate the exchange rate between the surface ocean and the interior (figure 2.3.9). The mechanisms of ventilation are presumed to be Ekman pumping, horizontal inflow (which is referred to as induction), and lateral mixing. These are represented by a simple exchange term $1/\tau$. The equation solved was:

$$\frac{\partial [^3\text{H}]_{interior}}{\partial t} = \frac{1}{\tau} \cdot ([^3\text{H}]_{surface} - [^3\text{H}]_{interior}) - \lambda \cdot [^3\text{H}]_{interior} \quad (2.3.10)$$

where λ is the decay constant of tritium (yr^{-1}), equal to ln 2 over the half-life. The surface tritium concentration was taken from *Dreisigacker and Roether* [1978]. The model was run with various values of τ until the 1972 value of the interior tritium concentration was correctly predicted. The results are shown in figure 2.3.10. Also shown in the figure are some revised calculations of ventilation time scales by *Doney and Jenkins* [1988]. The main result of these calculations is that the thermocline ventilation time scales are of the order of one or two decades or less and that the ventilation rate exceeds the Ekman pumping by a factor of 5. Horizontal induction is the largest contributor to the ventilation by far

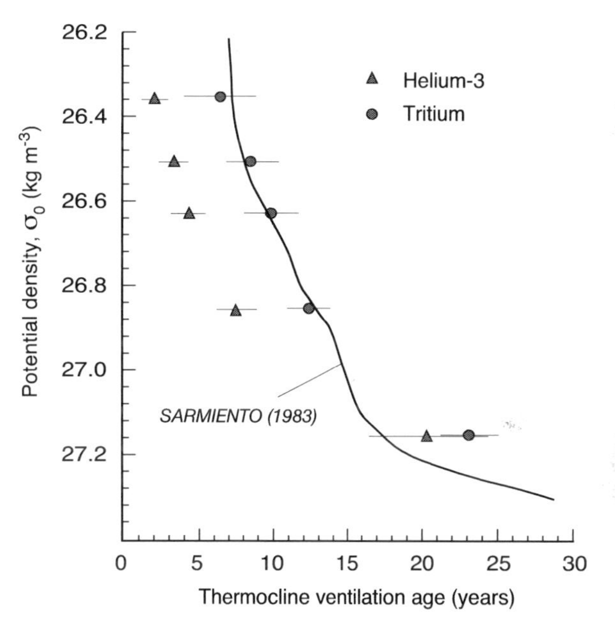

FIGURE 2.3.10: Thermocline ventilation rates determined from North Atlantic tritium observations by *Sarmiento* [1983] and *Doney and Jenkins* [1988], and from the helium-3 data set by *Doney and Jenkins* [1988]. Figure adapted from *Doney and Jenkins* [1988].

[e.g., *Huang*, 1990; *Speer and Tziperman*, 1992; *Marshall et al.*, 1993].

A powerful application of combined parent tritium/daughter helium-3 observations is the determination of an age since the water was last at the surface. Helium-3 equilibrates rapidly with the atmosphere by gas exchange. It builds up by decay of the parent once the water is isolated from the surface. The isolation time can be estimated from:

$$\tau_{\text{THe}} = \frac{\ln\left[\frac{[^3\text{H}]+[^3\text{He}]}{[^3\text{H}]}\right]}{\lambda} \quad (2.3.11)$$

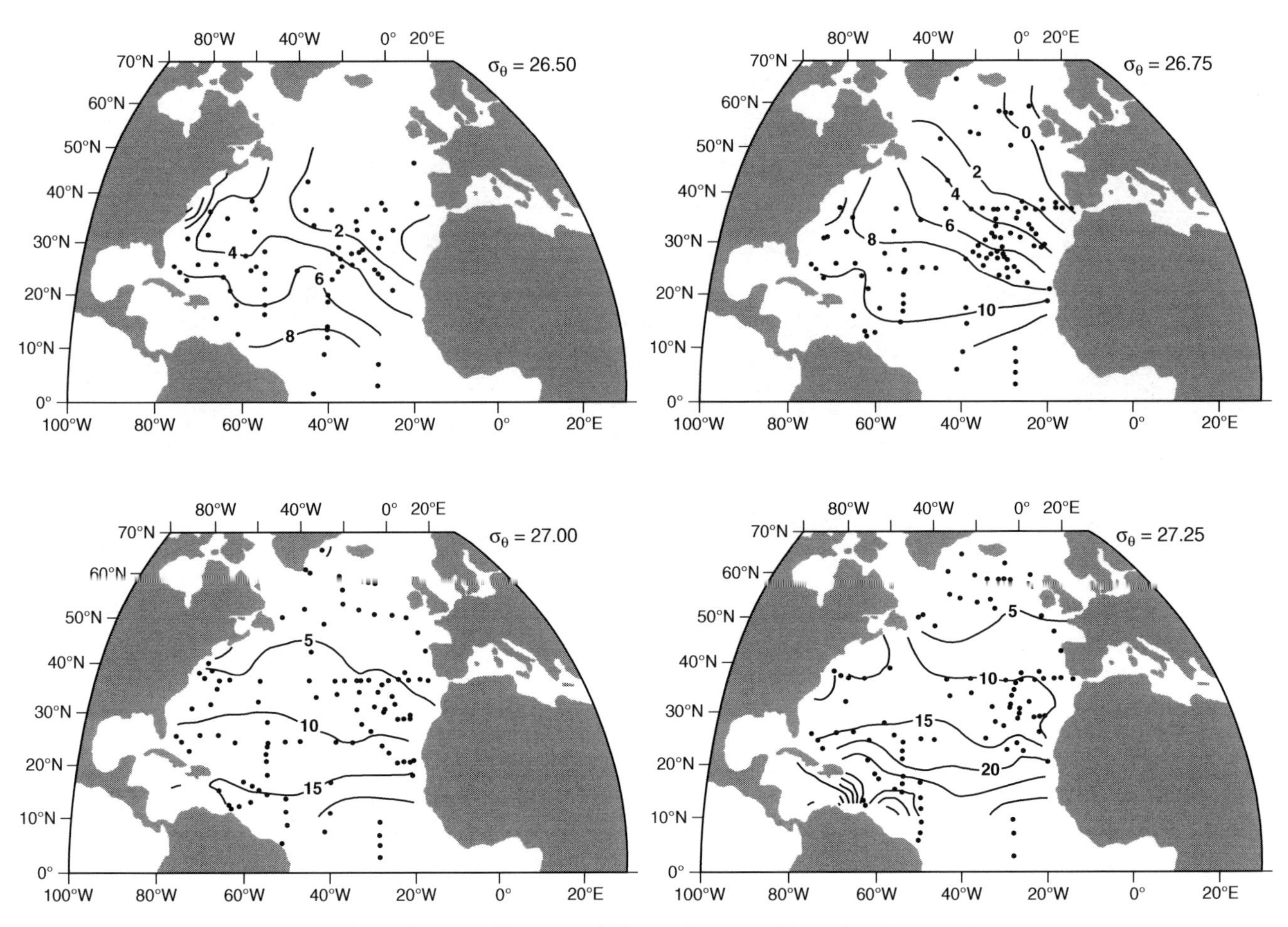

FIGURE 2.3.11: Isopycnal maps of ages determined by tritium/helium-3 dating [*Jenkins and Wallace*, 1992].

Figure 2.3.11 shows age contours on four isopycnal surfaces in the North Atlantic [*Jenkins and Wallace*, 1992]. The 15°N front is only hinted at due to the sparseness of observations in this region. However, the general patterns of gyre ventilation, and the increasing timescale with depth, are all evident in these figures. *Jenkins* [1998] has used the two tracers in combination to estimate the actual magnitude of the advection and diffusivity in the western side of the North Atlantic subtropical gyre.

In concluding this section, we show in figure 2.3.12 a map of potential vorticity [cf. *McPhaden and Zhang*, 2002] and CFC-11 age (John Bullister, personal communication, 2004) from the North Pacific Ocean (cf. also the tritium distributions analyzed by *Fine et al.* [1987]). The CFC-11 age is determined by assuming that the concentration of this gas is saturated at the surface of the ocean, which makes it possible to determine the age of any water parcel simply by comparing its concentration directly with the temporal history of the surface concentration shown in figure 2.3.7. *Thiele and Sarmiento* [1990] discuss the effect of mixing on this age determination, as well as tritium/helium-3 age estimates. These two tracers show the same patterns of the ventilated zone, shadow zone, and recirculation region of the main thermocline that we have seen in the Atlantic Ocean.

INSIGHTS FROM THE THERMAL WIND RELATIONSHIP

In regions where the geostrophic balance obtains, one should be able, in principle, to calculate velocities from (2.3.1) and (2.3.2). However, the horizontal pressure gradients cannot be measured with sufficient accuracy to do this. We get around the difficulty of determining the horizontal pressure gradient from observations by recasting the geostrophic equations in terms of the horizontal density gradient (the so-called *thermal wind relationship*). We do this using the *hydrostatic equation* (2.2.1c), which gives the vertical pressure gradient in terms of the density:

$$-\rho g = \frac{\partial p}{\partial z} \tag{2.3.12}$$

Note that we cannot represent density by its mean, ρ_0, in this equation, even though it is still safe to do so in the geostrophic equations (2.3.1) and (2.3.2). Without density variations, the vertical pressure gradient in (2.3.12) would have to be constant everywhere, which is a poor assumption. We eliminate the pressure gradient terms from (2.3.1) and (2.3.2) by differentiating them

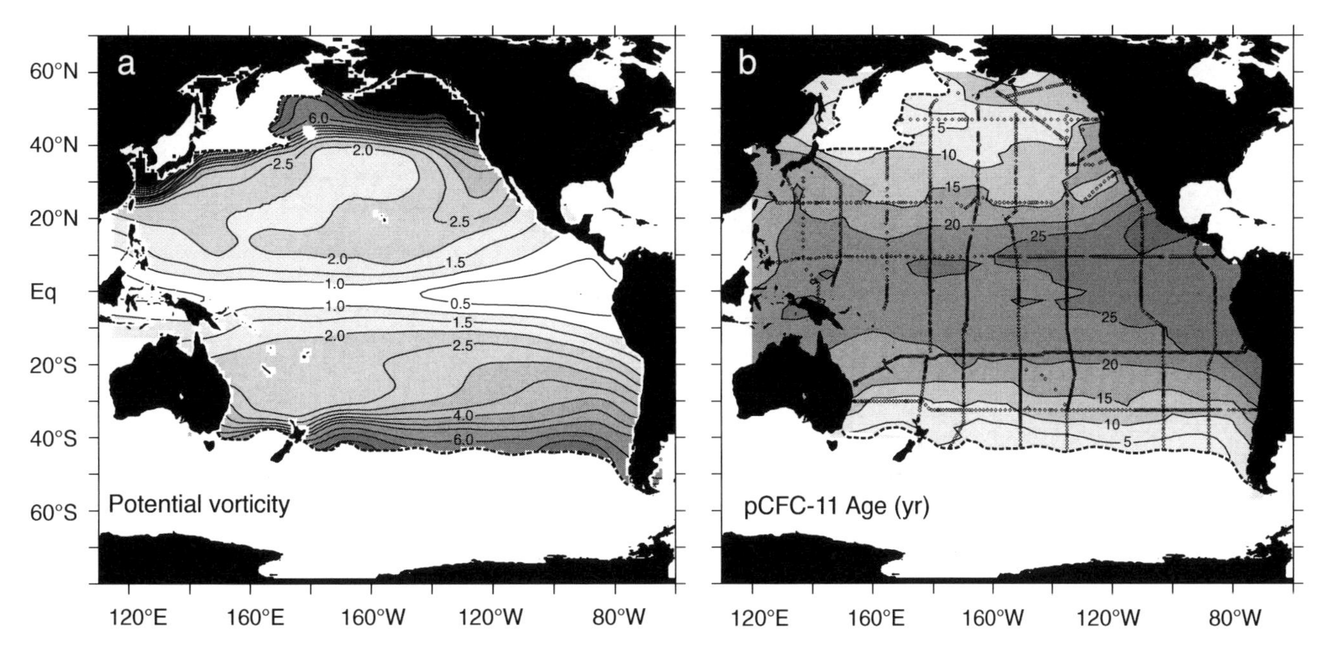

FIGURE 2.3.12: (a) Potential vorticity on the $\sigma_\theta = 26.4$ surface calculated from individual CTD and bottle profiles for the period 1950–present, and then objectively mapped. Heavy dashed contours mark the winter time outcrop of this isopycnal surface (Donxiao Zhang, personal communication, 2004). The contours are in units of $10^{-10}\ \mathrm{m}^{-1}\ \mathrm{s}^{-1}$, with intervals of 0.5 between 0 and 5, and 1 for values greater than 5. (b) Ages calculated from CFC-11 on the $\sigma_\theta = 26.4$ surface. The locations of the data are shown on the map. The data are mostly from the WOCE period, centered in the early-to-mid-1990s, but also including data from three sections, TPS-10, TPS-24, and TPS-47, from the mid-to-late 1980s. (John Bullister, personal communication, 2004.)

with respect to z and subtracting the differential of (2.3.12) with respect to x and y, respectively:

$$\begin{aligned} \frac{\partial(\rho_0 f v)}{\partial z} + \frac{\partial(\rho g)}{\partial x} &= \frac{\partial}{\partial z}\left(\frac{\partial p}{\partial x}\right) - \frac{\partial}{\partial x}\left(\frac{\partial p}{\partial z}\right) = 0 \\ -\frac{\partial(\rho_0 f u)}{\partial z} + \frac{\partial(\rho g)}{\partial y} &= \frac{\partial}{\partial z}\left(\frac{\partial p}{\partial y}\right) - \frac{\partial}{\partial y}\left(\frac{\partial p}{\partial z}\right) = 0 \end{aligned} \tag{2.3.13}$$

We now have:

$$\begin{aligned} \frac{\partial}{\partial z}(\rho_0 f v) &= -\frac{\partial}{\partial x}(\rho g) \\ \frac{\partial}{\partial z}(\rho_0 f u) &= \frac{\partial}{\partial y}(\rho g) \end{aligned} \tag{2.3.14}$$

The vertical gradient of f and the horizontal gradient of g are negligible, and the vertical gradient of ρ_0 is 0. Taking these terms out of the differentials gives the *thermal wind relationship*:

$$\begin{aligned} \frac{\partial v}{\partial z} &= -\frac{g}{\rho_0 f}\frac{\partial \rho}{\partial x} \\ \frac{\partial u}{\partial z} &= \frac{g}{\rho_0 f}\frac{\partial \rho}{\partial y} \end{aligned} \tag{2.3.15}$$

The thermal wind equations give us the vertical gradient of the horizontal velocity as a function of the horizontal density gradient. In order to find the absolute velocity, we must integrate this equation vertically from some depth at which we know the velocity. This requirement of determining the velocity at some depth level is known as the *reference level problem*. The historical practice has been to pick a depth between oppositely flowing water masses where it seems reasonable to assume that the velocity should be close to 0. In recent years there have been attempts to combine the geostrophic constraint with mass conservation to estimate absolute velocities and transports of heat, salt, and tracers [e.g., *Roemmich*, 1980; *Wunsch*, 1984; *Rintoul and Wunsch*, 1991; *Martel and Wunsch*, 1993; *Wunsch*, 1994; *Ganachaud*, 2003]. Here we apply the thermal wind relationship qualitatively in a couple of locations of particular interest.

Figure 2.3.13 gives a section across the narrow Gulf Stream off the North American coast. For simplicity, we treat the section as though it were east-west, impinging upon a north-south coastline, although in reality the orientation of the coastline is rotated towards the east-west direction. The density and temperature both show a very sharp front, with colder and denser waters near the coast and warmer and less dense waters in the interior. The horizontal gradient of density in the x direction across the front is negative (density decreases with increasing x). From (2.3.15) we see that this implies that the north-south velocity v must increase with increasing z (upwards). If we integrate this equation upward from some reference depth (e.g., 1500

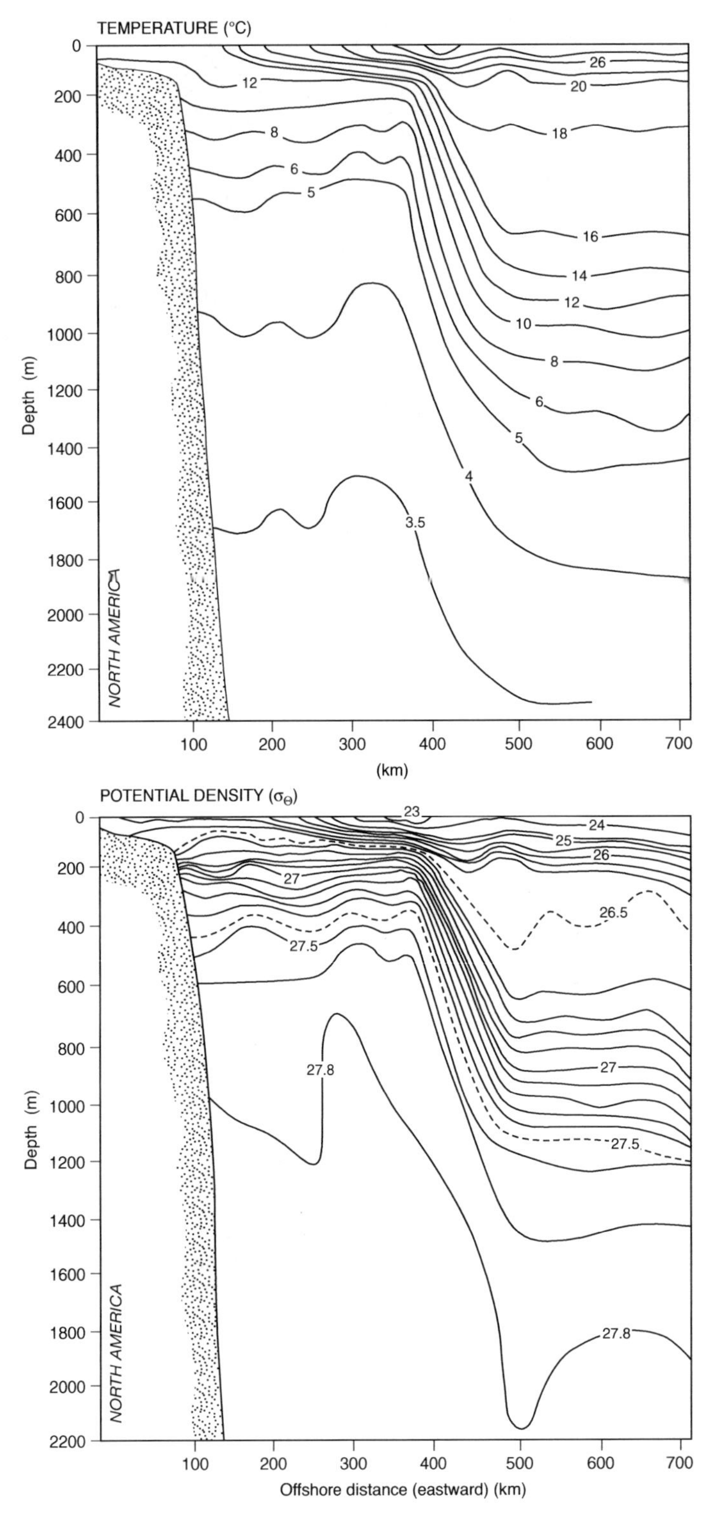

FIGURE 2.3.13: Vertical sections of potential temperature and potential density across the Gulf Stream [*Knauss*, 1997].

meters) we see that the velocity in the frontal region is towards the north. The intense current defined by the strong horizontal gradient region is quite narrow, of order 100 kilometers. This narrow boundary current, which is the Gulf Stream, closes the interior Sverdrup transport in Stommel's simple model of the gyre (cf. figures 2.2.5 and 2.2.9). Thus, we see that the large-scale flow features defined by the simple wind-driven theories of section 2.2 manifest themselves very clearly in the density structure of the ocean.

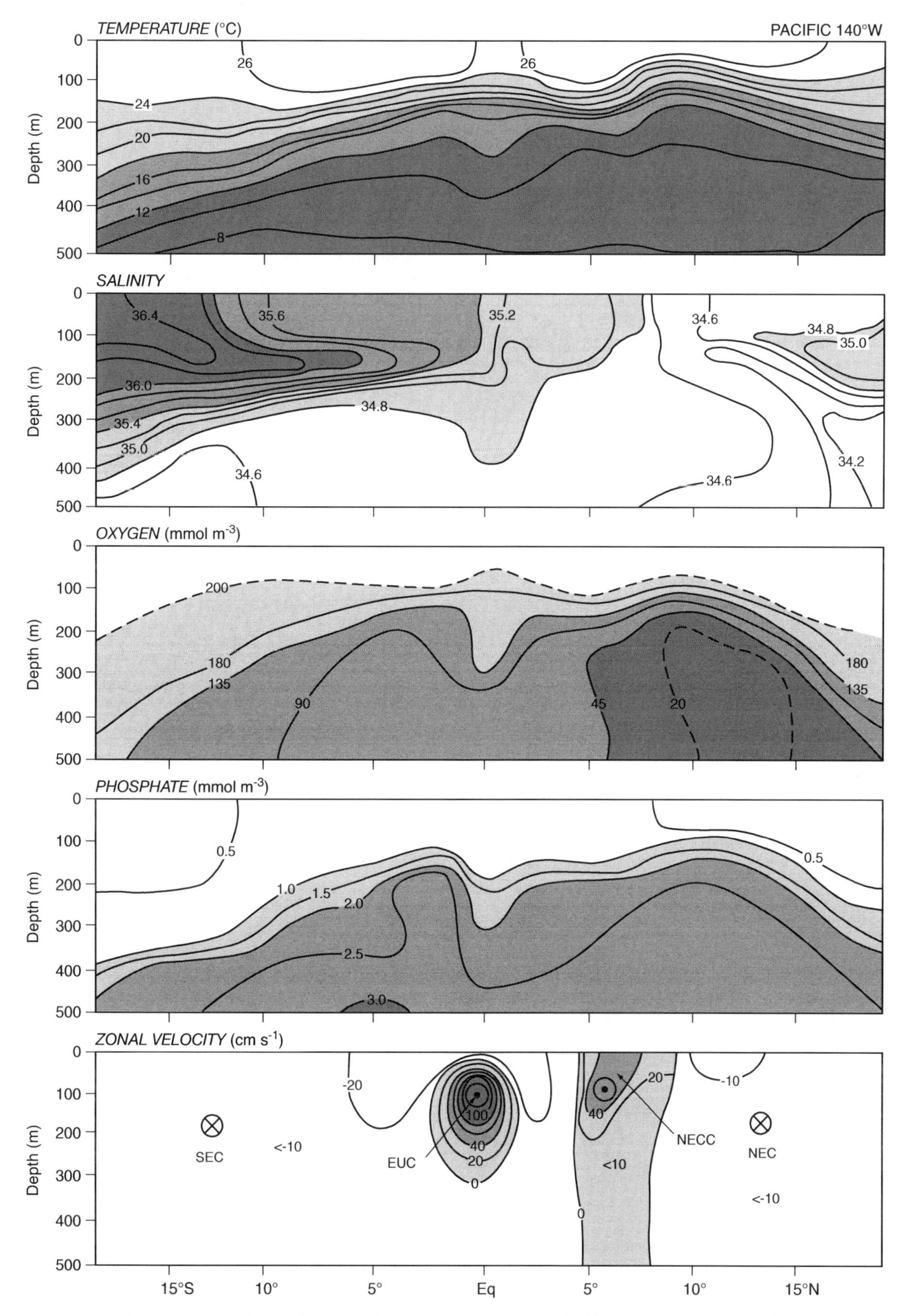

FIGURE 2.3.14: A north-south vertical section across the equator at 140°W in the Pacific Ocean of temperature, salinity, oxygen, phosphate, and velocity estimated by the geostrophic method [*Knauss*, 1997].

Another region of considerable interest is the tropics. We see from figure 2.2.9 that the large-scale trend of flow on the equatorward side of the subtropical gyres is towards the west (the North and South Equatorial Currents; cf. Also figure 2.2.5). However, figure 2.2.5 also shows an eastward flow just north of the equator. How do these features manifest themselves in the density structure and geostrophic velocity? Figure 2.3.14 shows north-south sections of various properties across the Pacific Ocean. We use the temperature to obtain a measure of the density. This section shows low-temperature (high-density) waters at shallow depths at the equator, and at about 8°N. The meridional density gradient is negative throughout most of the northern hemisphere (density decreases with increasing y) and positive in the southern hemisphere (density increases with increasing y). From (2.3.15) we see that the choice of a 0 velocity reference level at some depth below the thermocline implies a flow in the negative x direction in both hemispheres (recall that f changes sign from positive in the northern hemisphere to negative in the southern). This is as predicted by the Sverdrup transport solution. Between 5°N and 8°N, the flow is reversed to form the North Equatorial Countercurrent, a feature which is also predicted by the gyre circulation theories in analyses with realistic wind stresses. The only feature of the transport shown in the bottom panel of figure 2.3.14 that is not predicted by the theories we have developed so far is the Undercurrent right at the equator where the Coriolis force is 0. A simple pressure gradient force drives this current from the west towards the east.

2.4 Deep Ocean Circulation

There do not exist any generally accepted theories of the deep ocean circulation. Indeed, even the term *thermohaline circulation*, which is often used to describe this flow, is controversial because of its incorrect implication that the main source of energy for the deep ocean mass flux is buoyancy (heat and water fluxes) rather than mechanical energy provided by the wind and tides (cf. *Munk and Wunsch* [1998] and *Wunsch* [2002], who recommend using the term *meridional overturning circulation* or MOC instead). Buoyancy actually does play a role in the meridional overturning circulation, but mostly in determining where surface water sinks to the abyss. In this section, we will confine our attention primarily to a description of the main features of the meridional overturning circulation as inferred from observations and simulated by ocean general circulation models. In a final subsection, we will propose a synthesis.

Observations

We use the salinity and radiocarbon distributions to characterize the main features of the meridional overturning circulation and to estimate the timescale of deep ocean ventilation by this circulation. Figure 2.4.1 shows salinity along the sections illustrated in figure 2.3.3a, figure 2.4.2 shows zonal mean radiocarbon in the Atlantic and Pacific Oceans produced from gridded data as described in the figure caption, and figure 2.4.3 shows the age of radiocarbon at 3500 m determined by inverting the decay equation

$$C = C_0 \cdot \exp(-\lambda t) \qquad (2.4.1)$$

[*Matsumoto and Key*, 2004]. The ages are calculated with respect to the preindustrial atmosphere (0‰). Radiocarbon has a decay constant of $\lambda = 1.21 \times 10^{-4}\,\mathrm{yr}^{-1}$. Thus, for example, a radiocarbon concentration of −100‰ implies an age of 871 years. Before discussing these figures, we need to say a little about the interpretation of the radiocarbon distributions and ages.

Natural radiocarbon produced by cosmic rays in the atmosphere has long been the tracer of choice for estimating the timescales of the deep ocean meridional overturning circulation. Panel 1.2.1 explains how natural radiocarbon is formed, how it enters the ocean, and how it is measured. The natural radiocarbon concentration has fluctuated over time due to variability in the cosmic ray flux [cf. *Stuiver and Quay*, 1981]. However, on the timescale of the ocean circulation, these fluctuations are small relative to the contamination by the input of radiocarbon-free fossil fuel CO_2 that began with the start of the industrial revolution in the nineteenth century (called the *Suess effect* after *Suess* [1955]); and to the production of radiocarbon by nuclear bomb tests in the 1950s and 1960s, which nearly doubled the atmospheric concentration by 1964 (see figure 2.3.7).

The surface ocean radiocarbon concentrations before the Suess effect and nuclear bomb tests were about −40‰ in the low latitudes [cf. *Toggweiler et al.*, 1989], dropping to about −67‰ in the North Atlantic and −140‰ in the surface waters of the Weddell Sea in the Southern Ocean [*Broecker et al.*, 1998]. The surface North Atlantic and Weddell Sea concentrations correspond to ages of 570 years and 1250 years, respectively, and even −40‰ gives an age of 340 years. One would expect the concentrations of these surface waters to be close to atmospheric. The reason that the surface ocean does not equilibrate with the atmosphere is because the air-sea equilibration time for isotopic ratios of carbon is about 5 years for a 40 m-thick layer of the ocean (see chapter 3), and water does not spend that much time at the surface. In the Weddell Sea, this exposure time at

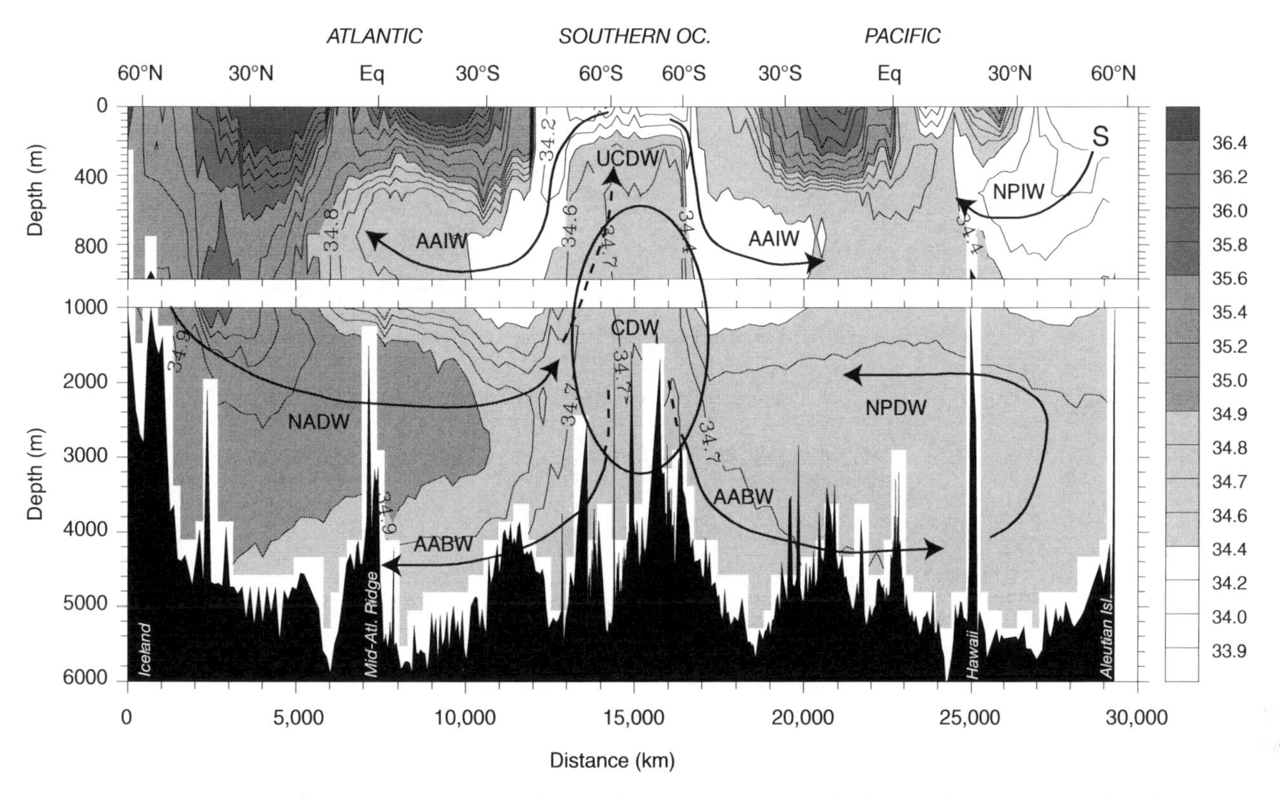

FIGURE 2.4.1: Vertical section of salinity (in the practical salinity scale) along the WOCE tracks shown in figure 2.3.3a. See also color plate 6. NADW is North Atlantic Deep Water, AABW is Antarctic Bottom Water, AAIW is Antarctic Intermediate Water, CDW is Circumpolar Deep Water, UCDW is Upper CDW, NPDW is North Pacific Deep Water, and NPIW is North Pacific Intermediate Water.

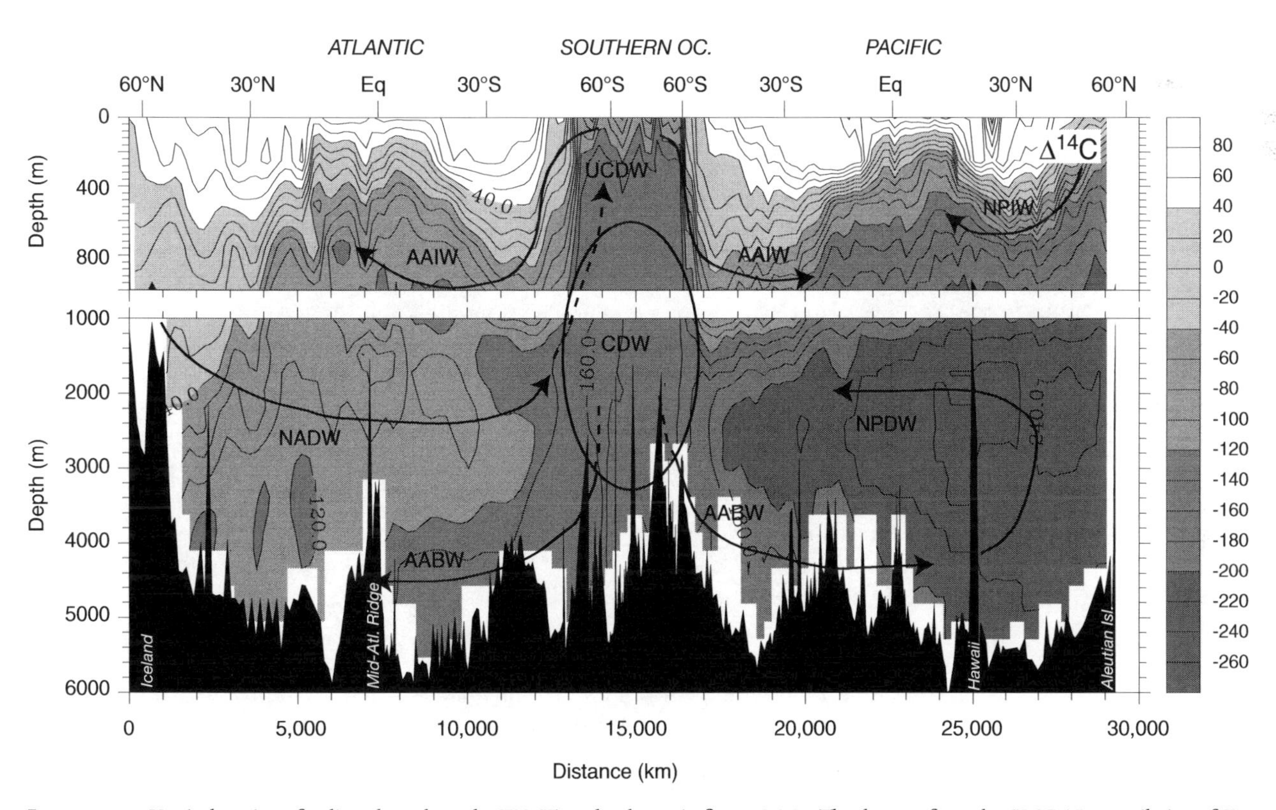

FIGURE 2.4.2: Vertical section of radiocarbon along the WOCE tracks shown in figure 2.3.3a. The data are from the GLODAP compilation of *Key et al.* [2004]. The data have not been corrected for the Suess effect and bomb radiocarbon input. See also color plate 6.

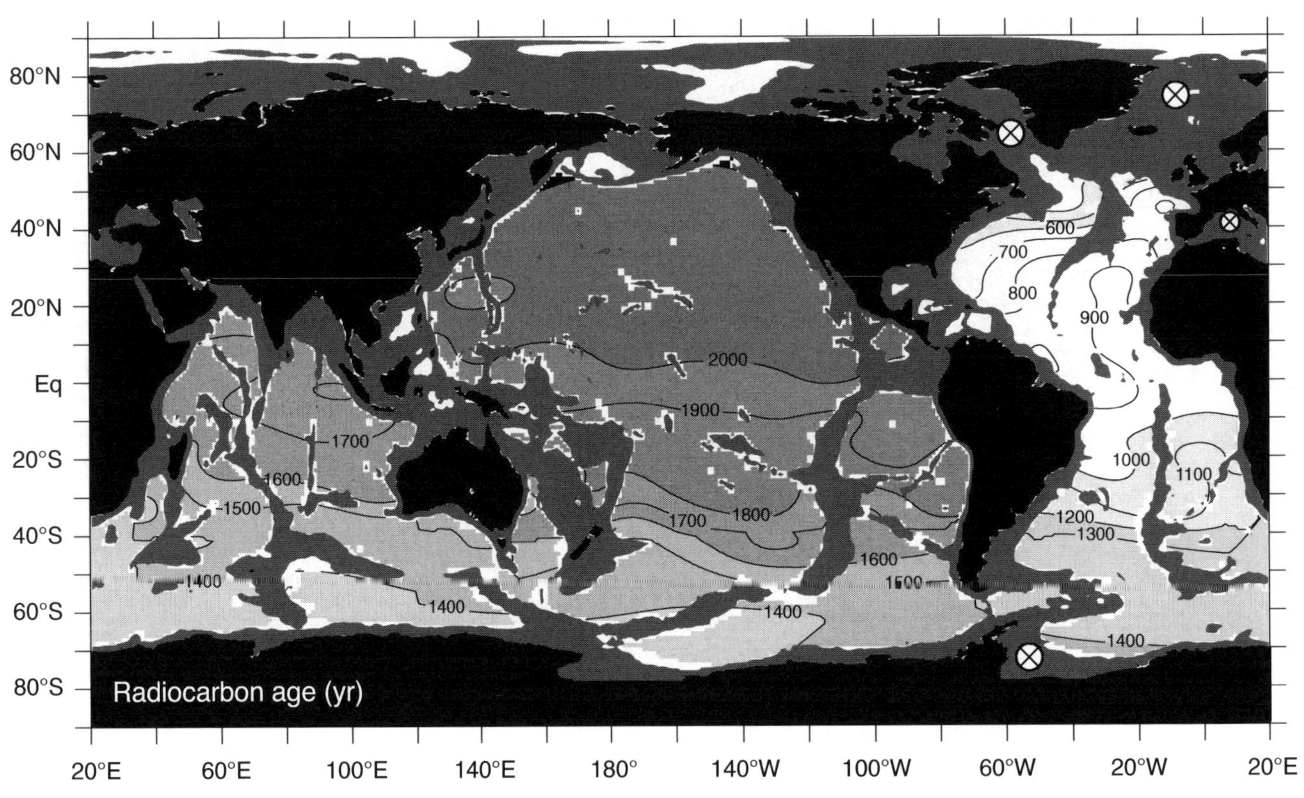

FIGURE 2.4.3: Map of radiocarbon-based age estimates in years with respect to the preindustrial atmosphere at a depth of 3500 m, based on the GLODAP gridded data set of *Key et al.*, [2004].

the surface is so short that essentially no radiocarbon is able to enter from the atmosphere and the surface concentration essentially reflects the deep ocean concentration in that region [*Weiss et al.*, 1979].

The observed radiocarbon concentrations in the main thermocline of the Atlantic and Pacific Oceans shown in figure 2.4.2 are clearly well above the surface values, indicating that the nuclear bomb test contamination has penetrated throughout the main thermocline to a depth of about 1000 m. We can infer from the distribution of CFC-11 and tritium in the North Atlantic that there is also contamination of bomb radiocarbon at the bottom of the ocean west of the mid-ocean ridge from Iceland all the way down to the equator and possibly as far as 5°S to 10°S [*Smethie et al.*, 2000]. This means that the age estimates for this region of the ocean shown in figure 2.4.3 are too young. However, the rest of the deep ocean at 3500 m has little if any such contamination.

The salinity distribution (figure 2.4.1) shows a layered structure with distinct minima and maxima at different depths. Since there are no significant internal sources or sinks of salinity, these minima and maxima can only originate at the surface of the ocean. Their pattern thus implies the existence of a large-scale meridional overturning circulation involving *deep water formation* (sinking from the surface into the deep ocean) in the North Atlantic and around the Antarctic, and penetration of these water types into the Atlantic, Indian, and Pacific Oceans. We use the salinity and radiocarbon distributions to identify the deep water masses that form by this sinking, beginning first in the North Atlantic and then progressing through the Southern Ocean into the Pacific.

The North Atlantic has a thick layer of relatively salty *North Atlantic Deep Water* (NADW; see arrows in figure 2.4.1). This water mass originates in the deep basins to the north of Iceland, and also includes shallower components that form in the Labrador and Mediterranean Seas, the latter having particularly high salinity (see circles with X in figure 2.4.3 for formation locations). The radiocarbon content of NADW is the highest in the deep ocean (figure 2.4.2), giving the youngest ages (figure 2.4.3). This would be true even without the bomb contaminant. The age map in figure 2.4.3 shows that the youngest ages of the NADW are along the western boundary of the basin, suggesting that it flows out of the North Atlantic predominantly as a western boundary current.

At the bottom of the Atlantic Ocean, mostly south of the equator, is found the relatively fresh (i.e., less salty) *Antarctic Bottom Water* (AABW; see figure 2.4.1). The predominant source of the AABW within the Atlantic is the Weddell Sea (see circle with X in figure 2.4.3), though there is some input from other formation regions around the Antarctic, as well as some entrainment of the so-called *Circumpolar Deep Water* that fills the deep Southern Ocean. As the AABW penetrates to the north, its influence gradually erodes due to mixing with salty NADW above. The AABW has lower radiocarbon than the NADW (figure 2.4.2). The 3500 m age map in figure 2.4.3, which lies at the top of the AABW, gives an age of 1400 years near the Antarctic continent. We would like to be able to use the radiocarbon timescales in figure 2.4.3 to estimate the ventilation rate of the Atlantic Ocean by the NADW and AABW. However, to do so we need to isolate that part of the aging that is due to actual decay of radiocarbon from that which is due to mixing of old AABW with young NADW and contamination by bomb radiocarbon in the North Atlantic. The most recent attempt to do this, which we describe here, is that of *Broecker et al.* [1998].

Consider a water parcel somewhere in the deep Atlantic that consists of some fraction f_n of NADW, and a corresponding fraction $(1-f_n)$ of AABW. The measured radiocarbon content of this water parcel will equal f_n time the initial radiocarbon concentration of the NADW at the time it left the surface, plus $(1-f_n)$ times the initial radiocarbon concentration of AABW, minus radiocarbon that has been lost to decay (the part that interests us), plus the radiocarbon contamination effects. Ignoring the contamination effects for now, we see that we can determine the decay if we know f_n and the initial radiocarbon concentrations of NADW and AABW. *Broecker et al.* [1998] estimate the initial radiocarbon concentrations of NADW and AABW (−67‰ and −140‰, respectively) by using observations as in *Broecker et al.* [1991]. They remove the bomb contamination from NADW by plotting radiocarbon versus tritium (also produced both by cosmic rays and by bomb tests) and extrapolating the relationship to the presumed pre-bomb tritium concentration of near 0. The AABW is assumed to be free of bomb contamination.

Determining f_n requires a conservative tracer that is clearly different in the NADW than in AABW. Salinity is conservative, but the separation into NADW and AABW is ambiguous because the four water masses that combine to form the NADW have different salinities. *Broecker et al.* [1991] proposed a new tracer, PO_4^*, that has nearly uniform values within the AABW and within the four components of the NADW (1.95 ± 0.07 and $0.73 \pm 0.07\ \mu$mol kg^{-1}, respectively; *Broecker et al.* [1998]). PO_4^* consists of the weighted sum of two tracers used by organisms during photosynthesis and respiration, namely phosphate and oxygen. As noted in chapter 1, observations suggest that deep ocean organisms remineralize organic matter in a fixed stoichiometric ratio of about 1 phosphate molecule released for every 170 oxygen molecules taken up. In other words, for every phosphate molecule added to the water by remineralization, 170 oxygen molecules are removed. Thus, if we were to add the phosphate concentration to the oxygen concentration divided by 170, the resulting tracer would be conserved except at the ocean surface, where gas exchange can affect the O_2 concentration. We discuss the underlying concepts in more detail in chapter 5. The specific definition for PO_4^* proposed by *Broecker et al.* [1991] is

$$[PO_4^*] = [PO_4^{3-}] + \frac{[O_2]}{175} - 1.95\ \mu\text{mol kg}^{-1} \qquad (2.4.2)$$

which has a slightly higher stoichiometric ratio of 175. The constant of 1.95 μmol kg^{-1} in (2.4.2) is arbitrary. Given the aforementioned NADW and AABW concentrations, the fraction of northern component water in a water sample can then be calculated from

$$f_n = \frac{1.95 - [PO_4^*]}{1.95 - 0.73} \qquad (2.4.3)$$

Figure 2.4.4 shows the average PO_4^* distribution at 3000 m in the world ocean. The separation between NADW and AABW in the Atlantic is clear. Interestingly, the remainder of the deep waters in the Indian and Pacific Oceans have a nearly uniform PO_4^*, with a mean of $\sim 1.4\ \mu$mol kg^{-1}. Plugging this into (2.4.3) gives that the Indian and Pacific Oceans are about 45% NADW and 55% AABW. Globally, including the Atlantic, this means that about half the deep water is NADW and half AABW. This implies that the rate of AABW formation must be about the same as the rate of formation of NADW [*Broecker et al.*, 1998]. So what is the rate of formation of NADW?

Once we know the fraction of northern and southern component waters from (2.4.3), we can estimate the change in radiocarbon concentration due to decay from:

$$\begin{aligned} \Delta(\Delta^{14}C) &= \Delta^{14}C_{initial} - \Delta^{14}C_{observed} \\ \Delta^{14}C_{initial} &= f_n(0.933) + (1 - f_n)(0.860) \end{aligned} \qquad (2.4.4)$$

Here 0.933 is R^* for the NADW component and 0.860 is R^* for the AABW component (R^* is defined by equation (2) of panel 1.2.1). Figure 2.4.5 shows the radiocarbon deficiency due to decay calculated from (2.4.3) and (2.4.4) overlaid with age contours. Recall that this radiocarbon deficiency and therefore the age are with respect to the initial radiocarbon of the NADW and AABW. They are thus a measure of the timescale of ventilation of the deep ocean with respect to the surface deep water formation regions. The analysis in figure 2.4.5 uses the vertically averaged data over water depths greater than 2000 m. The map shows very young water along the western boundaries, consistent with figure 2.4.3 and our earlier inference that the southward flow of NADW is confined primarily to western boundary currents. The observed distribution of chlorofluorocarbon and tritium provide strong confirmation of this western boundary undercurrent in the North Atlantic [*Jenkins and Rhines*, 1980; *Weiss et al.*, 1985; *Olsen et al.*, 1986; *Pickart et al.*, 1989; *Rhein et al.*, 1998; *Smethie et al.*, 2000].

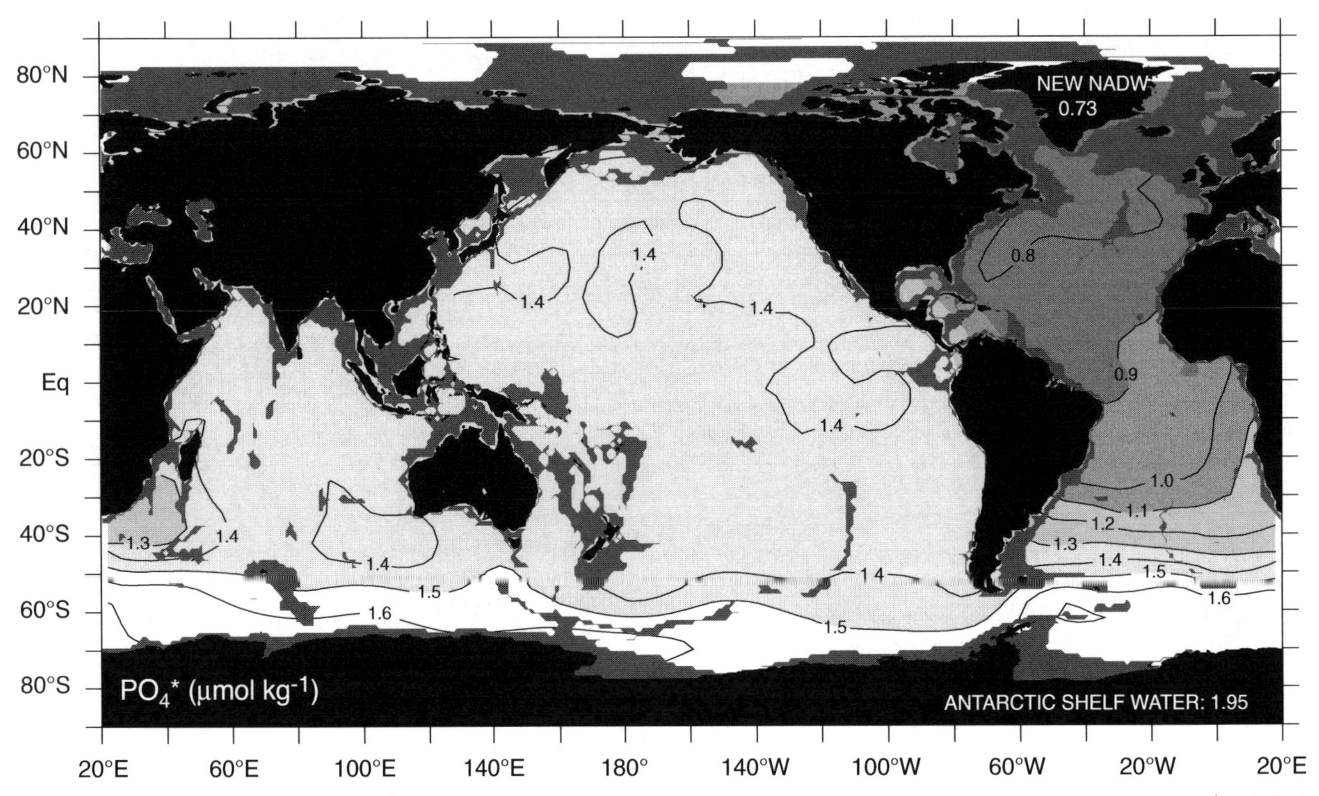

FIGURE 2.4.4: A map of average PO_4^* between 2000 and 4000 m depth based on the GLODAP gridded data set of *Key et al.* [2004]. PO_4^* is defined by equation (2.4.2).

The average radiocarbon deficit obtained from the data shown in figure 2.4.5 is 18‰, corresponding to a residence time of 161 years. After correcting for variations in the initial conditions due to the cosmic ray flux variability, *Broecker et al.* [1998] estimate the residence time as 180 years. For the deep Atlantic below 2000 m, this gives a deep water input of 22 Sv (recall that $1\,\text{Sv} = 10^6\,\text{m}^3\,\text{s}^{-1}$). Other studies suggest about 4 to 7 Sv of this is from AABW [cf. *Broecker et al.*, 1998], giving an NADW input of 15 to 18 Sv, with a corresponding input of AABW based on the global PO_4^* analysis of about 15 Sv. If we take the volume of $8.2 \times 10^{17}\,\text{m}^3$ for the ocean basins below 2000 m, this gives a ventilation timescale of ~870 years.

Returning to our analysis of the salinity distribution, we see that northward-flowing *Antarctic Intermediate Water* (AAIW) lies above the NADW (figure 2.4.1). AAIW is relatively fresh and has quite high radiocarbon (figure 2.4.2a). The AAIW originates in open ocean areas of the Southern Ocean. The upper part of the AAIW is actually a separate water mass identified as the *Subantarctic Mode Water* or SAMW, also flowing towards the north (cf. chapter 7, figures 7.3.3 to 7.3.5). Above the SAMW lies the main thermocline.

The circumpolar region, or Southern Ocean, that rings the Antarctic is characterized by a relatively uniform deep water mass called the *Circumpolar Deep Water* or CDW. This water mass is a crossroads of the deep ocean that blends together NADW, and corresponding deep water masses from the Indian and Pacific Oceans, as well as including deep water that forms around the Antarctic continent. It is remarkably uniform in the vertical, and the radiocarbon distributions shown in figure 2.4.2 suggest that this water is drawn toward the surface by upwelling, with a downward return flow toward the south that becomes AABW, and a northward flow that becomes AAIW as well as SAMW.

The Pacific Ocean has an AABW tongue along the bottom that erodes as the water penetrates to the north (figure 2.4.1). This AABW is often referred to as CDW to distinguish it from the slightly different AABW that flows into the Atlantic. The Pacific AABW has a strong component of salty NADW that it picks up around the Antarctic, making it the saltiest water in the deep Pacific [cf. *Reid and Lynn*, 1971]. Note from the salinity section in figure 2.4.1 that there is no evidence of deep water formation in the North Pacific. The radiocarbon observations shown in figure 2.4.2b suggest instead that the bottom water upwells and returns to the south at intermediate depths. This low-radiocarbon water mass is known as the *North Pacific Deep Water* (NPDW; see figure 2.4.1). Above the NPDW in the South Pacific can be seen the fresher signature of AAIW penetrating to the north (figure 2.4.1). A corresponding intermediate water mass formed in the North Pacific is referred to as the *North Pacific Intermediate Water* (NPIW).

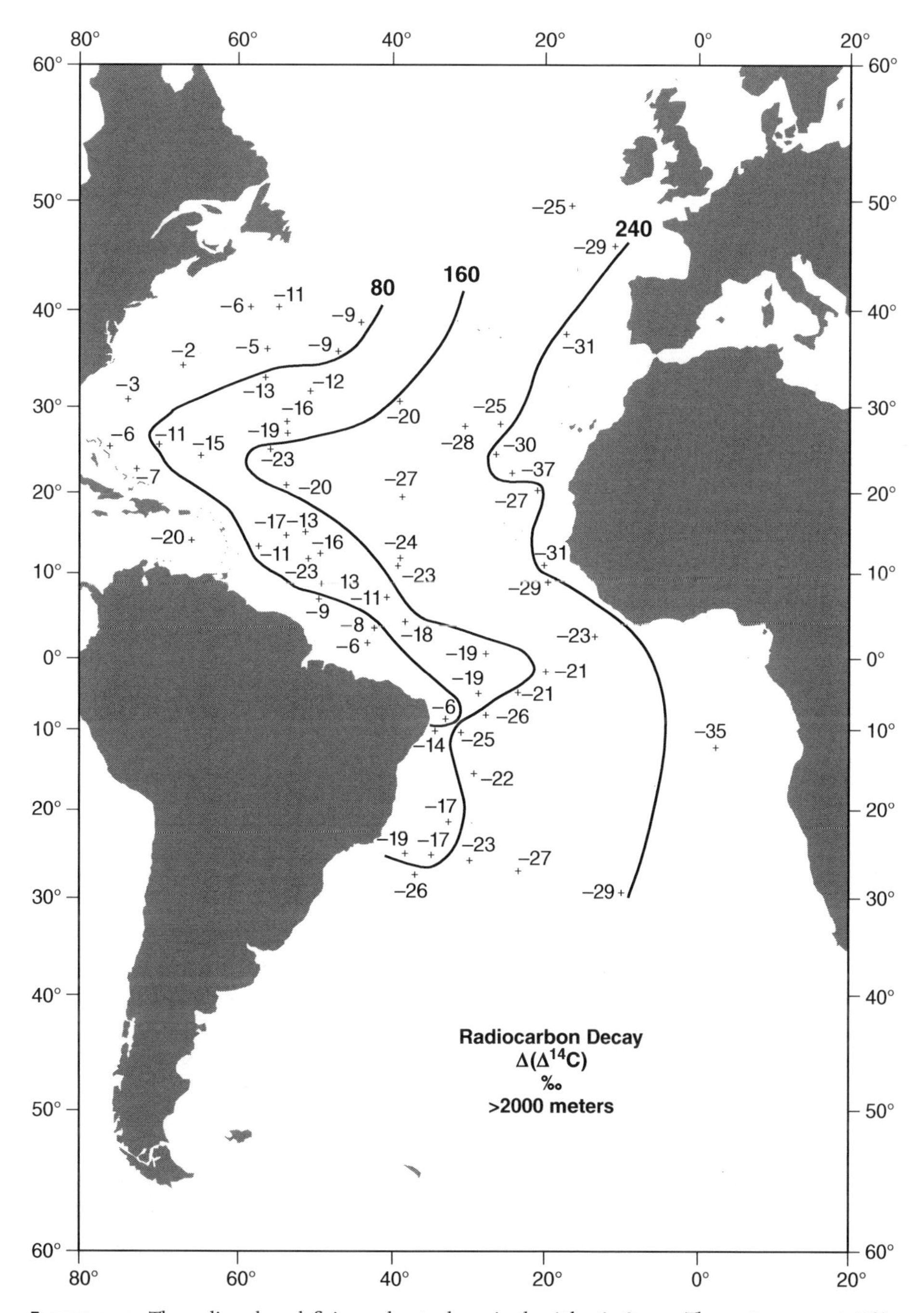

FIGURE 2.4.5: The radiocarbon deficiency due to decay in the Atlantic Ocean. The contours are at 10‰ intervals, and are labeled by age using 80 years per 10‰ interval (taken from *Broecker et al.* [1991, 1998]).

The Indian Ocean, which is not shown in figure 2.4.1, looks similar to the Pacific except that a modest amount of salty deep water sinks down from the Red Sea and forms a water mass referred to as the *Red Sea Deep Water* (RSDW).

We conclude our discussion of observations with a brief summary of an important box model analysis of the deep radiocarbon distribution by *Stuiver et al.* [1983]. Their box model is illustrated in figure 2.4.6. Their estimate of NADW formation is specified *a priori*, so we will not discuss it any further. Their flow diagram for the Indian and Pacific Oceans has water entering from the circumpolar region and upwelling across 1500 m. Our analysis of observations is inconsistent with this depiction of the flow. It appears instead that most of the outflow is in the deep waters well below 1500 m. However, the view represented by this diagram does correspond to the way that oceanographers used to view the deep ocean circulation, and it is interesting to examine its implications.

In order to determine the magnitude of the inflow from the Southern Ocean into the Indian and Pacific

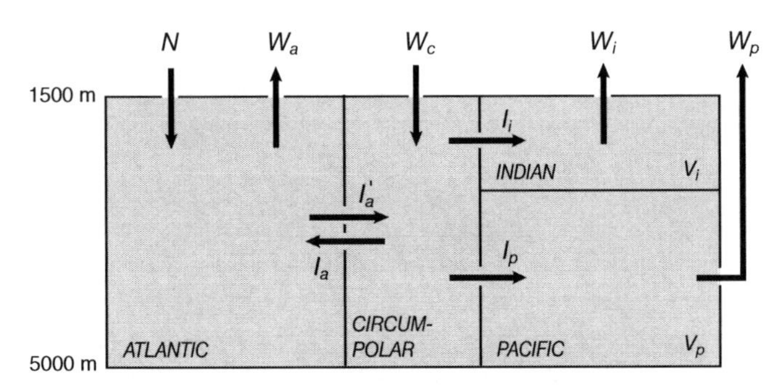

FIGURE 2.4.6: Box model diagram of the deep ocean circulation below 1500 m according to *Stuiver et al.* [1983].

Oceans in the *Stuiver et al.* [1983] box model we need to know the radiocarbon content of the inflow and outflow water. The radiocarbon content of the inflowing circumpolar water is −158‰, corresponding to an R^* of 0.842 (see definition in panel 1.2.1, equation (2)). The water at 1500 m has a radiocarbon concentration of −181‰ in the Indian Ocean, and −207‰ in the Pacific, corresponding to R^*'s of 0.819 and 0.793, respectively. Decay is equal to the decay constant times the deep ocean volume and the mean deep ocean concentration of −184‰ in the Indian Ocean, and −217‰ in the Pacific, corresponding to R^*'s of 0.816 and 0.783, respectively. A solution is obtained from the radiocarbon balance in combination with the mass balance constraint that inflows from the circumpolar region I balance the vertical upwelling W:

$$\begin{aligned} &\text{Indian Ocean:} \qquad I_i = W_i \\ &I_i(0.842) = W_i(0.819) + V_i\lambda(0.816) \\ &\text{Pacific Ocean:} \qquad I_p = W_p \\ &I_p(0.842) = W_p(0.793) + V_p\lambda(0.783) \end{aligned} \tag{2.4.5}$$

The volume of the Indian Ocean north of 50°S and below 1000 m is 1.6×10^{17} m^3 and that of the Pacific is 4.1×10^{17} m^3. Solution of these equations gives 20 Sv for the Indian Ocean inflow and 25 Sv for the Pacific. The ventilation timescales and 1500 m upwelling rates corresponding to these inflows are given in table 2.4.1. We discuss these calculations further in the following two subsections.

TABLE 2.4.1
Ventilation rates and timescales for deep water below 1500 m estimated from radiocarbon observations by *Stuiver et al.* [1983]

	Ventilation Rate (Sv)	*Ventilation Timescale (yr)*	*Upwelling Velocity (m yr^{-1})*
Atlantic*			
NADW	14	275	4
Circumpolar	−4		
Circumpolar	41	85	
Indian	20	250	10
Pacific	25	510	5
World	55	500	

*The NADW component of the Atlantic was specified by *Stuiver et al.* [1983]. The Atlantic upwelling velocity is based on the net inflow of 10 Sv. The ventilation time scale is based on the NADW inflow of 14 Sv summed to an estimated input of 7 Sv from the circumpolar region, which is balanced by a loss of 11 Sv to the circumpolar region, plus the upwelling of 10 Sv.

MODELS

Given an appropriate specification of boundary conditions at the surface and bottom of the ocean (i.e., wind forcing, heat and water fluxes, and friction), it is possible in principle to solve the three equations of motion, the equation of mass continuity, the equation of state relating ρ to T and S, and the conservation equations for heat and salt. This gives a total of seven equations, which match the number of unknowns: u, v, w, p, ρ, T, and S. However, there is no analytical solution for the full equations under realistic conditions. The solution is therefore approximated by breaking the ocean down into boxes of finite size, doing a volume integral around each box as we did in equation (2.1.13), and then expressing the resulting integrals in finite difference form and solving them numerically on a computer. The models that result from this are called Ocean General Circulation Models or OGCMs.

Any ocean circulation processes that are smaller than the resolution of the boxes in an OGCM have to be parameterized, usually as eddy diffusivity (see panel 2.1.2). We would thus prefer to use as large a number of boxes as is necessary to resolve major features of the ocean circulation that we believe to be important. However, as illustrated in figure 2.4.7, present computational resources are not up to the task, particularly for global circulation models such as those we discuss in this section, which correspond to the area denoted "T" in figure 2.4.7. In particular, these models are unable to resolve mesoscale eddies (shaded area 1 of the figure) such as are discussed in the following section. Typically, what is done to examine the role of mesoscale eddies is to develop slow-running models with higher resolution that barely permit eddies (area denoted "E" in figure 2.4.7) or a limited-area model that truly resolves the eddies. Another major problem with models is that the boundary conditions of wind stress, heat, and water fluxes are difficult to determine. Taking all these problems together, we must conclude that the numerical solutions of the flow that are obtained are all approximations to the real flow in the ocean. Nevertheless, this powerful approach is being used with increasing frequency to study ocean circulation, supplemented in studies such as the ECCO project [cf. *Stammer et al.*,

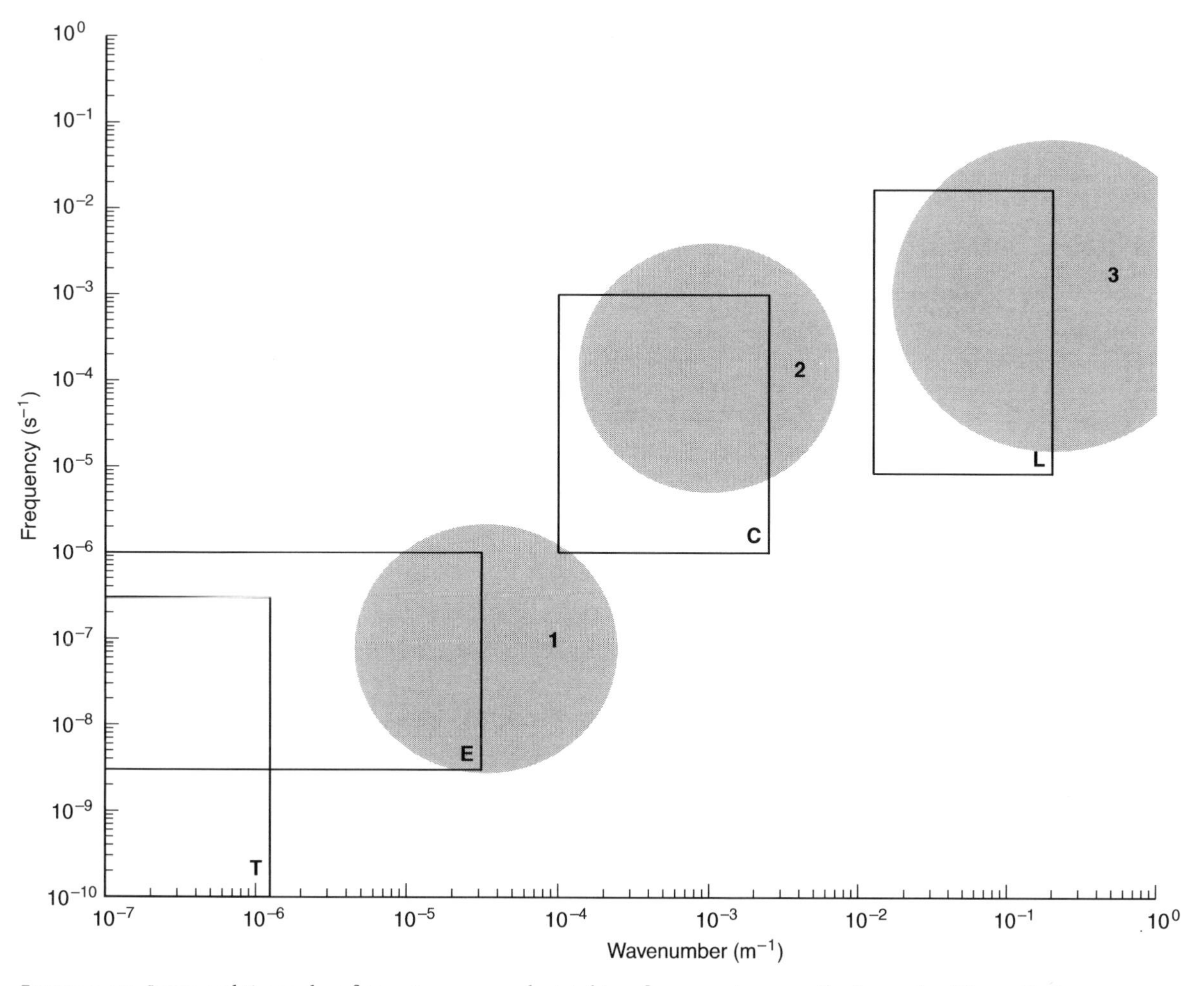

FIGURE 2.4.7: Space- and timescales of oceanic processes depicted in a frequency (one over the timescale of the motion) versus wave number (2π over the length-scale) diagram. Short time- and space-scale processes are in the upper right-hand corner of the diagram, whereas longer time- and space-scale processes are in the lower left-hand corner of the diagram. The shaded areas depict some of the relevant features of ocean circulation such as: (1) eddies, fronts, and western boundary currents; (2) deep-ocean convective overturning; and (3) turbulent motions including the very small-scale vertical motions. Present OGCMs of the large-scale ocean circulation cover approximately the area depicted by "T." Global eddy-resolving models cover the area depicted by "E," and small-scale models of ocean convection and eddy simulations of the ocean mixed layer cover the areas depicted by "C" and "L," respectively. The point of this diagram is that present OGCMs are not capable of resolving major features of the ocean circulation that are thought to be of importance. This figure is adapted from *Willebrand and Haidvogel* [2001].

2002] by increasingly sophisticated assimilation of observations (see figures 2.2.3 and 2.2.5).

The value of tracers such as radiocarbon and chlorofluorocarbon in helping to constrain the meridional overturning circulation predicted by OGCMs is illustrated by the diagrams shown in figure 2.4.8 from *Matsumoto et al.* [2004], which compare 19 different models (indicated by small diamonds) with each other and with observations of these tracers (indicated by dots and crosses giving estimates of the uncertainty). The abscissa of all these diagrams is the radiocarbon content of the Circumpolar Deep Water (CDW). Some of the conclusions that can be drawn from these diagrams include the following [cf. *Matsumoto et al.*, 2004]:

1. The radiocarbon in the CDW has a wide range in the models, with most of them being well outside the 2σ uncertainty shown in the figure. Only four models are within the uncertainty. The data are thus clearly indicating that most models are unable to simulate the radiocarbon content of the CDW correctly.

2. The CDW radiocarbon correlates linearly with the NPDW radiocarbon content, with a slope of almost 1:1 (see shaded ovals in figure 2.4.8a), but is independent of the NADW radiocarbon content. *Matsumoto et al.* [2004] conclude from this that most models have roughly the same circulation timescale between the Southern Ocean and North Pacific. The wide range of NPDW radiocarbon contents between the models is due primarily to the differing ability of the models to simulate the rate at which fresh radiocarbon is fed into the Southern Ocean either by gas exchange from above or laterally by input of young NADW.

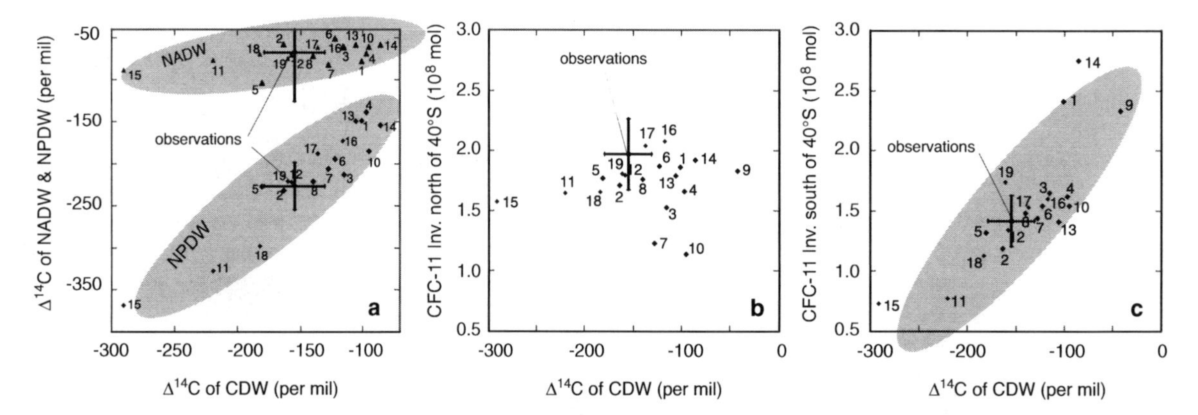

FIGURE 2.4.8: Radiocarbon and chlorofluorocarbon simulations by 19 different OGCMs [*Matsumoto et al.*, 2004] that participated in phase 2 of the Ocean Carbon Model Intercomparison Project (OCMIP-2; [*Dutay et al.*, 2001]). The P2A model we discuss in the text is identified by the number 19 in this figure. Also shown are observations indicated by a dot and crosses indicating 2σ error bars for radiocarbon and 15% uncertainty for the CFC-11 inventory estimate. The various water types are defined as follows: North Atlantic Deep Water (NADW) is equator–60°N, 1000–3500 m; North Pacific Deep Water (NPDW) is equator–60°N, 1500–5000 m; and Circumpolar Deep Water (CDW) is 90°S–45°S, 1500–5000 m. These boundaries have been applied to both observation and models, except that in models, a smaller NADW depth range of 1500–2500 m was used, because most models produce too shallow a NADW. (a) shows both NADW and NPDW on the vertical axis. The NADW trend is at the top of the figure (i.e., younger water) and the NPDW trend is at the bottom. (b) and (c) show the CFC-11 inventory in two regions of the ocean.

3. The CFC-11 inventories in the Southern Ocean have basically the same pattern as the CDW radiocarbon inventory on a model-by-model basis. Since CFC-11 has only been in the ocean for a few decades (see figure 2.3.7) and thus mostly reflects upper ocean processes, the CFC-11 results of figure 2.4.8c imply that the models have a wide range of surface to deep ventilation rates in the Southern Ocean. This is most likely the primary cause of the wide range in CDW radiocarbon simulations.

In what follows, we present a few results obtained by the model identified by 19 in figure 2.4.8. The general picture that emerges from this particular model is broadly consistent with a range of other studies [e.g., *England*, 1993; *Maier-Reimer et al.*, 1993; *Hirst et al.*, 1996; *Hirst and McDougall*, 1998]. This simulation was carried out at Princeton/GFDL based on the code originally developed by *Bryan* [1969] as modified by *Pacanowski and Griffies* [1999]. The model has a nominal resolution of 4° in the horizontal and 24 levels in the vertical ranging in thickness from 25 to 450 m, with highest resolution in the upper part of the water column. The particular version of the model we discuss here is identified as P2A in *Gnanadesikan et al.* [2004], where it is described. The small-scale processes in the conservation equation that are not resolved are represented by eddy diffusion terms with diffusivities of $0.15\,cm^2\,s^{-1}$ in the vertical increasing to $1.3\,cm^2\,s^{-1}$ at depth, and a lateral diffusivity along isopycnals of $1000\,m^2\,s^{-1}$. The boundary conditions used to force the model at the ocean surface include the momentum flux, equal to wind stress determined from observations; and heat and water fluxes specified also according to observations. In addition, the surface temperature and salinity are forced towards observations (which is equivalent to an additional heat or water flux). No momentum, heat, or salinity flux is permitted across the ocean bottom. The lateral walls have a no-slip boundary condition (i.e., 0 velocity) and no heat or salinity flux.

Figure 2.4.9a shows the global meridional overturning in the P2A model, and figure 2.4.9b and c show a breakdown of this into the Atlantic and Indo-Pacific Oceans. The contour interval is 2.5 Sv, i.e., the amount of water flowing between two contour lines is equal to 2.5 Sv, and the total flow between any two contour lines is equal to the difference between them. For example, the flow between a contour labeled 10.0 and one labeled -2.5 is 12.5 Sv. The motion is clockwise around maxima in the stream function, and counterclockwise around minima in the stream function. The primary features of the meridional overturning circulation outside the Southern Ocean are as follows:

1. There is a massive amount of upwelling in the upper 100 to 200 m at the equator. This upwelling is fed almost entirely from the upper few hundred meters of the subtropical gyre thermocline. The equatorial upwelling feature is consistent with what we inferred from our analysis of the Ekman transport.

2. The circulation in the deep main thermocline outside the Southern Ocean is characterized by a northward flow (figure 2.4.9a) that is confined primarily to the Atlantic Ocean (figure 2.4.9b), where it contributes to the formation of NADW. We have identified this as AAIW on the diagrams, as its behavior in the model, including the fact that it feeds into NADW formation, corresponds to what we believe the AAIW does in the real ocean [e.g., *Sloyan and Rintoul*, 2001].

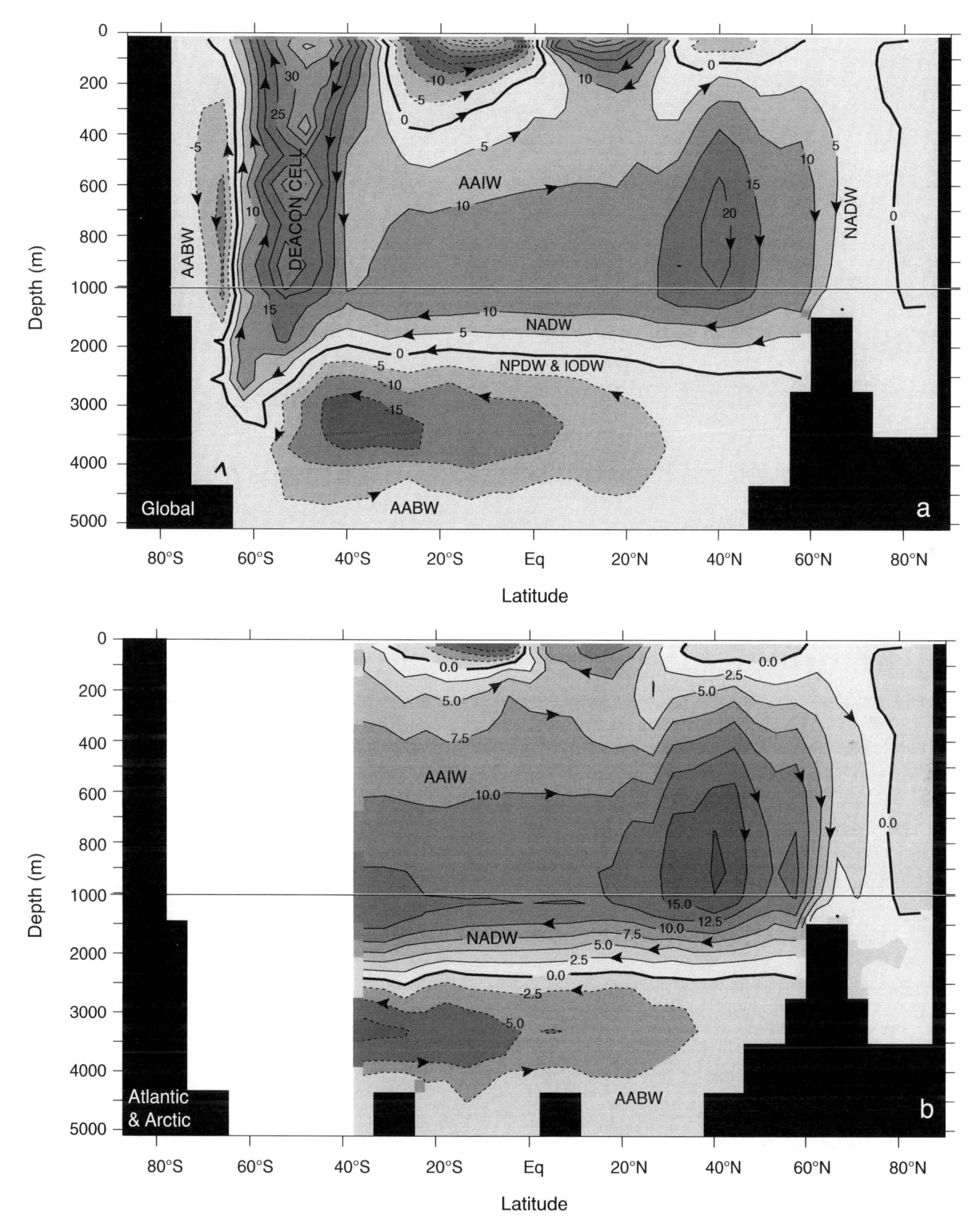

FIGURE 2.4.9: (a) Global meridional overturning stream function in the P2A OGCM described in the text. The contours are in Sverdrups ($1\,\mathrm{Sv} = 10^6\,\mathrm{m}^3\,\mathrm{s}^{-1}$). North is to the right. (b) and (c) give a basin breakdown of the global meridional stream function for the Atlantic and Indo-Pacific Oceans, respectively. See figure 2.4.1 caption for an explanation of the water types. IODW is Indian Ocean Deep Water.

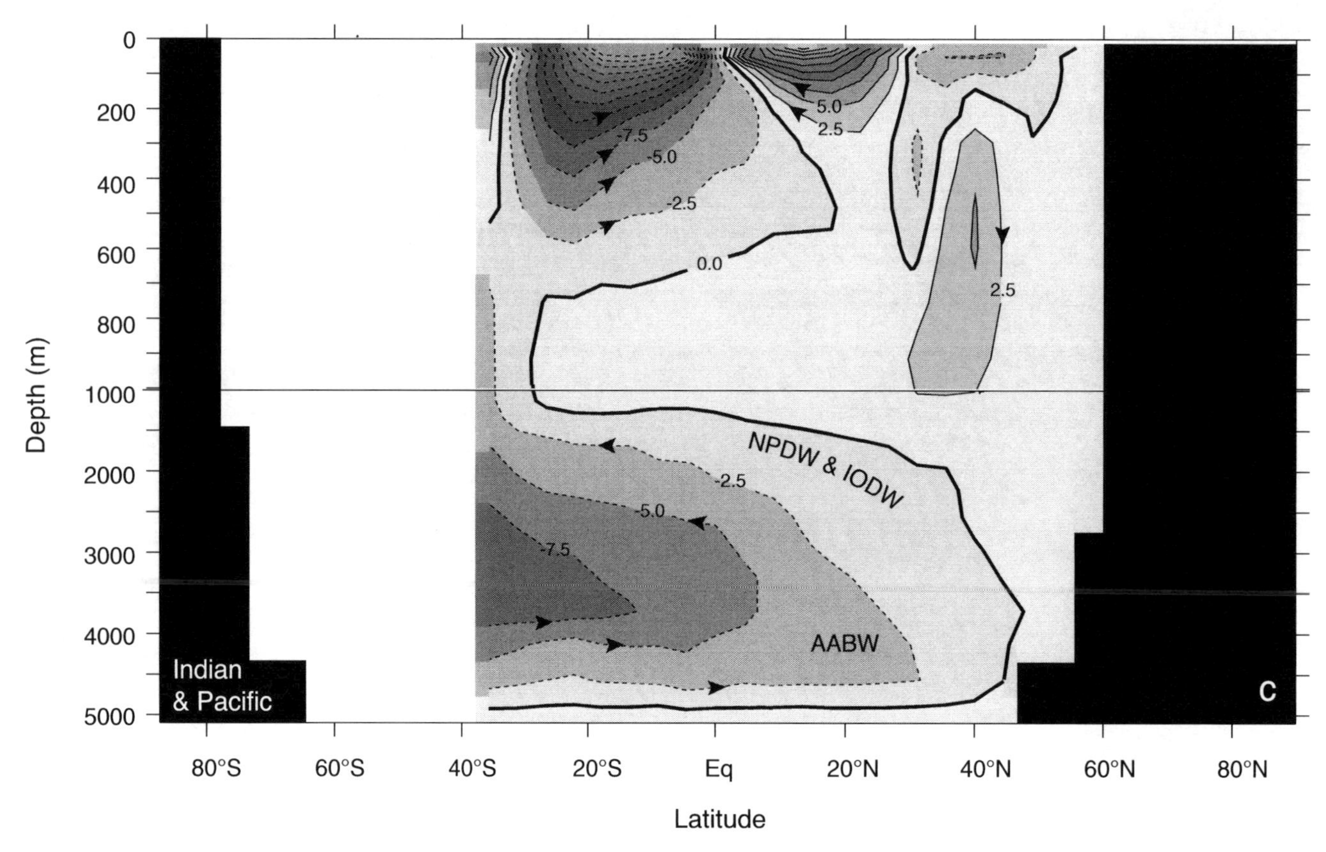

FIGURE 2.4.9: (Continued.)

3. The southward flow of deep water at intermediate depths of 1000 to about 3500 m occurs in all the ocean basins (figure 2.4.9). It has a contribution of about 20 Sv from NADW (figure 2.4.9b), with 5 to 10 Sv of additional flow due to NPDW and the corresponding *Indian Ocean Deep Water* (IODW; figure 2.4.9c). The Indo-Pacific deep waters are formed primarily by upwelling of AABW in the northern part of these basins, whereas the Atlantic deep waters are formed primarily by sinking from above of waters from the main thermocline.

4. There is a northward flow of AABW below about 3000 to 3500 m in all the ocean basins.

The model thus succeeds in qualitatively reproducing the main features of the deep ocean circulation that we inferred from observations in the previous subsection. It also gives about the same NADW formation rate as estimated by *Broecker et al.* [1998]. However, the bottom water inflow of ~8 Sv in the Indo-Pacific is much smaller than the 45 Sv obtained by *Stuiver et al.* [1983] (table 2.4.1), despite the fact that the OGCM radiocarbon simulation agrees quite with the observations (see figure 2.4.8a). The difference between the [*Stuiver et al.*, 1983] box model analysis and the OGCM is due in part to the fact that in the OGCM, most of the outflow of deep water from the Indian and Pacific Oceans occurs horizontally below 1500 m rather than vertically across 1500 m as assumed, most likely incorrectly, in the box model. Furthermore, inflows at one depth level are often balanced by outflows at another location but at the same depth level so that they do not show up in the zonally integrated meridional overturning. In addition, lateral mixing, which is not reflected in the stream function but is parameterized as advection in the box model, is an important contributor to the transport of radiocarbon out of the circumpolar region into the ocean basins [cf. *Gnanadesikan et al.*, 2004].

We conclude our discussion of the meridional circulation in the model with a brief description of the large overturning cell centered at about 50°S in the Southern Ocean, which is referred to as the *Deacon cell* [cf. *Speer et al.*, 2000], also commenting on the smaller and deeper cell centered at about 60°S to 55°S. An observationally based analysis of the circulation in this region is presented in chapter 7 (cf. figures 7.3.3 to 7.3.5) and by *Speer et al.* [2000]. Many aspects of the circulation shown in figure 2.4.9a are consistent with observational analyses, such as the large northward Ekman transport at the surface, which the model gives as about 40 Sv centered at 50°S to 45°S, whereas at least one observational analysis suggest it is more like 25 Sv [*Speer et al.*, 2000]; and the upwelling of deep water toward the surface between about 65°S and 50°S, which contains some NADW (see figure 2.4.1) but consists mostly of shallower Upper Circumpolar Deep Water (UCDW). Also consistent with our understanding of the deep ocean circulation is that some of the northward Ekman transport subsequently sinks and flows to the north as part of the intermediate water (10 Sv in the model). The downward limb of the Deacon cell of about 20 Sv that occurs at ~40°S is primarily an artifact of the way that the stream function is calculated along surfaces of constant depth rather than along surfaces of constant density [*Döös and Webb*, 1994; *Hirst et al.*, 1996].

The deep overturning cell of 10 Sv between 60°S and 50°S in figure 2.4.9a likely corresponds to the feature inferred from observations of downward sinking of water along the Antarctic continent to form AABW which subsequently flows to the north (cf. figure 2.4.1). However, the meridional circulation in the model shows this cell as being isolated rather than connected to the AABW.

The difficulties of model P2A and most models in simulating the Southern Ocean circulation correctly are due to many different problems that are difficult to resolve, including the various ways that eddy mixing is parameterized in models and the difficulty in correctly representing buoyancy forcing at the surface of the ocean (i.e., heat and water fluxes).

Summary of Deep Ocean Circulation

One of the most influential attempts to synthesize the deep ocean meridional overturning circulation views it as a "great ocean conveyor" starting with sinking of NADW, which then travels to the circumpolar region, where it is briefly brought back to the surface and cooled again, before continuing on its way into the deep Indian and Pacific Oceans, there to upwell all the way to the thermocline and surface ocean. The great ocean conveyor is closed by the return of Pacific Ocean surface water through the Indonesian Straits, where it joins Indian Ocean surface water to flow around South Africa [cf. *Broecker*, 1991a]. This view grows in part out of the remarkable "tour-de-force" model of *Stommel* [1958], which was concerned primarily with deep water formation and the flow paths that deep water takes after it leaves the deep water formation regions, but had also to deal with the problem of how to get rid of old deep water. *Stommel* [1958] imagined that the removal of old deep water occurred by upwelling that was due to heating of deep water by vertical mixing, and that this deep ocean upwelling joined up with the thermocline upwelling discussed by *Robinson and Stommel* [1959]. More elaborate views of the meridional overturning circulation, exemplified by the work of *Schmitz* [1995], include the various deep and intermediate water types we have identified above and confine the deep upwelling primarily to the North Pacific. These depictions tend to focus one's attention on the formation of NADW as the starting point for the meridional overturning circulation, and the importance of vertical upwelling to close the thermohaline circulation.

An alternative view of the meridional overturning circulation that has begun to emerge in the last decade focuses primarily on the closure of the circulation (that is to say, the removal of old deep water from the deep ocean) as the starting point for understanding the deep circulation. The main organizing principle is the removal of deep water through wind-driven upwelling that occurs in the Southern Ocean (see figures 2.4.10 and 2.4.11, based on *Toggweiler and Samuels* [1993b] and *Gnanadesikan and Hallberg* [2002], respectively). Southern Ocean upwelling is driven by the strong northward Ekman transport at the surface of the ocean due to the westerlies (cf. [*Speer et al.*, 2000]). A significant fraction of the upwelling gets drawn from great depths for reasons that have to do with the dynamics of the Antarctic Circumpolar Current (ACC) in this unique area of the ocean, where there is a completely open zonal path around the world [cf. *Toggweiler and Samuels*, 1993a, 1993b; *Döös and Webb*, 1994; *Gille*, 1997; *Gnanadesikan*, 1999b; *Hallberg and Gnanadesikan*, 2001; *Lee and Coward*, 2003]. As a consequence of this deep upwelling, the main feature of the deep ocean circulation becomes a mid-depth flow to the south, and very little vertical mixing is required, since the winds in the Southern Ocean do most of the work required to bring deep water to the surface. Buoyancy forces help determine where deep water formation occurs, but do not drive the deep circulation as they do in the old theory.

The specific consequences that follow from the fact that upwelling in the Southern Ocean appears to be the primary return path for deep waters to the surface of the ocean include the following (cf. figure 2.4.10):

1. The observational and model analyses suggest that there is a massive deep water flow to the south in all three ocean basins. Presumably, this is due to the drawing up of deep water by the Ekman upwelling in the Southern Ocean.

2. The deep water inflow to the Southern Ocean is fed by two circulation loops. The first is a shallow loop that exists mainly in the Atlantic and involves northward flow through the main thermocline (with input also from the Indian and Pacific Oceans) and formation of NADW in the North Atlantic. The second is a deep loop that involves AABW formation in the Southern Ocean, and northward flow of AABW in all three basins that then upwells and returns as deep water (NADW, NPDW, and IODW in the Indian Ocean [cf. *Schmitz*, 1995]).

3. Because almost all of the upwelling of deep water to the surface of the ocean appears to occur in the Ekman divergence region of the Southern Ocean, the meridional overturning circulation loops associated with it are asymmetric. The deep overturning loop that results from

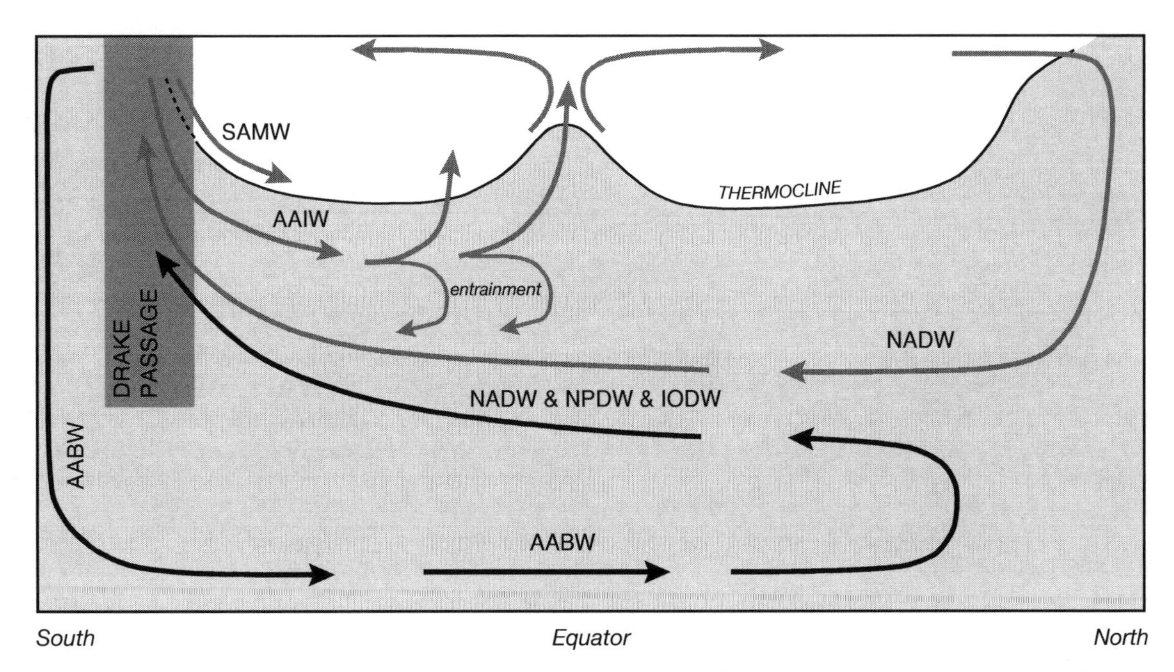

FIGURE 2.4.10: A schematic depiction of the meridional overturning circulation based on the analysis of *Toggweiler and Samuels* [1993b]. See figures 2.4.1 and 2.4.9 for an explanation of the water types. SAMW is Subantarctic Mode Water.

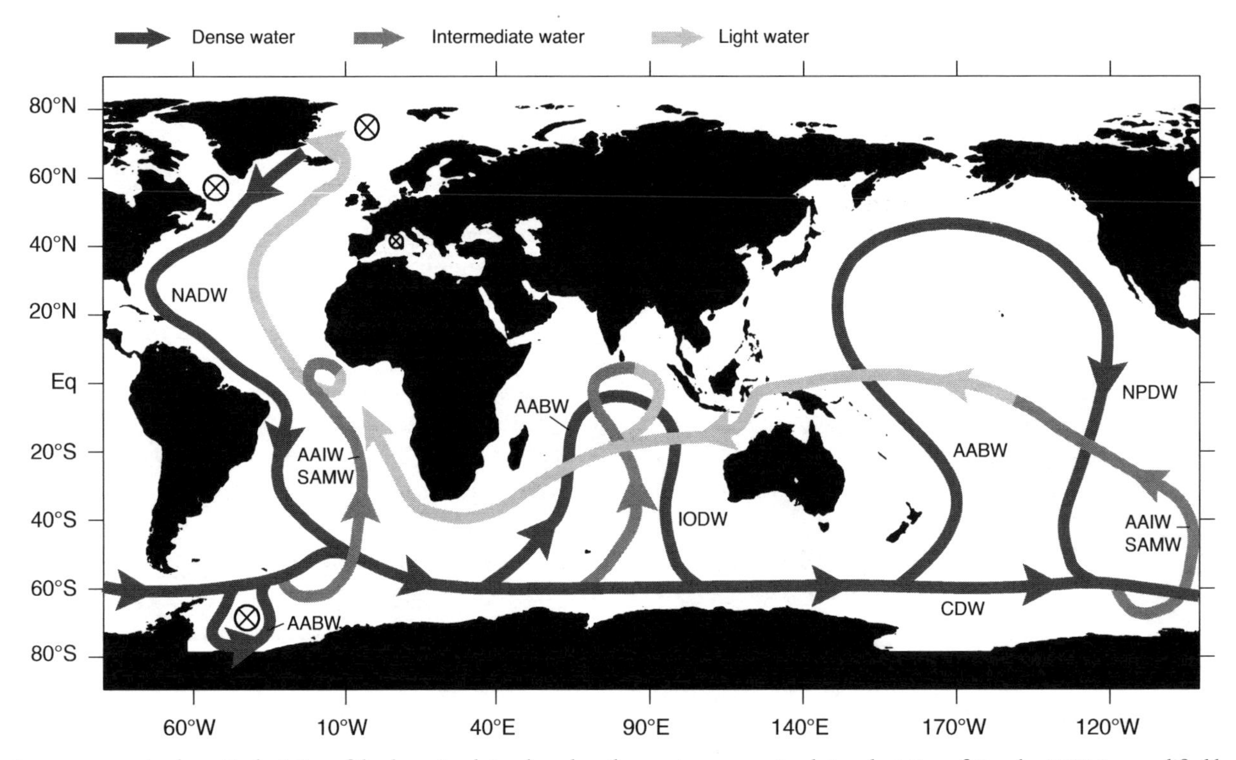

FIGURE 2.4.11: A schematic depiction of the deep circulation based on the great conveyor circulation depiction of *Broecker* [1991a] as modified by *Gnanadesikan and Hallberg* [2002].

southward flow at the surface and formation of AABW occurs very close to where the upwelling is. Thus this water spends very little time at the surface before sinking again. This explains why AABW does not have time to take up any radiocarbon from the atmosphere, and very likely plays a role in explaining why the surface nutrients are so high in this region, a feature we will discuss in chapter 4.

4. By contrast, the shallow overturning loop results from northward flow of AAIW and SAMW in the main thermocline

and surface waters, which then sinks in the North Atlantic to form NADW. The upper branch of this loop spends a great deal of time in the main thermocline and surface waters. Thus, the NADW that forms from this upper ocean water has relatively high radiocarbon and also tends to have lost most of its nutrients before it is cooled and sinks to the abyss.

We show in figure 2.4.11 a revision by *Gnanadesikan and Hallberg* [2002] of the Great Conveyor Belt Circulation diagram of *Broecker* [1991a] which attempts to capture the main features of the above thermohaline circulation in a map view. In this diagram, the AABW and deep water (NADW, NPDW, and IODW) flows are depicted as horizontal dense water circulation loops within each basin in which there is no conversion of "dense" to "intermediate" or "light" waters. We note that the original Great Conveyor Belt Circulation diagram of *Broecker* [1991a] had the bottom water upwelling all the way to the thermocline within the Indian and Pacific Oceans in association with vertical mixing as proposed by *Robinson and Stommel* [1959]. This is not supported by the observations and model we have discussed above [cf. *Wunsch et al.*, 1983]. The conversion of dense (NADW, NPDW, and IODW) to intermediate (AAIW and SAMW) water masses occurs instead within the Southern Ocean, where the strong upwelling associated with a large northward Ekman transport brings deep waters to the surface [cf. *Toggweiler and Samuels*, 1995; *Gnanadesikan*, 1999b]. It is these intermediate waters which flow to the north and are converted to light thermocline waters that eventually return to the North Atlantic to form NADW [cf. *Sloyan and Rintoul*, 2001], thereby closing the upper meridional overturning circulation loop. The Pacific light waters flow through the Indonesian Straits into the Indian Ocean, where they are joined by light waters formed in the Indian Ocean, and thence around South Africa. Note that NADW also includes an input of intermediate water that comes from the circumpolar current flowing around South America.

The remarkable "tour-de-force" model of the deep ocean circulation by *Stommel* [1958] is worth mention, as it is still highly influential in our thinking about deep ocean circulation. This model is based on the geostrophic vorticity equation (2.3.6) together with the assumption that interior mixing drives heat down into the ocean and leads to upwelling (as in the diffusive theories of the thermocline). This upwelling requires a compensatory input of dense plumes of water sinking into the deep ocean from the high-latitude regions where they form. Because the interior upwelling increases as z increases (z is positive upwards), the vertical velocity gradient $\partial w/\partial z$ is positive. From (2.3.6), we see that this requires a flow toward the pole in both hemispheres, directly toward the deep water formation regions! *Stommel* [1958] postulated that the flow away from the source regions must therefore be in narrow, intense, deep western boundary currents analogous to those at the surface. Indeed, one of the great successes of this model was the first observation of such currents after *Stommel* [1958] predicted their existence (cf. North Atlantic in figures 2.4.3 and 2.4.5).

The geostrophic vorticity constraint provides powerful insights on the flow of the deep ocean in the ocean interior away from the boundaries. However, in its original conception, the "tour-de-force" model had water upwelling across density surfaces all the way from the deep ocean to the thermocline before returning from there to the deep water formation regions. The observations we have described are inconsistent with this, suggesting instead that most of the flow in the deep ocean is nearly horizontal, involving only modest interior upwelling.

We conclude our discussion of the meridional overturning circulation with a brief discussion of the influence of geothermal heating on the ocean circulation. Examinations of the impact of a single plume or a series of plumes of hydrothermal waters along a mid-ocean ridge show that the effect is mostly local [e.g., *Stommel*, 1982; *Speer*, 1989] and can probably be ignored in global ocean circulation models. (The release of excess mantle helium-3 by the hydrothermal plumes in the Pacific Ocean provides a remarkable illustration of the deep ocean circulation in this basin [*Lupton*, 1998]). However, a more recent study of a spatially uniform geothermal heat flux shows that it can change the meridional overturning in the ocean by several Sverdrups [*Scott et al.*, 2001], which is quite significant compared to the deep ocean thermohaline circulation rates on the order of a few tens of Sverdrups obtained from the radiocarbon measurements.

2.5 Time-Varying Flows

We began our discussion of ocean circulation by examining systematically the balances between the various forces in the equations of motion (2.2.1) assuming that the ocean circulation is in steady state (i.e., by setting the net acceleration to zero). Inclusion of time dependence gives rise to a rich set of additional behavior including waves, turbulent motions, eddies, convection, large-scale ocean-atmosphere interactions, and many other such processes leading to variability on all time- and space-scales (see figure 2.4.7). This variability is a fundamental property of the ocean and the climate system, and therefore the ocean circulation can never be regarded as being in a true steady state. In this section, we discuss variability on two spatial and temporal scales.

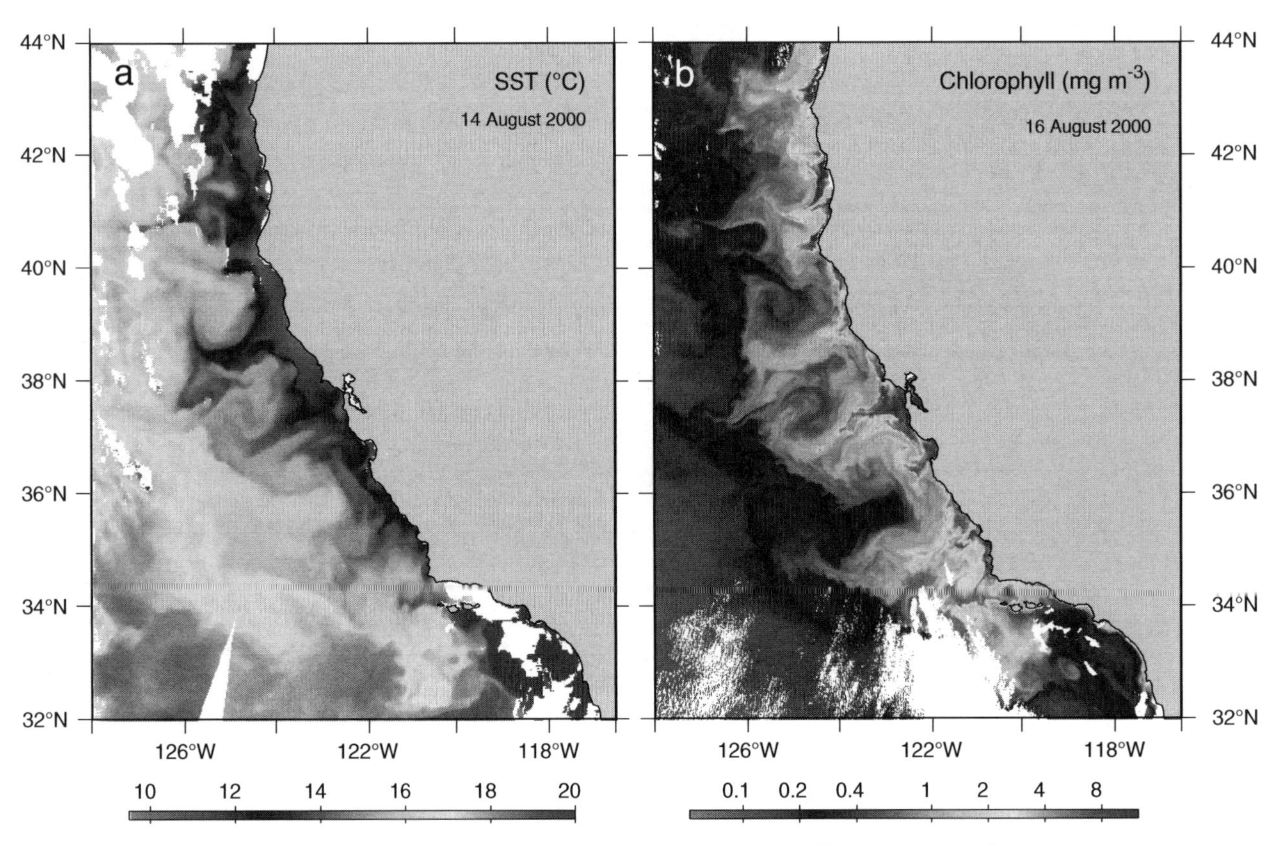

FIGURE 2.5.1: Satellite images of (a) sea-surface temperature (SST) and (b) chlorophyll-*a* off the West Coast of North America showing the influence of mesoscale eddies and other turbulent features on both properties. Image produced by John Ryan from the Monterey Bay Aquarium Research Institute (personal communication, 2005) using SST data from the Advanced Very High-Resolution Radiometer (AVHRR) sensor (obtained from Dave Foley, NOAA/PFEL) and chlorophyll a data from the Sea-viewing Wide Field-of-view Sensor (SeaWiFS; obtained from NASA/Goddard).

We first talk briefly about the role of mesoscale variability in the ocean, which is among the most important contributors to oceanic variability in phytoplankton productivity [see, e.g., *Doney et al.*, 2003]. We then proceed to a discussion of variability on interannual and decadal timescales. We will focus primarily on an interannual phenomenon in the tropics, the El Niño–Southern Oscillation (ENSO), which has received much attention in the public during recent years. We will then proceed to a discussion of extratropical variability patterns that have emerged in various recent analyses. *Enfield* [1989], *Philander* [1990], and *Cane* [1992] provide more detailed treatments on ENSO and its influence on world climate, while *Stocker* [1996] gives an overview of climatic variability on decadal and longer timescales.

Mesoscale Variability

The ocean is turbulent in nature. Any instantaneous image of sea-surface temperature or surface chlorophyll reveals a rich spectrum of spatial variability with a particular concentration of energy at the mesoscale, i.e., on spatial scales of tens to a few hundred kilometers and at timescales of a few days to weeks (figure 2.5.1). The turbulent nature of the ocean is a result of instabilities that occur because the full momentum equations characterizing the flow have acceleration terms that depend on the velocity of the flow. As the flow increases, the influence of these nonlinear terms grows, destabilizing the large-scale flow. Starting from a laminar flow regime, where all paths along which a given particle follows are parallel, the destabilization is often characterized first by the appearance of meanders. After some critical threshold is reached, the flow becomes progressively more unstable, first shedding *vortices* or circular flow features, and eventually becoming fully turbulent.

There are many processes that create oceanic turbulence. Many energetic currents in the ocean, such as the western boundary currents and the Antarctic Circumpolar Current (ACC, also known as the West Wind Drift; cf. figures 2.2.3, 2.2.5, and 2.2.9), are well beyond the critical threshold of instability, explaining their intense meandering and eddy shedding. In this case the instability is driven by the large-scale flow, with the energy mostly coming from the kinetic energy of this flow (*barotropic instability*). Instabilities can also draw their energy from the potential energy contained in

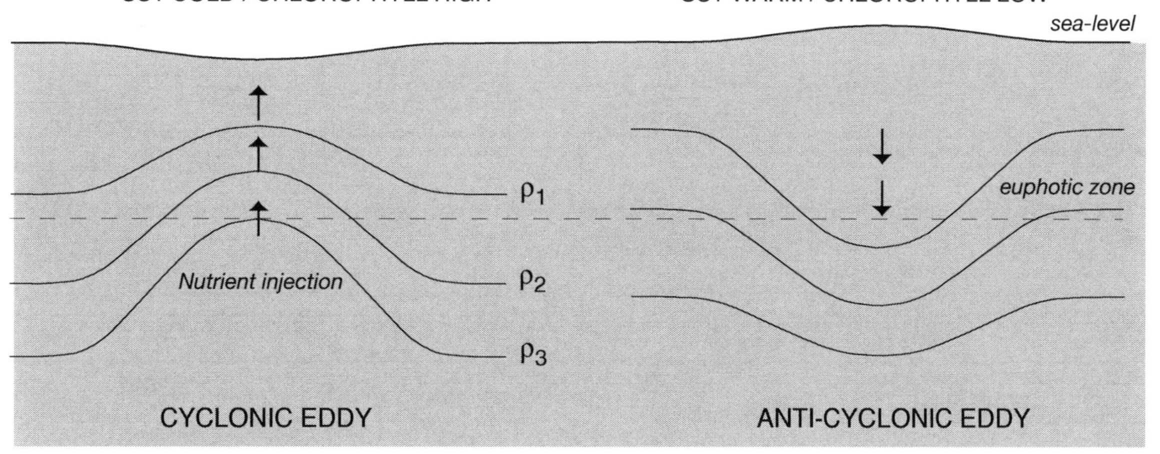

FIGURE 2.5.2: Schematic representation of the impact of (a) cyclonic and (b) anticyclonic eddies on surface ocean productivity. Adapted from *McGillicuddy et al.* [1998].

tilted isopycnals associated with the flow (*baroclinic instability*). Other sources of instability include flow past obstacles, such as islands, and other interactions with complex topography, as well as external forces such as wind and buoyancy fluxes.

Mesoscale eddies are a particularly prominent form of turbulence in the ocean because (a) theirs is the scale that tends to be preferentially formed in unstable flows, and (b) they tend to have a longer lifetime than most other turbulent phenomena. The scale of mesoscale eddies is a few hundred kilometers in the tropics, of order a hundred kilometers in the subtropics, and a few tens of kilometers in the high latitudes [*Chelton et al.*, 1998]. Mesoscale eddies tend to be geostrophically balanced, i.e., the pressure gradient force balances Coriolis force (see equations (2.3.1) and (2.3.2)), explaining in part their relatively long lifetime. The mesoscale eddies in the ocean are equivalent to the high- and low-pressure systems that characterize most of the atmospheric state at any given moment.

The most important effect of eddies on oceanic properties is to act as a mixing agent, thereby homogenizing gradients. However, it is now also well established that eddies can induce a mean transport in certain areas, such as the Antarctic Circumpolar Current or in the equatorial regions, and that this transport can be of leading order importance in determining the mean state of the system [e.g., *Stammer*, 1998; *McWilliams and Danabasoglu*, 2002]. Evidence that eddies and other meso- and sub-mesoscale phenomena can also fundamentally alter the distribution and dynamics of marine ecosystems started to accumulate in the late 1970s [e.g., *Denman*, 1976]. But it is only in the last 20 years, with the advent of satellite and autonomous *in situ* observation capabilities as well as high-resolution modeling, that such variations have come to be recognized as important components of marine biology and biogeochemical cycles. While earlier work focused on the role of mesoscale dynamics on phytoplankton patchiness (see review by *Denman and Dower* [2001]), much research in the past decade has studied the role of mesoscale processes in enhancing surface ocean productivity [e.g., *Falkowski et al.*, 1991; *McGillicuddy et al.*, 1998; *Oschlies and Garcon*, 1998; *Abraham et al.*, 2000].

The main reason for the enhancing effect of mesoscale eddies is that cyclonic eddies, i.e., those rotating around a depression of sea level, have upward-tilting isopycnals in the interior that may bring deeper isopycnal surfaces with elevated nutrients into the euphotic zone (figure 2.5.2). In nutrient-limited ecosystems, this may lead to an enhancement of biological productivity. Anticyclonic eddies, i.e., those rotating around an elevation of sea level, have downward-tilting isopycnal surfaces in the interior, leading to little expected change in productivity (figure 2.5.2). Given this differential response, it has been suggested that the net effect of eddies in such nutrient-limiting systems is to increase biological productivity. The magnitude of the influence of eddies on biological productivity is an issue of current debate [*McGillicuddy et al.*, 1998; *Oschlies*, 2001], but there is little doubt that mesoscale phenomena are among the dominant features in the spectrum of chlorophyll variability [*Doney et al.*, 2003].

INTERANNUAL TO DECADAL VARIABILITY

TROPICAL VARIABILITY

The single most prominent signal in year-to-year variability in the atmosphere-ocean system is the El Niño/Southern Oscillation (ENSO), which has its center of action in the tropical Pacific Ocean. Before we look at the ENSO phenomenon in more detail, it is instructive to discuss first the climatological mean state of the tropical Pacific.

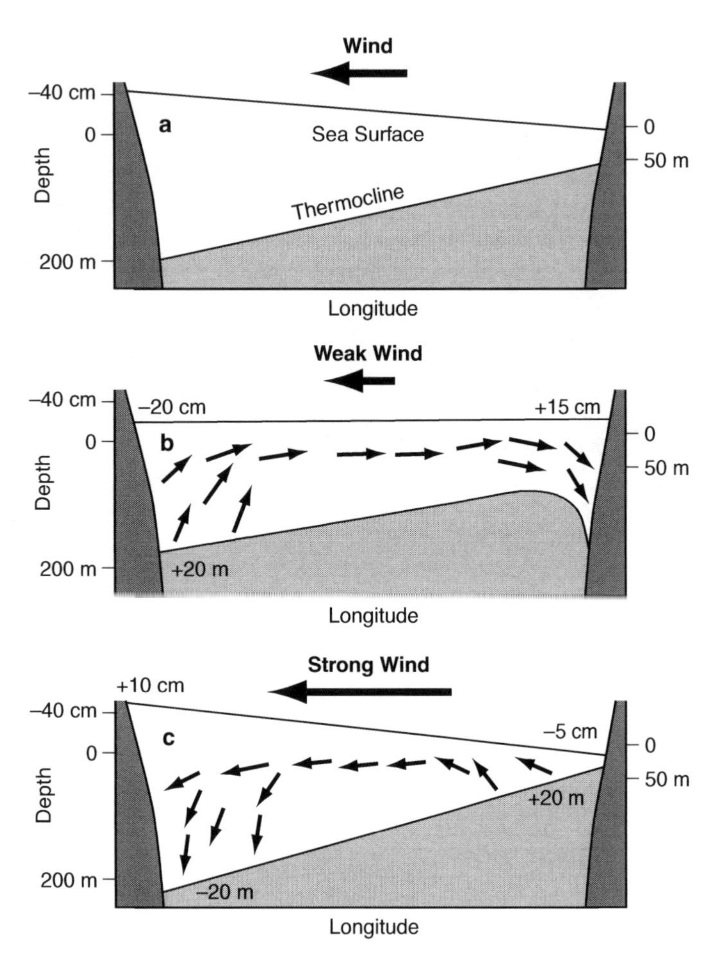

FIGURE 2.5.3: Response of the thermal structure of the equatorial Pacific to changes in surface winds [*Peixoto and Oort*, 1992]. Depth is positive downward. (a) Under normal easterly trade wind conditions, sea level rises to the west and the thermocline deepens. (b) When the trade winds relax, equatorial Kelvin waves propagate eastward, which leads to a rise in sea level and a deepening of the thermocline near the South American coast (El Niño conditions). (c) The normal situation is amplified during strong trade winds (La Niña conditions).

The tropical Pacific is characterized by warm surface water (29–30°C) in the west but much cooler temperatures in the east (22–24°C). The body of warm water in the west in known as the *Pacific warm pool.* It is associated with the Indonesian Low, a region of atmospheric heating, rising air, and therefore intense rainfall. This rising region constitutes one branch of a global zonal circulation, the so-called *Walker circulation.* A region of broad sinking over the eastern Pacific closes the Pacific cell of the Walker circulation. The cooler sea-surface temperatures in the eastern Pacific are the result of cold waters upwelling from below, forced by the divergent Ekman transport (see figure 2.2.8a) due to the trade winds. The westward-blowing trade winds also cause a deepening of the thermocline in the west, and a sea level difference of about 40 cm between the east and the west (figure 2.5.3). It is the pressure gradient resulting from this sea-surface height difference that gives rise to the Equatorial Undercurrent (EUC) shown in figure 2.3.14. Collectively, these east-west gradients and the associated westward-blowing trade winds constitute a state of quasi-equilibrium for the ocean-atmosphere system.

Every few years, this state of equilibrium breaks down, giving rise to an ENSO event. As the name suggests, ENSO consists of two components. The first (mainly oceanic) component, El Niño, has historically been associated with a warm, weak north-to-south current appearing annually around December off the coasts of Ecuador and Perú [*Philander*, 1990]. During El Niño conditions, the current is more intense than normal, and flows further to the south, greatly expanding the region of warm waters and leading to widespread mortality of fish and other marine wildlife. Very heavy rains in this region also accompany it. More recently it has been realized that these anomalies are connected to a much larger scale phenomenon in which the normally cold waters of the entire eastern tropical Pacific show dramatic warming. Therefore, the term "El Niño" is now used in a much broader sense to describe the warm state of the entire eastern tropical Pacific.

The second (mainly atmospheric) component of ENSO, the Southern Oscillation, is associated with large east-west shifts of mass in the tropical atmosphere between the Australian-Indonesian regions (the locus of the Indonesian Low) and the southeastern tropical Pacific (the locus of the South Pacific High). *Walker* [1924] first described this oscillation; however, it was not until the 1960s that the connections between the El Niño and the Southern Oscillation phenomena were discovered. Today these phenomena are considered two aspects of one global-scale oscillation in the combined ocean-atmosphere system.

The tight coupling of the oceanic and atmospheric anomalies is evident in figure 2.5.4, which compares an index of the Southern Oscillation with sea-surface temperature anomalies in the tropical Pacific. This index is based on the normalized pressure difference between Tahiti and Darwin, Australia, and therefore measures the strength of the Indonesian Low versus the South Pacific High. Figure 2.5.4 also shows that El Niño events, i.e., the warm excursions of sea-surface temperature in the tropical Pacific, are just one phase of the oscillation. The complementary phase, i.e., the cold anomaly, is often referred to as La Niña [*Philander*, 1990].

Cane [1992] and *Philander* [1990] have summarized the typical evolution of a warm ENSO event. In the early stages of an El Niño, large shifts in atmospheric mass occur so that the surface pressure decreases in the central and southeastern tropical Pacific and increases over the western Pacific and Indian Oceans, resulting in a weakening of the trade winds. The onset of an El Niño is then typically marked in the western Pacific by a series of prolonged bursts of winds from the west [*Webster and Peterson*, 1997]. They persist for one to three weeks and

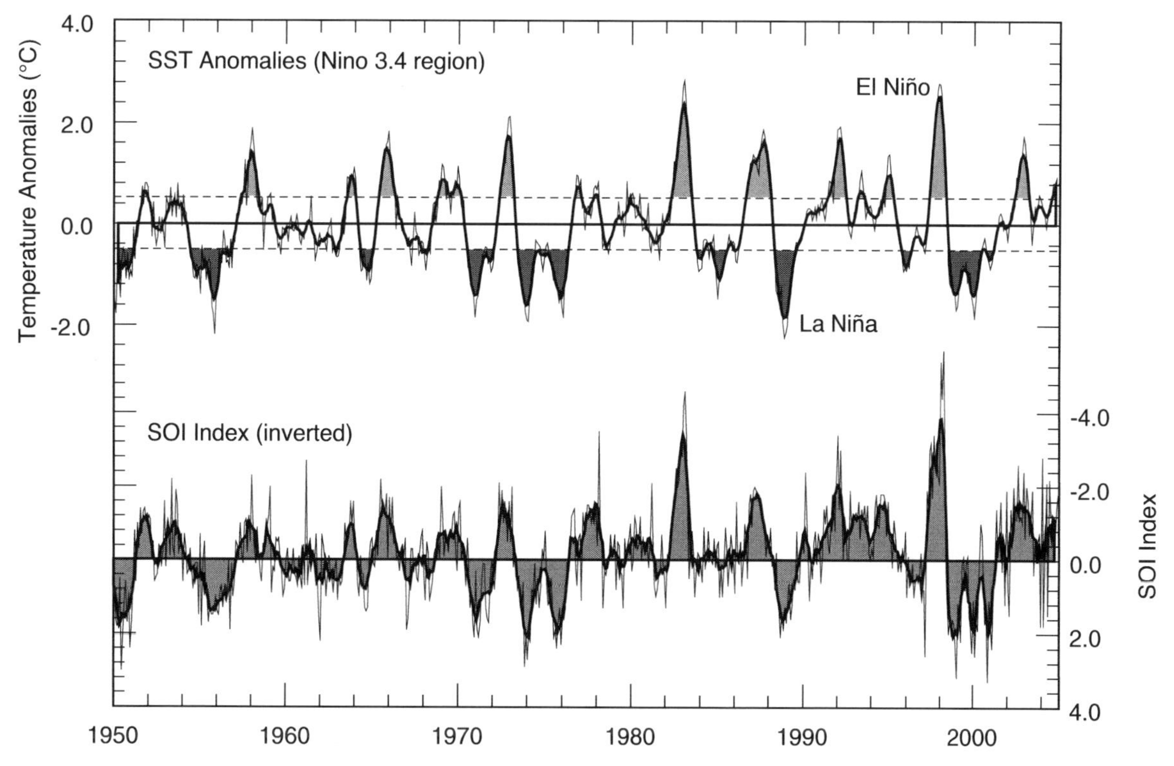

FIGURE 2.5.4: (*Top*) Time series of sea-surface temperature anomalies in the Niño 3.4 region in the Pacific region (120°W–170°W, 5°S–5°N). The anomalies have been calculated relative to a base climatology of 1950–1979 using data from *Trenberth and Hoar* [1997]. (*Bottom*) Time series of the *Southern Oscillation Index* (SOI). This index has been computed using monthly mean sea level pressures at Tahiti and Darwin, Australia. The thin lines indicate monthly values, while the thick lines depict the six-month running mean. Arrows indicate major El Niño events as defined by *Trenberth and Hoar* [1997].

replace the normally weak winds from the east over the warm pool. The most important impact of these wind bursts is to severely perturb the upper ocean and excite the eastward propagation of large-scale so-called *planetary waves*. By analogy with large surface waves in the ocean, where the restoring force for the disturbance to the surface that is caused by the waves is gravity, planetary waves are waves for which the restoring force is the Coriolis force. The particular type of planetary wave found at the equator is a so-called equatorial *Kelvin wave*. These waves have their largest amplitude in the thermocline and have wavelengths of thousands of kilometers. They are trapped at the equator (until they hit the continents on the eastern side of the basin) and owe their existence to the change of sign of the Coriolis force at either side of the equator. The main effect of these Kelvin waves is to suppress the upwelling of cold water in the eastern basin and to deepen the thermocline (figures 2.5.3b and 2.5.5, middle panel). This leads to an apparent eastward spreading of the warm water pool. However, this spreading is not a consequence of advective transport, but rather the result of the aforementioned changes in the vertical structure of the thermocline associated with the Kelvin wave.

The oceanic response to the change in atmospheric forcing is not passive, however. The eastward spreading of the warm water pool increases the heat flux from the ocean to the atmosphere in the central portions of the Pacific. The intense convective zone in the atmosphere over the western Pacific shifts eastward, further weakening the intensity of the trade winds in the central Pacific (figure 2.5.5). This positive feedback intensifies the oceanic anomalies. In the mature state of an El Niño, the warm sea-surface temperature anomalies extend across the entire eastern tropical Pacific from 20°S to 20°N, and even further poleward along the coast of the Americas (figure 2.5.6). A precursor of the end of an El Niño event is the appearance of cold surface waters in the eastern equatorial Pacific by mechanisms that are only poorly understood. These low sea-surface temperatures spread westward and inaugurate the opposite phase of the ENSO, La Niña. The average duration of an El Niño episode is on the order of 18 months.

It is important to note that ENSO is linked to the seasonal cycle of sea-surface temperatures in the tropical Pacific. Instead of cooling early in the northern spring, the eastern Pacific continues to warm. This observation has led to the speculation that the atmosphere-ocean instability that produces ENSO results from the inability of the warm pool to export enough heat each year, so that its heat content increases with time [*Webster and Peterson*, 1997]. The warm pool

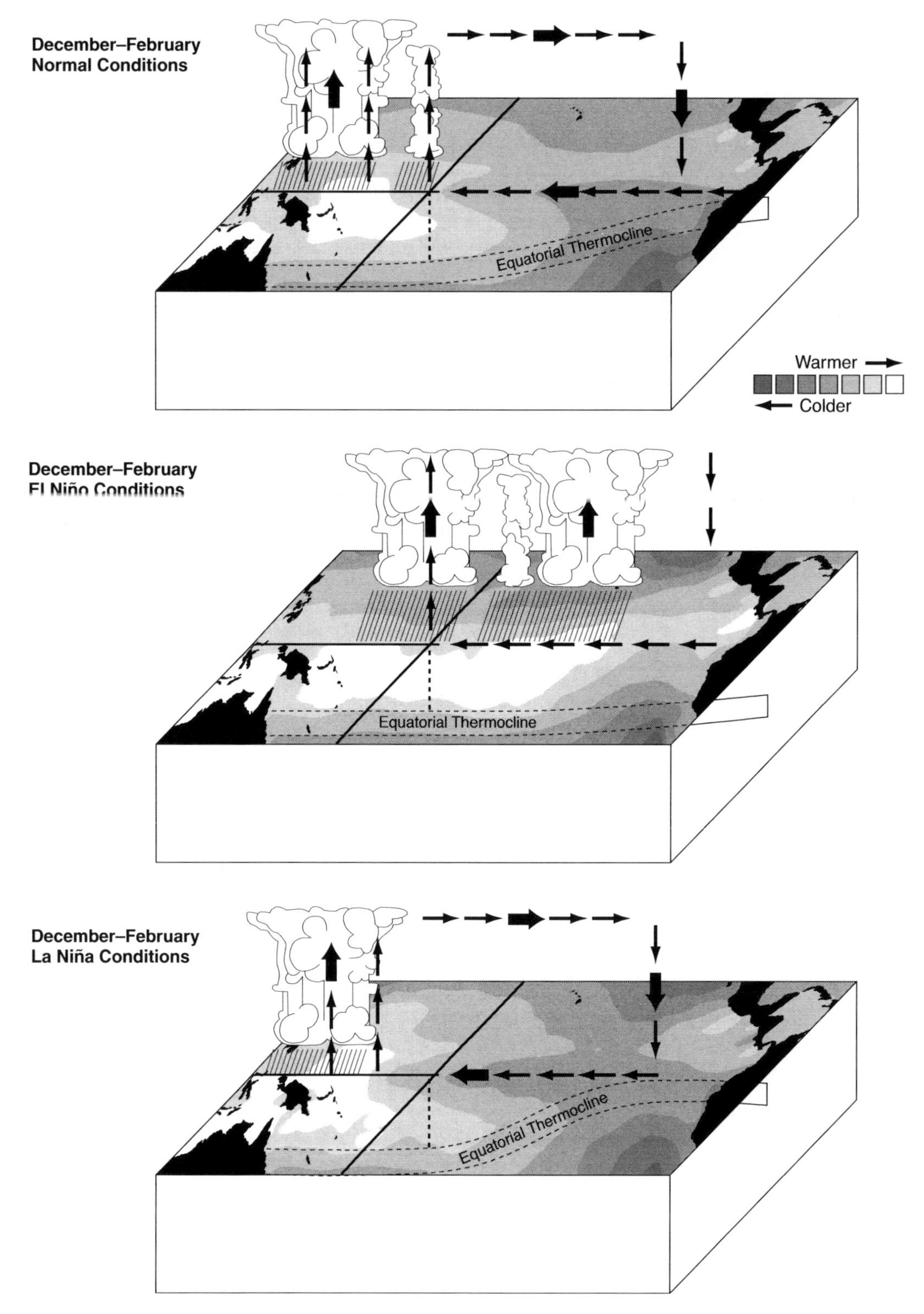

FIGURE 2.5.5: Schematic view of sea-surface temperature and tropical rainfall in the equatorial Pacific Ocean during normal, El Niño, and La Niña conditions. The sea-surface temperature is shaded, with dark being cold and light being warm. The arrows indicate the direction of air movement in the atmosphere: upward arrows are associated with clouds and rainfall and downward arrows are associated with a general lack of rainfall. The depth of the thermocline under the equator is depicted by the box labeled "Equatorial Thermocline." Adapted from http://www.cpc.noaa.gov/products/analysis_monitoring/ensocycle/enso_cycle.html.

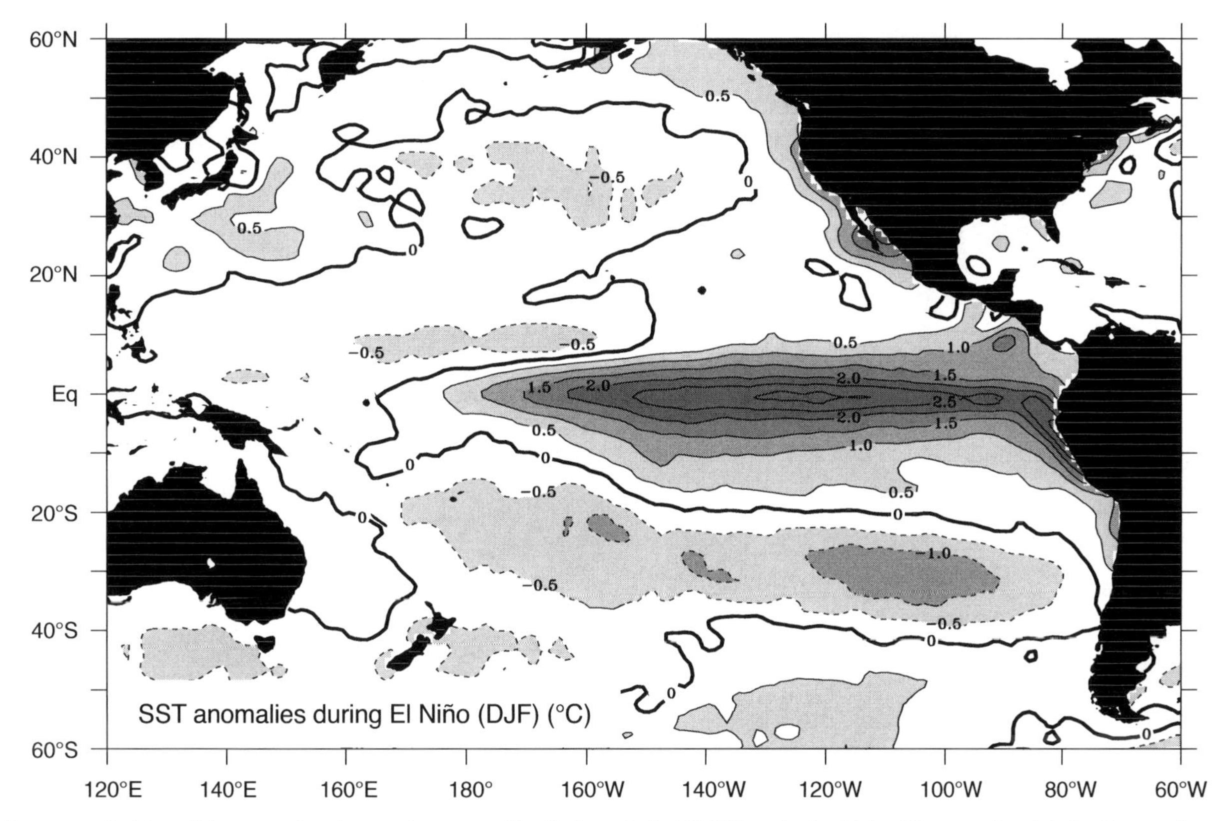

FIGURE 2.5.6: Map of the sea-surface temperature anomalies during a typical El Niño episode obtained by averaging data for the months December through February for the three major El Niño episodes between 1982 and 2001, i.e., 1982–83, 1987–88, and 1997–98. Data prepared by Holger Brix (personal communication, 2005) on the basis of the Reynold's optimum interpolated SST data provided by the NOAA-CIRES Climate Diagnostics Center, Boulder, Colorado (http:www.cdc.noaa.gov/).

is then preconditioned for El Niño, and the bursts of winds from the west may serve as triggers to release the stored energy.

The large changes in the ocean circulation, particularly the strength and source of upwelling, lead to dramatic changes in ocean biogeochemistry [*Feely et al.*, 1999; *Keeling and Revelle*, 1985], biology [*Barber and Chavez*, 1983; *Chavez et al.*, 1999; *Lehodey et al.*, 1997], and ecosystem structure [*Karl et al.*, 1995]. The impact of ENSO on the oceanic carbon cycle and atmospheric CO_2 will be discussed in detail in chapter 10.

ENSO is a regional phenomenon, but its impact is global. As a direct consequence of the large shift of atmospheric weather patterns over the Pacific, the warm phases of ENSO are usually accompanied by severe drought over Australia and Indonesia, together with weakened summer monsoon rainfall over South Asia [*Ropelewski and Halpert*, 1987]. Conversely, catastrophic flooding often occurs along the Pacific coast of South America. Moreover, coupling between ENSO and the large-scale atmospheric circulation may affect the climate of the extratropical regions remote from the Pacific [*Wallace and Gutzler*, 1981]. Such coupling tends to change the probability of certain weather regimes in a particular region. For example, the chance of strong winter storms over the southwestern and southern United States is significantly enhanced, leading, on average, to more precipitation.

A phenomenon quite similar to El Niño exists in the tropical Atlantic [*Philander*, 1990; *Zebiak*, 1993], but with a much smaller amplitude, likely a direct result of the fact that the width of the Atlantic is far smaller than that of the Pacific. In the tropical Pacific, interannual variations in the east and west are negatively correlated, whereas in the tropical Atlantic, they tend to be in phase and relatively uniform in the east-west direction. On the other hand, analyses of sea-surface temperatures in the tropical Atlantic have revealed that an important mode of variability exists in the meridional direction that appears to be symmetrical about the equator [*Chang et al.*, 1997; *Nobre and Shukla*, 1996]. The fluctuations associated with this quasi-dipole are of much smaller amplitude than those in the Pacific, but nevertheless have a pronounced effect on rainfall variations in northeastern Brazil and the Sahel region. Although there is still considerable debate over the exact structure of this dipole-like variability [*Enfield and Mayer*, 1997], it is very likely that it represents, much like ENSO, an unstable coupled mode of the ocean-atmosphere system [*Chang et al.*, 1997; *Zebiak*, 1993].

Extratropical Variability

Are there areas outside of the tropics that exhibit similar coherent variability on the interannual to decadal time-scale? In order to address this question, *Kawamura* [1994] performed a so-called empirical orthogonal function (EOF) analysis of global sea-surface temperature (SST). Such an analysis attempts to split the temporal variance in the observations into sets of spatially correlated patterns (the EOFs) and their associated magnitude of variability (the principal components or PCs). The EOFs that account for a large fraction of the variability are, in general, considered physically meaningful and connected with important centers of action.

Figure 2.5.7 shows the spatial pattern of the first four rotated EOF patterns and figure 2.5.8 shows the corresponding time series of the principal components. The spatial structure of EOF 1 shows strong positive values over the entire eastern tropical Pacific, extending toward the north along the coast of the Americas. The temporal variability of this pattern has a quasi-periodicity of 2–5 years with little inter-decadal variability. From the resemblance of the spatial pattern with the SST anomalies seen in figure 2.5.6 and from the similarity of the time-evolution of the EOF with the SO index shown in figure 2.5.4, this mode can clearly be identified with the ENSO phenomenon. This confirms the notion that ENSO is the most important mode of oceanic variability on the interannual to decadal timescale.

The spatial structure of EOF 2 shows positive values over the tropical and subtropical Indian Ocean, while pronounced negative values exist over the temperate North Pacific. This variability appears to be associated with the Pacific Decadal Oscillation (PDO) described below. The time series of the principal component of EOF 2 is drastically different from that of EOF 1. The time series of EOF 2 exhibits a long-term trend, with some short-term interannual variability overlaid.

By contrast with EOF 1 and EOF 2, the EOF 3 and EOF 4 patterns are confined to the Atlantic Ocean: EOF 3 to the North Atlantic, and EOF 4 to the tropical Atlantic. EOF 4 is clearly related to the tropical Atlantic variability described above, and EOF 3 is associated with the North Atlantic Oscillation (NAO) discussed below. Both time series have a strong decadal variability.

In summary, the analysis of *Kawamura* [1994] reveals two centers of action for extratropical variability: the North Pacific, where variability is linked to the PDO, and the North Atlantic, where variability is dominated by the NAO. These two patterns will now be discussed briefly.

The Pacific Decadal Oscillation pattern is the leading mode of sea-surface temperature variability over the North Pacific ([*Mantua and Hare*, 2002]; see also EOF 2, figures 2.5.7b and 2.5.8b). The PDO is associated with a deepening and southwestward shift of the Aleutian Low, which causes an intensification of the westerlies over the central North Pacific, and enhanced southerly winds along the coast of North America. These changes lead to a cooling over much of the temperate and subpolar central and western Pacific, whereas the eastern North Pacific (particularly the Gulf of Alaska) experiences anomalously warm waters. These SST changes in the temperate and subpolar central and western Pacific can be explained by increased heat loss caused by higher winds and deeper mixing, which brings cold waters to the surface. The surface anomalies propagate into the interior of the ocean and can be seen down to depths of over 400 m [*Deser et al.*, 1996]. The mechanisms causing the PDO are only poorly understood [*Mantua and Hare*, 2002]. On the interannual timescale, the PDO seems to be connected with the ENSO phenomenon. However, on longer timescales, the PDO is currently believed to represent an independent mode of variability.

The North Atlantic Oscillation (NAO) is the primary source of climatic variability on interannual to decadal timescales over the North Atlantic region (see also EOF 3; figures 2.5.7c and 2.5.8c). The atmospheric component of this phenomenon was discovered early in this century when meteorologists noticed that year-to-year fluctuations in wintertime air temperatures on either side of Iceland were often out of phase with one another [*Walker and Bliss*, 1939]. More recent research has demonstrated that the NAO can be understood as the North Atlantic expression of a whole Arctic phenomenon, dubbed the *Northern hemisphere Annular Mode* (NAM) [*Thompson and Wallace*, 2000]. This name originates from the observation that the primary mode of variability over the poles is zonally symmetric, and is associated with variations in the strength of the polar vortex. As the NAO and NAM are essentially the same phenomenon and our focus is on the North Atlantic, we continue to use here the historical term NAO. An index of the NAO has been defined as the normalized pressure difference between the Icelandic low and the Azores high, i.e., the meridional atmospheric pressure gradient over the North Atlantic.

A spectral analysis of the time history of the NAO index reveals that it contains significant variability at periods of 2 to 3 years and at periods of 6 to 10 years. These are significantly different from the ENSO periods. When the NAO index is high, the Icelandic low is anomalously deep and the Azores high exceptionally high. This causes a strengthening of the surface winds from the west across the Atlantic and increased southward flow of cold polar air over the Labrador Sea. These anomalous winds lead to a drastic warming over the European continent, whereas northeastern Canada and the Labrador Sea experience anomalously cold conditions.

The anomalous patterns of the NAO must leave an imprint in the ocean. On interannual timescales, sea-surface temperature anomalies across the entire At-

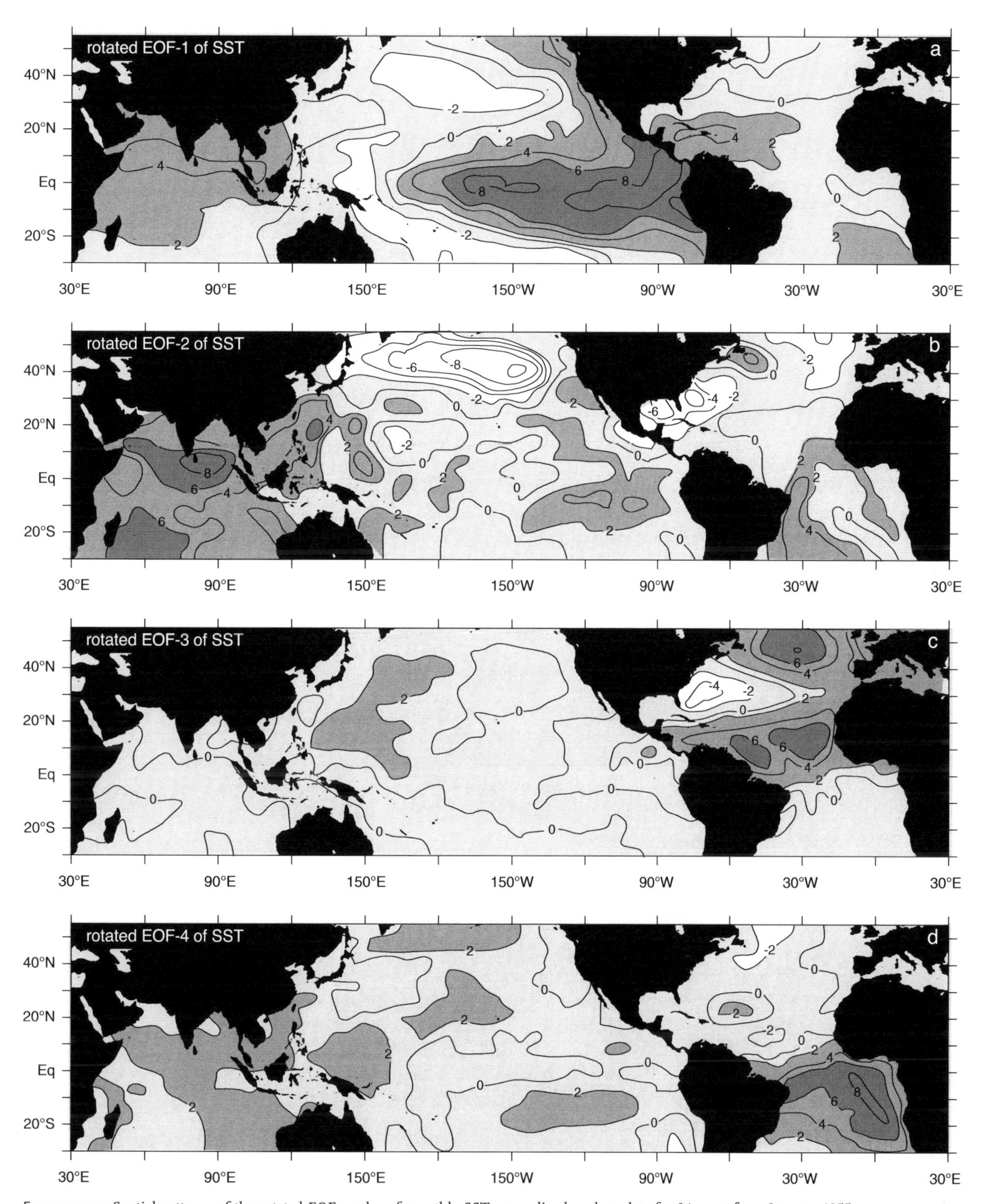

FIGURE 2.5.7: Spatial patterns of the rotated EOF modes of monthly SST anomalies based on data for 34 years from January 1955 to December 1988. Contour interval is 2.0 in relative units. (a) through (d) show the first through fourth modes, respectively. These figures are adapted from *Kawamura* [1994].

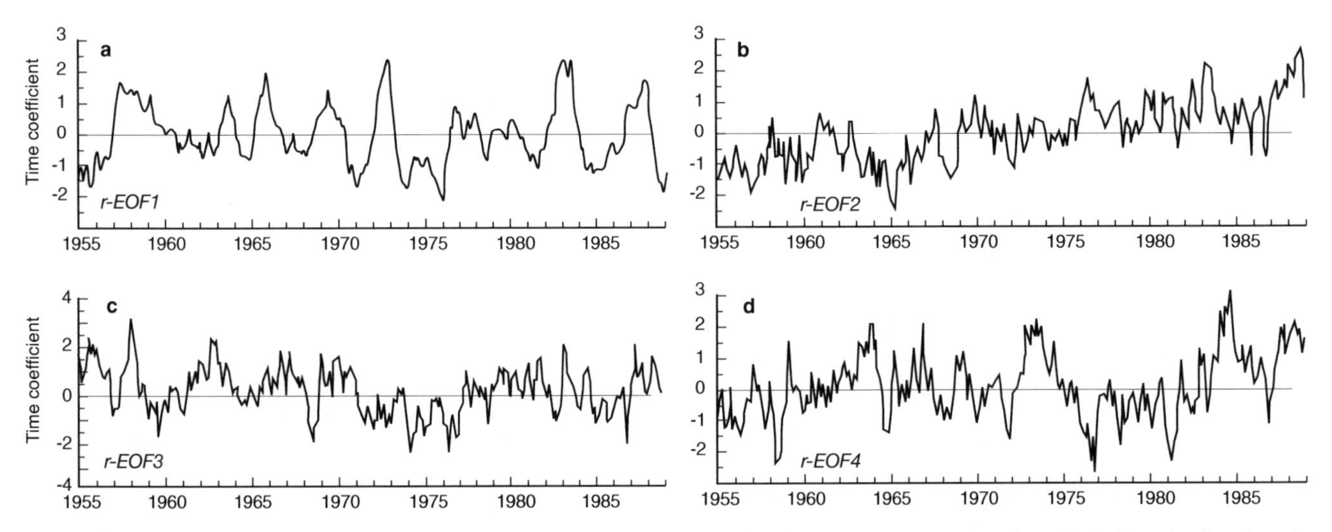

FIGURE 2.5.8: Time series of the rotated EOF modes of monthly SST anomalies shown in Figure 2.5.7. (a) through (d) show the first through fourth modes, respectively. These figures are adapted from *Kawamura* [1994].

lantic correlate very well with the NAO. Periods of high NAO lead to warm conditions in the western subtropical gyre and in the Norwegian Sea, whereas they lead to cold SST anomalies in the Labrador Sea. On these shorter timescales, the ocean response is probably just a passive response to the atmospheric forcing. However, variations associated with lower frequencies in the NAO will be reflected in subsurface water masses, because the ocean has a memory for winter conditions and integrates them over many years. Indeed, subsurface observations provide a clear depiction of the large changes that occurred in the North Atlantic region between the late 1960s, when NAO was mostly negative, to the early 1990s, when NAO was at a record high. In the late 1960s, convection in the Labrador Sea was tightly capped, whereas deep convection occurred in the Greenland Sea. By the early 1990s, conditions had completely reversed. This evolution is reflected in a substantial cooling and freshening of the mean properties of the Labrador Sea Water while the deep waters in the central Greenland sea have warmed and become saltier. The changes are not restricted to the high latitudes. For example near Bermuda, a substantial freshening of the subtropical mode water was detected over this period [*Dickson et al.*, 1996]. Furthermore, wintertime mixing near Bermuda is also strongly affected by the state of the NAO, with positive phases usually associated with shallow winter mixed layers, and vice versa. We will discuss the impact of these variations on the upper ocean carbon cycle and biological productivity in chapter 10.

We conclude with a brief mention of the modes of variability in the Southern Ocean. *White and Peterson* [1996] recently suggested that there exists a coherent large-scale variability pattern in sea-ice extent, sea-surface temperature, wind speed, and atmospheric pressure in the Southern Ocean. Analysis of observations over the last 20 years suggest that there are two such disturbances, one with a period of oscillation of about 3.3 years that is driven by sea-air interactions within the Southern Ocean; and a second with a period of about 5 years that may be forced by ENSO [cf. *Venegas*, 2003; *Turner*, 2004]. The disturbances propagate eastward around Antarctica in a wave-like pattern. These variability patterns have been dubbed the *Antarctic Circumpolar Wave* (ACW). The putative ACW owes its existence to the fact that the Southern Ocean is the only oceanic domain encircling the globe, and it is dominated by the strong eastward flow of the Antarctic Circumpolar Current. An alternative mode of Southern Ocean variability is thought to be associated with the Southern hemisphere Annular Mode (SAM), which is the counterpart to the NAM discussed above. Although this mode is the dominant mode of variability in the southern hemisphere in the atmosphere [*Thompson and Wallace*, 2000], relatively little is known about its impact on the ocean. *Hall* [2002] demonstrated on the basis of an analysis of a long integration of a fully coupled climate model that this mode and not the ACW explains most of the variations in oceanic properties.

Problems

2.1 A rocket launched at 10°S in the direction of the South Pole appears to be deflected to the left. Explain why.

2.2 Describe the pathway a rocket would takc if it were launched from New Zealand in the direction of the North Pole.

2.3 Name two regions where upper ocean Ekman transport induced by wind leads to downwelling and two regions where this effect leads to upwelling. Explain the reasons for your choices.

2.4 Suppose the Earth were shaped like a cylinder, and rotated around its own axis at the same angular velocity as Earth. Would a moving body on the surface of this cylinder be subject to the Coriolis force?

2.5 During La Niñas, the trade winds in the eastern equatorial Pacific are particularly strong. Discuss the implications of this observation for upper ocean circulation and biology during these periods. Draw a latitude-depth diagram ($\sim$15°S to $\sim$15°N, 300 m to surface) that shows the direction of the zonal mean winds, as well as the mean zonal and meridional currents. Then discuss how the strengthening of the trades changes this picture as well as impacts on upper ocean biology.

2.6 A long-term atmospheric pressure anomaly has led to a persistent strengthening of the westerlies and to a slackening of the trade winds by the same amount.

a. Discuss what happens to the strength of the equatorial upwelling and the strength of the subtropical downwelling.

b. How is the strength of the subtropical gyre affected?

2.7 Explain the term *geostrophy*. Name at least one situation in the ocean where this term is appropriate to use.

2.8 During El Niño years, the prevailing winds off the coast of Perú (southern hemisphere) change from being northward to being southward. Draw and explain the coastal circulation pattern before and during such years. Discuss the implications for the transport of nutrients from the deeper waters to the surface.

2.9 In 2092, nearly 100 hundred years after the first extra-solar planets were observed, two astronomers in Geneva find a solar system that rotates in the opposite direction from our solar system. They are particularly excited by the finding of a planet that looks almost identical to Earth, with oceans and continents, but which rotates in the opposite direction from our Earth.

a. In which directions would you expect the trade winds to blow on this planet?

b. Would you expect upwelling or downwelling on this planet near its equator?

c. Assuming that the strength of the trade winds is similar on this planet to that on Earth, how strong is the equatorial up- or downwelling on this planet relative to that on Earth?

d. In which direction would you expect the subtropical gyres to spin?

e. Would we find strong boundary currents (such as the Gulf Stream) in the oceans on this planet? If yes, on which side of the oceans basin would you expect them to be found?

2.10 Many decades before the astonishing finding of the astronomers from Geneva, a team of U.S. astronomers identify a planet that rotates twice as fast as Earth, yet in the same direction. In addition, they find that this planet has a very similar distribution of oceans and continents as Earth. Suppose that the strength and direction of the winds on this planet are similar to those on Earth.

a. How does the strength of the coastal upwelling and downwelling on this planet compare with that on Earth?

b. Would you expect a stronger or weaker subtropical gyre? Explain your answer.

c. How strong do you expect the western boundary currents to be on this planet relative to the Earth?

2.11 Draw a schematic latitude by depth section of salinity in the Atlantic Ocean and identify the following major water masses:

- North Atlantic Deep Water
- Antarctic Intermediate Water
- Antarctic Bottom Water

2.12 What happens to the density of a seawater parcel

a. if freshwater is added?

b. if a strong wind cools it?

c. if it is moved from the surface to 3000 m depth?

2.13 What is meant by the expression "stratification"? What determines stratification, and how does stratification relate to the vertical exchange of material (e.g., nutrients, etc.)?

2.14 Referring to the Atlantic and Pacific Ocean *PV* and age contours shown in figures 2.3.5, 2.3.11, and 2.3.12, illustrate and discuss the basic structure of *PV* contours on a typical isopycnal surface in the thermocline of the subtropical gyre, and discuss the flow associated with these contours, including the ventilated, recirculation, and shadow zones.

2.15 Suppose that all of the bomb-produced radiocarbon and all 550 kg of the bomb-produced tritium were introduced into the atmosphere in the major nuclear bomb tests of 1962.

a. How much of that tritium was left on Earth in the year 2005?

b. What fraction of the bomb radiocarbon was left in the year 2005?

2.16 Redraw figure 2.1.1 for a flow in the y-direction and show step-by-step how to obtain

$$\frac{1}{V}\frac{\partial C}{\partial t} = -\frac{\partial}{\partial y}(C \cdot v)$$

2.17 Consider an ocean in which surface waters sink into the deep ocean in the low latitudes, with the return flow from the deep ocean to the surface ocean occurring in the high latitudes. Use the integral form of the continuity equation to consider how a two-box model for such an ocean should be constructed.

a. What concentration should be used in the time derivative for the deep box?

b. What concentrations should be used in calculating the inflows to the deep box and outflows from the deep box?

c. All else being equal (e.g., surface ocean nutrients are depleted in low latitudes, but not in high latitudes), how would the deep ocean oxygen concentration of this ocean compare with that of the present ocean?

2.18 The Mediterranean is a closed basin except for exchange at the Straits of Gibraltar, where the inflowing water has a salinity of 36.20 and the outflowing water a salinity of 38.45. The net flux of freshwater into the Mediterranean by rivers and precipitation minus evaporation is $-0.040 \times 10^6\ m^3\ s^{-1}$ per second (i.e., it loses more water through evaporation than it gains by rivers and precipitation). Its volume is $3.8 \times 10^6\ km^3$.

a. What is the flow of water in and out at the sill?

b. What is the residence time of water in the Mediterranean Sea with respect to the inflow from the open ocean?

2.19 Consider a north-south continental boundary on the western side of an ocean basin in the southern hemisphere.

a. Draw an x-by-z section that intersects the boundary and projects out into the ocean. Schematically illustrate the ocean current pattern if the wind is blowing at a constant rate from the north. Include a depth scale (i.e., show the approximate depth range of the Ekman layer).

b. What is the total Ekman transport in Sverdrups across a 100 km long north-south section some distance from the shore? Assume a latitude of 20°S, and a τ_0^y of $-0.4\ g\ cm^{-1}\ s^{-2}$ at the position of the north-south section.

2.20 It is observed that contours of constant PV on isopycnal surfaces can traverse 20° or more of latitude from, for example, 30°N to 10°N. By what fraction would the thickness of the isopycnal have to decrease in going from 30°N to 10°N if all of the flow continued straight south?

Problems 2.21 to 2.23 refer to the illustration of a basin with an east-west dimension of 6000 km and a north-south dimension of 3300 km that extends from 20°N to 50°N (figure 1). The surface wind stress over this basin, also shown in the figure, is given by

$$\tau_0^x = -\tau_0 \cos\left(\pi \times \frac{y}{3300\,\text{km}}\right)$$

$$\tau_0^y = 0$$

where y is the distance in kilometers from the southern edge of the basin at 20°N, i.e.,

$$y = (\phi - 20°\text{N}) \times 110\,\text{km}$$

with ϕ being the latitude in degrees. (*Note*: This equation applies *only* to the portion of the basin shown in the figure, not the world as a whole. As part of solving this problem, you need to calculate the y derivative of f. Do this by using the chain rule to obtain

$$\frac{\partial f}{\partial y} = \frac{\partial f}{\partial \phi}\frac{\partial \phi}{\partial y}$$

and using the circumference of the Earth to determine the second derivative.) The value of the constant τ_0 is

$$\tau_0 = 1\frac{\text{g}}{\text{cm s}^2}$$

Figure for problems 2.21 to 2.23

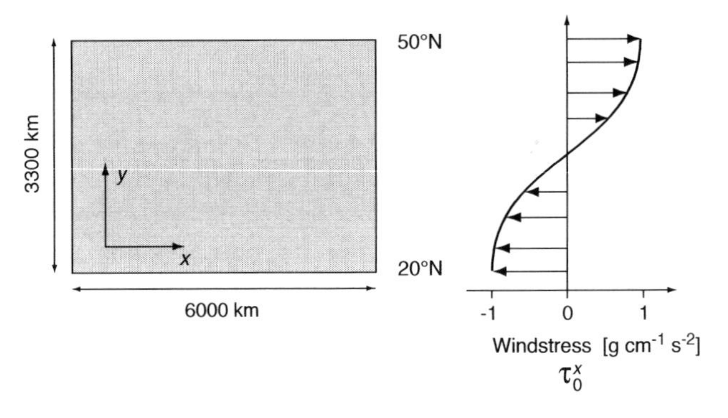

2.21 What is the total Ekman transport across the width of the basin at 30°N? Give your answer in Sverdrups.

2.22 What is the vertical velocity due to the Ekman convergence at 30°N?

2.23 Estimate the relative contributions of the terms containing the meridional wind stress gradient and meridional Coriolis parameter gradient to the upwelling/downwelling driven by the wind stress at 30°N.

CHAPTER 3

Air-Sea Interface

Having introduced the tracer conservation equation and discussed ocean transport, we now consider the processes at the air-sea and ocean bottom interfaces and biogeochemical processes internal to the ocean that need to be specified to solve for the tracer distributions. This chapter discusses gas exchange at the air-sea interface. This exchange has a major impact on the distribution of gases within both the ocean and the atmosphere. The first section of this chapter introduces the major gases in the atmosphere and ocean. We then discuss the solubility of gases in seawater, and the processes at the air-sea interface that control the air-sea exchange. A final section illustrates the application of the gas exchange model to several important problems.

3.1 Introduction

The gaseous composition of our atmosphere is unique in the solar system because of the impact of biological processes on all the major constituents except the noble gases: argon, neon, helium, krypton, and xenon (table 3.1.1). Our discussion in this chapter will focus primarily on three of these gases, O_2, CO_2, and N_2O, which are produced and consumed in large quantities by biogeochemical processes within the ocean. Two of them, CO_2 and N_2O, are among the gases that play an important role in controlling temperature at the surface of the Earth through the greenhouse effect of trapping long-wave radiation emitted from the surface of the Earth (see chapter 10). Both gases have been increasing at an unprecedented rate in the atmosphere due primarily to human impacts (for example, deforestation and fossil fuel burning in the case of CO_2) and are believed to be major contributors to global warming (see chapter 10). Concern about global warming, as well as our interest in the contribution of greenhouse gases to longer term climate variations, are among the major motivations for continuing interest in understanding what controls the air-sea flux of these and other gases.

The net air-sea flux of a gas A is directly proportion to the partial pressure difference across the air-sea interface or its corresponding concentration *anomaly*, defined as

$$\Delta p^{A} = p_w^{A} - p_a^{A} \quad \text{and} \qquad \Delta[A] = [A]_w - [A]_{equilibrium} \tag{3.1.1}$$

respectively. In these equations, [A] is concentration in mmol m^{-3}, p^A is the partial pressure in atmospheres, and the subscripts w and a refer to water and air, respectively. Partial pressure is defined such that the sum of the partial pressures of all gases is equal to the total atmospheric pressure. According to Henry's law, the partial pressure of a gas above a liquid with which it is in thermodynamic equilibrium is directly proportional to the concentration in the liquid. We find it convenient to state this equilibrium condition as

$$[A]_{equilibrium} = S_A \cdot p^A \tag{3.1.2}$$

where S_A is the solubility of A in units of mmol m^{-3} atm^{-1}. The solubility is a function of temperature and salinity. If the ocean is not in equilibrium with the atmosphere, as is commonly the case, then $[A]_w$ is not equal to $[A]_{equilibrium}$, nor will the partial pressure equivalent of $[A]_w$ calculated from (3.1.2) (i.e., $p_w^A = [A]_w / S_A$) be equal to the partial pressure of the gas in the air, p_a^A. Returning to the anomaly definitions (3.1.1), we see that a *saturated* or equilibrium solution will have an anomaly of 0. *Supersaturated* solutions are those that have higher partial pressures or concentrations in seawater than in the air, and thus positive anomalies. Supersaturated solutions will tend to lose gas A from the ocean to the air. *Undersaturated* solutions have negative anomalies and tend to gain A from the air. Figures 3.1.1 to 3.1.4 show concentration anomalies of O_2 and partial pressure

TABLE 3.1.1
Atmospheric composition given in units of the ratio of moles of each gas to the total number of moles of gas in dry air (the mixing ratio)

(a) Atmospheric Composition	
N_2	78.084 ± 0.004%
O_2	20.946 ± 0.002%
Ar	0.934 ± 0.001%
CO_2	365 ppm
Ne	18.18 ± 0.04 ppm
He	5.240 ± 0.004 ppm
CH_4	1.745 ppm
Kr	1.14 ± 0.01 ppm
H_2	0.55 ppm
N_2O	0.314 ppm
Xe	0.087 ± 0.001 ppm
CO	0.05–0.20 ppm
O_3	0.01–0.50 ppm

Sources: Weast and Astle [1982] for all gases except CO_2, CH_4, and N_2O, which are 1998 concentrations taken from *Ramaswamy, et al.* [2001].

(b) Influenced by Chemical and Biological Processes	
No	*Yes*
Ar	N_2
Ne	O_2
He	CO_2
Kr	CH_4
Xe	H_2
	N_2O
	CO
	O_3

anomalies for CO_2 and N_2O. We outline the main features of each of these figures in turn.

Figure 3.1.1 shows that oxygen is supersaturated over a large fraction of the surface ocean. The measurements summarized in figure 3.1.2, which are from the World Ocean Circulation Experiment (WOCE), show that during this expedition the supersaturation was commonly of order 3%. The supersaturation implies a source of oxygen from the ocean to the atmosphere. Where does this oxygen come from? By contrast, figure 3.1.1 shows that the high latitudes of the Southern Ocean, the north and eastern equatorial Pacific, and the region surrounding Indonesia, are all undersaturated. What causes these areas to be different from the rest of the world?

The disequilibrium between oceanic and atmospheric CO_2 shown in figure 3.1.3, which is for a nominal year of 1995, shows a much greater range than that of O_2. The units of this figure are the partial pressure of CO_2 in μatm. The atmospheric partial pressure of CO_2 in air in 1995 was about 360 μatm, so a disequilibrium of order 40 μatm, such as occurs in the equatorial Pacific, is equivalent to about 11% supersaturation, as contrasted with the values of about 3% observed for O_2. There are some similarities in the patterns of the disequilibrium of CO_2 to the concentration anomalies of O_2, such as the presence of a large disequilibrium at the equator, but the CO_2 anomalies tend to have much broader horizontal scales than O_2 and are often opposite in sign. For example, the equatorial supersaturation implies a source of CO_2 to the atmosphere in the same general regions where the O_2 anomalies imply a sink. What mechanisms might cause these two gases to have opposite anomalies? And why is the region of O_2 undersaturation much more tightly confined to the equator than the corresponding band of CO_2 supersaturation? Later we shall see that other measurements suggest that the anomalies of the two gases tend in fact to have the same sign over much of the ocean. This clearly is inconsistent with the comparison of figures 3.1.1 and 3.1.3. If there is a problem with one of these two data sets, it is more likely to be with the oxygen, which has an anomaly that is much smaller and thus more difficult to measure, and which has not been sampled nearly as frequently and carefully as CO_2.

An important consideration for CO_2 is that the atmospheric partial pressure of 365 μatm that was observed in 1998 represents quite a large increase over the partial pressure of ~280 μatm that existed prior to the beginning of the industrial revolution (see chapter 10). This increase is due to fossil fuel burning and land use changes. A large amount of this anthropogenic CO_2 has dissolved in the ocean and continues to do so. Detecting this invasion is the primary reason that so much more effort has been put into measuring CO_2 partial pressure differences than O_2 anomalies. One of the things we will do in the final section of this chapter is to estimate the net CO_2 flux into the ocean implied by the data of figure 3.1.3.

Finally, we see in figure 3.1.4 that the N_2O partial pressure is close to atmospheric equilibrium in most latitude bands except for a very tight band near the equator, where it is supersaturated. This implies an oceanic source to the atmosphere. Where does this nitrous oxide come from, and why is the band of supersaturated N_2O so narrow?

In what follows, we will address these queries starting with a discussion of the solubilities that are required in order to calculate the concentration anomalies, and then proceeding to a discussion of how we can translate the anomalies into an estimate of the actual flux of the gas across the air-sea interface. Armed with the ability to quantitatively estimate the gas fluxes that are implied by the anomalies, we turn in the final section to a discussion of the actual fluxes and what these imply about the processes that are causing the anomalies.

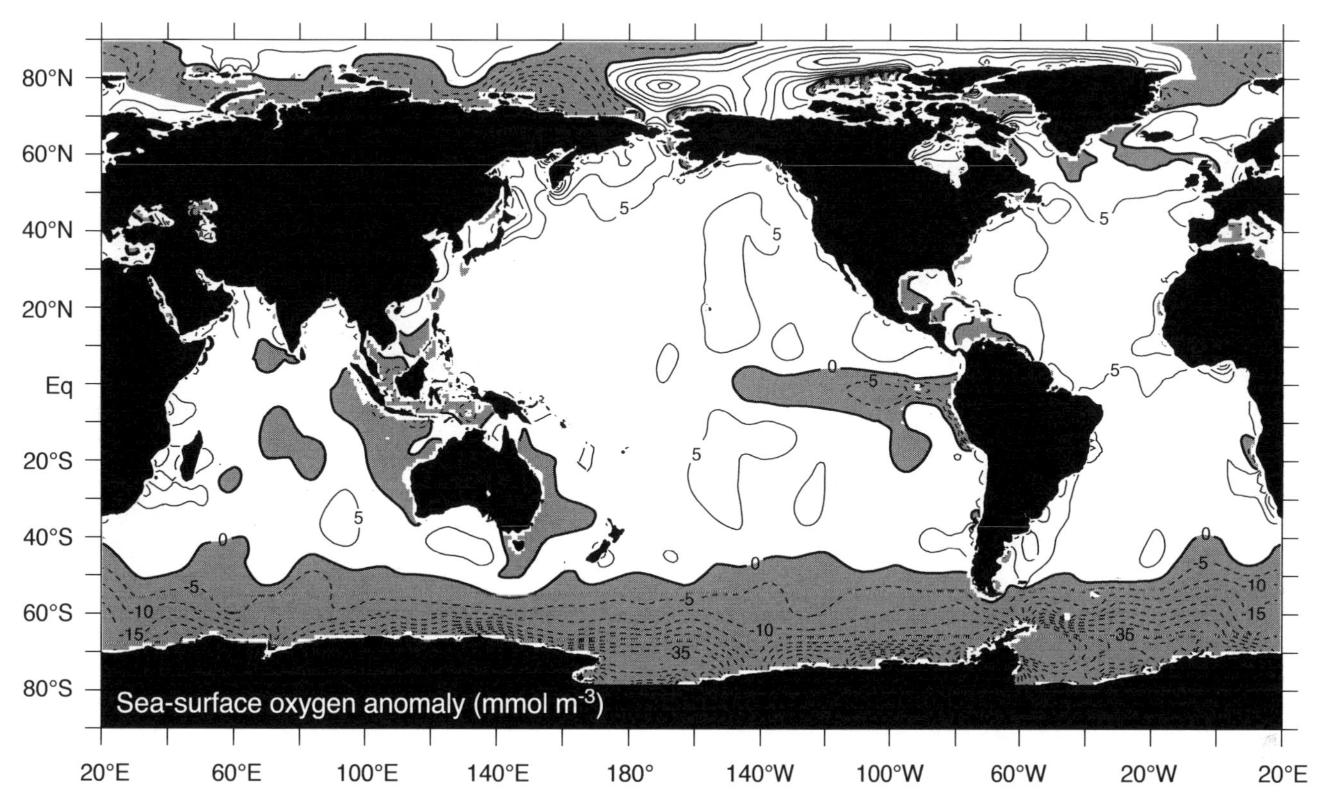

FIGURE 3.1.1: Map of the annual mean dissolved oxygen concentration anomaly in the surface ocean based on data from the World Ocean Atlas 2001 [*Conkright et al.*, 2002]. Supersaturation is positive and undersaturation is negative. As discussed in section 3.4, the globally integrated flux of oxygen across the air-sea interface is close to zero, which means that areas of oxygen undersaturation where the ocean is gaining oxygen must balance areas of oxygen supersaturation where oxygen is being lost from the ocean.

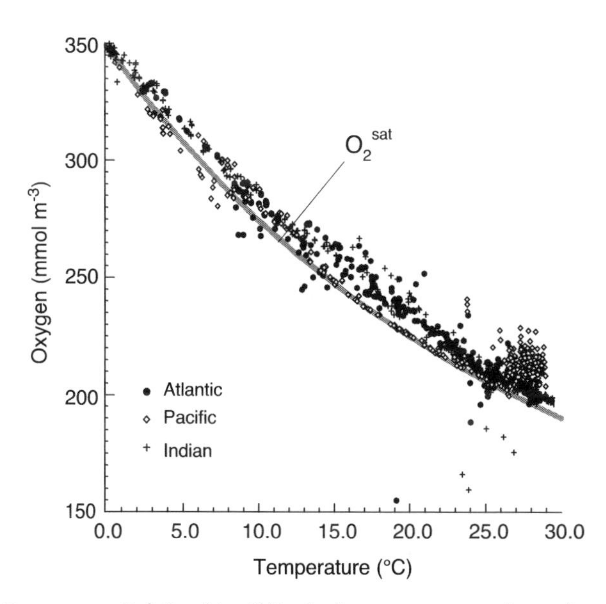

FIGURE 3.1.2: Relationship of dissolved oxygen concentration and temperature in the near-surface ocean measured throughout the global ocean during a few selected cruises of the WOCE program (A16, I8/9, P16). The grey line shows the saturation concentration of oxygen in equilibrium with the atmosphere.

3.2 Gas Solubilities

The most convenient of the numerous definitions for gas solubilities, and the one we will use here, is our restatement of Henry's law, equation (3.1.2). The solubility S_A is equivalent to the K' of *Warner and Weiss* [1985]. However, solubilities are commonly reported in the literature as the Bunsen coefficient β or the solubility function F, so it is necessary to specify how to convert from these solubility parameters to the parameter S_A. Panel 3.2.1 explains how to do this. Table 3.2.1 summarizes the necessary relationships derived in panel 3.2.1, and table 3.2.2 gives equations and coefficients for calculating Bunsen coefficients and solubility functions for several important gases.

Table 3.2.3 and figure 3.2.1 summarize the solubilities of various gases as a function of temperature. The solubility decreases with increasing temperature for all the gases. The upper panel of the table shows gases composed of a single element whose behaviors follow a progression from less to more soluble as their molecular weight increases. The lower panel of the table shows gases made of two or more elements whose solubility falls outside the overall trend of the upper

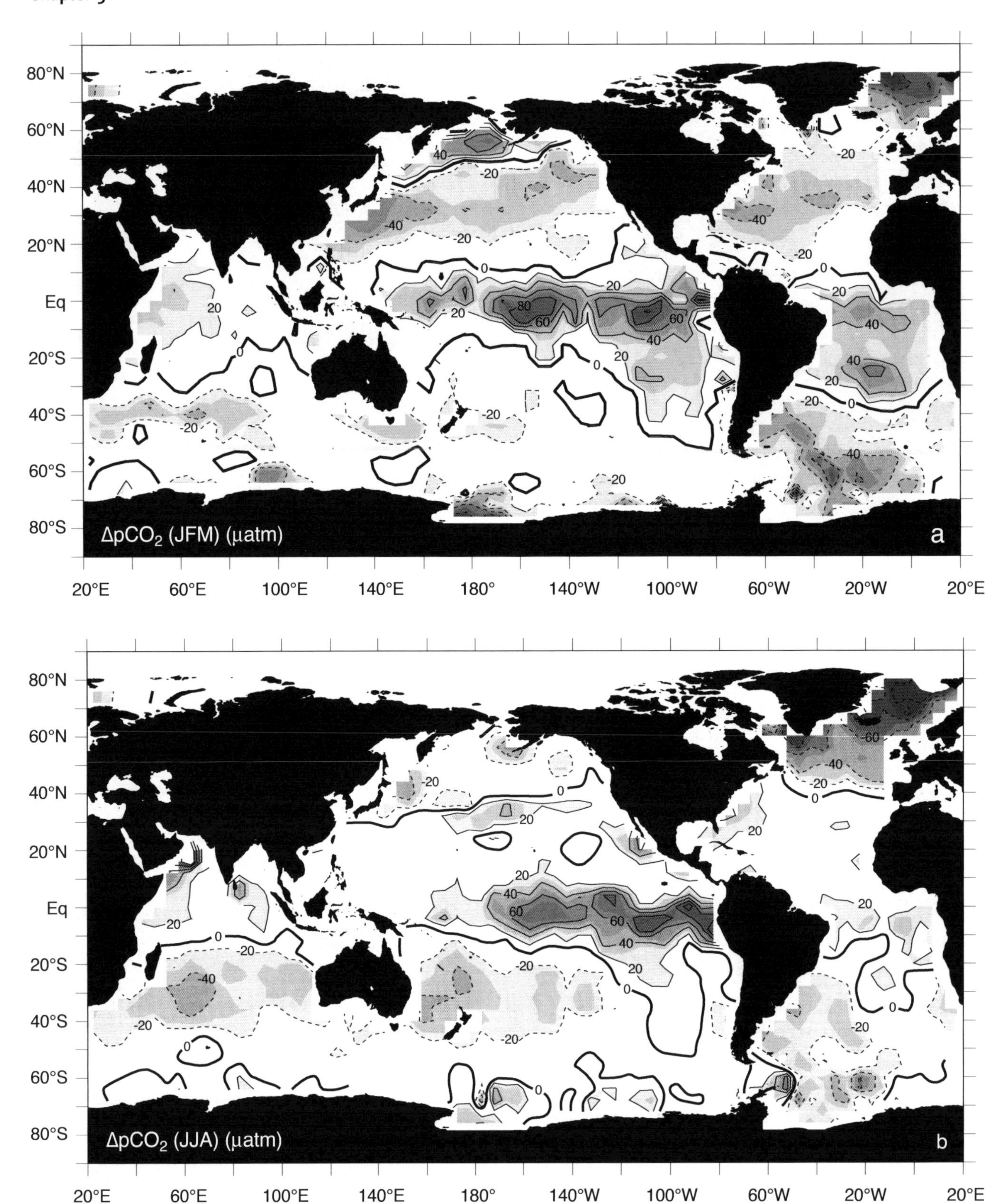

FIGURE 3.1.3: Maps of the climatological carbon dioxide partial pressure difference (ΔpCO_2) across the air-sea interface, with supersaturation indicated by positive numbers. (a) Average ΔpCO_2 for the northern hemisphere winter months (January through March); (b) Average ΔpCO_2 for the northern hemisphere summer months (June through August). Based on the climatology of *Takahashi et al.* [2002].

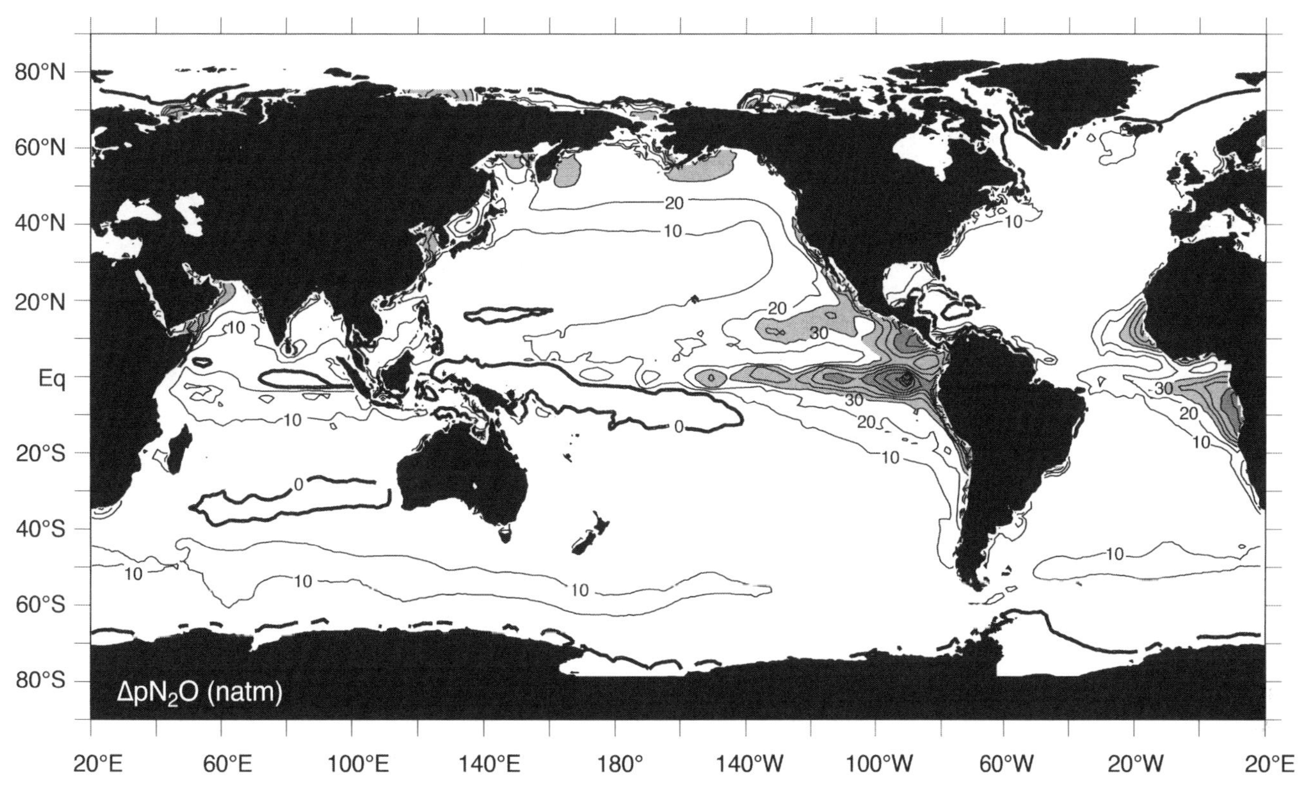

FIGURE 3.1.4: Annual mean map of the nitrous oxide partial pressure difference across the air-sea interface, based on data from *Suntharalingam and Sarmiento* [2000]. Supersaturation is positive.

Panel 3.2.1: Definitions and Conversions of Solubility Parameters

The Bunsen coefficient β is defined per *Weiss* [1974] as the volume of a gas at STP (standard temperature and pressure, i.e., 0°C and 1 atm) that dissolves in one unit volume of solution at temperature T with a total pressure of 1 atm and a fugacity of 1 atm, i.e.,

$$\beta_A = \frac{V_A}{V_{solution} \cdot f^A} \tag{1}$$

where β_A has units of l gas · (l solution)$^{-1}$ · (atm gas)$^{-1}$. The *fugacity*, f^A, which is a thermodynamic quantity analogous to the activity of dissolved substances, is related to the actual partial pressure of gas A, p^A, by

$$f^A = p^A \cdot \exp\left[\frac{P \cdot (\bar{V}_A - \bar{V}_{ideal})}{R \cdot T}\right] \tag{2}$$

The gas constant R has a value of 0.082053 l atm K^{-1} mol^{-1}, $\bar{V}_A$ is the molar volume of A at standard temperature and pressure, $\bar{V}_{ideal}$ is the molar volume of an ideal gas, which is equal to 22.4136 l mol^{-1}, and T is temperature in degrees Kelvin. The exponential term corrects the partial pressure for the nonideal behavior of the gas. The definition of solubility is

$$[A] = \frac{\text{moles of A}}{V_{solution}} = \frac{V_A/\bar{V}_A}{V_{solution}} \tag{3}$$

Combining (1) and (3), we find that

$$[A] = \left[\frac{\beta_A}{\bar{V}_A} \cdot f^A\right] \times 10^6 \, \frac{\text{mmol m}^{-3}}{\text{mol l}^{-1}} \tag{4}$$

The factor of 10^6 is required to obtain [A] in units of mmol m^{-3}.

We obtain the solubility S_A from the Bunsen coefficient by substituting f^A from (2) into (4) and setting this equal to (3.1.2). This gives

$$S_A = \frac{\beta_A}{\bar{V}_A} \cdot \exp\left[\frac{P \cdot (\bar{V}_A - \bar{V}_{ideal})}{RT}\right] \times 10^6 \, \frac{\text{mmol m}^{-3}}{\text{mol l}^{-1}} \tag{5}$$

Gases with near-ideal behavior have $\bar{V}_A - \bar{V}_{ideal} \approx 0$, giving

$$S_A \approx \left(\frac{\beta_A}{\bar{V}_A}\right) \times 10^6 \, \frac{\text{mmol m}^{-3}}{\text{mol l}^{-1}} \tag{6}$$

An important additional complication is that the partial pressures of gases are normally reported as p^A_{dry} with respect to dry air at a total pressure P of 1 atm. This avoids having to deal with the problem that the moisture content of air varies by quite a large amount as a function of temperature and the relative humidity. However, the concentration of a gas in seawater is determined by its partial pressure in moist air, p^A_{moist}, with

TABLE 3.2.1
Conversion of Bunsen coefficient (atm^{-1}) and solubility function ($mol\,l^{-1}$) parameters to solubility parameter S_A for use in calculating the seawater concentration from $[A] = S_A \cdot p^A$

The units of [A] are $mmol \cdot m^{-3}$, those for S_A are $mmol\,m^{-3}\,atm^{-1}$, and those for p^A are atm. The conversion from Bunsen coefficient to S_A ignores the correction for nonideality. This assumption is good to about 0.1% or better for most of the gases. Other gases with larger nonideal behavior (CO_2, N_2O, and the CFCs) have the correction included in the solubility function F.

Solubility Parameter	*Symbol*	*Conversion to S_A*
	For use with $p^A_{moist} = p^A_{dry} \cdot (1 - p^{H_2O}/P)$	
Bunsen coefficient	β_A	$S_A \approx \left(\frac{\beta_A}{\bar{V}_A}\right) \times 10^6 \frac{mmol\,m^{-3}}{mol\,l^{-1}}$
Solubility function	F_A	$S_A = \left\lfloor \frac{F_A}{(P - p_{H_2O})} \right\rfloor \times 10^6 \frac{mmol\,m^{-3}}{mol\,l^{-1}}$
	For use with p^A_{dry}	
Bunsen coefficient	β_A	$S_A \approx \left[\frac{\beta_A}{\bar{V}_A} \cdot \left(1 - \frac{p^{H_2O}}{P}\right)\right] \times 10^6 \frac{mmol\,m^{-3}}{mol\,l^{-1}}$
Solubility function	F_A	$S_A = \left(\frac{F}{P}\right) \times 10^6 \frac{mmol\,m^{-3}}{mol\,l^{-1}}$

Panel 3.2.1: (*Continued*)

respect to moist air at a total pressure P (usually of 1 atm). This can be obtained from the following relationship:

$$\begin{aligned} p^A_{moist} &= \chi^A_{dry} \cdot (P - p^{H_2O}) \\ &= \chi^A_{dry} \cdot P \cdot \left(1 - \frac{p^{H_2O}}{P}\right) \\ &= p^A_{dry} \cdot \left(1 - \frac{p^{H_2O}}{P}\right) \end{aligned} \tag{7}$$

where the mixing ratio, χ^A, is the number of moles of constituent A per unit mole of air, p^{H_2O} is the partial pressure of the water vapor, and P is the total atmospheric pressure in atmospheres. In deriving the above equation, we make use of the relationship between the mole ratio and partial pressure, $p^A = \chi^A \cdot P$. The water vapor pressure is generally assumed to be at saturation in the vicinity of the air-sea interface. The saturation water vapor pressure is given by *Weiss and Price* [1980] as

$$\frac{p^{H_2O}}{P} = \exp\left[24.4543 - 67.4509(100/T) - 4.8489 \ln(T/100) - 0.000544S\right] \tag{8}$$

where T is the absolute temperature in K, and S is the salinity. The value of $(p^{H_2O}/P) \cdot 100$ for $P = 1$ atm ranges between ~0.5% and 4.0% over the sea surface (see figure 1).

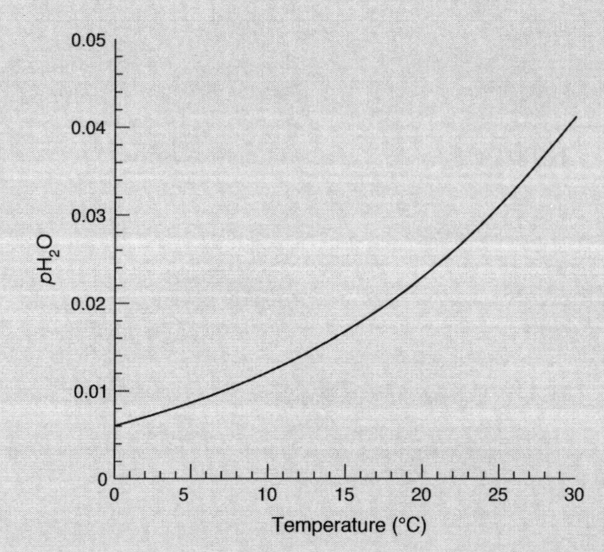

FIGURE 1: Plot of the partial pressure of water in saturated air (p^{H_2O}/P) as a function of temperature. Computed from (8).

TABLE 3.2.2
Coefficients for the fit of solubility to the following equation:

$$\ln(\ ) = A_1 + A_2(100/T) + A_3 ln(T/100) + A_4(T/100)^2 + S \cdot [B_1 + B_2(T/100) + B_3(T/100)^2]$$

T is the temperature in K and S is the salinity, ( ) = β is the Bunsen solubility coefficient in units of l gas (l solution)$^{-1}$ atm^{-1}, and ( ) = F is the volumetric solubility function in units of mol l^{-1} atm^{-1}.

Gas	( )	A_1	A_2	A_3	A_4	B_1	B_2	B_3	Estimated Accuracy
He	β	−34.6261	43.0285	14.1391	—	−0.042340	0.022624	−0.0033120	0.5%
Ne	β	−39.1971	51.8013	15.7699	—	−0.124695	0.078374	−0.0127972	0.5%
N_2	β	−59.6274	85.7761	24.3696	—	−0.051580	0.026329	−0.0037252	0.4%
O_2	β	−58.3877	85.8079	23.8439	—	−0.034892	0.015568	−0.0019387	0.4%
Ar	β	−55.6578	82.0262	22.5929	—	−0.036267	0.016241	−0.0020114	0.4%
Kr	β	−57.2596	87.4242	22.9332	—	−0.008723	−0.002793	0.0012398	0.4%
Rn	β	−11.95	31.66	—	—	—	—	—	—
CH_4	β	−68.8862	101.4956	28.7314	—	−0.076146	0.043970	−0.0068672	1%
CO_2	F	−160.7333	215.4152	89.8920	−1.47759	0.029941	−0.027455	0.0053407	0.3%
N_2O	F	−165.8806	222.8743	92.0792	−1.48425	−0.056235	0.031619	−0.0048472	0.14%
CCl_2F_2 (CFC-12)	F	−218.0971	298.9702	113.8049	−1.39165	−0.143566	0.091015	−0.0153924	1.5%
CCl_3F (CFC-11)	F	−229.9261	319.6552	119.4471	−1.39165	−0.142382	0.091459	−0.0157274	1.5%
SF_6	F	−80.0343	117.232	29.5817	—	0.0335183	−0.0373942	0.00774862	2%
CCl_4	F	−148.247	227.758	62.5557	—	−0.400847	0.265218	−0.0446424	2.5%

Sources: He and Ne, *Weiss* [1971]; N_2, O_2, and Ar, *Weiss* [1970]; Kr, *Weiss and Kyser* [1978]; Rn, *Hackbusch* [1979]; CH_4, *Wiesenburg and Guinasso* [1979]; CO_2 and N_2O, *Weiss and Price* [1980]; CFC-11 and CFC-12, *Warner and Weiss* [1985]; SF_6, *Bullister et al.* [2002]; CCl_4, *Bullister and Wisegarver* [1998].

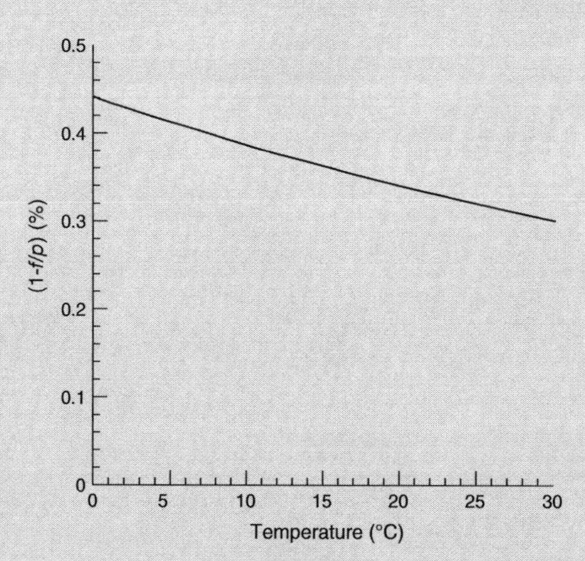

FIGURE 2: Plot of the temperature dependence of the correction factor that takes into account the nonideal behavior CO_2.

The correction for nonideal behavior, $\exp[(\overline{V}_A - \overline{V}_{ideal}) \cdot P/(R \cdot T)]$, is important for gases whose behavior departs significantly from ideality, such as CO_2, N_2O, and the chlorofluorocarbons. For example, figure 2 shows $\exp[(\overline{V}_A - \overline{V}_{ideal}) \cdot P/(R \cdot T)]$ for CO_2 as obtained with the *Weiss* [1974] binary mixture expression for $\overline{V}_A - \overline{V}_{ideal}$. The correction is approximately 0.4%, equivalent to a p^{CO_2} of about 1.5 μatm. *Weiss* [1974] incorporates the correction for nonideal gases in the solubility function F_A, defined by

$$[A] = [F_A \cdot \chi^A_{dry}] \times 10^6 \frac{\text{mmol m}^{-3}}{\text{mol l}^{-1}} \tag{9}$$

where

$$F_A = (P - p^{H_2O}) \cdot \frac{\beta_A}{\overline{V}_A} \cdot \exp\left[\frac{P \cdot (\overline{V}_A - \overline{V}_{ideal})}{R \cdot T}\right] \tag{10}$$

has units of mol l^{-1}. Combining (3.1.2) and (7) with (9) and (10) gives

$$S_A = \left[F_A \cdot \frac{\chi^A_{dry}}{p^A_{moist}}\right] \times 10^6 \frac{\text{mmol m}^{-3}}{\text{mol l}^{-1}} = \left[\frac{F_A}{(P - p^{H_2O})}\right] \times 10^6 \frac{\text{mmol m}^{-3}}{\text{mol l}^{-1}} \tag{11}$$

TABLE 3.2.3

Solubility parameters S_A (mmol m^{-3} atm^{-1}) at a salinity of 35.0

These are calculated from the Bunsen solubility coefficients and solubility functions given in table 3.2.2, and converted for use with moist air partial pressures as explained in the upper panel of table 3.2.1. The molar volume used for the Bunsen coefficient conversions is the ideal gas value of 22.4136 l mol^{-1}.

		T (°C)						
	Molecular Weight	*0*	*5*	*10*	*15*	*20*	*25*	*30*
He	4.0	349.4	342.8	338.0	334.7	332.9	332.3	332.9
Ne	20.2	448.6	430.5	415.1	402.0	390.7	381.1	372.9
N_2	28.0	818.8	731.8	661.6	604.5	557.9	519.6	488.2
O_2	32.0	1,725	1,524	1,363	1,232	1,126	1,039	967.6
Ar	39.9	1,879	1,665	1,492	1,351	1,236	1,142	1,064
Kr	83.8	3,820	3,281	2,854	2,515	2,241	2,019	1,838
Rn	222	31,150	25,290	20,690	17,040	14,130	11,790	9,892
CH_4	16.0	1,984	1,732	1,532	1,371	1,241	1,136	1,050
CO_2	44.0	64,400	53,350	44,880	38,300	33,110	28,970	25,650
N_2O	44.0	47,840	39,380	32,900	27,860	23,870	20,690	18,120
CCl_2F_2 (CFC-12)	120.9	6,686	5,086	3,963	3,157	2,566	2,124	1,789
CCl_3F (CFC-11)	137.4	27,380	20,090	15,140	11,700	9,242	7,459	6,138
SF_6	146.1	425.2	340.2	277.7	231.0	195.8	169.0	148.4
CCl_4	153.8	97,114	69,946	51,696	39,139	30,307	23,971	19,346

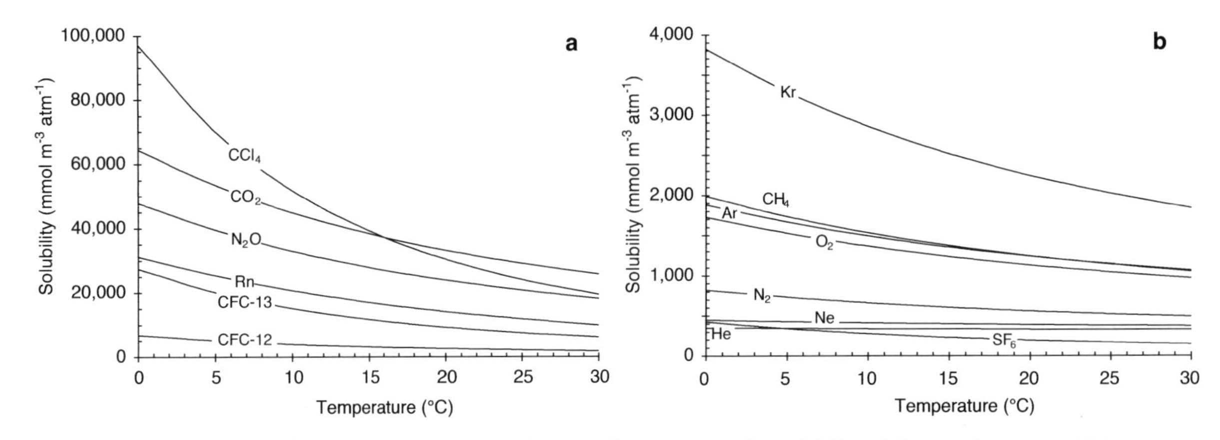

FIGURE 3.2.1: Plot of the solubility of various gases as a function of temperature. The solubility of all gases decreases with increasing temperature, but there exist large differences in the solubility of the different gases for a particular temperature. These differences can be understood in terms of their molecular weight (ideal gas) or other factors, such as molecular interactions between the gas and the water (nonideal gas). Note the change in vertical scale between panels (a) and (b). Based on the empirical functions listed in table 3.2.2.

part of the table; all of these except SF_6 have greater solubility than one would infer from the molecular weight trend in the upper panel of the table. Table 3.2.4 gives the equilibrium concentration for several gases from table 3.1.1 arranged in order of their oceanic concentration.

3.3 Gas Exchange

The fact that gases such as those shown in figures 3.1.1 to 3.1.4 have a nonzero anomaly over most of the ocean means that they are out of equilibrium with the atmosphere. Here we discuss how we can determine the magnitude of the fluxes that result from this disequilibrium. Our starting point is the basic concept that turbulent vertical motions in both the atmosphere and ocean are strongly suppressed near the air-sea interface. A variety of models have been developed to explain the effect of this phenomenon on the transfer of gases across the interface. The following discussion is based primarily on *Liss and Slater* [1974] and *Liss and Merlivat* [1986].

TABLE 3.2.4
Saturation concentration of gases in seawater at a salinity of 35.0 calculated from $[A] = S_A \cdot p^A_{moist}$

The solubilities are taken from table 3.2.3. The moist partial pressure is calculated from the dry air atmospheric mole fractions given in table 3.1.1 corrected to moist air using (7) from panel 3.2.1 and the saturation water vapor pressures given below and assuming a total atmospheric pressure of 1 atm. The saturation water vapor pressures were calculated with (8) from panel 3.2.1.

T (°C)	*0*	*5*	*10*	*15*	*20*	*25*	*30*
	Saturation Water Vapor Pressure $\left(\frac{p_{H_2O}}{P} \times 100\%\right)$						
	0.59%	*0.84%*	*1.19%*	*1.65%*	*2.26%*	*3.07%*	*4.11%*
	Concentrations in mmol m^{-3}						
N_2	635.6	566.6	510.5	464.2	425.7	393.3	365.6
O_2^a	359.1	316.5	282.1	253.9	230.5	211.0	194.4
	357.7	315.9	282.0	254.2	231.1	211.6	194.9
CO_2	23.37	19.26	16.09	13.60	11.61	10.00	8.66
Ar	17.44	15.42	13.77	12.41	11.29	10.34	9.53
	Concentrations in μmol m^{-3}						
N_2O	14.84	12.16	10.09	8.46	7.16	6.10	5.23
Ne	8.11	7.76	7.46	7.19	6.94	6.72	6.50
Kr	4.33	3.71	3.22	2.82	2.50	2.23	2.01
CH_4	3.44	3.00	2.64	2.35	2.12	1.92	1.76
He	1.81	1.77	1.74	1.72	1.70	1.68	1.67

[a]Two results are given for oxygen. The upper one uses solubilities from table 3.2.3. The lower one is a direct calculation of the saturation concentration using the more accurate equation of *Garcia and Gordon* [1992], $O_2^{saturation} = (1000/22.3916) \cdot e^{\ell}$, where $\ell = A_0 + A_1 \cdot T_s + A_2 \cdot T_s^2 + A_3 \cdot T_s^3 + A_4 \cdot T_s^4 + A_5 \cdot T_s^5 + S(B_0 + B_1 \cdot T_s + B_2 \cdot T_s^2 + B_3 \cdot T_s^3) + C_0 \cdot S^2$ and $T_s = \ln[(298.15 - T)/(273.15 + T)]$, with T in °C. The constants are:

A_0	A_1	A_2	A_3	A_4	A_5
2.00907	3.22014	4.05010	4.94457	−0.256847	3.88767

B_0	B_1	B_2	B_3	C_0
-6.24523×10^{-3}	-7.37614×10^{-3}	-1.03410×10^{-2}	-8.17083×10^{-3}	-4.88682×10^{-7}

Note that this equation uses a molar volume of 22.3916 l mol^{-1}, rather than the ideal volume used in the Bunsen coefficient calculations. The erroneous $A_3 \cdot T_s^2$ term in the original reference is left out.

STAGNANT FILM MODEL

We begin with a simple model of gas exchange which assumes that the suppression of turbulence at the air-sea interface leads to the formation of two stable stagnant films of finite thickness, one in the atmosphere and one in the ocean, through which gases can diffuse only by molecular diffusivity ε (see figure 3.3.1; cf. *Whitman* [1923]). Alternative more realistic but more complex models that will be discussed at the end of this section include the micrometeorological boundary layer model [*Deacon*, 1977], which assumes a more gradual transition from interior turbulence to molecular diffusion at the interface; and the film replacement model [*Danckwerts*, 1951; *Higbie*, 1935], which assumes the existence of a film that is periodically replaced by bulk fluid.

The fluxes through the stagnant films in figure 3.3.1 can be obtained from Fick's first law, $\Phi = -\varepsilon\partial[A]/\partial z$, where Φ has units of mmol m^{-2} s^{-1}, and ε in m^2 s^{-1} is the molecular diffusivity in the air or water, depending on which film we are talking about. If there are no sources or sinks of A in the stagnant film, the concentration gradient in the film will be linear and we can represent the flux by its finite difference form $\Phi = -\varepsilon \cdot \Delta[A]/\Delta z$, where Δ[A] is the concentration difference across the stagnant film and Δz is its thickness. Henry's law allows us to convert this equation to one in terms of partial pressures. We thus arrive at the following expressions for the gas exchange through the atmospheric stagnant film:

$$\Phi_a = -k_a \cdot (p_a^A - p_a^{A,0}) = -\frac{k_a}{S_A} \cdot ([A]_a - [A]_a^0) \qquad (3.3.1)$$

where $k_a/S_A = \varepsilon_a/\Delta z_a$ and $[A]_a$ is equal to the saturation concentration $[A]_{equilibrium}$ of (3.1.1). For the oceanic stagnant film we have

$$\Phi_w = -k_w \cdot ([A]_w^0 - [A]_w) \qquad (3.3.2)$$

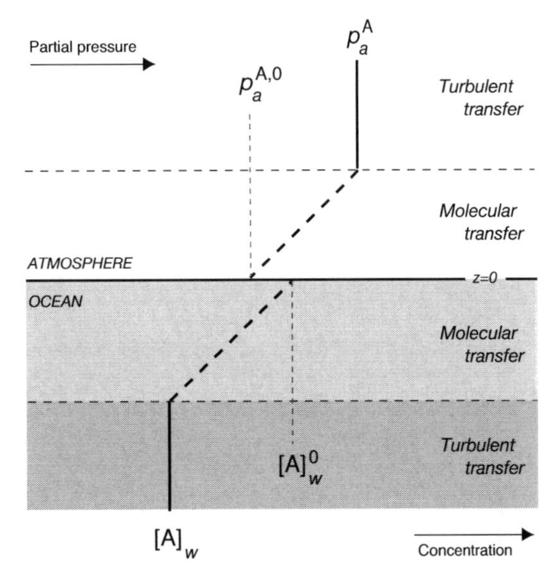

FIGURE 3.3.1: Schematic depiction of the stagnant film model. Shown is the air-sea interface ($z = 0$) with stagnant films in both the atmosphere and ocean and turbulent regions further away from the interface. Gas A is well mixed in the turbulent regions, resulting in a relatively uniform concentration $[A]_w$ in the water and partial pressure p_a^A in the air. However, the transfer across the stagnant films is assumed to be governed by molecular diffusion. Consequently, the concentration and partial pressure of this gas changes linearly in the stagnant layers toward the saturation levels at the interface, $[A]_w^0$ and $p_a{}^{A,0}$. The stagnant film model proposes that the time it takes for a gas to transfer across the air-sea interface is governed by how fast this gas passes through the two stagnant films.

where $k_w = \varepsilon_w/\Delta z_w$. The vertical coordinate is positive upwards. The partial pressures and concentrations are as defined in figure 3.3.1. The superscript 0 indicates that the concentration and partial pressures are at the air-sea interface (see figure 3.3.1). In these equations, k_a is a gas exchange coefficient with units of velocity times solubility, and k_w is a gas exchange coefficient with units of velocity. k_w is often referred to as the *transfer* or *piston* velocity, and we shall use that terminology here.

Measurements of partial pressures and concentrations are always made well above or below the air-sea interface, never at the interface. We therefore eliminate the interface partial pressure and concentrations from (3.3.1) and (3.3.2) by taking advantage of the requirement from continuity that without any sources or sinks in the stagnant film, $\Phi_a = \Phi_w = \Phi$ and $[A]_a^0 = [A]_w^0$. This gives

$$\Phi = -K \cdot ([A]_a - [A]_w) \qquad (3.3.3)$$

with

$$\frac{1}{K} = \frac{1}{k_w} + \frac{S_A}{k_a} \qquad (3.3.4)$$

For all gases of concern to us here, k_a/S_A is typically well in excess of 1000 cm hr^{-1}, as compared with values of order 10 cm hr^{-1} for k_w [*Liss and Slater*, 1974], allowing us to safely ignore the stagnant film in the atmosphere. This gives $K = k_w$ and

$$\Phi = -k_w \cdot ([A]_a - [A]_w) \qquad (3.3.5)$$

Equation (3.3.5) is also frequently expressed in terms of partial pressures as:

$$\Phi = -k_g \cdot (p_a^A - p_w^A) \qquad (3.3.6)$$

with $k_g = k_w \cdot S_A$. However, we mostly confine our attention here to the concentration form (3.3.5). Note that the gas exchange coefficient k_g, which is for the boundary layer in the water, is not to be confused with k_a, which has the same units but is for the boundary layer in the air.

Measurements of the gas transfer velocity have been made in wind tunnels (figure 3.3.2) and with the use of in situ tracers (figure 3.3.3) for a variety of gases under different wind and temperature conditions. These measurements and theory indicate that the gas exchange coefficient is controlled by the level of turbulence near the interface, which is primarily controlled by the wind stress at the boundary; and by the Schmidt number, $Sc = \nu/\varepsilon$, where ν is the kinematic viscosity of water, which influences the thickness of the film where molecular diffusivity dominates. Table 3.3.1 gives the parameters necessary to calculate the Schmidt number. Additional processes that may influence the gas transfer velocity are the type of wave at the surface, breaking waves and formation of air bubbles, and temperature and humidity gradients [cf. *Nightingale et al.*, 2000b], as well as organic films coating the surface of the ocean [e.g., *Frew*, 1997].

How are we to parameterize the influence of all these processes on the gas exchange coefficient? *Jähne et al.* [1987b] proposed the following relationship:

$$k_w = \gamma \cdot Sc^{-n} \cdot u_w^* \qquad (3.3.7)$$

where γ is a dimensionless constant which is the same for all gases. The friction velocity in the water, u_w^*, represents the wind stress effect. The exponent n in the Schmidt number is 1 for a stable stagnant film model such as depicted in figure 3.3.1, but can take on other values in different models of the boundary layer. In particular, the micrometeorological boundary layer model we mentioned earlier gives an n of 2/3, and the film replacement model gives an n of 0.5 [cf. *Liss and Merlivat*, 1986]. The value of n can be determined experimentally from measurements of gas transfer velocity with more than one gas, or with gas and heat at the same conditions:

$$n = \frac{-\log(k_{w_1}/k_{w_2})}{\log(Sc_1/Sc_2)} \qquad (3.3.8)$$

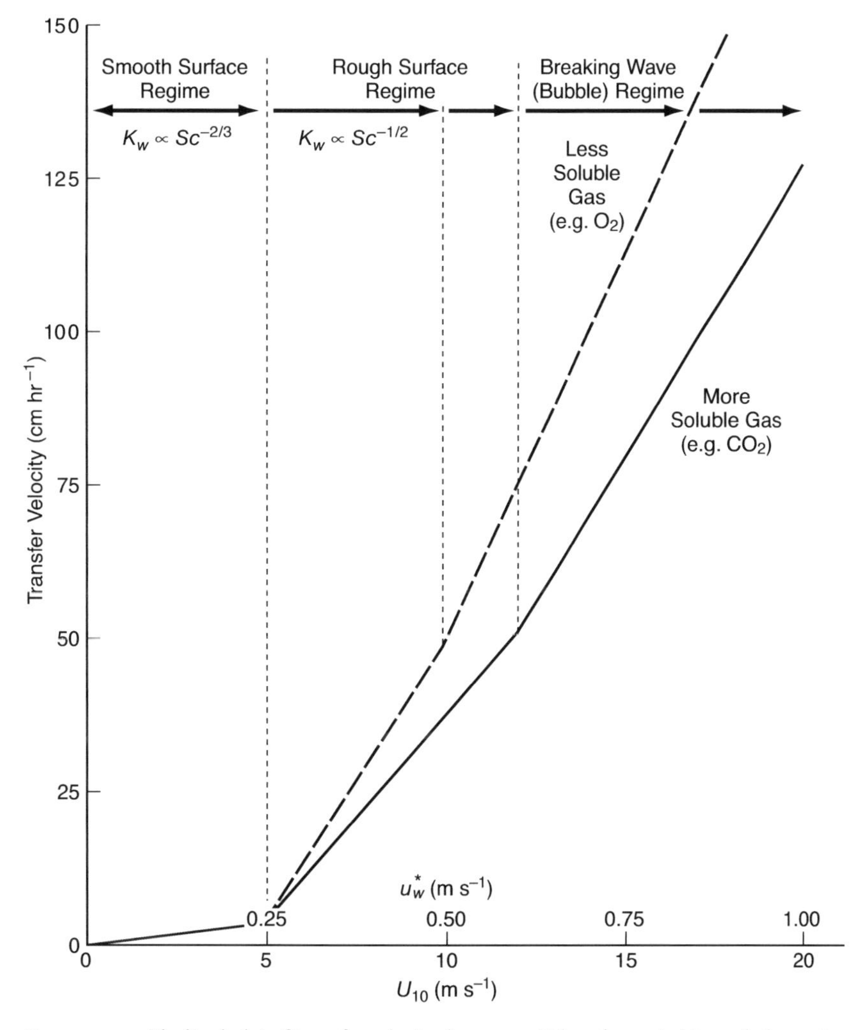

FIGURE 3.3.2: Idealized plot of transfer velocity, k_w, versus U based on wind tunnel observations [*Liss and Merlivat*, 1986]. The abscissa gives scales for both the friction velocity u_w^* and the 10 m wind velocity U_{10}.

Table 3.3.2 gives several commonly used formulations for the gas transfer velocity which are based on the approach of (3.3.7), but using the wind speed U at 10 m above sea level rather than the friction velocity, and also allowing the wind speed to be raised to a power of 2 or 3. Note that all of these are referenced to a Schmidt number of either 600 or 660 and need to be rescaled if applied to other temperatures and gases that have different Schmidt numbers. The global area weighted mean of the gas transfer velocities calculated with each of these models varies from a low of 10.2 cm hr^{-1} to a high of 17.3 cm hr^{-1} (table 3.3.2), and the sensitivity to the wind speed varies substantially from model to model (figure 3.3.3). In the following we first summarize observational constraints on the gas transfer velocity and then discuss the sources of disagreement between the models and which of them is more consistent with the measurements.

LABORATORY STUDIES

Wind tunnel studies typically estimate the gas transfer velocity by measuring the mass balance of a tracer in the water. The change in inventory over a given period is the flux. The gas transfer velocity is obtained by dividing the flux by the measured air-water concentration difference following (3.3.5). Such experiments show that transfer velocities vary as a function of wind speed U in a manner that was interpreted by *Liss and Merlivat* [1986] as suggesting the existence of three linear regimes separated by the transition from a smooth to a rough water surface, and from a rough surface to one where waves begin to break and form bubbles (figure 3.3.2). Micrometeorological boundary layer theory, with its prediction that the exponent n in (3.3.7) should be equal to 2/3 [*Deacon*, 1977], appears to apply in the smooth surface regime. The film replacement model,

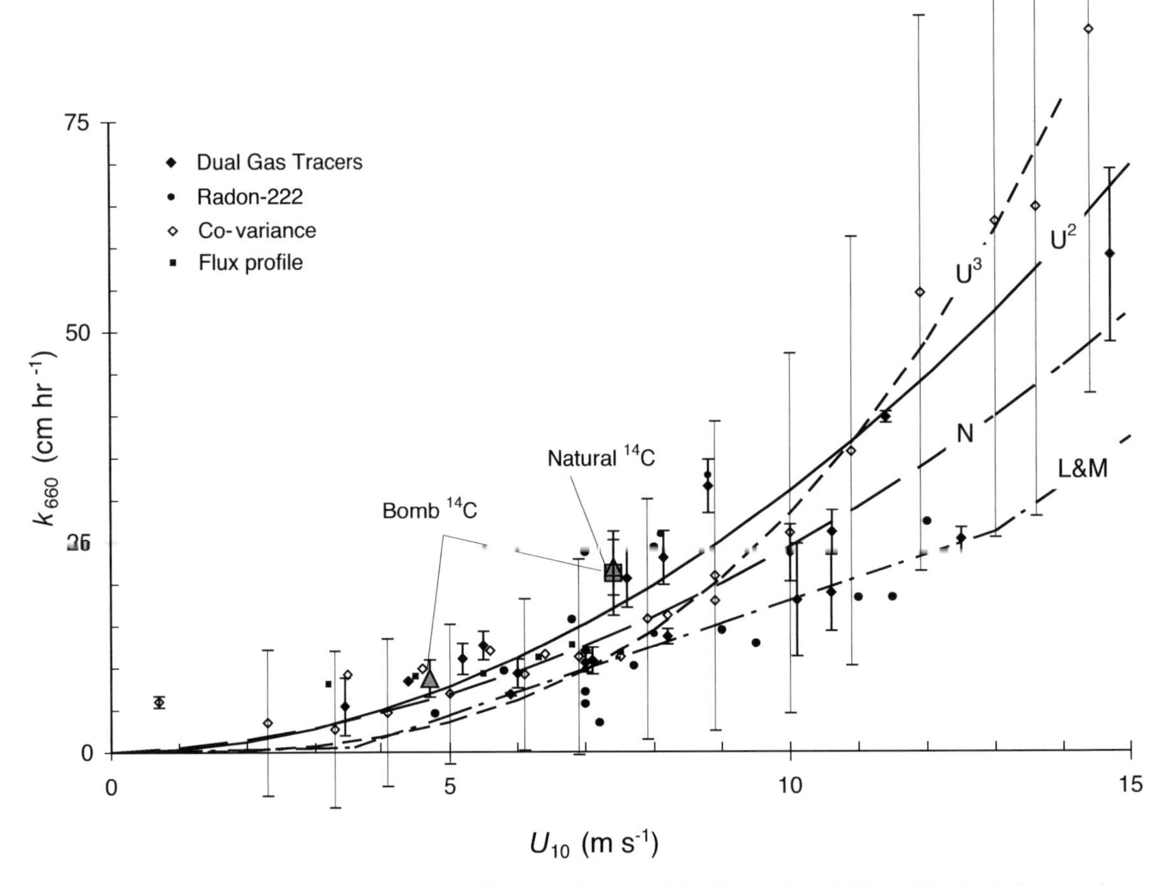

FIGURE 3.3.3: A summary of oceanic observations of k_w normalized to a Schmidt number of 660 (R. Wanninkhof, personal communication, 2004) plotted versus U_{10}, the wind speed at a height of 10 m. A Schmidt number of 660 is the value for CO_2 at 20°C. The dual gas tracer measurements are based on measurements of SF_6 and ^3He. They come from *Asher and Wanninkhof* [1998], *Wanninkhof et al.* [1997, 2004], and *Nightingale et al.* [2000a, 2000b]. The estimates based on radon-222 come from *Peng et al.* [1974, 1979], *Kromer and Roether* [1983], *Smethie et al.* [1985], *Glover and Reeburgh* [1987], and *Emerson et al.* [1991]. The estimates based on bomb and natural ^{14}C come from *Broecker et al.* [1985a, 1986] and *Cember* [1989] with the lower bomb ^{14}C estimate being for the Red Sea and the higher being for the world. The covariance and flux profile data are from W. McGillis (personal communication, 2004). The lines are based on the gas exchange models given in table 3.3.2. L&M is the *Liss and Merlivat* [1986] model, N is the *Nightingale et al.* [2000b] model, U^2 is the *Wanninkhof* [1992] model, and U^3 is the *Wanninkhof and McGillis* [1999] model.

with its prediction that $n = 1/2$ [*Danckwerts*, 1951; *Higbie*, 1935], appears to apply when the surface becomes rough, which occurs above a wind speed of 5 m s^{-1} in figure 3.3.2.

It is now understood that the first slope break in the gas transfer velocity-U relation, which represents the transition from a smooth to a rough surface, is in fact due to the presence of surfactant films that exist even in relatively clean experimental systems [*Frew*, 1997]. In experiments carried out in an annular wind-wave channel with surfactant present, Nelson Frew and colleagues confirmed the linear scaling of the gas transfer velocity and the kink in the line as the gas exchange shifts from the smooth to the rough surface regime (see summary by *Frew*, [1997]). By contrast, when the surface was clean, the gas transfer velocity was much higher for a given wind speed and increased smoothly with the *square* of the wind speed. The surfactant effect is found to increase with surfactant concentration; and in experiments done using natural waters it is found to increase linearly with various measures of the amount of organic matter in the water such as chlorophyll fluorescence, dissolved organic carbon concentration, and colored dissolved organic matter. The effect of surfactants is attributed primarily to their suppression of near-surface turbulence, inhibition of wave growth, and enhancement of the energy dissipation from small-scale waves.

Another finding of the wind tunnel experiments in the smooth and rough surface regimes is that the transfer velocities correlate strongly with the mean of the square of the wave slope $\langle s^2 \rangle$ over a wide range of wind tunnel systems and surfactant concentrations [e.g., *Jähne et al.*, 1987b; *Frew*, 1997]. The wave slope s is defined as the amplitude of the wave (half the wave height) times the wave number (2π over the wavelength). The correlation

TABLE 3.3.1
Coefficients for the calculation of the Schmidt number as a function of temperature in seawater of salinity 35 for temperatures between 0 and 30°C.
$Sc = A - BT + CT^2 - DT^3$, with T in °C.

Gas	*A*	*B*	*C*	*D*
He	410.14	20.503	0.53175	0.006011
Ne	855.1	46.299	1.254	0.01449
N_2	2206.1	144.86	4.5413	0.056988
O_2	1638	81.83	1.483	0.008004
Ar	1909.1	125.09	3.9012	0.048953
Kr	2205.0	135.71	3.9549	0.047339
Rn	3412.8	224.30	6.7954	0.08300
CH_4	2039.2	120.31	3.4209	0.040437
CO_2	2073.1	125.62	3.6276	0.043219
N_2O	2301.1	151.1	4.7364	0.059431
CFC-12	3845.4	228.95	6.1908	0.06743
CFC-11	3501.8	210.31	6.1851	0.07513
SF_6	3531.6	231.4	7.2168	0.090558
CCl_4	4295.8	281.52	8.7826	0.11025

Sources: The coefficients are all taken from *Wanninkhof* [1992] except for O_2 [*Keeling et al.*, 1998] and CFC-11 and CFC-12 (calculated by *Zheng et al.*, [1998] based on their measurements) and CCl_4 (R. Wanninkhof, personal communication, 2004). The coefficients for He, Ne, Kr, Rn, CH_4, and CO_2 are based on measurements by *Jähne et al.*, [1987a] and those for O_2 are based on measurements by *Jähne* [1980] as reported by *Liss and Merlivat* [1986]. The coefficients for N_2, Ar, N_2O, and SF_6, and CCl_4 are based on diffusivities calculated with the relationship of *Wilke and Chang* [1955] as discussed by *Wanninkhof* [1992].

TABLE 3.3.2
Selected gas transfer velocity formulations.
By convention, k_w is given in cm hr^{-1} and the wind speed, U, in m s^{-1}. The equations are for U measured at or interpolated to a nominal height of 10 m above sea level. The Schmidt number can be calculated using the coefficients given in table 3.3.1. The global average gas transfer velocity is obtained using the 41-year NCEP monthly mean wind speed climatology averaged over ice-free 4° × 5° areas corrected for variability as described in panel 3.3.1 and weighted by the ice-free area. The average gas transfer velocity for Liss and Merlivat [1986] is obtained using only the 3.6 to 13 m s^{-1} equation, as monthly mean U's are rarely outside this range.

Source	*Equation*	*Global Mean k_w at $Sc = 660^a$ (cm hr^{-1})*
Liss and Merlivat [1986]	$k_w = 0.17 \cdot U \cdot (Sc/600)^{-2/3}$, $U \leq 3.6$ m s^{-1}	
	$k_w = (U - 3.4) \cdot 2.8 \cdot (Sc/600)^{-0.5}$, $3.6 < U \leq 13$ m s^{-1}	11.2
	$k_w = (U - 8.4) \cdot 5.9 \cdot (Sc/600)^{-0.5}$, $U > 13$ m s^{-1}	
Nightingale et al. [2000b]	$k_w = (0.333 \cdot U + 0.222 \cdot U^2) \cdot (Sc/600)^{-0.5}$	14.9
Wanninkhof [1992]	$k_w = 0.31 \cdot U^2 \cdot (Sc/660)^{-0.5}$	20.0
Wanninkhof and McGillis [1999]	$k_w = 0.0283 \cdot U^3 \cdot (Sc/660)^{-0.5}$	18.7

[a] The transfer velocities are normalized to a Schmidt number of 660, which is the value for CO_2 at 20°C.

of k_w with $\langle s^2 \rangle$ suggests the possibility of parameterizing the gas transfer velocity as a function of the mean square wave slope, at least for the smooth and rough surface regimes. This is a very promising development, as the mean square wave slope can be related to measures of sea surface roughness obtained from the backscatter of radar measurements made from space [*Glover et al.*, 2002]. For the present, however, there are far more wind speed than mean square wave slope data available, and the relationship of the gas transfer velocity to the wind speed has been much more thoroughly explored, so all the models we will discuss here are wind speed–based.

Additional features of the schematic summary of laboratory studies in figure 3.3.2 are the second break in the lines at higher wind speeds, and the fact that the less soluble gas O_2 has a higher gas transfer velocity than the more soluble gas CO_2. Both of these features are due to the contribution of bubbles to the gas exchange process. Wave breaking, which occurs in wind tunnels at wind speeds above ~10 m s^{-1}, injects air bubbles into the water, which then exchange gas with the water. Eventually the bubbles will either dissolve completely or rise to the surface and burst back into the air. The net transfer of gas due to these processes is the difference between the gas injected into the water and the gas that is released back into the atmosphere, and can be in either direction. The flux of gas by bubbles is generally represented as consisting of two components, one due to bubble collapse or *injection*, the other due to *exchange* across the surface of the bubble in bubbles that do not dissolve completely. We represent these two processes by the following equation modified from *Keeling* [1993b], who derived his version by modifying the parameterization of *Fuchs et al.* [1987]:

$$\Phi_{bubble} = -k_{injection}\cdot[A]_a - k_{exchange}\cdot\left\{\left(1+\frac{\Delta P}{P}\right)\cdot[A]_a - [A]_w\right\} \tag{3.3.9}$$

The first term on the right-hand side represents the injection of gas A into the water by bubbles that dissolve completely. The second term represents gas exchange across the surface of the film. Both have an exchange coefficient k with units of velocity (cm hr^{-1}). The concentration of A that would be in equilibrium with the gas in the bubble is represented by the equilibrium concentration of air at 1 atm, $[A]_a$, corrected for the fractional increase in the total pressure of the bubble over the atmospheric pressure P (in atm) that results from surface tension and hydrostatic pressure (note: 10 m of water is equivalent to about 1 atm of pressure).

We can explain the fact that less soluble gases have greater bubble injection as follows: suppose air is being injected into the ocean at some volumetric rate per unit time per unit area as a consequence of bubble formation. The resulting injection rate v_{air} has units of velocity (e.g., cm hr^{-1}). We wish to determine Φ_A, the resulting flux of gas A in moles per unit area per unit time that results from this injection of air. Using the ideal gas law to convert the air volume into moles of gas A gives $\Phi_A = p_A\cdot v_{air}/R\cdot T$. However, the schematic illustration of data in figure 3.3.2 and the corresponding equations (3.3.5) and (3.3.9) all make use of gas transfer velocity formulations in terms of the concentration [A] rather than the partial pressure p_A. We convert p_A to [A] using Henry's law (3.1.2), which gives for the resulting bubble injection flux the expression $\Phi_A = \{v_{air}/(R\cdot T\cdot S_A)\}\cdot[A]$. The term in the curly brackets is essentially what is represented by the transfer velocities in (3.3.9) and (3.3.5). Its inverse dependence on solubility is what accounts for the higher gas transfer velocity for the less soluble gas, oxygen, in figure 3.3.2.

In regions where bubble injection is occurring, the total flux of gas across the air-sea interface will be equal to the bubble flux plus the flux across the air-sea interface $\Phi_{surface}$ given by (3.3.5), i.e.,

$$\begin{aligned}\Phi_{Total} &= \Phi_{surface} + \Phi_{bubble}\\ &= -k_{surface}\cdot([A]_a - [A]_w)\\ &\quad - k_{injection}\cdot[A]_a - k_{exchange}\\ &\quad \cdot\left\{\left(1+\frac{\Delta P}{P}\right)\cdot[A]_a - [A]_w\right\}\\ &= -(k_{surface}+k_{exchange})\cdot\{(1+\Delta_e)\cdot[A]_a - [A]_w\}\\ &= -k_w\cdot\{(1+\Delta_e)\cdot[A]_a - [A]_w\end{aligned} \tag{3.3.10}$$

where

$$\Delta_e = \frac{k_{injection} + k_{exchange}\cdot\dfrac{\Delta P}{P}}{k_{surface}+k_{exchange}} \tag{3.3.11}$$

The final form of (3.3.10) is equivalent to that of *Woolf* [1997]. In this equation, Δ_e is the equilibrium fractional supersaturation of the gas that results when input of the gas by bubble injection and exchange are balanced by loss across the surface of the water, i.e., when $\Phi_{Total} = 0$. Such supersaturation is commonly observed in the ocean and is generally considered to be due primarily to the influence of bubbles. Note from our previous discussion that Δ_e will be larger for gases with lower solubilities. Except for the Δ_e term, (3.3.10) is also equivalent to (3.3.5), but with the k_w now understood to include the effect of gas exchange across the bubble surface at higher wind speeds where waves begin to break.

Field Studies

The laboratory studies demonstrate that gas exchange is sensitive to a variety of conditions whose natural distribution is difficult to reproduce accurately in a laboratory situation. A number of techniques have been developed to study gas exchange under more realistic conditions in the field. Figure 3.3.3 summarizes results of such estimates obtained by three different techniques: the ocean radiocarbon inventory method, the radon deficit method, and the dual tracer method. Each of these is discussed in turn. We conclude with a brief discussion of recent work using the covariance between gas concentrations and vertical velocities to directly estimate the flux.

Note that the bubble supersaturation term Δ_e is not a significant contributor to the gas flux for the tracers discussed in this subsection. In the case of radon-222 and the tracers injected into the ocean in the dual tracer experiments, $[A]_a$ is negligible and the bubble supersaturation term in (3.3.10) thus drops out. The flux equation reduces to $\Phi = k_w\cdot[A]_w$, where k_w represents the combined effect of gas exchange across the air-sea interface as well as across the surface of the air bubbles. In the case of radiocarbon, there is an influence of bubbles on our estimate of the gas transfer velocity, but, as we shall see, its effect is quite small.

The ocean radiocarbon inventory method is based on the fact that the global ocean decay rate of pre-bomb/preindustrial radiocarbon must, in a steady state, be replaced by a flux from the atmosphere, where it is produced by cosmic ray spallation of nitrogen. The flux from the atmosphere divided by the preindustrial air-sea radiocarbon difference gives an estimate of the gas transfer velocity. The total concentration of dissolved inorganic radiocarbon equals the sum of all the species of carbon dioxide and its hydrolysis products in the ocean ($DI^{14}C = [^{14}CO_2]+[H^{14}CO_3^-]+[^{14}CO_3^{2-}]$). The total oceanic decay rate of radiocarbon that must be replaced by gas exchange with the atmosphere is thus $\int(\lambda \cdot DI^{14}C_{ocean})\, dV_{ocean} \approx V_{ocean} \cdot \lambda \cdot DI^{14}C_{ocean}$, where $DI^{14}C_{ocean}$ is the mean concentration of the entire ocean. The input from the atmosphere per unit area thus has to be:

$$\Phi = -\frac{V_{ocean}}{A_{ocean}} \cdot \lambda \cdot DI^{14}C_{ocean} \tag{3.3.12}$$

The negative sign is due to the fact that the radiocarbon flux is downward into the ocean in the negative z-direction. Following (3.3.5), the flux must also be equal to

$$\Phi = -k_w \cdot ([^{14}CO_2]_a - [^{14}CO_2]_{surface}) \tag{3.3.13}$$

The subscripts *a* and *surface* refer to the atmosphere and to the surface waters of the ocean. Note that for gas exchange we use the concentration of $[^{14}CO_2]$ rather than $DI^{14}C$ because this is the only component of the $DI^{14}C$ that is gaseous and can exchange across the air-sea interface.

Equating (3.3.12) to (3.3.13) gives

$$\begin{aligned} k_w &= \frac{V_{ocean} \cdot \lambda \cdot DI^{14}C_{ocean}}{A_{ocean} \cdot ([^{14}CO_2]_a - [^{14}CO_2]_{surface})} \\ &= H_{ocean} \cdot \lambda \cdot \frac{DI^{14}C_{ocean}}{([^{14}CO_2]_a - [^{14}CO_2]_{surface})} \end{aligned} \tag{3.3.14}$$

We now make use of equation (3) of panel 1.2.1 to rewrite (3.3.14) in terms of R^* and the total concentration of carbon as follows:

$$\begin{aligned} k_w &= H_{ocean} \cdot \lambda \cdot \frac{\alpha_{water} \cdot DIC_{ocean} \cdot R^*_{ocean}}{(\alpha_{air} \cdot [CO_2]_a \cdot R^*_a - \alpha_{water} \cdot [CO_2]_{surface} \cdot R^*_{surface})} \\ &\approx H_{ocean} \cdot \lambda \cdot \frac{DIC_{ocean}}{[CO_2]_{surface}} \cdot \frac{R^*_{ocean} \Big/ \left(\frac{\alpha_{air}}{\alpha_{water}} \cdot R^*_a\right)}{1 - R^*_{surface} \Big/ \left(\frac{\alpha_{air}}{\alpha_{water}} \cdot R^*_a\right)} \\ &\approx H_{ocean} \cdot \lambda \cdot \frac{DIC_{ocean}}{[CO_2]_{surface}} \cdot \frac{R^*_{ocean}/(1.017 \cdot R^*_a)}{1 - R^*_{surface}/(1.017 \cdot R^*_a)} \end{aligned} \tag{3.3.15}$$

The CO_2 concentration gradient across the surface film is much smaller than the radiocarbon gradient; therefore, we use the average surface ocean value $[CO_2]_{surface}$ to represent both the atmospheric and oceanic CO_2 concentration. The ratio $\alpha_{air}/\alpha_{water}$ is greater than one because the heavier molecule $^{14}CO_2$ is more soluble in seawater than its lighter counterpart $^{12}CO_2$. Because of this, the seawater concentrations of $^{14}CO_2$ and $^{12}CO_2$ that are in equilibrium with the atmosphere have a ratio that is about 1.7% higher than that in the air above the seawater. We account for this effect by using the atmospheric concentration of $^{14}CO_2$ in our equation but multiplying it by 1.017. The R ratios are averages for the entire world, i.e., $R^*_{surface}$ is the average over the entire surface of the ocean, and R^*_{ocean} is the average over the entire volume of the ocean.

We are now ready to estimate the global mean gas transfer velocity that is implied by the natural radiocarbon distribution. We use $H_{ocean} = 3800$ m for the mean ocean depth. The radiocarbon decay constant is 1/8267 years. We use a surface CO_2 concentration of 9.7 mmol m^{-3} obtained from the global mean $T = 17.64°C$ and salinity $S = 34.78‰$, and assuming a preindustrial pCO_2 of 280 µatm. We use the preindustrial pCO_2 because the radiocarbon balance we are interested in is that which was in place prior to the perturbation of the carbon balance by input of anthropogenic CO_2 to the atmosphere and input of radiocarbon by nuclear bomb tests. Anthropogenic CO_2 comes from fossil fuels and cement production, which have no ^{14}C. For DIC_{ocean} we use 2200 mmol m^{-3}. Finally, we have 0.95 for $R^*_{surface}/(1.017 \cdot R^*_a)$, and 0.84 for $R^*_{ocean}/(1.017 \cdot R^*_a)$. This gives for the global mean gas exchange a value of ~20 cm hr^{-1}. Note that this estimate is very sensitive to the exact value chosen for $R^*_{surface}/(1.017 \cdot R^*_a)$. For example, a slightly lower value of 0.94 would give a gas transfer velocity of 16.6. An independent check on the gas transfer velocity obtained using the bomb radiocarbon inventory agrees with this result, as long as one remembers to increase the surface CO_2 concentration to account for the fossil CO_2 increase [*Siegenthaler*, 1986; *Broecker et al.*, 1985a]. How do air bubbles affect our estimate of the gas transfer velocity? We show later an estimate that bubbles give a CO_2 equilibrium supersaturation of ~0.02%. Including this term reduces the gas exchange coefficient estimate by only about 0.4%, and even the much greater upper limit equilibrium supersaturation of 0.3% given by *Keeling*, [1993b] would yield only a 5.6% reduction in the gas transfer velocity estimate.

The radiocarbon method discussed here gives a global long-term average gas exchange without any spatial or temporal resolution. Its value lies primarily in its use as a benchmark against which various formulations of dependency of the gas transfer velocity on wind speed can be tested. We shall see in the following subsection

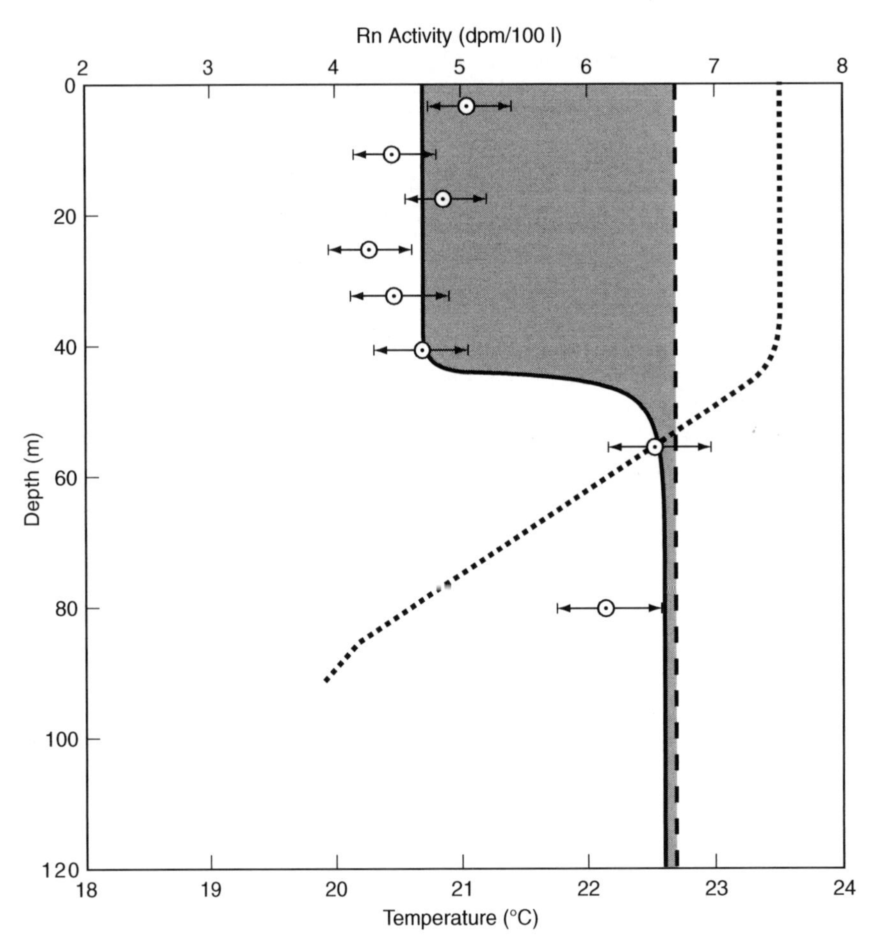

FIGURE 3.3.4: Radon-222 profile in the water column from *Broecker and Peng* [1982]. The dashed line is the estimated activity of radon-222 that would be in equilibrium with radium-226, if no radon-222 were lost to the atmosphere. The deficit relative to the expected amount is shaded. The dotted line is the vertical temperature profile, from which we can infer that the radon-222 deficit is confined to the mixed layer. This deficit is used to estimate the gas transfer velocity.

that its role in this regard, particularly that of the bomb radiocarbon constraint, is quite critical. Ongoing studies using models are exploring how one can use the natural and bomb radiocarbon constraint to obtain more information about spatial and temporal variability.

The radon deficit method is based on the distribution of radon-222 in the surface ocean. This 3.825 day half-life radioisotope is formed by decay of radium-226. We know from measurements of the parent concentration how much radon-222 is formed per unit of time. Measurements in the mixed layer generally show a deficit relative to the expected amount. The flux to the atmosphere can be estimated from this deficit and the gas transfer velocity found by dividing the flux by the air-sea concentration difference following (3.3.5). Figure 3.3.4 shows a typical example of a vertical profile of radon-222 in the surface ocean. The dashed line is the radon-222 distribution that the water would have if radon-222 were in radioactive equilibrium with radium-226. This is estimated by extrapolating the radon-222 concentration from the thermocline, where we know that it is in radioactive equilibrium with radium-226. This assumes that the loss of radium-226 from the mixed layer by biological uptake and export or adsorption onto sinking particles is negligible.

Before going through a sample calculation, we must briefly consider the behavior of a decay chain such as that involving radium-226 and radon-222. Radium-226 has a much longer half-life than radon-222 (1640 years). Thus, these two isotopes if left in isolation will achieve a *secular equilibrium*, in which the rate at which the daughter decays is equal to the rate at which the parent decays to form the daughter. It is thus convenient to express the concentration of these isotopes as the *activity* A, which is the decay constant λ times the number of atoms present per unit volume, e.g., $A_{\text{Rn-222}} = \lambda \cdot [^{222}\text{Rn}]$. The brackets represent concentration in atoms per unit volume. In the case of the isotopes under consideration here, the concentration is typically reported as dpm/100 kg, where dpm is disintegrations per minute.

The decay constant λ is simply $\ln(2)/t_{1/2}$, where $t_{1/2}$ is the half-life, 3.825 days. The expected equilibrium activity of radon-222 is thus $A_{\text{Rn-222}} = A_{\text{Ra-226}}$. The deficit of radon-222 with respect to this expected activity is that part of the radon-222 that has been lost to the atmosphere. Thus we have that the loss of radon-222 across the air-sea interface must be:

$$\Phi = h \cdot \{A_{\text{Ra-226}} - (A_{\text{Ra-222}})_w\} \quad (3.3.16)$$

where h is the thickness of the radon-222–depleted layer per figure 3.3.4, and the activities are the average for this layer.

The radon-222 gas exchange flux can also be calculated from (3.3.5) as follows:

$$\begin{aligned} \Phi &= -k_w \cdot ([^{222}\text{Rn}]_a - [^{222}\text{Rn}]_w) \\ &= -\frac{k_w}{\lambda} \cdot \{(A_{\text{Rn-222}})_a - (A_{\text{Rn-222}})_w\} \\ &\approx \frac{k_w}{\lambda} \cdot (A_{\text{Rn-222}})_w \end{aligned} \quad (3.3.17)$$

where we have taken advantage of the fact that the atmospheric radon-222 concentration is much lower than that in the ocean to drop this term. We now equate (3.3.16) and (3.3.17) to obtain

$$k_w = \lambda \cdot h \cdot \left(\frac{A_{\text{Ra-226}}}{(A_{\text{Rn-222}})_w} - 1 \right) \quad (3.3.18)$$

The profile in figure 3.3.4 has a radium-226–to–radon-222 activity ratio of 1.6 in the mixed layer and a mixed layer depth of 40 m, giving an estimate for the gas transfer velocity of

$$\begin{aligned} k_w &= \left[\frac{\ln(2)}{3.825\ \text{d}} \cdot 40\ \text{m} \cdot (1.6 - 1) \right] \\ &= 4.4\ \text{m d}^{-1} \\ &= 18\ \text{cm hr}^{-1} \end{aligned} \quad (3.3.19)$$

The radon-222 measurements shown as dots in figure 3.3.3 do not give any clear pattern with wind speed, although they do fall within the general range of the other measurements and models. Note that there is a large uncertainty in the measurement relative to the size of the signal (cf. error bars in figure 3.3.4). A further problem with radon-222 is that its distribution reflects an average of the air-sea gas exchange over a period of time of several days or weeks that it takes for this gas to equilibrate between the atmosphere and ocean. Some of the scatter in the measurements shown in figure 3.3.3 may be due to inadequate sampling of the wind speed over this period of time (cf. *Liss and Slinn* [1983] and discussion in panel 3.3.1).

The dual tracer technique purposefully injects two tracers into the ocean and traces their evolution in time. Most experiments to date have used ^{3}He and SF_6. The reason for using two tracers is to be able to separate the effect of horizontal and vertical transport within the ocean from the loss to the atmosphere. The evolution of the concentration ratio of the two gases $R = [^3\text{He}]/[\text{SF}_6]$ in a well-mixed layer of thickness h can be shown to be governed by the following equation [*Watson et al.*, 1991; *Wanninkhof et al.*, 1993]:

$$\begin{aligned} \frac{1}{R}\frac{dR}{dt} &= \frac{1}{[^3\text{He}]} \cdot \frac{d[^3\text{He}]}{dt} - \frac{1}{[\text{SF}_6]} \cdot \frac{d[\text{SF}_6]}{dt} \\ &= -\frac{(k_{\text{He}} - k_{\text{SF}_6})}{h} \end{aligned} \quad (3.3.20)$$

Observations of the behavior of R with time thus enable us to calculate the difference in the gas exchange constants but not their absolute value. We use (3.3.7) as an additional constraint that enables us to calculate the transfer velocities separately:

$$\frac{k_{\text{He}}}{k_{\text{SF}_6}} = \left(\frac{Sc_{\text{He}}}{Sc_{\text{SF}_6}} \right)^{-n} \quad (3.3.21)$$

The value for n that is typically used is 0.5 [*Nightingale et al.*, 2000b], though there is some concern about how well this number is known. Ideally, one would use a nonvolatile chemical to trace the water motion in order to avoid the problem of having to use (3.3.21). The use of bacterial spores, which are transported within the ocean but do not exchange across the air-sea interface nor sink significantly [*Nightingale et al.*, 2000b], is an interesting new development that gets around this problem. As figure 3.3.3 shows, the dual tracer measurements that are available so far tend to show a scatter comparable to that of the radon-222 measurements.

A recent promising development is the deployment of instruments that are capable of directly measuring the flux of gas in the atmospheric boundary layer. For example, the eddy correlation or covariance method simultaneously measures the vertical velocity and carbon dioxide concentration [*McGillis et al.*, 2001]. The air-sea flux of a gas A (CO_2, in this particular study) can be directly estimated from the cross-correlation of the gas concentration and the vertical velocity $\Phi_A = \overline{[A]' \cdot w'}$, where the prime quantities are as defined in panel 2.1.2. This assumes that net transport of air across the instrument is negligible compared to eddy exchange. The gas transfer velocity can be determined from the air-sea gradient by combining the above flux with (3.3.5) and dividing by the air-sea concentration anomaly. We shall return to a brief discussion of recent results from one such study in the following section.

Gas Transfer Velocity Models

Table 3.3.2 provides a summary of a variety of gas transfer velocity models based on the laboratory and field studies discussed above. The three linear features identified in the wind tunnel experiments have been

TABLE 3.3.3
Model- and observationally based estimates of cumulative bomb radiocarbon uptake between 1950 and the GEOSECS program.
The model based estimates use an ocean general circulation model with transfer velocities obtained using the climatic mean winds of Esbensen and Kushnir [1981] (B. Samuels, personal communication, 1990).

	Average k_w at Sc = 660 (cm hr^{-1})	*Oceanic Uptake ($\times 10^9$ atoms cm^{-2})*
Model using *Liss and Merlivat* [1986][a]	10.5	4.66
Model using *Broecker et al.* [1985a][b]	23.5	7.67
Radiocarbon observations[c]	21.9	8.24

[a] See table 3.3.2 for an explanation of the *Liss and Merlivat* [1986] gas transfer velocity formulation. The mean gas transfer velocity shown here is slightly lower than that shown in table 3.3.3 because of the use of a different wind speed data set.

[b] The *Broecker et al.* [1985a] gas exchange formulation is $k_g = 0.0118 \cdot (U - 2)$ mol m^{-2} yr^{-1} μatm^{-1}, which has a global mean of 0.065 mol m^{-2} yr^{-1} μatm^{-1}. This has been converted to k_w by dividing by the global mean CO_2 solubility of 3.44×10^{-5} mmol m^{-3} and multiplying by $(788/660)^{0.5}$ to normalize the gas transfer velocity to a Schmidt number of 660 from the global mean CO_2 Schmidt number of 788.

[c] The radiocarbon observational estimate is from *Broecker et al.* [1985a]. *Broecker et al.* [1995] give an updated bomb radiocarbon ocean column inventory estimate of 8.52×10^9 atoms cm^{-2}, while *Hesshaimer et al.* [1994] suggest it should be about 6.2×10^9 atoms cm^{-2}. The gas exchange coefficient estimate would change accordingly.

included by *Liss and Merlivat* [1986] in their model of gas transfer velocity given in the table and illustrated in figure 3.3.3. However, wind tunnel gas transfer velocities are generally different from those obtained from *in situ* field studies even at the same wind speeds (compare figures 3.3.2 and 3.3.3). Thus Liss and Merlivat calibrated their model with SF_6 gas exchange measurements made in small lakes [*Wanninkhof et al.*, 1985]. Many oceanic radon-222–based measurements as well as some of the recent oceanic dual tracer measurements (figure 3.3.3) support this calibration. However, when utilized in a three-dimensional ocean circulation model, the Liss and Merlivat function underpredicts the observed oceanic radiocarbon inventory between 1950 and the GEOSECS program of the 1970s by more than 40% (table 3.3.3).

An alternative model that is calibrated to radiocarbon measurements is that of *Wanninkhof* [1992], which has a quadratic dependence on wind speed of the form $k_w = a \cdot U^2$, where a is a constant determined from the observations that is related to the constants in (3.3.7) (see table 3.3.2 and figure 3.3.3). Wanninkhof notes that such a quadratic dependence does not have any physical significance, but it does give a reasonable fit to experimental results. In effect, his model replaces the three linear regimes of Liss and Merlivat with a dependence on the square of the wind speed and rescales the gas transfer velocity to match the radiocarbon observations. Wanninkhof also assumes that a square root dependence on the Schmidt number ($n = 0.5$) applies throughout the whole range of wind speeds. He ignores the smooth surface regime with its exponent of 2/3 on the Schmidt number, arguing that this regime seldom occurs in the ocean. The illustration in figure 3.3.3 shows that the Wanninkhof function sits well above that of Liss and Merlivat at all wind speeds.

A third model which is also calibrated with radiocarbon measurements is that of *Wanninkhof and McGillis* [1999], which has a cubic wind speed dependence. A theoretical justification for this is the proposal by *Monahan and Spillane* [1984] that gas transfer is proportional to whitecap coverage, which in turn scales approximately with U^3 [cf. *Monahan*, 2002]. This paper also summarizes a large set of new gas transfer velocity estimates obtained by the covariance method using CO_2. There is very good agreement between the data and their radiocarbon calibrated model.

We also show in table 3.3.2 the gas transfer velocity formulation of *Nightingale et al.* [2000b]. The global mean gas transfer velocity of this formulation is 13.7 cm hr^{-1} (table 3.3.2), which is intermediate to the *Liss and Merlivat* [1986] formulation on the low end, and to the models of *Wanninkhof* [1992] and *Wanninkhof and McGillis* [1999] at the high end. This formulation was obtained by fitting the dual tracer data set.

Which formulation for the gas transfer velocity should be used? First, it is important to note that for most gases the choice of gas transfer velocity formulation does not have a large effect on models of their concentration in the ocean. The air-sea exchange of CFCs, CH_4, and N_2O is so rapid relative to transport away from the interface that their surface concentrations are usually close to being in equilibrium with the atmosphere. CO_2 has the additional complexity that it reacts with water to form bicarbonate and carbonate ions. We see later that a consequence of this is that it takes ~20 times longer for CO_2 to equilibrate in the surface layer than it would without these chemical reactions. However, even anthropogenic CO_2 uptake models differ by at most a factor of 11% due to the factor of two uncertainties in the gas transfer velocity [*Sarmiento et al.*, 1992]; see table 3.3.4). On the other hand, the choice of gas transfer velocity is critical for modeling the uptake of radiocarbon (table 3.3.3) and for calculating the air-sea exchange from direct measurements of the air-sea gradient. In addition, the spatial structure of the air-sea gas exchange shows significant sensitivity to the gas transfer velocity, particularly in regions of upwelling and deep convection [*Sarmiento et al.*, 1992].

There is at present no universally accepted formulation for the gas transfer velocity. However, the bomb radiocarbon model simulation and observational results summarized in table 3.3.3 are very compelling in suggesting that the *Liss and Merlivat* [1986] function in particular cannot come close to predicting the observed radiocarbon distribution. Furthermore, the analysis of the oceanic oxygen distribution by *Keeling et al.* [1998] demonstrates that the *Liss and Merlivat* [1986] transfer velocities need to be increased by almost 150% in order to explain atmospheric oxygen observations that were calculated using fluxes estimated from the oceanic concentration anomaly. By contrast, the *Wanninkhof* [1992] relationship needed to be increased by less than 20% to explain the oxygen observations. In what follows we will therefore use primarily the bomb-calibrated formulations, specifically that of *Wanninkhof* [1992], which has been utilized in a wide range of studies.

We next discuss three processes that need to be taken into consideration when using the gas transfer velocity models to estimate the gas exchange from equation (3.3.5). The first of these is enhancement of gas exchange by chemical reactions in the boundary layer. This augmentation, known as *chemical enhancement*, increases the flux of gas by reducing the concentration within the boundary layer. One gas affected by chemical reactions in the boundary layer is CO_2, which reacts with water to form bicarbonate and carbonate ions when it dissolves in the ocean (see chapter 8). However, these carbon system reactions are slow in the absence of enzymes, and there is thus only minor chemical enhancement except at very low wind speeds. *Boutin and Etcheto* [1995] have studied the chemical enhancement effect on air-sea CO_2 exchange using the *Hoover and Berkshire* [1969] formalism. They found a noticeable effect in low-wind equatorial regions, but very little overall impact on a global scale. Their recommendation was that *in situ* experiments are needed, and the effect should be ignored in gas exchange models unless direct evidence for it is found. Enzymatic catalysis by carbonic anhydrase present in the water may play a role, but evidence for this has yet to be found [e.g., *Goldman and Dennett*, 1983]. One gas for which chemical enhancement is important is SO_2, which has extremely rapid hydration reactions, with a timescale of order 10^{-6} s. Because of this, the effective k_w for SO_2 is of order 34,420 cm h^{-1}. The rate limiting step for SO_2 gas exchange is thus gas transfer across the atmospheric boundary layer, which has a k_a/S_A of order 1600 cm h^{-1} [*Liss and Slater*, 1974].

The second process that must be taken into consideration when calculating gas exchange is the impact of cool skin temperatures on the solubility of atmospheric gases at the top of the boundary layer [*Robertson and Watson*, 1992]. The loss of heat at the surface of the ocean cools the upper 1 mm or so by a global mean ΔT of $\sim -0.3°$C relative to the temperature of the bulk water. *Van Scoy et al.* [1995] suggest that this should be viewed as a theoretical upper limit due to the

TABLE 3.3.4
Model-based estimates of cumulative oceanic uptake of CO_2 between 1750 and 1986.
Table uses an ocean general circulation model with transfer velocities obtained with the mean winds of Esbensen and Kushnir [1981], Sarmiento et al. [1992]. The units of k_g are mol m^{-2} yr^{-1} μatm^{-1}, and those of k_w are cm hr^{-1}.

	Average Gas Transfer Velocity[a]			*Oceanic uptake*	
Gas Transfer Velocity Formulation	k_g	k_w	*Normalized to BR85*	*Pg C*	*Normalized to BR85*
Liss and Merlivat [1986]	0.029	10.5	0.45	89.1	0.89
1.0x *Broecker et al.* [1985a][b] (BR85)	0.065	23.5	1.00	100.6	1.00
1.2x *Broecker et al.* [1985a]	0.078	28.2	1.20	103.1	1.02
2.0x *Broecker et al.* [1985a]	0.130	47.1	2.00	109.8	1.09

[a] The original reference by *Sarmiento et al.* [1992] reports only the value of k_g. This has been converted to k_w by dividing by the global mean CO_2 solubility of 3.44×10^{-5} mmol m^{-3} and multiplying by $(788/660)^{0.5}$ to normalize the gas transfer velocity to a Schmidt number of 660 from the global mean CO_2 Schmidt number of 788.

[b] The *Broecker, et al.* [1985a] gas exchange formulation is $k_g = 0.0118 \cdot (U-2)$ (cf. table 3.3.3).

fact that the *Hasse* [1971] parameterization used to estimate it tends to overestimate the magnitude of the skin temperature cooling at night [cf. *Kent et al.*, 1996]. The skin temperature effect has been taken into consideration by modifying the flux equation (3.3.5) to the form

$$\Phi = -k_w \cdot [(1+\zeta) \cdot [A]_a - [A]_w] \qquad (3.3.22)$$

where

$$\zeta = \frac{1}{S_A} \cdot \frac{dS_A}{dT} \cdot \Delta T \qquad (3.3.23)$$

Note that this assumes, as usual, that $[A]_a$, the saturation concentration of A in seawater, is calculated at the bulk temperature of the water. For CO_2, the temperature sensitivity term ζ is ~ -2.5 to -3.8% per °C, i.e., the

Panel 3.3.1: Influence of Spatiotemporal Variability on Flux Estimation

Tracer-based estimates of gas fluxes such as those used to obtain the transfer velocities summarized in figure 3.3.3 represent the average behavior over some region and period of time during which it is likely that all the parameters in the gas flux equations and gas transfer velocity formulations of table 3.3.2 are varying. For example, radon-222–based estimates represent the average flux over several radon-222 half-lives, i.e., one to two weeks. The dual tracer experiments are typically sampled at best twice in a 12-hour period, but usually represent the average behavior over a few days [e.g., *Nightingale et al.*, 2000b]. The bomb radiocarbon and natural radiocarbon tracers represent the globally averaged behavior over several decades to a millennium, respectively. The only observationally based gas exchange flux estimates that are arguably instantaneous are eddy correlation measurements such as those of *McGillis et al.* [2001], which are made at a frequency of several times a second and averaged over a period of 30 minutes (to reduce error). We discuss here how an improper representation of the variability in the parameters that determine the gas flux can affect our estimates of the magnitude of the gas flux.

Suppose we wish to determine the average flux of a gas A from measurements of its partial pressure difference defined by (3.1.1) using one of the gas exchange formulations given in table 3.3.2. A common way to do this would be to estimate the average wind speed over a timescale t that is comparable to the air-sea equilibration timescale of A and over an area that is representative of the region over which mixing and ocean circulation carried the water parcel that was sampled when we made the measurements of the partial pressure anomaly. This average wind speed would then be used to determine k_w from the gas exchange formulation, and the flux would be determined from (3.3.6) expressed as $\Phi^A = S_A \cdot k_w \cdot \Delta p^A$, where we have used the relationship $k_g = k_w \cdot S_A$. We use a simple example to show, by reference to figure 1, why such an approach would always give an underestimate of the gas flux with existing concave parameterizations of the gas transfer velocity as a function of the wind speed, and then go through a more detailed discussion of this and other issues that need to be dealt with to properly account for spatiotemporal variability. In our example, we use the *Wanninkhof* [1992] k_w function for a Schmidt number of 660, i.e., $k_w = 0.31 \cdot U^2$.

Panel figure 1 shows the gas transfer velocity as a function of wind speed. Suppose that over the timescale t and area S represented by our measurement of Δp^A, the wind blows half the time at a speed of 4 m s^{-1} and half the time at a wind speed of 11 m s^{-1}. The mean wind speed in such a case would be 7.5 m s^{-1}, with a corresponding gas transfer velocity of 17.4 cm hr^{-1}. However, if we instead were to separately calculate the gas transfer velocity corresponding to wind speeds of 4 m s^{-1} and 11 m s^{-1} (5.0 cm hr^{-1} and 37.5 cm hr^{-1}, respectively) and average the two transfer velocities together, which would be the right way to do this, we would obtain a gas transfer velocity of 21.2 cm hr^{-1}, which is 3.8 cm hr^{-1} larger than our first estimate.

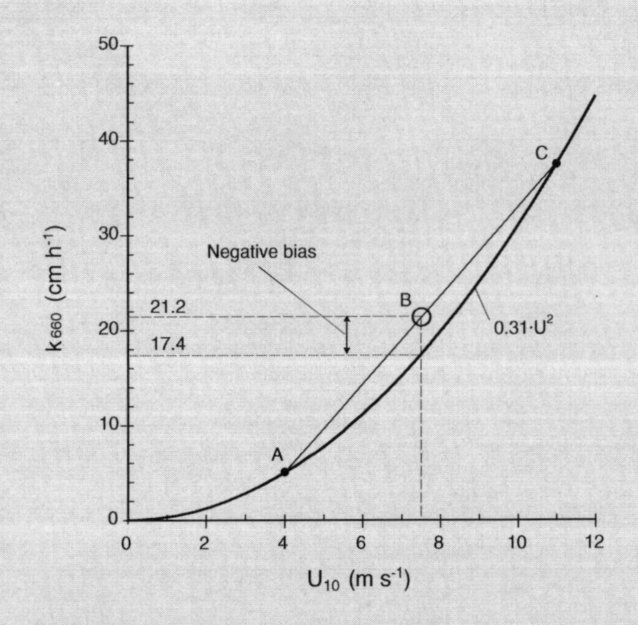

FIGURE 1: Schematic illustration of the influence of wind variability on the estimation of the mean gas exchange coefficient. Due to the nonlinear dependence of the gas exchange coefficient on the wind speed, simple averaging leads to biases.

solubility decreases with increasing temperature. Using the 1998 atmospheric CO_2 partial pressure of 365 μatm given in table 3.3.1, we can calculate that an average ΔT of 0.3°C would decrease the CO_2 partial pressure in equilibrium with water by 2.7 to 4.2 μatm. This correction is relatively modest compared to the peak air-sea partial pressure differences of order 40 μatm shown in figure 3.1.3, but is quite significant when considering, for example, that the total uptake of anthropogenic CO_2 by the ocean requires a global air-sea partial pressure difference of ~8 μatm. For oxygen, the temperature sensitivity term is −1.4 to −2.5% per °C for oceanic temperature ranges. We discuss the impact of this in section 3.4.

An issue that is not discussed in the papers that have attempted to estimate the impact of the skin tempera-

The negative bias introduced by using the average wind speed is characteristic of any measurements that are obtained over a span of time or over a region of space when the wind speed is varying. Note that the magnitude of the negative bias in figure 1 depends on how broadly the wind speed varies over time; if the variability is small, the bias will itself be small. The magnitude of the problem depends also on the concavity of the actual gas exchange relationship. Thus, the more concave *Wanninkhof and McGillis* [1999] U^3 relationship shown in figure 3.3.3 would have even larger biases than the *Wanninkhof* [1992] U^2 relationship; and the less concave *Nightingale et al.* [2000b] and *Liss and Merlivat* [1986] relationships would tend to have smaller biases. Only if the gas exchange formulation were a straight line with an intercept ≥ 0 would the bias disappear.

We now do a more rigorous analysis of how spatio-temporal variability may affect our flux estimate, beginning with the flux equation $\Phi^A = k_g \cdot \Delta p^A$. We average the flux equation in time and space as follows:

$$\overline{\Phi^A} = \overline{k_g \cdot \Delta p^A}, \text{ where}$$
$$\overline{(\;)} = \frac{\iint(\;)dS \cdot dt}{\iint dS \cdot dt} \tag{1}$$

S is the surface area under consideration, and t is time, as before. We expand the right-hand side of (1) in terms of an average $\overline{(\;)}$ and a time varying $(\;)'$ component using the approach described in panel 2.1.2. This gives

$$\overline{\Phi^A} = \overline{k_g} \cdot \overline{\Delta p^A} + \overline{k_g' \cdot \Delta p^{A'}} \tag{2}$$

[cf. *Keeling et al.*, 1998]. The second term on the right-hand side is the *covariance* between the gas exchange coefficient and pressure anomaly Δp^A. The magnitude of this and other covariance terms we will derive below can only be estimated properly by sampling sufficiently frequently to pick up all the timescales of variability of the parameters of interest. *Keeling et al.*, [1998] investigated the magnitude of this covariance term in the case of oxygen using models at two locations (Station Papa in the subtropical North Pacific and BATS near Bermuda) and an averaging time of one month. They found that using only the time average terms overestimated the flux by 0 to 10%. The covariance term reduces the flux due primarily to the fact that persistent periods of high gas exchange tend to drive the pressure anomaly to zero. This type of analysis has not been done for other gases that we are aware of. It is likely that the influence of the covariance term will be comparable for gases such as radon-222 that have similar air-sea equilibration times as oxygen, but smaller for gases such as CO_2 and radiocarbon that have much longer air-sea equilibration times and are thus slower to respond to perturbations in the gas transfer velocity.

We proceed by ignoring the covariance term between the gas exchange coefficient and pressure anomaly (as is usually done) and replacing the gas exchange coefficient by the product of the solubility and gas transfer velocity:

$$\begin{aligned}\overline{\Phi^A} &= \overline{k_g} \cdot \overline{\Delta p^A} \\ &= \overline{S_A \cdot k_w} \cdot \overline{\Delta p^A} \\ &= (\overline{S_A} \cdot \overline{k_w} + \overline{S_A' \cdot k_w'}) \cdot \overline{\Delta p^A}\end{aligned} \tag{3}$$

We are not aware of any attempts to estimate the covariance between the solubility and gas transfer velocity. The qualitative argument has been made that the temperature and salinity of the ocean, on which the solubility depends, vary more slowly than the winds over the surface of the ocean, and that the covariance term should thus be small. Assuming that this is the case gives:

$$\overline{\Phi^A} = \overline{S_A} \cdot \overline{k_w} \cdot \overline{\Delta p^A} \tag{4}$$

We arrive at the final step, in which we expand the gas transfer velocity $\overline{k_w}$ into average and varying components. We consider specifically the *Wanninkhof* [1992] U^2 dependency, although the following approach can be applied to any of the gas transfer velocity formulations. We assume that variability in the Schmidt number, which like solubility is a function of temperature and salinity, can be ignored. This gives for the average gas

ture effect on the gas exchange fluxes is the fact that the boundary layer for gases is much thinner than that for heat. We can get some idea of the thickness of the boundary layer for a gas from the stagnant film model, for which, as we noted earlier in this section, the transfer velocity k_w is equal to the molecular diffusivity ε divided by the stagnant film thickness Δz. For an average k_w of 20 cm hr^{-1} and a typical gas diffusivity of 10^{-5} cm^2 s^{-1}, this gives a stagnant film thickness of about 2×10^{-3} cm, i.e., about 0.02 mm, compared with the thermal film thickness of order 1 mm. Since the base of the gas stagnant film is well within the thermal boundary layer, the temperature difference between the top and bottom of the gas stagnant film should be much less than that between the skin temperature of the ocean and the bulk temperature of the mixed layer. Because of this, it may be that the skin temperature effect has been overestimated [*McGillis and Wanninkhof*, 2005].

The third process that needs to be taken into consideration when calculating gas exchange is the bubble-induced equilibrium supersaturation term Δ_e of (3.3.10), which is reproduced here for gas A:

$$\Phi = -k_w \cdot \{(1+\Delta_e^{\mathrm{A}}) \cdot [\mathrm{A}]_a - [\mathrm{A}]_w\} \quad (3.3.24)$$

Woolf and Thorpe [1991] have proposed that the Δ_e^{A} term can be represented for gas A by

$$\Delta_e^{\mathrm{A}} = 0.01 \cdot \left(\frac{U}{U_{\mathrm{A}}}\right)^2 \quad (3.3.25)$$

where U_{A} is the wind speed at which the supersaturation is 1%. The square dependence on the wind speed arises from balancing input of gas to the ocean by bubbles with loss of gas from the ocean. Bubble formation rate is related to the presence of foam, which is observed to scale approximately with U^3. Loss of gas from the ocean is assumed by them to scale as U. The

Panel 3.3.1: Continued

transfer velocity the expression:

$$\overline{k_w} = 0.31 \cdot (\overline{U} \cdot \overline{U} + \overline{U' \cdot U'}) \cdot \left(\frac{Sc}{660}\right)^{-0.5}$$
$$= 0.31 \cdot (\overline{U}^2 + \overline{U'^2}) \cdot \left(\frac{Sc}{660}\right)^{-0.5} \quad (5)$$

Note that the wind speed covariance term $\overline{U'^2}$ is equal to the standard deviation of the velocity, σ_U^2 [cf. *Boutin and Etcheto*, 1995]; this of course would not be the case if we used the U^3 relationship in (5). In our example of figure 1, the covariance term adds 0.8 m s^{-1} to the average wind speed of 7.5 m s^{-1}, giving an average gas transfer velocity of 21.2 cm s^{-1}.

We discuss in the text a recent attempt to estimate the contribution of wind speed variability to calculations of the air sea flux of CO_2 by *Wanninkhof et al.* [2002], updated by Rik Wanninkhof (personal communication, 2004; cf. tables 3.3.2 and 3.4.1). The transfer velocities are calculated using the 41-year (1958–1998) monthly average climatological wind speeds of the National Center for Environmental Prediction (NCEP) of NOAA corrected for the influence of wind speed variance calculated from the 6-hourly NCEP data set for the year 1995 (S. Doney and R. Wanninkhof, personal communication, 2004). The 41-year average wind speed is either squared or cubed according to the gas transfer velocity formulation, giving $(\overline{U_{41year}})^2$ or $(\overline{U_{41year}})^3$, respectively. The 6-hourly data are then used to calculate corrections consisting of $\overline{U^2_{6\,hour}} / (\overline{U_{6\,hour}})^2$ or $\overline{U^3_{6\,hour}} / (\overline{U_{6\,hour}})^3$, which are multiplied times the 41-year averages. The salient results from table 3.4.1 are the following:

1. In the case of the *Liss and Merlivat* [1986] and *Nightingale et al.* [2000b] calculations, the effect that one obtains by including the influence of wind speed variability is to increase the globally integrated flux of CO_2 by about 20%.

2. In the case of *Wanninkhof* [1992] and *Wanninkhof and McGillis* [1999], the column (a) and column (b) results use different equations, which are supposed to correct for the difference between using the long-term average winds (column a) and the instantaneous or short-term winds (column b; see footnote in table 3.4.1). The disagreement between the two columns, particularly in the case of the *Wanninkhof and McGillis* [1999] result, indicates either that the correction is not entirely successful, or that even 6-hourly winds are insufficient to capture the full range of wind variability.

The final point we would make with regard to the issue of the spatiotemporal variability of wind speeds is that the data shown in figure 3.3.3 should really be plotted versus mean wind speed velocities that take into consideration the functional form of the gas transfer velocity formulation with which they are being compared. Thus, if one wanted to compare the data with the *Wanninkhof* [1992] U^2 relationship, one should plot the data versus $\sqrt{\overline{U^2}}$ rather than versus $\overline{U}$, as is usually done.

value of U_A is given by *Woolf and Thorpe* [1991] for four gases:

$$\begin{aligned} U_{N_2} &= 7.2\,\mathrm{m\,s^{-1}} \\ U_{O_2} &= 9.0\,\mathrm{m\,s^{-1}} \\ U_{CO_2} &= 49\,\mathrm{m\,s^{-1}} \\ U_{Ar} &= 9.6\,\mathrm{m\,s^{-1}} \end{aligned} \qquad (3.3.26)$$

The fractional supersaturation for a typical wind speed of 7.5 $\mathrm{m\,s^{-1}}$ is relatively modest: 1.1% for nitrogen, 0.7% for oxygen, 0.02 % for CO_2, and 0.61% for Ar. Bubbles have a much smaller impact on CO_2 than the other gases because it is the most soluble gas of the four. In the final section of this chapter, we will use equations (3.3.24) to (3.3.26) to estimate the effect of bubbles on the supersaturation. However, it should be kept in mind that there is no consensus on how to represent the impact of bubbles either on the exchange rate k_w or on the supersaturation Δ_e. Many models have been proposed, and the observations are not consistent in indicating which does the best at representing the observations [e.g., *Emerson et al.*, 1995, 1997; *Farmer et al.*, 1993; *Keeling*, 1993b; *Schudlich and Emerson*, 1996; *Spitzer and Jenkins*, 1989; *Wallace and Wirick*, 1992; *Woolf*, 1997; *Woolf and Thorpe*, 1991].

A final point: in this chapter, as in most studies of gas fluxes, we are interested in determining the average gas flux over some period of time and region of space over which the wind speed and concentrations of the gas in the ocean and partial pressure in the air are varying in time. Until recently, the long-term average gas flux was commonly estimated by separately determining the long-term average wind speed and concentration anomaly and then plugging these into the corresponding gas transfer velocity and gas exchange formulations. The fluxes calculated by this incorrect procedure are biased low relative to what one obtains if one follows the correct procedure of using the instantaneous wind speed and concentration anomaly to determine an instantaneous gas flux, and then takes a long-term average of this instantaneous gas flux. Panel 3.3.1 provides a detailed explanation of this important problem and the difference that it makes to the answer that one obtains.

3.4 Applications

We now address the questions that were raised in the introduction to this chapter about the distributions of oxygen, carbon dioxide, and nitrous oxide. We begin with a brief overview of the biological and solubility processes that we need to be aware of and then proceed to a more detailed application of our gas exchange models to provide a quantitative analysis.

We saw in chapter 1 that the synthesis of organic matter in the ocean by phytoplankton can be represented by the reaction

$$106CO_2 + 16HNO_3 + H_3PO_4 + 78H_2O$$
$$\rightleftharpoons C_{106}H_{175}O_{42}N_{16}P + 150O_2.$$

Remineralization is the reverse of this reaction. The net synthesis of organic matter thus consumes CO_2 and increases O_2. By contrast, remineralization of organic matter consumes O_2 and increases CO_2. The impact of these processes on the air-sea gas fluxes will depend on which predominates in the surface ocean. One would generally expect organic matter synthesis to predominate over remineralization in the well-lit surface layers, and for remineralization to predominate in the thermocline and deep ocean. This being the case, biology should generally lead to uptake of CO_2 from the atmosphere and release of O_2 back to the atmosphere. However, there are many situations where rapid upward mixing or upwelling of thermocline and deep waters overwhelms the surface ocean and leads to CO_2 release to the atmosphere and O_2 uptake from the atmosphere.

Nitrous oxide N_2O is a special case of a gas that is produced only during the process of remineralization as a minor byproduct of the oxidation of organic nitrogen (see chapter 5). Photosynthesis neither produces nor consumes this gas. The ultimate fate of nitrous oxide is destruction in the atmosphere, which occurs on a timescale of order 120 years [*Prather and Ehhalt*, 2001].

The influence of solubility on the air-sea gas exchange depends primarily on the temperature sensitivity illustrated in figure 3.2.1. Heating of the ocean reduces the solubility, whereas cooling increases it. Thus, if the surface ocean is allowed to saturate with a gas at some temperature and then is heated, the water will become supersaturated and degassing will result. Conversely, cooling of the water will result in uptake from the atmosphere. We will refer to these solubility effects as the "thermal component." Changes in salinity by evaporation, rainfall, and river input also affect the solubility, but are much smaller than the thermal effects.

We turn now to a discussion of the oxygen supersaturation of order 3% observed over much of the ocean during the WOCE program (figure 3.1.2; cf. also map of figure 3.1.1). A full characterization of this supersaturation requires careful consideration of the spatial and temporal variability of all the parameters that affect the gas exchange. We show how this might be done by considering here a simple example using global mean values for all the relevant parameters. Two processes that we consider first are the skin temperature effect represented by (3.3.22) and (3.3.23), and the bubble effect represented by (3.3.24) and (3.3.25), both of which cause the ocean to be supersaturated with respect to the atmosphere. From the oxygen solubility we can estimate that a skin temperature cooler by 0.3°C will give a ζ of about 0.006. As noted earlier, this should be considered an upper limit. As regards the effect of bubbles, we have seen that the parameterization (3.3.25) gives an oxygen Δ_e of 0.007 for a typical wind speed of $7.5\,m\,s^{-1}$. In combination, these two processes could account for a supersaturation of ~0.013, or 1.3%. The remaining supersaturation of ~1.7% in the WOCE observations would then have to be due to the biological and thermal components discussed above. We can estimate the net impact of these processes by calculating the loss of oxygen to the atmosphere implied by the observed oxygen supersaturation.

We calculate the air-sea flux using (3.3.22) and (3.3.24) to correct for the bubble supersaturation and skin temperature effects:

$$\Phi = -k_w \cdot [(1+\zeta+\Delta_e) \cdot [O_2]_a - [O_2]_w] \qquad (3.4.1)$$

For the gas transfer velocity we use the *Wanninkhof* [1992] formulation given in table 3.3.2. Using a global mean temperature of 17.64°C gives for the *Sc* number of oxygen a value of 663. We use for the wind speed an average value of $7.5\,m\,s^{-1}$. This gives a gas transfer velocity of $17.4\,cm\,hr^{-1}$ from the Wanninkhof model. We use for the skin temperature and bubble supersaturation corrections the values of 0.006 and 0.007 we estimated previously. We replace the water concentration by 1.03 times the atmospheric concentration, as observed, i.e.,

$$[O_2]_w = 1.03 \cdot [O_2]_a \qquad (3.4.2)$$

We use for $[O_2]_a$ a value of $241\,mmol\,m^{-3}$, which is the saturation concentration at the global mean surface temperature of 17.64°C and salinity of 34.78. Converting the gas transfer velocity to units of $m\,yr^{-1}$ and substituting all the numbers into (3.4.1) yields

$$\begin{aligned}\Phi &= -(1520\,m\,yr^{-1}) \cdot [(1+0.006+0.007) \\ &\quad \cdot [O_2]_a - 1.030 \cdot [O_2]_a] \\ &= -(1520\,m\,yr^{-1}) \cdot (-0.017) \cdot 0.241\,mol\,m^{-3} \\ &= 6.2\,mol\,m^{-2}\,yr^{-1}\end{aligned} \qquad (3.4.3)$$

Based on the foregoing analysis we would conclude that about 40% of the oxygen supersaturation observed over much of the ocean may be due to bubble injection (the effect of which is highly uncertain) and the skin temperature effect (which should be considered an upper limit). Where would the remaining oxygen come from? It must be due to the net effect of other processes such as production of oxygen by biology, supersaturation induced by warming of the surface ocean, and vertical mixing. The loss rate of oxygen that we estimate is of the same order of magnitude as typical estimates of net biological oxygen production in the surface layer of order $5\,mol\,m^{-2}\,yr^{-1}$ [e.g., *Emerson*, 1987; *Emerson et al.*, 1997; *Jenkins and Goldman*, 1985; *Musgrave et al.*, 1988; *Peng et al.*, 1987; *Spitzer and Jenkins*, 1989; *Thomas et al.*, 1990]. However, such a comparison must be viewed with caution due to the nonnegligible contributions of warming and mixing discussed in these studies. One very interesting aspect of some of these studies is their use of inert noble gases to constrain the gas transfer velocity and bubble contribution. For example, *Spitzer and Jenkins* [1989] demonstrated that the low solubility helium-3 provided a powerful constraint on the bubble injection contribution.

A major issue that we have ignored up to now is that in a steady state the global climatic mean flux of oxygen across the air-sea interface has to be close to zero. Thus, losses of oxygen to the atmosphere such as implied by the supersaturation of figure 3.1.2 must on average be balanced by a gain either at other times of the year or in other locations of the ocean. Indeed, the annual mean concentration anomaly map shown in figure 3.1.1 shows that there are large areas of the ocean where oxygen is undersaturated. It just happens that a high percentage of the data in figure 3.1.2 were in regions (lower latitudes) and at times (warm seasons) when the ocean was supersaturated.

Figure 3.4.1 shows a recent estimate of the climatic mean air-sea flux of oxygen over various regions of the ocean implied by the interior distribution of oxygen. Ocean circulation models were used to estimate what gas fluxes are required in order to explain the interior distributions [*Gruber et al.*, 2001]. Downward bars represent uptake by the ocean; upward bars are degassing from the ocean. The figure shows that the ocean is generally losing oxygen to the atmosphere in the tropics between 36°S and 13°N, and gaining oxygen poleward of these latitudes except for the South Atlantic between 36°S and 58°S. The globally integrated net flux estimated by this method is quite small, as expected. Careful comparison of the signs of the fluxes in this figure with those that would be expected from the annual mean air-sea concentration anomaly map in figure 3.1.1 shows significant discrepancies, for example in the tropical Pacific where one would infer an oceanic uptake from the negative anomaly of figure 3.1.1, but this figure shows a degassing flux into the atmosphere. This inconsistency is due primarily to differences in the data sets used in the two analyses (surface concentration anomalies for figure 3.1.1 versus interior oxygen distributions in this figure) and is unlikely to be fully resolved without more high quality

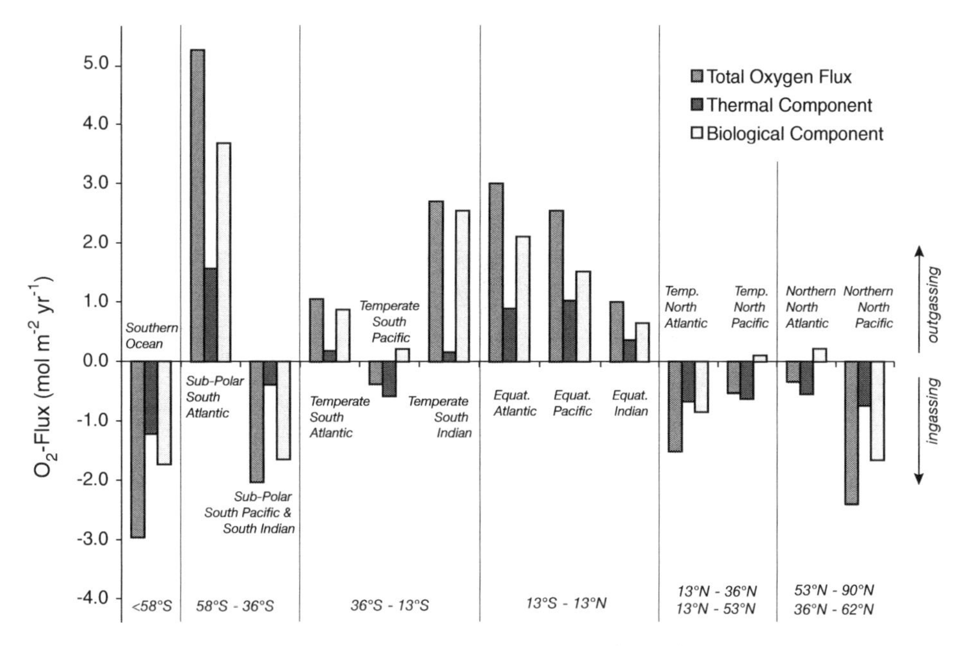

FIGURE 3.4.1: Oxygen flux estimates obtained from the study of *Gruber et al.* [2001]. Positive fluxes are out of the ocean. The breakdown of the fluxes into thermal and biological components is explained in the text.

measurements of the surface oxygen concentration anomaly.

Figure 3.4.1 also shows a separation of the total oxygen flux in each region into thermal and biological components. The thermal component is calculated from the relationship

$$\Phi_{thermal} = -(\partial S_A/\partial T) \cdot (Q/C_p) \tag{3.4.4}$$

where T is temperature, Q is the heat flux (positive for warming and negative for cooling of the ocean), and C_p is the specific heat of water [cf. *Keeling and Peng,* 1995]. The biological component is calculated by subtracting the thermal component from the total flux. One can see from figure 3.2.1 that the term $(\partial S_A/\partial T)$ is negative, thus a flux of heat into the ocean will give a positive gas flux, that is to say out of the ocean. In other words, heating makes gases less soluble and leads to a degassing flux, whereas cooling has the opposite effect. Equation (3.4.4) assumes that the air-sea equilibration of oxygen is extremely rapid, such that a heat flux will quickly be manifested as an oxygen flux. The breakdown into thermal and biological components shows that these two are almost always of the same sign, with biology predominating in most regions. The biological degassing flux in low latitudes and in the South Atlantic is due to a dominance of organic matter synthesis over upward mixing and upwelling of deep waters whose oxygen has been depleted by respiration. This dominance of organic matter synthesis occurs despite the influence of upwelling due to Ekman divergence at the equator and along the western margins of the continents (see chapter 2). Heating of the cold water that is brought up by the upwelling contributes an additional component to the degassing flux. The biological uptake flux of the higher latitudes indicates that the combined effect of respiration and the upward mixing of the low-oxygen waters of the thermocline and deep ocean dominates over organic matter synthesis. The uptake by the thermal component is due to cooling of warm waters from lower latitudes or brought up to the surface by convective overturning.

We conclude our discussion of oxygen by noting that there are significant seasonal variations of oxygen fluxes [e.g., *Louanchi and Najjar,* 2001; *Najjar and Keeling,* 1997, 2000], and that spatial and temporal variations of oxygen fluxes have a measurable impact on the atmospheric oxygen distribution that provides useful constraints on our understanding of oceanic processes [e.g., *Keeling et al.,* 1998; *Stephens et al.,* 1998]. In chapter 10, we discuss evidence that global warming appears to be already having a significant impact on the oxygen distribution of the ocean as well as the atmosphere.

We turn now to a discussion of CO_2. The thermal component of the air-sea CO_2 flux will have the same sign as the thermal component of the O_2 flux. However, the biological component will have the opposite sign because CO_2 is produced when O_2 is consumed and vice versa. Given that for oxygen the biological component dominates over the thermal component, we might expect that the CO_2 fluxes would generally be in the opposite direction from the O_2 fluxes. In fact, the CO_2 fluxes that we infer from the concentration anomalies shown in figure 3.1.3 indicate degassing in

TABLE 3.4.1
The globally integrated air-sea CO_2 flux based on the partial pressure difference maps of *Takahashi et al.* [2002] (figure 3.1.3).
The average gas transfer velocity for each quadrangle in that data set was obtained using the 41-year (1958–1998) monthly average climatological wind speeds of the National Center for Environmental Prediction (NCEP) of NOAA (column a) corrected in column b for the influence of wind speed variance calculated from the 6-hourly NCEP data set for the year 1995 as described in panel 3.3.1 (S. Doney and R. Wanninkhof, personal communication, 2004). The original wind speeds used by Takahashi et al. [2002] have been updated to adjust for their incorrect use of winds referenced to ~40 m above sea level, rather than 10 m (T. Takahashi, personal communication, 2004).

	Globally Integrated Air-sea CO_2 Flux ($Pg\ C\ yr^{-1}$)		
Gas Transfer Velocity Formulation	*(a) Without Wind Speed Variability*	*(b) With Wind Speed Variability*	*(b)/(a)*
Liss and Merlivat [1986]	−0.88	−1.06	1.21
Nightingale et al. [2000b]	−1.06	−1.25	1.19
Wanninkhof [1992][a]	−1.65	−1.58	0.96
Wanninkhof and McGillis [1999][a]	−2.35	−1.94	0.82

[a] The equations used for the *Wanninkhof* [1992] and *Wanninkhof and McGillis* [1999] estimates in column (a) are $k_w = 0.39 \cdot U \cdot (Sc/660)^{-0.5}$ and $k_w = (1.09 \cdot U - 0.33\ U^2 + 0.078 \cdot U^3) \cdot (Sc/660)^{-0.5}$, respectively. These are calibrated against the radiocarbon data in figure 3.3.3 and are thus representative of the behavior of the gas transfer velocity with respect to long-term average winds. The equations shown in table 3.3.2 and used in column (b) have been renormalized using a Rayleigh statistical distribution of the wind speed. See discussion in panel 3.3.1.

the tropics and uptake in high latitudes, the same as for O_2. How can this be? It would appear that for CO_2, unlike for O_2, the thermal component dominates over the biological component. That this indeed appears to be the case is confirmed by a set of model studies discussed in chapter 8 [*Murnane et al.*, 1999]. We believe that the reason the thermal component dominates over the biological component for CO_2 is because of the long timescale of air-sea equilibration for CO_2. We describe briefly why the air-sea equilibration time for CO_2 is so long (see chapter 8 for a full discussion) and then return to a discussion of how this may explain the dominance of the thermal component over the biological component.

Consider an example of oxygen in a 40-meter mixed layer at a temperature of 25°C with a wind speed of $7.5\ m\ s^{-1}$. The *Sc* number is 394, which gives a gas transfer velocity of $23\ cm\ hr^{-1}$ using the Wanninkhof U^2 gas exchange model. Converting this to units of $m\ yr^{-1}$ (1977) and dividing into the mixed layer thickness of 40 m gives an oxygen air-sea equilibration timescale of 7 days. For CO_2, we have a larger *Sc* number of 525, and therefore a slightly lower gas transfer velocity of $20\ cm\ hr^{-1}$. The residence time found by dividing the CO_2 mixed layer thickness by this gas transfer velocity is now 9 days. However, this is not the end of the story for CO_2, for which we must also consider the chemical reactions that it undergoes with water (see chapter 8). Only 0.5% of average surface seawater *DIC* is in the form of CO_2. The remainder is in the form of bicarbonate and carbonate ions. The ocean carbon chemistry is such that the majority of CO_2 molecules that exchange with the atmosphere come from bicarbonate ions via the reaction:

$$CO_2 + CO_3^{2-} + H_2O \rightleftharpoons 2HCO_3^-$$

At present, only 1 in every 20 CO_2 molecules that escapes to the atmosphere contributes toward reducing the concentration of CO_2 in the tropical regions. Thus, it takes 20 times longer for the CO_2 to equilibrate with the atmosphere than if it behaved like oxygen, i.e., 6 months! We note that the same analysis applies to any of the isotopes of carbon such as $^{13}CO_2$ and $^{14}CO_2$, but because of the way we usually report these measurements as isotopic ratios to $^{12}CO_2$, the multiplicative factor ends up being the ratio of DIC to CO_2, namely 200, which gives 60 months, i.e., 5 years as the air-sea equilibration time for isotopic ratios of carbon [cf. *Broecker and Peng*, 1982].

How does this long equilibration time affect the thermal versus the biological components for CO_2? The only reason there is a biological component to the air-sea fluxes of O_2 and CO_2 is because there are some regions of the surface where synthesis of organic matter is slow relative to upward mixing and advection. In such regions, there is a release of CO_2 to the atmosphere and an uptake of oxygen. In a steady state, these biological components must be balanced by equal and opposite biological components somewhere else in the ocean. However, these regions where strong upward mixing and advection overwhelm the biological uptake occur predominantly in the high latitudes, particularly of the Southern Ocean. The overturning rate in these regions is too rapid for much of the excess CO_2 to escape to the atmosphere, but not for O_2 uptake to occur. Thus the biological component for O_2 is quite large, whereas it is small for CO_2. By contrast, the thermal component is more widespread, and in regions where warming is occurring, the water column becomes stratified and water stays at the surface for a long time. Thus the thermal component is much more fully expressed for CO_2 than the biological component. Chapter 8 has a more complete discussion of these phenomena for CO_2.

One of the most important problems in ocean biogeochemistry today is to estimate the uptake of anthropogenic carbon by the ocean. This topic is covered in detail in chapter 10, but we summarize here a set of estimates of the net air-sea CO_2 flux based on the pCO_2 monthly average partial pressure difference data of *Takahashi et al.* [2002], which they mapped out over 4° north-south by 5° east-west quadrangles around the world. Their data set included approximately 940,000 measurements obtained since 1956, all of which were interpolated to 1995. Column (b) of table 3.4.1 gives results from the study of *Wanninkhof et al.* [2002] updated by Rik Wanninkhof (personal communication, 2004), which includes the influence of wind speed variability on the gas transfer velocity estimate (see panel 3.3.1). The range from the lowest anthropogenic carbon uptake estimate to the highest is almost a factor of 2. Three lines of evidence examined earlier strongly suggest that the lowest of these estimates based on the *Liss and Merlivat* [1986] function is unlikely to be correct: (1) comparison with empirical estimates of gas exchange using the radiocarbon observations (figure 3.3.3), (2) model simulations of bomb radiocarbon uptake by ocean general circulation models (table 3.3.3), and (3) comparison of model simulations with oxygen observations in the atmosphere [*Keeling et al.*, 1998]. The skin temperature effect has been estimated to increase the oceanic uptake by $0.4\ Pg\ C\ yr^{-1}$, i.e., about 25% of the total ([*Van Scoy et al.*, 1995], based on the radiocarbon calibrated linear gas transfer velocity model of *Tans et al.* [1990]). However, as noted earlier, this is likely to be an upper limit.

We conclude with a brief discussion of the nitrous oxide distribution shown in figure 3.1.4. One interesting aspect of this distribution is that nitrous oxide is supersaturated almost everywhere. This is exactly what we would expect from what we learned about its biogeochemistry at the beginning of this section. Another interesting aspect of this distribution is that the peak supersaturations are observed in the tropics of the Atlantic, Pacific, and Indian Oceans. We show in chapter 5 that the highest deep-ocean nitrous oxide concentrations are found in the thermocline waters of the low latitudes where the remineralization of organic matter is at a maximum. Since nitrous oxide is a byproduct of remineralization, this is exactly where we would expect the nitrous oxide to peak. It is the upwelling of these waters at the surface that we see manifested in the tropics in figure 3.1.4.

There have been numerous attempts to estimate the flux of nitrous oxide to the atmosphere using observations such as those shown in figure 3.1.4 as well as a range of models. The most recent observationally based estimate is that of *Nevison et al.* [2004], who give a global ocean N_2O degassing flux of 3.7–$4.3\ Tg\ yr^{-1}$ of N. This can be compared to a global "natural" (i.e., pre-anthropogenic) nitrous oxide production rate of 7 (range of 4 to 12) $Tg\ yr^{-1}$ of N for non-oceanic sources, mostly soils, given by *Prather and Ehhalt* [2001]. The oceanic source thus accounts for about 40% (range of 28% to 51%) of the total pre-anthropogenic input of nitrous oxide to the atmosphere.

Problems

3.1 You measure the oxygen content of a water sample and find that it has a concentration of 250 mmol m^{-3}. The temperature of this sample is 15°C and it has a salinity of 33.

a. Is this water sample super- or undersaturated with respect to current atmospheric oxygen concentrations (use table 3.2.4)? How large is the super- or undersaturation?

b. Over the hours since the sample was taken, its temperature rises from the initial 15°C to 20°C. Assuming that this sample has not gained or lost oxygen over this time, how large is the under- or supersaturation now?

3.2 If we were to increase atmospheric CO_2 fivefold by the burning of fossil fuels, how much would the atmospheric oxygen concentration change relative to what it is today? Assume that the burning of fossil fuels consumes 1.4 moles of O_2 per mole of CO_2 produced.

3.3 The analysis of an air sample has revealed that it has a dry atmospheric CO_2 mixing ratio of 370 ppm, and a total pressure of 1.020 atm, and is at saturation with regard to water vapor. The temperature at which this sample was taken was 20°C.

a. Compute the partial pressure of CO_2 for this air sample.

b. How would the partial pressure of CO_2 change if you warmed the sample isobarically to 30°C without changing its water vapor content?

c. Everything else being equal, how large would the partial pressure of CO_2 be if you happened to have sampled the air while a low pressure system was present, e.g., if the total pressure was 0.940 atm?

3.4 Explain step-by-step how to derive the flux equation $\Phi = -K \cdot ([A]_a - [A]_w)$, where $1/K = 1/k_w + S_A/k_a$, from (3.3.1) and (3.3.2).

3.5 We discussed in the text that one can use abundances of radiocarbon to constrain the global mean air-sea exchange coefficient. An important input parameter is the difference between the ^{14}C concentration in the atmosphere and the ocean, expressed as a ratio of the respective R^* values. Suppose now that the value of $R^*_{surface}/(1.017 \cdot R^*_a)$ i.e., the mean ratio of pre-bomb ocean surface to atmospheric ^{14}C including the fractionation correction, is 0.90 instead of 0.95 as assumed in the text (see equation (3.3.15)). What would the global mean gas transfer velocity be?

3.6 A recent study by *Peacock* [2004] suggests that the total ocean inventory of bomb radiocarbon is about 25% lower than that used by *Wanninkhof* [1992] to calibrate his air-sea gas exchange coefficient model. Discuss the implications of this finding for the calibration of the air-sea gas exchange coefficient.

3.7 How long does it take for a 100 m deep mixed layer to lose half of its initial oxygen supersaturation of 5 mmol m^{-3}? Assume a gas transfer velocity of 20 cm hr^{-1} and ignore any lateral loss or gain of oxygen or any exchange across the bottom of the mixed layer.

a. Derive the time-dependent mass balance equation for the oxygen anomaly, ΔC in this mixed layer.

b. Solve for the time dependence of the mixed layer oxygen anomaly.

c. Solve for the time τ required for $\Delta C(t)$ to drop to half of the initial oxygen anomaly, $\Delta C_{initial}$.

3.8 What equatorial upwelling rate would be required to maintain the oxygen deficit of about 4 mmol m^{-3} in the eastern equatorial Pacific (see figure 3.1.1) if there were no biological production? Assume that the upwelling water has an oxygen deficit of 40 mmol m^{-3} and that the deficit is zero in the waters at either side of the axis of upwelling. Assume a mixed layer of 40 m, and a gas transfer velocity of 10 cm hr^{-1}.

a. Draw a meridional/depth section of the upwelling system across the equator. Indicate the directions of the mass (water) fluxes and those of oxygen.

b. Derive a mass balance equation for oxygen in the mixed layer.

c. Assume steady state and solve for the upwelling velocity.

3.9 Observations from a mooring in the same region of the equatorial Pacific as in problem 3.8 indicate that the oxygen undersaturation was nearly constant for an entire month. You now would like to compute the ingassing of oxygen for this month. The information you have available is the monthly mean wind speed, i.e., 5 m s^{-1}, and its standard deviation, ± 2 m s^{-1}. Compute the ingassing of oxygen using the *Wanninkhof* (1992) parameterization (see table 3.3.2) for a case in which you take the covariance of the wind into account, and for a case in which this term is neglected. Discuss the reasons for the difference. Assume a Schmidt number of 660.

3.10 Analyses of the oxygen concentration in raindrops reveals that it is nearly always very close to oxygen saturation. This is surprising at first, since when a raindrop is formed, it starts with zero oxygen and the lifetime of a raindrop is only a few minutes. In contrast, we just learned that it takes weeks for a 50 m deep mixed layer in the ocean to equilibrate with the atmosphere. Explain how the effective equilibration timescale depends on the geometry of the problem, and demonstrate how it scales with the size of the raindrop. Assume that the raindrop is a perfect sphere. Note: Typical raindrops are $\sim$1 mm in diameter.

CHAPTER 4

Organic Matter Production

Were it not for the *biological pump*, the distribution of most chemicals in the ocean would be as uniform as that of salinity. Indeed, ocean circulation and mixing continually drive the distribution of chemicals toward just such a uniform distribution. The biological pump resists this tendency by stripping nutrients and carbon out of surface waters to form organic matter, by exporting this organic matter into the thermocline and deep ocean where the majority of it is remineralized, and by delivering the remainder to the sediments where most is remineralized and some is buried. At the surface of the ocean the physical processes of circulation and mixing are responsible for increasing the nutrients, and the biological pump is responsible for reducing them. In the deep ocean, the physical processes are responsible for reducing the nutrients and the biological pump is responsible for increasing them through remineralization of organic matter that comes from the surface mostly as particulate organic matter sinking under the influence of gravity. This chapter is about the balance of the physical and biological processes at the surface of the ocean and how these combine to maintain the nutrient concentrations at lower values than they would have if there were no biology. We discuss the water column and sediment processes in chapters 5 and 6, respectively.

4.1 Introduction

All organisms on Earth are made of five major elements, H, C, N, O, and P, that are found in approximately constant stoichiometric ratios in oceanic organisms. Oceanic organisms also contain relatively large amounts of S, and more than 50 trace elements [cf. *Williams*, 1981; *Falkowski et al.*, 2003]. Elements whose concentrations sometimes limit photosynthesis are referred to as nutrients, which we separate into macronutrients (N and P) and micronutrients (trace elements) according to their relative concentrations in the organisms. The mean concentration of macronutrients in the ocean is in the mmol m^{-3} range, and that of the micronutrients is in the μmol m^{-3} range and lower (e.g., iron). The remaining major elements, H, C, and O, are all generally present in abundance and therefore not usually considered to be limiting, as is the case for many of the trace elements as well.

The free-floating unicellular *phytoplanktonic* organisms that are responsible for photosynthesis in the ocean (see figure 4.1.1) need to take up these elements from seawater in order to form organic matter of the required stoichiometric composition. The distributions of nitrate (figure 4.1.2) and phosphate (figure 1.2.4) provide a remarkable illustration of the impact of this uptake on chemical concentrations at the surface of the ocean. Without the biological pump, the physical processes would take over and nitrate would have a mean surface concentration of $\sim$33 mmol m^{-3} and phosphate would have a mean concentration of $\sim$2.1 mmol m^{-3}, varying just by $\pm$10% as salinity does (see figure 4.1.3c). Instead, we find undetectably low concentrations over much of the ocean, and only rarely do nitrate and phosphate come anywhere close to the concentrations they would have if there were no biological pump. On the other hand, there are major regions of the ocean such as the eastern equatorial Pacific, the North Pacific, and vast areas of the Southern Ocean where nitrate and phosphate are present at very high concentrations under conditions that would otherwise seem ideal for total depletion by the phytoplankton.

The central goal of this chapter is to understand what controls the production and export of organic matter from the surface of the ocean and how this influences the surface nutrient concentration. The key observation that will guide our analysis is the observed nutrient distribution.

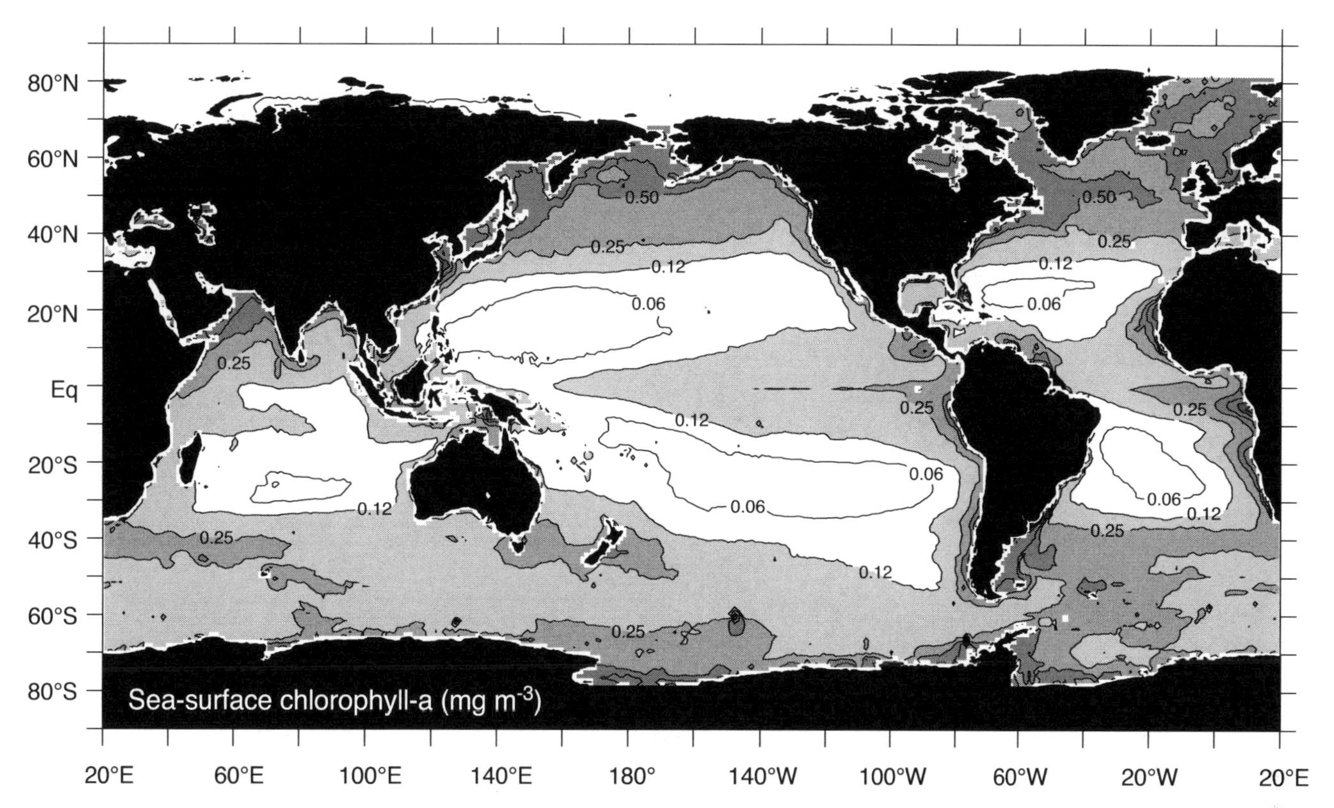

FIGURE 4.1.1: Map of the average surface chlorophyll-a in mg Chl m^{-3} for the period 1998–2002 estimated from satellite measurements of the water leaving radiance of visible light (i.e., ocean color) (see also color plate 3). The spectrum of the water leaving radiance signal is modified by chlorophyll absorption and scattering by plants in the upper ~ 10 meters of the water column. Chlorophyll is the most common index of phytoplankton abundance [e.g., *Cullen*, 1982], with phytoplankton carbon:chlorophyll mass ratios typically given as ~ 50, though measurements show a range of <10 to ~ 300 [cf. *Cloern et al.*, 1995] due to the down-regulation of chlorophyll by phytoplankton under high irradiance, referred to as *photoadaptation* [cf. *Geider et al.*, 1996]. The chlorophyll estimates shown in the figure were obtained from James Yoder and Maureen Kennelly (personal communication, 2004), who based their estimate on satellite data provided by the NASA Sea-viewing Wide Field-of-view Sensor (SeaWiFS) Project, NASA/Goddard Space Flight Center and ORBIMAGE. This satellite measures irradiance in eight different wavelengths, mostly coming from the atmosphere. The chlorophyll is estimated by its impact on the relative radiance of different wavelengths.

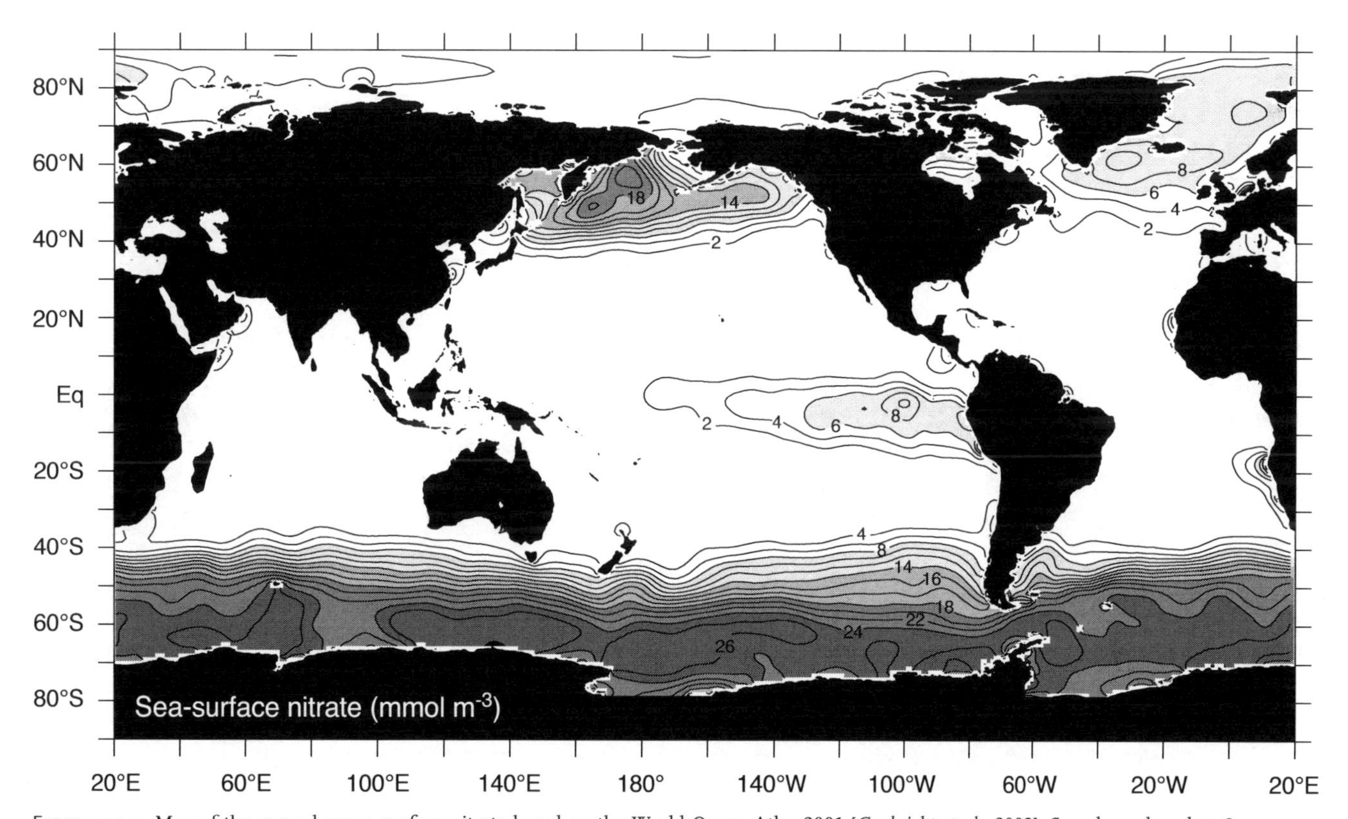

FIGURE 4.1.2: Map of the annual mean surface nitrate based on the World Ocean Atlas 2001 [*Conkright et al.*, 2002]. See also color plate 2.

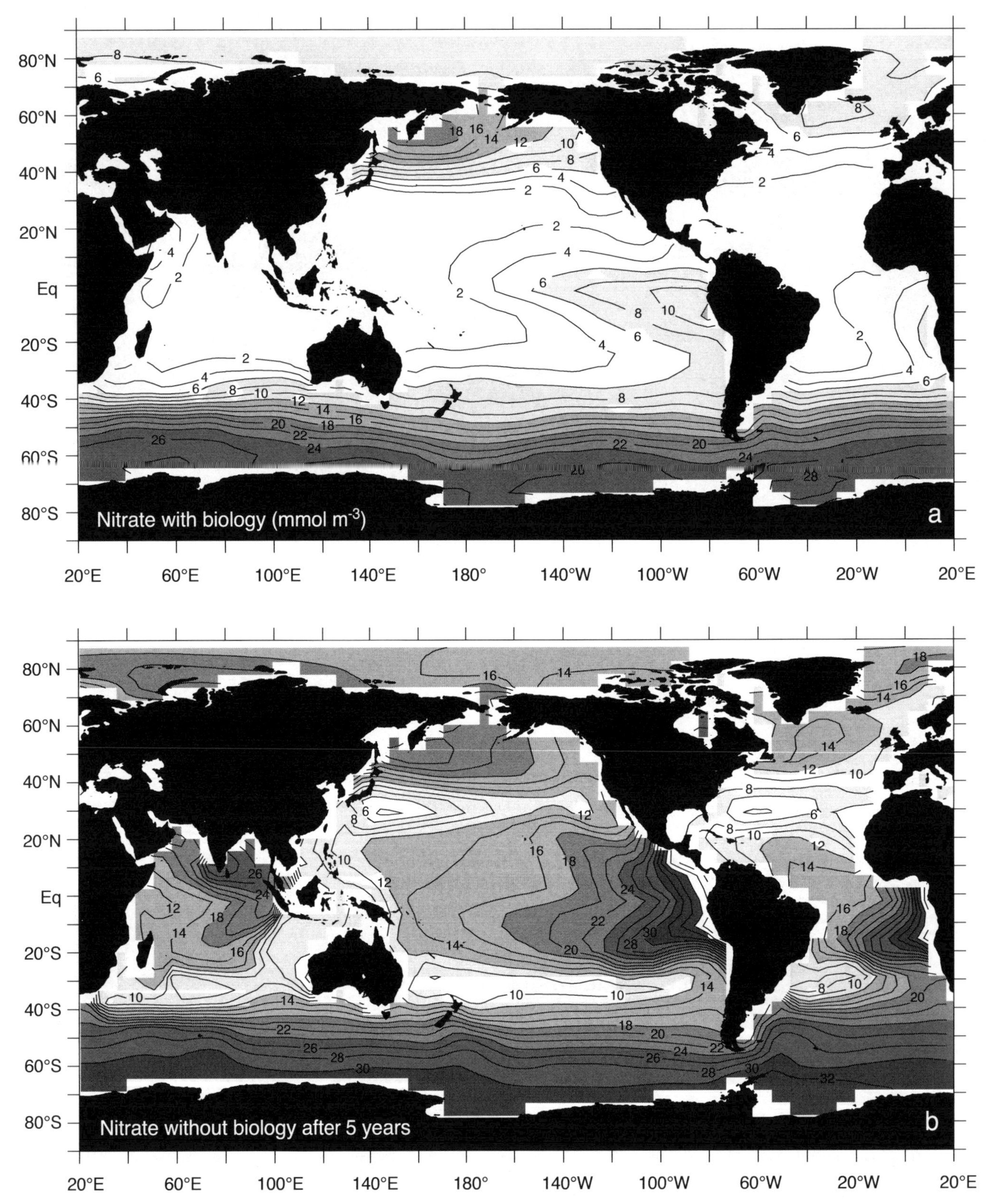

FIGURE 4.1.3: Maps showing model simulated distributions of surface nitrate to demonstrate the impact of biological uptake on the surface nitrate distribution. (a) Results from a model including biological processes. (b) Model simulated nitrate distribution five years after the biological pump was turned off. (c) Expected nitrate distribution after the biological pump has been turned off for several thousand years. Results are based on simulations using the ocean general circulation model described in chapter 2 and the so-called OCMIP protocols for nutrients described in chapter 5. The model was developed at Princeton University and GFDL/NOAA, and the simulations were carried out by Xin Jin at UCLA. The simulations were actually carried out using phosphate, and what is shown is the inferred nitrate distribution (16 · phosphate − 2.9 mmol m^{-3}), where the 2.9 mmol m^{-3} is a correction for the fact that the global mean nitrate is lower than that of phosphate. The results shown in (c) are approximated by using the salinity distribution multiplied by the ratio of global mean nitrate to global mean salinity.

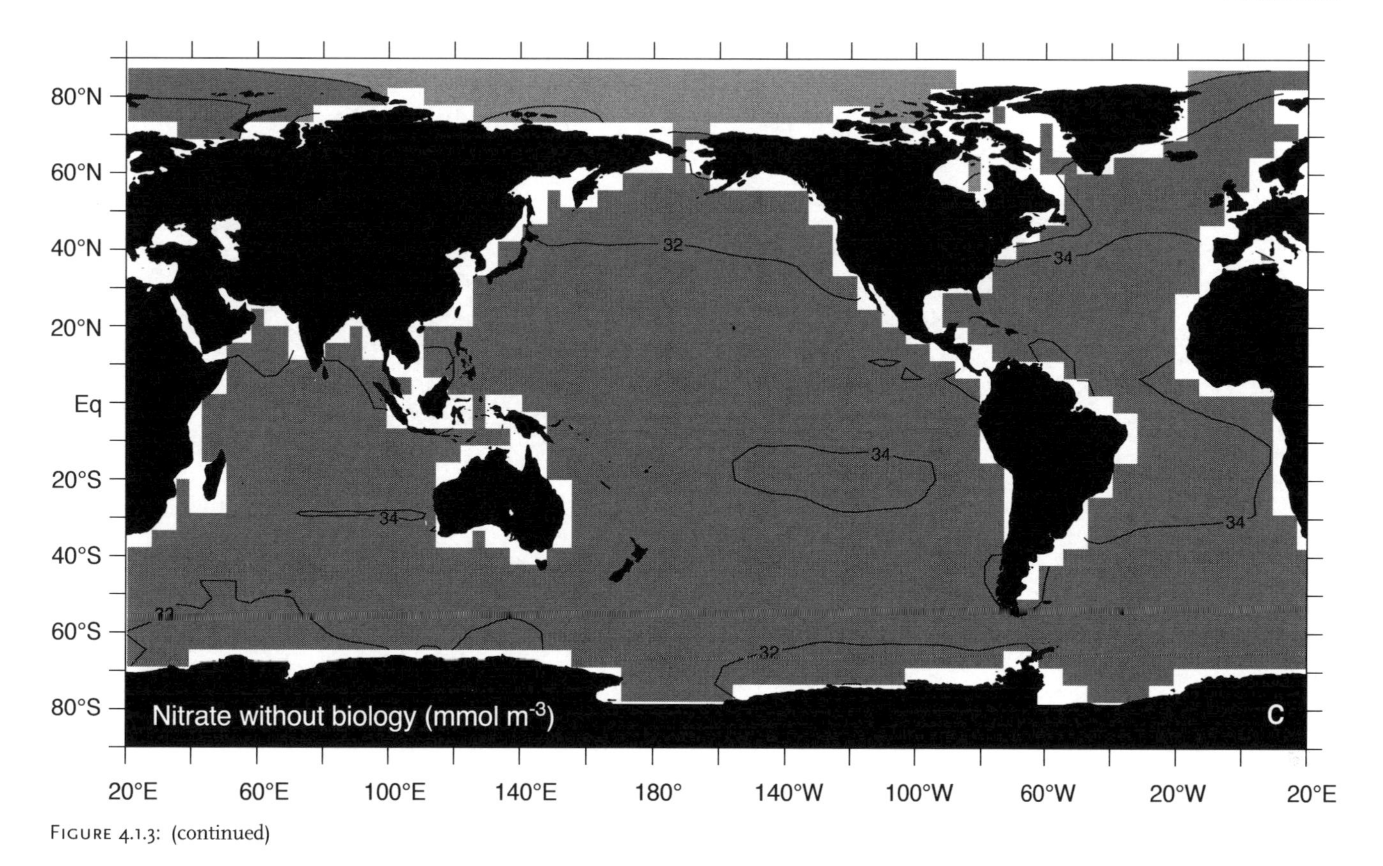

FIGURE 4.1.3: (continued)

A high proportion of the organic matter that is produced in the surface ocean is recycled there rather than being exported. A second goal of this chapter is to understand what controls this recycling efficiency and how this influences the surface nutrient concentration. One might expect that systems with higher recycling efficiency would also tend to have higher surface nutrients, and vice versa, but we will show that the observations indicate exactly the opposite over much of the ocean. How can we explain this counterintuitive result?

The decision tree chart in figure 4.1.4 summarizes the most important factors that are thought to play a role in determining the production and export of organic matter, the recycling efficiency of the ecosystem, and the influence of these on the surface nutrient concentration. These are: the nutrient supply by transport and mixing, the light supply, and the efficiency of the biological pump. We introduce each of these in turn. See panel 4.1.1 for a discussion of a reservoir and pool analogy that may help to understand some of the important concepts that we introduce here.

NUTRIENT SUPPLY

The starting point for the decision tree of figure 4.1.4 is the input of nutrients from the nutrient-rich deep ocean. This input comes primarily from the main thermocline by upwelling, by convective overturning (mostly as a result of wintertime cooling), and by vertical mixing, all processes that we discussed in chapter 2. It may also come by lateral transport from regions where vertical input occurs or by deposition from the atmosphere, which generally is small but can be locally significant.

We can use the vertical velocities of figure 4.1.5a and deep wintertime mixed layer depths of figure 4.1.5b to define a set of regions according to whether the supply of nutrients from below has the "potential" to be high (upwelling or deep wintertime mixing) or the likelihood of being low (downwelling and stably stratified). We use the term "potential" to describe the nutrient supply rate because the "actual" supply rate, even in the presence of strong upwelling and deep wintertime mixing, will depend on how depleted the surface nutrients are. It could even be 0 if the surface nutrients were equal to the deep nutrients. As an illustration of this important point, consider the simple two-box model of figure 1.2.3. In this case, the "actual" supply of nutrients is $\Phi_{nutrient} = v \cdot (C_d - C_s)$. This supply can range anywhere between a minimum of 0 when $C_s = C_d$ and a maximum of $\Phi_{nutrient} = v \cdot C_d$ when surface nutrients are completely depleted, i.e., $C_s = 0$. One can think of the term "potential nutrient supply" as denoting specifically the upper limit flux that one would obtain if surface nutrients were depleted, i.e., $\Phi_{nutrient} = v \cdot C_d$ in our two-box model.

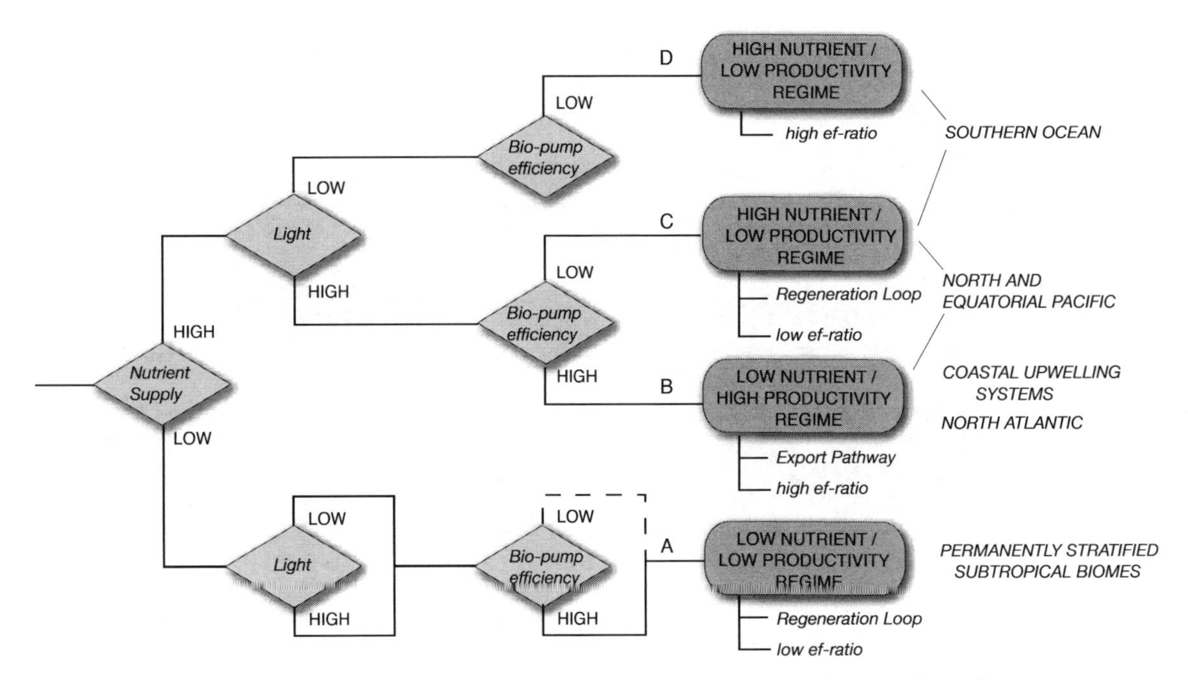

FIGURE 4.1.4: Schematic diagram representing the primary processes that are thought to determine the surface nutrient concentration and associated production and export of organic matter and recycling efficiency of the ecosystem. The different branches of the diagram indicate different regimes that are discussed in the text, with the final oval and branches off the oval representing the consequences of each regime for the surface nutrient concentration, biological productivity, ecosystem structure, and recycling efficiency. The *ef*-ratio is a composite indicator of the export efficiency of organic matter in the surface ocean as measured by the *f*- and *e*-ratios defined by equations (4.2.5) and (4.2.6). The recycling efficiency is thus $1 - ef$. The regeneration loop and export pathway are components of the surface ecosystem discussed primarily in section 4.4. The regeneration loop is made up primarily of the microbial loop consisting of heterotrophic bacteria and the picoplankton, and is very efficient at recycling organic matter. The export pathway usually exists as an add-on to the regeneration loop and consists of large phytoplankton and zooplankton that in combination are very inefficient at recycling organic matter.

Since the differences in nutrient supply mechanisms that we identified in the previous paragraph create distinctive environments that have a major impact on the types of organisms that live there, we refer to them as *biomes*. Figure 4.1.6 shows an attempt to classify the ocean into a set of biomes according to the nutrient supply mechanism. Following *Sarmiento et al.* [2004b], we define first an *equatorially influenced biome* between 5°S and 5°N where upwelling along the equator and rapid horizontal transport out of the upwelling band generally guarantees a high supply of nutrients. Poleward of the equatorially influenced biome are the subtropical gyres defined by a broad band of downwelling velocities. The equatorward half of the subtropical gyres has no significant deepening of the mixed layer in wintertime, so the nutrient supply is extremely low year around. We define it as the *permanently stratified subtropical biome*. The poleward half of the subtropical gyres, where deep wintertime mixing brings nutrients to the surface, defines the *seasonally stratified subtropical biome*. The subpolar gyres, where both upwelling and deep wintertime mixing bring ample supplies of nutrients to the surface, define the *subpolar biome*. Finally, we define a *low-latitude upwelling biome* to describe the nutrient-rich coastal upwelling bands around the margins of the subtropical gyres, and a *marginal sea ice biome* where the seasonal formation and melt-back of sea ice defines a unique environment that is rich in nutrients and has very high biological productivity. Note that only the permanently stratified subtropical biome has low nutrient supply. The other five biomes have the potential for high nutrient supply during at least part of the year.

Comparison of the geographical distribution of the biomes in figure 4.1.6 with the nitrate distribution in figure 4.1.2 shows a striking correspondence between biomes with the potential for high nutrient supply and observations of high nutrient concentrations, and between biomes with low nutrient supply and observations of low nutrient concentrations. The surface chlorophyll distribution shown in figure 4.1.1, which is the most commonly used index of phytoplankton biomass, also shows a strong correlation with regions of high nutrient supply. The conclusion that one might draw from this comparison is that the surface nutrient concentrations and phytoplankton biomass, and thus also the biological productivity based on phytoplankton, are controlled primarily by the transport of nutrients within the ocean. This accounts for the remarkable success of the early physically based depiction of oceanic productivity made by *Sverdrup* [1955] at a time when very few biological observations were available to back up his speculation [cf. *McGowan*, 2004]. However, one should not draw the mistaken conclusion from this physically based point of view that biology is not important in setting the pattern of biological production. The nutrient transport and concentrations would be completely different if there were no biology. As we have already noted, there would be no net input of nutrients to the surface if surface nutrients were not reduced by biological uptake.

A more subtle effect of biology is its control of lateral transport of nutrients. Consider the model simulation given in figure 4.1.3, which shows what happens in a global ocean circulation model when it is initialized by being forced toward the observed nitrate distribution at the surface (figure 4.1.3a) and then the biology is shut off for five years (figure 4.1.3b) and then for several thousand years (figure 4.1.3c). The transport of nutrients by ocean circulation and mixing is so rapid that surface nitrate begins to approach thermocline concentrations within just a few years. This is the case even in the equatorward half of the subtropical gyres. When the biological pump is shut off, the subtropical gyres are able to receive nutrients by lateral transport from regions where upwelling and deep mixing occur. There is also some vertical mixing of nutrients from immediately below. Thus we see that the low nutrient supply rate of the permanently stratified subtropical biome depends in part on biology stripping nutrients out of the surface near the upwelling and deep winter mixing source regions before lateral transport can carry them into low vertical supply areas. The final outcome of shutting off the biological pump for thousands of years gives a nearly uniform nitrate distribution, with surface concentrations equal to those in the deep ocean (figure 4.1.3c).

We depict the influence of nutrient supply in the flowchart of figure 4.1.4 by a control point (the diamond in the figure) labeled "Nutrient Supply." If the potential nutrient supply rate is low (lower branch) this will inevitably result in low surface nutrients (Regime A). These type A regions occur predominantly in the permanently stratified subtropical biome (figure 4.1.6), and are characterized not just by low nutrients (figure 4.1.2) but also, as might be expected, by low chlorophyll concentrations (figure 4.1.1) and relatively low biological productivity. The HOT time series station summarized in table 4.1.1, which has near-zero nitrate and relatively low chlorophyll, epitomizes the type of behavior normally seen in the permanently stratified subtropical biome. The BATS time series station also usually has the

Panel 4.1.1: A Semi-Infinite Reservoir and Pool Analogy for Nutrient Supply and the Biological Pump Efficiency

It is instructive to use the semi-infinite reservoir and pool shown in figure 1 as an analog to understand what controls the surface nutrient supply rate and concentration. In this analogy, the water level in the reservoir represents the nutrient concentration of the deep ocean, the diameter of the pipe connecting the reservoir to the pool represents the strength of the physical processes that carry nutrients from the deep ocean to the surface, the water level in the pool represents the surface nutrient concentration, and the flow of water between the reservoir and pool represents the nutrient supply rate. Finally, the diameter of the drainpipe out of the swimming pool represents the biological pump.

The flow rate from the reservoir to the pool, which is analogous to the nutrient supply, will depend not just on the diameter of the pipe connecting the pool and reservoir (i.e., on the physical processes), but also on how high the water level in the reservoir is relative to the water level in the pool (i.e., on the difference in nutrient concentration between the surface and deep water). Thus, a reservoir with a high water level (high deep nutrients) and wide diameter pipe (strong vertical exchange) has the potential to give a high water flux, but will only do so if the pool level (surface nutrients) is low. By contrast, if the pool is full, the flow, which depends on the height difference between the reservoir and the pool, will be small. Note that a low water level in the reservoir (low deep nutrient concentration) will give a low water flux into the pool, no matter how much water is in the pool.

In the infinite reservoir and pool analogy the type A behavior of figure 4.1.4 corresponds to a reservoir with a low water level (see figure 1) or one with a high level but extremely small diameter pipe connecting it to the pool. If the reservoir water level is low, the pool will always be near empty no matter how big the pipe that connects the reservoir and pool or how small the drain out of the pool. If the reservoir water level is high, but the pipe connecting it to the pool has a much smaller diameter than the drainpipe out of the pool, the pool water level will also be low. In this analogy, there is a potential nutrient supply rate that is represented by the flow from the reservoir to the pool if the pool is empty. This potential flow rate is small if deep nutrients are low.

The "type D" behavior of figure 4.1.4 corresponds in figure 1 to a situation where the reservoir is full (high deep nutrients) and connected by a large diameter pipe to the pool (strong vertical exchange), but there is no drain out of the pool (no biological pump). In this case, the flow rate of water from the reservoir to the pool will be 0, but the pool will be full (i.e., the surface nutrients will be high).

Panel 4.1.1: (*Continued*)

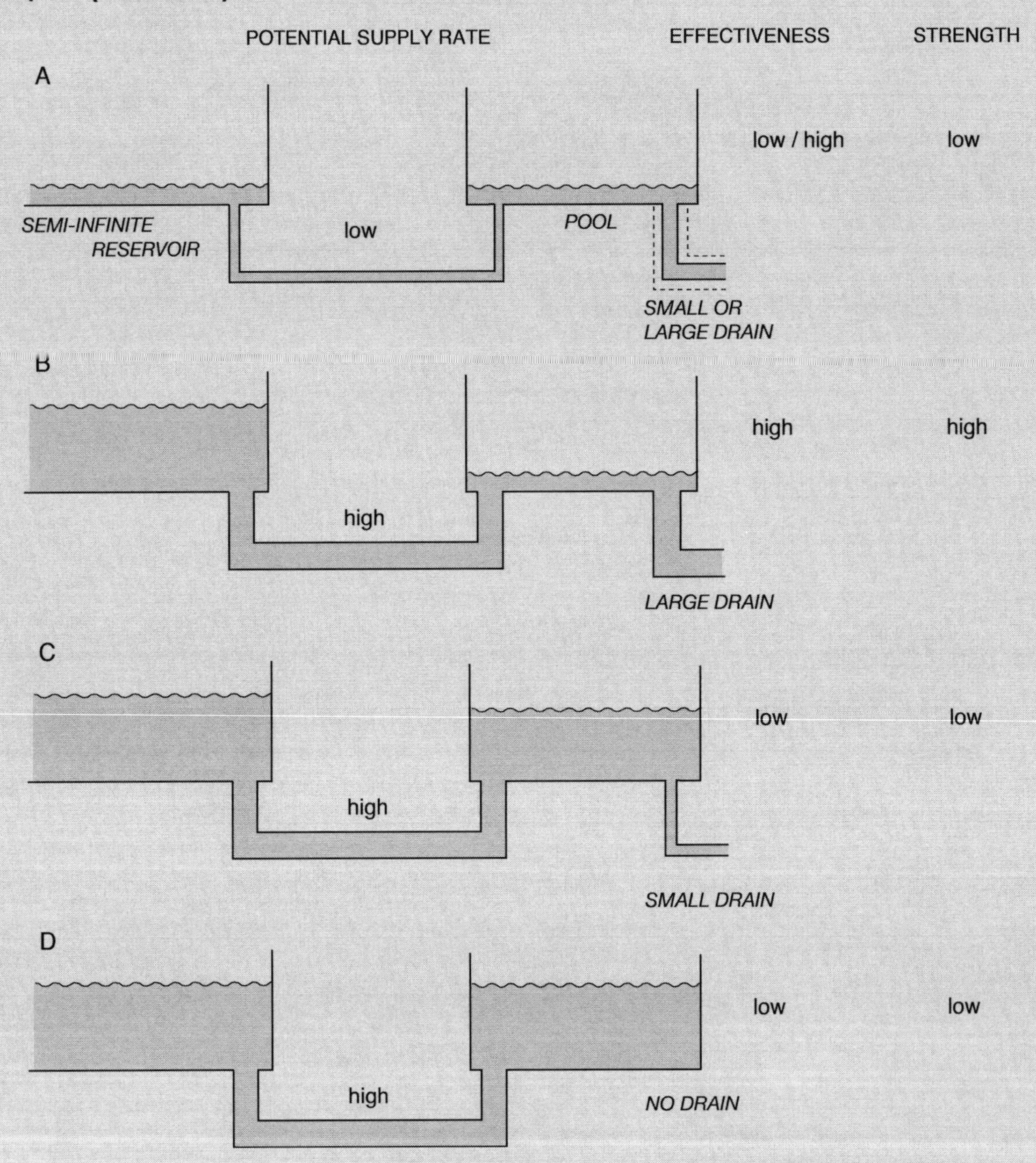

FIGURE 1: Schematic illustration of a semi-infinite reservoir-pool system connected by a pipe, with a drain out of the bottom of the pool. The type A through type D systems are intended as analogies to the type A through type D surface ocean systems depicted in figure 4.1.4.

The type B and C behaviors of figure 4.1.4 have a large pipe connecting the reservoir to the pool. However, in case B, the drain from the pool is so large that the pool level remains low even though the supply rate is very high, whereas in case C, the drain from the pool is so small that the pool becomes almost as full as it would be if there were no drain (type D behavior). In biogeochemical parlance, case B has a very efficient biological pump (one that empties the pool out), whereas case C has an inefficient biological pump (one that allows the pool to fill up).

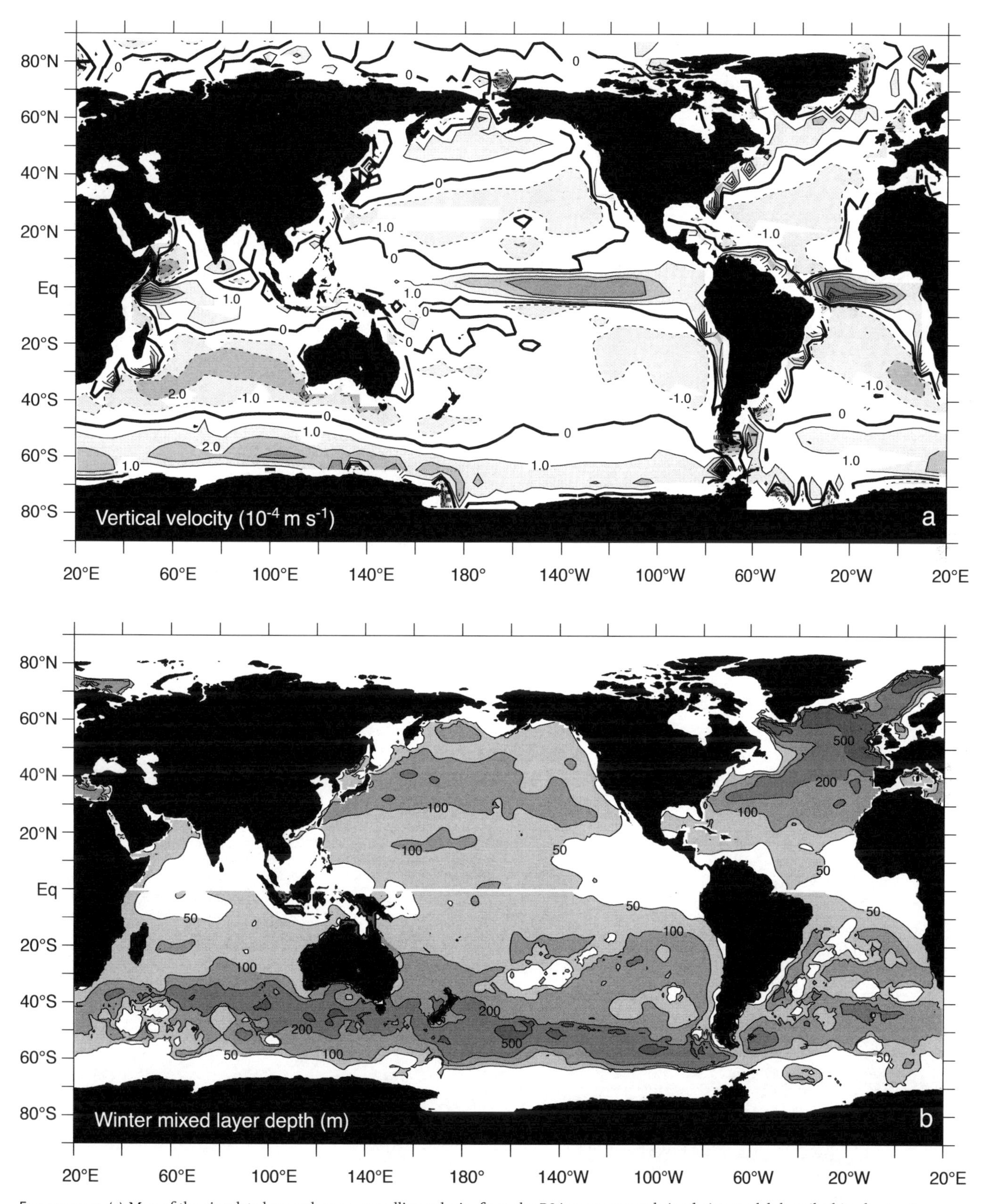

FIGURE 4.1.5: (a) Map of the simulated annual mean upwelling velocity from the P2A ocean general circulation model described in chapter 2. Upwelling is positive and downwelling negative. (b) Map of observed wintertime (JFM for the northern hemisphere and JAS for the southern hemisphere) mixed layer depth, based on the climatology of *Kara et al.* [2003].

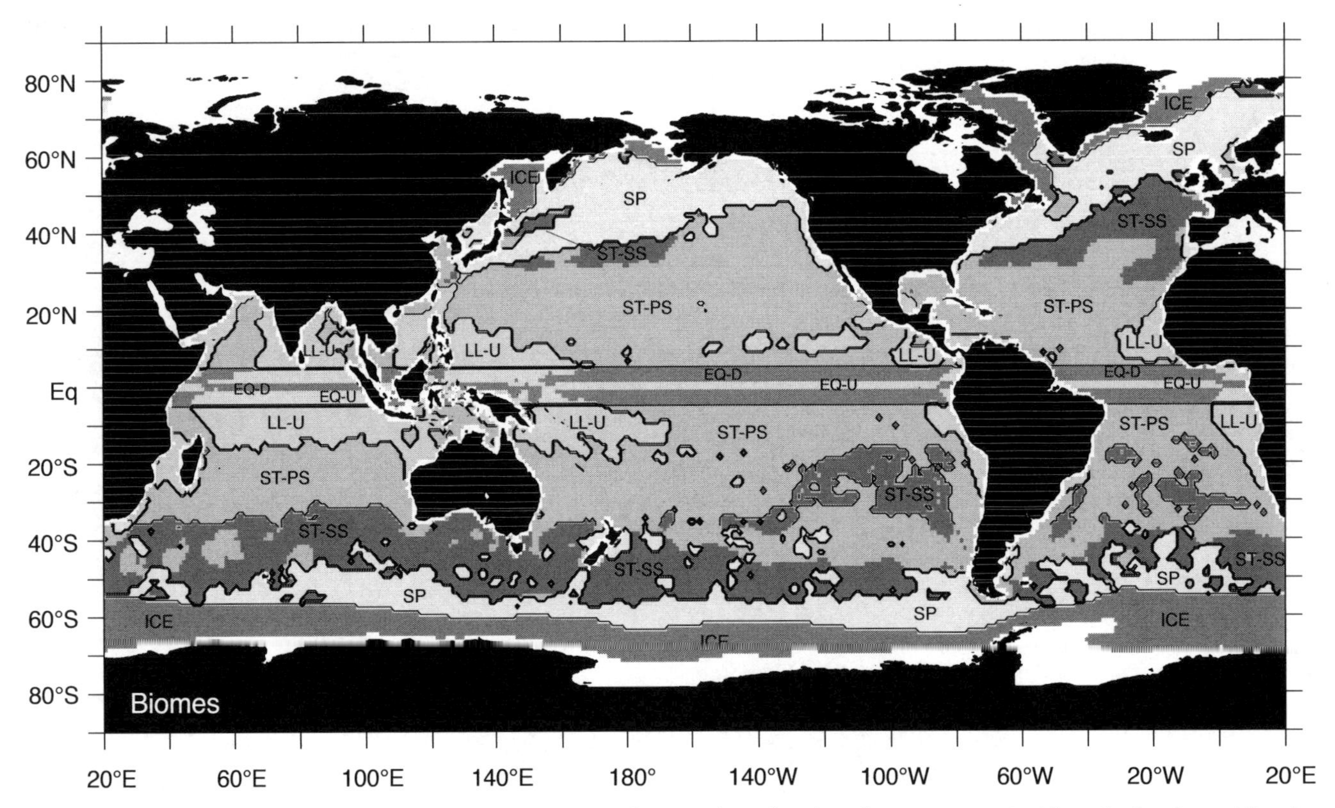

FIGURE 4.1.6: Map of global biome distribution using a classification scheme based on the wintertime mixed layer depth (where >150 m is considered "seasonally stratified," and <150 m is considered "permanently stratified") and whether the vertical velocity is up or down (see explanation in text). The symbols are as follows: Eq-D and Eq-U are the equatorially influenced biome split into downwelling and upwelling regions; ST-PS and ST-SS are the subtropics separated into a permanently stratified and seasonally stratified biome, respectively; LL-U is the low-latitude upwelling biome; SP is the subpolar biome; and Ice is the marginal sea ice biome. Adapted from *Sarmiento et al.* [2004].

TABLE 4.1.1

Mean properties at the four Joint Global Ocean Flux Study (JGOFS) time-series stations (BATS, HOT, OSP, and KERFIX) as compiled by *Kleypas and Doney* [2001] and also at OSI as compiled by M. Schartau (personal communication, 2004).

The standard deviations are standard deviations of the mean, not the measurement error. The nutrient concentrations (see especially nitrate) separate those regions identified as oligotrophic (low nutrients) from those identified as eutrophic (high nutrients and chlorophyll). Note: By the definitions given in table 4.4.2, Falkowski et al. [2003] would identify OSI, OSP, and KERFIX as mesotrophic because their chlorophyll is $<1\,mg\,m^{-3}$.

Station (Location; Dates of Analysis)	*Chlorophyll* ($mg\,m^{-3}$)	*Nitrate* ($\mu mol\,kg^{-1}$)	*Phosphate* ($\mu mol\,kg^{-1}$)	*Silicic Acid* ($\mu mol\,kg^{-1}$)
		Oligotrophic Regions		
BATS (31°40′N 64°10′W—Subtropical Atlantic; October 1998–December 1999)	0.10 ± 0.08	0.04 ± 0.11	0.01 ± 0.02	0.8 ± 0.3
HOT (22°45′N 158°W—Subtropical Pacific; October 1988–December 1998)	0.09 ± 0.04	0.00 ± 0.10	0.08 ± 0.03	1.3 ± 0.4
		Eutrophic Regions		
OSI (59°N 19°W—Subpolar Atlantic; 1970–74, 1995)	0.8 ± 0.5	9 ± 6		
OSP (50°N 145°W—Subpolar Pacific; 1959–1995)	0.4 ± 0.3	10 ± 5	1.0 ± 0.4	15 ± 8
KERFIX (50°40′S 68°25′E—South of Antarctic Polar Front; April 1990–March 1995)	0.3 ± 0.2	25 ± 2	1.8 ± 0.1	14 ± 4

*BATS stands for Bermuda-Atlantic Time Series, HOT for Hawaiian Time Series, OSI for India or Ocean Station India, OSP for Station P or Ocean Station Papa, and KERFIX for Kerguelen Fixed Station.

characteristics of the permanently stratified subtropical biome, though occasional deep wintertime mixing events modify its behavior.

Whether the high nutrient supply rates of all the other biomes, characterized by branches B, C, D in the chart, will result in high or low nutrients, and high or low organic matter export flux, depends on a series of other controlling factors that we outline next.

Light

The formation of organic matter from dissolved inorganic nutrients in the ocean is due almost entirely to photosynthesis, which requires sunlight. This defines the next decision point on the upper branch of figure 4.1.4. Obviously, regions where light supply is cut off, for example during the polar night, will have no photosynthesis. More subtle effects arise from the strong attenuation of sunlight by seawater, which results in a rapid decrease of the available light with depth. As a result, there is only enough light for photosynthesis to occur within a shallow layer near the surface whose depth depends on the amount of incoming sunlight and the clarity of the water. As we will discuss below, the depth of this well-lit layer, commonly referred to as the *euphotic zone*, is of order 100 m in the open ocean, but can be as shallow as a few meters in very turbid environments. If the thickness of the mixed layer is greater than the euphotic zone, which generally occurs only in wintertime, the phytoplanktonic organisms will spend at least part of their time in darkness even during the daytime. This is because phytoplanktonic organisms generally cannot control their vertical position in the water column against the overturning motion of the water within the mixed layer. Therefore, during at least part of the day, they are swept down to depths where there is not enough light for their growth to exceed respiration. If the mixed layer is deep enough, they will spend too much time outside the euphotic zone and will quickly die out.

If the light supply for photosynthesis is inadequate, photosynthesis will be less than respiration, and there will be no biological pump. As a consequence, the surface nutrient concentration will be high and the nutrient supply and resulting organic matter export will be low (regime D of figure 4.1.4). If both the potential nutrient supply rate and light supply are high (central branches), the chart of figure 4.1.4 takes us to the final decision point, where the efficiency of the biological pump is the critical process in determining whether surface nutrient concentrations are low (branch B) or high (branch C).

Efficiency of the Biological Pump

In the case of branch B of figure 4.1.4, the biological uptake and export of the supplied nutrients is so large that it outweighs the high supply rate. This leads to very high biological productivity with relatively small residual nutrient concentrations, i.e., a low-nutrient, high-productivity regime. In the case of branch C, biology might be elevated, but it is inefficient in stripping out and exporting the nutrients, so that we end up with a high-nutrient, and low- to intermediate-productivity regime.

We define the *biological pump efficiency*, E_{BP}, as a measure of the success of phytoplankton in maintaining low nutrient concentrations in the surface ocean. It is important to note that this measure is very different from the *strength* of the biological pump, which describes the magnitude of the downward flux of organic matter. The focus of this efficiency measure is the residual concentration of nutrients in the surface. For example, if the flux of organic matter out of the surface at a given location is large, but the residual surface concentration remains high, we would characterize that location as having a strong but inefficient biological pump. Consider the specific example of the two-box model of figure 1.2.3. For this case, we would define the biological pump efficiency as

$$E_{BP} = \frac{C_{deep} - C_{surface}}{C_{deep}} \qquad (4.1.1)$$

Thus a 100% efficient pump would be one where the surface concentration is zero, and a 0% efficient pump would be one where the surface concentration is equal to the concentration in the deep waters feeding into the surface. As a tool for understanding what controls the surface nutrient content, this concept is most useful when applied to regions with high nutrient and high light supply (branches B, C of figure 4.1.4). As we noted above, the type A regions will have low surface nutrient concentrations regardless of whether the biological pump has a high or low efficiency, and type D regions, which have inadequate light supply, will inevitably have a biological pump with a low efficiency. At the final decision point of branches B, C, the phytoplanktonic community has both adequate nutrient supply and light. We would thus expect the biological pump to be not only highly effective but also very strong (branch B of the diagram). Is it?

Figure 4.1.7a shows an estimate of E_{BP} based on an analysis of summertime nitrate observations in the 0 to 100 m depth range (surface nitrate) versus the 100 to 200 m depth range (deep nitrate). The overall pattern is one of extremely high efficiency in the permanently stratified subtropical gyre biome and extremely low efficiency in the Southern Ocean marginal sea ice, subpolar, and seasonally stratified subtropical biomes. Other regions such as the north and equatorial Pacific and north Atlantic, as well as many of the low-latitude upwelling biomes, have intermediate efficiencies (compare figure 4.1.7a with figure 4.1.6). The extremely low biological pump efficiencies of the Southern Ocean, and intermediate efficiencies of much of the rest of the

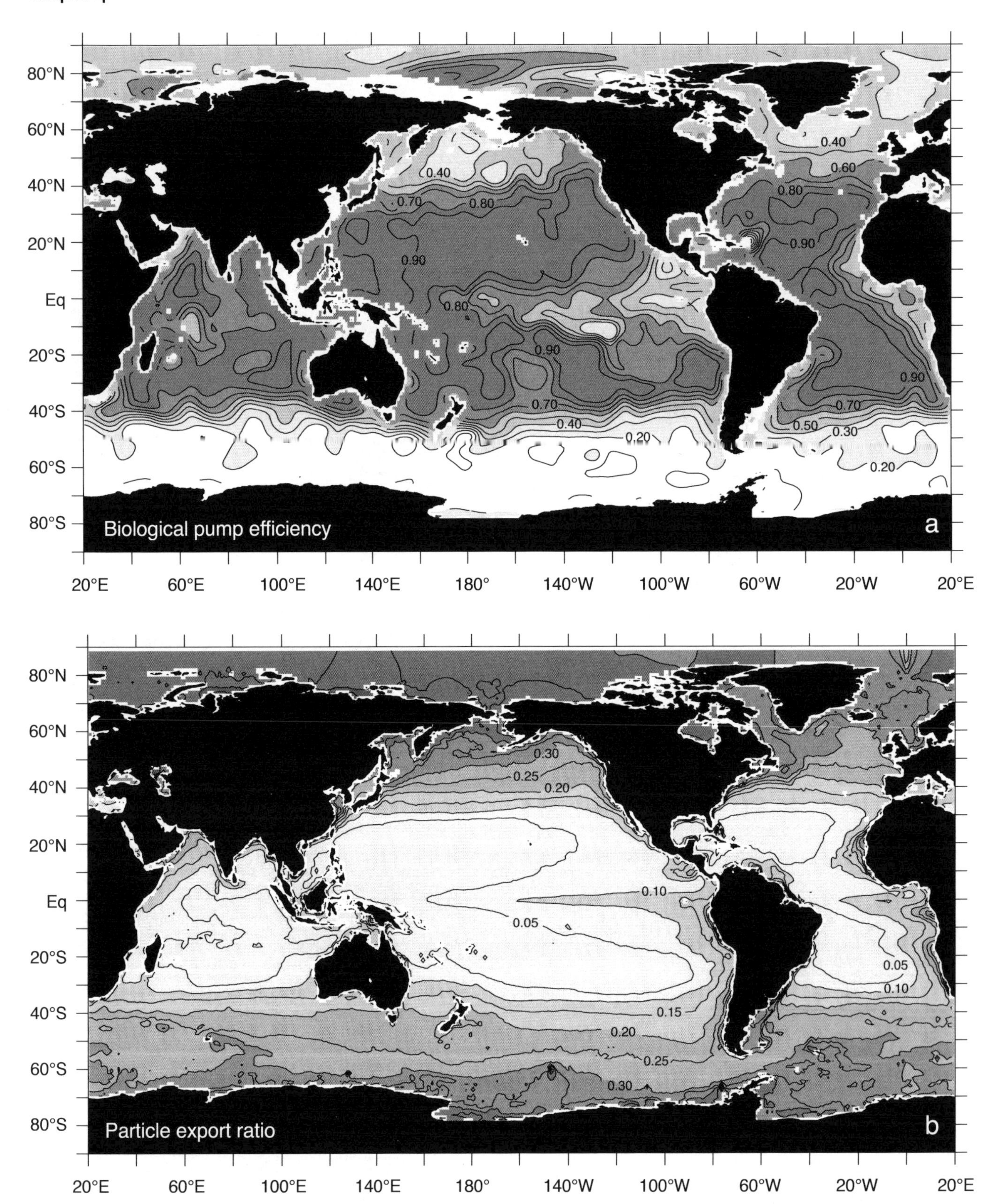

FIGURE 4.1.7: (a) Map of an observationally based estimate of the biological pump efficiency based on the mean nitrate concentration in the 100 to 200 m depth range versus that in the 0 to 100 m depth range. $E_{BP} = ([NO_3^-]_{deep} - [NO_3^-]_{surface}) / [NO_3^-]_{deep}$, as defined in the text. Based on data from the World Ocean Atlas 2001 [*Conkright et al.*, 2002]. (b) An estimate of the fraction of primary production that is exported as particulate organic matter from the surface using the chlorophyll concentration (*Chl*) and surface temperature (*T*) based on the empirical algorithm of *Dunne et al.* [2005b]: e_p-ratio $= \max[0.042, \min(0.72, -0.0078 * T + 0.0806 * \ln(Chl) + 0.433)]$.

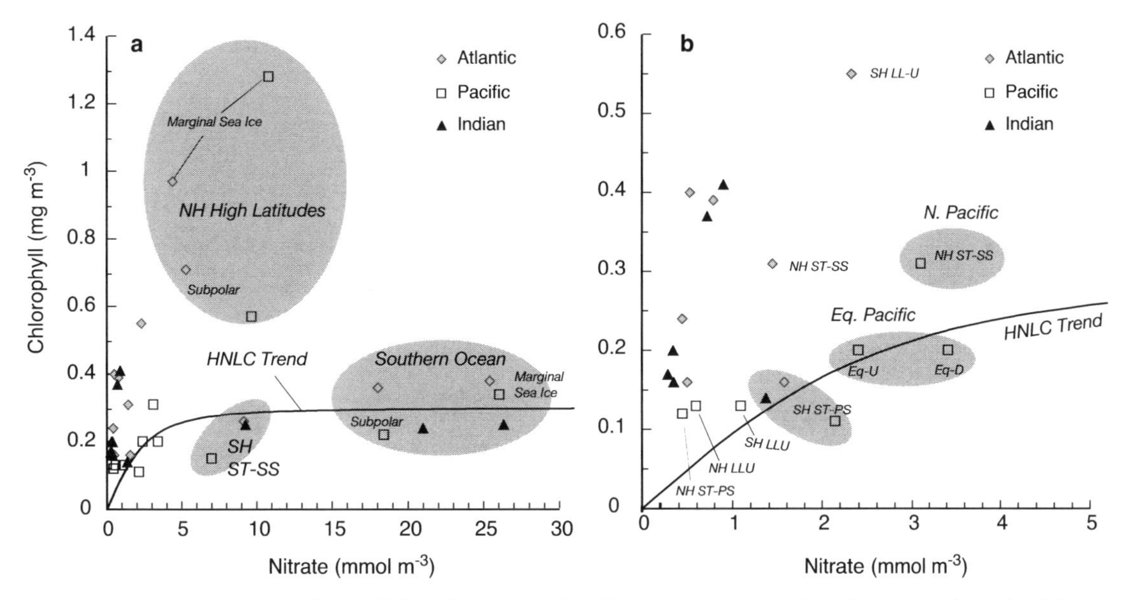

FIGURE 4.1.8: Geometric mean chlorophyll from figure 4.1.1 plotted versus mean nitrate from figure 4.1.2 for each of the biomes of figure 4.1.6, broken down into biogeographical regions by hemisphere (except for the equatorially influenced biome) and ocean basin. Panel (b) is a blow-up of panel (a). The trend lines are intended merely to guide the eye; they are not fits to the observations.

ocean outside the permanently stratified subtropical gyre biome, cannot easily be explained as being due to light. The equatorial Pacific and low-latitude upwelling biomes have plenty of light all year around; and the high-latitude regions of low E_{BP} have adequate light supply during most of the year. This implies that these regions may belong primarily to the high potential nutrient–high light supply–type C regions.

This finding that most of the high potential nutrient supply–high light regions have unexpectedly low biological pump efficiencies is one of the major puzzles of biogeochemical oceanography. Most of these regions correspond to the high nutrient–low chlorophyll (HNLC) regions that biological oceanographers have long identified as anomalous. The Ocean Station Papa (OSP) and KERFIX time series data summarized in table 4.1.1 provide a remarkable illustration of the HNLC phenomenon. The nutrient concentration at both locations is extremely high, being orders of magnitude greater than at the BATS and HOT time series stations. By contrast, the mean chlorophyll *a* concentration is only three to four times higher than that at the BATS and HOT oligotrophic time series stations. Coastal environments with nutrient concentrations such as those at OSP and KERFIX can have chlorophyll *a* concentrations in excess of $10\,\mathrm{mg\,m^{-3}}$, easily two to three orders of magnitude greater than those observed in these open ocean HNLC sites. Ocean Station India (OSI) in the North Atlantic also has higher chlorophyll, but what really distinguishes it from OSP and KERFIX is its seasonal behavior, discussed in section 4.3. Both OSP and KERFIX have only modest seasonal variability, whereas OSI shows frequent summertime episodes of a highly efficient biological pump and non-HNLC behavior, with chlorophyll *a* exceeding $1\,\mathrm{mg\,m^{-3}}$, and nitrate dropping well below $5\,\mathrm{mmol\,m^{-3}}$.

Figure 4.1.8 shows a plot of the geometric mean chlorophyll concentration versus mean surface nitrate concentration for each of the biomes identified in figure 4.1.6, broken down into biogeographical provinces by hemisphere and basin. The data that fall near the curve labeled HNLC are in biogeographical provinces that are generally considered to have anomalously low chlorophyll in conjunction with high nutrients. If we take a nitrate concentration cut-off of $\sim 2\,\mathrm{mmol\,m^{-3}}$ as a definition of "high nutrients," the HNLC biogeographical regions include: (1) the Southern Ocean, including the marginal sea ice zone, subpolar gyre, and seasonally stratified subtropics; (2) the equatorial Pacific; and (3) the northern hemisphere seasonally stratified biome of the North Pacific. The southern hemisphere permanently stratified subtropical gyre of all three ocean basins tends to have too low nutrients to qualify as HNLC by this definition, but they are close. Most of the rest of the ocean falls well above the HNLC trend line we have drawn in the figure. This is particularly so in: (1) the northern hemisphere and equatorial regions of the Atlantic Ocean, and (2) the northern hemisphere and equatorial regions of the Indian Ocean.

Why are the phytoplanktonic organisms in HNLC regions such as those represented by OSP and KERFIX unable to expand their population and strip out the nutrients? Explaining this enigma has been a major goal of oceanographic research over a long period of time. The leading hypotheses include: light limitation in the high latitudes, suppression of phytoplankton by herbivore grazing, an inadequate supply of iron or some other micronutrient, or some combination of these [cf. *Chisholm and Morel*, 1991; *Cullen*, 1996]. We will explore each of these in this chapter.

We now have a definition for the efficiency of the biological pump and a sense of why it is important to understand it, but what determines its magnitude? We turn again to our example of the two-box model of figure 1.2.3 to illustrate the kind of answer we will be seeking with the more sophisticated models we develop later in this chapter. To keep things simple, let us postulate that the net synthesis of organic matter in the surface ocean by phytoplankton, denoted *net primary production* (NPP), is directly proportional to the concentration of nutrients in the surface, i.e., $\mathrm{NPP} = \Lambda \cdot C_s$, where the nutrient assimilation rate Λ has units of $m^3\ s^{-1}$. Think of Λ as equivalent to the volume of water in the surface of the ocean whose nutrients are stripped out per unit time. Suppose, further, that a fraction e of the NPP is exported from the surface, the remaining $(1-e)$ fraction being recycled within the surface. In a steady state, the biological export must be balanced by the vertical flux of nutrients into the surface by the exchange rate v, which also has units of $m^3\ s^{-1}$ (see section 1.2), i.e., $e \cdot \mathrm{NPP} = e \cdot \Lambda \cdot C_s = v \cdot (C_d - C_s)$. Rearranging this equation and using the definition for E_{BP} given earlier yields $E_{BP} = e \cdot \Lambda/(e \cdot \Lambda + v)$. An efficient biological pump for this model is thus one in which the rate constant for formation of organic matter that is exported, $e \cdot \Lambda$, is much greater than the volumetric exchange rate coefficient v. The larger $e \cdot \Lambda$ is relative to v, the smaller the surface concentration will be.

It is tempting to make a connection between the export fraction e and biological pump efficiency E_{BP}; indeed, our two-box model result, $E_{BP} = e \cdot \Lambda/(e \cdot \Lambda + v)$, implies such a connection, with higher e giving higher E_{BP}, and vice versa. However, the observationally based estimate of a quantity closely related to the export fraction given in figure 4.1.7b, the particle export fraction e_p, shows almost exactly the opposite behavior from the biological pump efficiency of figure 4.1.7a over much of the ocean. For example, the permanently stratified regions of high E_{BP} in figure 4.1.7a, are all regions of extremely low e_p in figure 4.1.7b. Conversely, the Southern Ocean region of low E_{BP} in figure 4.1.7a corresponds to a high e_p region in figure 4.1.7b. Only the equatorial and coastal upwelling bands appear to have both low E_{BP} and low e_p. The North Pacific and North Atlantic have high e_p and intermediate values of E_{BP}. What can account for the surprising anti-correlation between these two quantities over so much of the ocean? We will seek an answer to this question in what follows.

Since the export fraction e goes the wrong way to explain the E_{BP} observations, we must look to Λ and v to explain what is happening. We have already seen that the magnitude of v is low in the high E_{BP} regions of the permanently stratified subtropical gyre, and high in the regions of low E_{BP}. This goes in the right direction to explain the E_{BP} observations, but as suggested above, biology, which we have parameterized here by Λ, must also certainly play a role.

Outline

The goals we have in this chapter are: (1) to understand what controls the production and export of organic matter from the surface of the ocean and how this influences the surface nutrient concentration; (2) to understand what controls the efficiency of organic matter recycling in the surface and how this influences the surface nutrient concentration; and (3) to understand why the recycling efficiency of organic matter in the surface ocean appears to be anti-correlated with surface nutrient concentrations, contrary to expectations. As indicated, the key observation that will guide our analysis is the surface nutrient distribution. Among the major features we would like to explain are the low-surface nutrient, low-productivity regime of the permanently stratified subtropical biome and the high nutrient, low to intermediate productivity regime of the HNLC regions. We have proposed the following explanations/hypotheses for these major features:

1. Nutrient supply controls the surface nutrients and the export of organic matter. We have argued that if the supply is low, the surface nutrients will also be low, but if it is high, then the light supply and/or biological pump must play a role in controlling surface nutrients. As regards the export of organic matter, we have seen that this must equal the nutrient supply rate, but since the nutrients supply rate itself depends on the extent to which the surface nutrients are depleted, this provides only a partial understanding of what controls organic matter export.

2. Light supply controls the surface nutrients and the export of organic matter. We have argued that if the nutrient supply is high but the light supply is low, surface nutrients will inevitably be high and the export of organic matter will be low. However, if the light supply is high, then the biological pump must play a role.

3. The efficiency of the biological pump controls the surface nutrients and the export of organic matter. We have noted that the efficiency of the biological pump makes the greatest difference when the nutrient and light supply are high, and that there are vast regions of the ocean commonly identified as HNLC where the biological pump efficiency is surprisingly low. In addition to light limitation, two fundamental hypotheses exist to explain the HNLC regions:

   a. The suppression of phytoplankton by herbivore grazing prevents them from expanding their population to where they can consume all the available nutrients.
   b. The ecosystem is limited by the supply of a micronutrient, with iron being the prime candidate.

There also exists a variant of these two HNLC hypotheses, the *ecumenical hypothesis* of *Morel et al.* [1991b], in which one or more components of the ecosystem may be grazing-limited but the ecosystem as a whole is iron-limited. Given the dependence of the nutrient supply rate on the surface nutrient concentration, we can readily infer that a more efficient biological pump will give a higher organic matter export, and vice versa.

One particularly intriguing aspect of the observations we analyzed in this section is that the efficiency of the biological pump is inversely related to the efficiency of the ecosystem in exporting organic matter from the surface. That is to say, the areas of the ocean with the least efficient biological pumps have ecosystems that are very efficient at exporting any organic matter that they produce, and vice versa, which is the opposite of what one might expect. The link between the ecosystem and surface nutrient concentrations clearly is a complex one.

Before we can evaluate the hypotheses we have proposed, we need to have a clear understanding of what we mean by each of them. We do this by building a set of models that illustrate the basic principles underlying them. In section 4.2, we review the basic components of the surface ecosystem and develop a toolkit for modeling them. In section 4.3, we combine the components in a suite of simple ecosystem models that illustrate the influence of light as well as the various ways we can combine ecosystem components to give systems that are either grazing-limited or iron-limited, or that have components that are grazing-limited while the system as a whole is iron limited. We also apply our understanding to interpret representative observations from different regions of the world. In section 4.4, we provide a synthetic overview of our present understanding of what controls surface nutrients based on these various insights.

4.2 Ecosystem Processes

Our objective in this and the next section is to develop a set of solutions of the conservation equation (2.1.11)

$$\frac{\partial C}{\partial t} = -\mathbf{U}\cdot\nabla C + \nabla\cdot(\mathbf{D}\cdot\nabla C) + SMS(C) \qquad (4.2.1)$$

for each component of the ecosystem. For convenience we lump the time rate of change and the negative of the ocean advection and mixing terms into a single term $\Gamma(C)$ such that

$$\Gamma(C) = \frac{\partial C}{\partial t} + \mathbf{U}\cdot\nabla C - \nabla\cdot(\mathbf{D}\cdot\nabla C) \qquad (4.2.2a)$$

and

$$\Gamma(C) = SMS(C) \qquad (4.2.2b)$$

We will only be dealing here with steady-state solutions, i.e., $\partial C/\partial t = 0$. Furthermore, for all components of the ecosystem except nutrients, we will assume that transport by advection and diffusion is negligible relative to the various components of the $SMS(C)$ term, i.e., we will assume $\Gamma(C) \approx 0$. The following four subsections describe how we model the four major components of the ecosystem: nutrients, phytoplankton, zooplankton (the herbivores and carnivores of the ecosystem), and bacteria. We discuss the modeling of dissolved and particulate organic matter and how they are lost from the surface in chapter 5.

Nutrients

In a steady state, the conservation equation for nutrients N will be (4.2.2) with $\partial N/\partial t = 0$, i.e., $\Gamma(N) = \mathbf{U}\cdot\nabla N - \nabla\cdot(\mathbf{D}\cdot\nabla N) = SMS(N)$. The $SMS(N)$ term will be determined entirely by the other components of the ecosystem. For example, for a system consisting of phytoplankton P, zooplankton Z, and bacteria B, we have that $\Gamma(N) = \mathrm{SMS}(N) = -SMS(P) - SMS(Z) - SMS(B)$. The P, Z, and B terms are all discussed in the following subsections. However, we do need to discuss in this subsection which nutrients we will model and how these nutrients are related to each other.

Rather than derive a separate equation for all the major and minor nutrients, our models will be designed around nitrogen as the master variable to which we key all components of the ecosystem. There are three reasons we use nitrogen for this:

1. As previously noted, organic matter is considered to have approximately constant stoichiometric ratios of $H:C:N:O_2:P$. The trace element ratios are more variable. Wherever this conjecture of constant major element stoichiometric ratios applies, it should be possible in principle to apply a model developed for any one of these major organic matter constituents to any one of the others simply by multiplying by the appropriate ratio.

2. There is evidence that whenever macronutrients limit growth, it is generally due to an inadequate supply of nitrate rather than phosphate. Later in the chapter we shall deal with the possibility of limitation by micronutrients such as iron.

3. Nitrogen in its various dissolved inorganic chemical forms is an excellent tracer of the different pathways that matter can flow in the surface ecosystem.

We discuss each of these in turn.

Composition of Organic Matter

Based on measurements of the composition of phytoplankton in the ocean, the traditional stoichiometric

TABLE 4.2.1
Stoichiometric ratios of phytoplankton organic matter and oxygen released during synthesis of the organic matter or consumed during remineralization.
The first line gives the traditional Redfield ratio. The second line is a revision of the hydrogen and oxygen content in the Redfield ratio based on a reevaluation of the range of composition of organic matter. The third line is based on an analysis of the remineralization ratio in the deep ocean below 400 meters.

	Organic Matter					*Oxygen*
	C	H	O	N	P	O_2
Redfield et al. [1963]	106	263	110	16	1	138
Anderson [1995]	106	164–186	26–59	16	1	141–161
Anderson and Sarmiento [1994]	117 ± 14	–	–	16 ± 1	1	170 ± 10

TABLE 4.2.2
Mean composition of primary components of marine phytoplankton.
From Anderson [1995]. Proteins and lipids were taken by him from the study of Laws [1991], carbohydrates from Strickland [1965], and nucleic acid from Adams et al. [1986]. Shown in parentheses are the mean compositions used by Hedges et al. [2002] in their study. Their carbohydrate composition was identical to Anderson [1995] and they did not include nucleic acid in their analysis.

Organic Matter Component	*Composition*	*H/C_{org} ratio*	*C_{org}/O ratio*
Carbohydrate	$C_6H_{10}O_5$	1.67	1.2
Lipid	$C_{40}H_{74}O_5$ ($C_{18}H_{34}O_2$)	1.85	8.0
Protein	$C_{3.83}H_{6.05}O_{1.25}N$ ($C_{106}H_{168}O_{34}N_{28}S$)	1.58	3.1
Nucleic acid	$C_{9.625}H_{12}O_{6.5}N_{3.75}P$	1.25	1.5

formula for the composition of marine phytoplankton organic matter is:

$$106\ CO_2 + 16\ HNO_3 + H_3PO_4 + 122\ H_2O \rightleftharpoons (CH_2O)_{106}(NH_3)_{16}(H_3PO_4) + 138\ O_2 \quad (4.2.3)$$

The stoichiometric ratios of C:N:P:O_2 of 106:16:1:−138 in this formula (see the first line of Table 4.2.1) are termed *Redfield ratios* in honor of the oceanographer A. C. Redfield, who was among the first to try to quantify them [*Redfield et al.*, 1963]. However, a number of more recent observations suggest that the oxygen and hydrogen contents of the organic matter proposed by *Redfield et al.*, [1963] are too high [e.g., *Takahashi et al.*, 1985; *Martin et al.*, 1987; *Hedges et al.*, 2002]. Reducing these has the overall effect of increasing the amount of oxygen required in order to oxidize the organic matter.

In his reexamination of the stoichiometry of organic carbon production in the surface ocean, *Anderson* [1995] begins with the mean composition of the primary components of phytoplankton organic matter summarized in table 4.2.2, and considers a variety of observational constraints on the relative concentration of the components and extreme values of the element ratios, assuming that the C:N:P ratio is equal to that of Redfield. The final range of estimates he arrives at is given in table 4.2.1. The new hydrogen and oxygen stoichiometry is lower than that proposed by Redfield and the oxygen required to remineralize the organic matter is higher. A stoichiometric formula for the "best guess" of the new ratios, which was that given in chapter 1, is:

$$106\ CO_2 + 16\ HNO_3 + H_3PO_4 + 78\ H_2O \rightleftharpoons C_{106}H_{175}O_{42}N_{16}P + 150\ O_2 \quad (4.2.4)$$

with a C:N:P:O_2 of 106:16:1:−150 (see the second line of table 4.2.1). The composition of this organic matter is 54.4% protein, 25% carbohydrate, 16.1% lipid, and 4.0% nucleic acid by dry weight. In a recent study, *Hedges et al.* [2002] carried out an observational analysis of phytoplankton at five contrasting sites that gave a composition of 65% protein, 19% lipid, and 16% carbohydrate (see table 4.2.2 for the compositions of these that they used in their study), with an overall stoichiometry of $C_{106}H_{177}O_{37}N_{17}S_{0.4}$ requiring 154 mol of O_2 for oxidation. They did not include nucleic acids in their study, which accounts for the missing phosphorus. Given the uncertainties in these estimates (cf. table 4.2.1), their results are in very good agreement with those of *Anderson* [1995].

We make full use of the Redfield ratio concept in this book. A major justification for doing this is the finding from analyses of water column measurements of nutrients (see third line of table 4.2.1) that the remineralization of organic matter exported from the surface is very similar to that given by (4.2.4). However, there are some interesting, often puzzling, and mostly unresolved discrepancies that arise from more detailed analyses of nutrient distributions and phytoplanktonic community compositions in the surface ocean. These are discussed in panel 4.2.1.

Limiting Nutrient

The concept of a limiting nutrient is used in two different ways by ecologists [cf. *Blackman,* 1905]. The *Liebig* concept is that the extent of growth, i.e., the stock of phytoplankton, will eventually be limited by the supply of a single cellular component such as a major nutrient or micronutrient. When the supply of that nutrient runs out, growth will cease. This is referred to as the law of the minimum and is attributed to von Liebig [*von Liebig,* 1840], although the attribution may not be entirely accurate [*de Baar,* 1994]. An alternative concept that focuses on the influence of nutrient concentration on the rate of photosynthesis, rather than on the extent of growth, is now referred to as the *Monod* concept [*Monod,* 1949]. The Monod concept can, in principle, apply to simultaneous limitation by two or more nutrients.

Biological oceanographers have historically tended to think of nitrogen in the form of nitrate as the major nutrient whose supply is most likely to limit photosynthesis [e.g., *Ryther,* 1969]. Clearly, photosynthesis cannot proceed without a supply of the necessary constituents such as CO_2 and the macronutrients nitrate and phosphate as well as essential micronutrients. This provides a first-order explanation for the observation that photosynthesis and biomass are strongly correlated with the presence of macronutrients in the surface ocean. The ambient concentration of CO_2 tends to be low relative to that required for the photosynthetic reaction to operate at maximum efficiency, but phytoplanktonic organisms have developed methods to concentrate CO_2 inside the cell so that it generally does not limit photosynthesis [cf. *Falkowski and Raven,* 1997]. Nitrate, on the other hand, becomes depleted, or nearly so, over large areas of the surface ocean. Usually nitrate becomes depleted before the other macronutrient, phosphate, as illustrated by the phosphate versus nitrate plot shown in figure 4.2.1. In addition, laboratory culture studies of marine phytoplankton strongly suggest that addition of nitrate in these nitrate depleted regions increases growth, while phosphorus additions do not (freshwater environments are generally the opposite [cf. *Valiela,* 1984]). It is thus nitrate that has tended to be used as the limiting macronutrient by biological oceanographers.

Geochemists traditionally regard phosphate as the limiting macronutrient. This is because organisms are able to directly modify the total amount of nitrogen accessible to them for synthesis of organic mater by *nitrogen fixation* (the ability to use gaseous nitrogen N_2 for organic matter synthesis), and *denitrification* (the use of nitrate as an alternative to oxygen for oxidation of organic matter when oxygen becomes depleted; see section 5.3), whereas the total oceanic supply of phosphorus is determined entirely by delivery from external sources. Despite the fact that nitrate may limit growth locally, it is considered that on larger spatial and temporal scales the nitrate sources and sinks adjust to keep the mean oceanic nitrate concentration at a relatively constant value that is determined essentially by how much phosphate is available (see discussion on the nitrogen cycle in section 5.3). *Tyrrell* [1999] uses the terms *proximate* versus *ultimate* limiting nutrient to distinguish between local short-term nitrogen limitation versus longer-term limitation. Our focus in this chapter is primarily on a smaller scale, where the use of nitrate rather than phosphate limitation is generally appropriate.

The limitation of photosynthesis by other nutrients has long been considered an important possibility. There is now strong evidence from a remarkable series of *in situ* fertilization experiments of an important role for iron in the control of photosynthesis (*Martin and Fitzwater* [1988]; see table 4.2.3 and discussions in sections 4.3 and 4.4). Iron is an important component of electron transport proteins involved in photosynthesis and respiration. It is also a component of enzymes required to utilize nitrate and nitrite (nitrate reductase and nitrite reductase) as well as for fixing nitrogen (nitrogenase). Reduced supplies of iron may lead to reduced rates of growth as well as reduced relative abundance of larger phytoplankton because of the lower efficiency of organisms with low surface area–to–volume ratios at gathering the required iron supplies through their cell wall. While iron is the only trace metal that has drawn a high level of attention and that we will discuss here, we note that there are other bioactive trace metals that are considered to be essential for the growth of phytoplankton and thought to influence phytoplankton productivity at some times and places either on their own or through synergistic or antagonistic interactions [e.g., *Morel et al.,* 1991a; *Bruland et al.,* 1991]. These include Mn, Co, Ni, Cu, Zn, and Cd, for which one set of studies gives an approximate plankton organic tissue composition of P:Fe:Zn:Cu, Mn, Ni, & Cd = 1:0.005:0.002:0.0004 (cf. *Bruland et al.* [1991]).

Paradigm of Surface Ocean Nitrogen Cycling

A fundamental paradigm of organic matter production in the surface ocean is that it consists of two components attributable to nutrients that are supplied either by recycling of organic matter within the surface ocean (referred to as *regenerated production*), or from external sources, mostly by upwelling or upward mixing of nutrients from the thermocline (referred to as *new production*). The distinction between new and regenerated

Panel 4.2.1: Is There Such a Thing as a Redfield Ratio?

The composition of phytoplankton organic matter produced in the surface ocean is highly variable, showing sensitivity to the rate of growth [e.g., *Goldman et al.*, 1979] and the concentration of nutrients [*Goldman et al.*, 1992] as well as taxonomy (e.g., *Arrigo et al.* [2000], who give a C:N:P ratio of $(94\pm20):(9.7\pm0.3):1$ for diatoms versus $(147\pm27):(19.2\pm0.6):1$ for *Phaeocystis antarctica* based on analyses of surface nutrient concentrations; *Sweeney et al.* [2000], who give a C:N:P ratio of $(80.5\pm2.3):(10.1\pm0.3):1$ for diatoms versus $(134\pm5):(18.6\pm0.4):1$ for *Phaeocystis antarctica* also based on analyses of surface nutrient concentrations; and *Quigg et al.* [2003], whose cultures show systematically higher ratios for the so-called red plastid superfamily than for the more recently evolved green plastid superfamily, with diatoms, which belong to the green superfamily, having a mean C:N:P ratio of $\sim$ 70:10:1). These results suggest that diatoms, which are responsible for a large fraction of global organic matter export, have much lower C:P and N:P than the Redfield ratios. By contrast, in regions where nitrogen fixation occurs, both the C:P and N:P ratios are higher than in areas where it does not (C:N:P $\sim$ 150:25:1, versus the standard Redfield stoichiometry of 106:16:1, per *Karl et al.* [2003]; see also chapter 5).

Other estimates of stoichiometric ratios of surface drawdown by more indirect methods are equally bewildering. Most estimates of stoichiometric ratios based on measurements of nutrient drawdown at the surface are much higher than Redfieldian:

1. Measurements at the Bermuda time series station show a large drawdown of surface dissolved inorganic carbon without any corresponding signal present in the nitrate distribution ([*Michaels et al.*, 1994]; see discussion in section 4.3). This has been partly attributed to nitrogen fixation by *Trichodesmium* [*Michaels et al.*, 1996a], but the problem of explaining this feature is largely unresolved.

2. *Karl et al.* [1997] estimate that nitrogen fixation accounted for a substantial fraction of biological production at the time series station ALOHA near Hawaii between 1989 and 1995, with correspondingly high C:P and N:P ratios.

3. *Sambrotto et al.* [1993] summarize a set of measurements of the reduction in surface nitrate and total dissolved inorganic carbon content at three locations in the Bering Sea, Gerlache Strait of the Antarctic, and North Atlantic [see also *Arrigo et al.*, 1999]. The air-sea flux of CO_2 was taken into consideration. The net photosynthetic uptake of nitrate and dissolved inorganic carbon imply a C:N ratio of about 10 to 12, considerably in excess of the ratio of 6.6 obtained from the Redfield ratio given in equation (4.2.3), or the range of 5 to 9 obtained from the summary in table 4.2.1. The fixation of gaseous nitrogen, which is largely confined to regions with water temperatures $\geq$25°C [cf. *Karl et al.*, 2002; *Staal et al.*, 2003], is not an important factor at the locations where these studies were carried out.

4. Similarly, *Kortzinger et al.* [2001] found that the C:N ratio of organic matter exported from the surface of the ocean in a north-south transect in the eastern Atlantic was very high in stations toward the south that were considered to be in post-spring bloom conditions. However, stations to the north that were considered to be in early-spring bloom conditions gave export ratios that were near to the Redfield stoichiometry. They also found that the suspended POM was close to Redfield stoichiometry at all times. They propose that the increase in export ratio to the south may be due to a change in the export mechanism, from removal of particles by mixing events in the early bloom conditions, to export of large sinking aggregates formed in part by high C:N organic matter in post-bloom conditions.

By contrast, an estimate of the ratios based on the relative rates at which the nutrients are supplied into the surface ocean from the thermocline tends to be below the Redfield ratio (table 1). On average, the supply ratios have to be balanced by a comparable ratio in the organic matter exported from the surface of the ocean, unless there is production of nutrients in the surface, such as is the case with nitrogen fixation, for example. This method has been used to estimate the $CaCO_3$-to-nitrogen and silicon-to-nitrogen supply ratios by *Sarmiento et al.* [2002 and 2004a, respectively] (cf. panel 7.2.1). The net N:P utilization ratios obtained by this same method are shown in table 1 [*Dunne et al.*, 2005c]. All of the ratios obtained this way, except those found in the North Atlantic subpolar and subtropical gyres and in the North Pacific subpolar gyre, are below the Redfield stoichiometry of 16:1. The difference from the Redfield stoichiometry is generally more modest than those discussed above, but is still relatively systematic. The deviation from Redfield stoichiometry tends to be particularly marked in the subtropical gyres. These deviations are likely due to deep ocean *denitrification*, which is the destruction of nitrate by bacteria that use it for respiration, generally in parts of the water column and sediments where oxygen is depleted or nearly so (*Dunne et al.* [2005c]; see also chapter 5). Presumably, nitrogen fixation in the

TABLE 1

N:P supply ratios between the top 100 m and the depth range of 100 and 200 m.

Calculated with the method of Sarmiento et al. [2002][a] (taken from Dunne et al. [2005c])

Region	Median N:P Ratio
Atlantic Ocean	
Subpolar Gyre (>45°N)	18.2 ± 0.3
Subtropical Gyre (15°N–45°N)	19.7 ± 1.5
Equatorial (15°S–15°N)	15.9 ± 0.2
Subtropical Gyre (45°S–15°S)	14.4 ± 0.3
Southern Ocean (<45°S)	15.5 ± 0.4
Indian Ocean	
Subtropical Gyre (15°N–45°N)	11.7 ± 0.3
Equatorial (15°S–15°N)	15.1 ± 0.1
Subtropical Gyre (45°S–15°S)	13.4 ± 0.2
Southern Ocean (<45°S)	13.5 ± 0.2
Pacific Ocean	
Subpolar Gyre (>45°N)	17.6 ± 0.1
Subtropical Gyre (15°N–45°N)	15.5 ± 0.2
Equatorial (15°S–15°N)	14.7 ± 0.1
Subtropical Gyre (45°S–15°S)	13.2 ± 0.3
Southern Ocean (<45°S)	15.0 ± 0.2

[a]The net N:P utilization ratios are obtained from the mean concentrations of NO_3^- and PO_4^{3-} using the following equation:

$$r_{N:P} = \frac{\Sigma(\langle[NO_3^-]\rangle_{100-200} - \langle[NO_3^-]\rangle_{0-100})}{\Sigma(\langle[PO_4^{3-}]\rangle_{100-200} - \langle[PO_4^{3-}]\rangle_{0-100})}$$

The surface value is represented by the mean over the top 100 m, and the thermocline value is represented by the mean in the 100–200 m range. The bootstrap method is used to estimate the nonparametric standard deviation of the mean. For this, 10,000 trial data sets were constructed from the original data set by random selection with replacement (that is, repeatedly selecting from the same set of observations). The most likely estimate of the ratio was then taken as the median of all trials, with the 95% confidence limits coming directly from the 2.5% and 97.5% tails of the distribution of the 10,000 trials [*Dunne et al.* 2005c].

surface ocean will counterbalance some or most of the nitrate deficit in the waters supplied from below.

What are we to make of these conflicting results? Much of the difference between the various estimates may be due simply to the spatiotemporal scale of the observations on which they are based. Most of the conflicting results summarized above come from measurements made on relatively small scales. From a biogeochemical point of view, the most important information we need is the larger scale mean behavior of the stoichiometric ratio of the organic matter that is exported from the surface. This must ultimately equal the stoichiometric ratio of the net supply of nutrients by transport from below and laterally, plus any nutrients added from the atmosphere or by river delivery or nitrogen fixation. To the extent that we continue to have confidence in the Redfield concept of constant stoichiometric ratios in organic matter exported from the surface, this comes primarily from what we have been able to infer from large-scale diagnostic studies of the impact of organic matter remineralization on the dissolved inorganic carbon, nitrate, phosphate, and oxygen concentrations in the main thermocline and deep ocean (cf. chapter 5). One such study by *Anderson and Sarmiento* [1994] given in the third line of table 4.2.1 found that the organic matter remineralized in the ocean below 400 m has a stoichiometric ratio very similar to the other estimates of marine phytoplankton organic matter shown in table 4.2.1 (including the original Redfield ratios), except that the organic carbon content and oxygen demand are slightly higher. This analysis thus suggests that below 400 m at least, all the processes that contribute to the complex results summarized above balance each other out so that the effective organic matter stoichiometric ratio comes back to about $C:N:P:O_2 = 117:16:1:-170$. Unfortunately, the analysis method of *Anderson and Sarmiento* [1994] could only be applied below about 400 meters depth and only in the Pacific and Indian Oceans. Above this depth and in the Atlantic Ocean, the data were too complex to be interpreted by their method. The stoichiometric ratios obtained by *Anderson and Sarmiento* [1994] are similar to a number of older studies of more limited regions of the ocean [*Takahashi et al.*, 1985; *Broecker et al.*, 1985b; *Peng and Broecker*, 1987], but disagree with others [*Minster and Boulahdid*, 1987; *Boulahdid and Minster*, 1989].

In the remainder of this book, we use primarily the stoichiometric ratios of *Anderson and Sarmiento* [1994] unless we need to know the hydrogen and oxygen components of the organic matter, in which case we use the stoichiometric ratios of *Anderson* [1995]. These two are in reasonable agreement with each other. The *Anderson and Sarmiento* [1994] ratios probably give the best idea of the

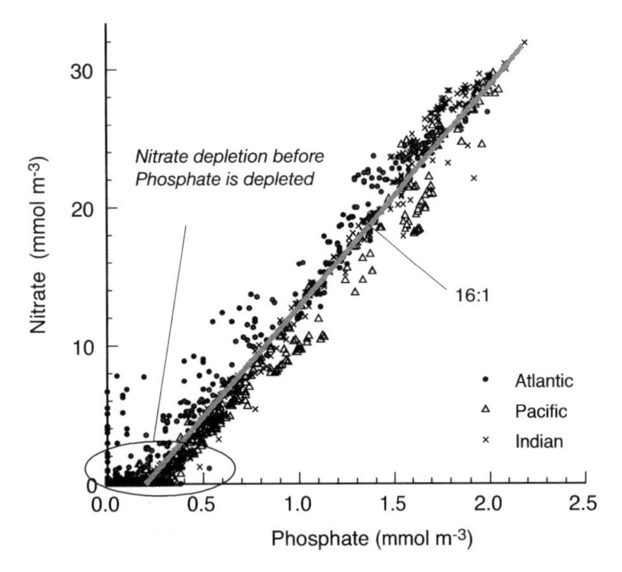

FIGURE 4.2.1: Plot of near-surface concentrations of phosphate versus nitrate from selected WOCE cruises (A16, I8/9, P16) and the full set of GEOSECS cruises. A grey line with a slope of 16:1 is added to emphasize that most data follow this trend. Note that in most regions of the world except the Atlantic, nitrate becomes depleted before phosphate.

production is a critical one for us because, in a steady state, the large-scale export of organic matter from the surface, referred to as *export production*, has to be equal to the large-scale new production; and export production is the main mechanism by which surface nutrient concentrations are reduced and prevented from climbing up to those of the deep ocean. In our example of the two-box model given in section 4.1, we saw that the export production and thus also the new production was equal to $e \cdot NPP = e \cdot \Lambda \cdot N_s$, where e is the export fraction. The regenerated production in this example would be $(1-e) \cdot NPP = (1-e) \cdot \Lambda \cdot N_s$. Nitrogen is distinctive among the nutrients in having several chemical forms that can be uniquely identified with new, regenerated, and export production.

The aspects of nitrogen chemistry that are relevant to our discussion are illustrated in Figure 4.2.2 (see section 5.3 for a more complete diagram and discussion), and the biological processes that lead to transformations of the nitrogen in the surface are illustrated in figure 4.2.3. Most phytoplanktonic organisms are able to use both nitrate and ammonium as a source of inorganic N for photosynthesis. However, ammonium costs less energy to convert to organic matter, and so phytoplanktonic organisms prefer it to nitrate. Because of this, the lifetime of ammonium in seawater is extremely short, and the vast majority of the ammonium that is present in the surface ocean is just that which is produced by local remineralization of organic matter. By contrast, there is little or no production of nitrate by remineralization in the well-lit surface ocean. This is because nitrifying bacteria, which carry out the nitrification reaction (the conversion of ammonium to nitrate that occurs in a series of steps as part of the remineralization of organic matter; see section 5.3), are inhibited by light [cf. *Zehr and Ward*, 2002]. Therefore, most of the nitrate used by phytoplankton must be supplied either by transport from below the euphotic zone or delivery through the atmosphere. Thus we see that ammonium uptake is identified uniquely with regenerated production, and nitrate uptake with new production. The method commonly used to separately measure nitrate from ammonium uptake is isotopic tagging with ^{15}N of ammonium and/or nitrate to trace the relative incorporation of these into phytoplankton, as proposed by *Dugdale and Goering* [1967].

Nitrate and ammonium are by far the most important sources of N for phytoplankton, but phytoplanktonic organisms are also able to use N_2 by fixation (as shown in both figure 4.2.2 and figure 4.2.3), as well as nitrite (which we normally sum to the nitrate pool), and also organic nitrogen compounds such as urea and amino acids, which are implicit in figure 4.2.2 as "organic nitrogen" and figure 4.2.3 as dissolved organic nitrogen, *DON*. Of these reactions, nitrogen fixation in particular can be locally quite important, but for now we will focus on just nitrate- and ammonium-based production.

With this background, we turn now to a discussion of the surface ocean nitrogen cycle depicted in figure 4.2.3 (see review by *Lipschultz et al.* [2002]). The basic concepts underlying the initial statement of the paradigm of new and regenerated production by *Dugdale and Goering* [1967] and *Eppley and Peterson* [1979] were two:

a. On average, the supply of nitrate to the surface ocean (primarily by pathway 1a in the figure) has to be balanced by export of organic nitrogen, which at the time was thought to be due primarily to sinking particles (pathway 6a in the figure).

Panel 4.2.1: (*Continued*)

composition of the net flux of organic matter exported from the surface to depths below 400 m. However, there is considerable uncertainty associated with the fact that it was not possible to do the analysis in waters shallower than 400 m, as well as the fact that nutrient removal from surface waters does not always appear to follow the Redfield stoichiometry. These issues will be addressed further, though not resolved, in the following chapter on remineralization. Clearly there is work to be done on this problem.

TABLE 4.2.3
In situ iron fertilization experiments carried out between 1993 and 2002.
This table is based on a summary by Francisco Chavez (personal communication, 2004). The iron fertilization experiments all show that iron addition leads to improved photosynthetic competency, with diatoms generally showing the greatest response except in the low-silicic acid waters of the northern SOFeX patch.

Experiment	*Location*	*Result*	*Nitrate Drawdown (mmol m^{-3})*
		North Pacific	
SEEDS, July 2001 (Japan)	Northwest Pacific (48.5°N, 165°E)	Resulted in an increase in chlorophyll to concentrations as high as 20 mg m^{-3} (from >1 mg m^{-3}), and large drawdowns in pCO_2 and nutrients. Perhaps the most dramatic result to date [cf. *Tsuda et al.*, 2003].	>15
SERIES, July 2002 (Canada & Japan)	Subpolar Pacific (50°N, 145°W)	An open ocean iron enrichment experiment with a large bloom observed from space. This is the first time series of the decline of an iron-induced bloom, which was observed to be associated with silicic acid depletion as well as iron exhaustion [cf. *Boyd et al.*, 2004].	>5
		Equatorial Pacific	
IRONEX I, Oct. 1993 (U.S.)	Equatorial Pacific (5°S, 90°W)	Perhaps the least dramatic of the experiments. Clear physiological response, small increase in chlorophyll, minor impact on nutrients and carbon dioxide, advection of the patch out of the euphotic zone [cf. *Martin et al.*, 1994].	None
IRONEX II, May 1995 (U.S.)	Equatorial Pacific (3.5°S, 104°W)	Clear physiological response, substantial bloom of diatoms, clear impact on nutrients and carbon dioxide, evidence of carbon removal from mixed layer [cf. *Coale et al.*, 1996].	~5
		Southern Ocean	
SOIREE, Feb. 1999 (New Zealand + others)	Southern Ocean (61°S, 140°E)	Revealed significant increase in phytoplankton biomass and production and a shift in planktonic community structure toward large diatoms. Resulted in modest drawdown of macronutrients and inorganic carbon. However, no significant impact on the magnitude of export production was evident [cf. *Boyd et al.*, 2000].	~3
EisenEx Nov. 2000 (Europe)	Southern Ocean (48°S, 21°E)	Showed that plankton growth in the Antarctic Circumpolar Current at 20°E is limited by iron availability and that addition of this element can lead to a quadrupling of biomass within a period of three weeks, despite heavy grazing and poor light conditions [cf. *Gervais et al.*, 2002].	<2
SOFeX, Jan–Feb. 2002 (U.S.)	Southern Ocean (56°S, 172°W and 66°S, 172°W)	Two patches, one in high-nitrate, low-silicic acid water, the other in high-nitrate, high-silicic acid water. Both resulted in increases in chlorophyll and diatoms. The high-nitrate, low-silicic acid response was the most noteworthy [cf. *Coale et al.*, 2004].	~2

Thus if one were able to measure either one of these pathways with sufficient time and space resolution, one would immediately know the magnitude of the other [*Eppley and Peterson*, 1979].

b. Phytoplankton production is based on uptake of nitrate (pathway 2a) that is supplied from the thermocline (pathway 1a) as well as ammonium (pathway 3) that comes from recycling of organic matter by zooplankton and bacteria

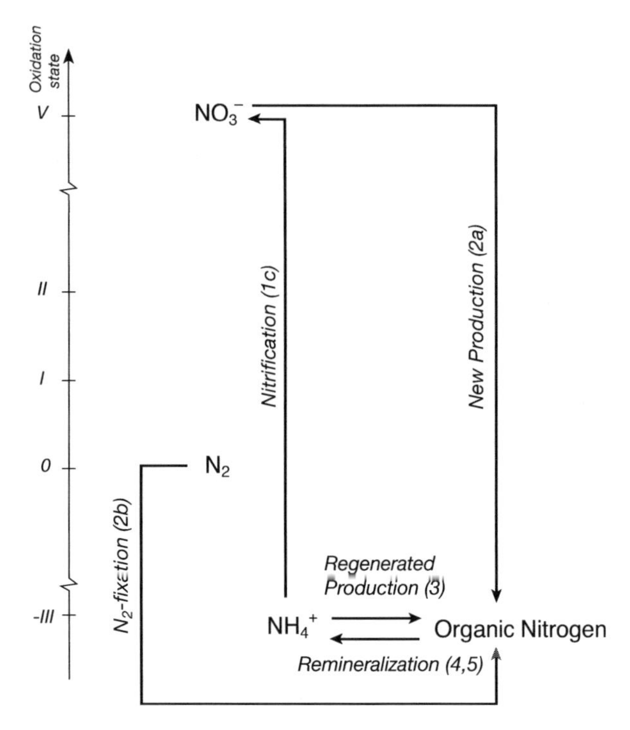

FIGURE 4.2.2: Schematic illustration of relevant aspects of nitrogen chemistry for surface ocean nitrogen cycling. Nitrogen species are plotted as a function of their oxidation state, as indicated by the vertical axis on the left-hand side. Nitrogen combined with O or H is referred to as *fixed*. See text for further discussion.

(pathway 4). It was *Dugdale and Goering* [1967] who recognized this distinction and used the terms new and regenerated production to describe pathways 2a and 3.

Subsequent work led to significant modifications in the initial view of the paradigm. With the development of an improved knowledge of the role of heterotrophic bacteria in the surface ocean, it came to be recognized that they are major contributors to the recycling of organic matter in the surface ocean [*Pomeroy*, 1974], and the concept of the *microbial loop* was born [*Azam et al.*, 1983]. The microbial loop starts with bacterial consumption of dissolved organic matter (DOM) produced by all components of the food web, and terminates with the consumption of bacteria by small zooplankton (nanoflagellates and small ciliates) as well as mucus net feeders such as larvaceans [e.g., *Azam et al.*, 1983; *Azam*, 1998; *Ducklow*, 2001]. Bacteria also produce ammonium and consume both ammonium and nitrate [e.g., *Caron*, 1994].

Globally, about half of the primary production is routed through DOM and processed by bacteria [cf. *Carlson*, 2002]. However, a significant amount of DOM escapes reprocessing and can be exported out of the euphotic zone by transport and mixing processes, such as subduction, convection, and diffusion. Estimates of the export of DOC from the surface ocean show it to be about $20 \pm 10\%$ of the total export of organic matter [cf. *Hansell*, 2002], with a range from 11% in the Ross Sea [*Carlson et al.*, 2000] to 52% in the subtropical North Pacific [*Emerson et al.*, 1997], and intermediate values of 33% in the subtropical North Atlantic and 19% in the tropical Pacific [*Carlson et al.*, 1994; *Hansell et al.*, 1997; *Dunne et al.*, 2000]. The new paradigm that emerged from this research thus included a significant role for dissolved organic matter and the microbial loop in the recycling of organic matter within the surface (pathway 5b in figure 4.2.3) and the export of organic matter from the surface (pathway 6b). With the addition of DOM to the ecosystem, the role of bacteria becomes essential because they are the primary consumers of DOM and also play a role in the production of DOM by excretion of enzymes that are able to break down particulate organic matter into smaller soluble molecules that they are able to consume [*Cho and Azam*, 1988].

Additional modifications to the original paradigm of surface ocean biological cycling are depicted in figure 4.2.3. These include:

a. The supply of nitrogen by fixation of N_2 (pathway 2b). *Dugdale and Goering* [1967] discussed this in their original paper, but thought that it was small. More recent estimates suggest that nitrogen fixation is about $130\,\mathrm{Tg\,N\,yr^{-1}}$ (see section 5.3), which would give a carbon export of about $0.8\,\mathrm{Pg\,C\,yr^{-1}}$, less than 10% of the global new production.

b. The input of nitrate and ammonium from the atmosphere (pathway 1b), estimated to be about $50\,\mathrm{Tg\,N\,yr^{-1}}$ (see section 5.3), which would give a carbon export of $0.3\,\mathrm{Pg\,C\,yr^{-1}}$ of carbon if all of it were used to form organic carbon with a C:N ratio of 117:16.

c. Nitrification occurring in the euphotic zone (pathway 1c). As noted earlier, the inhibition by light of the bacteria that carry out this reaction has generally led to this reaction being considered negligible in the euphotic zone, but there is evidence that most water column nitrification actually takes place in very low light levels at the base of the euphotic zone [cf. *Zehr and Ward*, 2002]. The implications of this for the new and regenerated production paradigm are unclear at this time.

The breakdown of biological production into new, regenerated, and export components suggests the following two definitions that have found considerable application in the literature: (1) the *f*-ratio:

$$
\begin{aligned}
f &= \frac{\text{New production}}{\text{Primary production}} \\
&= \frac{\text{New production}}{\text{New} + \text{regenerated production}} \qquad (4.2.5) \\
&= \frac{N_2\ \text{fixation} + NO_3^-\ \text{uptake (excluding nitrification)}}{N_2\ \text{fixation} + \text{uptake of}\left(NO_3^- + NH_4^+\right)}
\end{aligned}
$$

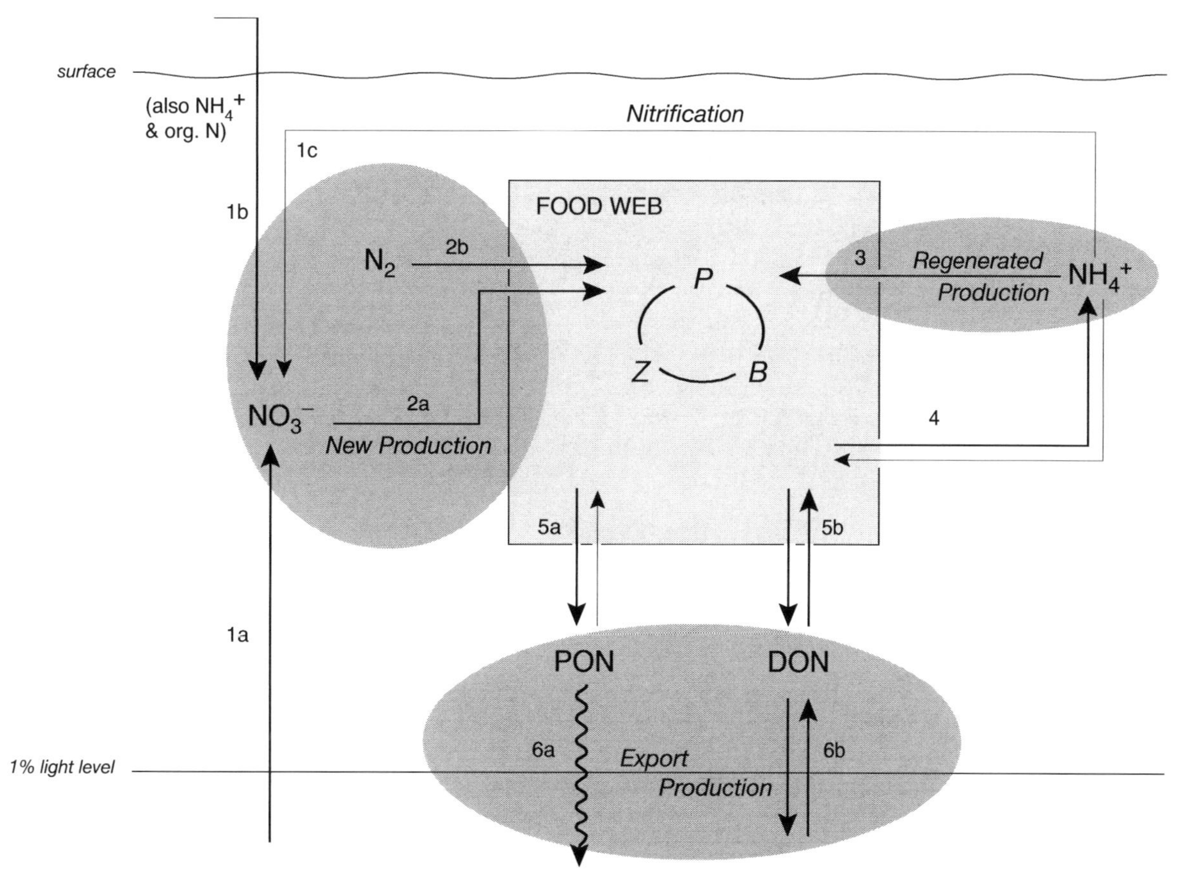

FIGURE 4.2.3: Schematic illustration of the paradigm for the cycling of nitrogen in the surface ocean. *P* is phytoplankton; *Z* and *B* are zooplankton and bacteria, respectively; *PON* and *DON* are particulate and dissolved organic nitrogen. New nitrogen is supplied to the system either from below (pathway 1a) or from the atmosphere (pathway 1b) as well as through nitrogen fixation (pathway 2b). Regenerated production is driven by ammonium (pathway 3), and export of organic matter from the surface ocean is in the form of *PON* (pathway 6a) as well as *DON* (pathway 6b), some of which is recycled back into the part of the ecosystem within the brackets before it can be exported (pathway 5). *DON* is both produced and recycled by the ecosystem (pathway 5b) as well as being exported from and possibly imported to the surface ocean (pathway 6b). Organic matter remineralization by the ecosystem can go all the way to nitrate rather than just to ammonium (nitrification; pathway 1c). Bacteria are also able to consume nitrate (not shown).

and (2) the export or *e*-ratio that we already mentioned in connection with our discussion of the two-box model:

$$e = \frac{\text{Export production}}{\text{Primary production}} \qquad (4.2.6)$$

The general expectation is that the *e*-ratio and *f*-ratio have to equal each other over large enough regions and timescales to account for lateral transport and temporal variability. In what follows, we will use the term *ef*-ratio when the distinction between these two is not relevant to the discussion.

PHYTOPLANKTON

A basic understanding of the organisms that comprise the ecosystem is useful to bound the range of possible biological dynamics in specific regions and to suggest how models should be formulated to capture these dynamics. That is the motivation for the following general discussion of the various ways in which we classify organisms in the ocean and the description in panel 4.2.2 of some of the dominant phytoplankton in open ocean ecosystems. After this, we discuss the distribution and productivity of phytoplankton and the specific processes that control photosynthesis and how we model these.

CLASSIFICATION OF ORGANISMS

The diverse marine organisms are classified by operational criteria (e.g., size), functional criteria (e.g., trophic status, impact on biogeochemical cycles, and life cycle), and taxonomic criteria. All of these have potential utility for understanding an ecosystem, and many of the operational and functional distinctions are basic to any modeling exercise. Marine organisms cover an extreme range of sizes, from submicron to many meters. Operationally, they are divided into five size categories at the small end of the spectrum: *picoplankton* (0.2–2.0 μm), *nanoplankton* (2.0–20 μm), *microplankton*

Panel 4.2.2: Dominant Phytoplanktonic Organisms

Just as on land, the oceanic food chain begins with photoautotrophic organisms (phytoplankton) that are able to utilize light energy to synthesize organic matter from inorganic chemicals (photosynthesis). The contribution of chemoautotrophic organisms, such as those that live off of energy supplied by the oxidation of reduced chemicals in hydrothermal waters emitted by mid-ocean ridges, is negligible compared to photosynthesis. The phytoplanktonic community is taxonomically diverse, with over 20,000 species in eight or more phyla. However, only a few species dominate over most of the open ocean. A few picoplanktonic groups dominate clear oceanic waters. In more productive and coastal waters, a few groups of larger, more conspicuous phytoplanktonic organisms also become quantitatively important components of the suspended biomass. Among these larger phytoplanktonic organisms are diatoms belonging to the class Bacillariophyceae; coccolithophorids, belonging to the Prymnesiophyceae; and in some cases, dinoflagellates, belonging to the class Dinophyceae.

The picophytoplanktonic community is comprised of *prokaryotes* (autotrophic bacteria) and *eukaryotes* (cells with internal organelles) that are less than a few microns in size. *Prochlorococcus* and *Synechococcus* are unicellular prokaryotes that dominate the picophytoplankton in low-nutrient open ocean regimes ([*Li et al.*, 1983; *Platt et al.*, 1983; *Chisholm et al.*, 1988]; cf. review by *McCarthy* [2002]). They are typically the dominant source of chlorophyll in the low-nutrient open ocean regions, though some poorly known autotrophic eukaryotes are also present and occasionally become important. Interestingly, *Prochlorococcus* and at least one strain of *Synechococcus* are unable to grow on nitrate [*Moore et al.*, 2002d], while all can grow on ammonium and urea. Thus many of these picophytoplanktonic organisms grow only by regenerated rather than new production. On a global scale, the picophytoplankton are responsible for about 40% of the primary production, despite constituting only about 24% of the biomass [cf. *Agawin et al.*, 2000a & b]. Their average turnover rate can be estimated from this to be $\sim 1/(5\,\mathrm{d})$, slightly faster than the average turnover rate of $\sim 1/(7\,\mathrm{d})$. Their fractional contribution to the total primary production increases with temperature due probably to higher metabolic rates; and it decreases with primary production. The decrease with primary production may be due in part to grazing limitation of the picophytoplankton and in part to the higher growth capacity of larger phytoplankton when nutrients are abundant (cf. [*Agawin et al.*, 2000]). Diatoms, for example, can have growth rates of up to $2.2\,\mathrm{d}^{-1}$. Thus picophytoplankton dominate in warm waters of low productivity ($\geq$50%),

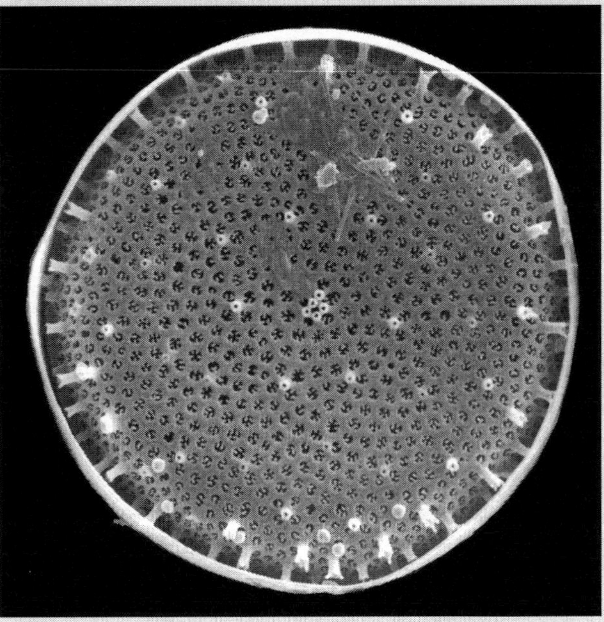

FIGURE 1: Photographs of typical diatoms. (a) *Fragilariopsis kerguelensis*, a Southern Ocean pennate diatom, and (b) *Thalassiosira gravida*, a Southern Ocean centric diatom. The white line in (a) is 10 μm. *T. gravida* is ~30 μm in diameter. Photographer: Fiona Scott, © Australian Antarctic Division 2002, Kingston, Tasmania 7050.

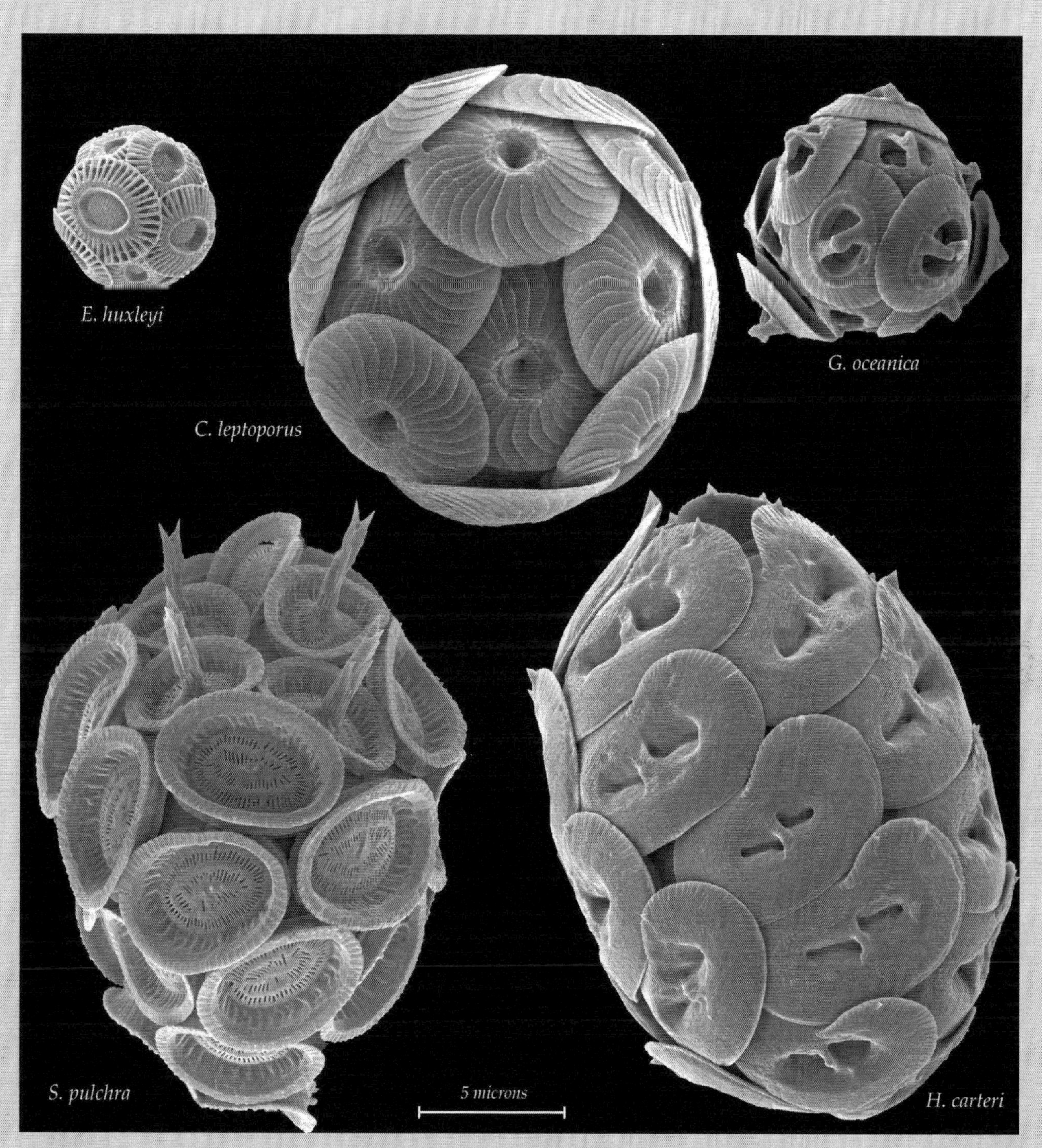

FIGURE 2: Examples of typical coccolithophorids, from *Stoll et al.* [2002].

(20–200 μm), *mesoplankton* (200–2000 μm), and *macroplankton* (>2000 μm). In the open ocean, most of the organic biomass is in the smallest of the size categories, while in highly productive coastal ecosystems, the micro-, meso-, and macroplankton can dominate the biomass. These size-based relationships have further importance as they determine the capability of an organism to sink (generally restricted to the microplankton and larger) and to consume other plankton (generally organisms eat particles that are smaller than they are).

Functional classifications of organisms are usually based on their trophic status, i.e., their position within an ecosystem with regard to the flow of energy and material. The primary distinction is between *autotrophs*, which can create organic matter directly from inorganic nutrients using solar or chemical energy, and *heterotrophs*, which are organisms that gain both energy and nutrition only from preexisting organic matter. The autotrophs we will discuss here are the microscopic floating marine plants, including bacteria, referred to as *phytoplankton*. The heterotrophs we will discuss here include small floating animals referred to as *zooplankton*, as well as bacteria. Zooplankton organisms are often further distinguished into herbivores, i.e., those whose food consists primarily of phytoplankton, and carnivores, i.e., those who prey primarily upon other zooplankton organisms. Some organisms known as *mixotrophs* can function as both autotrophs and heterotrophs. They do this either through the presence of both physiologies within one cell or through symbiotic relationships between dissimilar organisms, generally algal cells living within animal tissue. Both autotrophy and

Panel 4.2.2: (*Continued*)

and contribute less than 10% in highly productive polar regions.

The outstanding characteristic of diatoms is the beautiful external skeleton or *frustule* composed of two valves that they form from opal (e.g., figure 1; see chapter 7). The valves preserved in sediments provide evidence of the existence of diatoms since the early Jurassic period 185 million years ago. Diatoms are dominant in the high-latitude oceans and coastal and tropical upwelling regions where the supply of silicic acid, from which they make their opal frustules, is abundant. They are often the primary species responsible for the appearance of blooms. The ratio of silicic acid to nutrients in freshly upwelled water over most of the ocean is smaller than that required by diatoms (Si:N $\sim$1:1 [*Brzezinski*, 1985]), so when silicic acid becomes too low for diatoms they are succeeded by nonsiliceous phytoplankton that can live off the nutrients that still remain.

Coccolithophorids form an outer sphere of armor from small $CaCO_3$ plates known as coccoliths (see chapters 8 and 9). The coccoliths are formed inside the cell and are replaced as frequently as one every 15 minutes. *Emiliania huxleyi*, pictured in figure 2, is the most abundant coccolithophorid in the world. Indeed, it is the most productive lime-secreting organism on Earth [*Westbroek et al.*, 1984]. However, the dominant contributors to the export of $CaCO_3$ from the surface ocean and to its preservation in sediments are larger coccolithophorids with thicker coccoliths such as those also shown in the figure [*Stoll et al.*, 2002]. Of these, *Coccolithus pelagicus* is a significant contributor at high latitudes, whereas *Gephyrocapsa oceanica*, *Calcidiscus leptoporus*, and *Helicosphaera carterae* are significant contributors at lower latitudes. The coccoliths are preserved in sediments such as those that make up the white cliffs of Dover, providing a history of coccolithophorids since their first appearance some 170 million years ago. Although *E. huxleyi* is found almost everywhere except the polar seas, most coccolithophorids live only in warm waters. They tend to thrive in areas of reduced light intensity and can reach maximum abundance at depths of 100 m in clear tropical water. They also tend to appear later in phytoplankton blooms as the nutrients become depleted. Satellite images of coccolithophorid blooms show highest concentrations in the subpolar North Atlantic, North Pacific, and numerous low-latitude marginal seas, and shelf and slope regions [cf., *Brown and Yoder*, 1994; *Iglesias-Rodríguez et al.*, 2002b]. However, the production inferred from the distribution of nutrients and alkalinity in the water column (alkalinity is reduced by formation of $CaCO_3$; see chapter 8) implies a more widespread distribution than what is seen from satellites [cf. *Sarmiento et al.*, 2002]. *Phaeocystis* is a related species that lacks the $CaCO_3$ coccoliths, but also forms massive blooms. This species has been implicated in the flux of dimethylsulfide (DMS) to the atmosphere [cf. *Keller et al.*, 1989], which is thought to influence climate through providing seeds for the formation of cloud droplets [*Charlson et al.*, 1987].

Dinoflagellates (figure 3) are usually relatively uncommon, but are known to form large coastal blooms sometimes called "red tides." They differ from the nonmotile diatoms in that they have long whip-like structures referred to as *flagella* that are used in locomotion, and many are able to play the role of grazers as well as predators (mixotrophs or heterotrophs). Indeed, about

heterotrophy have a remarkable diversity in the bacteria, many of which are able to make a living on unique chemical transformations, usually in distinctive environments such as sediments where oxygen is totally depleted (see chapter 6). Within the zooplankton, functional feeding responses are often used to distinguish *raptorial feeders* (those that capture individual particles) from *filter feeders* (those that indiscriminately consume all particles retained in their feeding net). Another distinction is made between zooplankton that migrate vertically and those that do not.

Finally, organisms reach their highest level of identity in the sometimes confusing rules that govern the scientific naming and taxonomic relationships of each species. The diversity of similar but unique species within marine planktonic ecosystems is astounding. The recognition of this diversity generated a controversy over what processes could maintain such a large number of species (the Paradox of the Plankton [*Hutchinson*, 1961]), with both spatiotemporal variability of the environment and oscillatory and chaotic species interactions thought to play a role [cf. *Scheffer et al.*, 2003].

It is impractical to capture the remarkable degree of species diversity in the biological models that are used to study biogeochemistry. Nor is it clear that it is necessary to do so, since there are extensive overlaps in the ecological roles of different species. Instead, the focus in biogeochemistry has been on the *functional groups* or *biogeochemical guilds* concept of groupings of organisms that share similar metabolic pathways, e.g., the ability to fix gaseous N_2, or to precipitate $CaCO_3$ (calcite and aragonite) or SiO_2 (opal [cf. *Falkowski and Raven*, 1997;

FIGURE 3: Photographs of typical dinoflagellates. (a) *Protoperidinium latistriatum* and (b) *Protoperidinium defectum*, both found in the Southern Ocean. The white line in both figures is 10 μm. Photographer: Fiona Scott, © Australian Antarctic Division 2002, Kingston, Tasmania 7050.

half the dinoflagellates are obligate heterotrophs, unable to photosynthesize and living entirely by feeding on phytoplankton and other zooplankton. The majority of autotrophic dinoflagellates form a thick cellulose wall known as a theca. Dinoflagellates are capable of rapid division under favorable conditions, and when these motile cells aggregate in stratified surface layers, they can form very large blooms. Some species of dinoflagellates carry powerful neurotoxins that are responsible for the damage associated with some red tides and with the occasional toxicity of shellfish.

Although the picoplanktonic community dominates the low-nutrient open ocean regions of the ocean, there are some relatively large phytoplanktonic organisms or colonies of phytoplanktonic organisms that have found unique ways to make a living in nutrient-depleted environments. Among the most conspicuous of these is the marine prokaryote, *Trichodesmium*, which forms colonies comprised of multiple individuals. This phytoplanktonic organism has the capability to fix gaseous nitrogen. Gaseous nitrogen concentrations in the ocean are in excess of $\sim$400 mmol m^{-3} (table 3.2.4), compared with a globally averaged nitrate concentration of $\sim$30 mmol m^{-3}, making it by far the most abundant form of nitrogen in the ocean. The ability to fix nitrogen allows *Trichodesmium* and other nitrogen-fixing organisms, including nitrogen-fixing symbionts in diatoms [cf. *Karl et al.*, 2002], to flourish in environments such as the tropics, where the combined forms of nitrogen that other phytoplankton require, such as nitrate and ammonium NH_4^+, are in low abundance. We shall see in chapter 5 that the observed nutrient distribution gives evidence for a major role for nitrogen fixers in certain marine environments.

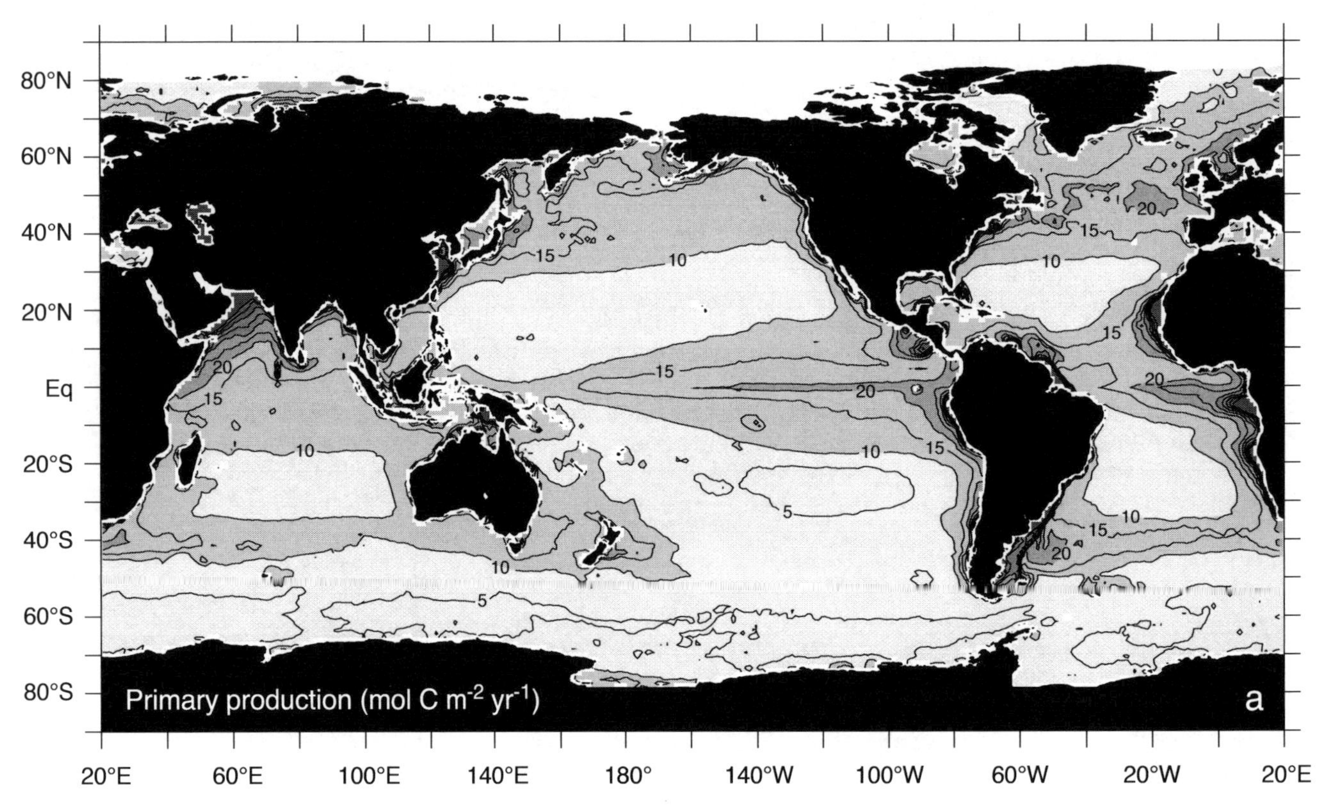

FIGURE 4.2.4: Maps of annual mean primary production and particle export. (a) Depth-integrated primary production estimates for the euphotic zone based on the SeaWiFS 4-year monthly chlorophyll climatology of figure 4.1.1 (see also color plate 3). This was converted to productivity using the average of three separate productivity algorithms as in *Dunne et al.* [2005c]. The productivity algorithms are those of J. Marra [*Marra et al.*, 2003; *Lee et al.*, 1996], *Carr* [2002] (but using the *Behrenfeld and Falkowski* [1997] definition of euphotic zone depth), and *Behrenfeld and Falkowski* [1997]. Primary production algorithms are typically calibrated with measurements of radiocarbon assimilation [*Steemann Nielsen*, 1952]. Radiocarbon-labeled bicarbonate is placed in two sets of water samples containing phytoplankton, one of which is left in the light, and the other of which is kept in darkness. This is done on the deck of a ship or by lowering the bottles *in situ*. The amount of radiocarbon taken up in the light bottle minus that taken up in the dark bottle over dawn-to-dusk incubations is assumed to give a measure of net photosynthesis, i.e., primary production [e.g., *Marra and Barber*, 2004]. There is a long history of methodological and interpretational questions with this technique [cf. *Karl*, 1999]. Trace metal contamination problems (oceanic waters have some of the lowest natural levels of trace elements) of older measurements may have led to systematic underestimates of production over much of the world oceans. Measurements with "clean" techniques [*Fitzwater et al.*, 1982] consistently yield rates 2 to 5 times the historical values [*Martin et al.*, 1987; *Lohrenz et al.*, 1992; *Welschmeyer et al.*, 1993; *Michaels et al.*, 1994; *Karl et al.*, 1996]. There are also unresolved questions about incubation conditions, time-space resolution, and the loss of radiocarbon to the dissolved organic pool during incubations. (b) Particle export at 100 m calculated using the chlorophyll concentration (*Chl*) and surface temperature (*T*) based empirical algorithm of *Dunne et al.* [2005b] for the ratio of particle export to primary production: e_p-ratio $= \max[0.042, \min(0.72, -0.0078*T + 0.0806*\ln(Chl) + 0.433)]$. See also color plate 4.

Boyd and Doney, 2002; *Iglesias-Rodríguez et al.*, 2002b]). A further subdivision of organisms that we will use in our models is according to size. The classification of organisms by size, which may cut across functional groups, is justified by the different ecological roles played by small versus large organisms, with the small organisms generally being highly efficient at recycling organic matter within the surface, and the larger organisms being highly efficient at exporting organic matter by generating large particles that can sink rapidly. We outline in panel 4.2.2 the dominant phytoplanktonic organisms that we will discuss in this and subsequent chapters.

Phytoplankton Distribution and Productivity

What do we know about the distribution and productivity of the phytoplankton community? The most common index of phytoplankton abundance is the distribution of chlorophyll shown in figure 4.1.1. Figure 4.2.4a shows the average primary production calculated from the chlorophyll map shown in figure 4.1.1 using three separate algorithms. *Primary production* (PP or NPP) is the net CO_2 uptake by phytoplankton, which is equal to the gross primary production (photosynthesis) minus phytoplankton respiration. The term *net community production* (NCP) is used to refer to PP minus the respiration by heterotrophs as well. Figure 4.2.4b shows an estimate of the particle export flux obtained by taking the product of the particle export ratio shown in figure 4.1.7 and the primary production of figure 4.2.4a. The correspondence with high levels of NPP is striking. Regions of low NPP recycle organic matter more efficiently (see figure 4.1.7) and thus tend to have

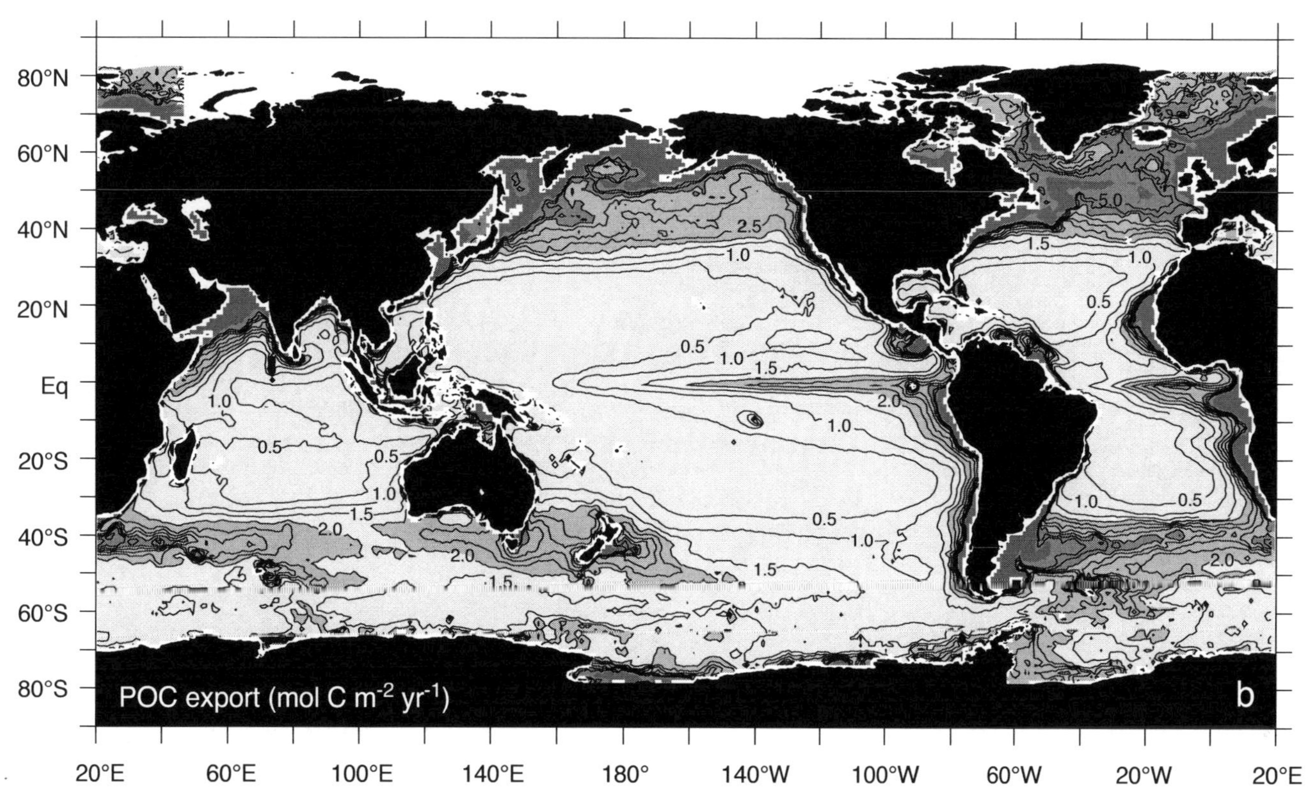

FIGURE 4.2.4: (continued)

extremely low export in general. The globally integrated particle export from this map is 10 ± 3 Pg C yr^{-1}. If we add to this the dissolved organic carbon export, which is estimated to be $20 \pm 10\%$ of the total, we obtain a total organic carbon export of 12 ± 4 Pg C yr^{-1}.

Both the surface chlorophyll map of figure 4.1.1 and the primary production map of figure 4.2.4a show a strong correspondence to the regions of high nutrient concentrations, as we saw in section 4.1 for chlorophyll. The nutrient-rich regions with high biological productivity are referred to as *eutrophic*. The vast subtropical gyre regions with low nutrients and low biological productivity are *oligotrophic*. We have already seen that there are some interesting surprises in these patterns. The most important from our point of view are the HNLC (high-nutrient/low-chlorophyll) regions in the equatorial Pacific, North Pacific, and Southern Ocean. About 40% of the Pacific Ocean is in one of these regions [*de Baar et al.*, 1994]. The Southern Ocean HNLC region is particularly important because it is one of the major regions of deep water formation, and thus the surface nutrient concentrations there play a central role in determining the deep ocean chemical properties. One of the major challenges we face is to explain how such features can exist.

We earlier introduced the functional groups and size concepts for representing those aspects of phytoplankton diversity that are most important for biogeochemical cycling. A powerful complement to these concepts is the putative "rule" of *universal distribution and local selection* articulated by *Falkowski et al.* [2003], according to which any given body of water has a seed population for all organisms, with the local environment determining which of these grows at that location. We use these concepts in combination to propose an allocation tree (figure 4.2.5). The idea of the allocation tree is to map out which functional group is expected to dominate for a given set of environmental conditions. In constructing the allocation tree, we assume that even with only a small advantage of one functional group over the others, that functional group will outcompete the less well-adapted functional groups. The allocation tree starts at the top with the total amount of chlorophyll present at a given location. The light and temperature determine the potential production of organic matter by this chlorophyll if the nutrient supply is unlimited. As we progress down the allocation tree, each branching point has successively tighter and tighter constraints. At the first branch, the only additional requirement is for phosphorus and iron. If these two nutrients are present while fixed nitrogen is not, N_2 fixers will find their optimal ecological niche and we would predict that they would become a dominant component of the upper ocean ecosystem.

The next branch in the allocation sequence of figure 4.2.5 occurs when the supply rate of phosphate and iron

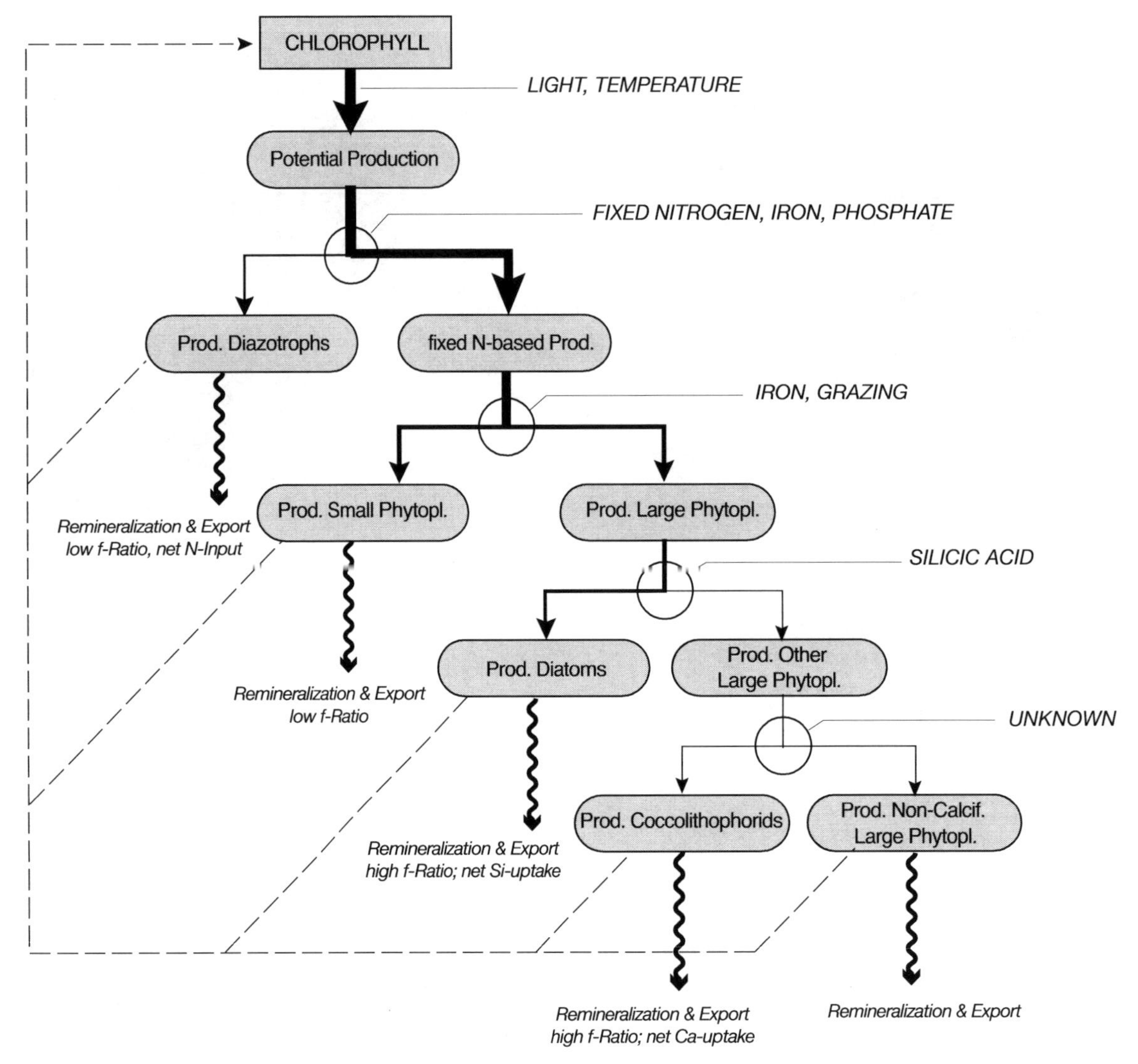

FIGURE 4.2.5: Schematic diagram showing the allocation of primary production to different phytoplankton functional groups. See discussion in text.

as well as fixed nitrogen are all sufficient for phytoplankton growth. Additional conditions at this junction determine whether production will occur primarily in small (pico) or large phytoplanktonic organisms. This branch is important because the picoplanktonic community is generally extremely efficient at recycling organic matter, giving very low *ef*-ratios, whereas the large phytoplanktonic organisms tend to form part of a community that has high *ef*-ratios, i.e., a more direct route from photosynthesis to export of organic matter. There are two controls on the small versus large phytoplankton branch. The first is microzooplankton, which keep the total amount of small phytoplankton in check by grazing. Thus, the capacity of small phytoplankton to take up nutrients has an upper limit, after which additional nutrients tend to go into large phytoplankton production, which is generally less subject to intense grazing. The second control on this branch is the supply of micronutrients, particularly iron. Even if there is an abundant supply of macronutrients, the shift to large phytoplankton will only occur if relatively abundant micronutrients are also present. This is because of the disadvantage that large phytoplanktonic organisms have in taking up iron from the water due to their low surface area-to-volume ratio as compared with picophytoplanktonic organisms.

However, the efficiency of large phytoplankton in exporting organic matter from the surface will depend on additional controlling factors that determine what type of large phytoplankton grow in a given area. We have added two additional branches to account for the processes that we believe will be most relevant. The first of these determines whether potential production will go primarily to opal-forming organisms such as diatoms or to other large phytoplankton. The paradigm is that production will go to diatoms if there is silicic acid present. Diatoms are generally considered very effective organic carbon exporters both because of their life cycle, which includes for some diatoms the formation of rapidly sinking aggregates when stressed (see chapter 7), and the fact that their frustules provide ballast (see chapter 5). In the absence of silicic acid, the remaining production potential will go either to calcifying organisms such as coccolithophorids or to noncalcifying large phytoplankton such as *Phaeocystis*. Because of their size, coccolithophorids are perhaps more properly considered members of the small phytoplankton group, but their ecological role vis-à-vis export efficiency leans more toward that of the large organisms. The controlling factors that determine the relative magnitude of allocations in this final decision tree are not well understood, though, as noted in Panel 4.2.2, coccolithophorids tend to appear late in a bloom as nutrients are depleted. Coccolithophorids, like diatoms, provide ballast for sinking of organic matter.

The focus in this chapter is on the allocation tree, beginning at the top and going down to the small versus large phytoplankton branch, with considerable attention also on diatoms because of their crucial role in the response of ecosystems to large inputs of macro- and micronutrients. The diatoms are discussed in more detail in chapter 7 and the coccolithophorids in chapter 9, and their important role as ballast for sinking organic matter is discussed in chapter 5.

Modeling Photosynthesis

Our goal is to develop a model of net growth of phytoplankton. This requires that we look at growth and loss. Most of the insights on ecosystem processes that we present here are based on laboratory studies of plankton physiology. It is important to realize that there is a major conceptual leap in translating inferences about plankton physiology from such studies to the behavior of a complex community in the ocean consisting of many species. The ability of different species to acclimate to a wide range of different environments, coupled with the fact that the behavior of a community with the potential for species succession will most likely be quite different from that of an individual species, must be taken into consideration as our models become more sophisticated.

The influence of light irradiance I ($\mathrm{W\,m^{-2}}$ or Einsteins $\mathrm{m^{-2}\,s^{-1}}$), nutrients N ($\mathrm{mmol\,m^{-3}}$), temperature T (°C), zooplankton grazing G ($\mathrm{mmol\,m^{-3}\,time^{-1}}$), and various other sinks on the source minus sink equation of phytoplankton P ($\mathrm{mmol\,m^{-3}}$ of nitrogen) are represented as a function of the form:

$$SMS(P) = V_P(T) \cdot \gamma_P(I,N) \cdot P - G(P) - \text{sinks} \qquad (4.2.7)$$

The first term on the right-hand side, $V_P(T) \cdot \gamma_P(I,N) \cdot P$, is equal to the net uptake and assimilation of inorganic nitrogen, and hence equivalent to NPP expressed in terms of nitrogen. The temperature-dependent maximum growth rate is $V_P(T)$ with units of divisions per unit time ($\mathrm{time^{-1}}$), and $\gamma_P(I,N)$ is a dimensionless function with a value from 0 to 1 that gives the combined influence of the light and nutrient supply on the growth rate. The functions $V_P(T) \cdot \gamma_P(I,N)$ in combination are the *specific growth rate* with units of $\mathrm{time^{-1}}$. Here we first examine the behavior of phytoplankton in the absence of grazing, i.e., $G(P) = 0$. We examine the nature of the grazing term $G(P)$ in the following subsection on the zooplankton. The "sinks" term includes losses by the slow escape of organic chemicals through the cell wall (*exudation*), and cell death due to, for example, breakdown of the cell wall (*lysis*), in which viruses probably play a role. It also includes losses due to sinking of phytoplankton, which would be represented by a term of the form $w_{sink} \cdot (\partial P/\partial z)$ (see section 5.4). However, we will ignore sinking of live organisms in this chapter. The sinks term is generally modeled as a first-order decay of the form "sinks" $= \lambda_P \cdot P$, where λ_P has units of $\mathrm{time^{-1}}$.

The effect of temperature on the maximum rate of photosynthesis has been studied by *Eppley* [1972], who gives the following empirical relationship based primarily on studies of diatoms:

$$\begin{aligned} V_P(T) &= a \cdot b^{cT} \\ a &= 0.6\,\mathrm{d^{-1}} \\ b &= 1.066 \\ c &= 1\,(^\circ\mathrm{C})^{-1} \end{aligned} \qquad (4.2.8)$$

The range from 0°C to 20°C is 0.6 to 2.2 $\mathrm{day^{-1}}$, with higher growth rates occurring in warmer waters. The number of doublings per day is equal to these numbers divided by ln(2), i.e., 0.9 to 3.1 doublings per day. The doubling rates are not only temperature sensitive, but also vary from species to species, so this equation should be considered as representative of an envelope response around a highly variable phenomenon. It must also be noted that the range of species studied, primarily diatoms, is quite limited. It is likely that the community sensitivity to temperature is smaller than the species sensitivity because of the adaptation of different organisms to a specific temperature range.

The light and nutrient sensitivity function $\gamma_P(I,N)$ determines how close to the maximum growth rate $V_P(T)$ the phytoplankton are able to grow. There are many situations in which light supply is clearly the major limiting factor. This can occur in high latitudes during the winter when the sun angle is extremely low; it occurs at depth, where light penetration is insufficient; and it can occur in deep mixed layers, where the average light supply encountered by phytoplankton as they are carried about by the turbulent motions is insufficient to maintain net growth. However, in the upper layers of most of the ocean, the light supply is

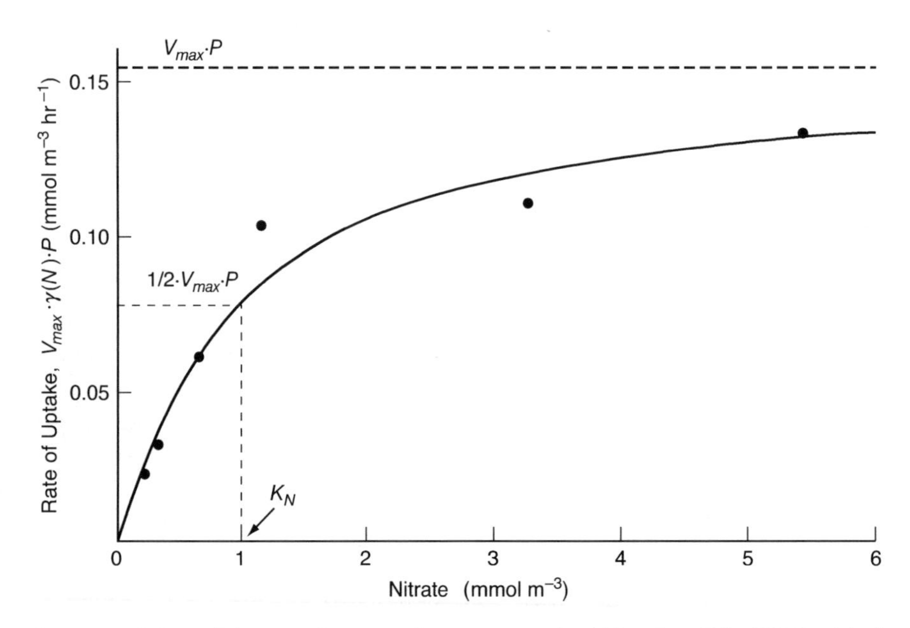

FIGURE 4.2.6: Rate of nitrate uptake versus nitrate concentration. Taken from *Dring* [1982]; original reference is *MacIsaac and Dugdale* [1969]. V_{max} is $V_P(T)$ and $\gamma_P(I) = 1$.

generally not a factor. In such situations the major concern is to identify the nutrients and/or micronutrients that are limiting the growth, i.e., the *limiting nutrients*.

The usual practice of modelers is to assume that $\gamma(I,N)$ is multiplicative, i.e., that it consists of the product of two independent functions of light and nutrient concentration:

$$\gamma_P(I,N) = \gamma_P(I) \cdot \gamma_P(N) \tag{4.2.9}$$

This is the approach we will use here. In what follows we discuss each of the terms on the right-hand side in turn. We focus on nitrate as the limiting nutrient, but more will be said later about the possibility of limitation by substances other than nitrate, as well as simultaneous limitation by several nutrients. An alternative formulation for $\gamma_P(I,N)$ is:

$$\gamma_P(I,N) = \min\{\gamma_P(I), \gamma_P(N)\} \tag{4.2.10}$$

Although we will not discuss this approach any further here, it has been used in a number of ocean ecosystem model studies such as those of *Kiefer and Atkinson* [1984] and *Hurtt and Armstrong* [1996].

There are both observational (figure 4.2.6) and theoretical grounds for assigning a hyperbolic functional form to $\gamma_P(N)$ [*Monod*, 1949]:

$$\gamma_P(N) = \frac{N}{K_N + N} \tag{4.2.11}$$

where K_N is the Monod constant equivalent to the concentration of N at which the growth rate is half its maximum value. Measurement-based estimates of this half-saturation constant for nitrate range from substantially less than 0.1 mmol m^{-3} up to values in excess of 3 mmol m^{-3} [e.g., *Eppley et al.*, 1969]. One must also take into consideration that laboratory experiments generally measure nitrate uptake, not growth. The uptake of nitrate and other substances by a cell occurs by transfer through the cell wall into the interior of the cell. We are interested in the growth of phytoplankton, i.e., the actual incorporation of the nutrients into organic matter. This involves a separate reaction from uptake and will not necessarily be equal to the instantaneous uptake rate. See *Morel* [1987] for a discussion of how the half-saturation constant for uptake and growth are related to each other. We use for the growth half-saturation constant K_N a value of 0.1 mmol m^{-3} (cf. values of 0.020 to 0.045 mmol m^{-3} given by *Lipschultz* [2001]), which corresponds to a half-saturation constant for uptake of $\sim$0.15 to 0.5 mmol m^{-3}.

What happens if more than one form of nitrogen is available? In the case of nitrogen, the two major forms utilized for photosynthesis are nitrate and ammonium. Ammonium is the form that requires the least amount of energy to utilize, and therefore it appears to be favored [e.g., *MacIsaac and Dugdale*, 1972; *MacIsaac*, 1978]. The main effect of ammonium thus appears to be to strongly suppress the uptake of nitrate. This has been represented in models as an additive function weighted toward ammonium:

$$\gamma_P(N) = \frac{NO_3^-}{K_{NO_3^-} + NO_3^-} \cdot \exp(-\Psi \cdot NH_4^+) + \frac{NH_4^+}{K_{NH_4^+} + NH_4^+} \tag{4.2.12}$$

[e.g., *Wroblewski*, 1977; *Fasham*, 1992], but can be represented in multiplicative and other functional forms as well [cf. *O'Neil et al.*, 1989]. In the model of *Fasham* [1992] both K's are assigned values of 0.5 mmol m^{-3} and

$\Psi = 1.5\ (\text{mmol m}^{-3})^{-1}$, though in what follows we will model only nitrate and use a $K_{NO_3^-}$ of 0.1 mmol m^{-3}. We will further discuss the role of ammonium in sections 4.3 and 4.4.

The case of limitation by multiple elements, as might occur for example if the supply of iron is inadequate, can also be dealt with by additive or multiplicative functions or by other functional forms [cf. *O'Neil et al.*, 1989]. This will be discussed further in section 4.3 in connection with the development of a multiple size class phytoplankton-zooplankton model, and later we give an example of a multiplicative function that has been found to work well for the combined effect of iron and silicic acid on diatom growth (see equation (7.2.1)).

We begin our discussion of $\gamma_P(I)$, the influence of light on photosynthesis, by discussing the processes that influence the irradiance I in the ocean. The light at the surface of the ocean is generally given as a function of the form

$$I_0 = f_{PAR} \cdot f(CL) \cdot f(\tau) \cdot I_n \tag{4.2.13}$$

I_n is the local clear sky irradiance at the surface of the ocean at noon either in energy units (W m^{-2}) or in terms of the number of quanta per unit time per unit area. Variations in I_n occur as a function of the latitude and time of the year as well as absorption of solar radiation in the atmosphere and reflection of light by the ocean surface. $f(\tau)$ is a dimensionless index with a range from 0 to 1 that defines how the irradiance curve varies with time of day at a given location. This can be represented as a triangular function, which has the advantage that it can be readily integrated to obtain a daily average. The impact of cloud cover CL on the light supply is represented by the dimensionless index $f(CL)$. The parameter f_{PAR}, which has a value of about 0.4 when I_n is in energy units, is the fraction of the total radiation that falls within the wavelength range that can be captured by chlorophyll and the other pigments used by phytoplankton.

The penetration of light into water is *attenuated* by selective absorption and scattering. The irradiance at a depth z can be represented by an equation of the form

$$\begin{aligned} I(z) &= I_0 \cdot \exp(-K \cdot z) \\ K &= k_w + k_P \cdot P + k_x \end{aligned} \tag{4.2.14}$$

The clear water attenuation coefficient k_w is of order 0.04 m^{-1}. The 1% light level typically chosen as the nominal base of the euphotic zone is at 115 m in clear water. The coefficient k_P in units of m^{-1} (mmol m^{-3})$^{-1}$ accounts for the effect of phytoplankton on the penetration of light, where P is the phytoplankton biomass in mmol m^{-3} of nitrogen. The above formulation assumes that k_P and P are uniform with depth. If they are not, a more appropriate representation for $k_P \cdot P$ would be $(\int_0^z (k_P \cdot P)dz)/z$. The coefficient k_x accounts for attenuation by other particulate and dissolved organic matter, which is generally ignored in models.

A wide range of values has been adopted for the phytoplankton attenuation coefficient. The model studies of *Fasham et al.* [1990] and *Sarmiento et al.* [1993], on which we base most of the parameter values given here, use $k_P \approx 0.03$ m^{-1} (mmol N m^{-3})$^{-1}$. The phytoplankton concentration obtained by the grazing-limited model discussed in the following section is 0.6 mmol N m^{-3}. Taking these numbers together and assuming that k_P and P are constant with depth gives a total attenuation coefficient of $K = k_w + k_P \cdot P = 0.04 + 0.03 \cdot 0.6 = 0.058$ m^{-1}, which gives a 1% light level at 79 m. However, this estimate is about half that obtained using the empirical equation proposed by *Morel* [1988]: $K = 0.121 \cdot CHL^{0.428}$, where CHL is the chlorophyll concentration in mg m^{-3}. The concentration of $P = 0.6$ mmol m^{-3} is equivalent to a chlorophyll concentration of ~1 mg m^{-3}, obtained using a carbon-to-chlorophyll ratio of 50 (see figure 4.1.1 caption), a C:N mole ratio of 117:16, and the atomic weight of carbon, 12 g mol^{-1}. Plugging this into the Morel equation gives $K = 0.12$ m^{-1}, which is consistent with the North Atlantic study of *Siegel et al.* [2002]. This gives a 1% light level depth of 38 m. In coastal waters where the concentration of chlorophyll, particles, and dissolved organic matter can be very high, the euphotic zone is often confined to less than this. Figure 4.2.7 schematically illustrates the penetration of light and maximum depth of various processes that depend on light.

The attenuation of light is wavelength-dependent. A number of studies have considered this in some detail [e.g., *Sathyendranath and Platt*, 1989; *Siegel and Dickey*, 1987a, 1987b], but the overall conclusion is that a simplified representation such as the one we go through here gives reasonable results for most purposes. Not surprisingly, the wavelength dependence of the pigments that are used by phytoplankton to capture light energy tends to match the wavelengths that penetrate most deeply.

We discuss now how the light is utilized by the phytoplankton. The uptake and utilization of light energy by phytoplankton is separated into two interrelated sets of processes, the light reactions and the dark reactions. The light reactions involve the initial absorbance of photons by pigments and their conversion to chemical energy in the form of reduced molecules like ATP and NADPH. During this process, water is consumed and oxygen is evolved. The dark reactions (called dark because they do not directly require photons; they do occur during the day) involve the utilization of ATP and NADPH to reduce CO_2 to organic carbon. All else being equal, the combination of the light and dark reactions gives an overall reaction with the stoichiometry $CO_2 + H_2O \rightarrow CH_2O + O_2$. However, some of the products of the light reactions are also used directly for

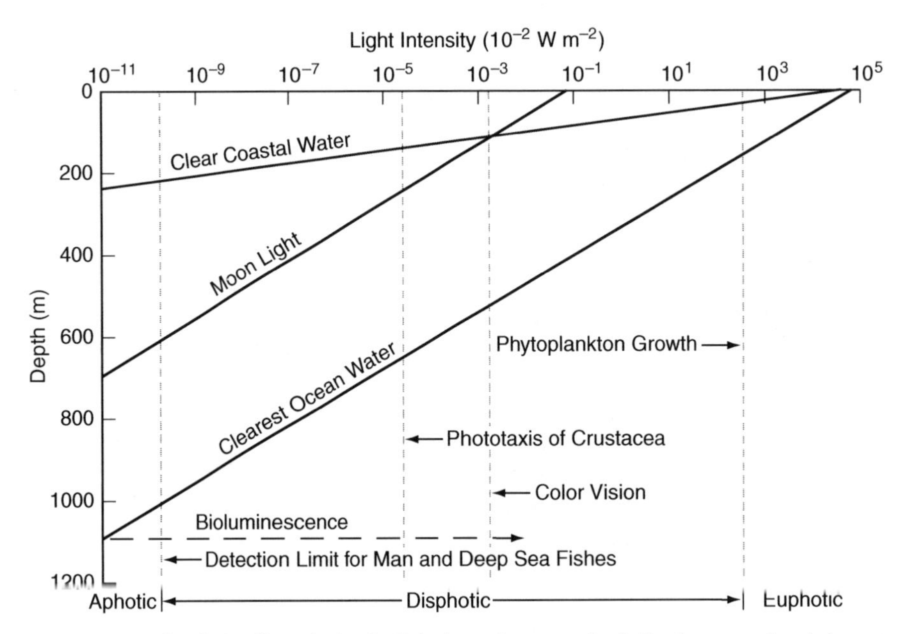

FIGURE 4.2.7: The sloping lines depict the light intensity versus depth for clear coastal and clearest open ocean waters, as well as for moonlight. Note that the horizontal scale is logarithmic. Also shown on the diagram are vertical lines depicting the light intensity cutoff for various biological processes. Taken from *Parsons et al.* [1984].

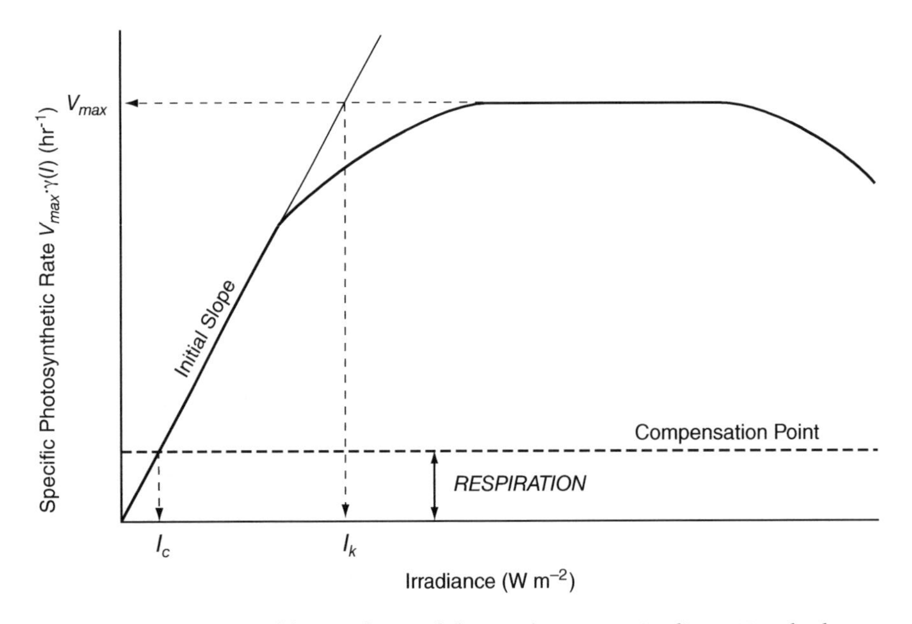

FIGURE 4.2.8: Schematic of the specific rate of photosynthesis versus irradiance. Note the drop-off at high irradiance due to photoinhibition. I_C is the compensation irradiance (see section 4.3).

assimilation of nitrate and other substances, and the proportion of the reductants used for other processes affects the photosynthetic quotient PQ defined as O_2 evolved/CO_2 incorporated. For example, production of nitrogen-poor compounds like carbohydrates results in low PQs, and production of nitrogen-rich compounds like amino acids and nucleic acids results in high PQs. This is because formation of organic nitrogen from nitrate releases oxygen. Traditional values of the photosynthetic quotient PQ are of order 1.1 to 1.8.

Figure 4.2.8 gives a schematic of the typical response of photosynthesis to irradiance when nothing else limits growth. This is known as a *P-I curve*. What this curve shows is the nutrient saturated specific growth rate $V_{max} \cdot \gamma_P(I)$, where

$$V_{max} = V_P(T) \tag{4.2.15}$$

and $\gamma_P(N) = 1$. Typical units for the specific growth rate are mg O_2 (mg chlorophyll a)$^{-1}$ hr^{-1}. Frequently photosynthesis is measured in carbon units, which recognizes that there are generally differences between the

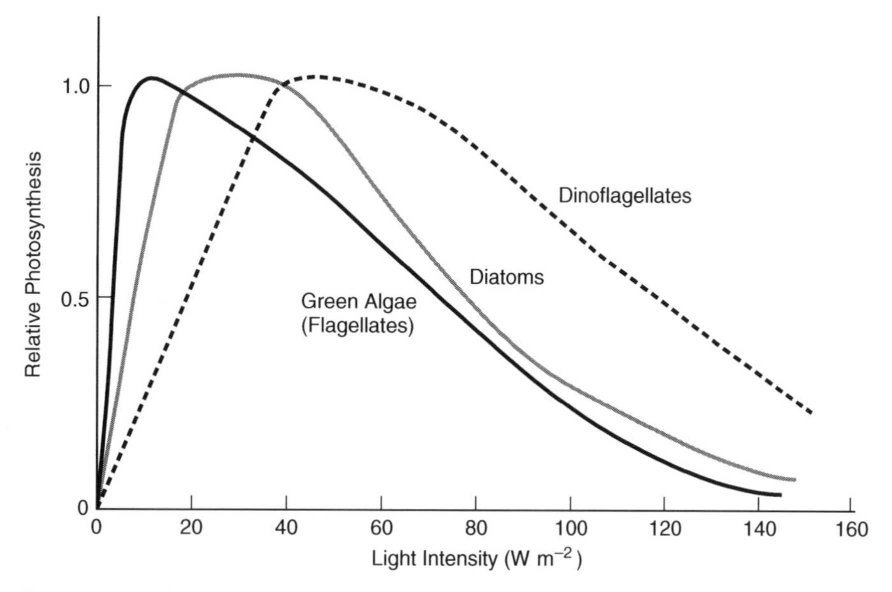

FIGURE 4.2.9: Schematic P-I curves for various major phytoplankton groups. Based on *Parsons et al.* [1984].

light and dark reaction rates that give a departure from the simple stoichiometry of $CO_2 + H_2O \rightarrow CH_2O + O_2$. For our models, we use N units. These are usually related to carbon uptake by the Redfield ratio, although there are differences between carbon and nitrogen uptake processes that occasionally become significant.

The maximum photosynthetic rate V_{max} of figure 4.2.8 is a function of the rate of transfer of reductants from the light reactions and the efficiency of their use in the dark reactions. These enzyme-dependent processes are the more temperature-sensitive part of the photosynthesis process. They are also related to the availability of nutrients both through the nutrient assimilation processes that affect PQ (see above) and through more complex interactions between carbon and nitrogen biochemistry in the cell [*Turpin*, 1991]. Photoinhibition generally sets in at higher irradiances.

The P-I curve in figure 4.2.8 has been represented by a number of analytical expressions. It looks somewhat like a Monod curve, which would give an expression for the $\gamma_P(I)$ term in $V_{max} \cdot \gamma_P(I)$ of $\gamma_P(I) = I/(I_k + I)$. However, this gives too steep a drop-off of the initial slope. A more accurate expression recommended by *Platt and Jassby* [1976] is:

$$\gamma_P(I) = \frac{I}{\sqrt{I_k^2 + I^2}} \qquad (4.2.16)$$

where the initial slope α of the P-I curve is:

$$\alpha = \frac{V_{max}}{I_k} \qquad (4.2.17)$$

A typical value for α is $0.025\,\mathrm{d^{-1}}\,(\mathrm{W\,m^{-2}})^{-1}$, but this and the magnitude of V_{max} have significant variations from one phytoplankton species to another that help explain how one species can dominate under certain circumstances. Figure 4.2.9 presents average relative P-I curves for a variety of phytoplankton. The figure shows green algae (flagellates) doing better at low light levels than diatoms, and dinoflagellates doing best at high light levels.

Zooplankton

The next ecosystem component is the zooplankton, which we discuss in panel 4.2.3, with an illustration of a few groups in panel 4.2.3, figure 1. Zooplankton grazing has three important effects on biomass in the upper ocean. First, herbivory has a direct influence on the amount of phytoplankton. Second, some proportion of the food ingested by zooplankton is excreted as particulate matter. Sinking of large detrital particles is one of the important mechanisms that removes organic matter from the surface waters. Third, an important product of zooplankton metabolism is ammonium. This form of inorganic nitrogen helps fuel additional photosynthesis over that which would occur if nitrate from the waters below the euphotic zone were the only nutrient supply.

In order to model the zooplankton, we must specify the grazing function $G(P)$ in equation (4.2.7). *Ivlev* [1945] determined from a study of a variety of fish that the amount of food eaten increased with the amount of food offered up to an upper limit. This is often represented by a Monod expression of the form

$$G(P) = g \cdot \frac{P}{K_P + P} \cdot Z \qquad (4.2.18)$$

with g (time^{-1}) being the maximum specific zooplankton grazing rate $G(P)/Z$, and K_P ($\mathrm{mmol\,m^{-3}}$) being the half-saturation constant for grazing of phytoplankton.

Usually a grazing threshold P_0 is included by replacing P with $P-P_0$, which helps prevent phytoplankton from going to 0 or even negative in a model. *Moore et al.* [2002c] solve this problem by using a quadratic dependence of the form $G(P)=g\cdot[P^2/(K_P^2+P^2)]\cdot Z$, which drops off more steeply as the food supply P supply drops below K_P. Another frequently encountered functional form for $G(P)$ is the Ivlev function:

$$G(P)=g\cdot[1-\exp(-k\cdot P)]\cdot Z \tag{4.2.19}$$

In what follows we will keep things simple by assuming we are in the linear portion of the feeding curve of the Monod function, i.e.,

$$G(P)=g\cdot\frac{P}{K_P}\cdot Z \tag{4.2.20}$$

Including death and other sink terms, the full $SMS(Z)$ expression might then be of the form

$$SMS(Z)=Z\cdot\left(\gamma_Z\cdot g\cdot\frac{P}{K_p}-\lambda_Z\right) \tag{4.2.21}$$

where γ_Z is an index with a value of 0 to 1 which measures the efficiency with which zooplankton assimilate organic matter into their biomass, and λ_Z is the zooplankton mortality term $\sim 0.12\,d^{-1}$ [cf. *Frost and Franzen*, 1992].

The specific growth rate, g, used in models is generally of order 1 d^{-1} [e.g., *Fasham et al.*, 1990] but can be larger for zooplanktonic organisms grazing on small phytoplanktonic organisms (e.g., 3 d^{-1}; [*Moore et al.*, 2002c]) and smaller for large zooplankton. The

Panel 4.2.3: Dominant Zooplanktonic Organisms

As with phytoplankton, zooplanktonic organisms come in a wide range of sizes (see figure 1). Generally zooplanktonic organisms consume prey that are smaller than their own size, and each step in a food chain will involve larger organisms than the previous step. Food chains that are dominated by large phytoplankton can result in significant populations of larger heterotrophs. Food chains that begin with small algae, as is common in most of the world's oceans, generally are dominated by very small heterotrophs. Higher levels of the food chain such as fish are generally thought not to be of importance in controlling the chemical composition of most of the surface ocean. The techniques used to study zooplankton have many shortcomings, as a consequence of which much still remains to be learned about their distribution. The traditional technique is to tow fine mesh nets through the water, but these damage many zooplankton and do not do a good job of sampling smaller sizes. Furthermore, many of the larger zooplankton have advanced visual capabilities and are good swimmers and thus are able to escape the nets.

The smaller zooplanktonic organisms are usually unicellular protozoa. These are always very abundant and they play an important role in most ecosystems. They range from zooflagellates (zooplankton with flagella) that can be almost picoplanktonic in size; through the ciliates (which are covered in short, dense hair-like structures) and heterotrophic dinoflagellates; to the large (up to a mm in diameter) shelled amoebas called acantharia, foraminifera, and radiolaria. Zooflagellates are the primary grazers on the picophytoplankton; they have rapid growth rates that can exceed the growth rates of their prey. The acantharia, foraminifera, and radiolaria have beautiful skeletons made of $SrSO_4$, $CaCO_3$, and opal, respectively. Many of these amoebas also contain symbiotic algae that allow them to be functionally autotrophic and heterotrophic at the same time.

Larger zooplanktonic organisms are often called net plankton in recognition of the dominant collection tool. About 70% of the zooplankton captured in net tows in coastal regions are Calanoida, which are Arthropoda (class Crustacea) belonging to the copepod subgroup. The calanoid copepods are found everywhere in the ocean and appear to play a major role as the larger zooplankton in many ecosystems. Other important members of the Crustacea are the euphausiids, one of which, *Euphausia superba*, is the krill of the Southern Ocean, a major food supply for large animals such as whales.

Two groups of the phylum Urochordata deserve mention because their unique feeding methods provide interesting links in food webs: larvaceans and salps. Larvaceans (or appendicularians) secrete spherical mucus "houses" that they use as filters to collect food. These houses have typical diameters of 0.5 to 4 cm, but one deep-dwelling species has a mucus house of up to 3 m in diameter. The filters can remove particulates as small as a few microns. After a period of use, the houses become clogged and are shed. This occurs as frequently as every few hours, contributing significantly to the formation of aggregates of particulate material known as *marine snow* that may play a major role in the flux of particulate organic matter from the upper ocean to the deep sea [*Alldredge and Silver*, 1988]. Salps are also filter feeders that indiscriminately capture particles down to a few microns in size. They are ubiquitous in the surface waters of the oceans and have a rapid form of asexual

temperature dependence of zooplankton growth rate can be extremely large. Observations on copepods have been fit with the function $g = 0.0445 \cdot \exp(0.111 \cdot T)\ \mathrm{d}^{-1}$, where T is in °C [*Huntley and Lopez*, 1992]. This equation gives a factor 16 range in g from $1/(22\,\mathrm{d}) = 0.0445\,\mathrm{d}^{-1}$ at 0°C to $1/(1.4\,\mathrm{d}) = 0.714\,\mathrm{d}^{-1}$ at 25°C. Such a large temperature dependence can have a major impact on the way the ecosystem functions [cf. *Laws et al.*, 2000]. In their model study, *Moore et al.* [2002c] used a more modest temperature dependence that gives a factor of 3.4 range over the same temperature span, whereas *Sarmiento et al.* [1993] assumed no temperature dependence at all.

Bacteria

Today we know that bacterial biomass is comparable to that of phytoplankton in the oligotrophic regions of the ocean, and that bacterial consumption of organic matter can be as much as a third or more of primary production and perhaps twice the zooplankton consumption [e.g., *Fuhrman*, 1992; *Azam*, 1998]. Bacteria are less dominant but still of considerable importance in eutrophic regions. Our interest in bacteria from the geochemical point of view stems primarily from their ability to compete with phytoplankton for uptake of dissolved inorganic nutrients such as phosphate, nitrate, and

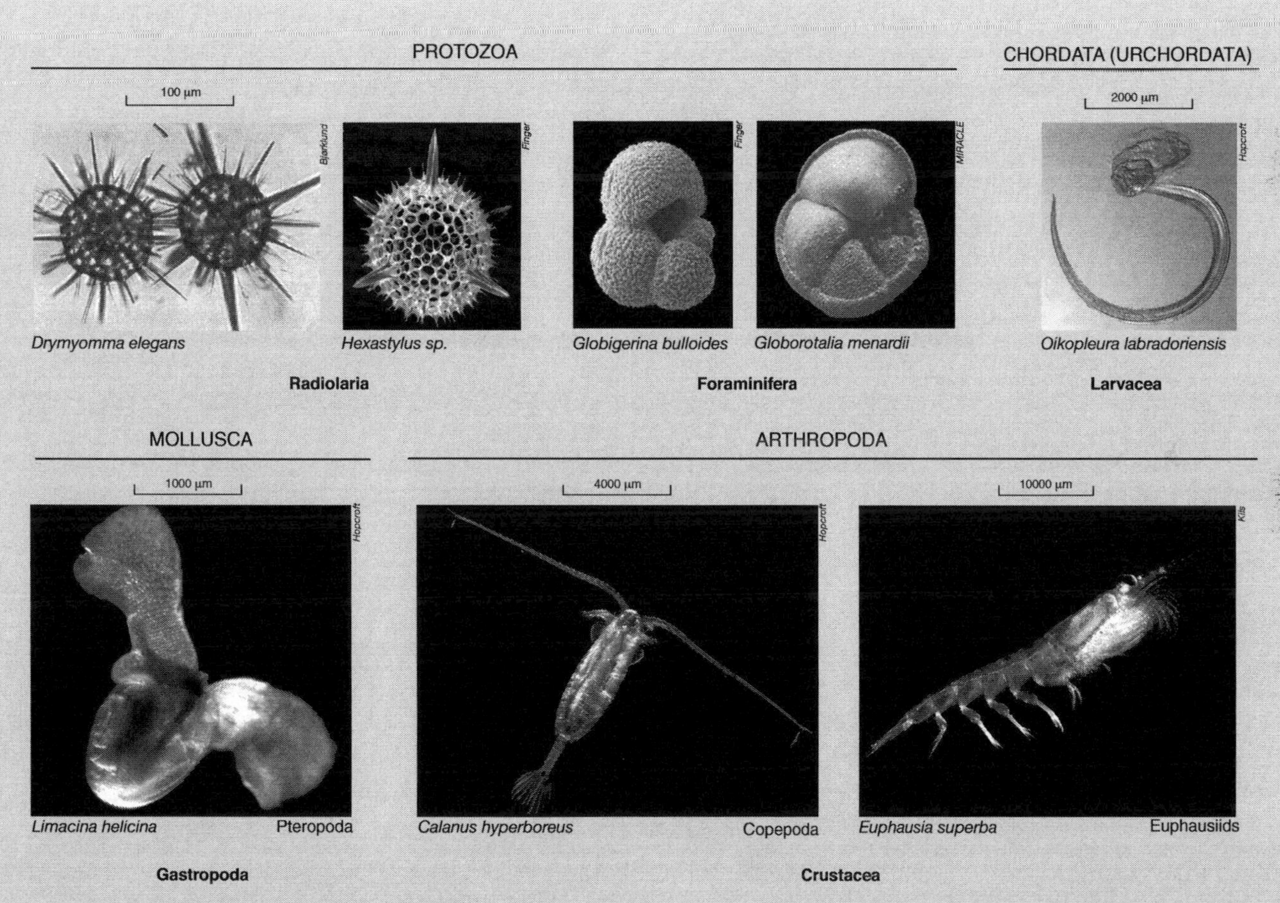

Figure 1: Images of representative zooplankton groups. Shown are two examples of radiolarians: *Drymyomma elegans* (image by Kjell R. Bjorklund) and *Hexastylus sp.* (image by Ken Finger); two examples of formanifera: *Globigerina bulloides* (image by Ken Finger) and *Globorotalia menardii* (image by Michael A. Kaminski and Paul R. Bown); one larvacea: *Oikopleura labradoriensis* (image by Russ Hopcroft) one pteropod: *Limacina helicina* (image by Russ Hopcroft); and two crustacea: *Calanus hyperboreus* (Russ Hopcroft); and *Euphausia superba* (Krill) (image by Uwe Kils).

reproduction that allows them to bloom in "salp swarms." When they become abundant they can rapidly deplete the ocean of nearly all small plankton. The animals themselves range from a few mm to many tens of cm in size. The large fecal wastes that they produce can have sinking speeds of thousands of meters per day.

ammonium [e.g., *Harrison et al.*, 1977; *Wheeler and Kirchman*, 1986; *Fuhrman et al.*, 1988; *Suttle et al.*, 1990; *Caron*, 1994], and the role they play in the recycling of organic matter via production and consumption of DOM, some of which is ultimately exported from the surface ocean and remineralized at depth.

The transport of organic matter across the cell walls of bacteria is limited to molecular masses of <500 Daltons (1 Dalton = 1/12 the mass of ^{12}C; cf. *Law* [1980], who studied only oligopeptides and proteins; the behavior of other compounds may be different), and thus the primary food source for heterotrophic bacteria is dissolved organic matter produced by other components of the ecosystem or by the solubilization of particulate organic matter by ectohydrolases produced by bacteria. The vast majority of bacteria are free-living, that is to say that they are suspended in the water rather than attached to particles and obtain their food by uptake of highly dilute DOM rather than from the concentrated organic matter found on particles [*Cho and Azam*, 1988].

The uptake of DOM and inorganic nitrogen by bacteria and their production of ammonium is generally modeled by including bacteria explicitly using a standard Monod equation for their uptake of DOM and nutrients and a first-order decay term for production of ammonium [e.g., *Fasham et al.*, 1990]; or more commonly, they are modeled implicitly by including decay terms from phytoplankton and other organic matter components of the ecosystem to account for uptake by bacteria, and conversion rates of organic matter to recycled nutrients (e.g., ammonium) to account for bacterial contributions to this process [e.g., *Hurtt and Armstrong*, 1996; *Steele*, 1998]. An example of an explicit formulation is that of *Fasham et al.* [1990], who give:

$$SMS(B) = V_B(T) \cdot \gamma_B(\mathrm{DON}, \mathrm{NH}_4^+) \cdot B - G(B) - \mathrm{sinks} \qquad (4.2.21)$$

Here $V_B(T)$ is the maximum bacterial growth rate, which *Fasham et al.* [1990] give as $2\,\mathrm{d}^{-1}$ and *Laws et al.* [2000] give as $1.2 \cdot \exp(0.0633 \cdot [T - 25])\,\mathrm{d}^{-1}$, with T in °C. The temperature-dependent function increases by almost a factor of 5 from 0°C to 25°C. The term $\gamma_B(\mathrm{DON},\mathrm{NH}_4^+)$ is the sum of two hyperbolic equations for the uptake of DON and ammonium. The relative uptake of DON and ammonium is fixed according to the C:N ratio of bacterial organic matter versus that of DOM. For a typical C:N ratio of 5 in bacteria [e.g., *Fenchel and Blackburn*, 1979] and 14 ± 3 in DON [*Bronk*, 2002], the DON:NH_4^+ uptake ratio would be 0.36. In effect, bacteria obtain all of their carbon and some of their nitrogen from DOM. The rest of the nitrogen they need in order to produce organic matter with a C:N ratio of 5:1 must come from ammonium. The $G(B)$ term is given by some version of (4.2.18) to (4.2.20), and the sinks term is simply a first-order decay $\lambda_B \cdot B$, where λ_B is given as $0.05\,\mathrm{d}^{-1}$ in the *Fasham et al.* [1990] study.

4.3 Analysis of Ecosystem Behavior

Now that we have some idea of how to model each ecosystem component, we can assemble a set of models that will permit us to explain exactly what we mean by the various hypotheses we listed at the end of section 4.1. We already discussed the physical processes that are responsible for the macronutrient supply in chapter 2 with a review in section 4.1. We thus begin here with a simple model of irradiance as a function of latitude and time of the year, cloud cover, and depth in the water column including the mixed layer thickness, and we use this to determine where and when the light supply may limit nutrient uptake.

Next, we discuss the efficiency of the biological pump by working our way through a hierarchy of increasingly complex ecosystem models that illustrate exactly what we mean by nutrient limitation, grazing limitation, and combined grazing and nutrient limitation, as well as the role of iron supply in all of this. Our discussion of ecosystem models is divided into two subsections, a first one on classical *N-P* and *N-P-Z* models as well as a more complex model including the microbial loop, and the second on multiple size class models. The first classic ecosystem model we consider includes just phytoplankton (P) and dissolved inorganic nitrogen (nitrate + ammonium) + DON, which we will treat as a single source of fixed nitrogen N. We do not explicitly differentiate between the various forms of dissolved nitrogen in the system, nor do we attempt to determine what fraction of the N comes from internal recycling versus external sources. We will see that the *N-P* system is able to consume all the N that is fed to it simply by expanding the phytoplankton population. The biological production is thus controlled entirely by the supply of N, which we refer to as *bottom-up limitation*. This system maintains surface N concentrations at extremely low values. It is thus extraordinarily effective as a biological pump.

The next classic ecosystem model we develop is for the *N-P-Z* system, which adds zooplankton to the *N-P* model. In this case we will find that the phytoplankton population is limited by zooplankton grazing, which we refer to as *top-down limitation*. Because the size of the phytoplankton population is now controlled by something other than nutrients, it is possible for the supply rate of nutrients to exceed the ability of the phytoplankton to consume it, and the N concentration in the water can climb to higher and higher values as the N

supply rate increases. This system is thus efficient as a biological pump if the supply rate is low enough, but becomes inefficient as the nutrient supply increases beyond the capacity of the fixed population of phytoplankton to take it up.

The third ecosystem model we develop illustrates the functioning of the microbial loop. It consists of an *N-P-Z-B* model, with *N* broken down into four explicit components: nitrate and a detrital pool of sinking particulate organic matter, and the two forms of fixed *N* that are consumed by bacteria, namely DON and ammonium.

The final model we develop allows multiple discrete size classes, with zooplanktonic organisms of each size class uniquely paired with phytoplanktonic organisms that are one size class lower, or with a range of smaller size classes of phytoplankton or zooplankton. This model illustrates how the ecosystem as a whole can be bottom-up limited at the same time that each phytoplankton-zooplankton pair, or all but the largest pair, is top-down limited. We will also discuss the role of iron supply in determining if larger phytoplanktonic organisms are present or not.

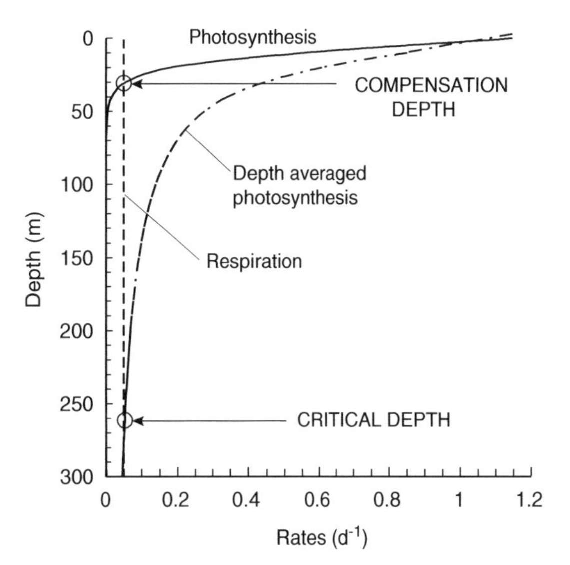

FIGURE 4.3.1: Schematic representation of the compensation and critical depth concepts as per *Sverdrup* [1953]. The curve labeled "photosynthesis" decreases exponentially with depth due to attenuation of the irradiance by adsorption and scattering. The respiration rate by plants and animals is assumed to be constant with depth. The compensation depth is the depth where photosynthesis and respiration are equal. In a stably stratified water column, this is the depth above which one would expect to find net photosynthesis occurring, and below which phytoplankton would not be able to survive. In a mixed layer that is deeper than the compensation depth, the phytoplankton would spend part of the time above the compensation depth and part of the time below this. If we assume that the rate of photosynthesis adjusts immediately as phytoplankton are mixed up and down in the mixed layer, we can calculate a critical mixed layer depth where the vertically integrated photosynthesis and respiration rates are equal to each other. The critical depth will always be much greater than the compensation depth. In the North Atlantic study of *Siegel et al.* [2002], the critical depth averaged about 7 times the compensation depth.

Role of Light Supply

Our primary goals in this subsection are three:

1. To discuss the minimum light irradiance at which phytoplankton photosynthesis is great enough to balance the community respiration. This is referred to as the *compensation irradiance* I_C (see figure 4.2.8). If the irradiance is below this, the phytoplankton and ecosystem community will die out. If it the irradiance is above this, the phytoplanktonic photosynthesis will exceed the community respiration, and the ecosystem will grow.

2. To estimate the *compensation depth* Z_C in the water column at which the irradiance is equal to the compensation irradiance. This is the depth above which the phytoplankton photosynthesis would exceed the community respiration if there were no mixed layer (see figure 4.3.1).

3. To estimate the thickness of the mixed layer over which the mean growth rate of the phytoplankton, as they sweep continuously in and out of the well-lit zone near the surface of the mixed layer, is large enough to balance the mean community respiration rate of the mixed layer (see figure 4.3.1). This thickness is known as the *critical depth* Z_{CR}. The critical depth always exceeds the compensation depth because the phytoplanktonic organisms that are at the compensation depth at any given time have spent at least part of their life above the compensation depth, where the light irradiance is great enough to give a net positive growth.

The commonly accepted definition for the base of the euphotic zone is the 1% light level [e.g., *Morel*, 1988], which is also used to calculate the thickness of the euphotic zone. In what follows, we will compare the 1% light level criterion with various estimates of the critical depth and compensation depth.

A necessary condition to have any net growth of the ecosystem community is that the irradiance at the surface of the ocean exceed the compensation irradiance. However, this is not a sufficient condition because if there is a mixed layer, as is usually the case, no net growth will occur unless the irradiance at the surface of the ocean is great enough to maintain a positive net photosynthesis when averaged over the entire mixed layer. Thus, the condition for community growth to occur is that the mixed layer be less than the critical depth. Determining exactly when this condition is met is central to understanding how light influences the uptake of nitrate by phytoplankton.

We begin with the phytoplankton conservation equation obtained by combining (4.2.2b), (4.2.7), and (4.2.9), i.e.,

$$\Gamma(P) = V_P(T) \cdot \gamma_P(I) \cdot \gamma_P(N) \cdot P - G(P) - \text{sinks} \qquad (4.3.1)$$

For purposes of this analysis, we make the following simplifications (see problems 4.14 to 4.16 for a more

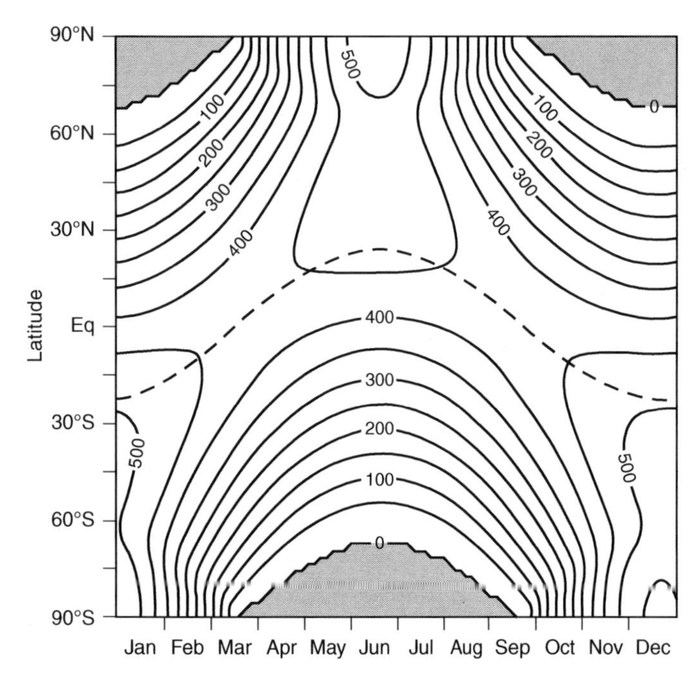

FIGURE 4.3.2: Mean daily irradiance at the top of the atmosphere in Wm^{-2} as a function of latitude and time based on *Hartmann* [1994].

complete analysis). We assume that advection and mixing of phytoplankton are negligible, which gives $\Gamma(P) = \partial P/\partial t$. Following 4.2.15 we define the maximum phytoplankton specific growth rate as $V_{max} = V_P(T)$. We use the *Platt and Jassby* [1976] function for $\gamma_P(I)$ (equation (4.2.16)), and assume that the nutrient supply is not limiting growth, i.e., $\gamma_P(N) = 1$. The final assumption we make is that the loss terms of the system, $G(P) + \text{sinks}$, are first order in P, which we represent by $\lambda_P \cdot P$. This gives

$$\frac{1}{P} \cdot \frac{\partial P}{\partial t} = V_{max} \cdot \frac{I}{\sqrt{I_k^2 + I^2}} - \lambda_P \tag{4.3.2}$$

The compensation irradiance, I_C, is defined as the irradiance where photosynthesis and the loss term are in balance, i.e., $\partial P/\partial t = 0$, which allows us to obtain the following solution of (4.3.2):

$$\begin{aligned} I_C &= I_k \cdot \sqrt{\frac{(\lambda_P/V_{max})^2}{1-(\lambda_P/V_{max})^2}} \\ &\approx I_k \cdot \frac{\lambda_P}{V_{max}} \end{aligned} \tag{4.3.3}$$

A typical value for λ_P would be $\sim 0.05\,\text{d}^{-1}$ and for V_{max}, $\sim 1.4\,\text{d}^{-1}$, such that $\lambda_P/V_{max} = 0.036$. This is $\ll 1$, which is what enables us to simplify the square root term, as shown. We substitute these values into (4.3.3), and use for I_k a value of $56\,\text{W m}^{-2}$ determined from equation (4.2.17) with a typical value of $0.025\,\text{d}^{-1}\,(\text{W m}^{-2})^{-1}$ for α. This gives $I_C = 2\,\text{W m}^{-2}$. How does our estimate of $I_C = 2\,\text{W m}^{-2}$ compare with others? *Siegel et al.* [2002] cite studies giving an order of magnitude range from 1 to $10\,\text{W m}^{-2}$. Their own quite interesting analysis, about which we shall say more, gives an estimate of $3.6\,\text{W m}^{-2}$.

We can now use I_C to calculate the compensation depth Z_C from the irradiance as a function of depth equation (4.2.14), i.e., $I_C = I_0 \cdot \exp(-K \cdot Z_C)$. Here K is the total extinction coefficient, for which we use the value of $0.12\,\text{m}^{-1}$ given in section 4.2. We estimate a surface irradiance of $I_0 = 80\,\text{W m}^{-2}$ from (4.2.13) using a noontime irradiance at the top of the atmosphere of $I_n = 1000\,\text{W m}^{-2}$, $f_{PAR} = 0.4$, $f(CL) = 0.8$, and $f(\tau) = 0.25$. The value for $f(\tau)$ is obtained by assuming that the sun shines 12 hours a day, with the irradiance increasing linearly from 6 AM to noon and decreasing linearly thereafter [cf. *Evans and Parslow,* 1985]. Our estimate for I_0 is comparable to the surface irradiance of $86 \pm 23\,\text{W m}^{-2}$ obtained by averaging the estimate of *Siegel et al.* [2002] for the irradiance at the time of the initiation of the spring bloom between 35°N and 75°N in the Atlantic (31 ± 8 mol photon $\text{m}^{-2}\text{d}^{-1}$ × conversion factor of 2.79 ± 0.14 W d (mol photon)$^{-1}$ for "blue-green" water obtained from *Morel and Smith* [1974]). The mean daily irradiance at the top of the atmosphere that we use is $I_n \cdot f(\tau) = 250\,\text{W m}^{-2}$, which occurs at a latitude of $\sim$30°N and S in the wintertime hemisphere (see figure 4.3.2). Plugging these values into (4.2.14) and solving for the compensation depth gives:

$$\begin{aligned} Z_C &= -\frac{\ln(I_C/I_0)}{K} \\ &= -\frac{\ln(2\,\text{W m}^{-2}/80\,\text{W m}^{-2})}{0.12\,\text{m}^{-1}} \\ &= 31\,\text{m} \end{aligned} \tag{4.3.4}$$

In section 4.2, we calculated that the 1% light level occurs at 38 m, as contrasted with the depth of 31 m given here. The irradiance at 31 m is 2.5% of the surface irradiance. Of course, our estimate will vary depending on what parameter values are specified in the above equation.

We turn now to the problem of estimating the critical depth. The notion underlying the *critical depth hypothesis* is that there is a maximum thickness of the mixed layer above which the average light supply experienced by phytoplanktonic organisms as they are swept in and out of the well-lit regions of the mixed layer is insufficient for them to maintain photosynthesis in excess of respiration. The critical depth hypothesis was first proposed by *Sverdrup* [1953], based on a conjecture by *Gran and Braarud* [1935], as an explanation for how the shallowing of the mixed layer in springtime determines the start of the spring bloom. It is one of the fundamental paradigms of biological oceanography. The critical depth is the thickness of a mixed layer Z_{CR} over

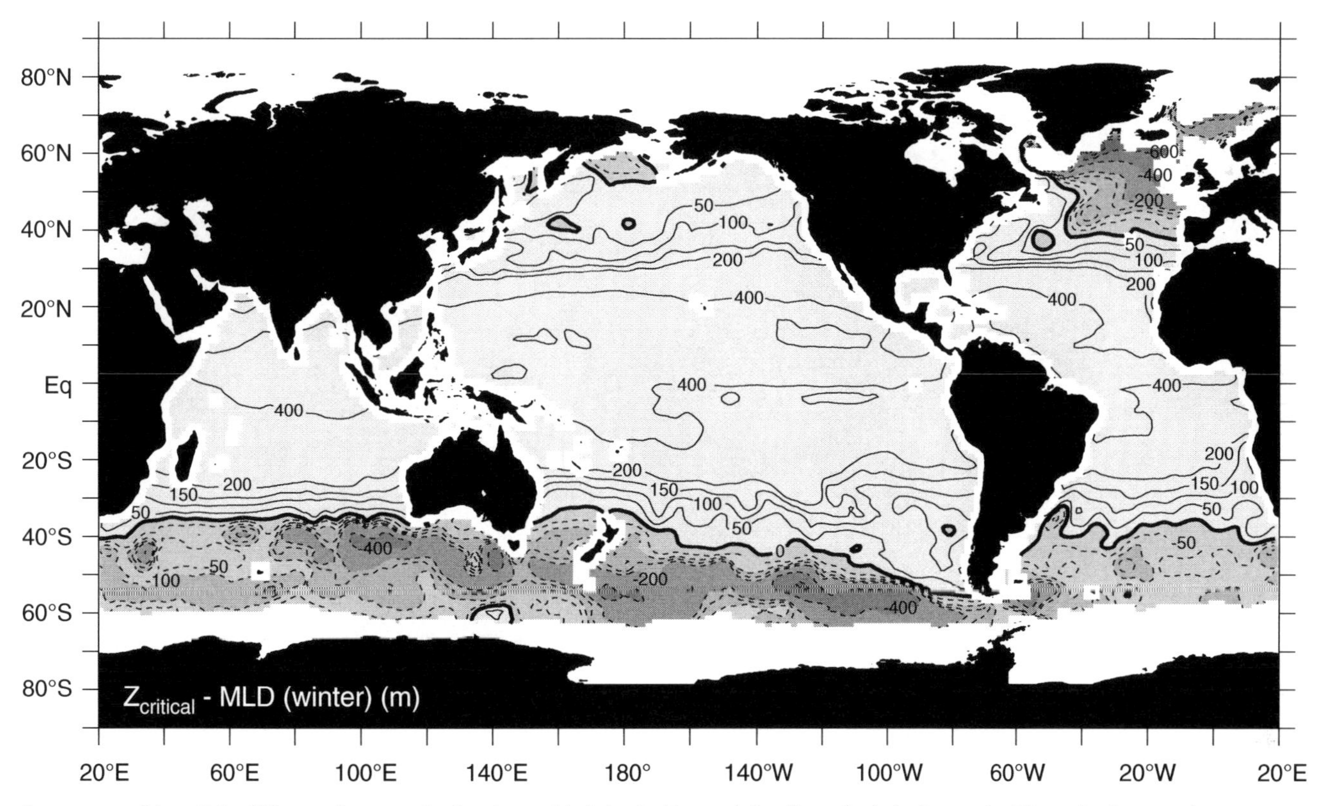

FIGURE 4.3.3: Map of the difference between the Sverdrup critical depth, Z_{CR}, and the climatological winter mixed layer depth, assuming a critical light intensity of 3.6 Wm^{-2} [*Siegel et al.*, 2002]. Positive values indicate that Z_{CR} is deeper than the winter mixed layer. Data provided by Colm Sweeney (personal communication, 2005).

which the vertically averaged specific growth rate balances the loss rate, i.e.,

$$V_{\max} \cdot \left[\frac{1}{Z_{CR}} \cdot \int_0^{Z_{CR}} \left(\frac{I}{\sqrt{I_k^2 + I^2}}\right) dz\right] = \lambda_P, \tag{4.3.5}$$

Here $I = I_0 \cdot \exp(-K \cdot z)$. We perform the averaging in a spreadsheet (see problem 4.16a) and obtain the vertically averaged specific growth rate shown in figure 4.3.1. This gives $Z_{CR} = 269$ m, much greater than the 1% light level depth of 38 m. However, our estimate of the critical depth is likely too high. The recent study by *Siegel et al.* [2002] gives an average critical depth of 170 ± 30 m for the North Atlantic from 35°N to 75°N based on the climatic mean thickness of the mixed layer on the date that the spring bloom was seen to begin in the SeaWiFS satellite ocean color imagery. Among the many possible problems with our calculation is that our phytoplankton "decay" rate may underestimate the magnitude of the community system respiration. A larger value for λ_P would reduce the critical depth. The change does not have to be large, given how steep the average phytoplankton growth rate curve is at that depth.

Figure 4.3.3 pulls together the information we have given above to generate an estimate of the critical depth minus the climatological mixed layer thickness for the hemispheric winter months. The caption explains how the critical depth is calculated. Positive numbers indicate that the mixed layer thickness is less than the critical depth so that light supply is adequate for photosynthesis to exceed community respiration; negative numbers indicate the mixed layer thickness is deeper than the critical depth so that these regions are light limited. The figure shows that there are large areas of both the North Atlantic and Southern Oceans that suffer from light limitation during the hemispheric winter when mixed layers are deep (cf. figure 4.1.5). The North Pacific suffers from light limitation in only a few isolated regions. The regions of wintertime light limitation extend about 25° to 35° closer to the equator than the polar night latitude that can be seen in figure 4.3.2. Clearly, the mixed layer thickness has a major influence in broadening the region of light limitation so that it covers a far greater area than can be accounted for by the polar night.

With the parameter values used in generating the plot in figure 4.3.3, the summer hemisphere never suffers from light limitation, including the Southern Ocean, contrary to what was found for that region by *Mitchell et al.* [1991]. However, the criterion for light limitation used by *Mitchell et al.* [1991], which is

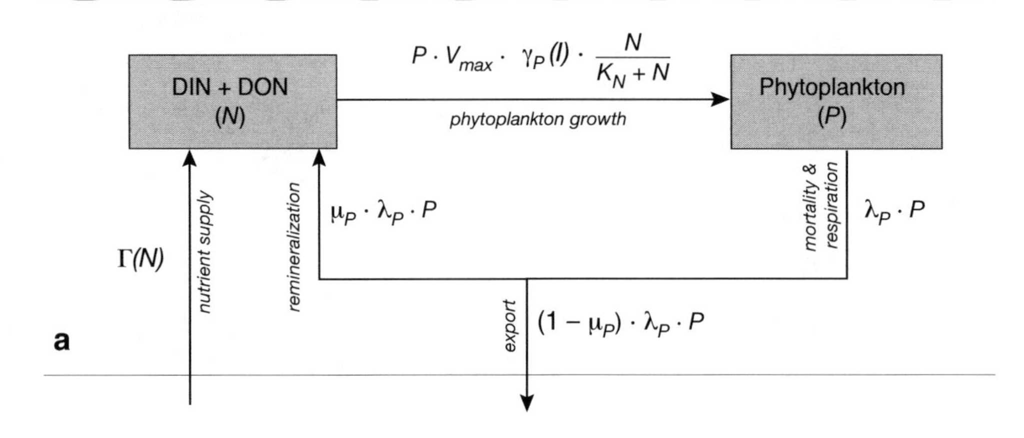

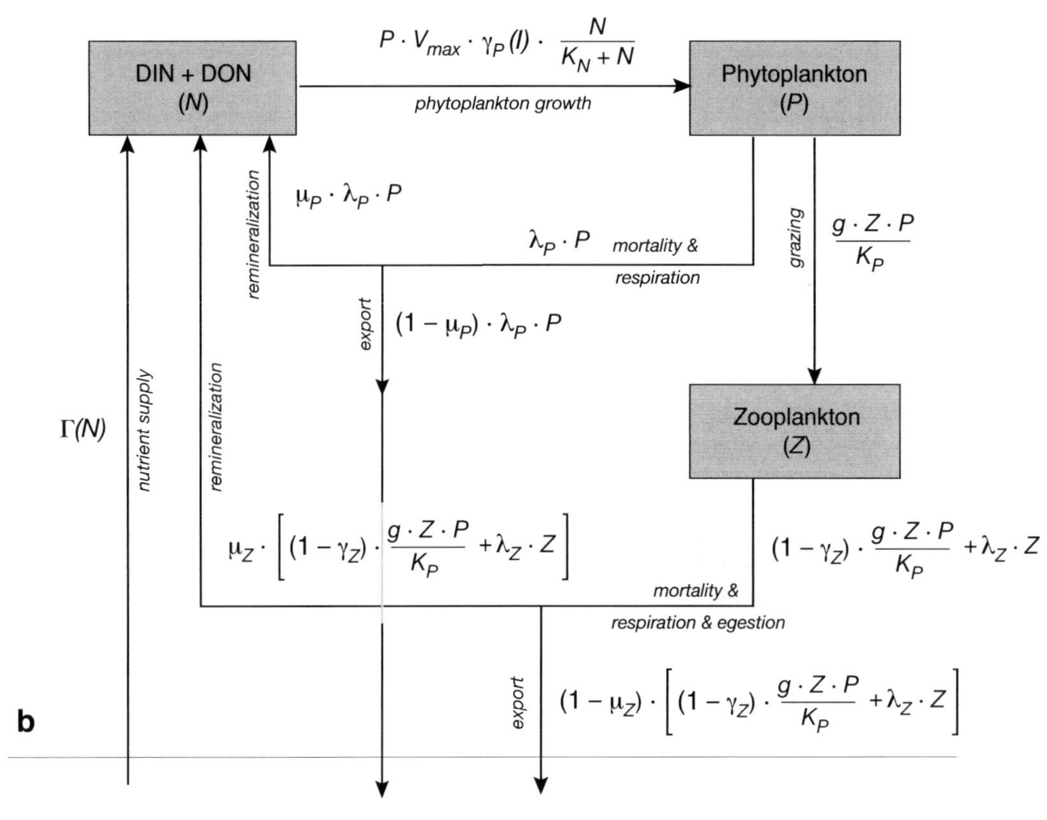

FIGURE 4.3.4: A schematic illustration of (a) a nitrate-phytoplankton model and (b) a nitrate-phytoplankton-zooplankton model. These models, discussed in the text, are used to determine how these particular ecosystems constrain the surface nitrate concentration.

"the potential for massive, nutrient-exhausting, phytoplankton blooms," is much stricter than the one we use here based on the threshold light level required for the spring bloom to commence in the North Atlantic observations ([*Siegel et al.*, 2002]; see figure 4.3.3 caption).

CLASSICAL ECOSYSTEM MODELS

N-P MODEL—BOTTOM-UP LIMITATION

The *N-P* model is illustrated in figure 4.3.4a. Its two components require two conservation equations. The conservation equation for phytoplankton is obtained by combining (4.2.2b), (4.2.7), and (4.2.9). In a steady state, and assuming that the contribution of ocean advection and mixing as well as sinking of organic matter is negligible, we have that $\Gamma(P) = 0$. We define a maximum phytoplankton specific growth rate as $V_{max} = V_P(T)$, we assume that the light supply is not limiting growth, i.e., $\gamma_P(I) = 1$ (see problems 4.14 to 4.16 for an analysis of a light-limited system), and we use the Monod growth function (4.2.11) for $\gamma_P(N)$. Since there are no zooplankton, the grazing term $G(P)$ is 0. The sinks such as phytoplankton exudation, lysis, and mortality (the "sinks" term of (4.2.7)) are assumed to be first order in P, i.e., sinks $= \lambda_P \cdot P$. This gives

$$\Gamma(P) = 0 = P \cdot \left(V_{max} \cdot \frac{N}{K_N + N} - \lambda_P \right) \qquad (4.3.6)$$

A steady state solution also requires that the supply of N by ocean advection and mixing plus decay of phytoplankton must balance the photosynthetic uptake, i.e.,

$$\Gamma(N) = P \cdot \left(-V_{max} \cdot \frac{N}{K_N + N} + \mu_P \cdot \lambda_P \right) \qquad (4.3.7)$$

where Γ is the sum of the time rate of change (assumed to be 0) and the negative of the transport terms (see (4.2.2a)), and μ_P is the fraction of the phytoplankton sink term that is remineralized back to dissolved inorganic nitrogen, DIN, within the system (see figure 4.3.4a). The fraction $(1 - \mu_P)$ is therefore the organic matter that is exported out of the upper ocean, which must be equal to the N supply term $\Gamma(N)$. We also find it convenient to define a third variable, the total amount of nitrogen in the two components of the ecosystem:

$$N_T = N + P \qquad (4.3.8)$$

The analytical solution to this set of equations defines two regimes:

1. In the first regime, which we will refer to as the N regime, the N concentration is so low that the phytoplankton growth term in (4.3.6) is smaller than the decay term, i.e., $V_{max} \cdot N/(K_N + N) < \lambda_P$, and any phytoplankton that are present will decay away faster than they can be replaced. In this regime, $P = 0$ and $N = N_T$.

2. In the second regime, which we shall refer to as the N-P regime, N climbs to the concentration obtained by solving (4.3.6) and stays there:

$$N = \frac{\lambda_P \cdot K_N}{V_{max} - \lambda_P} \qquad (4.3.9)$$

By analyzing (4.3.6), we can see that the solution (4.3.9) is the N concentration required for photosynthesis to balance mortality. If N is lower than this, the mortality rate will be greater than the photosynthesis rate and any phytoplankton present initially will die out, i.e., $P = 0$. If N is higher than this, the rate of photosynthesis will exceed mortality, and phytoplankton will increase until a new steady state is achieved with photosynthesis and mortality in balance. Note that this analysis ignores export by advective and diffusive transport of phytoplankton as well as sinking. If these terms contribute to the system, a steady-state solution will require that N be higher than (4.3.9) so that photosynthesis can exceed mortality by an amount equal to the export.

We can estimate the concentration of N required for P to survive by substituting in appropriate values for the parameters in (4.3.9). We use a value for K_N of 0.1 mmol m^{-3}, and for V_{max} a representative value of 1.4 d^{-1}. If we take a value for λ_P of order 0.05 d^{-1}, we arrive at

$$N = \frac{(0.05\,\mathrm{d}^{-1}) \cdot 0.1\,\mathrm{mmol\,m}^{-3}}{1.4\,\mathrm{d}^{-1} - 0.05\,\mathrm{d}^{-1}} = 0.0037\,\mathrm{mmol\,m}^{-3} \qquad (4.3.10)$$

In other words, the threshold nitrate concentration required for P to survive is near zero. Since our N-P system never permits the nitrate concentration to exceed the threshold value, the entire world ocean would have near 0 nutrient concentrations if this were the dominant ecosystem.

Another way of interpreting the solution (4.3.9) is that the nitrate concentration can never exceed the value specified by (4.3.10) because any excess will be converted into phytoplankton biomass. This can easily be understood by combining the solution (4.3.10) with (4.3.8) and solving for P, which gives $P = N_T - 0.004$ mmol m^{-3}. As N_T increases gradually from 0, initially all of it will go into nitrate ($P = 0$) until nitrate has crossed the threshold value defined by (4.3.9) for supporting P. Any nitrogen added to the system beyond this threshold value will then go into P. Figure 4.3.5 illustrates these relationships. This is considered to be a bottom-up limited system in the sense that the biological production can increase without limit to take up any additional nutrient that is added to it.

This simple phytoplankton-nitrate system can explain nutrient depletion to a concentration of $\sim$0.004 mmol m^{-3}, but not the existence of high-nutrient regions. Other factors must play a role in preventing the phytoplankton from consuming all the nitrate as in this model. The light supply can be of importance, but only over certain periods of the year when the sun angle is low or the mixed layer is very deep. We need more than this to explain the observed nitrate distribution, which only rarely if ever gets as low as 0.004 mmol m^{-3}. In the following subsection we add zooplankton and consider how this affects the ecosystem.

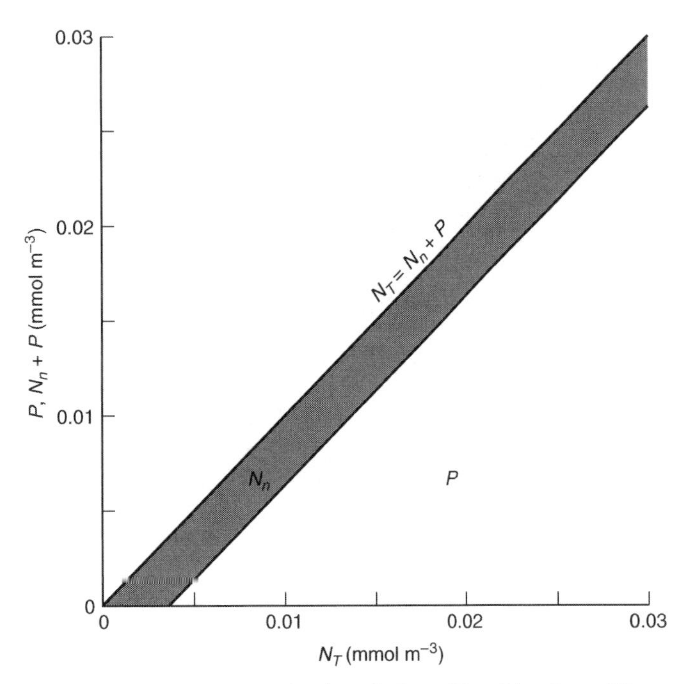

FIGURE 4.3.5: Nitrogen in phytoplankton (P) and in nitrate (N) as a function of total nitrogen in a two-component model. The vertical axis is concentration in each of the components plotted as the cumulative amount.

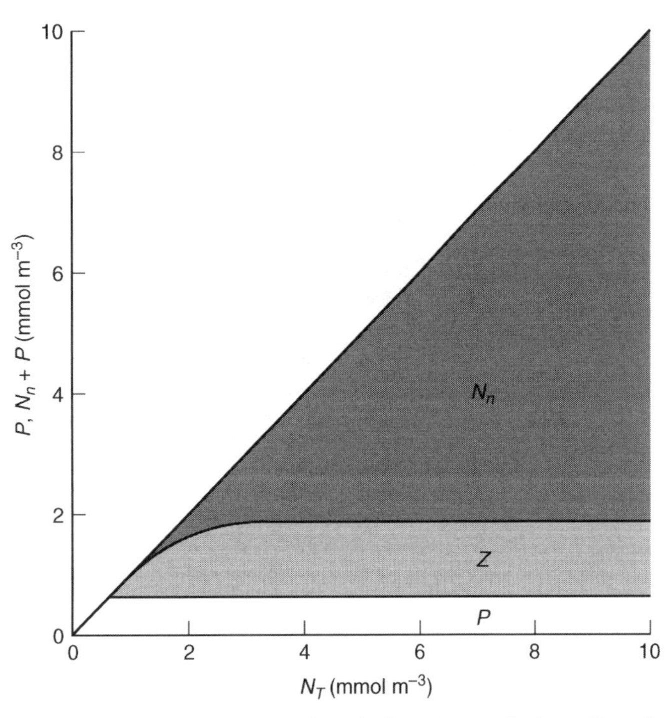

FIGURE 4.3.6: Nitrogen in phytoplankton (P), zooplankton (Z), and nitrate (N) as a function of total nitrogen. These results are from the three-component model of *Armstrong* [1994]. The vertical axis is concentration in each of the components plotted as the cumulative amount.

N-P-Z MODEL—TOP-DOWN LIMITATION

Figure 4.3.4b illustrates the N-P-Z model for which we will now obtain a solution. We add zooplankton grazing (4.2.20) to the phytoplankton equation (4.3.6). With $\gamma_P(I) = 1$, this gives for phytoplankton:

$$\Gamma(P) = 0 = P \cdot \left(V_{max} \cdot \frac{N}{K_N + N} - \lambda_P - \frac{g \cdot Z}{K_P} \right) \tag{4.3.11}$$

For zooplankton, we have growth due to grazing and also mortality $\lambda_Z \cdot Z$. The steady-state zooplankton equation, ignoring advection, diffusion, and swimming, is given by (4.2.21), reproduced here as

$$\Gamma(Z) = 0 = Z \cdot \left(\gamma_Z \cdot g \cdot \frac{P}{K_P} - \lambda_Z \right) \tag{4.3.12}$$

As previously noted, γ_Z is assimilation efficiency. The organic matter that is not assimilated gets excreted as fecal pellets and other material or metabolized as an energy source. The nitrate conservation equation for the N-P-Z ecosystem, equivalent to (4.3.7), is then

$$\Gamma(N) = P \cdot \left(-V_{max} \cdot \frac{N}{K_N + N} + \mu_P \cdot \lambda_P \right) + Z \cdot \mu_Z \cdot \left[(1 - \gamma_Z) \cdot g \cdot \frac{P}{K_P} + \lambda_Z \right] \tag{4.3.13}$$

where μ_Z is the fraction of zooplankton losses and unassimilated organic nitrogen that is remineralized back to dissolved inorganic nitrogen within the euphotic zone. Hence, the fraction $(1 - \mu_Z)$ represents the organic nitrogen exported out of the euphotic zone. Finally, by analogy with (4.3.8), we define the total nitrogen content of the system as the sum of dissolved inorganic nitrogen and the organic nitrogen in phytoplankton and zooplankton:

$$N_T = N + P + Z \tag{4.3.14}$$

This set of equations can be solved analytically for all three components of the ecosystem [e.g., *Wroblewski et al.*, 1988]. The parameters we use for phytoplankton are the same as those used in (4.3.10). The zooplankton parameters we employ are based on the study of *Frost and Franzen* [1992]. The maximum zooplankton growth rate g has a value of $1.4\,\text{d}^{-1}$, the half-saturation constant for zooplankton grazing of phytoplankton K_P is $2.8\,\text{mmol m}^{-3}$, the zooplankton mortality coefficient λ_Z is $0.12\,\text{d}^{-1}$, and the assimilation efficiency γ_Z is 0.4. Figure 4.3.6 shows a solution to the equations obtained by *Armstrong* [1994]. Three regimes are defined by these equations, the first two of which are identical to the two regimes in the N-P model:

1. In the first or N regime, N is too low to sustain P, and as a consequence both P and Z are 0. From the P equation

(4.3.11) with Z set to 0, we see that this regime applies when

$$V_{max} \cdot \frac{N}{K_N + N} < \lambda_P$$
$$\Rightarrow N < \frac{\lambda_P \cdot K_N}{V_{max} - \lambda_P} \quad (4.3.15)$$

Here we have that $N = N_T$. The threshold value of N defined by (4.3.15) is the concentration of 0.0037 mmol m^{-3} obtained in (4.3.10). This solution is the same as that obtained in the previous subsection and occupies the lower-left-hand region of the plot shown in figure 4.3.5.

2. The second N-P regime is also the same as that obtained in the previous subsection. The nutrient concentration N is sufficient to sustain P, but P is insufficient to support Z. From the Z equation (4.3.12) we see that the condition for this is:

$$\gamma_Z \cdot g \cdot \frac{P}{K_P} < \lambda_Z$$
$$\Rightarrow P < \frac{\lambda_Z \cdot K_P}{\gamma_Z \cdot g} \quad (4.3.16)$$

The solution in this case is exactly the same as (4.3.8) and (4.3.10) and is illustrated in the higher concentration region of figure 4.3.5 as well as in figure 4.3.6.

3. The third N-P-Z regime begins when P becomes large enough to sustain Z. From the Z equation (4.3.12) we see that this occurs when

$$P = \frac{\lambda_Z \cdot K_P}{\gamma_Z \cdot g} = \frac{0.12\,d^{-1} \cdot 2.8\,\text{mmol m}^{-3}}{0.4 \cdot 1.4\,d^{-1}}$$
$$= 0.60\,\text{mmol m}^{-3} \quad (4.3.17)$$

of nitrogen in phytoplankton. In fact, we see from an examination of the Z equation (4.3.12) that once P reaches the threshold value where Z comes into existence, P can never rise above this except temporarily, or if advective and diffusive transport or sinking removes zooplankton from the surface at a rate that is significant with respect to the rate of grazing and mortality. If P gets larger, the zooplankton uptake rate will rise above its mortality, and Z will expand. This will reduce P. The opposite happens if P goes too low. The only steady-state solution for P in this system is the one given by (4.3.17). The behavior in nature is almost certainly more complex than this, but it is nevertheless clear that herbivory exercises strong control over the population of the phytoplankton.

Note that the P concentration is equivalent to a chlorophyll-*a* concentration of ~ 1.0 mg m^{-3}, which is comparable to values observed in the high latitudes of the North Atlantic and North Pacific (see figure 4.1.1 and figure 4.1.8), but an order of magnitude higher than oligotrophic stations such as BATS and HOT (see figure 4.1.1, figure 4.1.8, and table 4.1.1). This is a consequence of the parameter values chosen. For example, if we use the parameter values of *Sarmiento et al.* [1993], i.e., $\lambda_Z = 0.05\,d^{-1}$, $K_P = 1$ mmol m^{-3}, $\gamma_Z = 0.75$, and $g = 1\,d^{-1}$, we obtain a threshold P concentration of 0.067 mmol m^{-3}, which translates into a chlorophyll concentration of 0.11 mg m^{-3}.

Now we see in figure 4.3.6 that as N_T increases beyond the minimum amount required to achieve (4.3.17), a small portion of it begins to go into an increase of N. This increases the phytoplankton uptake rate through the Monod function even though P remains the same. The increased phytoplankton growth rate feeds directly into Z, and the zooplankton population climbs. However, eventually the N concentration in this idealized system gets high enough that the Monod function for nutrient uptake asymptotes toward its maximum value of 1. At this point the nutrient uptake by phytoplankton and throughput as organic matter to zooplankton can no longer increase, and all new N_T begins to accumulate as N. This occurs at an N_T concentration of about 2 mmol m^{-3}. The dividing line is quite sharp. Below this we can think of the nitrate concentration as being dominated by biological uptake; above this the biological system is nitrate replete and most new nitrogen goes into nitrate (see problem 4.17 and discussion in *Sarmiento and Bender* [1994]). The biological pump in this case becomes inefficient, and the ecosystem is considered to be top-down limited in the sense that phytoplankton grazing by zooplankton keeps their population in check, thereby placing an upper limit on how much nutrient the ecosystem is capable of converting into organic matter.

In conclusion, we find that the addition of zooplankton makes it possible for the nutrient concentration to climb above the low levels predicted by the N-P model. This occurs because the zooplankton community does not permit the concentration of phytoplankton to climb above the threshold level determined by the zooplankton grazing and mortality parameters. The capacity of phytoplankton to take up nutrients in this ecosystem thus has an upper limit.

One can generalize the conclusion of this section by noting that the reason the N-P-Z model allows the nitrate concentration to grow is because zooplankton grazing depends on the phytoplankton concentration, which is referred to as *density-dependent* grazing. The solution to the zooplankton equation (4.3.12) is what fixes the phytoplankton concentration at the concentration given by (4.3.17), not allowing it to grow beyond this concentration once it achieves the threshold level for zooplankton to live. One can conceive of other density-dependent removal mechanisms, such as viral mortality, or flocculation and sinking of phytoplankton, which might give a similar outcome. For example, if the phytoplankton mortality rate λ_P in the N-P model equation (4.3.6) for P were replaced instead by a

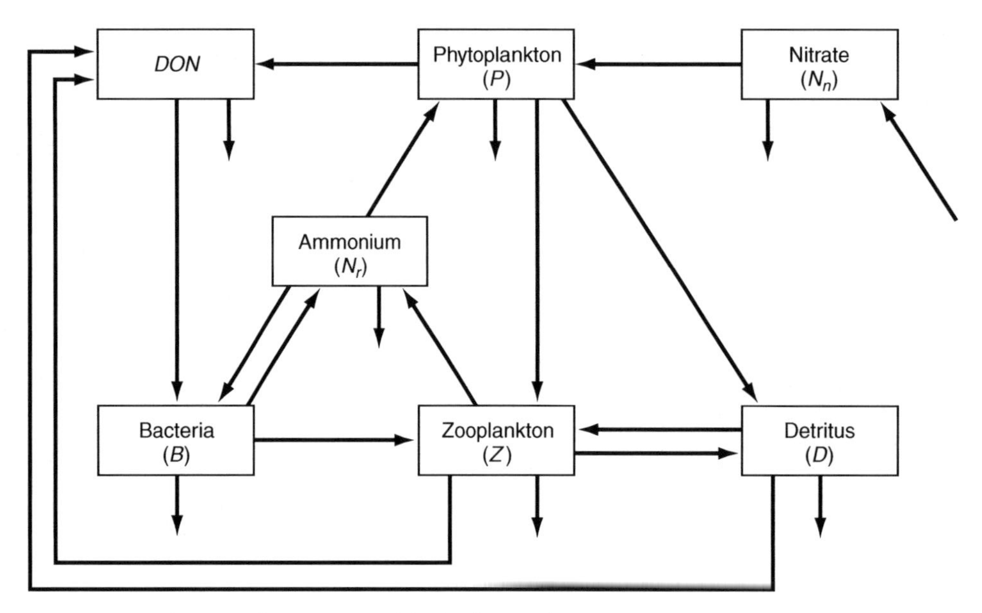

FIGURE 4.3.7: Schematic of ecosystem model of *Fasham et al.* [1990].

density-dependent mortality rate of the form $\lambda_Z \cdot P$, then the sink term in (4.3.6) would be $\lambda_Z \cdot P^2$, and the phytoplankton population would have a fixed concentration as in the *N-P-Z* model. Later we examine an additional mechanism, micronutrient limitation, which can also limit the growth of phytoplankton.

ADDING THE MICROBIAL LOOP

The basic structure of the *Fasham et al.* [1990] ecosystem model we will discuss here is illustrated in figure 4.3.7. The *N-P-Z* structure of the previous model is now expanded to include separate *N* sources for phytoplankton, i.e., nitrate and ammonium; and sinking particulate organic nitrogen (detritus) is modeled explicitly. The microbial loop is comprised of bacteria and dissolved organic matter, *DON*, and includes ammonium as a source of nitrogen and zooplankton grazing of bacteria. The model has seven equations, one for each compartment [see *Fasham et al.*, 1990; *Sarmiento et al.*, 1993].

The phytoplankton and zooplankton equations of the *Fasham et al.* [1990] model are similar to those of the *N-P-Z* model except that the phytoplankton component now has two separate sources of fixed nitrogen, nitrate and ammonium, whose uptake is governed by (4.2.12), including the exponential inhibition term of ammonium on nitrate uptake. In addition, the zooplankton grazing parameterization has to account for the preference between three different food sources: phytoplankton, bacteria, and detritus. Zooplankton excretion and mortality are the main source of ammonium in this model. A particulate detritus pool *D* consisting of waste material produced by phytoplankton and zooplankton has been added to the model to directly simulate the export of organic matter by sinking of particles as well as to serve as a source of food for bacteria via the production of *DON*. *DON* is also produced by zooplankton and phytoplankton decay.

The food source for bacteria is *DON* supplemented with ammonium. The *DON* simulated by this model includes only a short-lived highly *labile* component (i.e., easily degradable or transformed by organisms) that is not exported from the surface ocean in significant amounts (see chapter 5). It is produced by breakdown of detritus, which is probably bacterially mediated, and exudation and excretion by phytoplankton and zooplankton. As previously noted, bacterial uptake of ammonium is required so that bacteria can satisfy their need for a fixed nitrogen source to supplement the fixed nitrogen they obtain from dissolved organic matter (*DOM*). The $\gamma_B(DON, NH_4^+)$ term in the bacterial uptake equation (4.2.21) is the sum of two Monod equations for *DON* and ammonium, with a half-saturation constant of 0.5 mmol m^{-3}, and a complex functional form to determine the relative supply of *DON* and ammonium that is required for bacteria to be able to synthesize organic matter of the appropriate C:N ratio. The maximum bacterial uptake rate $V_B(T)$ is 2.0 d^{-1}, and bacterial mortality is given as a first-order decay with $\lambda_B = 0.05\ d^{-1}$. The microbial loop is closed by zooplankton grazing of bacteria and "decay" of bacteria to ammonium. Bacteria thus can be either a source of ammonium or a sink for it, though generally in this model the net flow is toward ammonium.

Figure 4.3.8 shows a series of equilibrium studies of this model with increasing total nitrogen. This is a top-down limited system just as the *N-P-Z* model. The three components that are grazed by zooplankton, i.e., phytoplankton, bacteria, and detritus (*ON* in the figure, which also includes *DON*), become limited by grazing as soon as zooplankton are able to survive, after which

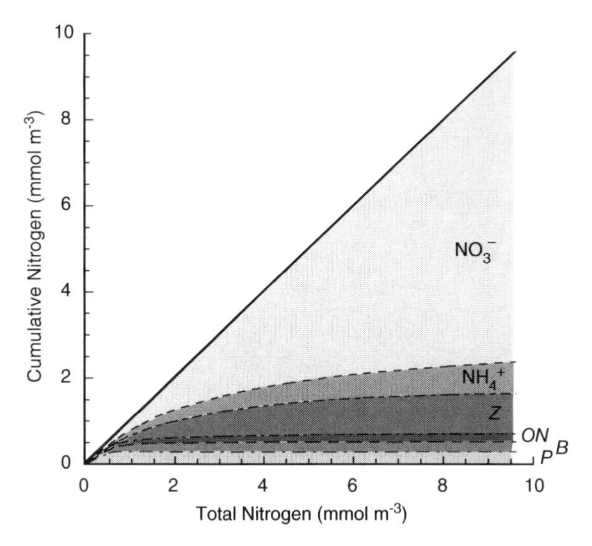

FIGURE 4.3.8: Steady-state nitrogen distribution as a function of total nitrogen in the components of the ecosystem model of *Fasham et al.* [1990] (Fasham, personal communication, 2004). The plot shows the cumulative distribution, i.e., the concentration of each component is the difference between its curve and the curve below it.

their concentration hardly changes in response to changing N_T. What is new about this ecosystem relative to that of the *N-P-Z* ecosystem is that as N_T increases beyond the zooplankton threshold, some of it goes into the recycled nitrogen pools of ammonium as well as *DON* rather than ending up in zooplankton or nitrate.

The most interesting behavior of the *Fasham et al.* [1990] model vis-à-vis the *N-P-Z* model is its response to perturbations such as the spring bloom that occurs in the North Atlantic upon shallowing of the mixed layer in the springtime [see *Sarmiento et al.*, 1993]. The early bloom is driven almost entirely by nitrate uptake, giving an extremely high *ef*-ratio. However, nitrate uptake drops precipitately once enough ammonium is generated to suppress nitrate uptake. The later part of the spring bloom is thus driven primarily by ammonium uptake with an *ef*-ratio of 0.3 or lower. This type of behavior would be difficult to simulate in an *N-P-Z* model.

Multiple Size Class Ecosystem Models

As we have noted, once the total nitrogen in the *N-P-Z* ecosystem increases beyond the threshold of $N_T = N + P = 0.004\ \text{mmol m}^{-3} + 0.60\ \text{mmol m}^{-3}$ that is required for zooplankton to survive, the zooplankton prevent the phytoplankton from climbing any higher no matter how much nitrogen is supplied to the system. However, while this top-down limitation may be characteristic of some regions of the ocean for part of the time, or of some components of a given ecosystem, it is also the case that one often observes ecosystems where the phytoplankton community, or at least some fraction of the community, increases essentially without limit so long as nutrients are available. For example, it is not unusual for coastal upwelling systems to have an order of magnitude greater chlorophyll than the concentration of $\sim$1 mg m^{-3} that we estimated with our *N-P-Z* model. One characteristic of such regions is that they tend to include multiple phytoplankton and zooplankton species, often varying quite markedly in size and other characteristics. In this section, we develop an ecosystem model that attempts to capture the essential behavior of such a system by multiple size classes of phytoplankton and zooplankton. What we find is that a system such as this can behave as though each phytoplankton size class is top-down limited, but because of the ability of larger size classes to take over when smaller size classes becomes top-down limited, the system as a whole can become bottom-up limited, that is to say, limited by the nitrate supply [e.g., *Hairston et al.*, 1960].

In addition to the insights multiple size class models offer on how top-down and bottom-up limitation might function in a more complex ecosystem, we use them to explore the role of micronutrient supply, specifically iron, in helping to determine the surface ecosystem structure. In particular, we will develop a model that permits only smaller phytoplankton such as Prochlorococcus to survive when iron is low, and larger organisms such as diatoms to grow when iron and macronutrients are abundant. We discuss the relevance of this to the *ef*-ratio in section 4.4.

The Model

The *Armstrong* [1994] model of multiple food chains that we discuss here is based on two key concepts: (1) the evidence briefly summarized above that although individual phytoplankton species may be limited by herbivores, the ecosystem system as a whole is limited by nutrient supply (or space); and (2) the observation that the total amount of phytoplankton (as measured by chlorophyll) in any given size class defined by logarithmic intervals is roughly constant [cf. *Chisholm*, 1992]. It appears that, as the total amount of phytoplankton increases, it does so by adding larger size classes (see figure 4.3.9, from *Raimbault et al.* [1988]). These two concepts taken together led *Armstrong et al.* [1994] to propose a simple multiple *P-Z* size class model, in which each successive size class is related to the previous size class by a logarithmic size progression. This allometric (size-based) modeling approach is grounded in the work of *Moloney and Field* [1989, 1991].

Armstrong et al., [1994] used nominal size classes of $L_0 = 1\,\mu$m, $L_1 = 4\,\mu$m, $L_2 = 16\,\mu$m, $L_3 = 64\,\mu$m, etc. Various interactions can be permitted between the different size classes of phytoplankton and zooplankton. The simplest is illustration 1 in figure 4.3.10, which permits zooplankton to graze only on phytoplankton in the size class immediately below it. In Armstrong's steady-state model based on this structure, the equations are exactly

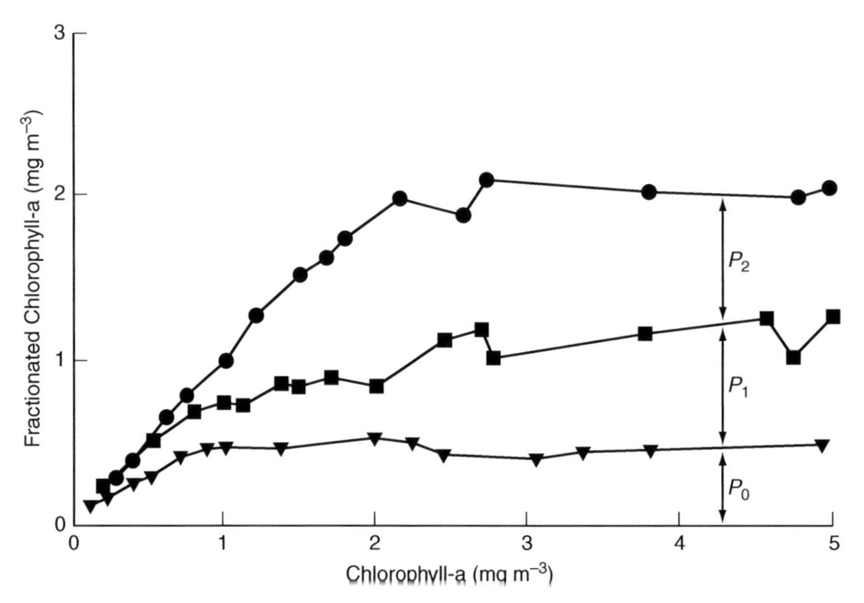

FIGURE 4.3.9: Size distribution of phytoplankton in the Mediterranean Sea. Each successively higher curve has the next larger size class of phytoplankton added on. P_0 is <1 μm, P_1 is between 1 and <3 μm, and P_3 is between 3 and <10 μm. Taken from *Chisholm* [1992]; original from *Raimbault et al.* [1988].

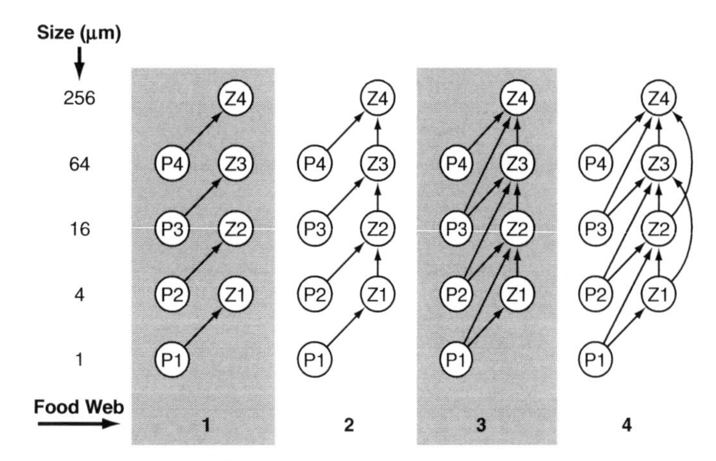

FIGURE 4.3.10: An illustration of various ways that various size classes of phytoplankton and zooplankton may interact with each other. Taken from *Davis and Steele* [1994].

the same as (4.3.11) for each phytoplankton size class, (4.3.12) for each zooplankton size class, and similar to (4.3.14) for the total amount of nitrogen present except that N_T now includes all size classes of phytoplankton and zooplankton. The values of the model parameters of the smallest size class in Armstrong's standard model are those given in the previous $1P$-$1Z$ model. All the parameters are kept the same for larger size classes except for the maximum phytoplankton growth rate, which follows a power law allometric relationship as a function of the growth rate of the smallest size class, $(V_{max})_0$:

$$(V_{max})_i = (V_{max})_0 \cdot (L_i/L_0)^{-\beta} \tag{4.3.18}$$

as in *Moloney and Field* [1991]. The parameter β is referred to as the *allometric coefficient*. Note that the exponent is negative, i.e., larger organisms have smaller specific growth rates than smaller ones. The suggestion that β is nearly constant for a variety of processes and among unrelated organisms has wide acceptance in biological sciences, but there is also strong evidence that the specific growth rate is not always sensitive to size (see review by *Chisholm* [1992]). For example, the salp is a large zooplankton that is capable of extremely rapid division, and *Goldman et al.* [1992] have shown that diatoms may also not follow typical allometric relationships.

In this multiple P and Z size class model an increase in nutrient concentration will lead to the addition of larger size classes such that, in principle, any amount of N_T can be accommodated by phytoplankton and zooplankton biomass, with extremely low nutrient N concentrations as in the N-P model. This is effectively a bottom-up controlled system, even though each P-Z pair, except for the largest, is top-down controlled. This is illustrated in figure 4.3.11a and b, where model results are shown with an allometric coefficient β of (a) 0.75, and (b) 0.4. The smaller allometric coefficient has a less rapid decrease in the maximum growth rate as size increases. Table 4.3.1 gives the minimum nutrient concentration that is required for a given size class of phytoplankton to invade the system per equation (4.3.10). We see from this that the smaller allometric coefficient of 0.4 permits a larger range of size classes for a given nutrient concentration, which accounts for the ability of this version to take up more of the total nutrient content into phytoplankton and

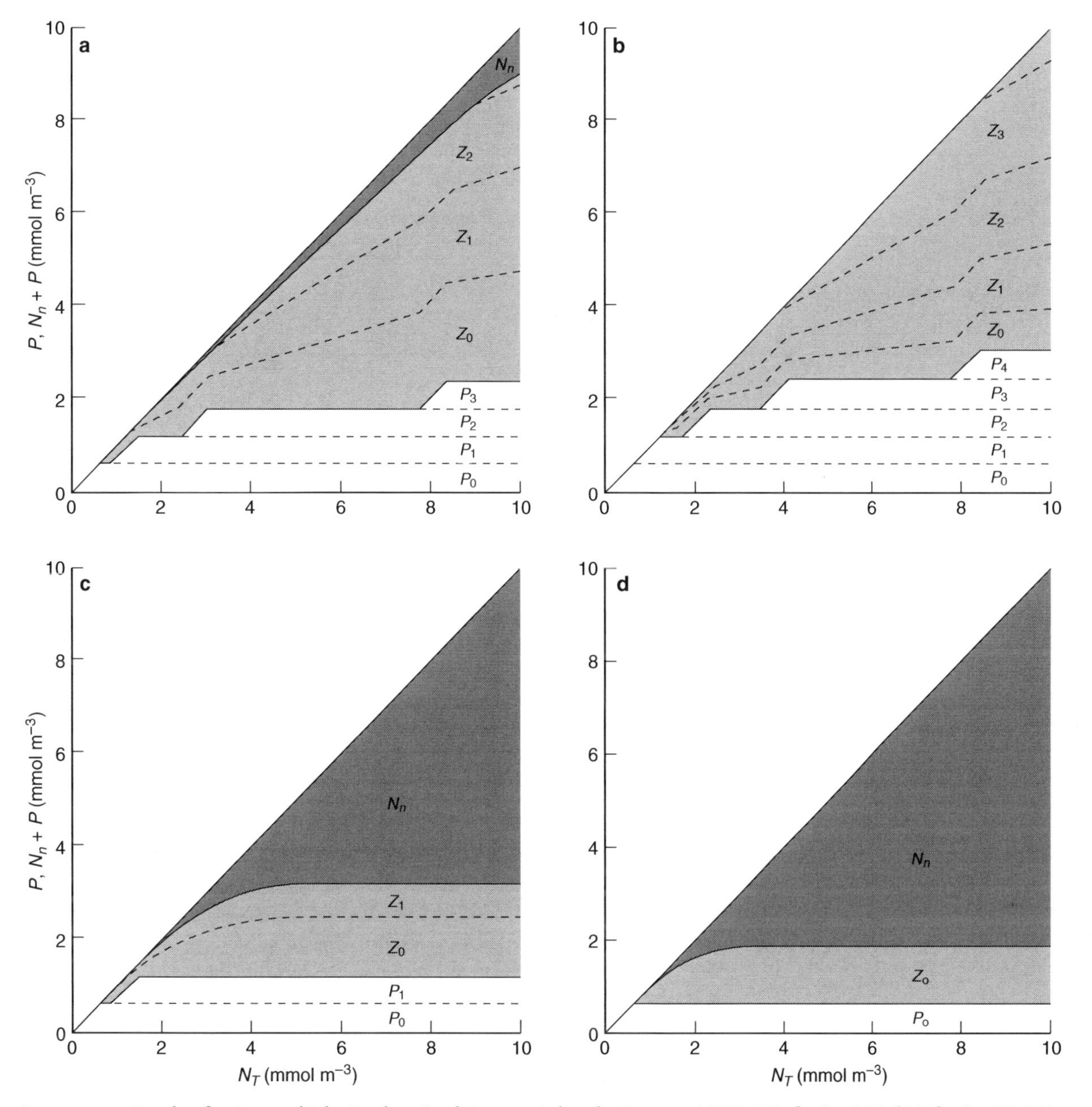

FIGURE 4.3.11: Results of various multiple size class simulations carried out by *Armstrong* [1994]. (a) is for $\beta = 0.75$, (b) is for $\beta = 0.4$, (c) is for $\beta = 1.0$ (transfer site limited), and (d) is for $\beta = 2.0$ (diffusion limited). See table 4.3.1 and discussion in text.

zooplankton than the example with an allometric coefficient of 0.75 (figure 4.3.11).

Armstrong [1994] also explored what happens when the zooplankton are allowed to graze on each other and on smaller phytoplankton size classes, as in illustrations 2, 3, and 4 of figure 4.3.10. Most (though not all) of these models have little or no nutrient N present as long as the allometric coefficient is sufficiently small. They tend to develop into systems with one or a few phytoplankton at the base of the food chain, and multiple zooplankton size classes. This may be of importance in controlling the export of organic matter from the surface, since the export tends to be greater with larger phytoplankton.

Influence of Micronutrients

The multiple size class model provides a way of exploring the potential impact of limited supplies of micronutrients such as iron. Iron limitation has been postulated to affect phytoplankton by reducing their growth rate, as well as making it more difficult for larger size classes to dominate the system. Experimental evidence supports this view (e.g., *de Baar* [1994]). Based on theoretical considerations, *Morel et al.* [1991a] proposed an allometric coefficient of 1 for cases where micronutrient uptake is limited by the density of sites on the cell surface that can transfer the micronutrient from seawater to the interior of the cell, and an allometric coefficient of 2 for cases where dif-

TABLE 4.3.1
Phytoplankton in multiple size class model of *Armstrong* [1994]

Case	*(a)*	*(b)*	*(c)*	*(d)*
$(V_{max})_0$ (d^{-1})	1.4	1.4	0.7	0.7
β	0.75	0.4	1.0	2.0
Nominal Size (μm)	Minimum Nitrate Concentration Needed for Invasion (mmol m^{-3})			
1	0.004	0.004	0.008	0.008
4	0.011	0.007	0.040	*
16	0.040	0.012	*	*
64	0.421	0.023	*	*
256	*	0.049	*	*
1024	*	0.133	*	*

* Outside the range of observed oceanic concentrations.

fusion of the micronutrient to the cell limits the uptake. With either of these mechanisms, small cells do better than large cells when there is iron limitation. Table 4.3.1 and figure 4.3.11 show that the two examples (c) and (d) with reduced growth rates and larger allometric coefficients give only a limited number of size classes and are thus less successful at taking up nutrients. Indeed, figure 4.3.11d is identical to the solution obtained with the 1*P*-1*Z* model of figure 4.3.6. Interestingly, the behavior of these two solutions with respect to nitrate is identical to the *N-P* ecosystem, which we earlier characterized as a top-down limited system. However, they are still bottom-up limited, but by iron rather than nitrate.

An alternative view of the impact of micronutrient limitation proposed by *Martin* [1991] is that it follows Liebig's law of the minimum, that is to say that the impact of micronutrient limitation is to control the total stock of phytoplankton. Growth of phytoplankton continues until all the iron is utilized, at which point the growth ceases. The evidence that is available, some of which has been summarized above, and which includes for example *in situ* work in the equatorial Pacific [*Kolber et al.*, 1993] and the iron fertilization studies summarized in table 4.2.3, suggests that the impact of iron limitation is primarily on the rates, as depicted in the model described above.

APPLICATIONS

Most of our discussion thus far has focused on the interpretation of annual mean observations and steady-state models. Yet we know from our discussion of the wintertime mixing illustrated in figure 4.1.5, and the influence of this and the seasonally varying sun angle on the light supply as manifested in the Sverdrup critical depth analysis of figure 4.3.3, that there are large seasonal variations in the forcing that are bound to have a major impact on the efficiency of the biological pump. The critical depth analysis shows that the biggest effects would be expected in the North Atlantic and Southern Ocean. Figure 4.3.12, showing the seasonal change in chlorophyll and nitrate, confirms this impression. There are also significant interannual variations that occur in response to the physical forcing discussed in section 2.5. We will not discuss those here, but note that there is clear documentation of quite dramatic interannual variability, for example in the North Pacific, which we mention briefly below [e.g., *Karl et al.*, 2001; *Emerson et al.*, 2001; *Chavez et al.*, 2003]; and in the equatorial Pacific due to El Niño [e.g., *Chavez et al.*, 1999].

Figure 4.3.13, showing the stations listed in table 4.1.1, provides a more detailed view of the seasonality at a few locations where long time series observations are available. The two oligotrophic time series stations (HOT and BATS) show a deepening of the mixed layer in winter, with generally higher nutrients and more chlorophyll during this time. By contrast, the nutrient-rich stations (OSI, OSP, and KERFIX), which also show a deepening of the mixed layer in winter, tend to show elevated chlorophyll and reduced nutrients in the summer. OSI, at 59°N, is particularly striking in this regard, with almost complete nutrient drawdown in the summer, and a rapid drop-off of chlorophyll at the end of the summer and increase at the end of winter. OSP and KERFIX, both near 50° latitude, have more modest seasonality. The contrast in seasonality between OSP and OSI, with particular focus on the failure of biology at OSP to draw down the nutrients, has been a subject of particular interest among oceanographers. In what follows, we will discuss in some detail the observations shown in figure 4.3.13 using the insights we have gained from the models we have developed. This discussion is based in part on the excellent reviews of time-series observations by *Michaels et al.* [1999] and *Karl et al.* [2003].

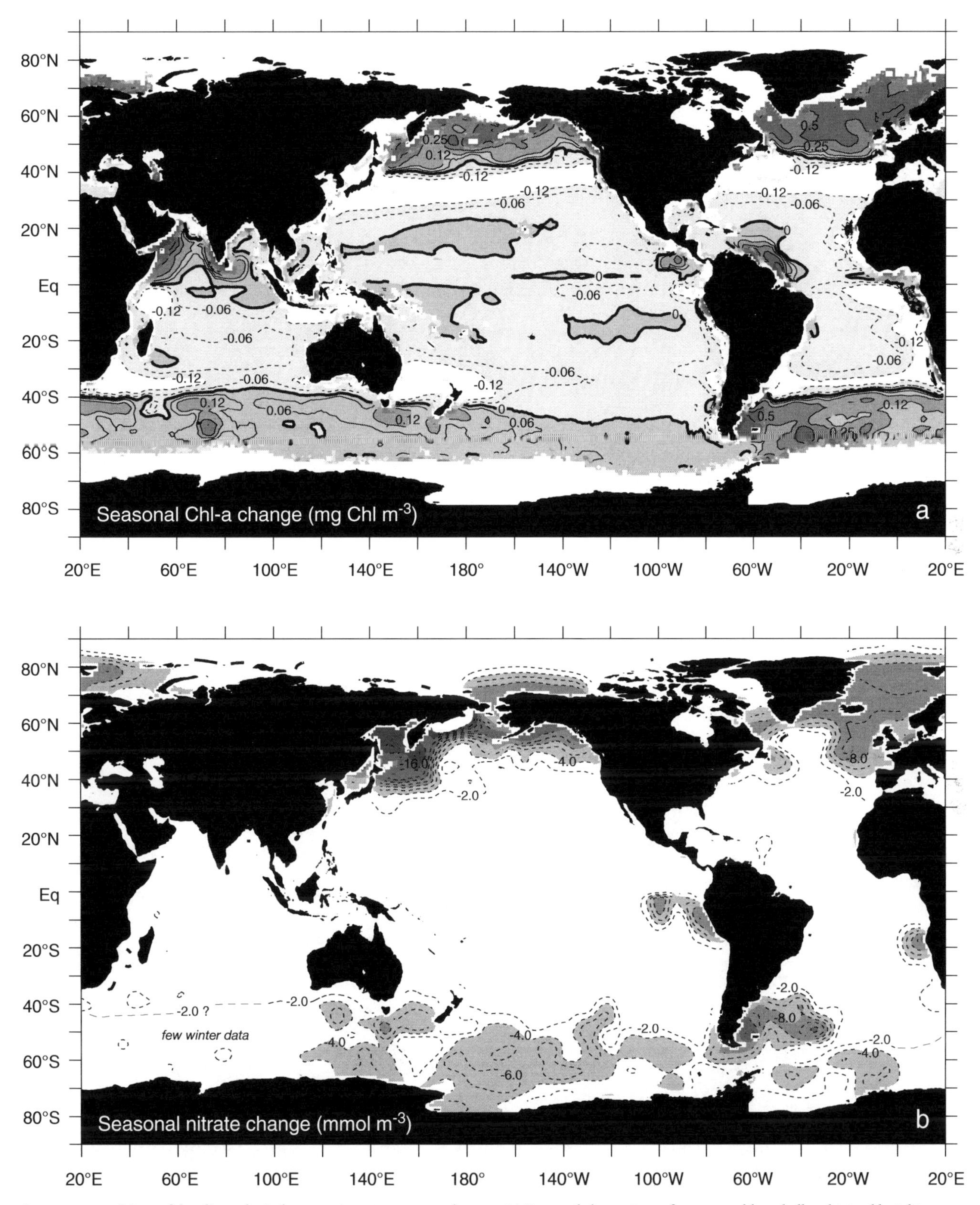

FIGURE 4.3.12: Maps of the climatological mean winter-to-summer changes. (a) Seasonal change in surface ocean chlorophyll-*a* obtained by taking the summer average (July through September for northern hemisphere, January through March for southern hemisphere) and subtracting the winter average (January through March for northern hemisphere, July through September for southern hemisphere). Positive values indicate an increase from winter to summer. Data are from the SeaWIFS sensor (cf. figure 4.1.1) (b) As (a), but for surface nitrate, using data from the World Ocean Atlas 2001 [*Conkright et al.*, 2002]. The seasonal nitrate drawdown in the Indian sector of the Southern Ocean is not well determined in the WOA01 climatology because of the lack of wintertime measurements in this region. A corresponding map of the seasonal phosphate drawdown suggests that the seasonal nitrate drawdown in the Indian sector of the Southern Ocean is similar to that observed in the other ocean basins.

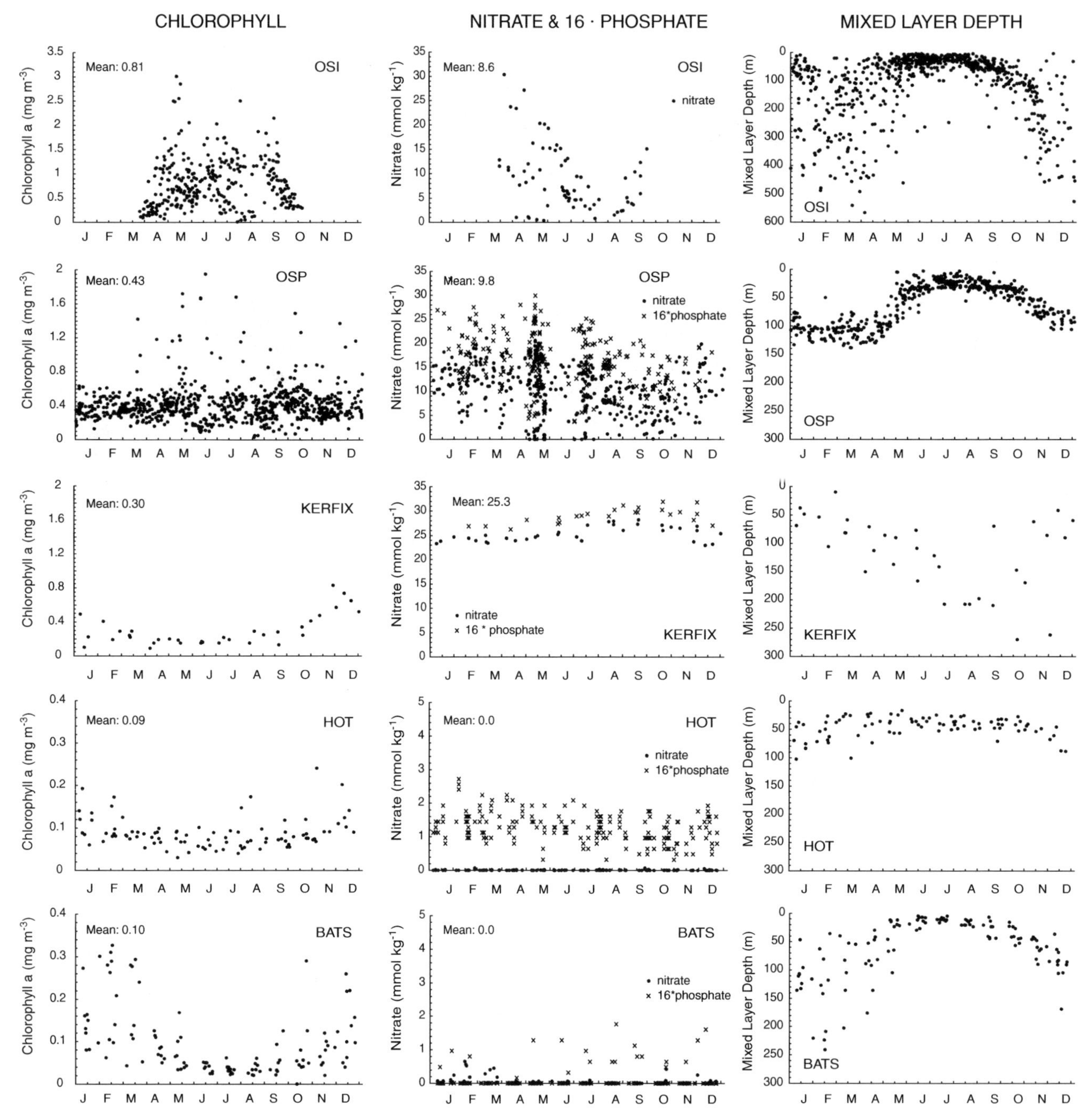

FIGURE 4.3.13: Composite time-series of chlorophyll-*a* (left column), of nitrate and 16 × phosphate (middle column), and mixed layer depth (right column) at the five time series stations described in table 4.1.1. Note the different vertical scales used for the different stations.

NORTH PACIFIC VERSUS NORTH ATLANTIC

Figure 4.3.14 gives a schematic illustration of the ecosystem behavior in the seasonally stratified subtropical gyres and subpolar regions of the Pacific and Atlantic. The illustration is based on observations at locations such as OSP in the North Pacific, and at OSI in the Atlantic (see table 4.1.1 and figure 4.3.13). The phenomenon that is of greatest interest from the geochemical point of view is that nutrients are depleted during much of the year in the North Atlantic but never in the North Pacific. Furthermore, figure 4.3.13 shows that chlorophyll-*a* at OSP is nearly constant year around at a mean concentration of 0.43 mg m^{-3}, though with a great deal of scatter around that mean, whereas at OSI the summer bloom is three or more times higher than this. In the decision tree of figure 4.1.4, these two stations could be identified as having high nutrient supply and, at least during the summer, high light supply. However, one of the stations follows the low biological pump efficiency type C pathway (OSP), whereas the

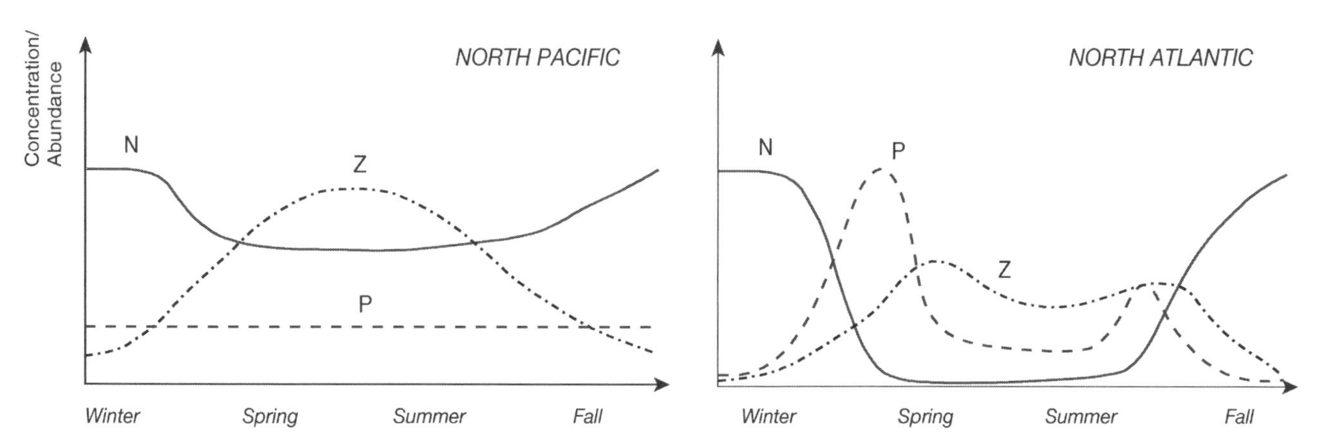

FIGURE 4.3.14: Schematic of typical seasonal behavior of the upper ocean ecosystem in (a) the North Pacific and (b) the North Atlantic. *N* is nitrate, *P* is phytoplankton, and *Z* is zooplankton. Taken from *Steele and Henderson* [1992].

other follows the high biological pump efficiency type B pathway (OSI). OSP is characterized as HNLC. Why are these two locations so different in their response to the availability of light and nutrients?

The models we have developed suggest two major hypotheses for the existence of HNLC regions: top-down control of phytoplankton by zooplankton grazing, and bottom-up control by a micronutrient (Fe). The key feature of the North Pacific is that the phytoplankton population is low and hardly changes with season. This suggests a major role for zooplankton grazing as in the results of figure 4.3.6. *Frost* [1987] examined the zooplankton grazing hypothesis using a combination of observations and model calculations. He first calculated the primary production PP and mixed layer nitrate concentration using an annual mean phytoplankton concentration equivalent to 0.3 mg m^{-3} of chlorophyll *a* and an estimate of the mean light supply in the mixed layer. The seasonal fluctuations in PP shown in figure 4.3.15a and b are due to changes in the sun angle, as well as changes and in the light supply resulting from variations in mixed layer depth (about 30 m in summer and >100 m in winter; see OSP data in figure 4.3.13). In a further set of models, Frost explored what population of herbivores would be required to prevent the phytoplankton population from expanding. An interesting result of this study was that the observed population of large zooplankton (copepods) was insufficient. Frost had to propose the existence of an herbivorous microzooplankton with a large seasonal fluctuation matching that of the primary production (figure 4.3.16).

The zooplankton grazing hypothesis thus appears capable of explaining the existence of the North Pacific HNLC region as manifested at OSP. The small phytoplankton that are present at OSP throughout the year are cropped by microzooplankton and never permitted to expand to the point where they would be able to remove all the nutrient that is present. However, the important question remains as to why larger phytoplankton such as diatoms are seldom seen at OSP, contrary to what might be expected from the observations of figure 4.3.9 and the models of figure 4.3.11a and b. This suggests that the supply of iron plays a significant role in explaining the HNLC conditions at OSP. The models of figure 4.3.11c and d and table 4.3.1 illustrate how the difficulty of larger cells to gather sufficient iron might limit the ecosystem to the smallest sizes.

The next question we examine is what makes the North Atlantic so different from the North Pacific and other HNLC regions. In the Atlantic there appears to be little or no limitation on the phytoplankton population during the early spring, with the consequence that the chlorophyll *a* expands considerably (see OSI data in figure 4.3.13). This behavior is the same as that obtained in the zooplankton-free model illustrated in figure 4.3.5 as well as the micronutrient-rich multiple size class models of figure 4.3.11a and b. It should be noted that the models we have developed are steady-state, whereas the spring bloom is a very dynamic feature. However, time-dependent versions of the models suggest that similar processes are occurring in the non–steady-state situation. The observations schematically illustrated in figure 4.3.14 suggest that the absence of zooplankton during the early part of the spring bloom is the crucial factor. In the absence of zooplankton grazing the phytoplankton population expands greatly during the spring bloom and reduces the nutrients to extremely low levels (see OSI data in figure 4.3.13).

However, observations of picoplankton in the North Atlantic near Bermuda [*Chisholm*, 1992] show that this small size class is present at near-constant biomass throughout the year. This near constancy implies the continuous presence of microzooplankton grazers, contrary to the schematic illustration in figure 4.3.14b. Thus, it appears that the spring bloom is superimposed on a near-constant background biological activity that is somewhat analogous to that in the North Pacific. The difference is that larger phytoplanktonic organisms are

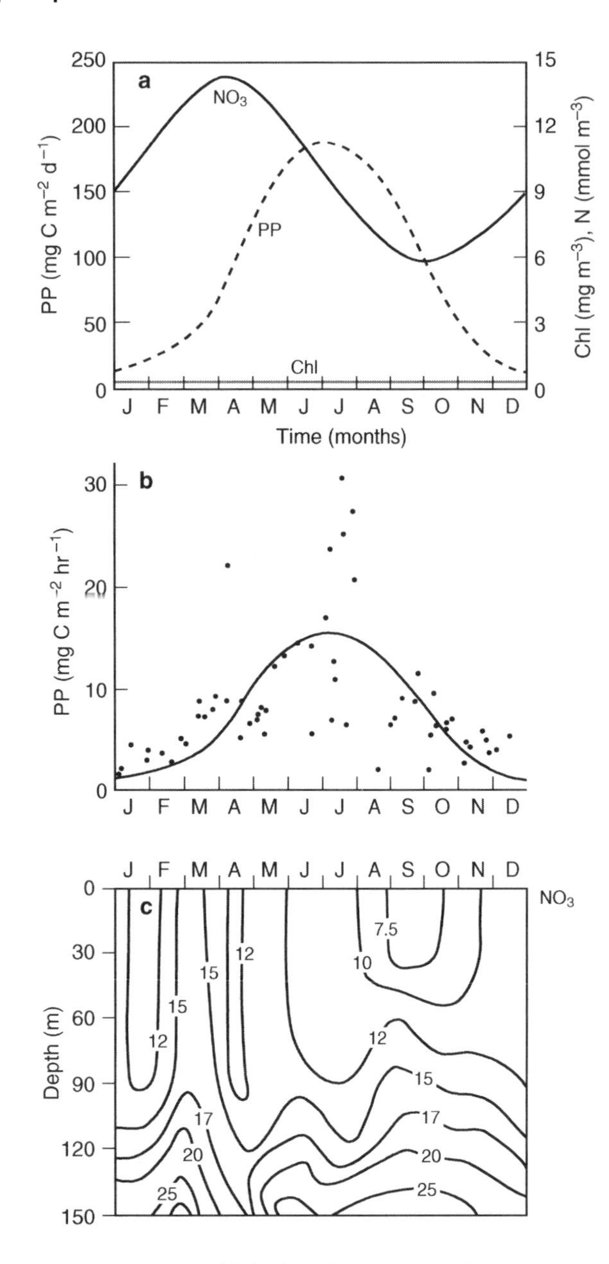

FIGURE 4.3.15: Seasonal behavior of various properties at Ocean Station Papa in the subarctic Pacific. (a) Schematic of the seasonal evolution of nitrate, chlorophyll, and primary production; (b) Composite seasonal evolution of observed primary production; and (c) composite seasonal cycle of nitrate in the upper 150 m. Adapted from *Frost* [1987].

able to grow in the Atlantic but not in the North Pacific. The model and observational analysis by *Taylor et al.* [1993] provides a nice illustration of how this works in the northeast Atlantic with a model that starts off the spring bloom with picophytoplankton that quickly become grazing-limited, followed by diatoms until silicic acid runs out, followed eventually by slow-growing dinoflagellates. They find no evidence of iron limitation in this region. We still have the problem, though, of explaining why the larger phytoplankton do not become grazing limited.

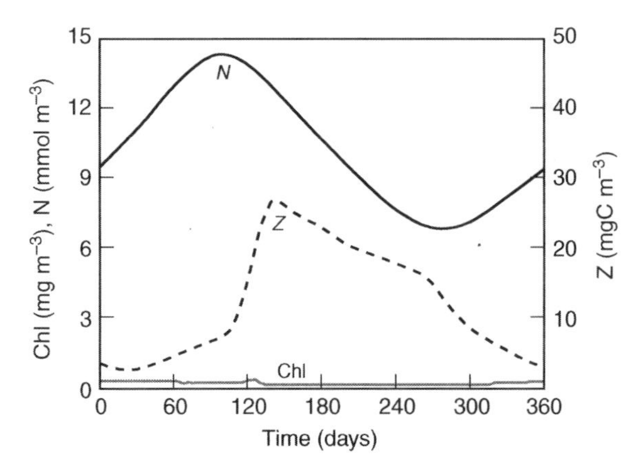

FIGURE 4.3.16: Model calculations of annual cycles of nitrate N, herbivorous zooplankton Z, and chlorophyll Chl at OSP. From *Frost* [1987].

Evans and Parslow [1985] provide a possible answer in their examination of the role of the annual cycle in determining the behavior of ecosystems. The North Atlantic has very deep mixed layers in the wintertime (see figure 4.1.5). As the mixed layer deepens and the critical depth is exceeded, the mean light supply is reduced to the point where there is little net production of organic matter. As a consequence, the zooplankton population plummets (figure 4.3.14) and their recovery in the spring is too slow to prevent the phytoplankton from first climbing to very high levels. The North Pacific is prevented from having such deep layers (figure 4.1.5) because of the stabilizing effect of low-salinity water at the surface of the ocean (figure 2.2.7b). An alternative or perhaps complementary view is that the life cycle of the zooplankton does not enable them to closely track the growth of the phytoplankton. The spring blooms often consist primarily of large diatoms and *Phaeocystis*. The large zooplanktonic organisms that feed on these are unable to respond rapidly to the growth of the phytoplankton.

An interesting illustration of the contrast between the North Pacific and North Atlantic as well as the potential role of competing phytoplankton species is figure 4.3.17, from *Parsons and Lalli* [1988]. Figure 4.3.17a schematically illustrates how the growth rate-irradiance curves of diatoms and flagellates are related to each other. The flagellates have a steeper response at low light levels, but diatoms have a higher growth rate. As a consequence, flagellates are favored in low light levels and diatoms in high light levels.

Figure 4.3.17b illustrates how this relative behavior would manifest itself at OSP in the North Pacific and OSI in the North Atlantic. The ordinate is solar radiation at 1 m depth. The abscissa is mixed layer depth. The light experienced by the phytoplankton is the average over the entire mixed layer. This assumes that mixing is sufficiently rapid within the mixed layer so that the phytoplanktonic organisms experience the av-

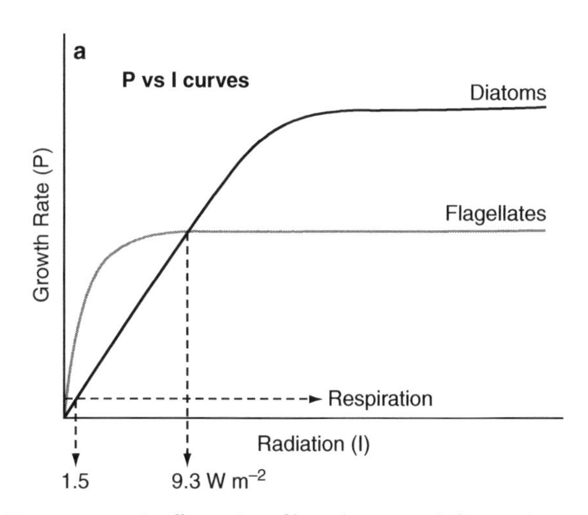

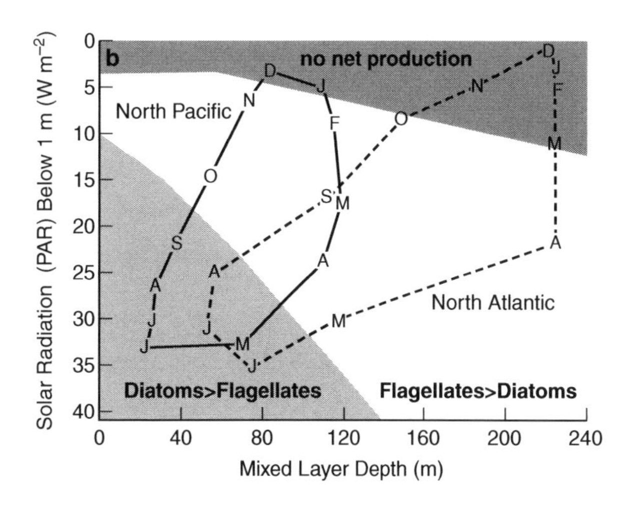

FIGURE 4.3.17: An illustration of how the seasonal changes in mixed layer depth and solar radiation combine to determine whether the critical depth is exceeded, and if not, whether diatoms or flagellates are likely to dominate based on the assumed growth rate-irradiance curves in the first panel. Based on *Parsons and Lalli* [1988].

erage light level over the entire layer. Two lines define three regimes: a first one in which light is insufficient to support net photosynthesis, a second one in which flagellates dominate over diatoms, and a third one in which diatoms are the dominant form. The two curves in Figure 4.3.17b depict the seasonal evolution of the North Atlantic and North Pacific in incident light–mixed layer depth space. The wintertime mixed layer depth of the North Atlantic is such that there are over four months during which no net production would be expected. This is the Sverdrup critical depth phenomenon that is at the root of the *Evans and Parslow* [1985] explanation for the absence of zooplankton during the spring bloom. The North Pacific is only marginally in the no-net-production zone during two months of the year. The behavior of the North Atlantic and North Pacific during the rest of the year is also interesting. The analysis suggests that flagellates would dominate in the early spring and late fall, and that diatoms would dominate in the summer, all else being equal. The period of diatom dominance in the North Pacific is about 5 months long. This makes it all the more interesting that diatoms do not seem to be present in the North Pacific except rarely. It appears that other factors are at play, such as iron limitation.

The KERFIX station in the Southern Ocean provides an interesting complement to the North Pacific–North Atlantic contrast (see table 4.1.1 and figure 4.3.13). The data set is more limited there, but the seasonal cycle can be readily observed. In their model and observational analysis of these data, *Pondaven et al.* [2000b] conclude that there are small-size classes of phytoplanktonic organisms that are top-down limited by microzooplankton grazing pressure. The spring bloom and associated nitrate drawdown are modest because of light limitation, and because the diatoms that are responsible for the bloom become limited by the supply of silicic acid as it gets drawn down to low levels during the summer. The model fit to the observations requires silicic acid drawdown by diatoms with very high Si:N ratios, which is generally taken to be indicative of iron stress of diatoms (see chapter 7).

OLIGOTROPHIC REGION

There are two important points we will make about the oligotrophic regions of the ocean using HOT and BATS as illustrations. The first is that the biological productivity in these regions is much greater than one might anticipate from the low surface chlorophyll concentrations illustrated in figure 4.1.1 and figure 4.3.13a. *Emerson et al.* [1997] give three independent estimates of the export production at HOT that have a modal value of $\sim 2\,\text{mol C m}^{-2}\,\text{yr}^{-1}$. *Carlson et al.* [1994] and *Gruber et al.* [1998] summarize several independent estimates of the export production at BATS. All but one of them give an export production of $\sim 4\,\text{mol C m}^{-2}\,\text{yr}^{-1}$. The higher export is probably due to greater input of nutrients from the deeper wintertime mixing at BATS than at HOT (figure 4.3.13). The one BATS estimate that disagrees with the others is based on the export of particulate and dissolved organic carbon, which together give $\sim 2\,\text{mol C m}^{-2}\,\text{yr}^{-1}$. The discrepancy between this estimate and the others may be due in part to undertrapping of particles by sediment traps (see chapter 5), but the reason for the disagreement is largely unresolved [cf. *Karl et al.*, 2003]. Based on extrapolation of the lower HOT result, *Emerson et al.* [1997] conclude that the subtropical regions of the ocean may account for $\sim 5\,\text{Pg C yr}^{-1}$ of organic carbon export, about half the global ocean total. Our more narrow definition for the permanently stratified subtropical biome (of which HOT is the prime example and BATS, with its occasional deep wintertime mixing, is at the margins) gives a global area of $145.3 \times 10^{12}\,\text{m}^2$ [*Sarmiento et al.*, 2004b]. The global

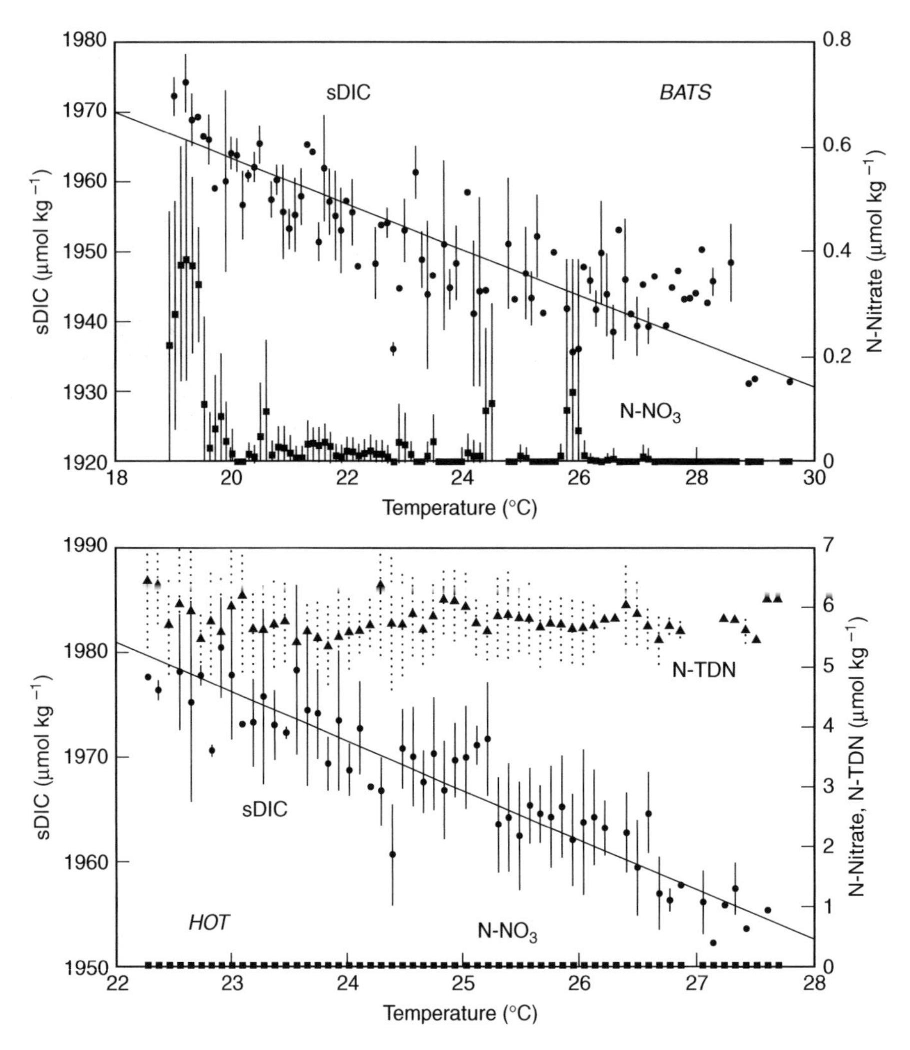

FIGURE 4.3.18: An illustration of the spring to summer drawdown in salinity normalized dissolved inorganic carbon (*sDIC*) in the absence of a measurable drawdown of fixed nitrogen, N-NO_3. Upper panel: Data from the Bermuda Atlantic Times-series Study (BATS) site in the subtropical North Atlantic with nitrogen represented by salinity normalized nitrate. Lower panel: Data from the Hawaii Ocean Timeseries (HOT) program in the subtropical North Pacific. N-TDN is total dissolved nitrogen, which includes nitrate and dissolved organic nitrogen. The data span multiple years and were first binned by temperature and then averaged. Adapted from *Karl et al.* [2003].

contribution from just this biome is thus 3.5 Pg C yr^{-1}, about one-third the global total.

In our discussion of the factors that control the supply of nutrients into the surface ocean, we identified the permanently stratified subtropical biome as an area that was likely to have low nutrient input and low biological productivity due to lack of upwelling and low wintertime mixing. BATS has significant wintertime mixing, but HOT does not, yet even HOT has a large enough export production to make the permanently stratified subtropical biome a major contributor to global export production. What did we miss in our earlier analysis? Part of the answer appears to lie in infrequent mixing events. *Jenkins* [1988] provided evidence from tracer observations at BATS, and in the North Atlantic more generally, that the upward supply of nitrate from deeper down is much larger than had previously been thought (see further discussion of these results in chapter 5). In a series of impressive modeling and observational studies, *McGillicuddy et al.* [1998] demonstrated that an important contributor to this supply at BATS may be the episodic supply of nutrients by mesoscale eddies. In their review of the time-series stations, *Michaels et al.* [1999] argue that episodic nutrient supply also occurs at HOT, where a number of nutrient intrusion events have been detected over time [cf. *Sakamoto et al.*, 2004].

Another part of the answer for the larger than expected productivities of the oligotrophic regions is likely related to the second important point we wish to make, which is that there is a large inconsistency between the nitrogen budget and the carbon budget at BATS and HOT [cf. *Karl et al.*, 2003] as well as more generally in the subtropics [e.g., *Lee et al.*, 2002]. This inconsistency is most likely due to nitrogen fixation. Figure 4.3.18 provides a very interesting illustration of the problem. The horizontal axis gives temperature and the vertical axis shows fixed nitrogen and dissolved inorganic carbon normalized to a constant salinity of 35‰. Think of the trajectory from cold temperatures on the left to warm temperatures on the right as the warming of the surface ocean during the spring and summer. What this figure shows is that there is a large drawdown of dissolved inorganic carbon during the spring and summer, with no discernable change in nitrate except for a few samples at BATS. The total dissolved nitrogen data at HOT indicate that the phytoplankton are not getting their fixed nitrogen from dissolved organic matter.

We illustrate the problem using the results of a diagnostic study by *Gruber et al.* [1998] of the processes affecting the mixed layer total carbon at BATS (cf. also studies by *Michaels et al.* [1994]; *Doney et al.* [1996]; *Marchal and Monfray* [1996]; *Bates et al.* [1996b]; and discussion in chapter 8). The biological uptake at Bermuda Station S of 33.3 mmol m^{-3} of carbon requires a nitrate supply of 4.4 mmol m^{-3}. Where does this come from? It certainly is not present at the beginning of the carbon drawdown. The outstanding feature of the BATS mixed layer nitrate plot in figure 4.3.13 is that it remains at near zero throughout the year. Bermuda has a deep mixed layer of up to 200 m in the winter. This occasionally raises the surface nitrate concentration to a few tenths of a mmol m^{-3}, but there is no indication that this deep mixing, which would also suppress biological production, brings anything like 4.4 mmol m^{-3} of nutrients to the surface. It also seems unlikely that the supply of nitrate by mixing through the thermocline can support the carbon uptake. If the nitrate were supplied by mixing through the thermocline, this would bring with it a large load of dissolved inorganic carbon, thus canceling out the large biological uptake reduction that is observed. Dust delivery through the atmosphere can supply only a small portion of the required nitrate [*Michaels et al.*, 1994].

The most likely explanation for the large carbon drawdown unsupported by nitrate is that the required nitrogen is being provided by fixation of gaseous N_2. *Lee et al.* [2002] estimate the total fixation in the tropical and subtropical nitrate-depleted regions ($T > 20°C$) to be about 0.8 Pg C yr^{-1}, accounting for 20% to 40% of the new production in these regions. Nitrogen fixation requires a high iron supply [e.g., *Kustka et al.*, 2003], and will of course consume phosphate. *Karl* [2002] reviews observations indicating that phosphate at BATS is drawn down much more than at HOT, due likely to a more abundant iron supply; and that HOT has seen a long-term reduction in the upper water column phosphate as a result of a long-term increase in nitrogen fixation that has occurred during the period of the time series observations [*Karl et al.*, 2001]. Other explanations that have been proposed for the carbon drawdown, though not now considered to be important, include the suggestion by *Marchal and Monfray* [1996] that the upper ocean is recycling nutrients but not carbon, i.e., that the effective C:N ratio of surface production is much higher than the traditional Redfield ratio. This suggestion is consistent with the results of *Sambrotto et al.* [1993], who, however, made their measurements in higher latitude regions, where nitrate was relatively abundant.

In conclusion, the time series observations at BATS and HOT, along with their forerunners, for example, at Bermuda Station S [cf. *Jenkins and Goldman*, 1985], have revolutionized our view of the oligotrophic regions of the ocean as represented most starkly by the permanently stratified subtropical biome. The biological productivity in these vast regions is much greater than had been thought prior to the time these observations were initiated. They contribute between one-third and half of global ocean export production. Furthermore, the contribution of nitrogen fixation to this export production is much greater than had been previously thought, fueling between about 20 and 40% of the new production and contributing between 5% and 10% of the global export production (see also discussion in chapter 5).

4.4 A Synthesis

The simple ecosystem models examined in the previous section provide three powerful insights that are at the core of most of the existing theories that have been developed to explain surface biological processes and their effect on surface nutrient concentrations. These are: (a) the concept of bottom-up limitation illustrated by the *N-P* ecosystem, where nutrients are shunted directly into phytoplankton and nitrate is always maintained at extremely low concentrations; (b) the concept of top-down limitation illustrated by the *N-P-Z* model, where the phytoplankton population is limited by zooplankton grazing and thus the capacity of the ecosystem to take up nitrate is also limited; and (c) the concept of simultaneous top-down limitation of individual *P-Z* components of an ecosystem, with bottom-up limitation of the ecosystem as a whole, as illustrated by the multiple size class model. The present view we have of how these basic concepts combine to explain the actual behavior of ecosystems in the ocean consists of the following five basic components listed in approximate historical order:

1. A high *ef*-ratio *N-P-Z* grazing food chain based on diatoms at the base of the food chain and going through large zooplankton and higher trophic levels to export of organic matter in the form of large particles.

2. A microbial loop consisting of heterotrophic bacteria that produce and consume dissolved organic matter as well as ammonium and are responsible for a large fraction of the recycling of organic matter in the surface ocean, thereby lowering the *ef*-ratio.

3. The picophytoplankton, e.g., *Prochlorococcus* and *Synechococcus*, many of which are unable to use nitrate, and are at the base of a food chain involving nano- and microzooplankton. This component of the ecosystem is viewed as having a low *ef*-ratio.

4. The role of iron availability in determining the physiological health of the phytoplankton, and whether only the picophytoplanktonic organisms are present (low iron) or if the diatom-based grazing food chain is also present (high iron in addition to adequate macronutrients and light).

5. The importance of nitrogen fixation in the oligotrophic gyres, and the role of iron supply in this.

We summarize our present understanding of how the first four of these ecosystem components fit together in the diagram shown in figure 4.4.1, which we discuss in detail below (see recent reviews of JGOFS studies by, e.g., *Karl* [1999, 2002]; *Landry* [2002]; *McCarthy* [2002]; *Ducklow* [2003]; *Falkowski et al.* [2003]; and *Karl et al.* [2003]). The discussion of nitrogen fixation is left largely to chapter 5, since many of the observational constraints that we will use to analyze it come from water column observations discussed there.

The Regeneration Loop

We use the term *regeneration loop* to refer to the low *ef*-ratio picophytoplankton-based ecosystem in the upper part of figure 4.4.1. The regeneration loop is fueled primarily by ammonium produced by recycling (recall that many picophytoplanktonic organisms are unable to use nitrate and even nitrite). Picophytoplanktonic organisms are generally top-down limited by zooplankton grazing. The zooplanktonic organisms that graze on picophytoplankton are small and rapidly growing, and they respond quickly to any perturbation and prevent the picophytoplankton from escaping predation and blooming. The regeneration loop also includes the microbial loop (recycling by heterotrophic bacteria). Export of organic matter from this system occurs primarily by downward transport of dissolved organic matter, and by sinking of larger particles produced by zooplankton such as mucus net feeders that are able to eat the tiny organisms in the regeneration loop.

The picophytoplanktonic organisms require iron to operate at full efficiency, but, with their high surface area-to-volume ratio, they are easily able to out-compete diatoms for iron if the supply is limited. The regeneration loop is a highly resilient component of the ecosystem that is present everywhere, though less abundant in colder waters, and quickly returns to its steady-state behavior if it is perturbed.

The Export Pathway

The classic textbook paradigm of the food web in the upper ocean is the *grazing food chain*, usually illustrated with the herring food web as proposed by *Hardy* [1924] for the east coast of England [cf. *Landry*, 2002]. The grazing food chain starts with large phytoplankton, then proceeds through grazing by herbivores to higher trophic levels, and ends with export from the surface ocean primarily as large rapidly sinking particulate organic matter. We use the term *export pathway* to refer to the grazing food chain. We depict it in the lower panel of figure 4.4.1 as an add-on to the regeneration loop. The export pathway is generally diatom-based, and the zooplanktonic organisms are large (e.g., copepods), with the food chain often proceeding to higher trophic levels including fish. The combined regeneration loop–export pathway ecosystem has recycling of organic matter involving both picophytoplankton and heterotrophic bacteria as well as ammonium uptake by diatoms, but the overall *ef*-ratio of this ecosystem is high, with loss both by aggregation and sinking of diatoms (see chapter 7) as well as formation of large rapidly sinking detritus particles. The export pathway requires abundant macronutrients as well as iron.

The diatoms of the export pathway may or may not be top-down limited by zooplankton grazing. This ecosystem usually kicks off in response to a major perturbation in the nutrient supply (with iron being a critical component) such as by deep wintertime mixing, upwelling, or an iron fertilization event in areas where the macronutrients are present in abundance. The large zooplanktonic organisms such as copepods that most often graze on diatoms have slow growth rates, particularly at low temperatures. This makes it difficult for them to respond quickly to diatom blooms. Thus diatom blooms often end in large export events due to aggregation and rapid sinking (see chapter 7) rather than achieving a stable top-down–limited equilibrium such as that of the regeneration loop ecosystem.

The export pathway of the surface ecosystem is opportunistic. It takes advantage of perturbation events in the nutrient supply that enable a large increase in biological productivity. However, it is not resilient; once the nutrient supply is exhausted, the export pathway quickly disappears and the ecosystem lapses back to the more resilient regeneration loop ecosystem. The export

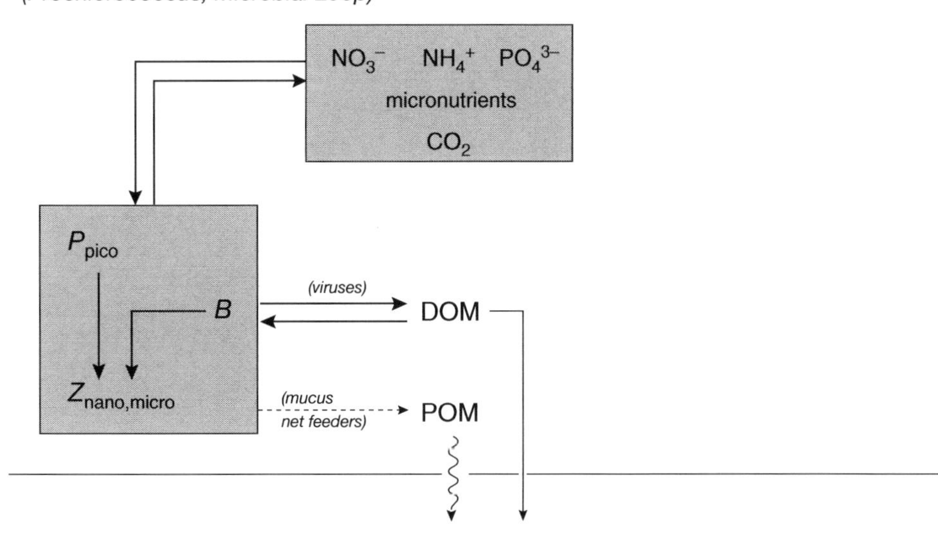

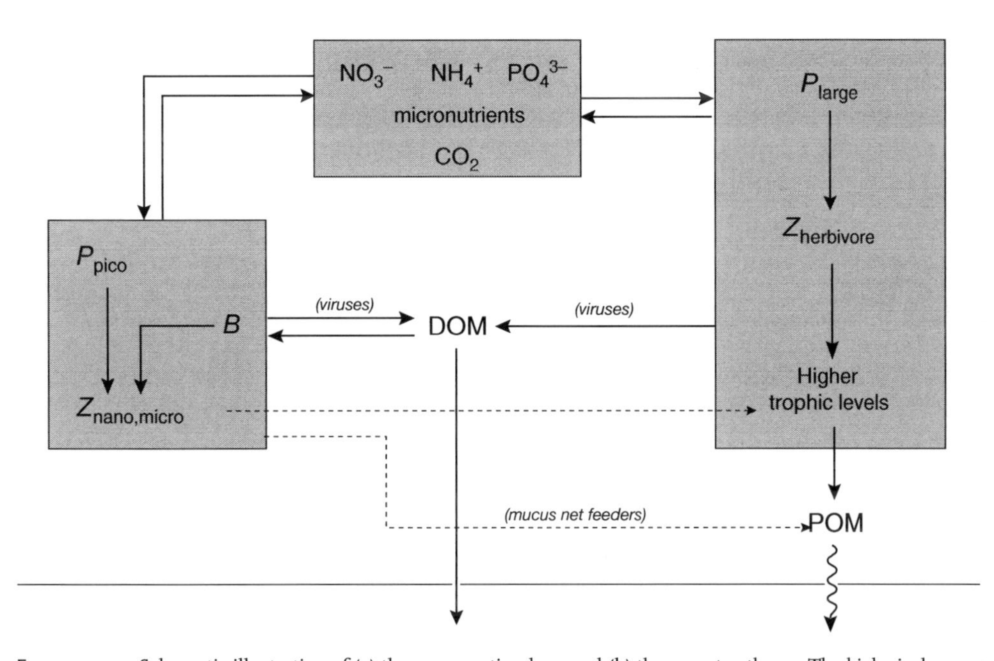

FIGURE 4.4.1: Schematic illustration of (a) the regeneration loop and (b) the export pathway. The biological pump with a low efficiency has just the regeneration loop present despite having both adequate light and high levels of nutrients. This is generally attributed to insufficient iron for large phytoplankton to survive. The high-efficiency biological pump has both the regeneration loop and export pathway. The export pathway is primarily based on diatoms and includes the grazing food chain. Viruses play a significant role in the formation of DOM. See text for further discussion.

pathway is also difficult to turn on. An interesting study of the contrasting response of the IRONEX I and IRONEX II iron fertilization experiments (see table 4.2.3) by *Pitchford and Brindley* [1999] in terms of an "excitable medium" model provides a possible explanation for the way that the export pathway might respond to a perturbation event. In this model, both diatoms and copepods are present at all times, but the grazing of diatoms by copepods is able to prevent them from blooming unless the perturbation to the system by iron fertilization is large enough and sustained over a long enough period of time for the diatoms to escape predation by the slowly growing copepods. Below this threshold, the system is highly resistant to perturbations. By this analysis, the IRONEX I experiment failed to bloom, despite showing enhanced phytoplankton growth rates, because the fertilization event was not sustained over a sufficiently long period of time for the diatoms to escape predation.

The Role of Iron

As spectacularly demonstrated by the *in situ* iron fertilization experiments summarized in table 4.2.3, iron limitation, or its converse of abundant iron supply, has a major effect on the physiology of all the phytoplankton, but preserves its most dramatic impact for larger phytoplanktonic organisms such as diatoms, which depend on there being sufficient dissolved iron for them to compete with the far more efficient picophytoplankton. The nitrate drawdown that occurs upon addition of iron in the fertilization experiments varies from 0 to greater than 15 $mmol\,m^{-3}$. The biggest effects were observed in the two North Pacific fertilization experiments (SEEDS and SERIES) and IRONEX II in the equatorial Pacific, when the iron addition lasted long enough for a bloom to occur. The Southern Ocean iron fertilization experiments all resulted in a relatively modest nitrate drawdown of 3 $mmol\,m^{-3}$ and less. In contrasting the modest Southern Ocean response at SOFeX with the massive North Pacific response at SEEDS, *Coale et al.* [2004] suggested that SOFeX may have been slower to respond because of lower temperatures (5–7°C in the northern patch, and -0.5°C in the southern patch, versus 9.5°C in SEEDS), and that it may have been suffering from light limitation due to deeper mixed layers (40 m in the northern patch and 45 m in the southern patch versus 10 m at SEEDS; cf. also *Mitchell et al.* [1991]).

What determines the natural supply of this most critical micronutrient? Our understanding of the iron cycle in the ocean is still rudimentary and will continue to be so until ongoing efforts to standardize measurement procedures [cf. *Measures and Vink,* 2001; *Bruland and Rue,* 2001], measure the global distribution of iron, and study the processes that control its distribution, begin to bear fruit [e.g., *Turner and Hunter,* 2001]. The following brief overview must therefore be considered provisional.

Iron exists in the ocean as Fe(II) and Fe(III). In the presence of oxygen and H_2O_2 in seawater, Fe(II), the more soluble form, has a half-life of seconds to a few hours, oxidizing to Fe(III), the relatively less soluble form [*Moffett,* 2001]. Typical [Fe(III)]/[Fe(II)] concentration ratios are expected to be $\sim 10^{10}$ [*Waite,* 2001]. Inorganic Fe(III) exists in the ocean almost entirely in hydrolyzed form as $Fe(OH)_2^+$ and $Fe(OH)_4^-$ [*Morel et al.,* 1991a; *Bruland et al.,* 1991; cf. *Waite,* 2001], with an overall solubility of $\sim 0.1\,\mu mol\,m^{-3}$ [*Waite,* 2001]. However, observations suggest that a large fraction of the iron in solution (typically defined as the total amount of iron that passes through a 0.4 μm pore-size filter, e.g., [*Johnson et al.,* 1997]) is bound to organic ligands [e.g., *Rue and Bruland,* 1997; *Bruland and Rue,* 2001; *Waite,* 2001]. Estimates of the solubility of iron including that bound to organic ligands are of order 0.2 $\mu mol\,m^{-3}$ in the surface ocean and 0.6 $\mu mol\,m^{-3}$ in the deep ocean, though this varies with the availability of organic ligands [e.g., *Millero,* 1998; *Nakabayashi et al.,* 2002; *Tani et al.,* 2003]. The increase in solubility with depth is attributed to lower temperatures [e.g., *Liu and Millero,* 2002] as well as to availability of humic-type fluorescent organic matter to which the iron is bound [e.g., *Tani et al.,* 2003]. Ocean model studies suggest that iron observations can be fit by a model in which iron exists in both the free and complexed form, with both forms available for biological uptake and remineralization [*Parekh et al.,* 2004; cf. also *Lefevre and Watson,* 1999; *Archer and Johnson,* 2000; *Christian et al.,* 2002; *Moore et al.,* 2002a]. However, the actual utilization of ligand-bound iron by phytoplankton likely involves an intermediate reduction step in which Fe(III) is converted to the soluble Fe(II) form either at the cell surface or by photochemical processes [cf. *Rue and Bruland,* 1997].

The Fe:C ratio of phytoplanktonic organisms ranges between $\sim 2\,\mu mol$ Fe $(mol\ C)^{-1}$ in areas that have low iron supply to $\sim 25\,\mu mol$ Fe $(mol\ C)^{-1}$ in coastal regions with high iron supply (see reviews by *Fung et al.* [2000] and *Sunda* [2001]). *Johnson et al.* [1997] propose a mean value of $\sim 5\,\mu mol$ Fe $(mol\ C)^{-1}$, which is equivalent to an Fe:N ratio of $\sim 40\,\mu mol$ Fe $(mol\ N)^{-1}$. Nitrogen fixers require many times this [*Kustka et al.,* 2003]. The mean ocean Fe concentration in seawater is $\sim 0.7\,\mu mol\,m^{-3}$ (see table 1.1.1), and the global mean nitrate concentration is $\sim 30\,mmol\,m^{-3}$, which gives a global mean Fe:N ratio in seawater of $\sim 20\,\mu mol$ Fe $(mol\ N)^{-1}$. This is half of what would be required to consume all the nitrate at an Fe:N ratio of 40 μmol Fe $(mol\ N)^{-1}$. However, observations such as from the iron fertilization experiments summarized in table 4.2.3 strongly suggest that iron is limiting growth primarily only in the North and equatorial Pacific and

Southern Ocean. Why these particular regions, and not others?

Iron is highly reactive and is removed swiftly from the ocean by biological uptake as well as scavenging by sinking particles. In chapter 1, we estimated a residence time of 40 years for iron with respect to river input. However, most of the river input of iron is removed locally in estuaries and on the shelves. The main sources of iron to the open ocean are thus by aeolian transport of dust particles [*Duce and Tindale*, 1991] and dissolution of iron that occurs in continental margin sediments, where high organic carbon contents consume all the oxygen and favor iron reduction from insoluble Fe(III) to soluble Fe(II) (*Elrod et al.* [2004]; see chapter 6).

Estimates of total aeolian delivery of iron range from 2.5×10^{11} mol Fe yr^{-1} [*Gao et al.*, 2001] to 5.7×10^{11} mol Fe yr^{-1} [*Duce and Tindale*, 1991], with a more recent estimate of 3.0×10^{11} mol Fe yr^{-1} and a range of 2.5 to 6.3×10^{11} mol Fe yr^{-1} given by *Jickells and Spokes* [2001]. However, only a fraction of this iron is soluble in seawater, with estimates ranging from a low of 1% determined by box model tuning [*Parekh et al.*, 2004], through values in excess of 30% measured for wet deposition, with a modal value of 17% determined from atmospheric observations and atmospheric model tuning [*Gao et al.*, 2003]. Ocean model studies typically use soluble fractions of 2% with sensitivity studies ranging up to 10% [e.g., *Fung et al.*, 2000; *Moore et al.*, 2002b]; but *Archer and Johnson* [2000] obtained values of 2% to 12% by ocean model tuning, depending on which atmospheric input function was used. In their recent analysis, *Jickells and Spokes* [2001] estimate that the iron solubility lies in the range of 0.3% to 6.8%, but that it is likely $\leq 2\%$. Assuming a total iron delivery of $\sim3\times10^{11}$ mol Fe yr^{-1} and that 10% of this is soluble, the input of soluble Fe to the ocean would be $\sim3\times10^{10}$ mol Fe yr^{-1}. If only 2% is soluble, the flux decreases to 0.6×10^{10} mol Fe yr^{-1}. The mean oceanic iron concentration of $\sim0.7\ \mu$mol m^{-3} gives a total oceanic inventory of 9×10^{11} mol. The larger flux estimate gives an oceanic residence time for iron with respect to the aeolian source of 30 years, whereas the smaller one gives 150 years. These numbers are consistent with estimates of the iron scavenging time in the deep ocean [e.g., *Bruland et al.*, 1994].

A recent estimate of the flux of soluble iron from continental shelf sediments gives 9×10^{10} mol Fe yr^{-1}, three or more times larger than the aeolian delivery of soluble iron [*Elrod et al.*, 2004]. The significance of this shelf source for the global iron distribution, the residence time of iron in the open ocean, and surface biological production, is still being evaluated, though there is strong evidence that its local influence on water column iron concentrations and continental margin biological production is very important.

Because of the short residence time of iron (~100 years) relative to the deep ocean circulation (~1000 years), the regional distribution of iron concentrations in the ocean will tend to reflect primarily the regional pattern of dust delivery overlain by the effects of the continental margin source and recycling of iron within each ocean basin. The dust delivery shown in figure 4.4.2 is generally higher in the northern than in the southern hemisphere. The tropics, primarily of the southern hemisphere, and the Southern Ocean are particularly low. In their recent summary of the iron concentration at 1000 m, *Parekh et al.* [2004] noted a general pattern that appears to be consistent with the pattern of the aeolian flux, with highest concentrations in the North Atlantic (0.6–0.8 μmol m^{-3}), and lowest in the Southern Ocean (0.2–0.3 μmol m^{-3}), and intermediate concentrations in the subtropical and tropical Pacific (0.4–0.5 μmol m^{-3}) as well as the North Pacific (0.6 μmol m^{-3}). However, more detailed sections based on measurements such as the new ones in the north Atlantic and Pacific Oceans shown in figure 4.4.3 (C. I. Measures and W. M. Landing, personal communication, 2004, and *Johnson et al.* [1997], respectively) suggest that these broad inter-basin generalizations will likely need to be refined [see also *de Baar and de Jong*, 2001]. In particular, it appears that the influence of continental margin sources extends well into the ocean interior, giving iron concentrations of ~0.8 μmol m^{-3} in the western North Pacific, also higher values of $>1.5\ \mu$mol m^{-3} in the Atlantic just north of the equator. Note that coastal waters can have concentrations as high as 100 μmol m^{-3} [*de Baar and de Jong*, 2001].

The vertical distribution of iron generally shows the typical behavior of a nutrient, with lower concentrations in the surface waters and higher concentrations below (figure 4.4.3). This implies that at least some of the iron removed from the surface ocean by biological uptake and adsorption onto particles is returned to solution in the deep water. The increase of iron concentration with depth means that ocean circulation and mixing will supply iron to the surface ocean [e.g., *Archer and Johnson*, 2000; *Fung et al.*, 2000; *Christian et al.*, 2002; *Moore et al.*, 2002a]. The model studies of *Archer and Johnson* [2000] and *Moore et al.* [2002a] suggest that on a global scale about three-quarters of the iron supply for new and export production comes from these processes rather than from dust input. However, as we have suggested above, the global pattern of oceanic iron that is available for transport to the surface is itself determined by the dust delivery (as well as by the continental shelf sources in some regions) and we thus find that two of the major regions where iron appears to be limiting growth by phytoplankton, the Southern Ocean and equatorial Pacific (see table 4.1.1), are also areas where the dust input is low (see table 4.4.1 and figure 4.4.2). The

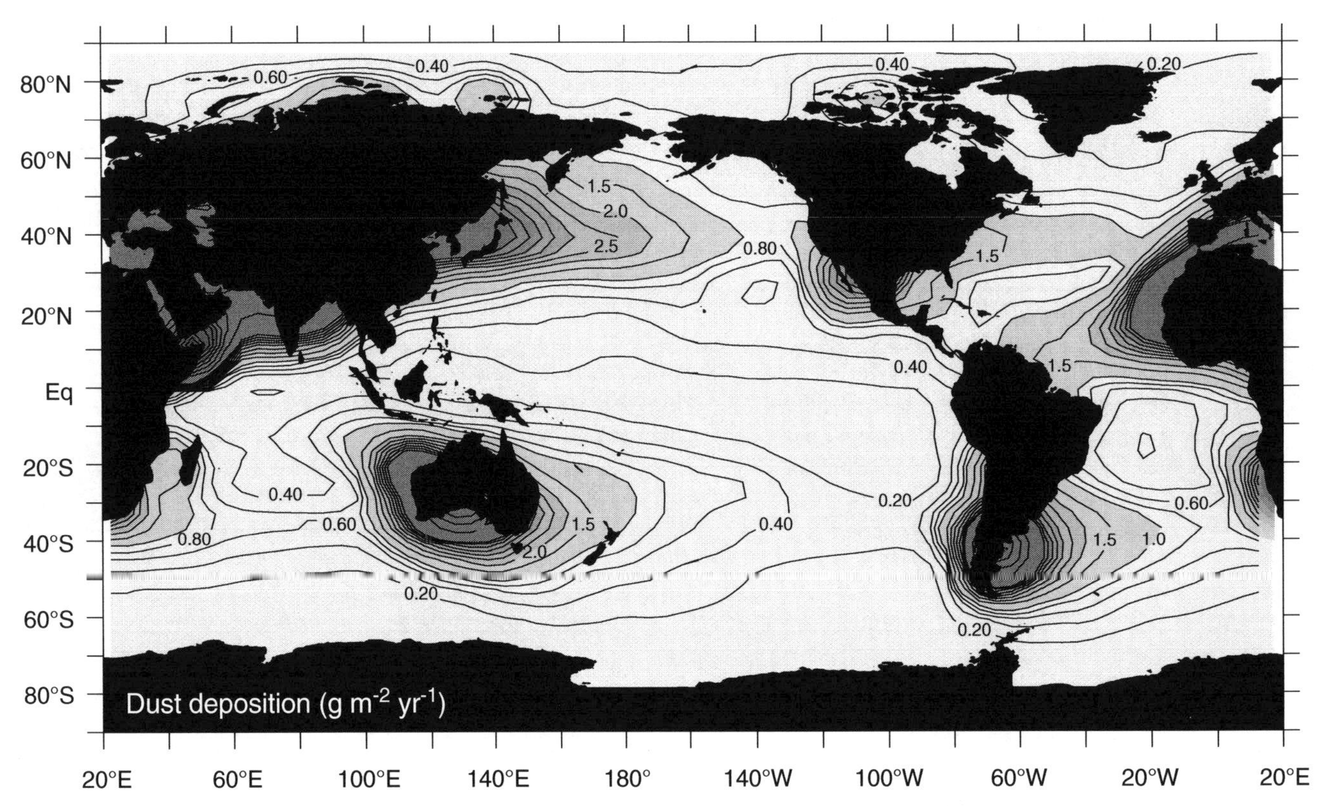

FIGURE 4.4.2: Map of the annual mean dust deposition flux onto the surface of the ocean. The flux estimates are model derived [*Tegen and Fung* 1994], and subject to considerable uncertainty.

North Pacific is another iron-limited region, but here it appears that low upward supply from below may be responsible, since the dust transport seems to be relatively high (see table 4.4.1 and figure 4.4.2).

CONCLUSIONS

We conclude this chapter by applying what we have learned to discuss the two goals we set for ourselves at the beginning of section 4.1, restated here as questions:

1. What controls the production and export of organic matter from the surface of the ocean, and how does this influence the surface nutrient concentration?

2. What controls the efficiency of organic matter recycling within the surface ocean, and how does this influence the surface nutrient concentration?

The first of these questions is basically about new and export production; the second is fundamentally about regenerated production; and both together are about primary production (see diagram in figure 4.2.3). We discuss each of these in turn.

A comparison of the nutrient distribution that would exist in a world without biology (figure 4.1.3c) with the actual nutrient concentrations shown in figure 4.1.2 (see also figure 4.1.3a) shows that biology is quite efficient at stripping nutrients out of the surface over most of the ocean. In other words, the zero order answer to the question of what controls the new and export production is that it is the potential nutrient supply, i.e., the supply rate of nutrients that would exist if nutrients at the surface were kept at a concentration of 0. The potential nutrient supply is controlled, in turn, by the thermocline nutrient concentration and the physical processes that transport nutrients from the thermocline to the surface.

In terms of the decision tree chart shown in figure 4.1.4, the observations of low surface nutrients in figure 4.1.2 tell us that most of the world falls into either the type A or type B regime, both of which have low surface nutrients. The type A regime has low nutrient supply and low productivity. The oligotrophic–permanently stratified subtropical biomes have mostly type A characteristics, with low nutrient concentrations due to low supply from below as well as laterally resulting from the fact that nutrients in the adjacent high-supply regions are stripped out before they can be carried horizontally into the oligotrophic regions. Although long viewed as deserts, a view strengthened by the near absence of surface chlorophyll, these oligotrophic regions have surprisingly high levels of new production, due partly to episodic nutrient supply by eddies and other mixing processes and partly to nitrogen fixation. The type B regime has high nutrient supply and high biological productivity. It corresponds primarily to areas of high

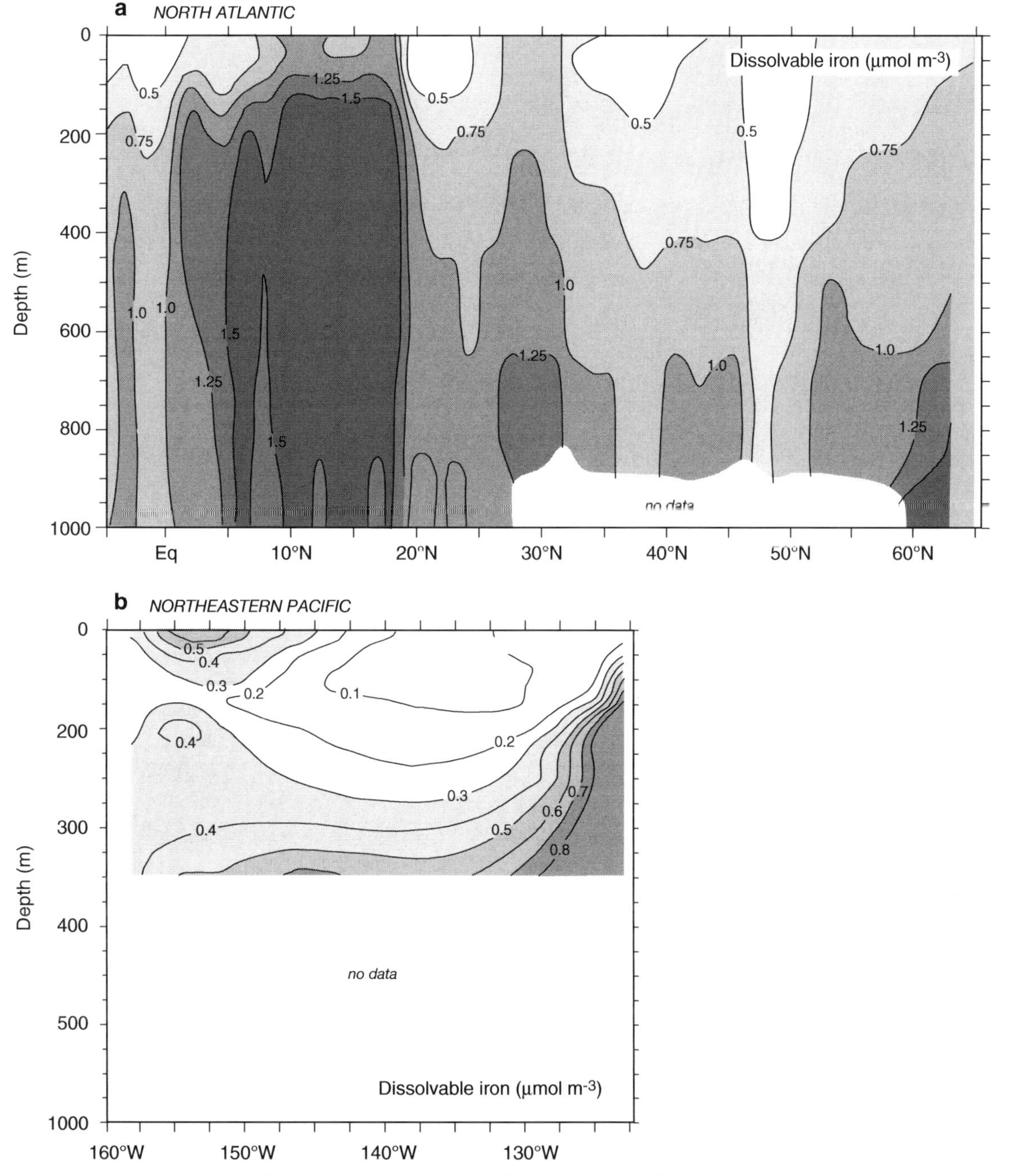

FIGURE 4.4.3: Vertical sections of dissolvable iron. (a) North-south section in the North Atlantic along about 20°W. Data are from C. I. Measures and W. M. Landing (personal communication, 2005). (b) East-west section in the eastern North Pacific going from Hawaii to California. Adapted from *Johnson et al.* [2002].

iron supply such as upwelling regions near the coast as well as the North Atlantic. Seasonality plays an important role in places like the North Atlantic, which shift back and forth between the type B regime in the summer and the light-limited type D regime in the winter. Episodic events, such as of strong upwelling, can also cause shifts between one regime and another.

However, there are also large regions of the world, primarily in the North and equatorial Pacific as well as the Southern Ocean, where nutrients are not drawn down to zero. In these cases, which correspond to the type C and D regimes of figure 4.1.4, the net supply rate of nutrients to the surface is lower than the potential nutrient supply rate. The new and export production are also less than would be predicted based on the potential nutrient supply rate. These regions correspond to those that biological oceanographers typically characterize as HNLC. Based on the model simulations and observational analyses we have discussed above, we propose the following hypotheses to explain the in-

TABLE 4.4.1
Estimated input of iron to the HNLC regions in nmol Fe m^{-2} d^{-1} [from *Gargett and Marra*, 2002; based on *de Baar et al.*, 1995, and *Landry et al.*, 1997; cf. *Watson*, 2001].
Note that the iron input to all three regions is comparable despite the North Pacific having ten times the aeolian input. The small upward transport in this region is due to a combination of low upwelling velocity and low Fe concentration in the upwelling water, per de Baar et al. [1995; cf. Watson, 2001].

	Upwelling and diffusion	*Aeolian*	*Total*
North Pacific	27	150[a]	177
Equatorial Pacific	130	15	145
Southern Ocean	145	15	160

[a]The estimated aeolian input to the North Pacific is based on *Duce and Tindale* [1991]. More recent estimates of the flux for the entire North Pacific by *Jickells and Spokes* [2001] and *Gao et al.* [2001] are about 20% to 40% of this (see summary by *Gao et al.* [2001]), which would reduce the aeolian contribution to 25 to 60 nmol Fe m^{-2} d^{-1}.

ability of new and export production to deplete the nutrients in these regions:

1. The high surface nutrient regions of the North and equatorial Pacific are due to iron limitation. In the North Pacific, this is primarily because of low vertical iron supply; in the equatorial Pacific it is primarily because of the low aeolian dust supply (table 4.4.1). When adequately fertilized with iron, both regions experience near depletion of nitrate (table 4.2.3). Iron fertilization should also lead to increased export of organic matter in these regions, though perhaps more in the high upwelling equatorial Pacific region than in the lower upwelling North Pacific. However, such an increase in export production with iron fertilization has yet to be clearly demonstrated. Analysis of the surface nutrient distributions in figure 4.1.2 and the biological export production in figure 4.2.4b leads us to classify these two regions as type B/C regimes with intermediate surface nutrients and intermediate export production.

2. The high surface nitrate region of the Southern Ocean is also iron limited due to low aeolian dust supply (table 4.4.1), but experimental application of iron in this region leads to only a modest reduction in nitrate compared with ambient concentrations (table 4.2.3). Possible explanations for this modest response include light limitation and low metabolic rates due to the cold temperatures [*Coale et al.*, 2004]. Although the summertime mixed layers are relatively shallow (e.g., ~40 m in the SOFeX experiment), the amount of light may be insufficient to draw down the high nitrate concentrations of this region [cf. *Mitchell et al.*, 1991]. If this analysis is correct and if it applies more broadly to the Southern Ocean as a whole, it would place this region in the light-limited type D regime. More likely, this is an area that oscillates between the type C regime due to iron limitation, and the type D regime due to periods of light limitation. The Southern Ocean is a region of potentially high organic matter export, but where the export would be much higher if biology were able to deplete the nutrients.

We conclude that new and export production are controlled by the potential nutrient supply in regions of nutrient depletion. In the HNLC regions, the actual nutrient supply is less than the potential nutrient supply because of the failure of organisms to deplete the nutrients due to a combination of iron and light limitation. The ecosystem role in determining the new and export production and the surface nutrient concentration can be explained in terms of the behavior of the two major ecosystem components that we have referred to as the regeneration loop and the export pathway. The regeneration loop is ubiquitous and stable, though less abundant in colder temperature waters than in warmer temperature waters; the diatom-based export pathway responds to perturbations in nutrient or light supply such as springtime shallowing, upwelling, and dust delivery or iron fertilization events. It is the export pathway that is responsible for drawdown of nutrients in high nutrient supply regions; and it is the relative absence of this component of the ecosystem that accounts

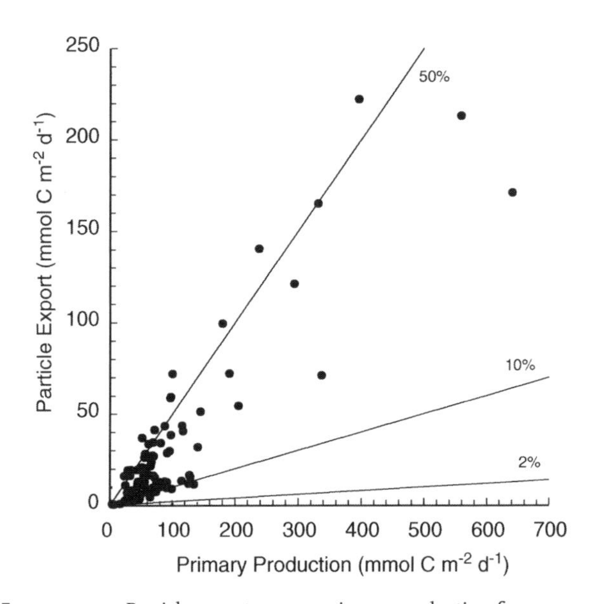

FIGURE 4.4.4: Particle export versus primary production from the data compilation of *Dunne et al.* [2005b]. The lines show slopes at e_P-ratios of 2%, 10%, and 50%.

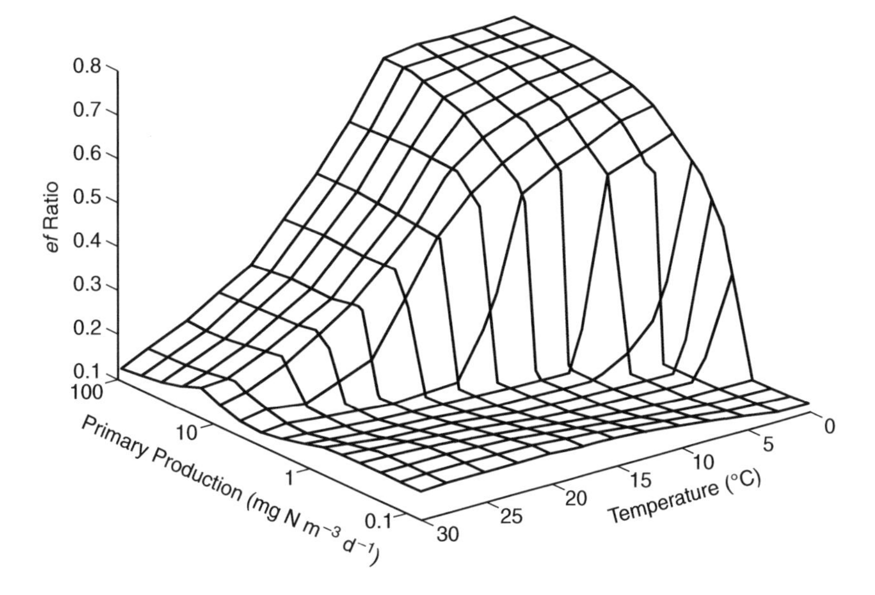

FIGURE 4.4.5: *ef*-ratio as a function of primary production (defined as N uptake) and temperature, from *Laws et al.* [2000]. The *Laws et al.* [2000] study on which this figure is based constructed a simple model with two size classes of phytoplankton that was able to account for 97% of the variance of averaged observations at 11 different sites. They explained the behavior of their system as resulting from the phytoplankton community being nutrient limited and the heterotrophic community (grazers and bacteria) being temperature limited. The primary production PP acts as a proxy for nutrient limitation, with high PP indicating abundant nutrients and resulting in surplus production that is exported (high *ef*-ratio), and low PP indicating nutrient limitation, with the heterotrophic community recycling a greater percentage of the organic matter (low *ef*-ratio). The effect of temperature, which captures 86% of the variance in their data set, works mainly through the high temperature sensitivity of the heterotrophs in their model, with recycling efficiency being much less at low than at high temperatures. The *Laws et al.* [2000] results are very appealing, but *Aufdenkampe et al.* [2001] showed with a large data set in the equatorial Pacific that using the *Laws et al.* [2000] mean *ef*-ratio approach to calculate new production in that region gives new production estimates with an error of ±50%, and the possibility of under- or overestimation by more than a factor of 3. *Dunne et al.* [2005b] used the much larger data set illustrated in figure 4.4.4 and a two-size-class phytoplankton model similar to that of *Laws et al.* [2000] to reexamine this issue. Their model accounts for 63% of the variance in their data set, as contrasted with 48% using the *Laws et al.* [2000] model. The mechanisms are different, however. The temperature sensitivity of their heterotrophs is much smaller than in *Laws et al.* [2000], and so the temperature effect is smaller though similar in direction (i.e., warmer temperature gives less recycling). The effect of nutrient limitation of the phytoplankton that *Laws et al.* [2000] identified in their analysis is explicitly explained in the *Dunne et al.* [2005b] analysis as being due to a shift from the low *ef*-ratio behavior associated with the small phytoplankton component of the food web, to a high *ef*-ratio associated with the large phytoplankton component of the food web as the primary production increases. *Tremblay et al.* [1997] also explored the importance of large phytoplankton in determining the *ef*-ratio.

for the inefficiency of the biological pump in iron- as well as light-limited regions.

We turn now to our second question of what controls the recycling efficiency of organic matter in the surface ocean. The recycling efficiency is defined as

$$\text{Recycling efficiency} = \frac{\text{Regenerated production}}{\text{Primary production}} \tag{4.4.1}$$

An important focus of our discussion in this chapter has been on the behavior of the f-ratio, which we defined in equation (4.2.5) as $f = $ New production/ Primary production, where Primary production = New production + Regenerated production. Rearranging these equations, it can readily be shown that the recycling efficiency is related to the f-ratio by

$$\text{Recycling efficiency} = (1 - f) \tag{4.4.2}$$

We can also show that regenerated production can be calculated from the new production by the equation:

$$\text{Regenerated production} = \frac{1-f}{f} \cdot \text{New production} \tag{4.4.3}$$

TABLE 4.4.2
Export production and *ef*-ratios calculated using the model of *Laws et al.* [2000] (from *Falkowski et al.* [2003]).

	Export (Pg C yr^{-1})	Mean *ef*-ratio
By Ocean Basin		
Pacific	4.3	0.19
Atlantic	4.3	0.25
Indian	1.5	0.15
Southern Ocean	0.62	0.28
Arctic	0.15	0.56
Mediterranean	0.19	0.21
Total (including Arctic and Mediterranean)	11.1	0.21
By Trophic Status		
Oligotrophic (chlorophyll $a < 0.1$ mg m^{-3})	1.0	0.15
Mesotrophic ($0.1 <$ chlorophyll $a < 1.0$ mg m^{-3})	6.5	0.18
Eutrophic (chlorophyll $a > 1.0$ mg m^{-3})	3.6	0.36

We thus see that understanding the magnitude of the recycling efficiency *per se* as well as the magnitude of the regenerated production essentially boils down to the problem of understanding the *f*-ratio. Because of the direct link between the *e*- and *f*-ratios over sufficiently large spatiotemporal scales, we have cast most of our discussion in terms of the joint *ef*-ratio.

Numerous studies have variously attempted to explain the *ef*-ratio distribution of the ocean as related to the primary production ([*Eppley and Peterson*, 1979]; see figure 4.4.4); the nitrate concentration [*Platt and Harrison*, 1985]; the nitrate and ammonium concentration [*Harrison et al.*, 1987]; a combination of ammonium limitation of nitrate uptake, grazing limitation of phytoplankton, and the recycling efficiency of the ecosystem including the rate at which detritus is exported [*Fasham et al.*, 1990; *Sarmiento et al.*, 1993]; or a combination of four variables consisting of either primary production or chlorophyll together with ammonium, nitrate, and temperature [*Aufdenkampe et al.*, 2001]. Our paradigm for the surface ecosystem as consisting of a regeneration loop system versus a system including both the regeneration loop and export pathway suggests the following hypotheses to explain the observed *ef*-ratio behavior:

1. The *ef*-ratio increases with increasing primary production primarily because of a shift from the regeneration loop system to a system including the export pathway.

The temperature sensitivity of remineralization of organic matter by bacteria, for example, suggests the following additional hypothesis to explain the observed *ef*-ratio behavior:

2. The *ef*-ratio decreases with increasing temperature because of more efficient recycling of organic matter by the heterotrophs (the microbial loop), which strengthens the regeneration loop.

Figure 4.4.5 depicts the results of a model based on similar principles to those enunciated above, though with some differences discussed in the caption. Table 4.4.2 shows a geographic breakdown of the *ef*-ratio calculated by the model depicted in figure 4.4.5, which illustrates the complex interplay of the temperature and nutrient limitation factors (e.g., the low *ef*-ratio in the Southern Ocean compared with the Arctic, both of which are cold, but one of which, the Southern Ocean, is less productive).

The final problem we need to address is to explain what link if any exists between the recycling efficiency and the surface nutrient concentration. We earlier defined the efficiency of the biological pump in depleting surface nutrients by equation (4.1.1). One might expect the biological pump efficiency (figure 4.1.7a) and *ef*-ratio (cf. figure 4.1.7b) to be directly correlated with each other. Instead, we find almost exactly the opposite over most of the ocean, with the biological pump being most efficient in regions where the *ef*-ratio is low (e.g., the permanently stratified subtropical gyres), and the biological pump being least efficient where the *ef*-ratio is high (almost everywhere else except for the equatorially influenced biome). We suggest the following hypotheses to explain this behavior:

1. The permanently stratified subtropical biome is easier to explain. The more modest supply of nutrients in this region is readily consumed by the phytoplankton, giving an efficient biological pump while at the same time favoring a picophytoplankton-based ecosystem with a low *ef*-ratio.

2. The high latitudes have very high nutrient supply rates, and most (except the North Atlantic) have an inadequate

iron supply to deplete the nutrients, exacerbated by inadequate light supply in wintertime and possibly even during the summer in some regions of the Southern Ocean. One would expect such low iron systems to be dominated by picophytoplankton, but in fact the observations show that the picophytoplanktonic community tends to be a much smaller component of the phytoplankton in the cold high-nutrient waters of the high latitudes than in the warm low-nutrient waters of the low latitudes, due perhaps to a temperature effect [*Agawin et al.*, 2000]. These systems thus behave like large phytoplankton–high *ef*-ratio systems, even though the iron supply is insufficient for the large phytoplankton to flourish.

We have intentionally characterized all our conclusions as hypotheses and have tried to keep them simple and general. While they are firmly grounded in the observational and model studies discussed in this chapter, we recognize that there are many alternative interpretations that have been proposed for all these phenomena, and that the ocean is very likely not as simple as we have depicted it. Hypotheses are meant to be tested and challenged, as we fully intend these to be.

• • •

Problems

4.1 What is meant by the terms *autotrophic*, *heterotrophic*, and *mixotrophic*? Associate the following groups of organisms with these terms: phytoplankton, zooplankton, fish, bacteria, diatoms, Prochlorococcus, dinoflagellates, coccolithophorids, picophytoplankton.

4.2 Figure 4.2.1 shows the observed surface ocean relationship between phosphate and nitrate.

a. What are the reasons that most observations lie close to a single line with a slope of 1:16?

b. What are the implications of the fact that there appear to be many observations where phosphate is still measurable, whereas nitrate is below detection limit?

c. Discuss possible reasons why the intercept of the phosphate versus nitrate relationship is not far from zero.

4.3 Define the terms *gross primary production, net primary production,* and *net community (ecosystem) production.* How are they related to each other and how do they relate to the transfer of energy/material through the upper ocean ecosystem?

4.4 What is the meaning of the terms *new production, regenerated production,* and *export production*? How are they related to net primary production and each other?

4.5 Discuss the reasons why one often finds a deep chlorophyll maximum at around 100 m in the centers of the subtropical gyres. Discuss also the implication of this observation with regard to detection of this chlorophyll from space (see figure 4.1.1).

4.6 Figure 4.1.1 and figure 4.2.4a show the annual mean distribution of surface chlorophyll and net primary production (NPP), respectively. Discuss and explain similarities and dissimilarities between the two plots.

4.7 The mean seasonal cycles of nutrients, phytoplankton, and zooplankton are very different between the North Atlantic and the North Pacific (figure 4.3.14). Discuss the hypotheses that have been put forward to explain this difference.

4.8 What is meant by the expressions *bottom-up* and *top-down control*? Which one appears to be more important in the ocean? Explain why you draw this conclusion.

4.9 List the major factors controlling phytoplankton production in the ocean. Elaborate briefly how these factors control phytoplankton growth.

4.10 Assume the composition of organic matter is $(CH_2)_{30}(CH_2O)_{76} \cdot (NH_3)_{16}(H_3PO_4)$.

a. Calculate the C:N:P stoichiometric ratio of this organic matter.

b. Calculate the amount of O_2 that would be required to oxidize this material if H_3PO_4, HNO_3, H_2O, and CO_2 are the oxidation products of phosphorus, nitrogen, hydrogen, and carbon, respectively. Give the full equation for the oxidation reaction. (Hint: Write down the full equation of organic matter reacting with oxygen to produce the above products, then balance each of the elements one at a time. You will have two unknowns, one for the amount of oxygen required, and the other for water. The oxygen and hydrogen atom balances will give you the two equations you need in order to solve for the unknowns.)

c. Suppose water upwelling to the surface has a total carbon concentration of 2000 mmol m^{-3}, an oxygen concentration of 160 mmol m^{-3}, a nitrate concentration of 5 mmol m^{-3}, and a phosphate concentration of 1 mmol m^{-3}. Which of these nutrients is likely to limit production if the light supply is adequate and there is no nitrogen fixation? Which of the elements will limit production if nitrogen fixation is allowed? In each case, calculate the concentration of the remaining nutrients after the limiting nutrient is exhausted.

4.11 Nitrate may serve as the terminal electron acceptor (i.e., oxidant) for the remineralization of organic matter if oxygen is not available. The nitrate loses its oxygen and is converted to dissolved N_2, in the process of which it gains electrons. This is referred to as *denitrification*.

a. Write a balance equation for the oxidation of the organic matter in problem 4.10 by denitrification. Assume that the organic matter reacts with nitrate in the form HNO_3, and that all the nitrogen present in both the organic matter and nitrate is converted to N_2. All other oxidation products are as in problem 4.10(b). See hint at the end of 4.10(b).

b. What fraction of the N_2 in (a) comes from nitrate?

4.12 Consider the surface nitrogen cycle illustrated in figure 4.2.3.

a. What combination of biological uptake fluxes should the primary production be equal to?

b. What fluxes are the new and regenerated production equal to?

c. Use the fluxes in the figure to define the traditional *f*-ratio based on the nitrogen cycle within the euphotic zone, and an *e*-ratio (the ratio of organic nitrogen exported from the surface to the primary production) based on the amount of nitrogen that is exported from the euphotic zone.

4.13 Derive *f*-ratio expressions for both the *N-P* and *N-P-Z* ecosystem models illustrated in figure 4.3.4 and discuss how the μ terms influence the magnitude of the *f*-ratio. Hint: In a steady state, losses must equal uptake. Define the *f*-ratio in terms of the loss terms rather than the uptake terms (e.g., input of nitrate = export of organic matter from the surface; uptake of nutrients by phytoplankton = mortality and grazing of phytoplankton).

4.14 This and the next two problems will guide you through the development of an ecosystem model that will illustrate how the light supply may control phytoplankton growth in the presence of a mixed layer. In this problem, you are to estimate the diurnally (24 hr) averaged light supply function $\gamma_P(I_0)$ at the surface of the ocean, which we will define as $\langle\gamma_P(I_0)\rangle$. Assume that I_n (the noontime clear sky irradiance at the surface of the ocean) at the equator is 1000 W m^{-2}, and that the diurnal variation of the irradiance function $f(\tau)$ is given as a triangular function that increases linearly from 0 at 6 AM to 1 at noon, then back to 0 at 6 PM. Do this in two steps:

a. Starting with (4.2.13), give an equation for the surface irradiance, I_0 for the first 6 hours of daylight in terms of the time t in hours, with t set to 0 at daybreak. Assume that the fraction of photosynthetically active radiation (PAR) is $f_{PAR} = 0.4$ and that the cloud cover coefficient is $f(C) = 0.8$.

b. Calculate $\langle\gamma_P(I_0)\rangle$. Use the Platt and Jassby formulation (4.2.16). To calculate I_k from (4.2.17), use for V_{max} the typical value of 1.4 d^{-1} given in the text, and the representative value for α of 0.025 d^{-1} (W m^{-2})$^{-1}$. Solve the problem analytically by stepwise integration over the 24 hours of the day, starting with the first 6 hours of daylight and using the integral form

$$\int \frac{x}{\sqrt{a^2+x^2}}dx = \sqrt{a^2+x^2}.$$

4.15 In this problem, you are to find the depth at which the diurnally averaged light supply $\langle\gamma_P(I(z))\rangle$ crosses the threshold necessary for phytoplankton to achieve the minimum concentration at which zooplankton can survive, 0.60 mmol m^{-3}. Use the temperature-dependent growth rate given by the Eppley relationship (4.2.8) for a temperature of 10°C, a mortality rate λ_P of 0.05 d^{-1}, and a nitrate half-saturation constant K_N of 0.1 mmol m^{-3}. Assume that the total nitrate concentration N_T is 10 mmol m^{-3}. Do this in two steps:

a. Find the minimum light supply function $\gamma_P(I)$ that is required in order for phytoplankton to cross the threshold concentration (assume zooplankton concentration $Z = 0$). Hint: Start by calculating N given $P = 0.60$ mmol m^{-3}, then use the steady-state equation for the phytoplankton distribution to solve for $\gamma_P(I)$.

b. Assuming that $\gamma_P(I)$ from (a) is equal to the diurnal average $\langle\gamma_P(I(z))\rangle$, at what depth H in the ocean will the diurnally averaged light supply function cross the threshold you estimated in (a)? Assume that P is constant with depth and use a total attenuation coefficient of $0.12\,m^{-1}$.

4.16 Suppose the phytoplanktonic organisms from problem 4.15 are living in a mixed layer of thickness h that has a total nutrient concentration N_T of $10\,mmol\ m^{-3}$. Do a cumulative plot of the three components of an N-P-Z ecosystem as a function of mixed layer thickness h. Assume that the light supply seen by the phytoplankton is the vertical average of $\langle\gamma_P(I(z))\rangle$ through the entire mixed layer. Do the plot only down to the depth where the vertically averaged light supply crosses the minimum threshold necessary to sustain P at the value of $0.60\,mmol\ m^{-3}$ required for Z to be present. Do this in two steps:

a. Calculate the vertically averaged magnitude of $\langle\gamma_P(I(z))\rangle$ in the mixed layer as a function of the thickness of the mixed layer h and plot it versus h. From the graph, pick the depth where the vertically averaged light function crosses the threshold you calculated in problem 4.15(a). Note: This problem can be solved numerically in a spreadsheet by solving for $\langle\gamma_P(I(z))\rangle$ every 5 m (for example) and then averaging these values from the surface to the depth h.

b. Assuming P is $0.60\,mmol\ m^{-3}$, obtain solutions for Z and N and then produce the cumulative distribution plot of P, $Z+P$, and $N+Z+P$. Use the zooplankton parameters given in the text (e.g., $g=1.4\,d^{-1}$, $K_P=2.8\,mmol\ m^{-3}$) and assume that the linear zooplankton feeding function (4.2.20) applies.

4.17 This problem illustrates how the nutrient supply and light supply work together to determine the biological productivity and surface nutrient concentration as the potential nutrient supply by vertical mixing increases. It requires you to add nitrate supply to the steady state N-P-Z ecosystem of problem 4.16. We do this by developing a two-box model. The top box of the model is the mixed layer with a depth of 50 m. The box below it has a uniform nitrate concentration of $10\,mmol\ m^{-3}$. The exchange rate between the mixed layer and the second layer is represented by an exchange coefficient v with units of velocity ($m\ yr^{-1}$). Assume that P and Z have active buoyancy and mobility that enable them to avoid being mixed out of the mixed layer; and that $\lambda_P \cdot P$, $\lambda_Z \cdot Z$, and $(1-\gamma_Z)\cdot g \cdot Z \cdot P/K_p$ are exported out of the mixed layer instantly.

a. Draw an illustration of the two-box model and put down the N, P, and Z equations governing the model.

b. Assume the temperature is 10°C. What is the phytoplankton concentration fixed at when zooplankton are present? How high does the nitrate concentration have to be in order for zooplankton to be present? For the remainder of the problem, assume that the phytoplankton concentration is fixed at this level.

c. Use the model to determine how the net input of nitrate ($mmol\ m^{-3}\ yr^{-1}$) to the mixed layer varies as a function of the exchange constant for a value of 0 to $1000\,m\ yr^{-1}$. Plot the nitrate input versus the exchange constant. Also, plot the surface nitrate concentration versus nitrate input.

d. Explain how and why the above system switches from one where the nitrate input increases with the exchange constant (a nitrate-deficient system) to one where the nitrate input remains constant with increasing exchange constant (a nitrate-replete system).

CHAPTER 5

Organic Matter Export and Remineralization

In the previous chapter we examined the uptake of nutrients to form organic matter by organisms in near-surface waters of the ocean. We saw that this uptake forces surface nitrate and phosphate to very low concentrations over almost the entire world ocean. It also reduces the surface concentration of total carbon and alkalinity, as well as having an impact on the surface concentration of micronutrients and many trace metals. We now examine the fate of organic matter that is exported from the surface. We will see that the majority of the organic matter exported is remineralized within the upper few hundred meters of the thermocline, with only a small but significant fraction making it into the deep ocean. We will also see that our understanding of the processes determining the remineralization of organic matter in this "twilight" zone is much less developed than our understanding of how organic matter is formed in the euphotic zone. Our approach will therefore be somewhat more descriptive in comparison to that we took in chapter 4.

5.1 Introduction

The vast majority of organic matter exported from near-surface waters is converted back into its inorganic constituents by heterotrophic organisms within the water column and sediments, a process referred to as *remineralization*. This process is the main energy source for these organisms, therefore providing a strong incentive for them to complete remineralization to the greatest extent possible. The subsequent upward flux to the surface of remineralized inorganic nutrients closes the nutrient cycling loop driven by the biological pump. These processes are depicted schematically in figure 5.1.1, which excludes remineralization processes and burial occurring in the sediments. Figure 5.1.1 separates the organic matter pool into *dissolved organic matter* (*DOM*) and *particulate organic matter* (*POM*), with the difference often defined operationally on the basis of what goes through a filter of a particular size and what does not. Functionally the main difference between these two pools is that *DOM* is too small to sink on its own; therefore its transport is primarily determined by water transport and mixing. In contrast, *POM* is large and heavy enough to sink, so in addition to water transport and mixing, it has a non-negligible vertical advection term. In this chapter, we examine the export of organic matter from the surface ocean, as well as remineralization in the water column. In chapter 6, we discuss remineralization in the sediments.

We begin this section with a first look at observations, and then discuss the two major remineralization reactions that occur in the water column and that determine the observed distributions. The first reaction uses oxygen as an oxidant, and is therefore termed *aerobic remineralization*. When oxygen is scarce, nitrate replaces oxygen as the oxidant, leading to a reaction termed *denitrification*. A major focus of this chapter will be on the influence of these reactions on the concentration of oxygen and nutrients, which we will show for the first time in this section. However, in order to study the impact of remineralization on the oxygen and nutrient distributions, we must first separate the remineralization component from the so-called *preformed* nutrient concentration transported in from the surface. The difference between the remineralized and preformed components is the third subject of this section. The final topic discussed in this section is the definition of particulate versus dissolved organic matter.

NUTRIENT AND OXYGEN DISTRIBUTIONS

To trace the impact of remineralization reactions on the distribution of nutrients and oxygen, we show in

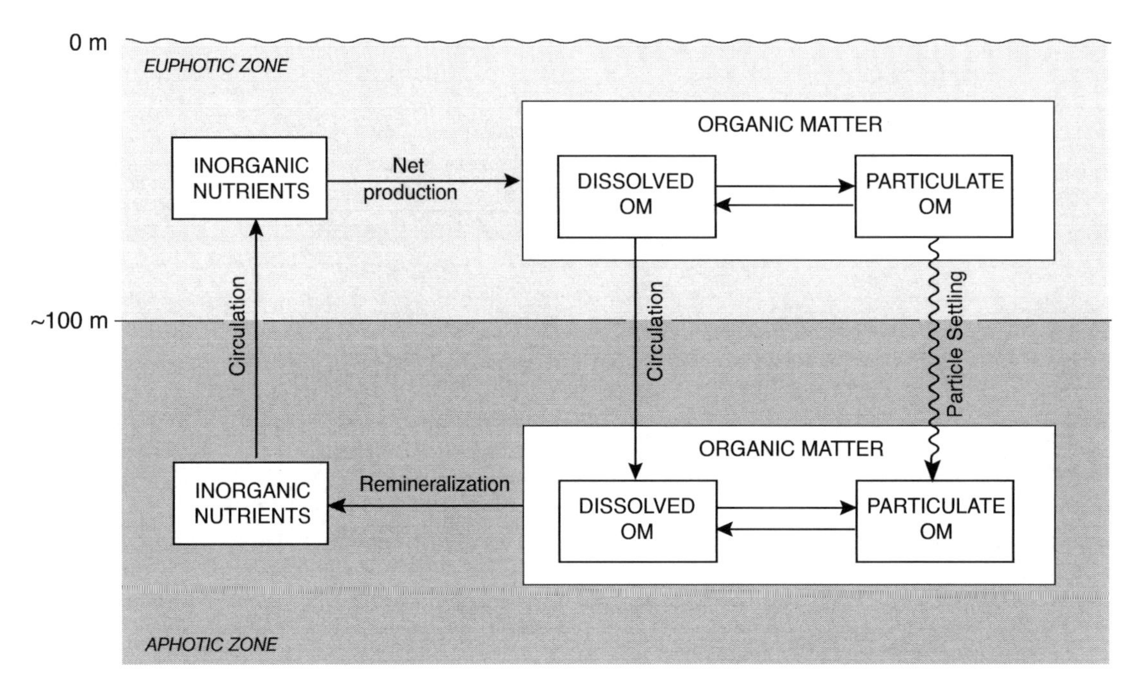

FIGURE 5.1.1: Schematic of the production and remineralization of organic matter in the ocean. Particulate organic matter (*POM*) and dissolved organic matter (*DOM*) are formed in the euphotic zone and then exported by sinking (*POM*) or by transport and mixing (*POM* and *DOM*). In the aphotic zone, *DOM* and *POM* are subject to remineralization, which returns the inorganic constituents back to seawater. Similar processes occur in the sediments, from where the solutes then diffuse back into the water column.

figures 5.1.2 and 5.1.3 the variations of these tracers along a global-scale section that is identical to that used earlier in this text for other tracers (see map in figure 2.3.3a). We can clearly identify the impact of ocean biology in the form of surface depletion and deep enrichment of nitrate and phosphate, and the corresponding decrease in deep ocean oxygen.

We can investigate this apparent coupling between the two nutrients and oxygen by plotting nitrate and oxygen versus phosphate (figure 5.1.4) using data from all depths. A tight correspondence exists between nitrate and phosphate (figure 5.1.4a), with a mean slope of about 16:1, very similar to the relationship that we observed for near-surface waters (figure 4.2.1). It thus appears that most of the remineralization of nitrate and phosphate can be understood simply as the reverse of the photosynthesis reaction (4.2.4). However, we also observe distinct trends in the various ocean basins, with the Atlantic Ocean, for example, having substantially higher nitrate concentrations relative to phosphate. In contrast, the Indian Ocean appears to be depleted in nitrate relative to the mean trend. Do these variations indicate inter-basin trends in the N:P stoichiometry of remineralization of organic matter, or are they driven by other processes? We will see that most of these nitrate "anomalies" are not driven by stoichiometric variations, but are caused by the removal of nitrate by denitrification and the addition of nitrate stemming from the remineralization of nitrogen-rich organic matter formed by N_2-fixers, i.e., organisms that can use dinitrogen gas (N_2) as a source of nitrogen.

Given the overall strong coupling between nitrate and phosphate, the absence of such a coupling between oxygen and phosphate in figure (5.1.4b) is at first surprising. If remineralization of organic matter were to dominate the interior ocean distribution of these two tracers, we would expect all points to lie along a line with a slope of about $-170:1$. This is clearly not the case. Is it the denitrification reaction mentioned above that decouples phosphate from O_2? The tight correspondence between nitrate and phosphate in figure 5.1.4a suggests that this process is unlikely the main cause. Therefore, other processes must be responsible for the decoupling. The only place where this decoupling can occur is the surface ocean, where oxygen can exchange with the atmosphere, while nutrients do not. These signals generated near the surface are then transported into the interior ocean as *preformed* concentrations. All subsequent changes arising from remineralization are changes relative to these preformed concentrations.

A useful reference point for the preformed concentration is a conservative tracer such as salinity that is influenced by the circulation but not affected by biological processes. Figure 2.4.1 shows salinity along the same section as displayed for nitrate and phosphate in figure 5.1.3a and b. The three tracers have almost identical patterns in the deep North Atlantic and in the deeper part of the southern hemisphere thermocline, but very

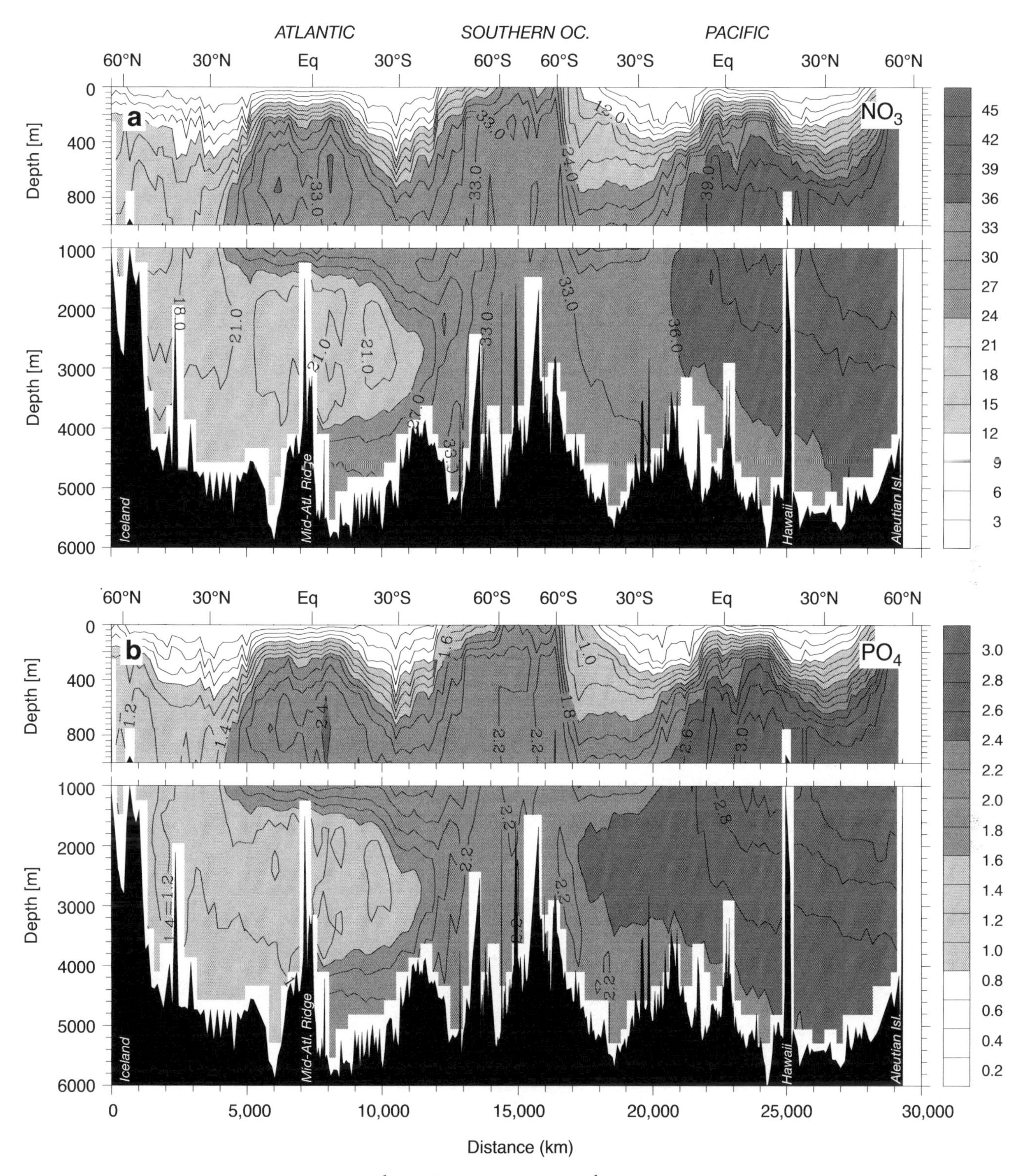

FIGURE 5.1.2: Vertical sections of (a) nitrate ($\mu mol\,kg^{-1}$) and (b) phosphate ($\mu mol\,kg^{-1}$) along the track shown in figure 2.3.3a. See also colorplate 6.

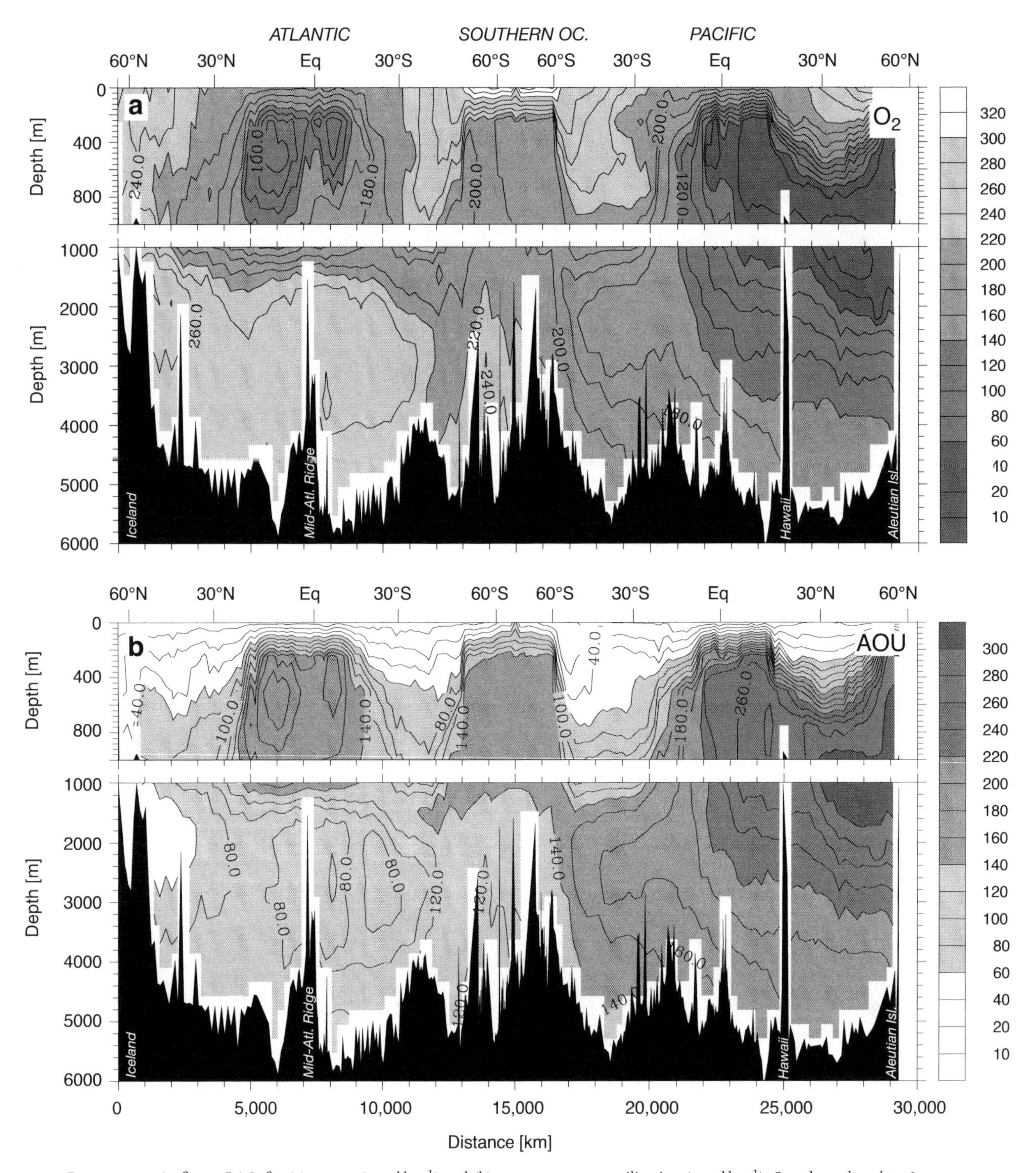

FIGURE 5.1.3: As figure 5.1.2, for (a) oxygen (μmol kg^{-1}) and (b) apparent oxygen utilization (μmol kg^{-1}). See also color plate 6.

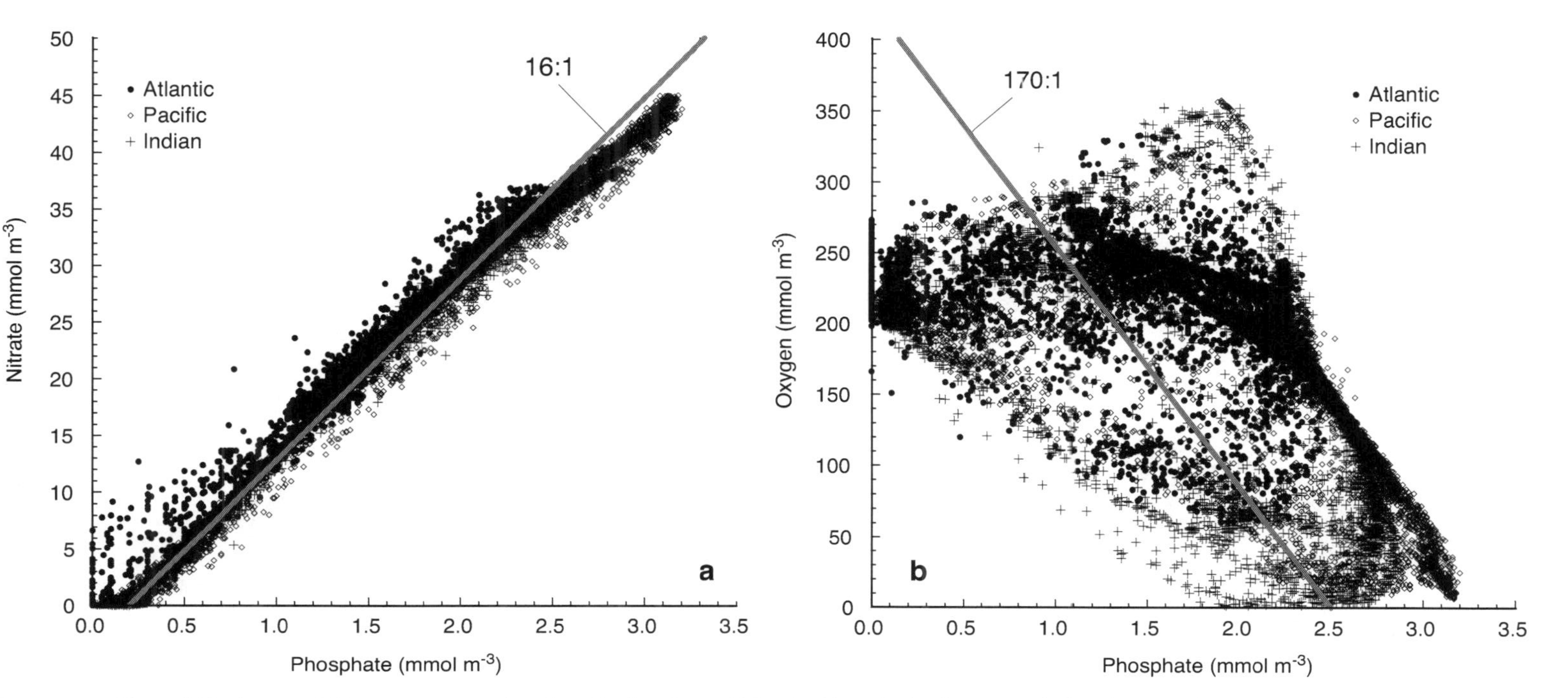

FIGURE 5.1.4: Plots of (a) nitrate and (b) oxygen versus phosphate for data from all depths. Lines depict the expected slope between the tracers as a result of photosynthesis and remineralization. Based on data from three long transects in the Atlantic (OACES NATL93, SAVE), Pacific (WOCE P16), and Indian (WOCE I9NI8S) Oceans.

different patterns in the deep Pacific and low-latitude thermocline. The strong similarity in the deep North Atlantic and the deeper parts of the southern hemisphere thermocline suggest that preformed concentrations play a major role in determining the overall nutrient and oxygen distributions in the interior ocean. If we are to understand the contribution of remineralization to the water column distributions, we need a way of separating its influence from that of the preformed concentration. A method exists to do this, but before we can explain it, we first need to understand more about the remineralization reactions.

Remineralization Reactions

Most of the water column remineralization of organic matter occurs by the reverse of the organic matter synthesis reaction at the surface of the ocean depicted by (4.2.4), i.e.,

$$C_{106}H_{175}O_{42}N_{16}P + 150\ O_2 \rightleftharpoons 106\ CO_2 + 16\ HNO_3 + H_3PO_4 + 78\ H_2O \quad (5.1.1)$$

However, the $C:N:P:O_2$ stoichiometry of the remineralization reaction below 400 m (which is the only depth interval where such estimates have been made on a global scale) appears to be somewhat different:

$$C_{organic}:N:P:O_2 = (117 \pm 14):(16 \pm 1):(1):(-170 \pm 10) \quad (5.1.2)$$

([*Anderson and Sarmiento* 1994]; see panel 4.2.1). Note that the O_2 in (5.1.2) refers only to the free dissolved oxygen that is removed from the water by the remineralization reaction. It does not include the oxygen that is bound up in organic matter. The technique used to obtain the estimate of remineralization stoichiometry given by (5.1.2) did not permit an estimate of the absolute amount of H and O in organic matter. This is because we are unable to detect the water produced by the full remineralization reaction such as illustrated by (5.1.1).

The conversion of nitrogen in organic matter to dissolved inorganic nitrate in (5.1.1) actually occurs in three distinct steps: *ammonification*, *ammonium oxidation*, and *nitrite oxidation*, where the latter two processes in combination are often referred to as *nitrification* (see figure 5.1.5). We explain these here, as we will be examining the nitrogen cycle in some detail in this chapter. Ammonification is the transformation of organic nitrogen to ammonium, NH_4^+. Since organic nitrogen is mostly at the same reduced oxidation state as ammonium, i.e., N(-III), this step usually does not involve a redox reaction, i.e., the transfer of an electron from one molecule to another. Nitrification refers to the oxidation of ammonium to nitrate. This step is done by nitrifying bacteria, which specialize in using the energy from this reaction to take up CO_2 and synthesize organic matter from it. They are therefore, like phytoplankton, autotrophic organisms, except that they use a chemical reaction rather than sunlight as a source of energy. The nitrifying bacteria are thus referred to as *chemoautotrophs*. The two steps of nitrification, i.e., ammonium oxidation (the oxidation of ammonium to nitrite, NO_2^-) and nitrite oxidation (the conversion of nitrite to nitrate) are most often done by distinct groups of organisms. The best known are *Nitrosomonas spp.*, an ammonium oxidizer, and *Nitrobacter spp.*, a nitrite oxidizer. Nitrification requires the presence of oxygen and tends to be inhibited by light, as noted in chapter 4. To summarize, the following three distinct reactions remineralize organic nitrogen to nitrate:

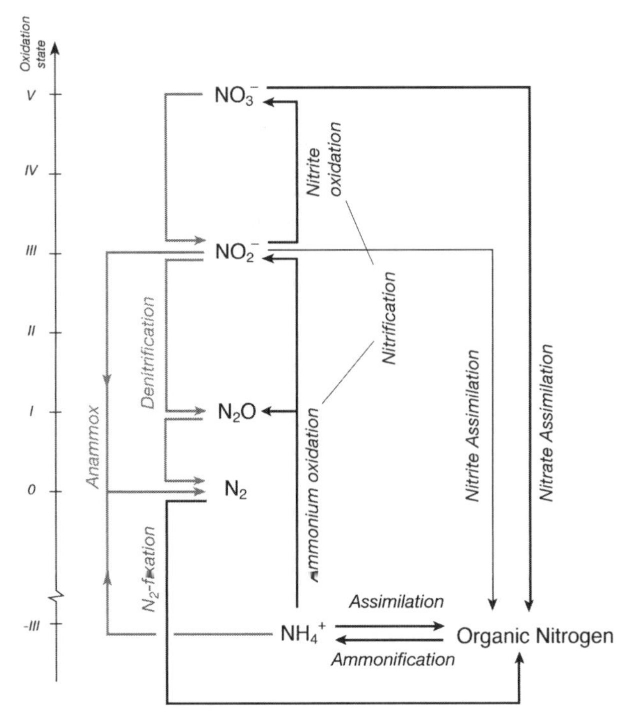

Figure 5.1.5: Transformation pathways of nitrogen in the ocean. Most of the fixed nitrogen atoms in the ocean follow a pathway that consists of NO_3^- and NH_4^+ uptake by phytoplankton, followed by *ammonification* (the conversion of organic N to NH_4^+). The resulting NH_4^+ is then oxidized in two reaction steps back to NO_3^-, collectively referred to as *nitrification*. Under anoxic conditions, NO_3^- gets used as an oxidant in bacterial respiration processes, termed *denitrification*, leading to its stepwise reduction to N_2. N_2 can also be *fixed* into organic matter by a few specialized organisms, called *diazotrophs*, balancing the loss of fixed nitrogen in the ocean. N_2O gets produced as a side-product in association with both nitrification and denitrification. N_2 can also be produced by the reaction of NH_4^+ with NO_2^-, a process called Anammox.

$$\text{Ammonification} : \text{org.N} \rightleftharpoons NH_4^+$$
$$\text{Ammonium oxidation} : 2NH_4^+ + 3O_2 \rightarrow 2NO_2^- + 4H^+ + 2H_2O$$
$$\text{Nitrite oxidation} : 2NO_2^- + O_2 \rightarrow 2NO_3^- \quad (5.1.3)$$

The remineralization reaction depicted by (5.1.1) requires oxygen and, except for many bacteria, is used by nearly all heterotrophic organisms, including all zooplankton, and fish. Most bacteria use this aerobic remineralization reaction as well, since the energy yield per unit of organic matter is higher in comparison to all other available oxidants. Most organisms are strict aerobes, i.e., they cannot switch to another oxidant for oxidizing organic matter to provide energy for their metabolism, and so they cannot live in the absence of free oxygen. However, many bacteria are more flexible and can use a wide range of other reactions to obtain energy from organic matter (see detailed discussion in chapter 6). The most common anaerobic process in the water column is *denitrification*. In the denitrification process, nitrate replaces oxygen as the oxidant, and is thereby reduced to N_2. The overall denitrification reaction for organic matter of composition (5.1.1) is:

$$C_{106}H_{175}O_{42}N_{16}P + 104\ HNO_3 \rightarrow 106\ CO_2 + 60\ N_2 + H_3PO_4 + 138\ H_2O \quad (5.1.4)$$

[*Gruber and Sarmiento*, 1997]. Unfortunately, it is not possible to work out the correct denitrification stoichiometry for organic matter with the deep ocean remineralization stoichiometry (5.1.2). This is because (5.1.2) does not include the correct organic O and H stoichiometry, as noted above.

Another anaerobic process that removes fixed nitrogen from the water is the *anaerobic ammonium oxidation* (anammox) process, in which bacteria combine ammonium and nitrite to form N_2 under anoxic conditions. This process, first discovered in wastewater bioreactors, appears to occur also in oceanic environments (see, e.g., *Kuypers et al.* [2003]). Its quantitative significance is not known yet on a global scale, but a recent study suggested that the anammox process could be responsible for most of the fixed nitrogen loss in the oxygen minimum zone of the Benguela upwelling system [*Kuypers et al.*, 2005]. We won't discuss this process further, but note that from a geochemical perspective, the anammox and denitrification processes play the same role in that they remove nitrate from the water while adding phosphate from the associated remineralization of organic phosphorus.

The loss of fixed nitrogen by denitrification and the anammox process requires a counterbalancing source elsewhere if the nitrate content of the ocean is to remain constant through time. The primary source is the N_2 fixation reaction discussed in the previous chapter. The N_2 fixation reaction converts N_2 to organic nitrogen, which ends up as nitrate upon subsequent reaction of the organic matter with oxygen. The extent to which the oceanic fixed nitrogen budget is actually in balance is controversial at present, as will be discussed below [e.g., *McElroy*, 1983; *Codispoti and Christensen*, 1985; *Gruber and Sarmiento*, 1997; *Codispoti et al.*, 2001; *Gruber*, 2004].

Not all nitrogen contained in organic matter in reactions (5.1.1) and (5.1.4) gets completely converted to nitrate or N_2. Some small fraction ends up as nitrous oxide, N_2O, part of which is then emitted into the atmosphere. We have seen in chapter 3 that N_2O, like CO_2, acts as a greenhouse gas; therefore variations of this gas in the atmosphere can lead to changes in Earth's temperature and climate. As we will see below, the oceanic production of N_2O constitutes a substantial fraction of the total emission of N_2O into the atmosphere. It therefore behooves us to understand the processes that form N_2O in the ocean and what controls them. In the case of aerobic remineralization, the formation of N_2O is associated with the oxidation of ammonium during nitrification (figure 5.1.5). Nitrous oxide is also formed during denitrification, as it represents an intermediate product during the reduction of nitrate to N_2 (figure 5.1.5). If some of this N_2O escapes the further reduction to N_2, denitrification can act as a source of N_2O. At the same time, denitrification can act as a sink for N_2O, with the overall balance generally believed to be positive, i.e., denitrification acting as a net source of N_2O. The relative importance of the two production pathways is still debated, but it appears as if the majority of the N_2O in the ocean is formed in association with nitrification [*Suntharalingam and Sarmiento*, 2000; *Nevison et al.*, 2003; *Jin and Gruber*, 2003]. However, given the much higher yield of N_2O at low oxygen concentrations, the production is biased toward the low-oxygen regions of the world ocean. We will discuss the impact of this on the oceanic distribution of N_2O, thereby permitting us to explain the distribution of surface N_2O we briefly covered in chapter 3.

Preformed and Remineralized Components

In the most general form, we can think of the nutrient or oxygen concentration of a water parcel in the interior of the ocean as consisting of two components. The first is the preformed concentration, $C_{preformed}$, that this water parcel had when it left the surface, and the second one is the change in concentration, ΔC_{remin}, that occurred since that time as a result of remineralization, i.e.,

$$C_{observed} = C_{preformed} + \Delta C_{remin} \quad (5.1.5)$$

We can use the three-box ocean model of figure 1.2.5 to provide a more detailed illustration of the difference between the preformed and remineralized components of deep ocean nutrient concentrations. If we assume that the low-latitude surface nutrient concentration C_l is 0, such as would be the case for nitrate or phosphate over much of the ocean, the flux balance of the low-latitude surface box gives for the low-latitude organic matter export and deep water remineralization term Φ_l the expression:

$$\Phi_l = T \cdot C_d \qquad (5.1.6)$$

where T is the rate of upwelling associated with the thermohaline overturning, and C_d is the deep ocean nutrient concentration. Recall that this model ignores sediment burial, so the flux of organic matter to the deep ocean is equal to the rate of remineralization in the deep ocean. This equation tells us that the export of organic matter from the low-latitude surface box is equal to the nutrient input by upwelling.

The high-latitude surface box also has an organic matter export flux and deep water remineralization equal to the net input of nutrients by ocean circulation. However, the expression is more complicated because the high-latitude nutrient concentration C_h is not zero:

$$\Phi_h = f_{hd} \cdot C_d - (f_{hd} + T) \cdot C_h \qquad (5.1.7)$$

The transport term f_{hd} represents the two-way exchange of surface high-latitude waters with the deep ocean. It brings deep ocean water to the surface and exports surface high-latitude water to the deep ocean in combination with the thermohaline transport term T.

Summing (5.1.6) and (5.1.7) and solving them for the deep ocean concentration gives the expression

$$C_d = C_h + \frac{\Phi_l + \Phi_h}{f_{hd} + T} \qquad (5.1.8)$$

From this we see that the deep ocean concentration is made up of the high latitude concentration C_h brought in from the surface outcrop region, which is the preformed concentration, $C_{preformed}$, and a second component due to the remineralization of organic matter divided by the total rate of deep water formation. This second term is equal to ΔC_{remin}. The first problem we discuss in section 5.2 is how to separate the preformed component from the remineralized component. The most useful tracer for this purpose is oxygen, because we are able to estimate its surface, i.e., preformed, concentration. We do this by assuming it is at or near equilibrium with the atmosphere.

The second problem we discuss in section 5.2 is how to estimate the rate of organic matter remineralization based on the rate of oxygen utilization. Rearranging (5.1.8) to solve for the sum of the two export fluxes, which in steady state are equal to the total interior ocean remineralization rate, gives the expression

$$\Phi_l + \Phi_h = (C_d - C_h) \cdot (f_{hd} + T) \qquad (5.1.9)$$

We can estimate the rate of remineralization if we know the portion of the concentration that is due to remineralization, i.e., the difference between *in situ* and preformed concentration, this being the solution to the first problem discussed in section 5.2, and if we know the rate of ocean circulation. The problem of estimating the remineralization rate from oxygen thus reduces to determining the rate of ocean circulation.

The three-box model is, of course, highly simplified. The actual distribution of nutrients in the deep ocean results from a complex interplay between the layered pattern of thermocline and deep ocean circulation discussed in chapter 2, and the remineralization of organic matter exported from the surface ocean. We use observations of tracers to help us determine ocean circulation rates.

Dissolved and Particulate Organic Matter

Remineralization cannot occur without a continuous supply of organic matter. Chemical oceanographers generally classify organic matter as either particulate (*POM*) or dissolved (*DOM*). *DOM* is operationally defined as the component of organic matter that passes through a filter of a given pore size, commonly 0.45 μm. However, this definition is not entirely satisfactory. For one thing, such a filter lets through many or most of the bacteria [*Lee and Fuhrman*, 1987]. Bacteria can only be removed by a 0.2 or 0.4 μm pore membrane filter. For another thing, even this smaller size range allows colloids through (see review by *Fuhrman* [1992]). Colloids have been operationally defined as particles with a molecular weight above 10,000 and having a size smaller than about 0.4 μm [*Baskaran et al.*, 1992]. To complicate matters, the measurement of organic matter, except in coastal environments, is now often made on water samples that have not been filtered. The measurement that is obtained is thus total organic matter (*TOM*). This is done so as to avoid problems with contamination that arise from handling the samples, and because in open ocean environments, the contribution from *POM* is very small, so that *DOM* is very nearly equal to *TOM* there [*Sharp et al.*, 1993; *Sharp*, 2002]. In coastal waters, where the contribution from *POM* is large, samples are usually filtered. To avoid confusion, we will continue to use the term *DOM*, even when the actual measurement was actually *TOM*.

Although the separation of organic matter into *DOM* and *POM* is largely driven by its operational simplicity, one of the primary merits is that the resulting *DOM* includes all organic matter that is either truly dissolved or in particulate form, but nonsinking. We can therefore regard *DOM* as that component of organic matter that is transported by ocean circulation and mixing, whereas *POM* is that component for which sinking is important. Another argument for this separation is that the pool sizes of these two components are very different. In terms of carbon, the dissolved pool comprises about 700 Pg C (1 Pg $= 10^{15}$ g), which is larger than the amount of CO_2 contained in the atmosphere in preindustrial times (600 Pg C) and is about equal to the living biomass on land. The *DOM* pool is also two orders of magnitude larger than the particulate pool, which has been estimated to be a few Pg C only. Although the simple separation into *DOM* and *POM* characterizes already much of the fundamentally different dynamics of these two pools, it is important to develop a better understanding of the various components within these two pools. As we will see below, turnover times within the *DOM* pool range from a few days to centuries, so without a better characterization it will be very difficult to capture this enormous range in behavior. While substantial progress has been made over the last two decades (see the book edited by *Hansell and Carlson* [2002] for an excellent summary of the present status of the field), the majority of the *DOM* pool is still chemically uncharacterized [*Benner*, 2002]. We will discuss the dynamics of *DOM* and *POM* in section 5.4.

Outline

We proceed in section 5.2 to examine the effect of remineralization on the distribution of oxygen, and in section 5.3 to a discussion of the phosphate and nitrate distributions and their cycles. In section 5.4, we turn to an examination of the particulate and dissolved organic matter cycling processes that give rise to the observed oxygen and nutrient distributions. The chapter ends in section 5.5 with a discussion of models of the cycling of organic matter. These models enable us to examine the connection between the organic matter processes discussed in section 5.4 and the oxygen and nutrient distributions discussed in sections 5.2 and 5.3.

5.2 Oxygen

The distribution of oxygen is the best place to start our discussion of the impact of remineralization on tracer distributions because of the possibility that exists for easily separating the preformed from the remineralized component. This separation will be the subject of the first subsection, followed in subsequent subsections with a discussion of estimates of the rate of oxygen utilization based on the combined use of oxygen and observations of tracers of ocean circulation.

Separation of Preformed and Remineralized Components

In order to study the impact of aerobic remineralization on the ocean interior distribution of oxygen, we would like to estimate its remineralized component i.e., the change in O_2 that occured since a water parcel was last in contact with the atmosphere:

$$\Delta[O_2]_{remin} = [O_2]_{observed} - [O_2]_{preformed} \tag{5.2.1}$$

In the case of oxygen, we are aided by the fact that, at the surface of the ocean, oxygen is generally close to its saturation concentration (cf. chapter 3), i.e., $[O_2]_{preformed} = [O_2]_{sat}$. The main theoretical argument in support of this assumption is that the exchange of O_2 across the air-sea interface is fast relative to perturbations of this balance, so that it is difficult to maintain large super- or undersaturations for a substantial amount of time. Since the saturation concentration can be computed for each water parcel from the observed potential temperature, θ, and salinity, S, i.e., $[O_2]_{sat} = f(\theta_{observed}, S_{observed})$, we can thus estimate the remineralized component of oxygen, for which the term *apparent oxygen utilization* (*AOU*) is commonly used:

$$AOU = [O_2]_{sat} - [O_2]_{observed} \tag{5.2.2}$$

where $AOU = -\Delta[O_2]_{remin}$. Note that *AOU* does not include the effect of remineralization that is due to denitrification. Also, it should be kept in mind that the actual preformed oxygen concentration is usually not exactly at saturation. The oxygen concentration over most of the low-latitude ocean is slightly supersaturated by an average of about 7 mmol m^{-3}(figure 3.1.2) even in the wintertime [e.g., *Jenkins and Goldman*, 1985; *Emerson et al.*, 1991; *Carlson et al.*, 1994]. These supersaturated waters ventilate the upper thermocline. Deeper density surfaces, at the base of the thermocline and below, outcrop in high-latitude regions, where the surface waters can be undersaturated by in excess of 100 mmol m^{-3} during the wintertime [e.g., *Peng et al.*, 1987] (see figure 3.1.1). It turns out that there are very few places where such large undersaturations occur, rendering the problem less dramatic. Nevertheless, the absolute magnitude of deep ocean *AOU* should be viewed with caution, although gradients in *AOU* would be less affected by this problem.

Figure 5.1.3b shows *AOU* computed from the observed O_2 data shown in figure 5.1.3a. The surface concentrations of *AOU* are near 0 or slightly negative due to the previously noted supersaturation. The high *AOU* values of the thermocline are well in excess of any uncertainties that arise from the assumptions inherent in the definition of *AOU*. One of the important questions we must address is the relative contributions of the rate of circulation versus the rate of remineralization to the signals that we see. In particular, are the higher *AOU* signals in the thermocline a result of larger rates of remineralization or slower circulation? We expect the former, given the fact that thermocline ventilation is very rapid (chapter 2). But why is the *AOU* signal in the equatorial thermocline so much larger than that in the subtropical thermocline? Is it because of high export of organic matter from the surface in the tropical regions causing a high rate of remineralization, or is it because of a more sluggish ventilation of the tropical thermocline in contrast to the extratropical thermocline? Finally we also see a large deep Pacific remineralization signal (the increase in concentration from south to north), and an indication that the deep Atlantic has very little increase in AOU along the pathway of the NADW (see chapter 2 for water mass nomenclature). Is less organic matter reaching the deep Atlantic in comparison to the Pacific, or has this difference something to do with ocean circulation and mixing?

In order to address these questions, we need to combine AOU with information about the rates of ocean transport and mixing. In particular, we will use tracer observations to estimate the equivalent of the f_{hd} and T terms in (5.1.9), permitting us to calculate the time rate of change of AOU, i.e., the oxygen utilization rate (OUR) associated with the remineralization of organic matter,

$$OUR = \frac{dAOU}{dt} \tag{5.2.3}$$

We do this first for the deep ocean, and then proceed to a discussion of the thermocline.

Deep Ocean Oxygen Utilization Rates

An estimate of the OUR in the deep Pacific can be obtained from the $25\times10^6\,\mathrm{m^3\,s^{-1}}$ ventilation rate estimated from GEOSECS radiocarbon observations by *Stuiver et al.* [1983] for the region north of 50°S and below 1500 m depth (see chapter 2). In the model of *Stuiver et al.* [1983], this flow represents water entering from the circumpolar region at 50°S and upwelling over the entire Pacific across the 1500 m depth contour. The average AOU concentration of the inflowing water at 50°S is about 150 ± 10 mmol m^{-3}, and that of the outflowing water at 1500 m is 225 ± 48 mmol m^{-3}. The steady-state AOU balance equation analogous to (2.4.5) is thus:

$$I_{Pacific} \cdot AOU_{in} + V_{Pacific} \cdot OUR = W_{Pacific} \cdot AOU_{out} \tag{5.2.4}$$

where $I_{Pacific} = W_{Pacific} = 25$ Sv is the ventilation rate of the deep Pacific (see chapter 2), and $V_{Pacific} = 0.41\times10^{18}\,\mathrm{m^3}$ is its volume. Solving (5.2.4) for OUR gives:

$$\begin{aligned} OUR &= \frac{I_{Pacific}(AOU_{out} - AOU_{in})}{V_{Pacific}} \\ &= \frac{(25\cdot10^6\mathrm{m^3s^{-1}})(225-150)\mathrm{mmol\ m^{-3}}}{0.41\times10^{18}\mathrm{m^3}} \\ &= 0.14\,\mathrm{mmol\ m^{-3}yr^{-1}} \end{aligned} \tag{5.2.5}$$

Including the uncertainties, the OUR in the deep Pacific is thus 0.14 ± 0.09 mmol m^{-3} yr^{-1}. In other words, in one century, the oxygen concentration decreases by about 14 mmol m^{-3} due to remineralization in the deep Pacific below 1500 m depth.

The deep Atlantic is more difficult to analyze because of the mixing of different water types, as discussed in chapter 2. In particular, as we follow the pathway of NADW from the North Atlantic toward the south, some of the increase in AOU is due to mixing from the AOU-rich AAIW above and AABW below. We have to remove this mixing effect in order to identify the effect of oxygen utilization by remineralization. Furthermore, because of uncertainties in the actual preformed oxygen concentration, the use of $[O_2]_{sat}$ to represent the preformed oxygen concentration is not sufficiently accurate. *Broecker et al.* [1991] have accomplished the job of identifying the effect of mixing by using the PO_4^* tracer defined by (2.4.2) to separate northern and southern water type components for Atlantic waters below 2000 m per (2.4.3). They analyze observations to estimate an initial O_2 concentration of 288 mmol m^{-3} for the northern component water, and 257 mmol m^{-3} for southern component water. This gives

$$\begin{aligned} &[O_2]_{initial} \\ &= f_n \cdot (288\,\mathrm{mmol\ m^{-3}}) + (1-f_n)\cdot(257\,\mathrm{mmol\ m^{-3}}) \end{aligned} \tag{5.2.6}$$

for the expected initial, i.e., preformed, oxygen concentration. f_n is the fraction of northern component water from (2.4.3). The oxygen deficit due to remineralization, i.e., $\Delta[O_2]_{remin}$, is then given by (5.2.1). These equations can be compared with the analogous equations (2.4.4) for radiocarbon.

The resulting map of $\Delta[O_2]_{remin}$ (figure 5.2.1) mirrors the radiocarbon age map (figure 2.4.5). The oxygen deficit $\Delta[O_2]_{remin}$ is largest in the eastern basin of the Atlantic, where the radiocarbon age is greatest. The lowest $\Delta[O_2]_{remin}$ and youngest ages are found along the western boundary, where the inflow of deep water is strongest.

One can calculate an average OUR simply by dividing the $\Delta[O_2]_{remin}$ by the radiocarbon age. In the Caribbean, $\Delta[O_2]_{remin}$ is ~60 mmol m^{-3}, and the radiocarbon decay of 20%, equivalent to 160 years, gives for OUR a value of 38 mmol m^{-3} per century, 2.7 times the deep Pacific OUR. However, the Caribbean appears to be an exception in this regard. Figure 5.2.2 shows a plot of the oxygen deficit versus radiocarbon decay data from the Atlantic Ocean. A linear regression of these data, corrected for contamination of radiocarbon by the bomb transient, gives an average OUR of about 12 mmol m^{-3} per century.

We thus find that the OURs estimated for the deep Atlantic below 2000 m and the deep Pacific below 1500 m are essentially the same: 12 versus 14 mmol m^{-3} per century. The difference in AOU signals between these two basins is due to the much slower rate of circulation in the Pacific Ocean. On the other hand, the Caribbean has

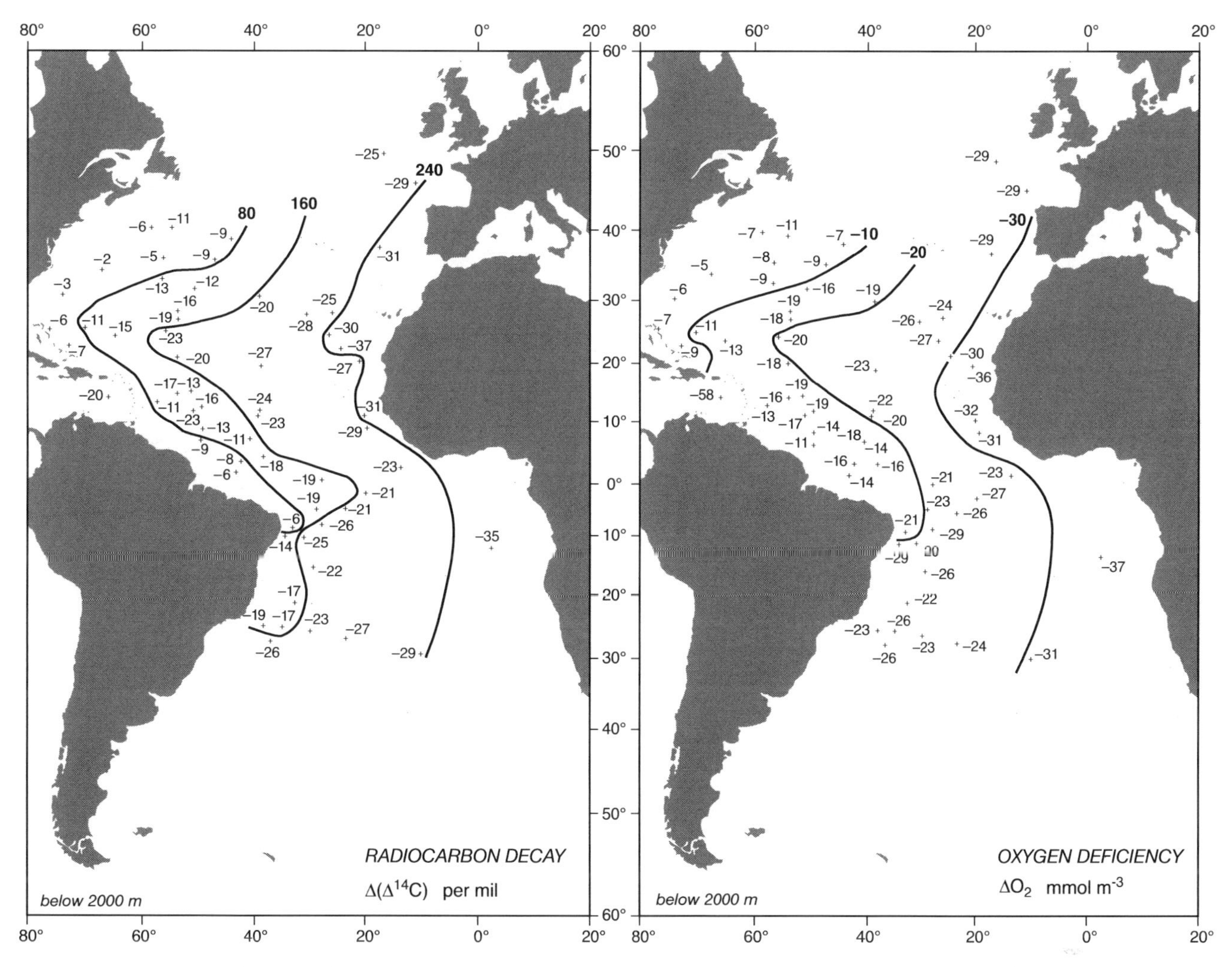

FIGURE 5.2.1: Maps of the deficiency of (a) radiocarbon and (b) oxygen in the deep Atlantic due to decay of the former and biological consumption of the latter as estimated by *Broecker et al.* [1991]. The numbers shown are vertical averages below 2000 m.

an exceptionally large OUR of 38 mmol m^{-3} per century, implying an organic matter flux about 3 times greater than the Atlantic and Pacific averages. This could be due to a higher export of organic matter from the surface, or a less efficient remineralization of organic matter above 2000 m.

Thermocline Oxygen Utilization Rates

In chapter 2, we studied the thermocline circulation by considering the distribution of various properties along isopycnal surfaces, primarily in the North Atlantic. One of the tracers we studied was the distribution of age determined by the tritium/helium-3 technique (figure 2.3.11). Figure 5.2.3 shows the distribution of AOU along the same set of surfaces. As with the deep Atlantic radiocarbon age and oxygen deficit maps of figure 5.2.1, the AOU plots of figure 5.2.3 show basically the same pattern as the tritium/helium-3 ages of figure 2.3.11. The AOU is low near the outcrop regions, and high in the shadow zone south of about 15°N.

One can estimate the thermocline OUR by dividing the AOU by the tritium/helium-3 age. Figure 5.2.4 shows OUR results obtained by *Jenkins* [1987] using earlier tritium/helium-3 age estimates and oxygen observations from the Atlantic Ocean "Beta Triangle" region in the vicinity of 30°N 30°W. Figure 5.2.4 also shows results from a number of additional Atlantic Ocean OUR estimates obtained by other techniques about which more shall be said below.

The first point to note about the results in figure 5.2.4 is that the Atlantic Ocean thermocline OUR is of order 10 mmol m^{-3} yr^{-1}, two orders of magnitude greater than the deep ocean OURs summarized in table 5.2.1. However, the depth range of the thermocline is less than that of the deep ocean. Table 5.2.2 summarizes the vertically integrated thermocline remineralization obtained from all the data of figure

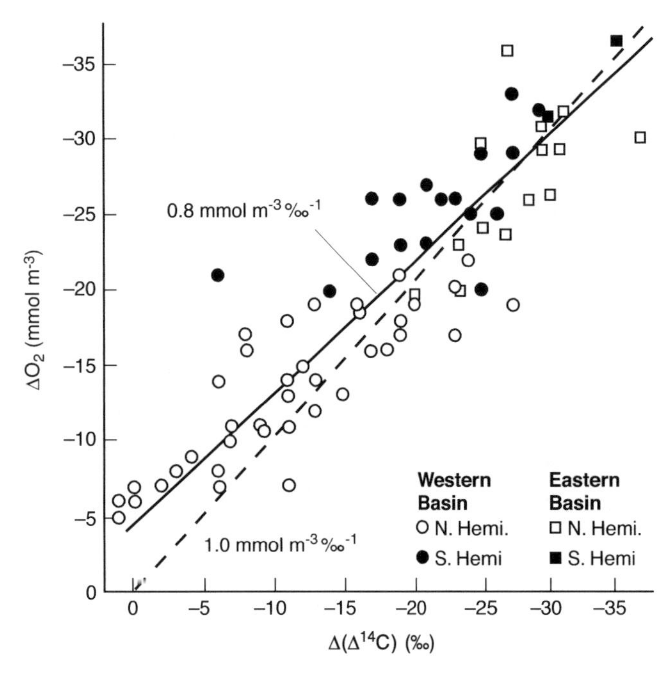

FIGURE 5.2.2: Plot of oxygen deficiency versus radiocarbon deficiency using data from figure 5.2.1. The slope of this line yields the oxygen utilization rate (*OUR*) [*Broecker et al.*, 1991].

5.2.4. The average vertically integrated thermocline remineralization for the depth range from 100 to 750 m is 5.5 mol O_2 m^{-2} yr^{-1}. If we use a depth range of 2000 m to 6000 m for the deep ocean, we obtain a vertically integrated remineralization of 0.4 mol O_2 m^{-2} yr^{-1}, about 7% of the total. Clearly, the great majority of the remineralization occurs within the main thermocline. Note that our estimate of 7% remineralization in the Atlantic below 2000 m is an upper limit, since our calculation does not include any constraint on the *OUR* in the depth range between 750 m and 2000 m.

A second point to note about figure 5.2.4 is the tendency for *OUR* to be higher in the upper thermocline than deeper down. This tendency, which is not uniform across all the data sets shown, is as expected from the fact that organic matter is produced in the upper ocean above about 100 m, and thus would be fresher and more abundant in the upper thermocline. Direct observations of both the particulate and dissolved organic matter pools confirm that they are indeed more abundant in the upper thermocline (see section 5.4). As this organic matter settles or is transported downward, one can assume that it becomes not only less abundant but also

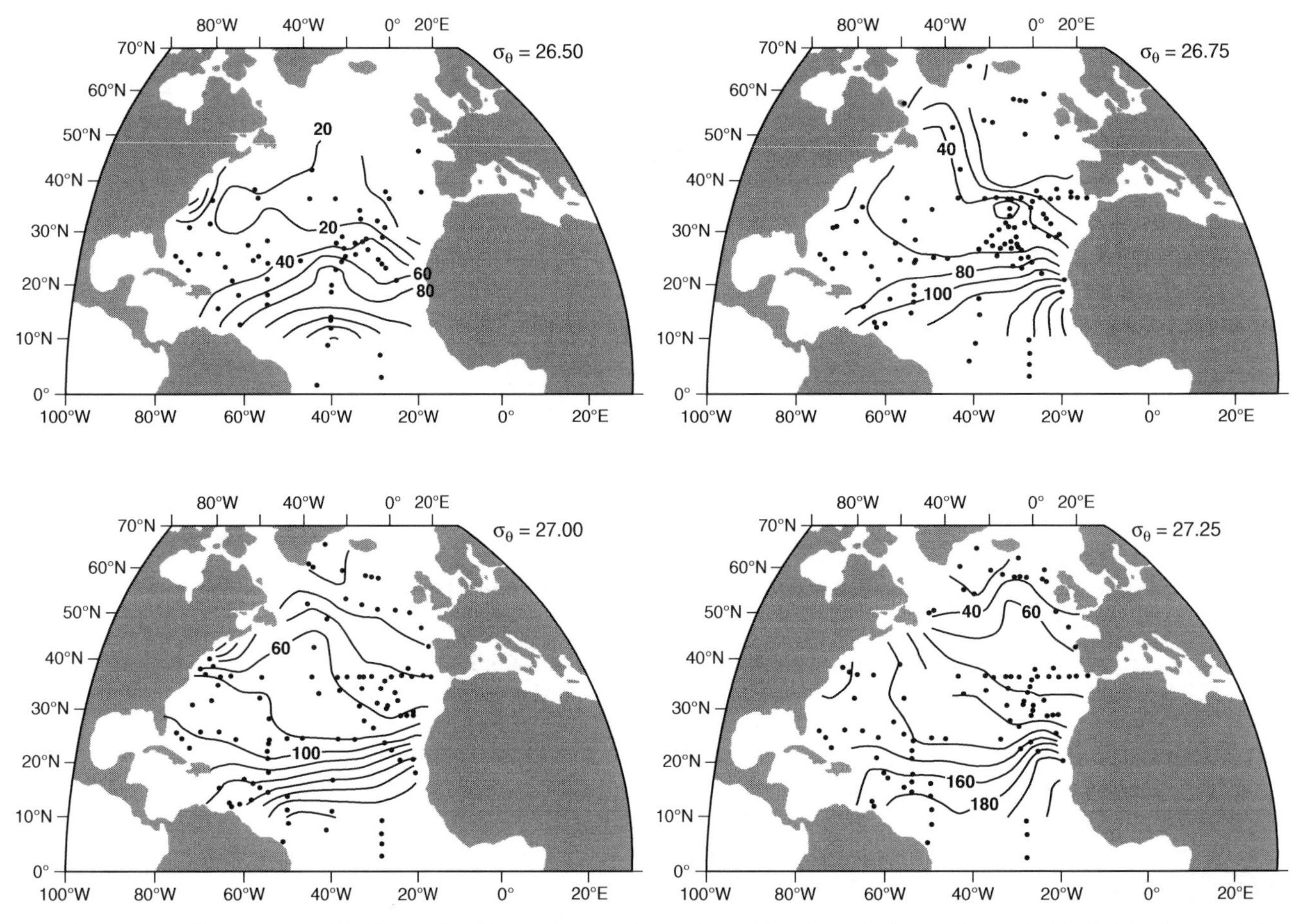

FIGURE 5.2.3: Contour maps of *AOU* (mmol m^{-3}) on various density surfaces within the main thermocline of the North Atlantic [from *Jenkins and Wallace*, 1992].

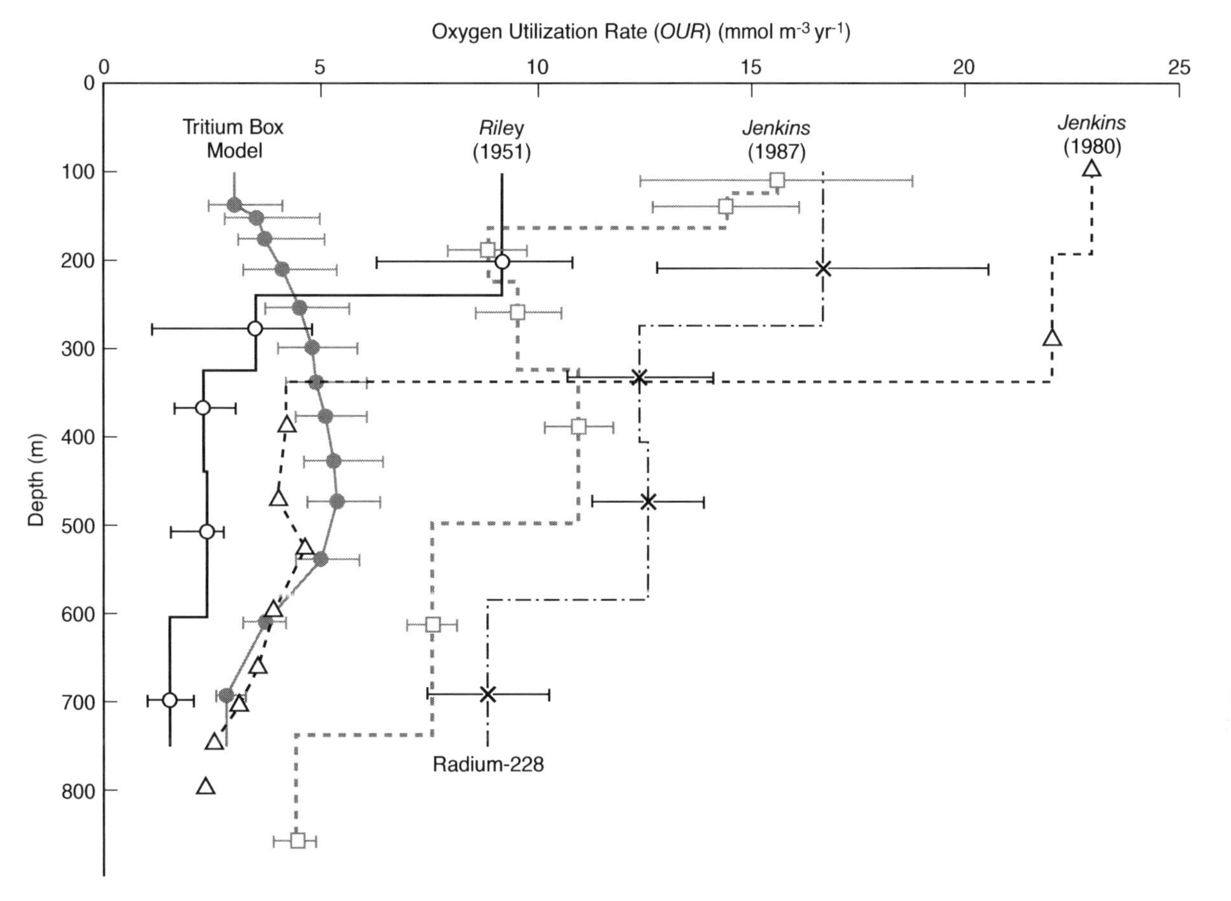

FIGURE 5.2.4: A summary of *OUR* estimates from the thermocline of the North Atlantic (from *Sarmiento et al.* [1990]).

more refractory. Both effects, then, decrease the rate of remineralization. We will discuss this in greater detail in section 5.4 below.

Figure 5.2.4 and table 5.2.2 show *OUR*s obtained by a variety of techniques other than the use of tritium/helium-3 ages. The tritium box model [*Sarmiento et al.*, 1990; *Jenkins*, 1980] results were obtained by the box model technique described in chapter 2 (equation (2.3.10) and figure 2.3.9). The *Riley* [1951] estimate is based on a model of geostrophic transport and mixing. The radium-228 technique [*Sarmiento et al.*, 1990] obtains a timescale from the distribution of the 5.75-year half-life radium-228 tracer. This tracer forms in shelf sediments by decay of its parent thorium-232. The presumption is that it enters the density surfaces primarily via the outcrops, just as does oxygen. If so, and if the mixing of this tracer is small relative to advection, one can use the decay of radium-228 on the isopycnal surface in order to obtain an age estimate.

Some of the differences between the results shown in figure 5.2.4 stem also from uncertainties in estimating the aging of waters from the various tracers. One particular issue is associated with the degradation of the age tracer resulting from mixing [*Jenkins*, 1987], since most traditional age tracers assume that flow occurs advectively with minimal dispersal. In the case of the ages estimated from tritium/helium or radium-228, the degradation occurs because the decay of the parent and ingrowth of the daughter are exponential with time rather than linear [cf. *Sarmiento et al.*, 1990; see also *Robbins et al.*, 2000]. For example, *Doney and Jenkins* [1988] have demonstrated that the tritium box model ages are probably too old (see figure 2.3.10), which would tend to give an underestimate for *OUR*, particularly in shallower waters. *Jenkins and Wallace* [1992]

TABLE 5.2.1
Deep ocean oxygen utilization rate (*OUR*) estimates

Region	*Reference*	*OUR* (*mmol* m^{-3} *per century*)
Pacific below 1500 m		14
Atlantic below 2000 m	[*Broecker et al.*, 1991]	12
Carribean below 2000 m	[*Broecker et al.*, 1991]	36

TABLE 5.2.2
Estimates of the vertically integrated remineralization in the thermocline of the North Atlantic from 100 to 750 m depth [*Sarmiento et al.*, 1990] in units of oxygen consumption

Source	*Technique*	*Region*	*Remineralization (OUR) (mol O_2 m^{-2} yr^{-1})*
Riley [1951]	Geostrophic transport and mixing model	54°S to 45°N	2.5 (1.6 to 3.1)
Jenkins [1980]	Tritium/He-3 box model	Sargasso Sea	7.2
Jenkins [1987]	Tritium/He-3 age versus *AOU*	Beta Triangle (~30°N 30°W)	6.4 ± 0.3
Sarmiento et al. [1990]	Tritium box model	15°N to $\sigma_\theta = 27.45$ outcrop	2.8 (2.4 to 3.5)
Sarmiento et al. [1990]	Radium-228	15°N to $\sigma_\theta = 27.45$ outcrop	8.5 ± 0.8

conclude that the overall effect of mixing is to give an uncertainty of about 20% in the *OUR* estimates (see also *Waugh* [2003] for a recent assessment of uncertainties). One way to decrease the uncertainties arising from these age estimates is to use multiple tracers with different timescales, which permit computation of a full age-spectrum (see, e.g., *Hall and Haine* [2002]).

5.3 Nitrogen and Phosphorus

We examine the primary impact of the remineralization of organic matter on the oceanic distribution of nitrate and phosphate by keying ourselves to the impact of this remineralization process on oxygen. In other words, we estimate the effect of remineralization on the nitrate and phosphate distribution using the stoichiometric ratios of these two to oxygen. The stoichiometric ratios are discussed in the following subsection, after which we turn to a discussion of phosphate, then nitrate. A phenomenon of particular interest in this analysis is the behavior of nitrate, which differs fundamentally from that of phosphate because of the ability of oceanic organisms to consume nitrate through the process of denitrification, and produce it from N_2 through the mechanism of N_2 fixation followed by nitrification of the resulting nitrogen-rich organic matter.

STOICHIOMETRIC RATIOS

Table 4.2.1 and panel 4.2.1 of chapter 4 provide a summary of a variety of estimates of the stoichiometry of organic matter. The most recent global-scale study of the stoichiometry of remineralization given in table 4.2.1 is that of *Anderson and Sarmiento* [1994]. The technique used by them was a nonlinear inverse fit of observations from the GEOSECS program to a one– or two–end-member mixing model along neutral surfaces as defined by *McDougall* [1987]. Similar techniques have been used more recently on the basis of data from the WOCE program, but in limited regions or depth horizons so far [e.g., *Hupe and Karstensen*, 2000].

Generally, an end-member mixing model assumes that a given water parcel consists of a mixture of n distinct water types, with end-member properties denoted by the subscripts i, which contribute fractions f_i to a given water parcel. Therefore, the observed potential temperature θ, salinity S, phosphate PO_4^{3-}, nitrate NO_3^-, oxygen O_2, dissolved inorganic carbon *DIC*, and alkalinity *Alk*, are equal to the sum of the respective end-members multiplied by the appropriate mixing fraction. An additional term is needed to account for the effect of remineralization. The final mixing model equations are:

$$\begin{aligned}
1 &= \sum_{i=1}^{n} f_i \\
\theta_{observed} &= \sum_{i=1}^{n} f_i \theta_i \\
S_{observed} &= \sum_{i=1}^{n} f_i S_i \\
[PO_4^{3-}]_{observed} &= \sum_{i=1}^{n} f_i [PO_4^{3-}]_i + \Delta[PO_4^{3-}]_{remineralized} \\
[NO_3^-]_{observed} &= \sum_{i=1}^{n} f_i [NO_3^-]_i + r_{N:P}\Delta[PO_4^{3-}]_{remineralized} \\
[O_2]_{observed} &= \sum_{i=1}^{n} f_i [O_2]_i + r_{O_2:P}\Delta[PO_4^{3-}]_{remineralized} \\
DIC_{observed} &= \sum_{i=1}^{n} f_i DIC_i + (r_{I:P} + r_{C:P})\Delta[PO_4^{3-}]_{remineralized} \\
Alk_{observed} &= \sum_{i=1}^{n} f_i Alk_i + (2r_{I:P} - r_{N:P})\Delta[PO_4^{3-}]_{remineralized}
\end{aligned} \quad (5.3.1)$$

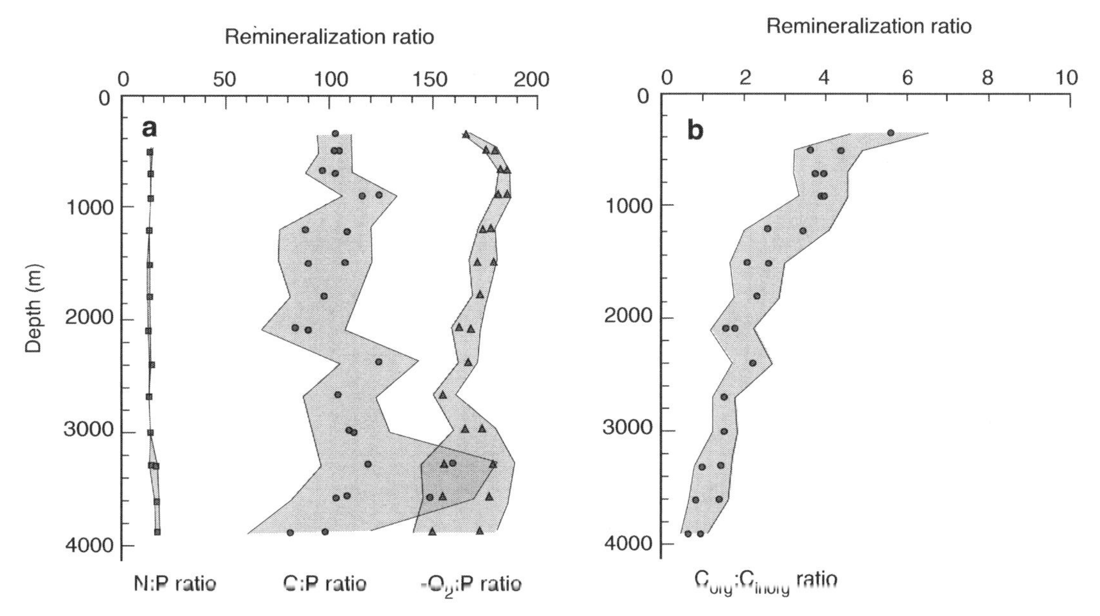

FIGURE 5.3.1: Estimates of the stoichiometric ratios of remineralization as a function of depth in the Pacific Ocean, taken from *Anderson and Sarmiento* [1994]. (a) N:P, C:P, and $-O_2$:P remineralization ratios of organic matter. (b) Organic-to-inorganic carbon remineralization ratio. The range of results denoted by the shading is an estimate of uncertainty based on a series of sensitivity studies.

The effect of remineralization is keyed to the addition of phosphate due to remineralization, $\Delta[PO_4^{3-}]_{remineralized}$, by means of the appropriate stoichiometric ratio r. The definition of the stoichiometric ratios is self-explanatory except for $r_{I:P}$, which is the stoichiometric ratio of inorganic carbon to phosphorus for the carbon removed by formation of $CaCO_3$ or added to the water by dissolution of $CaCO_3$. The organic matter component of carbon is represented by $r_{C:P}$. The cycling of $CaCO_3$ also affects the alkalinity, which is a measure of the charge balance of the water due to strong acids and bases (see chapter 8). It does so through the two positive charges of the Ca^{2+} ion. The alkalinity also includes a negative effect due to the formation of nitrate during remineralization. Equations (5.3.1) represent a nonlinear system of equations that require careful attention to obtain reasonable solutions.

Figure 5.3.1 shows the stoichiometric ratios obtained by *Anderson and Sarmiento* [1994] for the Pacific Ocean for neutral surfaces with a depth at the equator of 400 m to 4000 m. The average of all these estimates, plus similar estimates from the Atlantic and Pacific Oceans, gives the stoichiometric ratio C_{org}:N:P:O_2 = $(117 \pm 14):(16 \pm 1):(1):(-170 \pm 10)$ that has been discussed previously. One of the surprising results of this work is that there is no clear indication in the vertical profiles of C:P and $-O_2$:P ratios of a preferential remineralization of organic nitrogen and phosphorus shallower in the water column than organic carbon, such as might be expected based on sediment trap observations [e.g., *Martin et al.*, 1987]. This work suggests, instead, that the composition of the remineralized organic matter is relatively constant with depth. A recent global-scale investigation of the composition of sinking particulate organic matter collected by sediment traps by *Schneider* et al. [2003] supports this conclusion in part. They found that, after a correction was made for the contribution of lithogenic material, the C_{org}:N ratio of sinking *POM* is about (114 ± 2):16 in near-surface waters and increases with depth at a rate of about 0.2 (mol C):(mol N) per 1000 m. For example, they infer a mean C_{org}:N ratio of *POM* of about (117 ± 3):16 at 1000 m, and a ratio of (129 ± 6):16 at 5000 m. This increase of the C_{org}:N ratio with depth was found to be significant, and suggests a somewhat faster recycling of nitrogen over carbon from sinking particles. However, this increase with depth is relatively small and well within the uncertainties given by *Anderson and Sarmiento* [1994].

Recent studies by *Hupe and Karstensen* [2000] and *Li and Peng* [2002], however, suggest that the stoichiometric ratios of remineralization are substantially more variable than inferred by *Anderson and Sarmiento* [1994] and reported by *Schneider et al.* [2003]; (cf. also *Shaffer*, [1996]). For example, *Hupe and Karstensen* [2000] conclude on the basis of their analysis of data from the Arabian Sea that nutrients are recycled much faster than carbon, with the C_{org}:P ratios of the remineralized organic matter increasing from (99 ± 8):1 in the mid-thermocline to (132 ± 11):1 below 2000 m, i.e., suggesting a rate of increase with depth of about 0.9 (mol C):(mol P) per 1000 m—nearly 5 times as large as the depth increase found by *Schneider et al.* [2003].

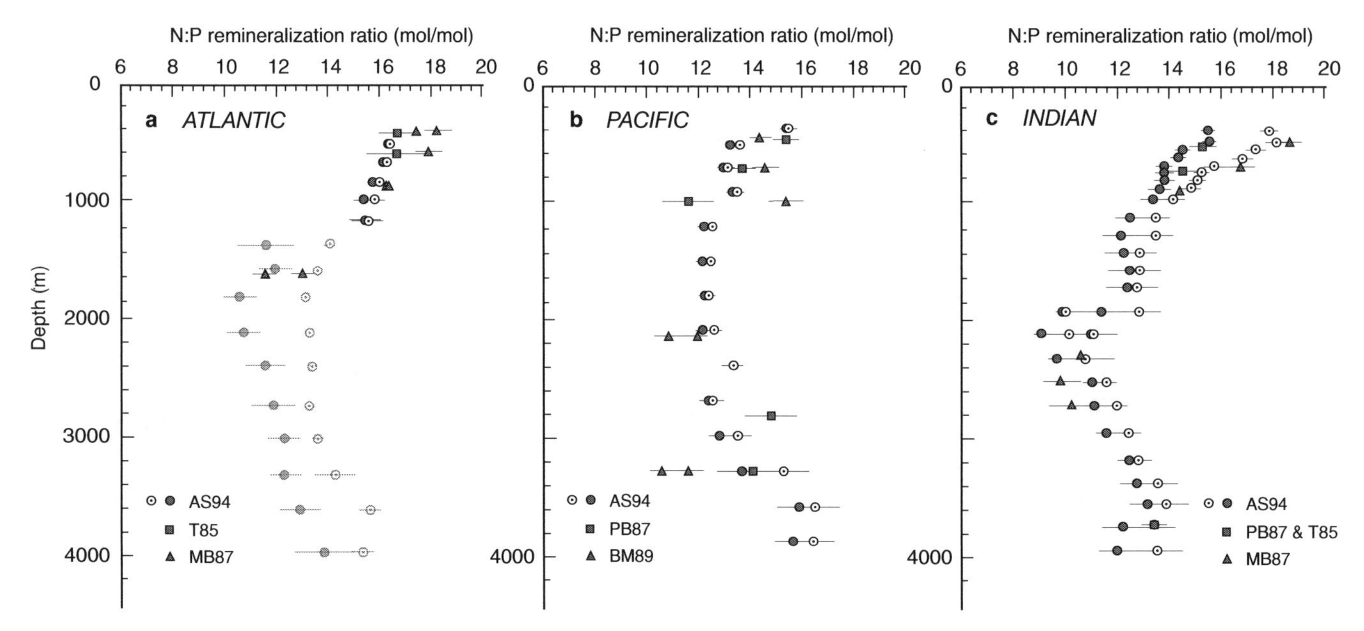

FIGURE 5.3.2: N:P stoichiometric ratios of remineralization estimated for (a) the South Atlantic, (b) the Pacific, and (c) the Indian Oceans. Open and closed circles are the results of *Anderson and Sarmiento* [1994] using either a one- or a two-end-member mixing model. Also shown are the estimates of *Takahashi et al.* [1985] (T85), *Minster and Boulahdid* [1987] (MB87), *Peng and Broecker* [1987] (PB87), and *Boulahdid and Minster* [1989] (BM89). Figure adapted from *Anderson and Sarmiento* [1994].

Given the presence of strong water column denitrification in the Arabian Sea, which is difficult to correctly account for in (5.3.1), it is currently not possible to assess the robustness of these results and whether they are applicable to the other ocean basins. A further problem exists in their using a linear method to solve the nonlinear problem given by (5.3.1), which can give rise to spurious results. We also question the results of *Li and Peng* [2002], in part because they implicitly assume very strong diapycnal mixing, which is in contradiction to the mostly isopycnal flow in the interior of the ocean (see chapter 2). Further analyses of the rich new data sets from the WOCE/JGOFS programs will hopefully resolve these discrepancies.

While there is considerable debate with regard to the stoichiometric ratios for organic carbon, *Anderson and Sarmiento* [1994] and *Hupe and Karstensen* [2000] agree on the finding of a significant depth trend in the stoichiometric profile of organic to inorganic (i.e., $CaCO_3$) carbon remineralization. This shows that the remineralization of organic carbon occurs shallower than the dissolution of $CaCO_3$. A discussion of the reason for this feature, which requires understanding the complex chemistry of the oceanic carbon system, is deferred until chapter 9.

Another surprising result of the work of *Anderson and Sarmiento* [1994] is the picture it gives of nitrogen cycling. Figure 5.3.2 shows the N:P stoichiometry for the South Atlantic, Indian, and Pacific Oceans in more detail. The N:P ratio that would be expected for fresh organic matter is 16:1 (see table 4.1.2). The magnitude and near constancy of the C:P and $-O_2$:P stoichiometric ratios are consistent with relatively fresh organic matter, but N:P shows a deficit of about a quarter of the nitrate in waters between about 1 and 3 km depth. *Anderson and Sarmiento* [1994] argue that the nitrate is being converted to N_2 by denitrification. The denitrification probably occurs in *anoxic* sediments (see chapter 6). The condition of *anoxia* (lack of biologically available oxygen) required for denitrification to occur is rarely met in the water column of the world's ocean. However, we will see below that there are a few distinct regions where the water column concentration of O_2 goes below detection, and denitrification is observed to occur. In what follows, we analyze first the behavior of a "normal" nutrient, i.e., phosphate. We then use the phosphate to infer the "normal" component of the behavior of nitrate. The difference between the expected normal behavior of nitrate and the actual behavior provides information on the impact of denitrification and nitrogen fixation.

PHOSPHATE

Figure 5.1.2b shows the distribution of phosphate in the ocean. Our analysis of the oxygen distribution and the stoichiometric ratios of organic matter remineralization now gives us tools we can use to separate the observed phosphate into a preformed and remineralized component. We estimate the remineralized phosphate component from the remineralized component of oxygen, i.e., the negative of AOU:

$$\Delta[PO_4^{3-}]_{remin} = -r_{P:O_2} AOU \qquad (5.3.2)$$

Note that this estimate misses the denitrification component of phosphate remineralization, since denitrification converts organic phosphorus to phosphate without consuming oxygen. However, this should not be an important factor over most of the ocean, as we shall see in the following section. An estimate of the preformed phosphate component is given by:

$$[PO_4^{3-}]_{preformed} = [PO_4^{3-}]_{observed} - \Delta[PO_4^{3-}]_{remin} \qquad (5.3.3)$$

The two components of the phosphate distribution are shown in figure 5.3.3. The remineralized component is exactly the same as AOU, which has been discussed previously. The preformed phosphate is essentially a conservative water mass tracer of ocean circulation. Slight variants of the preformed phosphate definition (5.3.3) have been exploited for studies of deep ocean circulation by *Broecker* [1974, 1979] and *Broecker et al.*, [1985b, 1991]. The separation of deep Atlantic water masses by (2.4.3) is based on one variant of the preformed phosphate definition. The main reason that preformed phosphate is useful to separate northern from southern component waters in the Atlantic is because of the lack of utilization of this phosphate in the surface waters of the Southern Ocean, possibly because of iron limitation (see discussion in chapter 4). By contrast, surface phosphate gets reduced to much lower levels in the North Atlantic, leading to lower preformed phosphate concentration, which then gets transported into the ocean's interior. We will see in chapter 7 that the transport of the high preformed nutrient concentrations from the Southern Ocean northward constitutes a major pathway for providing nutrients to the low-latitude thermocline. The importance of this transport can be inferred by simply comparing the magnitude of the preformed with the remineralized components (compare figure 5.3.3a with figure 5.3.3b). In much of the low-latitude thermocline, the majority of the phosphate present is preformed phosphate.

It is interesting to compare the preformed phosphate of figure 5.3.3 with the salinity shown in figure 2.4.1. The deep ocean distributions are similar, with the low-salinity, high-preformed phosphate waters of Southern Ocean origin being clearly distinguishable from the salty, low-preformed phosphate waters of North Atlantic origin. The end-member concentrations of preformed phosphate of these two water masses are about 0.8 mmol m^{-3} for the North Atlantic, and about 1.8 mmol m^{-3} for the Southern Ocean. As these two end-members are the primary end-member water masses for the waters filling up the deep Indian and Pacific, we can infer from the observed preformed concentration of about 1.3 mmol m^{-3} that the Southern Ocean and the North Atlantic each contribute about half to this water mass [see also *Broecker et al.*, 1999b]. By contrast to the deep ocean, the thermocline distribution of salinity and preformed phosphate differs significantly due to the fact that phosphate is stripped out of low-latitude surface waters by biological uptake. This gives a relatively large low-latitude-to-high-latitude surface gradient in preformed phosphate.

The Nitrogen Cycle

The most widespread remineralization reaction utilizes oxygen as the oxidant. We have seen above that the portion of this reaction that converts ammonium to nitrate is referred to as *nitrification*. If nitrification were the only remineralization reaction affecting nitrate, nitrate would have a distribution identical to that of phosphate, with a remineralized-nitrate-to-remineralized-phosphate ratio of 16:1. The distribution of nitrate would thus be of no additional interest relative to phosphate. However, figure 5.3.4 shows that although nitrate and phosphate are indeed strongly related to each other with a slope of 16:1, there are systematic deviations from this trendline. For example, many points in the Atlantic Ocean seem to lie above the mean trend, while there are many points in the Indian Ocean that appear to be below the mean trend. What causes the deviations from this trend?

We know that the distribution of nitrate is affected by denitrification that occurs when oxygen reaches very low levels. The nitrate-to-phosphate ratio for denitrification, as depicted by (5.1.4), is −104:1 (104 nitrate molecules are consumed for every phosphate molecule that is produced). The distribution of nitrate is also changed by N_2 fixation, which converts N_2 into N-rich organic matter, which is subsequently remineralized to nitrate. As we will see below, the N:P ratio of this N-rich organic matter formed by the organisms capable of N_2 fixation, a group called *diazotrophs*, is on the order of 40:1 to 125:1 [*Karl et al.*, 2002]. We turn now to a discussion of denitrification and N_2 fixation.

N as a Tracer of Denitrification*

We examine the effect of denitrification and N_2 fixation by defining a new tracer that uses phosphate to correct nitrate for that portion of its distribution due to the nitrification reaction. Variations in this tracer, called N^* [*Gruber and Sarmiento*, 1997], illustrate the combined effect of denitrification and remineralization of nitrogen-rich organic matter from N_2-fixing organisms. Our interpretation will be greatly aided by the fact that these two processes are often spatially well separated, so that we can interpret the signals as either coming from one or the other process.

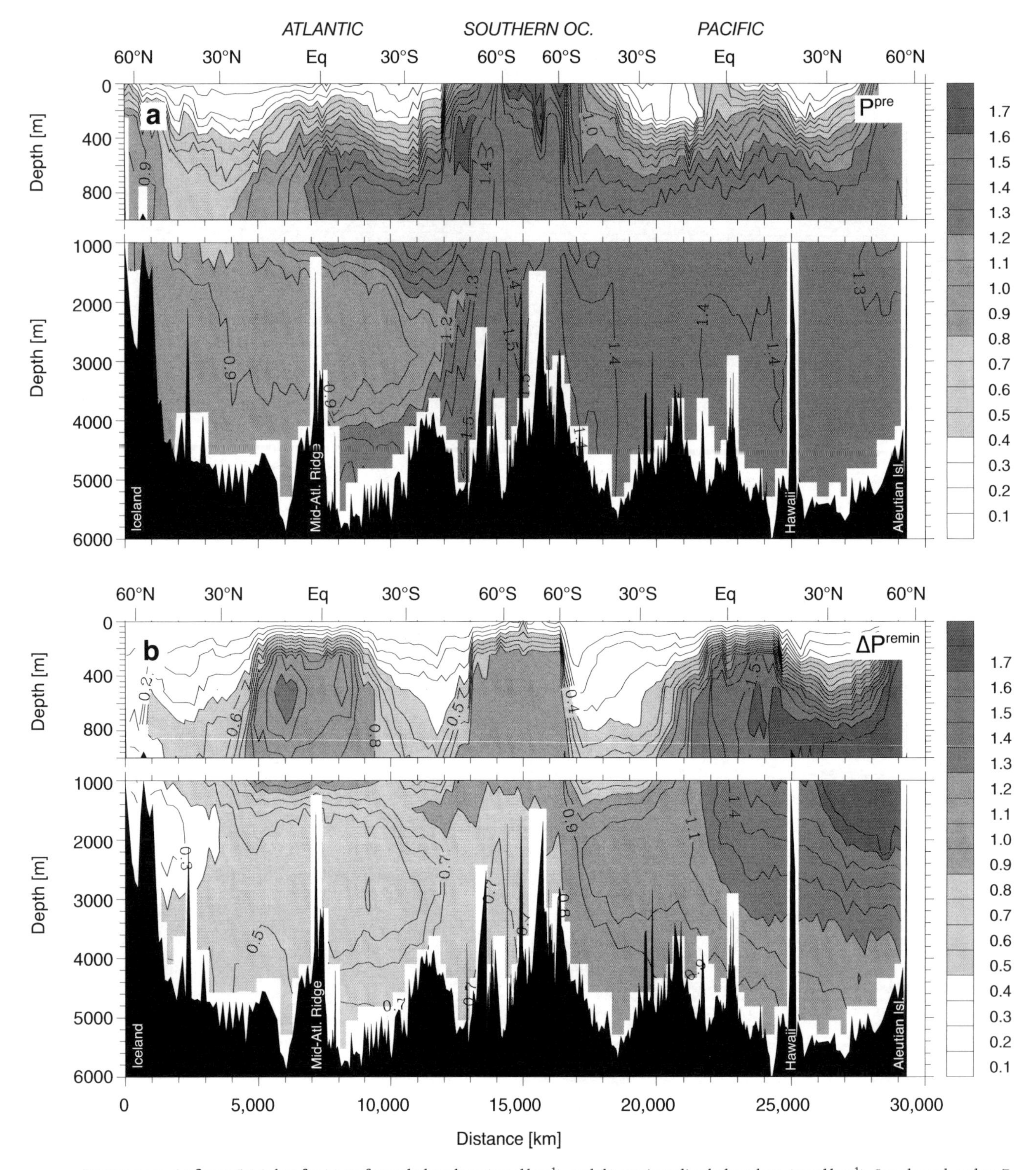

FIGURE 5.3.3: As figure 5.1.2, but for (a) preformed phosphate (μmol kg^{-1}), and (b) remineralized phosphate (μmol kg^{-1}). See also color plate 7. The two components of observed phosphate (see figure 5.1.2) have been separated on the basis of AOU (figure 5.1.3) and a $r_{P:O_2}$ ratio of 1:−170. See text for details.

We begin with the conservation equations for nitrate, N, and phosphate, P, considering only the effects of nitrification and denitrification, i.e., assuming initially that the effect of N_2 fixation is going to be manifested mainly in the surface ocean:

$$\begin{aligned}\Gamma(N) &= SMS_{nitrification}(N) + SMS_{denitrification}(N)\\ \Gamma(P) &= SMS_{nitrification}(P) + SMS_{denitrification}(P)\end{aligned} \tag{5.3.4}$$

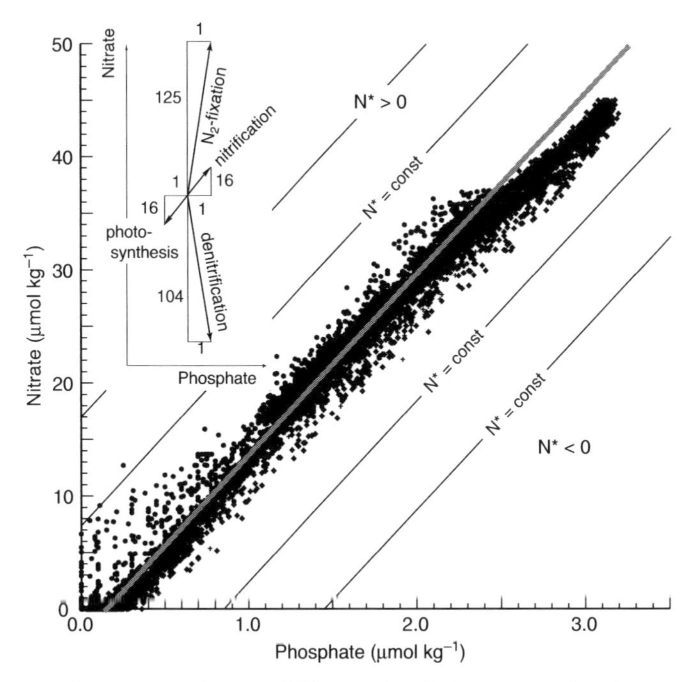

FIGURE 5.3.4: Concept of N^* on a plot of nitrate versus phosphate from all depths and from selected WOCE cruises in all ocean basins (A16, P16, I8N, I9S). The inset shows how the various processes influence the NO_3^- versus PO_4^{3-} distribution. The solid line represents the mean ocean trend with a slope of 16:1, while the thinner lines show trends of constant N^* (see text for definition).

where Γ is the transport and time rate of change operator defined by (4.2.2):

$$\Gamma(C) = \frac{\partial C}{\partial t} + U \cdot \nabla C - \nabla \cdot (\mathrm{D} \cdot \nabla C)$$

The *SMS* terms for nitrate and phosphate are related to each other by the stoichiometric ratios for nitrification and denitrification (see reactions 5.1.3 and 5.1.4):

$$\begin{aligned} SMS_{nitrification}(N) &= 16 \cdot SMS_{nitrification}(P) \\ SMS_{denitrification}(P) &= -(1/104) \cdot SMS_{denitrification}(N) \end{aligned} \qquad (5.3.5)$$

Substituting (5.3.5) into (5.3.4) gives

$$\begin{aligned} \Gamma(N) &= 16 \cdot SMS_{nitrification}(P) + SMS_{denitrification}(N) \\ \Gamma(P) &= SMS_{nitrification}(P) - (1/104) \cdot SMS_{denitrification}(N) \end{aligned} \qquad (5.3.6)$$

We can eliminate the nitrification term from the nitrate equation by subtracting 16 $\Gamma(P)$ from it. This gives

$$\begin{aligned} \Gamma(N) - 16 \cdot \Gamma(P) &= [1 + (16/104)] \cdot SMS_{denitrification}(N) \\ &= 1.15 \cdot SMS_{denitirification}(N) \end{aligned} \qquad (5.3.7)$$

In the final step we take advantage of the fact that both the nitrate and phosphate equations are linear, which means that the differential terms $\Gamma(N) - 16 \cdot \Gamma(P)$ can be combined to $\Gamma(N - 16P)$. This permits us to define a new tracer N^* whose interior distribution is affected only by transport and denitrification:

$$\Gamma(N^*) = 1.15 \cdot SMS_{denitrification}(N) \qquad (5.3.8a)$$

$$N^* = N - 16 \cdot P + 2.9 \text{ mmol m}^{-3} \qquad (5.3.8b)$$

(after *Gruber and Sarmiento* [1997] using the slightly revised definition of *Deutsch et al.* [2001]). The concentration of 2.9 mmol m^{-3} was added to bring the global mean of N^* to about zero. Note that $SMS_{denitrification}$ (N) is probably 0 over most of the ocean, in which case (5.3.8) describes the behavior of a conservative tracer with a global mean of 0 whose structure is set by the low extrema in regions where denitrification occurs. As we will see below, these deficits will be compensated by high extrema reflecting areas where nitrogen-rich organic matter from diazotrophs is remineralized.

Figure 5.3.5 shows vertical profiles of N^* obtained from WOCE observations in the Atlantic, Pacific, and Indian Oceans and averaged into 45° latitude bands. The deepest waters in the southern hemisphere below about 4000 to 4500 m are roughly constant at a value of 0 within the noise. However, the overall trend for waters above this depth is one of high values in the Atlantic, intermediate values in the Southern Ocean and Pacific, and lowest values in the Indian Ocean. This suggests that the Atlantic is a major source of nitrate via N_2 fixation, and that the Indian Ocean is a major sink relative to the other basins. This impression is reinforced by figure 5.3.6, which shows how N^* decreases progressively from the North Atlantic on the left-hand side of the figure to the North Pacific on the right-hand side. It also shows the extensive area of the N^* maximum in the thermocline of the North Atlantic as well as a well-defined N^* minimum in the thermocline of the subtropical North Pacific. Does this pattern coincide with our expectations based on direct studies of these processes?

The major areas of anoxia where water column denitrification is known to occur are in the thermocline of the Arabian Sea and eastern equatorial Pacific (figure 5.3.7) [cf. *Broecker and Peng*, 1982]. Both areas belong to what are known as the shadow zones of the thermocline. The waters in these shadow zones are not directly ventilated from the outcrop of these density surfaces, because there exists no direct pathway to them that permits conservation of potential vorticity (see discussion in chapter 2 and figure 2.3.12). As a result, these shadow zones are ventilated only in a diffusive manner, leading to relatively long residence times and a strong drawdown of the dissolved oxygen concentrations.

The impact of these anoxic regions on the distribution of N^* is very prominently displayed in figure 5.3.8, which shows the global distribution of N^* on two

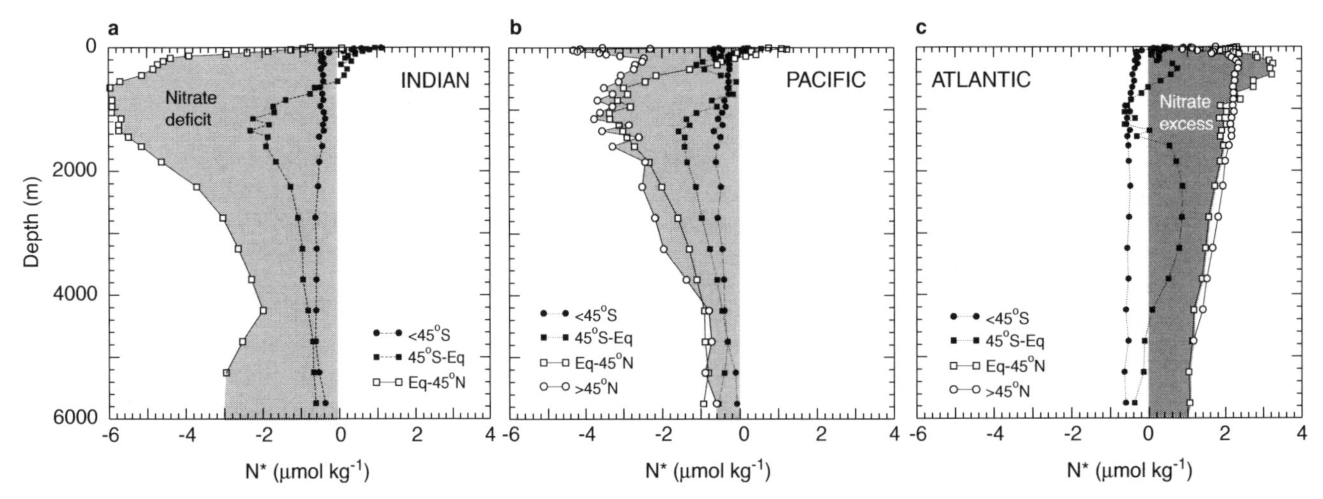

FIGURE 5.3.5: Horizontally averaged profiles of N^* in (a) the Indian, (b) the Pacific, and (c) the Atlantic Oceans. The profiles were computed by averaging all data within the ocean basin and the four latitude bands considered. The nitrate and phosphate data come from the World Ocean Circulation Experiment (WOCE) and selected pre-WOCE cruises and were individually corrected to common deep water trends by the Global Ocean Data Analysis Project (GLODAP).

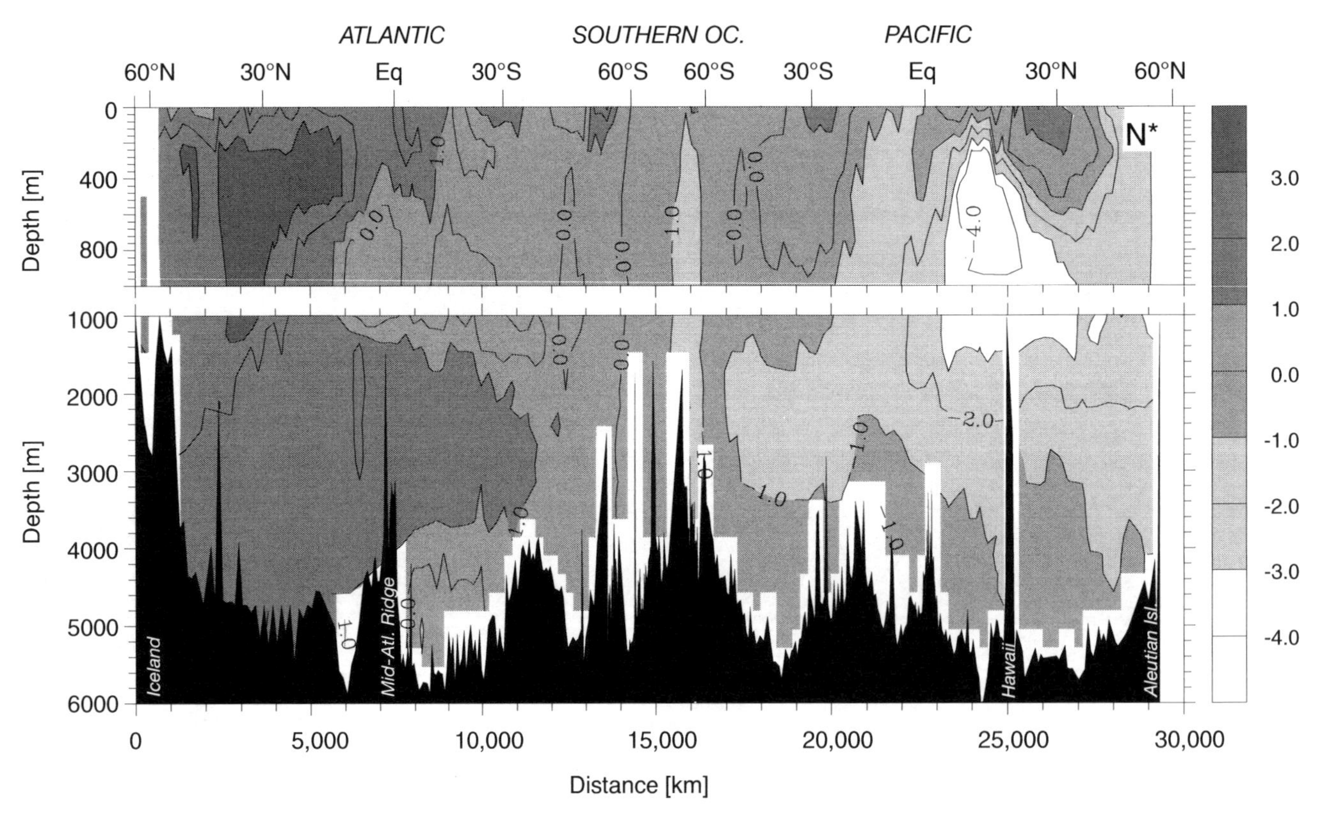

FIGURE 5.3.6: As figure 5.1.2, except for N^* (μmol kg^{-1}). See also color plate 7.

representative isopycnal surfaces in the thermocline. Figure 5.3.8a shows N^* on $\sigma_\theta = 26.50$, which reflects subtropical mode waters and lies at a mean depth of 200 to 400 m. Two tongues of strongly negative N^* can be observed in the eastern tropical Pacific corresponding directly to the two oxygen minimum regions in figure 5.3.7. The northern N^* tongue explains also the low N^* values that we have seen in the North Pacific thermocline in figure 5.3.6. The N^* minimum in the section located at about 160°W represents the westward advection out of the O_2 minimum zone of waters that have lost a significant amount of nitrate due to

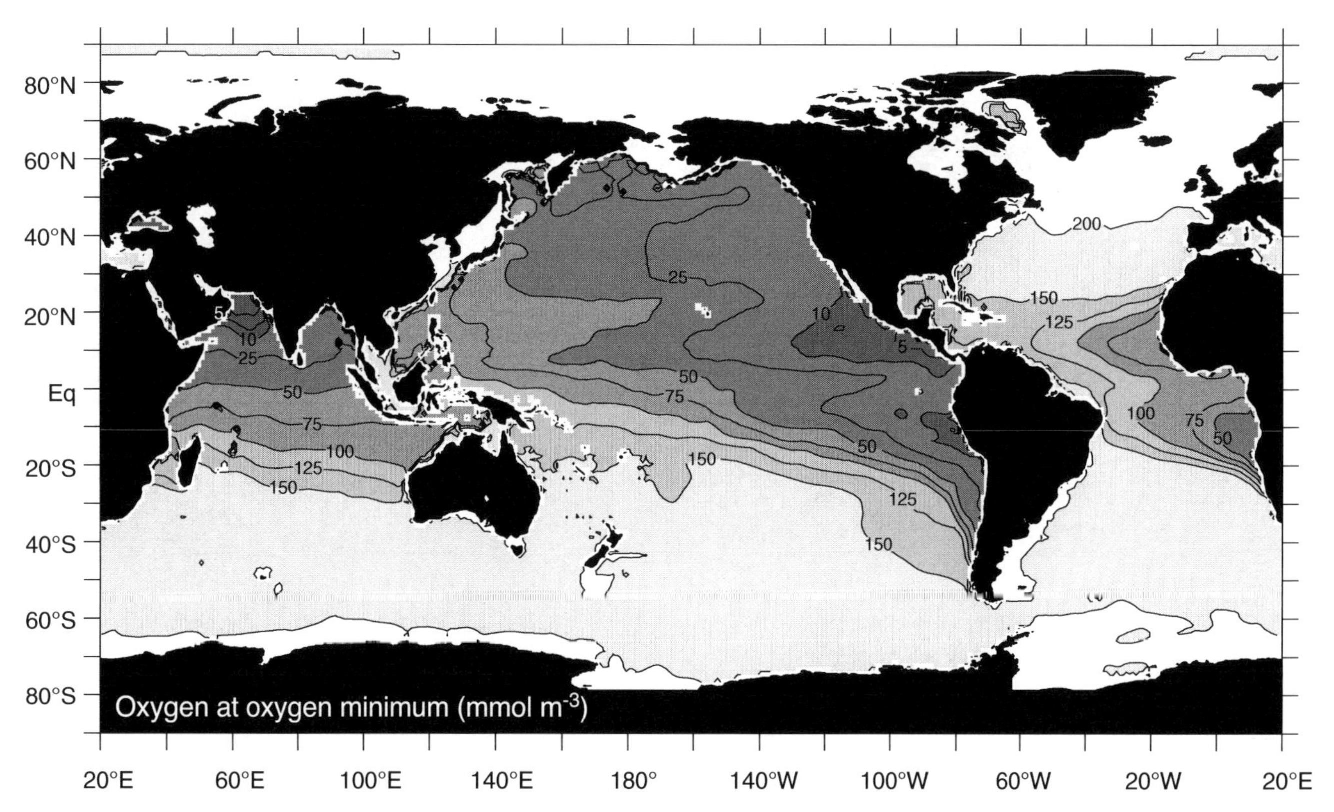

FIGURE 5.3.7: Oxygen concentration at the depth of the vertical oxygen minimum. Data are from the World Ocean Atlas 2001. [*Conkright et al.*, 2002]

denitrification. A similar correspondence between minima in N^* and oxygen can be found in the Arabian Sea. The oxygen minimum zones of the eastern tropical North Pacific (ETNP) and the Arabian Sea extend down to the $\sigma_\theta = 27.10$ surface, which lies at an average depth of about 450 to 800 m. By contrast, the oxygen minimum zone of the eastern tropical South Pacific (ESTP) is restricted to depths above this isopcynal surface [cf. *Deutsch et al.*, 2001]. This difference between the ETSP and ETNP is also well reflected in the distribution of N^* on this isopycnal surface (figure 5.3.8b), with the ETNP continuing to display very low N^* values, whereas the N^* minimum in the ETSP is already much weaker. We can therefore conclude that the observed nitrate deficits can nearly always be traced back to regions where denitrification is known to occur, as illustrated with the following example.

There is a large difference in the N^* of the intermediate and deep waters below the main thermocline of the Atlantic versus the Pacific and Indian Oceans (figure 5.3.6). The Atlantic Ocean has N^* concentrations about 2 to 3 mmol m^{-3} higher than the Pacific and Indian Oceans. We noted earlier that the stoichiometric analysis of *Anderson and Sarmiento* [1994] indicates that about a quarter of the nitrate due to remineralization appears to be missing in the deep waters below 1 km. The suggestion was made that this deficit is due to denitrification. The expected nitrate difference between the Atlantic and Pacific Oceans can be estimated from the remineralized phosphate component shown in figure 5.3.3. The remineralized phosphate component is approximately 0.4 mmol m^{-3} in the deep Atlantic and 1.1 mmol m^{-3} in the deep Pacific, giving a difference of 0.7 mmol m^{-3} for phosphate, and an expected difference of 11 mmol m^{-3} in nitrate. The N^* definition would cancel this difference out completely if there were no denitrification in the intermediate waters. If, on the other hand, a quarter of the remineralized nitrate is denitrified, then the N^* definition would give a decrease of $(11/4) = 2.7$ mmol m^{-3} between the Atlantic and Pacific. This is just as observed (figure 5.3.6), thus supporting the independent conclusion of *Anderson and Sarmiento* [1994] that there is a great deal of denitrification going on in intermediate and deep waters despite the absence of anoxia in these waters. It seems likely that the sediments are the major site for this denitrification, as discussed below and in the following chapter.

Deutsch et al. [2001] used the information from N^* in the ETSP and ETNP in conjunction with ventilation age estimates from CFCs (cf. figure 2.3.12) to estimate denitrification rates in these two regions. They arrived at estimates of 22 Tg N yr^{-1} (1 Tg $= 10^{12}$g) for the ETSP and of about 26 Tg N yr^{-1} for the ETNP, in good agreement with an extrapolation from direct rate measurements. Similar approaches have been employed in the Arabian Sea. *Howell et al.* [1997] used "NO," a geochemical tracer

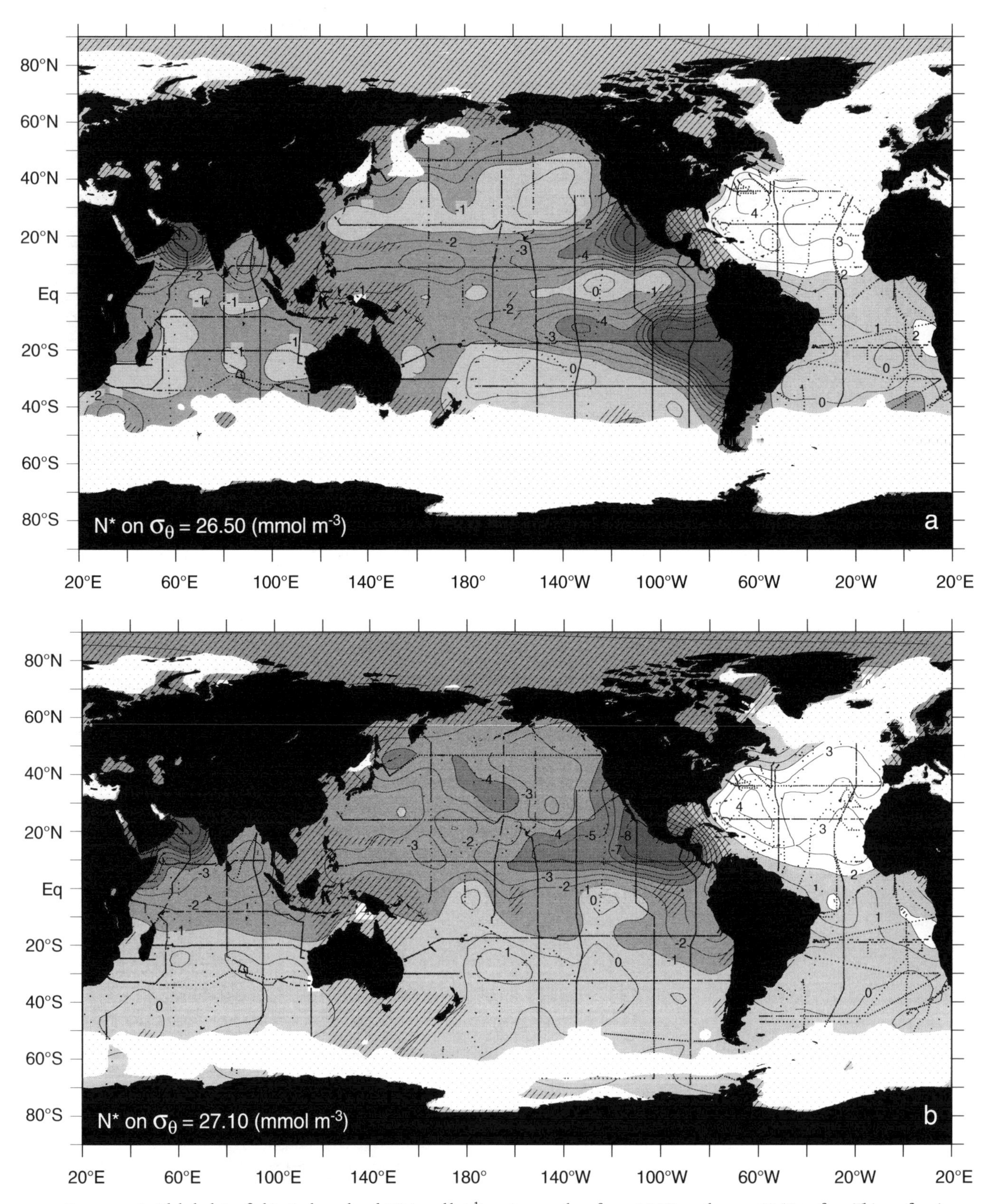

FIGURE 5.3.8: Global plots of objectively analyzed N^* (μmol kg^{-1}) on isopycnal surfaces. (a) N^* on the $\sigma_\theta = 26.50$ surface. This surface is characteristic of subtropical mode waters, such as the 18°C water in the North Atlantic. (b) N^* on the $\sigma_\theta = 27.10$ surface, representing subpolar mode waters. Circles denote the stations; stippling denotes areas where the waters of this potential density are not present in wintertime. Cross-hatched areas are regions where the estimated error in objectively analyzed N^* is greater than 0.7 μmol kg^{-1}. Data are from the WOCE program augmented by selected pre-WOCE cruises, and were adjusted by GLODAP.

TABLE 5.3.1
Present-day global marine nitrogen budgets of *Codispoti et al.* [2001] and *Gruber* [2004]

Process	*[Codispoti et al., 2001] Tg N yr^{-1}*	*[Gruber, 2004] Tg N yr^{-1}*
	Sources	
Pelagic N_2 fixation	117	120 ± 50
Benthic N_2 fixation	15	15 ± 10
River input (DON)	34	35 ± 10
River input (PON)	42	45 ± 10
Atmospheric deposition	86	50 ± 20
Total Sources	294	265 ± 55
	Sinks	
Organic N export	1	1
Benthic denitrification	300	180 ± 50
Water column denitrification	150	65 ± 20
Sediment burial	15	25 ± 10
N_2O loss to atmosphere	6	4 ± 2
Total sinks	482	275 ± 55

See *Gruber* [2004] and *Codispoti et al.* [2001] for details.

similar to N^*, to arrive at an integrated denitrification rate of about 21 Tg N yr^{-1} for this area. This gives a total loss of fixed nitrogen by denitrification of the order of 70 Tg N yr^{-1} (table 5.3.1; fixed nitrogen represents all chemical forms of nitrogen that can be used by common phytoplankton as a source of nitrogen, i.e., primarily nitrate, but including also dissolved organic nitrogen and ammonium). Denitrification not only occurs in the water column, but also in the sediments (see discussion in chapter 6). Most recent estimates put the sediment denitrification loss at about 180 Tg N yr^{-1}, (table 5.3.1). Adding sediment burial (25 Tg N yr^{-1}), loss of N_2O to the atmosphere (4 Tg N yr^{-1}), and export of fixed nitrogen by fisheries, the ocean appears to be losing about 280 Tg N yr^{-1}, with an uncertainty of at least ± 30% (table 5.3.1; *Gruber* [2004]). If this loss were not compensated by a gain of fixed nitrogen, the ocean would lose all of its fixed nitrogen within a few thousand years.

N^* as a Tracer of N_2 Fixation

We noted during our inspection of the distribution of N^* (see figures 5.3.5, 5.3.6, and 5.3.8; cf. also *Fanning* [1992]) that the North Atlantic appears to be one of the most important regions where fixed nitrogen is added to the ocean, helping to balance the large loss of fixed nitrogen stemming from water column and sedimentary denitrification. The Atlantic Ocean is known to be an important area of nitrogen fixation [*Carpenter and Romans*, 1991]. Several authors have suggested that this may be due to the high supply of iron by dust in this region [*Michaels et al.*, 1996a; *Gruber and Sarmiento*, 1997; *Wu et al.*, 2000], as recently experimentally demonstrated [*Mills et al.*, 2004]. The potential role of iron for limiting oceanic N_2 fixation is believed to be caused by the observation that N_2-fixing organisms have a higher iron demand than most other oceanic organisms [*Guider and Roche*, 1994; *Karl et al.*, 2002].

The fact that the North Atlantic N^* maximum appears to be a subsurface phenomenon violates our implicit initial assumption that the only effect of nitrogen fixation is as a boundary condition at the surface of the ocean. The organic matter produced by nitrogen fixation has a high N:P ratio. The addition of this high N:P organic matter to the model requires separating the nitrification term in (5.3.4) into two components: one for "normal" organic matter with N:P = 16, and one for high N:P organic matter associated with N_2-fixing organisms, i.e., diazotrophs [see *Gruber and Sarmiento*, 1997]:

$$SMS_{nitrification}(N) = SMS^{normal}_{nitrification}(N) + SMS^{diazotrophs}_{nitrification}(N)$$

$$SMS_{nitrification}(P) = SMS^{normal}_{nitrification}(P) + SMS^{diazotrophs}_{nitrification}(P) \quad (5.3.9)$$

We further have to consider that N_2-fixers will be taking up P in the surface ocean without a corresponding uptake of N, requiring the addition of a $SMS^{diazotrophs}_{uptake}(P)$ term to the phosphate equation in (5.3.4). With these modifications, the tracer conservation equation for N^* (5.3.8a) changes to

$$\Gamma(N^*) = 1.15 \cdot SMS_{denitrification}(N) + SMS^{diazotrophs}_{nitrification}(N) - 16 \cdot \left(SMS^{diazotrophs}_{nitrification}(P) + SMS^{diazotrophs}_{uptake}(P)\right) \quad (5.3.10)$$

with N^* defined as before in (5.3.8b). We can simplify (5.3.10) if our main concern is the impact of N_2 fixation on the distribution of N and P in the thermocline, as this permits us to set $SMS^{diazotrophs}_{uptake}(P)$ to zero. We can further simplify it by specifying the N:P ratio of organic matter of diazotrophic organisms. If we adopt the ratio of 125:1 that has been observed by *Karl et al.* [1992] during a *Trichodesmium* bloom in the Pacific near Hawaii, i.e.,

$$SMS^{diazotrophs}_{nitrification}(P) = (1/125) \cdot SMS^{diazotrophs}_{nitrification}(N) \quad (5.3.11)$$

then (5.3.10) becomes

$$\Gamma(N^*) = 1.15 \cdot SMS_{denitrification}(N) + \left(1 - \frac{16}{125}\right) \cdot SMS^{diazotrophs}_{nitrification}(N) = 1.15 \cdot SMS_{denitrification}(N) + 0.87 \cdot SMS^{diazotrophs}_{nitrification}(N) \quad (5.3.12)$$

where the exact value of the factor in front of $SMS^{diazotrophs}_{nitrification}(N)$ will depend on the N:P value of diazotrophic organic matter. *Karl et al.* [2002] report N:P ratios from recent laboratory studies that indicate that this ratio might be as low as 40:1. In this case the factor 0.87 becomes 0.60.

Figure 5.3.4 shows how we can interpret variations in N^* in relationship to N and P variations. The uptake or remineralization of organic matter with an N:P ratio of 16:1 will lead to changes in the N and P concentrations that move up and down along a line of constant N^*. By contrast, the removal of N and addition of P during denitrification moves a point toward the lower right, i.e., toward lower N^* values, with a slope of -104:1. The addition of N and P during the remineralization of N-rich organic matter from N_2-fixers moves a point toward the upper right, i.e., toward higher N^* values, with a slope of between 40:1 and 125:1. It is important to note, however, that the absolute value of N^* cannot be interpreted as a direct indication of active denitrification and/or N_2 fixation. It is rather the nonconservative change of N^* from one place to another that bears the information about the rates of these two processes. For example, let's start with a water parcel that has N^* value of $-2\ \mathrm{mmol\ m^{-3}}$. If we add NO_3^- and PO_4^{3-} to this water parcel by remineralizing $1\ \mathrm{mmol\ m^{-3}}$ of N-rich material with N:P ratio of 125:1, the N^* value of this water parcel will increase by $0.87\ \mathrm{mmol\ m^{-3}}$ (see (5.3.12)), resulting in a final N^* concentration of $-1.13\ \mathrm{mmol\ m^{-3}}$. If we mix this water parcel with another water parcel having N^* concentration of 2 mmol $\mathrm{m^{-3}}$, the final mixture will have N^* concentration $0.43\ \mathrm{mmol\ m^{-3}}$. Although the final concentration is positive, only that part of the N^* increase that is due to the addition of N-rich material, i.e., the nonconservative part bears direct information on N_2 fixation, while the second increase, i.e., the conservative mixing part, needs to be disregarded.

In places where denitrification is very small or negligible, we can use a nonconservative increase of N^* as a direct measure of the addition of N-rich material from N_2-fixers. Figure 5.3.8 shows that N^* has a tendency to increase from the outcrop areas in the North Atlantic toward the center of the subtropical gyre. These N^* variations along isopycnal surfaces can then be used in conjunction with information about the aging of waters (using ^{3}H-^{3}He, for example; see figure 2.3.11) to estimate the rate of addition of diazotrophic organic matter, i.e., $SMS^{diazotrophs}_{nitrification}(N)$. Once this is done for all isopycnal surfaces that show such a nonconservative increase in N^*, the rates can be integrated vertically and spatially to arrive at a basin-wide N_2 fixation rate. *Gruber and Sarmiento* [1997] followed this approach and obtained a North Atlantic N_2 fixation rate of about 28 Tg N $\mathrm{yr^{-1}}$. This estimate is relatively uncertain, as highlighted by the large discrepancy between similar geochemical estimates (compare the North Atlantic N_2 fixation estimate of *Michaels et al.* [1996b], 50–90 Tg N $\mathrm{yr^{-1}}$, with that of *Hansell et al.* [2004], 4 Tg N $\mathrm{yr^{-1}}$).

It is interesting to note that the North Atlantic areal average N_2 fixation rate determined by *Gruber and Sarmiento* [1997] is of about the right magnitude to provide the nitrogen source that fuels the biologically driven spring-to-summer drawdown in dissolved inorganic carbon typically observed near Bermuda, which occurs in the absence of a corresponding change in nitrate (see figure 4.3.18). Although this agreement in magnitude is enticing, the lack of high *in situ* N_2 fixation rates by *Trichodesmium* [*Orcutt et al.*, 2001] and the unexplained phosphorus source make this suggestion tentative. Vertical migration and large-scale horizontal transport of dissolved organic phosphorus have been suggested as possible mechanisms to supply the missing PO_4^{3-} (see discussion in *Karl and Bjorkman* [2002]), but none has been demonstrated to be of quantitative importance.

Deutsch et al. [2001] extended the N^* approach to the Pacific, and estimated a basin-wide N_2 fixation rate of about 59 Tg N $\mathrm{yr^{-1}}$. Using this estimate, and the *Gruber and Sarmiento* [1997] estimate for the North Atlantic, and then extrapolating the implied rates to the Indian Ocean, *Gruber* [2004] obtained a global N_2 fixation rate of about 120 Tg N $\mathrm{yr^{-1}}$, with uncertainties of about $\pm 40\%$ (table 5.3.1). This number is substantially smaller than the estimated total loss of fixed nitrogen of about 280 Tg N $\mathrm{yr^{-1}}$. Is the oceanic fixed nitrogen budget out of balance as suggested originally by *Codispoti and Christensen* [1985] and recently argued again by *Codispoti et al.* [2001]?

The Oceanic Nitrogen Budget

In order to put in perspective the input of fixed nitrogen into the ocean by pelagic N_2 fixation, the other source terms of the oceanic nitrogen budget need to be considered. Table 5.3.1 shows two recent budgets that list all the additional sources. Of importance are primarily the input of organic nitrogen from rivers (both DON and PON, for a total of about 80 Tg N $\mathrm{yr^{-1}}$) and the atmospheric deposition of organic forms of nitrogen (another $\sim$50 Tg N $\mathrm{yr^{-1}}$). Together with pelagic and benthic N_2 fixation, this brings the total input of fixed nitrogen to 265 ± 55 Tg N $\mathrm{yr^{-1}}$. If we adopt the total fixed N losses tabulated by *Gruber* [2004], we arrive at a nearly balanced oceanic N cycle for the present-day ocean, albeit with large uncertainties (table 5.3.1).

However, *Codispoti et al.* [2001] recently challenged many previous estimates of water column denitrification by arguing that most techniques have drastically underestimated these rates. They suggested that water column denitrification has to be at least twice as large as has been reported before. Since an oceanic mass balance of the isotope ^{15}N requires sedimentary denitrification to be about two to three times as large as water column denitrification [*Brandles and Devol*, 2002; *Deutsch et al.*, 2004], this elevated water column denitrification estimate requires a higher rate of sedimentary denitrification as well. Summing up, *Codispoti et al.* [2001] suggest that the total loss of fixed nitrogen in the ocean might be as large as 480 Tg N yr^{-1} (table 5.3.1), leading to a very large imbalance of over 100 Tg N yr^{-1}!

It is unlikely that imbalances of this magnitude have persisted over a long time, as this would have reduced the oceanic inventory of fixed nitrogen quite rapidly. With a mean concentration of about 30 mmol m^{-3}, the oceanic nitrate inventory is about 3.9×10^{16} mol (541,800 Tg N). Therefore, an imbalance of this magnitude would have depleted the oceanic fixed N inventory completely within about 5000 years. However, imbalances of this magnitude can not *a priori* be excluded if they are transient in nature, e.g., associated with recent human-induced changes, as argued for by *Codispoti et al.* [2001]. At the moment, our ability to quantify the rates of the different processes shown in table 5.3.1. is too poor to reject either the balanced hypothesis put forward by *Gruber* [2004] or the large imbalance hypothesis of *Codispoti et al.* [2001]. However, there is clear agreement that the marine nitrogen cycle is very dynamic, with N_2 fixation and denitrification playing a substantial role in altering the spatial distribution of nitrate and phosphate. As a result, the mean residence time of fixed nitrogen is less than 3000 years, which contrasts dramatically with that of P, which is primarily determined by river input and burial and is estimated to be about 30,000 to 50,000 years (*Delaney* [1998]; see also table 1.1.1).

The short residence time of fixed N immediately raises the question how the marine nitrogen cycle is capable of balancing the input and output of fixed nitrogen over long timescales. It appears at first that the sink and the source processes of fixed nitrogen are unrelated to each other, rendering it very difficult to stabilize the marine N cycle. The nature and magnitude of the negative feedback mechanisms responsible for the stabilization of the marine N cycle are not well understood yet. It appears, however, that the oceanic nitrate-to-phosphate ratio is the key control parameter, as it can link denitrification and N_2 fixation to each other and thereby ensure a long-term homeostasis (see *Tyrell* [1999] and *Gruber* [2004] for further discussion). The basic tenet of this model is that denitrification creates a deficit of fixed nitrogen relative to phosphorus in the ocean that then provides an ecological "niche" for N_2-fixers in the surface ocean, as they will be able to take advantage of the available excess phosphate, while normal phytoplankton are nitrogen limited. The resulting net addition of nitrogen to the ocean will therefore tend to eliminate the nitrate deficit created by denitrification. N_2 fixation is held in check by competition with normal phytoplankton, while the extent of denitrification is governed by total oceanic productivity, which is ultimately controlled by the amount of phosphorus in the ocean.

Nitrous Oxide

Although the loss of about 4 Tg N_2O per year from the ocean into the atmosphere is a relatively small part of the fixed nitrogen budget (table 5.3.1), this flux represents a substantial fraction of the total N_2O sources for the atmospheric N_2O budget. We mentioned in chapter 3 that the total source of N_2O to the preindustrial atmosphere is estimated to have been about 11 (range of 8 to 16) Tg N yr^{-1} [*Prather and Ehhalt*, 2001], which suggests that the oceanic source accounts for about 40% (range of 25% to 50%) of the total pre-anthropogenic input of N_2O to the atmosphere. We next discuss in greater detail the mechanisms that produce N_2O in the ocean and how they are expressed in the interior ocean distribution of N_2O as well as in the surface ocean. Our interest in N_2O is fueled by the role of N_2O as a greenhouse gas (as discussed in greater detail in chapter 10) and the observation that atmospheric N_2O has undergone substantial natural variations in the past 100,000 years [cf. *Fluckiger et al.*, 2004]. Atmospheric N_2O has also increased markedly in the last 200 years in response to human-induced changes in the global nitrogen cycle [cf. *Prather and Ehhalt*, 2001]. We start with the surface ocean and then work our way down into the ocean's interior.

Figure 3.1.4 showed that the efflux of N_2O out of the ocean is mostly concentrated in upwelling areas and high latitudes. This can be readily understood from our above discussion of the mechanisms that produce N_2O in the ocean. We have seen in section 5.1 that N_2O is formed in association with nitrification processes and, to a lesser extent, with denitrification. As nitrification is inhibited by light, most of the N_2O in the ocean is formed in the aphotic zone. The N_2O produced in the interior of the ocean can then come to the surface ocean and escape into the atmosphere only in those areas where waters from the aphotic zone are transported or mixed upward, such as occurs in upwelling regions, or areas of deep convection.

We next discuss the formation mechanisms for N_2O in more detail, as this has a substantial influence on how the marine N_2O cycle reacts to perturbations [cf. *Jin and Gruber*, 2003]. The meridional sections of N_2O

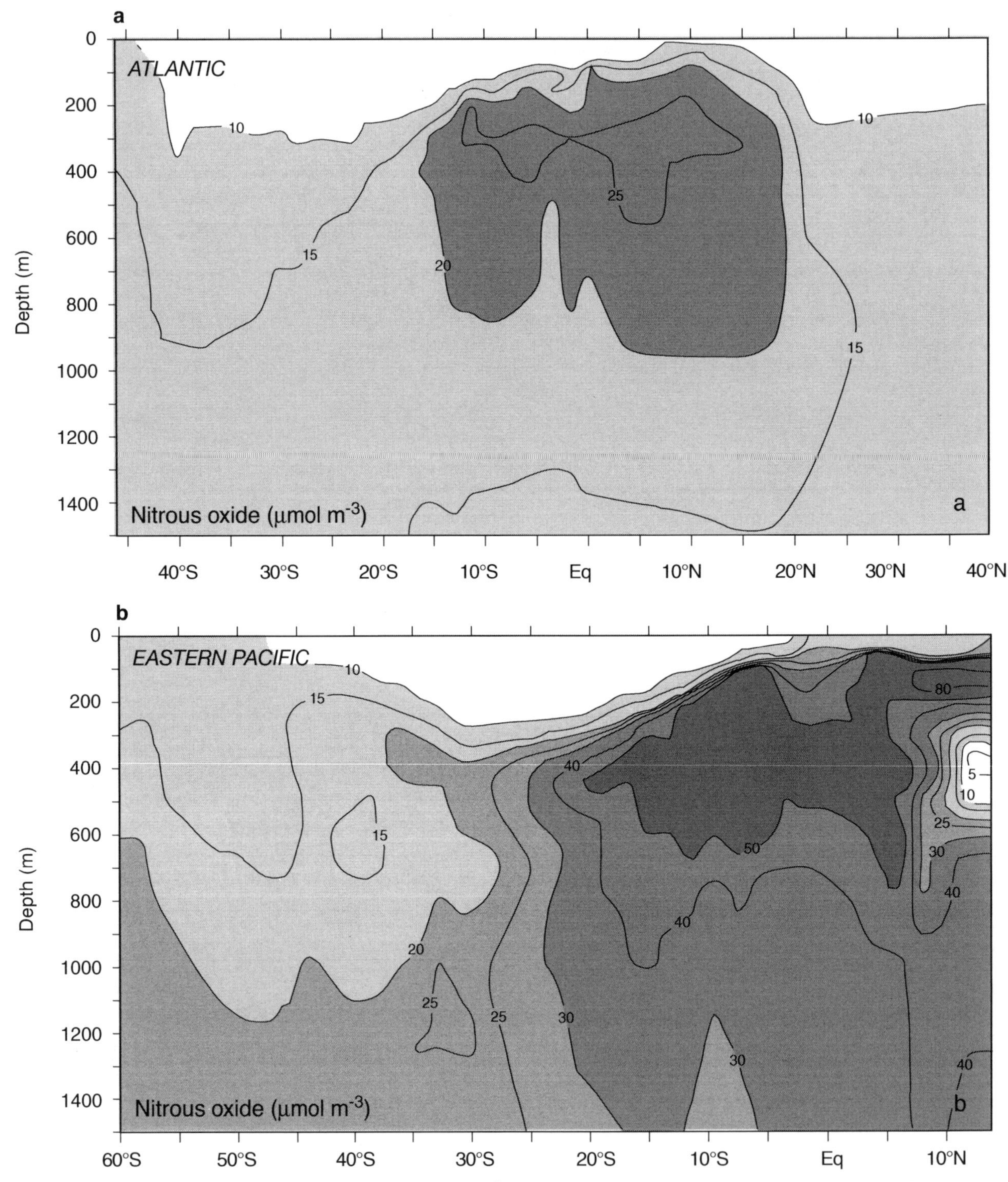

FIGURE 5.3.9: Meridional sections of the N_2O concentration (nmol l^{-1}): (a) Atlantic (along 20°W–57°W, BLAST II) and (b) Pacific (along 105°W, RITS89). Figure from *Nevison et al.* [2003].

in figure 5.3.9 show a substantial increase in N_2O between the thermocline of the Atlantic and that of the Pacific. A fraction of this difference in N_2O can be attributed to the difference in mean thermocline *AOU* between the two basins (see figure 5.1.4), as one expects N_2O to increase as *AOU* increases. This is because the latter is a direct measure of the amount of nitrification that has occurred since a water parcel left the surface. But the difference in N_2O between the two basins is larger than expected from *AOU* alone. In particular, the distribution of N_2O in the Pacific bears some inverse resemblance to N^* shown in figure 5.3.6, confirming

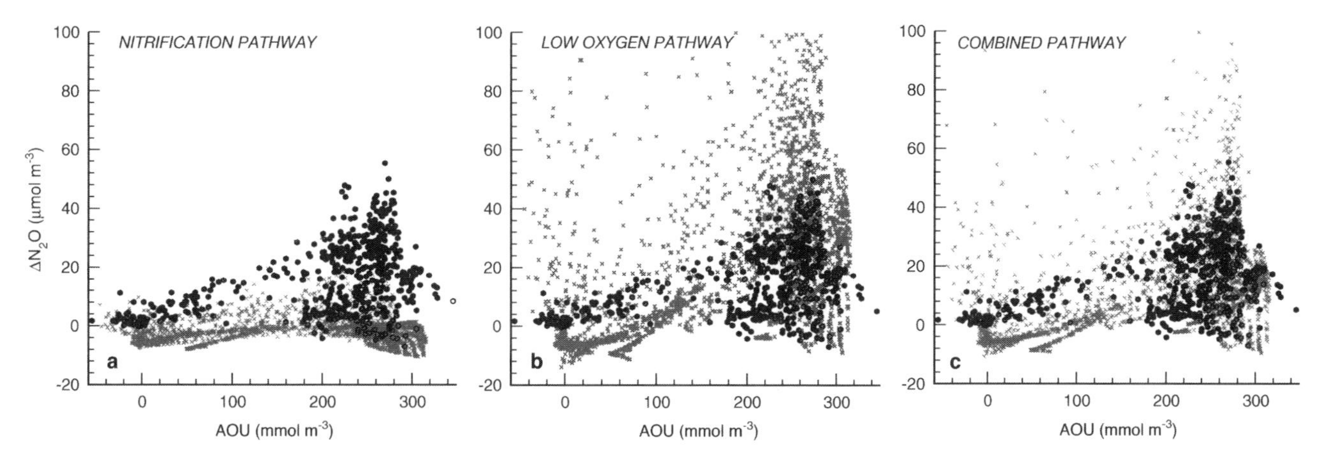

FIGURE 5.3.10: Plots of model simulated relationship between N_2O and AOU (shown as x) in comparison to observed relationships in the low-oxygen regions of the eastern tropical Pacific and Arabian Sea [*Bange et al.*, 2001; *Cohen and Gordon*, 1978] (observations shown as dots). (a) Results if N_2O is produced by the nitrification pathway; (b) results if only the low-oxygen pathway is active; and (c) results if half of the total oceanic N_2O production stems from the low-oxygen pathway and the other half from the nitrification pathway. The best agreement between the model and the data is found for the combined case. From *Jin et al.* [2005a].

the hypothesis that processes associated with denitrification may produce N_2O as well.

A convenient way to investigate the impact of the different pathways producing N_2O is to plot N_2O versus AOU, as shown in figure 5.3.10. If N_2O production was tied only to nitrification and the N_2O yield of this process was constant, this plot would show a linear relationship, with a slope that is equal to the yield. This is not the case, as the data indicate an increase in the slope toward higher AOUs, and then a rapid drop-off at very high AOUs. This suggests that the yield of N_2O must increase as AOU increases (or O_2 decreases), as experimentally demonstrated by *Goreau et al.* [1980], and that N_2O gets consumed at very low O_2 levels. The plotted data do not allow us to separate these effects very easily, as mixing and transport tend to eradicate the unique signatures from these processes.

Also shown in figure 5.3.10 are predictions from three model scenarios. In the first case, all N_2O in the ocean is produced by nitrification with a constant yield. In the second case, all N_2O is produced by a "low oxygen" mechanism, which is believed to include an interaction of nitrification and denitrification, and whose N_2O yield is highly dependent on the *in situ* oxygen concentration. In both cases, N_2O is denitrified inside anoxic regions. The third case is a combination of the first two cases, with each pathway contributing half of the total production of N_2O. In all three cases, the total production is tuned such that the total oceanic emissions match the observed emissions of about 4 Tg N yr^{-1} [*Suntharalingam and Sarmiento*, 2000; *Nevison et al.*, 2003]. Comparison with the observations indicates that both mechanisms contribute to the observed trends in N_2O and that a half/half contribution appears to give the best fit [see also *Suntharalingam et al.*, 2000; *Nevison et al.*, 2003]. We can therefore write the tracer continuity equation for N_2O below the euphotic zone as the sum of two production pathways (nitrification and low oxygen) and one consumption pathway associated with denitrification:

$$\Gamma(N_2O) = SMS_{nitrification}(N_2O) + SMS_{low\ oxygen}(N_2O) + SMS_{consumption}(N_2O) \quad (5.3.13)$$

where the individual source minus sink terms are given by

$$\begin{aligned} SMS_{nitrification}(N_2O) &= -\alpha \cdot SMS_{nitrification}(NH_4^+) \\ SMS_{low\ oxygen}(N_2O) &= -\beta \cdot f(O_2) \cdot SMS_{nitrification}(NH_4^+) \\ SMS_{consumption}(N_2O) &= -k_{consumption}[N_2O] \end{aligned} \quad (5.3.14)$$

where $SMS_{nitrification}(NH_4^+)$ is the sink term for NH_4^+ due to nitrification. Note that we write the nitrification term in terms of the reactant NH_4^+ rather than the product NO_3^- (e.g. (5.3.4)), because we no longer can assume that $SMS_{nitrification}(NH_4^+) = -SMS_{nitrification}(NO_3^-)$. The parameter α and the function $\beta \cdot f(O_2)$ represent the N_2O yield (mmol N_2O/mol NH_4^+) of the nitrification and low-oxygen pathways, respectively, and $k_{consumption}$ denotes a first-order consumption rate of N_2O in anoxic regions. Estimates for α and $\beta \cdot f(O_2)$ indicate that at oxygen concentrations of less than ~10 mmol m^{-3}, nearly every NH_4^+ molecule gets converted to N_2O, and that the yield decreases relatively rapidly to a more or less constant value of a little bit less than 1 molecule N_2O per 1000 molecules of NH_4^+ at O_2 concentrations above 50 mmol m^{-3} [*Jin et al.*, 2005a]. Similar parameterizations have been proposed by *Suntharalingam et al.* [2000] and *Nevison et al.* [2003].

5.4 Organic Matter Cycling

We turn now to a discussion of the organic matter cycling processes that determine the oxygen and nutrient distributions described above. We need to understand the processes that control the export of organic matter from the surface ocean, the processes that determine the relative proportions of particulate and dissolved organic matter in that export, and the processes that determine its elemental composition. We further need to understand how particulate and dissolved organic matter are transported to the deep ocean, and how they are remineralized there. The discussion will center on the *POC* and *DOC* production and consumption terms that determine the tracer continuity equation for particulate and dissolved organic matter:

$$\Gamma(POC) = SMS_{production}(POC) + SMS_{consumption}(POC) - w_{\text{sink}} \cdot \frac{dPOC}{dz} \qquad (5.4.1)$$

$$\Gamma(DOC) = SMS_{production}(DOC) + SMS_{consumption}(DOC)$$

where w_{sink} is the sinking velocity of *POC*.

The first topic we discuss in this section is the cycling of particulate organic matter. Historically, this has received more attention than the cycling of dissolved organic matter, both because of the early emphasis on the dominant role of particulate organic matter in export of organic matter from the surface, and because of the importance of particulate organic matter in determining the sedimentary record of ocean chemistry and climate.

The second topic we discuss is dissolved organic matter, which is now regarded as playing a role in organic matter cycling that is comparable in importance to that of particulate organic matter, particularly in the main thermocline. For example, *Doval and Hansell* [2000] showed that 21% to 47% of the *AOU* in the thermocline of the South Pacific is due to the oxidation of *DOC*. They also demonstrated that this large contribution is restricted to the upper 500 m. Below that depth, the changes in *AOU* were mainly driven by the oxidation of particulate organic matter.

Particulate Organic Matter

Overview

We begin our discussion of *POM* with a synopsis of what is known about the nature of particulate matter in general based on the summary of *Clegg and Whitfield* [1990]. Organic carbon constitutes 30% to 40% of the particulate matter above 100 to 200 m, but decreases to less than 10% at depth (figure 5.4.1). The remainder of the particulate matter is made up of biogenic $CaCO_3$, silicate, and other materials, including *aeolian* (i.e., windborne) and riverine particles that originate on the continents. As we will see below, these components of particulate matter play an important role in determining how organic matter is transported downward with sinking particles. The concentration of *POC* in near-surface waters ranges from 3 to 10 mmol C m^{-3}, and tends to decrease substantially with depth, with waters below 100 m having typical *POC* concentrations of around 3 mmol C m^{-3} (figure 5.4.2). A comparison of this concentration with typical *DOC* concentrations (see the *DOC* section along the same line in figure 5.4.14 below) shows that these open ocean *POC* concentrations are typically more than one order of magnitude smaller than those of *DOC*. Most variations along the section are associated with differing levels of marine productivity. For example, elevated *POC* levels are found between 5°N and about 15°S, associated with this transect crossing the productive eastern equatorial Pacific cold tongue area. Elevated *POC* concentrations are then found again further south in the more productive waters of the Southern Ocean.

Particles have a wide range of sizes in the ocean, but the size spectrum is relatively smooth (figure 5.4.3), suggesting that there are processes that convert particles from one size to another. Without such processes, the size distribution would tend to have peaks reflecting

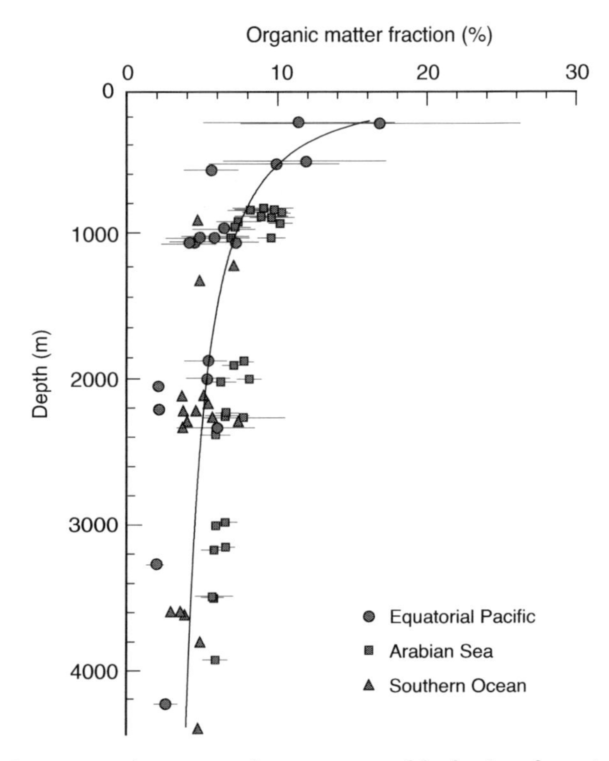

FIGURE 5.4.1: A summary of measurements of the fraction of organic carbon in particulate organic matter [*Clegg and Whitfield*, 1990].

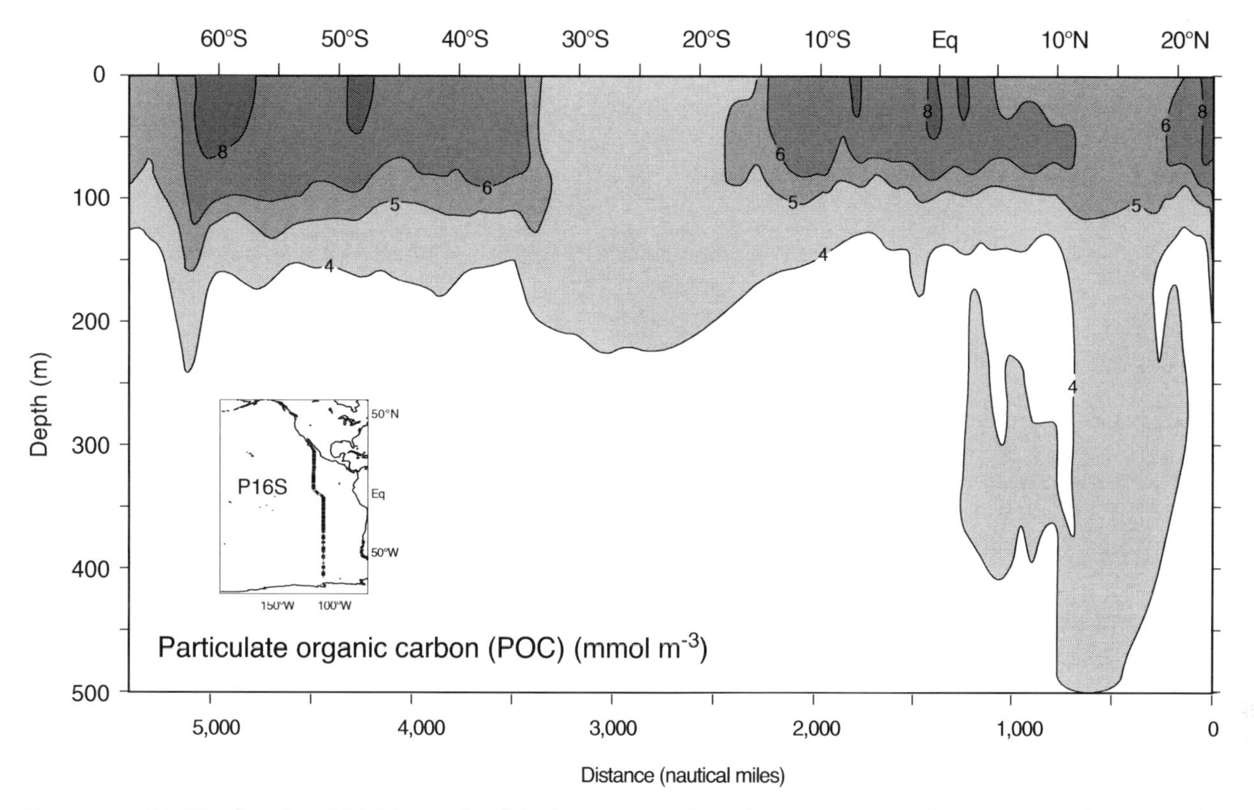

FIGURE 5.4.2: Meridional section of *POC* (in mmol m^{-3}) in the eastern South Pacific along approximately 105°W (WOCE line P18S). The *POC* concentration was estimated from measurements of beam attenuation using an empirical calibration. Data are from W. Gardner (personal communication, 2005).

the sources of the particles. One of the problems we must resolve in order to understand the contribution of particles to the sinking flux of organic matter and other constituents is how particles of different sizes interact with each other.

Only the largest particles of diameter >100 μm, which comprise roughly 1 particle in every 10^{10}, sink at rates that are sufficiently rapid to contribute significantly to the flux of *POM* from the surface to the abyss. The basic reason for this is that the sinking velocity is a very sensitive function of particle size, usually expressed as an equivalent particle radius, i.e., the radius of a sphere that has the same volume as the particle under consideration. Stokes's law for the sinking of spheres states that the sinking velocity, w_{sink}, is proportional to the square of the equivalent particle radius, r:

$$w_{sink} = \frac{2g \cdot r^2 \cdot (\rho_1 - \rho_2)}{9\mu} \quad (5.4.2)$$

where ρ_1 (kg m^{-3}) is the density of the particle falling through a liquid of density ρ_2 and dynamic viscosity μ (N s m^{-2}), and where g is the gravitational acceleration (9.81 m s^{-2}). The viscosity of a fluid is a measure of its internal friction. Equation (5.4.2) suggests therefore that the settling velocity is determined by the balance between gravitational acceleration of the particle (g times density excess, $\rho_1 - \rho_2$) and the drag acting upon the particle in relationship to its surface area (r^2) and the friction of the fluid (μ).

The sinking speed is therefore also proportional to the density difference between the particle and seawater. The density of organic matter is very close to that of water ($\rho_{org\,matter} = 1060$ kg m^{-3} [*Logan and Hunt*, 1987]). By contrast, $CaCO_3$, lithogenic material, and opal have densities that are at least twice as large ($\rho_{CaCO_3} = 2710$ kg m^{-3}; $\rho_{lithogenic} = 2710$ kg m^{-3}; $\rho_{opal} = 2100$ kg m^{-3} [*Klaas and Archer*, 2002]). Using these densities, a density of seawater of 1027 kg m^{-3}, and a viscosity of seawater at 15°C of about 1.25×10^{-3} N s m^{-2}, from (5.4.2) we estimate sinking velocities for a spherical particle with an equivalent radius of 50 μm (diameter of 100 μm) of about 12 m day^{-1} for a particle consisting of only organic matter, about 600 m day^{-1} for a particle made out of only $CaCO_3$ or lithogenic material, and about 400 m day^{-1} for a particle made out of opal. Assuming a depth of about 4000 m, a 100 μm organic matter particle would take nearly a year to reach the bottom of the ocean, whereas the heavy $CaCO_3$ or lithogenic material particle would take only about a week. A particle with an equivalent diameter of 10 μm would take nearly 100 years to reach the bottom if it were made out of organic matter alone, and even if it were made out of $CaCO_3$ or lithogenic material, it still

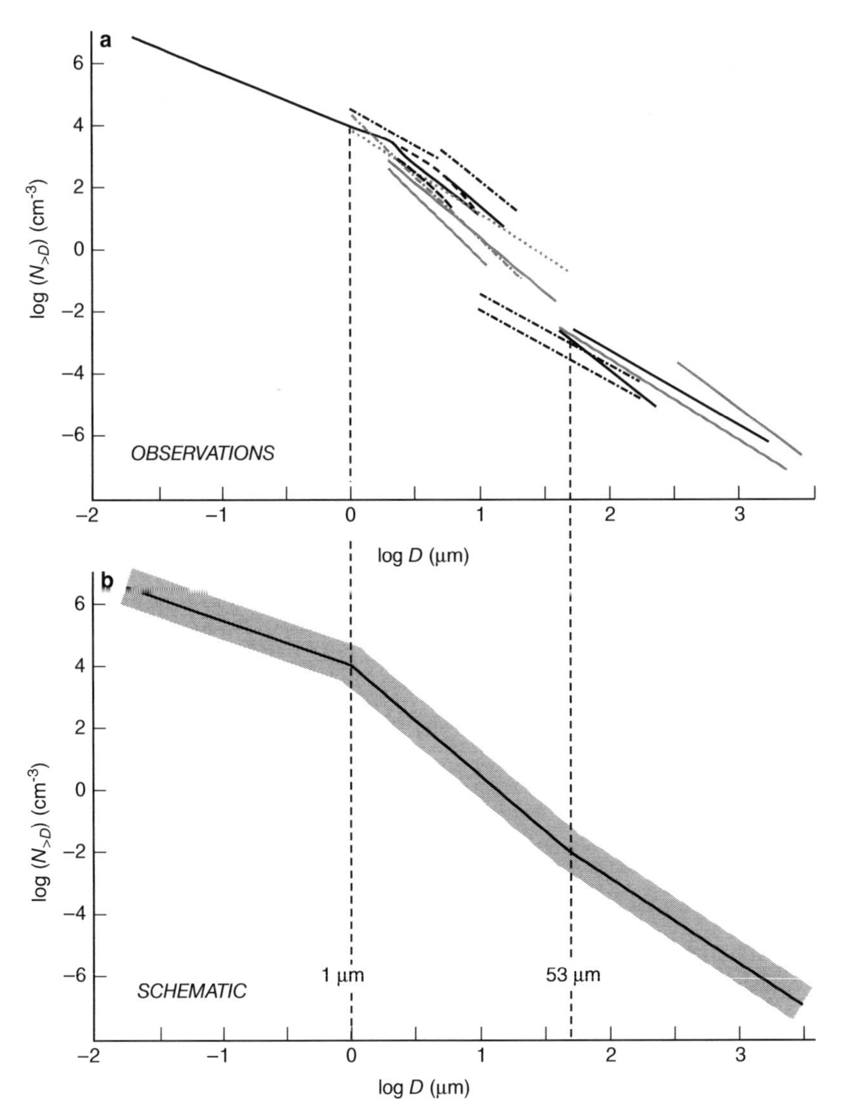

FIGURE 5.4.3: A summary of particle size spectra as a function of diameter. Upper panel: Measured spectra for a variety of different sites plotted as log ($N_{>D}$), where $N_{>D}$ is the number of particles of diameter greater than D (μm) per cm^{-3}. Lower panel: Schematic spectrum based on the data shown in the upper panel. From *Clegg and Whitfield* [1990].

would take nearly 2 years. Remineralization and dissolution processes are, on average, much faster than this timescale, so that particles of this size class never make it far. The importance of size and density in determining the settling velocity and consequently the fate of particles becomes therefore rather obvious, though we need to keep in mind that these estimates are only approximations, since none of the particles involved are spherical. Most of these particles are also porous, tending to slow down sinking rates.

Figure 5.4.4 shows a compilation of actual and theoretical sinking velocity estimates plotted versus particle size. The convolution of this information with the size spectrum of figure 5.4.3 gives the frequency distribution of mass flux as a function of particle size summarized in figure 5.4.5.

Also shown in figure 5.4.5 are the size distributions of mass and surface area. The remarkable finding from this type of analysis is that the vast majority of the mass and surface area is concentrated in smaller particles that are completely distinct from the large particles that contribute to the sinking flux. A proper representation of particle dynamics thus requires at the very least two distinct particle sizes, one of which is essentially nonsinking, and the other of which provides essentially all the sinking flux. Figure 5.4.5 also shows how these two pools change with depth. There is an indication of an increase in mean particle size of the sinking flux with depth, which would suggest an increase in sinking speed. *Berelson* [2002] demonstrated recently that such an increase of mean particle sinking velocities with depth may indeed exist. By analyzing sediment trap

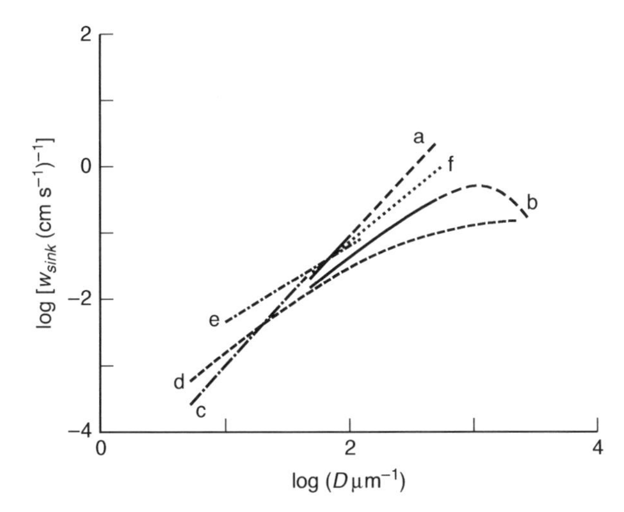

FIGURE 5.4.4: A summary of particle sinking rates as a function of diameter by *Clegg and Whitfield* [1990].

data from the Arabian Sea and the equatorial Pacific, he found mean particle sinking velocities in the upper ocean of the order of a few tens of m day^{-1}, whereas they appear to increase to values of 100 and 200 m day^{-1} at 1500 m and to even higher at around 3000 m.

We now proceed to a more detailed discussion of particles, beginning with particle flux and decomposition and concluding with a discussion of models of particle interactions.

PARTICLE FLUX

The major tool that has been used to characterize the nature and magnitude of sinking particles is the sediment trap, which consists of a cylindrical or conical container with one open end that is placed on a mooring or left to float in the water for a period of time. The bottom of the container into which the particles settle is poisoned so as to prevent microbial breakdown of organic matter. The ability of trap samples to represent the actual sinking flux of particles continues to be a matter of substantial controversy [e.g., *Buesseler*, 1991], centering on issues such as: (1) biases introduced by the hydrodynamics of processes such as the flow of water across a trap (e.g., conical traps lose much of their material due to the nature of flow induced within the trap), (2) biased collection due to trapping of swimming organisms that are killed by the poison layer, and (3) biases in the preservation of materials within the trap due to problems such as selective inhibition of decomposition or solubilization by the poison [e.g., *Jahnke*, 1990]. Because of this, one must be careful about using trap information for anything other than qualitative information about the particle sinking flux. The problem is much more acute for sediment traps deployed in the upper 1000 m, where lateral flows are much stronger, and swimmers are much more abundant.

Several approaches have been developed in the last decade to address some of these problems. The first

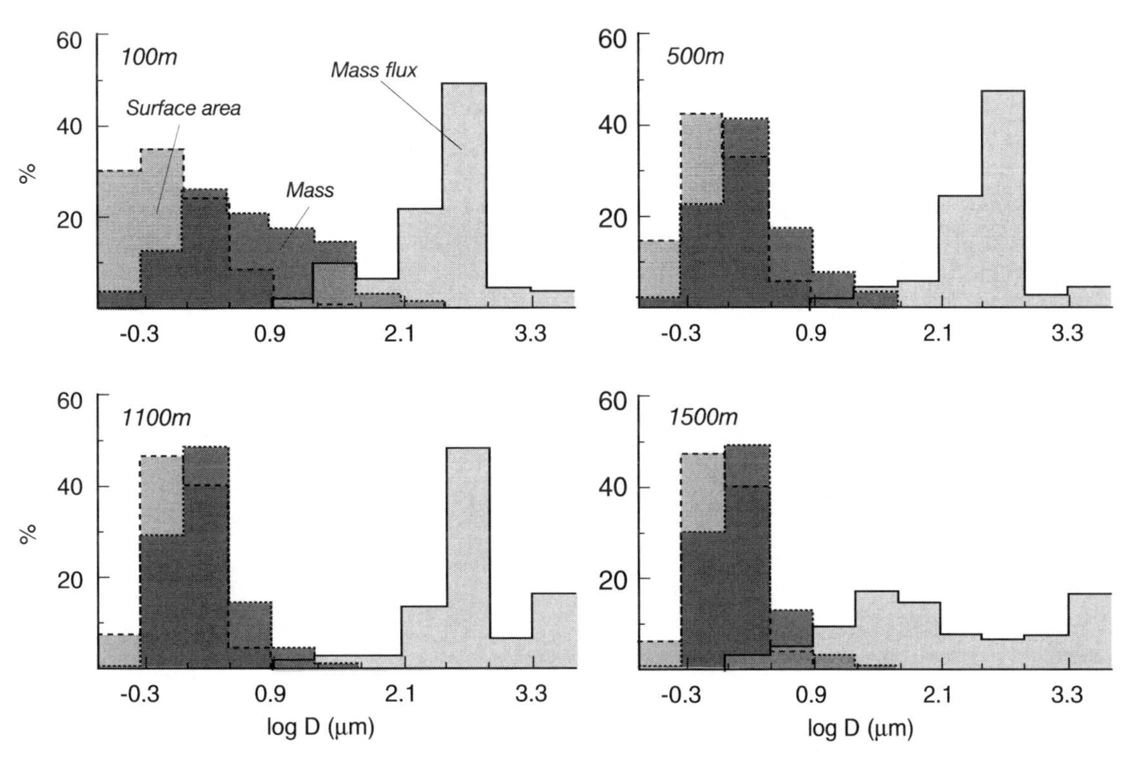

FIGURE 5.4.5: The frequency distribution of mass flux, mass, and surface area as a function of particle size at four depths. Shown is the relative contribution of each size bin to the total in percent. Figure adapted from *Clegg and Whitfield* [1990].

one is the design of neutrally buoyant sediment traps [e.g., *Buesseler et al.*, 2000], which drift with the currents and thus eliminate most of the hydrodynamic effects that plague tethered traps. Another important advance was the development of a thorium-protactinium technique to estimate the trapping efficiency and apply correction terms to the observed particle flux [e.g., *Yu et al.*, 2001; *Scholten et al.*, 2001]. *Yu et al.* [2001] demonstrated on the basis of this approach that the fluxes estimated from deep moored sediment traps (depths below 1500 m) are accurate, but that upper ocean traps tend to under-collect the vertical fluxes by up to 60%.

A third advance was the development of the thorium-234 technique, which permits an estimate of the export of particles from a particular water column on the basis of concentration measurements only, i.e., without resorting to sediment traps [see, e.g., *Buesseler et al.*, 1992]. This elegant method is based on the fact that thorium-234 is very particle-reactive and therefore tends to be rapidly scavenged by particles that are present in the water column. As we will discuss in more detail below, thorium-234 (half-life = 24.1 days) is produced in the oceans via the decay of its long-lived and highly soluble "parent," uranium-238. If there were no physical removal of either of these radioisotopes, the activities of them would be in secular equilibrium (i.e., identical) in the water column. However, the adsorption of thorium-234 onto particles followed by their vertical sinking leads to a deficit of thorium-234, the magnitude of which is proportional to the removal rate of thorium-234 on sinking particles from the upper ocean. By also measuring the ratio of particulate organic carbon to thorium-234 on sinking particles, one can then empirically determine the export fluxes of *POC*. This method not only avoids most of the issues associated with sediment traps but also has the advantage that it tends to integrate export fluxes over larger spatial and temporal scales. As this method requires extensive water sampling, it has been applied so far only at a few locations and with a strong focus on the near-surface ocean [e.g., *Buesseler*, 1998; *Buesseler et al.*, 1992].

Despite the substantial concerns regarding trap observations, they have been the primary means to observe and quantify the flux of sinking particles and as such provide one of the few data sets we have for developing hypotheses about the role of sinking particles in ocean biogeochemistry. Perhaps the most influential summary of trap results is that of *Martin et al.* [1987], based on measurements made in the northeast Pacific. The organic carbon fluxes as measured by *Martin et al.* [1987] show a sharp drop-off with increasing depth, implying that most of the sinking flux is either remineralized or broken down into nonsinking particles or *DOM* within the upper few hundred meters of the water column. *Martin et al.* [1987] fit all their organic carbon flux

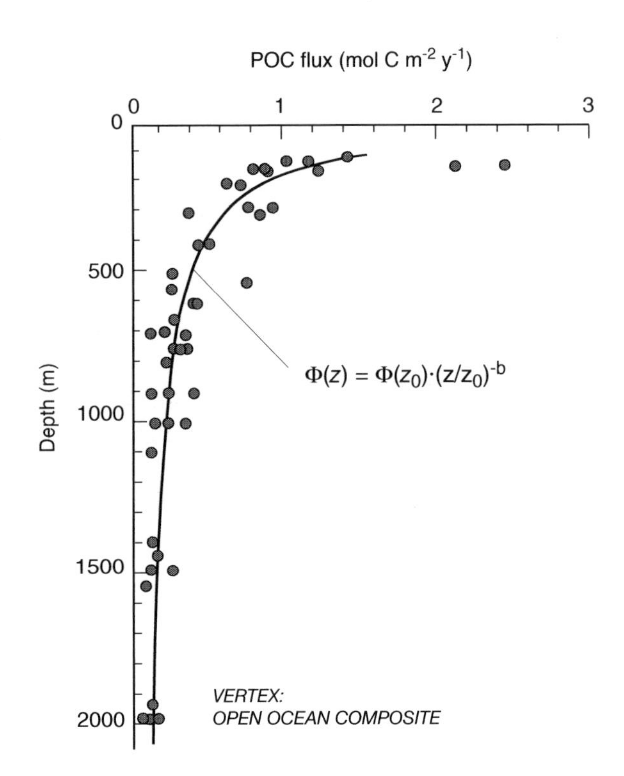

FIGURE 5.4.6: Particulate organic matter flux as a function of depth, from the sediment trap measurements of *Martin et al.* [1987]. The curve represents a fit with the power law function given in (5.4.3) in the main text.

observations (figure 5.4.6) to a simple power law as a function of depth and the flux at 100 m depth:

$$\Phi^{org}(z) = \Phi^{org}(z_0)\left(\frac{z}{z_0}\right)^{-b}, \; z_0 = 100\,\text{m} \qquad (5.4.3)$$

where z is depth (positive downward) and b is a dimensionless scaling factor determined to be 0.858. *Berelson* [2001] recently reviewed a large number of sediment trap studies from the JGOFS program and found that the power law function provides overall a good fit to the data, but that the scaling factor b varied quite substantially from region to region (figure 5.4.7), with a mean of 0.82 ± 0.16. *Berelson* [2001] noted also that a substantial fraction of the variability in the factor b can be explained by variations in the magnitude of the flux, with regions of higher fluxes having higher factors b. Similar variations of this factor b were found by *Schlitzer* [2002] analyzing data from the Southern Ocean. *Lutz et al.* [2002] also noted the inability of the *Martin et al.* [1987] curve with a single parameter b to provide a good fit to a large collection of sediment trap data. They suggested instead that one should use a double exponential curve, one with a relatively short remineralization length-scale representing labile organic matter, and one with a longer length-scale, representing more refractory organic matter.

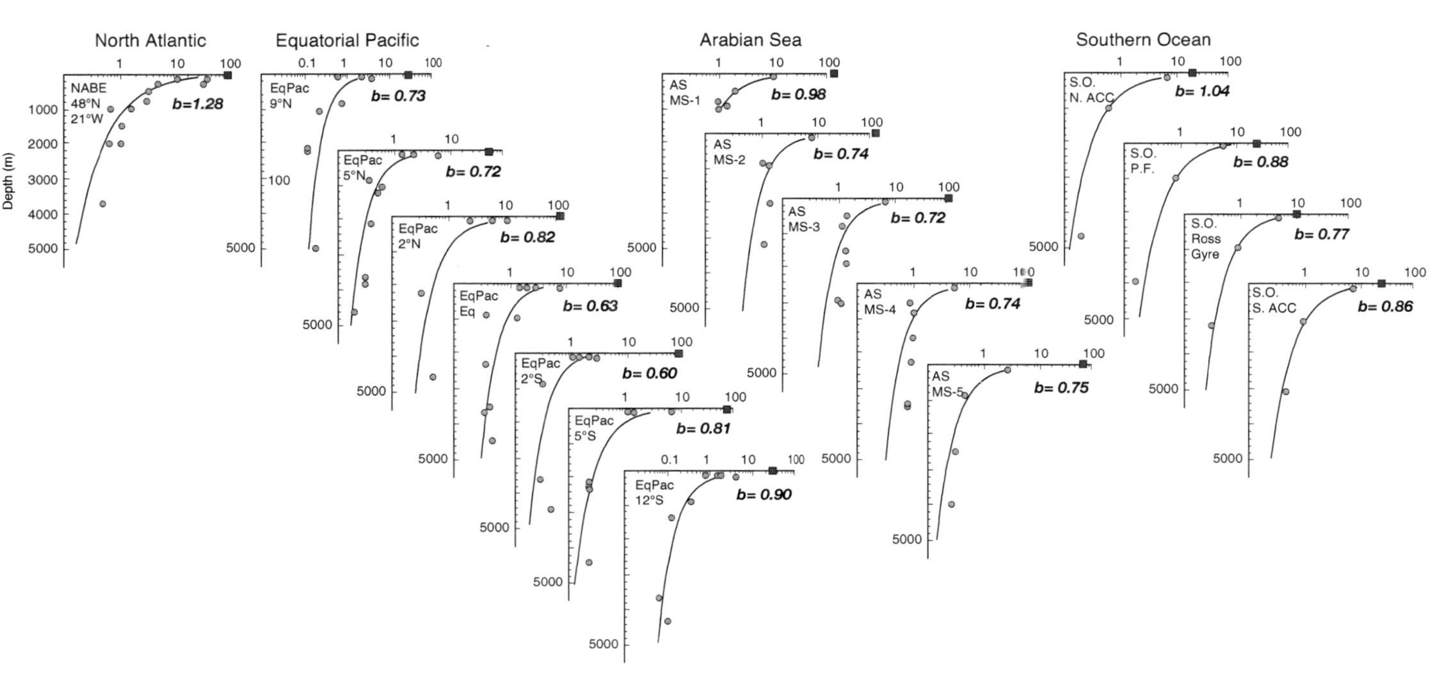

FIGURE 5.4.7: Plots of *POC* flux vs. depth for all 17 stations in the U.S. JGOFS regional studies conducted in the North Atlantic, equatorial Pacific, Arabian Sea, and Southern Ocean. Squares identify estimates of primary productivity, and solid circles are estimated flux values at various depths. The deepest points for the equatorial Pacific, Southern Ocean, and North Atlantic are defined by benthic recycling and sediment burial fluxes. Note the log scale for the flux (mmol C m^{-2} d^{-1}) and some variability in this scale. An average value of *POC* flux at 100 m was determined, and that value was used as an anchor point in defining the best fit given by the power law function equation (5.4.3). Equations of the fits to all the observed data for each of the study sites and the calculated fitting parameter *b* are shown. From *Berelson* [2001].

The role of Ballast

The variability of the factor b, the observation that particle fluxes tend to asymptote in the deep ocean, and the fact that organic carbon fluxes tend to cluster tightly when normalized to the total particulate mass flux, led *Armstrong et al.* [2002] to propose a new paradigm for the vertical flux of organic matter. They posit that the downward flux of organic matter in particles can be separated into two classes, one that is "quantitatively associated" with ballast minerals, and is perhaps protected from remineralization, and one that is unassociated and/or unprotected (see figure 5.4.8). The downward flux of organic matter below the euphotic zone is written as

$$\Phi^{org}(z) = \Phi^{org}_{ballast}(z) + \Phi^{org}_{unassociated}(z) \tag{5.4.4}$$

Armstrong et al. [2002] argued further that both components of the organic matter flux can be represented by a single exponential curve with the ballast-associated organic flux asymptoting to a nonzero value, $\Phi^{org}_{ballast}(\infty)$, that is protected from remineralization, while the unassociated organic carbon flux asymptotes to zero, thus:

$$\Phi^{org}_{ballast}(z) = \Phi^{org}_{ballast}(\infty) + \left[\Phi^{org}_{ballast}(z_0) - \Phi^{org}_{ballast}(\infty)\right] \cdot \exp\left(-\frac{z - z_0}{z^*_{ballast}}\right) \tag{5.4.5}$$

$$\Phi^{org}_{unassociated}(z) = \Phi^{org}_{unassociated}(z_0) \cdot \exp\left(-\frac{z - z_0}{z^*_{unassociated}}\right)$$

where z is positive downward, z_0 is 100 m, and $z^*_{ballast}$ and $z^*_{unassociated}$ are the length-scales of the remineralization/dissolution of ballast and unassociated organic matter, respectively. From fits to sediment trap results in the Arabian Sea and equatorial Pacific (see figure 5.4.8), *Armstrong et al.* [2002] found for both length-scales the same value of $z^* = 480$ m. This suggests that in these two tropical regions, the component of ballast that is removed as particles sink, e.g., by dissolution of opal or $CaCO_3$, disappears about as quickly as unassociated organic matter. This might not be the case in temperate or polar regions, as the mechanisms controlling dissolution of $CaCO_3$ and opal are quite distinct from each other (see chapters 7 and 9). Inserting (5.4.5) into (5.4.4) and using the equivalence of $z^*_{ballast}$ and $z^*_{unassociated}$ gives

$$\Phi^{org}(z) = \Phi^{org}_{ballast}(\infty) + \left[\Phi^{org}(z_0) - \Phi^{org}_{ballast}(\infty)\right] \exp\left(-\frac{z - z_0}{z^*}\right) \tag{5.4.6}$$

The fits to observations reveal that the asymptotic organic matter flux associated with ballast represents a fraction f of about 0.054 mg C_{org} (mg ballast)$^{-1}$ of the total asymptotic ballast flux, where

$$\Phi^{org}_{ballast}(\infty) = f \cdot \Phi^{total}_{ballast}(\infty) \tag{5.4.7}$$

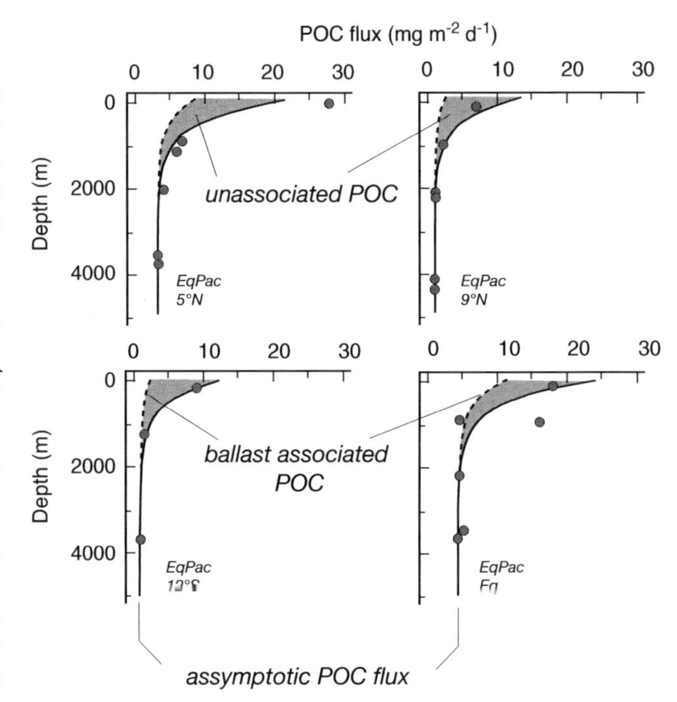

Figure 5.4.8: *POC* fluxes vs. depth are depicted for four latitudes in the equatorial Pacific with measured *POC* fluxes shown as filled circles. The dashed lines are the fitted ballast fluxes, while the solid lines are the fitted total fluxes. The shaded areas between these curves, therefore, represent estimates of fluxes of unassociated *POC*, which has virtually disappeared by 2000 m. Adapted from *Armstrong et al.* [2002].

The fits also show that the asymptotic flux of ballast, $\Phi^{total}_{ballast}(\infty)$ represents approximately 40% of the ballast flux leaving the surface ocean, $\Phi^{total}_{ballast}(z_0)$. *Armstrong et al.* [2002] compared the quality of fits of equations (5.4.6) and (5.4.7) to sediment trap data with the quality of fits obtained by using the power law function (5.4.3) with a fixed b, and concluded that the ballast model, despite having only a few more parameters, provided a substantially better fit.

The ballast model therefore provides a natural explanation for why the factor b in the results of *Berelson* [2002] varied from region to region. The flip side is that the ballast model requires not only knowledge about the remineralization of particulate organic matter, but also information about the production of ballast in the surface ocean and its fate as it sinks through the water column. *Armstrong et al.* [2002] noted in their study that different types of ballast, namely $CaCO_3$, opal, and lithogenic material (primarily dust), appear to have different associated fractions f. *Klaas and Archer* [2002] used a large synthesis of deep ocean sediment traps to analyze this hypothesis further. We restrict our discussion to their results using data from below 2000 m, because it is safe to assume that the unassociated organic matter fraction is virtually zero at and below this depth.

With this assumption, *Klaas and Archer* [2002] expanded (5.4.7) to

$$\Phi^{org}_{ballast}(\infty) = f_{carbonate} \cdot \Phi^{carbonate}_{ballast}(\infty) + f_{opal} \cdot \Phi^{opal}_{ballast}(\infty) + f_{lithogenic} \cdot \Phi^{lithogenic}_{ballast}(\infty) \quad (5.4.8)$$

and then used the sediment trap data to determine the protected fractions, f_i. They found that f appears to vary moderately between the different ballast materials, with $CaCO_3$ having the largest protected fraction and opal the smallest ($f_{carbonate} = 0.075 \pm 0.011$ mg C_{org} (mg $CaCO_3)^{-1}$, $f_{opal} = 0.029 \pm 0.004$ mg C_{org} (mg opal)$^{-1}$, and $f_{lithogenic} = 0.052 \pm 0.018$ mg C_{org} (mg lithogenic)$^{-1}$). *Klaas and Archer* [2002] computed also the contribution, $f_i \cdot \Phi^i$, of each of the different ballast fluxes, and found that about 80% of the organic matter fluxes to the ocean below 2000 m is driven by organic matter associated with $CaCO_3$ ballast. The contribution from opal ballast was found to be only about 14%, and that of lithogenic material, only about 6%. This is a puzzling result. Why is $CaCO_3$ much more important in carrying organic matter to the deep ocean than opal or lithogenic material?

Revisiting Stokes's law above (5.4.2) may provide part of the answer. We have seen that $CaCO_3$ is the densest ballast material, and has a sinking velocity that is about 50% higher than that of opal for the same equivalent particle radius. Thus, organic matter may simply have less time to be remineralized when it is sinking in association with $CaCO_3$. A second and perhaps more important reason is the spatial distribution of opal and $CaCO_3$ production and export. As we will discuss in chapters 7 and 9 in detail, most of the opal production and export occurs in a few regions, such as the Southern Ocean and the North Pacific. In contrast, $CaCO_3$ production and export is spatially much more uniform, and tends to scale with total organic matter export. Therefore, as the large sinking particles are being formed in the upper ocean, a larger fraction of global organic matter export can become quantitatively associated with $CaCO_3$.

A third reason is that the transfer efficiency, i.e., the fraction of organic matter exported from the euphotic zone arriving at 2000 m, appears to be higher for $CaCO_3$ than for opal [*Francois et al.*, 2002]. These authors used the density argument as a possible explanation, but hypothesized also that the organic matter exported from regions dominated by opal production (high latitudes) might be easier to break down by bacteria than the organic matter exported from regions dominated by $CaCO_3$ production (low latitudes). *Francois et al.* [2002] proposed that the more seasonal nature of the high latitudes may cause this difference in organic matter "quality." Another explanation is the observation that the $CaCO_3$-dominated low latitudes have a much lower *e*-ratio than the opal-dominated low latitudes (figure 4.1.7), so that the organic matter exported out of the euphotic zone of regions characterized by low *e*-ratios is expected to be more refractory and thereby more likely to be transported to depths below 2000 m.

Particle Remineralization

We next use the information from the flux equations to learn something about particle decomposition. With z positive downward, the vertical differential of $\Phi(z)$ is equal to the sinking term $-w_{sink} \cdot dPOC/dz$ in (5.4.1). In steady state, and assuming no net POC formation in the aphotic zone, the sinking term is equal to the remineralization term, $SMS_{consumption}(POC)$. Therefore, the flux parameterizations discussed above provide constraints on these two terms:

$$\frac{d\Phi^{org}(z)}{dz} = SMS_{consumption}(POC) = -w_{sink} \cdot \frac{dPOC}{dz} \quad (5.4.9)$$

where z is positive downward. For example, we can test various parameterizations for the consumption term. A first natural choice would be to assume a first-order remineralization reaction of the form

$$SMS_{consumption}(POC) = -k_{remin} \cdot POC \quad (5.4.10)$$

Inserting (5.4.10) into (5.4.9) and solving the second and third terms for $POC(z)$, and assuming that the sinking velocity is constant gives an exponential solution of the form

$$POC(z) = POC(z_0) \cdot \exp\left(-\frac{k_{remin}}{w_{sink}}(z - z_0)\right) \quad (5.4.11)$$

where z_0 is the depth of the euphotic zone. Recognizing that $\Phi(z) = -w_{sink} \cdot POC$, and hence also $\Phi(z_0) = -w_{sink} \cdot POC(z_0)$, we obtain a prediction of the organic matter flux in the case of a first-order remineralization of organic matter (5.4.10):

$$\Phi^{org}(z) = \Phi^{org}(z_0) \cdot \exp\left(-\frac{k_{remin}}{w_{sink}}(z - z_0)\right) \quad (5.4.12)$$

We thus find an exponentially decreasing flux, $\Phi(z)$, with an exponential scale length determined by k_{remin}/w_{sink}. This is analogous to the flux form assumed by *Armstrong et al.* [2002] for the unassociated organic matter flux (5.4.5), for which they found a scale length z^* of 480 m. The knowledge of the scale length permits us, for example, to determine k_{remin} for this unassociated organic matter:

$$k_{remin} = \frac{w_{sink}}{z^*} \quad (5.4.13)$$

Assuming a typical upper ocean particle sinking velocity of 10 to 50 m day^{-1} [*Berelson*, 2001], we find a first-order remineralization rate constant, k_{remin}, of 0.02 to 0.1 day^{-1}.

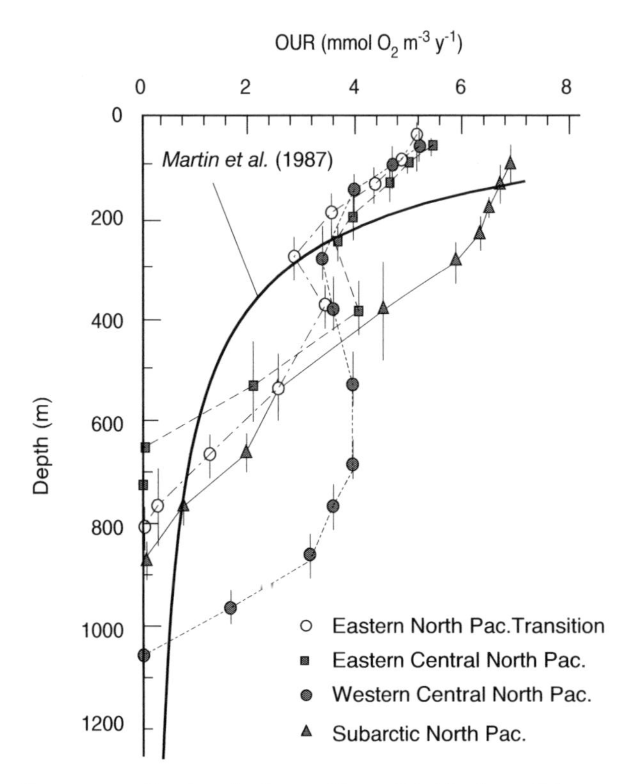

FIGURE 5.4.9: Remineralization profile predicted by the particle flux model of *Martin et al.* [1987] compared with a recent compilation of OUR estimates for the North Pacific by *Feely et al.* [2004b].

A second way to use the information contained in the flux formulations is to compare the consumption terms computed from the vertical differential of $\Phi(z)$ (first two terms in 5.4.9) with other estimates of organic matter remineralization. Figure 5.4.9 compares the oxygen loss rate implied by the *Martin et al.* [1987] trap data with recent *OUR* calculations from the North Pacific by *Feely et al.* [2004b]. The figure shows a reasonable agreement between these two very different approaches, despite the many possible shortcomings of traps and our neglecting of the *OUR* contribution from *DOC* remineralization.

One of the interesting results from trap observations is the large spatial and temporal variability on annual and interannual timescales (figure 5.4.10). This is generally attributed to variation in surface primary production, which has been shown to correlate with the flux of material caught in traps (e.g., *Suess* [1980]; see review by *Bruland et al.* [1989]). The equations (5.4.3) and (5.4.6) take this into account by normalizing the flux to the flux at 100 m.

We conclude our discussion of sediment traps with a summary of the type of material that is caught in them. Early results with conical traps suggested that sinking particles were composed primarily of fecal pellets. However, this turned out to be an artifact of the conical trap design, which tends to lose lighter material by currents generated within the trap by ambient water flow past the trap. Analysis of samples from cylindrical traps shows that the contribution from fecal pellets varies widely between different sites. *Le Borgne and Rodier* [1997] found for the equatorial Pacific a contribution of nearly 100%. In contrast, data from the JGOFS time series stations BATS (near Bermuda) and HOT (near Hawaii) suggest a more moderate contribution [*Roman et al.*, 2002].

Another important component of the sinking material caught in traps is marine snow [*Azam and Long*, 2001]. Marine snow consists of a diversity of macroscopic aggregates of detritus [*Alldredge and Gotschalk*, 1989]. Some aggregates are produced directly by organisms [cf. *Jackson and Burd*, 2002]. These include the cast-off mucus-feeding structures of zooplankton such as appendicularians, fecal pellets with no membrane, and mucus sheaths excreted by phytoplankton around their colonies. Another source of marine snow is the biologically mediated physical aggregation of particles such as phytoplankton, microorganisms, fecal pellets, inorganic matter, and microaggregates produced by processes such as spontaneous formation from *DOM* or adsorption of dissolved organic matter onto colloids [cf. *Jackson*, 2001]. Differential settling in the presence of shear plays a role, as does the enhancement of particle stickiness by organisms. Diatom blooms have been observed to flocculate into large mats that sink rapidly when they become nutrient stressed.

Other common particles found in traps are small, intact algae and protozoa such as radiolarians and foraminifera, many of which were alive when trapped [*Silver and Gowing*, 1991]. *Silver and Gowing* [1991] suggest that some of the decrease in trap flux with depth may simply reflect the ambient population of organisms.

To summarize, the processes forming large particles are poorly understood, but a dominant theme of what we do know is the importance of biological processes. The same is thought to be true of the primary mechanisms for the decomposition of particles. The two major categories of organisms that play a role in the decomposition of particles are bacteria and zooplankton. Some evidence suggests that bacteria do not colonize particles to a significant extent [e.g., *Taylor*, 1991]. However, the organic matter consumption rate of free-living bacteria is extremely large, accounting for nearly all of the carbon remineralization inferred from sediment trap flux measurements (e.g., [*Cho and Azam*, 1988]). To the extent that the bacteria are living off of the particle flux rather than the vertical flux of *DOC*, this implies that the sinking particles must be broken down into small particles and possibly *DOC* [e.g., *Azam and Long*, 2001]. However, there is little evidence for a substantial deep source of *DOC* from *POC* degradation [*Hansell and Ducklow*, 2003]. Zooplankton also have large carbon uptake needs that can account for most of the particle

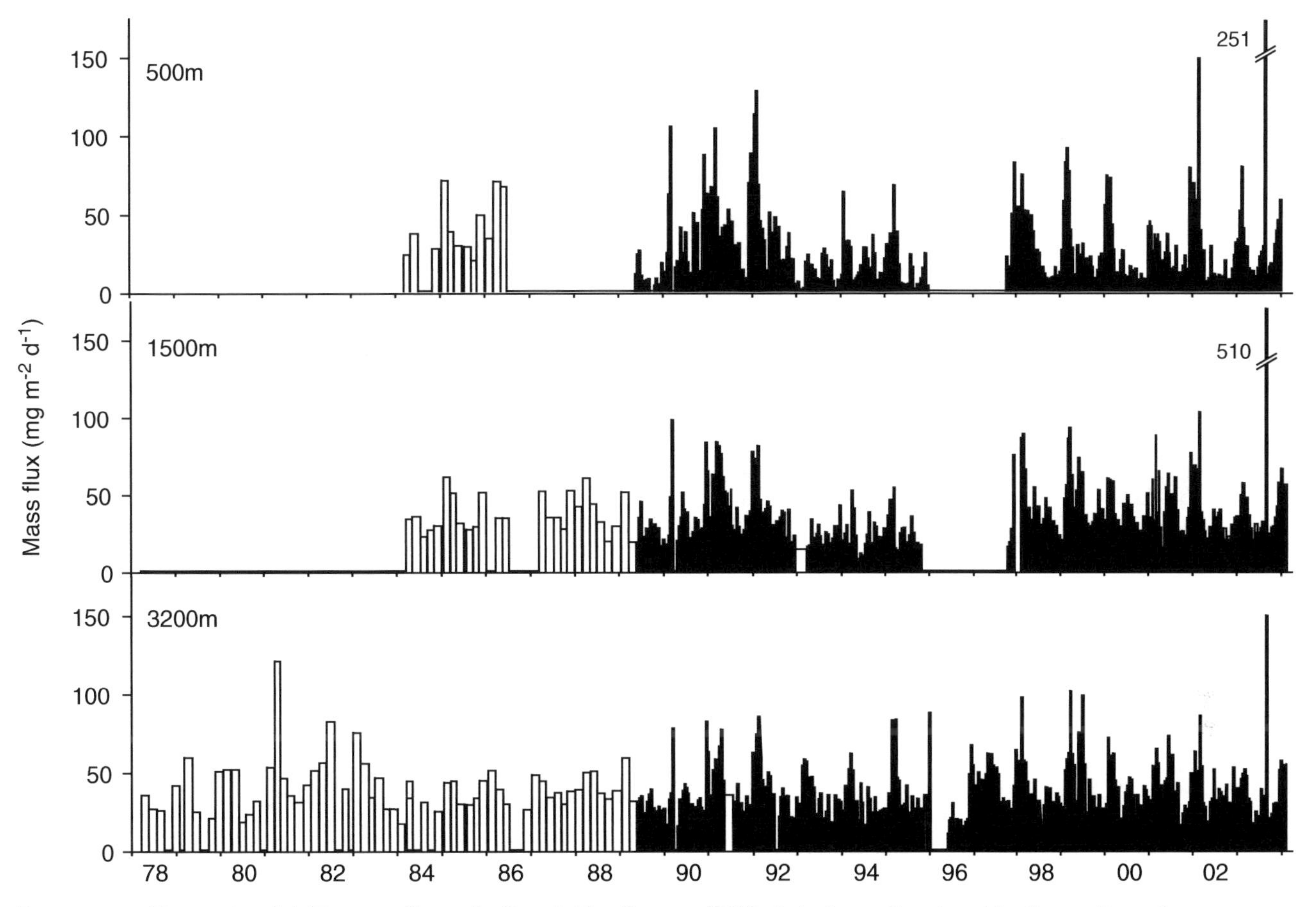

FIGURE 5.4.10: Time series of sinking mass flux at the Oceanic Flux Program (OFP) site in the northwestern Atlantic near Bermuda. Shown are the flux estimates from three depths with the horizontal extent of each bar representing the time period a particular measurement pertains to. In the second half of the record, the sampling frequency was increased from about bimonthly to biweekly, yielding a more finely resolved history of the flux variations. Updated from *Conte* [1998] (M. Conte, personal communication, 2005).

flux divergence, leaving little for microzooplankton and bacteria [e.g., *Bishop et al.*, 1987; *Lampitt*, 1992]. This discrepancy serves as a good illustration of our current lack of mechanistic understanding of the processes controlling the remineralization of *POC*.

Models of Particle Interactions

The observations discussed above give us two principal constraints on the processes that must be included in a model of particle dynamics: (1) the minimum representation of the particle size distribution is two sizes, one that sinks, and another that is nonsinking; (2) the formation and breakdown of particles is due primarily to biological processes. These two concepts can be represented by a simple model of particle interactions in the aphotic zone such as that shown in figure 5.4.11. The upper panel of the figure shows two boxes that represent small nonsinking particles, *S*, and large sinking particles, *L*, both of biogenic origin. The interaction between these two particle pools is assumed to be first order in the concentration of the particles with rate reactions, β_i [s^{-1}]. Remineralization is assumed to act only on the small particles, *S*, and also assumed to be first order. Therefore, large particles first have to decay to small particles before they can be remineralized to inorganic nutrients. No net formation of particles is included in this model, as we assume that all particles are formed in the surface ocean and supplied to the interior ocean by sinking. The tracer continuity equations for *S* and *L* are therefore:

$$\Gamma(L) = \beta_2[S] - \beta_{-2}[L] - w_{sink}\frac{d[L]}{dz}$$
$$\Gamma(S) = -\beta_2[S] + \beta_{-2}[L] - \beta_{-1}[S] \qquad (5.4.14)$$

We next have to find ways to determine the rate constants. Direct measurements are very difficult to undertake in natural settings. However, thorium provides an excellent tool to study particle dynamics, as it is highly particle-reactive, and its well-known decay rate gives a timescale that can be used to constrain the model rate constants [e.g., *Bacon and Anderson*, 1982]. Three isotopes of thorium are produced in the water column by soluble parents (see figure 5.4.12). The ur-

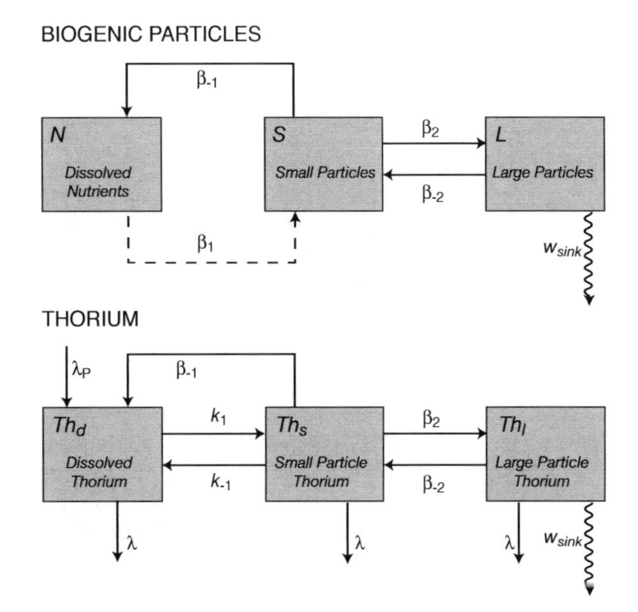

FIGURE 5.4.11: Schematic thorium and particle cycling model adapted from *Murnane et al.* [1994].

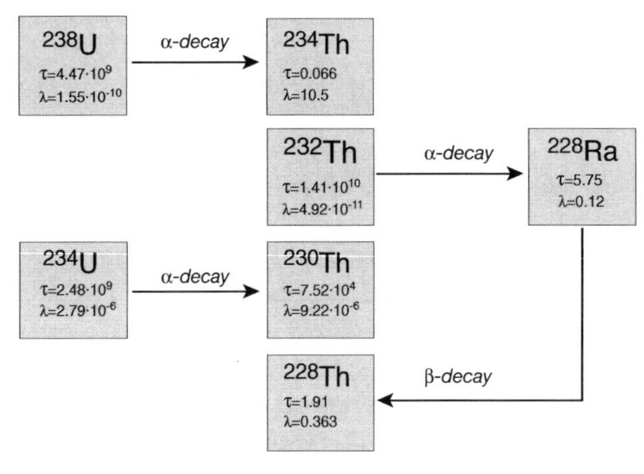

FIGURE 5.4.12: Uranium and thorium decay series. Also shown are the half-lives (τ in yr) and decay constants (λ in yr^{-1}) of the different isotopes.

anium parent isotopes ^{238}U and ^{234}U are very long-lived and nonreactive. Their distribution in the ocean is essentially uniform, and the rate of production of their daughters, ^{234}Th and ^{230}Th respectively, can be assumed to be the same everywhere within the error of the measurements. ^{228}Ra has a more complex distribution because its parent, ^{232}Th, is found only in the sediments, and the short half-life of ^{228}Ra (5.75 years) does not permit it to spread uniformly through the water column from its source in the sediments. The high concentrations of ^{228}Ra at the surface and bottom reflect the shelf and abyssal plain sedimentary sources, respectively. After production by decay of the dissolved parent, thorium adsorbs onto particles and remains with them as they become agglomerated and broken down [*Fisher et al.*, 1987].

We next develop a model for thorium that is parallel to that of the particles (figure 5.4.11). The equations are going to be similar, except that we have to consider the radioactive decay of dissolved uranium that produces Th and the radioactive decay of Th itself (figure 5.4.12). We also have to consider the absorption and desorption reactions between dissolved Th and the small particles (see, e.g., *Murnane et al.* [1994] for more details). The usual procedure for estimating the parameters of the particle cycling model from thorium observations is to assume steady state, and to solve the equations analytically using multiple isotopes of thorium with different half-lives.

Table 5.4.1 summarizes rate constants estimated for the biologically mediated processes of agglomeration of small particles, deagglomeration of large particles, and breakdown of small particles. The timescale of deagglomeration of large particles, $1/\beta_{-2}$, ranges from about a day to a year, with most estimates being at the low end of this range. The timescale of agglomeration, $1/\beta_2$, is typically on the order of a month to a few years. Thus, large particles are broken down more rapidly than they are being formed, which is consistent with the rapid drop-off in sediment trap fluxes. Finally, we have remineralization timescale estimates of $1/\beta_{-1}$ that range from less than a day to several years. Although highly unconstrained, the order of magnitude of this thorium-derived estimate of first-order particle remineralization is similar to that we inferred for unassociated organic matter above (5.4.13), which gave a timescale of 10 to 50 days.

The model illustrated in figure 5.4.11, with the addition of colloids, is among the more sophisticated models that exist today based on analysis of actual observations of particles and particle tracers (see also the work by *Jackson and Burd* [2002]). The recent recognition of the importance of ballast material in controlling the remineralization profile certainly calls for an extension of this model to include a pool for ballast-associated organic matter with higher sinking velocities. Clearly, much work remains to be done with regard to *POC* dynamics, particularly in the twilight zone, where most of the organic matter exported from the euphotic zone is remineralized.

A powerful independent constraint on particle cycling dynamics is radiocarbon (see chapter 2 for a general discussion on the usefulness of radiocarbon for estimating rates). Figure 5.4.13 shows results from one set of measurements made by *Bauer et al.* [1992] in the north central Pacific. Both the suspended and sinking *POC* show very high radiocarbon concentrations, indicating contamination with the bomb transient (figure 2.3.7). This is consistent with the relatively rapid turnover time for the particles inferred from the thorium isotope measurements. The increase in age with depth requires incorporation of old organic matter into the

TABLE 5.4.1

Rate constants determined by thorium observations for the particle cycling model illustrated in figure 5.4.11. The rate constant k_1 refers to a model where the reaction kinetics of the adsorption of thorium onto particles is assumed to be second order, whereas the rate constant k_1^* refers to a model, where this adsorption kinetics is assumed to be first order.

Taken from Murnane et al. [1994].

Source	*Site*	k_1 (10^4 m^3 kg^{-1} yr^{-1})	k_1^* (yr^{-1})	k_{-1} (yr^{-1})	β_{-1} (yr^{-1})	β_2 (yr^{-1})	β_{-2} (yr^{-1})
Nozaki et al. [1987]	Western Pacific		0.20 ± 0.27 to 0.44 ± 1.2	0.9 ± 1.2 to 1.9 ± 5.1		0.11 ± 0.03	16 ± 9
Bacon et al. [1989]	Arctic Ocean	5 to 8.7	0.16 to 0.47	2.6 to 9.8			
Bacon and Anderson [1982]	Panama Basin		0.2 to 1.3	1.3 to 6.3			
Nozaki et al. [1981]	Pacific Ocean		1.5	6.3			
Clegg et al. [1991]	Equatorial Pacific	~7 to ~70	1 to 4	2.5 ± 1.0	<0.4 to ~75	<0.1 to ~50	<50 to ≥365
Clegg et al. [1991]	North Pacific		3 to 70			<1 to ~40	~65
Murnane et al. [1994]	North Atlantic	5.0 ± 1.0		3.1 ± 1.5	4 to 1000	8 to 18	580 to 3000

particle pool. *Druffel et al.* [1992] suggest that this is due to incorporation of *DOC* into the particle pool. The cycling of *DOC* is the subject of the next subsection.

DISSOLVED ORGANIC MATTER

The primary elements making up *DOM* are dissolved organic carbon *DOC*, dissolved organic nitrogen *DON*, and dissolved organic phosphorus *DOP*. Considerably less is known about the latter two than about *DOC*, and little is known about the compounds that comprise the bulk of *DOM*. The concentrations of *DOC* and *TOC* are determined by oxidizing them to CO_2 and subsequently measuring the evolved CO_2 gas. The techniques for measuring *DOC* have undergone major improvements that began with a challenge by *Sugimura and Suzuki* [1988] to the traditional wet chemical oxidation techniques that had been used up to the mid-1980s (see *Hedges* [2002] for an excellent review of the historical development of research in *DOC*). *Sugimura and Suzuki* [1988] made use of a high-temperature oxidation method and obtained concentrations more than twice those measured by other techniques. Although these measurements were subsequently withdrawn by *Suzuki* [1993] because of his inability to reproduce them, and his recognition of a number of problems such as inadequate blank correction, this work spurred a rapid improvement of measurement techniques that now appear to be giving consistent results [e.g., *Sharp*, 2002]. The new measurements are consistent with the old wet chemical methods except that the precision has been greatly improved.

Figure 5.4.14 shows a north-south section of *TOC* in the eastern South Pacific along approximately 105° to 110°W (same section as shown for *POC* in figure 5.4.2). The high surface concentrations of order 70 mmol m^{-3} in the low latitudes and the lower surface concentrations of about 40 to 50 mmol m^{-3} in the high latitudes are representative of several other locations where measurements have been made (see *Hansell* [2002] for a recent summary). Also very typical is the drop over several hundred meters to deep ocean concentrations of ~30 to 40 mmol m^{-3}. Quite remarkable are the large differences in *TOC* of different water masses, suggesting that some of the distribution is driven also by variations in near-surface concentrations of *DOC* in the different water mass source regions, which are then transported into the interior ocean. As we have seen in figure 5.4.2, the concentration of *POC* is on the order of a few (1 to maximally 10) mmol m^{-3} in open ocean surface waters and rapidly drops to below 3 mmol m^{-3} with depth. The open ocean *TOC* is thus composed almost entirely of *DOC*. However, surface particulate organic carbon loads in eutrophic margin regions can exceed 20 mmol m^{-3}. It should be noted that since the vast majority of particles measured in the ocean do not contribute significantly to the vertical flux (figure 5.4.5), the *TOC* distribution probably gives quite a good measure of that portion of the organic carbon that is transported primarily by ocean circulation and mixing. Another remarkable feature of the oceanic *DOC* distribution is shown in figure 5.4.15, which depicts the slow reduction in the deep ocean *DOC* concentration from the North Atlantic (~48 mmol m^{-3}) to the North Pacific (~34 mmol m^{-3}).

A recent summary of measurements of *DON* gives values of order 6 ± 2 mmol m^{-3} in the surface ocean and about 4 ± 2 mmol m^{-3} in the deep ocean [*Bronk*, 2002]. These measurements give C:N ratios for *DOM* of about 14 in the surface ocean and 16 in the deep ocean. A similar summary for *DOP* gives typical surface concentrations of about 0.1 to 0.4 mmol m^{-3} and values below 0.1 mmol m^{-3} in the deep ocean [*Karl and Bjorkman*, 2002]. This yields very high C:P ratios of *DOM*, of the order of 200 to 600 [*Benner*, 2002].

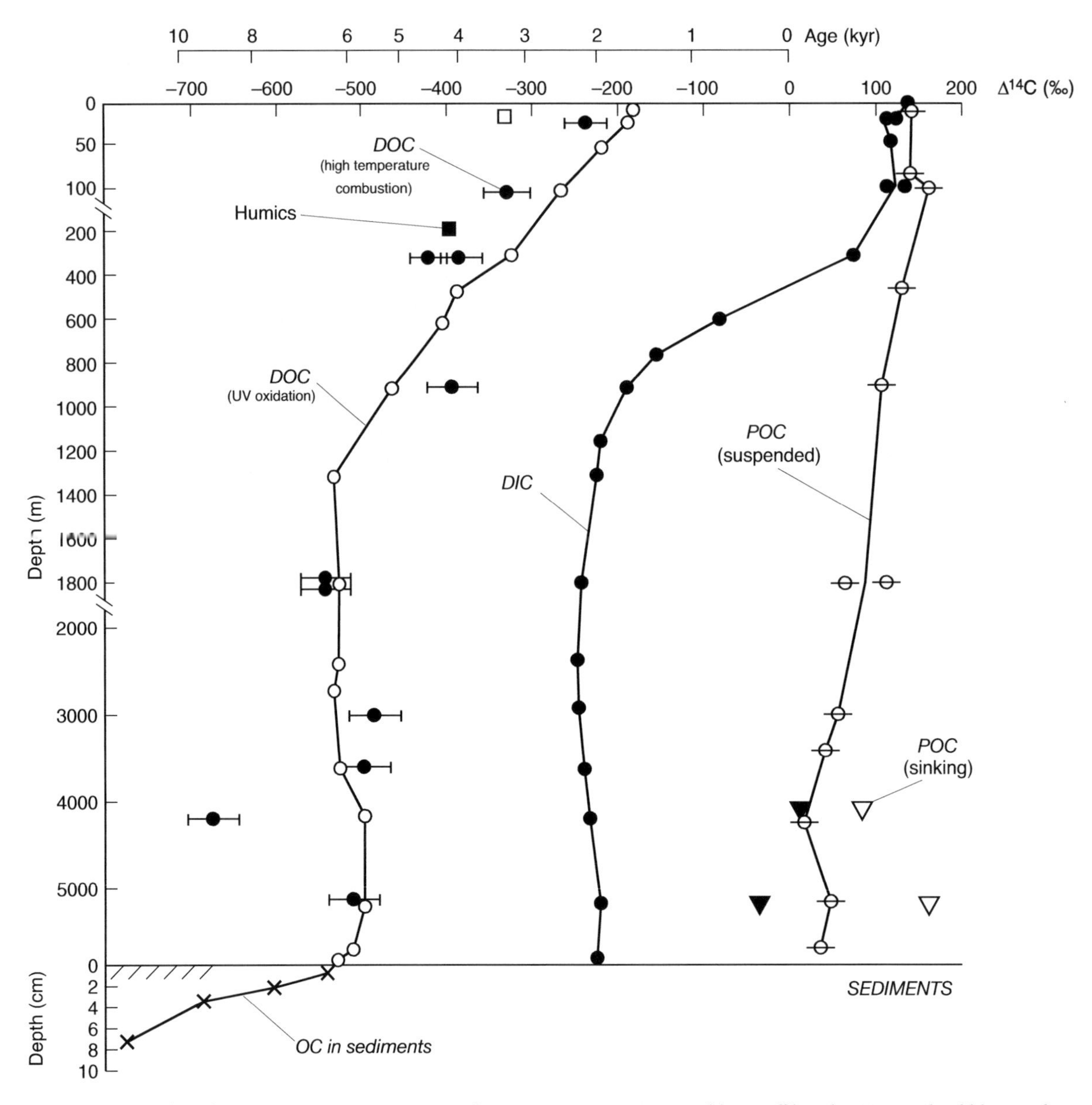

FIGURE 5.4.13: Radiocarbon measurements in various carbon components as measured by *Druffel et al.* [1992]. It should be noted that the radiocarbon measurements shown here were made before some of the *DOC* measurement issues were clarified. The results thus need to be confirmed.

A combination of measurements suggest the existence of three major pools of *DOC*: a refractory component, i.e., a component whose biological degradation is very slow, leading to a lifetime of several thousand years; a highly labile component, i.e., a component whose biological degradation is extremely rapid, resulting in generally low concentrations and a lifetime of a few hours to days; and a semilabile component with a sufficiently long lifetime to be transported downward well into the thermocline (figure 5.4.16). One piece of evidence supporting the existence of these components is the age of *DOC* as measured by radiocarbon (e.g., figure 5.4.13). The radiocarbon age of deep ocean *DOC* is ~4000 years in the Atlantic, aging to ~6000 years in the Pacific, consistent with deep ocean circulation. *Druffel et al.* [1992] estimate that these radiocarbon ages indicate that about 80% of the refractory deep *DOC* is returned unchanged each time the deep ocean water is replaced from the surface. The inferred removal of 20% of the deep *DOC* pool is consistent with the *DOC* reduction between the deep North Atlantic and the deep North Pacific (figure 5.4.15). The radiocarbon age of surface *DOC* is in excess of 1500 years [*Druffel et al.*, 1992; *Bauer et al.*, 1992]. This great age is due to the refractory component of ~ 40 mmol m^{-3} that comes from the deep ocean (figure 5.4.15).

The impression one gets from the deep ocean *DOC* observations in connection with their radiocarbon

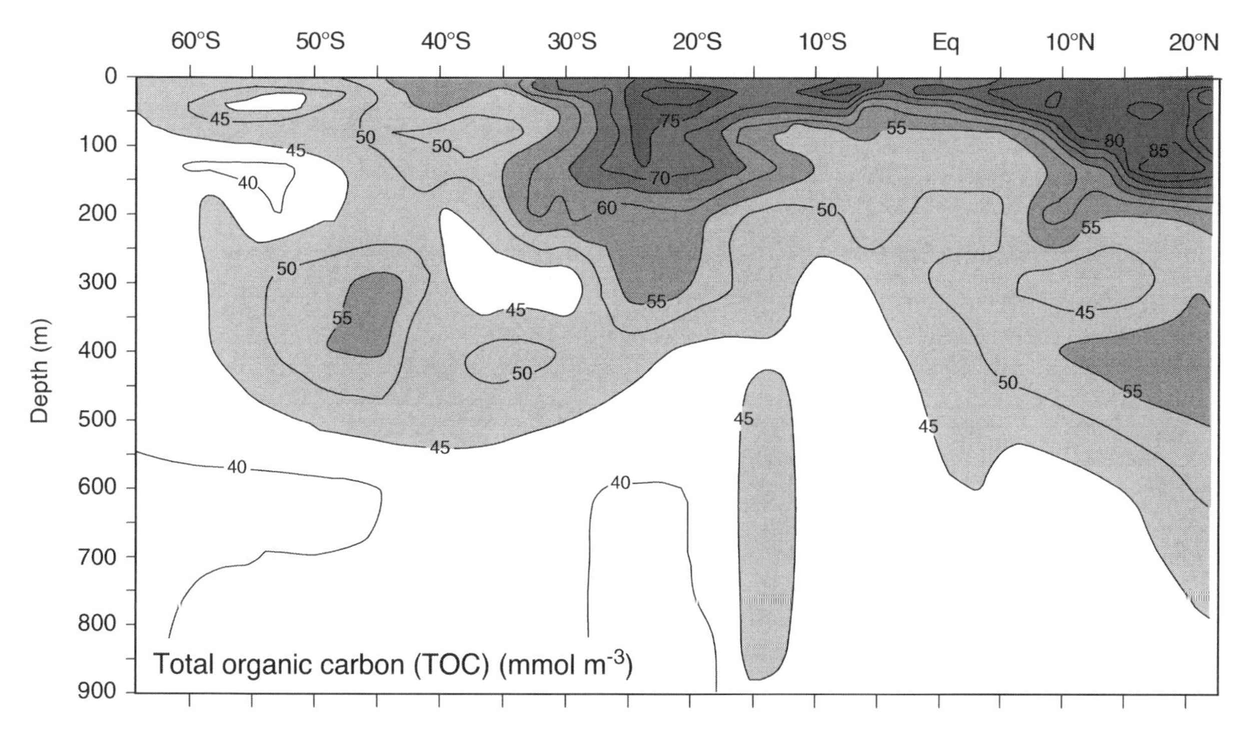

FIGURE 5.4.14: Meridional section of total organic carbon (*TOC*) in the eastern South Pacific along about 105°W (WOCE line P18*S*). Note the rather different *TOC* levels associated with different water masses. From *Hansell and Waterhouse* [1997].

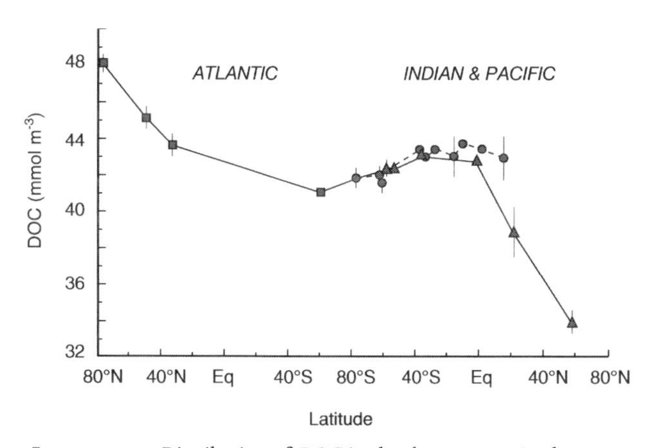

FIGURE 5.4.15: Distribution of *DOC* in the deep ocean. Analogous to the global vertical sections shown in figure 5.1.2 and elsewhere, the horizontal axis represents a combination of a southward section in the Atlantic, and northward sections in the Indian and Pacific Oceans. From *Hansell and Carlson* [1998].

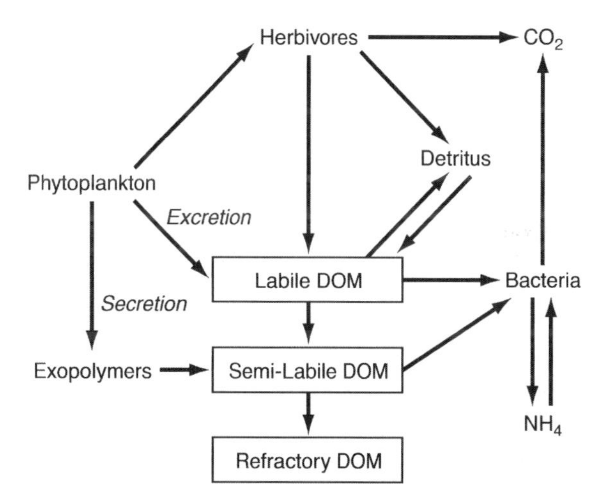

FIGURE 5.4.16: Schematic *DOC* cycling diagram modified from *Kirchman et al.* [1993]. This shows the labile, semi-labile, and refractory components referred to in the text as well as possible mechanisms for converting from one to the other.

ages is that almost all *DOC* entering the deep ocean comes from the North Atlantic, and that only a negligible amount of *DOC* is added along the deep branch of the global overturning circulation. The only other advective pathway for a substantial addition of *DOC* to the deep ocean is the Southern Ocean, but measurements summarized by *Hansell* [2002] indicate that this input appears to be very small. A second source of deep *DOC* are the sediments, but work summarized by *Burdige* [2002] suggests that this does not appear to have a significant impact on water column *DOC*, perhaps because it is rapidly remineralized near the sediment-water interface.

We must also consider the possibility that *DOC* may be produced by solubilization of *POC* in the water column. Measurements showing temporal variability of *DOC* in the North Pacific have been interpreted as possibly resulting from variations in *POC* solubilization due to changes in *POC* fluxes to the deep sea [*Bauer et al.*, 1998], but such variability is not supported by work elsewhere (e.g., in the Sargasso Sea [*Hansell and Carlson*, 2001]); even under conditions of high flux variability (e.g., Arabian Sea [*Hansell and Peltzer*, 1998]). The preponderance of evidence thus favors the interpretation that the large-scale *DOC* removal observed between the North Atlantic and North Pacific is driven primarily by input of *DOC* from the North Atlantic. *Bauer and Druffel* [1998] argued that this view is too simplistic, and the cycling of *DOC* in the deep ocean is more dynamic. They suggested that a significant amount of older *DOC* from continental margins is added throughout the ocean, requiring much higher rates of *DOC* turnover. However, this result is not supported by the study of *Hansell et al.* [2002]. If the view put forward by *Hansell* [2002] is correct, the apparent rate of deep *DOC* removal (about 30 pmol m^{-3} yr^{-1}) is so miniscule that it would hardly permit any organism to grow solely on this carbon source. *Hansell* [2002] therefore suggested that inorganic processes, such as adsorption onto particles, may play a role in the removal of deep *DOC*.

The *DOC* in the upper ocean that is in excess of the background deep ocean concentration is found all the way through the thermocline. This observation, coupled with the observation of a parallel increase in the radiocarbon age of *DOC* between surface and deep ocean values, suggests that a portion of the *DOC* must be semilabile, i.e., subject to bacterial breakdown. A simple mass balance calculation with radiocarbon measurements shows that the semilabile component is modern. In the central North Pacific, *Druffel et al.* [1992] found a surface *DOC* concentration of 84 mmol m^{-3} and a radiocarbon concentration of −163‰ (figure 5.4.13). If one takes the sample at 903 m (*DOC* = 38 mmol m^{-3}, and radiocarbon = −447‰) as representative of the refractory fraction, one finds that the semilabile fraction has a radiocarbon value of 95‰ at the surface. This high value indicates contamination with radiocarbon produced by nuclear bomb tests, which began entering the ocean mostly after 1960 (figure 2.3.7).

A final piece of evidence supporting the existence of a labile component of *DOC* are measurements by *Carlson et al.* [1994] and *Carlson and Ducklow* [1995] of near-surface *DOC* variability. These measurements suggest that there is a component of semi-labile *DOC* that has a residence time on the order of a season up to a year, and an additional small component of <5 mmol m^{-3} that has turnover times of hours to a few days [*Carlson and Ducklow*, 1995; see also *Amon and Benner*, 1994; *Hansell et al.*, 1995]. The seasonal and annual component is probably related to the semilabile component we see in the thermocline. The highly labile component with turnover times of hours to days is of importance only in surface biological processes. The export of carbon from the surface would be determined primarily by production of the semilabile component.

The chemical composition of *DOC* has not yet been well characterized (see review by *Benner* [2002]). Ultrafiltration and solid phase extraction methods are typically used to isolate and characterize *DOC*. The ultrafiltration method makes it possible to analyze two size fractions of the *DOC* pool, one that passes through an ~1 nm pore filter and one that is retained by this filter. The former is usually referred to as low molecular weight (*LMW*) *DOC* and the latter as high molecular weight (*HMW*) *DOC*. The cutoff between these two size classes is about 1000 atomic mass units or daltons (a dalton is 1/12 the weight of a ^{12}C atom). In the surface ocean, *DOC* is about 60% *LMW*, whereas this fraction increases to about 75–80% in the deep ocean [*Benner*, 2002].

The high molecular weight *DOC* was found to have a high carbohydrate content (about half in shallow samples and a quarter in deep samples). Surprisingly, this portion of the *DOC*, which is too large to enter the cell without further breakdown, was utilized quite rapidly by bacteria, whereas the remaining fraction containing smaller molecules was not [*Amon and Benner*, 1994]. The *HMW DOC* tends to have a C:N ratio of the order of 15 in surface waters, increasing to 20 in deep waters. This is much larger than typical bacterial C:N ratios of 4 to 5. It suggests that bacteria must obtain some nitrogen from other sources (see discussion on microbial loops in chapter 4). *Amon and Benner* [1994] observed a net uptake of dissolved inorganic nitrogen during bacterial incubations with the high molecular weight component as substrate. Incubations with the low molecular weight (< 1000 daltons) component gave the opposite result: net nitrogen regeneration. Amon and Benner conclude from this that the bioreactive *LMW* component has a low C:N ratio, i.e., that it is nitrogen-rich. This is consistent with its amino acid composition.

To summarize, surface *DOC* can be characterized as consisting of a highly labile pool of <5 mmol m^{-3} with extremely rapid turnover times of a few hours to days, a semilabile pool of order 20 mmol m^{-3} with a turnover time of order a season to 10 years, and a refractory pool of order 40 mmol m^{-3} with a turnover time of millennia [cf. *Kirchman et al.*, 1993; *Carlson and Ducklow*, 1995]. The lability of *DOC* is reflected in both the measured rate of uptake by bacteria [e.g., *Amon and Benner*, 1994; *Carlson and Ducklow*, 1995; *Hansell et al.*, 1995] and the distribution of *DOC* within the ocean (figures 5.4.14 and 5.4.15). The actual distribution of *DOC* pools is probably more continuous than the breakdown into three pools suggests. The semilabile pool is exported from the surface to the thermocline before being remineralized [*Hansell*, 2002]. The most labile component of *DOM* has high molecular weights and a C:N ratio of 15 to 20 [*Benner et al.*, 1992]. The labile component of the remaining low molecular weight *DOM* appears to be composed primarily of nitrogen-rich amino acids, explaining the observation that the C:N ratio of total dissolved organic matter is lower than that of *HMW DOM*.

Five mechanisms of *DOM* formation have been identified [*Carlson*, 2002]: (a) excretion and secretion directly from phytoplankton, (b) spontaneous autolysis and cell lysis resulting from viral attack, (c) sloppy feeding by zooplankton, (d) bacterial transformation and release, and (e) degradation of fecal material and other detritus. In his review of phytoplankton excretion of *DOM*, *Williams* [1990] concludes that despite the great amount of work that has been done in this area, understanding remains very poor. Excretion was shown in earlier studies to be <10% of the primary production [*Lancelot*, 1984], but a more recent study by *Bronk et al.* [1994] shows *DON* excretion rates more like a third of the primary production. The mechanisms *Williams* [1990] discusses include damage resulting from high light intensity, diffusion across the cell wall, either passive or due to overflow of excess organic matter during periods of high photosynthesis; and the impact of nutrient deficiency. He argues that loss of metabolic control resulting from environmental stresses is a common pattern in most of these mechanisms.

Kirchman et al. [1993] differentiate excretion, which consists of low molecular weight material that can pass through the cell wall, from secretion of extracellular material such as the polysaccharides making up the gelatinous matrix of colonial algae. He suggests that these may be the polysaccharides observed by *Benner et al.* [1992]. The viral infection of phytoplankton, which has been shown to be very common [e.g., *Fuhrman*, 1992], results in the production of *HMW* material by cell lysis. Spontaneous lysis, which may occur when phytoplankton are stressed by nutrient deficiencies at the end of a bloom, may also be of importance as a source of *HMW* dissolved organic matter. Certain zooplankton such as copepods that ingest plankton by filtering or use of raptorial appendages may damage cells and release dissolved organic matter in this way. This is referred to as "sloppy feeding." Some zooplankton may also excrete *DOM* as urea and amino acids. Finally, the fecal material produced by zooplankton can be degraded to *DOM* by hydrolytic enzymes produced by bacteria [e.g., *Smith et al.*, 1992]. Sinking particles, however, appear to be a relatively unimportant source of *DOC* [*Hansell and Ducklow*, 2003].

The primary consumers of *DOC* are bacteria, as noted in chapter 4. It is of importance to know the efficiency with which the organic matter taken up by bacteria is converted to biomass, and the rate at which bacteria convert *DOC* to *DIC*. In the Sargasso Sea, *Carlson and Ducklow* [1996] find bacterial growth efficiencies of the order of $14 \pm 6\%$. Somewhat higher (9–38%) but still small bacterial growth efficiencies were found in the Ross Sea [*Carlson et al.*, 1999]. It thus appears that a large portion of the *DOM* is going into respiration and is not used for forming bacterial biomass. The bacterial growth efficiency is almost certainly a function of food quality [*Ducklow and Carlson*, 1992; *Biddanda et al.*, 1994]. It appears to increase in more productive coastal and near-shore waters.

5.5 Models

The basic elements required for an ocean general circulation model of nutrient and oxygen cycles are: (1) a model of ocean circulation, (2) a model of gas exchange, (3) a model of organic matter production in the surface ocean, (4) a model for the transport and remineralization of *POM*, (5) a model for the transport and remineralization of *DOM*, and (6) information on the stoichiometry of the organic matter, which we use to key the cycling of nutrients and oxygen to each other. We will continue to ignore the sediments until the next chapter. Each of these model components has been discussed. Now it is time to put them together.

There have been many attempts over a period of several decades to develop box models of the ocean nutrient and oxygen cycles such as those discussed in earlier chapters. A major motivation for this work has been the attempt to understand the impact of the ocean on atmospheric CO_2 levels (see also the discussion in chapter 10). An early effort to develop such simulations in ocean GCMs is reported by *Sarmiento et al.* [1988a], who used an idealized sector model of the ocean circulation. This was followed by more realistic ocean model simulations carried out by *Bacastow and Maier-Reimer* [1990] and *Najjar* [1990]. Over the last ten years, this field has vastly expanded (see *Doney et al.* [2001] for a recent review). This section will summarize the major issues that one faces in developing such models, as well as some of the lessons that have been learned from the work that has been done to date. The focus will be on global-scale applications, recognizing that there are many coupled physical/ecological/biogeochemical modeling studies of note that focus on regional problems or that investigate certain phenomena in more idealized settings.

Model Development

Nearly all published global-scale ocean biogeochemical model studies are based on relatively coarse-resolution ocean general circulation models (OGCMs), mainly because of computational restrictions associated with the long integrations necessary to achieve steady-state solutions over the whole water column. Given the several hundred years turnover of the deep ocean circulation, such models typically need to be integrated over several thousand years. As a result, fundamental scales of ocean variability, particularly those at the mesoscale, which we have seen to contribute substantially to oceanic variability and presumably also to the mean state (section 2.5), are not explicitly resolved. Various parameterizations have been proposed to capture the effect of these mesoscale variations on the mean-state of ocean circulation [e.g., *Gent and McWilliams*, 1990], but no such parameterization exists for the ecological and biogeochemical components of these models. Chapter 2 shows and discusses the simulated circulation of one such OGCM, developed at Princeton University/Geophysical Fluid Dynamics Laboratory. Detailed comparisons of the physical fields and the uptake of CFCs and bomb radiocarbon with observations and other OGCMs demonstrate that this model is very typical in its strengths and deficiencies [see *Dutay et al.*, 2001; *Matsumoto et al.*, 2004]. We will use variants of this OGCM for many of our illustrative discussions here.

The modeling of the formation and export of organic matter in the surface ocean has taken essentially two directions (see also *Doney* [1999]). One class of models are the ecosystem models first implemented into OGCMs by *Fasham et al.* [1993], *Kurz* [1993], and *Sarmiento et al.* [1993]. These ecosystems models are structurally similar to those discussed in chapter 4. They recently have become increasingly more sophisticated and detailed by including, for example, several phytoplankton functional groups and multiple limiting nutrients [cf. *Moore et al.*, 2004]. We will not discuss such ecosystem models further here, as many of the very fundamental insights arising from the coupling of biogeochemical cycles and ocean circulation can already be captured by the other class of ecosystem models, which use highly parameterized formulations of biological production and export. A classical example of this second class of models is that of *Bacastow and Maier-Reimer* [1990], who estimated export production of organic matter directly as a function of light and nutrient concentration. An even simpler formulation was proposed by *Sarmiento et al.* [1988b] and later implemented in a large number of global models as part of the Ocean Carbon-Cycle Intercomparison Project (OCMIP) [cf. *Najjar et al.*, 2001]. This formulation estimates the surface ocean production and export of organic phosphorus, $SMS_{soft}(PO_4^{3-})$, by restoring the model-simulated phosphate concentration, $[PO_4^{3-}]_{simulated}$, toward the observed concentration, $[PO_4^{3-}]_{observed}$, i.e.,

$$SMS_{soft}(PO_4^{3-}) = -\frac{1}{\tau}([PO_4^{3-}]_{simulated} - [PO_4^{3-}]_{observed}),\ z \leq z_{euphotic} \qquad (5.5.1)$$

where $z_{euphotic}$ is the depth of the base of the euphotic zone and τ is a restoring timescale, typically of order of a few weeks to months. Equation (5.5.1) is only applied if $[PO_4^{3-}]_{simulated} > [PO_4^{3-}]_{observed}$, otherwise $SMS_{soft}(PO_4^{3-})$ is set to zero. In a steady state, this equation removes phosphate from the surface ocean at the same rate it is transported in laterally or from below. It is assumed that a fixed fraction, σ, of the total phosphate uptake is converted to semilabile *DOP* (the only form of *DOP* considered here), and the rest is converted to particulate organic phosphorus (*POP*). *DOP* is permitted to be advected and diffused, and is remineralized everywhere with a first-order rate constant, k_{DOP}. The tracer continuity equation for *DOP* is therefore

$$\Gamma(DOP) = \sigma \cdot SMS_{soft}(PO_4^{3-}) - k_{DOP} \cdot [DOP],\ z \leq z_{euphotic}$$

$$\Gamma(DOP) = -k_{DOP} \cdot [DOP],\ z > z_{euphotic} \qquad (5.5.2)$$

The flux of *POP* exported at the base of the euphotic zone is given by the vertical integral over the euphotic zone of the fraction of SMS_{soft} (PO_4^{3-}) that goes into *POP*, i.e.,

$$\Phi^{POP}(z_{euphotic}) = (1-\sigma)\int_{z=0}^{z=z_{euphotic}} SMS_{soft}(PO_4^{3-})dz \qquad (5.5.3)$$

In most global studies, the transport and remineralization of *POM* has been modeled implicitly by assuming that any particles produced in the surface are remineralized instantaneously immediately below where they are produced. A typical practice is to use the vertical derivative of the *Martin et al.* [1987] function (5.4.3) to obtain an equation for that portion of the remineralization due to *POP*. Adding the source term from the remineralization of *DOP*, the tracer continuity equation for PO_4^{3-} in the aphotic zone becomes

$$\Gamma(PO_4^{3-}) = \Phi^{POP}(z_{euphotic}) \cdot \left[\frac{z}{z_{euphotic}}\right]^{-(1+b)} + k_{DOP} \cdot [DOP],\ z > z_{euphotic} \qquad (5.5.4)$$

where b is the scaling parameter. For completeness we add the tracer continuity equation for PO_4^{3-} in the euphotic zone:

$$\Gamma(PO_4^{3-}) = -SMS_{soft}(PO_4^{3-}) + k_{DOP}[DOP],\ z \leq z_{euphotic} \qquad (5.5.5)$$

Since no explicit sediments are considered here, any particulate organic matter reaching the bottom of the ocean is assumed to be remineralized in the lowest layer of the model. Although (5.5.4) is by far the most often used formulation for remineralization of organic matter in the aphotic zone, recent models have begun to implement the ballast model of *Armstrong et al.* [2002] [e.g. *Moore et al.*, 2004; *Dunne et al.*, 2005a]. To our knowledge, however, no coupled biogeochemical-physical model resolving the global ocean has yet implemented more sophisticated particle cycling models such as those suggested by figure 5.4.11, although regional models have [e.g., *Gruber et al.*, 2005]. This problem will need to be addressed if we wish to use the thorium tracers to constrain the particle dynamics in OGCM simulations.

Other nutrients and oxygen are keyed to (5.5.1) through (5.5.5) using the stoichiometric ratios of remineralization discussed previously. Alternatively, similar restoring formulations can be used for the other nutrients and alkalinity, resulting in export estimates of organic nitrogen, opal, and $CaCO_3$ [see, e.g., *Gnanadesikan*, 1999a; *Deutsch et al.*, 2005; *Jin et al.*, 2005b].

Our simple biogeochemical modeling approach has the advantage of being considerably less costly in computation time than an explicit ecosystem model. It is also guaranteed to accurately reproduce the surface nutrient content. Its main disadvantage is that it is not readily adaptable to making predictions for situations where the ocean biology changes, such as occurs as a result of interannual variations in the ocean, associated with ENSO, or as a result of future climate change.

Sensitivity Studies

The ecosystem model given by (5.5.1) through (5.5.5) has a number of parameters that need to be determined before we can apply the model. A first parameter is the fraction of organic matter going into semilabile *DOP*, σ, a second one is the first-order rate constant for semilabile *DOP* remineralization, k_{DOP}, and a third one is the remineralization length scale of *POP* remineralization, b. A fourth parameter is the restoring timescale τ, but since uncertainties in this parameter have little influence on the overall model results, we do not have to worry about the the error introduced by this parameter. A relatively straightforward approach to determine the values of the remaining three parameters is to run the model to equilibrium with many separate sets of parameters and then determine which set provides results that are in best agreement with observations. An alternative approach would be to undertake a formal parameter optimization, whereby the model parameters are automatically adjusted until the disagreement with the observations is minimized. This latter approach requires sophisticated numerical modifications of the code and is computationally expensive. As a result, such optimization studies have not been used extensively (see e.g., *Schlitzer* [2002]).

We limit our discussion here to the first approach, as it also permits us to investigate directly the sensitivity of various model fields to changes in the parameters. The most extensive parameter sensitivity studies to date have been undertaken by *Anderson and Sarmiento* [1995] and *Yamanaka and Tajika* [1996, 1997]. All three studies used an ecosystem model that is structurally identical to the simple ecosystem model presented above. Some differences do arise between the work of *Anderson and Sarmiento* [1995], who used a circulation model developed at Princeton University/Geophysical Fluid Dynamics Laboratory (GFDL), and the work of *Yamanaka and Tajika* [1996, 1997], who used their own circulation model. In most situations, we expect relatively modest dependence of the parameter sensitivity on the underlying circulation model, so that we can transfer the results from one model to another.

Figure 5.5.1 shows the model-simulated distribution of semilabile *DOP* in response to variations in the fraction, σ, of total organic phosphorus production that is allocated to *DOP*, and in the lifetime of the produced *DOP*. As the lifetime of the semilabile *DOP* pool increases, both the vertical penetration (figures 5.5.1a and c) and the surface concentration (figure 5.5.1b) increase. Figure 5.5.1c actually shows the vertical penetration depth, which is defined as the vertical integral of the *DOP* concentration divided by the surface concentration. Since figure 5.5.1a depicts *DOP* concentrations normalized by the surface concentrations, the penetration depths can be readily inferred from this plot by multiplying the mean normalized concentration over the water column with the depth of the water column, i.e., in the case of a 10-year decay time for *DOP*, the average normalized *DOP* concentration is about 0.4, which multiplied with the depth of the water column of 800 m gives a penetration depth of 320 m, very close to the actually computed value of 327 m (see figure 5.5.1c).

Changing the *DOP* allocation fraction σ produces *DOP* distribution variations that are different from those computed for changes in the lifetime. As σ increases, the surface concentration of *DOP* increases, but the vertical penetration depth of *DOP* decreases. This is because a higher allocation to *DOP* does not lead to substantial increases in deep *DOP* concentrations, a result of the relatively slow ventilation of the ocean's interior. Consequently, with increasing values of σ, the vertical integral goes up less quickly than the surface concentrations, leading to a decrease in the vertical penetration depth.

Given this differential response to changes in these two parameters, the available *DOM* observations permitted *Yamanaka and Tajika* [1997] to separately estimate optimal values for the two parameters. As can be seen from figure 5.5.1b and c, a *DOP* lifetime of $1/k_{DOP} = 0.5$ yr and an allocation factor $\sigma = 0.67$ gives the best fit to *DOM* observations. It is important to realize, however, that the fraction of the net downward flux of organic matter across the depth of the euphotic zone that is in dissolved form is less than σ. This is because *DOM* is also remineralized within the euphotic zone (equation 5.5.2). For example, with $\sigma = 0.67$, *Yamanaka and Tajika* [1997] found that, on a global scale, *DOC* accounted for less than 30% of the total downward flux of organic matter across a depth of 100 m.

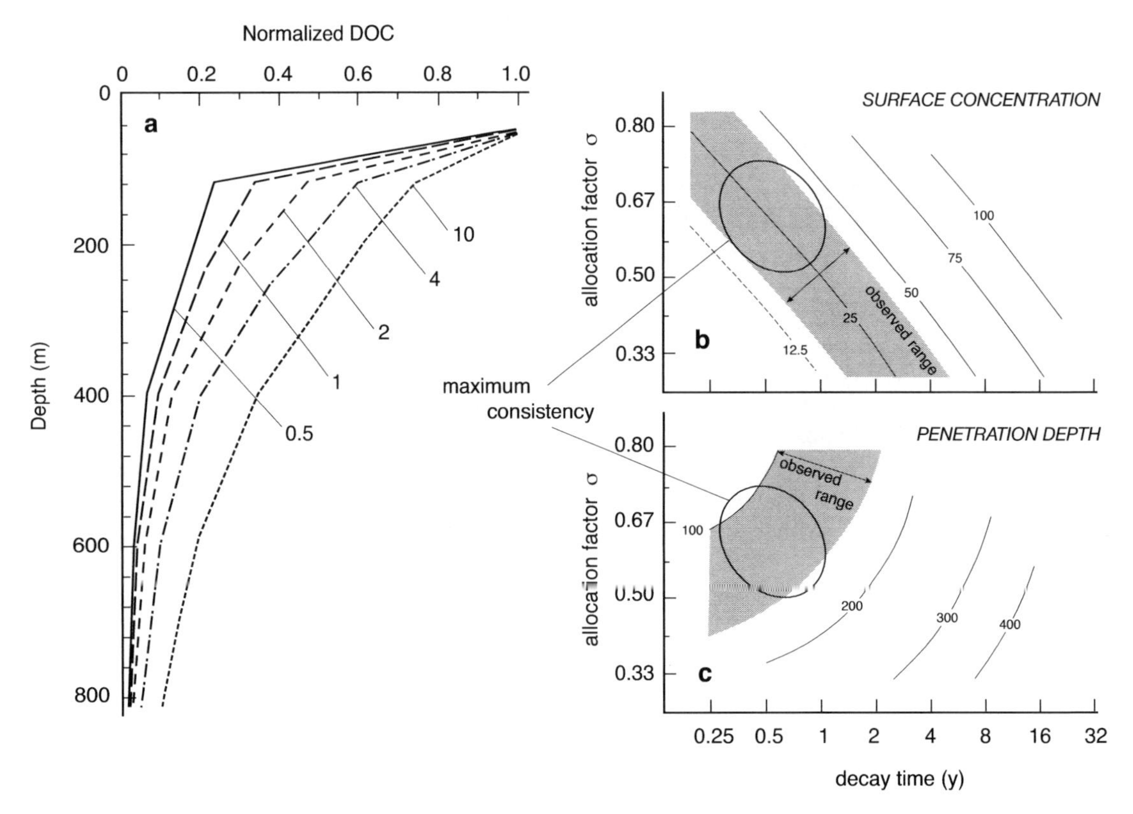

FIGURE 5.5.1: Model-simulated distribution of semilabile *DOP* in response to variations in the allocation factor σ, i.e., the fraction of organic matter going into the dissolved pool, and the lifetime of the semilabile *DOM* pool. (a) Global mean profiles of semilabile *DOP* in response to variations in the lifetime of *DOP* (numbers indicate the lifetime given as $1/k_{DOP}$ in years). Results are shown for $\sigma = 0.67$. (b) Mean surface *DOP* concentration as a function of σ and $1/k_{DOP}$. (c) Mean vertical penetration depth of *DOP* as a function of σ and $1/k_{DOP}$. Observational ranges are given by the shaded areas. Adapted from *Yamanaka and Tajika* [1997].

Such a high value for the allocation of total production into the dissolved organic matter pool deserves further discussion. At the time when the first coupled physical-biogeochemical GCM models began to be developed, it was thought that the export of organic matter from the surface occurred primarily by sinking of particles. However, it soon became clear that the particle parameterization given by (5.5.4) led to an unrealistically large accumulation of nutrients in regions of upwelling and strong convection [*Najjar et al.*, 1992]. The source of the difficulty lies in the shallow remineralization of (5.5.4), and the fact that the remineralization of particles occurs directly below where they are produced. The nutrient concentration below regions of high surface productivity must build up to exceptionally high levels before the large input of inorganic nutrients by remineralization can be balanced by export.

Historically, this nutrient trapping problem has been dealt with by either modifying the parameterization of particulate organic matter, for example, by lengthening the depth scale of the particle remineralization, or by adding the semilabile component of dissolved organic matter to the model. The latter was the procedure adopted by *Sarmiento et al.* [1988b], and many subsequent studies [e.g., *Bacastow and Maier-Reimer*, 1990; *Najjar*, 1990; *Anderson and Sarmiento*, 1995]. *DOM* avoids the problem of nutrient trapping by permitting the organic matter to be transported laterally and vertically over great distances before being remineralized. The sensitivity studies of *Yamanaka and Tajika* [1997] therefore corroborate this approach to solving the nutrient trapping problem. However, *Matear and Holloway* [1995] and, later, *Aumont et al.* [1999] pointed out that the nutrient trapping is also a consequence of the relatively poor resolution of the ocean physics by these models in many upwelling areas, particularly in the tropics. The coarse-resolution models typically used for such global-scale studies tend to simulate excessive upwelling, and the upwelled waters tend to have too high nutrient concentrations, therefore providing a mechanism to worsen any existing nutrient trapping.

Anderson and Sarmiento [1995] made an interesting observation when they varied the mean *DOP* concentration in their model between 0.02 and 0.2 mmol m^{-3} by adjusting the *DOP* production parameters. The observed

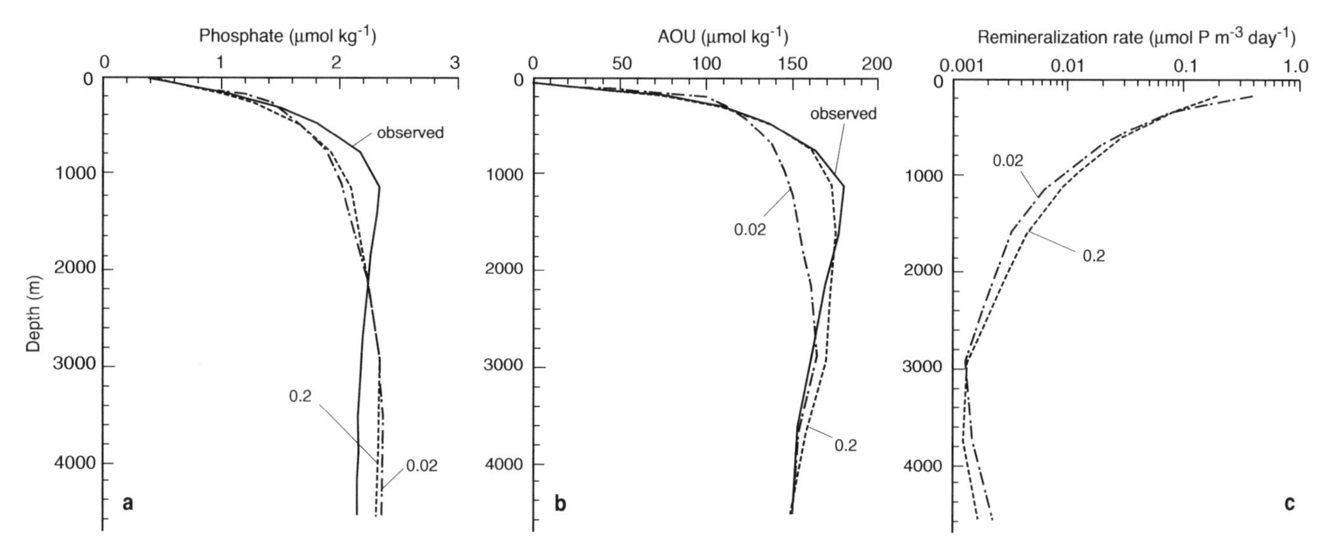

FIGURE 5.5.2: Global mean vertical profiles of (a) phosphate, (b) *AOU*, and (c) phosphorus remineralization rate in response to varying global mean *DOP* concentrations. Shown are two simulations with global mean *DOP* of 0.2 and 0.02 μmol kg^{-1}, respectively. Note in (c) that the low-*DOP* case has a higher remineralization rate in the upper thermocline than the high-*DOP* case, but that the situation reverses below about 300 m. Based on simulations reported by *Anderson and Sarmiento* [1995].

mean *DOP* is about 0.1 mmol m^{-3} [*Karl and Bjorkman*, 2002]. While the mean *DOP* concentration had very little influence on the model-simulated phosphate concentration (figure 5.2.2a), *AOU* changed quite considerably (figure 5.2.2b). The higher the *DOP*, the higher the *AOU*, which means lower oceanic oxygen inventory. These changes in oxygen inventory are not due to changes in export production, which are relatively small: the global mean new production in carbon units changed only from 2.98 mol C m^{-2} yr^{-1} in the simulation with a global mean *DOP* of 0.02 μmol kg^{-1} to 2.26 mol C m^{-2} yr^{-1} in the simulation with *DOP* = 0.2 μmol kg^{-1}. Instead, the cause of the oxygen inventory difference is most likely the change in remineralization that occurs in the upper few hundred meters. The phosphate remineralization rate profile in figure 5.5.2c shows that, in the upper 300 m, the low-*DOP* simulation has a higher remineralization rate, i.e., higher O_2 demand than the high-*DOP* simulation. Since export in the two simulations is about equal, the larger remineralization in the upper thermocline in the low-*DOP* case results in a lower remineralization rate in the middle thermocline and below, ie., smaller O_2 demand. The low-*DOP* simulation will thus have less oxygen in the upper thermocline, and consequently establishes a stronger vertical gradient to draw oxygen in from the surface ocean. This enhanced O_2 supply plus the lower O_2 demand in the middle and lower thermocline act together and lead then to the surprising phenomenon that the low-*DOP* case has a substantially lower overall *AOU* than the high-*DOP* case, despite the two having only small differences in simulated phosphate and export production.

The third parameter of interest is the remineralization length-scale of the *Martin et al.* [1989] curve, *b*.

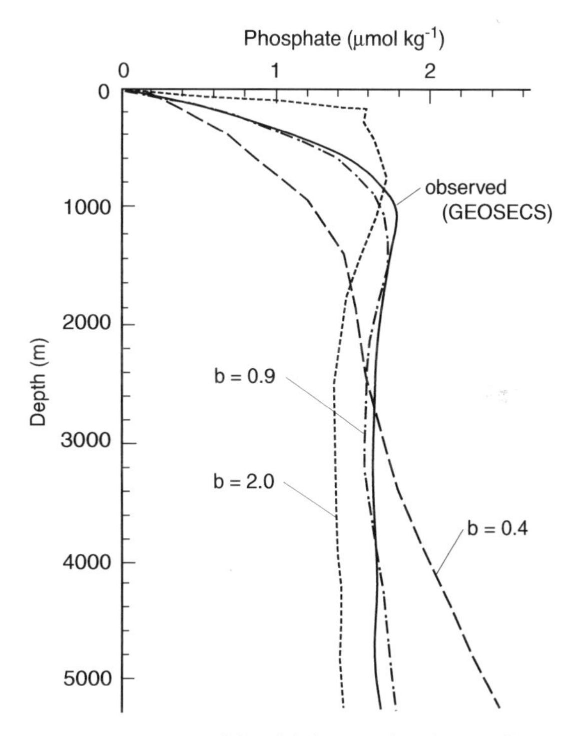

FIGURE 5.5.3: Sensitivity of the global mean phosphate profile for three GCM simulations with differing values of the remineralization length parameter *b*, i.e., the exponent of the power law function for vertical POM flux proposed by *Martin et al.* [1989] (see equation (5.4.3)). An exponent $b = 0.9$ gives the best fit to the observed phosphate profile, in agreement with the original value of $b = 0.858$ derived from sediment traps. Adapted from *Yamanaka and Tajika* [1996].

Figure 5.5.3 from *Yamanaka and Tajika* [1996] demonstrates that the model-simulated phosphate distribution responds very sensitively to variations in this parameter. The larger the parameter *b*, the faster or-

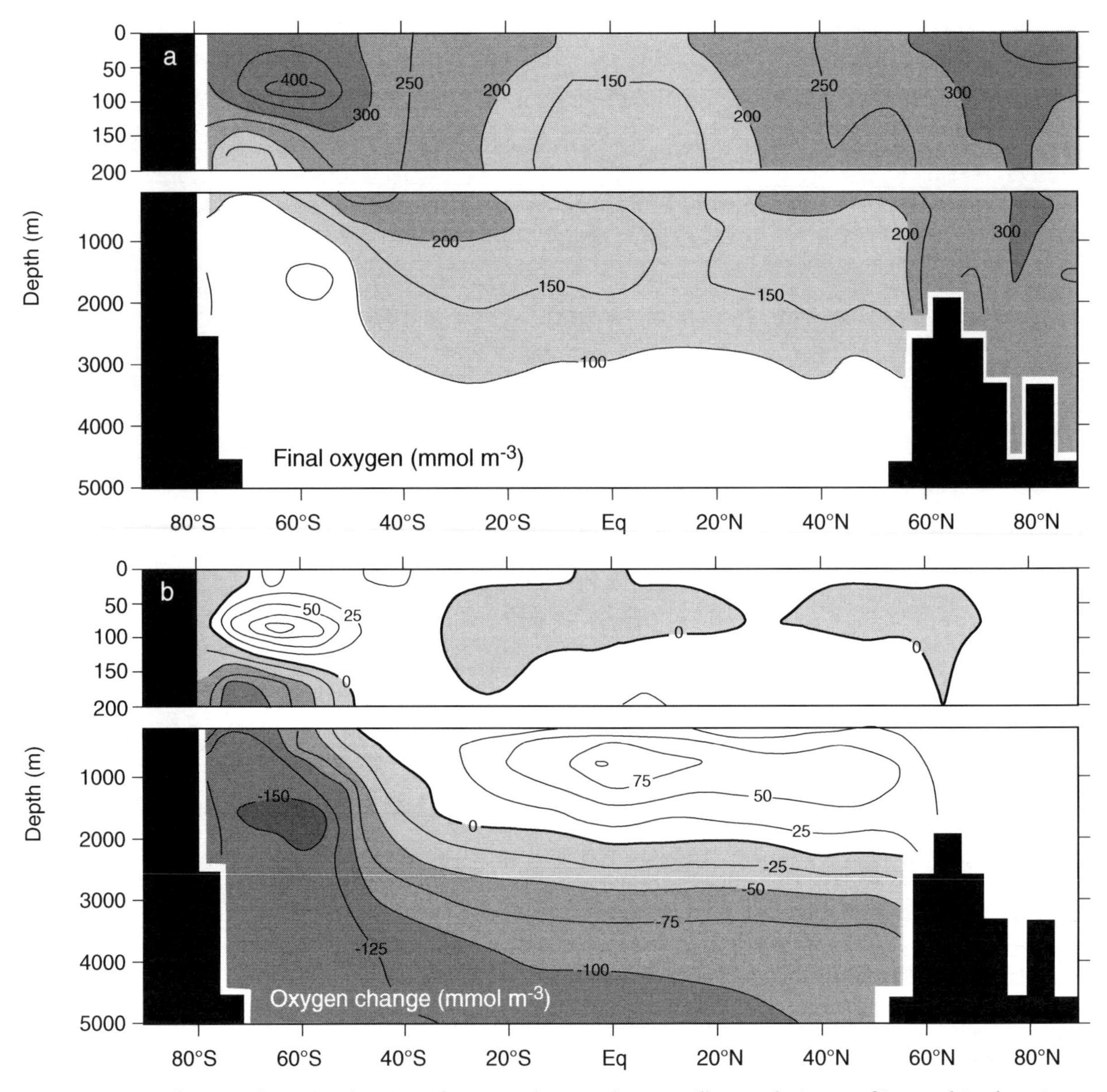

FIGURE 5.5.4: Zonally averaged meridional sections of oxygen and oxygen changes to illustrate the impact of increased Southern Ocean biological productivity and organic carbon export. (a) Equilibrium oxygen concentration in a case where biological productivity in the Southern Ocean was increased to the level where surface phosphate became zero. (b) Difference in oxygen concentration shown in (a) from a case where biological production in the Southern Ocean was approximately as today, achieved by forcing model simulated phosphate toward the observed concentration. Note the reduction in oxygen in the abyssal ocean, but the increase in the thermocline and the upper parts of the deep ocean. Based on simulations performed by X. Jin at UCLA (X. Jin, personal communication, 2005).

ganic matter is remineralized, leading to a very rapid initial increase of phosphate with depth, and much slower subsequent increase deeper in the water column. Conversely, if b is small, particulate organic matter gets exported very deeply, depleting the main thermocline of nutrients, while enriching the deep ocean. The best agreement with the observed profile of phosphate was found for $b = 0.9$, very close to the original value of $b = 0.858$ proposed by *Martin et al.* [1989]. Changes in the depth of remineralization have huge implications for surface ocean productivity. In the case of very shallow remineralization, exported nutrients are quickly resupplied to the surface ocean, stimulating renewed export. Conversely, the deep remineralization removes the nutrients from the upper ocean, leaving upwelling in high latitudes as the only pathway of nutrients back to the surface. *Yamanaka and Tajika* [1996] found for $b = 2.0$ a global export production that was more than three times larger than that for $b = 0.9$, which itself was three times larger than the estimated global export in the case of $b = 0.4$. In addition, large spatial changes in export production were simulated, with the low latitudes greatly diminished in importance in the case of deep remineralization, while the low latitudes dominated in the shallow remineralization case. Given the observation that variations in the remineralization length-scale tend to be governed by variations in the flux of ballast material into the interior ocean, one can hypothesize that a change in the delivery or production of ballast material in the open ocean could lead to a change in ocean interior remineralization and, subsequently, changes in surface productivity.

Applications: Control of Oceanic Oxygen

We end this chapter with an illustration of how an ocean GCM can lead to quite different insights on the biogeochemical behavior of the ocean than the box models that have generally been used for such purposes in the past. We saw at the end of chapter 1 that the three-box model of figure 1.2.5 predicts total deep ocean anoxia when the high-latitude phosphate content is reduced to a low value. We can calculate from (1.2.8) that the deep ocean would become anoxic if the high-latitude surface phosphate dropped to 0.33 mmol m^{-3} from its present estimated value of 1.3 mmol m^{-3}:

$$[O_2]_d = [O_2]_h - 170 \cdot ([PO_4^{3-}]_d - [PO_4^{3-}]_h)$$
$$= 301 - 170 \cdot (2.1 - 0.33) \text{ mmol m}^{-3}$$
$$= 0 \text{ mmol m}^{-3}$$

This control of ocean chemistry by the high-latitude surface nutrient content has been proposed as an explanation for the global episodes of anoxia that occurred during the Cretaceous [e.g., *Sarmiento et al.*, 1988a]. The equivalent equations for the carbon system have been used to propose a mechanism for reductions of atmospheric CO_2 during the last ice age [e.g., *Knox and McElroy*, 1984; *Sarmiento and Toggweiler*, 1984; *Siegenthaler and Wenk*, 1984]. However, the absence of observational support for the associated low levels of oxygen predicted by the models has been taken as evidence that this mechanism is incorrect.

Figure 5.5.4 shows results of an ocean GCM simulation in which the high-latitude phosphate was reduced all the way to 0 [*Sarmiento and Orr*, 1991]. Based on the box model results, we would expect the entire deep ocean to be anoxic. In fact, the only place in the GCM model where anoxia occurs is a small region at mid-depths south of Africa near the Antarctic coast. The zonal mean oxygen section shown in figure 5.5.4 gives no indication of the total anoxia predicted by the three-box model. The global mean oxygen content drops by less than 40 mmol m^{-3} from the present global mean of 162 mmol m^{-3}.

Why does the GCM model give such a different sensitivity to the high-latitude nutrients than the three-box model? The primary reason is the much higher resolution of the GCM model. The removal of phosphate from the high-latitude surface ocean is accomplished by increasing the biological production. In a steady state, the oxygen produced by the increased photosynthesis must be taken up by the deep ocean, to be utilized there for remineralization of the organic matter. In other words, a large increase in the flux of oxygen to the deep ocean is required in order to compensate for the increased remineralization due to the increased organic matter flux. This increase in oxygen flux requires an increase in the vertical gradient between the surface waters and ocean interior. Because surface ocean O_2 is determined primarily by equilibrium with the atmosphere, the only way such a gradient can be achieved in the three-box model is by reducing the oxygen in the entire deep ocean box. The GCM model accomplishes most of the task locally in the high latitudes [*Sarmiento and Orr*, 1991], and, because of its more sophisticated circulation, does not require anoxia to do this except in a small region of the high latitudes.

In concluding this brief discussion of model results, we note that they illustrate how all the processes discussed in previous chapters can be brought together to develop a sophisticated model of the oceanic distribution of nutrients and oxygen. They also provide an illustration of how our gradually improving understanding of water column remineralization has had quite a large impact on model simulations. The interaction between model development and measurement efforts should continue to prove fruitful in years to come.

Problems

5.1 What two major biochemical pathways are responsible for the remineralization of most organic matter in the ocean? Explain under which conditions each of these two pathways occur and why.

5.2 Define and explain the terms *apparent oxygen utilization* (AOU) and *oxygen utilization rate* (OUR). How can these two terms be estimated on the basis of observations?

5.3 Explain the difference between the preformed and remineralized components of phosphate. What assumptions are made if one wants to compute these two components based on observations?

5.4 A water sample taken in the deep Pacific has a potential temperature of 1°C, a salinity of 34, an oxygen concentration of 120 mmol m^{-3}, and a phosphate concentration of 2.6 mmol m^{-3}.

a. Compute the apparent oxygen utilization (AOU) of this water sample.

b. Compute the preformed and remineralized components of phosphate for this water sample (assume an oxygen-to-phosphate stoichiometry of $-170{:}1$).

c. Discuss the deep water circulation and remineralization processes that determine the concentrations given in (b).

5.5 The oxygen utilization rate in the thermocline is found to be at least an order of magnitude greater than that found in the deep ocean below 2000 m. Discuss and explain this observation.

5.6 Estimates of vertically integrated oxygen utilization rates (below 100 m) range between 3 and 9 mol $O_2\,m^{-2}\,yr^{-1}$.

a. Using a mean OUR of 6 mol $O_2\,m^{-2}\,yr^{-1}$, how large is the globally integrated OUR? (The global surface area of the ocean is $360\times10^{12}\,m^2$.)

b. How does the global OUR compare to estimates of global net primary production (50 Pg C yr^{-1})? Explain why the two numbers are different.

5.7 We would like to compute the oxygen utilization rate of the deep Indian Ocean and compare it with estimates of OUR in the deep Atlantic and deep Pacific. The deep Indian Ocean can be considered to be ventilated from a single source in the Southern Ocean with an outflow at mid-depths. This allows us to represent the deep Indian Ocean as a single box with one inflow and one outflow. The inflowing waters are found to have an *AOU* of 140 mmol m^{-3}, and the outflowing waters an *AOU* of 200 mmol m^{-3}. The flow rate of the inflowing waters has been determined from a ^{14}C budget and estimated to be about 5 Sv (5×10^6 m^3 s^{-1}). The volume of the deep Indian Ocean is about 8.0×10^{16} m^3.

a. Draw a schematic of the circulation of the deep Indian Ocean.

b. Derive a mass conservation equation for *AOU* in the deep Indian Ocean.

c. Assume steady state and solve the mass conservation equation for the oxygen utilization rate (*OUR*) in units of mmol m^{-3} y^{-1}.

d. Discuss the result found in (c) and compare it with the estimates of *OUR* in the deep Atlantic and deep Pacific given in table 5.2.1.

5.8 Figure 5.3.8 shows very low N^* values in the thermocline of the northern Indian Ocean and in the thermocline of the eastern tropical Pacific.

a. Define and discuss the meaning of N^*.

b. What process is most likely responsible for generating these low N^* values?

c. Discuss the occurrence of these low N^* values in relation to what you know about large-scale ocean circulation and ocean remineralization.

5.9 Observations show that the oxygen content of the main thermocline of the Pacific Ocean has decreased in recent decades. Global climate change simulations indicate that the buildup of greenhouse gases in the atmosphere may lead to a slowdown of the ventilation of thermocline waters, and it has been suggested by some scientists that this may be the cause of the oxygen decrease. Answer the following questions assuming that such a slowdown of the thermocline ventilation has indeed occurred. Base your answer on the thermocline ventilation diagram shown in figure 2.3.9, i.e., assume that you can represent an isopycnal layer as a single slab with a single outcrop.

a. Suppose that the export of organic matter from the surface to the isopycnal surfaces of the main thermocline remains unaffected by the ventilation slowdown. What kind of change would you expect for the oxygen and *AOU* concentrations in the thermocline?

b. Now suppose that the biological export production decreased as a result of this slowdown, because of the smaller vertical supply of nutrients to the surface euphotic zone. What kind of effect do you expect now for oxygen and *AOU*?

c. Finally, global climate change not only affects the thermocline ventilation, but also leads to a warming of the waters. How does the warming of the waters affect oxygen and *AOU*?

5.10 Iron fertilization has been proposed as a way to increase the export production of organic matter out of the surface ocean, with the goal of thereby inducing an uptake of CO_2 from the atmosphere. Given what you have learned about the cycling of N_2O in the ocean, discuss the potential implications of an increase in export production on the marine production of N_2O.

5.11 This problem asks you to separate the phosphate concentration into remineralized and preformed concentrations and explain the resulting distributions. Use the global mean vertical profile given at the end of the problem set.

a. Do a plot of temperature versus salinity and label the major deep water types: AAIW (Antarctic Intermediate Water), NADW (North Atlantic Deep Water), and AABW (Antarctic Bottom Water). Can you also identify the Subtropical Mode Water (a salinity maximum commonly found in the upper thermocline)?

b. Plot *AOU* versus depth and discuss its major features, including their association with water types. Also explain why the maximum in *AOU* occurs at about 1000 m.

c. Calculate preformed and remineralized phosphate and plot both of them versus depth (preformed alone, and total = preformed + remineralized). Discuss the major features of their distributions. Note, in particular, that most of the phosphate in the ocean is preformed.

5.12 The opal in diatoms grown with adequate light and nutrients is observed to have a Si:N ratio of $Si(OH)_4$ about 1. Suppose we use this observation to define a *Si** tracer as $Si^* = Si(OH)_4 - NO_3^-$. Use the global horizontal mean data at the end of the problem set to calculate *Si** and plot this versus depth. What conclusions can you draw about the cycling of silicate versus nitrate from the distribution of *Si**?

5.13 A detailed analysis of sediment trap observations from the Arabian Sea and a few other places has suggested that the sinking velocity of particles may increase with depth. Derive analytically the shape of the *POC* flux as a function of depth, i.e., $\Phi(z) = -w_{sink} \cdot POC$ (analogous to equation (5.4.12)) given the assumptions that the sinking velocity is a linear function of depth, i.e., $w_{sink} = b \cdot z$; that the remineralization of *POC* is a first-order reaction with a rate constant k_{remin}; and that there is no net *POC* production. Discuss the implications of this result for the *POC* flux equation (5.4.3) versus (5.4.12).

5.14 Preformed inorganic phosphorus: You saw in problem 5.11(c) that most of the phosphate in the ocean is preformed. Preformed phosphate is conserved except at the surface of the ocean, where it can be taken up by phytoplankton and converted to organic phosphorus.

a. Write down the vertical one-dimensional conservation equation for the steady-state distribution of preformed inorganic phosphorus. Assume here and in the remainder of these problems that advection and diffusion are constant with depth.

b. Give a solution to the equation in (a) using as boundary conditions that the surface concentration is $PIP(0)$ and the concentration at $z = H$ is $PIP(H)$. Use the symbol z^* for the exponential scale depth K/w.

c. Use a spreadsheet to calculate and plot the preformed phosphate given by your solution in (a) for the depth interval from 0 to −500 m. Use the observed preformed phosphate from problem 5.11(c) to specify the boundary concentrations at $z = 0$ and $H = -500$ m. Add the data from problem 5.11(c) to your plot and convince yourself that a value of $z^* = 2000$ m does a reasonable job of fitting these data (which have a kink in the middle due mostly to lateral processes).

PROBLEMS 5.14 TO 5.16:

These problems form a sequence that will take you through the development of a simple vertical one-dimensional advection (*w*) diffusion (*K*) model of the distribution of preformed inorganic phosphorus (*PIP*), organic phosphorus (*OP*), and remineralized inorganic phosphorus (*RIP*). Assume horizontal exchange is negligible. These problems will give you insight into the processes that control the vertical distribution of nutrients in the ocean.

d. Add two curves to your figure from part (c) to show how values of $z^* = 1000$ m and 4000 m affect the shape of the preformed phosphate curve. Typical values for the vertical diffusivity in models such as this are of order 2 $cm^2\,s^{-1}$ (see chapter 2 for a discussion of how this reflects primarily lateral input along isopycnals rather than vertical mixing). How does w vary if you keep K fixed, and how does this help to explain the results of this sensitivity test?

5.15 Organic phosphorus distribution: Assume that OP is particulate organic matter that sinks with a vertical sinking velocity w_{sink} of order -100 m d^{-1} (as observed). This is large enough that the advective and diffusive transport terms w and K can be ignored. Assume furthermore that you can model consumption of particulate organic matter as first-order decay with a rate constant k_{remin} (s^{-1}), and that the distribution of OP is in steady state.

a. Write down the conservation equation for OP.

b. Give a solution for $OP(z)$ using the boundary condition that $OP = OP(0)$ at $z = 0$. Use the symbol z^*_{OP} for the exponential scale depth w_{sink}/k_{remin}.

5.16 Biogenic (i.e., remineralized) phosphate distribution:

a. Write down the conservation for remineralized inorganic phosphorus, RIP.

b. Give an analytical solution to the RIP equation subject to the boundary conditions that $RIP = 0$ at $z = 0$ and $RIP(H)$ at $z = H$.

c. Use a spreadsheet to calculate and plot the remineralized phosphate given by your solution in (b) for the depth interval from 0 to -500 m. Use the observationally determined remineralized phosphate from problem 5.11(c) to specify the boundary concentration at $z = H = -500$ m. Add the data from problem 5.11(a) to your plot and convince yourself that a value of $z^*_{OP} = 300$ m and the previously determined value for $z^* = 2000$ m do a good job of fitting these data.

d. Add two curves to your figure from part (c) to show how values of $z^*_{OP} = 150$ m and 600 m affect the shape of the remineralized phosphate curve. How does k_{remin} vary if you keep w_{sink} fixed at its typical value of order 100 m day^{-1}, and how does this help to explain the results of the sensitivity test?

5.17 This problem is on the use of ^{230}Th to study the cycling of particles in the ocean. Note from figure 5.4.12 that the radioactive parent of ^{230}Th is ^{234}U. ^{234}U is well mixed throughout the ocean, so its activity, the rate at which it produces ^{230}Th by decay ($A(^{234}U) = \lambda \cdot {}^{234}U$), is constant everywhere. Its activity is 2800 dpm m^{-3}, i.e., it has a decay rate (activity) of 2800 disintegrations per minute per m^{-3} of seawater. If its daughter ^{230}Th were *not* scavenged from the water column, it would have an activity of 2800 dpm m^{-3} as well. This is because the daughter half-life is very short relative to the parent half-life. Because of this, after a few half-lives, the activity of the daughter would build up until you achieved a "secular equilibrium" where its decay rate equaled the rate at which it was being formed by decay of the parent. Water column observations of ^{230}Th are much lower than this expected activity. Observations also show an approximately linear increase of ^{230}Th with depth.

Global horizontal mean hydrographic and nutrient properties from the *Levitus et al.*, [1998] data set

Rescaled Depth (m)	*Temp. (°C)*	*Salinity*	O_2	O_2^{sat}	AOU	*Phosphate*	*Silicate*	*Nitrate*
			(mmol m^{-3})					
0	17.26	34.89	239	243	5	0.6	8.4	6.2
−50	15.09	35.01	218	254	36	0.82	11.3	9.4
−100	13.30	35.03	197	263	66	1.01	14.3	12.5
−150	11.89	34.98	183	271	88	1.17	17.2	14.8
−250	9.98	34.86	167	282	115	1.43	22.2	19.0
−350	8.57	34.76	159	291	133	1.64	27.0	22.6
−450	7.39	34.68	154	299	146	1.83	32.5	25.3
−550	6.44	34.63	150	306	157	1.97	38.4	27.8
−750	5.13	34.59	141	316	174	2.18	51.1	30.9
−950	4.21	34.61	139	323	184	2.26	63.0	32.0
−1450	2.97	34.70	148	332	185	2.25	85.3	32.2
−1950	2.32	34.74	161	338	176	2.18	95.9	31.4
−2450	1.96	34.75	171	341	170	2.15	101.5	31.0
−2950	1.72	34.74	178	343	165	2.15	106.3	30.8
−3950	1.37	34.73	190	346	156	2.13	112.3	30.8

Note: The rescaled depth has been reset to 0 at 50 m, the nominal base of the mixed layer, i.e., the actual depth is that given in the table minus 50 m.

a. Draw a diagram analogous to the model shown in figure 5.4.11 for the "irreversible" scavenging of ^{230}Th with only one particle size (large). The model should include formation of dissolved thorium from the parent, direct adsorption of dissolved ^{230}Th onto large particles (k_1), sinking of large particles, and decay of both dissolved and large-particle thorium. Leave out all the biological terms, the small-particle box, and also the desorption term k_{-1}. Write down and solve the equations for dissolved ^{230}Th and particulate ^{230}Th. Assume that k_1 is constant with depth. Keep in mind that $A = \lambda \cdot C$ and express all your answers in terms of activities. What should the vertical profiles of dissolved and particulate thorium look like and how do they compare with the linear increase found in observations? Hint: Be sure to consider the relative magnitudes of all the terms (e.g., $\lambda \ll k_1$) and simplify the equations where appropriate.

b. Now add the desorption term k_{-1} to the above diagram and solve the equations. How do the equations compare to the vertical profile? This analysis is the way that oceanographers first realized that trace metals were being desorbed from particles.

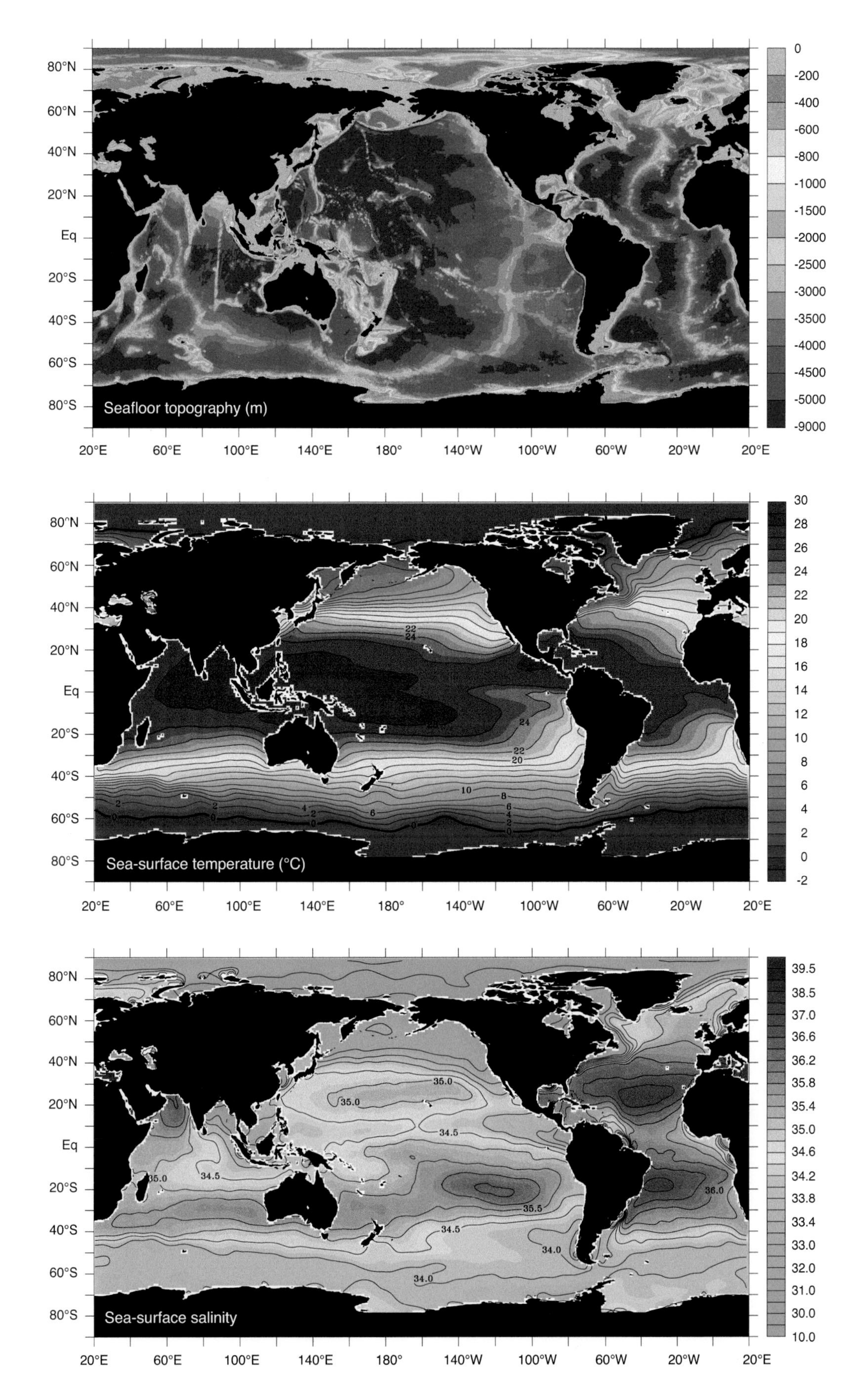

COLOR PLATE 1: Maps of seafloor topography taken from the National Geophysical Data Center [2001], sea surface temperature (see figure 2.2.7a), and sea surface salinity (figure 2.2.7b)

Color Plate 2: Maps of sea surface nitrate (figures 4.1.2 and 7.1.1b), phosphate (figure 1.2.4), and silicic acid (figure 7.1.1a)

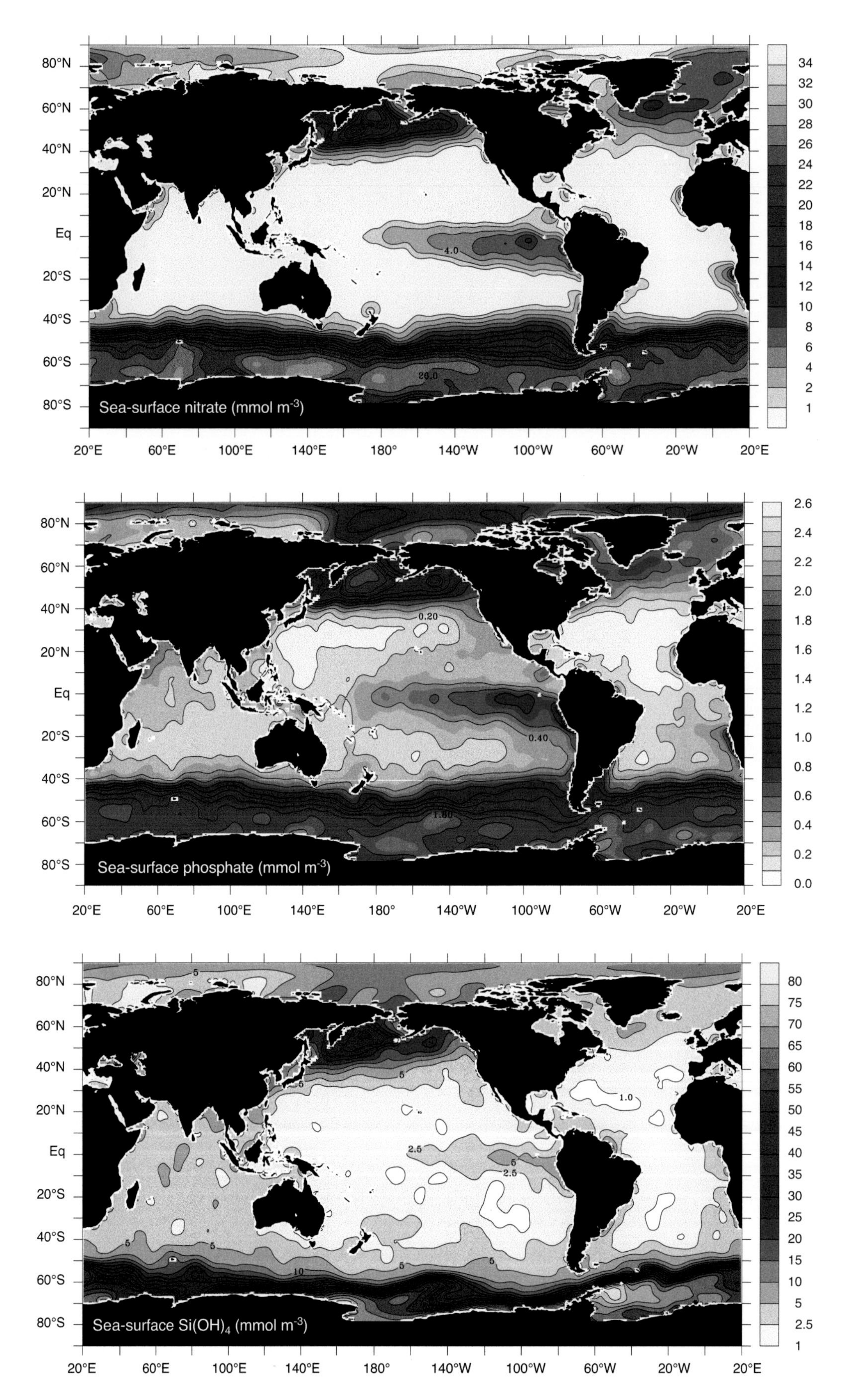

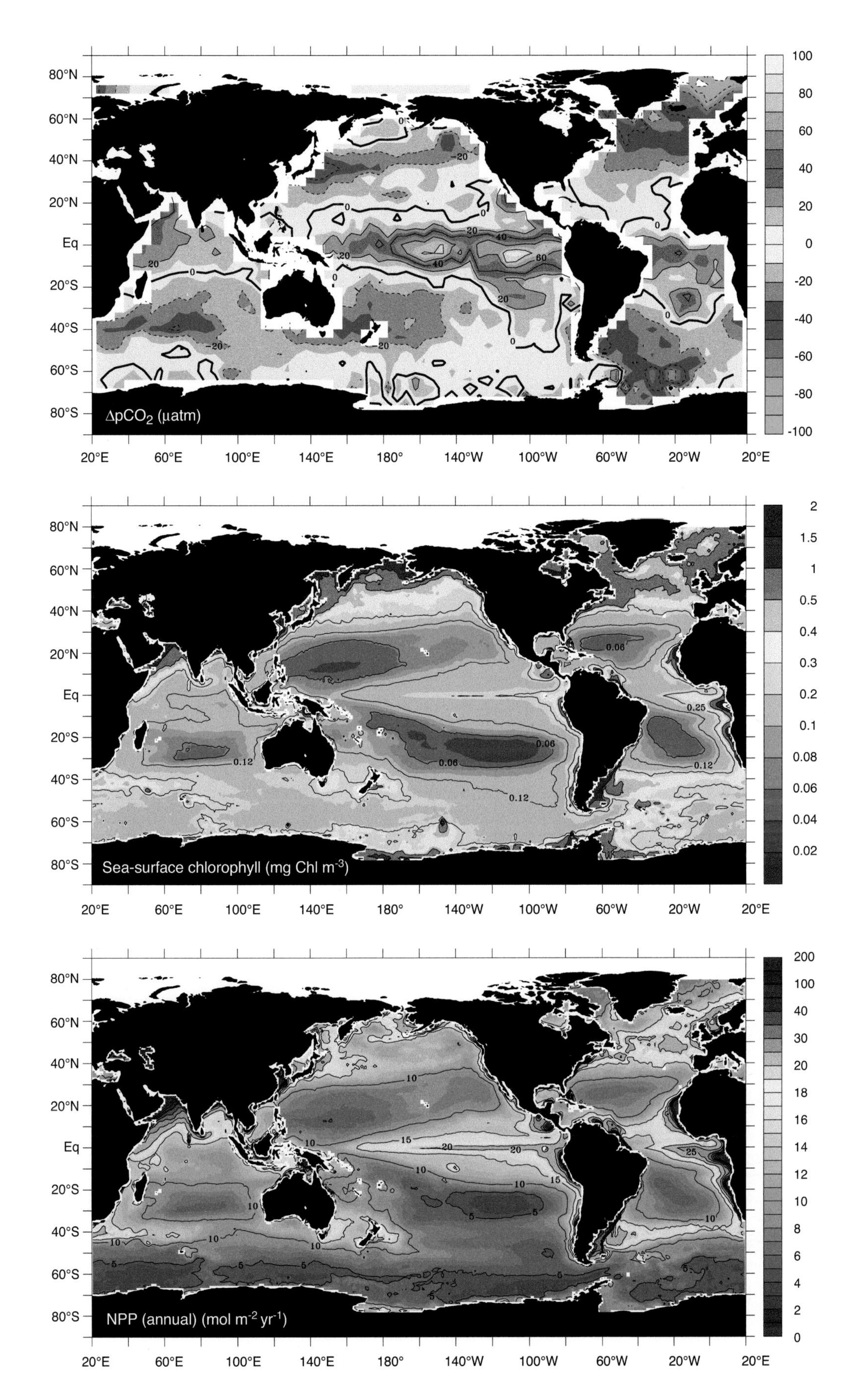

COLOR PLATE 3: Maps of sea-air pCO_2 difference (figure 8.1.1), sea surface chlorophyll-a (figure 4.1.1), and primary production (figure 4.2.4a)

COLOR PLATE 4: Maps of particulate organic carbon (*POC*) export (figures 4.2.4b and 7.2.5a), $CaCO_3$ export (figure 9.2.2), and opal export (figure 7.2.5b)

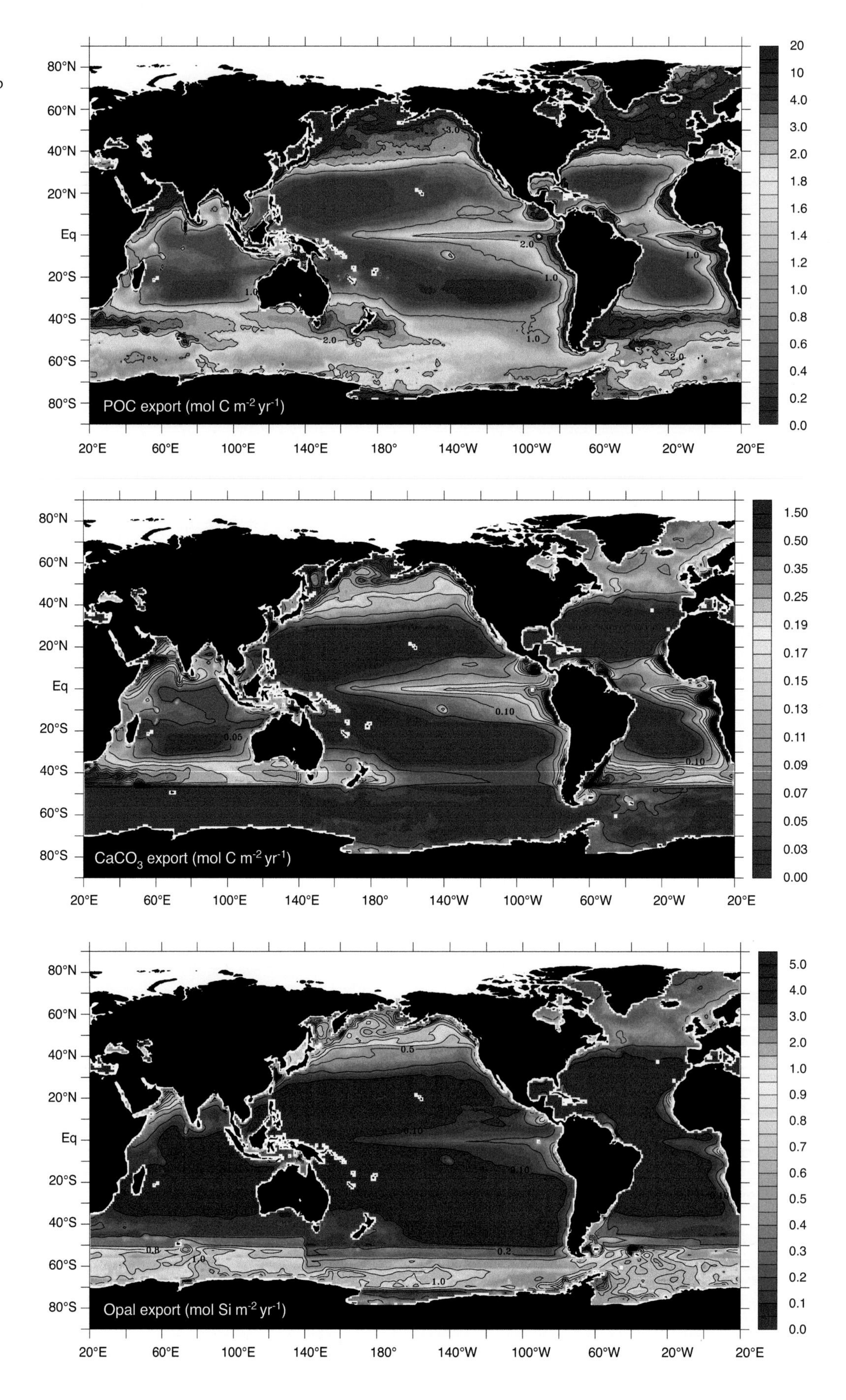

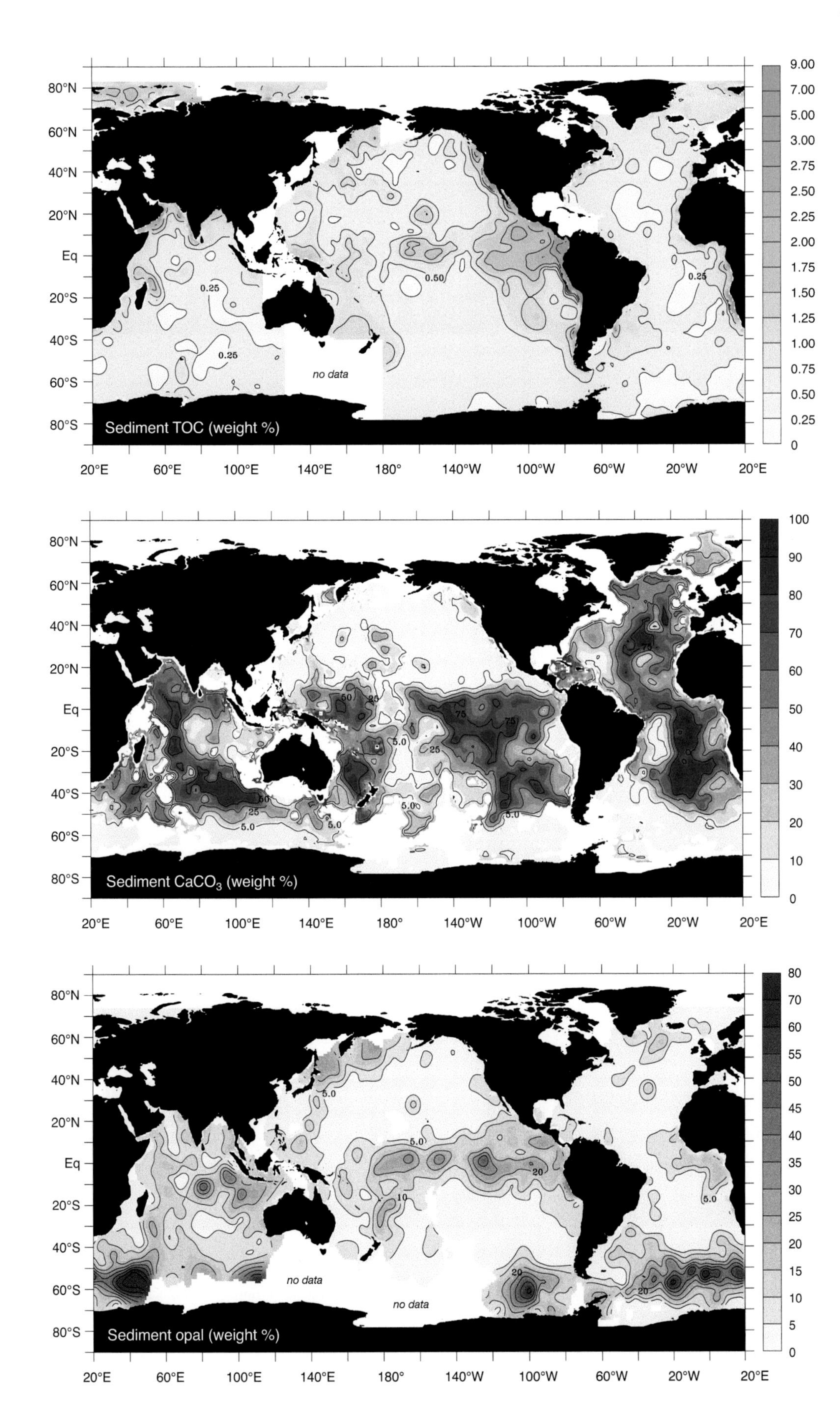

COLOR PLATE 5: Maps of the sediment distribution of total organic carbon (*TOC*; figure 6.1.4), $CaCO_3$ (figure 9.1.3), and opal (figure 7.1.3), all given in percent dry weight.

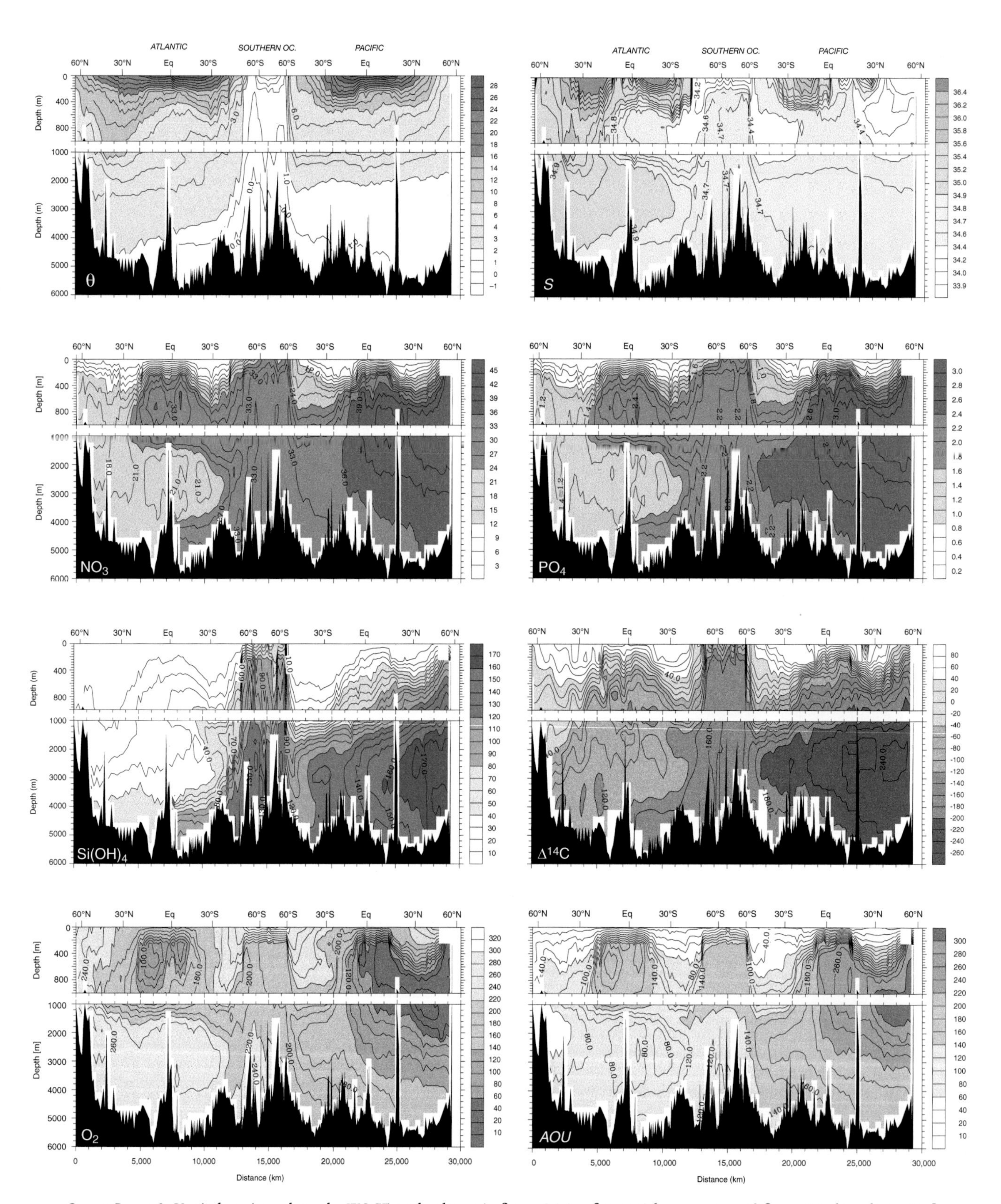

Color Plate 6: Vertical sections along the WOCE tracks shown in figure 2.3.3a of potential temperature (θ figure 2.3.3b), salinity (S; figure 2.4.1), nitrate (figures 5.1.2a and 7.1.2b), phosphate (figure 5.1.2b), silicic acid (figure 7.1.2a), radiocarbon ($\Delta^{14}C$; figure 2.4.2), oxygen (figure 5.1.3a), and apparent oxygen utilization (AOU; figure 5.1.3b)

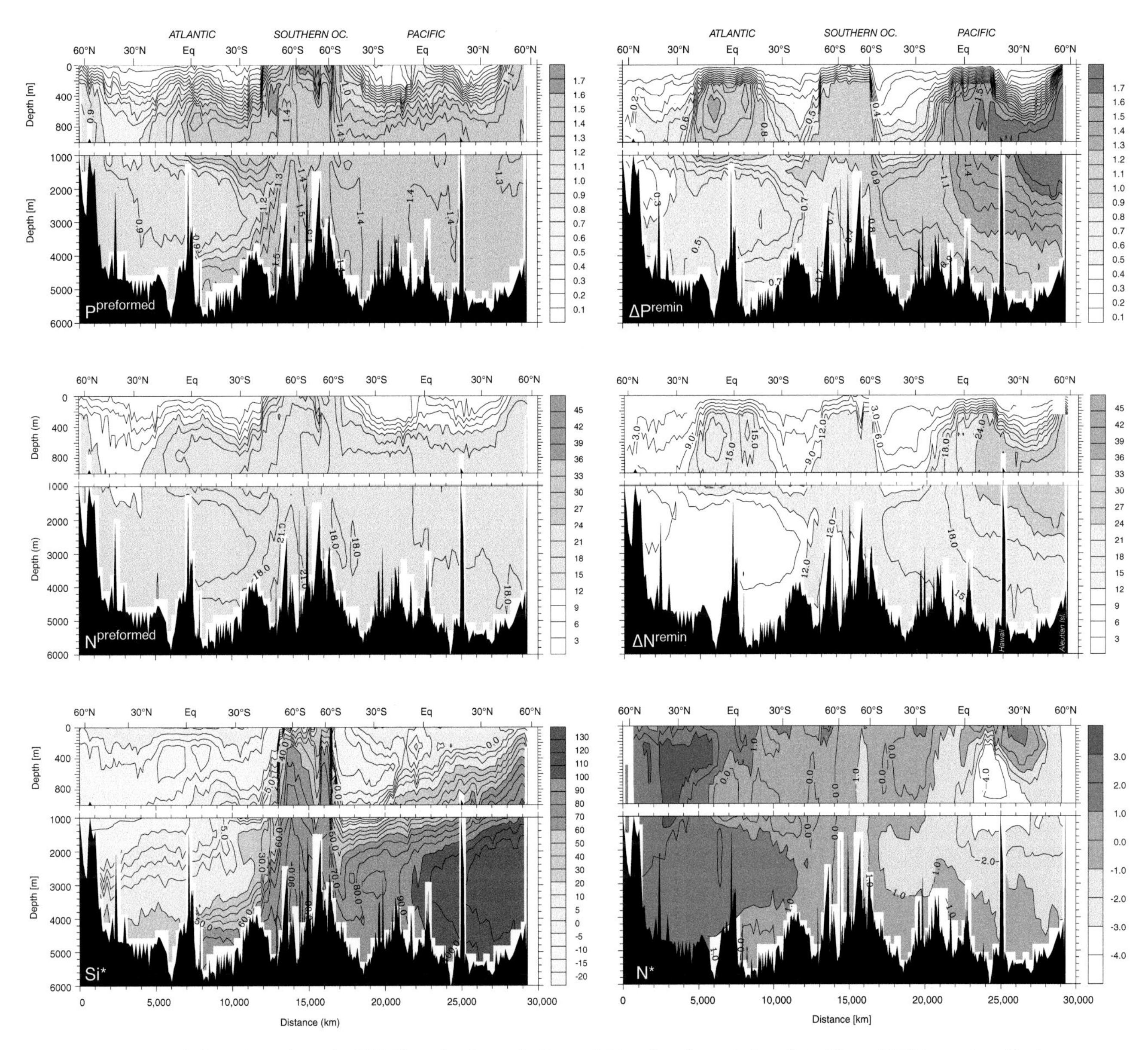

COLOR PLATE 7: Vertical sections along the WOCE tracks shown in figure 2.3.3a of preformed phosphate (figure 5.3.3a), remineralized phosphate (figure 5.3.3b), preformed nitrate (figure 7.3.2a), remineralized nitrate (figure 7.3.2b), Si* (figure 7.3.6), and N* (figure 5.3.6).

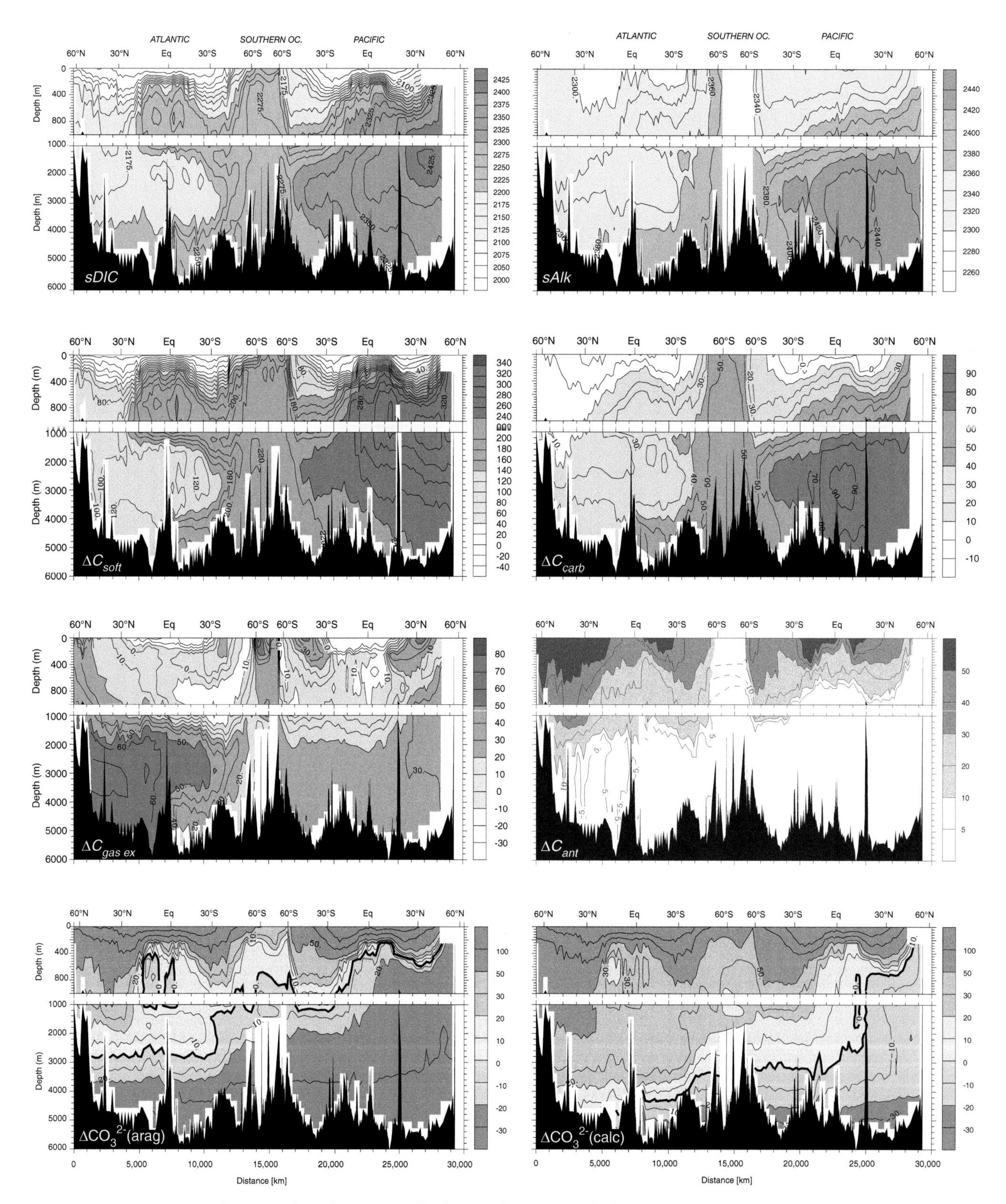

COLOR PLATE 8: Vertical sections along the WOCE tracks shown in figure 2.3.3a of salinity-normalized dissolved inorganic carbon (*sDIC*; figure 8.1.3a), salinity-normalized alkalinity (*sAlk*; figure 8.1.3b), the soft tissue pump component of *DIC* (ΔC_{soft}; figure 8.4.3a), the carbonate pump component of *DIC* (ΔC_{carb}; figure 8.4.3b), the gas exchange pump component of *DIC* ($\Delta C_{gas\ ex}$; figure 8.4.3c), the estimated anthropogenic carbon component of *DIC* (C_{ant}; figure 10.2.7), the carbonate ion concentration with respect to aragonite solubility (ΔCO_3^{2-} (arag); figure 9.3.2a), and the carbonate ion concentration with respect to calcite solubility (ΔCO_3^{2-} (calc); figure 9.3.2b)

Chapter 6

Remineralization and Burial in the Sediments

About a quarter of the organic matter that is exported from the surface of the ocean escapes remineralization in the water column and rains onto the sediments. Our objective in this chapter is to examine the processes in the sediments that determine the fate of this organic matter, whether it is remineralized or lost to long-term burial, and how these sediment processes affect and are affected by the distribution of properties in the water column.

More than 90% of the organic matter that rains onto the sediments is remineralized. We will see that we have a good understanding of how this occurs, or at least of the chemical reactions and physical processes that are involved. However, explaining the burial of the remaining organic matter has proven to be more of a challenge, probably involving some complex mixture of processes such as the presence or formation of highly refractory carbon, the diverse efficiencies of different oxidants or combinations of oxidants, and protection of organic carbon by interaction with mineral surfaces. Finally, while the focus of this book is primarily on "open" ocean processes away from the continental margins, continental margin sediments are a crucial part of the story for sediment cycling of organic matter, and we will discuss them here.

Over time, biogeochemists have learned to measure a wide range of chemical properties in the sediments while minimizing the disturbance to them, and to do a remarkably good job of modeling the distribution of these chemicals. The treatise on sediment *diagenesis* by *Berner* [1980] (diagenesis is the term we use to refer collectively to all the processes that determine the properties of sediments) summarized the existing knowledge and set a standard for quantitative interpretation of the chemical distributions that still stands today [cf. *Boudreau*, 1997]. However, many challenges to our understanding still remain.

6.1 Introduction

Sediments consist of solid particles that form a matrix of spaces filled with *pore water* (figure 6.1.1). The particles consist primarily of clay, $CaCO_3$, silica $SiO_2 \cdot nH_2O$, and organic matter. The particles enter the sediments mostly by sinking through the water column rather than being formed *in situ*. Typical sediment accumulation rates for the deep ocean range between about 0.1 and 1.0 cm per kyr^{-1} including the pore water (1 kyr = 1000 yr; figure 6.1.2). A measure of the fraction of the total sediment volume that is liquid is the *porosity* ϕ, defined as

$$\phi = \frac{V_l}{V_s + V_l} \tag{6.1.1}$$

In this equation, V is volume, and the subscripts l and s refer to liquid and solid, respectively. As figure 6.1.3 shows, the porosity of ocean sediments is typically about 80% to 90%, and gets larger in the less consolidated sediments near the sediment-water interface. In the following subsection, we begin our investigation of sediment processes with a first look at the solid and pore water concentrations of several of the properties we will be discussing in this chapter.

Observations

Figure 6.1.4 shows the organic carbon content of surface sediments in the world ocean, as percent dry weight. The dominant feature of this map is the high concentrations adjacent to the continental margins. Organic carbon in the sediments comes mainly from delivery of organic carbon to the sediment-water interface, which occurs primarily by sinking of particulate organic matter. Comparison of this distribution with a global map of export of particulate organic matter (figure 4.2.4b) shows good correspondence along the margins, but not in the ocean

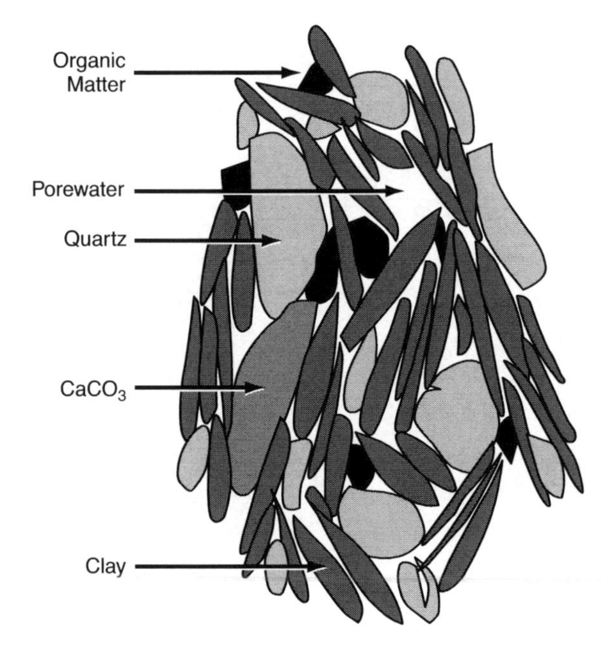

FIGURE 6.1.1: Schematic of interstitial waters and particles. From *Boudreau* [1997].

interior, where the organic matter content of sediments is almost featureless, while oceanic productivity is highly structured. Furthermore, comparison of this map with maps of total sediment accumulation rate (figure 6.1.2) and percent dry weight of $CaCO_3$ and silica (mostly in the form of amorphous silica, i.e., opal formed by diatoms; see chapters 7 and 9 and color plate 5) show that all are quite distinct from each other. Biogenic opal is dominated by preservation in deep ocean regions of high biological productivity, although with an important contribution from continental margins; and biogenic $CaCO_3$ preservation occurs principally along shallow topographic features, mainly mid-ocean ridges. What sets the large-scale pattern of organic matter concentration in sediments, and why does this differ from other sediment constituents? In addition to their intrinsic interest, these are important issues for paleoceanographers who would like to be able to use the organic matter content of sediments deposited at earlier times as an indicator of past oceanic conditions.

Figure 6.1.5 shows representative vertical profiles of the top 5 cm of weight percent organic content in dry sediments and oxygen concentrations in pore waters. These measurements were made at two locations in the Pacific Ocean, the top pair on the continental margin, and the bottom pair in the deep ocean. The origin of the vertical coordinate at $z = 0$ is at the sediment-water interface, with depth increasing downward. The top pair of panels is typical of continental margin sediments. Here the oxygen concentration drops from a value of almost 140 mmol m^{-3} at the sediment-water interface to 0

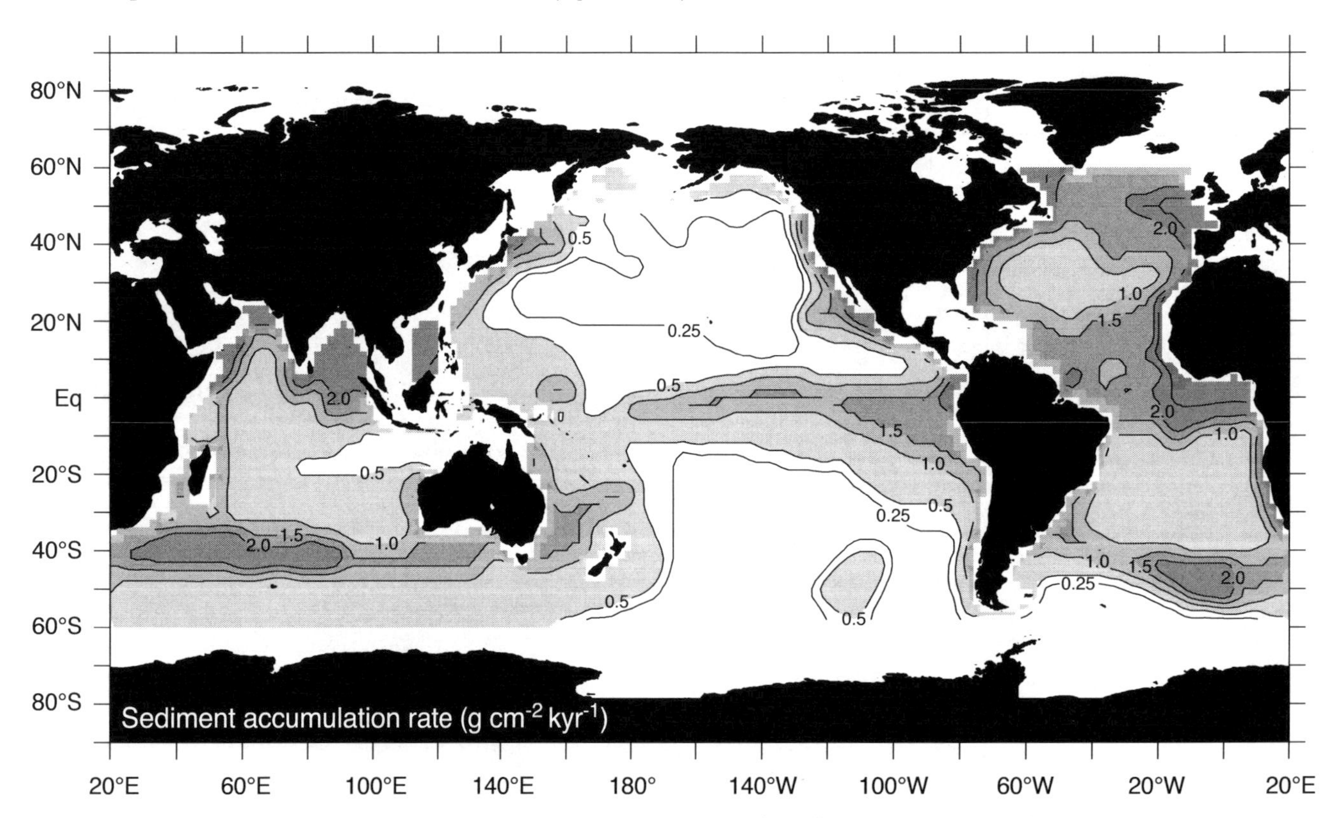

FIGURE 6.1.2: Global map of sedimentation rate from *Jahnke* [1996] in g cm^{-2} kyr^{-1}. To convert to cm kyr^{-1}, divide by the density of the dry sediments multiplied by $(1 - \phi)$, where ϕ is the porosity illustrated in figure 6.1.3. The density of dry sediments is typically of order 2.1 to 2.7 g cm^{-3} [cf. *Emerson*, 1985; *Reimers et al.*, 1992]. For a porosity of 0.8 and density of 2.5 g cm^{-3}, the sedimentation rate in g cm^{-2} kyr^{-1} should be multiplied by 2 to convert to cm kyr^{-1}. If the porosity were 0.6, the conversion factor would be 1 g cm^{-2} kyr^{-1} = 1 cm kyr^{-1}.

within the top couple of centimeters of the sediments. The organic carbon content drops only modestly. The lower pair of panels is typical of deep ocean sediments, where oxygen is reduced significantly within the top couple of centimeters, but not depleted; and the organic matter content is reduced to lower levels than in the continental margin sediments.

Particulate organic carbon and oxygen concentrations decrease with depth in sediments such as those of figure 6.1.5 because of remineralization. Among the questions we will address in this chapter are: What is the rate of remineralization of organic carbon in sediments such as these, and what processes determine the vertical distribution of the organic carbon? Oxygen, whose concentration at the sediment-water interface is set by the concentration in the overlying water column, decreases in the sediments because it is being utilized to oxidize the organic carbon. What are the rate-limiting steps for diffusion of oxygen into the sediments from the overlying water column, and what sets the shape of its profile? What impact does this oxygen flux have on the water column oxygen content, and how does oxygen influence the rate of remineralization as well as organic carbon burial in the sediments?

There are many differences between the two locations in figure 6.1.5 that are puzzling. The upper profile from a continental margin site behaves as if it were oxygen-limited, with abundant organic carbon left over in the sediments after oxygen runs out. By contrast, remineralization of organic carbon in the lower profile from an open ocean site appears to stop at a depth where there is still plenty of oxygen and organic carbon left. There is nothing about these profiles that provides an obvious hint as to why organic carbon should be consumed at one location until all the oxygen is gone, while at the other location, remineralization appears to stop while there is plenty of both oxygen and organic carbon. Indeed, the organic carbon and oxygen concentrations at the sediment-water interface differ only modestly at both stations.

Figure 6.1.6 shows a schematic distribution of a range of chemicals in pore water such as might be typical of continental margins (cf. upper panel of figure 6.1.5). The remineralization of organic matter by aerobic respiration consumes oxygen and proceeds through a series of steps that produce several chemicals, including nitrate (see equation (5.1.3)). The production of nitrate causes the concentration of nitrate in the pore water to increase with depth in the region where oxygen decreases. This will result in a flux of nitrate from the sediments into the water column. The water column concentration of nitrate is, as with oxygen, represented by the concentration at the sediment-water interface. However, at about the depth in the sediments where oxygen runs out, the nitrate concentration begins to decrease. Furthermore, as we go deeper in the sediments, we see the appearance of a whole range of other chemicals in pore water, including manganese, ammonia, iron or sulfide ions, and finally methane. What is happening with increasing depth in sediments that gives rise to this changing chemistry? In the previous chapter we talked about the consumption of nitrate by denitrification in regions where oxygen is depleted. We will see later that the availability of other chemicals in pore water permits a series of additional remineralization reactions to take place, breaking organic matter down even in the absence of oxygen or nitrate.

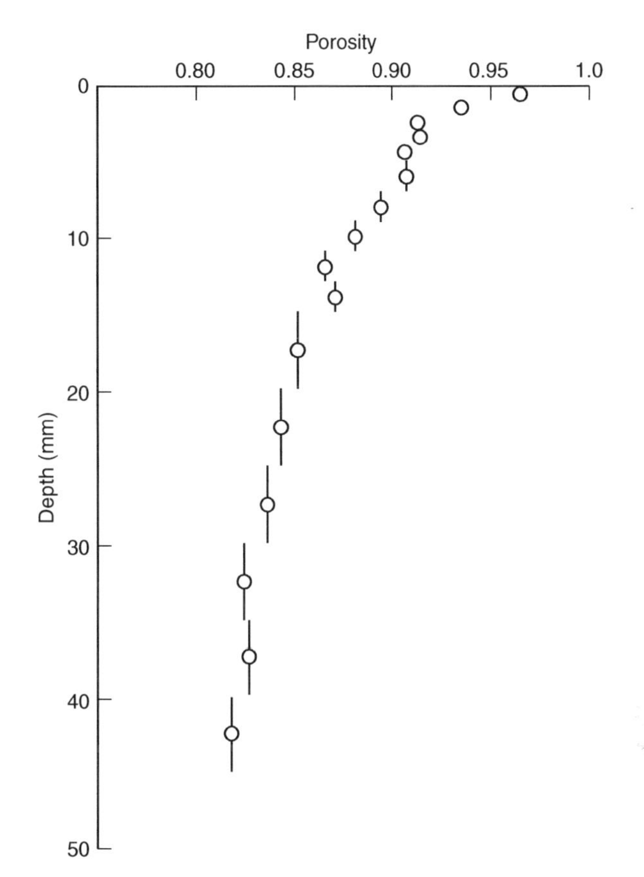

FIGURE 6.1.3: A typical vertical profile of porosity [*Reimers et al.*, 1992].

In the remainder of this section, we provide an initial overview of the processes that are thought to play a role in determining the remineralization and burial of organic carbon in the sediments. We discuss first the processes that affect pore waters, then those that affect the solids in sediments, and finally the range of remineralization reactions that occur in sediments. We then proceed in the following section to the development of a set of equations governing sediment diagenesis.

Sediment Properties and Processes

Pore water is seawater trapped in sediments as particles accumulate. Once buried, its properties are altered by remineralization of organic matter and dissolution or precipitation and adsorption or desorption of chemicals.

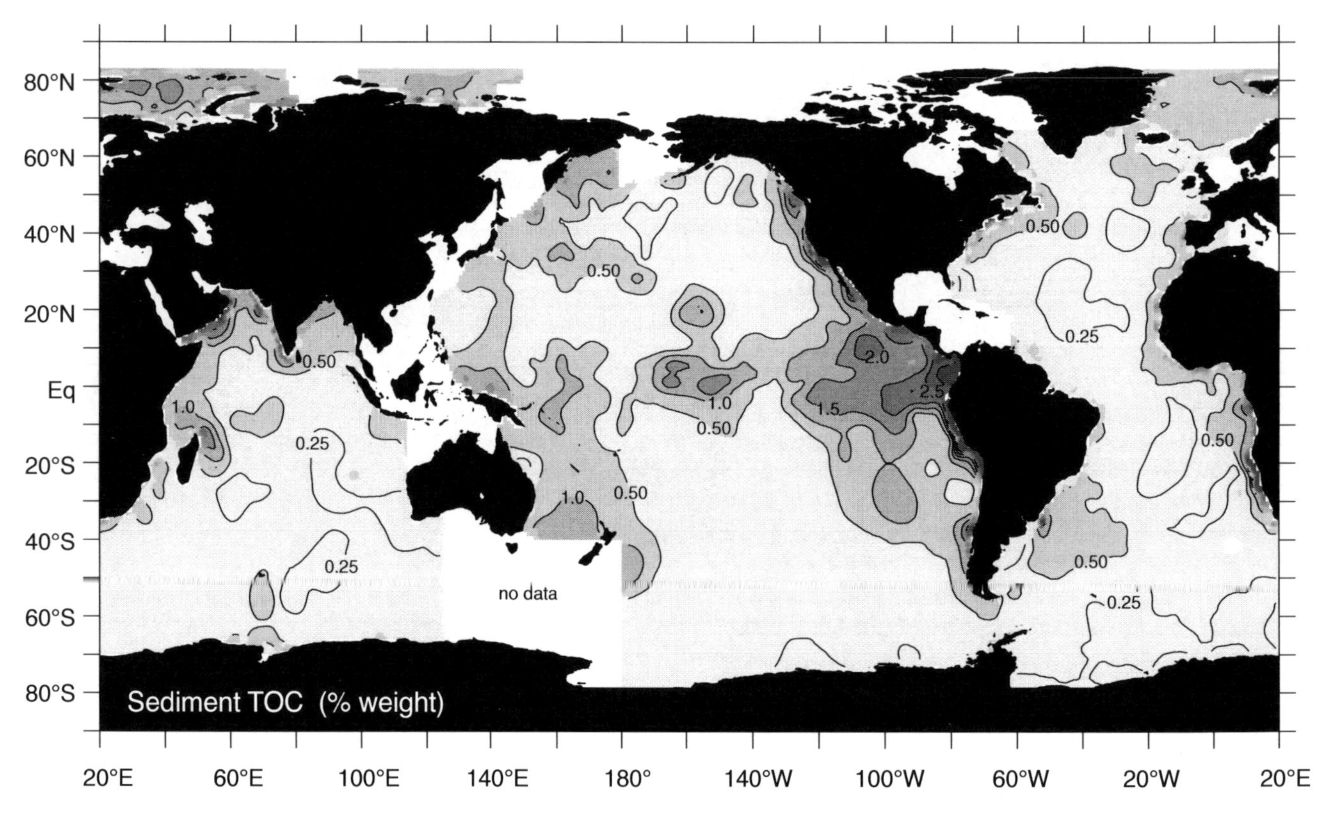

FIGURE 6.1.4: Global map of total organic carbon content (TOC) in dry sediments, as a percentage of dry weight, taken from *Seiter et al.* [2004]. See also color plate 5.

These chemical and biological processes create large concentration gradients between the pore water and the overlying seawater. The gradients result in diffusion of solutes to and from the overlying water column, as well as to and from different regions of the sediments.

The mechanism by which solutes diffuse in pore water is molecular diffusivity. However, the molecules cannot follow straight paths, but rather must detour around the particles. These detours decrease the effective diffusivity over that in pure water because of the reduction in the cross-sectional area over which diffusion occurs, by lengthening the mean path the molecules must follow, and by reducing the effective gradient that the molecule sees, since horizontal gradients are generally much smaller than vertical gradients.

The flux of solutes between the sediments and the water column is also affected by the existence of a *diffusive boundary layer* in the water column just above the sediments. This boundary layer results from the suppression of turbulent motions adjacent to the sediment-water interface, a feature that we also had to consider at the air-sea interface (chapter 3). Oxygen microelectrodes with a spatial measurement resolution of a few tens of μm provide dramatic illustrations of the existence of such layers. Figure 6.1.7 shows one such set of measurements from the coast of Denmark [*Gundersen and Jorgensen*, 1990]. The oxygen concentration at the sediment-water interface is substantially lower than in the overlying water column, a factor that we will need to consider when we develop a model of sediment diagenesis.

An additional process that may affect pore water is advection, the net motion of water moving up or down in the sediments. Advection in coastal and shelf environments may result from the direct impact of the orbital motion of waves in contact with the seabed or indirectly from pressure oscillations caused by the waves, which cause water to move up in one region and down in another [e.g., *Precht and Huettel*, 2003]. It may also result from interactions of tidal or bottom currents with bedform topography [e.g., *Huettel et al.*, 1996]. In the deep ocean, pore water advection can be caused by pressure gradients induced by flow across the top of sediments, or convective currents associated with heat loss from the oceanic crust [e.g., *Rudnicki et al.*, 2001]. Organisms such as deposit feeders may move water from one depth to another. This pore water *irrigation* can be quite important in coastal settings. We will not discuss it further here other than to point out that is has been modeled as advection but also in other ways. Another component of advection results from compaction of sediments at depth, which squeezes pore water upwards. An additional component is the virtual velocity that results from the fact that the origin of the vertical axis is always kept at the sediment-water interface even as sediments accumulate or are eroded away. Pore water advection is gen-

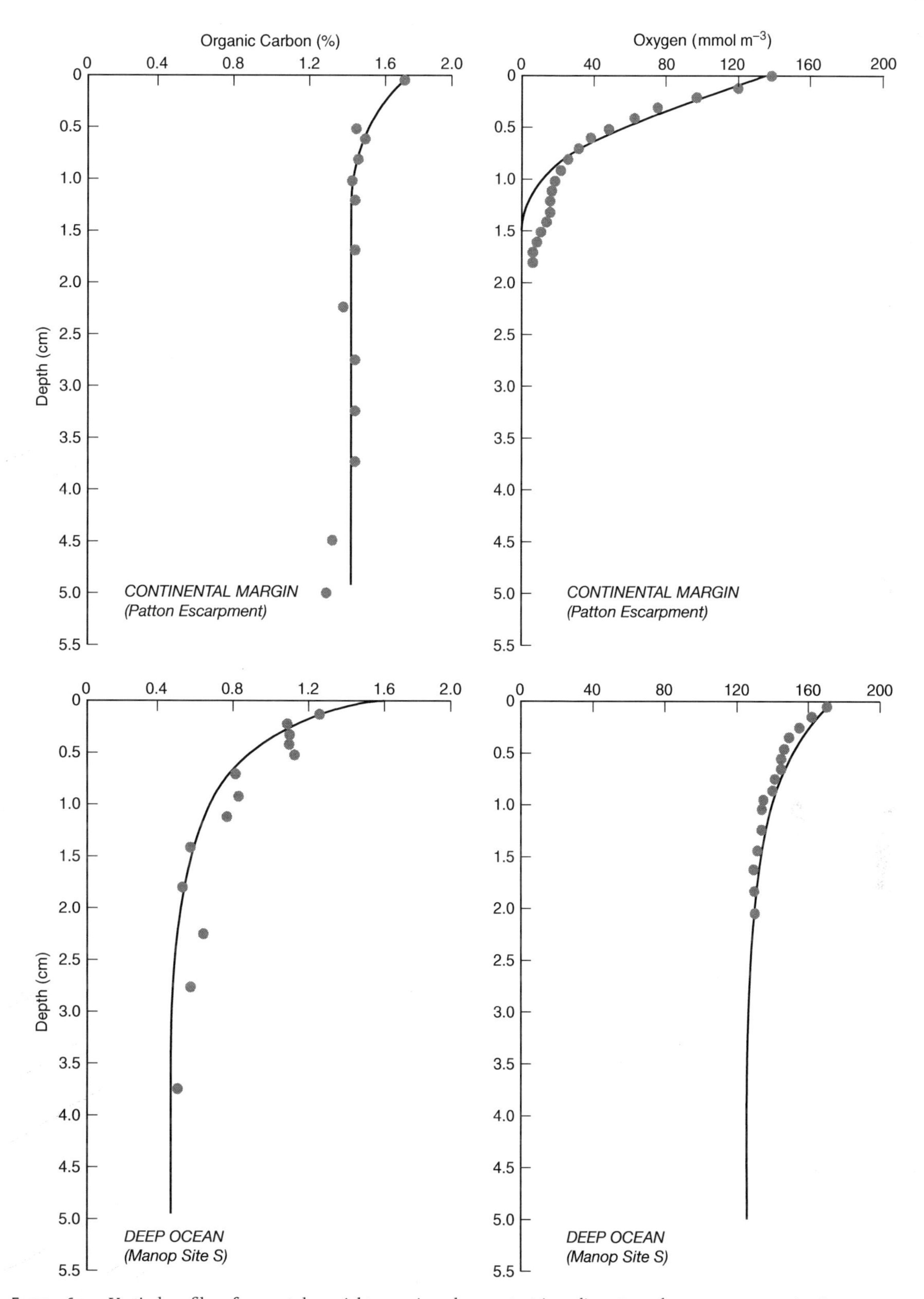

FIGURE 6.1.5: Vertical profiles of percent dry weight organic carbon content in sediments, and oxygen concentration in pore water [*Rabouille and Gaillard*, 1991a]. The 0 oxygen level at 1.5 cm in the Patton Escarpment core was detected in a separate core at the same site. The Patton Escarpment samples are from 32° 19.9' N, 120° 38.5' W at a depth of 3730 m. The MANOP Site S samples are from 11° 2' N, 140° W at a depth of 4910 m.

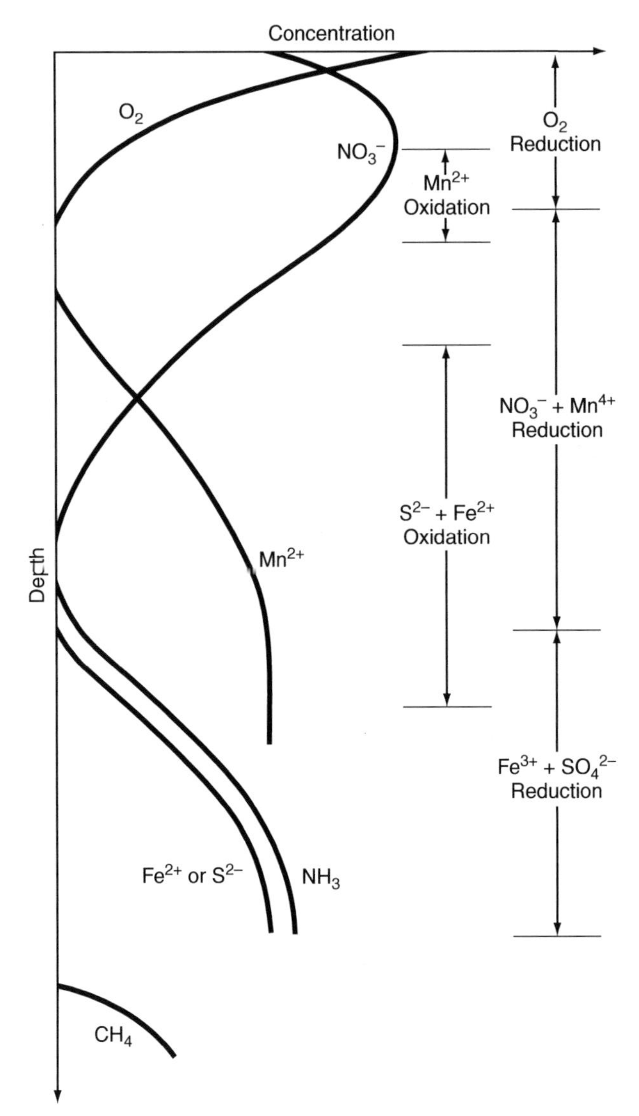

FIGURE 6.1.6: Schematic vertical profiles of reactants and products of remineralization reactions in sediments. Also shown are the reduction zones where the various reactions of table 6.1.1 occur, as well as zones where the soluble products of reduction reactions are reoxidized as they diffuse upward from their formation regions deeper in the sediments. Ammonium is produced by reactions (4) through (6) of table 6.1.1. Note that actual measurements rarely show the full sequence of reactants and products [*Emerson and Hedges,* 2003]. Deep sediments with high bottom water oxygen and low organic matter flux run out of reactive organic matter before sulfate reduction can occur; and near-shore sediments often have very thin zones of aerobic respiration, denitrification, and manganese reduction.

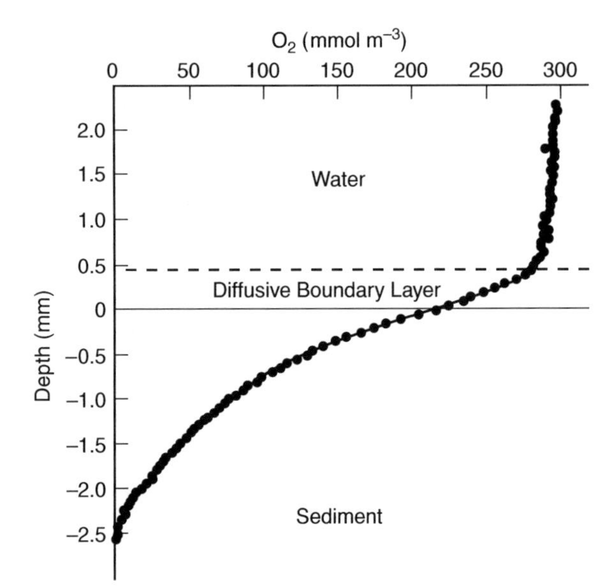

FIGURE 6.1.7: A vertical profile of oxygen at the sediment-water interface as measured by oxygen microelectrodes with a spatial resolution of 25 to 50 μm [*Gundersen and Jorgensen,* 1990]. The sediment-water interface is detected by a break in the vertical gradient of the profile that occurs in the transition from the sediments to the diffusive boundary layer. The top of the boundary layer is defined as the break that occurs in the transition from the linear gradient in the diffusive boundary layer to the uniform distribution in the overlying water column.

erally thought to be relatively unimportant in areas of thick sediments overlying older crust such as are characteristic of the abyssal plains [cf. *Sayles and Martin,* 1995; *Archer and Devol,* 1992; *Berelson et al.,* 1987, 1990; *Reimers and Smith,* 1986].

Solids in the sediments undergo chemical and biological reactions that complement those of the pore water. Thus, remineralization resulting from *aerobic respiration* (which uses oxygen) reduces the amount of particulate organic matter in the solid phase at the same time it increases the dissolved inorganic concentrations of the products of remineralization and reduces the oxygen concentrations in pore water. Material that is lost from pore water due to chemical precipitation or adsorption or biological uptake becomes part of the solids.

In addition to the foregoing nonconservative processes, solids in sediments are subject to physical processes that are somewhat analogous to advection and diffusion. The continued addition of particles at the sediment-water interface might be expected to move the interface upward in the water column. However, there is also a downward component of motion due to consolidation of deeper sediments, the contraction of the lithosphere underlying the sediments due to cooling as newly formed crust moves away from the mid-ocean ridges, and down-warping of the crust in response to the weight of the sediments. The traditional way of thinking about sediments is to ignore any vertical motion of the sediment-water interface relative to the earth's geoid and fix the origin of the z-coordinate ($z = 0$) at the sediment-water interface. The z-coordinate increases with depth into the sediments (e.g., figure 6.1.5). The addition of solids thus results in burial and a downward advection of solids with respect to the sediment-water interface. The process of sediment accumulation is regarded as advection with respect to the sediment-water interface. The sediments are like a movie strip in which any given signal, such as might be left by volcanic ash from an eruption, for example, slowly moves downward and away

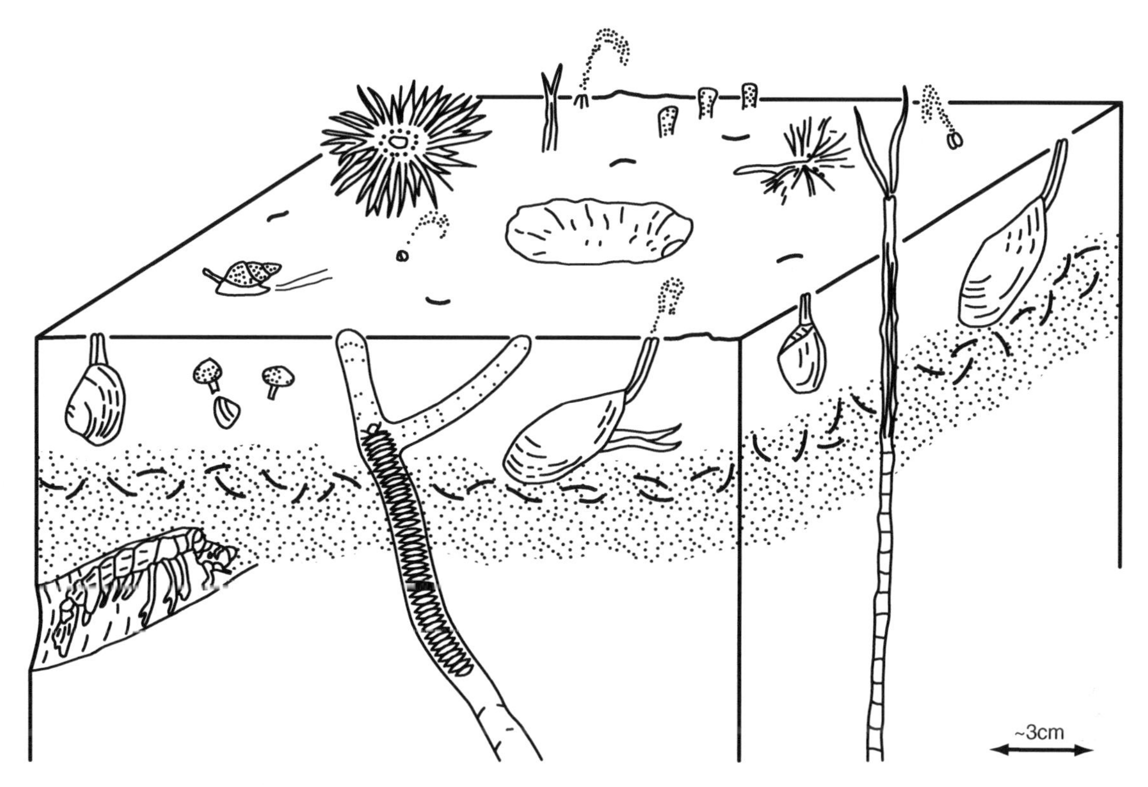

FIGURE 6.1.8: Schematic of sediment bioturbation processes. From *Berner* [1980].

from the sediment-water interface. One can think also of the vertical coordinate as a timescale with increasing age toward the bottom.

An additional important process for particles in sediments is *bioturbation*, the active movement of particles by burrowing fauna such as clams and worms (figure 6.1.8). Bioturbation encompasses a wide range of processes such as ingestion, defecation, and burrow construction. Organisms can be highly selective as to which particles they eat or utilize for their burrows, and thus many of the bioturbation processes are not random. Despite these issues, bioturbation is usually modeled as a kind of diffusion that results in mixing of the sediments. Measurements and models both show that bioturbation has a major impact on sediment diagenesis.

Finally, we note that resuspension and lateral transport of sediment particles by ocean currents and other phenomena is common in shallow seas and regions of steep topography.

Remineralization Reactions

As we saw in chapter 5, organisms obtain energy by remineralization of organic matter. The process of remineralization involves transfer of electrons from an *electron donor* (organic matter) to an *electron acceptor* (e.g., oxygen, nitrate, iron (III), manganese (IV), sulfate, or organic matter in the case of methane fermentation). During the process of transferring electrons, the electron donor is *oxidized* and the electron acceptor is *reduced*. We also refer to electron donors as *reductants*, because they reduce the electron acceptor; and we describe electron acceptors as *oxidants* because they oxidize the electron donor. Thus oxygen might be alternatively referred to as an electron acceptor or oxidant, and the transformation it undergoes would be described as reduction. Additional terminology that we will use here includes the term *oxic* to refer to oxygenated sediments and *aerobic* to refer to the respiration reaction involving oxygen; and the term *anoxic* to refer to oxygen-free sediments and *anaerobic* to refer to the reactions that take place in anoxic environments. We also use the term *substrate* to refer to the food source for organisms, i.e., organic matter.

Table 6.1.1 summarizes the most important reactions organic matter may undergo after it arrives in the sediments. Large organisms such as worms and clams as well as bacteria carry out aerobic respiration. Bacteria carry out all the other reactions shown in the table. The standard view of these reactions is that they do not occur simultaneously, but rather that they tend to follow the sequence shown in the table [*Froelich et al.*, 1979]. The sequence is determined by the relative amounts of free energy released by each reaction (table 6.1.2). For example, the reduction of oxygen releases more energy than the reduction of nitrate, manganese, and other oxidation reactions. Because of this, and because most of the enzymes that carry out reactions (2) to (6) do not function in the presence of oxygen, the standard view is

TABLE 6.1.1

Organic matter remineralization reactions based on organic matter *OM* with the composition $OM = C_{106}H_{175}O_{42}N_{16}P$, which is the median of that proposed by *Anderson* [1995] (cf. table 4.2.1).

We use this composition rather than that of Anderson and Sarmiento *[1994] because the latter does not provide an estimate of the hydrogen and oxygen component of the organic matter. See equation (5.1.3) for a breakdown of the nitrification component of reaction (1). The denitrification reaction was originally given in equation (5.1.4).*

Reaction	*Stoichiometry*
(1) Aerobic respiration	$OM + 150\ O_2 \rightarrow 106\ CO_2 + 16\ HNO_3 + H_3PO_4 + 78\ H_2O$
(2) Denitrification	$OM + 104\ HNO_3 \rightarrow 106\ CO_2 + 60\ N_2 + H_3PO_4 + 138\ H_2O$
(3) Manganese reduction	$OM + 260\ MnO_2 + 174\ H_2O \rightarrow 106\ CO_2 + 8\ N_2 + H_3PO_4 + 260\ Mn(OH)_2$
(4) Iron reduction	$OM + 236\ Fe_2O_3 + 410\ H_2O \rightarrow 106\ CO_2 + 16\ NH_3 + H_3PO_4 + 472\ Fe(OH)_2$
(5) Sulfate reduction	$OM + 59\ H_2SO_4 \rightarrow 106\ CO_2 + 16\ NH_3 + H_3PO_4 + 59\ H_2S + 62\ H_2O$
(6) Methane fermentation (methanogenesis)	$OM + 59\ H_2O \rightarrow 47\ CO_2 + 59\ CH_4 + 16\ NH_3 + H_3PO_4$

that denitrification and manganese reduction do not occur until aerobic respiration removes all or most of the oxygen. Similarly, reduction of iron and sulfate begin after the nitrate and MnO_2 are no longer freely available.

Vertical profiles of relevant chemicals in the sediments reflect the sequencing of the remineralization reactions (see schematic in figure 6.1.6). The decrease in oxygen concentration at the top of the sediments indicates the zone of aerobic respiration. The formation of nitrate in the aerobic respiration zone leads to an increase of nitrate concentration with depth in the upper part of this zone. However, at greater depth, the nitrate concentration first peaks, then declines due to the influence of denitrification. Manganese generally exists in the insoluble MnO_2 form in the presence of oxygen. When oxygen runs out, the reduction of MnO_2 to form soluble $Mn(OH)_2$ becomes energetically favorable, and one begins to see the appearance of dissolved Mn^{2+} in the pore waters of the sediments.

The aerobic respiration reaction (reaction (1) in table 6.1.1) produces nitrate by nitrification (cf. chapter 5, equation (5.1.3)), but denitrification consumes it. The effect of these sets of reactions often overlaps in sediments; some of the nitrate produced by the nitrification in the aerobic respiration zone diffuses downward into the zone where denitrification is going on and contributes to the total amount of denitrification. This is referred to as coupled nitrification-denitrification. Figure 6.1.6 shows how these overlapping processes affect the shape of the nitrate profile. The increase with depth and the peak in the upper part of the nitrate profile are due to nitrification in the aerobic respiration zone. Below this peak, nitrate drops, eventually going to zero in the denitrification zone. The shape of the profile is consistent with a downward flux of nitrate into the denitrification zone. The concentration of nitrate at the sediment-water interface reflects the water column concentration corrected for the effect of the diffusive boundary layer. The peak in the nitrate profile shown in the figure thus indicates that in this particular case some of the nitrate produced by aerobic respiration will also diffuse out of the sediments into the water column.

There are a wide range of energetically favorable reactions between the components of remineralization that create the potential for additional coupling of the remineralization reactions [cf. *Hulth et al.*, 1999]. Figure

TABLE 6.1.2

Free energy changes for remineralization reactions [*Morel and Hering*, 1993].

Note that the stoichiometry of these reactions is different from that of Anderson *[1995] used in table 6.1.1. The reactions in table 6.1.1 involve a complex mixture of organic matter molecules (see table 4.2.2) with complicated energetics. CH_2O represent sucrose.*

Reaction	*Free Energy Change (kJ mol^{-1} of CH_2O)*
$CH_2O + O_2 \rightarrow CO_2 + H_2O$	−476
$5CH_2O + 4\ NO_3^- \rightarrow 2\ N_2 + 4\ HCO_3^- + CO_2 + 3\ H_2O$	−452
$CH_2O + 3\ CO_2 + H_2O + 2\ MnO_2 \rightarrow 2\ Mn^{++} + 4\ HCO_3^-$	−388
$CH_2O + 7\ CO_2 + 4\ Fe(OH)_3 \rightarrow 4\ Fe^{++} + 8\ HCO_3^- + 3\ H_2O$	−187
$2\ CH_2O + SO_4^= \rightarrow H_2S + 2HCO_3^-$	−82
$2CH_2O \rightarrow CO_2 + CH_4$	−71

TABLE 6.1.3
Indirect oxidation reactions based on organic matter *OM* with the composition $OM = C_{106}H_{175}O_{42}N_{16}P$ proposed by *Anderson* [1995].
The top three rows depict the coupling of methane fermentation with anaerobic oxidation of methane by sulfate to produce an overall reaction with the stoichiometry of the sulfate reduction reaction. The lower three rows depict the coupling of sulfate reduction with the reaction of sulfide and ammonium with oxygen to give an overall reaction that has the same stoichiometry as aerobic respiration.

Reaction	*Stoichiometry*
Methane fermentation	$OM + 56\ H_2O \rightarrow 47\ CO_2 + 59\ CH_4 + 16\ NH_3 + H_3PO_4$
Anaerobic oxidation of methane	$59\ H_2SO_4 + 59\ CH_4 \rightarrow 59\ CO_2 + 118\ H_2O + 59\ H_2S$
Overall reaction	$OM + 59\ H_2SO_4 \rightarrow 106\ CO_2 + 16\ NH_3 + H_3PO_4 + 59\ H_2S + 62\ H_2O$
Sulfate reduction	$OM + 59\ H_2SO_4 \rightarrow 106\ CO_2 + 16\ NH_3 + H_3PO_4 + 59\ H_2S + 62\ H_2O$
Oxidation of sulfide and ammonium	$59\ H_2S + 16\ NH_3 + 150\ O_2 \rightarrow 59\ H_2SO_4 + 16\ HNO_3 + 16\ H_2O$
Overall reaction	$OM + 150\ O_2 \rightarrow 106\ CO_2 + 16\ HNO_3 + H_3PO_4 + 78\ H_2O$

6.1.6 shows evidence for three of these in the upward disappearance of CH_4, Fe^{2+}, and Mn^{2+}, above the depths of methanogenesis and iron and manganese reduction, respectively. The existence of the anaerobic methane oxidation reaction (see table 6.1.3) was first proposed based on observations of the methane distribution in sediments [*Barnes and Goldberg*, 1976; *Martens and Berner*, 1977; *Reeburgh*, 1976], and appears to be widespread in deep sediments wherever methanogenesis occurs [cf. *D'Hondt et al.*, 2002; *Reeburgh*, 2003]. The upper part of table 6.1.3 shows how the coupling of the anaerobic methane oxidation reaction with methanogenesis gives an overall reaction that is identical to sulfate reduction. See also discussion of Anammox reaction in chapter 5.

Vertical profiles of sulfide (cf. model study by *Boudreau* [1991]) show a decrease above the sulfate reduction zone that can occur by a variety of oxidation reactions [cf. *Hulth et al.*, 1999]. Ammonia also decreases above the iron reduction zone [*Reimers et al.*, 1992] due to oxidation. The lower part of table 6.1.3 depicts a coupled reaction in oxic sediments in which the products of sulfate reduction are ultimately oxidized by oxygen, giving an overall reaction that has the stoichiometry of aerobic respiration. Similarly, the soluble Mn^{2+} and Fe^{2+} produced by the manganese and iron reduction reactions can be oxidized back to their insoluble Mn^{4+} and Fe^{3+} forms in a sequence of reactions that ultimately sum up to the aerobic respiration stoichiometry [cf. *Hulth et al.*, 1999; *Reimers et al.*, 1992]. Indeed, as pointed out by *Canfield* [1993], the overall flux of oxygen into sediments gives a good approximation to the total carbon oxidation rate under many circumstances, even in cases where much or most of the oxidation is by anaerobic processes. The only remineralization it misses is that associated with reduced species that are buried in sediments (e.g., the reduced Mn and Fe phases as well as sulfide in the form of pyrite, FeS) as well as denitrification, which results in the formation of N_2 gas that escapes the sediments without being oxidized.

The mixing of sediments by bioturbation and physical processes such as estuarine circulation, tidal and other currents, and waves, which are particularly strong in deltaic environments, can boost the remineralization rate by (1) increasing the rate of reoxidation by reactions such as those described above; (2) leading to repetitive reaction zone successions, in which a given set of sediments may switch from one reaction zone to another repeatedly; (3) leading to exchange of reaction components between one zone and another; and (4) mixing in of fresh organic matter (e.g., marine plankton debris) with refractory organic matter (e.g., refractory terrestrial components [*Aller*, 1998]). The decomposition of organic matter in such systems is exceptionally efficient and can follow rather surprising pathways. For example, *Hulth et al.* [1999] describe a set of coupled reactions involving the N and Mn cycles in which the coupled nitrification-denitrification cycle is replaced by one in which nitrification occurs anaerobically by reaction of ammonia with MnO_2:

$$4\ MnO_2 + NH_4^+ + 6\ H^+ \rightarrow 4\ Mn^{2+} + NO_3^- + 5\ H_2O$$

Once nitrate is produced, it can be quickly denitrified by any one of numerous potential reductants in addition to organic matter, such as ammonia, Fe^{2+}, and HS^-.

In a related study by *Aller et al.* [1996], remineralization in Amazon shelf sediments has been shown to be about 75% anaerobic, but whereas the primary anaero-

TABLE 6.1.4
Contribution of various sediment remineralization processes to global oxidation rate

Remineralization Reaction	*Estimate of Global Organic Carbon Oxidation Rate*	
	Pg C yr^{-1}	*% of Total*
Aerobic respiration	1.6	65.0%
Denitrification	0.16 ± 0.05	6.5%
Manganese reduction	$0.01 - 0.02$	$\sim$0.6%
Iron reduction	$0.003 - 0.010$	$\sim$0.3%
Sulfate reduction	0.44	17.9%
Methanogenesis	0.24	9.8%
Total	2.5	

The aerobic respiration and sulfate reduction estimates are from *Canfield* [1993]. The denitrification estimate is from *Dunne* et al. [2005c] using the metamodel of *Middelburg et al.*, [1996]. *Gruber* [2004] obtains an identical answer. Estimates of the role of manganese and iron reduction are based on chemical gradients of these trace metals in the sediments [*Bender and Heggie*, 1984; *Heggie et al.*, 1987; *Bender et al.*, 1989]. The methanogenesis estimate is from *Hinrichs and Boetius* [2002]; earlier estimates by *Canfield* [1993] and *Reeburgh et al.*, [1993] are about a quarter of the value shown.

bic reaction in most shelf sediments is sulfate reduction, the dominant anaerobic reaction on the Amazon shelf is iron reduction. The periodic exposure to oxygen through physical reworking of the sediments, which, in the case of iron, converts it from Fe(II) to Fe(III), is an essential part of the process by which reactions such as those involving iron and manganese can become an important contributor to anaerobic remineralization [cf. *Wang and Cappellen*, 1996]. In regions where the seabed is stable, which is the case for most of the ocean, the role of Mn and Fe reduction is generally negligible, and sulfate is the dominant anaerobic oxidant, as we shall see in a later section.

Table 6.1.4 summarizes various estimates of the global remineralization rate due to each of the oxidants. Aerobic respiration is by far the dominant process, accounting for 65% of the total remineralization. Sulfate reduction is next, followed by methanogenesis and denitrification. The reduction of trace metals (manganese and iron) makes a negligible contribution overall. Later we shall see that deep ocean sediment remineralization is primarily due to aerobic remineralization, with sulfate reduction and methanogenesis being confined almost exclusively to the continental margins, and about 81% of the denitrification occurring in sediments with water depths less than 1000 m.

We conclude this prefatory overview by noting that the sediment processes that have the greatest influence on the oceanic distribution of oxygen and dissolved organic carbon, as well as nitrate, phosphate, and other chemicals used by organisms such as silicic acid and $CaCO_3$, all occur within the top few centimeters of the sediments. This chapter will thus be concerned only with what is referred to as *early diagenesis* occurring in the top $\sim$10 cm of the sediments (cf. figure 6.1.5), with occasional excursions in anoxic sediments to depths of $\sim$20 cm. We also note that in the deep ocean, oxygen is by far the most important electron acceptor in early diagenesis, with nitrate playing a distant secondary role. The vast majority of the sulfate reduction and methanogenesis shown in table 6.1.4 occurs on continental margins, where organic carbon fluxes and sedimentation rates are much higher than in the deep ocean and the reactions progress further down through the reaction sequence of table 6.1.1. The chief focus of our more detailed discussions below will be on aerobic respiration, although we will also provide an assessment of anaerobic processes.

The remainder of this chapter is organized as follows. In section 6.2, we develop the tools we need in order to model sediment diagenesis, and in section 6.3 we use those tools and an analysis of observations to investigate the remineralization of organic matter. We also show that, while the models provide valuable insights into how remineralization functions, they are not able to explain what controls burial of organic matter. In section 6.4, we turn to a discussion of the processes that have been proposed to control burial. Finally, in section 6.5, we pull together all that we have learned in this and the previous two chapters to put together a budget that will enable us to assess how sediment processes contribute to the cycles of carbon, oxygen, and nitrate in the ocean as a whole.

6.2 Sediment Diagenesis Models

This section develops the building blocks (*constitutive equations*) for putting together a model of sediment diagenesis. Specifically, we consider the mass balance equation analogous to (4.2.1) for a constituent C which is dissolved in pore water, and a particulate constituent B which is part of the solid particles. In order to keep the derivation general while minimizing the number of symbols required, we will use the symbol C to refer to both the concentration in the pore water and the molecule itself. Likewise, we will use the symbol B to refer to both the concentration in the solids and the molecule itself. We thus have

$$\frac{\partial \phi \cdot C}{\partial t} = -ADV(C) + DIFF(C) + SMS(C) \qquad (6.2.1a)$$

and

$$\frac{\partial(1-\phi)\cdot B}{\partial t} = -ADV(B) + DIFF(B) + SMS(B) \qquad (6.2.1b)$$

The form of the advection ADV and diffusion $DIFF$ terms will be specified below. SMS is *in situ* sources minus sinks due to biological and chemical processes. In these equations, the concentration C has units of mmol m^{-3} of pore water only, and is multiplied by porosity to convert it to units of mmol m^{-3} of pore water plus solids. The concentration B has units of mmol m^{-3} of solid with no porosity and is multiplied by $(1-\phi)$ to convert it to units of mmol m^{-3} of pore water plus solids. Note that particulate constituents in solid sediments are normally measured as the ratio f_B of the mass of the constituent per unit mass of the total solid, and then converted to the concentration B in mmol m^{-3} of solid with no porosity. Panel 6.2.1 explains how to do this conversion.

We start by considering the physical, chemical, and biological processes that determine the properties of pore water, then the processes that determine the properties of the solid component of the sediments. We consider only the vertical coordinate of the conservation equation, as the lateral gradients of sediment properties are normally very small and the contribution from lateral processes is generally thought to be insignificant.

Pore Waters

We consider first the contribution of molecular diffusion to the $DIFF(C)$ term of (6.2.1a). We begin with the vertical (z) component of Fick's first and second laws for molecular diffusion introduced in section 2.1 as equations (2.1.3) and (2.1.4). These equations need to be modified in order to take into account that the sediments include particles as well as pore water. We express Fick's first law for a porous medium as

$$\Phi^C(z) = -\phi\cdot\varepsilon_s\cdot\frac{\partial C}{\partial z} \qquad (6.2.2)$$

where $\Phi^C(z)$ is the flux per unit area of solids plus pore water (mmol per m^2 of liquid plus solid) and ε_s is the effective molecular diffusivity in the sediments (m^2 s^{-1}). Panel 6.2.2 gives a detailed discussion of the form of this equation and the various other forms that are found in the literature as well as the factors that contribute toward making ε_s smaller than the molecular diffusivity in seawater, ε. From Fick's second law, we obtain that the contribution of molecular diffusion to (6.2.1a) is

$$DIFF(C) = -\frac{\partial\Phi^C(z)}{\partial z} = \frac{\partial}{\partial z}\left(\phi\cdot\varepsilon_s\cdot\frac{\partial C}{\partial z}\right) \qquad (6.2.3)$$

Panel 6.2.1: Unit Conversions for Solid Constituents

The dried solids of sediments are often partitioned into four components: particulate organic carbon POC, opal SiO_2, and calcium carbonate $CaCO_3$, all of which are measured directly; and nonbiogenic detritus, which is defined as whatever is left over after measuring the first three of these. The concentrations of each component A is measured as the percent dry weight ratio to solids, $100\%\cdot f_A$. We explain here how to convert concentrations in mmol m^{-3} of solids with no porosity into dry sediment weight ratios, which can also be inverted to convert weight ratios to concentrations.

In order to convert f_{POC} to concentration $[POC]$ in mmol m^{-3} of solids, we need to know the molecular weight of carbon ($MW_C = 12$ g carbon per mole of carbon), the mass ratio of organic matter to carbon ($\alpha = 2.7$ g organic matter per gram organic carbon), the density of organic matter ($\rho_{OM} = 1.1$ g organic matter per cm^3 of organic matter), and the density of the nonorganic sediment material ($\rho_{sediment} = 2.7$ g of sediment per cm^3 of sediment with no porosity; values from *Emerson* [1985]). Define $[POC]^+ = [POC]\cdot MW_C \times 10^{-9}$ g carbon per cm^3 of solid with no porosity, where the factor 10^{-9} comes from converting mmol to mol (10^{-3}) and m^3 to cm^3 (10^{-6}). Then $f_{POC} = [POC]^+/\{[POC]^+\cdot\alpha + (1 - [POC]^+\cdot\alpha/\rho_{OM})\cdot\rho_{sediment}\}$ [cf. *Emerson*, 1985]. The first term in the denominator is the grams of organic matter per unit volume of solid with no porosity. The second term is the grams of nonorganic matter solids per unit volume of solid with no porosity.

By analogy with POC, conversion of f_{opal} to $[SiO_2]$ in mmol m^{-3} of solid with no porosity requires knowing the molecular weight of opal, $MW_{SiO_2} = 60$ g opal per mole of opal (this ignores the water content of the opal; the density of opal ($\rho_{opal} = 2.1$ g opal per cm^3 of opal); and the density of non-opaline sediment material given above. Define $B^+ = [SiO_2]\cdot MW_{SiO_2} \times 10^{-9}$ g opal per cm^3 of solid, as above. Then $f_{opal} = B^+/[B^+ + (1 - B^+/\rho_{opal})\cdot\rho_{sediment}]$. The first term in the denominator is the grams of opal per unit volume of solid with no porosity. The second term is the grams of non-opaline solids per unit volume of solid with no porosity.

Finally, conversion of f_{CaCO_3} to $[CaCO_3]$ in mmol m^{-3} of solid with no porosity requires knowing the molecular weight of $CaCO_3$ ($MW_{CaCO_3} = 100$ g $CaCO_3$ per mole of $CaCO_3$); the density of $CaCO_3$ ($\rho_{CaCO_3} = 2.7$ g opal per cm^3 of opal); and the density of the remaining sediment material, which is the same as that of $CaCO_3$. Define $B^+ = [CaCO_3]\cdot MW_{CaCO_3} \times 10^{-9}$ g $CaCO_3$ per cm^3 of solid with no porosity. Then $f_{CaCO_3} = B^+/[B^+ + (1 - B^+/\rho_{CaCO_3})\cdot\rho_{sediment}]$. Since the $CaCO_3$ and sediment densities are equal, this reduces to $f_{CaCO_3} = ([CaCO_3]\cdot MW_{CaCO_3}/\rho_{CaCO_3}) \times 10^{-9}$.

We summarize in table 6.2.1 empirical equations for estimating the molecular diffusivity ε of nitrate and oxygen in seawater, and another set of equations for estimating the ratio of these to the effective molecular diffusivity in the sediments, ε_s.

The contribution of advection is represented by

$$ADV(C) = \frac{\partial}{\partial z}(\phi \cdot w_{PW} \cdot C) \quad (6.2.4)$$

where w_{PW} is pore water flow in m s^{-1} and porosity converts the concentration to mmol m^{-3} of solid plus liquid to

Panel 6.2.2: Diffusion in Sediments

The literature on how to represent diffusion in porous media such as sediments and soils is formidable and can be quite confusing. The following is an attempt to shed some light by going back to some of the earlier studies on which geoscientists have based their analyses. We begin with the following modified representation of Fick's first law (6.2.2) for a porous medium, in which the porosity is folded into a newly defined effective diffusivity ε_e:

$$\Phi^C(z) = -\varepsilon_e \cdot \frac{\partial C}{\partial z} \quad (1)$$

The ratio between the effective diffusivity and the molecular diffusivity ε in seawater is referred to as the *diffusibility* Q [*van Brakel and Heertjes*, 1974], i.e.,

$$\frac{\varepsilon_e}{\varepsilon} = Q \quad (2)$$

There exist theoretical derivations of expressions for the diffusibility for certain kinds of porous media [e.g., *Petersen*, 1965; *Whitaker*, 1967; cf. *Boudreau*, 1997], but these provide only marginal insight and are generally not easily, much less widely, applicable. We prefer the empirical approach of *Petersen* [1965] and *van Brakel and Heertjes* [1974] in which experimental values are related to the diffusibility by the following expression:

$$Q = \frac{\phi \cdot \delta}{\tau} \quad (3)$$

where

ϕ = porosity, as before

δ = *constriction factor*

τ = *tortuosity factor*

The impact of each of these three parameters can be understood qualitatively as representing the effective reduction in the area available for the diffusion of a species in a porous medium (porosity), the fact that diffusing molecules must follow tortuous paths that deviate around particles rather than being able to diffuse directly in the z-direction (tortuosity), and the fact that the channels through the sediments are not uniform in cross-section (constrictivity [*Petersen*, 1965]). We see by comparing the above relationships with (6.2.2) that $\varepsilon_s/\varepsilon$, where ε_s refers to the definition in the text, is equal to δ/τ.

In their study, *van Brakel and Heertjes* [1974] provide arguments for relating the tortuosity factor to the square of the *tortuosity* θ, which is a fractional measure of the path length followed by the molecules defined as:

$$\theta = \frac{dL}{dz} \quad (4)$$

Here, dL is the distance actually traveled as the molecule detours around the particles, and dz is the vertical linear distance [cf. *Berner*, 1980]. This gives for the diffusibility the expression

$$Q = \frac{\phi \cdot \delta}{\theta^2} \quad (5)$$

We note that there is considerable confusion in the literature resulting from the failure to differentiate clearly between the tortuosity factor τ and the tortuosity θ. Equation (3) is a more general expression for the diffusibility in which no particular functional form is assumed for the tortuosity factor. Expressions (4) and (5) assume a specific form for the tortuosity factor.

Most geoscientists take the additional step of ignoring the constrictivity or, more commonly, assuming it is folded into the tortuosity. The diffusibility is then given as

$$Q = \frac{\phi}{\theta^2} \quad (6)$$

[e.g., *Berner*, 1980; *Boudreau*, 1997], where it is understood that the constrictivity is folded into the tortuosity term. However, we find this form of the equation to be misleading, in that the derivation by *van Brakel and Heertjes* [1974] of the tortuosity squared relationship given in (5), with tortuosity defined as in (4), assumes that the constrictivity is considered separately. We therefore prefer not to use expression (6).

The confusion occasioned by the various expressions for the diffusibility term, tortuosity factor, and tortuosity generally end out being of minor concern in biogeochemical applications. We usually use empirical relationships based on measurements of diffusivity in pure seawater and sediments or measurements of the *formation factor* f in order to estimate the appropriate diffusivity to use in the sediments. The formation factor is the ratio of the specific electrical resistivity of bulk sediments to the resistivity of pore water alone [*Boudreau*, 1997].

TABLE 6.2.1
Empirical equations for pore water transport parameters

Parameter	*Equation*
$\varepsilon\,(O_2)$ Molecular diffusivity for O_2 ($m^2\ s^{-1}$)	$(0.2604 + 0.006383 \cdot (T/\mu)) \times 10^{-9}$
$\varepsilon\,(NO_3^-)$ Molecular diffusivity for nitrate ion ($m^2\ s^{-1}$)	$(9.5 + 0.388 \cdot t) \times 10^{-10}$
μ dynamic viscosity (centipoise; 1 centipoise $= 10^{-3}$ N m^{-2})	$1.7910 - 6.1440\times10^{-2} \cdot t + 1.4510\times10^{-3} \cdot t^2 - 1.6826\times10^{-5} \cdot t^3 - 1.5290\times10^{-4} \cdot P + 8.3885\times10^{-8} \cdot P^2 + 2.4727\times10^{-3} \cdot S + t \cdot (6.0574\times10^{-6} \cdot P - 2.6760\times10^{-9} \cdot P^2) + S \cdot (4.8429\times10^{-5} \cdot t - 4.7172\times10^{-6} \cdot t^2 + 7.5986\times10^{-8} \cdot t^3)$
$\varepsilon\ /\ \varepsilon_s$ (unitless)	$1 - 2 \cdot \ln(\phi)$

Equations are taken from *Boudreau* [1997]. The temperature T is in degrees Kelvin and t is in degrees Celsius, P is pressure in bars, and S is salinity in practical salinity units. The dynamic viscosity is equal to the kinematic viscosity times the density. A typical deep ocean value for the dynamic viscosity for an ocean temperature of 4°C at 5000 db pressure and a salinity of 34.7 is 1.61 centipoise, which is equal to 1.61×10^{-3} N m^{-2}. The units for μ in the oxygen diffusivity equation are centipoises. Thus, in our example, the value for dynamic viscosity that should be used to calculate the oxygen solubility is 1.61, not 1.61×10^{-3}.

match the remainder of the equation. We neglect the small contribution from sediment accumulation.

The $SMS(C)$ term will depend on the process we are talking about. In this chapter we discuss specifically the remineralization of particulate organic matter by reaction with an electron acceptor. We use the negative of the Monod growth function used previously for photosynthesis, (4.2.11), multiplied by the maximum potential organic matter remineralization rate R_{max} in mmol $m^{-3}\ s^{-1}$ of organic matter in solid sediments. It is frequently the case that both the organic matter and the electron acceptor limit the rate of remineralization. Sediment geochemists generally assume that the joint dependence of the reaction rate on two substrates is multiplicative, in which the Monod hyperbolic function for the electron acceptor (C in this case) is multiplied by the hyperbolic function for organic matter (B in this case) to obtain the disappearance rate of C:

$$SMS(C) = -(1-\phi) \cdot r_{C:B} \cdot R_{\max}^{B,C} \cdot \gamma(B,C) \qquad (6.2.5a)$$

where

$$\gamma(B,C) = \left(\frac{B}{K_B + B}\right) \cdot \left(\frac{C}{K_C + C}\right) \qquad (6.2.5b)$$

(cf. *O'Neil et al.* [1989] for a discussion of alternative approaches). The constant $r_{C:B}$ is the stoichiometric ratio with which organisms take up C when remineralizing B (obtained from table 6.1.1). The half-saturation constant K_B has units of mmol m^{-3} of organic carbon in the solid sediments to match the units of B; and the half-saturation constant K_C has units of mmol m^{-3} of pore water to match the units of C. Note that equation (6.2.5) does not include an explicit dependence on the population of bacteria or other organisms. In effect, we assume that the the reaction is never limited by their availability. We have used a negative sign on the right-hand side to represent consumption of the electron acceptor C. The sign would be positive if C were produced by the remineralization reaction. Panel 6.2.3 discusses a series of simplifications to equation (6.2.5b) that have been employed frequently in the literature.

The final pore water diagenetic equation for a dissolved constituent C is obtained by combining the constitutive equation for transport obtained by inserting (6.2.3) through (6.2.5) into (6.2.1):

$$\begin{aligned}\frac{\partial \phi \cdot C}{\partial t} &= -ADV(C) + DIFF(C) + SMS(C) \\ &= -\frac{\partial}{\partial z}(\phi \cdot w_{PW} \cdot C) + \frac{\partial}{\partial z}\left(\phi \cdot \varepsilon_s \cdot \frac{\partial C}{\partial z}\right) \\ &\quad - (1-\phi) \cdot r_{C:B} \cdot R_{\max}^{B,C} \cdot \left(\frac{B}{K_B + B}\right) \cdot \left(\frac{C}{K_C + C}\right)\end{aligned} \qquad (6.2.6)$$

It is common practice in solving sediment diagenesis models to use as a boundary condition that the concentration at the top of the sediments, C_0, is equal to the bottom water concentration C_{BW} measured several meters above the sediments. This approach is fine for most situations unless the reactions in the sediments occur very near the sediment-water interface, such as is often the case with aerobic respiration (cf. figure 6.1.7). Because the diffusive boundary layer, which is where the concentration transitions from C_0 to C_{BW}, is only a few mm thick, direct measurement of the interface concentration is generally impractical.

The usual way of dealing with the influence of the diffusive boundary layer in sediment diagenesis models is to represent it as consisting of a stagnant film in which molecular diffusivity predominates, and an interior water column layer in which the diffusivity has its full interior value, with the concentration at the top of the stagnant film being equal to the interior bottom water concentration, C_{BW} (see the stagnant film model discussion in chapter 3). In the absence of significant remineralization within the boundary layer, the flux of C through the boundary layer has to equal $\Phi^C_{z=0}$, the flux of C in or out of the sediments at $z=0$. The finite difference form of Fick's first law (6.2.2) in the water column is thus $\Phi^C_{z=0} = -\varepsilon_s \cdot \Delta C/\Delta z$, with $\Delta C = C_0 - C_{BW}$. The stagnant film is set at the effective thickness δ_e that can account for the retarding effect of the full diffusive boundary layer given the expression:

Panel 6.2.3: Remineralization Kinetics

Equation (6.2.5) gives the multiplicative remineralization kinetics expression for two reactants. In practice, this expression can almost always be simplified based on analyses of the magnitudes of K_B and K_C relative to the substrate concentrations B and C. We briefly review a few of the more important and useful simplifications that have been employed in the literature.

The value of K_C for oxygen is ~3 mmol m^{-3} [cf. *Rabouille and Gaillard*, 1991b]. The water column and the top part of deep ocean sediments typically have oxygen concentrations that are much higher than this, i.e., $[O_2] \gg K_{O_2}$. Under conditions such as this, the consumption of oxygen by remineralization of organic matter B, or any other electron acceptor C that satisfied this condition, would be

$$SMS(C) = -(1-\phi) \cdot r_{C:B} \cdot R^B_{max} \cdot \frac{B}{K_B + B} \tag{1}$$

Another common assumption is that the half-saturation constant for organic matter is much larger than the organic matter concentrations, i.e., $K_B \gg B$. A further simplification of (1) is then

$$SMS(C) = -(1-\phi) \cdot k \cdot B \tag{2a}$$

where

$$k = r_{C:B} \cdot \frac{R^B_{max}}{K_B} \tag{2b}$$

with units of time^{-1}. Alternatively, for an electron acceptor that did not satisfy the condition $C \gg K_C$, but where $K_B \gg B$, we have

$$SMS(C) = -(1-\phi) \cdot k \cdot B \cdot \left(\frac{C}{K_C + C}\right) \tag{3a}$$

where

$$k = r_{C:B} \cdot \frac{R^{B,C}_{max}}{K_B}, \tag{3b}$$

also with units of time^{-1}.

The simplest expression, equation (2a), is the form first proposed by *Berner* [1964] for sulfate, and is similar to that used for oxygen by *Jahnke et al.* [1982a]. As noted, the condition for oxygen that $C \gg K_C$ is typical in the water column and in areas where the flux of organic matter to the sediments is very low, and is the reason we generally model water column remineralization of organic matter as in equation (2) (see phytoplankton and zooplankton equations (4.3.6) and (4.3.12) in chapter 4, and particulate and dissolved organic matter equations (5.4.10) and (5.5.2) in chapter 5). This form of the equation should not be used when oxygen drops down to values near or below 3 mmol m^{-3}.

Nitrate, whose K_C is estimated to be about 20 mmol m^{-3} of pore water [*Rabouille and Gaillard*, 1991b], generally does not satisfy the condition that $C \gg K_C$, and the appropriate equation to use in that case is (3). Indeed, in their study of the sediment diagenesis of organic matter, *Rabouille and Gaillard* [1991b] used (3) for both oxygen and nitrate.

As we noted when discussing dissolved organic matter in the water column in chapter 5, organic matter consists of a wide range of constituents that differ considerably in how labile they are. This has typically been dealt with using the *G-type kinetics* approach of *Berner* [1980]. The G-type kinetics model assumes that the organic matter in the sediments can be broken up into discrete decomposable components, G_i in mmol m^{-3} of dry sediment each of which has its own distinct rate constant. The total organic carbon is

$$G_T = \sum_{i=1}^{n} G_i \tag{4}$$

If we adopt kinetic expression (2a), the total source minus sink would be:

$$SMS(G_T) = -(1-\phi) \cdot \sum_{i=1}^{n} k_i \cdot G_i \tag{5}$$

An alternative to the discrete G-type kinetic approach, which requires determining a separate reaction rate constant for each component, is to treat the organic matter as consisting of a continuum of constituents with smoothly varying reactivities such as implied by the data in figure 6.4.1. The gamma distribution function proposed by *Boudreau and Ruddick* [1991] requires only two fitting parameters.

$$\Phi_{z=0}^{C} = -\frac{\varepsilon}{\delta_e} \cdot (C_0 - C_{BW}) \tag{6.2.7}$$

The sign convention in (6.2.7) is the same as for the sediments (i.e., z is positive downwards). The thickness of the effective stagnant film may be found by empirical relationships such as that given by *Shaw and Hanratty* [1977]:

$$\delta_e = \frac{\varepsilon \cdot Sc^{0.704}}{0.0889 \cdot u_*} \tag{6.2.8}$$

Empirical equations for the Schmidt number Sc are given in chapter 3. The friction velocity u_* can be estimated from velocity measurements in the water column above the sediments [e.g., *Wimbush*, 1976]. This equation predicts a diffusive sublayer thickness of 1 mm for a friction velocity of 0.1 cm s^{-1}, which would be obtained with a velocity above the boundary layer of order 2 to 3 cm s^{-1}. This is comparable to estimates such as those that were made based on the rate of dissolution of alabaster plates [*Santschi*, 1991].

An important issue with the stagnant film approach is that the roughness of the sediment surface will lead to variations in the structure of the boundary layer. A remarkable microprobe survey by *Gundersen and Jorgensen* [1990] of a 10 mm^2 area of sediments revealed a boundary layer ranging in thickness from less than 0.3 mm to more than 0.9 mm. The sediment area was increased by 88% due to the roughness, and that of the diffusive boundary layer in contact with the overlying water column was increased by 35%. The authors estimate that the roughness they observed would increase the flux across the diffusive boundary layer by 2.5 times that obtained from equation (6.2.7). See *Boudreau* [1997] for a further discussion of this and related issues.

Fortunately, the stagnant film is not rate-limiting for most pore water constituents, as we shall see, so we generally do not have to concern ourselves with the impact of this on pore water processes.

Solids

The processes that affect a particular chemical in the solid particles are: (1) the slow downward movement with respect to the sediment-water interface caused by deposition of fresh particles on the surface of the sediments, (2) bioturbation, and (3) remineralization reactions. The first two of these processes are generally treated as advection and diffusion. By analogy with (6.2.3) and (6.2.4) we have

$$DIFF(B) = \frac{\partial}{\partial z}\left((1-\phi) \cdot D_B \cdot \frac{\partial B}{\partial z}\right) \tag{6.2.9}$$

and

$$ADV(B) = \frac{\partial}{\partial z}((1-\phi) \cdot S \cdot B) \tag{6.2.10}$$

where B is, as above, the concentration of the solid phase of the chemical in units of mmol m^{-3} of solids, D_B is the bioturbation coefficient in $m^2\,s^{-1}$, and S is the sedimentation rate in m s^{-1}.

Remineralization reactions are complementary to those for the pore waters. For example, the constitutive equation for the reaction represented by (6.2.5) would be

$$SMS(B) = \frac{SMS(C)}{r_{C:B}} = -(1-\phi) \cdot R_{max}^{B,C} \cdot \left(\frac{B}{K_B + B}\right) \cdot \left(\frac{C}{K_C + C}\right) \tag{6.2.11}$$

The sign of the term is the same as in (6.2.5) because, for the case of B representing organic carbon and C representing oxygen (or, for example, nitrate in the denitrification zone of the sediments), both disappear with remineralization (see table 6.1.1). The product of this reaction is dissolved inorganic carbon.

Combining the constitutive equations for advection, diffusion, and sources minus sinks gives the final diagenetic equation for the solid component of the sediments:

$$\frac{\partial(1-\phi) \cdot B}{\partial t} = -ADV(B) + DIFF(B) + SMS(B) = -\frac{\partial}{\partial z}((1-\phi) \cdot S \cdot B) + \frac{\partial}{\partial z}\left((1-\phi) \cdot D_B \cdot \frac{\partial B}{\partial z}\right) - (1-\phi) \cdot R_{max}^{B,C} \cdot \left(\frac{B}{K_B + B}\right) \cdot \left(\frac{C}{K_C + C}\right) \tag{6.2.12}$$

The advection term in this equation is relatively straightforward, and the remineralization reaction term has been discussed previously. The one term in this equation that has yet to be discussed is bioturbation, which we cover next.

Bioturbation is the mixing of sediments by organisms. The actual processes by which this occurs are poorly understood and difficult to parameterize (see *Berner* [1980] and *Boudreau* [1997] for a full discussion), but their influence on the property distributions in the solid component of the sediments is beyond doubt. We gain insight into the impact of bioturbation by working our way through two tracers that illustrate various aspects of how this process affects sediment properties. The first bioturbation tracer that we consider is age calculated from radiocarbon measured in $CaCO_3$ particles. Figure 6.2.1 shows a vertical profile of age from a sediment core taken in the Indian Ocean [*Peng et al.*, 1977]. The solid line in figure 6.2.1 illustrates what the age profile would look like if there were

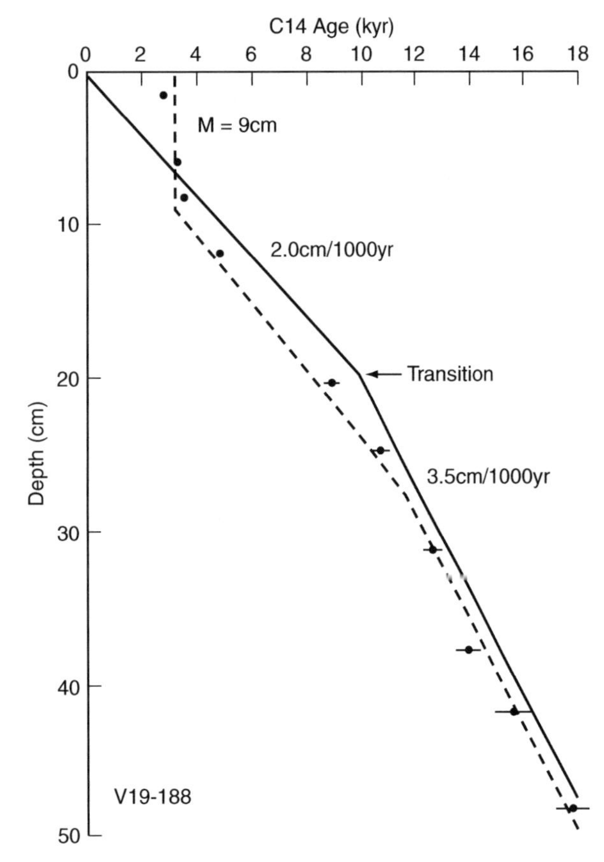

FIGURE 6.2.1: Radiocarbon age versus depth in a sediment core from the Indian Ocean [*Peng et al.*, 1977].

no mixing of the sediments. The age would be zero at the surface of the sediments (this is with respect to the surface ocean waters where the $CaCO_3$ is produced) and would increase linearly with depth, with a slope determined by sedimentation rate S. The higher sedimentation rate at depth in the core occurred during the last ice age.

The actual radiocarbon data shown in figure 6.2.1 differ considerably from the idealized curve. The shallowest sample in the sediments is almost 3000 years old due to upward mixing of old $CaCO_3$ from below. Furthermore, the measurements clearly indicate that bioturbation is not constant with depth. The large organisms that cause bioturbation do not penetrate more than 10 or 20 cm below the sediment-water interface (cf. figure 6.1.8). In the data shown in figure 6.2.1, it looks like bioturbation is confined to about the upper 9 cm, where the top three samples have almost the same age.

We can construct a simple model of the radiocarbon observations in figure 6.2.1 if we assume that a well-mixed layer of thickness L can represent the effect of bioturbation. By analogy with the discussion in panel 1.2.1, we represent radiocarbon by the product of the ratio $R^* = (\Delta^{14}C \times 10^{-3} + 1)$ times the appropriate $CaCO_3$ concentration. In a steady state, the loss of radiocarbon, which is due to decay within the mixed layer and flux out the bottom, must balance the input of radiocarbon through the top of the box. The input of $CaCO_3$ at the sediment water interface is $S \cdot [CaCO_3]$. We assume in this instance that there is no loss of $CaCO_3$ by dissolution in the sediment mixed layer. In such a case, the loss of $CaCO_3$ out the bottom of the mixed layer is equal to the input from above. This then gives us the following steady-state radiocarbon balance:

$$
\begin{aligned}
&R^*_{Surface\ Ocean} \cdot S \cdot [CaCO_3] \\
&= R^*_{\text{MixedLayer}} \cdot S \cdot [CaCO_3] + \lambda \cdot R^*_{MixedLayer} \cdot L \cdot [CaCO_3]
\end{aligned} \quad (6.2.13)
$$

i.e., input to the top of the bioturbated layer = output from the bottom of the bioturbated layer + radiocarbon decay over the thickness of the bioturbated layer. We assume that the $CaCO_3$ is formed at the surface of the ocean and thus has the radiocarbon concentration of surface water. Solving the above for the thickness of the mixed layer, and noting that $R^*_{Mixed\ Layer} = R^*_{Surface\ Ocean} \cdot \exp(-\lambda\tau)$, where τ is the radiocarbon age of the mixed layer, gives

$$
\begin{aligned}
L &= \frac{S}{\lambda}\left(\frac{R^*_{Surface\ Ocean} - R^*_{Mixed\ Layer}}{R^*_{Mixed\ Layer}}\right) \\
&= \frac{S}{\lambda}(\exp(\lambda\tau) - 1)
\end{aligned} \quad (6.2.14)
$$

In the case of figure 6.2.1, the sedimentation velocity is 2.3 cm kyr^{-1} and the age of the mixed layer is 3000 years, giving a mixed layer thickness of 8 cm. *Peng et al.* [1977] applied (6.2.14) to a large number of other cores for which they had surface radiocarbon age and sedimentation rate estimates. The mixed layers ranged from 6 to 18 cm in thickness.

The above analysis illustrates one important aspect of bioturbation: that it is largely confined to about the upper 10 to 20 cm of the sediments. The lead-210 activity profiles shown in figure 6.2.2 illustrate another important aspect of bioturbation: that the rate of mixing in the mixed layer is finite. Lead-210 is a radioisotope with a half-life of 22.3 years. It is part of the naturally occurring uranium-238 decay series. Production of lead-210 as part of this decay series occurs in both the water column and the sediments. The vertical line in figure 6.2.2 passes through the background lead-210 activity due to production in the sediments. The excess lead-210 in the upper part of the sediments has been stripped out of the water column by adsorption onto sinking particles. Without bioturbation, and in sediments with typical sedimentation rates of 1 cm kyr^{-1},

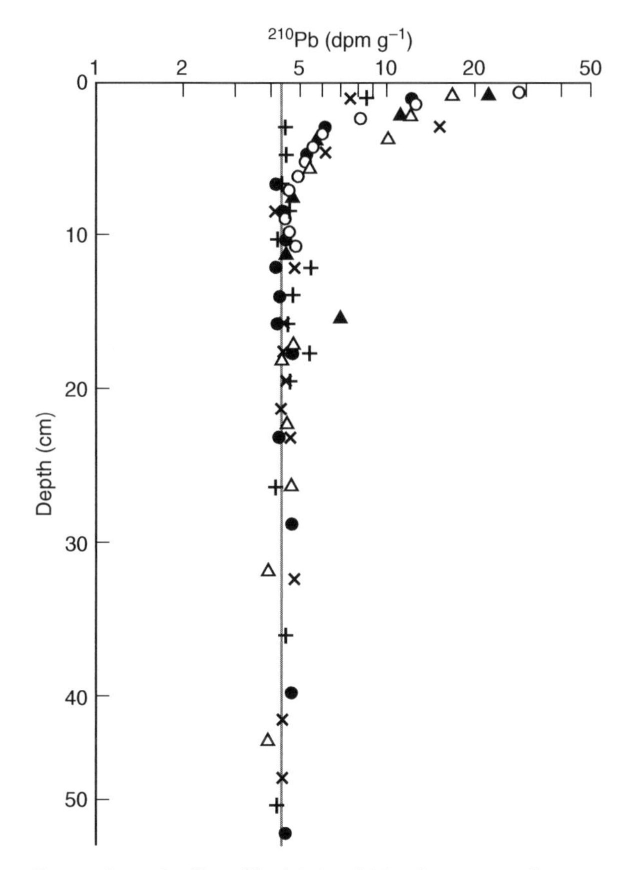

FIGURE 6.2.2: Profiles of lead-210 activities from cores taken on the Madeira Abyssal Plain [*Thomson et al.*, 1988]. The concentration units are activity, $A_{Pb-210} = \lambda_{Pb-210} \cdot [^{210}Pb]$, reported as disintegrations per minute per gram of solid sediment dpm g^{-1}).

this excess lead-210 would decay within the top mm or less of the sediments.

One can use excess lead-210 activity calculated from profiles such as those of figure 6.2.2 to estimate the bioturbation rate within the upper few cm of the sediments. We start with (6.2.1b) with *DIFF* defined by (6.2.9), then assume steady state and ignore the advection term, which is negligible on the timescale of the decay of lead-210. We add decay of lead-210 to obtain

$$0 = \frac{\partial}{\partial z}\left[(1-\phi)\cdot D_B \cdot \frac{\partial[^{210}Pb]}{\partial z}\right] - \lambda_{Pb-210}\cdot(1-\phi)\cdot[^{210}Pb]$$
$$= \frac{1}{\lambda_{Pb-210}}\cdot\frac{\partial}{\partial z}\left[(1-\phi)\cdot D_B\cdot\frac{\partial A_{Pb-210}}{\partial z}\right] - (1-\phi)\cdot A_{Pb-210} \quad (6.2.15)$$

where $[^{210}Pb]$ is the concentration of excess lead-210. We have made use of the definition of activity (see figure 6.2.2. caption) to arrive at the second form of the equation. The activity $A_{Pb\text{-}210}$ refers to the excess lead-210 to the right of the line in Figure 6.2.2. The effect of bioturbation is represented by the diffusion coefficient D_B. If porosity and bioturbation are constant with depth, the above equation simplifies to

$$0 = D_B \cdot \frac{\partial^2 A_{Pb\text{-}210}}{\partial z^2} - \lambda_{Pb\text{-}210}\cdot A_{Pb\text{-}210} \quad (6.2.16)$$

We take as boundary conditions that $A_{Pb\text{-}210} = (A_{Pb\text{-}210})_0$ at $z = 0$, and that $A_{Pb\text{-}210} \rightarrow 0$ as $z \rightarrow \infty$. This gives the solution

$$A_{Pb-210} = (A_{Pb-210})_0 \cdot \exp -\frac{z}{z^*}$$
$$z^* = \sqrt{\frac{D_B}{\lambda}} \quad (6.2.17)$$

This can be readily fit to the excess lead-210 calculated from profiles such as those in figure 6.2.2 in order to estimate the bioturbation rate D_B. The estimated bioturbation rate from the particular set of profiles shown in figure 6.2.2 is 80 to 180 cm^2 kyr^{-1} [*Thomson et al.*, 1988], which is equal to 2.5 to 5.7×10^{-13} m^2 s^{-1}, more than three orders of magnitude smaller than the effective molecular diffusivity in pore water.

The importance of the sedimentation rate S relative to the bioturbation can be evaluated using a nondimensional number, $S \cdot L/D_B$, where L is a scale-length over which the sedimentation is being compared with the bioturbation rate. This nondimensional number is directly analogous to a Peclet number, which indicates the relative contribution of advection versus diffusion in fluid media. We will use $L = z^*$. In section 6.3, we provide estimates for all three of these quantities at the typical oxygen-rich deep ocean site shown in figure 6.1.5. The values we specify or obtain by fitting the observations are $S = 1$ cm kyr^{-1}, $L = 0.9$ cm, and $D_B = 25$ cm^2 kyr^{-1}. Together, these give a very small Peclet number of 0.04, indicating that the bioturbation rate is of much greater importance than the sedimentation rate in determining the properties of the solids in the sediment.

Modeling approaches similar to that used for lead-210 have been used to estimate the bioturbation rate from a number of other radioisotopes as well as other tracers such as microtektites, which are small glass spherules of extraterrestrial origin. Figure 6.2.3 shows a compilation of such bioturbation rate estimates, and figure 6.2.4 shows estimates of the mixing layer depth over which bioturbation rates such as those of figure 6.2.3 are thought to apply. The bioturbation rate ranges over more than six orders of magnitude. Bioturbation rates are generally observed to increase with the supply of organic matter [e.g., *Pope et al.*, 1996], as might be expected, but this is far from enough to explain all the variation that is observed.

Part of the problem in characterizing the behavior of bioturbation is the remarkable observation that the bioturbation coefficient tends to be different for differ-

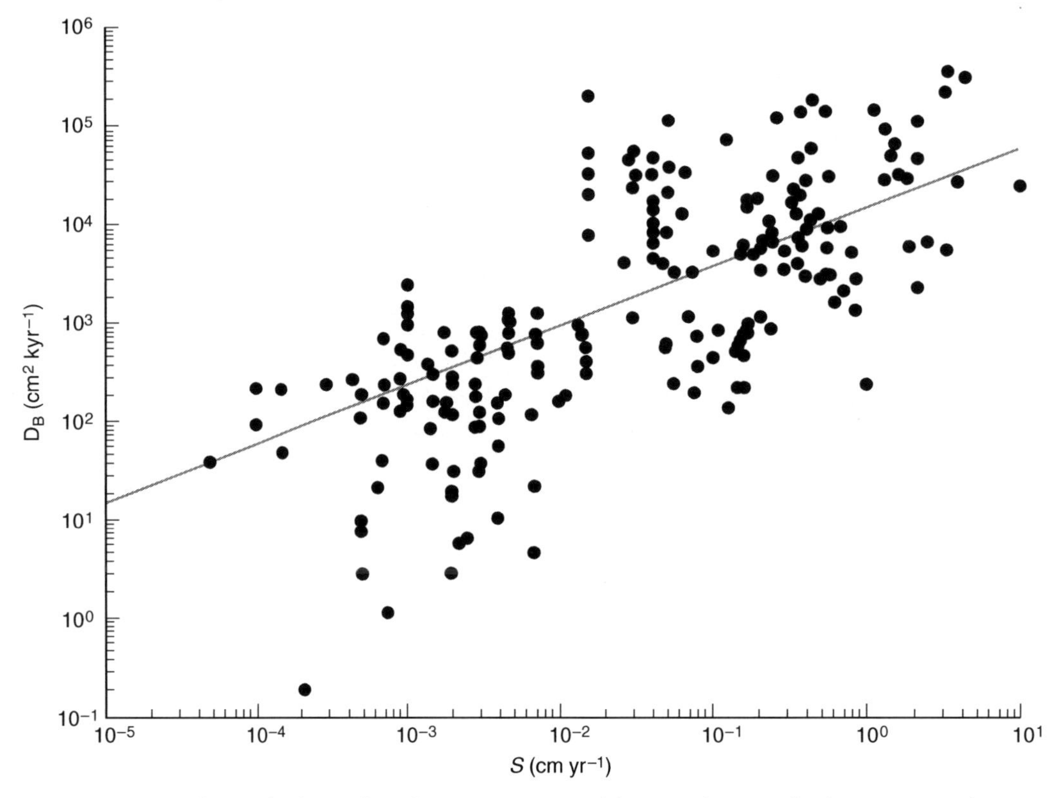

FIGURE 6.2.3: Compilation of sediment bioturbation rates estimated from a wide range of radioisotopes [*Boudreau*, 1997. This figure is from a study seeking correlations between the bioturbation rate and sedimentation rate [*Boudreau*, 1996]. The best fit line is $D_B = 15.7 \cdot S^{0.69}$. It explains only 25% of the variance.

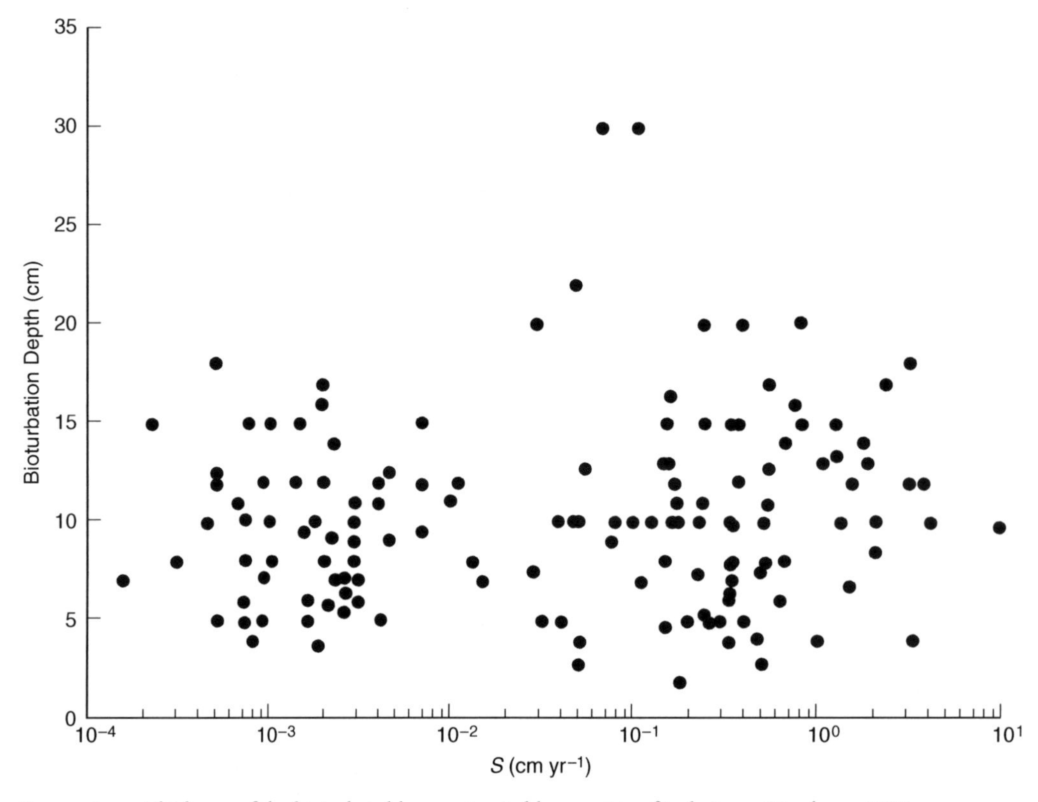

FIGURE 6.2.4: Thickness of the bioturbated layer estimated by a variety of techniques [*Boudreau*, 1997].

TABLE 6.2.2
Diagenetic equations for a dissolved constituent C in pore water, and a solid constituent B in the solid sediments.
C has units of mmol m^{-3} of pore water, and B has units of mmol m^{-3} of dry solids.

$$\frac{\partial\phi\cdot C}{\partial t}=-\frac{\partial}{\partial z}(\phi\cdot w_{PW}\cdot C)+\frac{\partial}{\partial z}\left(\phi\cdot\varepsilon_s\cdot\frac{\partial C}{\partial z}\right)-(1-\phi)\cdot r_{C:B}\cdot R_{\max}^{B,C}\cdot\left(\frac{B}{K_B+B}\right)\cdot\left(\frac{C}{K_C+C}\right)$$

$$\frac{\partial(1-\phi)\cdot B}{\partial t}=-\frac{\partial}{\partial z}((1-\phi)\cdot S\cdot B)+\frac{\partial}{\partial z}\left((1-\phi)\cdot D_B\cdot\frac{\partial B}{\partial z}\right)-(1-\phi)\cdot R_{\max}^{B,C}\cdot\left(\frac{B}{K_B+B}\right)\cdot\left(\frac{C}{K_C+C}\right)$$

ent tracers, even in the same core. Differences in the timescale represented by the different tracers probably explain part of this difference [e.g., *Smith et al.*, 1993]. For example, tracers that reside permanently in the sediments, such as microtektites, might experience occasional deep mixing episodes that a short-lived radioisotope would miss. Microtektites would thus tend to give deeper mixing layer depths but lower bioturbation rates, as in the summary of *Schink and Guinasso* [1977]. There also appear to be complex feeding effects in which, for example, particles containing isotopes produced by nuclear bomb tests (plutonium and cesium) are mixed more deeply and rapidly than particles containing lead-210 [e.g., *Cochran*, 1985]. The speculation is that benthic organisms are selectively moving certain types or sizes of particles. A particularly interesting result along these lines comes from the study of *Soetaert et al.* [1998] using oxygen, nitrate, ammonium, and organic carbon profiles. They found that the bioturbation rates estimated from these data were an order of magnitude larger than those estimated from lead-210 at the same location.

In concluding this section, we would mention one additional point about the nature of bioturbation, which is that it is carried out by actual organisms that do not behave like turbulent motion in the sea. A frequently observed phenomenon that results from the way that organisms feed in the sediments is the existence of subsurface maxima of surface source radioisotopes (e.g., *Smith and Shafer* [1984]; see also the 17 cm data point in figure 6.2.2).

The final diagenetic equations for both pore water and sediment constituents in dry solids are summarized in table 6.2.2. We turn now to applications of these equations to understanding the remineralization of organic matter.

6.3 Remineralization

In this section, we examine the remineralization of organic matter in sediments. We begin with a model of aerobic respiration in oxic sediments. This model shows that aerobic respiration is extremely rapid and will consume all labile particulate organic matter within the top 5 to 10 cm of the sediments as long as the supply of oxygen is sufficient. The only way there can be preservation of organic carbon in the presence of oxygen is if there is some component of the organic carbon that is highly resistant to aerobic degradation. This model also shows what controls the oxygen concentration in sediments and allows us to calculate a threshold organic matter rain rate above which the supply of oxygen is insufficient to remineralize all the labile organic matter and the sediments go anoxic.

The second topic we cover is anaerobic remineralization, which takes over when the sediments go anoxic. We discuss briefly how we estimate the contribution of the various anaerobic reactions to sediment diagenesis, and then show where in the ocean anaerobic remineralization occurs and discuss the relative contribution of the various anaerobic reactions to organic matter diagenesis in different regions of the ocean. We demonstrate that anaerobic remineralization tends to be less efficient than aerobic remineralization, but note that there are always enough oxidants available in sediments that there should be no organic matter preserved in the sediments. The fact that there is preservation indicates, as with aerobic sediments, that there is some component of the organic carbon that is highly resistant to anaerobic degradation. The efficiency of organic matter preservation relative to organic matter rain onto the sediments is found to increase with decreasing exposure to oxygen, indicating that aerobic remineralization is more efficient than anaerobic degradation.

Before turning in section 6.4 to a discussion of various processes that may contribute to making organic matter resistant to remineralization, we conclude this section with a discussion of the solubilization of particulate organic matter to produce dissolved organic matter that is able to escape into the water column. This loss of organic carbon from the sediments can be quite a large component of the organic carbon budget in some places.

Oxic Sediments

We begin with a model of particulate organic carbon *POC*, then a model of the oxygen distribution, both of which we apply to the deep ocean MANOP Site S observations shown in figure 6.1.5. The governing equation for solids in sediments is (6.2.12). We make the following assumptions in order to simplify the equation to a form that can be solved analytically:

1. The profiles shown in figure 6.1.5 represent a steady-state distribution, allowing us to eliminate the time derivative.

2. We ignore vertical variations in the porosity such as those shown in figure 6.1.3, enabling us to pull the $(1 - \phi)$ term out of the derivatives and cancel it out of the equation.

3. We ignore vertical variations in the bioturbation rate, which enables us to move the bioturbation rate out of the derivative. The observations in figure 6.1.5 and the model we derive below show that organic matter remineralization is confined to the top few cm of the sediments. The bioturbation depth is generally deeper than this (figure 6.2.4).

4. Based on the sediment Peclet number analysis we did earlier, we assume that in oxic sediments such as these, the bioturbation is dominant over the sedimentation rate, and we therefore leave this term out of the equation.

5. For the remineralization kinetics expression we use equation (2) of panel 6.2.3, divided by $r_{C:B}$. This *SMS* expression gives a remineralization that is independent of the concentration of the electron acceptor and linearly dependent on the concentration of the organic matter.

An additional complication that we must deal with in considering how to model the observations shown in figure 6.1.5 is that they strongly imply the existence of a more labile component confined mostly to the top 2 to 3 cm of the sediments, and a less labile or even totally refractory component below this. As long as there is oxygen present and as long as we assume that the organic carbon remineralization rate k is constant with depth, any reasonable solution of the diagenetic equations will remineralize essentially all the organic carbon in the sediments within the top few centimeters, which clearly is not consistent with the observations. We deal with this problem by assuming as a boundary condition that the organic matter deep in the sediments approaches a refractory nonzero asymptotic value POC_∞ which has negligible decomposition, and that the only portion of the particulate organic carbon that is susceptible to bacterial breakdown is thus $POC - POC_\infty$. This is equivalent to a two-component *G-type kinetics* model (see equation (5) of panel 6.2.3). The final expression for particulate organic carbon diagenesis is then

$$0 = D_B \cdot \frac{\partial^2 POC(z)}{\partial z^2} - k \cdot (POC(z) - POC_\infty) \tag{6.3.1}$$

where $k = R^{POC}_{max}/K_B$ has units of time^{-1}.

Equation (6.3.1) is a standard second-order partial differential equation with a solution of the form

$$POC(z) = \gamma_1 \cdot \exp(\lambda_1 z) + \gamma_2 \cdot \exp(\lambda_2 z) + \gamma_3$$
$$\text{where } \lambda_1, \lambda_2 = \pm\sqrt{\frac{k}{D_B}} \tag{6.3.2}$$

Note that λ_1 is positive, whereas λ_2 is negative. The λ's have units of m^{-1}. The integration constants γ_1, γ_2, and γ_3, which have units of mmol m^{-3} of solid, are determined by the specified boundary conditions. The usual practice for a boundary condition at the top of the sediments ($z = 0$) is to require that the flux of organic carbon downward from the sediment-water interface be equal the flux of organic matter arriving from above by sinking through the water column, $\Phi^{POC}_{z=0}$, i.e.,

$$\Phi^{POC}_{z=0} = -(1 - \phi) \cdot D_B \cdot \frac{\partial POC(z)}{\partial z} \tag{6.3.3}$$

The flux of organic matter at the sediment-water interface is often measured by sediment traps near the interface, or else inferred from the distribution of properties within the sediments. The other boundary condition we employ is $POC(z) \to POC_\infty$ as $z \to \infty$. This lower boundary conditions require that $\gamma_1 = 0$ and $\gamma_3 = POC_\infty$. Using boundary condition (6.3.3) to determine γ_2 gives the following solution we will employ in the analysis to follow:

$$POC(z) - POC_\infty = \gamma \cdot \exp(\lambda \cdot z) \tag{6.3.4a}$$

where

$$\gamma = -\frac{\Phi^{POC}_{z=0}/(1 - \phi)}{D_B \cdot \lambda} = \frac{\Phi^{POC}_{z=0}/(1 - \phi)}{\sqrt{k \cdot D_B}} \tag{6.3.4b}$$

Note that we have substituted $\gamma = \gamma_2$ and $\lambda = \lambda_2$. Since λ is negative, this defines a concentration profile that decreases exponentially with increasing depth in the sediments with an e-folding length-scale $z^* = -1/\lambda$, over which $POC(z) - POC_\infty$ decreases to 36.8% of its value at the sediment-water interface.

There are four parameters that define the shape of the organic carbon profile in the model we have derived: (1) the bioturbation rate D_B, (2) the organic remineralization rate constant k, (3) the total organic carbon flux at $z = 0$, $\Phi^{POC}_{z=0}$, and (4) the concentration of the refractory carbon POC_∞. The first two of these constants determine the e-folding length-scale $z^* = -1/\lambda$. The bioturbation rate is sometimes, but not always, estimated independently by mea-

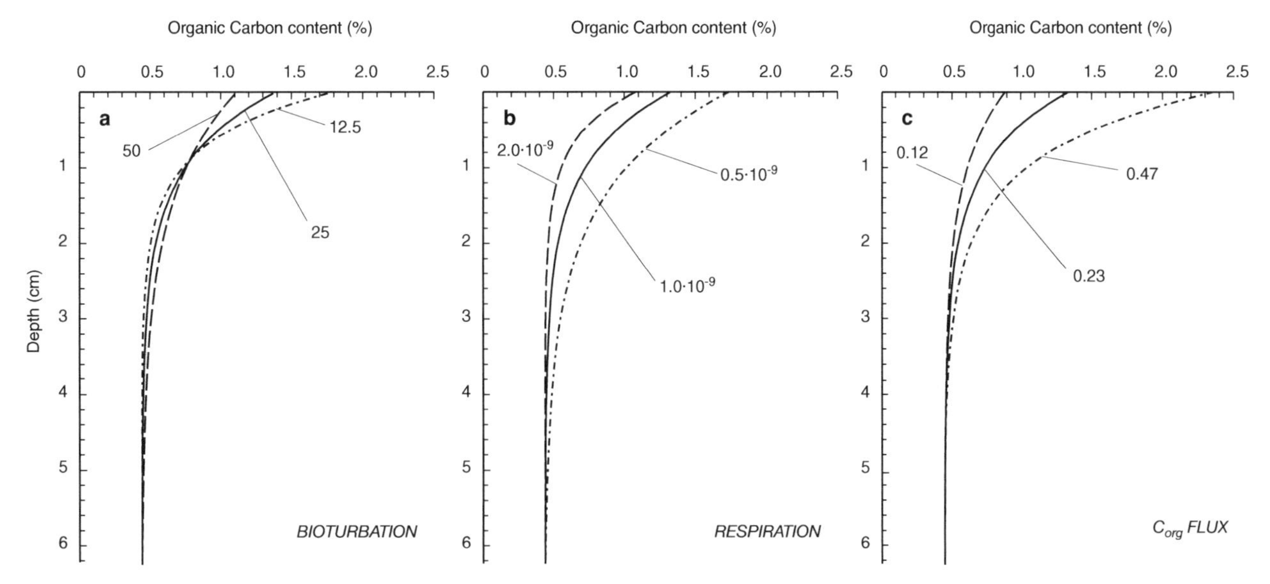

FIGURE 6.3.1: Influence of (a) bioturbation ($cm^2\ kyr^{-1}$), (b) respiration (s^{-1}), and (c) organic matter flux to the sediments ($mmol\ C\ m^{-2}\ d^{-1}$) on the profile of percent dry weight organic carbon in the top 10 cm. This is from the model fit to the deep ocean MANOP Site S data discussed in the text.

surements such as those discussed in the previous section. The remaining constants are usually determined by fitting the observed profiles. Note that the flux of organic matter out of the base of the bioturbated layer is equal to $(1-\phi) \cdot S \cdot POC(z)$. The sedimentation rate S is usually estimated directly using radiocarbon measurements or marker horizons that can be correlated with corresponding horizons in other cores that have been dated.

In our example of the deep ocean MANOP Site S, the organic carbon profile suggests that the concentration of the refractory carbon POC_∞ is 0.45 percent dry weight (see lower panel of figure 6.1.5). We estimate the remaining three constants by fitting the observations, including oxygen observations, using the model discussed in the next subsection. The bioturbation rate we obtain is 25 $cm^2\ kyr^{-1}$, which falls within the group of data below the line in figure 6.2.3. The organic carbon remineralization rate we obtain is $1.0 \times 10^{-9}\ s^{-1}$. This is equivalent to a remineralization timescale $\tau = 1/k$ of 32 years. The labile organic carbon component of the sediment is quite clearly far less labile than the particulate organic carbon that is remineralized in the water column, where typical sinking rates of order 50 $m\ d^{-1}$ and disappearance of ~90% of the organic carbon by 1000 m imply decay timescales of a few weeks at most. The total organic carbon flux that fits the data best is about 0.23 $mmol\ C\ m^{-2}\ d^{-1}$, of which 1.8% is refractory and the remaining 98.2% labile. All these values are comparable to those obtained for the same data set with a more sophisticated numerical model by *Rabouille and Gaillard* [1991a]. Now that we have an estimate of all the parameters, and using a sedimentation rate estimate of 1 $cm\ kyr^{-1}$ [cf. *Rabouille and Gaillard*, 1991a], we can check whether our assumption that the sediment Peclet number, $S \cdot L/D_B$, is small was correct. For L we use the e-folding length $1/\lambda$, for which we get 0.9 cm. The sediment Peclet number we get is 0.041, which supports our original assumption that it would be small.

The sediment POC remineralization rate of $0.982 \times 0.23 = 0.23$ $mmol\ C\ m^{-2}\ d^{-1}$ that we obtain is equivalent to an oxygen utilization rate (OUR) of 41 μmol $O_2\ m^{-3}\ yr^{-1}$ if the sediments draw oxygen from a 3000 m thick water layer above the sediments, and 122 μmol $O_2\ m^{-3}\ yr^{-1}$ if they draw oxygen from a 1000 m thick layer. In chapter 5 we estimated the deep ocean OUR as ~140 $\mu mol\ m^{-3}\ yr^{-1}$ below ~1500 m in the Pacific (table 5.2.1), so the organic carbon flux we infer from the deep ocean MANOP Site S can account for quite a substantial fraction of the oxygen utilization in the water column above it. We shall return to a further discussion of the contribution of sediment remineralization to the total deep ocean remineralization in section 6.5.

Figure 6.3.1 shows how the various parameters of the model control the shape of the organic carbon profile. The first thing to note in all these figures is that the labile organic carbon is gone below a depth of 5 to 10 cm. However, above this depth, there is a considerable range in behavior. Panel a of the figure shows that increasing the bioturbation rate leads to greater penetration of the labile organic carbon into the sediments, with lower surface concentrations and higher deep concentrations. Panel b shows that increasing the remineralization rate constant k leads to reduced labile organic carbon content at all depths. Panel c shows the influence of the organic carbon flux on the sediment organic carbon content. The higher this flux is, the higher the organic carbon content. The e-folding length-scale z^* is about 0.9 cm for the "standard" model, drops to about

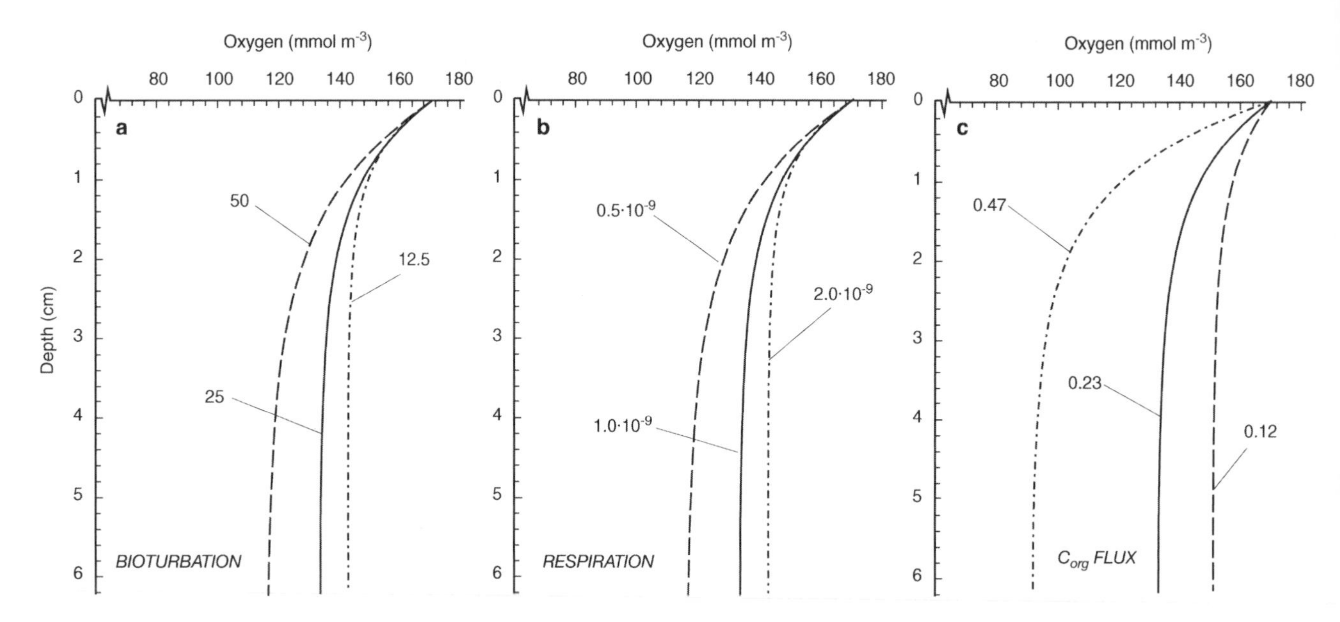

FIGURE 6.3.2: Influence of (a) bioturbation ($cm^2\ kyr^{-1}$), (b) respiration (s^{-1}), and (c) organic matter flux ($mmol\ C\ m^{-2}\ d^{-1}$) to the sediments on the oxygen profile in the top 10 cm. This is from the model fit to MANOP Site S data discussed in the text.

0.6 cm with either a halving of the bioturbation or a doubling of the respiration rate, and increases to about 1.3 cm with either a doubling of the bioturbation rate or halving of the respiration rate.

The ratio of the burial flux to the total delivery of organic carbon to the sediments, which is referred to as the *burial efficiency*, is 1.8% at the deep ocean MANOP Site S. This is less than a quarter of the global mean of $7.6 \pm 3.7\%$ that we obtain with the ocean carbon budget discussed later in section 6.5, which includes continental margin sediments that have much higher burial efficiencies. That the remineralization of organic matter should be so effective in the sediments is not immediately evident from examining the sediment profiles in figure 6.1.5, which have surface organic carbon concentrations that are only 3 to 4 times higher than the asymptotic organic carbon concentration deeper in the sediments. The reason that the refractory carbon concentration is so high, despite being only a small fraction of the total flux of organic carbon to the sediments, is because it keeps accumulating in the sediments until it is advected away by the sedimentation velocity S, which is only 1 cm per thousand years.

A final point we would emphasize about the burial efficiency in our model is that it depends entirely on our specification of a refractory pool of POC. Without this pool, all the POC would be remineralized. Our model thus tells us that burial of organic carbon in the presence of oxygen requires a pool of POC that is inaccessible to enzyme attack, but it does not tell us what processes make the organic carbon refractory and protect it from enzyme attack. We return to this problem in section 6.4.

We now develop an analytical model of oxygen parallel to our model of particulate organic carbon. The governing equation for oxygen is (6.2.6). We start with the following assumptions in addition to those made for the organic carbon model:

1. that advection of pore water is negligible relative to diffusivity, enabling us to eliminate the advection term, and
2. that the diffusivity ε_s is constant with depth, which enables us to take this out of the derivative.

This gives

$$0 = \phi \cdot \varepsilon_s \cdot \frac{\partial^2 [O_2]}{\partial z^2} - (1-\phi) \cdot r_{O_2:C} \cdot k \cdot (POC(z) - POC_\infty) \tag{6.3.5}$$

which can be rearranged to give

$$\frac{\partial^2 [O_2]}{\partial z^2} = k' \cdot (POC(z) - POC_\infty) \qquad k' = \frac{(1-\phi)}{\phi} \cdot \frac{r_{O_2:C}}{\varepsilon_s} \cdot k \tag{6.3.6}$$

Plugging in the solution (6.3.4) for $POC(z) - POC_\infty$ and integrating with respect to z gives:

$$[O_2] = k' \cdot \frac{\gamma}{\lambda^2} \cdot \exp(\lambda \cdot z) + \gamma_3 \cdot z + \gamma_4 \tag{6.3.7}$$

We use as a boundary condition at $z = 0$ that the concentration of oxygen at the sediment-water interface is $[O_2]_0$, which is equal to the bottom water concentration $[O_2]_{BW}$ corrected for the diffusive boundary layer effect. The profiles in figure 6.1.5 have been corrected for this already [cf. *Rabouille and Gaillard*, 1991a]. This gives that $\gamma_4 = [O_2]_0 - k' \cdot \gamma/\lambda^2$. Our second boundary condition is that the vertical gradient of oxygen approaches 0 as depth approaches infinity, which gives that $\gamma_3 = 0$. The final solution is then

$$[O_2] = [O_2]_0 - \left\{k' \cdot \frac{\gamma}{\lambda^2}\right\} \cdot [1 - \exp(\lambda \cdot z)] \quad (6.3.8a)$$

$$= [O_2]_0 - \{[O_2]_0 - [O_2]_\infty\} \cdot [1 - \exp(\lambda \cdot z)] \quad (6.3.8b)$$

The shape of the oxygen profile predicted by (6.3.8) is defined by seven parameters, the first three of which, D_B, k, $\Phi^{POC}_{z=0}$, also contribute to defining the *POC* profile, and have already been discussed above; and ϕ, $r_{O_2:C}$, ε_s, and the oxygen concentration at the sediment-water interface, $[O_2]_0$. We obtain the "standard" oxygen profile shown in figure 6.3.2 using a porosity of 0.85, the observed deep ocean oxygen to organic carbon remineralization ratio of 170:117 = 1.45 from *Anderson and Sarmiento* [1994], an oxygen diffusivity of $0.975 \times 10^{-9}\ m^2\ s^{-1}$, an oxygen concentration at the sediment-water interface of $170\ mmol\ m^{-3}$, and the parameters specified above. Note that the oxygen to organic carbon ratio of 150:106 = 1.42 from table 6.1.1 (cf. also [*Hammond, et al.*, 1996]) is very similar to the one we use, which comes from table 4.2.1. The oxygen diffusivity is obtained from table 6.2.1 at a temperature of 4°C, a dynamic viscosity of $1.61 \times 10^{-3}\ N\ m^{-2}$ given in the footnote of the table, and a porosity of 0.85.

The parameter sensitivity studies shown in figure 6.3.2 correspond to those shown in figure 6.3.1. They show that the asymptotic value of oxygen in the deep sediments is determined essentially by how much of the labile organic carbon escapes the rapid remineralization very near the sediment-water interface and penetrates into the sediments. Thus, whenever organic carbon is high below 1 cm, the oxygen is low, and vice versa. An additional parameter in the oxygen equation is the oxygen at the sediment-water interface, but this is just a constant added onto the depth-dependent term in the equation. Thus, in this model, changing the oxygen at the sediment-water interface shifts the concentrations by exactly the same amount at all depths, while preserving the shape of the oxygen profile. In fitting the oxygen and organic carbon data of figure 6.1.5, it was found that the surface concentration of organic carbon and the difference in oxygen concentration between the sediment-water interface and the asymptotic value at depth in the sediments were the two key constraints in estimating the parameters.

To summarize, the main insights that arise from our model analysis of open-ocean oxygen-rich sediments such as those at MANOP Site S are the following:

1. All labile organic carbon appears to be remineralized within the top few centimeters, with a timescale of a few decades.

2. The depth of penetration of labile organic carbon in the upper 5 to 10 cm increases with increasing bioturbation, decreasing respiration rate, and increasing organic carbon flux. The depth penetration is insensitive to the sedimentation rate because of the slowness of this rate relative to bioturbation.

3. The asymptotic oxygen content is directly related to the bottom water concentration and inversely related to the depth of penetration of labile organic carbon. The deeper the penetration of labile organic carbon, the lower the deep oxygen content, and vice versa.

4. The only way that oxic sediments such as these can preserve any organic carbon for burial is if there is a fraction of the organic carbon that is highly refractory or inaccessible to enzymes.

One can infer from equation (6.3.8) and figure 6.3.2 that the factors that might contribute to oxygen depletion include low bottom water oxygen concentration, and any processes that increase the labile organic carbon concentration in the deeper part of the sediments, such as a high organic matter flux to the sediments and increased bioturbation or reduced organic matter oxidation rates [cf. *Emerson*, 1985]. We can estimate an upper limit to the amount of oxygen that can flux into the sediments using an equation obtained by combining Fick's first law, equation (6.2.2), with our oxygen solution (6.3.8b):

$$\Phi^{O_2}_{z=0} = -\phi \cdot \varepsilon_s \cdot \lambda \cdot ([O_2]_0 - [O_2]_\infty) \quad (6.3.9)$$

Assuming a mean deep ocean oxygen concentration of $168\ mmol\ m^{-3}$, ignoring the influence of the stagnant film, assuming an asymptotic deep sediment oxygen concentration of $0\ mmol\ m^{-1}$ in order to get a maximum estimate of the flux, and using the ε_s and ϕ given above and for λ the value of $-1/(0.9\ cm)$ we obtained from the MANOP Site S data, we obtain an oxygen flux of $1.34\ mmol\ m^{-2}\ d^{-1}$. This is sufficient to remineralize $0.92\ mmol\ m^{-2}\ d^{-1}$ of organic carbon.

How does this "upper limit" or "threshold" oxygen flux and its equivalent carbon remineralization rate compare with actual remineralization rates in the sediments and the relative contribution of oxygen to them? We have seen that MANOP Site S has an organic carbon remineralization rate of $0.23\ mmol\ m^{-2}\ d^{-1}$, well below this threshold. Figure 6.3.3a shows data from the continental margins of northwest Mexico and the state of Washington in the United States that mostly exceed this threshold. There is only one measurement below the threshold carbon remineralization rate, and it has a significant amount of anaerobic remineralization

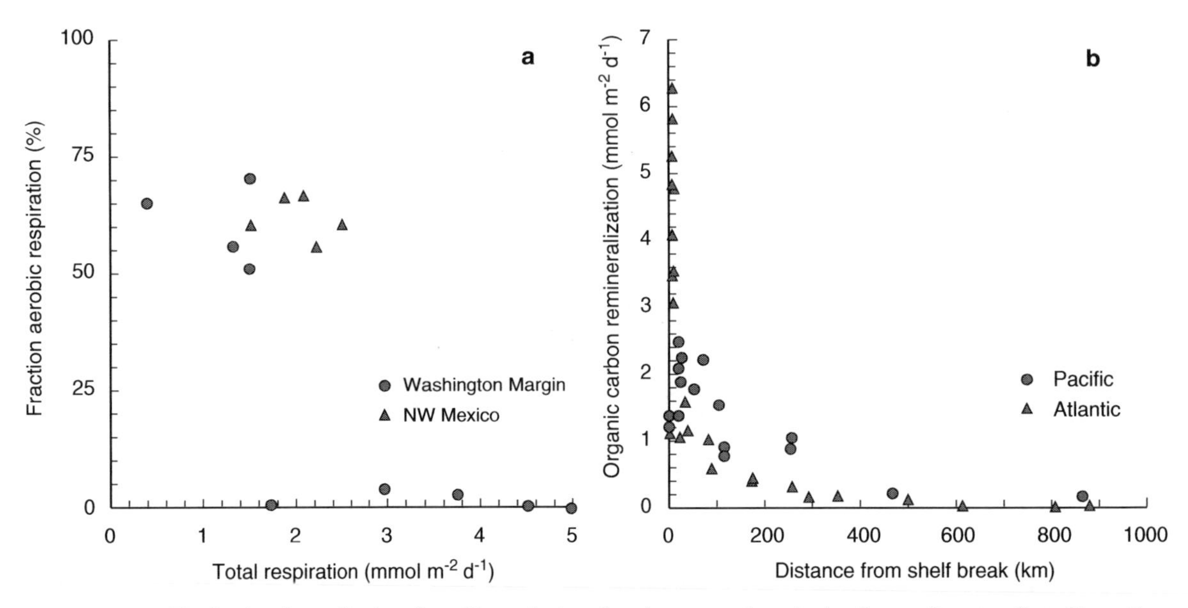

FIGURE 6.3.3: (a) The fractional contribution of aerobic respiration plotted versus total respiration from sediments collected in sections perpendicular to the continental margin off northwest Mexico and the Washington State. The data are from the study of *Hartnett and Devol* [2003] and include only measurements in water depths below 200 m, that is to say off the shelf, and where the bottom water oxygen is greater than 5 mmol m^{-3}. (b) Organic carbon remineralization versus distance from the shelf break, from measurements made in the Atlantic along a section approximately between Woods Hole and Bermuda, and in the Pacific along a section extending out approximately from Monterrey. The data are from the compilation of *Jahnke and Shimmield* [1995].

taking place. However, this location has low bottom water oxygen, which can explain most of the discrepancy. This figure shows an interesting feature of several data points with finite oxygen in the water column, but no aerobic respiration. All the oxygen that is going into the sediments is being used to oxidize the reduced products of anaerobic oxidation reactions. Figure 6.3.3b shows data from two sections perpendicular to the continental margin off the Atlantic and Pacific coasts of North America. Most sediments within ~50 to 100 km of the shelf break exceed the threshold oxygen flux we calculated. This boundary where organic carbon remineralization exceeds the upper limit oxygen availability would be pushed much further into the open ocean if we used a lower, more realistic, bottom water oxygen concentration. Thus, we see that it is only in the open ocean that aerobic respiration is likely to be dominant, and that near the continental margins, oxygen is unable to keep up with the organic matter flux and anaerobic remineralization begins to take over.

Anoxic Sediments

What happens if oxygen runs out in the sediments while there is still abundant labile organic carbon available, such as appears to be the case in the Patton Escarpment profile at the edge of the continental slope off California (see figure 6.1.5)? And how does the depletion of oxygen affect the fraction of organic carbon that is remineralized? The presumption is that if oxygen runs out, new consortia of bacteria will continue to remineralize organic matter via the anaerobic reactions in table 6.1.1. The main goal of this subsection is to describe the relative contribution of each of the possible remineralization reactions.

We begin by discussing the measurements that are used to estimate the vertically integrated carbon remineralization rate of each reaction in table 6.1.1 [cf. *Canfield*, 1993]. The vertically integrated remineralization rate has units of mmol m^{-2} d^{-1} and is exactly equivalent to a flux. The remineralization rate can be determined either directly, for example, by sediment incubations (e.g., the *Hartnett and Devol* [2003] sulfate utilization experiments), or indirectly from the fluxes of the reactants into or products out of the zone in the sediments where a given remineralization reaction is taking place. Fluxes are determined by measuring vertical pore water concentration profiles and using Fick's first law to calculate the flux from the pore water gradients, or by putting benthic chambers over the sediments and measuring the concentration changes versus time of the water that overlies the sediments in the chamber [cf. *Reimers et al.*, 1992; *Hartnett and Devol*, 2003]. The fluxes for any given compound can be converted to carbon remineralization rate using the stoichiometric ratios given in table 6.1.1. These methods should all agree unless pore water advection or bioirrigation (the exchange of pore water by animals) is a significant con-

TABLE 6.3.1
The relative contribution of various oxidants to organic matter remineralization at representative sites.
Based on a similar table by Middelburg et al. *[1993], with the addition of Central California data from* Reimers et al. *[1992] and Washington State and northwest Mexico data from* Hartnett and Devol *[2003]. The latter three data sets have been averaged, with sediment cores grouped by those shallower than 1500 m and deeper than or equal to 1500 m.*

	Percentage of Total				
Site	O_2	NO_3^-	*Mn & Fe*	SO_4^{2-}	CH_4
	Coastal Sediments				
Danish coast	40	3	—	57	—
Cape Lookout Bight	—	—	—	68	32
Saanich Inlet	—	—	—	76	24
	Continental Slope and Rise (<1500 m)				
Central California (3 cores)	29	57	0.8	14	—
Washington State (19 cores)[a]	10	35	—	55	—
NW Mexico (19 cores)[a]	0	45	—	55	—
	(≥1500 m)				
Central California (5 cores)	71	17	1.8	10	—
Washington State (4 cores)[a]	62	26	—	12	—
NW Mexico (5 cores)[a]	60	30	—	9	—
	(Other)				
Indian Ocean	87	12	0.3	0.6	—
Hatteras Cont. Rise	76	8	1.7	14	—
Bermuda Rise	78	12	1.4	9	—
Savu Basin	61	25	6.5	7.2	—
	Deep Sea				
Manop H	99	0.8	0.4	—	—
Manop M	91	6.9	0.4	—	—
Hatteras Abyssal Plain	96	4	—	—	—

[a]In their study of Washington State and Northwest Mexico, *Hartnett and Devol* [2003] assumed that iron and manganese reduction are negligible.

tributor, in which case benthic chamber measurements, which include the effect of bioirrigation, should exceed pore water gradient flux calculations, which they do not.

Table 6.3.1 shows the relative contribution of various oxidants to organic matter remineralization at a range of representative locations that provide a regional breakdown of the global totals in table 6.1.3. There is abundant evidence from these data and from model studies that remineralization of organic matter continues to occur even after oxygen runs out [e.g., *Dhakar and Burdige*, 1996; *Jahnke et al.*, 1982a, 1982b, 1989; *Haeckel, et al.*, 2001; *Soetaert et al.*, 1998; *Bender and Heggie*, 1984; *Reimers et al.*, 1992]. However, over the vast reaches of the deep ocean, where oxygen levels are generally high and the flux of organic carbon to the sediments is relatively modest, aerobic respiration generally accounts for >90% of the remineralization, with denitrification accounting for most of the remainder (table 6.3.1 [cf. *Jahnke et al.*, 1982a; *Bender and Heggie*, 1984; *Rabouille and Gaillard*, 1991b]). It is only in coastal and continental margin sediments that anaerobic reactions contribute significantly to the total remineralization. Coastal sediments are the only regions where early diagenesis is sometimes found to proceed past sulfate reduction to include a substantial amount of methanogenesis, particularly if oxygen is absent as in the Saanich Inlet observations off Victoria Island in Canada. Interestingly, Cape Lookout Bight underlies oxygen-containing waters, but the organic matter flux here is extremely high and overwhelms the

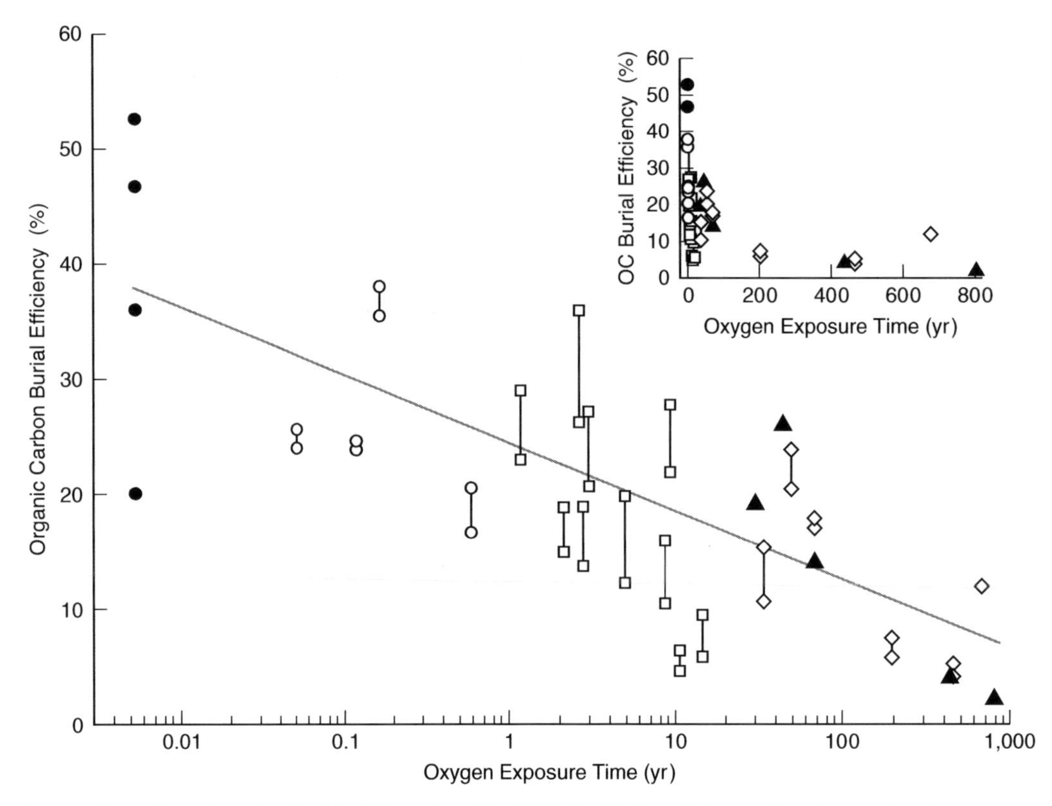

FIGURE 6.3.4: Organic carbon burial efficiency (total burial divided by burial plus total oxidation) plotted versus oxygen exposure time. Oxygen exposure is calculated by dividing the thickness of the oxic layer by the sedimentation rate. The plot is taken from *Hartnett et al.* [1998] and includes data from the Mexican slope in regions where water column oxygen is deficient (filled circles, plotted at an arbitrary oxygen exposure time of 0.003 yr); Mexican shelf (open circles); Washington shelf and upper slope (squares); Washington lower slope (diamonds); and California margin (triangles). The inset shows the same plot with a linear-linear scale, which shows the exponential nature of the relationship.

oxygen supply. We note that this table does not include results from studies such as those of *Aller et al.* [1996] and *Hulth et al.* [1999], which have shown that iron and manganese can be the dominant oxidants in regions where physical reworking of sediments is substantial (see section 6.1).

The continental slope and rise represent a transition between coastal and deep ocean sediments, with a negligible contribution from methanogenesis, but a substantial contribution from denitrification and sulfate reduction (table 6.3.1). Manganese and iron reduction contribute only a few percent (note that Mn and Fe reduction were not measured at the Washington State and northwestern Mexico margin sites). The most notable feature of these measurements is the major transition at about 1500 m, with aerobic respiration playing a secondary role above this depth, and being the primary oxidant below it.

How efficient are anaerobic processes in remineralizing organic matter? Figure 6.3.4 summarizes a large number of estimates of organic carbon burial efficiency from the Washington State margin, the northwestern Mexican margin, and the California margin (these are the same sites summarized in table 6.3.1). One minus the burial efficiency is the fraction of organic matter that is remineralized, i.e., small numbers on the ordinate mean high remineralization efficiencies. The abscissa of figure 6.3.4 is the *oxygen exposure time*, a measure of the potential role of aerobic respiration proposed by *Hartnett et al.* [1998], which they calculate by dividing the thickness of the layer within which oxygen is found (which is generally only a few millimeters in the sediments studied by *Hartnett et al.* [1998]) by the sedimentation rate. This figure reveals that the remineralization efficiency for most of these margins sediments is still high, of order 70% or greater, although clearly not as high as the MANOP Site S profile in the deep ocean (~98%). Secondly, there is a strong tendency for sediments with shorter oxygen exposure times to have lower remineralization efficiencies. However, the data are quite noisy, and it is very likely that other factors are at play as well [cf. *Reimers*, 1998].

The organic matter raining onto shelf and slope sediments has experienced less water column reminer-

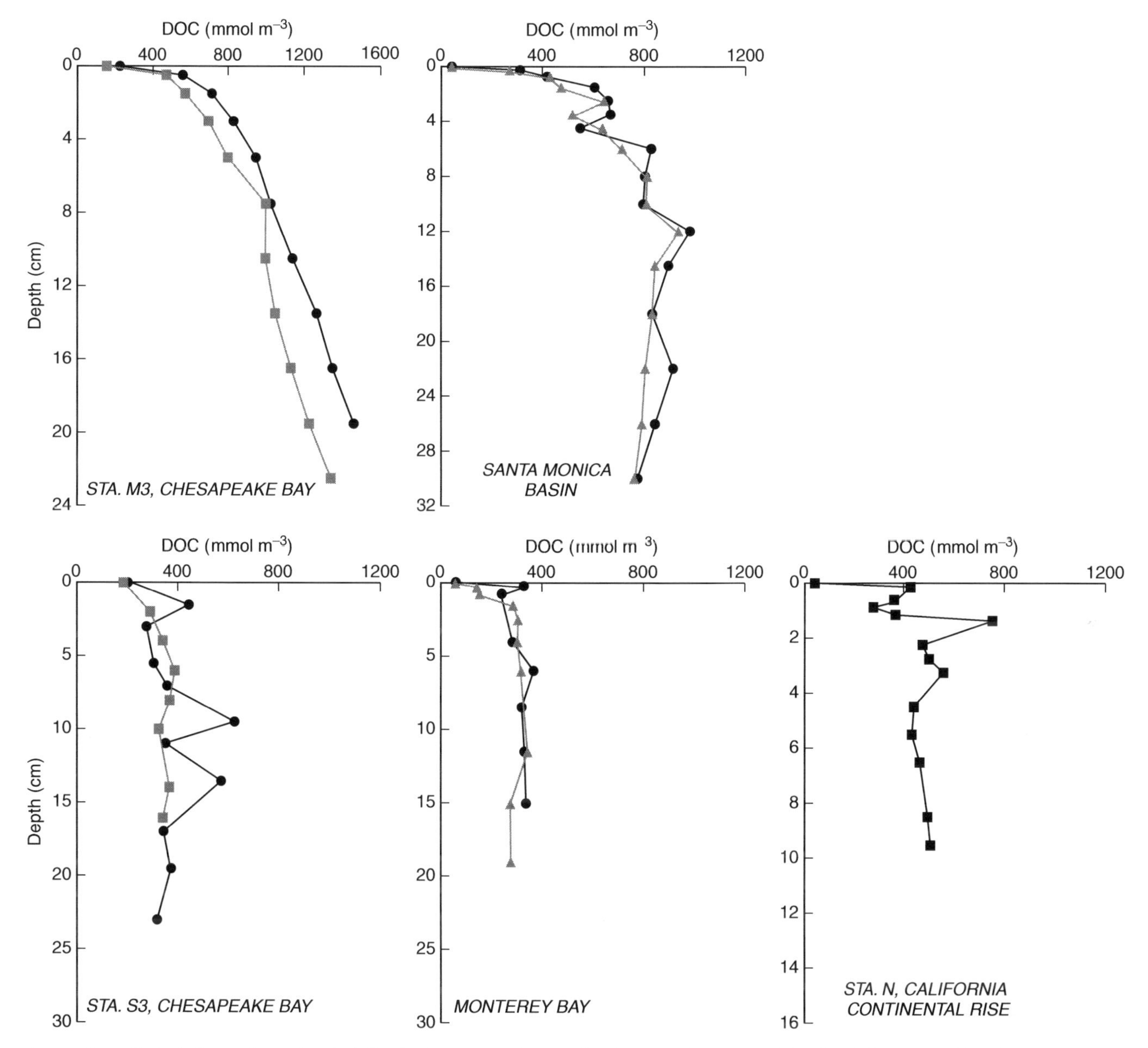

FIGURE 6.3.5: Typical observations of vertical dissolved organic carbon profiles in anoxic (upper panels) and oxic (lower panels) sediments. Taken from *Burdige* [2002].

alization than that which rains out onto deep ocean sediments, and should therefore have a much higher labile organic carbon content. Therefore, one would expect in principle that shelf sediments would have lower burial efficiencies than deep ocean sediments, and yet the opposite is the case. This clearly is not for lack of oxidants, as there is generally plenty of sulfate in pore water (see problem 6.12), and if that runs out, the methane fermentation reaction can take over. Understanding what controls remineralization and preservation efficiencies, and how to explain the contrasts that we see in figure 6.3.4, is a central focus of our discussion in section 6.4. First, however, we conclude this subsection with a brief discussion of dissolved organic carbon, which is an important intermediate in the remineralization of particulate organic carbon and can represent a major loss term of organic carbon from the sediments to the water column.

Dissolved Organic Carbon

We saw in chapter 5 that typical deep ocean *DOC* concentrations are $\sim 40\,\mathrm{mmol\,m^{-3}}$. The maximum *DOC* content of the anoxic sediments shown in the upper panels of figure 6.3.5 are 20 to 30 times as high as this. Oxic sediments such as those shown in the lower panels of figure 6.3.5 are about 10 times as high. This implies that *DOC* is being produced in the sediments

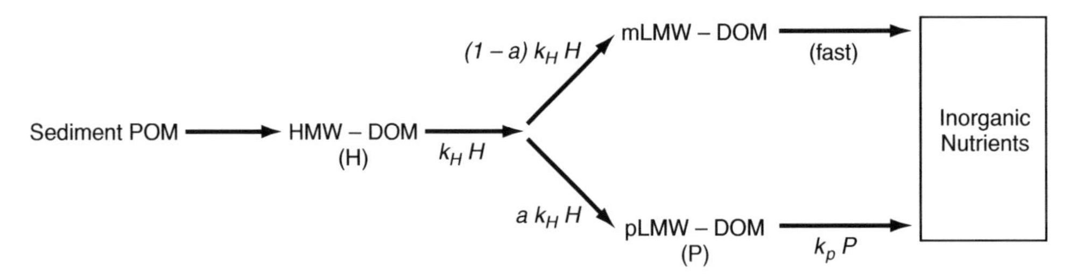

FIGURE 6.3.6: A conceptual model for the remineralization of dissolved organic matter, taken from *Burdige* [2002]. See discussion in text. High molecular weight (HMW) dissolved organic matter includes biological polymers such as dissolved proteins and polysaccharides. Monomeric low molecular weight (mLMW) dissolved organic matter includes acetate, other small organic acids, and individual amino acids. Polymeric LMW dissolved organic matter (pLMW) consists of humic substances.

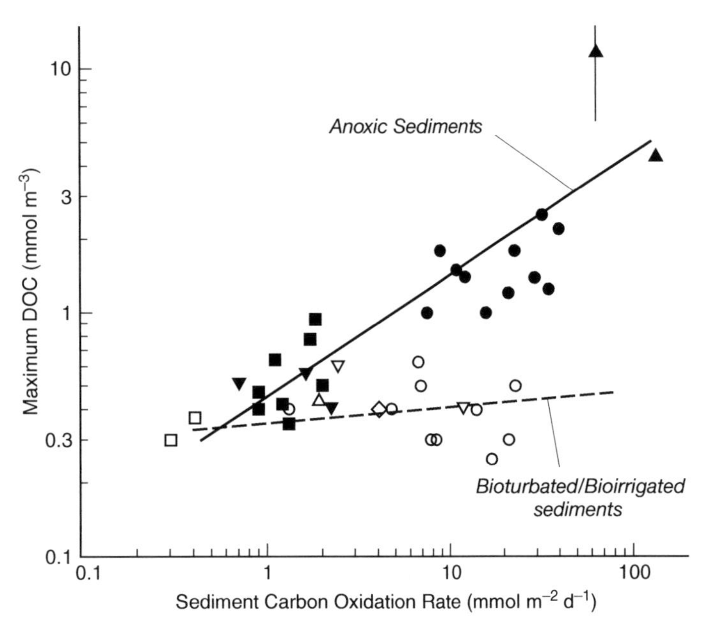

FIGURE 6.3.7: The maximum *DOC* concentration in the top 20 to 30 cm of sediments plotted versus the total oxidation rate in sediments, R_{ox}. The closed symbols represent data from anoxic sediments, while the open symbols are from sediments that have oxygen in the overlying water and experience bioturbation and, in some cases, bioirrigation. Taken from *Burdige* [2002].

and that it must be fluxing out of them into the water column. Estimates of the *DOC* flux in sediments with extremely high organic matter oxidation rates in excess of $\sim 5\,\mathrm{mmol\,m^{-2}\,d^{-1}}$ (which are found on the shelf, but are uncommon in slope, rise, and deep ocean environments; see figure 6.3.3b), show that the *DOC* flux is only about $1.6 \pm 0.2\%$ of the oxidation rate [cf. *Alperin et al.*, 1999]. However, the *DOC* flux does not drop off linearly with the oxidation rate and can become a major contributor to the sediment organic carbon budget in regions where the oxidation rate is lower. For example, at one site at 3560 m in the eastern North Atlantic that had an organic carbon oxidation rate of $0.8\,\mathrm{mmol\,m^{-2}\,d^{-1}}$, *Papadimitriou et al.* [2002] found that the *DOC* flux out of the sediments was 57% of the organic carbon oxidation rate. These fluxes are large enough that they should have a substantial impact on the oceanic *DOC* content. However, there is no evidence in the deep ocean *DOC* content of a significant deep ocean source (see chapter 5), which suggests that the *DOC* coming out of the sediments into the water column may be consumed fairly rapidly near the sediment-water interface.

Burdige et al. [1999] did an empirical fit of the ratio of *DOC* flux, Φ^{DOC}, to organic total carbon oxidation rate in the sediments, R_{ox}, as a function of the organic carbon oxidation rate, using data compiled from coastal and continental margin sediments. He obtained $\Phi^{DOC}/R_{ox} = 0.36 \cdot (R_{ox})^{-0.71}$ (R_{ox} in units of $\mathrm{mmol\,m^{-2}\,d^{-1}}$) with a rather low but statistically significant r^2 of 0.25. The data compilation they used does not include any deep ocean measurements away from continental margins and should not be extrapolated to such low-R_{ox} regions.

Where does the high *DOC* content of sediments come from? As we discussed in chapter 5, the remineralization of particulate organic matter by bacteria requires that it first be split into molecules that are small enough for them to ingest. Bacteria do this using extracellular enzymes that break down the particulate organic matter by a series of hydrolysis and/or fermentation steps involving dissolved organic matter as an intermediate [*Amon and Dacy*, 1996]. Figure 6.3.6 shows a conceptual model by *Burdige* [2001] of how this process might work in sediments. The model is based on the size-reactivity continuum model of *Amon and Benner* [1996] and ideas of *Alperin et al.* [1994]. The initial hydrolysis and/or fermentation step in this model produces high molecular weight (HMW) dissolved organic matter, which is rapidly broken down into low molecular weight (LMW) dissolved organic matter, some of which is very labile and is consumed rapidly by bacteria (mLMW), and some of which is refractory and consumed slowly (pLMW, which consists of humic substances [cf. *Burdige*, 2002]). The observed increase in *DOC* concentration with depth in the sediments is correlated with humic-like fluorescence [cf. *Burdige*, 2002], indicating that the increase is due primarily to an accumulation of pLMW.

The profiles in figure 6.3.5 and more detailed analysis in figure 6.3.7 show that there is a large difference in the maximum sedimentary *DOC* concentration between non-bioturbated anoxic sediments (upper panels of figure 6.3.5) and sediments that are oxygenated and bioturbated and/or bioirrigated (lower panels of figure 6.3.5). Furthermore, the maximum *DOC* concentration shows a tendency to increase with the vertically integrated organic carbon oxidation rate in anoxic sediments, but not in oxygenated sediments (figure 6.3.7; cf. *Burdige* [2002] for a discussion of the possibility of measurement artifacts). The difference in behavior between anoxic and oxygenated sediments suggests that there is some fraction of the products derived from HMW dissolved organic matter that is resistant to anaerobic remineralization but not to aerobic remineralization (cf. *Middelburg et al.* [1993] and discussion in following subsection). Other sediment processes that can influence the DOC concentration are bioirrigation (the pumping of pore water by macrofauna) and bioturbation, with larger values of these giving lower *DOC* concentrations [*Burdige*, 2002].

6.4 Burial

The simple models and data analysis that we discussed in section 6.3 have given us an understanding of how remineralization occurs in sediments, and what determines the vertical profiles of particulate and dissolved organic carbon and dissolved oxidants in the sediments. However, except for the need in our model to specify a C_∞, we still do not have an explanation for what determines the burial of organic matter in sediments. We will now discuss three major factors that are thought to play a central role: (1) the influence of organic matter composition (i.e., the substrate), in particular evidence from data such as those shown in figure 6.4.1 for organic matter becoming more recalcitrant as it degrades; (2) the implications from various types of observations that diagenesis in sediments depends on the oxidant, in particular observations suggesting that organic matter in sediments exposed to oxygen are more efficiently degraded that those that experience only anoxia (see figure 6.3.4); and (3) the possible role of inorganic minerals in protecting organic matter from degradation. The synthesis at the end of the section relates these three factors to others such as primary production and sedimentation rate that are often identified as the controlling variables for organic carbon preservation.

The Substrate

One of the iconic images of sediment diagenesis is the log-log plot shown in figure 6.4.1 of the reaction rate constant k versus the timescale over which the rate constant was determined [*Hedges and Keil*, 1995; cf. also *Emerson and Hedges*, 1988]. These data indicate that the remineralization rate constant decreases with increasing timescale. The implication is that organic matter consists of a range of constituents, with more labile components disappearing during early stages of diagenesis, and increasingly recalcitrant components being left behind as more and more of the organic matter is degraded [cf. *Middelburg et al.*, 1993; *Hedges et al.*, 1999a; *Dauwe et al.*, 1999].

An important issue raised by the data of figure 6.4.1 is that it would not in general be appropriate to characterize organic matter as consisting of a single constituent that can be modeled with a single degradation rate constant. We dealt with this issue at MANOP Site S by assuming that the organic matter there consisted of two

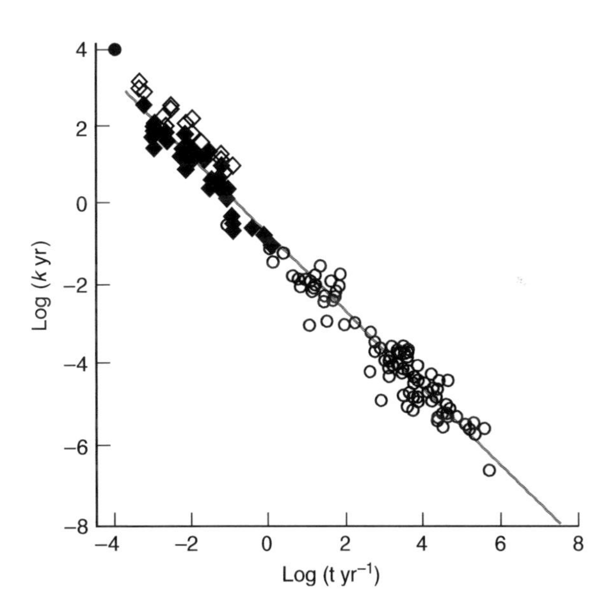

FIGURE 6.4.1: A log-log plot showing various estimates of the remineralization rate constant for organic matter, k, versus the time interval of the experimental measurement, t. The closed diamonds are from laboratory studies, the open diamonds from water column measurements, the open circles from "well-dated" sediment cores, and the closed circle from zooplankton gut passage. The figure is taken from *Hedges and Keil* [1995], who based it on *Middelburg et al.* [1993], who in turn based it on *Middelburg* [1989]. The sediment core calculations were made using the equation

$$k = -\frac{1}{\Delta z} \cdot S \cdot \ln(B_2/B_1)$$

$$\Delta z = z_2 - z_1$$

This assumes that bioturbation is negligible, which leads to an underestimate of the rate constant for sediments that are overlain by oxygen and experience bioturbation (see problem 6.5). The line in the plot represents a linear fit that corresponds to $k = 0.21 \cdot t^{-0.985}$.

constituents. The G-type kinetics model introduced in panel 6.2.3, equations (4) and (5), allows for any number of discrete constituents, with the appropriate number depending primarily on the time scale being considered: 2 or 3 for decadal time scales such as at MANOP Site S, up to about 8 for time scales of 10^4 to 10^7 years [cf. *Middelburg*, 1989]. Despite numerous efforts over many years [e.g., *Middelburg et al.*, 1993; *Henrichs*, 1992], only recently has there begun to be some success in linking specific types of organic matter with different degradation rate constants [e.g., *Dauwe et al.*, 1999].

An alternative to the discrete G-type kinetic approach, which requires determining a separate reaction rate constant for each component, is to treat the organic matter as consisting of a continuum of constituents with smoothly varying reactivities such as implied by the data in figure 6.4.1. The gamma distribution function proposed by *Boudreau and Ruddick* [1991] which we mention in panel 6.2.3 requires only two fitting parameters, and is consistent with the data of figure 6.4.1 if one assumes that the slope of the data is exactly equal to -1, which is close to what it actually is (cf. figure 6.4.1 caption; *Middelburg et al.* [1993]).

The data summarized in figure 6.4.1 imply that given enough time, all organic matter should eventually be remineralized. However, ancient shales have organic carbon contents of $\sim$1% that are very similar to those of their modern fine-grained continental margin precursors. The implication is that remineralization of organic carbon gets turned off at some point. In their insightful synthesis of organic matter preservation, *Hedges and Keil* [1995] point out that remineralization and preservation of organic matter are opposite processes and that the key to understanding the ultimate extent of remineralization in sediments is perhaps more likely to be found in asking why some organic matter escapes the determined efforts of bacterial communities rather than by focusing on what controls remineralization.

A traditional view has been that the rate-limiting step in remineralization is the initial breakdown of particulate organic matter by extracellular enzymatic hydrolysis. In her synthetic overview of this topic, *Arnosti* [2004] points out that bacteria are capable of producing extracellular enzymes on timescales of hours to days. She suggests that the slowing or cessation of organic matter degradation may result instead from a wide range of possible mismatches between the organic matter and the bacteria that are present or the enzymes that they deploy. Below we discuss the roles that aerobic versus anaerobic bacterial communities might play in leading to problems such as these. We also discuss the role that protection by mineral sorption might play in preservation.

A final point we would make regarding figure 6.4.1 is that the "sediment core" degradation rate constants shown in the figure (open circles) were determined using a simple model that ignores bioturbation [see *Middelburg*, 1989]. Middelburg (personal communication, 2004) says that he only used data below the bioturbated layer for these calculations, but our analysis of MANOP Site S would suggest that this would miss most of the remineralization in oxygenated sediments, which occurs primarily within the bioturbated zone. One can show that within the bioturbated zone, and for rates of bioturbation and sedimentation that are characteristic of deep ocean locations, ignoring bioturbation can lead to an underestimate of the degradation rate constant by as much as one to two orders of magnitude (see problem 6.15). Our conclusions regarding the decrease in remineralization rate with increasing age of the organic matter are unlikely to be affected by this omission. However, rate constants for bioturbated sediments (i.e., those underlying oxygenated waters) might, we suggest, fall above the line shown in the figure if the calculations were performed in the bioturbated layer with the effect of bioturbation included. In the following subsection we discuss further evidence suggesting that remineralization rates in oxic sediments do indeed appear to be higher than those in anoxic sediments.

The Oxidant

Our discussions so far have given several indications that organic matter subjected to aerobic respiration and associated bioturbation tends to degrade more rapidly and completely than organic matter subjected only to anaerobic remineralization (see figure 6.3.4 and discussion in connection with figure 6.4.1). This finding is consistent with what we understand about the nature of organic matter and the chemical processes by which it gets broken down [e.g., *Middelburg et al.*, 1993; *Canfield*, 1994; *Kristensen et al.*, 1995; *Sun et al.*, 2002]. In particular, aerobic decomposition by bacteria involves a wide range of enzymes, many of which attack specific classes of compounds, with oxygen playing a role not only as an electron acceptor but also as a reactant in the efficient splitting of oxygen-poor–hydrolysis-resistant organic molecules (e.g., lignin, lipids, and photosynthetic pigments; cf. *Hedges and Keil* [1995]) by oxygen-requiring enzymes. By contrast, anaerobic bacteria (e.g., sulfate reducers and methanogens) are not able to break apart most polymeric organic compounds and must rely instead on hydrolytic and fermentative bacteria to break down particulate and HMW dissolved organic matter into LMW compounds that they can ingest and metabolize. Other reasons given for why anaerobic remineralization may be less efficient are the lower energy yields of anaerobic remineralization reactions; the buildup of toxic byproducts such as H_2S from anaerobic remineralization, and the reduction in bioturbation, bioirrigation, and grazing of bacteria by macrofauna that is a consequence of the inability of aerobic organisms to survive in anoxic environments [cf. *Hedges and Keil*, 1995].

Despite expectations based on the above, and observations such as those alluded to at the beginning of this subsection, the role of aerobic degradation in determining the degree of remineralization and burial efficiency of organic matter in sediments has not been easy to demonstrate convincingly either in the laboratory or in the field. There is evidence from laboratory experiments such as those of *Kristensen et al.* [1995] and *Kristensen and Holmer* [2001] [cf. *Hedges and Keil*, 1995] that fresh organic matter decomposes at the same rate in aerobic and anaerobic cultures, but that old and refractory organic matter is much more resistant to degradation in anaerobic cultures. However, laboratory experiments have the limitation that they fall far short of the typical remineralization timescales of sediments, which are mostly centuries to many millennia. The most direct field evidence in support of the importance of oxygen exposure comes from studies in initially well-mixed deep sea turbidites (deposits that result from underwater slumping of sediments) where a portion of the sediment has been exposed to oxygen and a portion has not for many tens of thousands of years (see review by *Emerson and Hedges* [2003]). Such deposits show far greater remineralization of both marine and terrestrial organic matter (e.g., pollen) in the oxygenated portion of the sediments.

One argument used by those who question the central role of oxygen in organic matter remineralization and burial efficiency is the rather poor correlation that exists between sediment organic carbon concentration or burial efficiency and bottom water oxygen concentrations (cf. review by *Hedges and Keil* [1995]). However, *Hedges and Keil* [1995] point out that bottom water oxygen concentration can be a rather poor indicator of exposure of sediments to oxygen, and propose instead using the oxygen exposure time which was defined in connection with figure 6.3.4. While noisy, figure 6.3.4 strongly suggests that longer oxygen exposure times are associated with a more efficient remineralization of organic matter.

We note that the oxygen exposure time calculation used in generating figure 6.3.4 ignores the effect of bioturbation and bioirrigation. In addition to modifying the effective oxygen exposure times, these processes may lead to oscillations between oxic and anoxic conditions (*redox oscillation*), which *Aller* [1994] has proposed may be a distinct environment that is functionally different from either of the oxic or anoxic end-members, and which may promote efficient remineralization of organic matter. Another situation where the juxtaposition of oxidants may lead to enhanced remineralization is in the horizon where methane is oxidized by sulfate. Berelson (personal communication, 2005) observes that this interface shows evidence of fostering bacterial interactions that efficiently process large amounts of labile organic carbon that survive the sulfate reduction zone [cf., also, *D'Hondt et al.*, 2002].

Protection by Mineral Adsorption

An important development of the past decade has been the increasingly detailed understanding of the character of the physical association between fine-grained, high-surface-area minerals and organic carbon (cf. reviews by *Hedges and Keil* [1995] and *Emerson and Hedges* [2003]). An important early study was that of *Mayer* [1994], who found a linear correlation between asymptotic refractory organic carbon and mineral surface area in sediments. Most of the mineral surface area was found to be concentrated in slit-shaped openings called *mesopores* (size = 2–50 nm), and the average of 0.9 mg m^{-2} of organic carbon per unit surface area that they obtained was noted to be equivalent to a *monolayer coating* (a coating consisting of a single molecular layer) of organic matter. *Keil et al.* [1994] also found a monolayer equivalent of organic matter in sediments, and further demonstrated that a majority of gently desorbed organic matter (OM) was remineralized on a timescale of days.

While the concept of a monolayer equivalent of organic matter continues to be valuable as a measure of the amount of organic matter associated with minerals, it has become evident that most mineral surfaces are in fact free of organic matter [e.g., *Mayer*, 1999]. The majority of the organic matter in margin sediments appears, instead, to be concentrated in discrete clusters [e.g., *Ransom et al.*, 1998; *Arnarson and Keil*, 2001]. There are indications that particular types of particles have more organic carbon associated with them than others [e.g., *Ransom et al.*, 1998; *Arnarson and Keil*, 2001; *Kennedy et al.*, 2002], and that this organic matter is significantly less degraded [*Arnarson and Keil*, 2001]. Smectite, a clay mineral, is particularly noteworthy in this regard [*Ransom, et al.*, 1998; *Arnarson and Keil*, 2001; *Kennedy et al.*, 2002], with the minerals quartz and feldspar dominating in sediment fractions with less organic matter, and biogenic minerals such as diatom fragments and calcite and other minerals playing minor roles [*Arnarson and Keil*, 2001]. Furthermore, it appears that the monolayer-equivalent relationships are confined primarily to continental margin sediments, with sub-monolayer equivalency (organic carbon present in less than the monolayer-equivalent concentration) predominating in oxygenated, low-productivity deep ocean regions as well as in deltaic sediments (where redox oscillation may play an important role in efficient remineralization); and super-monolayer equivalency (organic carbon present in

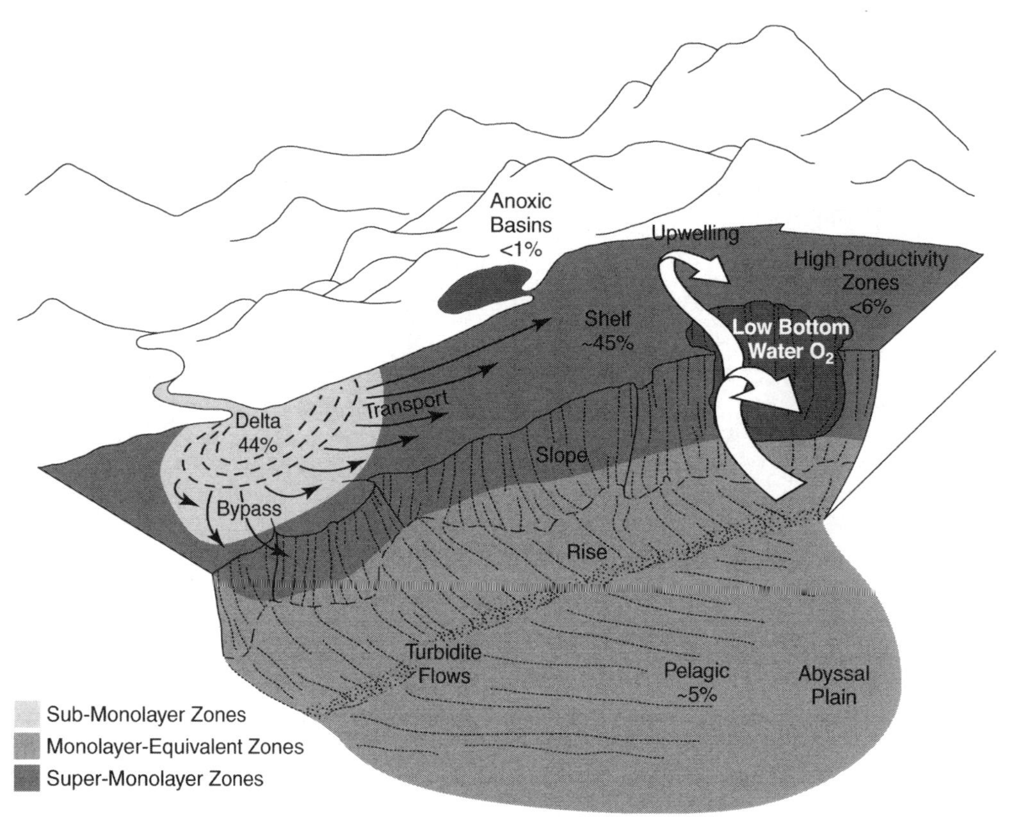

FIGURE 6.4.2: Idealized diagram depicting various oceanic regimes according to whether the organic preservation is monolayer-, sub-monolayer-, or super-monolayer-equivalent. From *Hedges and Keil* [1995]. The numbers shown for each environment represent the percentage of the global total organic C burial that occurs in that particular environment.

greater than the monolayer-equivalent concentration) predominating in anoxic basins and high-surface-productivity, low-bottom-water oxygen zones (figure 6.4.2; cf. *Hedges and Keil* [1995]).

Dissolved organic matter may play an important role as an intermediate in the processes by which organic matter becomes associated with minerals and preserved [cf. *Burdige*, 2002]. Two of the common conceptual models for how the association between minerals and organic matter develop are geopolymerization (abiotic polymerization) and sorption of dissolved organic matter in mesopores, both of which have dissolved organic matter intermediates. In these schemes, the more efficient preservation of organic matter in anoxic sediments might be explained as resulting from the higher dissolved organic matter concentrations alluded to earlier. However, there are other mechanisms that do not have dissolved organic matter as an intermediate, such as selective preservation of large particulate molecules. The role of dissolved organic matter in preservation is thus the subject of continued research [*Burdige*, 2002].

SYNTHESIS

Having in mind that almost any assertion one can make about the cycling of organic matter in sediments continues to be controversial, we nevertheless conclude this section by proposing a simple conceptual model for the primary controls on the extent of organic matter diagenesis. Our model is based largely on that proposed by *Hedges et al.* [1999a], with additional insight from other more recent studies that have been discussed above. The starting point for the model is the conclusion we came to in discussing the remineralization rate versus timescale plot of figure 6.4.1: that these data imply the existence of some fraction of the organic matter in sediments that is simply not accessible to degradation. Many ideas exist as to what might cause the organic matter to become inaccessible to degradation. We focus on two for which there appear to be strong evidence: (1) the nature of the oxidant to which the organic matter is exposed, with oxygen being particularly effective at degrading organic matter (figure 6.3.4) and redox oscillation also appearing to be particularly efficient; and (2) the association of organic matter with minerals, with smectite associations appearing to be particularly effective at protecting organic matter from degradation.

Our conceptual model implies that sediments with low exposure to oxygen and high mineral surface areas (probably with a focus on particular types of minerals such as smectite) will have higher organic preservation efficiencies than sediments with high exposure to oxygen and low mineral surface areas. *Hedges et al.* [1999a] si-

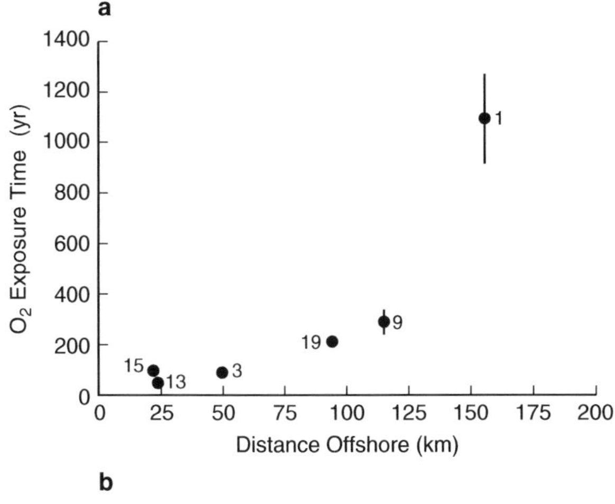

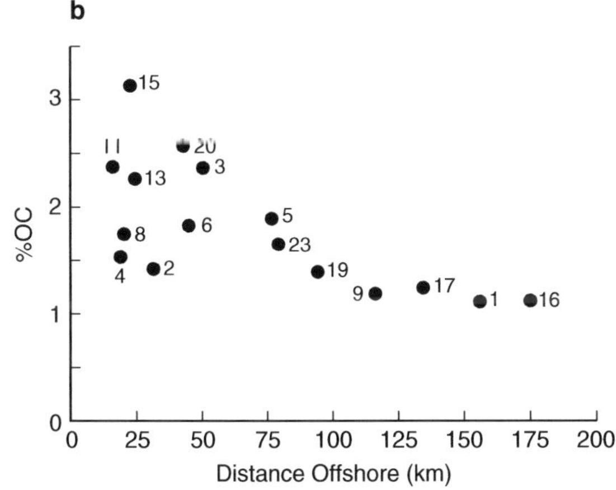

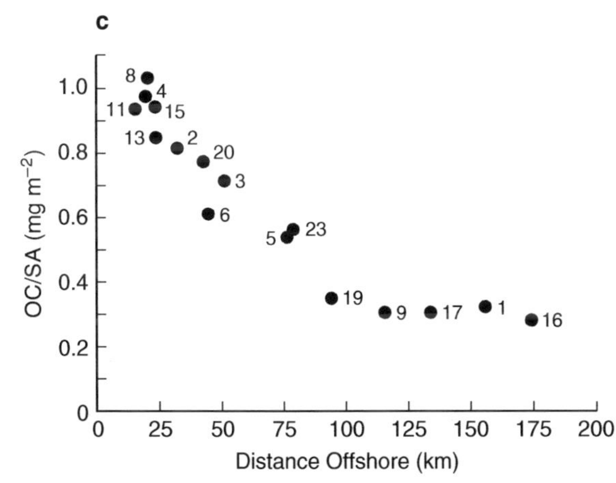

FIGURE 6.4.3: Plots of measurements made on cores from the southern Washington State continental margin by *Hedges et al.* [1999a]. (a) Oxygen exposure time versus distance offshore. See figure 6.3.4 caption for an explanation of how oxygen exposure time is calculated. (b) Percent dry weight organic carbon versus distance offshore. (c) The ratio of organic carbon to mineral surface area, *OC/SA*, plotted versus distance offshore.

multaneously measured all the relevant properties except mineralogy in one set of cores off the southern Washington State margin. They characterize the influence of oxidant using the log of the oxygen exposure

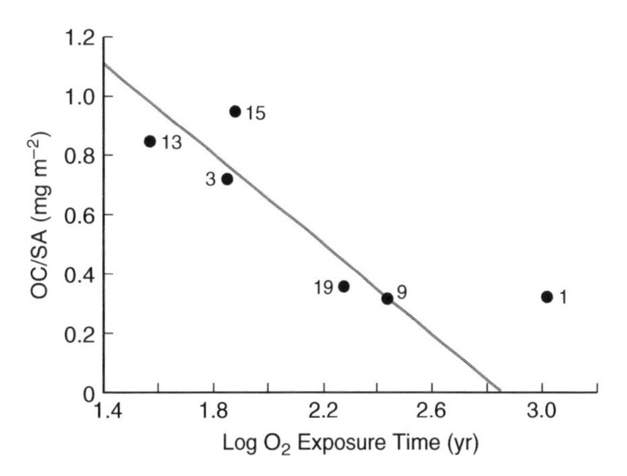

FIGURE 6.4.4: The ratio of organic carbon to mineral surface area *OC/SA* plotted versus the log of the oxygen exposure time

time calculated as in figure 6.3.4. The oxygen exposure time increases with distance offshore (figure 6.4.3a) as the sedimentation rate decreases and oxygen penetration depth increases. They characterize the influence of mineral surface area by considering the variable *OC/SA*, where *OC* is the mass of organic carbon per unit volume of sediments, and SA is the mineral surface area of the sediments per unit volume in units of mg m^{-2}. The plot of the percent dry weight of organic carbon shown in figure 6.4.3b does not show a consistent pattern with distance offshore, but when replotted in figure 6.4.3c as *OC/SA*, the scatter disappears. What emerges is a decrease in surface area normalized preservation efficiency with increasing distance offshore down to a lower threshold *OC/SA* of about 0.3 mg m^{-2}. Figure 6.4.4 pulls all these results together in a single plot of *OC/SA* versus log of oxygen exposure time but with less resolution due to the more limited oxygen exposure time data set.

Over time, many empirical relationships have been found between organic matter preservation and primary production, organic carbon rain rate to the sediments, sedimentation rate, organic carbon degradation rate, and bottom water oxygen concentration [cf. *Hartnett and Devol*, 2003]. All of these can be related in one way or another to our conceptual model. Higher primary production would generally be associated with a higher organic carbon rain rate to the sediments. We saw in our discussion of the aerobic oxidation model we derived that both of these and a reduced organic carbon degradation rate will reduce the oxygen concentration in the sediments, thereby reducing the oxygen exposure time. Sedimentation rate will have a direct impact on oxygen exposure time.

We conclude by noting that there are many obvious steps that could be taken to refine the conceptual model, including an improved definition of oxygen exposure time that takes bioturbation into account, a more refined analysis of the influence of oxidants other than just oxygen [cf. *Reimers*, 1998], and a more nuanced parameterization of the surface area effect that, for example, takes the mineral type into consideration. We note, for example, that there are some interesting correlations between the map of percent dry weight organic carbon (figure 6.1.4) and maps of the smectite content of sediments [e.g., *Griffin et al.*, 1968], particularly in the tropical Pacific, where both are high. Undoubtedly there will continue to be many future developments in this field that will require modifications to the conceptual model, including, for example, improved understanding of the role of biological processes such as passage through the guts of macrofauna in associating organic matter with minerals [e.g., *Berner*, 1995], the role of dissolved organic matter [e.g., *Burdige*, 2002], and structural effects such as those alluded to by *Arnosti* [2004].

6.5 Organic Matter Budget

The main goal of this final section is to examine the components of the sediment organic matter budget and to put them in the context of the full oceanic budget depicted schematically in figure 6.5.1. Table 6.5.1 and figure 6.5.2 show a *POC* budget for the whole ocean, which we will use to discuss the vital role of continental margins for the oceanic *POC* budget as a whole [cf. *Berner*, 1982; *Smith and MacKenzie*, 1987; *Brink et al.*, 1995; *Tromp et al.*, 1995; *Middelburg et al.*, 1997; *Rabouille et al.*, 2001; *de Haas et al.*, 2002; *Andersson et al.*, 2004; *Chen*, 2004]. Carbon cycling in deep ocean sediments is relatively unimportant to the global ocean carbon budget, but has a major impact on the deep ocean carbon budget [cf. *Jahnke and Jackson*, 1987;

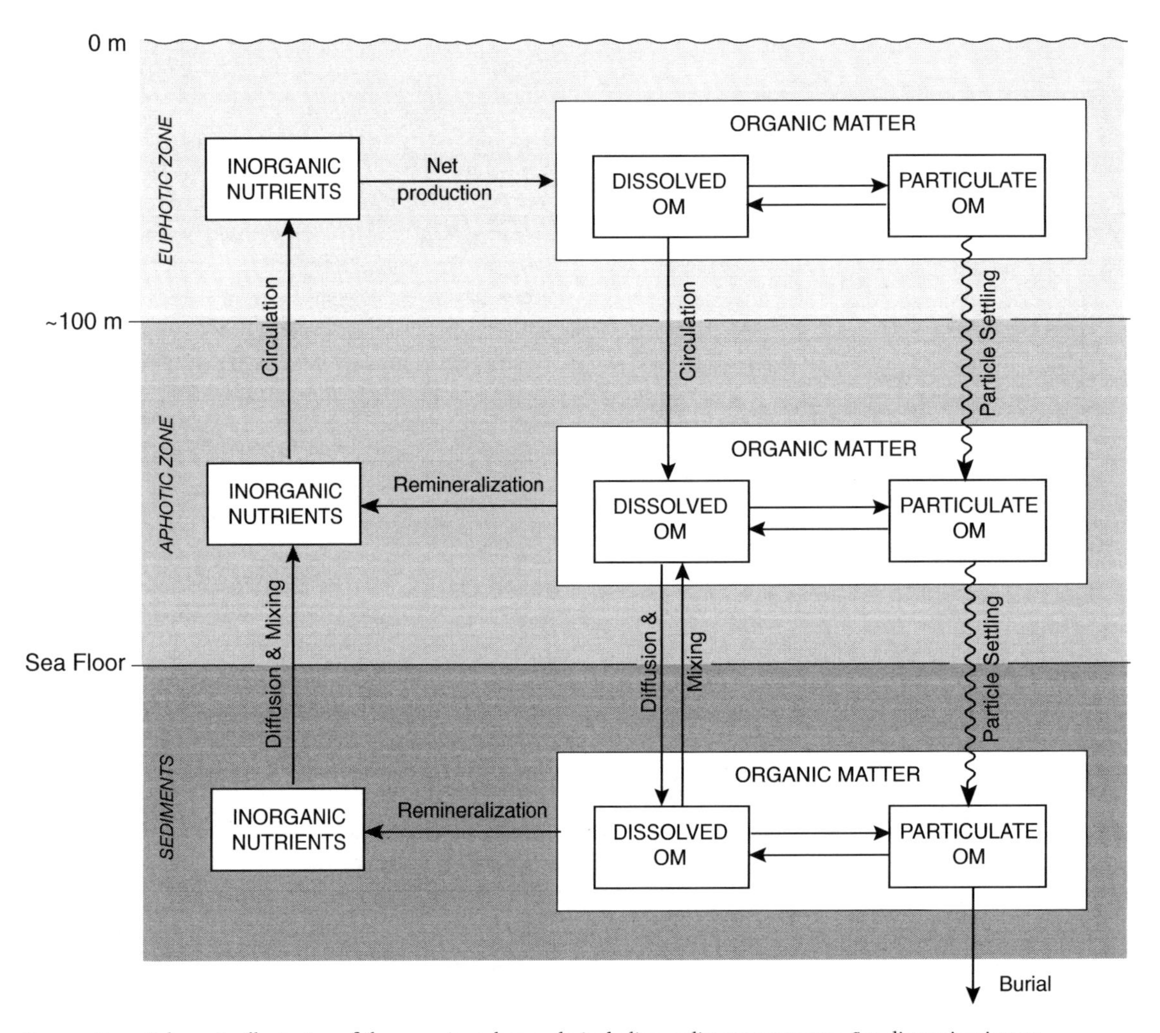

Figure 6.5.1: Schematic illustration of the oceanic carbon cycle including sediment processes. See discussion in text.

TABLE 6.5.1

The global ocean particulate organic carbon (*POC*) budget partitioned into continental margin and open ocean components.

The "continental margin" is defined as all regions where the floor of the ocean is <1000 m deep. Units are Pg C yr^{-1}. The % is with respect to the global total for each row. This table is based on Dunne et al. *[2005c].*

	Deep Ocean		*Continental Margin*		*Global*
Ocean area ($\times 10^{14}$ km^2)	3.221	(92%)	0.276	(8%)	3.497
			Water column POC budget		
Production[a]	45 ± 7	(84%)	8.6 ± 0.8	(16%)	53 ± 8
Export (100 m)[b]	7.0 ± 2.4	(71%)	2.8 ± 0.9	(29%)	9.8 ± 3.3
Remineralization[c]	6.6 ± 2.4	(91%)	0.63 ± 1.13	(9%)	7.3 ± 3.4
Flux to seafloor[d]	0.34 ± 0.14	(14%)	2.2 ± 0.7	(86%)	2.5 ± 0.8
			Sediment POC budget		
Remineralization[e]	0.24 ± 0.14	(11%)	1.9 ± 0.7	(89%)	2.1 ± 1.3
Burial[f]	0.018 ± 0.016	(9%)	0.174 ± 0.066	(91%)	0.19 ± 0.07
			Other sediment processes		
DOC flux to water column[g]	0.080 ± 0.033	(40%)	0.120 ± 0.039	(60%)	0.20 ± 0.07
Denitrification[h]	0.032 ± 0.014	(19%)	0.133 ± 0.043	(81%)	0.165 ± 0.054

[a] The primary production is the average of three estimates calculated from satellite color observations with the algorithms of *Behrenfeld and Falkowski* [1997], *Carr* [2002], and *Marra et al.* [2003], as described in chapter 4. The uncertainty was calculated by taking the standard deviation of the average of the three different primary production algorithms.

[b] The *POC* export is the product of the primary production times a particle export ratio estimated with the empirical model of *Dunne et al.* [2005b]. The uncertainty was calculated by error propagation using the error in the primary production and including an estimated uncertainty of 35% in the particle export ratio. The total export of organic carbon includes *DOC* as well, which *Hansell* [2002] estimates as contributing $20 \pm 10\%$ of the total on a global scale (i.e., *DOC* export $=$ *POC* export $\cdot\, 0.25 = 2.5$ Pg C yr^{-1}).

[c] *POC* remineralization in the water column is obtained by subtracting the bottom flux from the 100 m flux.

[d] The *POC* flux at the ocean floor is the average of 33 separate POC flux estimates obtained by combining the three *POC* primary production algorithms of *Behrenfeld and Falkowski* [1997], *Carr* [2002], and *Marra et al.* [2003], with the particle export ratio model of *Dunne et al.* [2005b] and 11 sediment trap models of *Lutz et al.* [2002]. Note that all these estimates are very sensitive to the choice of sediment trap scaling, which is highly uncertain, as discussed in chapter 5. This is why we have chosen to estimate the flux independently using all 11 of the *Lutz et al.* [2002] sediment trap models and to present the flux as an average with a standard deviation [*Dunne et al.*, 2005c]. This calculation is done with a horizontal resolution of 1° by 1°, and thus includes the effects of variability of surface production and differing depths of the ocean.

[e] *POC* remineralization in the sediments calculated by subtracting *DOC* loss from *POC* flux to the bottom.

[f] *POC* burial at water depths greater than 1000 m is estimated at 0.015 Pg C yr^{-1} by *Jahnke* [1996] from maps of the organic carbon content (figure 6.1.4) and sediment accumulation rate (figure 6.1.2), with no uncertainty given. We obtain a similar estimate of 0.018 ± 0.016 based on a new algorithm developed by *Dunne et al.* [2005c]:

$$\Phi^{POC}_{burial} = \Phi^{POC}_{bottom} \cdot \left[0.0055 + 0.76 \cdot \left(\frac{\Phi^{POC}_{bottom}}{5.3 + \Phi^{POC}_{bottom}} \right)^2 \right]$$

where Φ^{POC}_{bottom} is the flux of organic carbon reaching the sediments and Φ^{POC}_{burial} is the organic carbon accumulation rate, both in units of mmol m^{-2} d^{-1} of organic carbon. This algorithm was found to explain 65% of the variance in the data. Sediment burial on the continental margin is the mean of the *Berner* [1982] global estimate of 0.126 Pg C yr^{-1} minus the deep ocean burial of 0.018 Pg C yr^{-1}, and the *Wollast* [1993] shelf and slope estimate of 0.240 Pg C yr^{-1}. The "standard deviation" of the margin burial represents the range of the two estimates.

[g] *DOC* fluxes above 2000 m are obtained by multiplying the bottom fluxes of *POC* by $0.14/(1+0.14) = 0.12$. Below 2000 m the multiplication factor is $0.36/(1+0.36) = 0.26$. These are based on the observations of *Papadimitriou, et al.* [2002].

[h] We use the metamodel of *Middelburg et al.* [1996] to estimate denitrification as a function of the organic carbon flux to the sediments:

$$\log(\Phi^{Denitrification}_{bottom}) = -0.9543 + 0.7662 \cdot \log(\Phi^{POC}_{bottom}) - 0.2350 \cdot [\log(\Phi^{POC}_{bottom})]^2$$

where $\Phi^{Denitrification}_{bottom}$ is the denitrification rate in the sediments, and Φ^{POC}_{bottom} is the flux of *POC* to the sediments, both in units of μmol C cm^{-2} d^{-1}.

Jahnke, 1996]. Figure 6.5.3 shows a carbon budget for the ocean below 1000 m, which we will use to discuss this. We conclude this section and the chapter with a brief discussion of how the organic carbon budget of the ocean is linked to that of oxygen, and how both are kept in homeostasis.

Table 6.5.1 and figure 6.5.2 provide a side-by-side summary of the *POC* budgets of the continental mar-

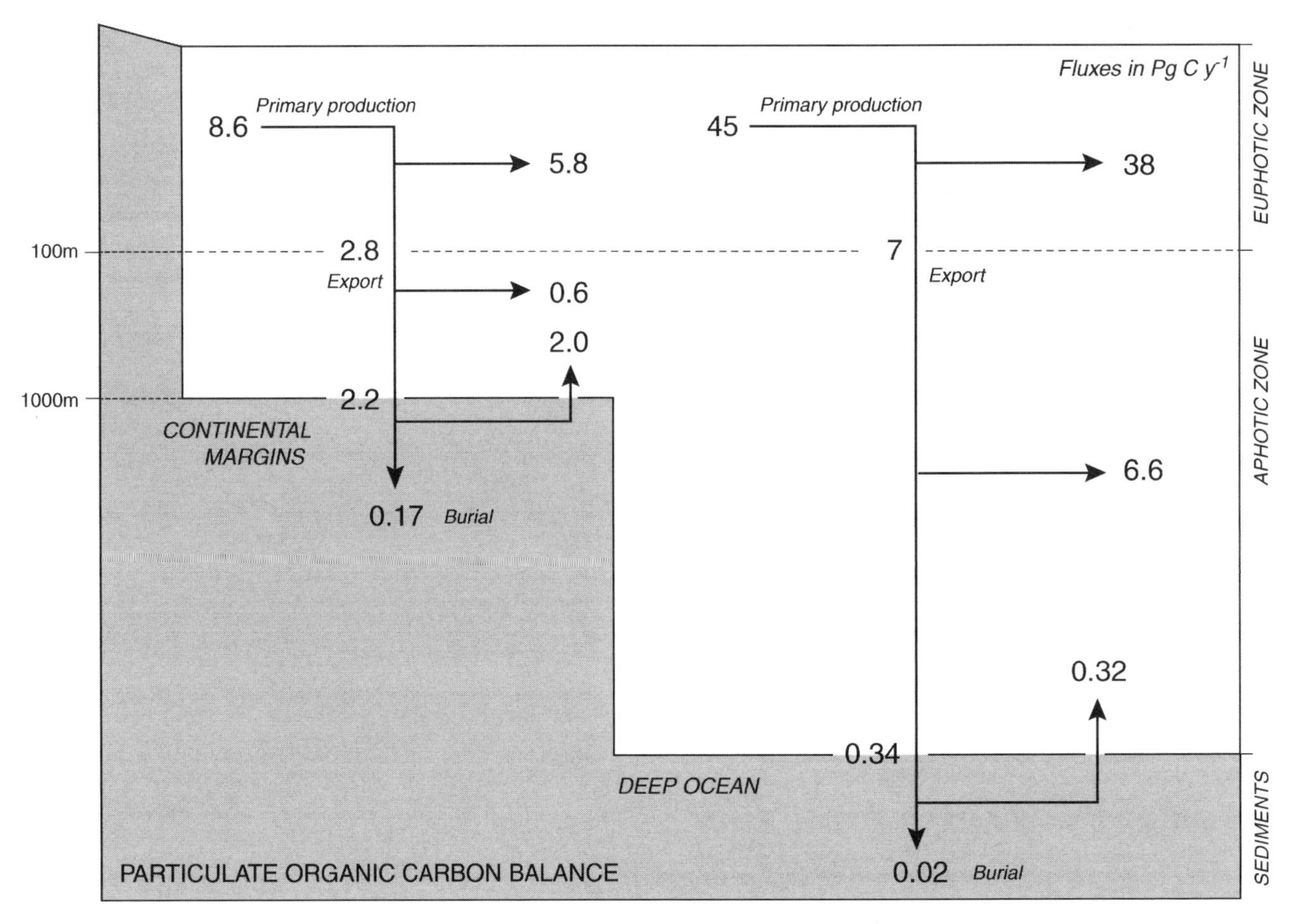

FIGURE 6.5.2: The global ocean particulate organic carbon budget based on table 6.5.1. Units are Pg C yr^{-1}. Downward arrows represent *POC* flux and horizontal arrows to the right represent remineralization including *DOC* generation. The initial numbers at the top of the water column are primary production estimates. The difference between the primary production and *POC* export from the surface ocean is represented by an arrow going off to the right. Most of this is remineralized in the surface ocean, but a fraction is exported as *DOC*. *Hansell* [2002] estimates the *DOC* export as equal to about $20 \pm 10\%$ of the total export on a global scale (i.e., *DOC* export $\sim 0.25 \cdot$ *POC* export).

gin, which we define as those areas of the ocean where the seafloor is shallower than 1000 m, and the "open" ocean, which we define as those areas of the ocean where the seafloor is deeper than 1000 m. On a global scale, a quarter of the *POC* exported from the surface of the ocean escapes water column remineralization and rains out onto the sediments, but almost 90% of this occurs in the continental margins. Why are the continental margins so dominant?

The continental margins account for 16% of the global production of POC by net primary production despite occupying only 8% of the total ocean area. The per unit area POC net primary production is thus 2.2× higher on the margins than in the open ocean. This relative importance of the shelves increases as the *POC* progresses through the carbon cycle from particle export all the way through sediment preservation. The euphotic zone on the margins is much less efficient in recycling *POC* than that in the open ocean. Thus, the fraction of *POC* net primary production that is exported from the surface is about 32% in the continental margins, as opposed to 16% in the open ocean. Furthermore, because the ocean floor is much shallower on the margins, only a small fraction of the POC that is exported from the surface is remineralized in the water column. As a consequence, almost 80% of the export arrives at the sediments on the continental margins, versus only 5% in the open ocean.

Sediment burial is also more efficient on the margins, where $7.9 \pm 3.9\%$ of the *POC* arriving at the sediments is buried, by contrast with the deep sediments, where we calculate an average deep ocean burial efficiency of $5.3 \pm 5.2\%$ using the numbers in table 6.5.1. The deep ocean burial efficiency drops to ~3.1% if we use the organic matter remineralization rate of 0.49 Pg C yr^{-1} of figure 6.5.3b together with the *DOC* loss and carbon burial fluxes of figure 6.5.3a to estimate a new rain rate to the sediments of $0.49 + 0.08 + 0.018 = 0.59$ Pg C yr^{-1}. Overall, the continental margins account for about 80% of the global sediment burial of *POC*. The estimated loss of organic carbon from the sediments as *DOC* is relatively unimportant on the shelves, though it is comparable to the burial rate. However, in the open ocean, our budget shows that it accounts for 24% of the particle flux arriving at the sediments and that it is more than 4 times the burial rate, although this number is not well constrained.

Most of the *POC* exported from the surface in the open ocean is remineralized within the upper 1000 m, as we saw in chapter 5. The flux across 1000 m is just 11% of the global export of organic matter in the open ocean, ~8% if we include the export over the continental margins. What is the fate of the organic carbon that survives this upper ocean remineralization? We see from the sediment trap–based analysis of figure 6.5.3a that 43% of the *POC* that sinks past 1000 m rains out onto the deep ocean sediments. Once in the sediments, only ~5% of the organic carbon is buried. The remainder is either remineralized or solubilized and lost from the sediments as *DOC*. The ratio of sediment organic carbon remineralization to total deep ocean remineralization of both *POC* and *DOC* based on the organic carbon budget of figure 6.5.3a is $(0.24\ \text{Pg C yr}^{-1})/(0.85\ \text{Pg C yr}^{-1}) = 0.28$ [cf. *Andersson et al.*, 2004]. On the other hand, the oxygen and nitrate budgets of figure 6.5.3b imply a sediment remineralization contribution of $(0.49\ \text{Pg C yr}^{-1})/(1.13\ \text{Pg C yr}^{-1}) = 0.43$. In either case, it is clear that sediment remineralization is a major contributor to the deep ocean organic carbon budget.

In chapter 5 we saw that organic carbon remineralization in the water column as inferred from sediment traps or other information follows a rather smoothly decreasing function of depth that is often fit with an exponential or power law function. How might the large amount of remineralization going on in sediments affect this? Figure 6.5.4 shows plots of several empirical models of the deep ocean remineralization rate plotted versus depth for the water column below 1000 m (taken from *Jahnke and Jackson* [1987]). The vertical distribution of remineralization that one obtains from a consideration of sediment remineralization processes is quite different than those obtained from standard exponential or power law functions such as the other curves shown on this plot. This is because a large fraction of the sediment area and also the areas of higher organic carbon flux to the sediments are concentrated in the slope regions of the continental margins. Clearly, a model of oceanic remineralization that used only a power law or exponential function of remineralization in the water column without including remineralization at the ocean floor would not be able to properly represent the spatial distribution of oceanic remineralization.

We conclude by discussing the fascinating problem of how the organic carbon burial in sediments is linked to the global biogeochemical cycle of other elements, and what keeps the combined system in check (i.e., homeostasis). Organic carbon is formed by reactions with the following simplified stoichiometry: $CO_2 + H_2O \rightarrow CH_2O + O_2$ (cf. aerobic respiration reaction in table 6.1.1). Thus, any time organic matter is buried in the sediments and lost to the atmosphere-ocean-land system, there is an associated reduction in CO_2 and increase of O_2 in the atmosphere and ocean. Once buried, organic carbon becomes part of the geologic cycle of plate tectonics. The crustal overturning that occurs in association with this cycle eventually exposes the organic matter on the land surface, where weathering reactions reverse the above reaction, thereby consuming O_2 and releasing CO_2. This combination of organic matter burial and weathering provides the backbone of processes that are thought to keep the global oxygen budget in check and contribute also to the carbon dioxide budget (cf. chapters 8 through 10 for a discussion of the carbon system). Much has been written about how these processes are linked by feedbacks to the atmospheric oxygen content in such a way as to maintain oxygen homeostasis [e.g., *Berner*, 1982; *Garrels and Lerman*, 1984; *Holland*, 1984; *Kump and Garrels*, 1986; *Kump*, 1988; *Berner and Canfield*, 1989; *Lasaga*, 1989; *Betts and Holland*, 1991; *Lovelock*, 1995; *Van Cappellen and Ingall*, 1996; *Colman et al.*, 1997; *Hedges et al.*, 1999b; *Berner et al.*, 2000; *Lenton and Watson*, 2000b; *Berner*, 2001; *Lasaga and Ohmoto*, 2002]. The following brief overview is based principally on the analysis of *Lenton and Watson* [2000b], supplemented by the perspectives of *Hedges et al.* [1999b], which grow out of a view of sediment burial that is the basis for the synthesis we proposed in the previous section.

The starting point for our analysis is a discussion of timescales. The burial of organic carbon in ocean sediments that we gave above is $0.19\ \text{Pg C yr}^{-1}$. Converting this to mol C yr^{-1} and using a 1:1 mole ratio of carbon dioxide uptake to oxygen release, as is normally assumed to be the case for organic carbon burial, gives an oxygen production rate of $16\ \text{Tmol yr}^{-1}$ ($1\ \text{Tmol} = 10^{12}\ \text{mol}$). The combined atmosphere-ocean inventory of oxygen is 3.8×10^7 Tmol [cf. *Van Cappellen and Ingall*, 1996], thus the residence time of oxygen in the atmosphere-ocean reservoir with respect to organic carbon burial is 2,300,000 years. By contrast, crustal overturn takes on the order of 100,000,000 years, during which the atmosphere-ocean oxygen reservoir would have been replaced about 40 times. One of the feedbacks that a system such as this would have is between the amount of organic carbon buried in sedimentary rocks, and the amount of organic carbon in the rocks that is available for being weathered on the land surface. However, unless the crustal overturn is somehow short-circuited, this feedback would operate on a timescale of ~100,000,000 years, and wide variations of atmosphere-ocean oxygen would be possible on shorter timescales. Do such variations occur?

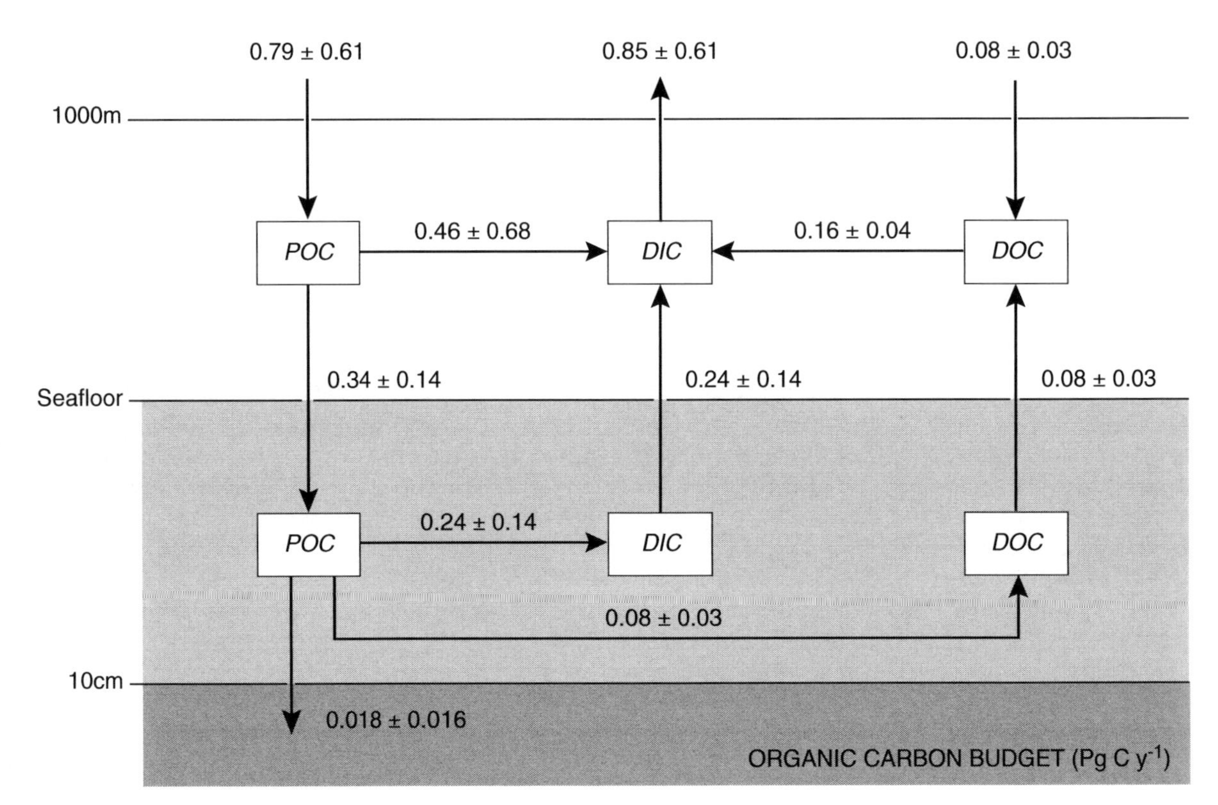

FIGURE 6.5.3: Deep ocean (>1000 m) and sediment (top 10 cm) carbon budget inferred from (a) constraints on the particulate and dissolved organic carbon fluxes and (b) oxygen and nitrate utilization rates using the stoichiometric ratios given in table 6.1.1. The budget terms are calculated as follows:

1. All the *POC* flux estimates, the *DOC* flux out of the sediments, and the denitrification in the sediments, are either from table 6.5.1 or calculated by the same approach described in that table. Denitrification in the water column is assumed to be negligible at these depths.

2. The *DOC* input to the deep ocean is estimated from

$$\Phi_{1000}^{DOC} = \frac{\Delta DOC\ (\text{mmol m}^{-3}) \times 10^{-3} (\text{mol mmol}^{-1})}{\Delta\tau} \cdot V_{>1000} \cdot 12 \times 10^{-15} \frac{\text{Pg C}}{\text{mol C}}$$

where ΔDOC is the difference in concentration between the North Atlantic and North Pacific, $\Delta\tau$ is the age difference between these two locations estimated from radiocarbon measurements, and $V_{>1000} = 1.0 \times 10^{18}$ m^3 is the volume below 1000 m. *Hansell and Carlson* [1998] give a *DOC* concentration of 44 mmol m^{-3} at 32°N in the Sargasso Sea (1000–4000 m). In the North Pacific, they give a concentration of 39 mmol m^{-3} at 23°N (900–4750 m) and 34 mmol m^{-3} at 58°N (1000–1500 m). *Druffel et al.* [1992] give an age difference of 1540 ± 70 years between the deep Atlantic and the deep Pacific. This gives a *DOC* utilization rate of 0.05 to 0.10 Pg C yr^{-1}, which we show as 0.08 ± 0.03 Pg C yr^{-1} in the figure (cf. the *Hansell* [2002] estimate of 0.12 to 0.13 Pg C yr^{-1}, which, however, includes intermediate waters at or above our 1000 m cutoff).

3. Total deep oxygen metabolism (sediments plus water column) is estimated from the deep Pacific and Atlantic *OUR* estimates of table 5.2.1 as follows:

$$\Phi_{1000}^{O_2\,\text{based}} = OUR\ (\text{m}^{-3}\,\text{yr}^{-1}) \times 10^{-3} (\text{mol mmol}^{-1}) \cdot V_{>1000} \cdot r_{C:O_2} \cdot 12 \times 10^{-15} \frac{\text{Pg C}}{\text{mol C}}$$

We use an *OUR* of ~0.13 mmol m^{-3} yr^{-1}.

4. The flux of oxygen into the sediments comes from *Jahnke* [1996], who bases it on a global extrapolation of a wide range of measurements of sediment-water oxygen fluxes and gradients in the sediments.

5. The remaining terms are all estimated by requiring mass balance.

In fact, the evidence suggests that oxygen has remained within the relatively narrow bounds of ~17–~30% mole ratio of oxygen to total air [cf. *Lenton and Watson*, 2000b], compared to the present mole ratio of 21% (table 3.1.1). What are the oceanic feedbacks that might contribute to keeping the oxygen at homeostasis on the timescale of the 2,300,000-year residence time of oxygen in the atmosphere-ocean? The models that have been developed to explore this issue all start from the basic premise that the burial of organic carbon in the ocean is related in some way to the export of organic carbon from the surface of the ocean. The relationship

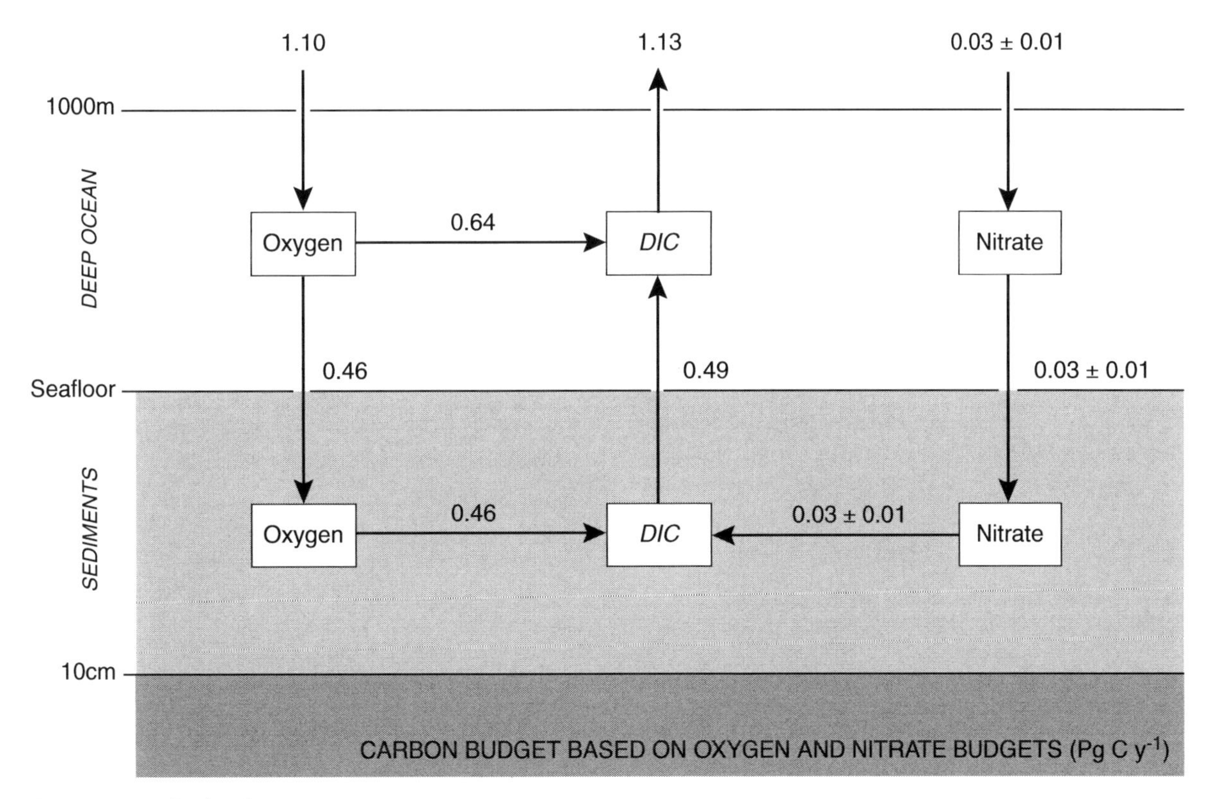

FIGURE 6.5.3: (*Continued*)

assumed in these studies is usually linear, but *Lenton and Watson* [2000b] propose a square dependence. The export production, in turn, is assumed to be a function of the supply rate of nitrate and/or phosphate to the euphotic zone of the ocean. A negative feedback would thus be one where an increase in oxygen would somehow trigger a decrease in the supply rate of nutrients to the surface ocean, usually by reducing the oceanic inventory of the nutrients. This would reduce organic carbon burial and thus reduce oxygen release to the atmosphere. If the uptake of oxygen by weathering on land continues at the same rate as before the perturbation (which we assume in the thought experiments to follow), the reduction in oxygen production will lead to a reduction in the atmosphere-ocean oxygen reservoir. A positive feedback would be one where an oxygen increase would trigger an increase in the supply rate of nutrients to the surface ocean. In their study, *Lenton and Watson* [2000b] include the following four intra-ocean mechanisms for varying the oceanic nutrient inventory that pretty much cover the range of possibilities that have been proposed:

1. Enhanced denitrification under low-oxygen conditions (cf. chapter 5; *Falkowski* [1997]; *Lenton and Watson* [2000a]). Denitrification occurs in low oxygen or anoxic environments. If oxygen in the atmosphere-ocean reservoir were to drop, water column anoxia would increase. This would lead to increased denitrification, which would reduce oceanic nitrate, reduce biological productivity, and reduce burial of organic carbon in the sediments (the source of oxygen to the atmosphere-ocean reservoir). Denitrification is thus a destabilizing positive feedback in which reduced oxygen leads to a further reduction in oxygen, and vice versa.

2. Enhanced phosphorus burial due to adsorption onto iron(III) hydroxides, which form only in oxic sediments [e.g., *Van Cappellen and Ingall*, 1996; see also *Colman et al.*, 1997]. This is a stabilizing negative feedback, in which, for example, an oxygen decrease will decrease phosphorus burial, increase the oceanic phosphorus inventory, and thus increase export production and burial of organic matter. This would increase the atmosphere-ocean oxygen inventory.

3. Enhanced phosphorus burial due to enhanced accumulation of phosphorus by aerobic bacteria [cf. *Van Cappellen and Ingall*, 1994], also a negative feedback.

4. Enhanced organic burial under anoxic conditions (see section 6.4), also a negative feedback.

The primary conclusion that *Lenton and Watson* [2000b] arrive at regarding the strength of these oceanic-based nutrient inventory feedbacks is that, while they certainly play a central role, they are not sufficient on their own to keep the oxygen within the required bounds. They examined the response to a doubling or halving of oxygen uptake by weathering on land, which

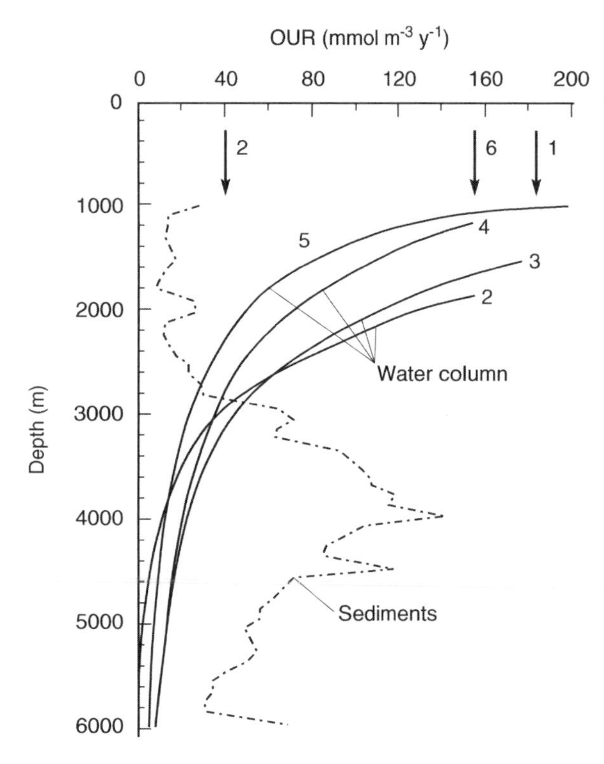

FIGURE 6.5.4: Organic carbon remineralization rates in the North Pacific. The sediment remineralization rates are from *Jahnke and Jackson* [1987]. The water column remineralization rate estimates numbered 1 and 2 are from the models of water column processes of *Craig* [1971] and *Fiadeiro and Craig* [1978], respectively. Those labeled 3 through 5 are empirical fits to sediment trap observations by *Betzer and al.* [1984], *Suess* [1980], and *Pace et al.* [1986], respectively. The arrow labeled 6 is the average deep ocean remineralization rate estimate of *Hinga* [1985].

is approximately within the range inferred from strontium isotope and phosphorus records in sediments [cf. *Lenton and Watson*, 2000b]. Their base model with feedbacks (1) and (2) predicted oxygen stabilization at 8.7% at the low end, but was unable to stabilize oxygen at the high end because the ocean became completely oxic. Adding feedback (3) to the base model reduced the response to the range 12.7% to 26.5%, while adding feedback (4) to the base model reduced the range to 15.1% to 22.8%. Nevertheless, *Lenton and Watson* [2000b] conclude from their analysis that oceanic feedbacks are likely insufficient to keep the oxygen within the required bounds, particularly at the high end, where a runaway increase can occur when the ocean becomes completely oxic. They show that land-based feedbacks, such as enhanced fire frequency under high oxygen conditions, the inhibition of C-3–based photosynthesis by oxygen, and a pair of mechanisms whereby terrestrial processes can alter the phosphorus supply to the ocean, contribute significantly toward keeping the oxygen in check.

The question arises as to how the conceptual model of sediment cycling we developed in section 6.4 would fit into the framework developed by *Lenton and Watson* [2000b]. *Hedges et al.* [1999b] have made the intriguing suggestion that the long-term link of organic carbon burial in ocean sediments with weathering on land is due to the association of organic matter with minerals. If weathering of organic matter increases, the flux of minerals into the ocean will also increase, and burial of organic matter will become more efficient. The effect of high oxygen levels in leading to severe aerobic degradation, such as in the deep ocean, would serve as a moderating effect on this long-term process. Specifically, *Hedges et al.* [1999b] suggest that the line between sediments with "monolayer-equivalent" loadings on the shelf and slope, and sediments with "sub-monolayer-equivalent" loadings on the rise and abyssal plains (figure 6.4.2; see also the break in the aerobic contribution to remineralization that occurs at ~1500 m in the data shown in table 6.3.1), may be controlled by the amount of oxygen in the water column. This line, which they refer to as the *organic carbon compensation depth*, would move up in response to increased oxygen, thus decreasing organic carbon burial and reducing oxygen production. The reverse would happen if oxygen decreased.

A final point: the residence time of CO_2 in the ocean-atmosphere that we obtain from the sediment organic carbon burial term divided into the ocean-atmosphere inventories shown in figure 10.1.1 is an order of magnitude smaller than for oxygen, 210,000 years versus 2,300,000. If we add the ~0.2 Pg C/yr burial rate of $CaCO_3$ to the carbon budget [*Sarmiento and Sundquist*, 1992], we get instead 100,000 years as the CO_2 residence time for the ocean-atmosphere reservoir with respect to sediment burial. With such a short residence time, we might expect that the atmospheric CO_2 would be much more sensitive to perturbations, which indeed it is (see chapter 10).

Problems

6.1 Define and explain the following terms:

a. porosity

b. diffusive boundary layer

c. pore water

d. burial efficiency

6.2 Within the sediments, what are the basic physical processes that you need to consider for a chemical that is

a. dissolved in pore water?

b. a mineral component?

6.3 Discuss the possible reasons for why methanogenesis is probably the most important contributor to anaerobic organic matter oxidation in freshwater swamps, while sulphate reduction is twice as important as methanogenesis in marine sediments (see table 6.1.4).

6.4 Explain why in a typical pore water profile, dissolved Mn^{2+} starts to increase at much shallower depth than dissolved Fe^{2+}.

6.5 Among the energetically favorable reactions that occur in sediments, anaerobic methane oxidation (AMO; see table 6.1.3) and anaerobic oxidation of ammonium (Anamox) have recently received a lot of attention by the scientific community. This is primarily because it appears that these processes are mediated by distinct groups of microorganisms that take advantage of the net energy gain associated with these reactions. Explain these reactions and their significance for the marine carbon and nitrogen cycles.

6.6 Assume that all the nitrate produced by nitrification during aerobic respiration is subsequently consumed by denitrification. What would the combined stoichiometry of these two reactions be for organic matter of the composition given in table 6.1.1? How does the oxygen consumption of the combined reactions compare with the oxygen consumption required if there were no contribution from denitrification?

6.7 Discuss the relative roles of (a) the bioturbation rate, (b) the organic matter remineralization rate constant, and (c) the total organic carbon flux at the sediment surface, in controlling the sediment penetration depth of labile organic matter. Assume that all organic matter is oxidized aerobically.

6.8 Using the analytical model for the sediment distribution of oxygen (equation (6.3.8a)) and the definition for γ (equation (6.3.4b)) and k' (equation (6.3.6)), plot and discuss the asymptotic concentration of oxygen (i.e., for $z \rightarrow \infty$) as a function of the organic carbon flux arriving at the surface of the sediments (i.e., for $z = 0$). Assume that the bottom water concentration of oxygen is 168 mmol m^{-3} and use the same values for the other parameters as given in the text. In particular, address the following questions:

a. At what value of the organic carbon flux does the asymptotic oxygen concentration become zero? (Assume a constant aerobic oxidation rate constant).

b. Given that aerobic oxidation starts to slow down at oxygen concentrations below a few mmol m^{-3} (see discussion in panel 6.2.3), reconsider your answer to (a).

c. In what way would you have to change your model if your organic carbon flux arriving at the sediment surface exceeds this threshold?

6.9 List the most important factors that determine the burial efficiency of organic carbon in sediments and briefly discuss them.

6.10 Discuss why about 90% of the global organic carbon burial occurs on the continental margins, despite the fact that they cover only about 8% of the global surface area.

6.11 In section 6.4, we discussed evidence that a reduction of the bottom water oxygen concentration leads to an enhancement of the organic matter burial. Discuss this process in the context of the question of what controls the atmospheric oxygen concentration on timescales of a few million years. In particular, address whether this is a positive or a negative feedback, and explain why. Discuss this assuming two scenarios: (a) the burial of phosphorus is tightly linked to the burial of organic carbon, and (b) the burial of phosphorus is unaffected by the organic matter burial.

6.12 Assume an average concentration for deep ocean seawater of 200 mmol m^{-3} of oxygen, 35 mmol m^{-3} of nitrate, 0.0002 mmol m^{-3} of manganese ion, 0.0007 mmol m^{-3} of iron ion, and 28,000 mmol m^{-3} of sulfate. Prepare a table with five columns, the first two of which consist of the name of the oxidant and its concentration in seawater. Use the remaining columns to add the following:

a. What is the ratio of organic carbon to oxidant for the remineralization reaction given in table 6.1.1?

b. Suppose that 1 m^3 of seawater were isolated and enough carbon added to it to react with all the above oxidants present. How much of the organic carbon would get used by each reaction? (Note: Add all nitrate from nitrification to the denitrification budget.)

c. How does the relative contribution of the different oxidants compare to the actual relative contributions shown in table 6.1.4? Why are they different?

6.13 Suppose a flux Φ_{Total} of organic matter arrives at the sediments with half of it being oxidized by aerobic processes, the magnitude of which is given by the solution to problem 6.8, the other half by anaerobic processes. How does the relative contribution of aerobic to anaerobic oxidation change if you halve the bottom water oxygen concentration? How would this ratio change if you kept the bottom water oxygen concentration constant and doubled the flux of labile organic matter (assume that the flux of refractory organic matter remains the same)? Explain the basis for your answers.

6.14 Compare the thickness of the stagnant film at the sediment water interface (which we estimated was typically about 1 mm) with the e-folding length of oxygen in the sediments (the depth at which oxygen concentration drops by 1/e over its surface value) for the solution given by equation (6.3.8).

a. What is the e-folding length for the deep ocean MANOP Site S set of parameters?

b. Assuming that bioturbation rate remains the same, what would the reaction rate have to be in order for the e-folding length to be the same as the stagnant film thickness?

6.15 In this problem, we would like to estimate the remineralization rate k from measurements of organic matter in a core [cf. *Middelburg*, 1989].

a. Given the bioturbation rate and sedimentation rate and measurements of organic carbon at two depths, show how you can estimate the remineralization rate.

b. Using the values from deep ocean MANOP Site S, estimate the relative contributions of bioturbation and the sedimentation rate to the estimate of k.

CHAPTER 7

Silicate Cycle

In previous chapters, we discussed biogeochemical processes at the air-sea interface, in the euphotic zone, in the interior of the ocean, and in the sediments. As we worked our way from the top to the bottom of the water column, we examined the impact of the processes occurring in each of these zones on the distribution of organic matter, oxygen, nitrogen, and phosphorus. In this chapter and the next two we apply the basic tools we have developed to a system analysis of the entire oceanic cycle of the two remaining major elements involved in biogeochemical processes, namely silicon and carbon.

7.1 Introduction

Silicon exists in the ocean as *silicate*, which is a compound or ion containing SiO_4 tetrahedra either in the solid phase (referred to as *silica*) or in solution. It exists in seawater primarily as *silicic acid*, $Si(OH)_4$, with about 5% of it present as its dissociated anion, $SiO(OH)_3^-$. We will use the term silicic acid to refer to the aggregate of silicic acid and its dissociation products. There are several organisms that take up silicic acid to form various types of structures, including sponges, radiolaria, silicoflagellates, and diatoms. Of these, the most important by far, and thus the primary focus of this chapter, are the diatoms. Diatoms utilize silicic acid (specifically $Si(OH)_4$ [*Del Amo and Brzezinksi*, 1999; *Wischmeyer et al.*, 2003]) to form their cell wall or *frustule*. The frustule is composed of amorphous silica $SiO_2 \cdot nH_2O$, which is referred to as *biogenic silica* or *biogenic opal*. Amorphous silica can also be formed by inorganic processes, for which the terms *lithogenic silica* or *lithogenic opal* are used. Most of the lithogenic opal in the ocean comes from the land. In this chapter, we will use the word *opal* to refer to biogenic opal or biogenic silica, and we will use the expression *lithogenic silica* to refer to silica formed by inorganic processes.

The behavior of opal in the ocean, with formation and recycling at the surface, as well as sinking and dissolution in the water column, has many analogies to that of organic matter. The analogy extends even to the tendency for the stoichiometric proportions of elements in diatoms to be nearly constant. As we shall see, diatoms with adequate light and nutrients generally contain Si and N in a 1:1 ratio [*Brzezinski*, 1985]. On the other hand, there are several ways in which the silicon cycle in the ocean differs quite significantly from the other nutrient cycles: (a) silicic acid is utilized almost exclusively by diatoms, whereas nitrate and phosphate are utilized by all phytoplankton; (b) siliceous material is not passed up the food chain to a significant degree; (c) regeneration of opal to silicic acid is due to dissolution, not biological (i.e., metabolic) degradation (although biological processes play an important role in generating or destroying the organic matrix that protects the opal surface from dissolution by undersaturated water); and (d) Si exists in seawater only in its inorganic form, primarily as silicic acid, whereas N and P (especially N) exist in several inorganic and organic forms [*Brzezinski and Nelson*, 1989].

Diatoms are relative newcomers to the evolutionary tree [cf. *Baldauf*, 2003]. They first appear in the geologic record during the early Jurassic (ca. 185 million years ago), though it is believed that they may have evolved earlier [*Wiebe et al.*, 1996; cf. *Harwood and Gersonde*, 1990]. Intense spreading of diatoms into new habitats occurred during the Cretaceous period [*Harwood and Gersonde*, 1990], which spans 144 to 65 million years ago, and they were abundant in the ocean by the beginning of the Eocene 55 million years ago [*Siever*, 1991]. There are more than 10,000 recognized diatom species, each of them with a different cell wall structure [cf. *Pickett-Heaps et al.*, 1990]. Diatoms are extremely efficient at taking up silicic acid from seawater, and *Siever* [1991] has argued that diatom evolution was associated with a massive drop in the mean oceanic concentration of silicic acid during the late Cretaceous and into the early Tertiary. Today they continue to exercise a dominant influence on the oceanic silicate cycle.

Diatoms are unusually important in oceanic biological productivity. *Nelson et al.* [1995] say diatoms may account for as much as one-third to half the primary production. Diatoms are also disproportionately important in the export of organic matter from the surface ocean [cf. *Buesseler*, 1998; *Kemp et al.*, 2000].

What are the consequences of this astonishing dominance of diatoms for the distribution of silicic acid in the water column and of opal in ocean sediments? And how does the resulting distribution of silicic acid in the water column affect diatom productivity? These are the underlying themes of this chapter. We begin by first looking at measurements of silicic acid in the water column and then opal and silicic acid in the sediments.

Water Column Observations

Figure 7.1.1a shows the silicic acid concentration at the surface of the ocean and figure 7.1.2a shows it at depth. As a reference point against which to compare these observations, we discuss in panel 7.1.1 the solubility of opal in the ocean. Comparison of the data in figures 7.1.1 and 7.1.2 with the solubilities in figure 2 of the panel shows that both surface and deep silicic acid concentrations are highly undersaturated with respect to opal. How can such extreme undersaturation be maintained? Clearly, diatoms must be capable of precipitating opal from highly undersaturated water; and the dissolution rate, which has an average timescale of 1/(23 days) at the surface but is affected by a variety of environmental factors as well as variations in the properties of the opal (see discussion in panel 7.1.1), must be slow enough for the opal to survive rapid dissolution.

The surface concentrations of figure 7.1.1a are highest in the Southern Ocean, but are also high in the North and equatorial Pacific. Silicic acid is very nearly depleted in other regions such as the subtropical gyres. Given the very similar distributions of nitrate (figure 7.1.1.b) and silicic acid, and given that diatoms cannot grow without silicic acid, one would infer from the observations in figure 7.1.1 that diatoms would tend to have relative abundances at the surface of the ocean that are very similar to those of other phytoplankton.

However, there are some striking differences between the nitrate and silicic acid concentrations at the surface that have proven difficult to explain. The most remarkable of these is the band around the Southern Ocean between about 40°S and 60°S, where nitrate is abundant in a region where silicic acid is about as low as it gets anywhere (figure 7.1.1a and b). This is the only major region of the ocean where such a phenomenon exists. What is special about this region? Does this phenomenon simply reflect the ratio of silicic acid to nitrate in the water supplied to the surface in the Southern Ocean? Or are diatoms and/or other phytoplankton different in this band than elsewhere?

Figure 7.1.2 shows vertical sections of silicic acid and nitrate along a track that goes from the North Atlantic through the Southern Ocean and into the North Pacific (figure 2.3.3). The deep concentrations are much higher than surface concentrations. The silicic acid distribution is much like that of nitrate, with low values in the North Atlantic Deep Water and increasing concentrations along the pathway of the conveyor belt circulation all the way to the North Pacific. The highest values are found in the North Pacific Deep Water. However, there are some significant differences between the vertical nitrate and silicic acid distributions. The most important of these is that silicic acid appears to be virtually absent from the main thermocline except in the North Pacific. Also, the nitrate maximum in the North Pacific is much shallower than that of silicic acid. What are the processes that give rise to the high silicic acid concentrations in the deep ocean, and what prevents the high silicic acid in the deep ocean from getting into the thermocline and surface waters?

A careful examination of the Southern Ocean silicic acid and nitrate distributions reveals another puzzle. Here we see that nitrate is vertically well mixed except for reduced concentrations within the upper 100 m or so, whereas the silicic acid gradient penetrates down to 1000 m and deeper. Furthermore, most of the silicic acid appears to be trapped in the Southern Ocean, that is to say, there is little apparent transport of silicic acid out of the Southern Ocean except perhaps at the very bottom. *Nelson and Gordon* [1982] noted that the enrichment of deep Southern Ocean water with silicic acid is the most pronounced anomaly in the global silicon cycle, but that its origin is not understood. Have we learned anything new since then that might shed some light on the causes of this anomaly?

Sediment Observations

Figure 7.1.3 shows the distribution of opal at the surface of the sediments in weight percent on a calcite-free basis, i.e., relative to the dry weight of the sediments once all calcite has been removed (see *Archer et al.* [1993] for a discussion of possible issues introduced by removing the calcite fraction). The pattern of opal concentration, with high values in the equatorial Pacific and Southern Ocean, and intermediate values in the equatorial Atlantic, Indian Ocean, and North Atlantic, is unlike that of any other biogenic substance. Organic carbon (figure 6.1.4) is confined primarily to the continental margins, while $CaCO_3$, which will be discussed in chapter 9, is confined primarily to shallow topography, including the mid-ocean ridges. What determines the pattern of opal in

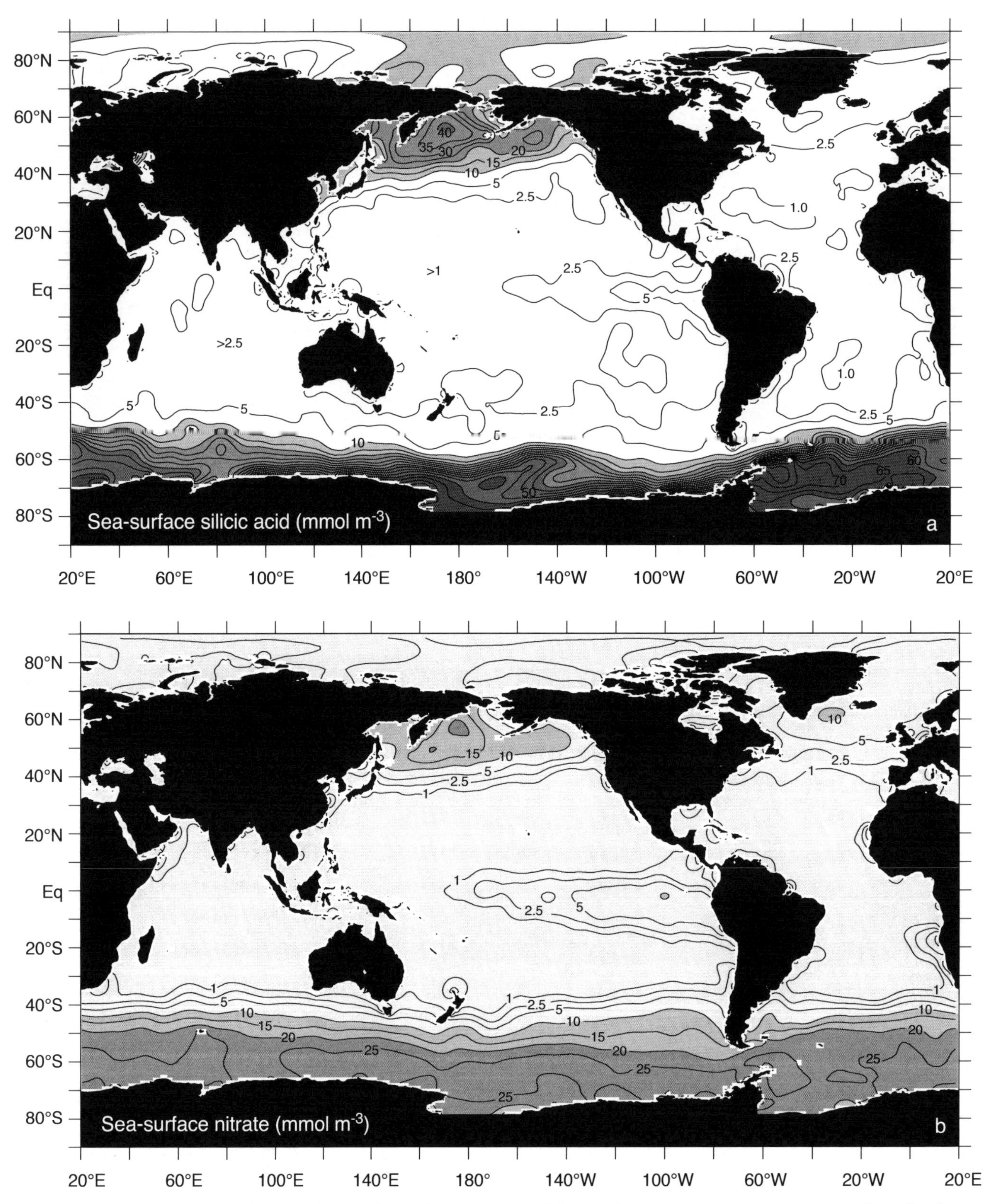

FIGURE 7.1.1: Surface maps of (a) annual mean silicic acid concentration and (b) annual mean nitrate concentration in units of mmol m^{-3}. See also color plate 2. Data are from the World Ocean Atlas 2001. [*Conkright et al.*, 2002].

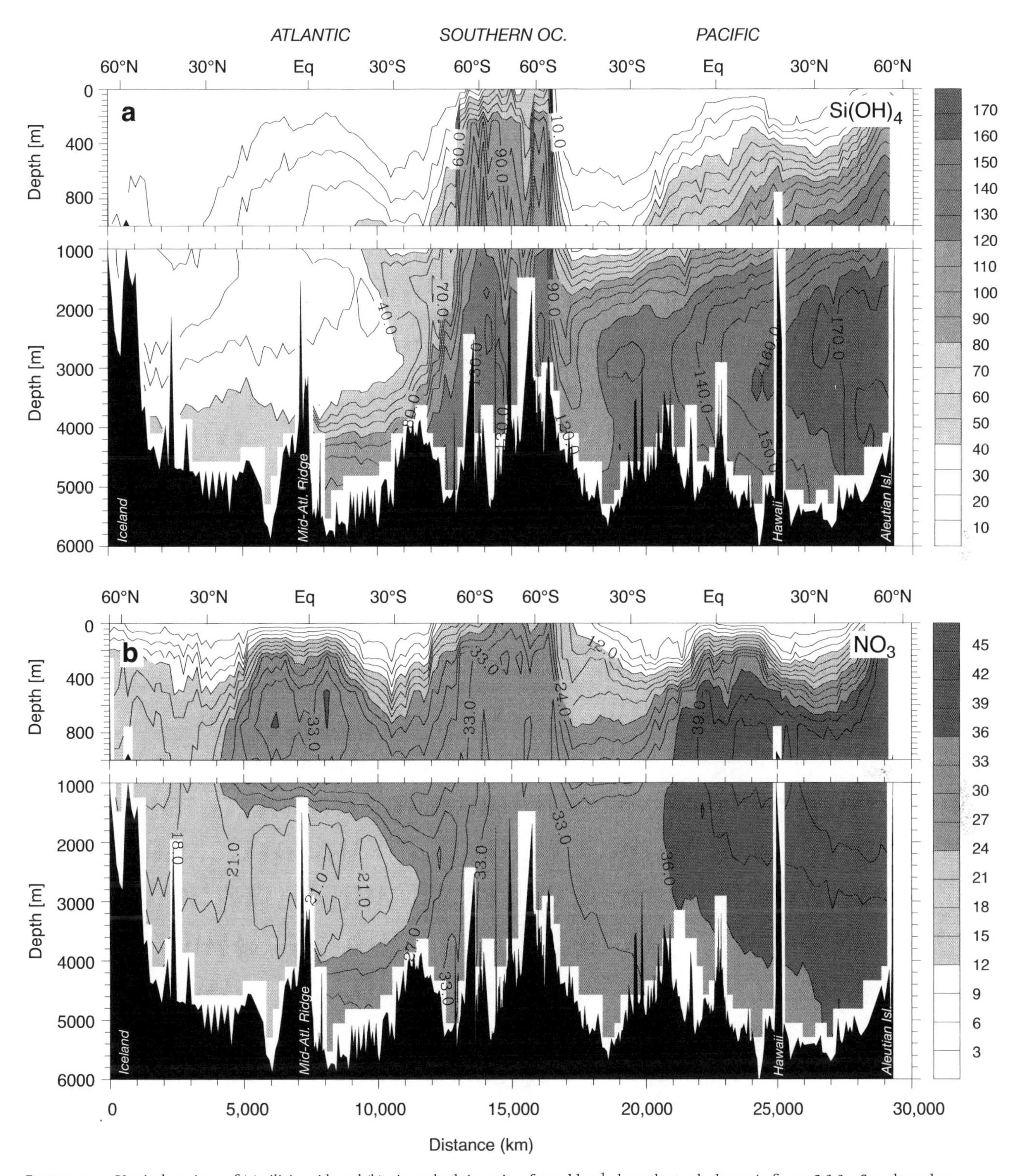

FIGURE 7.1.2: Vertical sections of (a) silicic acid, and (b) nitrate both in units of $\mu mol\ kg^{-1}$ along the track shown in figure 2.3.3a. See also color plate 6.

the sediments, and why is the pattern so different from that of other biogenic substances?

Figure 7.1.4 shows the silicic acid concentration in pore waters at representative locations from around the world. All of the profiles show an increase with depth to an asymptotic value that ranges from < 200 to $> 800\ mmol\ m^{-3}$. The fact that these concentrations are all higher than water column concentrations implies a flux from the sediments to the water column. What is the consequence of that flux for the con-

Panel 7.1.1: Opal Solubility and Dissolution Kinetics

The solubility of opal, $[Si(OH)_4]_{sat}$, is defined as the concentration of silicic acid in seawater that is in equilibrium with opal. It has been shown to be sensitive to temperature, the *specific surface area* (*SSA*, defined as the surface area of the opal per unit mass), pressure, pH, and the incorporation of Al in the silica matrix. For the solubility of opal in the water column, the most important terms are the specific surface area, pressure, and temperature, while for the solubility of opal in sediments the incorporation of Al into the matrix plays a dominant role. This incorporation can occur at the time the opal is formed [e.g., *Gehlen et al.*, 2002] as well as by reprecipitation in sediments with high Al(III) concentration in pore water. We focus here on water column solubility, and return to the effect of Al incorporation in section 7.4.

Table 1 gives an equation for the solubility of opal as a function of temperature, surface area, and pressure. The pH dependence for oceanic ranges is relatively small [cf. *Van Cappellen and Qiu*, 1997a] and will not be discussed any further. Comparison of this function with measured solubilities for cultured and surface-dwelling diatoms, all normalized to a surface area of zero, shows generally good agreement, though the cultured and surface-dwelling diatoms show somewhat smaller temperature dependence (figure 1). In order to consider the influence of specific surface area (*SSA*), we need to know typical values for opal. *Dixit and Van Cappellen* [2003] show that typical surface ocean diatoms have *SSA* ~100 $m^2\ g^{-1}$ (cultured diatoms range from 50 to above 200 $m^2\ g^{-1}$), while sediments tend to be lower, around 25 $m^2\ g^{-1}$. This decrease in SSA is likely a consequence of the aging of biogenic opal that reduces the surface area by more rapidly eroding diatom frustules with high surface area and removing delicate structures. If we adopt a high value for SSA of 250 $m^2\ g^{-1}$, the solubility shown in figure 1 increases by an average of 19% (20% at 0°C to 18% at 30°C).

The effect of temperature and incremental effects of pressure and surface area are illustrated in figure 2 using the observed global horizontal mean temperature profile. Pressure increases the solubility, and decreasing surface area decreases it, giving a vertical solubility profile that is quite different in shape from what one would obtain by considering temperature alone. Table 2 shows how the solubility changes as opal transitions from typical low-latitude surface water conditions to typical core top

TABLE 1

Solubility of opal.

The first equation from Van Cappellen and Qiu [1997a] gives the solubility as a function of temperature at 1 atm of total pressure. It is based on samples from an opal-rich Antarctic core with an average surface area of ~1500 $m^2\ mol^{-1}$ (25 $m^2\ g^{-1}$ times the molecular weight of SiO_2, 60 $g\ mol^{-1}$). The second equation from [Dixit et al., 2001] shows how the surface area of the opal modifies solubility. The third equation, also from Dixit et al. [2001], shows how pressure modifies the solubility. The final equation combines the first three, assuming that the pressure dependence is not sensitive to surface area, and that the surface area dependence is not sensitive to the pressure. T is absolute temperature in K, P is the total pressure in atmospheres, and SSA is the specific surface area in $m^2\ mol^{-1}$ (i.e., 60 $g\ mol^{-1}$ × the SSA in the commonly used units of $m^2\ g^{-1}$).

Temperature Dependence (SSA = 25 $m^2\ g^{-1}$)

$$[SiOH_4]_{SAT}(T, P = 1\ \text{atm},\ SSA = 1500\ m^2\ mol^{-1}) = 2.754 \times 10^6 \cdot \exp\left(\frac{-2229}{T}\right)\ \text{mmol m}^{-3}$$

Surface Area Dependence

$$\frac{[SiOH_4]_{SAT}(T, P = 1\ \text{atm},\ SSA)}{[SiOH_4]_{SAT}(T, P = 1\ \text{atm},\ SSA = 0\ m^2\ mol^{-1})} = \exp\left(3.688 \times 10^{-3} \cdot \frac{SSA}{T}\right)$$

Pressure Dependence

$$\frac{[SiOH_4]_{SAT}(T, P,\ SSA)}{[SiOH_4]_{SAT}(T, P = 1\ \text{atm},\ SSA)} = \exp\left[\frac{1}{T} \cdot \left(0.20 \cdot P - 2.7 \times 10^{-4} \cdot P^2 + 1.46 \times 10^{-7} \cdot P^3\right)\right]$$

Combined Equation

$$[SiOH_4]_{SAT}(T, P,\ SSA) = 2.754 \times 10^6 \cdot \exp\left[\frac{1}{T} \cdot \begin{pmatrix} -2229 \\ -3.688 \times 10^{-3} \cdot 1500 + 3.688 \times 10^{-3} \cdot SSA \\ +0.20 \cdot P - 2.7 \times 10^{-4} \cdot P^2 + 1.46 \times 10^{-7} \cdot P^3 \end{pmatrix}\right]$$

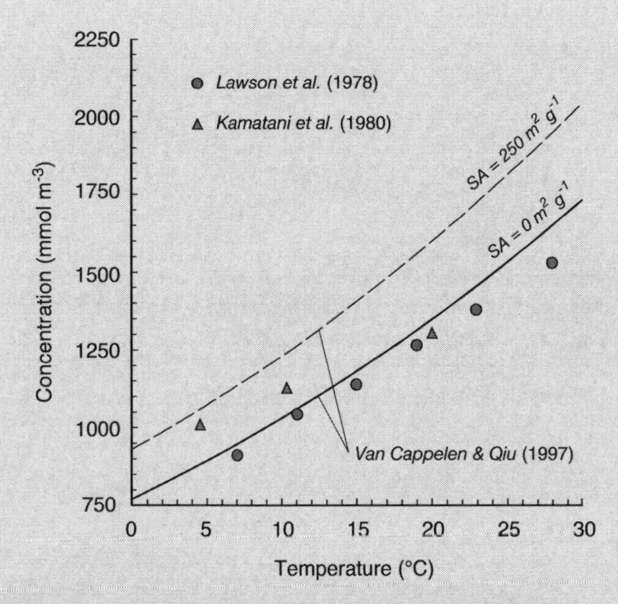

FIGURE 1: Opal solubility as a function of temperature at a total pressure of 1 atm. The curve and all the data on this plot have been renormalized to a surface area of 0 using the second equation in table 1. The solid line is based on the first equation in table 1, renormalized from a surface area of 25 $m^2\ g^{-1}$. The *Lawson et al.* [1978] and *Kamatani et al.* [1980] data were renormalized using the measured surface areas as summarized by *Dixit et al.* [2001], which average 108.5 $m^2\ g^{-1}$ for the *Lawson et al.* [1978] data and 258 $m^2\ g^{-1}$ for the *Kamatani et al.* [1980] data. The *Lawson et al.* [1978] data were obtained using surface plankton collected in Kaneohe Bay, Hawaii, while those of *Kamatani et al.* [1980] were obtained using cultured diatoms. Also shown on the figure is the *Van Cappellen and Qiu* [1997a] line renormalized to a surface area of 250 $m^2\ g^{-1}$.

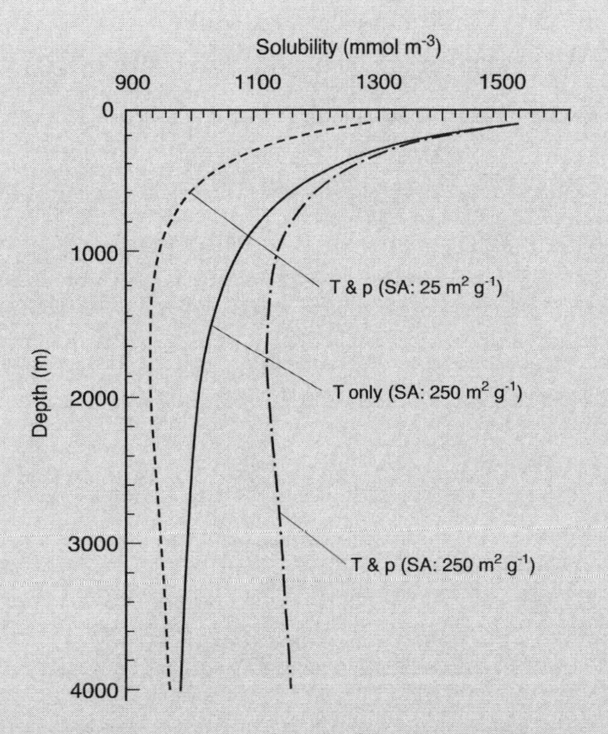

FIGURE 2: Opal solubility as a function of depth using global horizontal average potential temperatures, which range from a high of 17.3°C at the surface to a low of 1.4°C at 4000 m (cf. table in chapter 5 problem set; note: one should use *in situ* temperatures for this calculation, though the difference is small). The solid line is for a pressure of 1 atm and a surface area of 250 $m^2\ g^{-1}$; the dash-dot line includes the pressure dependence of the solubility; and the short dashed line is for a surface area of 25 $m^2\ g^{-1}$ including also the pressure dependence.

conditions in the deep ocean at 5000 m depth and an *in situ* temperature of 2°C. The total change is from a solubility of 1850 mmol m^{-3} at the surface to 1004 mmol m^{-3} at the ocean floor. The highest silicic acid concentrations observed in the ocean are $\sim$170 mmol m^{-3}, which means that the water column is highly undersaturated everywhere.

Because silicic acid in the ocean is highly undersaturated with respect to opal, one would expect there to be extensive opal dissolution in the water column. However, the extent of dissolution depends not just on the degree of saturation, but also on the kinetics of dissolution relative to the residence time of diatom opal in the water column. The dissolution rate of opal has been shown to depend on the temperature, the total surface area of opal present per unit volume of the solution, the fractional extent of undersaturation $1-[Si(OH)_4]/[Si(OH)_4]_{SAT}$, the density of reactive sites on the opal surface and various processes that can modify this, and the presence of organic coatings on diatom frustules. Pressure and pH have only a minor impact in the ocean. The rate of dissolution of opal is usually specified as a function of the form

$$\frac{\partial(1-\phi)\cdot[SiO_2]}{\partial t} = -k\cdot\{(1-\phi)\cdot[SiO_2]\}\cdot\left(1-\frac{[Si(OH)_4]}{[Si(OH)_4]_{SAT}}\right) \quad (1)$$

[cf. *Schink et al.*, 1975]. Here k is the dissolution rate constant with units of time^{-1} and $(1-\phi)\cdot[SiO_2]$ is the amount of opal present in mmol m^{-3} ($[SiO_2]$ is the opal concentration in mmol m^{-3} of solids, and ϕ is the porosity). The $(1-\phi)\cdot[SiO_2]$ term is intended to represent the effect of surface area on the solubility, assuming that surface area is directly proportional to the opal present in the water. However, as we have seen above, the surface area per unit mass of opal, i.e., the specific surface area *SSA*, is quite variable. The impact of variability in *SSA* on k is taken into account as follows:

$$\left.\frac{k}{k_{reference}}\right|_{T=constant} = \frac{SSA}{SSA_{reference}} \quad (2)$$

[cf. *Hurd and Birdwhistell*, 1983]. For a typical specific surface area range of 25 to 100 $m^2\ g^{-1}$, this results in a dissolution rate constant that varies by a factor of 4. The

centration of silicic acid in the water column above the sediments? Opal profiles are mostly constant with depth within the measurement uncertainty. The fact that the silicic acid concentrations approach an asymptotic value while there is still opal present in the sediments (see concentrations in figure 7.1.3) would normally be taken as implying that the silicic acid is in equilibrium with the opal in the sediments. Is this the case, and if so, how can one explain the fact that even the highest of these concentrations is below the saturation concentration of $\sim$1004 mmol m^{-3} that we calculate in panel 7.1.1 for 2°C at a depth of 5000 m; and what is the explanation for the wide range in the putative equilibrium concentrations? If the concentration of $Si(OH)_4$ is not in equilibrium with the opal in the sediments, how does one explain the fact that $Si(OH)_4$ apparently achieves a steady concentration while there is still opal present?

The long-term burial of opal in the sediments is the only sink for silicon in the ocean. If the oceanic silicic acid concentration is to be maintained at a steady state, the loss of silicon by burial must be equal to the sources, which are primarily rivers, with some input from dust-borne delivery and weathering of ocean floor basalts. As we have seen in chapter 1, an important requirement for maintaining a steady-state oceanic concentration of a given chemical is that there be a link between its concentration and the processes that

Panel 7.1.1: *(Continued)*

TABLE 2

An illustration of how opal solubility might change as it transitions from typical low-latitude surface water conditions for an opal particle with a relatively high surface area (*SSA*) of 250 m^{-2} g^{-1} of opal to a typical core top sample at 5000 m depth with a surface area of 25 m^{-2} g^{-1} of opal.

Parameters	*Solubility (mmol m^{-3})*
T=25°C, P=1 atm, SSA=250 m^2 g^{-1}	1850
T=2°C, P=1 atm, SSA=250 m^2 g^{-1}	1004
T=2°C, P=500 atm, SSA=250 m^2 g^{-1}	1203
T=2°C, P=500 atm, SSA=25 m^2 g^{-1}	1004

effect of temperature on k is given by the Arrhenius equation

$$\left.\frac{k}{k_{reference}}\right|_{SSA=constant} = \exp\left[\frac{E_a}{R}\left(\frac{1}{T_{reference}} - \frac{1}{T}\right)\right] \tag{3}$$

where E_a is the experimental activation energy of opal dissolution ($\sim 50 \pm 15$ kJ mol^{-1}, based on the data summarized by *Van Cappellen et al.* [2002]); T is absolute temperature in degrees K; and R=8.3144 J mol^{-1} K^{-1} is the universal gas constant. For a typical surface temperature range from 0°C to 25°C, the dissolution rate would increase by a factor of 6. Organic coatings have been shown to reduce the dissolution rate by as much as a factor of 10 [*Bidle and Azam*, 1999]. In addition, temperature influences the rate at which bacteria break down the organic coatings and may sometimes contribute inadvertently to measurements of the Arrhenius temperature effect [*Bidle et al.*, 2002]. The influence on opal solubility of the density of reactive sites on the opal surface is particularly influential in sediments and will be discussed in section 7.4.

The observed dissolution rate constant k of opal at the surface of the ocean varies widely from 10 to 100 yr^{-1} (i.e., a $1/k$ timescale of 4 to 40 days; cf. compilation by *Nelson et al.* [1991]) with a global average of 16 yr^{-1} ($1/k$ timescale of 23 days [cf. *Van Cappellen et al.*, 2002]). Over most of the water column the silicic acid concentration is so highly undersaturated, i.e., $[Si(OH)_4] \ll [Si(OH)_4]_{sat}$, that we can simplify (1) to

$$\frac{\partial[SiO_2]}{\partial t} = -k \cdot [SiO_2] \tag{4}$$

(assuming also that porosity is constant with time). Equation (4) has the solution

$$[SiO_2] = [SiO_2]_{t=0} \cdot \exp(-k \cdot t) \tag{5}$$

From (5), we can calculate that if diatoms sank at a rate of 100 m d^{-1} over a depth range of 5000 m, 89% of the opal would dissolve if k remained constant at 1/(23 d). The range of k of between 1/(4 d) and 1/(40 d) gives a dissolution fraction of about 100% to 70% over 5000 m. However, given the effect of temperature and other parameters in reducing the dissolution rate, these are likely to be upper limits to the extent of dissolution.

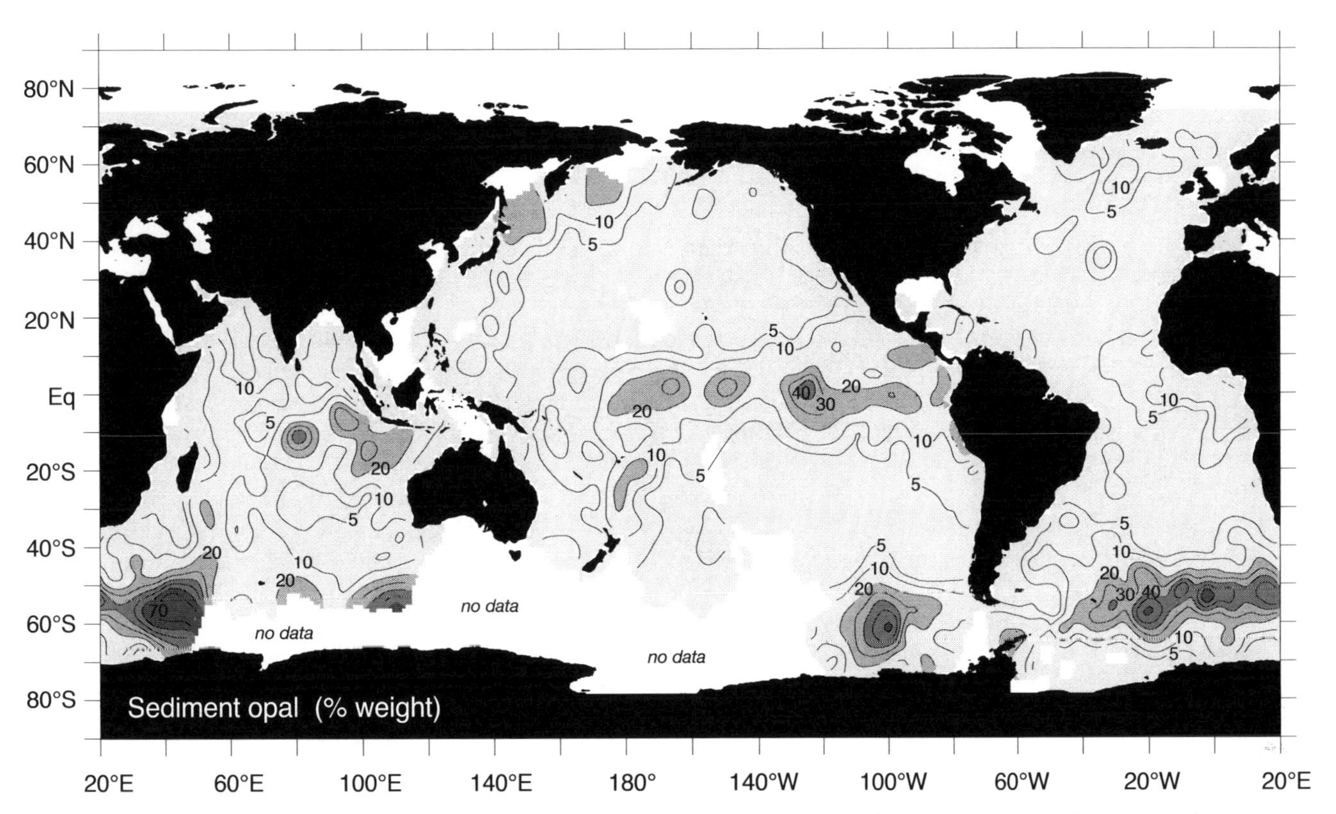

FIGURE 7.1.3: Global map of opal weight % (mass opal to mass total solid dry sediments) in sediments from *Seiter et al.* [2004]. See also color plate 5.

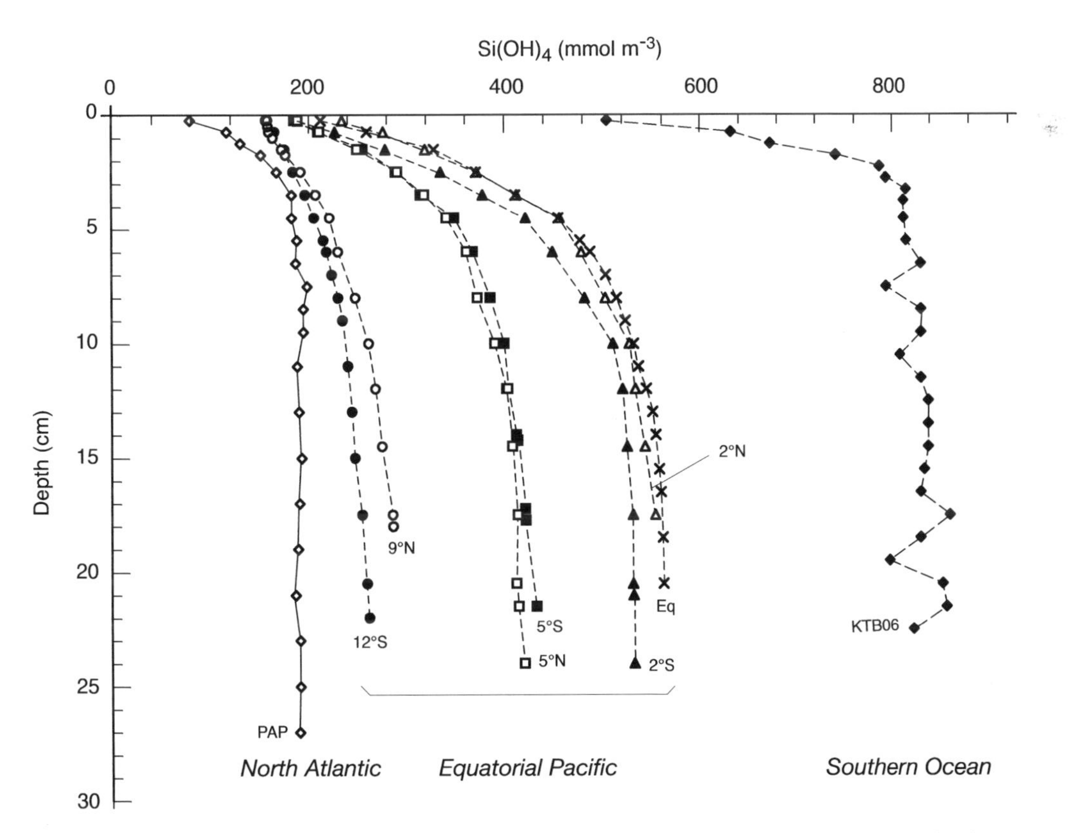

FIGURE 7.1.4: Silicic acid concentration in sediment pore waters from representative locations in the North Atlantic [*Ragueneau et al.*, 2001], equatorial Pacific [*McManus et al.*, 1995], and Southern Ocean [*Rabouille et al.*, 1997]. Data were provided by Olivier Ragueneau (personal communication, 2004).

remove it from the water column. What are the feedback processes between the water column silicic acid concentration and opal burial in the sediments?

Outline

The remainder of this chapter is aimed at answering the queries that have been raised during our brief survey of water column and sediment observations. We begin in section 7.2 with the surface layer. This is the euphotic zone, where diatoms fabricate opal frustules from silicic acid, and where about 50% of the opal is returned to solution and the remainder is exported. In section 7.3, which is on the water column, we consider the silicon cycle within the main thermocline and deep waters. Our main concern will be to understand how silicon that is exported from the surface of the ocean as opal is returned back to the surface as silicic acid in order to continue to fuel diatom production. We will see that about 38% of the opal originally produced in the surface (77% of the exported opal) dissolves directly in the water column, and the remaining 12% (25% of the exported opal) rains onto the sediments. In section 7.4, which is on sediments, we consider the cycling of opal within the sediments, where about four-fifths of the opal rain (~10% of the surface opal production) dissolves and is returned to the water column, and the remainder (~3% of the opal production) is buried and eventually becomes sedimentary rock.

We emphasize that our discussion in this chapter is focused primarily on open ocean processes, not including continental margins. The recent study of *DeMaster* [2002] suggests that the continental margins may be much more important to the marine silica budget than had been previously thought. In his budget, the burial of opal in continental margins accounts for ~40% of the total oceanic sediment burial, as contrasted with earlier studies that had given estimates more like 15%. This has significant implications for how we view the global silica budget. In particular, as we shall see, the open ocean observations indicate that the silica cycle and organic carbon cycle tend to be decoupled, whereas continental margin sediments show a much tighter coupling between these two budgets than implied by the open ocean observations [cf. *DeMaster*, 2002]. We shall return to a brief consideration of continental margins in the concluding section.

7.2 Euphotic Zone

The outstanding feature of diatoms vis-à-vis other phytoplankton is their frustule. We begin in the first subsection with a discussion of the functions that frustules are thought to serve for diatoms, and how frustules contribute to the success of diatoms vis-à-vis other phytoplankton. This is followed by a discussion of the metabolism of silicon in diatoms. The second subsection discusses opal production and export, and the contribution of diatoms to biological productivity and the export of organic matter from the surface of the ocean.

Diatoms

The diatom frustule is composed of two half shells (*valves*) that fit together like a pillbox, with a series of girdle bands that are deposited after the new valves are formed. The outer shell is referred to as the *epitheca*, and the inner as the *hypotheca* (see figure 7.2.1 and discussion by *Kröger and Wetherbee* [2000]). When a diatom initiates asexual reproduction, it forms two silica deposition vesicles that gradually expand to form two new *hypovalves* (i.e., inner valves) within the original cell. During cell division, the two new half shells are always manufactured to fit inside the old half shells. This leads to smaller and smaller cells over time. Diatoms restore their cells to full size during sexual reproduction, which occurs infrequently.

The long-standing theory regarding frustules is that they protect diatoms from predators. Indeed, a recent set of experiments by *Hamm et al.* [2003] shows that the structure of diatom shells makes them remarkably resistant to breakage. Another important suggestion is that a cell wall made of silica requires less energy to build than a cell wall made of other materials, thus potentially leaving more energy for other purposes [*Martin-Jézéquel et al.*, 2000; *Raven*, 1983]. *Milligan and Morel* [2002] suggest that silica may also enhance the ability of diatoms to acquire inorganic carbon (CO_2) from seawater. Diatoms, like many other organisms, have extracellular carbonic anhydrase that can catalyze the conversion from bicarbonate ion (HCO_3^-), which is abundant in seawater, to CO_2, which is much less abundant but is the form of inorganic carbon required for photosynthesis. Opal is an effective pH buffer that can enhance carbonic anhydrase catalytic rates.

The complex details of frustule morphology may serve other functions as well. For example, laboratory experiments of the flow of suspended particles past diatom shells [*Hale and Mitchell*, 2001] indicate that the structure of the shell channels particles away from the pores of the frustule. This keeps the pores free of particles, thus allowing for more efficient diffusion of nutrients into the cell. *Hale and Mitchell* [2001] also suggest that the flow regime may contribute toward breaking up

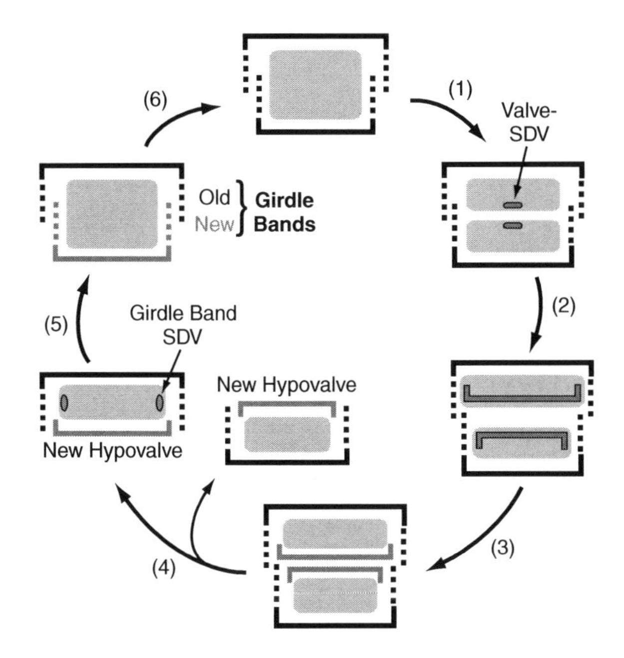

FIGURE 7.2.1: Diatom frustules and asexual cell reproduction cycle [per *Kröger and Wetherbee*, 2000].

of colloids and making the nutrients in them more readily available to the diatom.

Diatom frustules thus likely play a central role in the success of diatoms relative to other phytoplankton, but many other factors have been proposed as well. The success of one phytoplankton over another requires that it be better able to take advantage of existing resources (i.e., more efficient production) and/or that it be less susceptible to zooplankton grazing, viral infections, and other causes of mortality. Frustules make it harder for zooplankton to get at diatoms, but *Smetacek* [1999] points out that there do exist predators, such as copepods with silica-edged mandibles, and euphausiids with gizzards lined with sharp teeth, that are able to break the shells of diatoms [cf. *Hamm et al.*, 2003]. There is also evidence of other predators that are able to feed on diatoms by penetrating them with a stalk-like appendage called a peduncle, or enveloping them in a digestive sack deployed around the cell [*Smetacek*, 1999]. Thus the ability of diatoms to successfully escape predation requires further explanation. *Smetacek* [1999] argues that grazing of diatom-feeding zooplankton by other zooplankton may play an important role. He further proposes that bulky diatoms are not easy for zooplankton to cope with, and that the plasma in a given cell provides a relatively small amount of food for the amount of effort. He also cites evidence that diatoms may inhibit fertility in copepods [cf. *Miralto et al.*, 1999], although a recent survey of field data at a large number of locations shows no such effect [*Irigoien et al.*, 2002].

Other theories for the success of diatoms focus more on the production side. Diatoms are able grow faster under low light conditions and are able to take up and store nutrients at a more rapid rate than the picophytoplankton that are responsible for recycled production [*Smetacek*, 1998]. *Brzezinski et al.* [1998] argue that the unusually large half-saturation constants of many diatoms with respect to the silicic acid concentration may give diatoms an advantage in areas of episodic nutrient supply. The slow replication rate of diatoms in waters with low silicic acid concentration may help to keep their metabolic machinery intact and poised to respond rapidly when a pulse of nutrients arrives.

The final suggestions that we mention are two strategies that enable diatoms to access nutrients in regions where the concentrations are extremely low. Diatoms have large vacuoles in them that allow them to regulate their buoyancy [*Round et al.*, 1990], aided perhaps by the high density of opal and the presence of protrusions such as spines that may help modify their buoyancy. It has been observed in some oceanic and freshwater diatoms that they reduce their buoyancy when they are nutrient-depleted [cf. *Richardson et al.*, 1996; *Titman and Kilham*, 1976]. It has been suggested that buoyancy regulation might be a means by which a few of the very largest diatoms that are able to undergo vertical migration of significant magnitude might access nutrients that are deeper in the water column [e.g., *Richardson et al.*, 1996; *Villareal et al.*, 1993]. In an interesting analysis, *Titman and Kilham* [1976] use a simple model to show how sinking of phytoplankton in the presence of Langmuir circulation could function to optimize the ability of diatoms to access nutrients deeper in the water column while preventing them from sinking out of the euphotic zone. In regions where diatoms can access phosphate but not nitrate by this or other methods, those with endosymbiotic nitrogen-fixers (e.g., *Rhizosolenia* and *Hemiaulus*) may flourish [cf. *Karl et al.*, 2002; *Villareal*, 1991].

To summarize, there is as yet no emerging single theory for the success of diatoms over other phytoplankton. It seems likely that the frustule plays an important role, both by helping to provide some resistance to predation, and by increasing the metabolic efficiency of diatoms in several different ways. Numerous other suggestions have been made, including largely anecdotal evidence for the relative absence of viral infections in diatoms (although see the recent study by *Nagasaki et al.* [2004] for evidence to the contrary).

The uptake of silicic acid by diatoms follows the hyperbolic Monod saturation function (see discussion in section 4.2 and figure 7.2.2 from *Leynaert et al.* [2004]). However, the maximum uptake rate V_{max} and the half-saturation constant $K_{Si(OH)_4}$ are highly variable. *Martin-Jézéquel et al.* [2000] summarize results from a large number of batch and continuous cultures from different diatoms under a variety of conditions. The maximum uptake rate obtained in these experiments ranges from a low of 0.02 to a high of 0.12 h^{-1}, and $K_{Si(OH)_4}$ ranges

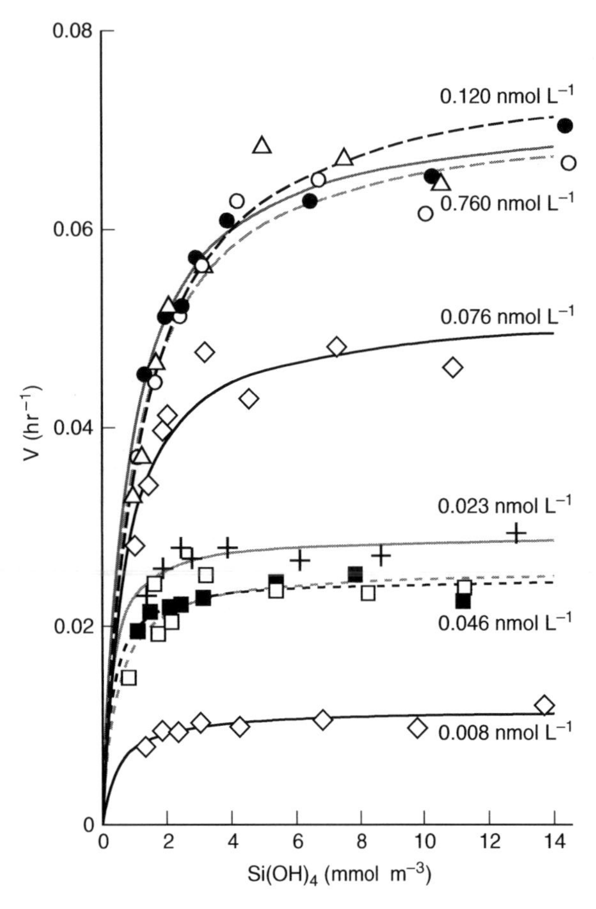

FIGURE 7.2.2: Diatom culture measurements of silicic acid uptake rates plotted here versus silicic acid concentration for different iron concentrations. The iron concentrations are shown adjacent to the lines [*Leynaert et al.*, 2004].

from a low of 0.2 to a high of 97 mmol m^{-3}. *Martin-Jézéquel et al.* [2000] discuss evidence indicating that there are multiple silicic acid transport proteins involved in transporting silicic acid across the cell wall [e.g., *Hildebrand et al.*, 1998], which may help explain why there is such great variability between diatoms, and even within a single diatom during different growth phases.

Trace metals such as iron and zinc can also have an influence on silicic acid uptake kinetics [cf. *De La Rocha et al.*, 2000]. Figure 7.2.2 shows silicic acid uptake estimates from a recent series of laboratory culture experiments with the diatom *Cylindrotheca fusiformis* where silicic acid concentrations were varied systematically in cultures with different iron concentrations [*Leynaert et al.*, 2004]. One can estimate a separate maximum silicic acid uptake rate V_{max} and the half-saturation constant $K_{Si(OH)_4}$ for each of the experiments with a different iron concentration. From the figure, one can see that generally V_{max} increases and $K_{Si(OH)_4}$ decreases as the iron concentration increases. The authors tested a variety of kinetic expressions to represent the codependence of the growth rate on iron and silicic acid. They found that a multiplicative model of the form

$$V = V_{max} \cdot \frac{[\mathrm{Fe}]}{K_{\mathrm{Fe}} + [\mathrm{Fe}]} \cdot \frac{[\mathrm{Si(OH)_4}]}{K_{\mathrm{Si(OH)_4}} + [\mathrm{Si(OH)_4}]} \tag{7.2.1}$$

(see discussion of uptake kinetics in chapter 4) fit their data quite well, with a V_{max} of 0.077 h^{-1}, a $K_{Si(OH)_4}$ of 0.80 mmol m^{-3}, and a K_{Fe} of 0.031 μmol m^{-3} (i.e., nmol l^{-1}). The concentration [Fe] in this equation refers to total dissolved iron. *Leynaert et al.* [2004] show that the uptake kinetics of silicic acid is strongly dependent on cell size, which in turn is influenced by the iron concentration, suggesting that the iron works by influencing the surface-to-volume ratio of the diatoms.

As shown in panel 7.1.1, the solubility of opal for surface ocean and cultured diatoms ranges between about 900 and 2000 mmol m^{-3} as temperature increases from 0°C to about 28°C. Comparison with the observed concentrations in figure 7.1.1a shows that surface waters are highly undersaturated with respect to opal. How do diatoms manage to precipitate opal? The concentration of silicic acid inside diatom cells exceeds that of seawater by two to three orders of magnitude, ranging from 19,000 to 340,000 mmol m^{-3} in a small number of studies summarized by *Martin-Jézéquel et al.* [2000]. The existence of such high concentrations in the cell clearly indicates active transport of silicic acid across the cell wall. As previously noted, it appears that there are multiple types of transporter proteins in diatoms involved in this transport. With such high cellular concentrations, a problem for diatoms is actually how to keep the silicic acid in the cell from polymerizing spontaneously. This suggests the existence of silica-binding organic material in the cell [cf. *Martin-Jézéquel et al.*, 2000].

Once in the cell, the precipitation of silicic acid to form opal occurs only during the part of the cell life cycle just prior to cell division [cf. *Claquin et al.*, 2002; *Martin-Jézéquel et al.*, 2000]. Opal formation is decoupled from the C and N metabolisms and continues to occur even in cells suffering from slow growth due to limitation of major nutrients, iron, or light. *Claquin et al.* [2002] and *Martin-Jézéquel et al.* [2000] suggest that the continuation of opal formation in slow-growing cells may explain why diatoms that are suffering from stress have higher Si/C and Si/N elemental ratios. Specifically, cells that are grown with adequate nutrients and light tend to have a Si:N ratio of order 1:1 [*Brzezinski*, 1985], whereas those grown under iron or other limitation have Si:N ratios that can be many times higher than this [cf. *Claquin et al.*, 2002; *Hutchins and Bruland*, 1998; *Martin-Jézéquel et al.*, 2000; *Takeda*, 1998; *Watson et al.*, 2000; *Leynaert et al.*, 2004].

OPAL PRODUCTION AND EXPORT

Table 7.2.1 summarizes average biological opal production rates from a variety of regions around the world

TABLE 7.2.1
Estimates of opal production in the world ocean based on the silicon isotope method.
Summary taken from Ragueneau et al. [2000].

Region	*Number of Profiles*	*Production rate* ($mmol\ m^{-2}\ d^{-1}$ *of Si*) *Low*	*Mean*	*High*
Coastal upwelling	77	2.3	90	1140
Other coastal	55	0.2	15	131
Deep ocean	84	0.2	2.3	12
Southern Ocean	74	0.9	15	93

[cf. *Ragueneau et al.*, 2000]. The mean opal production values range from a low of 2 $mmol\ m^{-2}\ d^{-1}$ in regions outside the Southern Ocean and coastal areas, to a high of 90 $mmol\ m^{-2}\ d^{-1}$ in coastal upwelling regions. One can get some sense of the magnitude of these opal production rates by noting that over a 100 m thick euphotic zone and a period of 100 days, they would reduce the silicic acid concentration by 2 to 90 $mmol\ m^{-3}$. It is interesting to note by reference to figure 7.1.1.a that if there were no resupply of silicic acid to the surface ocean, these removal rates would be more than enough to deplete the surface layer within the 100-day timescale we chose. In an earlier study, *Nelson et al.* [1995] used data such as that shown in table 7.2.1 to estimate that the global mean opal production was 1.6 to 2.1 $mmol\ m^{-2} \cdot d^{-1}$, which translates into a global opal production of 200–280 Tmol Si yr^{-1}. What is the fate of this opal?

As we have already noted, the concentration of silicic acid in the surface ocean is well below the saturation concentration of opal. With the global mean dissolution timescale of 23 days that we give in panel 7.1.1, one might expect that a substantial fraction of the opal produced by diatoms would dissolve over the period of several weeks that a typical bloom might last (cf. chapter 4). The large temperature sensitivity of opal solubility suggests that the fraction of opal dissolution should be much greater in warm than in cold waters. On the other hand, diatoms have a vested interest in preventing their frustules from dissolving, both for their own purposes, and also to preserve the quality of the inherited frustule from generation to generation. Indeed, culture experiments indicate that opal dissolution in healthy diatoms is negligible [*Milligan et al.*, 2004] and measurements at sea indicate that the ratio of the dissolution rate of opal to the uptake rate of silicic acid declines from about 70% during non-bloom conditions to about 20% during bloom conditions, with a modest tendency for the ratio to decrease with increasing opal production rate [*Brzezinski et al.*, 2003b]. The usual reason given for the resistance of opal to dissolution in healthy diatoms is that the opal is encased in an organic matrix. It has been found that the breakdown of this matrix by bacterial action on diatom detritus accelerates the rate of opal dissolution by more than an order of magnitude [*Bidle and Azam*, 1999], and that the breakdown rate as well as the dissolution rate are highly temperature-sensitive [cf. *Bidle et al.*, 2002].

Figure 7.2.3 shows measurement-based estimates of the ratio of export to opal production at nine different sites, summarized in the review of *Ragueneau et al.* [2002]. There is a wide range, from a low of 11% in the warm waters of the equatorial Pacific (89% of the opal dissolves in the surface layer), to a high of 84% (16% of the opal dissolves in the surface layer) in the cold waters of the Pacific sector of the Southern Ocean, with an overall average of 47%. This is consistent with the temperature sensitivity of the dissolution rate. Globally, the average export efficiency given by *Treguer et al.* [1995] is about 50%. Given the starting point of opal production of 200–280 Tmol yr^{-1}, this means that the export would be about 100 to 140 Tmol yr^{-1}.

Another way to estimate the opal export is from the organic nitrogen export $\Phi^{Organic\ N}$ multiplied by the Si:N export ratio calculated as described in panel 7.2.1 (see figure 7.2.4):

$$\Phi^{Opal} = \Phi^{Organic\ N} \cdot \left(\frac{Si}{N}\right) \qquad (7.2.2a)$$

where

$$\Phi^{Organic\ N} = (NPP \cdot p_e) \cdot \left(\frac{\Phi^{Organic\ C}}{\Phi^{POC}}\right) \cdot r_{N:C} \qquad (7.2.2b)$$

The first term in parentheses on the right-hand side of (7.2.2b) is the particulate organic carbon flux Φ^{POC}, which is calculated at a horizontal resolution of 1° by 1° using the satellite-based estimate of net primary production *NPP* previously discussed in chapter 4, times the particle export ratio $p_e = \Phi^{POC}/NPP$, from *Dunne et al.* [2005b] (see figure 7.2.5a). The next term in parentheses is the ratio of total organic carbon export to

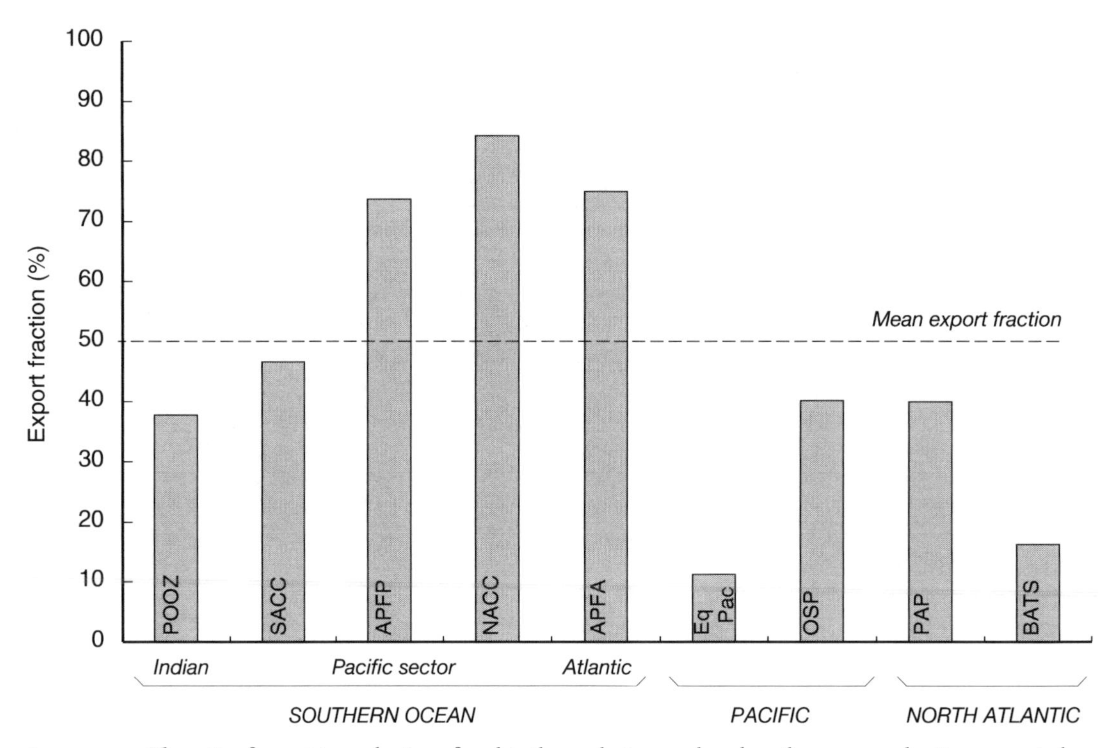

FIGURE 7.2.3: The ratio of export to production of opal in the euphotic zone based on the summary by *Ragueneau et al.* [2002]. These measurements were obtained at 9 representative locations in the Southern Ocean, Pacific, and North Atlantic. The mean of the data shown in the diagram is 47%, which is very close to the mean of 50% given by *Treguer et al.* [1995] in their study of the global ocean silicon cycle.

particulate organic export. We use a global mean value of 1.25 ± 0.12, which we base on the estimate by *Hansell* [2002] that dissolved organic carbon export is $20 \pm 10\%$ of total organic carbon export. For the stoichiometric $r_{N:C}$ ratio we use the value of 16:117 from *Anderson and Sarmiento* [1994]. Figure 7.2.5.b and table 7.2.2 show the opal export obtained by the above method but without the $\Phi^{Organic\,C}/\Phi^{POC}$ correction, since this is not available with regional resolution.

The global mean particulate organic carbon export and corresponding opal export we obtain without the $\Phi^{Organic\,C}/\Phi^{POC}$ correction are 9.8 ± 3.3 Pg C yr^{-1} and 104 Tmol Si yr^{-1}, respectively. Including this correction gives an organic carbon export of 12.2 ± 4.3 Pg C yr^{-1} and an opal export of 130 Tmol Si yr^{-1}, which is within the range of estimates given by *Treguer et al.* [1995]. As regards the regional distribution of opal export, table 7.2.2 and figure 7.2.5b reveal that the highest fluxes occur in three regions of the world: the equatorial and North Pacific (33% of the global total), and the Southern Ocean (40% of the global total south of 45°S, despite accounting for only 17% of the world area). The correlation with the organic carbon export is strong, of course, but as figure 7.2.5 demonstrates, there are major differences between the patterns of the two that remain to be explained.

One of the remarkable characteristics of diatoms that we have previously alluded to is the central role that they appear to play in the export of organic matter from the surface of the ocean [e.g., *Buesseler*, 1998]. We can now assess this contribution of diatoms. As an upper-bound estimate, we assume that diatoms export Si and N with a

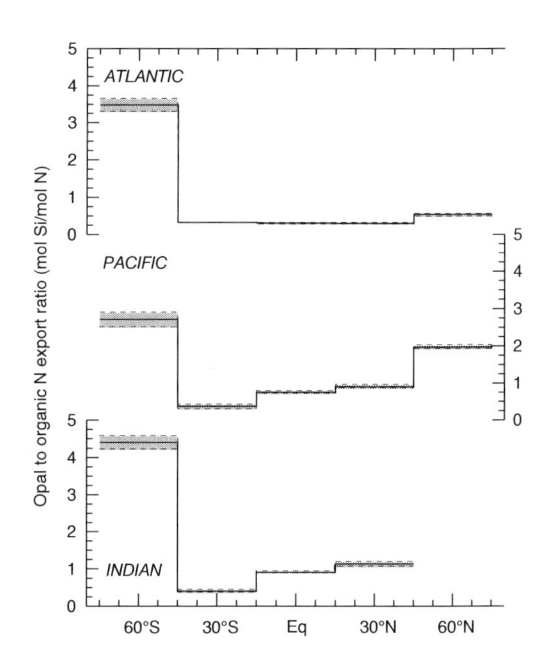

FIGURE 7.2.4: The zonal mean of the opal to organic nitrogen export ratio estimated from the vertical gradient of silicic acid divided by that of nitrate between the surface 100 m and the next 100 m (data taken from *Sarmiento et al.* [2004a]; cf. explanation in panel 7.2.1).

"healthy diatom" ratio of 1:1. If we further assume that the organic matter that is quantitatively associated with diatoms during export has an organic C-to-N ratio of 117:16, we find an upper-bound organic matter export driven by diatoms of about 950 Tmol C yr^{-1}, or about 11 Pg C yr^{-1}. As a lower-bound estimate, we assume a Si to N export ratio of 4:1 based on the export ratios in the Southern Ocean (see figure 7.2.4 and discussion in section 7.3), which gives an organic matter export driven by diatoms of about 3 Pg C yr^{-1}. Earlier, we saw that primary production and export production of organic carbon are 53 ± 8 Pg yr^{-1} and 12 ± 4 Pg yr^{-1} of carbon, respectively. Thus diatoms appear to be driving between 20% and 90% of organic matter export, with a central estimate of about 55%, supporting these previous claims. This is a remarkable contribution by a single group of phytoplankton. Is this simply a consequence of their contribution to ballasting organic matter, or are other characteristics also at play?

It has been observed that diatoms often terminate their blooms with the formation of algal mats that sink rapidly out of the euphotic zone. *Smetacek* [1985] has suggested that this represents a transition from a growing to a resting stage in the life cycle of diatoms, although the evidence for this is inconclusive [*Alldredge and Gotschalk*, 1989]. Observational studies by *Alldredge and Gotschalk* [1989] led them to conclude that mass flocculation likely involves differential settlement as a

Panel 7.2.1: Opal to Nitrogen Export Ratio

The export of opal and organic nitrogen from the surface ocean must, in a steady state, be balanced by biological uptake of silicic acid and nitrate, which in turn must be supplied by transport. Thus, if we can estimate the supply ratio of silicic acid to nitrate, we will know the export ratio [cf. *Sarmiento et al.*, 2002]. The conservation equation for a steady state that expresses the balance between transport and biological uptake is

$$SMS(C) = \mathbf{U} \cdot \nabla C - \nabla \cdot (D \cdot \nabla C) \tag{1}$$

where $SMS(C)$ is biological uptake, U is the velocity vector, D is the eddy diffusivity tensor, and C is tracer concentration. We consider the tracer balance in a box in the surface ocean over which we perform a volume integral:

$$\int SMS(C)dV = \int [\mathbf{U} \cdot \nabla C - \nabla \cdot (D \cdot \nabla C)]dV \tag{2}$$

The right-hand side of (2) can be converted into a surface integral by the Gauss divergence theorem, giving

$$\int SMS(C)dV = \oint (\mathbf{U} \cdot C - D \cdot \nabla C)dA \tag{3}$$

where A is the surface area of the box.

We consider here only the simplest case in which the horizontal gradient of tracer is negligible, i.e., the supply is predominantly due to vertical velocity and diffusion. Equation (3) then takes the form

$$\int SMS(C)dV = \left[w(C_{deep} - C_{surface}) - D_z\left(\frac{C_{surface} - C_{deep}}{\Delta z}\right)\right] \cdot A \tag{4}$$

where w (which can be either positive or negative depending on whether the water is upwelling or downwelling) is the vertical input of water with a concentration C_{deep}, or loss of water with a concentration $C_{surface}$. An equal and opposite lateral flux of water must balance this input or loss by vertical transport. Note that any net input or loss of water that takes place due to evaporation, precipitation, or river runoff will modify the actual volume of water gained or lost. We account for this effect by using salinity-normalized tracer concentrations. The second term accounts for the vertical eddy flux of tracer, where D_z is the vertical component of the diffusivity and the term in the parentheses is the vertical gradient of C. Equation (4) can be rearranged to give

$$\overline{SMS(C)} = \frac{A}{V} \cdot \left[w + \frac{D_z}{\Delta z}\right] \cdot (C_{deep} - C_{surface}) \tag{5}$$

where $\overline{SMS(C)}$ is the average uptake over the entire volume of the box. The export ratio of opal to organic nitrogen is then

$$\left(\frac{SMS(opal)}{SMS(organic\ nitrogen)}\right) = \frac{\sum\left(\langle[Si(OH)_4]\rangle_{100-200\ m} - \langle[Si(OH)_4]\rangle_{0-100\ m}\right)}{\sum\left(\langle[NO_3^-]\rangle_{100-200\ m} - \langle[NO_3^-]\rangle_{0-100\ m}\right)} \tag{6}$$

from which we see that the uptake ratio is directly related to the ratios of the vertical gradients of the tracers. The $\langle\,\rangle$ brackets symbolize the vertical average over the depth interval shown, and the sum is over the region of interest. The concentration gradient ratio is calculated by taking for the surface concentration the mean over the top 100 m, and for the deep concentration the mean over 100 to 200 m. As previously noted, the concentrations used in (6) are all normalized to constant reference salinity.

We avoid cases where there are large horizontal gradients and vertical supply is negligible essentially by considering only large regions where (6) would be expected to apply. The approach has been tested by *Sarmiento, et al.* [2002] using ocean model simulations and was found to work quite well for determining $CaCO_3$-to-organic nitrogen export ratios (see chapter 8).

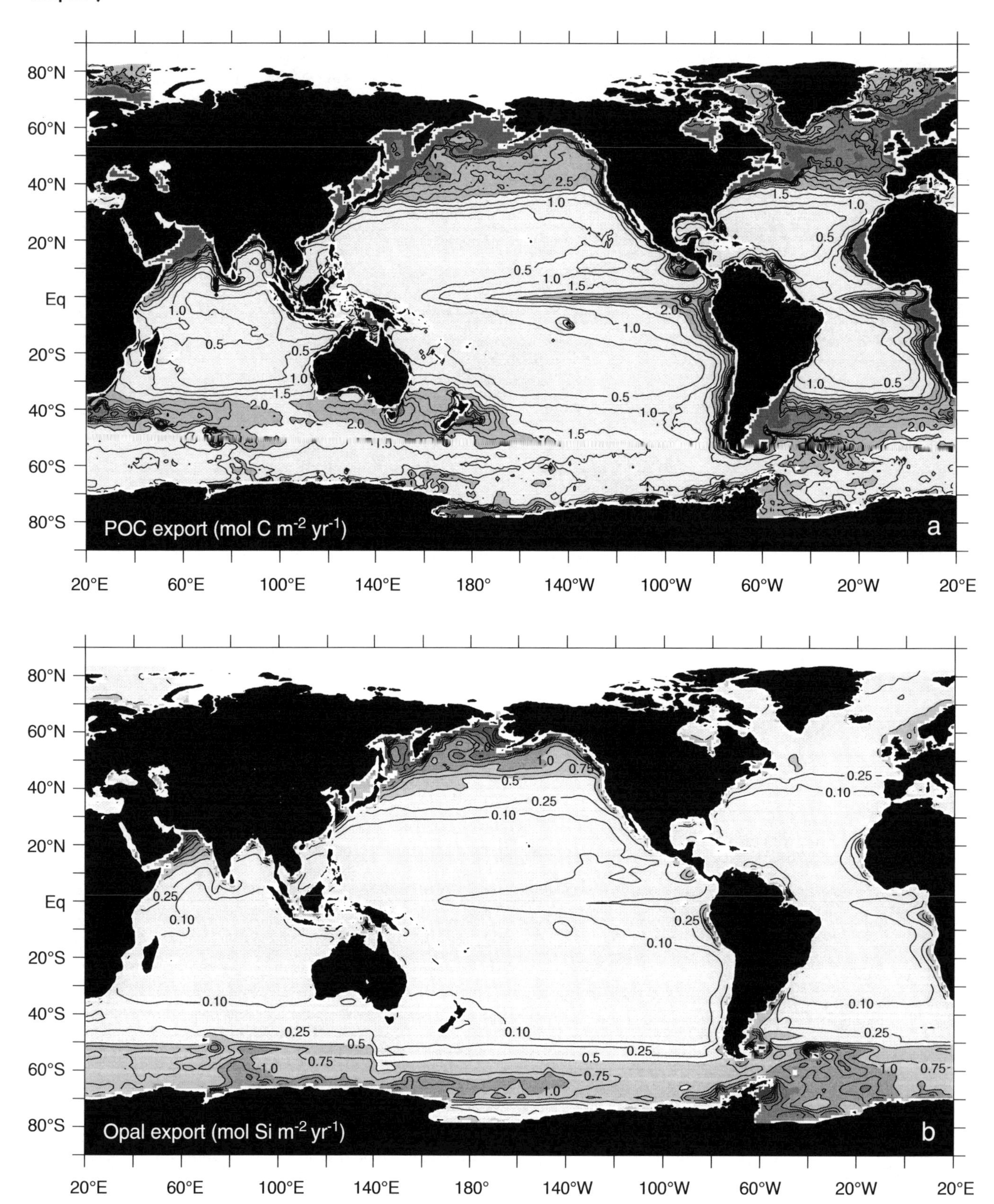

FIGURE 7.2.5: Global maps of annual particulate matter export: (a) Particulate organic matter export calculated from satellite color data and the particulate organic matter export ratio as described in the text (data from *Dunne, et al.* [2005b]; cf. figure 4.2.4b). (b) Opal export calculated by combining the *POC* export in 7.2.5a with an estimate of the opal to *POC* export ratio. Data are from *Dunne et al.* [2005c]. See also color plate 4.

TABLE 7.2.2
Estimates of particulate organic carbon export and associated opal from the surface of the ocean, from *Dunne et al.* [2005b] obtained as described in the text and panel 7.2.1.
The final line shows the exports multiplied by a factor of 1.25 to account for the contribution of DOC to the organic carbon export (see discussion in text).

Region	*Area* ($\times 10^{13}$ m^2)	*Organic C* (Pg yr^{-1})	*Opal* (Tmol yr^{-1})
	Atlantic Ocean		
Subpolar Gyre (45°N–80°N, 70°W–20°E)	1.096	0.81	5.8
Subtropical Gyre (15°N–45°N)	2.471	0.73	2.5
Equatorial (15°S–15°N)	1.987	0.73	2.6
Subtropical Gyre (45°S–15°S)	2.111	0.74	2.8
Southern Ocean (<45°S)	1.477	0.47	16.0
	Indian Ocean		
Subtropical Gyre (15°N–45°N)	0.281	0.32	4.1
Equatorial (15°S–15°N)	2.474	0.53	5.5
Subtropical Gyre (45°S–15°S)	2.932	0.55	2.5
Southern Ocean (<45°S)	1.711	0.34	13.2
	Pacific Ocean		
Subpolar Gyre (>45°N)	0.962	0.69	15.5
Subtropical Gyre (15°N–45°N)	3.907	0.86	11.0
Equatorial (15°S–15°N)	5.714	0.92	7.9
Subtropical Gyre (45°S–15°S)	4.310	0.58	2.4
Southern Ocean (<45°S)	2.501	0.56	12.7
	Global Ocean (excluding Arctic)		
	33.93	9.78	103.7
Including correction for *DOC* export		12.22	129.6

mechanism for bringing diatoms together, and surface features such as spines and mucus layers that facilitate aggregation.

There is now clear evidence of two distinct seasonal episodes of diatom export events (figure 7.2.6) at a wide range of locations [*Kemp et al.*, 1999, 2000]. The first of these commonly follows termination of the spring bloom [*Alldredge and Gotschalk*, 1989; *Smetacek*, 1985]. The second "fall dump" is made up of larger diatoms that *Kemp et al.* [2000] propose may be "shade flora," diatoms adapted to low light conditions that might regulate their buoyancy to access deep nutrients as well as using symbionts to fix nitrogen. These shade flora appear to be very sensitive to stratification. The breakdown of stratification that typically occurs in fall/winter leads to mass sedimentation. There is some evidence from analysis of laminated sediments at a few locations that the fall dump may sometimes account for a higher fraction of the flux of organic matter to the sediments than the spring bloom [*Kemp et al.*, 2000].

Another interesting difference between the cycling of silicon and organic carbon in the surface ocean is their export ratios. The global mean organic carbon export ratio is (12.2 Pg C yr^{-1})/(53 Pg C yr^{-1}) = 0.23, compared with the mean opal export ratio of 0.50 in figure 7.2.3. The more efficient recycling of organic matter (77%) relative to dissolution of opal (50%) in the surface layer persists all the way down the water column, as we shall see in the following section.

7.3 Water Column

We have seen that about 50% of the opal that is produced in the surface survives dissolution and sinks out. In this section we discuss the fate of this opal export and how silicic acid is returned back to the surface to feed continued diatom production. We begin with a discussion of opal fluxes estimated by sediment traps and then proceed to a discussion of the silicic acid distribution in the water column. We shall see that about three-quarters of the opal that leaves the surface dissolves in the water column, with the remaining 25% sinking all the way to the sediments. We also will see that the most important pathway for resupplying silicic acid to the upper ocean appears to be upwelling of silicic acid–rich deep waters in the Southern Ocean followed by formation of Subantarctic Mode Water or SAMW. The silicic acid content of SAMW is unusually low relative to nitrate because it

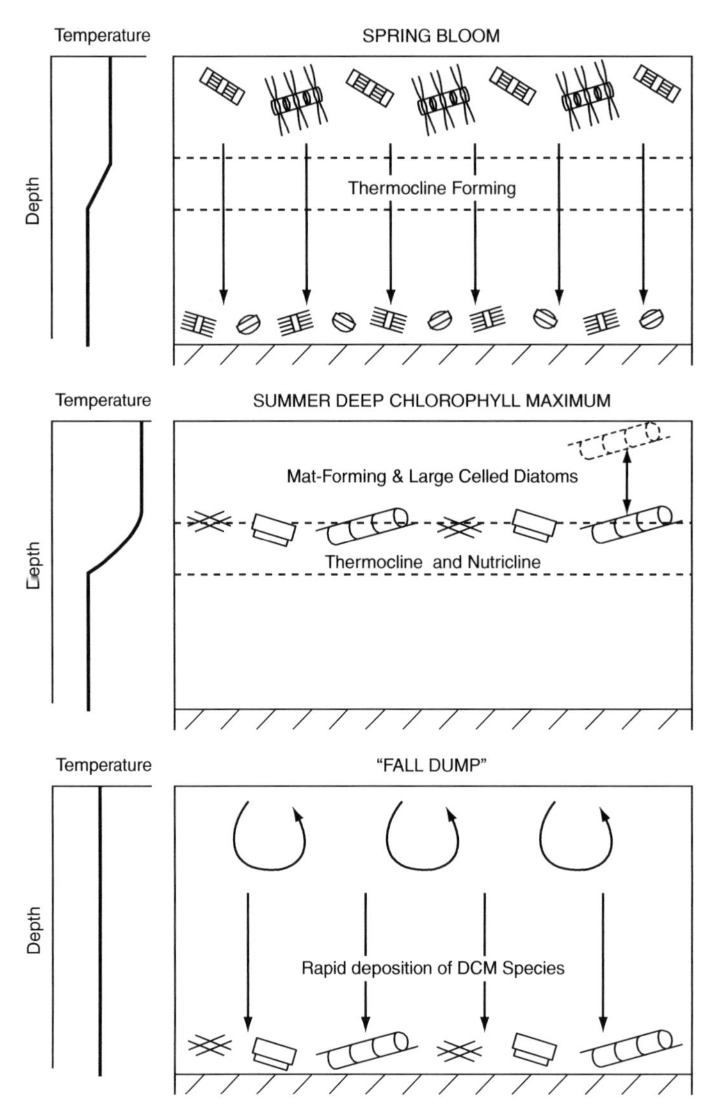

FIGURE 7.2.6: Schematic of biannual diatom flux events as depicted by *Kemp et al.* [2000]. The upper panel illustrates a spring bloom or upwelling episode lasting a few days to weeks and terminated with rapid aggregation and sinking. The middle panel illustrates the summer growth of diatoms that are adapted to stratification. This lasts several months and may involve vertical migration. The fall/winter "fall dump" occurs when the breakdown of the summer stratification leads to mass sedimentation of the diatoms that have grown during the summer.

forms in a region where lack of iron appears to lead to exceptionally high Si:N ratios in the diatoms that are exported from this region. Were it not for the preferential stripping out of silicic acid in this region, it is likely that diatoms would be even more dominant in the world oceans than they are today.

OPAL

We earlier used equation (5) of panel 7.1.1 to get a preliminary estimate that only 11% of the opal leaving the surface of the ocean would arrive at the sediments given a sinking rate of 100 m d^{-1} over 5000 m and a dissolution rate constant k of 1/(23 d). Since the temperature decreases with depth, we would expect the dissolution rate constant to decrease as well, thus our estimate of how much opal survives dissolution is likely a lower limit. On the other hand, the sinking rate of particles tends to be more like 10 to 50 m d^{-1} in the upper ocean [*Berelson*, 2001], which means that our opal survival estimate may be too high. How does our estimate compare with sediment trap observations?

Figure 7.3.1 shows the ratio of opal rain arriving at the bottom of the ocean to opal export leaving the surface of the ocean at the same nine locations as in figure 7.2.3 [*Ragueneau et al.*, 2002]. The estimates on

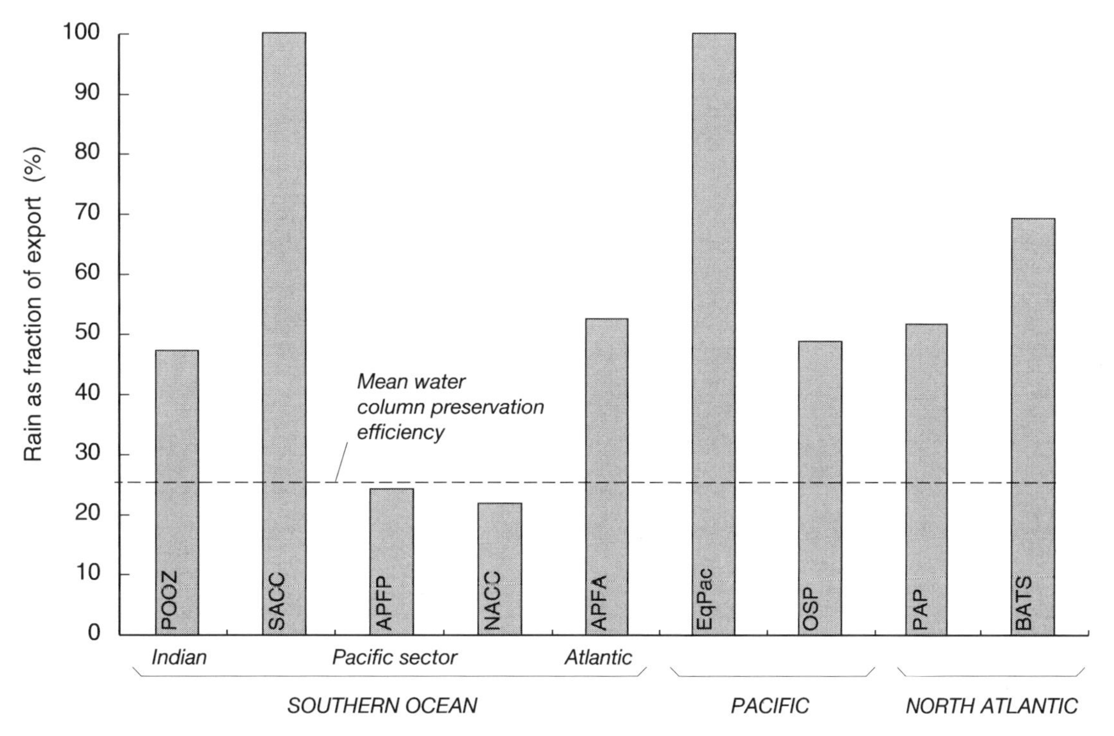

FIGURE 7.3.1: The ratio of opal rain arriving at the sediments to opal export from the surface of the ocean, based on the summary by *Ragueneau et al.* [2002]. These measurements were obtained at nine representative locations in the Southern Ocean, Pacific, and North Atlantic. The mean of the data shown in the diagram is 59%, which is quite a bit higher than the estimate of 25% given by *Treguer et al.* [1995] in their study of the global ocean silicon cycle. This is due to just two of the nine locations.

which this figure is based include sediment trap measurements, as well as model-based estimates of opal export from the surface of the ocean and opal rain to the sediments. The estimates range from a low of just over 20% in two regions of the Pacific Southern Ocean to a high of 100% in the Equatorial Pacific and another sector of the Pacific Southern Ocean, with a mean of 59%. This rather high average value is strongly influenced by the data at two locations. A much lower area estimate of 25% was obtained by *Treguer et al.* [1995]. Over the 5000 m depth range of our sample calculation in the previous paragraph, we can calculate from equation (5) of panel 7.1.1 that 75% dissolution would imply a mean $1/k$ dissolution timescale of 36 days, which is only 1.6 times the global mean surface $1/k$ timescale of 23 days. If we use a lower sinking rate of $50\,\mathrm{m\,d^{-1}}$ instead of $100\,\mathrm{m\,d^{-1}}$, the mean remineralization $1/k$ timescale obtained from equation (5) of panel 7.1.1 doubles from 36 days to 72 days, 3.1 times the surface mean. There is no obvious pattern to the variation from location to location, which is likely due as much to temporal variability as to any fundamental difference between these regions.

It is interesting that despite the evidence for the importance of diatoms in the export of organic matter from the surface of the ocean that we noted at the end of section 7.2, the strongest correlations of open ocean organic matter fluxes in sediment traps further down the water column is with $CaCO_3$. We have already discussed the data that support this and explanations for why this might be the case in section 5.4.

How does the opal dissolution rate compare with organic carbon remineralization in the water column? We estimated in section 5.4 that for a typical upper ocean sinking rate of 10 to $50\,\mathrm{m\,d^{-1}}$, the organic carbon remineralization rate constant is $k_{remin} \sim 1/(10\,\mathrm{d})$ to $1/(50\,\mathrm{d})$. This is quite close to the surface ocean opal dissolution rate constant of $1/(23\,\mathrm{d})$, but the comparison between these two numbers needs to be made with care, since the former is for the thermocline, whereas the latter is for the surface. In fact, sediment trap observations clearly show that opal dissolution in the water column is less efficient than organic matter remineralization, with the consequence that the Si:C ratio in traps increases with depth, and the silicic acid distribution in the water column is displaced downward with respect to dissolved nutrients. *Ragueneau et al.* [2002] summarize a set of sediment trap measurements of the opal and organic carbon fluxes as a function of depth that they fit with the following equation:

$$\frac{\Phi^{Opal}(z)}{\Phi^{Organic\ C}(z)} = \frac{Prod(Opal)}{NPP} \cdot z^{0.41} \qquad (7.3.1)$$

where *Prod(Opal)* is the production of opal in the euphotic zone prior to dissolution, and the other terms have all been previously defined. For example, the flux at 1000 m has a Si:C ratio that is 2.6 times greater than the export flux at 100 m. By 5000 m, the Si:C ratio has increased to 5 times that at 100 m. One can use this equation in combination with one of the organic matter remineralization relationships in chapter 5 (e.g., equation (5.4.3)) to estimate the actual opal flux as a function of depth:

$$\begin{aligned}\Phi^{Opal}(z) &= \frac{Prod(Opal)}{NPP}\cdot\Phi^{Organic\,C}(z)\cdot z^{0.41}\\ &= \frac{Prod(Opal)}{(\Phi^{Organic\,C}(z_0)/p_e)}\cdot\left[\Phi^{Organic\,C}(z_0)\cdot\left(\frac{z}{z_0}\right)^{-0.858}\right]\cdot z^{0.41}\\ &= Prod(Opal)\cdot p_e\cdot z_0^{0.858}\cdot z^{-0.448}\end{aligned} \tag{7.3.2}$$

where $z_0 = 100$ m. From this particular equation, we can calculate that the flux at 1000 m is 35.6% of the flux at 100 m and that the flux at 5000 m is 17.3% of that at 100 m, which is somewhat lower than the observed rain:export ratios in figure 7.3.1 but certainly within the uncertainty of the measurements and model.

Silicic Acid

In chapter 1, we used a three-box model of the ocean to gain powerful insights into the cycling of phosphate and oxygen in the deep ocean. We focused particularly on the importance of preformed phosphate and oxygen concentrations in enabling us to obtain a reasonable solution for the deep ocean concentration. We begin this subsection with a return to the three-box model to analyze the distribution of silicic acid in the water column relative to nitrate and to estimate the opal export. The rather surprising (and incorrect) results we obtain are used as a springboard for a more detailed examination of how and why the silicic acid distribution differs from that of nitrate, and what the consequences of this are for diatom production in the ocean.

The steady-state solution for the deep box of the model shown in figure 1.2.5 is

$$\Phi_l + \Phi_d = (f + T)(C_d - C_h) \tag{7.3.3}$$

We calculate the opal-to-organic nitrogen export ratio by taking the ratio of (7.3.3) for the silicon balance to that for the nitrogen balance, i.e.,

$$\frac{(\Phi_l+\Phi_d)_{Si(OH)_4}}{(\Phi_l+\Phi_d)_{NO_3^-}} = \frac{([Si(OH)_4]_d - [Si(OH)_4]_h)}{([NO_3^-]_d - [NO_3^-]_h)} \tag{7.3.4}$$

Substituting in mean deep ocean concentrations (below 100 m) and mean surface concentrations for the Southern Ocean south of 40°S and shallower than 100 m gives a putative export ratio of

$$\frac{(\Phi_l+\Phi_d)_{Si(OH)_4}}{(\Phi_l+\Phi_d)_{NO_3^-}} = \frac{(84 - 20\text{ mmol m}^{-3})}{(29.7 - 17.1\text{ mmol m}^{-3})} = 5.1 \tag{7.3.5}$$

where the concentrations are in mmol m^{-3} and the ratio is in units of mol/mol.

We can use the ratio in (7.3.5) to estimate the flux of opal from the surface of the ocean based on the two-box model estimate of 3.2 Pg yr^{-1} of carbon export we obtained in chapter 1 as follows:

$$\begin{aligned}\Phi^{opal} &= 3.2\times10^{15}\,\frac{\text{g C}}{\text{yr}}\cdot\frac{1\text{ mol C}}{12\text{ g C}}\cdot\frac{16\text{ mol N}}{117\text{ mol C}}\cdot\frac{5.1\text{ mol Si}}{\text{mol N}}\\ &= 186\times10^{12}\,\frac{\text{mol Si}}{\text{yr}}\end{aligned} \tag{7.3.6}$$

Here we have used the C/N stoichiometric ratio of 117/16 given in chapter 5 to convert carbon export to nitrogen export. The Si export we obtain is 186 Tmol yr^{-1} of opal. However, if we were to use the more realistic organic carbon export estimate of 12 ± 4 Pg C y^{-1} that we obtained above, the Si export would increase to 710 Tmol yr^{-1}. Observations suggest that the export is smaller than either of these estimates, of order 130 Tmol yr^{-1} (see table 7.2.2).

Why does the three-box model give us predictions that are so inconsistent with direct observational constraints on the opal export flux? Our earlier look at the silicic acid and nitrate distributions in the water column (figure 7.1.2) suggests the reason why. Silicic acid is strongly fractionated toward the deep ocean relative to nitrate. When we use the mean deep ocean concentration of silicic acid in (7.3.4) and (7.3.5), we are effectively assuming that the water returning from the deep ocean to the surface in low as well as high latitudes can be represented by the mean deep ocean concentration. While figure 7.1.2c suggests that this is perhaps not unreasonable for nitrate (the concentration of nitrate in the main thermocline and vicinity of the Southern Ocean outcrop is close to the oceanic mean of ~30 mmol m^{-3}), it is quite clear that it is not remotely correct for silicic acid, which is below 10 mmol m^{-3} over almost the entire thermocline, compared with its oceanic mean of ~84 mmol m^{-3}. Another way of thinking about this is to consider the integral form of the conservation equation (2.1.13), where we see that the exchange between one box and another is calculated as an integral around the surface of the box. In other words, the appropriate concentration to use in modeling exchange between one box and another is a value that is representative of the vicinity of the boundary of the box, rather than its interior.

The usual explanations for the deep fractionation of silicic acid relative to nitrate are that opal dissolution occurs at greater depths than remineralization of organic matter (see equation (7.3.1); cf. *Broecker and Peng* [1982]); and that opal produced in the surface ocean is recycled less efficiently than nutrients, resulting in a preferential export of opal to the deep ocean (the silica

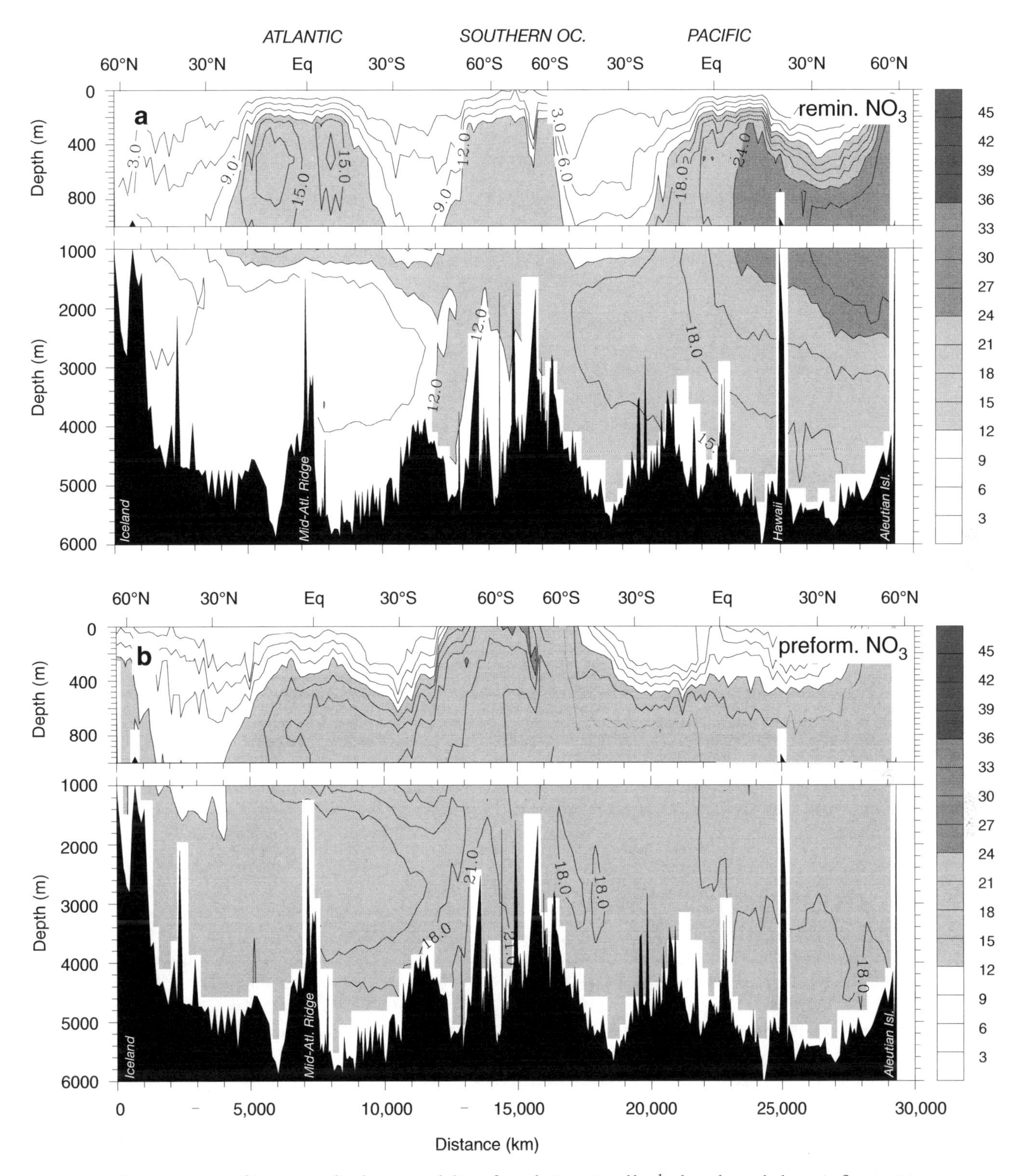

FIGURE 7.3.2: Vertical sections of (a) remineralized nitrate and (b) preformed nitrate ($\mu mol\,kg^{-1}$) along the track shown in figure 2.3.3a. Compare to figure 7.1.2b. See also color plate 7.

pump hypothesis of *Dugdale et al.* [1995]). While these mechanisms definitely play a central role, the near absence of silicic acid in the main thermocline (figure 7.1.2b) suggests that something else is at play as well. We can gain insight into the unexpectedness of the thermocline deficit in silicic acid by considering the sections of preformed and remineralized nitrate shown in figure 7.3.2. We calculate remineralized nitrate using AOU and the stoichiometric $N:O_2$ ratio of remineralization of organic matter, and then subtract this from the total nitrate to calculate the preformed concentration, as discussed in chapter 5 for phosphate (equation (5.3.3) and figure 5.3.3). Recall that the preformed concentration is that component of the nutrients that is transported directly in from the surface of the ocean, having never participated directly in biological processes since leaving the surface.

The remarkable thing about the nitrate distribution in the main thermocline is that most of it is preformed; in other words, except for a few regions (notably the tropical Atlantic and Pacific as well as North Pacific), the nitrate that is present in the main thermocline comes mainly from water that sinks down from the surface of the ocean (figure 7.3.2b) rather than from remineralization of organic matter (figure 7.3.2a). Viewed from this perspective, the almost complete absence of silicic acid in figure 7.1.2b implies that the silicic acid in waters entering the main thermocline from the surface of the ocean must be extremely low. We saw in figure 7.1.1 that the only major region of the surface ocean where silicic acid is low and nitrate high is in a band between about 40°S and 60°S. We show next that the high-nitrate, low-silicic acid waters of the main thermocline are formed in this region. We discuss why these waters have high nitrate and low silicic acid content how they come to have such an overwhelming influence on the global thermocline (except in the North Pacific, where silicic acid is high), and what the consequences of this are for diatom productivity throughout the world.

To understand what is happening in the Southern Ocean we need first to understand the meridional overturning circulation pattern schematically illustrated in figure 7.3.3, which is based on a similar depiction by *Speer et al.* [2000]. Circumpolar Deep Water (CDW) upwells to the surface of the ocean in the Antarctic. The upper component of the upwelled CDW is carried to the north by the wind-driven Ekman transport, where some of it sinks to form Antarctic Intermediate Water (AAIW) in the vicinity of the so-called Antarctic Polar Front (APF), and more sinks to form Subantarctic Mode Water (SAMW) in the vicinity of the Subantarctic Front (SAF). The key to our story is the SAMW, which we shall see is the main conduit for nutrients from the deep ocean into the base of the main thermocline [cf. *Toggweiler et al.*, 1991; *Sarmiento et al.*, 2004a]. The SAMW is a pycnostad (a thick layer of relatively homogeneous density) that originates in the deep wintertime mixed layers found in a ring around the Southern Ocean (figure 7.3.4b) and spreads northward at the base of the main thermocline of the subtropical gyres of the southern hemisphere [*McCartney*, 1977].

The upwelling of deep waters and deep mixing in the Antarctic brings extremely high nutrients to the surface of the Antarctic Circumpolar Current to the south of the APF (figure 7.1.1a and b). As the Ekman transport carries these waters to the north, organisms utilize the nitrate [*Sigman et al.*, 1999b] and silicic acid as well as other nutrients (lower panel of figure 7.3.3). However, the drawdown of silicic acid is more efficient than that of nitrate, with the consequence that nitrate concentrations are still high in a band where silicic acid concentrations are extremely low (lower panel of figure 7.3.3; figure 7.1.1a and b). The preferential removal of silicic acid within the Antarctic and PFZ is a well-known phenomenon [cf. *Kamykowski and Zentara*, 1985; *Pondaven et al.*, 2000a; *Smith et al.*, 2000; *Brzezinski et al.*, 2001] generally attributed to the influence of iron limitation in dramatically increasing the Si:N content of diatoms in this band [*Brzezinski et al.*, 2003a; *Franck et al.*, 2000].

This preferential removal of silicic acid from the surface ocean is what traps silicic acid in the Southern Ocean, preventing it from entering into the main thermocline with the SAMW. The downward displacement of the silicic acid relative to nitrate distribution within the Southern Ocean (figure 7.1.2b and c) is likely due in part to this preferential removal, but also to the deeper dissolution of opal relative to remineralization of organic matter.

A powerful tool to examine the relationship between nitrate and silicic acid is the tracer $Si^* = Si(OH)_4 - NO_3^-$ [cf. *Brzezinski et al.*, 2002; *Sarmiento et al.*, 2004a]. Si* will be positive in regions where silicic acid is in excess of the amount that would be required for diatoms to use up all of the nitrate in the 1:1 ratio that we saw earlier is characteristic of healthy diatoms, and negative in areas where silicic acid is in deficit relative to this ratio. Figure 7.3.4a shows clearly the band of silicic acid deficit around the Southern Ocean as a negative anomaly of $-10\ \mu mol\ kg^{-1}$ to $-15\ \mu mol\ kg^{-1}$. The belt of low Si* is confined primarily to the region spanning the SAF, where the deep mixed layers that form SAMW are found (figure 7.3.4). The particular combination of properties in this region are unique; there is nowhere else at the surface of the ocean that has such low Si* concentrations.

We can use the unusually low Si* values of the Southern Ocean to trace the influence of the SAMW at the base of the main thermocline. Figure 7.3.5a shows Si* concentrations mapped on the $\sigma_\theta = 26.8$ (potential density $= 1.0268\ kg\ m^{-3}$) surface, which is the median of the range of densities that make up the SAMW ($\sigma_\theta = 26.5$ in the western Atlantic, increasing toward the east to $\sigma_\theta = 27.1$ in the southeast Pacific *Hanawa and Talley*, 2001). The depth of this surface, shown in figure 7.3.5b, places it at about the base of the main thermocline. *Sarmiento et al.* [2004a] give evidence that Si* is nearly conserved on this density surface except in the North Pacific. In other words, by coincidence, the depth interval of this isopycnal surface is one where the remineralization of organic nitrogen, which happens preferentially in shallow waters, and the dissolution of opal, which is biased toward deeper waters relative to organic matter

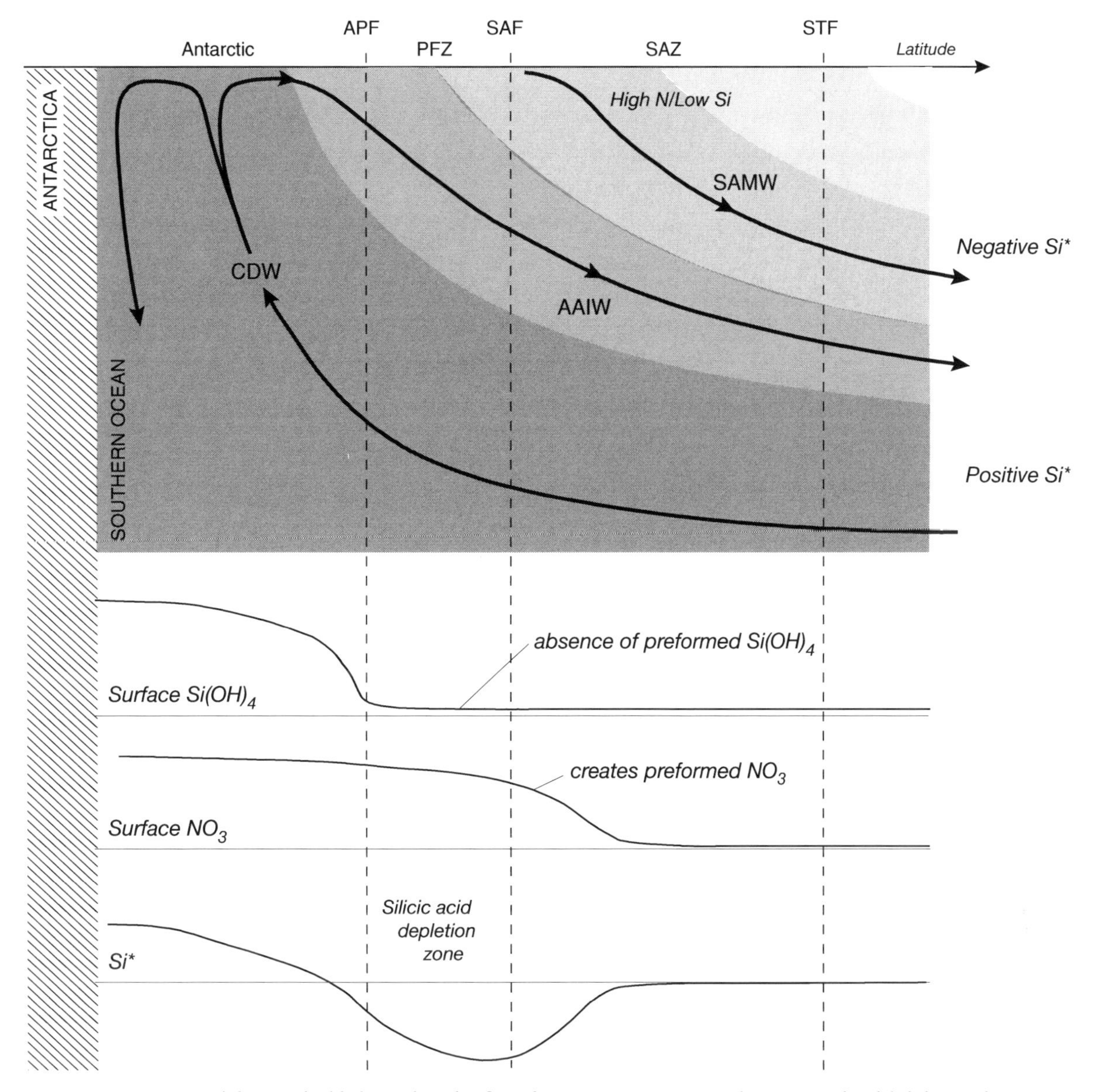

FIGURE 7.3.3: Conceptual diagram highlighting the role of Southern Ocean processes in determining the global thermocline nutrient properties [adapted from *Sarmiento et al.*, 2004a]. APF is the Antarctic Polar Front, PFZ is the Polar Front Zone, SAF is the Subantarctic Front, SAZ is the Subantarctic Zone, and STF is the Subtropical Front, which often is split into a southern and northern component (see figure 7.3.4).

remineralization (see equation (7.3.1)), occur with a ratio of ~1:1. The low Si* SAMW extends not only throughout the southern hemisphere subtropical gyres [*McCartney*, 1977, 1982; *Reid*, 1997], but also into the North Atlantic, where it forms part of the upper water return flow of the global meridional overturning circulation [*Rintoul and Wunsch*, 1991; *Schmitz*, 1995; *Sloyan and Rintoul*, 2001]. The vertical section shown in figure 7.3.6 shows clearly how this negative Si* water fills the main thermocline, except in the North Pacific.

The North Pacific Ocean is an important exception to the dominance of the SAMW influence. The deep waters of this ocean have exceptionally high silicic acid values that result essentially from the fact that this is the dead end of the global thermohaline circulation during which the water has been continuously accumulating silicic acid (as well as other nutrients [*Broecker and Peng*, 1982; cf. also *Ragueneau et al.*, 2000]). However, as we have previously noted, the oceanic stratification generally creates an almost impenetrable barrier between deep waters and the main thermocline. How, then, does the main thermocline of the North Pacific tap into the high-nutrient waters of the deep ocean in this region, as indicated by figures 7.1.1, 7.3.5, and 7.3.6 [cf. *Tsunogai*, 2002]?

The $\sigma_\theta = 26.8$ surface shown in figure 7.3.5 is the median density of upper North Pacific Intermediate Water (NPIW [*Talley*, 1997]), the formation of which is

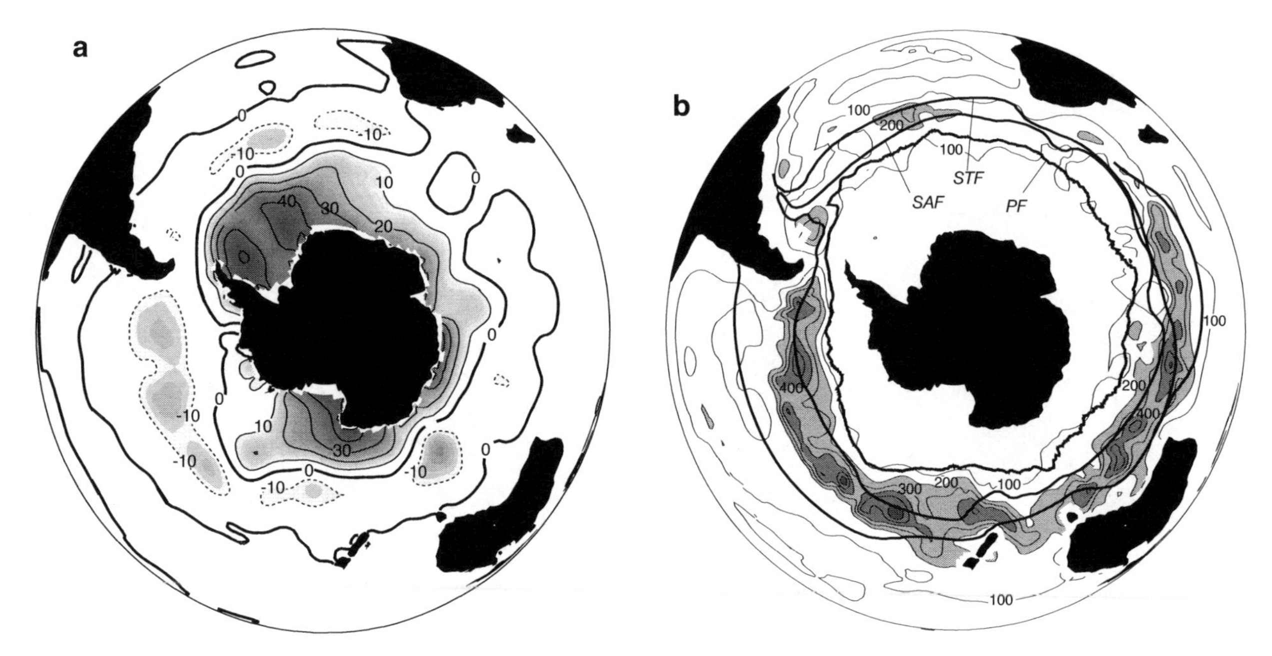

FIGURE 7.3.4: Polar stereographic maps of the Southern Ocean showing (a) annual mean $Si^* = SiOH)_4 - NO_3^-$ computed from the World Ocean Atlas 2001 [*Conkright et al.* 2002] and (b) wintertime mixed layer thickness (July through September) taken from *Kara et al.* [2003]. Also shown in (b) are the mean positions of the major hydrographic fronts: the southernmost black line indicates the position of the Antarctic Polar Front. In sequence from south to north, the remaining lines indicate the position of the Subantarctic Front, then Southern Subtropical Front, and the Northern Subtropical Front.

the result of a complex series of processes occurring in the Sea of Okhotsk [*Talley*, 1991; *Yasuda et al.*, 2002] and the "mixed water region" between the Kuroshio and Oyashio Currents [*Talley*, 1993, 1997; *Yasuda et al.*, 2001]. Strong vertical mixing driven by tidal interactions in the Kurile Islands is thought to be an important contributor to the processes that determine the properties of NPIW [*Nakamura et al.*, 2000; *Talley*, 1991]. *Sarmiento et al.* [2004a] suggest that this strong vertical mixing acting across the base of the relatively shallow thermocline in this region is what brings deep nutrients up into the thermocline both at the depth of the NPIW and above it.

Thus, it may be possible that the high silicic acid concentrations of the North Pacific thermocline are a consequence of an unusually large amount of vertical mixing in this region. The dominance of SAMW elsewhere in the world occurs only because the vertical mixing is too small to tap into the deep ocean high–silicic acid waters. It is interesting to consider why the silicic acid is not stripped out at the surface of the ocean in the NPIW formation region, as it is in the SAMW zone. This may perhaps have something to do with a difference in iron supply (more abundant iron in the region where NPIW forms), but *Sarmiento et al.* [2004a] speculate that it is more likely due to the fact that the nutrients are apparently not exposed at the surface for any length of time during formation of the NPIW.

The influence of NPIW extends over the entire North Pacific to the equator (figure 7.3.5). The high silicic acid North Pacific water is critical for diatom production in the Equatorial Pacific, where it accounts for 70% of the silicic acid supply to the region despite the fact that half the supply of nitrate for biological production comes from the southern hemisphere [*Dugdale and Wilkerson*, 1998]. The influence of this water type can be seen at the surface of the ocean throughout the North and equatorial Pacific, where the opal-to-organic nitrogen export ratios are higher than almost anywhere else except the Southern Ocean (figure 7.2.4). There is also a modest amount of high Si* water in the western equatorial Indian Ocean and Bay of Bengal (figure 7.3.5a), with corresponding high opal-to-organic nitrogen export ratios (figure 7.2.4). This may be generated by strong vertical mixing in the Indonesian Straits or leakage from the North Pacific. The influence of these high Si* waters is confined to a small region of the Indian Ocean.

The foregoing analysis of the Si* distribution shows that nutrient input to the base of the main thermocline occurs primarily in association with the formation and lateral spreading of SAMW, with an important contribution of NPIW. This result has far-reaching consequences for what controls biological production of organic matter and opal in the ocean. The present export of organic matter across the $\sigma_\theta = 26.8$ surface is large enough to deplete the thermocline of nutrients within <60 years. The return of nutrients from the deep ocean to the thermocline is thus critical to maintaining biological productivity. Model simulations of

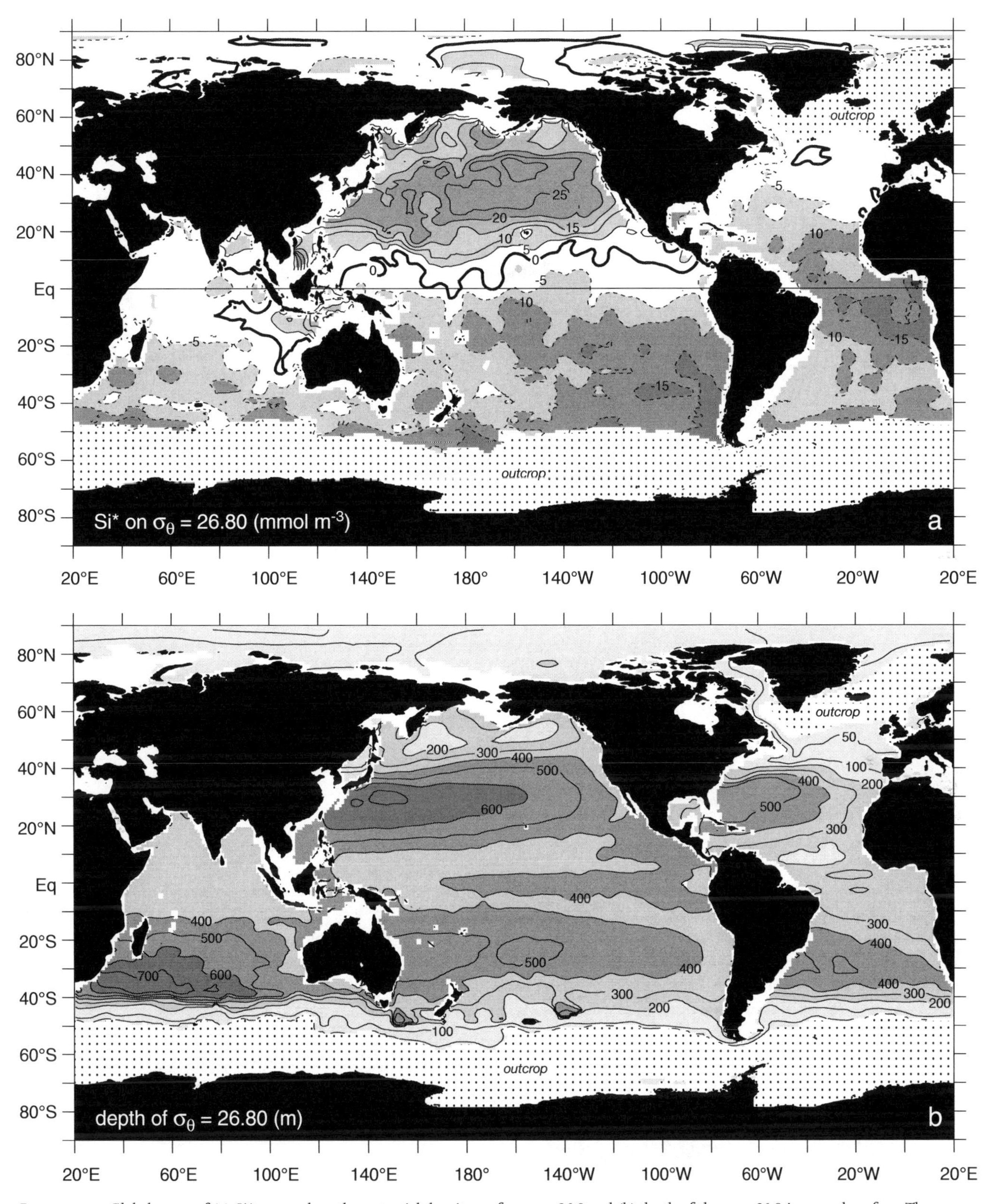

FIGURE 7.3.5: Global maps of (a) Si* mapped on the potential density surface $\sigma_\theta = 26.8$ and (b) depth of the $\sigma_\theta = 26.8$ isopycnal surface. The data are from the World Ocean Atlas 2001 [*Conkright et al.*, 2002].

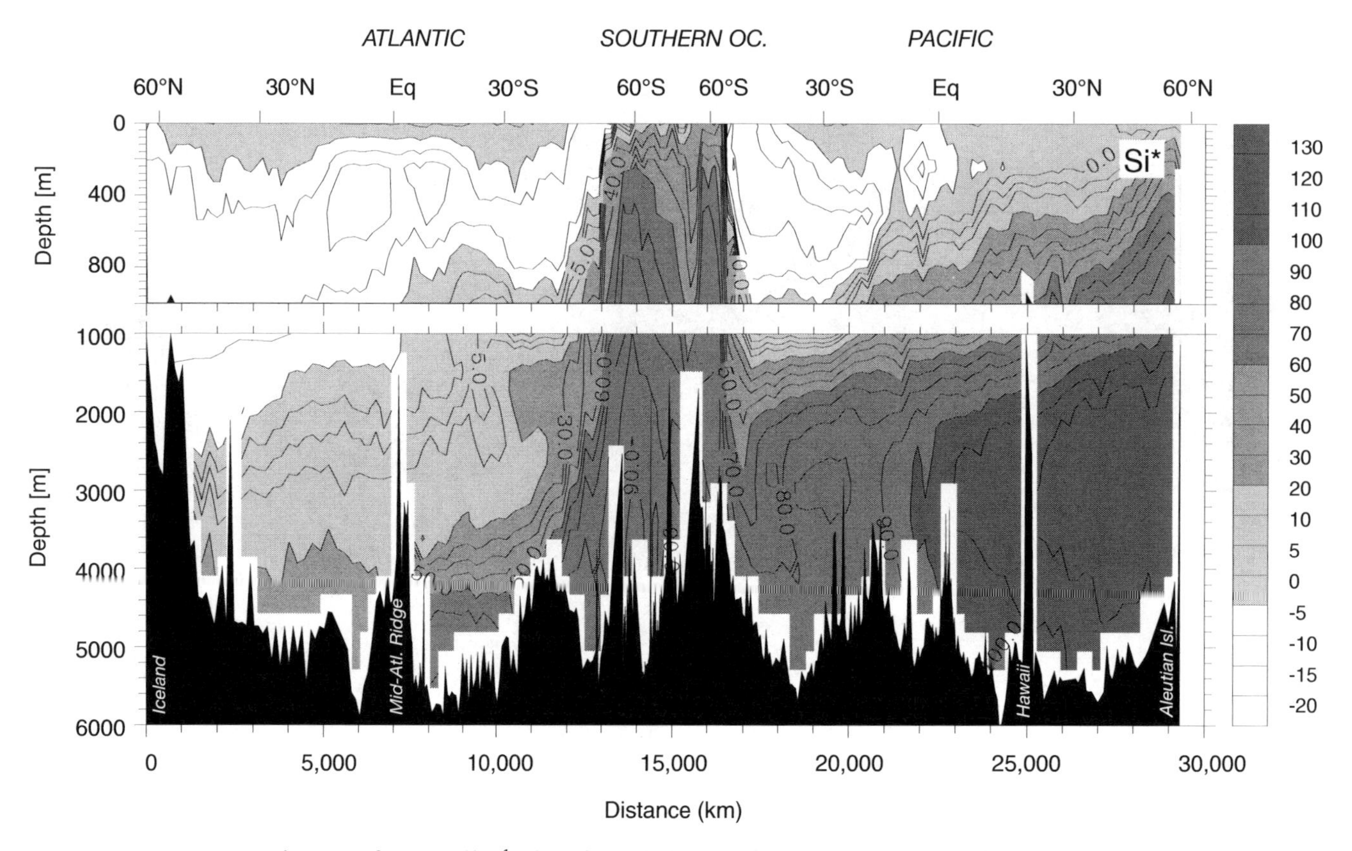

FIGURE 7.3.6: Vertical section of Si* (μmol kg^{-1}) along the track shown in figure 2.3.3a. Compare to figure 7.1.2a. See also color plate 7.

Southern Ocean nutrient depletion such as that described by *Sarmiento et al.* [2004a] demonstrate the importance of the Southern Ocean for biological production in the rest of the world. Biological productivity in their model is predicted by forcing nutrients toward observations. In order to examine the influence of high-latitude nutrients on biological productivity in the rest of the global ocean, a simulation is performed where nutrients are forced toward zero over the entire Southern Ocean south of 30°S latitude. This closes off the return flow of nutrients from the deep ocean to the thermocline, except for that supplied by vertical mixing. The nutrient content of the thermocline plummets, and biological productivity north of 30°S drops by about 75% (see chapter 4). In other words, nutrient input from the Southern Ocean accounts for about 75% of biological production north of 30°S.

In addition to being a powerful water mass tracer, Si* has the further benefit of being an indicator of the nutrient status vis-à-vis the requirements of diatoms. The ubiquity of waters with a low silicic acid-to-nitrate ratio, i.e., negative Si*, in the main thermocline (figure 7.3.5a) provides an explanation for why the silicic acid-to-nitrate supply ratio to surface waters shown in figure 7.2.4 is so low throughout most of the ocean, and the more general observations of low diatom production over much of the ocean ([*Ragueneau et al.*, 2000]; see table 7.2.1 and 7.2.2). The importance of the high silicic acid content of North Pacific water to diatom production in the equatorial Pacific was discussed in the previous subsection.

We conclude this section by noting that there are many features of the oceanic cycling of silicic acid and nitrate that we cannot yet explain or for which the explanations we have offered must be considered preliminary. For example:

1. Why does Si* go to zero almost exactly at the Polar Front and reach a minimum at the Subantarctic Front? The coincidence of these features suggests the possibility of a physical rather than biological mechanism.

2. Why does Si* begin to increase to the north of the SAF? This is probably due to the influence of subtropical waters depleted in both silicic acid and nitrate (which gives Si* = 0) that are transported from the north into the Subantarctic Zone SAZ [*Speer et al.*, 2000].

3. What are the pathways by which nutrients are transported from the base of the thermocline into the upper waters of the thermocline? Possible mechanisms include upwelling and vertical mixing in the tropics and lateral mixing in intense western boundary current regions.

4. Why does the North Pacific have a high input of nutrients from the deep ocean? That this is an area of unusually high vertical mixing due to tidal forcing remains to be proven.

5. What is different about the northwest Pacific relative to the Southern Ocean that negative Si* waters are not generated from the high nitrate and silicic acid waters found at the surface of the ocean?

7.4 Sediments

In the previous section, we learned that three-quarters of the opal that leaves the euphotic zone dissolves in the water column. In this section we discuss the fate of the remaining 25% that arrives at the sediments. Opal raining onto the sediments will generally either dissolve or be buried. Lateral transport of sediments is important in some places, but we will stay away from them in our discussion. Our objective in this section is thus to understand what controls dissolution and burial. The discussion will be built around four fundamental observations and ideas which form a conceptual framework that captures our present understanding:

1. Opal dissolution is highly efficient, accounting for an average of ~80% of the rain rate (figures 7.4.1 and 7.4.2a). The first-order control on the flux of silicic acid out of the sediments is thus how much opal rains onto the sediments. The opal rain is large enough in high-deposition regions that one can readily observe the impact of the resulting silicic acid flux on water column silicic acid concentrations.

2. Analysis of the flux equation for silicic acid in sediments suggests that there is an upper limit to the loss of silicic acid from the sediments defined by the thermodynamic properties of opal and physical properties of the sediments. If the opal rain onto the sediments exceeds this threshold, there should inevitably be burial of the excess rain above the threshold. If the opal rain falls below this threshold, it becomes difficult for preservation of opal to occur. This threshold coincides with an observed *opal preservation discontinuity*, which occurs at an opal rain rate of ~2 mmol m^{-2} d^{-1} (cf. figure 7.4.2b). Observations above the discontinuity show opal burial efficiencies of order 30%. Observations of sediments below the discontinuity show two quite surprising characteristics: (a) there is still opal present, although the average opal preservation efficiency of ~5% is 1/6th what one would expect by extrapolating the behavior of sediments from above the discontinuity (cf. figure 7.4.2b); and (b) the asymptotic concentration of silicic acid in these sediments is far lower than the concentration of ~1000 mmol m^{-3} that one would expect if the pore water were in equilibrium with opal, even though opal is still present (see, for example, the North Atlantic and equatorial Pacific profiles in figure 7.1.4).

3. Laboratory studies indicate that the low asymptotic concentration of opal in the sediments is due to a variety of factors, of which the most important is the formation of authigenic aluminosilicates (*authigenic* means generated or formed in place) with lower solubilities. These authigenic aluminosilicates, when combined with the opal that is still present, give a low apparent solubility and contribute to the observed low asymptotic silicic acid concentration. Lower apparent solubilities reduce the flux of silicic acid from the sediments to the water column, but all else being equal, the conversion rate of relatively fresh opal to aluminosilicates is fast enough that all the opal should still disappear.

4. The only way to preserve opal in sediments that fall below the opal preservation discontinuity is thus if the dissolution rate of opal becomes slow enough to prevent this from happening. Laboratory studies once again provide the key insight, which is that the dissolution rate constant of opal does indeed decrease with age as the number of reactive sites on the diatom surface declines and they become increasingly blocked by contaminants such as aluminum.

We emphasize that all of these concepts are subject to uncertainty and that the second through fourth ones, in particular, should be considered working hypotheses rather than proven facts. We discuss each of these topics, and the evidence that supports them, in turn.

Opal Dissolution and Burial

The rate of opal dissolution is generally estimated from the flux of silicic acid out of the sediments, calculated with Fick's first law:

$$\Phi_{z=0}^{Si(OH)_4} = -\varepsilon_s \cdot \{\partial(\phi \cdot [Si(OH)_4])/\partial_z\}_{z=0} \tag{7.4.1}$$

as well as by placing benthic chambers over the sediments. Figure 7.4.1 shows the ratio of the opal dissolution rate in sediments to the total rain of opal onto the sediments, estimated either from deep sediment traps above the sediments, or by summing the silicic acid flux out of the sediments to the opal burial rate, which is equal to the sedimentation rate multiplied by the asymptotic opal concentration $S{\cdot}[SiO_2]_\infty$. These data are from the same nine sites as in figure 7.2.3 [*Ragueneau et al.*, 2002]. The dissolution efficiency at these nine sites ranges from 60% to 96%, with an average of 86%. Figure 7.4.2a shows a larger selection of data of opal dissolution plotted versus calculated rain rate. The data fall approximately into two groups, one with a dissolution efficiency of about 70% and the other with a dissolution efficiency of about 95%. Averaged over the world as a whole, the mean dissolution efficiency of opal is about 80% [*Treguer et al.*, 1995].

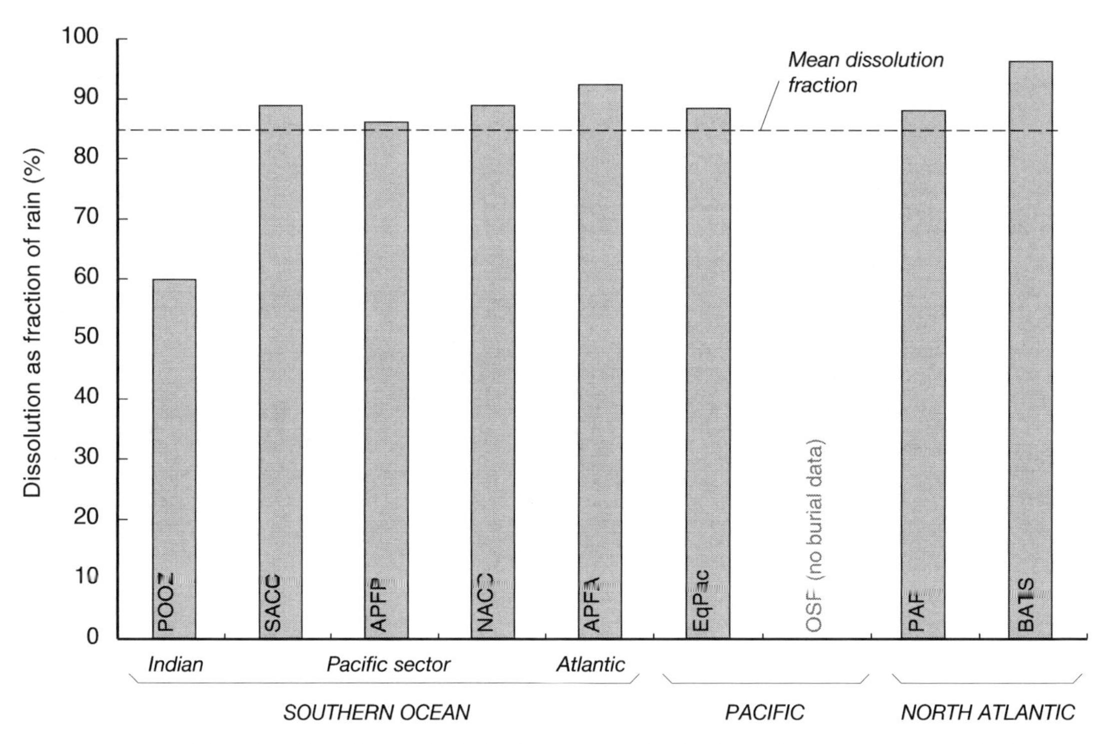

FIGURE 7.4.1: Dissolution of opal in sediments divided by the rain rate of opal arriving at the sediments. The data shown are from the same nine locations as figure 7.2.3 [*Ragueneau et al.*, 2002].

With such a high dissolution efficiency, and relatively modest variations around the mean, this means that, the dissolution of opal and flux of silicic acid out of the sediments will be determined primarily by the pattern of opal rain to the sediments, which in turn is determined by the opal export. Below we show that the silicic acid flux out of the sediments is in the range of -0.09 to $-2.4\,\text{mmol}\,\text{m}^{-2}\,\text{d}^{-1}$. How would fluxes such as these affect the water column concentration of silicic acid? If we take a 1000 m-thick water column, these fluxes would increase the concentration by an average of 4–110 $\text{mmol}\,\text{m}^{-3}$ over a typical ocean basin residence timescale of ~100 years. From figure 7.2.5b, we would expect to observe the highest silicic acid fluxes out of the sediments in the Southern Ocean, North Pacific, and some of the continental margins and equatorial Pacific. The analysis of deep ocean silicic acid observations by *Edmond et al.* [1979] shows anomalies in the middle of the range of our estimated concentration increase in three of these regions—the Weddell-Enderby Basin (~50 $\text{mmol}\,\text{m}^{-3}$; figure 7.4.3a), as well as other Antarctic basins (~10 $\text{mmol}\,\text{m}^{-3}$), the North Pacific (~10 $\text{mmol}\,\text{m}^{-3}$), and the northern Indian Ocean (~20 $\text{mmol}\,\text{m}^{-3}$; figure 7.4.3b)—all of which they attribute to fluxes of silicic acid out of the sediments in areas where the opal content is relatively high (figure 7.1.3).

One can infer from the correspondence between the opal production rate in figure 7.2.5 and the burial rate in figure 7.1.3 that the opal rain arriving at the sediments is also the primary determinant of the large-scale distribution of the opal burial. However, there are quite large variations in the burial efficiency, as shown in figure 7.4.2b. Understanding these variations is particularly important for paleoceanographers, who are interested in using opal burial as a proxy for paleoproductivity. The most remarkable feature of figure 7.4.2b is the separation into two groups that occurs at an opal rain rate of about $2\,\text{mmol}\,\text{m}^{-2}\,\text{day}^{-1}$. Above this threshold, the opal burial efficiency is high, up to 30%. Below this threshold, the opal burial efficiency is low, only about 5%. As suggested above, this separation of the preservation efficiency into two distinct groups occurs at an opal rain rate that corresponds approximately to a threshold defined by the upper-limit silicic acid flux out of the sediments. In order to better understand what controls opal burial efficiency, we solve the diagenetic equation for silicic acid and substitute this into (7.4.1). The solution we obtain allows us to examine the basic parameters that control the flux of silicic acid into the water column, and to estimate the actual magnitude of the threshold flux.

Panel 7.4.1 shows how we obtain a solution to the diagenetic equation for the distribution of $[Si(OH)_4]$ as a function of depth in the sediments. Substituting equation (6) from this panel into (7.4.1) gives

$$\Phi_{z=0}^{Si(OH)_4} = -\phi \cdot \varepsilon_s \cdot \lambda \cdot ([Si(OH)_4]_0 - [Si(OH)_4]_{SAT}) \qquad (7.4.2)$$

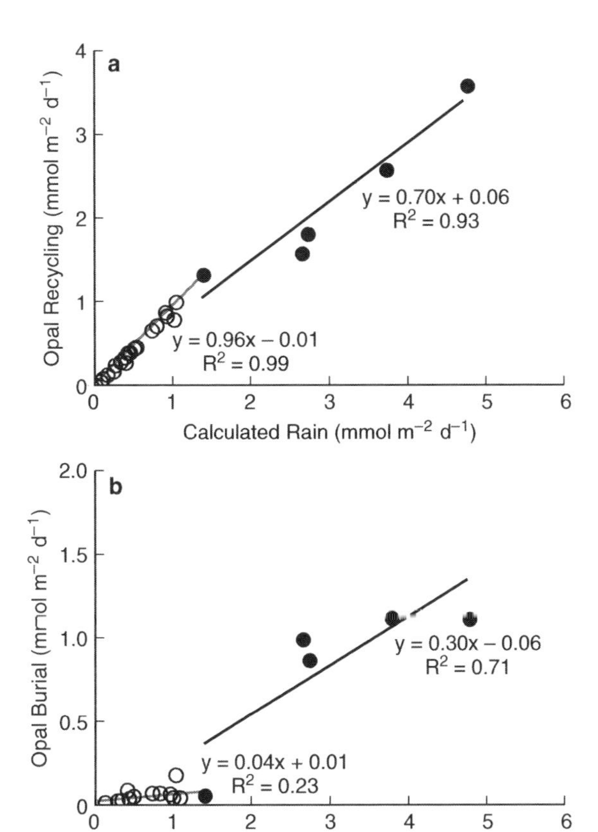

FIGURE 7.4.2: (a) Flux of silicic acid from the sediments to the water column plotted versus the rain rate of opal arriving at the sediment-water interface [*Ragueneau et al.*, 2000]. The rain rate is calculated by summing the flux of silicic acid out of the sediments to the burial rate of opal. The burial rate of opal is calculated by multiplying the measured asymptotic opal content times the sedimentation rate S. The data are from a wide range of locations around the world. (b) Burial rate of opal plotted versus rain rate. The regions with low opal rain rate, which are representative of most of the deep ocean, typically have about a 5% preservation efficiency, whereas the high rain rate locations, which are generally from the high latitudes in the Bering Sea and Southern Ocean, have about a 30% preservation efficiency.

where $\lambda = -\sqrt{k'/\varepsilon_s}$, and the modified dissolution rate constant $k' = ((1-\phi)/\phi) \cdot k \cdot ([SiO_2]/[Si(OH)_4]_{SAT})$ depends on the dissolution rate constant, k, the $Si(OH)_4$ saturation concentration, the opal content of the sediments, and the porosity. Note from the discussion in panel 7.4.1 that λ is negative, and since $[Si(OH)_4]_0 - [Si(OH)_4]_{SAT}$ is also always negative (see figure 7.1.4), the flux will be in the negative direction, that is to say out of the sediments. From this equation, we see that the flux is a function of the porosity, the pore water diffusivity, the dissolution rate constant of opal, the opal content in the sediments, and the difference between the silicic acid concentration at the sediment-water interface and the saturation concentration in the sediments. Neither the porosity nor the diffusivity differs very much from one region to another, so it is really only the dissolution rate constant, opal content, and silicic acid concentration that cause the flux to vary from one location to another.

We are now ready to estimate the silicic acid fluxes out of the sediments for profiles such as those shown in figure 7.1.4. For porosity, we use a value of 0.85, which is typical for the upper part of deep ocean cores (cf. figure 6.1.3 and *Dixit and Van Cappellen* [2003]). We estimate the effective molecular diffusivity, ε_s, based on the free water-silicic acid molecular diffusivity of $\varepsilon = 5.5 \times 10^{-10}\,m^2\,s^{-1}$ given by *Wollast and Garrels* [1971] for deep water conditions. For a porosity of 0.85, the equation in table 6.2.1 yields an $\varepsilon/\varepsilon_s$ of 1.33. This gives $\varepsilon_s = 4.1 \times 10^{-10}\,m^2\,s^{-1}$ for silicic acid. Typical values for $[Si(OH)_4]_0 - [Si(OH)_4]_{SAT}$ are about -100 to $-700\,mmol^{-3}$ (cf. figure 7.1.4 and summary by *Dixit and Van Cappellen* [2003]), assuming for now that the observed asymptotic concentration of silicic acid in sediments is indeed representative of the saturation concentration, and using a representative $[Si(OH)_4]_0$ concentration of $\sim 100\,mmol\,m^{-3}$ (cf. figure 7.1.2).

In order to proceed, we need a quantitative estimate for λ, which we obtain by analyzing silicic acid profiles in the sediments such as those shown in figure 7.1.4. We note that the e-folding length scale of our solution to the silicic acid pore water equation is $z^* = -1/\lambda = \sqrt{\varepsilon_s/k'}$. The significance of the e-folding length-scale in this particular equation can perhaps be more easily understood if we rearrange equation (6) of panel 7.4.1 to the following form:

$$\frac{[Si(OH)_4] - [Si(OH)_4]_0}{[Si(OH)_4]_{SAT} - [Si(OH)_4]_0} = 1 - \exp\left(-\frac{z}{z^*}\right) \qquad (7.4.3)$$

When $z = z^*$, the right-hand side of this equation is 0.632, i.e., z^* is the depth over which the profile increases by almost two-thirds of the difference between the surface concentration and deep asymptotic concentration. From the observations in figure 7.1.4, we see that typical e-folding length-scales are 1 to 4 cm, i.e., $\lambda = -0.25$ to $-1\,cm^{-1}$. Note that these length scales are much greater than the typical deep-ocean stagnant film thickness of order 1 mm (see section 6.2), which means that $[SiOH_4]_0$ is essentially identical with the bottom water concentration $[SiOH_4]_{BW}$.

Plugging the above parameters into (7.4.2) gives a range of fluxes between -0.1×10^{-5} and -2.4×10^{-5} $mmol\,m^{-2}\,s^{-1}$, which is equivalent to diurnal flux of -0.08 to $-2.1\,mmol\,m^{-2}\,d^{-1}$. Comparisons of the fluxes obtained by this method with fluxes estimated *in situ* with benthic chambers generally show excellent agreement within the uncertainty of the measurements [e.g., *Berelson et al.*, 1987; *McManus et al.*, 1995]. The range of our estimates encompasses most of the flux estimates summarized by *Dixit and Van Cappellen* [2003], which were obtained either by benthic chambers or more sophisticated models than represented by equation (6) of

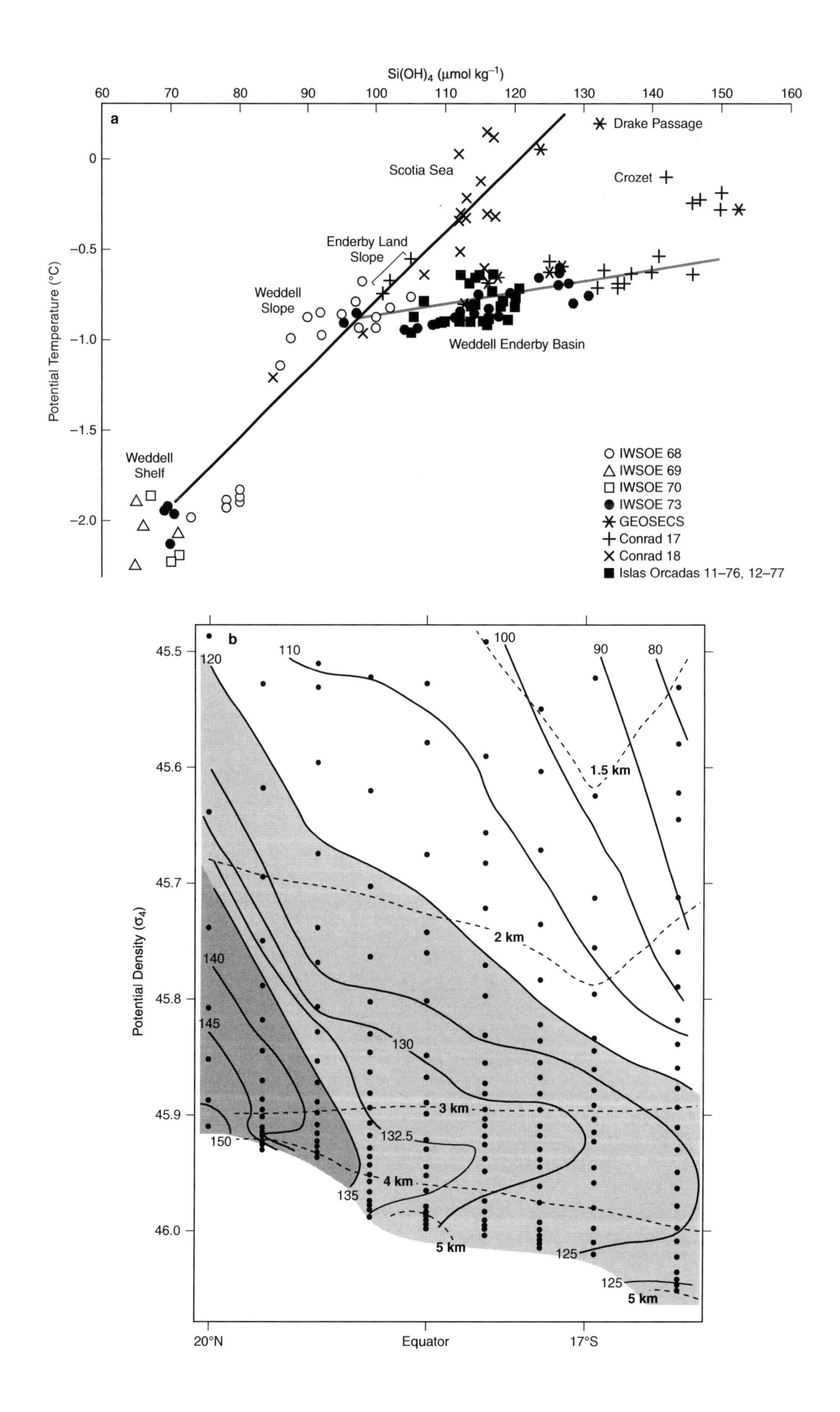

Si(OH)4 (μmol kg−1)
Potential Temperature (°C)
a
Drake Passage
Scotia Sea
Crozet
Enderby Land
Slope
Weddell
Slope
Weddell Enderby Basin
Weddell
Shelf
IWSOE 68
IWSOE 69
IWSOE 70
IWSOE 73
GEOSECS
Conrad 17
Conrad 18
Islas Orcadas 11–76, 12–77
b
Potential Density (σ4)
1.5 km
2 km
3 km
4 km
5 km
20°N
Equator
17°S

panel 7.4.1 (cf. also figure 7.4.2b). Only some of the Southern Ocean and most of the Scotia Sea estimates are higher (-2.9 to $-5.6\,\text{mol m}^{-2}\,\text{d}^{-1}$), and only one estimate in each of the Norwegian Sea and Somalia Basin are lower ($-0.05\,\text{mol m}^{-2}\,\text{d}^{-1}$).

We now have the tools and information we need in order to estimate an upper limit threshold flux. We use the flux equation (7.4.2) with the values of porosity and silicic acid diffusivity given above. For λ we use the largest estimate we obtained from examination of the observed profiles, $\sim 1\,\text{cm}^{-1}$; and for $[Si(OH)_4]_0 - [Si(OH)_4]_{SAT}$ we use an upper limit of $\sim 900\,\text{mmol m}^{-3}$ obtained using a representative bottom water concentration of $100\,\text{mmol m}^{-3}$, and the opal saturation concentration of $1004\,\text{mmol m}^{-3}$ we calculated for bottom water in panel 7.1.1. This gives an upper-limit silicic acid flux out of the sediments of $\sim 3\,\text{mmol m}^{-2}\,\text{d}^{-1}$.

What is the significance of this upper limit flux? From mass conservation, and in the absence of lateral transport of sediments, we have in a steady state that:

Opal burial = opal rain – silicic acid flux out of the sediments – (burial of silicic acid + conversion of opal to authigenic aluminosilicates)

Observations suggest that the contribution of the final two terms in parentheses is small compared to the others. Thus, all else being equal, if the opal rain exceeds the threshold silicic acid flux out of the sediments, any opal rain above this flux should lead to burial. If the opal rain is less than the threshold flux, then the burial should be negligible, unless the asymptotic concentration or some other property of the sediments changes.

Within the uncertainty of our estimate of the threshold flux, we see that it corresponds quite well with the opal preservation discontinuity of $\sim 2\,\text{mmol m}^{-2}\,\text{d}^{-1}$ that we infer from figure 7.4.2. However, the data shown in figure 7.4.2 show that there is opal present in most of the sediments below the preservation discontinuity, and that the burial rate averages about 5% of the opal rain rate. We also observe that the pore water silicic acid concentration generally asymptotes toward a level that is well below saturation with opal, even in sediments that are rich in opal (e.g., figures 7.1.4 and 7.4.4). In other words, the opal preservation discontinuity certainly appears to be a real phenomenon, and it matches quite well the threshold flux we estimated above, but we appear to still be missing an important part of the puzzle that can help us to understand the behavior of the system below the opal preservation discontinuity. The next section on opal chemistry discusses laboratory experiments that provide possible answers.

Opal Chemistry

As we have seen, the asymptotic concentration of pore water silicic acid varies widely between $< 200\,\text{mmol m}^{-3}$ and $> 800\,\text{mmol m}^{-3}$. In panel 7.1.1, we noted that opal has a pressure and surface area–corrected solubility of $\sim 1004\,\text{mmol m}^{-3}$ at typical deep ocean temperatures and pressures, and with a typical sediment opal surface area of $25\,\text{m}^2\,\text{g}^{-1}$. This suggests that most sediments are greatly undersaturated with respect to opal, but with the degree of undersaturation varying by a factor of ~ 4 from about 200 to $800\,\text{mmol m}^{-3}$. A wide variety of mechanisms have been proposed to explain how silicic acid concentrations can remain so low relative to saturation with fresh diatom opal, while the opal concentrations in the sediments remain high. These all fall into two fundamental categories: (a) equilibrium mechanisms, in which the sediment silica is assumed to have a lower saturation concentration than fresh diatom opal, with the asymptotic concentration in the sediments being equal to the saturation concentration so that net dissolution ceases while there is still abundant opal; or (b) kinetic mechanisms, in which the dissolution rate decreases with age, becoming negligible at some depth in the sediments where there is still abundant opal and the silicic acid concentration is below saturation. The outcome of model comparison studies of these mechanisms, such as that of *McManus et al.* [1995], is that it is difficult to discriminate between the proposed mechanisms by analysis of the observed silicic acid and opal distributions. The major new insights that have made it possible to begin to distinguish between these hypotheses and make some progress in our understanding have come from laboratory studies under controlled conditions, such as those described next.

The laboratory solubility and dissolution kinetics measurements we shall discuss here are made in both batch and flow-through reactors. In batch experiments, a sediment sample is exposed to a solution that initially has little or no dissolved silicic acid. The solution is monitored over

FIGURE 7.4.3: Evidence of the flux of silicic acid from the sediments in the bottom waters in: (a) the Weddell-Enderby and Crozet basins in the Southern Ocean, which correspond to the band of high opal deposits in a band across the South Atlantic and into the Indian Ocean (see figure 7.1.3); and (b) the western Indian Ocean [*Edmond et al.*, 1979]. Panel (a) shows a plot of potential temperature versus silicic acid. Most of the oceanic data in this region fall along the straight line connecting the Weddell shelf with the Scotia Sea. The branch labeled Weddell-Enderby Basin and grouping of data labeled Crozet Basin are as much as 40 to $50\,\mu\text{mol kg}^{-1}$ (~ 40 to $50\,\text{mmol m}^{-3}$) higher in concentration than other water types in this region, due to the high flux of silicic acid from the sediments. Panel (b) shows a vertical section with very high concentrations in the high sediment silica region of the northwestern Indian Ocean. These high silica waters are undercut by low silica waters further to the south and thus form a maximum above the bottom.

Panel 7.4.1: Solution to Silicic Acid Diagenesis Equation

The diagenetic equation for silicic acid in pore water based on equation (6.2.6) is

$$\frac{\partial \phi \cdot [Si(OH)_4]}{\partial t} = -\frac{\partial}{\partial z}(\phi \cdot w \cdot [Si(OH)_4]) + \frac{\partial}{\partial z}\left(\phi \cdot \varepsilon_s \cdot \frac{\partial [Si(OH)_4]}{\partial z}\right) + SMS(Si(OH)_4) \quad (1)$$

Here, $[Si(OH)_4]$ (mmol per m^3 of pore water) is silicic acid concentration, and the source-minus-sink term $SMS(Si(OH)_4)$ (mmol s^{-1} per m^3 of pore water plus bulk sediment) represents the net production rate of silicic acid. We simplify the equation by making the following assumptions:

1. that the silicic acid distribution is in steady state, allowing us to eliminate the time derivative;
2. that vertical variations in the porosity such as those shown in figure 6.1.3 can be ignored, enabling us to take ϕ out of the derivatives (see *McManus et al.* [1995] for a discussion of how variable porosity can affect the model predictions);
3. that advection of pore water is negligible relative to diffusivity, allowing us to eliminate the advection term; and
4. that the diffusivity ε_s is constant with depth, which enables us to take this out of the derivative.

This gives

$$0 = \phi \cdot \varepsilon_s \cdot \frac{\partial^2 [Si(OH)_4]}{\partial z^2} + SMS(Si(OH)_4) \quad (2)$$

The source-minus-sink term in (2) represents the chemical dissolution and precipitation of opal and any other reactive silica phases that may be present in the sediments. This is commonly expressed in the form given by equation (1) of panel 7.1.1,

$$SMS(Si(OH)_4) = k \cdot \{(1-\phi) \cdot [SiO_2]\} \cdot \left(1 - \frac{[Si(OH)_4]}{[Si(OH)_4]_{SAT}}\right) \quad (3)$$

Recall that $[SiOH_4]_{SAT}$ is the saturation concentration of silicic acid in equilibrium with the silica in the sediments, and the term in parentheses is a fractional measure of the departure of the actual concentration from the saturation concentration. The dissolution rate constant k has units of s^{-1}. Plugging (3) into (2) and dividing by the porosity gives

$$0 = \varepsilon_s \cdot \frac{\partial^2 [Si(OH)_4]}{\partial z^2} + k' \cdot ([Si(OH)_4]_{SAT} - [Si(OH)_4]) \quad (4a)$$

where

$$k' = \frac{(1-\phi)}{\phi} \cdot k \cdot \frac{[SiO_2]}{[Si(OH)_4]_{SAT}} \quad (4b)$$

is a revised dissolution rate constant also with units of s^{-1}.

In what follows, we assume that k' is constant with depth, which allows us to obtain a simple analytical solution of the form

$$[Si(OH)_4] = [Si(OH)_4]_{SAT} + \gamma_1 \cdot \exp(\lambda_1 z) + \gamma_2 \cdot \exp(\lambda_2 z), \quad (5a)$$

where

$$\lambda_1, \lambda_2 = \pm\sqrt{\frac{k'}{\varepsilon_s}} \quad (5b)$$

Note that λ_1 is positive and λ_2 is negative. The λ's have units of m^{-1}. The integration constants γ_1 and γ_2, which have units of mmol m^{-3} of pore water, are determined by the boundary conditions. We take as a boundary condition at the sediment-water interface ($z = 0$) that $[Si(OH)_4]$ equals the bottom water concentration corrected for the stagnant film effect, $[Si(OH)_4]_o$. At the bottom, we assume that $[SiOH_4]$ approaches $[Si(OH)_4]_{SAT}$ as z approaches infinity (recall that the sign convention we use in sediments is positive z downwards) so that the dissolution of opal goes to zero. The boundary condition at infinite depth requires us to set γ_1 to zero because the exponential term with positive λ_1 becomes infinite as z approaches infinity. The final solution is

$$[Si(OH)_4] = [Si(OH)_4]_{SAT} + ([Si(OH)_4]_0 - [Si(OH)_4]_{SAT}) \cdot \exp(\lambda \cdot z) \quad (6)$$

where $\lambda = \lambda_2 = -\sqrt{k'/\varepsilon_s}$. The exponential term $\exp(\lambda \cdot z)$ ranges from a value of 1 at the sediment-water interface to 0 at infinite depth. The equation thus defines an exponential profile that starts with $[SiOH_4]_o$ at the sediment-water interface and increases asymptotically to $[SiOH_4]_{SAT}$ at infinite depth. This solution is identical to that used, for example, by *Archer et al.* [1993], *McManus et al.* [1995], and *Dixit and Van Cappellen* [2003], and provides a good fit to pore water profiles such as those shown in figure 7.1.4. Where studies such as these and others differ widely from each other is in the interpretation of $[SiOH_4]_{SAT}$ and k'. For example, *McManus et al.* [1995] assume that k' varies exponentially with depth, which would be the case if porosity, the rate constant k, and the concentration of opal $[SiO_2]$ either were all constant or themselves varied exponentially with depth. Their solution has exactly the same form as equation (6), but with a different interpretation of λ.

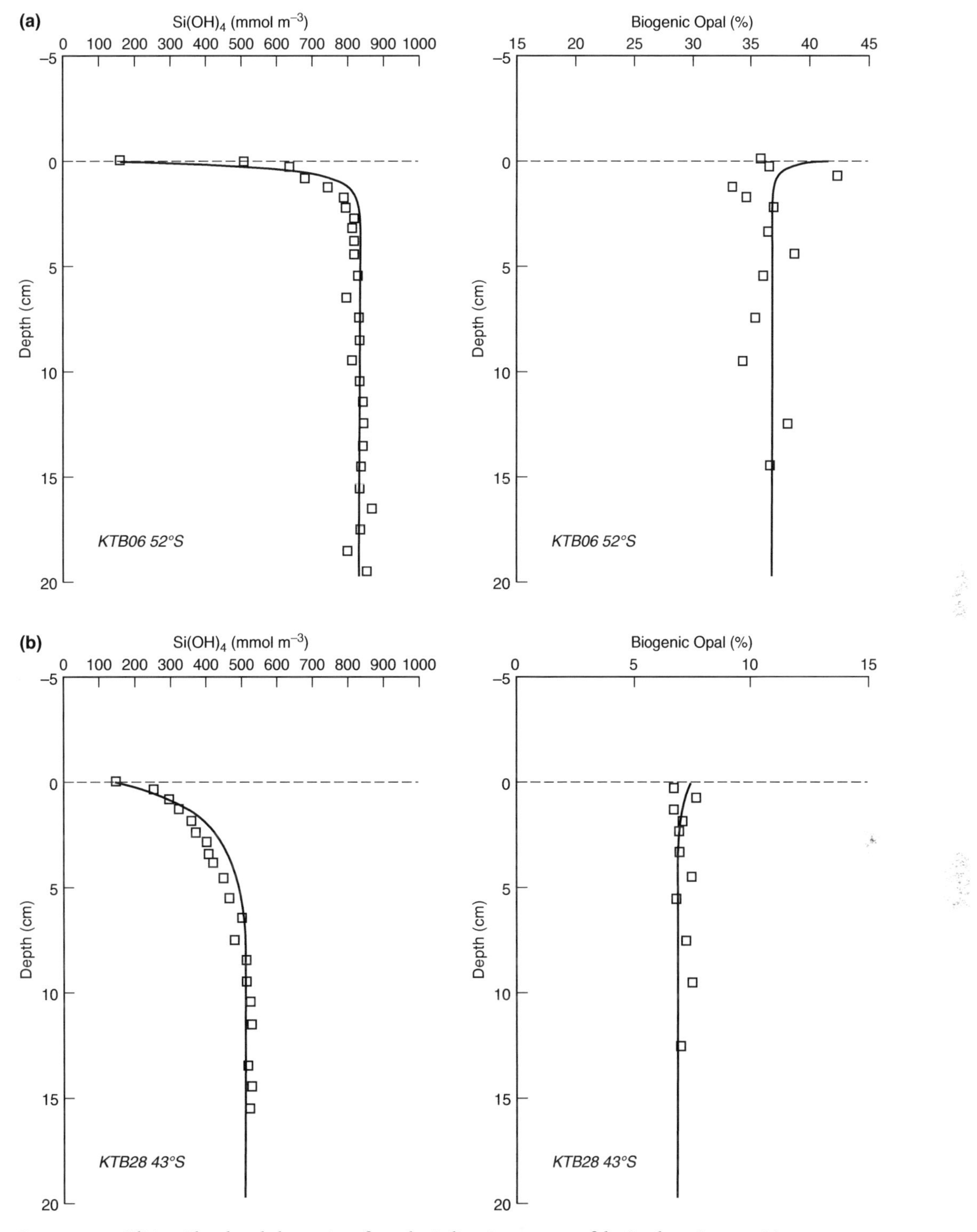

FIGURE 7.4.4: Silicic acid and opal observations from the Indian Ocean sector of the Southern Ocean at (a) an area of opal rich "siliceous oozes" at 52°S and 61°E, where both the silicic acid and opal concentrations are high and detrital concentrations (11%) are low, and (b) an area at 43°S and 58°E where the opal and silicic acid concentrations are much lower and nonbiogenic detritus concentration (76%) is high [*Rabouille et al.*, 1997]. The data are shown as open squares. The lines are model fits to the data (see caption to figure 7.4.12 for a discussion of the model). Note that at both locations, the opal is relatively uniform with depth.

time until it stabilizes at a concentration that is taken to represent the solubility of the reacting silica. Flow-through reactors have a constant inflow of a solution with a known silicic acid concentration $[Si(OH)_4]_{in}$ (mmol m^{-3}) that is varied from one experiment to the next and then run until the outflow stabilizes at a composition $[Si(OH)_4]_{out}$, which may be greater or less than $[Si(OH)_4]_{in}$, depending on whether silica is dissolving or precipitating in the

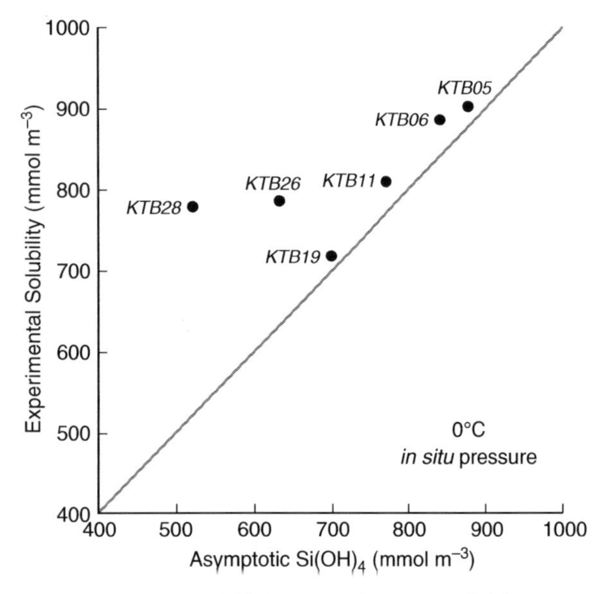

FIGURE 7.4.5: Experimentally determined apparent solubilities plotted versus the observed asymptotic pore water silicic acid concentration [*Dixit et al.*, 2001]. The apparent solubilities were determined using flow-through reactors. All the experimental solubilities are higher than the asymptotic silicic acid concentrations. However, except at cores KTB26 and KTB28, which were taken in detritus-rich sediments, the two are strongly correlated with each other. The cores were taken between 55°S (KTB05) and 43°S (KTB28) across the Polar Front in the Indian Ocean sector of the Southern Ocean. Cores KTB26 and KTB28 were taken in detritus-rich regions and are much more highly undersaturated. Vertical profiles of opal and silicic acid from two of the cores, KTB06 and KTB28, are shown in figure 7.4.4.

vessel. The *apparent solubility* is defined as the concentration of silicic acid when the outflow concentration equals the specified inflow concentration. The term "apparent" is used because the solubility thus obtained represents the average behavior of the heterogeneous silica mixture contained in the sediment sample weighted toward those phases that are more reactive.

Figures 7.4.5 and 7.4.6 from *Dixit et al.* [2001] and *Gallinari et al.* [2002], respectively, compare measured apparent solubilities with observed asymptotic concentrations at several locations around the world. The apparent solubilities are much lower than the value of $\sim 1004\,\mathrm{mmol\,m^{-3}}$ we estimated as representative of opal in the deep ocean. Furthermore, except for the cores labeled KTB26 and KTB28 in figure 7.4.5, there is a strong correlation between the asymptotic concentration and apparent solubilities over a range from a low of $\sim 200\,\mathrm{mmol\,m^{-3}}$ in the North Atlantic (figure 7.4.6) to $\sim 900\,\mathrm{mmol\,m^{-3}}$ in the Southern Ocean (figure 7.4.5). These results imply that solubility does exert control on the asymptotic silicic acid concentration, but the fact that these two are not identical at any location suggests that the kinetic mechanism may be involved as well.

The large range in apparent solubilities and the fact that they are so much lower than those we obtain from the equations in panel 7.1.1, table 1 is remarkable and baffling. As noted in section 7.1, laboratory experiments and theoretical studies indicate that the solubility of silica depends on a wide range of parameters including the temperature, pH, pressure, surface area of the silica per unit mass, the inclusion of trace metal contaminants in the silica crystal, and the presence or authigenic formation of other silica phases in the sediment. We already discussed all of these except the last two and found that they give us a deep ocean saturation concentration of $1004\,\mathrm{mmol\,m^{-3}}$ with only a modest variability of a few 10's of percentage that cannot come close to explaining the wide range in the observations. Experiments show that the principal culprit in the remaining two mechanisms appears to be Al(III), which can be incorporated into diatoms at the time of their formation (referred to as *primary uptake*), or during early diagenesis (*secondary uptake*) and is also linked to the formation of authigenic aluminosilicate phases in sediments.

The observations summarized in figure 7.4.7 show a large drop in observed asymptotic concentrations of silicic acid in pore water, $[\mathrm{Si(OH)_4}]_\infty$, as the ratio of nonbiogenic detritus to opal in sediments increases. Combined with the indication from observations that the pore water Al(III) concentration increases with percentage of detritus, this is consistent with Al playing a role in the apparent solubility of silica in the sediments. However, these observations are not sufficient in themselves to prove that Al is important, nor do they provide any insight into the mechanisms by which Al might be influencing the apparent solubility. The influence on the apparent opal solubility of Al incorporation into the opal and the formation of authigenic silica phases, has been widely investigated [e.g., *Lewin*, 1961; *Iler*, 1973; *Mackin*, 1987; *van Bennekom et al.*, 1989, 1991; *Van Cappellen and Qiu*, 1997a; *Dixit et al.*, 2001; *Gallinari et al.*, 2002]. The most important findings from laboratory investigations are as follows:

1. Direct incorporation of Al into opal, probably due mostly to secondary uptake, gives a range of ~7 in the Al/Si ratio (see abscissa of figure 7.4.8). However, the solubility of the opal drops by only 15% to 25% even with these large changes in Al/Si ratio (see figure 7.4.8, taken from *Dixit et al.*, [2001]; and *Gallinari et al.* [2002]). If we take the pressure and surface area–corrected solubility of $\sim 1004\,\mathrm{mmol\,m^{-3}}$ calculated above, a correction of 25% gives an upper limit reduction in the solubility to $\sim 753\,\mathrm{mmol\,m^{-3}}$, which is within the range of the Antarctic cores of figure 7.4.5, but well above the equatorial Pacific and North Atlantic cores of figure 7.4.6.

2. In a remarkable series of experiments, *Dixit et al.* [2001] were able to demonstrate the formation of authigenic aluminosilicate minerals in flow-through reactors when both Al(III) and silicic acid were present in the inflowing water. Figure 7.4.9 shows steady-state silicic acid concentrations

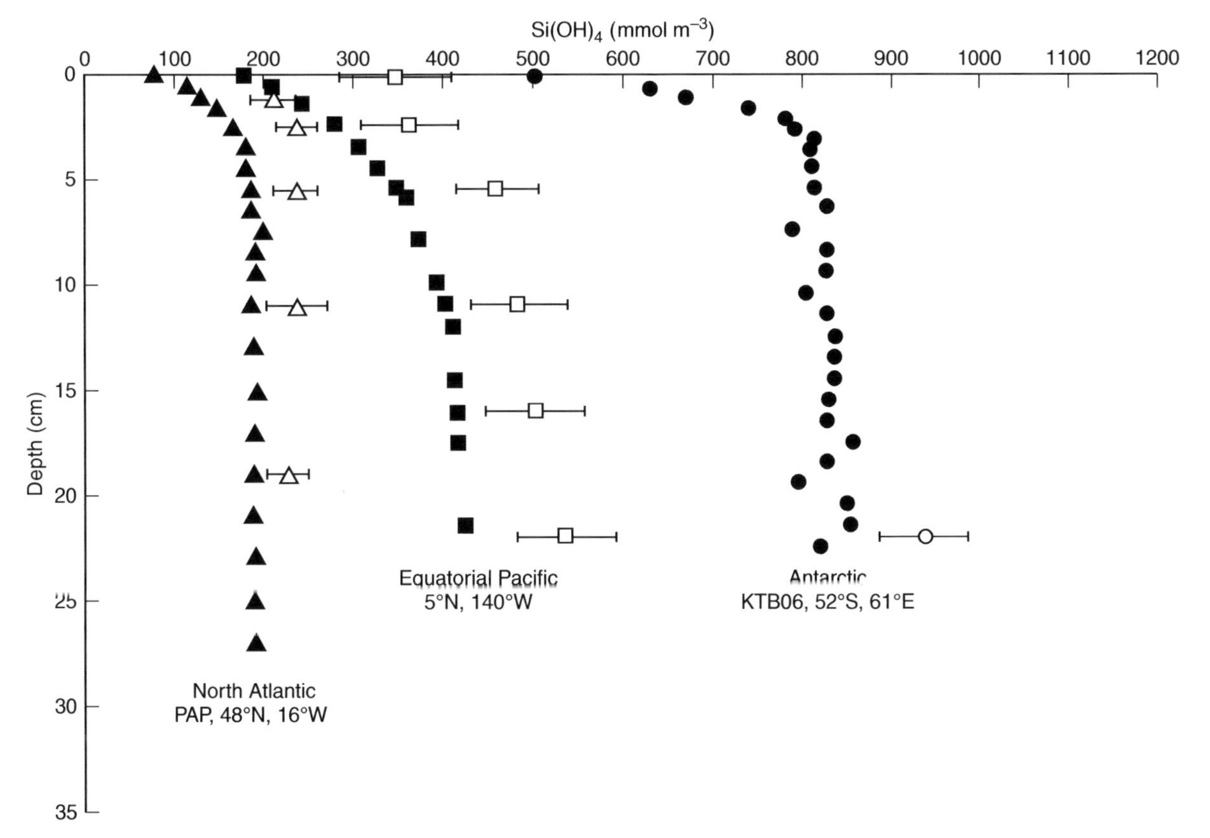

FIGURE 7.4.6: Vertical profiles of pore water silicic acid concentration at three representative locations identified in the figure (filled-in symbols; figure taken from *Gallinari et al.* [2002]). Also shown are apparent solubilities determined experimentally with flow-through reactors (open symbols with measurement uncertainties). As with figure 7.4.5, the measured solubilities are higher than the pore water silicic acid concentrations, but there is a clear relationship between the two.

from a series of batch solubility experiments that were carried out using sediments with varying nonbiogenic detritus-to-opal ratios. The range in apparent solubilities is comparable to the observed range shown in figure 7.4.7. We have seen that secondary uptake of Al in the biogenic opal crystal can account for < 25% of this variability. Taken together, all these data strongly suggest that the formation of aluminosilicates is the primary cause of the large drop in apparent solubility.

In the case where there is a biogenic opal phase $[SiO_2]_a$ and an authigenic aluminosilicate phase $[SiO_2]_b$, and the dissolution kinetics for both is assumed to be first order (that is to say, linear in the degree of undersaturation), the source-minus-sink term might take the form

$$\frac{SMS(Si(OH)_4)}{\phi} = k_a' \cdot ([Si(OH)_4]^a_{SAT} - [Si(OH)_4]) + k_b' \cdot ([Si(OH)_4]^a_{SAT} - [Si(OH)_4])$$
$$k_a' = \frac{(1-\phi)}{\phi} \cdot k_a \cdot \frac{[SiO_2]_a}{[Si(OH)_4]^a_{SAT}}$$
$$k_b' = \frac{(1-\phi)}{\phi} \cdot k_b \cdot \frac{[SiO_2]_b}{[Si(OH)_4]^b_{SAT}} \qquad (7.4.4)$$

which can be rearranged to give

$$\frac{SMS(Si(OH)_4)}{\phi} = k' \cdot ([Si(OH)_4]_{SAT} - [Si(OH)_4]) \qquad (7.4.5a)$$

where

$$k' = k_a' + k_b' \qquad (7.4.5b)$$

and

$$[Si(OH)_4]_{SAT} = \frac{k_a' \cdot [Si(OH)_4]^a_{SAT} + k_b' \cdot [Si(OH)_4]^b_{SAT}}{k_a' + k_b'} \qquad (7.4.5c)$$

(D. Hammond, personal communication, 2004). We thus see that, in this case, the apparent dissolution rate constant would be the sum of the opal and authigenic silica rate constants, and the apparent solubility would be the mean of the two silica phases weighted by their reactivities.

Our discussion of apparent solubilities has been framed in terms of a putative solubility of ~1004 mmol m^{-3} in deep waters at 2°C and the ways in which this biogenic opal might be altered in order to modify its apparent solubility. We conclude our discussion of solubility by noting that the response of the apparent

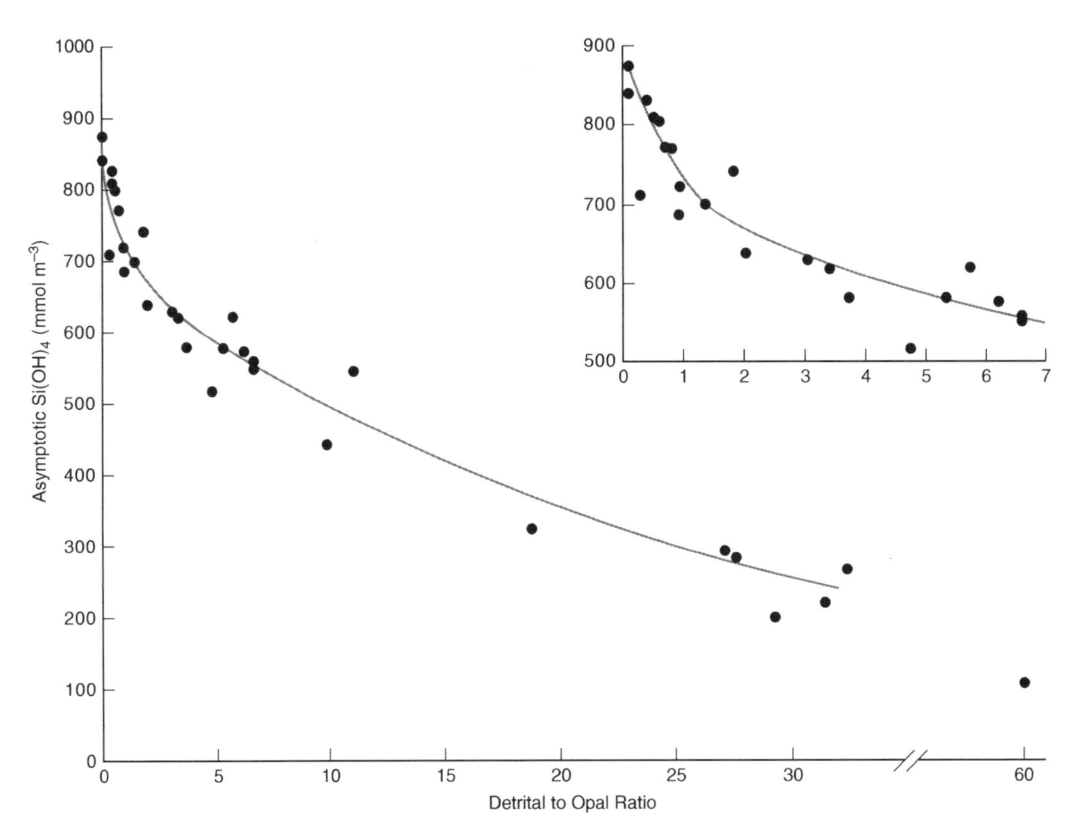

FIGURE 7.4.7: A summary of observed asymptotic concentrations of silicic acid in pore water, $[SiOH_4]_\infty$, plotted versus the ratio of nonbiogenic detritus to opal in sediments (taken from *Dixit and Van Cappellen* [2003]). The data are from the following locations: Indian sector of the Southern Ocean, Somalia Basin, Scotia Sea, Peru Basin, Juan de Fuca Ridge, Arabian Sea, Northeastern Atlantic, and Norwegian Sea. *Dixit and Van Cappellen* [2003] have fitted the data with two exponentials:

$$[\mathrm{Si(OH)_4}]_\infty = 811 \cdot \exp\left(-0.16 \cdot \frac{\text{detrital}}{\text{opal}}\right), \quad \frac{\text{detrital}}{\text{opal}} \leq 2$$

$$[\mathrm{Si(OH)_4}]_\infty = 693 \cdot \exp\left(-0.04 \cdot \frac{\text{detrital}}{\text{opal}}\right), \quad \frac{\text{detrital}}{\text{opal}} > 2$$

The reduction in asymptotic silicic acid concentration with increasing nonbiogenic detritus to opal ratio is hypothesized to result from reduced apparent solubility due to formation of authigenic aluminosilicates by reprecipitation of opal.

solubilities to increases in the ratio of nonbiogenic detritus to opal can differ depending on the type of detrital material (figure 7.4.9) and the organism that produced the opal [e.g., *Gallinari et al.*, 2002]. There appear to be considerable temporal and spatial fluctuations in the properties of the opal arriving at the sediments [e.g., *Ragueneau et al.*, 2001; *Gallinari et al.*, 2002]. *Gallinari et al.* [2002] show that there is a significant offset between the equatorial Pacific and Antarctic asymptotic concentrations when plotted versus the ratio of nonbiogenic detritus to opal. The authors offer no explanation for the differences between these two regions, but one is left to wonder whether they might be due to differences in the original biogenic opal (e.g., higher radiolarian content in the equatorial Pacific), or perhaps to differences in the history of the opal alteration. Clearly, there continues to be much to be learned about what controls apparent opal solubility.

Thus far we have ignored the role that dissolution kinetics plays in the observed diagenetic behavior of opal. Figures 7.4.5 and 7.4.6 show that, while the apparent solubility of silica in the sediments correlates strongly with the asymptotic concentration, the asymptotic concentrations still appear to be undersaturated. Furthermore, the apparent solubility obtained from two phases is actually the result of a dynamic process whereby silicic acid is continuously dissolving from the more soluble phase (opal) and precipitating onto the less soluble phase (authigenic aluminosilicate). If the dissolution rate is fast enough, the opal should eventually disappear.

We can estimate the magnitude of the dissolution rate constant k' in the sediments from our earlier estimates of λ and ε_s and the relationship $k' = \lambda^2 \cdot \varepsilon_s$. From this, we find that k' is $1/(3\,\mathrm{d})$ to $1/(40\,\mathrm{d})$. However, we need to convert this to the so-called *specific dissolution rate* k by

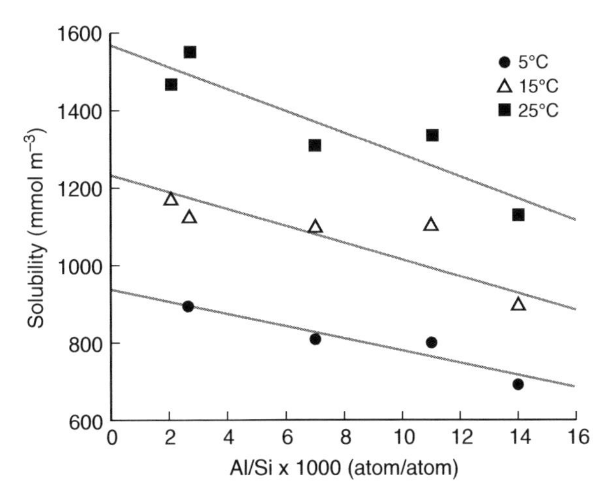

FIGURE 7.4.8: Apparent solubilities measured at different temperatures on core top samples from the Southern Ocean plotted versus the mole ratio of Al/Si in diatom frustules from those sediments (taken from *Dixit et al.* [2001]). The samples are from the same set of data shown in figure 7.4.5. This figure shows that Al(III) in diatom frustules is only able to account for a < 25% range in opal solubility.

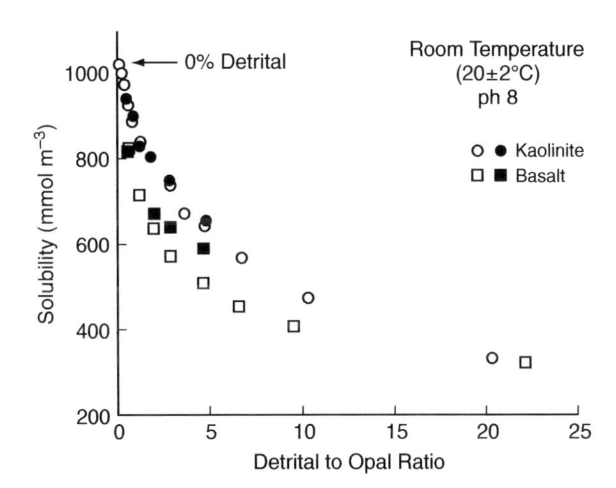

FIGURE 7.4.9: Apparent solubilities determined experimentally plotted versus the nonbiogenic detritus to opal ratio, which is an index of Al(III) availability in the pore water [*Dixit, et al.*, 2001]. The detrital-to-opal ratio was varied in these experiments using either kaolinite or basalt as detritus. The large reduction in apparent solubility with increasing detrital-to-opal ratio is hypothesized to result from the authigenic formation of low-solubility aluminosilicates by reprecipitation of opal. See text for discussion.

inverting equation (4b) of panel 7.4.1. We use representative values for ϕ and $[Si(OH)_4]_{SAT}$ of 0.9, and 500 mmol m^{-3} of pore water, respectively. The concentration of opal $[SiO_2]$ is more difficult to specify because, as with organic carbon, measurements of opal are usually reported as the mass ratio of opal to total sediments, f_{opal} (see figure 7.4.4). Panel 6.2.1 explains how to make the conversion. By inverting the equations given there, we find that a typical f_{opal} of 0.10 is equivalent to an $[SiO_2]$ concentration of 4.4×10^6 mmol m^{-3} of

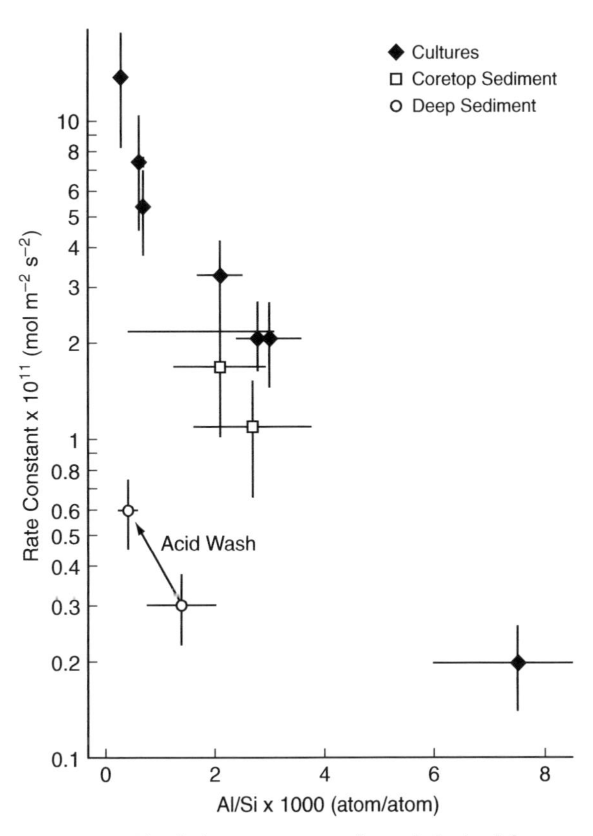

FIGURE 7.4.10: Dissolution rate constant for opal obtained from batch reactor experiments plotted versus the Al/Si ratio in the opal (figure taken from *Van Cappellen et al.* [2002]). The rate constants are normalized to the surface area of the opal in the reactor. The vertical displacement of data from cultured diatoms at the top, to core top sediments in the middle of the diagram, to deep sediments in the lower part of the diagram, shows the influence of aging in reducing the dissolution rate, presumably due to a reduction in reactive silanol sites (see discussion in text). The reduction in dissolution rate with increasing Al/Si ratio shows the influence of increasing Al incorporation in the diatoms. The influence of Al has been parameterized by *Dixit and Van Cappellen* [2003] as a function of the nonbiogenic detritus to opal ratio based on fitting experimental data that were obtained using sediment opal samples and diluting them with increasing amounts of kaolinite:

$$k_{DVC} = 0.0648 \cdot \left(\frac{\text{detritus}}{\text{opal}}\right)^{-0.32}$$

where k_{DVC} has units of nmol s^{-1} g^{-1} of opal. This can be converted to the units in the figure by multiplying by 10^{-9} mol nmol^{-1}, and dividing by the surface area, for which a typical ediment value might be 25 m^2 g^{-1} of opal. Conversion to k as defined by equation (1) of panel 7.1.1 in units of s^{-1} requires multiplying by 6×10^{-8} g nmol^{-1} of opal.

dry solids. We thus find that typical values for k are of order 1/(10 yr) to 1/(80 yr). This is two to three orders of magnitude slower than the average dissolution rate of 1/(23 d) at the surface of the ocean! However, it is so rapid compared with the timescales of sediment accumulation and bioturbation that there should not be any

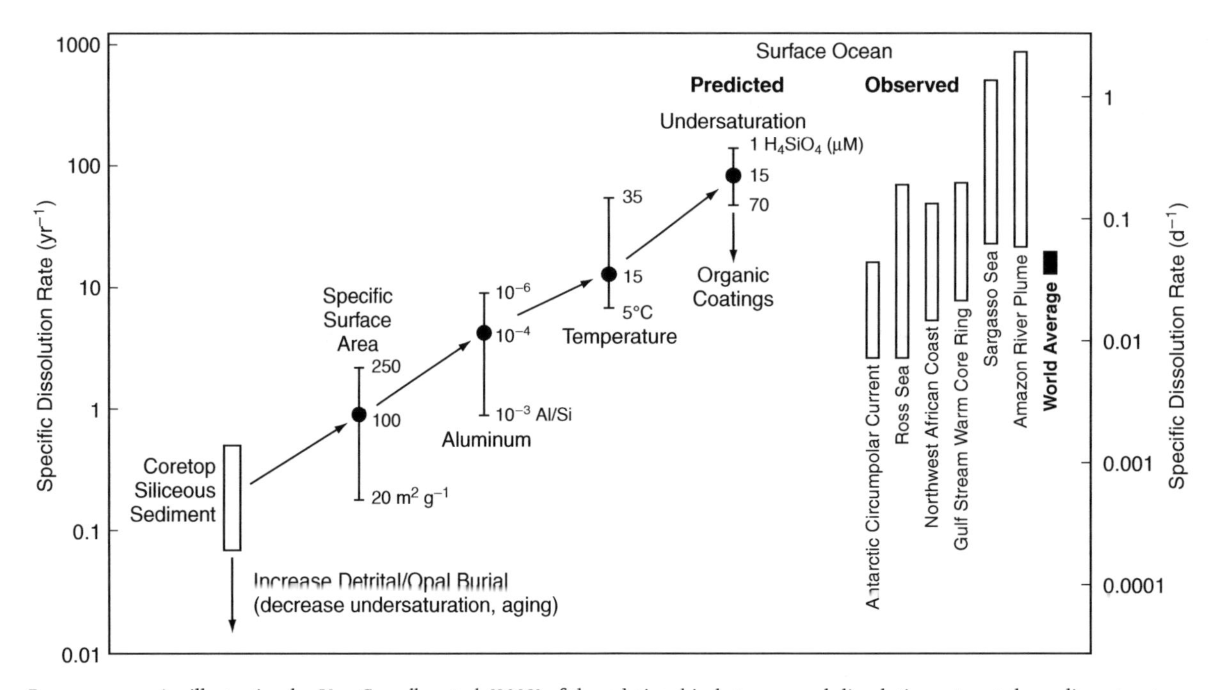

FIGURE 7.4.11: An illustration by *Van Cappellen et al.* [2002] of the relationship between opal dissolution rates at the sediment-water interface ("core top") and at the surface of the ocean. The diagram starts with core top measurements at the bottom left and adds first the influence of specific surface area (4.5×), then Al (5×), then temperature (3×), and finally the degree of undersaturation (6.5×) (cf. table 7.4.1). On the far right of the diagram are observed estimates of opal dissolution rates at the surface of the ocean. The extrapolated surface ocean dissolution rate is higher than the observed surface ocean dissolution rate, due most likely to the influence of organic coatings (∼0.1×).

TABLE 7.4.1
Material properties and environmental conditions affecting the rate of dissolution of opal in the ocean [from *Van Cappellen et al.*, 2002].
The core top sediments are based on siliceous sediments from the Southern Ocean, and the surface ocean properties are world averages. The rate increase is for surface ocean with respect to siliceous sediments.

	Siliceous Sediments	*Surface Ocean*	*Rate Increase*
Specific surface area (m^2 g^{-1})	22	100	4.5×
Temperature (°C)	0	15	3×
Al/Si ratio (atom/atom)	0.0025	0.0001	5×
Water depth (m)	>3000	0–200	Minor
Undersaturation[a]	0.40	0.985	6.5×

[a]The degree of undersaturation is defined as $1-[Si(OH)_4]/[Si(OH)_4]_{SAT}$, with $[Si(OH)_4]=$ 500 mmol m^{-3} in sediments and 15 mmol m^{-3} in the surface ocean.

NOTE: For the conditions given, the dissolution rate equation of *Hurd and Birdwhistell* [1983] gives a rate increase due to the specific surface area effect of 5.3×, and a rate increased due to the temperature effect of 11×.

opal left in the sediments below a meter or so (cf. *Rabouille et al.* [1997]). These observations suggest that the slowdown in opal dissolution that is seen in the transition from surface ocean to the upper few centimeters of the sediments must continue with depth in the sediments. What causes the slowdown?

We saw in section 7.1 that the dissolution rate of opal is affected by the temperature, the total surface area of opal present per unit volume of the solution, the fractional extent of undersaturation, the presence of reactive sites on the opal surface and various processes that can modify these, and the presence of organic coatings on diatom frustules. The only one of these that we did not discuss in section 7.1 is the influence of reactive sites on the opal surface. Figure 7.4.10 from *Van Cappellen et al.* [2002] summarizes a large number of dissolution rates obtained from flow-through and batch reactors. The rate constants in figure 7.4.10 have been normalized by the surface area of the opal in the reactor. The main messages of figure 7.4.10 are two:

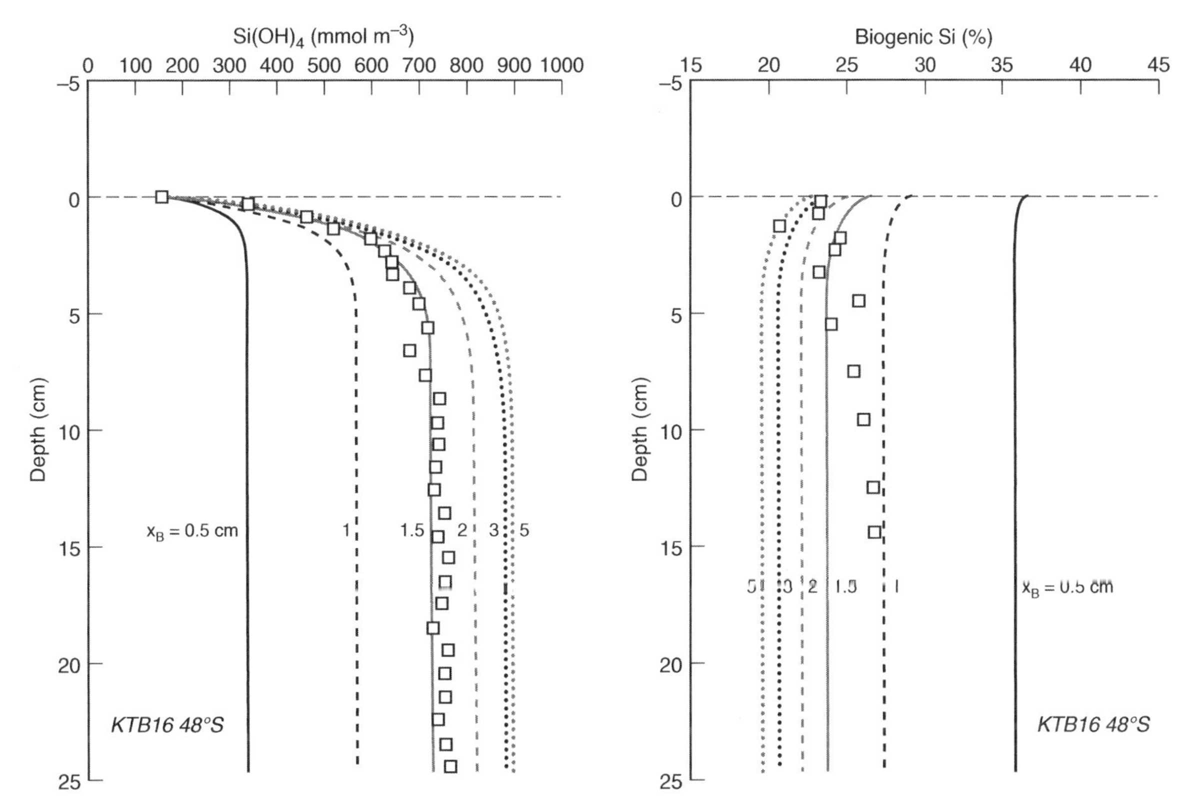

FIGURE 7.4.12: Silicic acid and opal profiles in sediments from the Indian Ocean sector of the Southern Ocean at 48°S and 56°E. Measurements are shown as open squares, and model simulations are shown as lines. The figure, including model simulations, is from *Rabouille et al.* [1997]. The distinctive feature of this model is that the dissolution rate constant is assumed to decrease exponentially from a value of k_0 at the sediment water interface to a value of k_∞ as the depth increases to infinity according to $k = k_\infty + (k_0 - k_\infty) \cdot \exp(-z/x_B)$. The various model simulations shown in the figure are for different scale lengths of the exponential decrease x_B, as labeled in the figure. As the scale length increases, from 0.5 cm to 5 cm, the opal concentration decreases and the asymptotic silicic acid increases. The best fit to the data are for a length-scale of 1.5 cm. Fits to a wide range of profiles such as those shown in figure 7.4.4 give x_B's ranging from 1 to 10 cm. In particular, the siliceous ooze core shown in figure 7.4.4a has a scale length of 10 cm and a k_0 of 1/(9 yr); and the detritus-rich core shown in figure 7.4.4b has a scale length of 1.8 cm and a k_0 of 1/(60 yr). The influence of detritus on both the scale length and initial dissolution rate is quite clear from these and the other cores in the study.

that the per unit area rate constant decreases (a) with increasing Al incorporation, and (b) with aging of the opal. Experimental measurements suggest that both of these work by modifying the availability of reactive sites on the crystal surface. The critical factor appears to be the availability of silanol groups $>SiOH^0$ (where > symbolizes the underlying lattice) the presence of which can be detected by potentiometric titrations and aluminum and cobalt adsorption experiments [cf. *Dixit and Van Cappellen*, 2002]. Such measurements show clearly that both Al and increasing age reduce the density of such reactive sites.

Table 7.4.1 and figure 7.4.11 show a summary by *Van Cappellen et al.* [2002] relating the dissolution rate on a core top sample from the Southern Ocean to dissolution rates obtained in opal samples from the surface ocean at a variety of locations around the world. In addition to the influence of the surface area (4.5×), which was normalized out of the data in figure 7.4.10, and the influence of Al (5×), this figure includes the influence of temperature (3×) calculated with the Arrhenius equation (3) of panel 7.1.1, and degree of undersaturation (6.5×). Aging effects are not included. Pressure effects are minor. The extrapolated dissolution rate at the surface of the ocean is somewhat higher than the observations, as shown in figure 7.4.11, and would be even higher if the aging effect were taken into consideration. The authors propose that the remaining differences between the extrapolated rate constants and the measured rate constants may be due to the influence of organic coatings on the surface ocean samples, which do not appear to be present in deep ocean sediments [*Van Cappellen and Qiu*, 1997b]. If we turn around and go from the top of the core down into the sediments, figure 7.4.10 indicates that the decreasing trend of the dissolution rate with age continues into the sediments.

To summarize, the behavior of relatively unaltered opal in sediments suggests that there should be no opal preservation below a threshold rain rate of ~2 to 3 mmol m^{-2} d^{-1}, and yet the observations indicate that there is. Furthermore, the asymptotic silicic acid concentration in these sediments with low rain rates is highly undersaturated with respect to opal solubility even though there is opal present. It is only with the recent series of laboratory experiments described above that biogeochemists have been able to move beyond the realm of conjecture in their attempts to explain these surprising observations. The story that has begun to emerge has many threads, but a critical one to both the dissolution rate and undersaturation conundrums appears to be the presence of Al(III) in pore waters (which comes from lithogenic detritus). A major cause of the decrease in the apparent solubility of opal appears to be the formation of low-solubility authigenic aluminosilicate minerals. A major cause of the slowdown of dissolution rate appears to be the blocking of reactive silanol sites by Al adsorption.

Several model studies presaged how opal might behave if the dissolution rate and apparent solubility of opal were allowed to vary [e.g., *Archer et al.*, 1993; *McManus et al.*, 1995; *Rabouille et al.*, 1998]. Figure 7.4.12 from one such study by *Rabouille et al.* [1997] shows how preservation in the sediments is affected by varying the way that the dissolution rate constant decreases with depth (see discussion in figure caption). As one might expect, the preservation increases as the dissolution rate constant decreases.

A final point: There have been many attempts over time to relate the burial of opal to parameters such as sedimentation rate or bioturbation. For example, *Ragueneau et al.* [2000] use a large set of data to show that opal preservation efficiency increases exponentially with sedimentation rate [cf. *Sayles et al.*, 2001]; and *Rabouille et al.* [1997] argue that the inefficient opal preservation of low rain rate samples may be due to a longer residence time of opal in the reactive zone. In another study, *Ragueneau et al.* [2001] argue by a process of elimination that variations in preservation efficiency in a set of cores they took in the northeast Atlantic may be due to changes in the rate of bioturbation. A possible mechanistic explanation for a link of the burial to the sedimentation rate or bioturbation is that there exists a reactive zone of high dissolution within the top few centimeters of the sediments, with burial rate increasing if opal spends less time in this reactive zone due to high bioturbation rate D_B and/or high sedimentation rate S. We are not aware of any modeling studies that have demonstrated how preservation might be influenced by bioturbation or sedimentation rate in light of our emerging new understanding of opal diagenesis and the influence on it of Al(III) and other processes.

7.5 Conclusion

In concluding this chapter, we first review the most important lessons we learned while discussing the marine Si cycle. We then proceed to propose a marine Si budget, and finally discuss possible controls on the long-term marine Si cycle.

Overview

In sections 7.1 and 7.2, we discussed a range of processes from local ecology to global silicic acid cycling that help explain the dominance of diatoms in many regions of the ocean and their relative absence in others as documented by figure 7.2.5b and the summary of this figure in table 7.2.2. There have been numerous efforts in recent years to add diatoms to euphotic zone ecosystem models in both one-dimensional physical models and OGCMs that incorporate insights such as those we discussed in section 7.1 (e.g., 1-D model studies by *Pondaven et al.* [1999], *Lancelot et al.* [2000], *Chai et al.* [2002], and *Yool and Tyrrell* [2003]; and OGCM studies by *Moore et al.* [2000a, 2001] and *Chai et al.* [2003]; the model by *Gnanadesikan* [1999a] includes diatoms implicitly by forcing the model silicic acid toward the observed silicic acid at the surface of the ocean as in equation (5.5.1)). The OGCM study by *Heinze et al.* [2003] is particularly notable in that it includes an end-to-end model of silicate cycling, from the production of opal by diatoms at the surface of the ocean to the dissolution of opal in the water column and sediments and the burial of opal in the sediments, and the cycling of silicic acid throughout. On the other hand, the euphotic zone biology in this model is much simpler than that used in most of the other studies mentioned, including the fact that there is no iron dependence on opal production.

The complexities of the interactions between various phytoplankton types in the euphotic zone ecosystem models, all requiring nitrate, phosphate, and iron, with only diatoms requiring silicic acid, make for very intricate models with a large number of poorly known parameters and thus a great deal of uncertainty. Nevertheless, these studies have begun to yield valuable insights about the role of silicic acid, macronutrients (nitrate and/or phosphate), and iron, as well as competition by other phytoplankton, in determining the predominance of diatoms in specific regions of the ocean, the Si:N ratio of diatom production and export, and the effect of perturbations such as iron fertilization on the ecosystem. We have previously discussed some salient results of these models in section 4.4.

As regards the large-scale distribution of silicic acid within the ocean, until recently there have been two major hypotheses that have framed our view of this distribution and how it relates to the global cycles of nutrients and carbon. The basic premises of both are that (1) organic matter is remineralized at depths shallower than where dissolution of opal occurs, and (2) the downward shift in silicic acid relative to nutrients that results from this is a global-scale phenomenon, that is to say, it is a consequence of the global biological productivity. The two hypotheses are the vertical trapping/global conveyor belt circulation mechanism explored most eloquently by *Broecker and Peng* [1982], in which silicic acid becomes trapped in the deep waters of the thermohaline circulation; and the silica pump hypothesis of *Dugdale et al.* [1995], in which even if there is a high input of silicic acid to the surface, the preferential surface recycling of nutrients will lead to a rapid preferential depletion of the silicic acid.

Sarmiento et al. [2004a] proposed an alternative hypothesis that focuses on the preferential stripping out of silicic acid in a specific region, the Southern Ocean, which then feeds high Si:N waters into the deep ocean, and low Si:N into the upper ocean. The North Pacific is a special case where strong vertical mixing brings high Si:N to the upper ocean. In other words, their hypothesis focuses more on the importance of the unique conditions of production in the Southern Ocean (rather than remineralization), and geographically on the Southern Ocean (as well as the North Pacific), rather than on the world. Specifically, their alternative to the *Broecker and Peng* [1982] vertical trapping–horizontal circulation mechanism is that silicic acid is rich in deep sea and poor in the thermocline primarily because it is stripped out preferentially into deep waters in the Southern Ocean and thus cannot get into the main thermocline. And by contrast with *Dugdale et al.* [1995], they emphasize the importance of low Si:N in supply waters as determining and limiting diatom production, rather than preferential recycling of N, as proposed by *Dugdale et al.* [1995].

The foregoing analysis has important implications for how we attempt to understand the potential impact of climate change on the global nutricline, biological productivity, and the carbon cycle. Many biogeochemists have drawn attention to the importance of the Southern Ocean to the global carbon cycle and biological productivity [cf. *Caldeira and Duffy*, 2000] and the role of these in climate change [*Sarmiento et al.*, 1998; cf. *Sigman and Boyle*, 2000]. Most recently, it has been suggested that modifications of the silicic acid cycle in the Southern Ocean during glacial times [*Brzezinski et al.*, 2002] could have had global impacts on low-latitude productivity that might even be able to explain the decreased atmospheric carbon dioxide of the ice ages [*Matsumoto et al.*, 2002]. What the above analysis suggests is that it is specifically the processes determining the properties of SAMW, and the formation of SAMW in the Subantarctic Zone (SAZ; see figures 7.3.3 and 7.3.4), that would be most crucial in determining how Southern Ocean processes affect the low-latitude productivity not just of diatoms, but also of other phytoplankton through the regulation by SAMW of the supply of nitrate and phosphate. It is likely that global warming will have a substantial impact in the SAZ. The SAZ is a region of deep mixed layers which models suggest are formed by a combination of cooling of subtropical gyre water and northward wind-driven Ekman transport *Ribbe* [1999]. Both of these are likely to change with global warming. Other mechanisms have been proposed for the regulation of climate change by the silica cycle [e.g., *Froelich et al.*, 1992; *Harrison*, 2000]. The role of the ocean in the global carbon cycle changes accompanying climate change is discussed further in chapter 10.

As regards the sediments, we have seen that a new paradigm is beginning to emerge that is finally able to offer experimentally based explanations for the unusual behavior of silicic acid and opal in the sediments. These explanations focus primarily on the effect of Al on both the dissolution rate of opal and the apparent solubility of silica. The consequences of these processes for what actually controls the preservation of opal in the sediments are still a bit muddled, though there are some interesting pointers to what might be happening. From the data that are available, there appears to be a divide between sediments that have high opal rain rates in excess of $\sim 2\,\mathrm{mmol\,m^{-2}\,day^{-1}}$, and also high opal preservation efficiencies of about 30%, on average; and those that have lower rain rates and low opal preservation efficiencies of about 5% on average (see figure 7.4.2). As might be expected from this analysis, most of the regions of high opal burial documented in figure 7.1.3 coincide with the regions of high opal export shown in figure 7.2.5.

The simple opal model we derived in section 7.3 suggests that the divide in opal preservation efficiency occurs at about the point where opal rain rates exceed a putative upper limit to the silicic acid flux out of the sediments. Above the opal preservation discontinuity, the opal rain rate exceeds the upper-limit diffusive flux out of the sediment, and burial is inevitable. Below the discontinuity, the opal rain rate is less than the potential diffusive flux out of the sediment, from which we might expect that all the opal should dissolve. The fact that the opal does not dissolve completely suggests that other processes come into play that limit the dissolution of opal, a hypothesis that is strongly supported by the modeling and experimental studies discussed in this chapter.

Marine Si Budget

We now combine all the information we have discussed in this chapter to put together an overall silicate budget

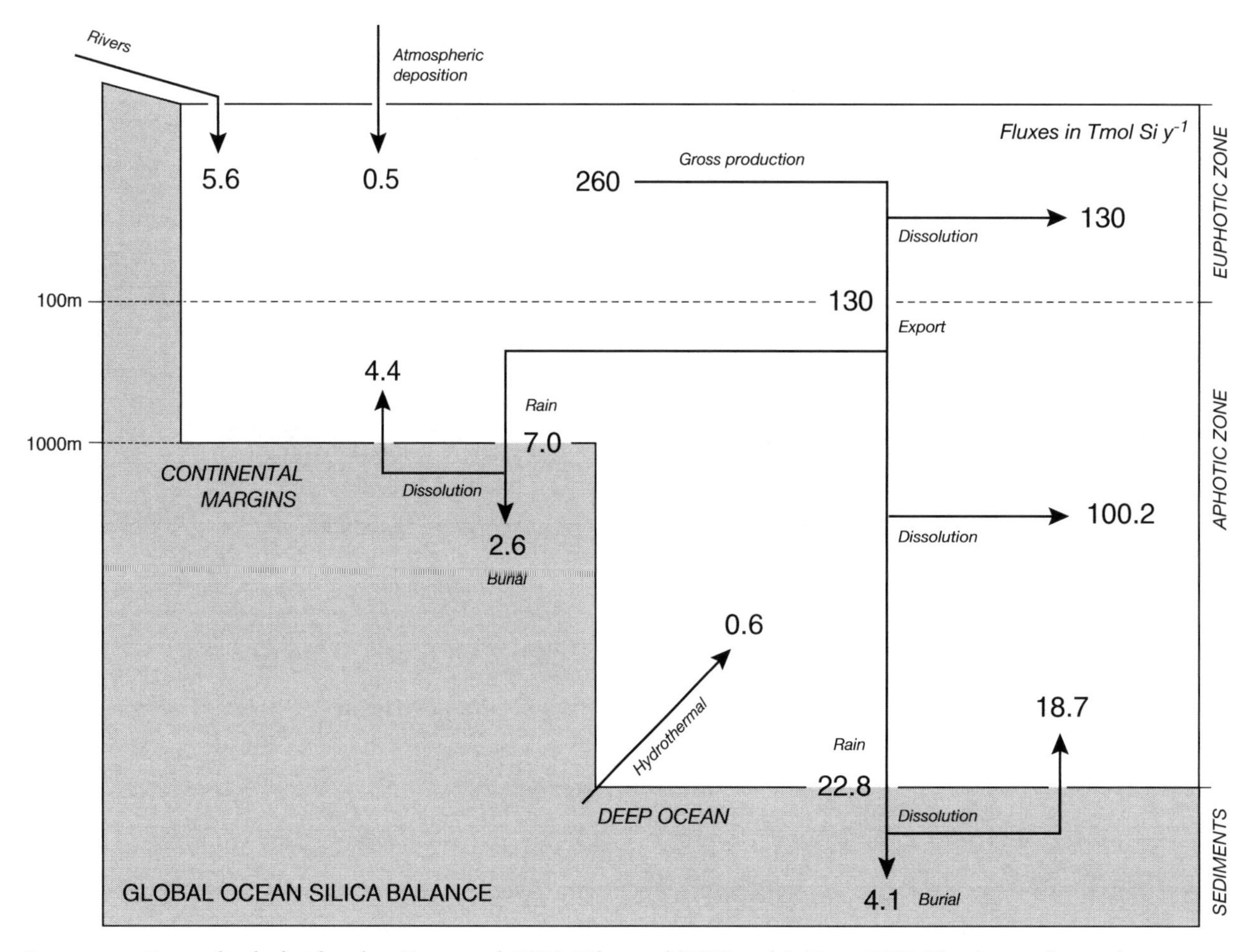

FIGURE 7.5.1: Ocean silica budget based on *Treguer et al.* [1995], *Nelson et al.* [1995], and *DeMaster* [2002]. The river, aeolian, and hydrothermal inputs, as well as the continental margin and deep ocean rain rates, all come from *Treguer et al.* [1995]. The opal burial rates are from *DeMaster* [2002]. The global opal export rate of 130 Tmol Si y^{-1} is from table 7.2.2 [*Dunne et al.* 2005c], and the gross production is twice this number per the estimates by *Treguer et al.* [1995] and *Nelson et al.* [1995] that half of the opal produced in the surface ocean dissolves within the surface ocean. The remaining numbers are all found by requiring mass conservation. Arrows pointing the right symbolize dissolution. The export flux of *Dunne et al.* [2005c] is available on a global scale only and thus this number and the water column remineralization of 100.2 Tmol Si y^{-1} are not broken down into separate continental margin and deep ocean components.

based on that of *Treguer et al.* [1995] and *Nelson et al.* [1995], as modified by *DeMaster* [2002] and *Dunne et al.* [2005c] (see figure 7.5.1). The new component of our budget for the interior of the ocean is the export across 100 m of 130 Tmol yr^{-1} of opal obtained from *Dunne et al.* [2005c], and the gross opal production of 260 Tmol yr^{-1} of Si that is obtained from this number by using the 50% recycling efficiency estimate of *Treguer et al.* [1995] and *Nelson et al.* [1995]. Our estimate is very close to that of *Nelson et al.* [1995], and in the midrange of a variety of other estimates using box models and OGCMs [e.g., *Lerman and Lal*, 1977; *Gnanadesikan*, 1999a; *Usbeck*, 1999; *Moore et al.*, 2002a; *Heinze et al.*, 2003; *Yool and Tyrrell*, 2003].

The sediment burial and dissolution flux estimates in our silicate budget are from *DeMaster* [2002] and *Treguer et al.* [1995], respectively; and the river, aeolian, and hydrothermal inputs are all from *Treguer et al.* [1995]. In this budget, about 40% of the global opal burial occurs on continental margins with a mean burial efficiency (burial divided by rain) of about 37%. Three-quarters of the deep ocean burial occurs in the Southern Ocean, which therefore accounts for about 45% of the global total, with the remaining ~15% occurring almost entirely in the North Pacific, Bering Sea, and Sea of Okhotsk [*DeMaster*, 2002]. Overall, the deep oceans have a burial efficiency of 18%. This seems rather high given that much of the deep ocean has a preservation efficiency of more like 5%, but is probably reasonable if weighted by where most of the opal rain occurs.

Given a mean oceanic silicic acid concentration of 84 mmol m^{-3} and a deep ocean volume of 1.29×10^{18} m^3 (cf. chapter 1), which gives an oceanic inventory of ~108,000 Tmol of Si, the residence time of silicic acid in the ocean with respect to the external source and sink total of 6.7 Tmol yr^{-1} of Si is about 16,000 years. With a

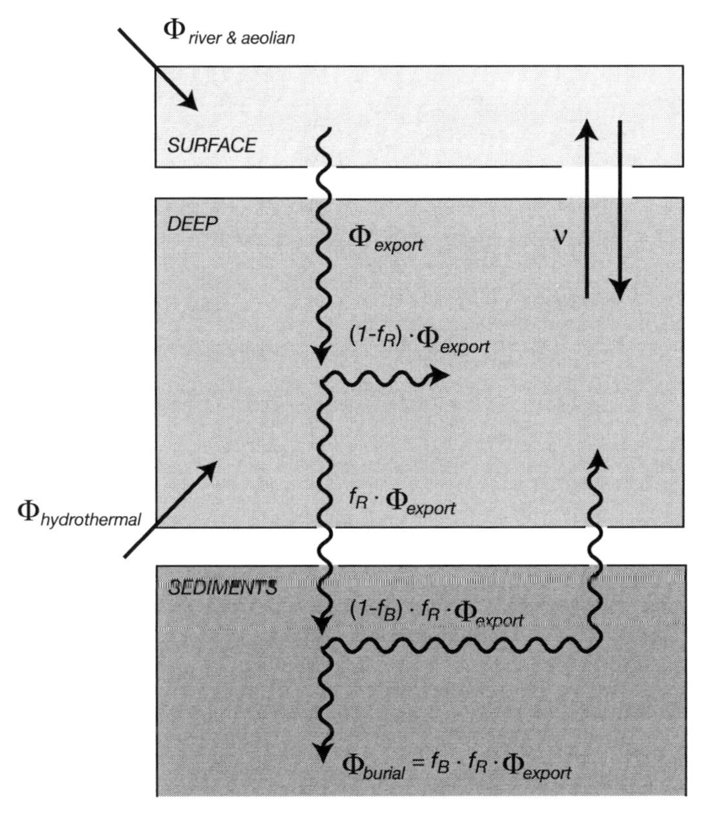

FIGURE 7.5.2: Two-box model used to examine the silicate cycle in the ocean. This model is based on figure 1.2.3, modified to include the river input and sediment loss terms of figure 1.1.2. The inputs to the ocean include the combined river and aeolian input $\Phi^{Si(OH)_4}_{river+aeolian}$ and the hydrothermal input $\Phi^{Si(OH)_4}_{hydrothermal}$. Removal from the ocean by loss to the sediments is symbolized by Φ^{Opal}_{burial}. The term f_R is the fraction of the opal exported from the surface that rains onto the sediments, and f_B is the fraction of the rain that is buried in the sediments. All the other terms are the same as in figure 1.2.3.

gross production of 260 Tmol yr^{-1} of Si, the cycling time of silicic acid through biogenic opal is about 420 years, which means that silicon goes through the biogenic opal phase an average of $\sim$40 times before being removed from the ocean [cf. *Treguer et al.*, 1995].

Long-Term Homeostasis

We conclude this chapter with a discussion of the one question we asked in the introduction that we have not yet addressed: What are the feedback processes between the oceanic silicic acid concentration and opal burial in the sediments that maintain the oceanic silicic acid concentration in *homeostasis*? That is to say, what prevents the silicic acid from wandering off into some extreme value of, for example, total depletion, or complete saturation with opal? Clearly chemical equilibrium with opal does not play a direct role, since the oceanic silicic acid content is highly undersaturated with respect to opal. This suggests that the controls are kinetic. In particular, we suggest that the oceanic silicic acid content is controlled by its influence on the export of opal from the surface ocean in combination with the processes that determine the ultimate efficiency with which this opal is buried in the sediments. Despite the limitations we discussed in section 1.2 and above, the two-box model of figure 7.5.2 provides a useful context in which to examine these feedbacks (cf. chapter 1 and *Yool and Tyrrell* [2003]).

From the model in figure 7.5.2, we can readily see that in a steady state the loss of opal by burial is

$$\Phi^{Opal}_{burial} = f_B \cdot f_R \cdot \Phi^{Opal}_{export} \quad (7.5.1)$$

where f_R is the fraction of the opal exported from the surface that rains onto the sediments, f_B is the burial efficiency, and Φ^{Opal}_{export} is the export of opal from the surface. The burial rate of opal will increase if any one or more of these three terms increases. If such an increase in opal burial were to occur starting from an initial steady state where the burial and input were equal to each other, it is easy to see that the oceanic silicic acid content would have to decrease if the input remained constant. The reverse is obviously true as well, that is to say, if burial decreased starting from a balanced initial steady state, the silicic acid content of the ocean would have to increase.

What is it that keeps the oceanic silicic acid content in homeostasis? From our discussions in this chapter, there is no obvious direct feedback between the oceanic silicic acid concentration and the f_B and f_R terms, since the ocean is so highly undersaturated with respect to opal that it would take extreme changes in silicic content to have a significant impact on these terms. We are thus left with the opal export term Φ^{Opal}_{export} as the likely candidate for a feedback mechanism. We obtain a solution for Φ^{Opal}_{export} from the steady-state balance equation for the surface box in figure 7.5.2:

$$\Phi^{Opal}_{export} = v \cdot ([\mathrm{Si(OH)_4}]_d - [\mathrm{Si(OH)_4}]_s) + \Phi^{\mathrm{Si(OH)_4}}_{river+aeolian} \tag{7.5.2}$$

From this, we can readily see that a decrease in the oceanic silicic acid content (and thus $[\mathrm{Si(OH)_4}]_d$) such as would occur if opal burial increased, would lead to a decrease in opal export production. From equation (7.5.1), we see that this decrease in opal export would lead, in turn, to a reduction in opal burial. Here we have the negative feedback mechanism that links the oceanic silicic acid concentration to opal burial.

It is of considerable interest to ask what insights this model can provide about how the oceanic silicic acid content might change in response to perturbations. We will consider only steady-state solutions. We begin by deriving a new equation obtained by inserting (7.5.2) into (7.5.1), noting that, in steady state, $\Phi^{Opal}_{burial} = \Phi^{\mathrm{Si(OH)_4}}_{river+aeolian} + \Phi^{\mathrm{Si(OH)_4}}_{hydrothermal}$. Solving the resulting equation for $[\mathrm{Si(OH)_4}]_d$ gives

$$[\mathrm{Si(OH)_4}]_d = [\mathrm{Si(OH)_4}]_s + \frac{\Phi^{\mathrm{Si(OH)_4}}_{hydrothermal} + \Phi^{\mathrm{Si(OH)_4}}_{river+aeolian} \cdot (1 - f_B \cdot f_R)}{f_B \cdot f_R \cdot v} \tag{7.5.3}$$

[*Harrison*, 2000] and *Treguer and Pondaven* [2000] have explored a scenario where $\Phi^{\mathrm{Si(OH)_4}}_{river}$ increased during the last ice age to the point where diatoms were able to outcompete coccolithophorids. The impact of reduced coccolithophorid production on atmospheric CO_2, which has been explored by *Archer et al.* [2000b], is discussed further in chapter 10. We are intrigued by the possibility that f_R, and more particularly f_B, might change under different climates (cf. sensitivity studies by *Yool and Tyrrell* [2003]). We have seen that the burial efficiency f_B in high opal rain areas of the deep ocean tends to be about 30%, whereas in low opal rain areas of the deep ocean it tends to be about 5%. Over the entire deep ocean, which accounts for about 60% of the total opal burial, the burial efficiency averages about 18%. By contrast, the burial efficiency in continental margins, which accounts for about 40% of all burial today, is about 37%. What would happen if circumstances were to modify the balance between these various regions? One can readily conceive of such scenarios. For example, a lowering of sea level and exposure of the shelves might modify the continental margin–deep ocean balance. Or an increase of iron flux in the high-latitude bands of high opal export might reduce the Si:N ratio in diatoms and lead to reduced export of opal in these regions and enhanced opal export in regions of lower burial efficiency. Another possibility is that the efficiency of opal remineralization in the water column, $(1 - f_R)$, might change in response to temperature. *Sigman et al.* [2004] published observations from the North Pacific and Southern Ocean that show a large drop in opal preservation occurring 2.7 million years ago, which they attribute to a change in oceanic stratification leading to lower biological productivity. Such a drop may have reduced the burial efficiency f_B and led to an increased deep ocean silicic acid concentration.

Taking all the above possibilities together, noting from (7.5.2) that the deep ocean silicic acid content is also inversely related to the mixing rate , and recalling from our discussion of the SAMW and NPIW formation processes that the return of silicic acid from the deep ocean to the thermocline follows a complex and potentially quite sensitive pathway, we leave the reader with the thought that the silicate balance in the ocean has the potential to be quite dynamic.

Problems

7.1 Define and explain the following terms:

a. silicic acid

b. opal

c. silicate

d. lithogenic silica

e. diatom

7.2 Name and discuss at least three hypotheses that have been put forward to explain the success of diatoms in today's ocean.

7.3 Discuss the relationship between opal production and export. Contrast the processes determining this relationship with those that control the relationship between net primary production and export of organic carbon.

7.4 Explain why we can use the vertical gradient ratio of nitrate and silicic acid to infer the export ratio of organic nitrogen to opal (see panel 7.2.1). In particular, discuss and evaluate the assumptions underlying this approach.

7.5 Show how the normalized silicic acid uptake rate V/V_{max} given by equation (7.2.1) varies with iron and silicic acid concentrations.

a. Do a contour plot of V/V_{max} on a diagram of normalized iron concentration $[Fe]/K_{Fe}$ on the ordinate versus normalized silicic acid concentration $[Si(OH)_4]/K_{Si(OH)_4}$ on the abscissa. Use a range of 0 to 5 for $[Fe]/K_{Fe}$ and $[Si(OH)_4]/K_{Si(OH)_4}$.

b. Suppose that $[Fe]/K_{Fe}$ and $[Si(OH)_4]/K_{Si(OH)_4}$ have a value of 1. If $[Fe]/K_{Fe}$ were to decrease to 0.5, by how much would $[Si(OH)_4]/K_{Si(OH)_4}$ have to increase to compensate and maintain the same growth rate?

7.6 Compare the fraction of estimated global organic carbon production that takes place south of 45°S with the fraction of estimated opal production that occurs in the same region. Explain why the two are so different.

7.7 Based on the information given in panel 7.1.1, how does the dissolution rate constant of opal, k, change if you

a. increase temperature by 10°C from 0°C?

b. decrease the specific surface area by a factor of two?

c. change the silicic acid concentration dissolved in seawater from 1 to 100 mmol m^{-3}?

7.8 Compute the solubility of opal with a specific surface area, *SSA*, of 250 m^2 g^{-1} for two conditions:

a. Surface of the ocean (T = 18.5°C)

b. At 1000 m depth ($P \sim$ 100 atm and T = 4°C)

7.9 Suppose that opal sinks at a rate of 100 m d^{-1} and that 35.6% of it survives between 100 m and 1000 m.

a. What is the mean dissolution rate k of the opal between 100 and 1000 m?

b. Suppose that the water temperature increases from an average of 9°C to 12°C. How would the mean dissolution rate in (a) change?

c. Given the dissolution rate in (b), what fraction of the 100 m opal flux would survive at 1000 m?

7.10 Considering the effect of temperature on both the solubility of opal and the kinetics of opal dissolution, how would you expect a 10°C warming of the deep ocean to affect silicic acid concentration in the water column? Explain your answer.

7.11 In this chapter we discussed the relationship between the export of organic matter and that of opal.

a. What is the global mean opal-to-organic nitrogen (Si:N) export ratio? Use the export fluxes given in table 7.2.2 to calculate this.

b. What would the deep concentration of silicic acid have to be in the three-box model (figure 1.2.5) in order to give the observed export ratio? Assume that the nitrate concentrations and the surface high-latitude silicic acid concentration are as observed. Further assume that the low-latitude concentrations of these two nutrients are zero. Use the values given in the text for the remaining parameters.

c. Suggest how you might modify the structure of the three-box model in order to give a more realistic simulation of both the silicic acid and nitrate distributions. Specifically, where would you add another box and why?

7.12 Explain why silicic acid is low in the thermocline whereas nitrate concentrations are high. Discuss the consequences of this for low-latitude diatom production.

7.13 Compare and contrast the various hypotheses for why the deep ocean has high Si*.

7.14 Investigate whether you can capture the impact of preferential remineralization of organic nitrogen over the dissolution of opal with a stacked three-box model, i.e., where a thermocline box (t) is added in between the surface (s) and deep box (d) of the two-box model (see figure). Assume a thickness of 75 m for the surface box, a thickness of 525 m for the thermocline box, and a thickness of 4400 m for the deep box. The exchange between the boxes is determined by the terms $v_1 = 11 \times 10^6$ m^3 s^{-1} and $v_2 = 31 \times 10^6$ m^3 s^{-1} [see *Sarmiento et al.*, 1988].

a. Calculate the fractions γ_{Corg} and γ_{opal} of the flux out of the surface box of organic carbon and opal that gets remineralized/dissolved in the thermocline box, assuming an exponential length-scale for organic carbon of $z^*_{Corg} = 500$ m (see section 5.4) and for opal of $z^*_{opal} = 2000$ m.

b. Calculate the thermocline and deep concentrations of silicic acid and nitrate given a global mean concentration of 29 mmol m^{-3} for nitrate and 83 mmol m^{-3} for silicic acid. Assume that the surface concentrations of these two tracers are zero. (Hint: Set up the conservation equations for a tracer C in the surface and deep boxes and assume steady state. Use the surface box to solve for the opal flux, then substitute this into the deep box equation to obtain the thermocline concentration in terms of the deep concentration. Use the equation for the mean concentration as an additional constraint.)

c. Compute Si* for the thermocline and deep boxes from your solutions in (b) and discuss your results. In particular, how does the Si* value you obtain compare with the observed thermocline concentrations in figure 7.3.6?

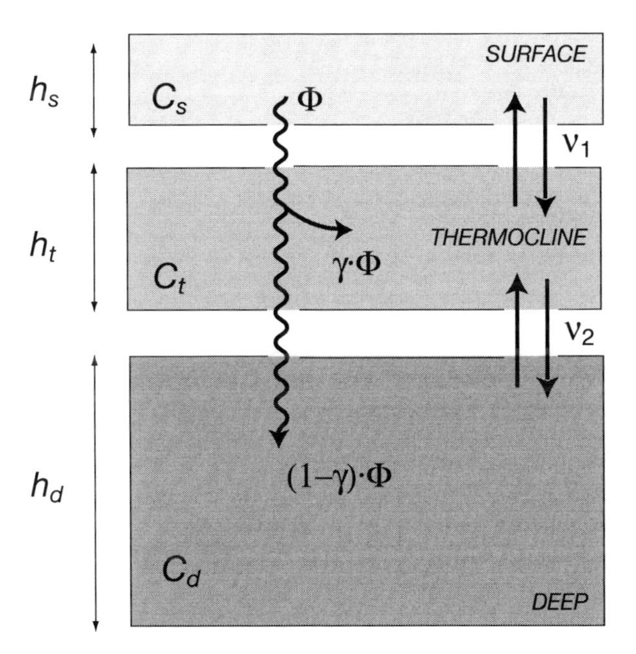

Figure for problem 7.14

7.15 An alternative explanation for the Si* distribution in the thermocline is that it primarily reflects differences in preformed concentrations of silicic acid and nitrate. We attempt to capture this by augmenting the above stacked three-box model with an outcrop box for the thermocline (o), and by extending the deep box all the way to the surface as well (see figure). We introduce two additional exchange terms, v_3 and v_4, to describe the exchange between the outcrop box and the thermocline box and between the deep box and the outcrop box. In addition, we consider an overturning circulation, T, which transports material properties from the outcrop into the thermocline box, upwells into the surface box, and finally transports material properties from the low-latitude surface box back to the high-latitude outcrop box. In our application here, we focus just on the thermocline and surface boxes, and fix the concentrations in the deep and outcrop boxes.

a. Set up the conservation equations for the surface, outcrop, and deep boxes. Assume steady state and that the surface concentration is 0. Use the three equations to eliminate the opal flux terms and then solve the resulting equation for the thermocline Si* concentration.

b. The solution to problem 7.14 showed that it is difficult to obtain a low thermocline Si* with realistic values of γ_{Corg} and γ_{opal}. Referring to the solution to part (a) of this problem, discuss other ways in which one can obtain a more realistic thermocline concentration for Si*.

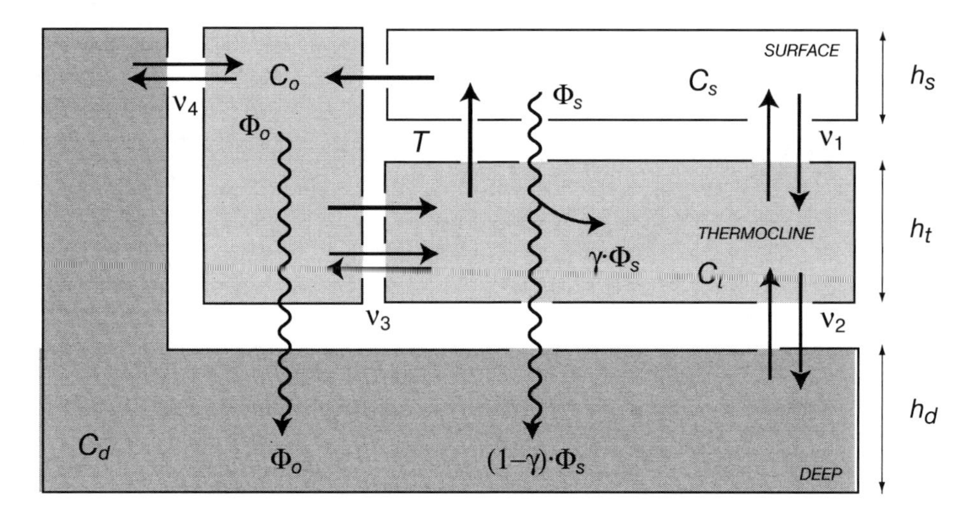

Figure for problem 7.15

7.16 List and discuss the main factors that control the preservation of opal in sediments.

7.17 Our global oceanic silicon budget suggests that about 40% of the global Si burial occurs on continental margins. Discuss the reasons for this high burial rate, and contrast this to the fraction of organic carbon that is buried on the margins.

7.18 Flux of silicic acid out of the sediments:

a. What is the flux of silicic acid out of the sediments if the e-folding length is 1 cm, the asymptotic silicic acid concentration is 400 mmol m^{-3}, and the ambient water column concentration is 100 mmol m^{-3}?

b. How much would the silicic acid concentration in a 1000 m-thick water column above the sediments in (a) increase over 100 years?

7.19 Consider the sediments described in problem 7.18 (a).

a. What is the solubility rate constant k'?

b. Suppose that the solubility of opal is 1000 mmol m^{-3} and that there is an authigenic aluminosilicate phase with a solubility of 200 mmol m^{-3}. What is the solubility rate constant k_a' for the opal and k_b' for the aluminosilicate phase?

c. Given the opal solubility rate constant in (b), and assuming that the opal concentration in the sediments is 5%, how many years would it take for 50% of the opal to disappear? Hint: Use equation (5) of panel

7.1.1, but first convert from k' to k using the definition of k' given after equation (7.4.2). You will need to use the unit conversion equation given in panel 6.2.1 to convert the opal fraction to opal concentration.

7.20 How would the oceanic silicic acid concentration change if the burial of opal on continental margins were to cease, as might occur if sea level dropped? Use the model of figure 7.5.2 with an exchange parameter $v = 1.2 \times 10^{15}$ m^3 yr^{-1} as obtained in panel 1.2.1.

a. Use the present silica balance in figure 7.5.1 to determine the values of all the parameters in the model. Use the total rain rates arriving at the sediments and the total sediment dissolution fluxes, i.e., combine the values for the margin and the deep ocean.

b. Use the parameters in (a) to estimate the deep ocean concentration required by the model of figure 7.5.2, assuming that the surface concentration is 0 mmol m^{-3}. (Note that the concentration obtained with this model is higher than the observed concentration of 84 mmol m^{-3}.)

c. Now assume that the sea level drops and continental margin opal deposition disappears. Use equation (7.3.2) to determine f_R for a mean deep ocean depth of 3800 m, and use the data in figure 7.5.1 to determine the burial efficiency f_B in the deep ocean. Using these two new values for the rain rate and burial efficiencies, estimate the new deep ocean silicic acid concentration, assuming all else in the model remains the same.

d. Calculate the expected opal export given the deep concentration computed in part (c). Compare this result with the opal export in the present ocean.

CHAPTER 8

Carbon Cycle

Having completed our first grand tour of ocean biogeochemical cycles by reviewing and discussing the oceanic cycling of silicate, we are now ready to assess the marine carbon cycle, one of the most important and also most fascinating biogeochemical cycles in the ocean. In discussing this cycle, we will need to employ all the concepts that have been introduced and discussed in the previous chapters. In this chapter we will study how the marine carbon cycle works and how this cycle is connected with the atmospheric CO_2 concentration. We will see that the oceans contain approximately 60 times more carbon than the atmosphere and that the ocean therefore exerts a dominant control on atmospheric CO_2. We will introduce the concepts necessary to understand this fact and also discuss how biological and physical processes act together to reduce the surface concentration of inorganic carbon relative to the deep ocean. Without this reduction, atmospheric CO_2 would increase by more than 150 ppm, with the exact magnitude depending on the ocean carbon cycle model.

Our ability to understand and model the carbon cycle in the oceans has made large advances over the last twenty years since the Geochemical Ocean Section Study (GEOSECS) provided a first complete picture of the inorganic carbon distribution in the oceans. The recently completed Global Ocean CO_2 Survey, a joint effort between the Joint Global Ocean Flux Study (JGOFS) and the World Ocean Circulation Experiment (WOCE), provides us now with almost two orders of magnitude more data with improved precision and accuracy. Analysis and interpretation of this new data set has just begun, and it quite surely will provide us with many new insights on details of the marine carbon cycle. Over the last ten years, several global models of the carbon cycle in the oceans have been developed and successfully applied to many interesting questions. These models also increasingly incorporate the relevant processes prognostically, thus offering the opportunity to employ the models in conditions other than the current climate. However, there still exist large gaps in our understanding of several key issues.

Probably the largest puzzle in ocean carbon research is our inability to explain the large glacial-interglacial variations in atmospheric CO_2. As the oceans contain 60 times more carbon than the atmosphere and 17 times more carbon than is stored in the terrestrial biosphere (see figure 10.1.1), the causes for these variations must lie within the oceans. This lack of knowledge also hampers our ability to predict more reliably how the oceanic carbon cycle is going to change in the future as a result of human-induced climate change. The investigation of natural variability might offer some insight into how the marine carbon cycle responds to changes in the climatic state, but little is known regarding the role of the oceans on interannual to decadal timescales. While considerable advances have been made in pinning down the role of the oceans as a sink for anthropogenic CO_2, there still exists large uncertainties, especially regarding future uptake. We will deal with these outstanding issues in the last chapter of this book (chapter 10).

We start this chapter with an introduction that establishes a number of problems that provide a focal point for the discussion in the remainder of the chapter. We then construct the building blocks for answering most questions by introducing the inorganic carbon chemistry in seawater. Afterward we look at the processes that control surface ocean partial pressure of CO_2 and its seasonal variability. The last section extends the view into the water column by discussing the causes for the vertical variations of dissolved inorganic carbon and alkalinity.

8.1 Introduction

A major motivation to understand the ocean carbon cycle is its importance in controlling atmospheric CO_2, which, in turn, is an important factor of the climate system due its greenhouse gas properties (see chapter 10). How does the ocean exert this control? Are physical processes primarily responsible for controlling atmospheric CO_2, or does marine biology play the dominant role?

The first step to answering these questions is to look at the surface ocean distribution of pCO_2, because it is only the surface ocean that is in direct contact with the atmosphere and hence determines atmospheric pCO_2. Figure 8.1.1 shows a global map of the annual mean surface distribution of the sea-air difference in the partial pressure of CO_2 ($\Delta pCO_2 = pCO_2^{oc} - pCO_2^{atm}$; see also definition in equation (3.1.1)). The data show relatively wide bands of positive ΔpCO_2, i.e., supersaturated waters, in the tropical regions, with maximum values attained in the eastern equatorial Pacific and large bands of negative ΔpCO_2, i.e., undersaturated waters, in the mid-latitudes. The high latitudes in the south have near zero ΔpCO_2 values. The northern high latitudes exhibit a clear distinction between the North Pacific (positive ΔpCO_2) and the North Atlantic (negative ΔpCO_2). Since atmospheric CO_2 is relatively uniform over the globe, the variability in ΔpCO_2 is largely driven by variations in surface ocean pCO_2. What controls the distribution of surface ocean pCO_2?

As we will see in detail below, variations in surface ocean pCO_2 are determined by surface ocean temperature, salinity, dissolved inorganic carbon (*DIC*, the sum of all inorganic carbon species) and alkalinity (*Alk*, a measure of the excess of bases over acids). Variations in these properties, and hence pCO_2, stem from a complex interplay of chemical, biological, and physical processes. The transfer of CO_2 between the atmosphere and the surface ocean tends to eliminate any air-sea pCO_2 difference, but as we have seen at the end of chapter 3, it takes about 6 months to equilibrate a 40 m-thick surface layer with the atmosphere. This is long compared to the timescale of the biological and physical processes that tend to perturb the system. Why is the air-sea exchange of CO_2 so slow compared with other gases like oxygen or nitrogen? We have already briefly seen in chapter 3 that only about 1 out of 20 molecules that exchange with the atmosphere leads to a change in pCO_2. In order to fully understand this, we must first learn more about the reaction of CO_2 in seawater to

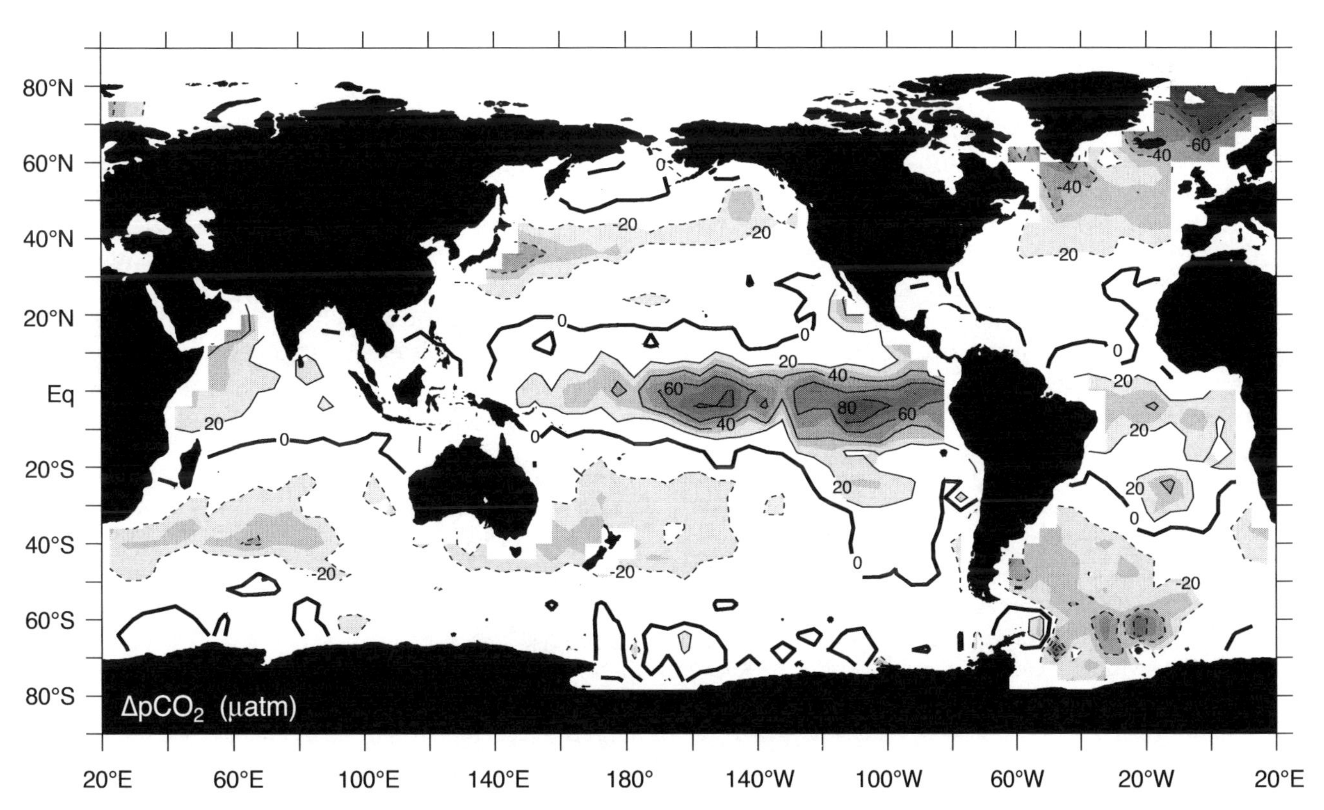

FIGURE 8.1.1: Map of the annual mean sea-air difference of the partial pressure of CO_2. See also color plate 3. Based on data from *Takahashi et al.* [2002].

form bicarbonate and carbonate ions. This topic will be covered in section 8.2. We will then investigate the controls on surface ocean pCO_2 in detail in section 8.3.

Although atmospheric CO_2 is controlled by surface ocean properties (i.e., temperature, salinity, *DIC*, and *Alk*), it is the deep ocean concentrations of *DIC* and *Alk* that ultimately determine the surface concentrations of these two chemicals and hence atmospheric CO_2. Figure 8.1.2 shows that the surface concentration of *DIC* is about 15% lower than deep ocean concentrations, while the gradient for *Alk* amounts to only about 5%. What are the mechanisms that maintain these gradients against the continuous action of ocean transport and mixing, which try to homogenize the ocean tracer distributions? Since these mechanisms operate against a gradient, they are often referred to as "pumps." The importance of these pumps becomes immediately clear when we turn them off and allow the ocean to homogenize completely (cf. also figure 4.1.3). The three-box model described in chapter 1 (figure 1.2.5) predicts that atmospheric CO_2 would rise from the preindustrial level of about 280 ppm to more than 420 ppm, i.e., an increase of about 50% [*Gruber and Sarmiento*, 2002]!

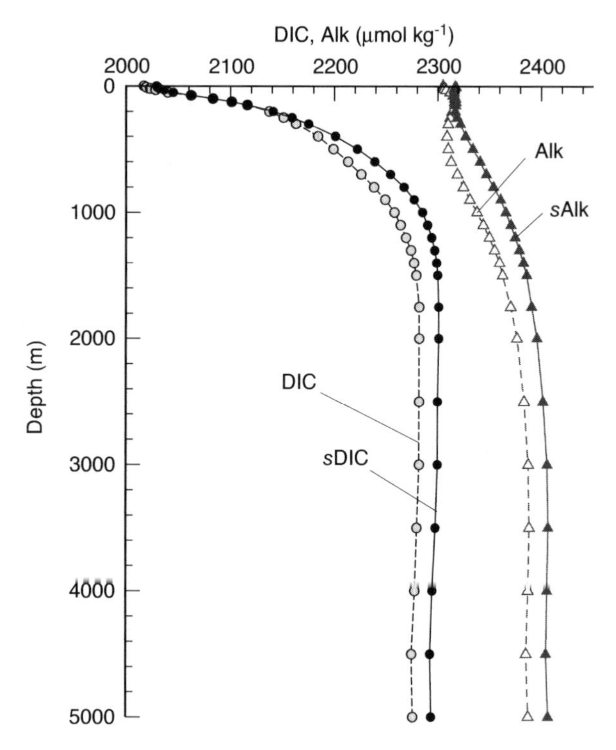

FIGURE 8.1.2: Horizontally averaged profiles of *DIC* and *Alk* in the global oceans. Also shown are the profiles of their salinity normalized concentrations, *sDIC*, and *sAlk*, computed by dividing the measured concentrations by the measured salinity and then multiplying the result with a standard salinity of 35. Based on the gridded climatological data from the GLODAP project [*Key et al.*, 2004].

Given the importance of carbon for organisms, we expect that ocean biology constitutes such a pump. This would imply that *DIC* and *Alk* should have a distribution similar to phosphate. This indeed appears to be the case when we compare the sections of *DIC* and *Alk* (figures 8.1.3) with those of nitrate and phosphate (figure 5.1.2). Note that we normalized *DIC* and *Alk* here to constant salinity in order to remove variations due to the addition and removal of freshwater by precipitation and evaporation. We will be using the symbols *sDIC* and *sAlk* to distinguish these salinity-normalized concentrations (see detailed discussion in the next section below). Is biology the whole answer to our question of what causes the vertical *DIC* variations? If biological processes were the only ones affecting phosphate and *sDIC*, then all oceanic observations would lie on a single line on a plot of *sDIC* versus phosphate. This is because plants appear to take up CO_2 in a nearly constant ratio relative to phosphate (see discussion in panel 4.2.1). During respiration and remineralization processes, both carbon and phosphate are then returned in dissolved inorganic form to the seawater in the same ratios (see discussion on stoichiometric ratios in chapter 5).

Figure 8.1.4a reveals that this is only partially the case. While most of the points indeed lie clustered around a line with a slope predicted from the stoichiometric ratio of photosynthesis and remineralization, the data also exhibit substantial systematic offsets from such a trend. In general, the data lie below the biological trend line for low concentrations and lie above the trend line for high concentrations. Clearly, processes other than photosynthesis, respiration, and remineralization play a role in creating the variations seen in *sDIC*. As shown in the inset, these processes are air-sea gas exchange and the formation and dissolution of mineral calcium carbonate, which affect *sDIC* but do not change phosphate. For example, given the strong relationship between *sDIC* and temperature (figure 8.1.4b), which follows roughly an expected trend defined by the temperature sensitivity of the *DIC* concentration in equilibrium with a fixed atmospheric CO_2, variations in temperature quite clearly must play a role as well. Determination of the relative role of these processes in controlling the distribution of *DIC* will be covered in section 8.4.

What about the processes controlling alkalinity? Given the much smaller variations in *sAlk* relative to the variations in *sDIC* (figures 8.1.2 and 8.1.3), one might think that these processes are not particularly important in controlling atmospheric CO_2. However, on timescales of millenia and longer, such as associated with the glacial-interglacial changes in atmospheric CO_2, the cycling of alkalinity within the ocean and its sediments emerges as a key component controlling atmospheric CO_2. The main processes controlling the cycling of alkalinity are the biological production of mineral calcium carbonates ($CaCO_3$) in the surface ocean and dissolution of these minerals in the water column and in the sediments. We will introduce the basic chemistry of these reactions in this chapter, but discuss the

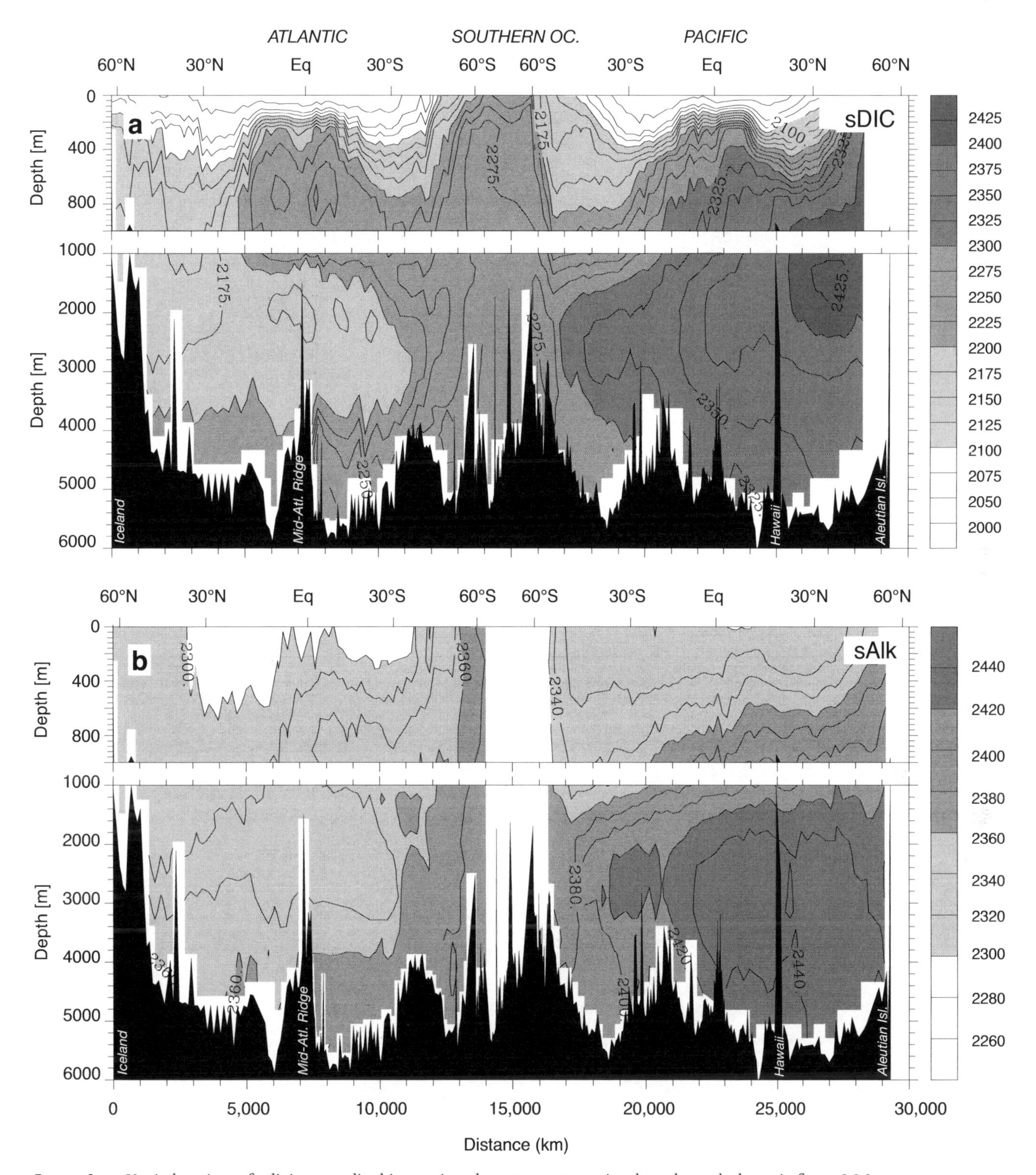

FIGURE 8.1.3: Vertical sections of salinity-normalized inorganic carbon system properties along the track shown in figure 2.3.3a. (a) Salinity-normalized *DIC* (μmol kg^{-1}). (b) Salinity-normalized *Alk* (μmol kg^{-1}). See also color plate 8.

cycling of $CaCO_3$ in chapter 9. The interaction of the oceanic $CaCO_3$ cycle with climate is taken up in chapter 10.

There is much about the ocean carbon cycle that is puzzling and sometimes even counterintuitive. For example, it is not self-evident why the ocean contains 98% and the atmosphere only 2% of the carbon stored in the combined reservoirs. For most other gases, like oxygen or nitrogen, this distribution pattern is actually reversed, despite the fact that all these gases have similar

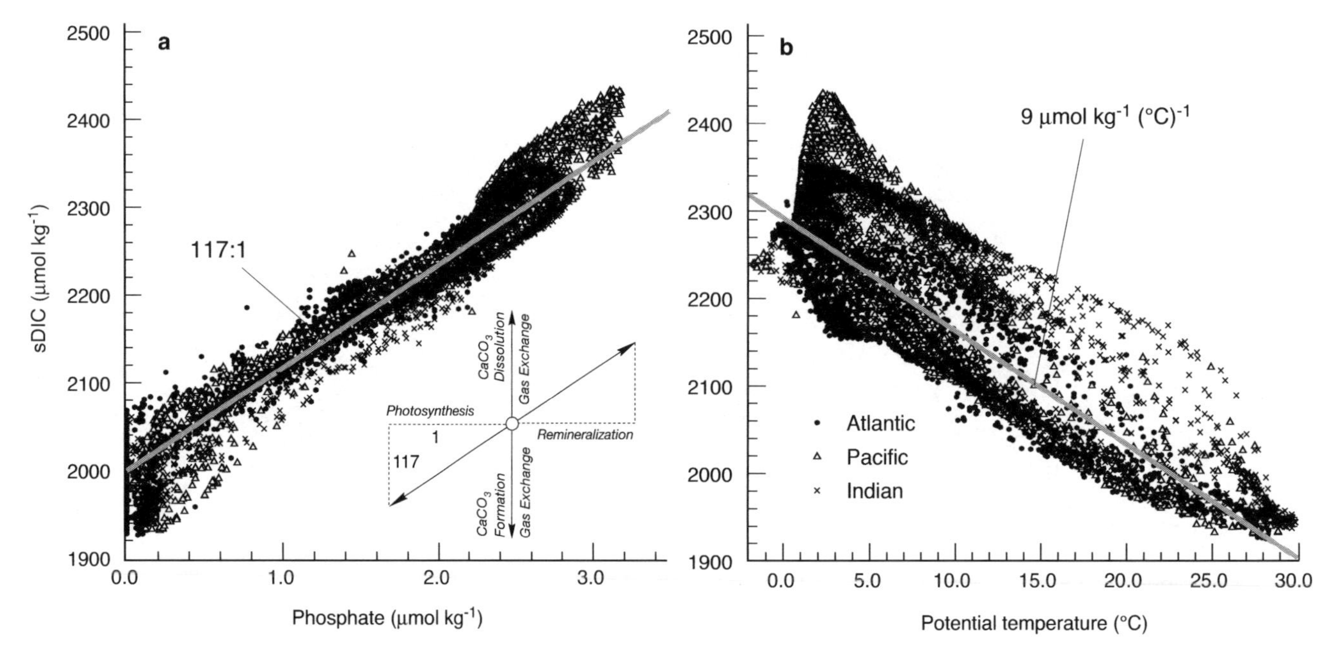

FIGURE 8.1.4: Plot of dissolved inorganic carbon, *DIC*, versus (a) phosphate and (b) potential temperature, using data from all depths. Both *DIC* and phosphate have been normalized to a constant salinity of 35, in order to remove the effects of evaporation and precipitation. The line in (a) shows the expected trend if all variations of *sDIC* were caused by biological processes. The inset shows the trends in *sDIC* and phosphate caused by photosynthesis, respiration, and remineralization, as well as the trends caused by air-sea gas exchange and the formation and dissolution of $CaCO_3$. The line in (b) shows the expected trend if all variations of *sDIC* were a result of differences in temperature. At constant oceanic pCO_2 and alkalinity, this trend is nearly 9 μmol kg^{-1} per degree Celsius [*Gruber et al.*, 1996]. Data are from WOCE legs A16, P16, and I9NI8S.

solubilities. The key to explaining this fact and almost all of the puzzles of the marine carbon cycle that we have posed in this introduction, is to first gain an understanding of the complex chemistry of CO_2 in seawater.

8.2 Inorganic Carbon Chemistry

The solution chemistry of carbon dioxide sets it apart from that of most other gases in that CO_2 not only dissolves in seawater but also acts as an acid (proton donor) that reacts with water to form free protons (H^+), and the conjugate bases bicarbonate (HCO_3^-) and carbonate (CO_3^{2-}). In this section we will introduce the basic concepts of these reactions and how they can be quantitatively described. We also attempt to establish some simple tools to aid in understanding the nonlinearities of the oceanic CO_2 system. For the purpose of this presentation we will employ a number of simplifications. The reader interested in a more thorough discussion is referred to *Dickson and Goyet* [1994].

When gaseous CO_2 dissolves in seawater, it first gets hydrated to form aqueous CO_2 ($CO_{2\,(aq)}$), which reacts with water to form carbonic acid (H_2CO_3). It is difficult to distinguish analytically between the two species $CO_{2(aq)}$ and H_2CO_3. As a result, it is usual to combine these two species and express the sum as the concentration of a hypothetical species, $H_2CO_3^*$ [*Stumm and Morgan*, 1981; *Dickson and Goyet*, 1994]. This hypothetical acid (proton donor) dissociates then in two steps to form bicarbonate (HCO_3^-) and carbonate (CO_3^{2-}) ions. These reactions are very fast, so that for all practical purposes we can assume that thermodynamic equilibrium between the species is established. These three reactions can be summarized as follows:

$$CO_{2\,(gas)} + H_2O \rightleftharpoons H_2CO_3^* \tag{8.2.1}$$

$$H_2CO_3^* \rightleftharpoons H^+ + HCO_3^- \tag{8.2.2}$$

$$HCO_3^- \rightleftharpoons H^+ + CO_3^{2-} \tag{8.2.3}$$

The equilibrium relationships between these species are given by

$$K_0 = \frac{[H_2CO_3^*]}{pCO_2} \tag{8.2.4}$$

$$K_1 = \frac{[H^+][HCO_3^-]}{[H_2CO_3^*]} \tag{8.2.5}$$

$$K_2 = \frac{[H^+][CO_3^{2-}]}{[HCO_3^-]} \tag{8.2.6}$$

where pCO_2 is the partial pressure of CO_2 in the air and where the brackets denote total concentrations. Strictly speaking, these equations should be expressed as activities rather than concentrations, but we neglect this small difference here. In particular, the CO_2 concentration in air is sometimes expressed as a fugacity, fCO_2, which is a thermodynamic property analogous to the activity of a species dissolved in water (see panel 3.2.1). Since fugacity takes the nonideality of CO_2 in air into account, its value is somewhat lower than the pCO_2. Equation (2) in panel 3.2.1 shows how the fugacity can be calculated from pCO_2. Here, we adopt pCO_2 as our primary quantity expressing the concentration of CO_2 in air.

In the literature, the equilibrium constants K_1 and K_2 are usually expressed in units of moles per kilogram of solution rather than in units of milimoles per cubic meter as adopted generally in this text. For consistency with the published literature, we use the symbol K_0 for the CO_2 solubility in units of $\text{mol kg}^{-1}\,\text{atm}^{-1}$, rather than S_{CO_2}, which is in units of $\text{mmol m}^{-3}\,\text{atm}^{-1}$ (see chapter 3).

The CO_2 system so far consists of 5 unknowns (pCO_2, $[H_2CO_3^*]$, $[HCO_3^-]$, $[CO_3^{2-}]$, and $[H^+]$) and 3 equations (8.2.4, 8.2.5, and 8.2.6). In order to determine the system, we have to specify any two of the five unknowns. The partial pressure of CO_2 in equilibrium with the water (pCO_2) and pH ($\text{pH} = -\log_{10}[H^+]$) are frequently measured, but these two parameters are ill suited to use in modeling, because neither is conservative with respect to changes in state (i.e., temperature, salinity, and pressure). In order to avoid these complications, the two parameters used in models are dissolved inorganic carbon (DIC) and total alkalinity (Alk), defined as

$$DIC = [H_2CO_3^*] + [HCO_3^-] + [CO_3^{2-}] \tag{8.2.7}$$

$$Alk = [HCO_3^-] + 2[CO_3^{2-}] + [OH^-] - [H^+] + [B(OH)_4^-] + \text{minor bases} \tag{8.2.8}$$

where $[OH^-]$ is the concentration of the hydroxide ion, and $[B(OH)_4^-]$ is the concentration of the borate ion. Both DIC and Alk are conservative with respect to changes in state. The total alkalinity is a measure of the excess of bases (proton acceptors) over acids (proton donors), and is operationally defined by the titration with H^+ of all weak bases present in the solution (see *Dickson* [1981] for an exact definition of alkalinity). Alternatively, total alkalinity can be viewed as the charge balance of all strong acids and bases unaffected by this titration, i.e.,

$$Alk = [Na^+] + [K^+] + 2[Mg^{2+}] + 2[Ca^{2+}] + \text{minor cations} - [Cl^-] - 2[SO_4^{2-}] - [Br^-] - [NO_3^-] - \text{minor anions} \tag{8.2.9}$$

The contribution of the minor bases like phosphate, silicate, and sulphate to variations in Alk are usually well below one percent, and we are therefore going to neglect them.

The definition of dissolved inorganic carbon and total alkalinity introduces four new unknowns (DIC, Alk, $[OH^-]$, $[B(OH)_4^-]$) for a total of 9, but only two new equations for a total of 5. Additional constraints on the system of equations are needed. The hydroxide ion originates from the self-dissociation of water

$$H_2O \rightleftharpoons H^+ + OH^- \tag{8.2.10}$$

with the dissociation constant

$$K_w = [H^+][OH^-] \tag{8.2.11}$$

Borate is formed by the dissociation of boric acid by the reaction

$$H_3BO_3 + H_2O \rightleftharpoons H^+ + B(OH)_4^- \tag{8.2.12}$$

with the dissociation constant

$$K_B = \frac{[H^+][B(OH)_4^-]}{[H_3BO_3]} \tag{8.2.13}$$

The total boron concentration is very nearly conservative within the ocean, and its concentration changes only with net exchange of water at the sea surface by evaporation and precipitation. Since these processes control salinity, the total boron concentration can be assumed to be proportional to the salinity, S,

$$[B(OH)_4^-] + [H_3BO_3] = c \cdot S \tag{8.2.14}$$

Equations (8.2.11), (8.2.13), and (8.2.14) add three new equations and 1 new unknown, H_3BO_3. We therefore end up with 8 equations and 10 unknowns (see table 8.2.1). As was the case above, the specification of two unknowns completely determines the inorganic carbon system. Models commonly employ DIC and Alk as state variables of the carbon system, whereas at sea, often a combination of pCO_2, DIC, and Alk are measured. With the introduction of a method for the high-precision determination of pH [*Clayton and Byrne*, 1993], this quantity is being increasingly determined as well. This system of eight equations cannot be solved analytically, because a higher order polynomial equation has to be solved. Iterative methods are commonly used (see table 8.2.1).

Table 8.2.2 lists the empirical equations for calculating the dissociation constants. Note that these constants are given in units of moles per kilogram of solution. The exact values for the dissociation constants, K_1 and K_2, of carbonic acid and bicarbonate, respectively, are currently still subject to considerable debate [*Millero et al.*, 1993; *Lee and Millero*, 1995; *Lee et al.*, 1997; *Lueker*, 1998]. We have adopted here the dissociation constants of *Mehrbach et al.*

TABLE 8.2.1
The CO_2 system in seawater

Unknowns:

$CO_{2(gas)}$, $H_2CO_3^*$, HCO_3^-, CO_3^{2-}, H^+, OH^-, $B(OH)_4^-$, H_3BO_3, *DIC*, *Alk*

Reactions:

$$CO_{2(gas)} + H_2O \overset{K_0}{\rightleftharpoons} H_2CO_3^* \quad (1)$$

$$H_2CO_3^* \overset{K_1}{\rightleftharpoons} H^+ + HCO_3^- \quad (2)$$

$$HCO_3^- \overset{K_2}{\rightleftharpoons} H^+ + CO_3^{2-} \quad (3)$$

$$H_2O \overset{K_w}{\rightleftharpoons} H^+ + OH^- \quad (4)$$

$$H_3BO_3 + H_2O \overset{K_B}{\rightleftharpoons} H^+ + B(OH)_4^- \quad (5)$$

Equilibrium constants:

$$K_0 = \frac{[H_2CO_3^*]}{pCO_2} \quad (6)$$

$$K_1 = \frac{[H^+][HCO_3^-]}{[H_2CO_3^*]} \quad (7)$$

$$K_2 = \frac{[H^+][CO_3^{2-}]}{[HCO_3^-]} \quad (8)$$

$$K_w = [H^+][OH^-] \quad (9)$$

$$K_B = \frac{[H^+][B(OH)_4^-]}{[H_3BO_3]} \quad (10)$$

Concentration definitions†:

$$DIC = [H_2CO_3^*] + [HCO_3^-] + [CO_3^{2-}] \quad (11)$$

$$Alk = [HCO_3^-] + 2[CO_3^{2-}] + [OH^-] - [H^+] + [B(OH)_4^-] \quad (12)$$

$$TB = [B(OH)_4^-] + [H_3BO_3] = c \cdot S \quad (13)$$

Solution:
In models of the ocean carbon cycle, two cases are most commonly encountered: (a) specifying *DIC* and *Alk* and calculating pCO_2; (b) specifying pCO_2 and *Alk* and computing *DIC*.

(a) *DIC* and *Alk*:
The goal is to rewrite the equation for *Alk* (12) in terms of the known parameters *DIC* and *TB*, and then solve the resulting equation for unknown $[H^+]$. As a first step, we use (7), (8), and the *DIC* definition (11) to write expressions for $[HCO_3^-]$ and $[CO_3^{2-}]$ that contain only $[H^+]$ as an unknown. In the case of $[HCO_3^-]$, this is done as follows. Solve (7) for $[H_2CO_3^*]$ and insert into (11). Similarly, solve (8) for $[CO_3^{2-}]$ and insert into (11). This gives

$$DIC = \frac{[H^+][HCO_3^-]}{K_1} + [HCO_3^-] + \frac{K_2[HCO_3^-]}{[H^+]} \quad (14)$$

Next, we use (8), solve it for $[HCO_3^-]$, and insert it into (14). This gives a second equation for *DIC*:

$$DIC = \frac{[H^+]^2[CO_3^{2-}]}{K_1 K_2} + \frac{[H^+][CO_3^{2-}]}{K_2} + [CO_3^{2-}] \quad (15)$$

Solving (14) for $[HCO_3^-]$ and (15) for $[CO_3^{2-}]$, respectively, and inserting the results into (12) eliminates $[CO_3^{2-}]$ and $[HCO_3^-]$ from this latter equation. The concentrations of $[OH^-]$ and $[B(OH)_4^-]$ are similarly eliminated using (9), (10), and (13). This results in our final equation, which is fourth order in the unknown $[H^+]$. This equation can then be solved for $[H^+]$ using an iterative approach. Once $[H^+]$ has been calculated, pCO_2 can be calculated from

$$pCO_2 = \frac{[DIC]}{K_0} \frac{[H^+]^2}{[H^+]^2 + K_1[H^+] + K_1 K_2} \quad (16)$$

where this equation has been derived by solving (6) for pCO_2, and then using (7), (8), and (11) to eliminate the unknowns $[H_2CO_3^*]$, $[HCO_3^-]$, and $[CO_3^{2-}]$. See also (17)–(19) below.

(b) *Alk* and pCO_2:
Rewrite the equations for the three carbon species, i.e., (6), (7), and (8), in terms of $[H^+]$ and pCO_2, thus

$$[H_2CO_3^*] = K_0\, pCO_2 \quad (17)$$

$$[HCO_3^-] = \frac{K_0 K_1\, pCO_2}{[H^+]} \quad (18)$$

$$[CO_3^{2-}] = \frac{K_0 K_1 K_2\, pCO_2}{[H^+]^2} \quad (19)$$

Substitute these terms into the equation for *Alk* (12) and also recast the other species, i.e., $[OH^-]$ and $[B(OH)_4^-]$, in terms of $[H^+]$ and total concentrations. The resulting equation can again be solved using an iterative approach. Once $[H^+]$ is found, *DIC* is calculated using (11) and (17)–(19).

† The minor contribution of other weak bases to alkalinity are neglected here (see Dickson and Goyet [1994] for more details).

TABLE 8.2.2
Equations used to calculate seawater equilibrium constants
T: Temperatures in [K], S: Salinity on the practical salinity scale.

Equation		Source
Solubility of CO_2 ($\text{mol kg}^{-1}\,\text{atm}^{-1}$):		
$\ln K_0 = -60.2409 + 93.4517\left(\frac{100}{T}\right) + 23.3585\ln\left(\frac{T}{100}\right) + S\left(0.023517 - 0.023656\left(\frac{T}{100}\right) + 0.0047036\left(\frac{T}{100}\right)^2\right)$	(20)	[*Weiss*, 1974]
Dissociation constants of CO_2† (mol kg^{-1}):		
$-\log K_1 = -62.008 + \frac{3670.7}{T} + 9.7944\ln(T) - 0.0118\,S + 0.000116\,S^2$	(21)	[*Mehrbach et al.*, 1973] as refitted by *Dickson and Millero* [1987]
$-\log K_2 = +4.777 + \frac{1394.7}{T} - 0.0184\,S + 0.000118\,S^2$	(22)	[*Mehrbach et al.*, 1973] as refitted by *Dickson and Millero* [1987]
Dissociation constants of other species† [$(\text{mol kg}^{-1})^2$] for K_w and (mol kg^{-1}) for K_b:		
$\ln K_w = 148.96502 + \frac{-13847.26}{T} - 23.6521\ln(T) + S^{\frac{1}{2}}\left(-5.977 + \frac{118.67}{T} + 1.0495\ln(T)\right) - 0.01615\,S$	(23)	[*Millero*, 1995]
$\ln K_b = \frac{1}{T}(-8966.9 - 2890.53\,S^{0.5} - 77.942\,S + 1.728\,S^{1.5} - 0.0996\,S^2) + 148.0248 + 137.1942\,S^{0.5} + 1.62142\,S + 0.053105\,S^{0.5}\,T + \ln(T)\,(-24.4344 - 25.085\,S^{0.5} - 0.2474\,S)$	(24)	[*Dickson*, 1990]
Total boron equation ($\mu\text{mol kg}^{-1}$):		
$TB = 11.88 \cdot S$	(25)	[*Uppström*, 1974]

† All dissociation constants are given with respect to the seawater pH scale [*Dickson*, 1993].

[1973] as refitted by *Dickson and Millero* [1987] because they appear to produce internally consistent data when considering *DIC*, *Alk*, and pCO_2, as is typically the case in models [*Lee et al.*, 1997; *Lueker*, 1998; *Wanninkhof et al.*, 1999b; *Lueker et al.*, 2000]. However, the refitted dissociation constants of *Mehrbach et al.* [1973] appear to be problematic when pH or other parameters of the CO_2 system are considered [*Lee et al.*, 1997]. Typical values for the dissociation constants and the solubility at various temperatures are listed in table 8.2.3.

Solving the equations for global mean surface seawater properties yields

$$\begin{aligned} DIC &= [H_2CO_3^*] + [HCO_3^-] + [CO_3^{2-}] \\ &= \quad 0.5\% \qquad 88.6\% \qquad 10.9\% \end{aligned}$$

$$\begin{aligned} Alk &= [HCO_3^-] + 2[CO_3^{2-}] + [OH^-] - [H^+] + [B(OH)_4^-] \\ &= 76.8\% \qquad 18.8\% \qquad 0.2\% \qquad 4.2\% \end{aligned}$$

(see also table 8.2.4). This shows that only a very small fraction of the dissolved inorganic carbon exists as dissolved CO_2, and that the majority of the carbon exists as bicarbonate ion and a smaller amount in the form of carbonate ion. Therefore, for many purposes, we can approximate *DIC* as the sum of bicarbonate and carbonate ions only:

$$DIC \approx [HCO_3^-] + [CO_3^{2-}] \tag{8.2.15}$$

This computation also reveals that the contribution of the dissociation of water to variations in alkalinity is negligible and that the contribution of borate is of the order of a few percent. Therefore, *Alk* can often be reasonably well approximated by the carbonate alkalinity, i.e.,

$$Alk \approx Carb\text{-}Alk = [HCO_3^-] + 2[CO_3^{2-}] \tag{8.2.16}$$

TABLE 8.2.3
Numerical values of the equilibrium constants as a function of temperature for a salinity of 35

Temp. °C	$-\log_{10} K_0$ mol kg^{-1} atm^{-1}	$-\log_{10} K_1$ mol kg^{-1}	$-\log_{10} K_2$ mol kg^{-1}	$-\log_{10} K_w$ mol kg^{-1}	$-\log_{10} K_b$ (mol kg^{-1})^2	$-\log_{10} \frac{K_2}{K_1}$	$-\log_{10} \frac{K_2}{K_0 K_1}$ kg mol^{-1} atm
0	1.202	6.106	9.384	14.300	8.906	3.277	2.076
2	1.235	6.080	9.346	14.203	8.878	3.266	2.031
4	1.267	6.055	9.310	14.108	8.851	3.255	1.988
6	1.298	6.030	9.274	14.015	8.824	3.244	1.945
8	1.329	6.007	9.238	13.924	8.797	3.232	1.903
10	1.358	5.984	9.203	13.834	8.771	3.219	1.862
12	1.386	5.962	9.169	13.746	8.746	3.207	1.821
14	1.413	5.941	9.135	13.659	8.721	3.194	1.781
16	1.439	5.920	9.101	13.575	8.696	3.181	1.741
18	1.465	5.901	9.068	13.491	8.671	3.167	1.702
20	1.489	5.882	9.035	13.409	8.647	3.154	1.664
22	1.513	5.863	9.003	13.329	8.623	3.140	1.627
24	1.536	5.846	8.971	13.250	8.600	3.125	1.590
26	1.558	5.829	8.940	13.172	8.576	3.111	1.553
28	1.579	5.813	8.909	13.096	8.553	3.096	1.517
30	1.599	5.797	8.878	13.021	8.530	3.081	1.482

All dissociation constants are given with respect to the seawater pH scale [*Dickson*, 1993]. Sources: K_0: *Weiss* [1974]; K_1 and K_2: *Mehrbach et al.* [1973] as refitted by *Dickson and Millero* [1987]; K_b: *Dickson* [1990]; K_w: *Millero* [1995].

TABLE 8.2.4
Mean values of potential temperature, salinity, *sDIC*, *sAlk*, $H_2CO_3^*$, HCO_3^-, and CO_3^{2-} in the world oceans, based on data collected during the World Ocean Circulation Experiment

Region	θ °C	*S*	*sDIC* μmol kg^{-1}	*sAlk* μmol kg^{-1}	$[H_2CO_3^*]$ μmol kg^{-1}	$[HCO_3^-]$ μmol kg^{-1}	$[CO_3^{2-}]$ μmol kg^{-1}
Surface Ocean (0–50 m)							
Low and mid-lat.†	23.0	35.15	2003	2315	10	1772	221
Southern hemisphere high latitudes†	3.7	34.05	2119	2291	17	1977	125
Northern hemisphere high latitudes†	4.7	32.90	2049	2257	13	1889	147
Mean surface ocean	18.1	34.75	2026	2308	11	1815	200
Deep (>1200 m) and entire ocean							
Mean deep ocean	1.8	34.73	2280	2381	29	2164	86
Mean global ocean	3.6	34.73	2256	2364	28	2138	90

The carbon species have been calculated using the dissociation constants of *Mehrbach et al.* [1973] as given by *Dickson and Millero* [1987].
† Low and mid-latitudes: 45°S–45°N; High latitudes: south 45°S for the southern hemisphere, and 45°N–80°N for the northern hemisphere.

Combining these two equations ((8.2.15) and (8.2.16)) allows us to express the concentration of bicarbonate and carbonate in terms of *DIC* and *Alk* only:

$$[HCO_3^-] \approx 2 \cdot DIC - Alk \qquad (8.2.17)$$

$$[CO_3^{2-}] \approx Alk - DIC \qquad (8.2.18)$$

These two approximations are usually good to within about 10%. As we will see below, these approximation will provide a powerful tool for discussing many peculiarities of the CO_2 system in seawater.

In many studies, pH is used as a master variable. It is therefore instructive to investigate the inorganic carbon system in seawater also as a function of this parameter. Figure 8.2.1 shows how the concentration of the three inorganic carbon species varies as a function of pH for fixed *DIC*. Below $pK_1 = -\log K_1$, $H_2CO_3^*$ dominates. At pH $= pK_1$, $[H_2CO_3^*]$ is equal to $[HCO_3^-]$ by definition (see (8.2.5), whereas for $pK_1 <$ pH $< pK_2 = -\log K_2$, HCO_3^- is the species dominating *DIC*. At pH $= pK_2$, $[HCO_3^-] = [CO_3^{2-}]$, and at pH $> pK_2$, $[CO_3^{2-}]$ dominates. Since the ocean has a mean surface pH of slightly above 8, in

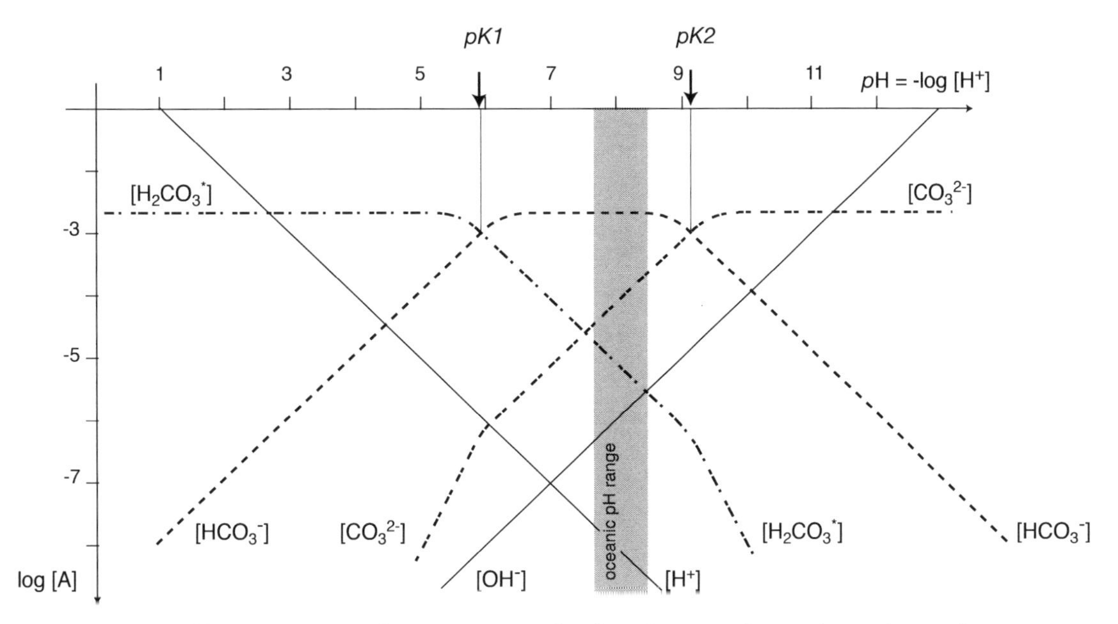

FIGURE 8.2.1: Plot of the concentrations of $H_2CO_3^*$, HCO_3^-, and CO_3^{2-} as functions of $pH = -\log_{10}[H^+]$. Note that the vertical axis is logarithmic to the base 10 as well. The concentrations are plotted for a *DIC* concentration of 2000 μmol kg^{-1}. By definition, the crossing point of an acid/base pair is equal to the *pK* of the corresponding dissociation constant.

between pK_1 and pK_2, we see immediately why HCO_3^- is the most important species present in seawater. Oceanic pH is also closer to pK_2 than to pK_1, so that it also becomes evident why CO_3^{2-} is the second most dominant inorganic carbon species, and why the concentration of $H_2CO_3^*$ is so small.

A note at the end: We usually normalize *DIC* and *Alk* to a constant salinity in order to remove the effect of freshwater fluxes. As described above, these normalized concentrations are denoted by the symbols *sDIC* and *sAlk*. The motivation of this normalization is that freshwater contains very little *DIC* and *Alk*, so that the addition of freshwater, for example, leads to a decrease of the *DIC* and *Alk* concentration, i.e., it dilutes the concentration of all chemical species present in seawater in direct proportion of the dilution of salinity. The opposite effect occurs if an excess of evaporation over precipitation leads to a net removal of freshwater from the surface ocean. Thus, net freshwater exchange at the surface leads to variations in *DIC* and *Alk* that can mask the "chemical" changes that we are mostly interested in, i.e., those changes driven by ocean biology, chemistry, and mixing/transport. For consistency, salinity normalization should be done for all other chemical properties that we investigate. However, since the concentration variations of most chemical properties of interest are large relative to their mean concentration, salinity normalization has little influence on these properties, and is therefore often neglected. By contrast, typical chemically or biologically induced variations in *DIC* and *Alk* are much smaller than their mean concentrations, so that freshwater flux variations have a much larger impact on these two properties. We therefore generally apply the salinity normalization to *DIC* and *Alk* observations.

8.3 The Surface Ocean

Having reviewed the most important aspects of marine carbon chemistry, we are now ready to tackle the problems we outlined at the beginning of this chapter. We start with the problem of what controls the annual mean spatial distribution of the surface pCO_2 and then proceed to a discussion of its seasonal variability.

ANNUAL MEAN DISTRIBUTION

The simplest answer to what controls surface pCO_2 is that it is determined by the concentration of $H_2CO_3^*$ in the water and the CO_2 solubility, K_0, i.e.,

$$pCO_2 = \frac{[H_2CO_3^*]}{K_0}. \qquad (8.3.1)$$

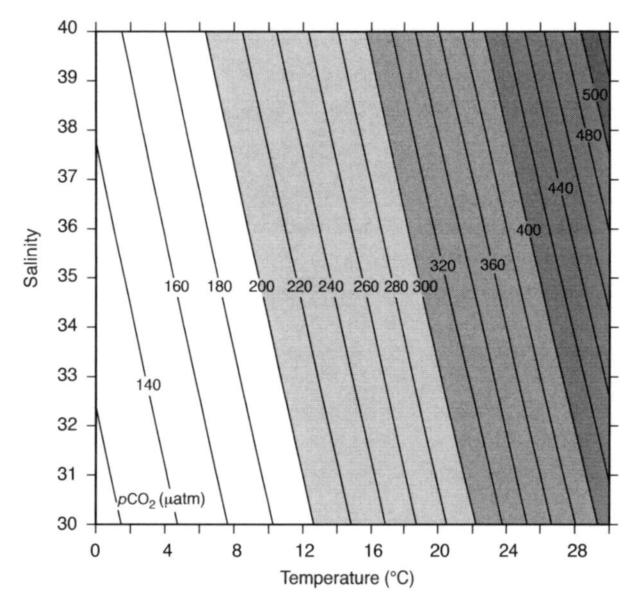

FIGURE 8.3.1: Plot of the partial pressure of CO_2 (pCO_2) as a function of temperature and salinity for constant DIC and Alk. Shown are the results for a typical surface water sample with an alkalinity of $2322\,\mu mol\,kg^{-1}$ and a DIC content of $2012\,\mu mol\,kg^{-1}$.

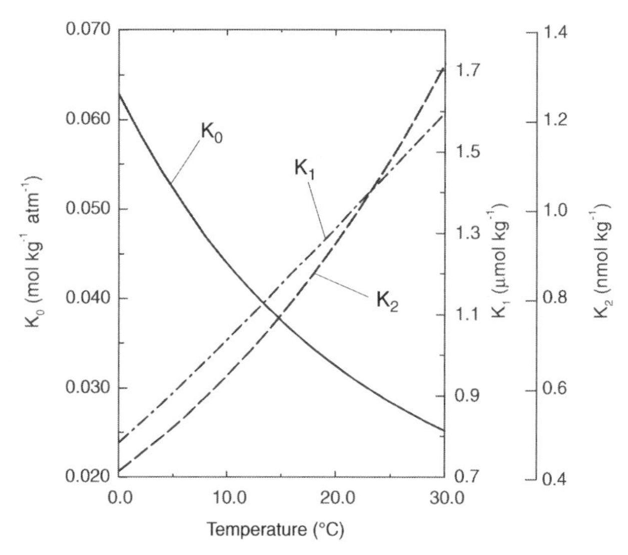

FIGURE 8.3.2: Plot of the CO_2 solubility (K_0), and of the first and second dissociation constants of carbonic acid (K_1 and K_2) as a function of temperature.

However, since $H_2CO_3^*$ constitutes only a very small fraction of the DIC pool, the bicarbonate and carbonate ions, rather than $H_2CO_3^*$, are ultimately controlling the pCO_2. To demonstrate this, we recast equations (8.2.4), (8.2.5), and (8.2.6) to relate the partial pressure of CO_2 in seawater to variations in carbonate and bicarbonate ions, and the three equilibrium constants K_0, K_1, and K_2,

$$pCO_2 = \frac{K_2}{K_0 \cdot K_1} \frac{[HCO_3^-]^2}{[CO_3^{2-}]} \tag{8.3.2}$$

In order to facilitate our discussion, we use approximations (8.2.17) and (8.2.18) to replace the bicarbonate and carbonate ion concentrations with DIC and Alk:

$$pCO_2 \approx \frac{K_2}{K_0 \cdot K_1} \frac{(2 \cdot DIC - Alk)^2}{Alk - DIC} \tag{8.3.3}$$

From consideration of (8.3.3), we see that the question of what controls the surface pCO_2 distribution is more complex than first suggested by (8.3.1). The best way to answer the question is to break it into three parts: (i) what controls the ratio of the equilibrium constants, $K_2/(K_0 \cdot K_1)$, (ii) what controls the DIC concentration, and (iii) what controls the Alk concentration.

We know from table 8.2.2 that the three equilibrium constants K_0, K_1, and K_2 are a function of temperature and salinity. The first of these questions can thus be readily answered by considering how the pCO_2 of a water parcel changes with temperature and salinity while keeping DIC and Alk constant. It turns out that the answer for the other two questions cannot be given independently of each other. The concentration of DIC is affected by air-sea gas exchange and by biological processes, whereas Alk is only affected by ocean biology. It it thus more convenient to separate the controls on surface ocean pCO_2 into physical processes and biological processes.

Physical Processes

We start our discussion by considering how the pCO_2 varies with temperature and salinity while keeping DIC and Alk constant. As shown in figure 8.3.1, the isolines of pCO_2 are almost vertical over the oceanic range of temperature and salinity, indicating a much greater sensitivity to temperature than to salinity variations. It turns out that about two-thirds of the temperature sensitivity of pCO_2 is a result of the strong temperature dependence of the solubility K_0 (figure 8.3.2), while the contribution of the ratio of the dissociation constants K_2/K_1 explains the remaining third. Since K_0 is relatively insensitive to salinity, about 70% of the salinity dependence of pCO_2 is governed by the ratio K_2/K_1.

Takahashi et al. [1993] provide a useful relationship that summarizes the temperature sensitivity of pCO_2 in a closed system, i.e., one where the concentration of DIC and Alk remain constant. They determined this sensitivity experimentally and found that a logarithmic dependence gave accurate results, thus

$$\frac{1}{pCO_2}\frac{\partial pCO_2}{\partial T} = \frac{\partial \ln pCO_2}{\partial T} \approx 0.0423^\circ C^{-1} \tag{8.3.4}$$

TABLE 8.3.1
Summary of the most important pCO_2 sensitivities in seawater

Parameter	*Definition*	*Mean Global*	*Mean High Latitudes*	*Mean Low Latitudes*
Temperature	$\frac{1}{pCO_2}\frac{\partial pCO_2}{\partial T}$	$0.0423°C^{-1}$	$0.0423°C^{-1}$	$0.0423°C^{-1}$
Salinity	$\gamma_S = \frac{S}{pCO_2}\frac{\partial pCO_2}{\partial S}$	1	1	1
DIC	$\gamma_{DIC} = \frac{DIC}{pCO_2}\frac{\partial pCO_2}{\partial DIC}$	10	13.3	9.5
Alk	$\gamma_{Alk} = \frac{Alk}{pCO_2}\frac{\partial pCO_2}{\partial Alk}$	−9.4	−12.6	−8.9
Freshwater[†]	$\gamma_{freshwater} = \frac{S}{pCO_2}\frac{\partial pCO_2}{\partial S}\Big\rvert_{freshwater} = \gamma_S + \gamma_{DIC} + \gamma_{Alk}$	1.6	1.7	1.6

[†] See chapter 10 for derivation.

Similarly, one finds for the salinity dependence of pCO_2,

$$\gamma_S = \frac{S}{pCO_2}\frac{\partial pCO_2}{\partial S} = \frac{\partial \ln pCO_2}{\partial \ln S} \approx 1 \qquad (8.3.5)$$

For example, if we take a water parcel with an initial pCO_2 of 300 μatm at 20°C and with a salinity of 35, a one-degree warming increases pCO_2 by approximately 13 μatm, whereas a salinity increase of 1 results in a pCO_2 increase of 9 μatm. Since temperature varies in the oceans by about 30°C, whereas salinity varies only by about 7, temperature rather than salinity has to be regarded as the dominant physical factor controlling pCO_2. A note of caution, however. The above salinity dependence is for constant *DIC* and *Alk*. Therefore, the CO_2 change given by (8.3.5) includes only the influence of salinity on the dissociation constants. Variations in salinity in the surface ocean, however, are mostly driven by changes in the balance between evaporation and precipitation. Therefore, if one is interested in using salinity as a tracer of the impact of the freshwater balance on pCO_2, one needs to take into account freshwater-induced *DIC* and *Alk* changes in addition to the direct salinity effect. We will show in chapter 10 how one can calculate this freshwater balance effect on pCO_2. We demonstrate there that the net effect of freshwater changes on the pCO_2 sensitivity is to increase the pure salinity-driven pCO_2 changes by about 60% (see table 8.3.1).

The next issue we consider is what happens to the carbon system if we permit gas exchange of CO_2 to occur while keeping *Alk* constant. Gas exchange will change the *DIC* concentration, but not *Alk*. The influence of air-sea gas exchange can be assessed with the following thought experiment taken from *Broecker and Peng* [1982] (see figure 8.3.3). We consider two extreme scenarios. In the first case, air-sea exchange is assumed to be sluggish compared to the residence time of water at the sea surface (figure 8.3.3a). Such a scenario is equivalent to the closed system that was discussed in the previous paragraph. Since air-sea exchange is severely restricted, *DIC* is nearly constant throughout the ocean and pCO_2 changes according to its temperature sensitivity of about 4% per degree centigrade. This results in the low latitudes having an oceanic pCO_2 twice as large as the high latitudes.

In the second scenario, air-sea gas exchange is assumed to be very rapid, so that oceanic pCO_2 comes close to equilibrium with atmospheric CO_2 everywhere (figure 8.3.3b). Since oceanic pCO_2 is now almost constant everywhere, oceanic *DIC* has to change in order maintain chemical equilibrium. What is the magnitude and direction of this change? We have learned that the factor $K_2/(K_0 \cdot K_1)$ increases with increasing temperatures. Therefore, in order to maintain the same oceanic pCO_2, we infer from (8.3.3) that *DIC* has to decrease with increasing temperature. Therefore, we find higher *DIC* values in the high-latitude ocean and lower *DIC* in the low latitudes.

Where in the range between these two extremes would we expect an ocean with realistic gas exchange to lie? This depends essentially on the timescale of air-sea gas exchange relative to the timescale of processes perturbing the local equilibrium. We have seen in chapter 3 that the air-sea gas exchange timescale for many gases is of the order of days to a few weeks. This is short relative to most perturbations, so that the oceanic partial pressure of these gases is usually very close to equilibrium with the atmosphere, as observed, for example, for oxygen (see figure 3.1.2). However, as we also saw in chapter 3, the gas exchange timescale of CO_2 is of the order of 6 months. Why is this the case? We were previously able to give only a partial explanation, but now we know enough about CO_2 chemistry to fill in this gap.

The time rate of change of any gas, A, in a surface mixed layer box that exchanges only with the atmosphere is given by

$$\frac{\partial[A]_w}{\partial t} = -\frac{\partial \Phi}{\partial z} = \frac{k_w}{z_{ml}}([A]_a - [A]_w) \qquad (8.3.6)$$

where k_w is the gas exchange coefficient (for CO_2 typically $20\,\text{cm hr}^{-1}$) and z_{ml} is the thickness of the mixed layer box (see also chapter 3). If we assume for the moment that the atmospheric concentration is fixed, (8.3.6) is a first-order differential equation in $[A]_w$. Its time-dependent solution is therefore an exponential with a timescale given by $(k_w/z_{ml})^{-1}$:

$$\tau = \frac{z_{ml}}{k_w} = \frac{z_{ml}}{20\,\text{cm hr}^{-1}} = 5\,\text{hr m}^{-1}\,z_{ml} \qquad (8.3.7)$$

For a 10 m deep surface mixed layer, the timescale amounts to about 8 days.

Because CO_2 needs to equilibrate with the entire *DIC* pool in the surface ocean and not just with the $H_2CO_3^*$ pool, the timescale for CO_2 is much longer. When CO_2 enters the ocean from the atmosphere, approximately 19 out of 20 molecules react with carbonate (the strongest base of the CO_2 system) to form two bicarbonate ions, i.e.,

$$H_2CO_3^* + CO_3^{2-} \rightleftharpoons 2HCO_3^- \qquad (8.3.8)$$

leaving behind only one molecule as $H_2CO_3^*$.

For CO_2, we therefore have to consider the time rate of change of all inorganic carbon species, i.e.,

$$\frac{\partial DIC}{\partial t} = \frac{\partial DIC}{\partial[H_2CO_3^*]}\frac{\partial[H_2CO_3^*]}{\partial t} = \frac{k_w}{z_{ml}}([H_2CO_3^*]_a - [H_2CO_3^*]) \qquad (8.3.9)$$

Solving for the time rate of change of $[H_2CO_3^*]$ results in

$$\frac{\partial[H_2CO_3^*]}{\partial t} = \left(\frac{\partial DIC}{\partial[H_2CO_3^*]}\right)^{-1}\frac{k_\omega}{z_{ml}}([H_2CO_3^*]_a - [H_2CO_3^*]) \qquad (8.3.10)$$

This equation is analogous to (8.3.6), except that the e-folding timescale for equilibration of a perturbation in the CO_2 system is given by

$$\tau = \frac{\partial DIC}{\partial[H_2CO_3^*]}\left(\frac{z_{ml}}{k_w}\right) \approx 20\frac{z_{ml}}{k_w} \qquad (8.3.11)$$

Here, we have used a partial derivative $\partial DIC/\partial[H_2CO_3^*]$ of 20 based on solving the carbon chemistry equations. This derivative indicates that only about 1 molecule of CO_2 in 20 molecules entering or leaving the ocean stays as $H_2CO_3^*$. Why is this the case? One might be tempted to argue that this partial derivative should be about equal to the concentration ratio of $DIC/[H_2CO_3^*] \approx 200$. However, as it turns out, variations in the concentration of the carbonate ion are the primary determinant of this partial derivative (see figure 8.3.4).

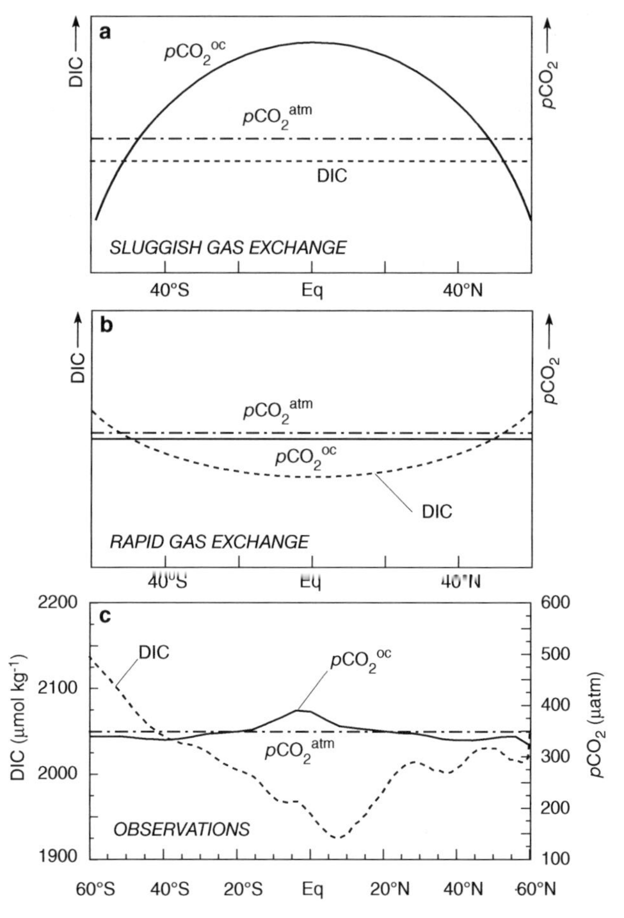

FIGURE 8.3.3: Hypothetical and observed meridional variations of pCO_2 and *DIC* in the surface ocean. (a) Hypothetical distribution in a case where gas exchange is very slow. It is assumed that no biological processes take place and that therefore *Alk* remains constant [*Broecker and Peng*, 1982]. (b) As in (a), but for a case with very rapid gas exchange. (c) Observed zonal mean variations of pCO_2 and *DIC*. Based on the pCO_2 climatology of *Takahashi et al.* [2003] and the GLODAP climatology of *Key et al.* [2004].

This can best be understood by reaction (8.3.8). The fraction of the total number of CO_2 molecules added to the ocean that are converted to HCO_3^- and consequently do not increase the $H_2CO_3^*$ pool depends on the availability of CO_3^{2-}. As the concentration of carbonate is about 15 to 20 times higher than that of $H_2CO_3^*$, the partial derivative is of the order of 20 rather than 200. A more rigorous derivation is given in panel 8.3.1.

We now have established that the characteristic timescale for air-sea exchange of CO_2 is of the order of 6 months, but still have not answered the question of where between the two extreme scenarios discussed above the real ocean lies. Are 6 months long or short compared to the average residence time of waters near the surface and the timescale of perturbations? The ΔpCO_2 observations in figure 8.1.1 suggest that the real ocean lies between the two scenarios: the pCO_2 is clearly out of equilibrium with the atmosphere, but

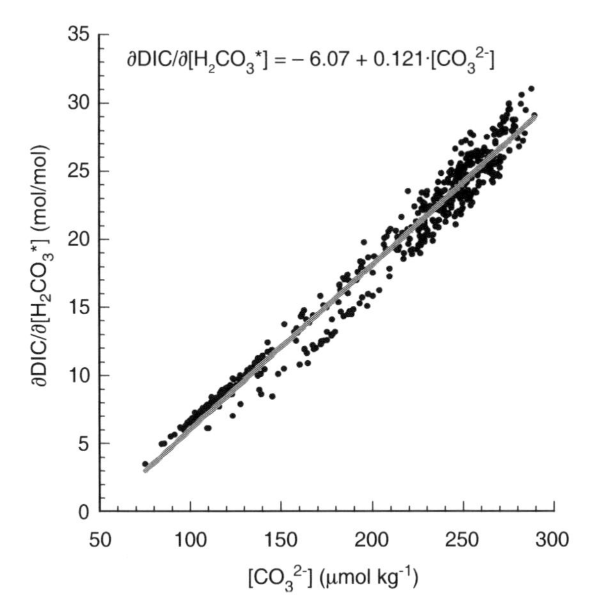

FIGURE 8.3.4: Plot of the partial derivative of *DIC* with respect to $[H_2CO_3^*]$. ($\partial DIC/\partial[H_2CO_3^*]$) versus the concentration of the carbonate ion $[CO_3^{2-}]$. Plotted are the surface ocean (< 50 m) data as obtained during the GEOSECS program.

nowhere near the factor of two between the coldest and warmest waters that would be expected from (8.3.4). In fact, observations of *DIC* suggest that the real ocean is closer to the rapid gas exchange case than it is to the sluggish one (figure 8.3.3c).

BIOLOGICAL PROCESSES

We now turn to a consideration of the influence of biology, which changes both the *DIC* and *Alk* distributions. The most important biological processes altering the concentration of *DIC* in the ocean are the photosynthetic uptake of CO_2 to form organic matter, and the reverse processes of respiration and remineralization (see reaction (4.2.4)):

$$106CO_2 + 16NO_3^- + HPO_4^{2-} + 78H_2O + 18H^+ \rightleftharpoons C_{106}H_{175}O_{42}N_{16}P + 150O_2 \tag{8.3.12}$$

We have recast the original reaction (4.2.4) to emphasize that NO_3^- and HPO_4^{2-} rather than HNO_3 and H_3PO_4, respectively, are the dominant chemical species present in seawater. Thus, in addition to decreasing the concentration of *DIC*, the formation of organic matter also decreases the concentration of the free protons, $[H^+]$, and therefore increases alkalinity (see definition (8.2.8)) [*Brewer et al.*, 1975]. Most of these protons are consumed by the assimilatory reduction of nitrate to organic nitrogen (see figure 5.1.5). The influence of the nitrate uptake on *Alk* can also be understood by considering the alternative definition of *Alk* (8.2.9), which shows that a decrease in the nitrate concentration increases *Alk*. While the dominant species at the pH of seawater is indeed HPO_4^{2-}, it is not well established whether this is actually the species taken up by phytoplankton or rather the uncharged H_3PO_4.

Panel 8.3.1: Derivation of the partial derivative $\partial DIC/\partial[H_2CO_3^*]$

We want to demonstrate the fact that the partial derivative $\partial DIC/\partial[H_2CO_3^*]$ is controlled by the CO_3^{2-} ion concentration and is about 20. We start by inserting the two approximations for HCO_3^- and CO_3^{2-} into (8.3.2) and by replacing $K_0 \cdot pCO_2$ with $[H_2CO_3^*]$. This gives

$$[H_2CO_3^*] \approx \frac{K_2}{K_1}\frac{(2 \cdot DIC - Alk)^2}{Alk - DIC} \tag{1}$$

The partial derivative of *DIC* with respect to $[H_2CO_3^*]$ is equal to the inverse of the partial derivative of $[H_2CO_3^*]$ with respect to *DIC*. This gives

$$\frac{\partial DIC}{\partial[H_2CO_3^*]} = \left(\frac{\partial[H_2CO_3^*]}{\partial DIC}\right)^{-1} \approx \left(\frac{K_2}{K_1}\frac{\partial}{\partial DIC}\left(\frac{(2 \cdot DIC - Alk)^2}{Alk - DIC}\right)\right)^{-1} \tag{2}$$

Computing the derivative and substituting the *DIC* and *Alk* approximations back into the resulting equation gives

$$\frac{\partial DIC}{\partial[H_2CO_3^*]} \approx \frac{K_1}{K_2}\frac{[CO_3^{2-}]^2}{4[HCO_3^-][CO_3^{2-}] + [HCO_3^-]^2} \tag{3}$$

Inserting typical surface ocean concentrations for bicarbonate and carbonate gives a partial derivative of 20.

It is, however, not yet evident why the carbonate ion content is the main controlling factor. In order to demonstrate this, we are going to simplify the equation even further. As the concentration of CO_3^{2-} is much smaller than the concentration of HCO_3^- (see table 8.2.4), the first part of the denominator is much smaller than the second, i.e., $4[HCO_3^-][CO_3^{2-}] \ll [HCO_3^-]^2$ and can thus be neglected. This results in:

$$\frac{\partial DIC}{\partial[H_2CO_3^*]} \approx \frac{K_1}{K_2}\frac{[CO_3^{2-}]^2}{[HCO_3^-]^2} \approx \frac{[CO_3^{2-}]}{[H_2CO_3^*]} \approx 20 \tag{4}$$

where we have made use of (8.3.2) to replace $(K_2/K_1) \cdot ([HCO_3^-]^2/[CO_3^{2-}])$ with $[H_2CO_3^*]$. Note from equation (2) in this panel that this partial derivative is largely independent of temperature, but is largely controlled by variations in the *Alk*-to-*DIC* ratio.

TABLE 8.3.2
Changes in pCO_2 as a typical water parcel from the deep sea is moved to the surface and adjusted to typical low-latitude surface values (see table 8.2.4 for values)

	(1) Deep Water	(2) Move to Surface	(3) Warm Up	(4) Formation of Organic Matter	(5) Formation of $CaCO_3$
Pressure (atm)	400	1	1	1	1
Temp (°C)	1.8	1.8	23	23	23
PO_4 (μmol kg^{-1})	2.1	2.1	2.1	0	0
DIC (μmol kg^{-1})	2298	2298	2298	2058	1994
Alk (μmol kg^{-1})	2400	2400	2400	2433	2305
pCO_2 (μatm)		510	1220	293	348

The changes in *DIC* and *Alk* due to the removal of organic matter and the loss of $CaCO_3$ were calculated using the stoichiometry of Anderson *and* Sarmiento [1994].

As we have discussed in detail in chapter 4, organic matter is produced in the uppermost sunlit layers of the ocean. A fraction of the organic matter is exported to the deeper layers through settling particles or advection of dissolved organic carbon. This leads to a net consumption of CO_2 in these upper layers. Upon remineralization of this organic matter in the deeper layers, this CO_2 is returned to the seawater. Thus these biological processes lead to a net transfer of inorganic carbon from the surface into the abyss. This process is often termed the "soft-tissue" pump [*Volk and Hoffert*, 1985]. In a steady state, this net downward flux has to be compensated by a net upward flux of inorganic carbon by transport processes.

The second biological reaction of great importance for carbon cycling is the biogenic formation and dissolution of calcite or aragonite:

$$Ca^{2+} + CO_3^{2-} \rightleftharpoons CaCO_3 \tag{8.3.13}$$

The formation and dissolution of mineral calcium carbonates changes *Alk* twice as much as it changes *DIC*. This is best understood by considering the charge balance definition of *Alk* (equation (8.2.9)), which shows that a 1 mol reduction in the concentration of Ca^{2+} results in a 2 mol change in *Alk*. Mineral calcium carbonate shells are formed in the upper layers of the ocean mainly by three groups of organisms: coccolithophorids, foraminifera, and pteropods. The first group are phytoplankton, whereas the second and third groups are zooplankton (see panels 4.2.2 and 4.2.3, respectively). Upon the death of these organisms, their shells sink and eventually dissolve, either in the water column or in the sediments, except for a small fraction that is buried permanently. The net effect of this process is a downward transport of *DIC* and *Alk* from the surface ocean into the abyss. This process is often dubbed the "carbonate pump."

The large influence of the biological pumps on pCO_2 can be dramatically demonstrated by the following thought experiment after *Broecker and Peng* [1982] (see table 8.3.2). We take a typical deep water parcel and bring it to the surface, adjusting its properties to typical low-latitude values (see table 8.2.4). After the sample is warmed from 1.8°C to 23°C, it has a pCO_2 in excess of 1200 μatm. Next, we remove all phosphate present and form organic matter, thus removing *DIC* from the water and adding *Alk* in proportion to the change in nitrate, assuming a constant nitrate to phosphate ratio of 16:1. What pCO_2 does this water parcel have after these changes? In order to compute the answer, we have to know the sensitivity of pCO_2 to changes in *DIC* and *Alk*.

We estimate the sensitivity of pCO_2 to changes in the *DIC* and *Alk* content of seawater by considering (8.3.3). This equation tells us that pCO_2 decreases when *DIC* decreases, but that pCO_2 increases for a decrease in *Alk*. The magnitude of these changes in pCO_2 depends strongly on the relative proportions of the *DIC* and *Alk* concentrations (see figure 8.3.5). To evaluate this more quantitatively, we investigate the following two dimensionless sensitivities:

$$\gamma_{DIC} = \frac{DIC}{pCO_2}\frac{\partial pCO_2}{\partial DIC} = \frac{\partial \ln pCO_2}{\partial \ln DIC} \tag{8.3.14}$$

$$\gamma_{Alk} = \frac{Alk}{pCO_2}\frac{\partial pCO_2}{\partial Alk} = \frac{\partial \ln pCO_2}{\partial \ln Alk} \tag{8.3.15}$$

We will refer to the sensitivity of pCO_2 to changes in *DIC* as the "buffer factor" or "Revelle factor" [*Takahashi et al.*, 1980] after Roger Revelle, who was among the first to point out the importance of this sensitivity for the oceanic uptake of anthropogenic CO_2. By analogy, we will refer to the alkalinity sensitivity, γ_{Alk}, as the alkalinity factor. We can estimate these factors by calculating the partial derivatives using the approximation (8.3.3). With some modifications we arrive at:

$$\gamma_{DIC} \approx \frac{3 \cdot Alk \cdot DIC - 2 \cdot DIC^2}{(2 \cdot DIC - Alk)(Alk - DIC)} \tag{8.3.16}$$

$$\gamma_{Alk} \approx -\frac{Alk^2}{(2 \cdot DIC - Alk)(Alk - DIC)} \tag{8.3.17}$$

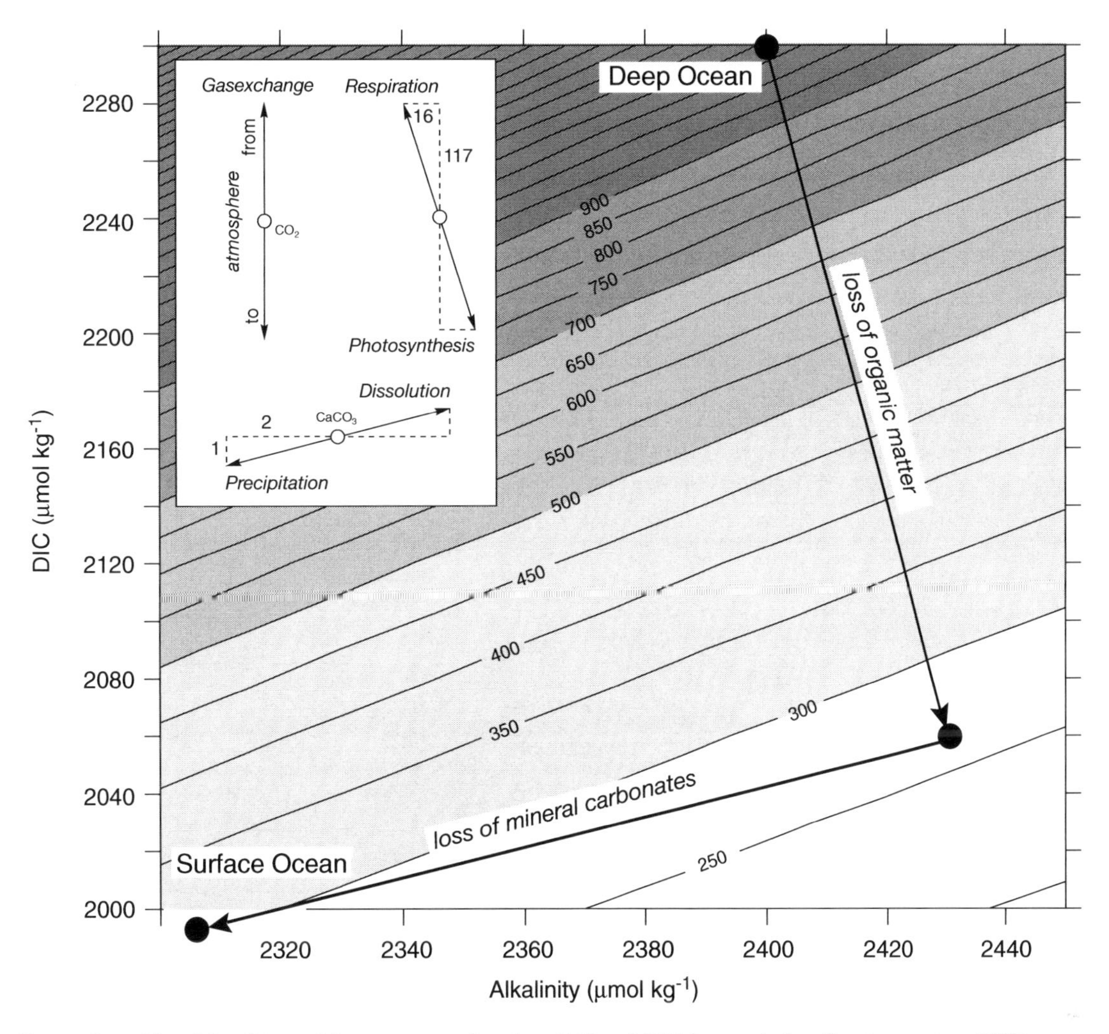

FIGURE 8.3.5: Plot of the CO_2 partial pressure as a function of *Alk* and *DIC* for a typical surface temperature (20°C) and salinity (35), computed from the equations given in table 8.2.1. Also shown is a simplified representation of the composition changes caused by biological processes. Water of average deep water composition is brought to the surface, and is warmed up to 20°C (point 1). It undergoes photosynthesis (point 1 to 2) and calcium carbonate precipitation (point 2 to 3). The inset shows the vectors of the different processes affecting *DIC* and *Alk*, and hence also pCO_2.

Inserting typical values for the surface ocean yields a buffer factor (γ_{DIC}) of about 12 and an alkalinity factor (γ_{Alk}) of about −10. These factors, computed with the full chemistry including borate, are about 10 and −9.4, respectively, meaning that pCO_2 increases by about 10% when *DIC* is increased by 1%, whereas pCO_2 decreases by about 9.4% when *Alk* is increased by 1%.

It is important to recognize that the magnitudes of these two factors are not a direct function of temperature. Instead, these factors are mainly determined by the relative concentrations of *DIC* and *Alk* as evident from (8.3.16) and (8.3.17) [*Takahashi et al.*, 1980]. Because *DIC* concentrations tend to correlate with temperature (see figure 8.1.4b), while surface *Alk* is spatially more homogeneous, the resulting correlation of the *DIC*-to-*Alk* ratio with temperature leads to a correlation of the buffer factor with temperature as well. This correlation between the buffer factor and temperature is a result of temperature causing some of the variations in *DIC* and not because of temperature directly driving variations in the buffer factor.

The gas exchange thought experiment that we used above (figure 8.3.3) provides a framework to illustrate this. We assume in both the rapid and sluggish gas exchange cases a spatially uniform alkalinity. In the sluggish gas exchange case, *DIC* would be spatially uniform as well, making the buffer factor essentially constant. In the rapid gas exchange case, the temperature variations are fully expressed in the surface concentration of *DIC*, leading to strong meridional variations in the *DIC*-to-*Alk* ratio. This also causes strong meridional variations in the buffer factor. The real ocean is somewhat closer to the fast gas exchange case, explaining the observed meridional variations in the buffer factor. As we will see

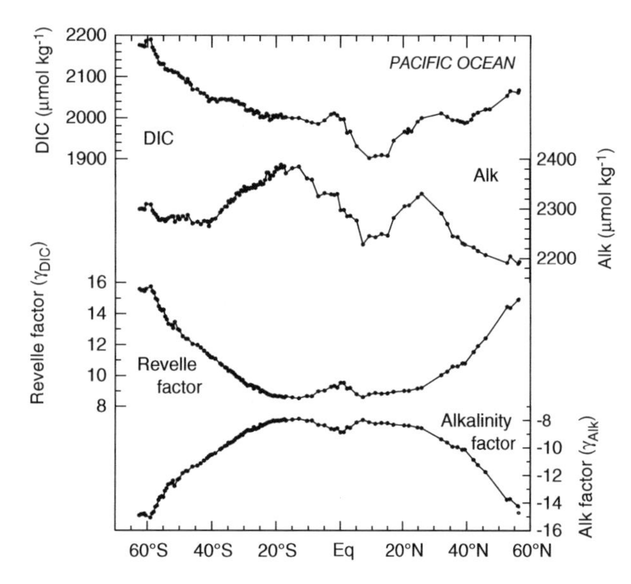

FIGURE 8.3.6: Meridional profiles of *DIC*, *Alk*, buffer factor (γ_{DIC}), and the alkalinity factor (γ_{Alk}) in the surface ocean of the Pacific. Based on data from the WOCE P16 meridional section along approximately 160°W.

in chapter 10, these spatial variations in the buffer factor have important implications for the uptake and storage of anthropogenic CO_2.

In summary, typical values for γ_{DIC} and γ_{Alk} in the surface oceans are

$$\gamma_{DIC} \approx 10 \quad \begin{matrix} 9.5 \text{ low latitudes} \\ 13.3 \text{ high latitudes} \end{matrix} \tag{8.3.18}$$

$$\gamma_{Alk} \approx -9.4 \quad \begin{matrix} -8.9 \text{ low latitudes} \\ -12.6 \text{ high latitudes} \end{matrix} \tag{8.3.19}$$

(see also table 8.3.1 and figure 8.3.6).

Having established the sensitivity of pCO_2 to *DIC* and *Alk*, we can now return to our thought experiment shown in table 8.3.2. In this experiment, we wanted to know how the biologically induced changes in *DIC* and *Alk* affect pCO_2. We estimate the change in pCO_2 by assuming that the effects of changes in *DIC* and *Alk* are independent of each other, and therefore additive. Solving both (8.3.14) and (8.3.15) for $d\ln pCO_2$ and then integrating the left and the right sides of the resulting equations gives:

$$\int d \ln pCO_2 = \gamma_{DIC} \int d \ln DIC + \gamma_{Alk} \int d \ln Alk \tag{8.3.20}$$

Integrating (8.3.20) from an initial state (i) to a final state (f) and inserting mean ocean values for γ_{DIC} and γ_{Alk} yields:

$$\ln pCO_2^f \approx \ln pCO_2^i + 10\,(\ln DIC^f - \ln DIC^i) - 9.4\,(\ln Alk^f - \ln Alk^i) \tag{8.3.21}$$

We estimate the final state f of *DIC* and *Alk* as a result of the formation of organic matter from the change in phosphate, $\Delta PO_4^{3-} = [PO_4^{3-}]^f - [PO_4^{3-}]^i$:

$$DIC^f = DIC^i + r_{C:P}\Delta PO_4^{3-} \tag{8.3.22}$$

$$Alk^f = Alk^i - r_{N:P}\Delta PO_4^{3-} \tag{8.3.23}$$

where $r_{C:P} = 117$ and $r_{N:P} = 16$ are the stoichiometric ratios of reaction (8.3.12). On the basis of approximation (8.3.21), we calculate that the removal of all phosphate present in this deep ocean water parcel, i.e., $\Delta PO_4^{3-} = -[PO_4^{3-}]^i$, lowers the pCO_2 by more than 900 μatm. Our approximate calculation is remarkably close to an estimate on the basis of a full carbon chemistry model, which predicts a decrease of 927 μatm (table 8.3.2).

As a last step, we adjust alkalinity to typical low-latitude surface values by forming mineral calcium carbonate. The change in pCO_2 can again be estimated using (8.3.21), but with DIC^f given by

$$DIC^f = DIC^i + \frac{1}{2}(Alk^f - Alk^i) \tag{8.3.24}$$

where we use the final values obtained after the formation of organic matter (8.3.22) and (8.3.23) as our initial values, DIC^i and Alk^i, respectively. We set Alk^f to the observed value for low-latitude surface waters (see table 8.2.4). The formation of carbonates increases pCO_2 slightly, such that a final pCO_2 value of about 350 μatm is computed, close to the atmospheric pCO_2 at the time when these *DIC* and *Alk* measurements were obtained (1988–1996).

VECTOR DIAGRAMS

Another way to depict the various processes that affect pCO_2 in the surface ocean is as vectors in a plot of *DIC* versus *Alk* (see inset in figure 8.3.5) [*Baes*, 1982]. The three major processes discussed form the following vectors:

1. *Gas exchange:* When CO_2 is transferred across the air-sea interface, the oceanic *DIC* changes proportionally, whereas *Alk* remains unaffected. Hence the vector is a vertical line with a length that is directly proportional to the number of molecules transferred.

2. *Soft-tissue pump:* The formation of organic matter reduces *DIC* and increases *Alk* proportional to the ratio of carbon to nitrogen, or about −117:16. Thus on a *DIC* versus *Alk* diagram, the vector forms a line with a slope of about −7.3.

3. *Carbonate pump:* When calcium carbonate is precipitated or dissolved, the water composition is changed along a line of slope ½ on the diagram. As we have seen above, this is because the carbonate ion contributes two moles to *Alk* for each mole of *DIC*.

A good example of the power of these vector diagrams is the depiction of our thought experiment of a deep water sample that was brought to the surface and adjusted to typical low-latitude values. Figure 8.3.5

shows a summary of the changes in pCO_2 that occur as *DIC* and *Alk* are changed.

Seasonal Variability

Summer-winter differences of surface ocean pCO_2 amount in many extratropical regions to more than 40 μatm (figure 8.3.7a) and are therefore comparable to the spatial variability of annual mean pCO_2 (figure 8.1.1). Figure 8.3.7a also reveals a distinct difference between the regions poleward of 40°, which generally show a decrease in surface ocean pCO_2 from winter to summer, and the subtropical latitudes, which experience an increase in surface ocean pCO_2 over the same period (cf. *Takahashi et al.* [2002]). What are the causes of these seasonal variations?

We learned from our discussion of (8.3.3) above that the processes that control surface ocean pCO_2 are temperature and salinity, which act through their influence on the equilibrium constants, $K_2/(K_0 \cdot K_1)$; and processes that affect the concentrations of *DIC* and *Alk*, such as gas exchange, biology, and lateral and vertical transport and mixing. We start our discussion by considering the seasonal variations in sea surface temperature. We neglect seasonal variations in salinity, as they are negligible.

We have seen that oceanic pCO_2 changes by about 13 μatm for each degree of warming or cooling. This permits us to estimate the effect of seasonal changes in SST on pCO_2 simply by multiplying the observed seasonal SST changes, ΔSST, by this factor, i.e.,

$$\Delta pCO_2\big|_{thermal} \approx pCO_2 \cdot 0.0423(°C)^{-1} \cdot \Delta SST \qquad (8.3.25)$$

The "thermally" forced pCO_2 amplitude, $\Delta pCO_2|_{thermal}$, reveals that the warming of the sea surface from winter to summer tends to increase surface ocean pCO_2 everywhere, with values reaching as high as nearly 200 μatm (figure 8.3.7b). A comparison of this thermally forced pCO_2 amplitude with the observed pCO_2 amplitude (figure 8.3.7a) shows that this component can explain the sign of the summer-minus-winter change in the subtropical gyres, but tends to overestimate the magnitude. In the high latitudes, the observed sign of the change is opposite to that forced by seasonal SST changes. Here, SST predicts an increase in pCO_2 from winter to summer, whereas the observations clearly show a decrease. This suggests that the effect of the seasonal SST changes on oceanic pCO_2 must be counteracted nearly everywhere by seasonal reductions in pCO_2 induced by changes in *DIC* and *Alk*, i.e., $\Delta pCO_2|_{DIC,Alk}$. We estimate this component simply by subtracting the thermal component from the observed summer-minus-winter pCO_2 difference, $\Delta pCO_2|_{observed}$ i.e.,

$$\Delta pCO_2\big|_{DIC,Alk} = \Delta pCO_2\big|_{observed} - \Delta pCO_2\big|_{thermal} \qquad (8.3.26)$$

As shown in figure 8.3.7c, the magnitude of these *DIC*- and/or *Alk*-driven pCO_2 changes is smaller than the thermally driven changes in the subtropical latitudes, while in the high latitudes, the *DIC*- and/or *Alk*-driven changes apparently outweigh the winter-to-summer increase in pCO_2 stemming from the warming of the sea surface. The negative sign of $\Delta pCO_2|_{DIC,Alk}$ implies a winter-to-summer drawdown in *DIC* or a winter-to-summer increase in *Alk*. How large are these changes?

We can estimate the magnitude of the required summer-minus-winter changes in *DIC* and *Alk*, ΔDIC and ΔAlk, by using the definitions of the buffer and alkalinity factors (equations (8.3.14) and (8.3.15) i.e.,

$$\Delta DIC \approx \frac{DIC}{pCO_2 \cdot \gamma_{DIC}} \Delta pCO_2\big|_{DIC,Alk} \qquad (8.3.27)$$

$$\Delta Alk \approx \frac{Alk}{pCO_2 \cdot \gamma_{Alk}} \Delta pCO_2\big|_{DIC,Alk} \qquad (8.3.28)$$

Figure 8.3.7c suggests a $\Delta pCO_2|_{DIC,Alk}$ of 40 to 80 μatm for the northern hemisphere subtropical gyres. Using low-latitude values for the buffer and alkalinity factors from (8.3.18) and (8.3.19), we compute that a winter-to-summer *DIC* decrease of about 20 to 40 μmol kg^{-1} or an *Alk* increase of 30 to 60 μmol kg^{-1} over the same time period would be required to explain the *DIC*- and/or *Alk*-driven winter-to-summer pCO_2 drawdown. In the high northern latitudes of the North Atlantic, $\Delta pCO_2|_{DIC,Alk}$ is well over 100 μatm. Given typical high-latitude buffer and alkalinity factors, the *DIC* drawdown must be more than 50 μmol kg^{-1}. Alternatively, *Alk* must increase from winter to summer by more than 60 μmol kg^{-1}.

As we will see below, seasonal variations in *Alk* are small, so that most of the $\Delta pCO_2|_{DIC,Alk}$ variations are, in fact, a result of a seasonal drawdown of *DIC*. But what, then, controls this *DIC* drawdown? Is it air-sea gas exchange, mixing, and/or biology? Since we know the air-sea difference in pCO_2, we can already assess the contribution of air-sea gas exchange. In the high northern latitudes, air-sea gas exchange can be excluded. This is because surface waters in this region are strongly undersaturated with respect to atmospheric CO_2 in summer (figure 3.1.3b), thus taking up CO_2 from the atmosphere. This leads to an increase in *DIC* rather than a decrease as required to explain the winter-summer pCO_2 drop. Therefore, in the high northern latitudes, it must be transport, mixing, and biology that dominate variations in *DIC*. In contrast, air-sea exchange cannot be excluded in subtropical regions, as these waters tend to be supersaturated in summertime (figure 3.1.3b). Therefore, some fraction of the required *DIC* reduction from winter to summer could stem from the loss of CO_2 across the air-sea interface.

We investigate in more detail three regions where different processes dominate seasonal pCO_2 variations. These regions are the subtropical gyres, where seasonal

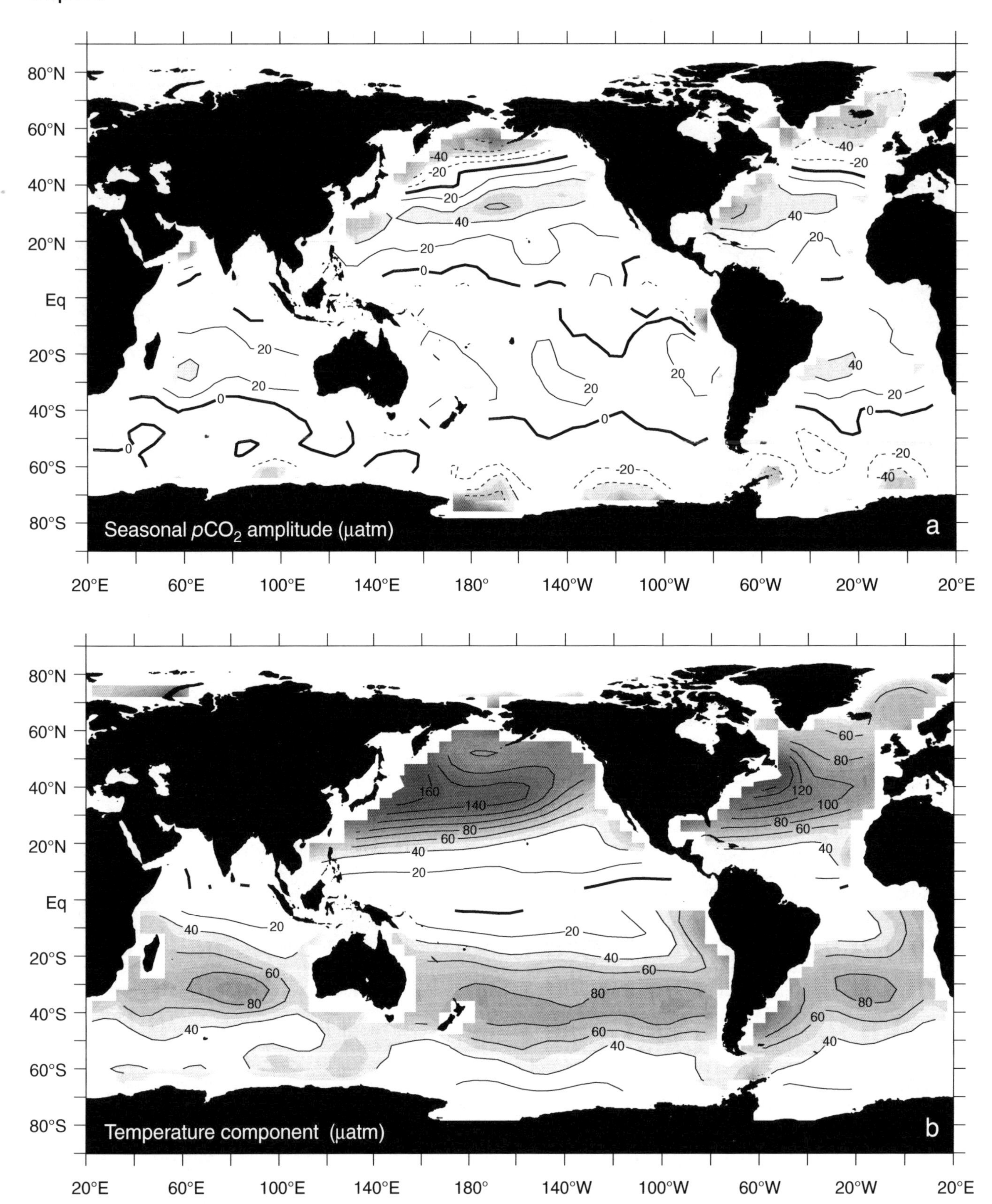

FIGURE 8.3.7: Maps of the seasonal amplitude of surface ocean pCO_2 and its driving forces. (a) summer-minus-winter difference of surface ocean pCO_2. (b) Sea-surface temperature-forced summer-minus-winter difference in surface ocean pCO_2 computed from the temperature sensitivity of pCO_2. (c) *DIC*- and *Alk*-forced summer-minus-winter difference in surface ocean pCO_2, computed by differencing (a) and (b). This change is primarily driven by seasonal variations in biology, and hence can be viewed as the "biological component," but gas exchange and transport/mixing also play a role. In the northern hemisphere, data from the months January through March (JFM) have been averaged to represent winter, and data from the months July through September (JAS) have been averaged for the summer period. In the southern hemisphere, JFM represent the summer season and JAS the winter season. Based on the climatology of *Takahashi et al.* [2002].

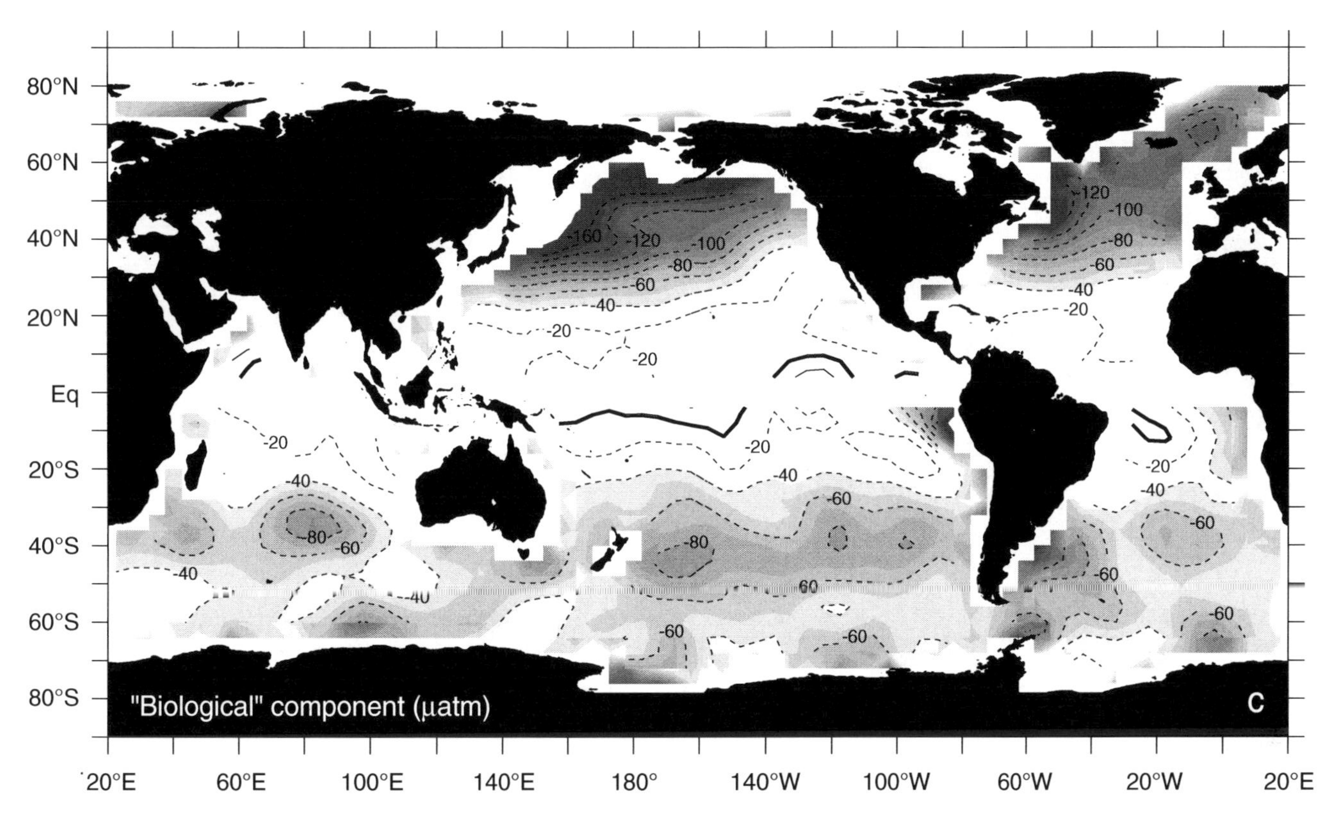

FIGURE 8.3.7: *(Continued)*

pCO_2 changes are primarily temperature controlled, the high latitude North Atlantic, where seasonal pCO_2 changes are dominated by biology and mixing, and the North Pacific, where all three mechanisms are important.

SUBTROPICAL GYRES

Figure 8.3.8 shows seasonally aggregated time series of temperature, *sDIC*, $^{13}C/^{12}C$ ratio of *DIC*, *sAlk*, and oceanic pCO_2 in the surface mixed layer at the Bermuda Atlantic Time-series Study (BATS) site, which lies near Bermuda in the northwestern Sargasso Sea [*Gruber et al.*, 2002]. Figure 8.3.9 shows a similar time series from the Hawaii Ocean Time-series (HOT) site, which is located north of the Hawaiian island chain [*Keeling et al.*, 2004]. Both sites exhibit well-defined seasonal cycles in *sDIC* and pCO_2 (see discussions by *Gruber* [1998], *Bates et al.* [1996b], *Quay and Stutsman* [2003], and *Keeling et al.* [2004]), but negligible variations in *sAlk*, with the exception of a few brief drawdowns occurring at BATS apparently linked to localized blooms of calcifying organisms (most likely coccolithophorids) [*Bates et al.*, 1996a]. The seasonal amplitude of *sDIC* at BATS is about 30 μmol kg^{-1}, approximately twice as large as that observed at HOT and in good agreement with our rough estimates based on (8.3.27).

The competing effects of changing temperature and *sDIC* on the seasonal evolution of pCO_2 at BATS are shown in more detail in figure 8.3.10a. If *sDIC* were kept constant, the seasonal cycle in temperature at this station would lead to a seasonal amplitude of more than 110 μatm, with a maximum occurring in mid-August. If temperature were kept at the annual mean value, the seasonal changes in *sDIC* would lead to a pCO_2 change opposite in phase and with an amplitude of about 60 μatm. The situation is similar but less pronounced in the subtropical North Pacific [*Keeling et al.*, 2004] (figure 8.3.10b). Thus, the detailed time-series data confirm the general trends for the subtropical gyres seen from the global maps very well, i.e., that in these regions, the thermally forced increase in pCO_2, $\Delta pCO_2|_{thermal}$, outweighs the negative $\Delta pCO_2|_{DIC,\,Alk}$ component, leading to a winter-to-summer increase in oceanic pCO_2 as displayed in figure 8.3.7a.

Having established the causes for the seasonal changes in pCO_2, what is driving the seasonal changes in *sDIC*? Of particular interest is the summer/fall period, when *sDIC* is drawn down almost continuously every year at both sites in the absence of measurable nitrate and phosphate concentrations (see also discussion in section 4.3). The processes that we need to consider are air-sea gas exchange (*ex*), net community production (*ncp*), i.e., net primary production minus community respiration, lateral transport (*trsp*), vertical diffusion (*diff*), and vertical entrainment (*ent*). The latter process describes the transport of *DIC* from the thermocline into the mixed layer whenever the surface mixed layer deep-

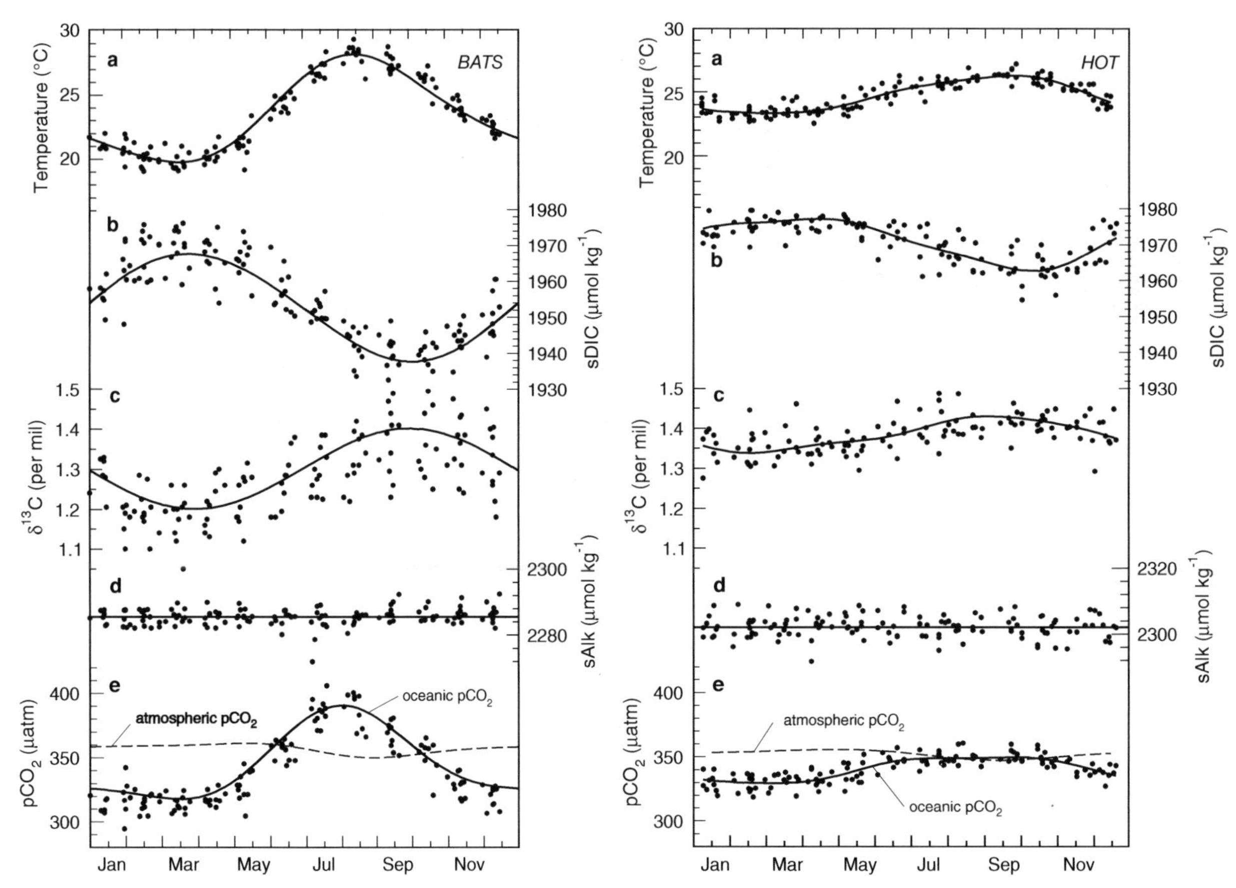

FIGURE 8.3.8: Annual composite time series of upper ocean quantities at the Bermuda Atlantic Time-series Station (BATS) in the subtropical Atlantic (31°50'N, 64°10'W): (a) temperature, (b) salinity-normalized *DIC*, (c) the reduced isotopic ratio ($\delta^{13}C$) of *DIC*, (d) salinity-normalized *Alk*, and (e) computed pCO_2. The filled circles represent the observations, whereas the smooth curve represents the results of a harmonic fit through the observations. Also shown in (e) is the estimated atmospheric pCO_2 near Bermuda. Data are for 1988–2001. Based on data reported by *Gruber et al.* [2002].

FIGURE 8.3.9: As 8.3.8 except for the Hawaii Ocean Time-series (HOT) ALOHA site in the subtropical North Pacific (22°45'N, 158°00'W). Based on data reported by *Keeling et al.* [2004].

ens. The time-rate of change for the mean *sDIC* concentration in the mixed layer can therefore be written as:

$$\frac{\partial sDIC}{\partial t} = \left.\frac{\partial sDIC}{\partial t}\right|_{ex} + \left.\frac{\partial sDIC}{\partial t}\right|_{ncp} + \left.\frac{\partial sDIC}{\partial t}\right|_{trsp} + \left.\frac{\partial sDIC}{\partial t}\right|_{diff} + \left.\frac{\partial sDIC}{\partial t}\right|_{ent} \tag{8.3.29}$$

Near Bermuda, air-sea gas exchange can explain a fraction of this drawdown, as oceanic pCO_2 is greater than atmospheric pCO_2 for most of this period (see figure 8.3.8e). However, estimates of the CO_2 evasion during this period are an order of magnitude too small to explain the drawdown [*Michaels et al.*, 1994, *Gruber et al.*, 1998]. Near Hawaii, oceanic pCO_2 is actually undersaturated for most of the year (see figure 8.3.9e), therefore air-sea gas exchange would tend to increase *sDIC* from spring to fall.

Therefore, other processes, either of biological origin or of a physical nature, such as transport and mixing, must create the seasonal reduction in *sDIC*. We learned in chapter 4, on the basis of several lines of evidence, that the *sDIC* drawdown near Bermuda is most likely caused by biology and not by physical processes. One such line of evidence is the concomitant observation of an increase in the $^{13}C/^{12}C$ ratio of *DIC* (figure 8.3.8c). Plants strongly prefer the light isotope ^{12}C over the heavier isotope ^{13}C during the photosynthetic uptake of CO_2, leading to isotopically light organic matter (low $\delta^{13}C$ values), whereas the remaining pool of inorganic carbon becomes isotopically heavier (high $\delta^{13}C$). This biological fractionation is substantially larger than the fractionation by any other process, giving the opportunity to use observations of $\delta^{13}C$ as an indicator of the biological uptake and release of CO_2.

Figure 8.3.11 shows the results of two studies that used the concurrent $^{13}C/^{12}C$ and *DIC* observations as well as other observations to diagnose the relative contribution of the different processes affecting the *sDIC*

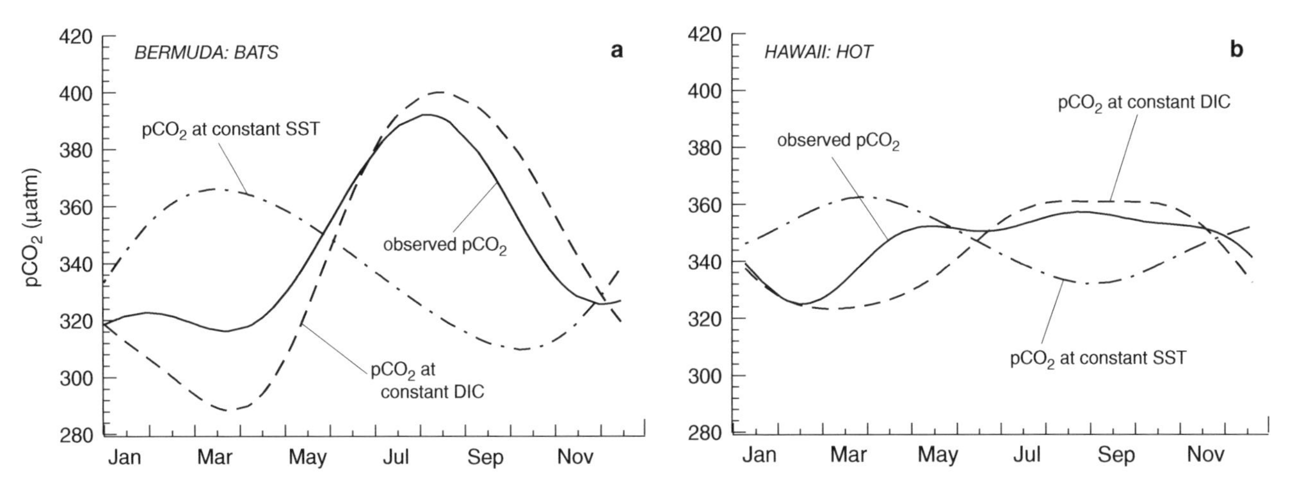

FIGURE 8.3.10: The role of variations in SST and DIC in driving seasonal variations in surface ocean pCO_2 in the subtropical gyres: (a) results for the BATS site, and (b) results for the HOT site. The solid line shows the results of a harmonic fit to the observed seasonal variations of pCO_2. The dashed line shows the expected pCO_2 variations if *sDIC* is held constant at its annual mean of 1954 μmol kg^{-1} (BATS) and 1971 μmol kg^{-1} (HOT). The dash-dotted line depicts the expected pCO_2 variations in case SST is held constant at its annual mean of 23.1°C (BATS) and at 24.8°C (HOT).

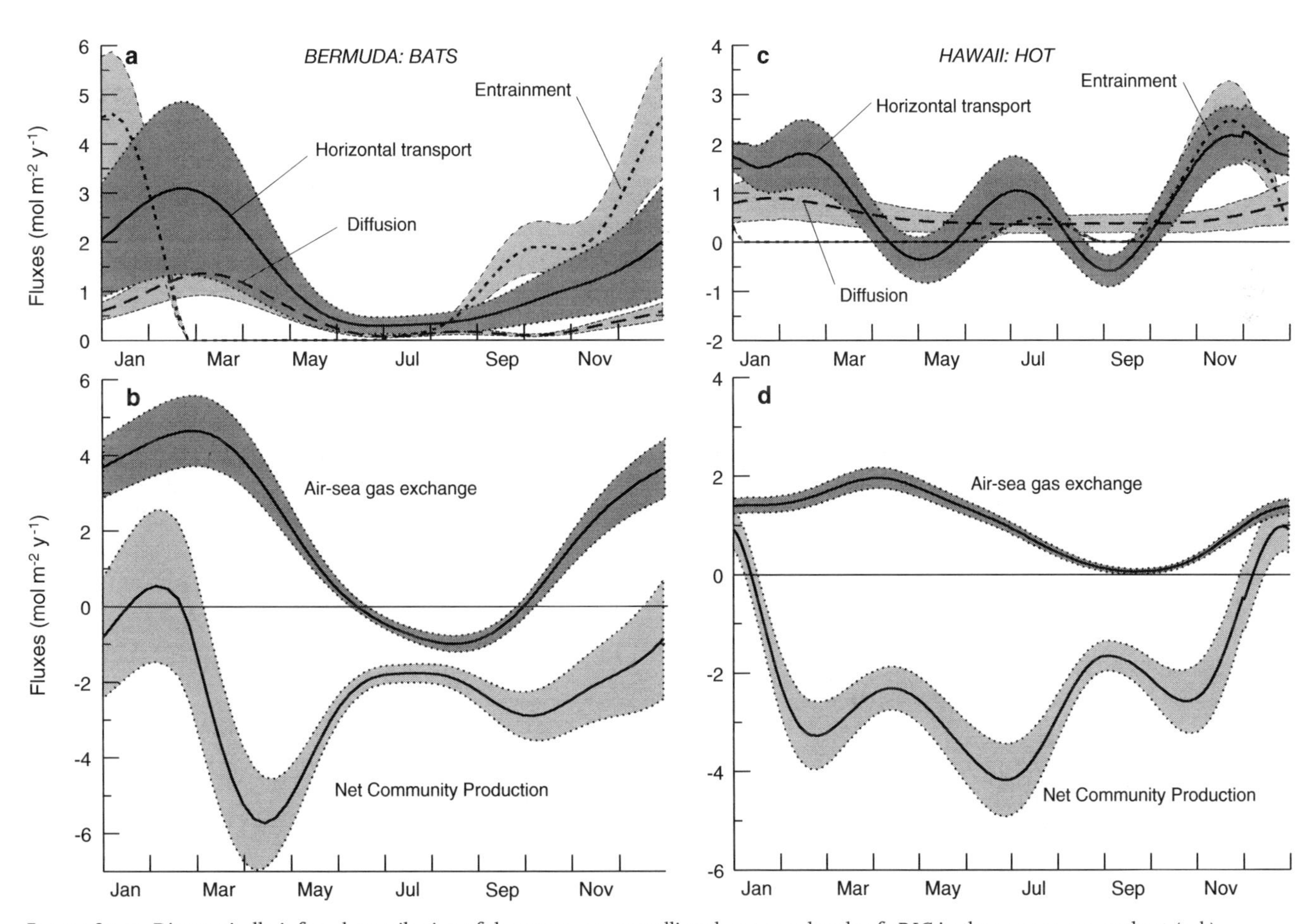

FIGURE 8.3.11: Diagnostically inferred contribution of the processes controlling the seasonal cycle of *sDIC* in the upper ocean cycle at (a–b) BATS and (c–d) HOT. The solid lines denote the computed curve for the standard set of parameters, and the dotted lines denote the upper and lower limit of the uncertainty intervals as evaluated from Monte Carlo simulations. Adapted from *Gruber et al.* [1998] and *Keeling et al.* [2004].

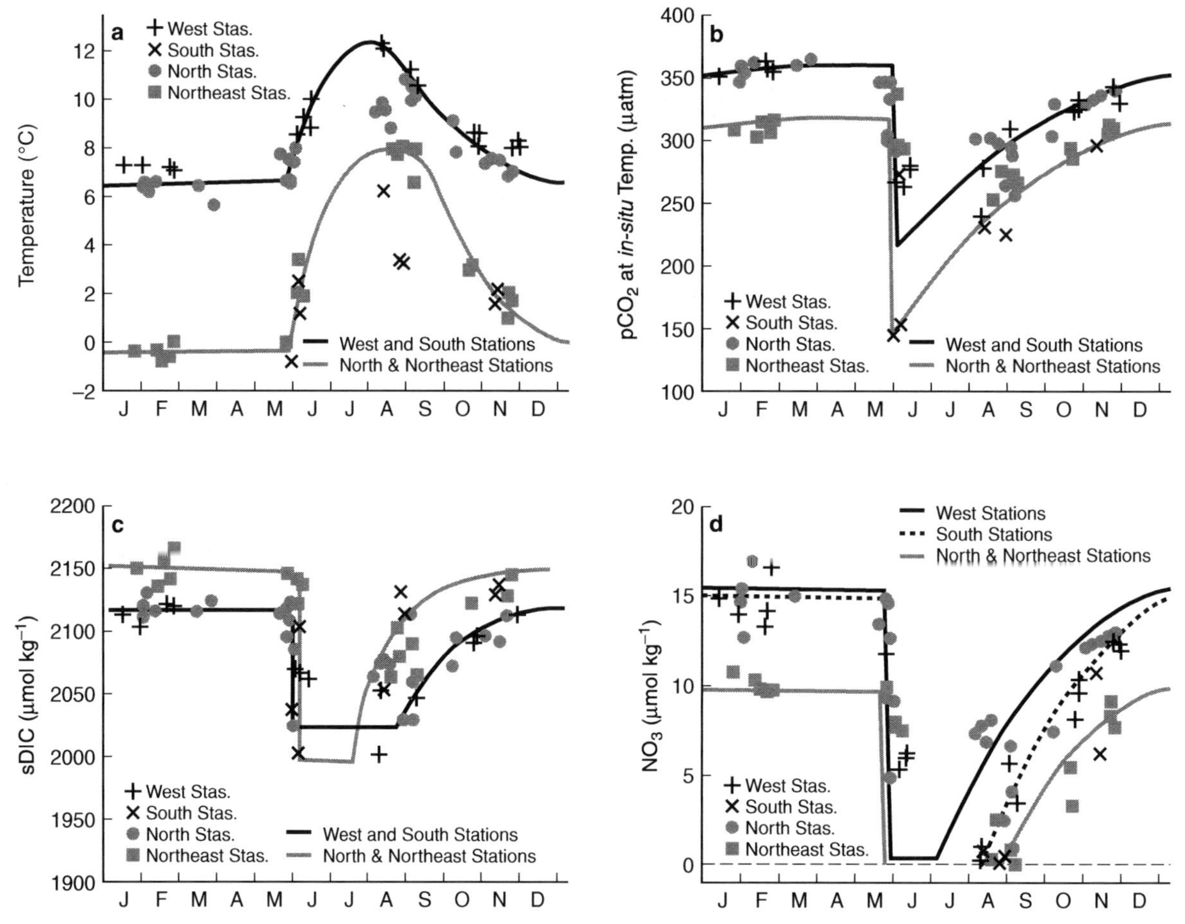

FIGURE 8.3.12: Seasonal observations of upper ocean quantities made at four groups of station around Iceland within approximately 120 miles from shore: (a) surface ocean temperature; (b) pCO_2 in surface water at *in situ* temperature; (c) *sDIC*; (d) nitrate. The curves indicate a general seasonal trend. Adapted from *Takahashi et al.* [1993].

evolution in the mixed layer at BATS and HOT; i.e., they determined each term in (8.3.29) [*Gruber et al.*, 1998, *Keeling et al.*, 2004] (see also studies by *Marchal et al.* [1996], *Bates et al.* [1996b] and *Quay and Stutsman* [2003]). According to these diagnostic analyses, the seasonal cycle of *sDIC* at both locations is a consequence of a complex interplay between the various processes, with different processes dominating at different times of the year. Both studies show also that the spring-to-summer drawdown of *sDIC* at both sites is primarily of biological origin, i.e., a result of net community production. The reasons why biology is able to have a positive net community production in the absence of any measurable nutrients is still somewhat of a mystery. In chapter 4, we suggested that N_2 fixation is the most likely candidate process (see also chapter 5), although we could not provide a satisfactory answer to the problem of where the required phosphorus would come from, particularly not in the case of the North Atlantic, where near-surface concentrations of phosphate are much lower than those of nitrate (see figure 5.1.4 and *Wu et al.* [2000]). Vertical migration has been suggested [*Karl et al.*, 1992], but has not yet been demonstrated to be of quantitative importance. This is a wonderful illustration of how much there is still to learn about ocean biogeochemistry.

NORTH ATLANTIC

The high latitudes of the North Atlantic show smaller winter-to-summer warming than the subtropical latitudes to the south. Nevertheless, if there were no seasonal changes in *DIC* and *Alk*, this warming would still lead to winter-to-summer increases in pCO_2 of more than 40 μatm (figure 8.3.7b). In contrast, the observations reveal that oceanic pCO_2 decreases from winter to summer. Since salinity-normalized alkalinity mostly behaves conservatively without a clearly defined seasonal cycle over the entire North Atlantic [*Brewer et al.*, 1986, *Millero et al.*, 1998], this decrease in pCO_2 must be caused by a sharp reduction of *sDIC*.

Indeed, data from time series stations around Iceland (figure 8.3.12) and from other stations in the North Atlantic reveal precipitous drops of *sDIC* occurring in spring [*Peng et al.*, 1987; *Takahashi et al.*, 1993]. At the Iceland sites, the observed *sDIC* drop of

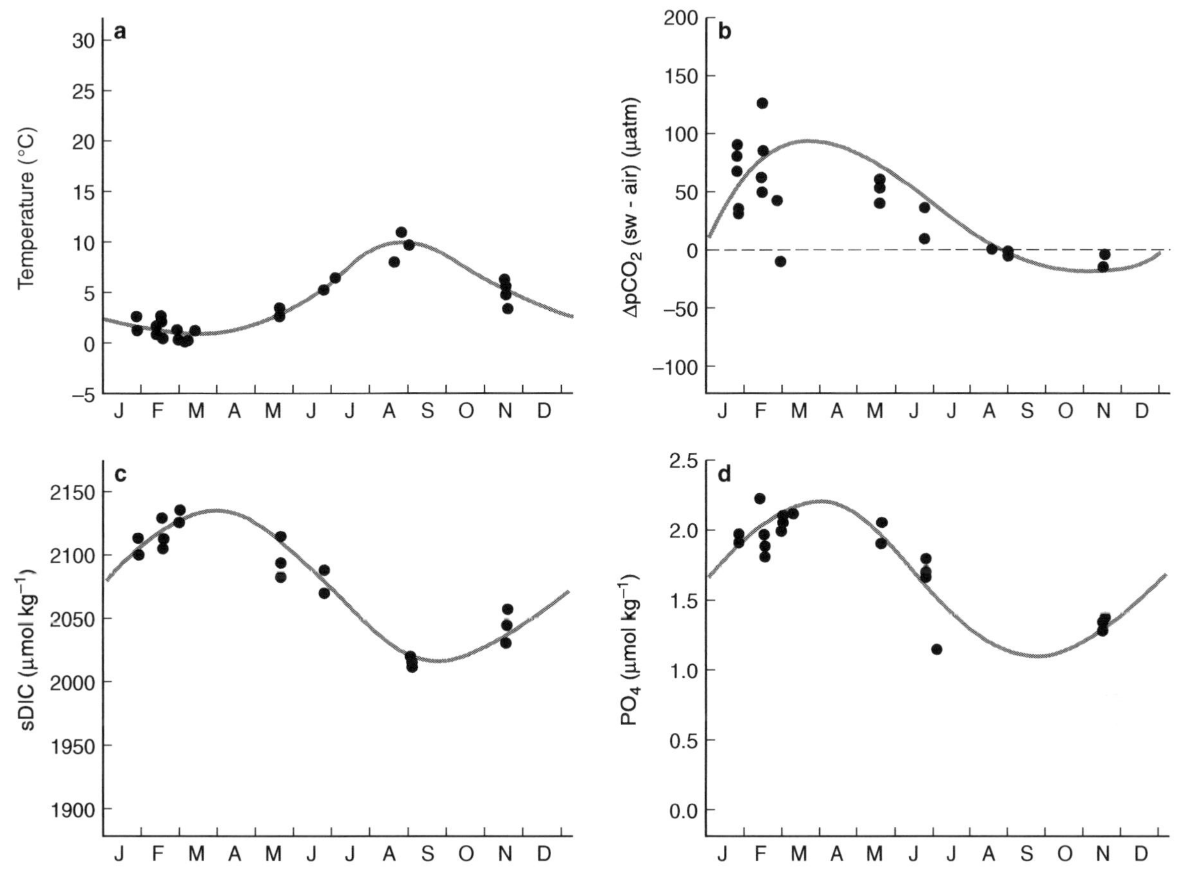

FIGURE 8.3.13: Seasonal changes of (a) temperature, (b) sea-air pCO_2 difference, (c) *sDIC*, and (d) phosphate observed in the surface mixed layer waters of the northwestern subarctic Pacific (50°N–54°N; 162°E–170°E). Circles represent actual observations, while the line was added to emphasize a general seasonal trend. Adapted from *Takahashi et al.* [1993].

>100 μmol kg^{-1} reduces pCO_2 by more than 150 μatm. This is enough to compensate for the warming effect. The drop of *sDIC* in spring is accompanied by equally precipitous nutrient decreases in nearly standard stoichiometric ratios. Thus, the *sDIC* drop likely can be attributed to the phytoplankton spring bloom typical of the northern North Atlantic (see discussion in chapter 4). In conclusion, in the high latitudes of the North Atlantic, the seasonal changes of *sDIC* and pCO_2 are dominated by biological processes, with temperature playing a lesser role.

North Pacific

Seasonal changes in SST are somewhat larger in the North Pacific than in the North Atlantic, largely because North Atlantic SSTs do not get as cold in winter due to the large northward heat transport in this basin. This leads to a stronger winter-summer thermal forcing on pCO_2, with values reaching well above 100 μatm (figure 8.3.7b) across a wide swath of the high-latitude North Pacific. Nevertheless, as was the case in the high-latitude North Atlantic, the surface ocean warming from winter to summer is accompanied by a reduction in surface ocean pCO_2.

Figure 8.3.13 shows the seasonal evolution of the upper ocean carbon cycle in the northwestern subarctic Pacific in more detail (region from 50°N to 54°N, 162°E to 170°E). Here, pCO_2 and the concentrations of *sDIC* and nutrients are highest in winter and lowest during summer. *Takahashi et al.* [1993] attributed this to upwelling and entrainment of deep waters during winter and biological uptake in summer. Therefore, as in the North Atlantic, the *sDIC* reduction by biological uptake in spring and summer is large enough to outweigh the warming effect on pCO_2.

It is intriguing to compare the seasonal pattern of *sDIC* in the subarctic North Pacific with that in the northern North Atlantic. The precipitous drop in *sDIC* and nutrients observed in springtime in the northern North Atlantic is absent in the subarctic Pacific. Why is this the case? Since we attributed the observed drawdowns in both regions to the biological uptake of *sDIC*, these differences must be caused by differences in the seasonal evolution of phytoplankton activity. We have seen in chapter 4 that the North Atlantic supports a large spring bloom, whereas the development of such a bloom is strongly attenuated in the North Pacific (see figure 4.3.14) for reasons not completely understood. We have mentioned zooplankton grazing and iron limitation as the two main hypotheses, with evidence accumulating that iron is the lead cause. This thus represents a remarkable example of the impact of iron limitation of phytoplankton on the upper ocean carbon cycle.

8.4 Water Column

We learned in the introduction that the surface concentration of *sDIC* is lower than the deep ocean concentration by about 15% (figure 8.1.2). This vertical gradient has major consequences for atmospheric carbon dioxide. If the ocean were to be mixed uniformly, models show that atmospheric CO_2 would climb by more than 50% as previously noted. It is thus of considerable importance to understand what "gradient makers" act to maintain the *sDIC* gradient in the face of ocean circulation and mixing, which are continuously trying to weaken and alter these gradients.

We have identified two biological processes that contribute toward creating the vertical gradient of *sDIC*: the *soft-tissue pump* and the *carbonate pump*. We have also seen that air-sea gas exchange, responding to changes in surface ocean pCO_2, which in turn are driven by changes in temperature, *DIC*, and *Alk* variations, can also lead to variations in oceanic *sDIC*. We will refer to this process as the *gas exchange pump*, although the word "pump" is not really appropriate here, since this process acts only in the surface ocean. It is, in fact, ocean circulation and mixing that takes these gas exchange–induced surface ocean variations and transports/mixes them into the ocean interior, leading then to vertical gradients in *sDIC*. Note that our definition of the *gas exchange pump* is different from the classical definition of the *solubility pump* by *Volk and Hoffert* [1985]. As we will demonstrate below, the solubility pump is a component of the gas exchange pump.

We expect on the basis of the strong covariation between *sDIC* and phosphate (figure 8.1.4a) that the soft-tissue pump likely plays a major role. However, figure 8.1.4b also reveals that variations in *sDIC* roughly track a trend versus temperature that is expected from the temperature sensitivity of the equilibrium *sDIC* concentration for a constant atmospheric CO_2 concentration. Clearly, both processes, soft-tissue pump and temperature variations driving the gas exchange pump, could by themselves explain the majority of the *sDIC* variations, but they cannot do so simultaneously. Which one is dominating, and what is the role of the third process, i.e., the carbonate pump?

Outline

We will tackle this question by breaking the *sDIC* observations down into individual components that can be identified with the three different pumps. We introduce this breakdown by going through a sequence in which we separately consider the distribution if only one or two of these processes are simultaneously active. We will see that we can associate the soft-tissue pump directly with variations in phosphate, and the carbonate pump with variations in nitrate-corrected alkalinity, giving us then the gas exchange component as a residual from the observed *sDIC* distribution once the anthropogenic CO_2 component has been subtracted.

We then discuss the distribution of each component in turn, noting that the distribution of the soft-tissue pump harbors few suprises, while the carbonate pump component reveals some interesting new features. The most interesting component is the gas exchange component, which shows a completely unexpected distribution. We attempt to unravel the reasons behind this puzzling pattern, first looking at the global distribution only, and then analyzing regional differences. We will see that the key to understanding this distribution is the fact that heating- and cooling-induced air-sea CO_2 fluxes and biologically induced air-sea CO_2 fluxes often oppose each other. The resulting CO_2 flux pattern is further modified by the relatively long timescale of air-sea exchange of CO_2, so that surface waters are seldom in equilibrium with atmospheric CO_2. This causes a substantial difference between the "potential" gas exchange pump, i.e., the oceanic *sDIC* pattern that would exist if surface waters were fully equilibrated, and the actually observed gas exchange pump. In section 8.5, we connect the ocean interior distribution of the gas exchange component with surface fluxes, permitting us to investigate in greater detail the consequences of the opposing trends of heat fluxes and biology on the gas exchange pump and how kinetic limitations alter the flux pattern.

Pump Components

We start our pump separation with a thought experiment using a two-dimensional, i.e., latitude-depth model of the ocean, but focus on the mean vertical gradient of *sDIC* only. We then use the insights from this thought experiment to define the pump components more rigorously. The approach we follow here is explained in more detail in *Gruber and Sarmiento* [2002].

Our first experiment considers an ocean that has no gas exchange and no biology, but a realistic temperature distribution. As the inhibition of air-sea gas exchange prevents the expression of the temperature-dependent solubility, such an ocean would have uniform *DIC* concentration, except for small variations induced by the addition or removal of freshwater (figure 8.4.1a). These latter variations are largely eliminated by our using salinity-normalized *DIC* for our analysis.

If we make the ocean isothermal, i.e., let it have constant temperature, and add biology to it, while still not permitting air-sea gas exchange to occur, *sDIC* will begin to show variations. In a case where we just have organic matter production by phytoplankton in the upper ocean followed by export and remineralization of this material at depth, *sDIC* would be directly proportional to the distribution of phosphate, with a slope that is equal to the C:P stoichiometric ratio of organic matter, $r_{C:P}$ (figure 8.4.1d). We can therefore identify the soft-tissue pump in the absence of gas exchange, ΔC_{soft}, by looking at deviations of phosphate relative to a constant reference value, $[PO_4^{3-}]^{ref}$, which we set equal to the mean phosphate concentration in the surface ocean:

$$\Delta C_{soft} = r_{C:P} \cdot ([PO_4^{3-}] - [PO_4^{3-}]^{ref}) \qquad (8.4.1)$$

Similarly, if the formation and dissolution of biogenic calcium carbonate were the only process affecting *DIC* in an isothermal ocean without air-sea gas exchange, variations of *sDIC* would be coupled entirely to variations in *sAlk*, with a slope of 1:2 (figure 8.4.1g). We can thus identify the contribution of the carbonate pump in the absence of gas exchange, ΔC_{carb}, by studying deviations of *sAlk* relative to a constant background value, $sAlk^{ref}$, chosen here to be the global mean *sAlk* in the surface ocean. In the real ocean, *sAlk* is also affected by the soft-tissue pump, and we therefore have to add a small correction that is proportional to nitrate:

$$\Delta C_{carb} = \frac{1}{2}(sAlk - sAlk^{ref} + [NO_3^-] - [NO_3^-]^{ref}) \qquad (8.4.2)$$

$$= \frac{1}{2}(sAlk - sAlk^{ref} + r_{N:P} \cdot ([PO_4^{3-}] - [PO_4^{3-}]^{ref})) \qquad (8.4.3)$$

where we made use of the very tight coupling between the cycling of nitrate and phosphate (see discussion in chapter 5) to write our separation in terms of phosphate only. Therefore, the term $r_{N:P} \cdot ([PO_4^{3-}] - [PO_4^{3-}]^{ref})$ accounts for the impact of the soft-tissue pump on *Alk*. The sign of this correction term is positive, since the remineralization of organic matter decreases *Alk*. *Brewer et al.* [1975] introduced the term "potential alkalinity" for this nitrate corrected alkalinity, i.e., $P_{Alk} = Alk + NO_3^-$. Therefore, except for the salinity normalization, ΔC_{carb} is equal to the difference between observed P_{Alk} and a constant reference P_{Alk}^{ref}.

Now, let us permit gas exchange with the atmosphere to occur. In our first case of an abiological ocean, the exchange of CO_2 across the air-sea interface permits the inorganic CO_2 system in the surface ocean to work toward an equilibrium with atmospheric CO_2. As a result, waters that cool as they are transported into higher latitudes will tend to take up CO_2 from the atmosphere, thereby getting enriched in *sDIC*. Waters that warm up as they are transported in the opposite direction lose CO_2 and consequently have lower *sDIC* concentrations. As the high latitudes constitute the source regions for the deep ocean, a vertical gradient of *sDIC* will develop (figure 8.4.1b). The magnitude of this vertical gradient depends on how rapidly air-sea gas exchange can supply the CO_2 necessary to bring high-latitude surface waters that sink into the abyss close to saturation (figure 8.4.1c).

The effect of air-sea gas exchange on the *sDIC* distribution in an isothermal, biological ocean is opposite to that in the abiological ocean discussed above (except for the small carbonate pump effect discussed later; figure 8.4.1e). Gas exchange will tend to remove carbon from the high-latitude outcrops of the carbon-rich deep waters, and add it to the lower-latitude surface waters [*Gruber and Sarmiento*, 2002; *Toggweiler et al.*, 2003]. As a result, the vertical gradient of *sDIC* produced by biological processes will be reduced in the presence of air-sea exchange. Again, the magnitude of this reduction in the gradient depends on how fast air-sea exchange is removing CO_2 from the surface relative to the removal by other processes, such as biological uptake (figure 8.4.1f).

Thus, the impact of air-sea exchange on the distribution of *sDIC* is twofold and includes a thermally driven as well as a biologically driven component. We refer to the sum of these two effects as the *gas exchange pump*:

$$\Delta C_{gas\ ex} = \Delta C_{gas\ ex}^{bio} + \Delta C_{gas\ ex}^{therm} \qquad (8.4.4)$$

Note that the commonly used term "solubility pump" [*Volk and Hoffert*, 1985] refers only to the thermal component of the gas exchange pump, i.e., $\Delta C_{gas\ ex}^{therm}$, i.e. those variations in *sDIC* that are driven only by temperature variations and ocean transport and mixing. Unfortunately, the gas exchange pump cannot be estimated directly from observations. Therefore, we estimate this pump by subtracting the two biological pumps from the observations, while taking into account that the invasion of anthropogenic CO_2 from the atmosphere has already altered the upper ocean distribution of *sDIC*. We have developed the means to estimate this contribution, C_{ant}, directly from observations [*Gruber et al.*, 1996] (see chapter 10), and subtract it from the observed *sDIC* distribution to reconstruct a preindustrial *sDIC* distribution. We thus estimate $\Delta C_{gas\ ex}$ by

$$\Delta C_{gas\ ex} = sDIC - sDIC^{ref} - C_{ant} - \Delta C_{soft} - \Delta C_{carb} \qquad (8.4.5)$$

where DIC^{ref} is a constant reference concentration chosen here as the mean surface ocean *sDIC* in preindustrial times. Note that this reference concentration and those used in the definitions of ΔC_{soft} and ΔC_{carb}

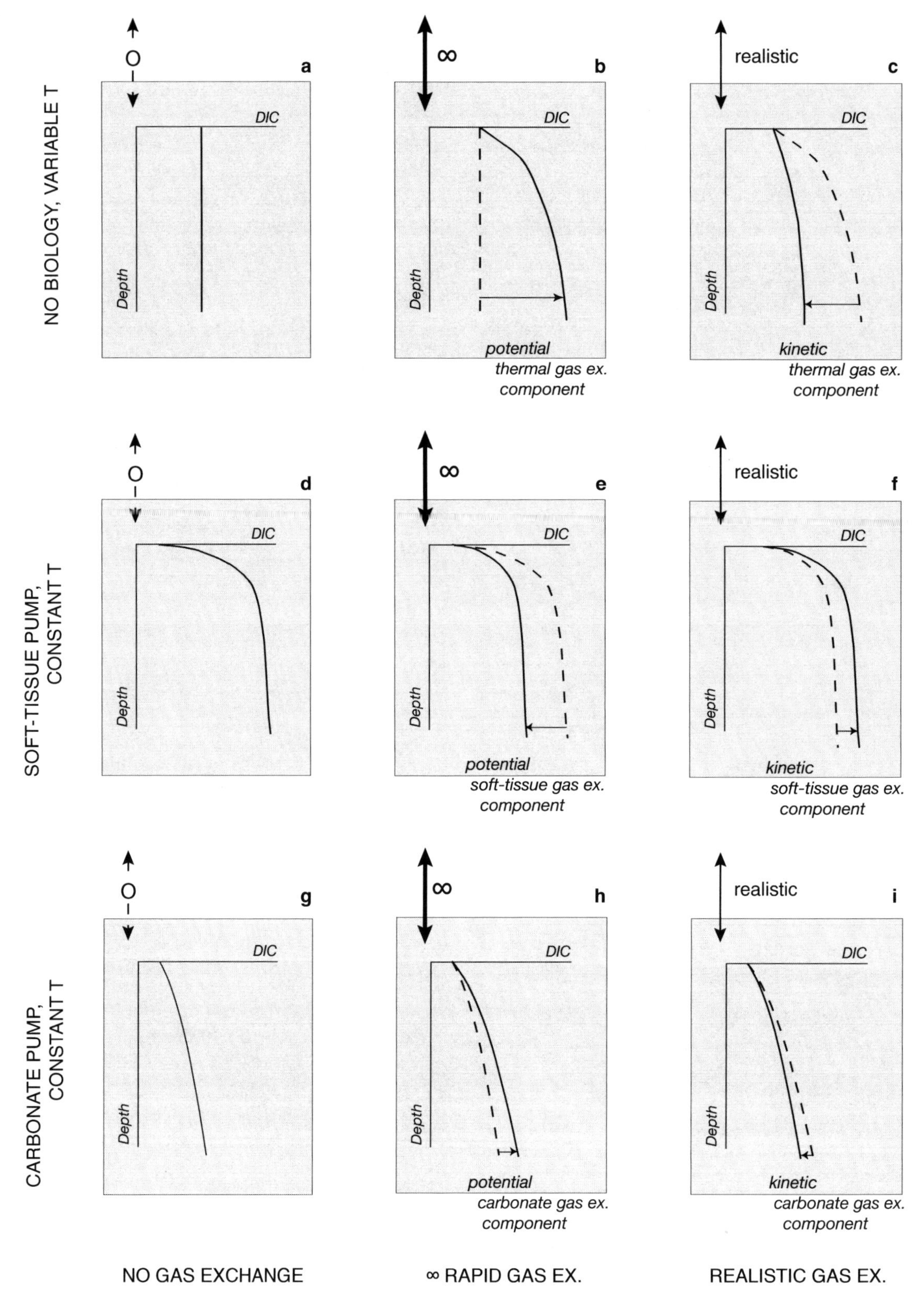

FIGURE 8.4.1: Hypothetical vertical profiles of *sDIC* in a series of thought experiments to elucidate the contributions of the various pump components. The first row depicts the *sDIC* distribution in an abiological ocean with realistic temperature distribution and various gas exchange scenarios. The second row shows the *sDIC* distributions in a biological ocean with just the soft-tissue pump operating and with fixed temperature distribution under various gas exchange scenarios. The third row is as the second, but with only the carbonate pump being active. The potential gas exchange pump components are the changes in the *sDIC* distributions that occur when one goes from an ocean without air-sea gas exchange to one with infinitely fast gas exchange. The kinetic gas exchange components are then the changes in *sDIC* that are caused by replacing the infinitely fast gas exchange with a realistic one. The main point to note is that the thermal and biological gas exchange pump components have a tendency to operate in opposite directions. This is the main reason why the gas exchange pump component is relatively small.

were only added for the convenience of setting the global mean surface concentration of each of the three pump components to zero. This is an arbitrary choice and has no influence on the interpretation, since the information of interest is contained in the gradients of these pump components from one place to another.

It needs to be emphasized that ΔC_{soft} and ΔC_{carb} refer only to the impact of ocean biology on *sDIC* in the absence of air-sea gas exchange. Therefore, these two components can be viewed as "potential" pumps. They differ from the common definition of these biological pumps, as the latter include also the impact of biology on the air-sea gas exchange component, i.e., $\Delta C^{bio}_{gas\ ex}$ [*Volk and Hoffert*, 1985; *Toggweiler et al.*, 2003a, b]. If we were able to estimate $\Delta C^{\mathrm{bio}}_{\mathrm{gas\ ex}}$, we could add it to ΔC_{soft} and ΔC_{carb} and determine the complete contribution of these two processes. However, this is unfortunately not possible on the basis of observations. Below, we will use model results to undertake this split of $\Delta C_{\mathrm{gas\,ex}}$ and discuss the implications. Another argument in favor of the pump separation adopted here is that it clearly separates the influence of air-sea gas exchange from the influence of processes that only redistribute *sDIC* internally within the ocean. This can be thought of as separating the closed system response (no exchange with the atmosphere) from the open system response (including exchange with the atmosphere) to processes that lead to perturbation in the oceanic carbon cycle.

We will now discuss the contribution of each of these pumps to the observed distribution of *sDIC* in turn, beginning with the soft-tissue pump. The major focus of the discussion will be on the contribution of the gas exchange pump to the *sDIC* distribution, which holds interesting surprises.

The Biological Pumps

Figure 8.4.2 reveals that the soft-tissue pump in the absence of air-sea exchange, ΔC_{soft}, is responsible for about 215 μmol kg^{-1} (70%) of the surface-to-deep gradient of preindustrial *sDIC* of about 305 μmol kg^{-1}. The carbonate pump, ΔC_{carb}, accounts for approximately 60 μmol kg^{-1} (20%) of the observed gradient, and the gas exchange pump, $\Delta C_{gas\ ex}$, for the remainder.

The dominance of the soft-tissue pump in creating the vertical gradient in preindustrial *sDIC* is not surprising, since we have seen this dominance already expressed in the strong covariation between *sDIC* and phosphate (figure 8.1.4a). The substantially smaller contribution of the carbonate pump will be confirmed in chapter 9, where we will show that the downward transport of carbon as $CaCO_3$ is about 10 times smaller than the downward transport of organic matter [*Sarmiento et al.*, 2002]. Figures 8.4.3a and b show the contribution of the biological pumps in the different ocean basins. Again, the soft-tissue and carbonate pump contributions exhibit

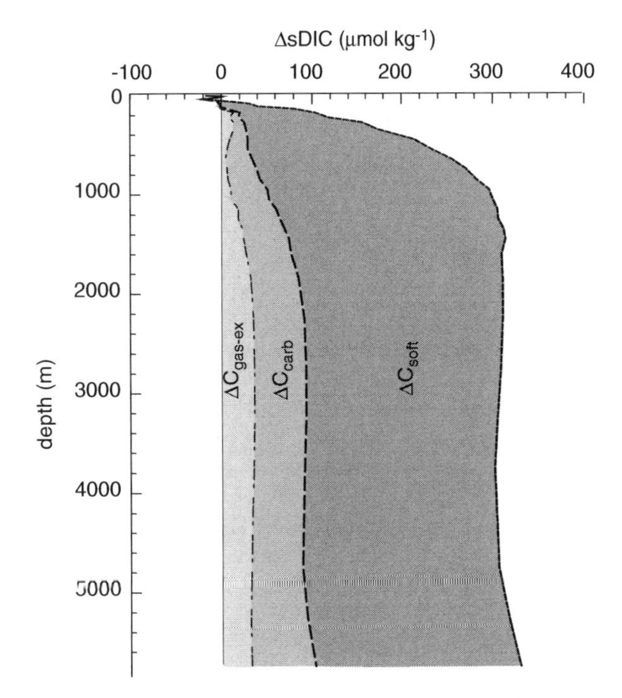

Figure 8.4.2: Global mean profiles of the three main carbon pumps. The data are plotted cumulatively. Based on GLODAP data [*Key et al.*, 1994].

the expected pattern of increasing contributions of these pumps in the deep waters along the flow path of the lower limb of the global-scale overturning circulation (figure 2.4.11).

A more detailed inspection of figure 8.4.3 reveals that the two biological pumps have quite different vertical structures. Almost the entire surface-to-deep gradient in the soft-tissue component, ΔC_{soft}, occurs in the main thermocline, with little additional contribution below 1000 m. By contrast, the contribution of the carbonate pump component, ΔC_{carb}, to the variability of preindustrial *sDIC* in the main thermocline is relatively small (<40μmol kg^{-1}). However, ΔC_{carb} increases slowly with depth below 1000 m and reaches a maximum at depths between 2000 to 6000 m. This is significantly deeper than the maximum of ΔC_{soft}, which occurs at about 1000 m.

The difference between ΔC_{soft} and ΔC_{carb} is particularly striking in the Atlantic basin, where ΔC_{carb} gradually increases all the way to the bottom, whereas ΔC_{soft} shows a pronounced maximum in intermediate waters at around 1000 m. The maximum in ΔC_{carb} shoals from the Atlantic to the Indian and Pacific Oceans but still occurs significantly deeper than the maximum in ΔC_{soft}. This indicates that the formation and dissolution of calcium carbonate is a process independent from the formation and remineralization of organic matter. We will discuss the cycling of calcium carbonate in more detail in the following chapter, but we can point out here already that the spatial distribution of ΔC_{carb} is

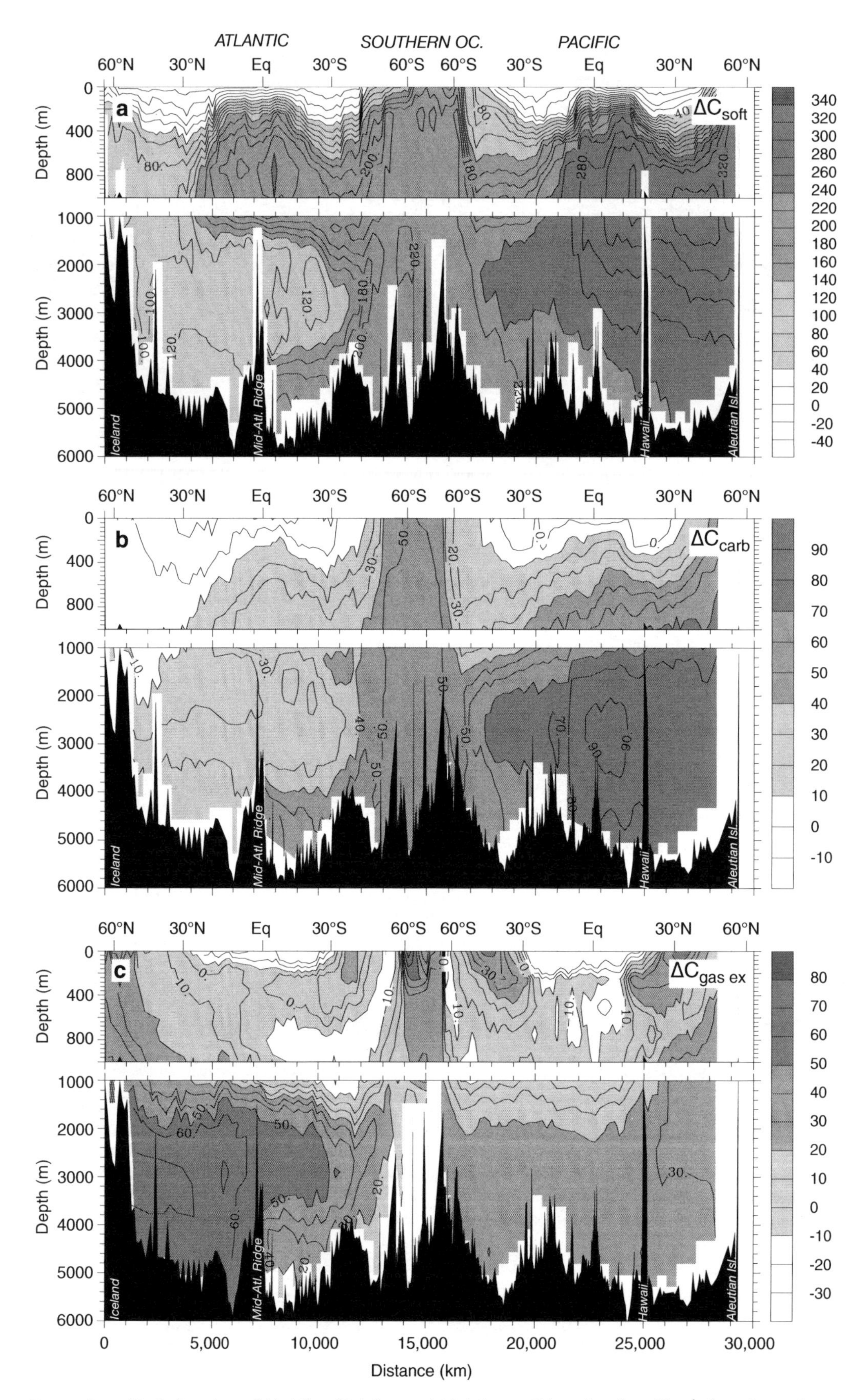

FIGURE 8.4.3: Vertical sections of (a) ΔC_{soft}, (b) ΔC_{carb}, and (c) $\Delta C_{gas\ ex}$, all in units of μmol kg^{-1} along the track shown in figure 2.3.3a. See also color plate 8.

consistent with the variation of the saturation horizons for the two major phases of calcium carbonate, calcite and aragonite. The saturation horizon for aragonite, the more soluble phase, occurs in the Atlantic at depths of about 3000 m, whereas it lies within the main thermocline in the Pacific and Indian Ocean [*Takahashi et al.*, 1981, *Broecker and Peng*, 1982, *Feely et al.*, 2002, *Chung et al.*, 2003].

THE GAS EXCHANGE PUMP

Let us turn now to the gas exchange pump. Figure 8.4.2 reveals that the contribution of the gas exchange pump to the global mean vertical profile of preindustrial *sDIC* amounts to only about 30 μmol kg^{-1} (10%). How does this compare to what one would expect on the basis of the surface-to-deep temperature gradient? We can estimate this temperature-induced gradient by combining the temperature sensitivity of pCO_2 (8.3.4) with the definition of the buffer factor (8.3.14) (see equation (10.2.31)). This gives, on average, a temperature sensitivity of *DIC* of about 8 to 9 μmol kg^{-1} per degree of temperature change (see slope in figure 8.1.4b). Therefore, with a surface-to-deep temperature gradient of about 18°C, one would expect a temperature-induced surface-to-deep gradient in *sDIC* of about 140 to 160 μmol kg^{-1}. This is several times larger than our estimate on the basis of $\Delta C_{gas\ ex}$. Such a small imprint of air-sea gas exchange on the distribution of *sDIC* is particularly surprising given the strong correlation of *sDIC* with temperature (figure 8.1.4b). Why is this case? Is this small $\Delta C_{gas\ ex}$ signal a result of the biological contribution, $\Delta C^{bio}_{gas\ ex}$, offsetting the thermal component, $\Delta C^{therm}_{gas\ ex}$, or is it because of kinetic limitations, i.e., the slow exchange of CO_2 not permitting the full temperature variations in the ocean to be reflected in *sDIC*?

Consideration of $\Delta C_{gas\ ex}$ in the different ocean basins reveals additional puzzles (figure 8.4.3c). Hidden in the small global mean surface-to-deep difference of $\Delta C_{gas\ ex}$, we find a very large imprint of the gas exchange pump in the deep Atlantic. This indicates that these waters must have taken up a substantial amount of CO_2 from the atmosphere before they descended into the deep Atlantic. The deep Pacific and deep Indian have smaller and vertically more uniform values of $\Delta C_{gas\ ex}$. Figure 8.4.3c also shows that Antarctic Intermediate Water (AAIW), which dominates the deeper thermocline in the southern hemisphere (see salinity minimum in figure 2.4.1), has comparatively low concentrations of $\Delta C_{gas\ ex}$, suggesting that the source waters for AAIW have lost CO_2 to the atmosphere before they descended into the ocean's interior. What are the reasons for these spatial variations in $\Delta C_{gas\ ex}$, and why does the deep Atlantic have such high concentrations of $\Delta C_{gas\ ex}$ in comparison to the deep Indian and Pacific?

The latter question is linked to an important research topic in global carbon cycle research. Since deep waters in the Atlantic flow southward, these high concentrations of $\Delta C_{gas\ ex}$ provide a conduit for transporting CO_2 taken up by the ocean in the northern hemisphere into the southern hemisphere, where it is released back into the atmosphere. In a preindustrial steady-state, such an oceanic southward transport of CO_2 must have been compensated by a northward transport of CO_2 in the atmosphere. This requires the existence in preindustrial times of a south-to-north concentration gradient in atmospheric CO_2. Accurate knowledge of such a preindustrial interhemispheric gradient in atmospheric CO_2 is of prime importance for atmospheric inverse studies that attempt to determine the present sources and sinks of anthropogenic CO_2 in the global carbon cycle [*Keeling et al.*, 1989; *Tans et al.*, 1990; *Ciais et al.*, 1995; *Fan et al.*, 1998; *Gurney et al.*, 2002; *Gloor et al.*, 2003]. This is because one requires accurate estimates of all natural processes causing atmospheric CO_2 variations, before the inversions can determine where anthropogenic CO_2 is taken up currently at the surface of the Earth.

We continue our discussion of the $\Delta C_{gas\ ex}$ distribution by addressing next the puzzle of the small global mean contribution of the gas exchange pump, and consider the Atlantic gas exchange signal afterwards. As we will see, however, the underlying cause for these two patterns is the same. It is the interaction of air-sea fluxes of heat with ocean biology and large-scale ocean circulation that determines how strongly the gas exchange pump influences the oceanic *sDIC* concentrations.

GLOBAL MEAN

The first key to understanding variations in $\Delta C_{gas\ ex}$ is to recall that both heat fluxes and biological processes control the exchange of CO_2 across the air-sea interface (see (8.4.4)). The second key is that the exchange of CO_2 across the air-sea interface is slow relative to the residence time of waters near the surface. This prevents full expression of the impact of both heat fluxes and biological changes on the respective gas exchange fluxes.

It is therefore instructive to separate the gas exchange pump even further into potential components ($\Delta C^{bio,\,pot}_{gas\ ex}$, $\Delta C^{therm,\,pot}_{gas\ ex}$) that would reflect the $\Delta C_{gas\ ex}$ distribution if gas exchange were infinitely rapid, and into kinetic components ($\Delta C^{bio,\,kin}_{gas\ ex}$, $\Delta C^{therm,\,kin}_{gas\ ex}$) that describe the changes from an infinitely rapid exchange to a case with realistic air-sea gas exchange:

$$\Delta C^{bio}_{gas\ ex} = \Delta C^{bio,\,pot}_{gas\ ex} + \Delta C^{bio,\,kin}_{gas\ ex} \tag{8.4.6}$$

$$\Delta C^{therm}_{gas\ ex} = \Delta C^{therm,\,pot}_{gas\ ex} + \Delta C^{therm,\,kin}_{gas\ ex} \tag{8.4.7}$$

The role of each of these four gas exchange components can be understood better by returning to the thought experiments we introduced at the beginning of this section to separate ΔC_{soft}, ΔC_{carb}, and $\Delta C_{gas\ ex}$ (see figure 8.4.1). At the same time, we will attempt to estimate the magnitude of each of these components, using a combination of observations and model results.

We start again with an ocean with a realistic temperature distribution, but without biology. We have seen that without air-sea gas exchange, this ocean would have a uniform *sDIC* distribution. We then turn on infinitely rapid air-sea exchange and let the ocean fully equilibrate with a fixed atmospheric pCO_2 of 280 μatm (figure 8.4.1b). The difference between this case and the no-gas-exchange case is the contribution of the potential $\Delta C_{gas\ ex}^{therm}$ pump (also called "potential solubility pump" [*Murnane et al.*, 1999]). The potential $\Delta C_{gas\ ex}^{therm}$ component can be estimated directly from data by computing the *sDIC* concentration in equilibrium with a preindustrial atmospheric pCO_2 of 280 μatm and assuming constant salinity-normalized total alkalinity. The resulting mean global surface-to-deep gradient of $\Delta C_{gas\ ex}^{therm,\ pot}$ amounts to about 155 μmol kg^{-1} (figure 8.4.4).

What happens if we then scale down air-sea gas exchange to realistic values? As it takes about 6 months to equilibrate *DIC* in a 40 m-deep surface layer with the atmosphere (see discussion in section 8.3 above), and typical residence times of surface waters are between a few days to a few years, the CO_2 system in surface waters generally does not achieve equilibration with the atmosphere. Therefore, the potential thermal gas exchange pump cannot be achieved fully (figure 8.4.1), resulting in a reduced vertical *sDIC* gradient (figure 8.4.4), with the difference being the kinetic $\Delta C_{gas\ ex}^{therm}$ effect. Unfortunately, we cannot estimate the kinetic $\Delta C_{gas\ ex}^{therm}$ component directly from observations, but simulations by *Murnane et al.* [1999] indicate that the slow kinetics of CO_2 exchange reduces $\Delta C_{gas\ ex}^{therm}$ by 90 μmol kg^{-1} (see figure 8.4.4), suggesting that the slow kinetics of air-sea gas exchange plays an important role. *Toggweiler et al.* [2003] pointed out, however, that this particular model may tend to overestimate the kinetic effect and that the true thermal $\Delta C_{gas\ ex}$ gradient might be somewhat closer to the potential thermal $\Delta C_{gas\ ex}$ gradient, $\Delta C_{gas\ ex}^{therm,\ pot}$.

We use our second and third thought experiments with an isothermal ocean and fully operational soft-tissue and carbonate pumps to elucidate the influence of biology on air-sea gas exchange (figure 8.4.1). We have seen above that, in the absence of air-sea gas exchange, the interior ocean distribution of *sDIC* would be directly proportional to either phosphate (soft-tissue pump) or potential alkalinity (carbonate pump). If we turn on infinitely rapid air-sea gas exchange, surface waters will become fully equilibrated with a fixed atmosphere of 280 μatm.

In the case of the soft-tissue pump, gas exchange in the high latitudes, where the elevated *sDIC* waters of the deep ocean come to the surface, would cause a loss of CO_2 to the atmosphere, depleting the waters in *sDIC* relative to a case without gas exchange. Conversely, gas exchange in the low latitudes would lead to an uptake of CO_2 from the atmosphere, as the biological fixation of CO_2 creates a deficit of CO_2 relative to the atmosphere. Both changes lead to a reduction of the surface-to-deep gradient. We can estimate the resulting surface-to-deep gradient in *sDIC* from O_2, as the cycling of CO_2 in the rapid gas exchange case would follow very closely that of O_2, except for a small deviation caused by surface variations in *sAlk*. The reason this works is because the surface concentration of O_2 is generally very close to saturation, as expected in a case of infinitely rapid air-sea gas exchange. In particular, in an isothermal ocean with biology and infinitely fast gas exchange, the interior distribution of *sDIC* would be directly proportional to the apparent oxygen utilization (*AOU*), with the proportionality given by the $C:O_2$ ratio of organic matter remineralization. Thus, multiplying the mean global surface-to-deep *AOU* gradient with $r_{C:O_2} = 117 : -170$ [*Anderson and Sarmiento*, 1994], and subtracting the result from ΔC_{soft}, i.e., the *sDIC* distribution that would exist in the absence of gas exchange, we arrive at a soft-tissue pump contribution to $\Delta C_{gas\ ex}^{bio,\ pot}$ of about -145 μmol kg^{-1}.

In the case of the carbonate pump, turning on infinitely fast gas exchange would lead to an effect opposite to that in the case of the soft-tissue pump. As the high latitudes in such an ocean would have very high concentrations of *sAlk* relative to *sDIC* due to the presence of a dissolution signal from the deep ocean, these waters would tend to take up CO_2 from the atmosphere. The low latitudes, in turn, would tend to lose CO_2 as the excess of $CaCO_3$ formation over dissolution drives oceanic pCO_2 above that of the atmosphere (see figure 8.3.5). Therefore, the presence of air-sea gas exchange enhances the surface-to-deep *sDIC* gradient in a case where the formation and dissolution of $CaCO_3$ is the only biological process. This enhancement can be approximated by calculating the difference between the *sDIC* concentration in equilibrium with the atmosphere for a constant salinity-normalized alkalinity and the *sDIC* equilibrium concentration for the observed preformed alkalinity (taken from *Gruber et al.* [1996]). As it turns out, this enhancement, i.e., the carbonate pump contribution to potential $\Delta C_{gas\ ex}^{bio}$, is almost an order of magnitude smaller than that of the soft-tissue pump, amounting to only about 15 μmol kg^{-1}. Adding the potential carbonate and soft-tissue pump components together, we arrive at a total contribution of $\Delta C_{gas\ ex}^{bio,\ pot}$ of -130 μmol kg^{-1} (figure 8.4.4).

When we use a realistic air-sea gas exchange instead, the gas exchange–induced reduction of the imprint of the biological pumps is smaller (figure 8.4.4). This leads to a vertical *sDIC* gradient from the biological pumps and the $\Delta C_{gas\ ex}^{bio}$ pump that is somewhere between the fully expressed biological pumps (figure

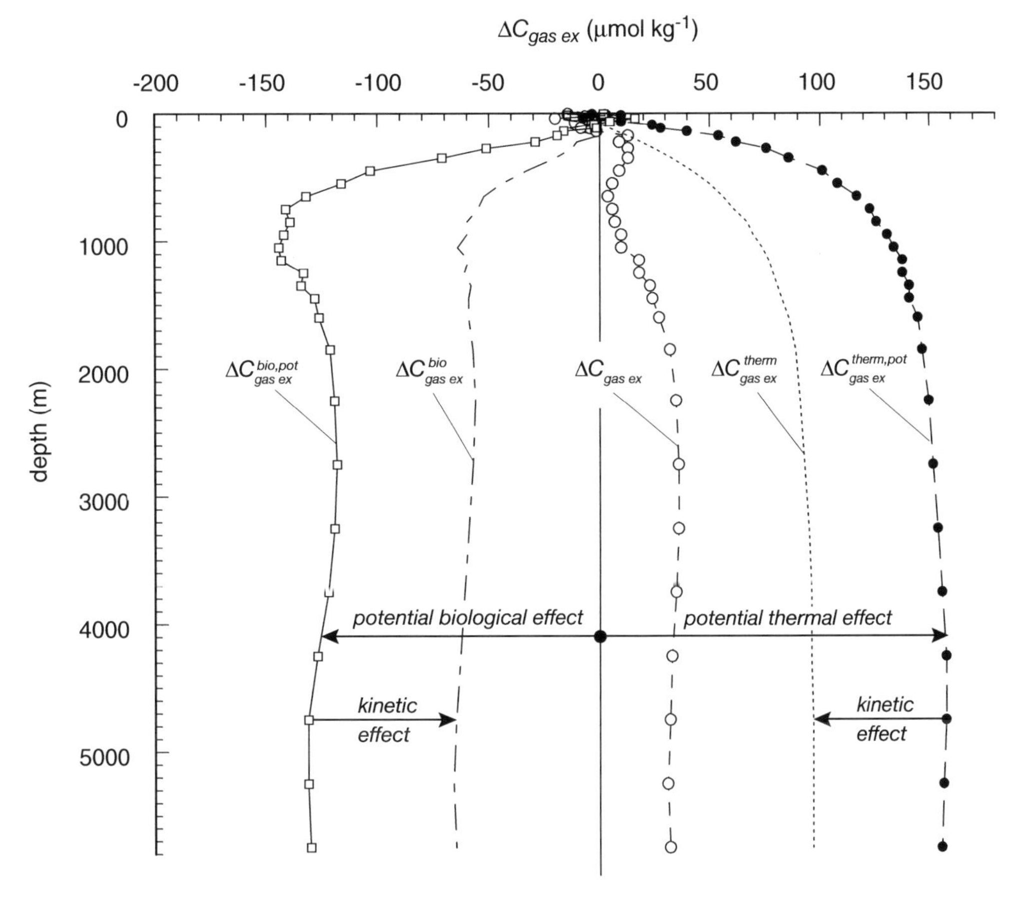

FIGURE 8.4.4: Global mean profiles of the two biological and the two thermal gas exchange pump components. The potential $\Delta C^{therm}_{gas\ ex}$ was calculated assuming that surface *sDIC* is in equilibrium with the atmosphere using a constant salinity-normalized total alkalinity. The potential $\Delta C^{bio}_{gas\ ex}$ was computed from *AOU* taking into account a change in the surface *Alk* concentration due to biology on the basis of preformed surface alkalinity estimates [*Gruber et al.*, 1996]. The kinetic $\Delta C^{therm}_{gas\ ex}$ profile is taken from the model of *Murnane et al.* [1999], and the kinetic $\Delta C^{bio}_{gas\ ex}$ component was calculated by difference.

8.4.1d) and the potential $\Delta C^{bio}_{gas\ ex}$ case (figure 8.4.1e). This means that the kinetic $\Delta C^{bio}_{gas\ ex}$ is positive, i.e., it increases the surface-to-deep gradient in *sDIC* (figure 8.4.4). As was the case for the kinetic $\Delta C^{therm}_{gas\ ex}$ contribution, it is currently not possible to estimate the kinetic $\Delta C^{bio}_{gas\ ex}$ contribution directly from data. However, if we adopt the estimated $\Delta C^{therm}_{gas\ ex}$ of *Murnane et al.* [1999], the kinetic effect on the biological component of $\Delta C_{gas\ ex}$ is substantial and amounts to around 60 μmol kg^{-1}. This results in a $\Delta C^{bio}_{gas\ ex}$ component of only about $-70\,\mu$mol kg^{-1} (figure 8.4.4).

While the exact values of the kinetic components are not known, the overall conclusion is very clear. The small imprint of air-sea gas exchange on the vertical *sDIC* distribution ($\Delta C_{gas\ ex}$) is a consequence of generally opposing tendencies of $\Delta C^{bio}_{gas\ ex}$ and $\Delta C^{therm}_{gas\ ex}$. The magnitude of these two components depends on the magnitude of biological fluxes and heat fluxes that create "potential" gas exchange fluxes, and the kinetics of air-sea exchange that determines to what degree these two potential pumps are realized. If the simulations by *Murnane et al.* [1999] are approximately correct, it appears that in the present ocean, a relatively small fraction of the potential gas exchange pumps are expressed.

This hypothesis was recently investigated in greater detail and confirmed by *Toggweiler et al.* [2003a, b]. They also highlighted that this kinetic limitation and how it is represented in different ocean models determines to a substantial degree how strongly atmospheric CO_2 responds to changes in upper ocean biology and chemistry, particularly when these changes occur in the high latitudes. We will come back to this topic in chapter 10, as it has strong implications on how we can explain the large atmospheric CO_2 variations associated with glacial-interglacial cycles in climate.

ATLANTIC VERSUS PACIFIC

The high $\Delta C_{gas\ ex}$ concentrations of the deep Atlantic in contrast to the low concentrations in the Pacific can be understood following arguments similar to those used to understand the global mean distribution of $\Delta C_{gas\ ex}$. However, it is this time more instructive to trace the

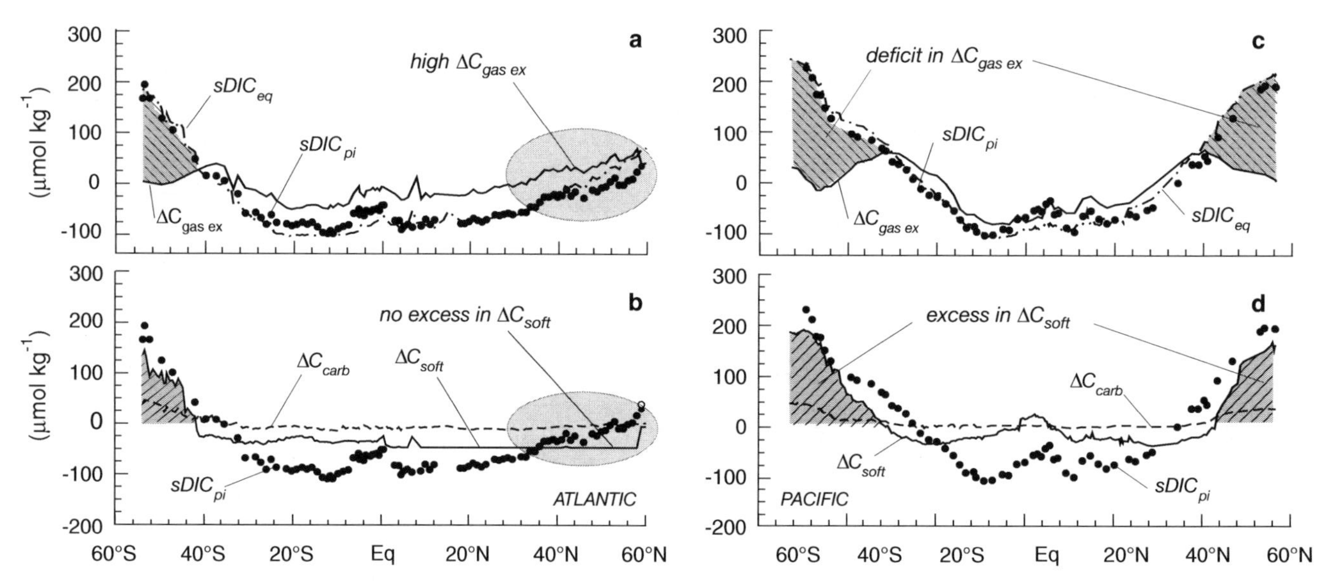

FIGURE 8.4.5: Meridional variations of the ocean carbon pumps in the surface waters (a–b) of the Atlantic and (c–d) of the Pacific. Data are from WOCE cruises A16 and P16.

different water masses back to the places where they were last in contact with the atmosphere and obtained their air-sea gas exchange signal.

The $\Delta C_{gas\ ex}$ signal in the deep Atlantic is associated with NADW, and since this water is formed in the northern North Atlantic, we have to look at surface water properties there. To better understand the particularities of the North Atlantic, it is helpful to contrast this basin with the surface distribution in the Pacific.

Figure 8.4.5 shows surface ocean concentrations of $\Delta C_{gas\ ex}$ as a function of latitude in the Atlantic and Pacific. Also shown are the equilibrium $sDIC$ concentrations and reconstructed preindustrial $sDIC$ concentrations ($sDIC_{pi}$). We subtracted a global mean surface concentration from the latter two in order to emphasize trends. The first point we note is that $\Delta C_{gas\ ex}$ shows more variations at the sea surface than we have seen in the water column. However, differences in $\Delta C_{gas\ ex}$ between the temperate latitudes, where most of the thermocline waters are formed, and the polar latitudes of the southern hemisphere, where most of intermediate and deep waters of the Indian and Pacific have their sources, are relatively small, e.g., maximally about 50 μmol kg^{-1}. This is consistent with our observation of a very small vertical gradient in $\Delta C_{gas\ ex}$, since the Indian and Pacific dominate the global mean profile.

By contrast to the Pacific, the $\Delta C_{gas\ ex}$ signal in the North Atlantic increases continuously toward the north, reaching higher levels than found in the North Pacific and in the Southern Ocean. While this explains where the signals observed in the interior come from, it does not explain why the different regions have such a different surface expression of $\Delta C_{gas\ ex}$.

If air-sea gas exchange were infinitely fast and there were no biology, $\Delta C_{gas\ ex}$ would follow the equilibrium $sDIC$ concentration $sDIC_{eq}$ shown in figure 8.4.5 (note that the equilibrium concentration is equal to $\Delta C_{gas\ ex}^{therm,\ pot}$). In such a case, waters that flow from low latitudes to high latitudes and vice versa would need to exchange large amounts of CO_2 with the atmosphere to account for the large temperature sensitivity of $sDIC$. The $\Delta C_{gas\,ex}$ in the North Atlantic Ocean has the same structure as $sDIC_{eq}$, suggesting that these waters do indeed pick up the required CO_2 from the atmosphere. However, $\Delta C_{gas\,ex}$ in the North Pacific behaves quite differently. It follows the $sDIC$ equilibrium concentration south of 40°N, but drops well below the $sDIC$ equilibrium trend to the north of this. Despite this failure to take up CO_2 from the atmosphere, the surface water $sDIC_{pi}$ is everywhere very close to $sDIC_{eq}$. Obviously, a process other than uptake of CO_2 from the atmosphere must increase the $sDIC_{pi}$ in the North Pacific. Note that the southern hemisphere high latitude waters also behave the same as the North Pacific.

The solution is revealed in the lower panels of figure 8.4.5, which show the contribution of the two biological pumps to surface ocean $sDIC$ variations. It turns out that these pumps provide the necessary mechanism to change surface ocean $sDIC$. The surface waters in the Southern Ocean contain large amounts of $sDIC$ that stem from the remineralization of organic matter and the dissolution of $CaCO_3$ (high ΔC_{soft} and ΔC_{carb}). This is a consequence of upwelling processes that bring intermediate and deep ocean waters with high ΔC_{soft} and ΔC_{carb} to the surface, and a biological pump that is relatively inefficient (see chapter 4) and therefore not able to remove these biological signals as rapidly as they are brought up.

When these waters are transported from the Southern Ocean to low latitudes, biological production removes

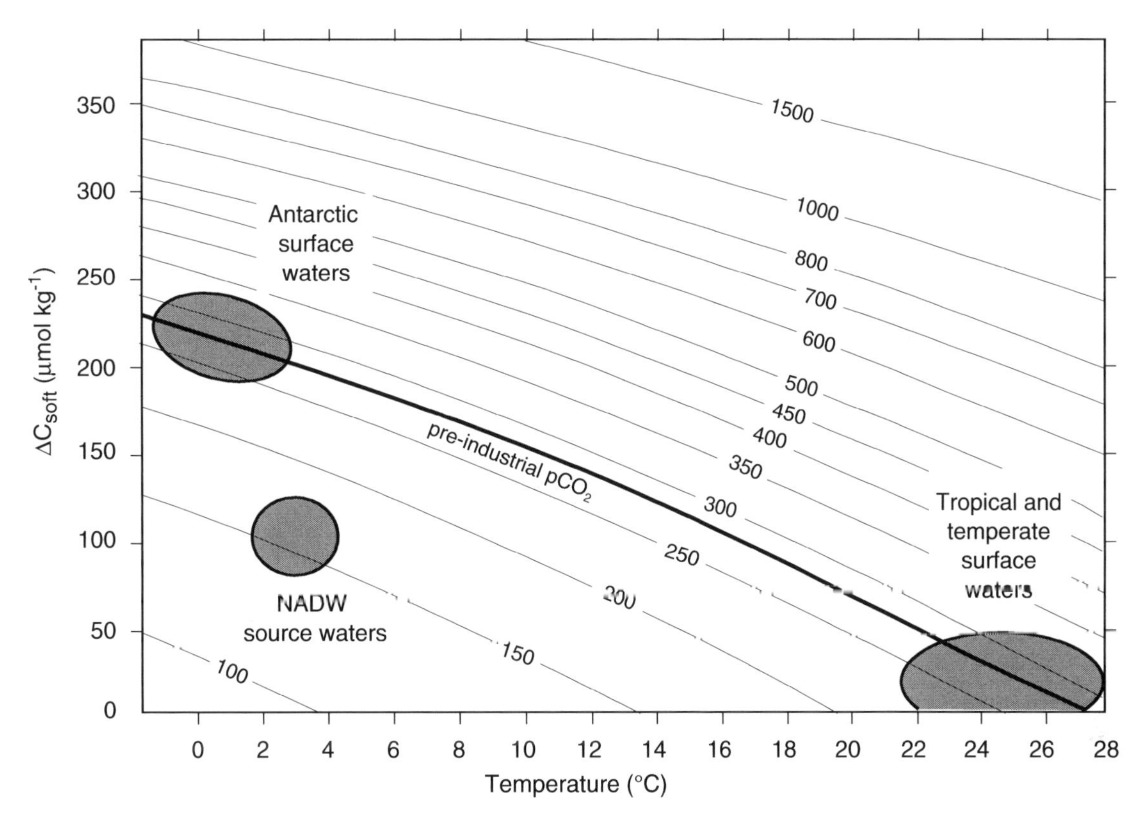

FIGURE 8.4.6: Contours of CO_2 partial pressure (in μatm) for a water sample with salinity of 35, a preindustrial *DIC* concentration of 1900 μmol kg^{-1}, and an alkalinity of 2310 μmol kg^{-1}, as a function of water temperature and phosphate content (and hence also ΔC_{soft} content). The heavy contour is that representing the preindustrial atmosphere (280 μatm). The important point here is that the high content of ΔC_{soft} compensates for the low temperature of Antarctic surface waters, giving them nearly the same pCO_2 as warm nutrient-free surface water. By contrast, because of their lower ΔC_{soft} content, surface waters in the northern Atlantic take up CO_2 given off to the atmosphere by other regions of the ocean. Adapted from *Broecker and Peng* [1992].

almost all of this "excess" ΔC_{soft} and ΔC_{carb}, reducing *sDIC* accordingly. This reduction is apparently of very similar magnitude to the change in *sDIC* required to accommodate the temperature change. Therefore, these waters have to exchange only limited amounts of CO_2 with the atmosphere to remain near equilibrium with the atmosphere.

In the North Atlantic, the contributions of ΔC_{soft} and ΔC_{carb} to surface *sDIC* variations are much smaller, indicating a lesser degree of biological compensation for the thermally driven CO_2 fluxes. This creates a strong tendency for the North Atlantic to take up CO_2 from the atmosphere. Figure 8.4.6 illustrates this difference between surface waters of the North Atlantic and Southern Ocean more explicitly by plotting the ΔC_{soft} for typical surface waters as a function of temperature. Also shown are isolines of calculated pCO_2 assuming a constant alkalinity and a fixed background *DIC* to which ΔC_{soft} was added according to the value given on the ordinate. Although North Atlantic surface waters are warmer than Antarctic surface waters, the much lower ΔC_{soft} concentration of the North Atlantic surface waters leads to their having much lower pCO_2 than that computed for Antarctic surface waters. This gives the North Atlantic surface waters a large potential to take up CO_2 from the atmosphere. By contrast, the Antarctic surface waters are near preindustrial pCO_2, as are typical tropical and temperate surface waters. These waters, therefore, have to exchange little CO_2 with the atmosphere as they are converted from one to another.

The reason for this different behavior of the North Atlantic is the global-scale meridional overturning circulation, which feeds low-nutrient surface waters from the subtropics into the NADW formation region. On the way, these waters are giving off large amounts of heat, leading to a strong cooling [*Schmitz*, 1996]. These waters remain near the surface for a sufficiently long time, so that the slow kinetics of air-sea gas exchange does not impede the uptake of substantial amounts of CO_2 from the atmosphere. This uptake from the atmosphere can occur because the cooling occurs without excessive entrainment of nutrients from below and with it respired CO_2 as is happening in the Southern Ocean. The lack of nutrient entrainment in the North

Atlantic is because the nutrient content of the thermocline is relatively low as a result of the nutrient export by NADW.

An additional factor contributing to the low ΔC_{soft} and ΔC_{carb} signals in the North Atlantic in comparison with the Southern Ocean may be the large difference in the bio-availability of iron. In the Southern Ocean, observational evidence is steadily increasing that the low availability of iron as a result of low atmospheric deposition is preventing the complete drawdown of nutrients, thus creating the largest high-nutrient, low-chlorophyll (HNLC) region [*Martin et al.*, 1990b, 1991; *de Baar et al.*, 1995; *Boyd et al.*, 2000] (see detailed discussion in chapter 4). The atmospheric delivery of iron to the North Atlantic is almost an order of magnitude higher [*Duce et al.*, 1991, *Tegen and Fung*, 1994] and appears to be sufficient to allow the near-complete drawdown of nutrients, creating low ΔC_{soft} and ΔC_{carb} concentration. This example illustrates how the interaction of biological processes with large-scale ocean circulation and possibly the biogeochemical cycles of iron can lead to quite unexpected results.

8.5 Carbon Pumps and Surface Fluxes

In this last section, we would like to close the great biogeochemical loop, and investigate the connection between the ocean carbon pumps and air-sea fluxes of CO_2. In particular, we will look at the biological and thermal components of the gas exchange pump, in order to understand what drives the CO_2 fluxes across the air-sea interface. Unfortunately, it is not possible to split the observed air-sea flux of CO_2 into these two components. We therefore use model results as an illustration.

Following *Murnane et al.* [1999], we ran two parallel simulations in an ocean biogeochemistry model, one with the thermal gas exchange pump only, and a second one that includes both biological pumps. The latter simulation is termed "combined pump." The difference in air-sea CO_2 fluxes between these two simulations is taken as the contribution of the biological gas exchange pump. The model employed here is very similar to that described in detail in chapter 5 except that a representation of the carbonate pump has been included as well. In order to arrive at a realistic atmosphere-ocean distribution of CO_2, all simulations were run to a quasi steady-state by setting atmospheric CO_2 to its pre-industrial value of 280 ppm and letting the ocean adjust to this concentration (see *Najjar and Orr* [1998] for details). This takes several thousand years of spin-up.

Figure 8.5.1a shows the meridional pattern of the zonally integrated air-sea gas exchange as predicted by this model. The patterns exhibited by the thermal gas exchange pump are related to the transport of heat and water within the ocean and its exchange with the atmosphere. In the high latitudes, surface waters lose heat to the atmosphere and consequently take up CO_2 from the atmosphere. In the equatorial regions, colder waters that are upwelled to the surface are warmed up, which leads to a loss of CO_2 to the atmosphere. The correlation of the air-sea exchange fluxes of CO_2 with the heat fluxes are modified, however, by the slow air-sea exchange of CO_2 relative to heat. This shows up particularly in the equatorial region, where the peak of the CO_2 gas exchange is much wider than would be predicted by the heat exchange alone.

The biological pump represents, to a first approximation, a closed loop within the ocean. Waters that are brought to the surface contain an excess of inorganic carbon stemming from the biological pump relative to these surface waters. At the same time, these waters also contain an excess of nutrients. Biological uptake at the surface is generally fast enough to strip out all or most of the nutrients and the excess CO_2 before CO_2 is lost to the atmosphere. Only in regions where the biological uptake is inefficient in reducing the nutrients and the associated biological inorganic carbon coming to the surface will CO_2 escape into the atmosphere. As we discussed in detail in chapter 4, a good diagnostic of such regions of low efficiency is high surface-nutrient concentrations, which occur in the Southern Ocean, in the equatorial Pacific, and in the North Pacific. The Southern Ocean, characterized by the highest surface ocean-nutrient concentration, also represents the region with the highest loss of CO_2 to the atmosphere associated with the biological pump. This biologically induced outgassing outweighs the uptake of CO_2 from the atmosphere as a consequence of the thermal gas exchange pump, leading to a small residual outgassing in the combined pump scenario in this region (figure 8.5.1a). This is exactly the same compensation that we alluded to before when trying to explain the low vertical $\Delta C_{gas\ ex}$ signal in the observations. The loss of CO_2 in the high latitudes due to the biological pump is compensated by a net uptake of CO_2 in the mid-latitudes, so that the global net exchange of CO_2 in association with the biological pump is zero as required. Simulations of the contribution of the carbonate pump to the variation in the biological pump show it to be generally small [*Murnane et al.*, 1999].

The pattern of the zonally integrated air-sea exchange due to the combined pumps represents the sum of the patterns generated by the thermal and biological pumps. In this particular model, the thermally forced CO_2 fluxes tend to dominate, except in the Southern Ocean. The air-

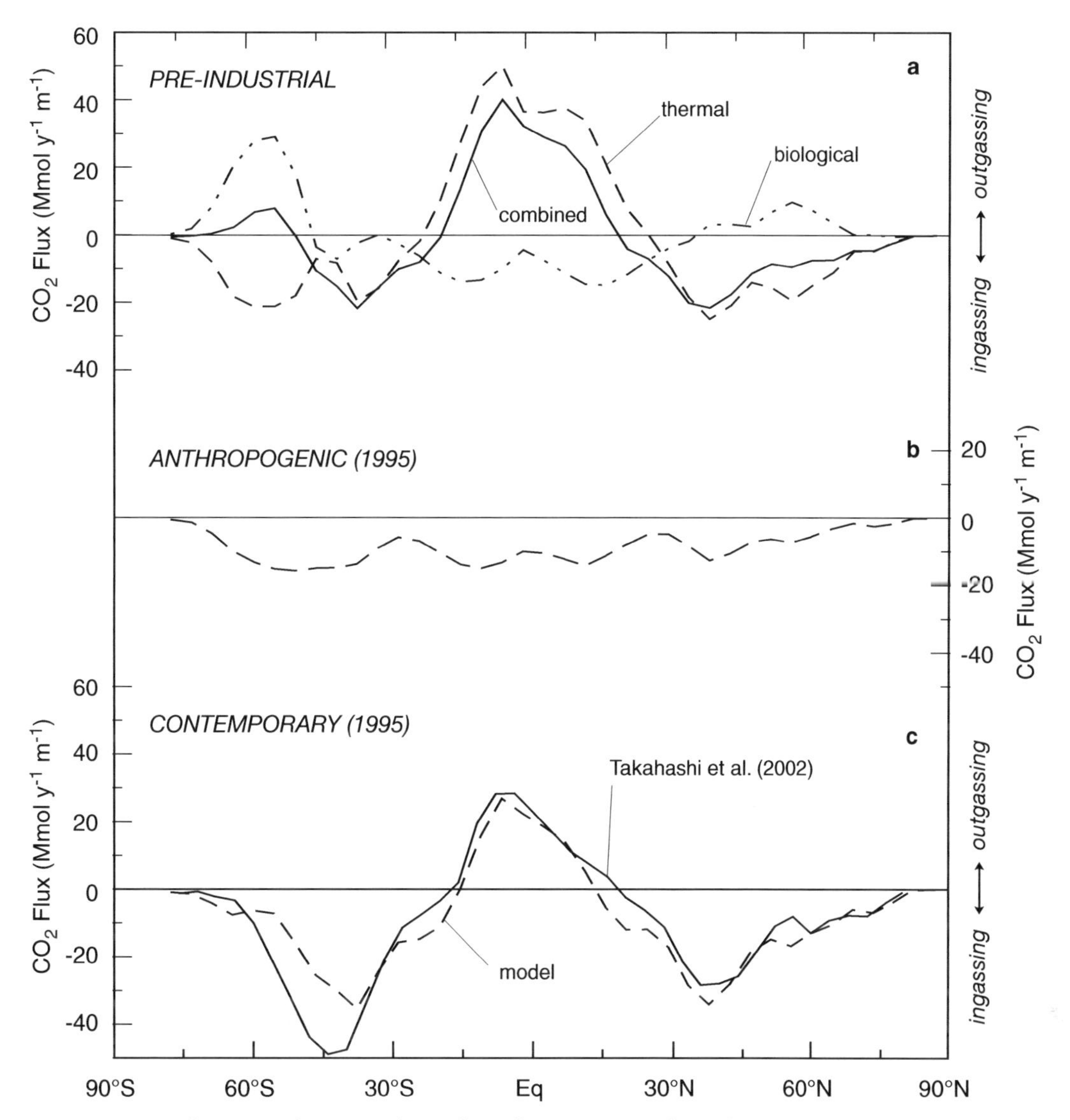

FIGURE 8.5.1: Zonally integrated sea-air exchange fluxes for CO_2 (positive fluxes denote outgassing from the ocean). (a) Model-simulated sea-air fluxes of CO_2 induced by the thermal gas exchange pump, the biological gas exchange pump, and the sum of these two pumps. (b) Model-simulated sea-air flux of anthropogenic CO_2 for 1995. (c) Model-simulated total sea-air flux for 1995 in comparison to observationally derived fluxes based on the ΔpCO_2 climatology of *Takahashi et al.* [2002] as estimated by *Wanninkhof et al.* [2002]. The model employed is the KVLOW-AILOW model as described by *Gnanadesikan et al.* [2002].

sea CO_2 fluxes induced by the two gas exchange pumps generally tend to oppose each other, similar to what we have seen for their influence on the ocean interior distribution of *sDIC* in section 8.4 above.

Our model-based conclusions about the relative roles of the two driving forces for the exchange of CO_2 across the air-sea interface would be strengthened if we could demonstrate that the model with the combined pumps compares well with observationally based flux estimates. The latter can be obtained by combining the observed air-sea pCO_2 difference (see figure 8.1.1) with an estimate of the gas exchange coefficient (see chapter 3). Before we can undertake this comparison, we have to take into account that the anthropogenically driven increase of atmospheric CO_2 has led to a substantial perturbation of the air-sea CO_2 fluxes, which cannot be separated from the observed fluxes. However, it is relatively straightforward to simulate these anthropogenic CO_2 perturbation fluxes (see chapter 10 below) in our model, as this simply requires the model to be run further forward in time, while changing the atmospheric CO_2 concentration according to its observed time history. Figure 8.5.1b shows that the anthropogenic perturbation flux is directed into the ocean everywhere, with largest magnitude in the regions where intermediate and deep waters come to the surface, i.e., in the tropics, along upwelling margins, and in mid- and high-latitude regions with intense convection.

If we add this anthropogenic perturbation flux to the CO_2 fluxes from the combined pump simulation, the resulting total flux compares reasonably well with the observationally based estimates (figure 8.5.1c). This suggests that the model is fundamentally able to represent the biological and thermally driven components. We therefore conclude that our model-based result of generally competing effects on air-sea fluxes of CO_2 arising from thermal and biological forcing applies also to the real world. This conclusion has major implications for how the ocean responds to changes in climate, as any change in meteorological forcing impacts both the thermal and biological components of air-sea gas exchange. We will show in chapter 10, for example, that future climate change will lead to a substantial thermally forced reduction in the uptake of anthropogenic CO_2 from the atmosphere. Climate-induced changes in the biological forcing, however, will lead to an enhancement of this uptake, albeit of smaller absolute magnitude than the thermally forced reduction, causing the total uptake to lie in between the thermally and biologically forced solutions. The exact magnitude of these climate change–induced changes in the uptake of anthropogenic CO_2 from the atmosphere is currently poorly understood and represents a major research issue.

Problems

8.1 Define and explain the terms *DIC*, *Alk*, and pCO_2. Discuss why these terms are important for describing the ocean inorganic carbon system.

8.2 Write down the inorganic carbon system reactions in seawater starting from the transfer of CO_2 from the air phase into the water phase. Define the main reaction constants and give their pK values for a temperature of 20°C.

8.3 Why is the ocean so important in the global carbon cycle?

  a. Compute the relative distribution of inorganic carbon between the atmosphere and the ocean. Use the ocean mean *DIC* concentration given in table 8.2.4 and use a preindustrial atmospheric mixing ratio of 280 ppm. The number of molecules in the atmosphere is about 1.8×10^{20} mol. The ocean mass is about 1.35×10^{21} kg.

  b. Explain and discuss the results of (a).

8.4 Describe how *DIC*, *Alk*, and pCO_2 change as a result of

  a. the formation of organic matter by photosynthesis,

  b. the formation of calcium carbonate shells, and

  c. the uptake of CO_2 from the atmosphere.

8.5 How does *Alk* change if you precipitate

  a. salt (NaCl)?

  b. dolomite ($MgCO_3$)?

  c. $MgSO_4$?

8.6 How does denitrification affect *Alk*? Use the denitrification reaction given in chapter 5 (reaction 5.1.4).

8.7 Compute the change in pCO_2 caused by the formation of 10 mmol $CaCO_3$ in a volume of 1 m^3. Assume that this volume originally had a *DIC* concentration of 2000 mmol m^{-3} and an *Alk* of 2300 mmol m^{-3}.

8.8 Precipitation (rainfall) and evaporation not only change surface salinity; these processes also alter surface ocean *DIC* and *Alk*. What is the overall sensitivity of pCO_2 in surface water to the addition or removal of freshwater? Compute this sensitivity in analogy to equation (8.3.5) in the text, i.e., (S/

$pCO_2) \cdot (\partial pCO_2/\partial S)$, but take into consideration how *DIC* and *Alk* change with *S*. Assume that precipitation has zero *DIC* and *Alk*.

8.9 During the last ice age, the surface ocean was both cooler and saltier. Calculate the change in mean surface ocean pCO_2 that results from these changes in surface properties by assuming that the cooling was 4°C and the salinity increase was 1.1. Use the surface mean concentrations given in table 8.2.4. Note: Use the salinity dependence computed in problem 8.8 rather than equation (8.3.5).

8.10 Discuss the equilibration timescale for the exchange of CO_2 across the air-sea interface:

a. Explain why CO_2 takes about 20 times longer than oxygen to equilibrate between the atmosphere and the surface ocean by air-sea gas exchange.

b. The invasion of anthropogenic CO_2 into the ocean leads to an increase in surface ocean *DIC*, while leaving surface *Alk* unchanged. How does this affect the air-sea equilibration time of CO_2? Hint: Consider equation (4) in panel 8.3.1 in conjunction with the approximations (8.2.17) and (8.2.18).

8.11 Consider a well-mixed surface ocean box with mean surface ocean properties. The only process considered is air-sea gas exchange. Investigate how this system responds to perturbations.

a. Write down the time-dependent conservation equation for the $H_2CO_3^*$ concentration in this surface ocean box (8.3.10), then convert this into a time-dependent equation for the anomaly of $H_2CO_3^*$ with respect to saturation with the atmosphere, $\Delta H_2CO_3^* = [H_2CO_3^*]_a - [H_2CO_3^*]_w$.

b. Solve the first-order differential equation for time-dependent behavior of $\Delta H_2CO_3^*$. Assume that atmospheric pCO_2 and the buffer factor, i.e., the partial derivative $\partial[DIC]/\partial[H_2CO_3^*]$, remain constant.

c. How long does it take to equilibrate the entire inorganic carbon system in this surface layer box to within 66%, 95%, and 99% of the atmospheric concentration after an instantaneous change in surface temperature? Use a gas exchange coefficient of 20 cm h^{-1} and a value of 20 for the partial derivative $\partial[DIC]/\partial[H_2CO_3^*]$, and assume that this surface box is 50 m deep.

d. How do the results in (c) change (i) if we increase the depth of the box to 100 m, or (ii) if we decrease the value of the partial derivative $\partial[DIC]/\partial[H_2CO_3^*]$ to 10?

8.12 Anthropogenic CO_2 emissions have led to a strong increase in atmospheric pCO_2 since the beginning of the industrial period. Much research over the last decades has focused on the role of the ocean as a sink for this anthropogenic CO_2 (see chapter 10 for more details).

a. The present rate of increase in atmospheric CO_2 is about 1.5 μatm yr^{-1}. Estimate the rate of increase in surface ocean *DIC* under the assumption that the surface ocean is able to keep up with the atmospheric perturbation, i.e., remains in equilibrium with the change in atmospheric CO_2. Assume a contemporary pCO_2 of 360 μatm.

b. Discuss this result considering that the measurement accuracy of *DIC* is about 2 μmol kg^{-1} and the precision about 0.5 μmol kg^{-1}. Consider also what you learned about seasonal variability of *DIC*. How often and how long do we have to repeatedly measure *DIC* in order to detect its increase in the surface ocean reliably?

8.13 What will be the ultimate fate of anthropogenic carbon in the atmosphere-ocean system?

a. Calculate the fraction of a perturbation added to the atmosphere that will be found in the ocean after full equilibration occurs. Assume that the buffer factor remains constant at its present value of 10, and use a pCO_2 of 280 μatm and a *DIC* of 2000 μmol kg^{-1} in your calculation.

b. In fact, the buffer factor does not remain constant. Use the approximation (8.3.16) to estimate the change in the buffer factor if *DIC* increases by 100 μmol kg^{-1}. How does this affect the fraction calculated in (a) above?

8.14 Perform a thought experiment similar to that shown in table 8.3.2. Instead of taking a deep water sample, take a sample from the waters feeding the equatorial upwelling with the following properties: $T = 20°C$, $[PO_4^{3-}] =$ 1.5 μmol kg^{-1}, $DIC = 2230$ μmol kg^{-1}, $Alk = 2320$ μmol kg^{-1}, and $S = 35$. For photosynthesis, assume constant stoichiometric ratios of 117:1 for carbon to

TABLE FOR PROBLEM 8.14

Oceanic $pCO2$ in μatm as a function of *Alk* (rows in μmol kg^{-1}) and *DIC* (columns in μmol kg^{-1}) for a temperature of 20°C and a salinity of 35.

		Alk (μmol kg^{-1})														
		2160	*2180*	*2200*	*2220*	*2240*	*2260*	*2280*	*2300*	*2320*	*2340*	*2360*	*2380*	*2400*	*2420*	*2440*
DIC	1900	349	320	295	273	253	235	219	205	191	179	168	158	149	141	133
(μmol kg^{-1})	1920	388	355	326	301	278	258	240	223	209	195	183	172	162	152	144
	1940	434	395	361	332	306	283	263	244	228	213	199	187	176	165	156
	1960	486	441	402	368	338	312	289	268	249	232	217	203	191	179	169
	1980	548	495	449	409	375	344	318	294	273	254	237	221	207	194	183
	2000	621	557	503	456	416	381	350	323	299	278	259	241	226	211	198
	2020	708	631	566	511	464	423	388	357	329	305	283	264	246	230	215
	2040	812	719	641	575	520	472	430	394	363	335	310	288	268	251	234
	2060	936	823	729	651	584	528	480	438	401	369	341	316	293	273	255
	2080	1086	949	835	740	660	593	536	487	445	408	376	347	321	299	278
	2100	1268	1100	961	846	750	670	602	545	495	452	415	382	353	327	304
	2120	1485	1282	1113	974	858	761	680	612	553	503	460	422	388	359	333
	2140	1744	1500	1297	1127	986	870	772	690	621	562	511	467	429	395	365
	2160		1759	1516	1311	1141	999	881	783	700	630	571	519	475	436	401
	2180			1775	1531	1325	1154	1012	893	794	711	640	579	527	482	443
	2200				1791	1546	1340	1168	1025	905	805	721	649	588	535	490
	2220					1806	1561	1354	1182	1037	917	816	731	659	597	544
	2240						1822	1576	1369	1195	1050	929	827	741	668	606
	2260							1837	1592	1383	1209	1063	941	838	751	678
	2280								1853	1607	1398	1223	1076	953	849	762
	2300									1869	1622	1413	1236	1089	965	860
	2320										1884	1637	1427	1250	1102	977
	2340											1900	1652	1442	1264	1115
	2360												1916	1668	1456	1278
	2380													1931	1683	1471
	2400														1947	1698

phosphate, and 16:1 for nitrate to phosphate. Compute the pCO_2 of this water sample after

a. it has been brought to the surface,

b. it is warmed to $T = 26°C$,

c. all phosphate is removed by net formation of organic matter, and

d. its *Alk* is adjusted to 2310 μmol kg^{-1} by the formation of $CaCO_3$ (remember to remove *DIC* also).

Hint: Use the lookup table to compute the oceanic pCO_2 for a given *DIC* and *Alk* rather than employing approximation (8.3.3). However, it is safe to use approximation (8.3.4) for making the temperature correction.

8.15 Figures 8.3.12 and 8.3.13 show the seasonal cycles of pCO_2 for the Atlantic and Pacific Oceans, respectively. Compare these two seasonal cycles and relate the differences to what you have learned about the seasonal dynamics of surface production in these two areas.

8.16 The seasonal cycle of pCO_2 in the subtropical gyres appears to be determined primarily by the seasonal variation of sea-surface temperature, with a partially compensating effect due to variations in *DIC* (see figure 8.3.10). Discuss how sea-surface pCO_2 in these regions would change in response to a positive *SST* anomaly of 1°C and a negative *DIC* anomaly of 1 mmol m^{-3}. (Hint: Use the temperature and *DIC* sensitivities of pCO_2 given by (8.3.4) and (8.3.14) to compute the change in pCO_2. Also use the mean concentrations of *DIC* and *Alk* for the low-latitude ocean given in table 8.2.4.)

8.17 Describe the main features of the ΔpCO_2 distribution shown in figure 8.1.1. Explain and discuss the reasons for this distribution. In particular, why are the tropical regions supersaturated and the high latitudes undersaturated with respect to atmospheric CO_2?

8.18 Discuss how ocean biology impacts the air-sea exchange of CO_2. In particular, address whether the strength or the efficiency of the biological pump is more relevant for controlling the air-sea flux of CO_2. Explain the basis for your choice.

8.19 Figure 8.1.3 shows the *sDIC* distribution along the meridional overturning circulation.

a. Describe the main features of this distribution.

b. Often, the term "pumps" is used to describe processes that affect the distribution of *sDIC*. Name the three major pumps and describe briefly how they work.

c. Relate the observations that you described in (a) to the action of the three pumps.

8.20 Speculate about the effects of an increase in the efficiency of nutrient utilization in the Southern Ocean on the vertical distribution of the gas exchange component $\Delta C_{gas\ ex}$.

8.21 During the Younger Dryas, a period of abrupt cooling in the northern hemisphere about 11 kyr ago, deep water formation in the North Atlantic reduced dramatically. Discuss what impact this might have had on *DIC* and the different pump components in the deep Atlantic.

Calcium Carbonate Cycle

9.1 Introduction

Mineral calcium carbonates ($CaCO_3$), which occur in the ocean mainly in the form of calcite, represent a significant fraction of deep ocean sediments, and belong to the most common minerals on Earth's surface. Most mineral $CaCO_3$ that is found at the bottom of the ocean is actually formed by organisms living in near-surface waters. The contribution of abiotic precipitation in today's ocean is virtually zero, despite the fact that near-surface waters are supersaturated with respect to mineral $CaCO_3$. The high relative contribution of $CaCO_3$ to open ocean sediments is a direct consequence of the observed fact that about half of the $CaCO_3$ that leaves the surface ocean arrives at the sediments, and that about a third of this $CaCO_3$ is buried, i.e., about 13% of the surface export of $CaCO_3$ is lost from the ocean every year (figure 9.1.1). It is very instructive to contrast this high burial efficiency with the open ocean budget of organic matter (see table 6.5.1), where 95% of the organic matter exported from the surface ocean is remineralized within the water column, and where only about 0.3% of surface organic matter export is buried.

The high burial of $CaCO_3$ leads to a significant loss of *Alk* from the ocean, which is replaced by riverine input of *Alk* stemming from continental weathering, and a small input from hydrothermal vents (see figure 9.1.1). The magnitude of these fluxes is two orders of magnitude smaller than the exchange of carbon between the atmosphere and ocean, and result in a residence time of *Alk* in the ocean of about 10^5 years [*Sundquist*, 1991]. These small fluxes and the overall small contribution of the carbonate pump to the distribution of *DIC* (see discussion in chapter 8) might lead to the conclusion that the oceanic cycling of mineral $CaCO_3$ is a relatively unimportant component of the ocean carbon cycle.

On timescales longer than a few hundred years, however, the cycling of $CaCO_3$ in the ocean emerges as a crucial component of the global carbon cycle. This is because the vast pool of $CaCO_3$ in the surface sediments becomes an active pool of the ocean carbon cycle on these timescales. It is conceivable that imbalances between the river input of *Alk* and the loss by burial might develop on such long timescales, as these two processes appear to be independent. Changes in mean ocean alkalinity resulting from the interaction with the ocean sediments and from an imbalance between river input and ocean burial could be quite substantial, possibly resulting in large changes in atmospheric CO_2. If such large changes in atmospheric CO_2 had ever occured, they would have threatened life on Earth either by overheating or by CO_2 starvation of terrestrial plants. It turns out, however, that the interaction of the oceanic carbon cycle with the underlying sediments leads to a negative, i.e., stabilizing, feedback, which forces the system toward a balance between the input of *Alk* by weathering and the loss by burial. This feedback involves a change in the mean ocean *DIC*-to-*Alk* ratio such that the magnitude of $CaCO_3$ burial on the seafloor is adjusted until any imbalance with the weathering input is removed, a process termed *calcium carbonate compensation* [*Broecker and Peng*, 1987]. Thus the ocean $CaCO_3$ system acts as a "homeostat" which on millennial timescales and longer maintains a steady state by linking deep sea burial rate to the balance of continental weathering minus shallow water deposition [*Broecker and Takahashi*, 1977a; *Archer*, 1996a].

On timescales of hundreds of thousands of years, a second feedback mechanism stabilizing atmospheric CO_2 has been proposed. It links volcanic outgassing of CO_2 and continental weathering of igneous rocks which consumes CO_2 (see reaction 10.2.12) and whose rates tend to depend on Earth's temperature [*Walker et al.*, 1981; *Berner et al.*, 1983; *Broecker and Sanyal*, 1998]. This feedback is thought to be provided through the police action of atmospheric CO_2. The hypothesis is that if the

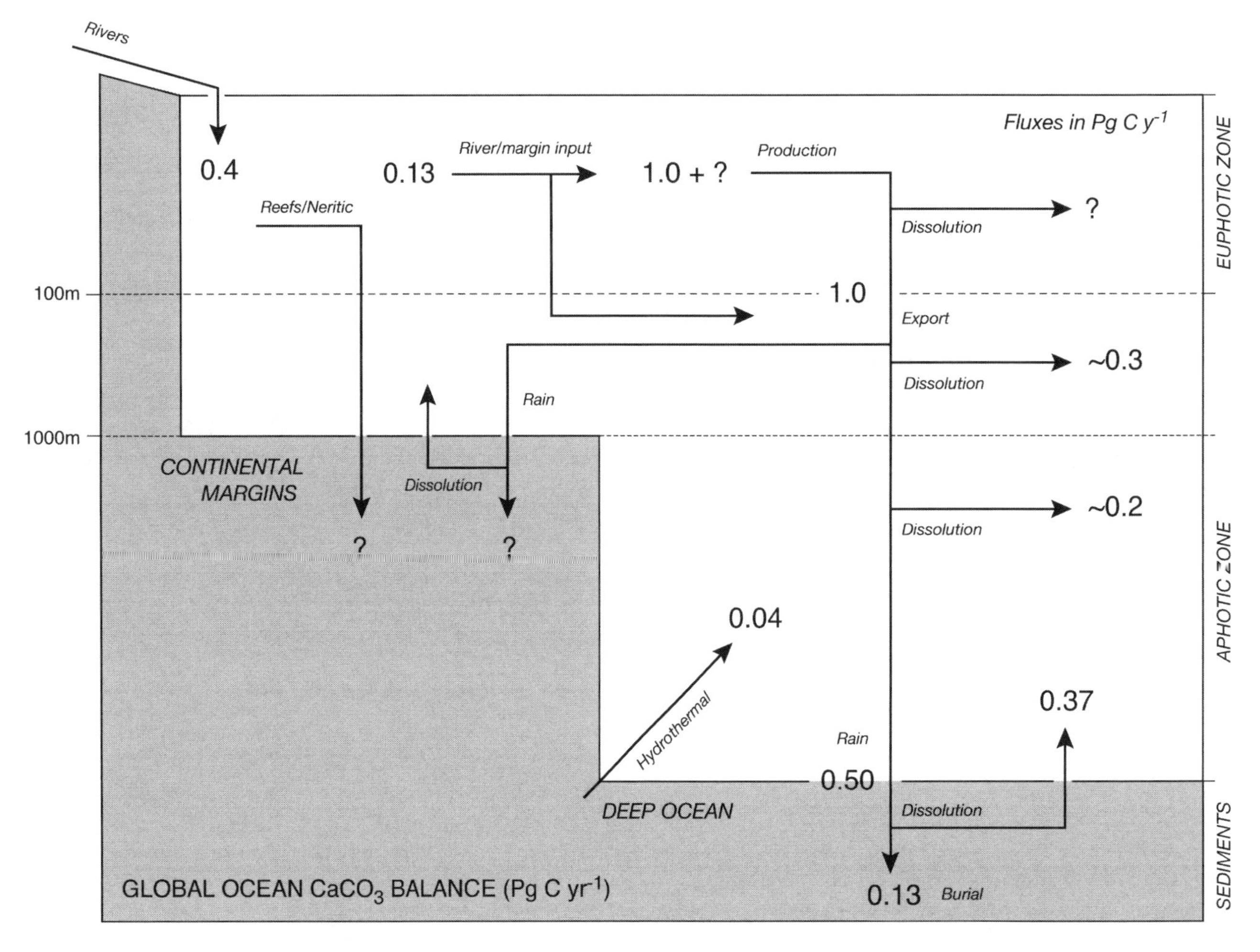

FIGURE 9.1.1: Budget of carbonate (alkalinity) of the global ocean in units of Pg C yr^{-1}. Modified from *Milliman and Droxler* [1996]. In particular, open ocean $CaCO_3$ production and export was increased to reflect more recent estimates (see text). The resulting increased rain rate of $CaCO_3$ onto deep ocean sediments results in an organic carbon-to-$CaCO_3$ rain ratio of about 0.8, in good agreement with estimates inferred from sediment modeling [*Archer*, 1996a].

consumption of atmospheric CO_2 by chemical weathering of igneous rocks were too small to match the input of CO_2 by volcanic outgassing, the CO_2 content of the atmosphere would rise, causing the temperature of Earth to rise, which in turn would increase the rate of chemical weathering until a new balance is achieved. Conversely, if the rate of CO_2 consumption by chemical weathering exceed volcanic outgassing, atmospheric CO_2 would be drawn down, causing Earth to cool, thereby reducing the rate of chemical weathering. This hypothesis is still being debated [see, e.g., *Broecker and Sanyal*, 1998; *Raymo and Ruddiman*, 1992], but, it provides a wonderful example of an interaction between tectonics, climate, and the carbon cycle.

We will not dwell on this second feedback any further (see *Broecker and Peng* [1998] for more details). However, we will discuss calcium carbonate compensation in detail, since this feedback plays an important role in many hypotheses trying to explain glacial-interglacial variations of atmospheric CO_2 [*Sigman and Boyle*, 2000]. Before we can address the question of how this feedback works, we first need to understand the individual components of the oceanic $CaCO_3$ cycle. This cycle starts with the formation of mineral $CaCO_3$ in near-surface waters, continues with the export of this material from the upper ocean and sinking through the water column, where it is subject to dissolution, and ends with diagenetic processes in the sediments that will determine how much will be dissolved versus how much will be buried.

As we go through these individual processes, we will encounter many puzzling questions, like why the highest CO_3^{2-} concentrations are found in places where CO_3^{2-} is actually consumed (surface ocean) and the lowest CO_3^{2-} concentrations are found in the places where CO_3^{2-} is being produced (deep ocean) (see figure 9.1.2). Another important question is what controls the preservation of $CaCO_3$ in the sediments. Figure 9.1.3 shows that the distribution of $CaCO_3$ in the near-surface sediments differs drastically from that of opal (figure 7.1.3) and

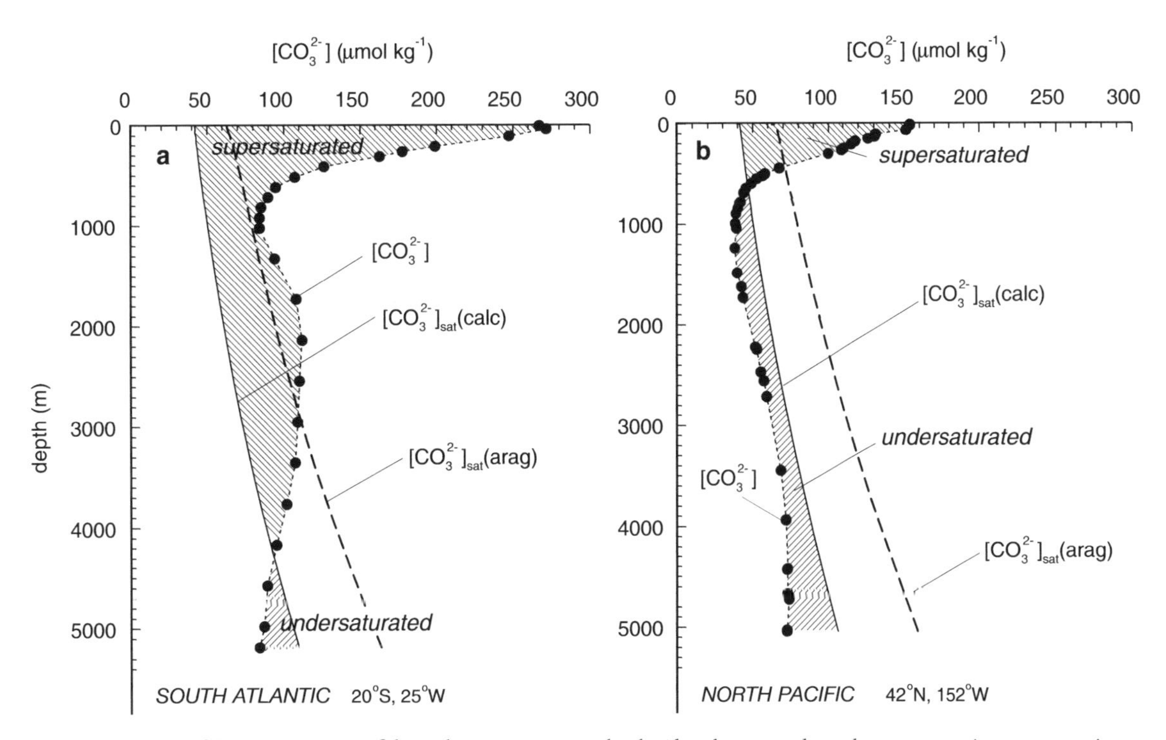

FIGURE 9.1.2: Plot of the concentration of the carbonate ion versus depth. Also shown are the carbonate saturation concentrations with regard to calcite (solid line) and aragonite (dashed line). (a) South Atlantic Ventilation Experiment station 334 in the South Atlantic at 20°S, 25°W. (b) WOCE P16 station 48 in the North Pacific at 42°N, 152°W. The region of supersaturation and undersaturation with regard to calcite are highlighted with cross hatches. The saturation concentrations were calculated using the solubility products of *Mucci* [1983] and the pressure dependence of *Millero* [1983]. For calcium, the salinity dependence of *Millero* [1982] was adopted.

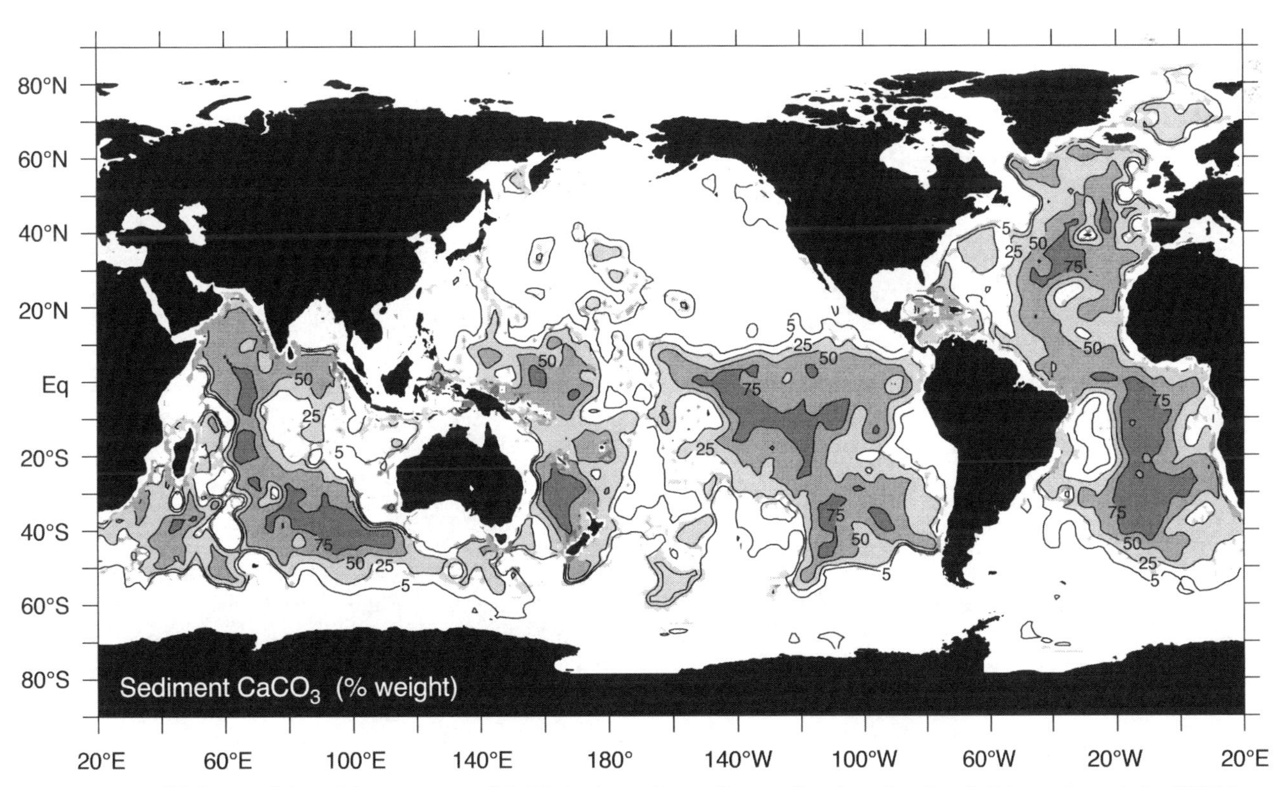

FIGURE 9.1.3: Gridded map of the weight percentage of $CaCO_3$ in the surface sediments. See also color plate 5. Data are from *Archer* [1996b].

organic matter (figure 6.1.4). Comparing this map with a bathymetric map reveals that the regions with high $CaCO_3$ content occur mostly on the oceanic ridges, whereas the deep ocean basins are almost completely devoid of $CaCO_3$. (*Broecker and Peng* [1982] compared this distribution to snowcapped mountains). What controls this burial pattern? Is it dissolution kinetics as is the case for opal, or is the burial pattern determined by the degree of saturation? Figure 9.1.2 suggests that it is the latter, as we find accumulation of $CaCO_3$ in shallow sediments, where the ocean is supersaturated with respect to calcite, the more stable form of $CaCO_3$, and no $CaCO_3$ in deep ocean sediments where the ocean is undersaturated.

Understanding the processes that lead to this preservation pattern are of great importance because variations in the $CaCO_3$ content of deep-sea sediments constitute one of the principal indicators of past changes in the global carbon cycle. Coupled with isotopic and chemical analyses, variations in the $CaCO_3$ content of the sediments provide important clues for assessing how marine systems responded to changing climate conditions in the past. In order to fully exploit the information stored in the sediments, it is imperative to develop a fundamental understanding of the processes that link production of $CaCO_3$ in the surface oceans with the resulting sedimentary record.

9.2 Production

In the open ocean, mineral calcium carbonates are formed almost exclusively by biological processes. The lack of inorganic precipitation of $CaCO_3$, despite the fact that these waters are supersaturated with respect to these minerals, is generally thought to be due to nucleation barriers; the necessary seed crystals may be absent, or magnesium ions, phosphate, or organic compounds may act as surface inhibitors for nucleation [*Stumm and Morgan*, 1981, p. 316]. Interestingly, the situation in lakes is reversed; almost all carbonates are formed as inorganic precipitates, and very little by organisms [*Kelts and Hsü*, 1978].

Organisms

Three groups of marine organisms are mainly responsible for the precipitation of $CaCO_3$ in the open ocean (see also chapter 4, panels 4.2.2 and 4.2.3). Coccolithophorids (see panel 4.2.2, figure 2) are a phytoplankton group that form an outer sphere of small calcite plates. The other two groups forming $CaCO_3$ shells are heterotrophic zooplankton. Foraminifera are amoebas that form beautiful skeletons made out of calcite (panel 4.2.3, figure 1). Finally, pteropods are a group of mollusks that produce crystals made out of aragonite rather than calcite. Between these two phases of mineral $CaCO_3$, calcite is the more stable phase at any pressure and temperature found in the oceans. Since surface waters are supersaturated with respect to both mineral phases, evolutionary history is likely the key determinant for which mineral phase organisms precipitate.

There are many reasons marine organisms may precipitate $CaCO_3$ shells. In the case of foraminifera, the reason is most likely protection from predation. The foraminifera shells consist actually of a series of hollow chambers, which are produced sequentially as the organism grows. When foraminifera reproduce, either sexually or asexually, they start again from a new initial chamber, although the chamber is much larger in the case of an asexually produced generation, in comparison to a sexually produced generation.

In the case of coccolithophorids, it has been hypothesized early on (*Paasche* [1962]; *Sikes et al.* [1980]) that calcification could be linked to photosynthesis, as the formation of $CaCO_3$ from HCO_3^- produces aqueous CO_2, which can then be used as a source of inorganic carbon during photosynthesis:

$$2HCO_3^- + Ca^{2+} \rightleftharpoons CO_2 + CaCO_3 + H_2O \qquad (9.2.1)$$

This would provide an advantage for coccolithophorids, as this strategy circumvents the need for CO_2 concentration mechanisms, which are employed by other phytoplankton groups to increase the intracellular concentration of aqueous CO_2, thereby increasing the efficiency of photosynthesis. However, the magnitude of the coupling between photosynthesis and $CaCO_3$ formation is still a topic of intense debate, with some experiments suggesting that a very large fraction of the aqueous CO_2 fixed during photosynthesis stems from calcification [e.g., *Buitenhuis et al.*, 1999] while other studies indicate a much smaller fraction (see recent review by *Paasche* [2002]). One reason against the CO_2 supply argument is that photosynthesis and calcification occur in very different locations within the cells of coccolithophorids. Photosynthesis occurs in the plastids, whereas the $CaCO_3$ platelets (called liths) are formed in special coccolith vesicles, both of which are separated by membranes from the rest of the cell. Once formed, the liths are then transported to the outside of the cell, where they stay within a coccosphere until they are shed individually after the death of the cell or sometimes earlier. It appears that the source of inorganic carbon for photosynthesis and $CaCO_3$ precipitation is only partially the intercellular medium, and that inorganic carbon can also be taken up directly from seawater. In all cases, the source of $CaCO_3$ is HCO_3^-, whereas photosynthesis is fueled by aqueous CO_2.

Although the strength of the coupling between photosynthesis and calcification remains unclear, experimental results in the laboratory show a very clear dependence of calcification rates by coccolithophorids on the availability of aqueous CO_2 [*Riebesell et al.*, 2000], with calcification rates plummeting as the aqueous CO_2 concentration increases. This strong calcification response has important implications for the future. The progressive increase in atmospheric CO_2 has already increased the surface ocean CO_2 concentration substantially (and correspondingly decreased the CO_3^{2-} concentration) and this increase will likely continue into the foreseeable future (see chapter 10). It therefore has been suggested that marine rates of calcification might strongly decrease in the future [*Riebesell et al.*, 2000], leading to a negative feedback for atmospheric CO_2 [*Zondervan et al.*, 2001]. This negative feedback arises because a decreasing rate of formation of $CaCO_3$ would tend to lower oceanic pCO_2 (see discussion in section 8.3 above), thereby increasing the uptake of CO_2 from the atmosphere and enhancing the oceanic mitigation of the anthropogenic perturbation of the global carbon cycle [*Sarmiento and Gruber*, 2002].

It is interesting to point out that most of the organisms that precipitate $CaCO_3$ in today's open ocean are relatively young from an evolutionary perspective. The fossil record suggests that coccolithophorids emerged in the mid-Triassic, i.e., around 200 million years ago [*Bown et al.*, 2004] and rose to prominence after the Cenomanian/Turonian boundary, about 90 million years before present [*Iglesias-Rodríguez et al.*, 2002b]. Similarly, planktonic foraminifera do not appear in the fossil record until the mid-Jurassic, i.e., about 180 million years ago. Before that time, there were essentially no open ocean planktonic organisms that precipitated $CaCO_3$. Only shallow water and benthic organisms formed biogenic $CaCO_3$.

Export Estimates

Global estimates of current open ocean $CaCO_3$ production and export are of the order of 1 $Pg\,C\,yr^{-1}$ (see summary by *Iglesias-Rodríguez et al.* [2002a]). From an inventory of alkalinity and the residence time of various water masses, *Milliman and Droxler* [1996] estimated a global open ocean export of mineral carbonates of about 0.7 $Pg\,C\,yr^{-1}$. *Lee* [2001], looking at the seasonal drawdown of potential alkalinity, found a lower-bound estimate for $CaCO_3$ export of about 1.1 $Pg\,C\,yr^{-1}$. Most of the open ocean $CaCO_3$ export is believed to be in the form of calcite, driven primarily by coccolithophorids, although some recent studies in the Arabian Sea and equatorial Pacific suggest a substantial contribution from foraminifera [*Iglesias-Rodríguez et al.*, 2002a]. In contrast, aragonitic pteropods seem to account for only about 10% of the total open ocean carbonate export [*Berner and Honjo*, 1981; *Fabry*, 1990].

Comparing these $CaCO_3$ export estimates with our estimate of organic matter export of about 12 $Pg\,C\,yr^{-1}$ (see chapters 4 and 5) reveals an inorganic-to-organic carbon export ratio of about 0.06 to 0.08, i.e., only 6 to 8% of the total carbon is exported out of the surface ocean in the form of $CaCO_3$. Assuming that $CaCO_3$-precipitating organisms have an inorganic-to-organic carbon ratio of about 1:1 (as implied by a tight connection between photosynthesis and calcification), this ratio suggests also that only around 6 to 8% of the total organic matter export is driven by $CaCO_3$-precipitating organisms. While the downward carbon transport in the form of $CaCO_3$ may be a small component of the total downward transport of carbon by the two biological pumps, we have seen in chapter 5 that $CaCO_3$ is singularly important as a ballast material, having the highest efficiency in transporting organic matter in particles through the thermocline and into the deep ocean.

The low contribution of carbonate-secreting organisms such as coccolithophorids to carbon export stands in stark contrast to the export of opal, for which we have seen that diatoms may account for half or more of the export of organic matter (chapter 7). We have also seen that the opal-to-organic matter export ratio varies substantially across oceanic regions (figure 7.2.4), with variations by more than a factor of five between the low latitudes and the Southern Ocean.

Inorganic-to-Organic Carbon Export Ratio

Is the inorganic-to-organic carbon export ratio as variable as the export ratio of opal to organic matter? There are only few and sparse sediment trap studies in the upper ocean that would permit us to draw a global picture. However, we can make use of the signals that the formation and export of organic matter and $CaCO_3$ leave on the distribution of tracers in the ocean, much as we used the tracer signals of the formation and export of opal and organic matter to infer the opal-to-organic matter export ratio (see panel 7.2.1 in chapter 7). *Sarmiento et al.* [2002] showed that the ratio of the upper ocean gradients of potential alkalinity and nitrate is equal to the inorganic-to-organic carbon export ratio if steady state is assumed and the data are averaged over large spatial scales. Recall that potential alkalinity, P_{Alk}, is equal to $Alk + NO_3^-$, and its variations are only governed by the formation and dissolution of $CaCO_3$ [*Brewer et al.*, 1975].

Let us consider the vertical balance between a surface box *s* and a thermocline box *th*. The horizontal extent of these boxes is of the order of several 1000 kilometers, such that the horizontal exchange time is of the order of many years, much longer than the vertical exchange time, which is of the order of one year or shorter. In steady state, the downward export of organic carbon and $CaCO_3$ needs to be equal to the upward supply of potential alkalinity and of nitrate multiplied with the organic carbon-to-nitrogen consumption ratio, $r_{C:N}$. Using the notation introduced in chapter 1 (see 1.2.2), we have

$$\Phi^{CaCO_3} = \frac{v}{2} \cdot ([P_{Alk}]^{th} - [P_{Alk}]^{s}) \quad (9.2.2)$$

$$\Phi^{C_{org}} = v \cdot r_{C:N} \cdot ([NO_3^-]^{th} - [NO_3^-]^{s}) \quad (9.2.3)$$

where v is an exchange term, and where the factor of 0.5 accounts for the fact that the formation and dissolution of $CaCO_3$ changes *Alk* by two moles for each mole of carbon transformed. In analogy to the derivation of the export ratio for opal (7.3.2), we can compute the $CaCO_3$-to-organic carbon export ratio by taking the ratio of (9.2.2) and (9.2.3):

$$R_{CaCO_3:C_{org}} = \frac{([P_{Alk}]^{th} - [P_{Alk}]^{s})}{14.6\ ([NO_3^-]^{th} - [NO_3^-]^{s})} \quad (9.2.4)$$

where we used an $r_{C:N}$ of 117:16 [*Anderson and Sarmiento*, 1994].

Figure 9.2.1 shows the $CaCO_3$-to-organic export ratio estimated from (9.2.4) across the different ocean basins. In sharp contrast to the opal-to-organic matter export ratio, the estimated $CaCO_3$-to-organic carbon export ratio is relatively constant, with a global mean of about 0.06 [*Sarmiento et al.*, 2002]. The $CaCO_3$-

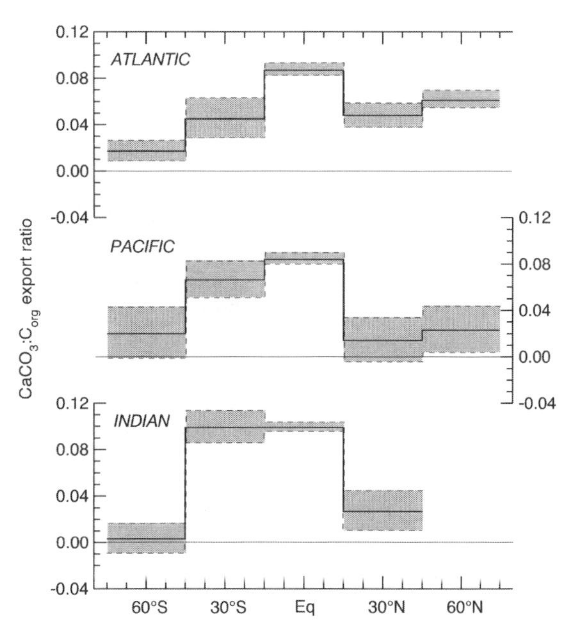

FIGURE 9.2.1: Plot of the $CaCO_3$-to-organic carbon export ratio estimated from observations. Each line segment represents the horizontal average over 30° latitude bands. Adapted from *Sarmiento et al.* [2002].

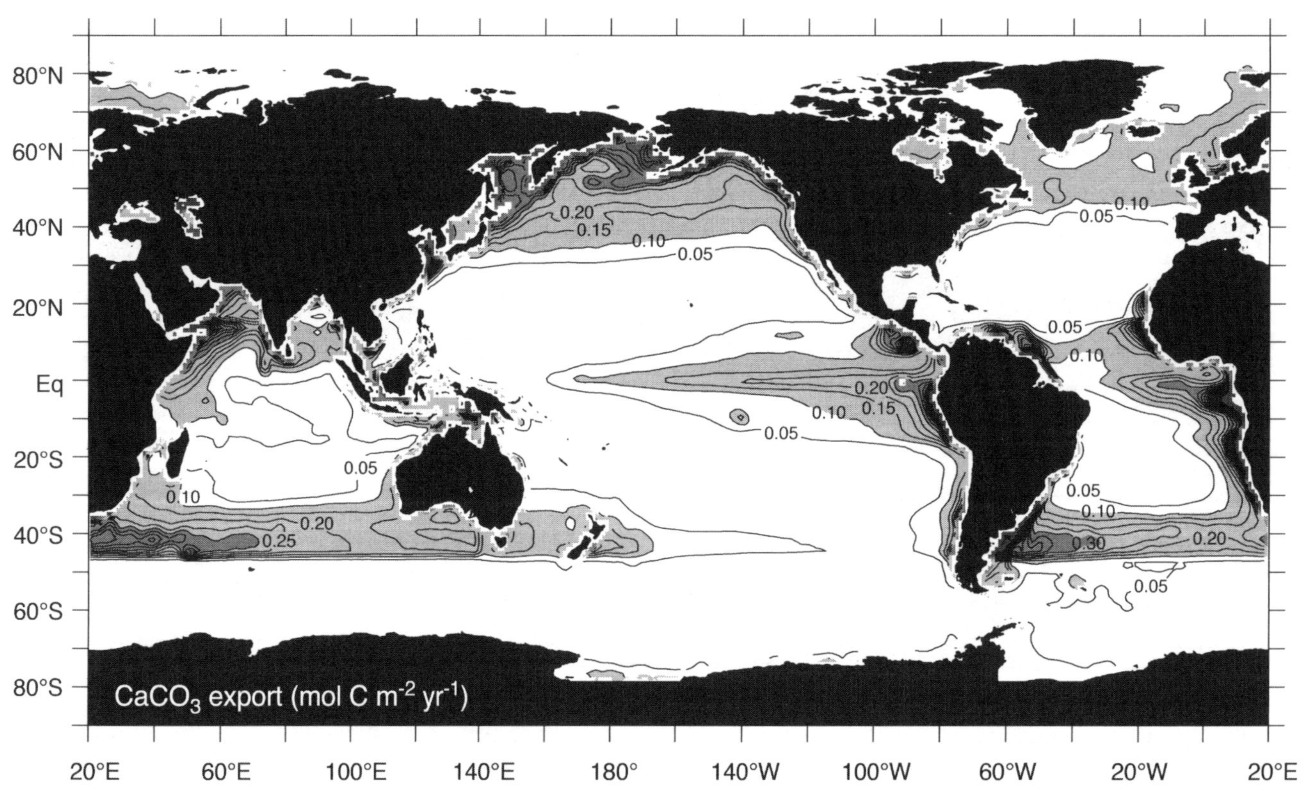

FIGURE 9.2.2: Global map of the export flux of $CaCO_3$ from surface waters. See also color plate 4. This map was derived by multiplying the $CaCO_3$-to-organic carbon export ratio shown in figure 9.2.1 with an estimate of the global organic matter export. The latter was estimated by combining primary production inferred from satellite observations of chlorophyll with the empirical export model of *Dunne et al.* [2005b].

to-organic carbon export ratio tends to be slightly higher in the low latitudes, with values around 0.10, and tends to be uniformly low in the Southern Ocean south of 45°S, with values around 0.02. These spatial variations might be a result of the fact that the coccolithophorids, which are overall dominating the $CaCO_3$ export, are adapted to relatively low turbulence regimes, such as the low latitudes, and tend to get outcompeted by diatoms in high-turbulence regimes [*Iglesias-Rodríguez et al.*, 2002b]. Another factor might be simply differences in the silicic acid concentrations, which permits coccolithophorids to become more prevalent only after diatoms have lost their competitive advantage by having exhausted all available silicic acid.

As a consequence of the relatively uniform $CaCO_3$-to-organic carbon export ratio, a map of the global export flux of calcium carbonate from surface waters looks similar to a map of the global export flux of organic carbon (compare figure 4.2.4 with figure 9.2.2). This $CaCO_3$ export map looks, however, rather different from that for opal. Opal export is particularly large in the high latitudes of the Southern Ocean where $CaCO_3$ export is negligible (figure 7.2.5).

We end with a cautionary note. While our focus here is on the open ocean realm, it is important to note that a substantial amount of $CaCO_3$ is also fixed in near-shore environments by coral and a few marine plants growing attached at the bottom of the ocean, such as *Halimeda spp. Milliman and Droxler* [1996] estimated that coral reefs and carbonate shelves produce about 0.3 Pg C yr^{-1} of $CaCO_3$ each year, which is about a third of the open ocean production. Most of this $CaCO_3$ is in the form of aragonite or high-magnesium calcite, which has a higher solubility than true calcite. Some of this material might get eroded and transported into the open ocean, but the magnitude of this transport is not well known. *Milliman et al.* [1999] suggested that this transport may be as large as 0.1 Pg C yr^{-1} (see figure 9.1.1).

9.3 Water Column Processes

Once the $CaCO_3$ particles leave the productive upper layers of the ocean, they are potentially subject to dissolution. The primary driver for dissolution of $CaCO_3$ particles is thought to be the state of saturation of the water with respect to the mineral phase. As we have seen in figure 9.1.2, the upper waters are generally supersaturated with respect to both mineral phases of $CaCO_3$, whereas deep waters are undersaturated. We therefore generally expect relatively little dissolution to occur in the upper ocean, and substantial dissolution only to take place once the $CaCO_3$ particles enter the undersaturated zones of the water column. Understanding the dynamics of this dissolution of $CaCO_3$ is not only important for the transport of $CaCO_3$ itself, but also for the transfer of organic matter from the surface into the ocean interior, since, as we have seen in chapter 5, $CaCO_3$ represents one of the most important ballast materials.

What determines the distribution of under- and supersaturation of the water column with respect to $CaCO_3$? This situation is obviously rather different from that of opal, with respect to which we have seen that waters are undersaturated almost everywhere. In order to answer this question, we first need to know the solubilities of calcite and aragonite, and then we have to understand the processes that control the concentrations of the CO_3^{2-} and dissolved calcium ions.

$CaCO_3$ Solubility

Let us start by defining the terms *saturation*, *undersaturation*, and *supersaturation*. The equilibrium relationship for the dissolution reaction of mineral $CaCO_3$ (see reaction (8.3.13)) defines the solubility product, $K_{sp}^{CaCO_3}$,

$$K_{sp}^{CaCO_3} = [CO_3^{2-}]_{sat}[Ca^{2+}]_{sat} \qquad (9.3.1)$$

where $[CO_3^{2-}]_{sat}$ and $[Ca^{2+}]_{sat}$ are the concentrations of the carbonate and dissolved calcium ions in equilibrium with mineral $CaCO_3$. Equations for $K_{sp}^{CaCO_3}$ with respect to calcite and aragonite are given in table 9.3.1, and numerical values are listed in table 9.3.3.

We define the degree of saturation, Ω, as the ratio of the product of the solutes (sometimes called ion activity product) over the product of the solutes at saturation (solubility product),

$$\Omega = \frac{[CO_3^{2-}][Ca^{2+}]}{[CO_3^{2-}]_{sat}[Ca^{2+}]_{sat}} = \frac{[CO_3^{2-}][Ca^{2+}]}{K_{sp}^{CaCO_3}} \qquad (9.3.2)$$

where $[CO_3^{2-}]$ and $[Ca^{2+}]$ are the observed carbonate ion and calcium concentrations. We can calculate the former from the measured *DIC* and *Alk* concentrations using the equations given in table 8.2.1. The latter is generally assumed to be constant (see below). When $\Omega > 1$, seawater is supersaturated with respect to mineral $CaCO_3$; conversely, when $\Omega < 1$, seawater is undersaturated.

Since dissolved calcium is a major constituent of seawater with a mean concentration of about 10,280 μmol kg^{-1} (table 1.2.1), its concentration is almost three orders of magnitude larger than that of CO_3^{2-}. In addition, we have found in section 8.4 above that *DIC* variations caused by the formation and dissolution of $CaCO_3$

TABLE 9.3.1
Equations used to calculate the solubility products of calcite and aragonite
T = temperature in K, S = salinity on the practical salinity scale, and K_{sp} is in (mol kg^{-1})^2.

Equation	*Source*
Solubility product of calcite for a total pressure of 1 atm:	[*Mucci*, 1983]
$$\ln K_{sp}^{calcite} = -395.8293 + \frac{6537.773}{T} + 71.595\ln(T) - 0.17959 \cdot T + \left(-1.78938 + \frac{410.64}{T} + 0.0065453 \cdot T\right) S^{1/2} - 0.17755 \cdot S + 0.0094979 \cdot S^{3/2} \quad (1)$$	
Solubility product of aragonite for a total pressure of 1 atm:	[*Mucci*, 1983]
$$\ln K_{sp}^{aragonite} = -395.9180 + \frac{6685.079}{T} + 71.595\ln(T) - 0.17959 \cdot T + \left(-0.157481 + \frac{202.938}{T} + 0.0039780 \cdot T\right) S^{1/2} - 0.23067 \cdot S + .0136808 \cdot S^{3/2} \quad (2)$$	

TABLE 9.3.2
Changes in molal volume, ΔV_i and compressibility, ΔK_i for calcite and aragonite in seawater.
T_c: Temperature in degree Celsius; S: Salinity on the practical salinity scale. The unit of ΔV_i is cm^3 mol^{-1} K^{-1}, and that of ΔK_i is cm^3 mol^{-1} K^{-1} bar^{-1}.

Equation	*Source*
Molal volume and compressibility for calcite:	[*Millero*, 1983]
$$\Delta V_{calcite} = -65.28 + 0.397 \cdot T_c - 0.005155 \cdot T_c^2 + (19.816 - 0.0441 \cdot T_c - 0.00017 \cdot T_c^2) . (S/35)^{1/2} \quad (3)$$	
$$\Delta K_{calcite} = 0.01847 + 0.0001956 \cdot T_c - 0.000002212 \cdot T_c^2 + (-0.03217 - 0.0000711 \cdot T_c + 0.000002212) \cdot (S/35)^{1/2} \quad (4)$$	
Molal volume and compressibility for aragonite:	[*Millero*, 1983]
$$\Delta V_{aragonite} = -65.50 + 0.397 \cdot T_c - 0.005155 \cdot T_c^2 + (19.82 - 0.0441 \cdot T_c - 0.00017 \cdot T_c^2) \cdot (S/35)^{1/2} \quad (5)$$	
$$\Delta K_{aragonite} = 0.01847 + 0.0001956 \cdot T_c - 0.000002212 \cdot T_c^2 + (-0.03217 - 0.0000711 \cdot T_c + 0.000002212) \cdot (S/35)^{1/2} \quad (6)$$	

in the ocean amount to maximally about 100 μmol kg^{-1}. This directly translates into maximum Ca^{2+} variations induced by these processes of the same magnitude, representing variations of less than 1%. As these changes in Ca^{2+} concentrations are very small relative to those of CO_3^{2-}, we can approximate $[Ca^{2+}]$ as being constant. This permits us to simplify the definition of Ω:

$$\Omega \approx \frac{[CO_3^{2-}]}{[CO_3^{2-}]_{sat}} \quad (9.3.3)$$

An often convenient measure of the saturation state is simply the difference between the observed CO_3^{2-} ion concentration and the saturation CO_3^{2-} concentration, thus

$$\Delta CO_3^{2-} = [CO_3^{2-}] - [CO_3^{2-}]_{sat} \quad (9.3.4)$$

Calculation of Ω and ΔCO_3^{2-} from *DIC*, *Alk*, temperature, and salinity at various depths requires that the pressure dependence of the various dissociation constants be taken into account. *Millero* [1995] showed that the pressure effect can be estimated from molal volume and compressibility:

$$\ln(K_i^p/K_i^o) = -\frac{\Delta V_i}{RT}P + \frac{0.5\Delta K_i}{RT}P^2 \quad (9.3.5)$$

TABLE 9.3.3
Numerical values of the $CaCO_3$ solubility products as a function of temperature and pressure for a salinity of 35

	Surface (0 dbar)		*Thermocline (1000 dbar)*		*Deep Ocean (6000 dbar)*	
Temp. °C	$-\log_{10} K_{sp}$ *Calcite* $(mol\,kg^{-1})^2$	$-\log_{10} K_{sp}$ *Aragonite* $(mol\,kg^{-1})^2$	$-\log_{10} K_{sp}$ *Calcite* $(mol\,kg^{-1})^2$	$-\log_{10} K_{sp}$ *Aragonite* $(mol\,kg^{-1})^2$	$-\log_{10} K_{sp}$ *Calcite* $(mol\,kg^{-1})^2$	$-\log_{10} K_{sp}$ *Aragonite* $(mol\,kg^{-1})^2$
0	6.368	6.166	6.282	6.086	5.894	5.724
2	6.367	6.166	6.283	6.088	5.903	5.734
4	6.366	6.166	6.285	6.090	5.912	5.743
6	6.366	6.167	6.286	6.092	5.921	5.753
8	6.366	6.168	6.287	6.094	5.929	5.762
10	6.365	6.169	6.288	6.097	5.937	5.771
12	6.365	6.170	6.290	6.100	5.944	5.780
14	6.365	6.172	6.291	6.103	5.952	5.789
16	6.366	6.174	6.293	6.106	5.958	5.797
18	6.366	6.177	6.294	6.110	5.965	5.806
20	6.367	6.180	6.296	6.114	5.971	5.814
22	6.368	6.183	6.298	6.118	5.977	5.822
24	6.369	6.187	6.300	6.123	5.983	5.830
26	6.371	6.191	6.302	6.127	5.988	5.837
28	6.372	6.195	6.305	6.132	5.993	5.845
30	6.374	6.200	6.307	6.138	5.998	5.853

Sources: The solubility products at 1 atm pressure are from *Mucci* [1983]. The pressure dependence is after *Millero* [1983].

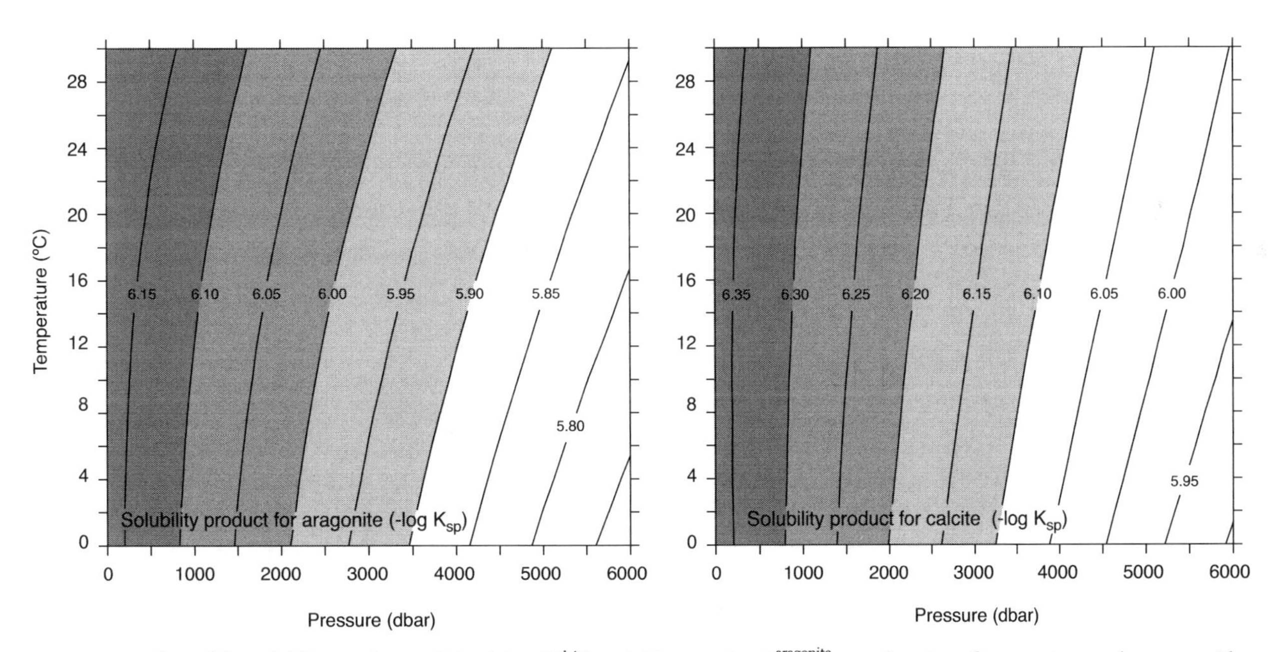

FIGURE 9.3.1: Plots of the solubility products of (a) calcite, $K_{sp}^{calcite}$, and (b) aragonite, $K_{sp}^{aragonite}$, as a function of temperature and pressure. The solubility products are shown as the negative logarithms to the base 10. The solubility products are from *Mucci* [1983], with the pressure dependence after *Millero* [1983].

where K_i^p is the value of the dissociation constant at pressure P in bar, K_i^o is the value of the dissociation constant at 1 atm pressure, T is temperature in Kelvin, R is the gas constant, and ΔV_i and ΔK_i are changes in molal volume and compressibility for the respective dissociation constants. Table 9.3.2 lists the equations for the molal volumes and compressibilities for aragonite and calcite in seawater.

Over the range encountered in the oceans, the solubility products of calcite and aragonite depend strongly on pressure and relatively little on temperature (figure 9.3.1). It is nevertheless interesting to note that calcite

TABLE 9.3.4
Saturation concentration of CO_3^{2-} as a function of temperature and pressure for a salinity of 35

	Surface (0 dbar)		*Thermocline (1000 dbar)*		*Deep Ocean (6000 dbar)*	
Temp. °*C*	$[CO_3^{2-}]_{sat}$ *Calcite* $\mu mol\ kg^{-1}$	$[CO_3^{2-}]_{sat}$ *Aragonite* $\mu mol\ kg^{-1}$	$[CO_3^{2-}]_{sat}$ *Calcite* $\mu mol\ kg^{-1}$	$[CO_3^{2-}]_{sat}$ *Aragonite* $\mu mol\ kg^{-1}$	$[CO_3^{2-}]_{sat}$ *Calcite* $\mu mol\ kg^{-1}$	$[CO_3^{2-}]_{sat}$ *Aragonite* $\mu mol\ kg^{-1}$
0	41.6	66.3	50.7	79.8	124.2	183.7
2	41.7	66.3	50.6	79.4	121.5	179.5
4	41.8	66.3	50.5	79.1	119.0	175.5
6	41.8	66.2	50.3	78.7	116.6	171.7
8	41.9	66.1	50.2	78.3	114.4	168.1
10	41.9	65.9	50.0	77.8	112.4	164.6
12	41.9	65.7	49.9	77.3	110.5	161.3
14	41.9	65.4	49.7	76.7	108.7	158.1
16	41.8	65.0	49.5	76.1	107.0	155.1
18	41.8	64.7	49.3	75.5	105.4	152.1
20	41.7	64.2	49.1	74.8	103.9	149.3
22	41.6	63.7	48.9	74.0	102.5	146.6
24	41.5	63.2	48.7	73.3	101.2	143.9
26	41.4	62.6	48.4	72.5	99.9	141.3
28	41.2	62.0	48.2	71.6	98.7	138.9
30	41.0	61.3	47.9	70.8	97.6	136.4

The solubility products at 1 atm pressure are from *Mucci* [1983]. The pressure dependence is after *Millero* [1983]. The calcium concentration was assumed to be 0.01028 μmol kg^{-1}.

and aragonite are unusual salts in that their solubilities decrease with increasing temperature (table 9.3.4). Table 9.3.3 and 9.3.4 show that the difference in the solubilities and CO_3^{2-} saturation concentrations between typical surface waters and typical deep ocean waters is about a factor of 3. These tables also show that aragonite has a 50% higher solubility than calcite.

Variations in Saturation State

Having established the CO_3^{2-} saturation concentrations, we can now compute the saturation state of the waters with respect to calcite and aragonite. We know already from figure 9.1.2 that surface waters are generally supersaturated and deep waters are generally undersaturated. However, the depth horizon that separates waters that are supersaturated from those that are undersaturated (called the *saturation horizon*) is far from uniform. A closer inspection of figure 9.1.2 reveals that the saturation horizon with respect to calcite lies at more than 4000 m depth in the South Atlantic, and shoals to less than 1000 m in the North Pacific.

Figure 9.3.2 shows the variations of ΔCO_3^{2-} with respect to calcite and aragonite solubility in more detail. Waters with a positive ΔCO_3^{2-} are supersaturated with respect to mineral $CaCO_3$, whereas the waters with a negative ΔCO_3^{2-} are undersaturated. In the latter case, the absolute value of ΔCO_3^{2-} is a measure of the tendency for the mineral $CaCO_3$ to dissolve. The saturation horizon ($\Delta CO_3^{2-} = 0$) with respect to calcite is deepest in the North Atlantic (more than 4500 m), intermediate in the South Atlantic and South Pacific (about 4500 to 3500 m), and shallowest in the North Pacific (less than 1000 m).

The saturation horizon for aragonite is located about 1000 to 1500 m shallower than that for calcite in the Atlantic, as expected from its metastability and hence higher solubility. In the Pacific, the aragonite saturation horizon lies within the main thermocline and almost reaches the surface in the North Pacific.

These spatial variations in the saturation horizons are mainly a consequence of variations in the CO_3^{2-} ion concentrations, since the saturation concentrations themselves vary little between the different ocean basins. Therefore, in order to understand the distribution of the saturation horizons, we need to understand what controls the distribution of the CO_3^{2-} ion.

Carbonate Ion Distribution

The distribution of the CO_3^{2-} ion shown in figure 9.1.2 seems counterintuitive and puzzling. Since carbonate is removed in the surface waters and added back to seawater mainly in the interior of the oceans, one would expect a nutrient-like distribution, with a minimum at the surface and a maximum in the deep ocean. Our discussion of ocean carbon chemistry in section 8.2 gives us the tools to understand this surprising

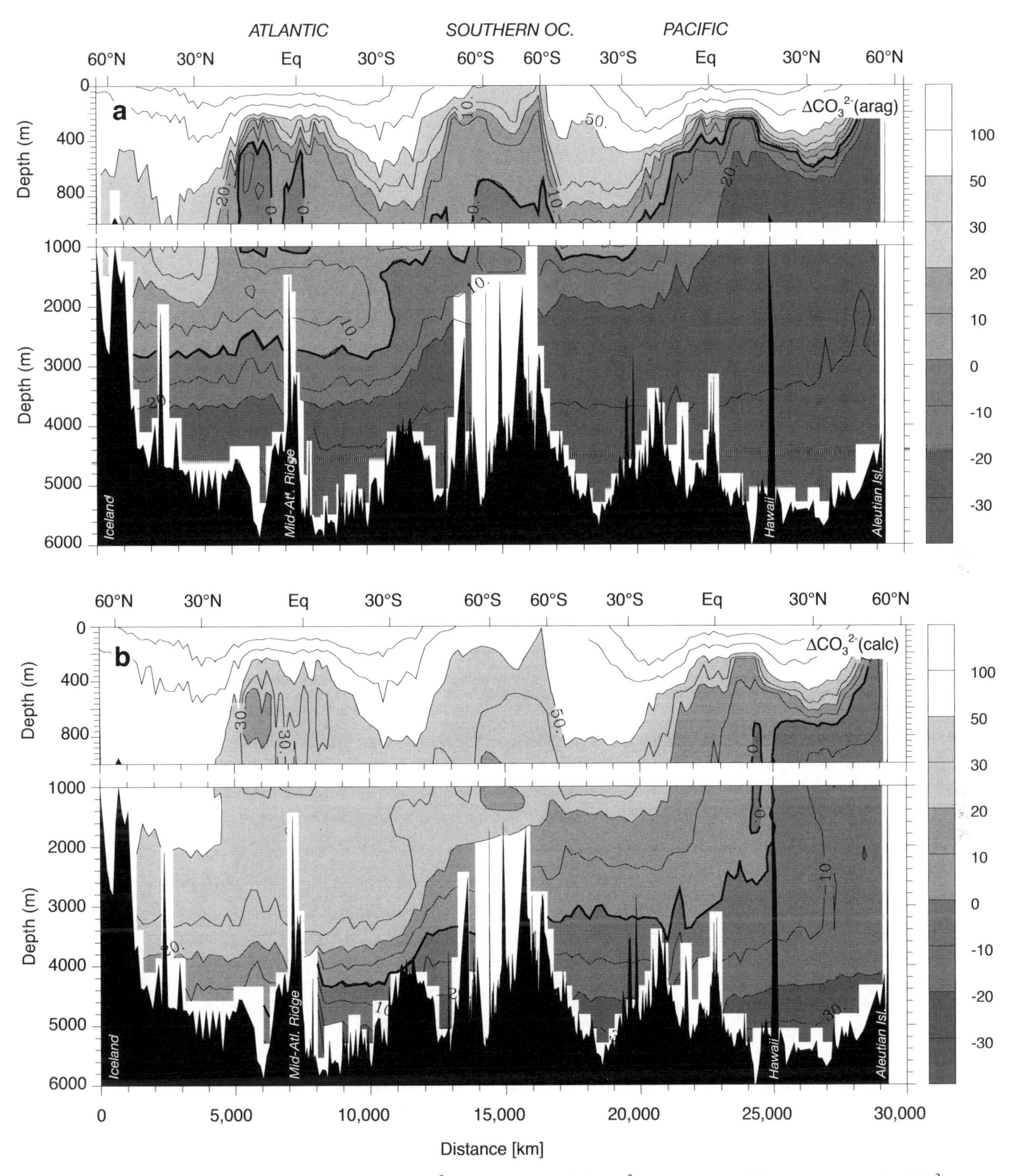

FIGURE 9.3.2: Vertical sections of the difference between the CO_3^{2-} concentration and the CO_3^{2-} saturation (a) with respect to aragonite, ΔCO_3^{2-} (aragonite), and (b) with respect to calcite, ΔCO_3^{2-} (calcite), along the track shown in figure 2.3.3a. See also color plate 8. The saturation concentration of CO_3^{2-} was calculated using the solubility product for calcite and aragonite from *Mucci* [1983], with the pressure dependence after *Millero* [1983]. The calcium concentration was assumed to be a function of salinity [*Millero*, 1982].

TABLE 9.3.5
Changes in DIC, Alk, $[HCO_3^-]$, and $[CO_3^{2-}]$ as a result of the remineralization of organic matter and dissolution of mineral $CaCO_3$

	Mineralization of Org. Matter	*Dissolution of $CaCO_3$*
ΔDIC (mol/mol)	+1	+1
ΔAlk (mol/mol)	−0.14	+2
$\Delta[CO_3^{2-}]$ (mol/mol)	−1.14	+1
$\Delta[HCO_3^-]$ (mol/mol)	+2.14	0

The changes in Alk due to the remineralization of organic matter were calculated using the nitrogen-to-carbon stoichiometry of *Anderson and Sarmiento* [1994]. The changes in $[CO_3^{2-}]$ and $[HCO_3^-]$ have been calculated using the approximations (8.2.17) and (8.2.18), thus $\Delta[CO_3^{2-}] \approx \Delta Alk - \Delta DIC$ and $\Delta[HCO_3^-] \approx 2 \cdot \Delta DIC - \Delta Alk$.

TABLE 9.3.6
Changes in DIC, Alk, HCO_3^-, and CO_3^{2-} as a typical surface water parcel from the low latitudes is moved first to intermediate depths and then to the deep ocean

	(1) Surface Water	*(2) Move to Intermed. Water*[a]	*(3) Move to Deep Water*[b]
DIC (μmol kg^{-1})[c]	2026	2246	2346
Alk (μmol kg^{-1})[c]	2308	2320	2413
$[CO_3^{2-}]$ (μmol kg^{-1})[d]	202	80	80
$[HCO_3^-]$ (μmol kg^{-1})[d]	1812	2123	2220

[a] Remineralizing 200 μmol kg^{-1} of organic matter and dissolving 20 μmol kg^{-1} $CaCO_3$.

[b] Remineralizing 50 μmol kg^{-1} of organic matter and dissolving 50 μmol kg^{-1} $CaCO_3$.

[c] The changes in DIC and Alk were calculated using the stoichiometries shown in table 9.3.5.

[d] The changes in $[HCO_3^-]$ and $[CO_3^{2-}]$ were calculated using a full chemistry model and employing the dissociation constants of *Mehrbach et al.* [1973] as refitted by *Dickson and Millero* [1987]. The temperature and salinity were assumed to remain constant at 24°C and 35.34.

distribution. Let us recall that we can approximate the CO_3^{2-} ion concentration by (8.2.18),

$$[CO_3^{2-}] \approx Alk - DIC \qquad (9.3.6)$$

We have seen in section 8.3 that both DIC and Alk are affected by the soft-tissue and carbonate pumps, but in different proportions (table 9.3.5). The remineralization of 1 mol of organic matter increases DIC by 1 mol, but decreases Alk by $16/117 = 0.14$ mol because of the oxidation of organic nitrogen to nitrate. The dissolution of 1 mol of $CaCO_3$ increases DIC by 1 mol and Alk by 2 mol. Using the above approximation, we can easily calculate that the remineralization of 1 mol organic matter results in a $[CO_3^{2-}]$ decrease of 1.14 mol, whereas the dissolution of 1 mol $CaCO_3$ leads to a 1 mol increase in $[CO_3^{2-}]$. Thus, the two biological pumps have competing effects on the CO_3^{2-} concentration, with nearly equal strength. Hence, the net effect depends on the relative contribution of the two biological pumps.

If the ratio of the contribution from the soft-tissue and carbonate pumps were the same everywhere, the $[CO_3^{2-}]$ distribution would follow some simple rules. For example, if the organic carbon-to-$CaCO_3$ remineralization ratio is substantially larger than one, the CO_3^{2-} ion content should be controlled by the remineralization of organic matter. Alternatively, if this ratio is substantially smaller than one, the CO_3^{2-} concentration should be controlled by the dissolution of $CaCO_3$.

We saw in chapter 5 (figure 5.3.1b) that the organic carbon-to-$CaCO_3$ remineralization ratio varies substantially with depth. In the upper thermocline, this ratio was found to be around 5:1, but decreased to values near 1:1 in the deep ocean. As this ratio is nearly always above 1:1, it appears that the CO_3^{2-} ion concentration in the oceans is mainly determined by the soft-tissue pump. It is only in the deep ocean, where the organic carbon-to-$CaCO_3$ input ratio approaches unity, that the carbonate pump is strong enough to exert a major influence. Since the remineralization of organic matter reduces the CO_3^{2-} content of seawater, the distribution of CO_3^{2-} is negatively correlated with the major nutrients, i.e., high near the surface and low in the deep ocean.

To examine the relative influence of the soft-tissue and carbonate pumps on the CO_3^{2-} concentration more specifically, let us perform a thought experiment similar to that shown in table 8.3.2. We start with a typical surface ocean water sample that has a DIC concentration of about 2026 μmol kg^{-1} and an alkalinity of 2308 μmol kg^{-1} (see table 8.2.4), giving a CO_3^{2-} content of about 202 μmol kg^{-1} (table 9.3.6). We then add carbon from the remineralization of organic matter and the dissolution of $CaCO_3$ in a 10:1 proportion as found in surface waters, i.e., about 200 μmol kg^{-1} from the soft-tissue pump and 20 μmol kg^{-1} from the carbonate pump. If the ocean were borate-free and therefore approximation (9.3.6) were accurate, table 9.3.5 tells us that this should lead to a decrease in the CO_3^{2-} concentration of about 208 μmol kg^{-1}. The real decrease is only about 122 μmol kg^{-1}, because of strong buffering by borate (table 9.3.6). In a second step, we add carbon from the remineralization of organic matter and the dissolution of $CaCO_3$ in a 1:1 proportion as found in the deep ocean, i.e., about 50 μmol kg^{-1} from the soft-tissue pump and 50 μmol kg^{-1} from the carbonate pump. Again, in a borate-free ocean, this would lead to a reduction in the CO_3^{2-} content of about 7 μmol kg^{-1}. Including borate, there is actually no reduction at all (table 9.3.6).

In summary, it is important to remember that the CO_3^{2-} ion is only a relatively small fraction of the fast-exchanging pool of total inorganic carbon. The CO_3^{2-}

concentration is set by all reactions that affect this pool, not only those that produce or consume CO_3^{2-} directly. As the soft-tissue pump dominates the interior ocean distribution of *DIC*, it also controls the concentration of the CO_3^{2-} ion.

Water Column Dissolution

The degree to which a settling $CaCO_3$ particle get dissolved in the water column once it enters undersaturated conditions depends on how fast the dissolution reaction is relative to the time it takes for the particle to settle to the bottom of the ocean. For that purpose, we need to have an estimate of the dissolution kinetics and of the settling speed.

Laboratory studies have demonstrated that the dissolution kinetics of $CaCO_3$ can be described by an expression depending on the state of $CaCO_3$ saturation, Ω. Following *Morse and Berner* [1972], we write the rate of $CaCO_3$ dissolution as

$$\frac{d[CaCO_3]}{dt} = -[CaCO_3]\, k_{CaCO_3} (1 - \Omega)^n, \text{ for } \Omega < 1 \qquad (9.3.7)$$

where $[CaCO_3]$ is the concentration of $CaCO_3$ in units of mmol m^{-3}, k_{CaCO_3} is the $CaCO_3$ dissolution rate constant in units of day^{-1}, and the exponent n represents an apparent "order" of the reaction. *Morse and Arvidson* [2002] recently presented an extensive review and synthesis of the dissolution kinetics of carbonates, and showed that the apparent order n is in most cases larger than 1, i.e., that the reaction is nonlinear. *Keir* [1980] suggested originally on the basis of his laboratory studies an exponent n of 4.5 for calcite and 4.2 for aragonite. However, *Hales and Emerson* [1997] recently reevaluated these data using newly calculated saturation states, Ω, and suggested that a first-order reaction, i.e., an exponent n of 1, represents a better fit to Keir's laboratory data. In addition, they argued that *in situ* measurements in sediments are also more consistent with a first-order reaction than with a 4.5-order reaction.

There is also little agreement with regard to the value of k_{CaCO_3}. Assuming a 4.5-order kinetics, *Keir* [1980] obtained the best fit to his data with a value of the order of 10 day^{-1}. There exist almost no measurements of dissolution kinetics in the water column, but determinations of k_{CaCO_3} based on sediment data generally have yielded much lower values than those determined in the laboratory. Measurements of *in situ* benthic fluxes [*Jahnke et al.*, 1994; *Berelson et al.*, 1994], *in situ* microelectrode pore water pH profiles [*Hales and Emerson*, 1996; *Archer et al.*, 1989; *Hales et al.*, 1994], or carbonate chemistry of pore waters collected with *in situ* samplers [*Martin and Sayles*, 1996], interpreted with models of pore water reactions and 4.5-order kinetics, give estimates of 5×10^{-5} to 1 day^{-1}. The much lower dissolution rates found in the sediments might be related to surface effects, including slower diffusive transport of solutes away from the surface, and possible chemical and physical inhibitions of the reactions at the surface of the mineral [*Morse and Arvidson*, 2002]. Theoretical models also suggest that the area-to-volume ratio of the particles plays a role [*Dreybrodt et al.*, 1996], explaining some of the differences between various sediment study locations [*Hales and Emerson*, 1997].

The corresponding values of k_{CaCO_3} for a first-order kinetics are numerically lower than those estimated by assuming a reaction order of $n = 4.5$. The reevaluation of Keir's laboratory data by *Hales and Emerson* [1997] with a first-order kinetics model gives a value of 0.38 day^{-1}, while the sediment data suggest values for k_{CaCO_3} of the order of 1×10^{-4} day^{-1}.

The $CaCO_3$ dissolution equation (9.3.7) is a homogeneous first-order differential equation, whose solution is an exponential decay function with an e-folding timescale of $\tau = 1/(k_{CaCO_3}(1-\Omega)^n)$. If we assume that the dissolution of $CaCO_3$ follows a first-order kinetics, i.e., $n = 1$, and adopt a value midway between the laboratory and the sediment data–based estimates of 0.01 day^{-1} for k_{CaCO_3}, the e-folding timescale for the dissolution reaction in the case of an undersaturation $\Omega = 0.9$ is about 1000 days.

How does this compare to the time it takes for $CaCO_3$ particles to settle through the water column? Because of their high density (see chapter 5), carbonate mineral particles are among the fastest settling particles, and sinking velocities of more than 100 m day^{-1} have been observed [*Berelson*, 2002]. Such rapid sinking translates into a residence time in a 5000 m-deep water column of about 50 days. If we consider, however, that the upper portion of the water column is supersaturated, the time $CaCO_3$ particles spend in the undersaturated portion may be only a few weeks. This is very short in comparison to the timescale of dissolution.

This argument, together with the observation that $CaCO_3$ particles caught in sediment traps show little corrosion [*Honjo*, 1978], has led to the paradigm that $CaCO_3$ behaves conservatively above the calcite saturation horizon [*Sverdrup et al.*, 1941; *Honjo*, 1978; *Deuser*, 1986; *Milliman*, 1993] and that dissolution in the water column below the saturation horizon is relatively small. However, there is increasing evidence that a considerable amount of $CaCO_3$ dissolves in the upper 1000 m of the oceans, even in waters that are supersaturated with respect to calcite. Integrating flux data from deep sediment traps over the entire ocean, *Milliman* [1993] estimated that about 0.30 Pg C yr^{-1} are leaving the upper 1000 m of the ocean, mainly in the form of calcite. Accepting a surface export of about 0.7 Pg C yr^{-1} suggests that perhaps more than 50% of the exported $CaCO_3$ gets dissolved before reaching 1000 m. This conclusion is relatively uncertain because of possibly

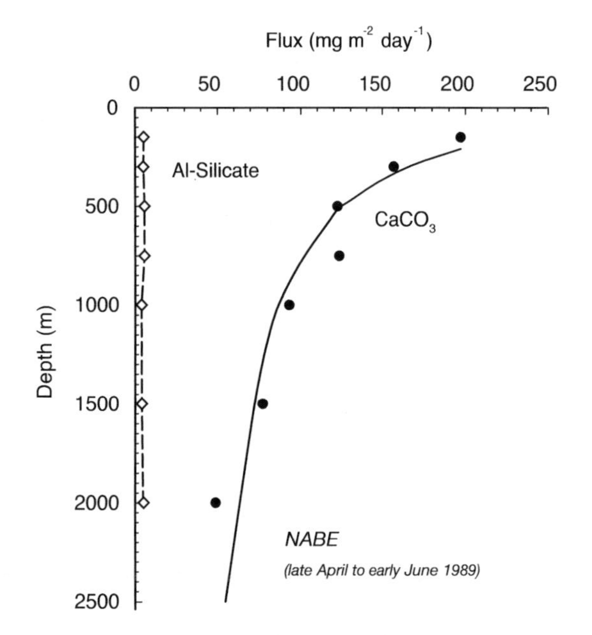

FIGURE 9.3.3: Plot of the vertical flux of $CaCO_3$ and aluminosilicates in the upper 2000 m during the North Atlantic Bloom Experiment (NABE; 47°30′N, 19°W) as measured by floating traps (total deployment period of six weeks). Carbonate flux decreases by 75% between 150 and 2000 m, but the aluminosilicate flux remains relatively constant (5.3 vs. 4.9 mg m^{-2} day^{-1}, respectively). From *Milliman et al.* [1999], based on data from *Martin et al.* [1993].

large errors in both flux estimates. However, as discussed by *Milliman et al.* [1999], there exists corroborating evidence.

Sediment trap data from the North Atlantic Bloom Experiment that resolve the vertical change in $CaCO_3$ fluxes in the thermocline show a large decrease of the flux of $CaCO_3$ between 150 and 1000 m [*Martin et al.*, 1993], i.e., in waters that are supersaturated with respect to both calcite and aragonite (figure 9.3.3). Other corroborating data come from the equatorial Pacific and the Arabian Sea, where surface calcification rates were on average several times larger than the sediment trap–measured $CaCO_3$ fluxes at 1000 m depth [*Balch and Kilpatrick*, 1996; *Milliman et al.*, 1999].

Furthermore, *Armstrong et al.* [2002] noted in their analyses of sediment traps from the equatorial Pacific that the exponential length-scale of ballast dissolution is only about 500 m and about equal to the length-scale of the remineralization of the unprotected fraction of organic matter (see equations (5.4.5) and (5.4.6)). Since $CaCO_3$ represents one of the most important components of ballast, this also suggests a substantial dissolution of $CaCO_3$ in the upper ocean.

Additional evidence comes from investigations of potential alkalinity changes in the various ocean basins by *Sabine et al.* [2002b], *Feely et al.* [2002], and *Chung et al.* [2003] as summarized by *Feely et al.* [2004a]. These authors studied the difference between measured potential alkalinity, P_{Alk}, and preformed potential alkalinity, $[P_{Alk}]_{preformed}$, in order to estimate the cumulative contribution of $CaCO_3$ dissolution to *Alk* from the time a water parcel has lost contact with the atmosphere. These authors multiplied this difference by a factor of 0.5 in order to express it in terms of changes in *DIC*. This tracer, ΔDIC^{CaCO_3}, is conceptually analogous to the apparent oxygen utilization, *AOU*, and is given by:

$$\begin{aligned}\Delta DIC^{CaCO_3} &= \frac{1}{2}([P_{Alk}] - [P_{Alk}]_{preformed}) \\ &= \frac{1}{2}(Alk + [NO_3^-] - (Alk_{preformed} + [NO_3^-]_{preformed}))\end{aligned} \qquad (9.3.8)$$

where $[NO_3^-]_{preformed}$ is preformed nitrate. Preformed nitrate can be estimated from *AOU* exactly the same way that we estimated preformed phosphate (see (5.3.2) and (5.3.3)), i.e.,

$$[NO_3^-]_{preformed} = [NO_3^-] + r_{N:O_2} AOU \qquad (9.3.9)$$

where $r_{N:O_2}$ is the N:O_2 ratio of organic matter remineralization, i.e., $r_{N:O_2} = 16:-170$ [*Anderson and Sarmiento*, 1994]. Inserting (9.3.9) into (9.3.8) gives

$$\Delta DIC^{CaCO_3} = \frac{1}{2}(Alk - Alk_{preformed} - r_{N:O_2} AOU) \qquad (9.3.10)$$

Various empirical relationships have been determined for preformed alkalinity by fitting near-surface measurements of *Alk* as a function of tracers that are conservative in the interior of the ocean, such as salinity, *S*, potential temperature, and $PO = [O_2] - r_{O_2:P}[PO_4^{3-}]$. We adopt here the empirical relationship of *Gruber et al.* [1996], as it pertains to the global ocean. The other empirical relationships give similar results.

$$Alk_{preformed} = (367.5 + 54.9 \cdot S + 0.074 \cdot PO)\ \mu\text{mol kg}^{-1} \qquad (9.3.11)$$

Figure 9.3.4 shows the distribution of ΔDIC^{CaCO_3}, resulting from $CaCO_3$ dissolution in the Atlantic and Pacific, as well as the saturation horizons for calcite and aragonite. This comparison reveals that the onset of a substantial impact of $CaCO_3$ dissolution–derived *Alk* lies at great depth in the North Atlantic and shoals to substantially less than 1000 m in the Pacific. The 10 μmol kg^{-1} isoline roughly follows the saturation horizon of aragonite, and is well above the saturation horizon for calcite. *Sabine et al.* [2002b], *Feely et al.* [2002], and *Chung et al.* [2003] used the distribution of ΔDIC^{CaCO_3} together with age tracers to estimate $CaCO_3$ dissolution rates and found that the dissolution between the aragonite and calcite saturation horizons represents a substantial fraction of the total water column dissolution.

This is at odds, however, with the finding that only about 10% of the total surface production and export of

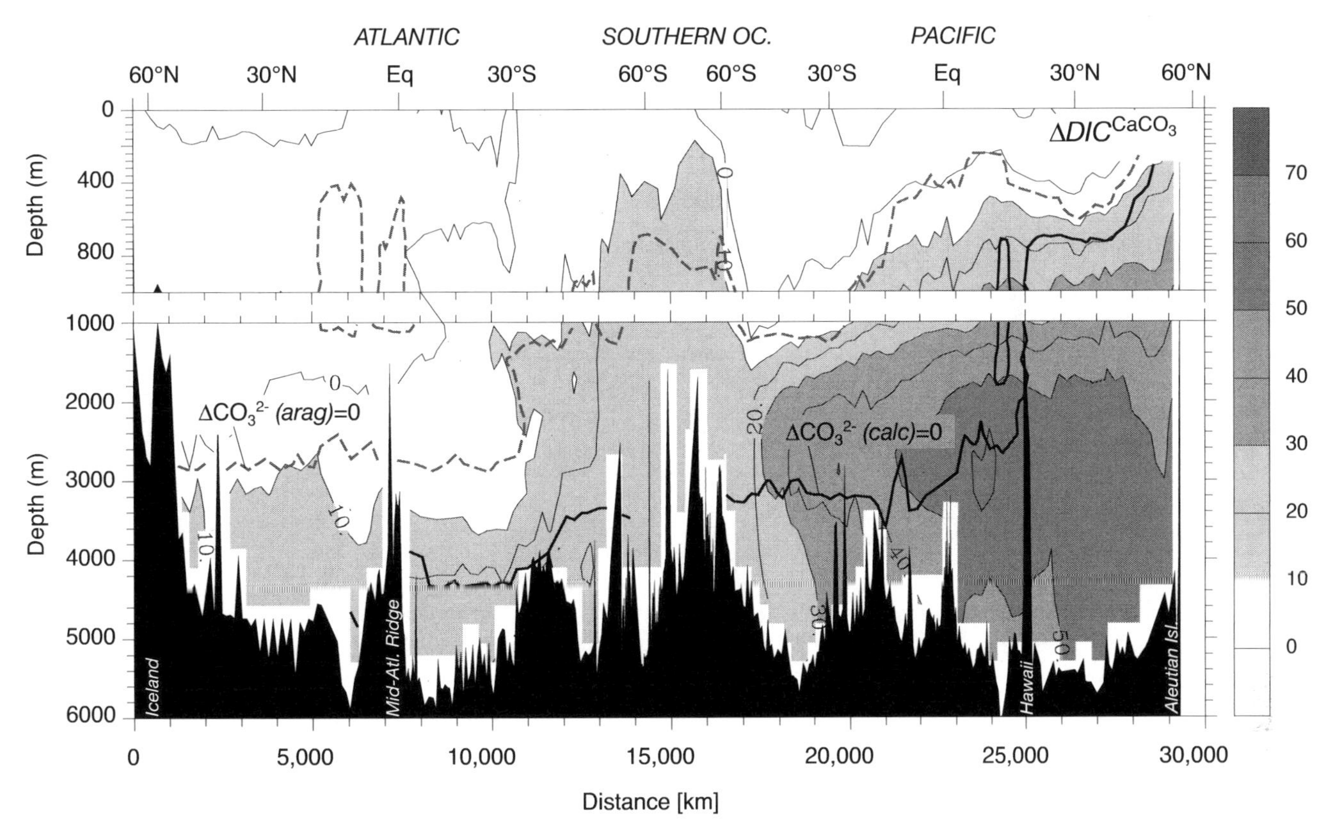

FIGURE 9.3.4: Vertical section of ΔDIC^{CaCO_3} along the track shown in figure 2.3.3a. Also shown are the saturation horizons for aragonite and calcite (i.e., the zero isolines of ΔCO_3^{2-}). Above these isolines, waters are supersaturated with respect to the respective mineral phases, whereas waters are undersaturated below these lines. The onset of a significant dissolution signal tends to coincide with the saturation horizon for aragonite.

$CaCO_3$ is in the form of aragonite [*Berner and Honjo*, 1981; *Fabry*, 1990]. Therefore, either a much larger fraction of the $CaCO_3$ exported from the surface ocean consists of aragonite or of a more soluble form of calcite, such as high-magnesium calcite [*Iglesias-Rodríguez et al.*, 2002a], or dissolution occurs through other mechanisms than just corrosion driven by the saturation state of seawater. *Fiadeiro* [1980] suggested that dissolution of $CaCO_3$ above the $CaCO_3$ saturation horizons may occur in connection with biological transformations. One such possible mechanism is dissolution in the acidic guts and feces of grazers. Another mechanism involves the microbial oxidation of organic matter in sinking material [*Alldredge and Cohen*, 1987]. This reaction produces CO_2, and hence creates an acidic micro-environment conducive to dissolution.

In summary, it appears that the cycling of $CaCO_3$ in the water column is much more dynamic than previously thought, and that the old paradigm of a conservative behavior of $CaCO_3$ above the calcite saturation horizon needs to be discarded. It is not yet clear what the new paradigm will be, as not much is really known about the processes that control the cycling of particles (organic and inorganic) in the main thermocline. A first step may be the separation of the sinking pool of $CaCO_3$ particles into a meta-stable pool, e.g., high-Mg calcite, that has a much higher solubilty and gets dissolved well above the saturation horizon for calcite, and into a second pool that consists of pure calcite, which settles to the bottom of the ocean without undergoing much dissolution at all.

Reconciling these uncertainties is of great importance for considering the impact of the decrease in saturation state of the upper ocean arising from the accumulation of anthropogenic CO_2 in the ocean (see also discussion in chapter 10). We can readily see from (9.3.6) that the increase in *DIC* caused by the invasion of anthropogenic CO_2 from the atmosphere decreases the CO_3^{2-} concentration, and hence lowers ΔCO_3^{2-}, i.e., the saturation state. *Feely et al.* [2004a] showed that the anthropogenic CO_2 that is present in the ocean (see Figure 10.2.7) has already lifted the saturation horizon for aragonite by up to several hundred meters in areas where anthropogenic CO_2 has penetrated to the depth of this horizon. As anthropogenic CO_2 is very likely to increase far above present levels in the ocean (see chapter 10), this change in the saturation state will likely become much larger, possibly having an impact on the dissolution processes governing the $CaCO_3$ dynamics in the water column. Given the present uncertainties in understanding the dissolution processes, accurate predictions of their response to future changes are challenging.

However, regardless of the uncertainties associated with the exact processes controlling water column dissolution of $CaCO_3$ particles, the difference from the remineralization of organic matter remains striking. For example, in a series of sediment traps distributed in the Arabian Sea, 98% of the surface-produced particulate organic carbon disappeared on average before reaching the traps at about 800 m, whereas only about 70% of the $CaCO_3$ disappeared [*Milliman et al.*, 1999]. This is consistent with our analyses of the different carbon pumps (see figure 8.4.2), which showed the much deeper dissolution signal for $CaCO_3$ in comparison to the shallower remineralization signal for organic matter. Consequently, $CaCO_3$ dissolution in the water column is often modeled as a simple exponential curve with a length scale on the order of 3500 m [*Najjar and Orr*, 1998; *Yamanaka and Tajika*, 1996], which leads to a much slower rate of $CaCO_3$ dissolution in the water column than achieved with the power law function commonly used for organic matter remineralization. The same conclusion holds if the newer double exponential remineralization parameterizations for organic matter are used (see chapter 5).

9.4 Diagenesis

Having followed the $CaCO_3$ cycle from the surface down the water column, we finally arrive at the sediments. Because of the much smaller role of water column processes for $CaCO_3$, sediments processes play a much more important role in the whole ocean budget of $CaCO_3$ than in the cycling of organic matter (although we have seen in chapter 6 that sediments play a major role for organic matter cycling below 1000 to 2000 m). On the other hand, as can be seen from a comparison of a deep ocean budget (>1000 m) for $CaCO_3$ with that of organic matter (table 9.4.1), a much bigger fraction of the $CaCO_3$ entering the ocean realm below 1000 m depth finally gets buried in the sediments. In the case of organic matter, this fraction is about 2%, whereas in the case of $CaCO_3$, this fraction is approximately 19%. In this section, we analyze what controls the amount of $CaCO_3$ that gets buried and how this is related to the $CaCO_3$ burial pattern that we have seen in figure 9.1.3.

TABLE 9.4.1
Deep ocean (>1000 m) budget for $CaCO_3$ and organic carbon.

	Organic Carbon[a] *Pg C yr*$^{-1}$	$CaCO_3$[b] *Pg C yr*$^{-1}$	*Ratio*[c]
Input across 1000 m depth horizon	0.87	0.70	1.2
Burial in sediments	−0.02	−0.13	0.2
Remineralization or dissolution	−0.85	−0.57	1.5

[a] Values are taken from Figure 6.5.3.

[b] Modified budget of *Milliman et al.* [1999]. The sinking flux of $CaCO_3$ across 1000 m has been increased to account for the observation that the organic carbon–to–$CaCO_3$ remineralization ratio in the deep ocean is about 1. The resulting higher dissolution flux from the sediments is within the range of estimates (0.30 to 0.48 Pg C yr^{-1}) given by *Archer* [1996a, 1996b].

[c] Organic carbon–to–$CaCO_3$ ratio.

Figure 9.4.1 shows in more detail the transition from the depth where $CaCO_3$ is well preserved in the sediments to the depth where $CaCO_3$ is completely absent from the sediments. The upper limit of the transition zone, where sediments are subjected to very little dissolution, is often called the *lysocline*, whereas the depth, where the sediments have lost virtually all their calcite due to dissolution, is referred to as the *calcite compensation depth* (CCD). At this depth the dissolution of $CaCO_3$ in the sediments matches exactly the deposition of $CaCO_3$ at the top of the sediments.

When the lysocline and the CCD are mapped throughout the oceans, an interesting pattern emerges (figure 9.4.2). Not only does the lysocline get shallower from the North Atlantic through the Indian to the North Pacific, but also the thickness of the transition zone between the lysocline and the CCD varies strongly from one place to another. We thus have two questions that we want to answer: What controls the depth of the lysocline, and what controls the thickness and shape of the transition zone between the lysocline and the CCD?

$CaCO_3$ Dissolution in Sediments

We have seen in chapter 7 that the dissolution of opal in the ocean depends on two major factors. First, it is influenced by the degree of undersaturation. Second, it is influenced by the degree of incorporation of aluminium into the opal lattice and the impact of this on opal solubility and dissolution kinetics.

One would expect that some of the same factors would control the dissolution of $CaCO_3$ in the sediments. While this conclusion is correct, there are several major differences between the dissolution of these two minerals. The most important difference has to do with

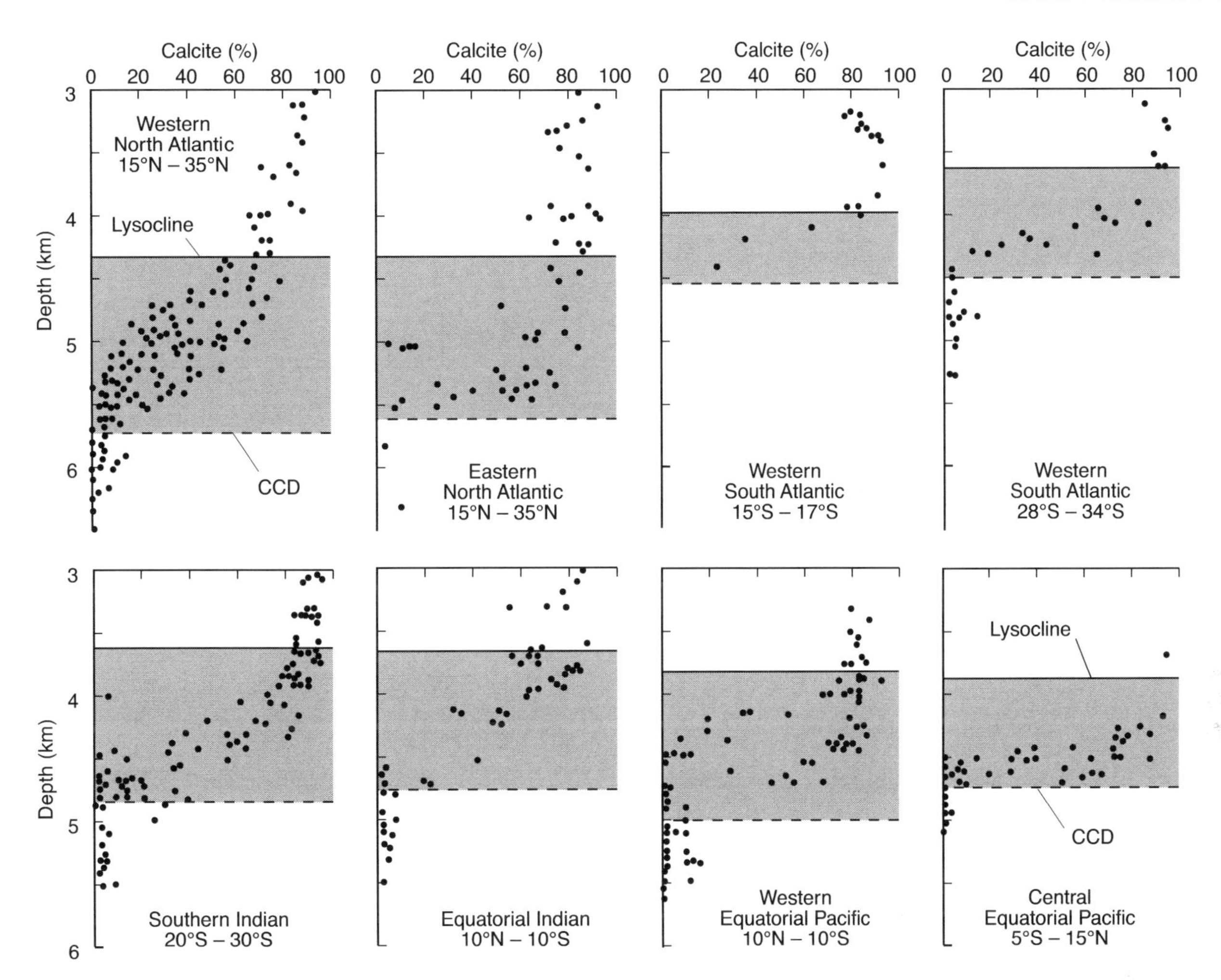

FIGURE 9.4.1: Plots of $CaCO_3$ content in surface sediments as a function of depth, separated into 8 different regions. From *Broecker and Peng* [1982].

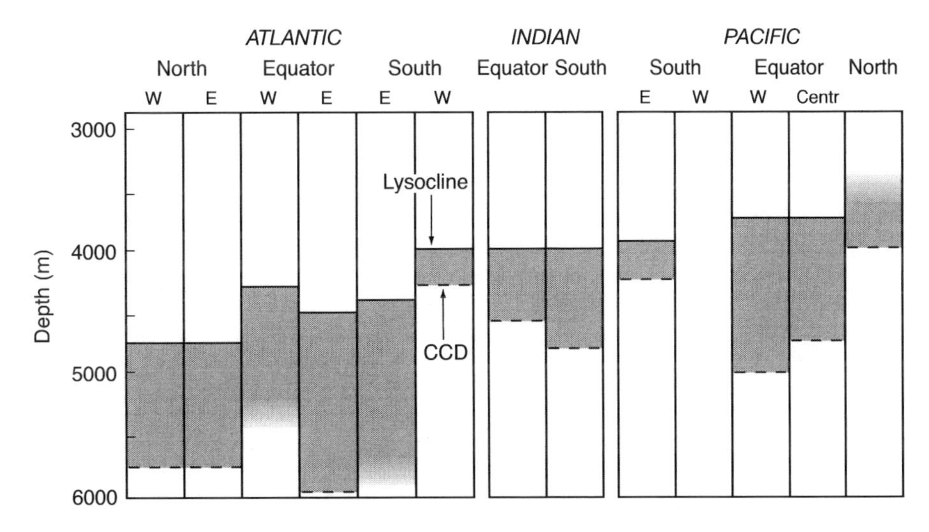

FIGURE 9.4.2: Water depths at the top (lysocline) and base (carbonate compensation depth, CCD) of the transition zone between sediments subjected to little dissolution and sediments that have lost virtually all their calcite to dissolution. From *Broecker and Takahashi* [1977b].

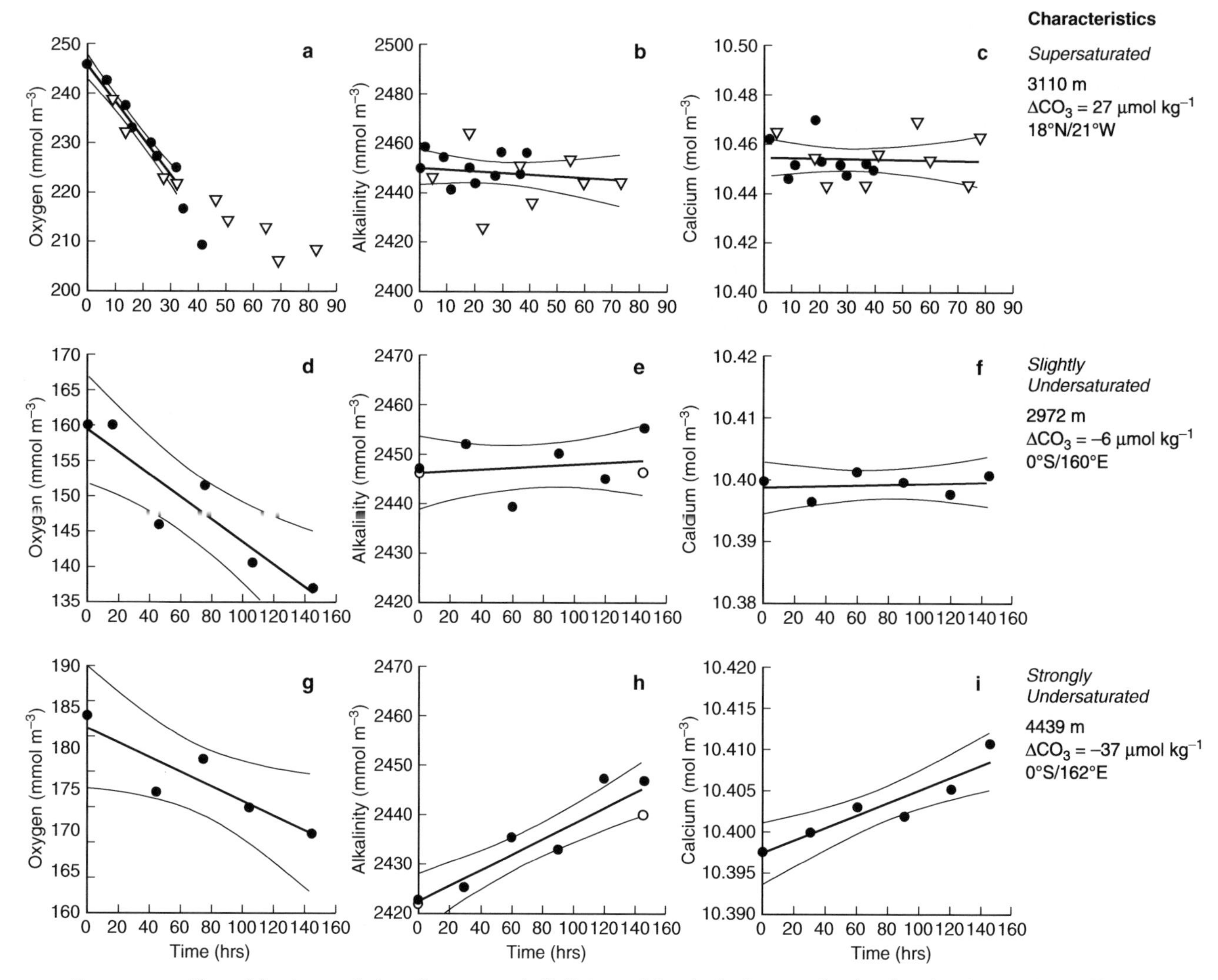

FIGURE 9.4.3: Plots of the time evolution of oxygen, total alkalinity, and dissolved calcium in benthic flux chambers at three different sites. (a–c) Atlantic site at 18°28′N, 21°02′W. Calcite saturation, $\Delta CO_3^{2-} = 27 \mu mol\ kg^{-1}$ (d–f). Pacific site at 0°00′S, 160°25′E. Calcite saturation, $\Delta CO_3^{2-} = -6 \mu mol\ kg^{-1}$. (g–i) Pacific site at 0°00′S, 162°41′E. Calcite saturation, $\Delta CO_3^{2-} = -37 \mu mol\ kg^{-1}$. The characteristics of each site are shown on the right side of each row. Straight lines represent the linear regression of the data; curved lines are the 95% confidence limits of the regression lines. Adapted from *Jahnke et al.* [1994].

the thermodynamic driving factor. Seawater is everywhere strongly undersaturated with respect to opal, resulting in a nearly uniform driving factor throughout the oceans. By contrast, the carbonate saturation state with respect to calcite varies substantially (figure 9.3.2). Therefore the variability in this driving factor must play a more important role in controlling the diagenesis of $CaCO_3$ in comparison with that of opal. We find strong evidence for this in the observation that the depth of the lysocline (figure 9.4.2) is broadly consistent with the depth of the saturation horizon (figure 9.3.2). This remarkable consistency suggests that the thermodynamic driving factor is the dominant factor determining $CaCO_3$ preservation.

One way to test this hypothesis is to look at results from benthic chambers that were deployed in waters of differing bottom water saturation [*Jahnke et al.*, 1994]. Figure 9.4.3 shows results from three such deployments. The first row depicts measurements from a site at 3100 m in the tropical North Atlantic, where bottom waters have a ΔCO_3^{2-} of 27 $\mu mol\, kg^{-1}$ with respect to calcite, i.e., they are supersaturated. The second and third rows depict measurements from two sites in the western equatorial Pacific, at depths of about 3000 m and 4400 m, respectively. The ΔCO_3^{2-} of the former site is $-6\ \mu mol\, kg^{-1}$, i.e., slightly undersaturated, whereas the ΔCO_3^{2-} of the latter is $-37\ \mu mol\, kg^{-1}$, i.e., strongly undersaturated. The bottom waters enclosed by the benthic chamber show a substantial decrease of oxygen at all three sites, reflecting the oxygen demand of the sediments as organic matter is remineralized. By contrast, large differences exist between the three sites in the sediment-water fluxes of *Alk* and Ca^{2+}. Whereas the strongly undersaturated site shows a strong efflux of *Alk*

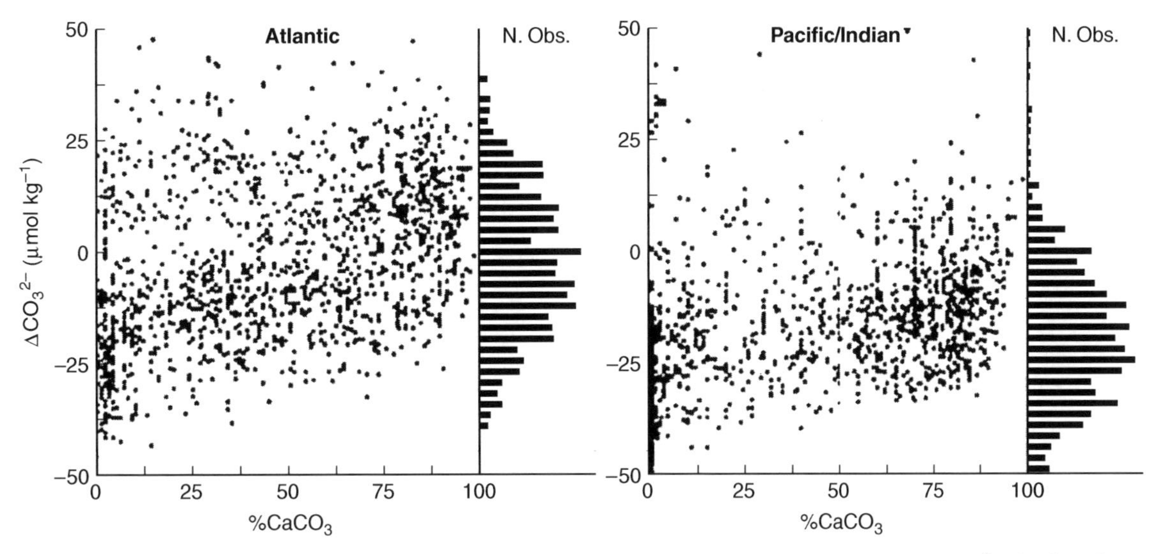

FIGURE 9.4.4: Carbonate content of sediments plotted as a function of the overlying water saturation state, ΔCO_3^{2-}, for the Atlantic and Pacific/Indian Oceans. Histograms on the right side of the plots show the frequency distribution of the observations. From *Archer* [1996a].

and Ca^{2+} from the sediments and hence an accumulation of these properties in the chamber, little change occurs at the supersaturated and slightly undersaturated sites. This suggests that dissolution of $CaCO_3$ does not begin in any substantial manner until bottom waters are significantly undersaturated, supporting the conclusion reached on the basis of the co-location of the lysocline and the calcite saturation horizon. Further support comes from the absence of significant changes in dissolution indices based on foraminiferal assemblages above the depth of the saturation horizon [*Broecker and Clark*, 2003].

If the saturation state were the only controlling factor, we would expect a strong correlation between the $CaCO_3$ content of the sediments and the saturation state of the overlying seawater. The absence of such a clear trend in figure 9.4.4 indicates that processes other than the saturation state of the overlying seawater must also play an important role. Two features of figure 9.4.4 are particularly noteworthy. First, there exist many sites where the overlying bottom waters are supersaturated with respect to calcite and nevertheless the $CaCO_3$ content of sediments is very small. Secondly, although masked by a substantial amount of scatter, *Archer* [1996b] showed that the average $CaCO_3$ content for a given level of water column saturation, ΔCO_3^{2-}, is about 10% lower in the Atlantic than in the Indian/Pacific.

What are the processes apart from the saturation state that control the preservation? Is it the rain rate of noncarbonate material or the rain rate of carbonate itself that influences the details in the preservation pattern? It turns out that while these two processes do influence diagenesis of $CaCO_3$, there exists a further major difference between the dissolution of opal and carbonate. Diagenetic reactions independent of the diagenesis of $CaCO_3$ can produce acid or base and thus change the carbonate saturation state of the pore waters in the sediments independently of the saturation state of the overlying waters.

Berger [1970] was the first to point out that CO_2 released into the pore water from the remineralization of organic matter acidifies the solution and can thus drive $CaCO_3$ dissolution even if the overlying bottom waters are supersaturated. Quantitative models of the influence of metabolically produced CO_2 on $CaCO_3$ dissolution have suggested that when the molar ratio of organic carbon to $CaCO_3$ of the deposited material is near 1:1 or greater, this process might have a large impact on $CaCO_3$ preservation [*Emerson and Bender*, 1981; *Archer*, 1991; *Hales*, 2003].

This can be understood by inspecting table 9.3.5, which shows that the remineralization of organic matter and the dissolution of $CaCO_3$ have nearly opposite effects on the CO_3^{2-} ion concentration. Thus, if we remineralize 1 mol of organic matter and use the CO_2 produced by the reaction to dissolve $CaCO_3$ according to

$$CO_2 + CaCO_3 + H_2O \rightleftharpoons 2HCO_3^- + Ca^{2+} \qquad (9.4.1)$$

we can dissolve 1.14 mol of $CaCO_3$ without changing the CO_3^{2-} saturation state. If the amount of organic carbon remineralized exceeds that of $CaCO_3$, one could expect complete dissolution.

As we have seen above, the ratio of organic carbon to $CaCO_3$ that is deposited on top of deep ocean sediments is of the order of 1:1 or higher. Since almost all organic matter gets remineralized, enough CO_2 is produced to potentially dissolve all $CaCO_3$ in the sediments. The fact that a substantial fraction of the deposited $CaCO_3$

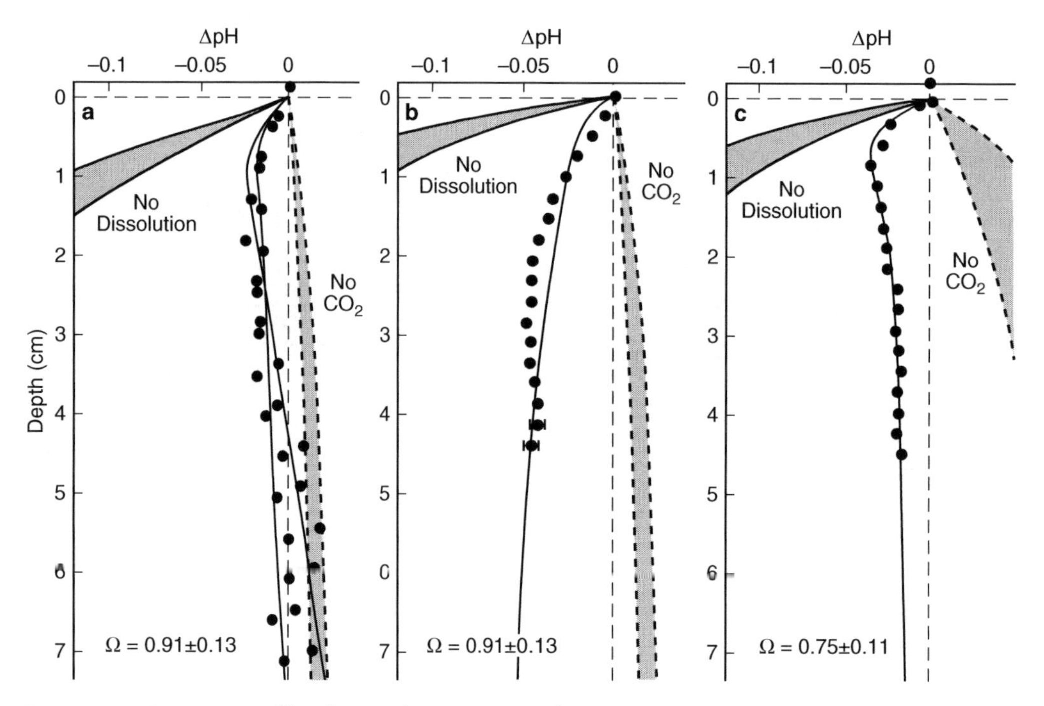

FIGURE 9.4.5: Pore water profiles of pH at three stations on the Ontong-Java Plateau in the western equatorial Pacific. (a) Station 2A at a depth of 2322m, with a calcite saturation $\Omega = 0.91 \pm 13$. (b) Station 2B at a depth of 2335 m, $\Omega = 0.91 \pm 0.13$. (c) Station 3 at 2966 m, $\Omega = 0.75 \pm 0.11$. Also shown are model-simulated pH curves for three scenarios: "no dissolution" (heavy solid lines), "no CO_2" generation from organic matter remineralization (heavy dashed lines), and the best fit to the data (light solid line). From *Hales and Emerson* [1996].

material gets permanently buried indicates that the coupling between remineralization of organic matter and dissolution of $CaCO_3$ is variable.

A powerful way to investigate the coupling between organic matter remineralization and $CaCO_3$ dissolution in sediments is to measure' vertical profiles of pore water constituents, especially $pH = -\log_{10}[H^+]$. Since the remineralization of organic matter produces free protons (see reaction 8.3.12), this reaction leads to a decrease in the pH of pore waters. In contrast, as the dissolution of $CaCO_3$ releases a base (CO_3^{2-}), which tends to consume a free proton to form HCO_3^-, this reaction leads to an increase in pH (see figure 8.2.1). Thus, pore water profiles of pH can give a qualitative indication of the relative role of organic matter remineralization versus $CaCO_3$ dissolution.

Three pH pore water profiles, obtained by carefully lowering a pH electrode into the sediments, are shown in figure 9.4.5. These profiles are from the Ontong-Java Plateau in the western equatorial Pacific. The three stations are located at water depths ranging from 2322 to 2966 m and are overlain with undersaturated waters with respect to calcite (Ω between 0.75 (station 3) and 0.91 (stations 2A and 2B), corresponding to ΔCO_3^{2-} between -5 and $-13\,\mu mol\,kg^{-1}$). All three pH profiles show decreasing pH with depth near the core top, indicating a small excess of organic matter remineralization over $CaCO_3$ dissolution. Thus, it indeed appears that the remineralization of organic matter may enhance the dissolution of $CaCO_3$.

If respiration is indeed a main driver for $CaCO_3$ dissolution, then why don't we see more $CaCO_3$ dissolution at shallower depths, where the rain of organic matter arriving at the sediments greatly exceeds that of $CaCO_3$, potentially providing more than enough free protons to dissolve $CaCO_3$ entirely?

One reason for the decoupling between remineralization of organic matter and the dissolution of $CaCO_3$ is that most of the organic matter is remineralized very quickly in the upper few centimeters of the sediments (see discussion in chapter 6 about the reactivity of organic matter), whereas the kinetics of $CaCO_3$ dissolution tends to be slower. This leads to a vertical separation in the sediments between the very near surface, where organic matter is remineralized, and the deeper layers, where most of the $CaCO_3$ dissolution might occur. As a result, a substantial fraction of the free protons produced by remineralization in the near-surface sediments diffuses out of the sediments and gets buffered by the bottom waters, decreasing the efficiency of respiratory-driven $CaCO_3$ dissolution [*Hales*, 2003]. The strength of the buffering depends on how fast HCO_3^- can diffuse into the sediments, while the degree of vertical decoupling between organic matter remineralization and $CaCO_3$ dissolution depends mainly on the accumulation rates and the differences in the kinetics. In shallower waters, the vertical separation tends to be larger, because of higher accumulation rates. The buffering of the protons freed by remineralization is more effective as well, mainly because of a generally higher degree of bottom water supersaturation. Both effects will tend to lead to a decrease in the efficiency of respiration to drive $CaCO_3$ dissolution in shallower waters [*Hales*, 2003]. In order to investigate this coupling between the different reactions and their interaction with diffusion and advection more quantitatively, we need a sediment model of early diagenesis, which we will develop and apply in the next section.

Modeling $CaCO_3$ Diagenesis

Chapters 6 and 7 provide us with most of the building blocks that we need to construct a model describing $CaCO_3$ diagenesis. Since we are especially interested in investigating the role of organic matter remineralization on the dissolution of $CaCO_3$, we start out with the model that we developed for organic matter diagenesis. We need to add the $CaCO_3$-specific reactions and equations, i.e., a reaction equation for mineral $CaCO_3$, and diffusion-reaction equations for all dissolved species of the carbonate system in seawater, i.e., CO_3^{2-}, HCO_3^-, and $H_2CO_3^*$. The three equations for CO_3^{2-}, HCO_3^-, and $H_2CO_3^*$ are coupled to each other by the dissociation reactions of the carbonate system in seawater (reactions 8.2.1 to 8.2.3). Since these dissociation reactions are very rapid, we can assume equilibrium for all practical purposes. Finally, we have to add conservation equations for *DIC* and *Alk*, as well as boundary conditions for *DIC* and *Alk*. The complete set of equations is given in table 9.4.2. These will now be discussed in more detail.

We begin with the equation describing the $CaCO_3$ content in the sediments. As is the case for organic matter, we have to take into account three processes: slow downward movement caused by the continuous deposition of fresh particles at the sediment water interface, bioturbation, and dissolution/remineralization reactions. The first two processes can be handled identically to the organic matter equation (6.2.1), but unlike the remineralization of organic matter, dissolution of mineral $CaCO_3$ is a purely chemical process. We adopt (9.3.7) for sediments by taking the porosity ϕ into account, and expressing the $CaCO_3$ concentration in terms of solids. This gives:

$$\frac{\partial(1-\phi)[\mathrm{CaCO_3}]}{\partial t} = -k_{\mathrm{CaCO_3}} \cdot [\mathrm{CaCO_3}](1-\phi)(1-\Omega)^n \quad (9.4.2)$$

where $[CaCO_3]$ is the concentration of $CaCO_3$ in units of mmol m^{-3} of solids.

We turn next to the solute equations for CO_3^{2-}, HCO_3^- and $H_2CO_3^*$. *Boudreau* [1987] pointed out that neglecting the reversible dissociation reactions between CO_3^{2-}, HCO_3^-, and $H_2CO_3^*$ results in a significant underestimation of the diffusive transport. This underestimation occurs because each inorganic carbon species is diffusing separately, experiencing a different spatial concentration gradient. For example, a fraction of the CO_3^{2-} produced by the dissolution of $CaCO_3$ goes into HCO_3^- and $H_2CO_3^*$, which are then transported away independently of the diffusion of the CO_3^{2-} ion.

Remineralization of organic matter in the sediments produces $H_2CO_3^*$. Hence the organic matter degradation term from the organic matter equation appears as a source in the reaction-diffusion equation for $H_2CO_3^*$ (see table 9.4.2). By contrast, the dissolution of $CaCO_3$ releases CO_3^{2-} ions. Therefore, the dissolution term from the $CaCO_3$ equation needs to be included in the CO_3^{2-} equation. In addition, the three governing equations for CO_3^{2-}, HCO_3^-, and $H_2CO_3^*$ are linked to each other by inorganic carbon system reactions in order to maintain chemical equilibrium. These exchange terms are denoted by R_{pH}.

To solve the system of equations of our diagenetic model, we need to specify dissociation constants, define *DIC* and *Alk*, and specify boundary conditions for all solutes and solids. In particular, we need to provide the rain rates of particulate organic matter and mineral $CaCO_3$ as boundary conditions.

Several similar $CaCO_3$ diagenesis models have been developed by various investigators [*Archer*, 1991, 1996a; *Archer et al.*, 1989; *Jahnke et al.*, 1994; *Hales and Emerson*, 1996] with varying details. The most important difference in the models is the absence or inclusion of anoxic remineralization of organic matter.

Model Applications

One of the goals of constructing this diagenetic model was to assess the importance of calcite dissolution driven by metabolic processes relative to that driven by bottom water undersaturation. This can best be assessed by simulating two extreme cases and then comparing them to the pore water pH profiles shown in figure 9.4.5. In the "no dissolution" case, the simulation includes only organic matter remineralization and no dissolution of $CaCO_3$ is permitted to occur. In the "no CO_2" case, there is no remineralization and calcite dissolves only in response to bottom water undersaturation.

The results of these two cases applied to the three sites on the Ontong-Java Plateau is plotted over the observations in figure 9.4.5. The two curves for the two scenarios represent upper and lower bounds of solutions when the input parameters are varied within their uncertainties. Neither of these cases can correctly reproduce the *in situ* pH observations. In the "no dissolution" case, the simulated pH decreases rapidly with depth, since the $H_2CO_3^*$ added by remineralization of organic matter does not get neutralized by $CaCO_3$ dissolution. In the "no CO_2" case, the modeled pH

TABLE 9.4.2
Governing equations for the $CaCO_3$ diagenesis model.
After Archer et al. [1989].

Solid equations:

$$(1-\phi)\frac{\partial[POC]}{\partial t} = D_B\frac{\partial}{\partial z}\left((1-\phi)\frac{\partial[POC]}{\partial z}\right) - k_{POC}[POC](1-\phi) \qquad (9.4.3)$$

$$(1-\phi)\frac{\partial[\mathrm{CaCO_3}]}{\partial t} = D_B\frac{\partial}{\partial z}\left((1-\phi)\frac{\partial[\mathrm{CaCO_3}]}{\partial z}\right) - k_{\mathrm{CaCO_3}}[\mathrm{CaCO_3}](1-\phi)(1-\Omega)^n \qquad (9.4.4)$$

Solute equations:

$$\phi\frac{\partial[\mathrm{H_2CO_3^*}]}{\partial t} = \frac{\partial}{\partial z}\left(\varepsilon_s^{\mathrm{H_2CO_3^*}}\frac{\partial[\mathrm{H_2CO_3^*}]}{\partial z}\right) + k_{POC}[POC](1-\phi) \pm R_{pH} \qquad (9.4.5)$$

$$\phi\frac{\partial[\mathrm{HCO_3^-}]}{\partial t} = \frac{\partial}{\partial z}\left(\varepsilon_s^{\mathrm{HCO_3^-}}\frac{\partial[\mathrm{HCO_3^-}]}{\partial z}\right) \pm R_{pH} \qquad (9.4.6)$$

$$\phi\frac{\partial[\mathrm{CO_3^{2-}}]}{\partial t} = \frac{\partial}{\partial z}\left(\varepsilon_s^{\mathrm{CO_3^{2-}}}\frac{\partial[\mathrm{CO_3^{2-}}]}{\partial z}\right) + k_{\mathrm{CaCO_3}}[\mathrm{CaCO_3}](1-\phi)(1-\Omega)^n \pm R_{pH} \qquad (9.4.7)$$

$$\phi\frac{\partial[\mathrm{O_2}]}{\partial t} = \frac{\partial}{\partial z}\left(\varepsilon_s^{\mathrm{O_2}}\frac{\partial[\mathrm{O_2}]}{\partial z}\right) - r_{\mathrm{O_2:C}}k_{POC}[POC](1-\phi) \qquad (9.4.8)$$

Equilibrium equations:

$$K_1 = \frac{[\mathrm{H^+}][\mathrm{HCO_3^-}]}{[\mathrm{H_2CO_3^*}]} \qquad (9.4.9)$$

$$K_2 = \frac{[\mathrm{H^+}][\mathrm{CO_3^{2-}}]}{[\mathrm{HCO_3^-}]} \qquad (9.4.10)$$

$$Alk = 2[\mathrm{CO_3^{2-}}] + [\mathrm{HCO_3^-}] \qquad (9.4.11)$$

$$DIC = [\mathrm{CO_3^{2-}}] + [\mathrm{HCO_3^-}] + [\mathrm{H_2CO_3^*}] \qquad (9.4.12)$$

Boundary conditions:
at $z = 0$:

$DIC = DIC^o$
$Alk = Alk^o$
$\mathrm{O_2} = \mathrm{O_2^o}$
rain rate of POC
rain rate of $CaCO_3$

at $z = Z_{max}$:

$$\frac{\partial(\text{solute})}{\partial z} = 0$$

$$\frac{\partial(\text{solid})}{\partial z} = 0$$

ϕ	=	sediment porosity
k_{POC}	=	organic degradation rate constant, s^{-1}
$k_{\mathrm{CaCO_3}}$	=	calcite, aragonite dissolution rate constant, day^{-1}
Ω	=	saturation state
n	=	dimensionless exponent for calcite and aragonite dissolution
$[POC]$	=	concentration of particulate organic material, mmol m^{-3} of total sediments
$[\mathrm{CaCO_3}]$	=	concentration of $CaCO_3$, mmol m^{-3} of total sediments
D_B	=	solid diffusion coefficient
ε_s^i	=	effective solute diffusion coefficient (see panel 6.2.2)
$r_{\mathrm{O_2:C}}$	=	stoichiometric oxygen-to-carbon ratio of organic matter
R_{pH}	=	addition or removal of carbon species in order to maintain equilibrium
K	=	dissociation constants of inorganic carbon system in seawater

increases with depth, since the CO_3^{2-} is added to the pore water without neutralization.

This shows that dissolution driven by metabolic CO_2 production plays a significant role in determining the pore water chemistry at these sites. *Hales and Emerson* [1996] also determined which combination of input parameters reproduced the observations optimally (light solid line in figure 9.4.5). They found from this statistical evaluation of the pore water pH data and the model that over 65% of the total dissolution is driven by metabolic CO_2. Assuming a 4.5-order kinetics, this optimization also constrains the calcite dissolution rate constants to values of about 5×10^{-5} to 1.6×10^{-3} day^{-1}, lower than most other field results, but consistent with the general trend of lower rate constants determined from field observations compared with laboratory studies [*Keir*, 1980; *Archer et al.*, 1989; *Hales et al.*, 1994; *Berelson et al.*, 1994; *Jahnke et al.*, 1994; *Martin and Sayles*, 1996; *Hales and Emerson*, 1997]. Similar conclusions about the importance of respiration-driven dissolution have been found from the North Atlantic [*Hales et al.*, 1994] and the eastern equatorial Pacific [*Berelson et al.*, 1994].

As outlined in the previous section, the importance of respiration-driven dissolution in the sediments of the deep Pacific and Atlantic is at odds with the absence of strong dissolution signals at shallower depths, as the higher rates of organic matter delivery should create an even larger tendency for respiration-driven dissolution. *Jahnke and Jahnke* [2004] recently suggested that this discrepancy could be reconciled by two possible mechanisms: buffering of metabolic CO_2 by overlying water or re-precipitation of $CaCO_3$ in the supersaturated surface sediments. *Hales* [2003] proposed recently that the buffering may be greatly enhanced by the fact that organic matter generally remineralizes faster than $CaCO_3$ is being dissolved, so that the two processes tend to be vertically separated in the sediments. As a result, the efficiency of organic matter remineralization driving $CaCO_3$ dissolution may be diminished, since many protons generated by organic matter remineralization diffuse into the overlying water column before they can dissolve any $CaCO_3$.

Figure 9.4.6 shows that the degree of the depth separation between organic matter remineralization and $CaCO_3$ dissolution, and consequently, also $CaCO_3$ dissolution fluxes and respiratory dissolution efficiencies, may vary substantially in response to bottom water undersaturation, Ω, and the assumptions being made about the kinetics of organic matter remineralization. Respiratory dissolution efficiency is defined, for a specific organic matter component, as the mole ratio of CO_2 used to dissolve $CaCO_3$ to total CO_2 generated. When *Hales* [2003] used first-order kinetics (i.e., $n=1$) for organic matter remineralization in a model with one organic component, about 80% of the $CaCO_3$ was dissolved by respiration-driven dissolution when waters were undersaturated, with this percentage decaying to about 50% when waters were supersaturated by 30%. However, when *Hales* [2003] assumed that organic matter in the sediments was separated into a fast and a slowly decaying pool (in analogy to the G-type kinetics that we introduced in chapter 6). The respiration-driven dissolution efficiency was substantially smaller, i.e., only about 50% in undersaturated conditions, dropping to about 20% in supersaturated conditions. The dissolution fluxes of $CaCO_3$ dropped correspondingly. The main reason for this difference is that in the case of the two-component model, the fast-decaying fraction of organic matter is being remineralized very close to the sediment-water interface so that most protons generated from this reaction are being buffered by the overlying water column before they can dissolve any $CaCO_3$ in the sediments. The slowly decaying organic matter fraction of the two-component model is also less efficient in dissolving $CaCO_3$ than the one-component model, but only when the bottom water is supersaturated. As the water becomes undersaturated, the buffer capacity drops, and the slowly decaying fraction becomes more efficient. Although there is so far little corroborating evidence for this being the correct mechanism, both this water-column buffering mechanism and the $CaCO_3$ re-precipitation mechanism of *Jahnke and Jahnke* [2004] provide a possible explanation for the observation that respiration-driven dissolution becomes less efficient as bottom waters become supersaturated.

Our model of $CaCO_3$ diagenesis permits us now to investigate in detail which parameters and processes are most important in determining the depth of the lysocline and the thickness of the transition zone. It turns out that the depth of the lysocline relative to the saturation state of the overlying water column is primarily determined by the ratio of the organic carbon to calcite rain. We show also that the thickness of the transition zone is controlled by the dissolution rate constant for calcite, k_{CaCO_3}, and the rate of delivery of noncarbonate refractory material, such as dust, to the sediments.

Unfortunately, neither the dissolution rate constant nor the rain ratio are well known from observations. We have discussed above that estimates for k_{CaCO_3} vary by several orders of magnitude, leaving this parameter virtually unconstrained. The situation is not as bad for the organic carbon–to–$CaCO_3$ rain ratio. Sediment traps below 2000 m measure ratios between 0.5 and 1.5, with a mean of about 0.8 ± 0.2 [*Klaas and Archer*, 2002]. The relatively small variations in the organic carbon–to–$CaCO_3$ rain ratio in the deep ocean have been interpreted as evidence of a close association between organic matter and $CaCO_3$, with the $CaCO_3$ providing the ballast to transport the organic matter efficiently through the water column (see discussion in chapter 5) [*Klaas and Archer*, 2002; *Armstrong et al.*, 2002; *Francois et al.*, 2002]. Nevertheless, this range is still too large to sufficiently constrain a diagenetic model that attempts to reproduce

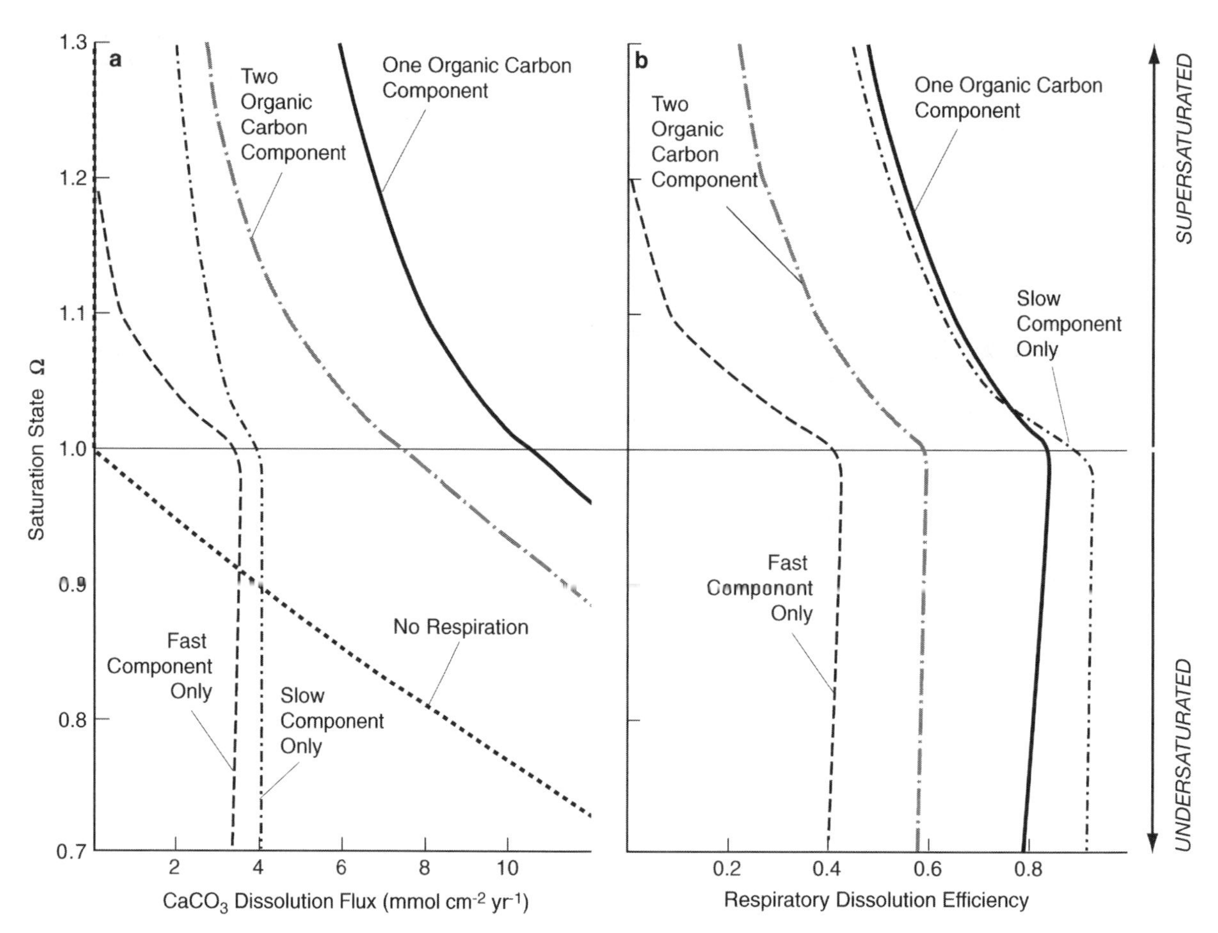

FIGURE 9.4.6: Calcite dissolution fluxes and efficiency of respiration-driven dissolution as a function of the degree of bottom water saturation, Ω. (a) Dissolution flux driven by various forcing components. The heavy dashed line represents the dissolution that would occur due to bottom-water undersaturation only. The dot-dashed line is the total calcite dissolution model, i.e., driven by bottom-water undersaturation and organic matter remineralization in the two-organic carbon component model. The two light dashed lines show the dissolution driven by each organic carbon component (fast or slow, as labeled) of that model. Note that the dissolution driven by the slow component is fairly constant above the saturation horizon ($\Omega = 1$) in comparison to that driven by the fast component, which is near zero down to a bottom-water saturation of about 1.1, and then increases rapidly until the saturation horizon is reached. The solid line shows, for comparison, the dissolution driven by a respiration formulation with a single-organic carbon component with a scale depth of 2 cm. This respiration rate drives more dissolution for the same organic carbon rain than the two-component formulation in all saturation cases, particularly above the saturation horizon. (b) Dissolution efficiency defined as the ratio of the dissolution flux driven by a particular process to the amount of CO_2 produced. Note that the efficiency of the two-organic carbon component is only about 0.2 above the saturation horizon, implying very small dissolution fluxes for typical oceanic organic carbon rain rates. The calculations were done for typical oligotrophic ocean-basin sediments with a calcite content of 50%, a first-order dissolution reaction (i.e., $n=1$) with a rate constant of 1×10^{-4} day^{-1}, and a total organic carbon respiration rate of 13 $\mu mol\, cm^{-2}\, yr^{-1}$. Modified from *Hales* [2003].

the observed lysocline distribution. However, given that we have good observations of the $CaCO_3$ distribution in sediments, one can turn the problem around and ask whether it is possible to constrain the ill-known parameters with these observations, i.e., solve the inverse problem.

In order to find the optimal set of parameters, we first need to establish how the modeled lysocline changes in response to variations of each parameter separately. Figure 9.4.7a shows the simulated shape of the transition between high and low calcite sediments as a function of the dissolution rate for calcite, k_{CaCO_3}, using a 4.5-order kinetics [*Archer*, 1991]. In this case an organic carbon–to–$CaCO_3$ rain ratio of 0.7 was used. As k_{CaCO_3} increases, both the "thickness" of the transition zone and the undersaturation at the CCD decrease. However, above a value of 3 day^{-1} the shape of the transition zone becomes relatively insensitive to the value of k_{CaCO_3}. These particular simulations were done with an exponent in the dissolution kinetics of $n = 4.5$ (see (9.4.2)). If an exponent $n = 1$ were used as argued for by *Hales and Emerson* [1997], the changes in the shape of the modeled lysocline in response to changes in k_{CaCO_3} would remain similar, but correspond to drastically different values of k_{CaCO_3}.

Variations of the organic carbon–to–$CaCO_3$ rain ratio induce large changes in the shape of the transition zone

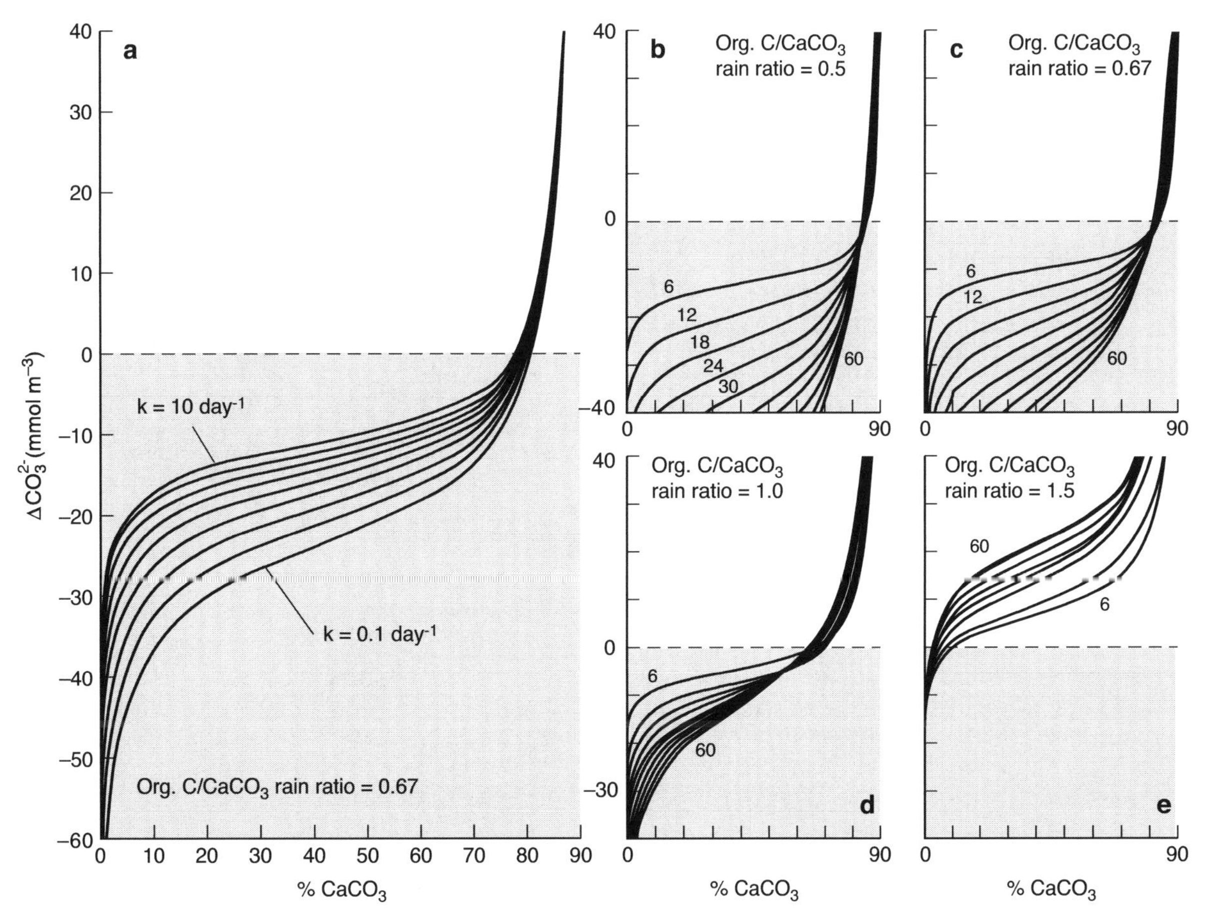

FIGURE 9.4.7: Sensitivity of the modeled transition zone between the lysocline and the carbonate compensation depth to changes in model parameters. (a) Model-predicted shape as a function of the dissolution rate constant for calcite, k_{CaCO_3}, using a 4.5-order kinetics. The rate constants used were 0.1, 0.18, 0.32, 0.56, 1.0, 1.8, 3.2, 5.6, and 10 day^{-1}. For all simulations, the rain rates of organic carbon and calcite were 12 and 18 μmol cm^{-2} yr^{-1}, respectively. The detrital mass flux was 10% of the total nonorganic carbon mass flux. (b–e) Model-predicted shape of the transition zone as a function of the organic carbon–to–$CaCO_3$ rain ratio and carbon rain rates. The rain rates of calcite were changed between 6 and 60 μmol cm^{-2} yr^{-1}. The organic rain is proportional to the calcite rain, using ratios of (b) 0.5, (c) 0.67, (d) 1.0, and (e) 1.5. The calcite dissolution rate constant was held constant at 1 day^{-1}. Adapted from *Archer* [1991].

between the lysocline and the CCD (Figure 9.4.7b–e) [*Archer*, 1991]. The influence of respiration becomes very apparent when the organic carbon–to–$CaCO_3$ rain ratio gets larger. At low rain ratios, both the thickness of the transition zone and the $CaCO_3$ undersaturation at the CCD are very large and very sensitive to variations in the $CaCO_3$ rain. This high sensitivity is mainly because bottom water undersaturation is the only factor driving dissolution at such low rain ratios. At high rates of $CaCO_3$ delivery to the sediments, the resulting higher downward advection of the sediments transports $CaCO_3$ more rapidly out of the reach of the bottom water undersaturation, resulting in increased preservation and hence a thicker transition zone.

As the rain ratio approaches 1 (figure 9.4.7d), both the lysocline thickness and the undersaturation at the CCD decrease substantially. At the same time, the undersaturation at the CCD becomes less sensitive to variations in the $CaCO_3$ rain. At organic carbon–to–$CaCO_3$ rain ratios above 1 (figure 9.4.7e), the CO_3^{2-} concentration at the CCD is above the $CaCO_3$ saturation concentration and largely independent of the calcite rain. However, these simulations were done with a single-organic matter component model, and use of a two-component model would lead to a substantially smaller sensitivity of the modeled lysocline to the rain ratio, as is illustrated in figure 9.4.6.

In summary, variations in the organic-to-inorganic carbon rain ratio determine the position of the lysocline relative to the saturation state of the overlying bottom waters, while variations in dissolution kinetics control mainly the thickness of the transition zone. As mentioned above, a second factor influencing the thickness of the transition zone is the rain of non-$CaCO_3$ material, such as dust particles and opal. Using a gridded data base of the sediment distribution of $CaCO_3$, *Archer* [1996a] attempted to find an optimal set of parameters that reproduces the observed variations. According to his model, a rain ratio of 0.5 leads to too strong preservation in the Atlantic, while a rain ratio of 1.0 leads to almost no preservation in the Pacific. A rain ratio of about 0.7 seems to best represent the observations. This value is in remarkably good agreement with the sediment trap–derived estimate of 0.8 ± 0.2 [*Klaas and Archer*, 2002]. *Archer* [1996a] also demonstrated that the main cause of the differences in the thickness of the transition zone between the Atlantic and Pacific is the rain of non-$CaCO_3$ material, with the Atlantic receiving much higher rates of such material than the Pacific.

Concluding Remarks

Before leaving this subject, we summarize the most important conclusions. The promotion of calcium carbonate dissolution by the degradation of organic matter plays a critical role in the dissolution of $CaCO_3$ in the sediments. Model results indicate that as one goes from supersaturated to undersaturated conditions in the overlying water column, already 40% to 70% of the $CaCO_3$ gets dissolved by this process when one reaches saturation. This large dissolution percentage does not necessarily show up as a large drop in the percent dry weight $CaCO_3$ content of the sediments for reasons that are discussed next.

The relationship between $CaCO_3$ rain, dissolution, and final $CaCO_3$ content in the sediments can be illustrated by computing the final $CaCO_3$ content, f_{CaCO_3}, as a function of the proportion of $CaCO_3$ to inert (e.g. terrigenous) material arriving at the sediment, R, and of the fraction of the deposited $CaCO_3$ that dissolves in the sediments, f_{diss}. This gives $f_{CaCO_3} = (1 - f_{diss})/(1 - f_{diss} + 1/R)$. If $R = 9$ and $f_{diss} = 0$, then $f_{CaCO_3} = 90\%$. If 50% of the $CaCO_3$ dissolves, i.e., $f_{diss} = 0.5$, the $CaCO_3$ content in sediments is still very high at 82%. Even with $f_{diss} = 0.9$, the sediment $CaCO_3$ drops to just 47%. Only if 99% of all $CaCO_3$ arriving at the sediments dissolves, will the final $CaCO_3$ sediment content fall below 10%. Thus, the $CaCO_3$ content of sediments is not a sensitive indicator of $CaCO_3$ dissolution fluxes until dissolution is very high, which also makes it difficult to detect the onset of dissolution from the depth of the lysocline.

9.5 Calcium Carbonate Compensation

One of the most exciting aspects of the marine $CaCO_3$ cycle is its possible role as a homeostat for regulating the *DIC* and *Alk* budgets of the ocean on millennial and longer timescales. Since the ocean carbon cycle controls atmospheric CO_2 on these timescales, this homeostat is also of great importance for controlling atmospheric CO_2. Without negative feedbacks, slight imbalances between the input and loss of alkalinity to the ocean would over time lead to large changes in oceanic alkalinity and hence atmospheric CO_2 concentrations. Such large excursions in atmospheric CO_2 do not appear to have occurred over the last tens of millions of years [*Prentice et al.*, 2001]. In order to have a stabilizing effect on atmospheric CO_2, the $CaCO_3$ homeostat in the ocean must somehow link the input of alkalinity with the loss of alkalinity in order to maintain a balance between the two. As implied in figure 9.1.1, the major inputs of alkalinity into the open ocean are export from the margins (mainly river input of weathering-derived alkalinity) and input from hydrothermal vents. The only loss term is the burial of $CaCO_3$ at the sea floor.

At first view, the input and loss processes seem to be completely uncoupled, so the existence of a strong coupling between these processes seems surprising. However, we have seen that the preservation of $CaCO_3$ in the sea-floor sediments is determined by a number of factors and most importantly by the degree of undersaturation of the seawater with respect to calcite. As the saturation state is governed by the concentration of the CO_3^{2-} ion, any process that changes the concentration of CO_3^{2-} in the deep ocean necessarily alters also the preservation pattern of $CaCO_3$ in the sediments. Because $[CO_3^{2-}] \approx Alk - DIC$, a change in the CO_3^{2-} ion content can be generated either by changes in the alkalinity supply to the open ocean or by changes in the ocean biological pumps that affect *DIC*. The first mechanism thus opens the possibility for maintaining a long-term "external" balance in the ocean between the input and loss of alkalinity. The second mechanism plays an important role in many scenarios that attempt to explain glacial-interglacial changes in atmospheric CO_2. It builds on the premise that any anomaly in the deep ocean CO_3^{2-} content generated by the processes that are invoked to explain the lower atmospheric CO_2 concentration needs to be removed in order to re-establish the balance between input and burial of *Alk*. This process, termed "$CaCO_3$ compensation" [*Broecker and Peng*, 1987], can either enhance or diminish the effect of the original process in changing atmospheric CO_2. We will first discuss the $CaCO_3$ homeostat and its role in maintaining the "external" balance between input and burial. We will discuss the second mechanism thereafter.

$CaCO_3$ Homeostat

In order to understand the mechanisms that link variations in the alkalinity input to variations in the burial of $CaCO_3$, let us turn to figure 9.5.1. In the present steady state, the alkalinity input by rivers and hydrothermal

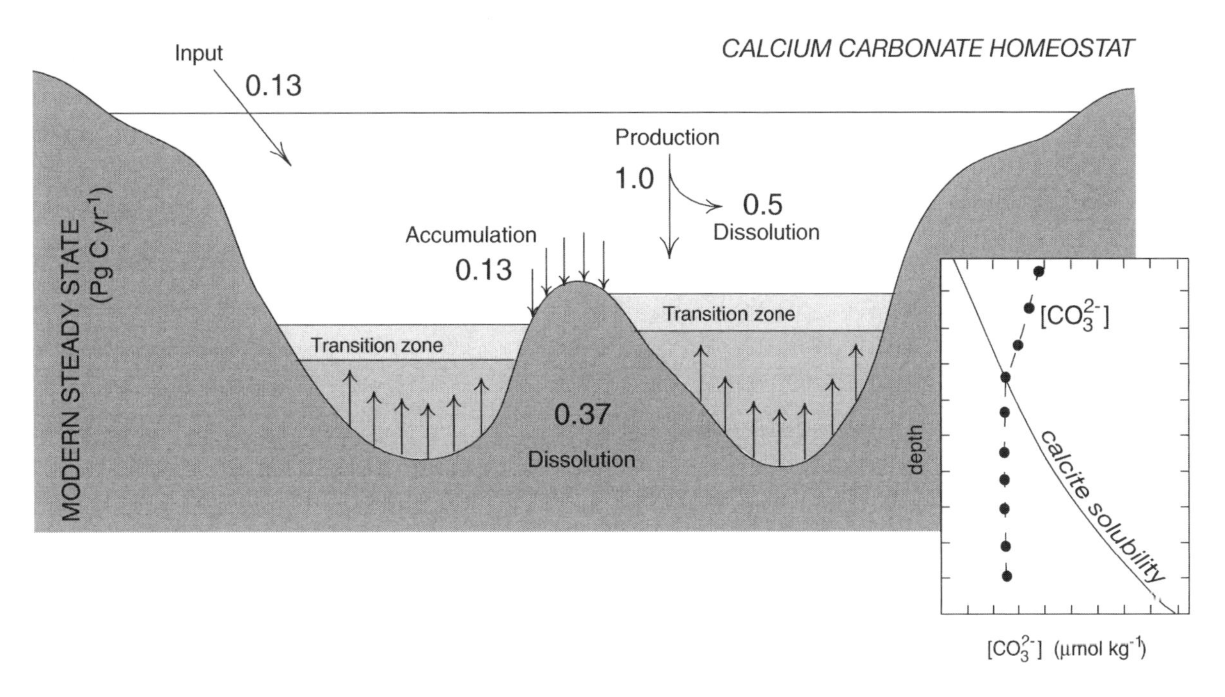

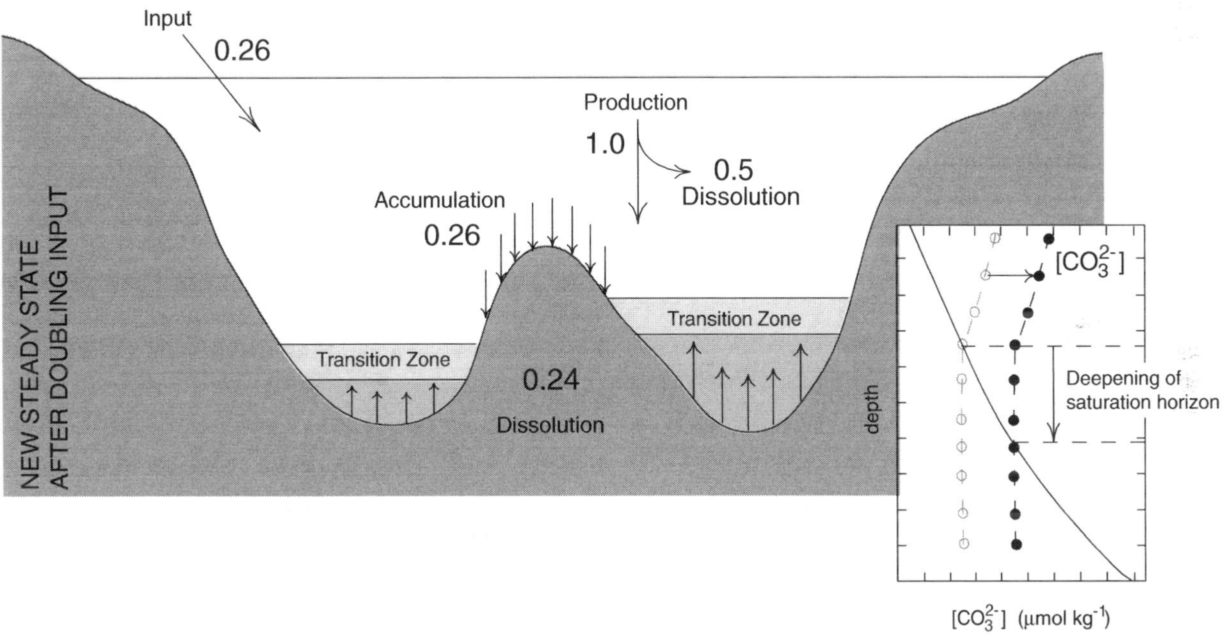

FIGURE 9.5.1: Illustration of the $CaCO_3$ homeostat. (a) Modern steady-state budget for $CaCO_3$ in the open ocean realm in units of Pg C yr^{-1} modified from [*Milliman et al.*, 1999]; see also figure 9.1.1). (b) New steady-state budget after the weathering input has been doubled from 0.13 Pg C yr^{-1} to 0.26 Pg C yr^{-1}. Insets show the CO_3^{2-} ion concentrations as a function of depth. The increase in the input of alkalinity leads to an increase in the CO_3^{2-} ion concentration in the entire ocean. This leads to a depression of the calcite saturation horizon, so that fewer sediments are exposed to undersaturated bottom waters. This process continues until the increased deposition of $CaCO_3$ in the supersaturated zone balances the input.

vents into the open ocean of about 0.13 Pg C yr^{-1} is balanced by deep sea burial (figure 9.5.1a). This balance is achieved by dissolving roughly 0.5 Pg C yr^{-1} of the surface $CaCO_3$ production of about 1.0 Pg C yr^{-1} in the water column and dissolving an additional 0.37 Pg C yr^{-1} in the sediments. If we now increase the input from terrestrial weathering, the CO_3^{2-} content of the entire ocean starts to increase (figure 9.5.1b). This results in a significant deepening of the saturation horizon, reducing the area of the sediments that are exposed to corrosive waters. The CO_3^{2-} concentration continues to increase and the saturation horizon continues to deepen until the burial of $CaCO_3$ in the deep sea sediments has reached the magnitude of the new input. In case of a doubling of the input, the new steady state requires that about 0.26 Pg C yr^{-1} are buried and 0.24 Pg C yr^{-1} are dissolved in the sediments. This is the feedback that keeps alkalinity in the ocean within rather tight bounds and hence provides a strong constraint on how much atmospheric CO_2 can vary over time. Using a multi-box model, *Keir* [1995] showed that a doubling of the weathering input leads to only a 25 to 35 ppm reduction in atmospheric CO_2. This small change is supported by the more recent results of *Archer et al.* [2000b] on the basis of a 3-D ocean biogeochemistry model that has been coupled to a sediment model similar to that described above.

The strength of this negative feedback depends on how strongly the saturation state of the bottom water determines the dissolution of $CaCO_3$ in the sediments. This $CaCO_3$ homeostat would be weaker if parameters other than the CO_3^{2-} ion concentration of the bottom waters were exerting a more dominant control. We have learned above that the remineralization of organic matter has a significant influence on the dissolution of $CaCO_3$ in the sediments. However, as long as the organic matter–to–$CaCO_3$ rain ratio arriving at the sediments is below 1:1, the saturation state of the bottom water remains the dominant factor. We have also seen that the efficiency of respiration-driven dissolution tends to decrease strongly if bottom waters are supersaturated, even in cases where the rain ratio is well above 1:1. There are thus substantial limits for how strongly the ocean's lysocline can deviate from the saturation horizon [*Sigman et al.*, 1998].

An important note remains to be added regarding the timescale of the $CaCO_3$ homeostat. One might be tempted to argue that the timescale is of the order of the residence time of alkalinity in the ocean, which is about 100 kyr. However, it turns out that the e-folding timescale for the oceanic adjustment is only about 5–10 kyr [*Archer et al.*, 1997, 2000b]. This rather surprising result is again a consequence of the peculiarities of the oceanic CO_2 system. Only about 5% of the *DIC* in the ocean is in the form of CO_3^{2-}. Thus the CO_3^{2-} content of seawater can adjust approximately 20 times faster than the 10^5 years it takes to change the entire *DIC* pool by the input from weathering processes and volcanic outgassing. The $CaCO_3$ homeostat is therefore able to interact with the global carbon cycle on timescales shorter than the 100 kyr cycle associated with the glacial-interglacial climate cycles of the last million years.

$CaCO_3$ Compensation

The balance between river input of alkalinity and burial can also be changed by ocean internal changes, such as alterations of the marine production of $CaCO_3$, or changes in the oceanic distribution of ΔCO_3^{2-}. We have seen above that the latter is mainly controlled by the soft-tissue pump. Therefore any change of the ocean's biological pumps will upset the ocean's alkalinity balance and require an adjustment of the ocean's carbonate cycle. However, the response of the latter to changes in the two biological pumps is fundamentally different.

In the case of the carbonate pump, a change in the production and export of $CaCO_3$ will lead directly to an imbalance between river input of alkalinity and burial, which is then adjusted by a change of the depth of the lysocline until a new balance is achieved. This carbonate compensation response is fundamentally similar to the external homeostat described above. Figure 9.5.2a shows how the various components of the oceanic system react to a sudden decrease in the $CaCO_3$ flux from the upper ocean, and how the new equilibrium of the ocean's alkalinity balance is achieved.

Initially, the unabated input of *Alk* by rivers, coupled with decreased burial, leads to a rapid increase in mean ocean alkalinity, which also causes ocean mean $[CO_3^{2-}]$ to go up. As a result, the lysocline starts to deepen, so that a greater fraction of the produced $CaCO_3$ can be buried. The process continues until the lysocline has reached a depth at which burial of *Alk* is equal to the river input of *Alk*. The increase in ocean mean *Alk* also decreases the oceanic buffer factor, leading to an imbalance of inorganic carbon between the atmosphere and ocean. The balance is restored by the ocean taking up CO_2 from the atmosphere, hence causing a drawdown of atmospheric CO_2. In addition to this "open system" response [*Sigman and Boyle*, 2000], atmospheric CO_2 is further drawn down by a "closed system" response, which is more direct and exists even in the absence of a need to balance the ocean's alkalinity input. This closed system response is simply a consequence of the fact that a reduction in the surface export of $CaCO_3$ will lead to an increase in surface ocean *Alk*, which will tend to lower oceanic pCO_2, and hence lead to an uptake of CO_2 from the atmosphere. Using a multi-compartment model of the ocean, *Sigman and Boyle* [2000] showed that the "open system" response to a halving of the $CaCO_3$ export lowers atmospheric CO_2 by about 20 ppm, whereas the "closed system" response leads to a change in atmospheric CO_2 of about 30 ppm. The two changes are cumulative, so that the total atmospheric CO_2 change is about 50 ppm. However, as we will discuss in chapter 10 in more detail, the exact magnitude of the atmospheric CO_2 response depends quite sensitively on the ocean model being employed [*Broecker et al.*, 1999a, *Archer et al.*, 2000a]. The open and closed system responses are also associated with quite different timescales. The closed system response has a timescale of maximally a few hundred years, primarily determined by the whole ocean overturning timescale. By contrast, the open system response is quite slow, of the order of hundreds to thousands of years, primarily because of the low rates of $CaCO_3$ burial and the slow kinetics of sediment dissolution.

The response of the oceanic carbonate system to changes in the soft-tissue pump are quite different. In this case, and disregarding the potential role of a change

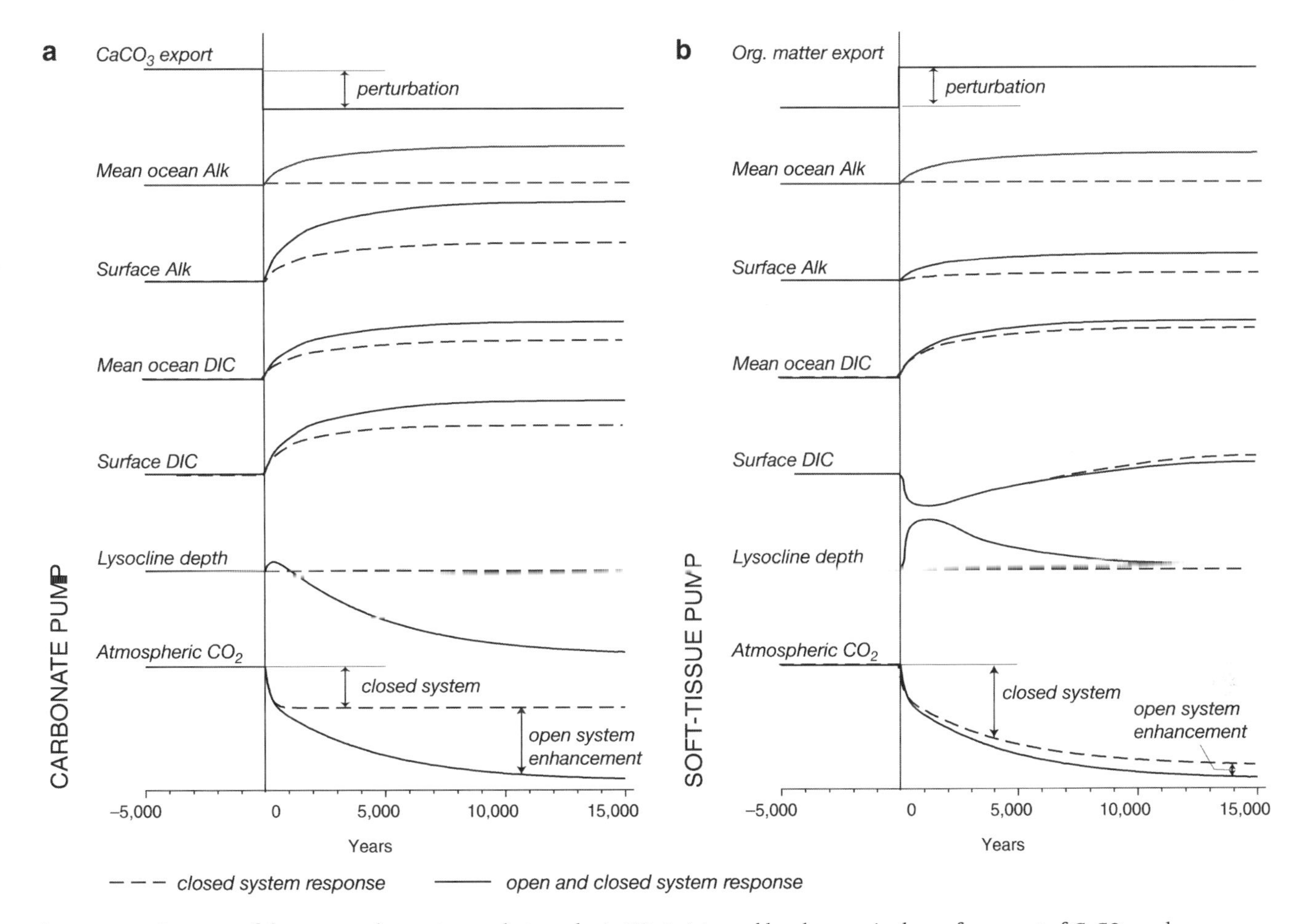

FIGURE 9.5.2: Response of the ocean carbon system and atmospheric CO_2 to (a) a sudden decrease in the surface export of $CaCO_3$, and (b) a sudden increase in the surface export of organic matter. The closed system responses are shown by the dashed lines, while the open system responses, i.e., those that include the river/burial balance of *Alk*, are shown as solid lines. The curves are approximate.

in the rain ratio on the decoupling of preservation from the saturation horizons, the main perturbation to the oceanic *Alk* balance consists of a change in the oceanic distribution of $[CO_3^{2-}]$, which changes the distribution of ΔCO_3^{2-}, and hence the depth of the saturation horizons. In the case of a sudden increase of the soft-tissue pump, the concentration of deep ocean CO_3^{2-} would fall quite strongly, leading to a rapid shoaling of the saturation horizon. Assuming that the depth of the saturation horizon is the main factor controlling preservation, this would lead to a large increase in $CaCO_3$ dissolution, which increases ocean mean *Alk* and hence $[CO_3^{2-}]$. As a consequence, the lysocline starts to deepen again until it has reached a level where the ocean's alkalinity balance is again met. This depth will be somewhat shallower than the original depth of the lysocline, with the exact location depending on the the intial $[CO_3^{2-}]$ distribution and where the additionally exported organic carbon gets remineralized in the water column.

As was the case for the carbonate pump, a change of the soft-tissue pump will lead to both open and closed system responses of atmospheric CO_2 (figure 9.5.2b). The closed system response is a result of the drawdown of surface ocean *DIC* induced by organic matter export, which will lead to an uptake of CO_2 from the atmosphere. The open system response is a consequence of the increase in ocean mean *Alk*, and will enhance the closed system response by removing additional CO_2 from the atmosphere. The exact magnitude of these two responses depends quite sensitively on the model being employed. For a 30% increase in the soft-tissue pump and using a multi-box model, *Sigman and Boyle* [2000] estimated a drawdown of 46 ppm, with most of this being driven by the closed system response (43 ppm), and the open system response adding only 3 ppm.

In summary, the ocean's need for establishing a long-term *Alk* balance (i.e., the open system response) leads to a tendency for carbonate compensation to enhance the initial (i.e., closed system) response of atmospheric CO_2 to changes in the ocean's biological pumps, i.e., acting as a positive feedback in the system. This has important consequences for discussing scenarios that attempt to explain the 80 ppm variations that occurred in atmospheric CO_2 between glacial and interglacial periods. We will return to this issue in chapter 10.

In chapter 10, we will also discuss a third type of $CaCO_3$ compensation, which occurs as a result of perturbations of atmospheric CO_2 external to the ocean, e.g., driven by imbalances in the terrestrial carbon cycle, or by the release of CO_2 from the burning of fossil fuel. In this case, the oceanic carbonate system acts as a negative feedback, neutralizing the carbon that has been added to the atmosphere-ocean-sediment system.

Problems

9.1 Define and explain the following terms:

a. Calcite and aragonite

b. Calcium carbonate saturation horizon

c. Calcium carbonate compensation depth

d. Lysocline

e. Rain ratio

f. Coccolithophorids and pteropods

9.2 What are the main processes responsible in the ocean for the production of aragonite and calcite? Where in the ocean are they produced, and what are the controlling processes?

9.3 Discuss why coccolithophorids precipitate $CaCO_3$.

9.4 Contrast the latitudinal distribution of the $CaCO_3$–to–organic carbon export ratio in figure 9.2.1 with the opal–to–organic nitrogen export ratio in figure 7.2.4. Discuss why the two distributions are so different.

9.5 Laboratory experiments have suggested that the invasion of anthropogenic CO_2 into the surface ocean will lead to a reduction of the calcification rate of corals and coccolithophorids.

a. Estimate how the invasion of anthropogenic CO_2 into the ocean changes the concentrations of $H_2CO_3^*$ and CO_3^{2-}. Take the mean surface *DIC* and *Alk* concentrations given in table 8.2.4 and compute the changes in $[H_2CO_3^*]$ and $[CO_3^{2-}]$ after you add 100 μmol kg^{-1} of anthropogenic *DIC* to this average surface water sample. Hint: Use the lookup table for problem 8.14 in chapter 8 for computing the oceanic pCO_2 for a given *DIC* and *Alk*. Then use this to estimate $[H_2CO_3^*]$.

b. Discuss how these changes in $[H_2CO_3^*]$ and $[CO_3^{2-}]$ change the saturation state of our average surface seawater sample with regard to calcite and aragonite.

c. Given these chemical changes, discuss the possible reasons why the calcification rates in corals and coccolithophorids are expected to decrease in response to the invasion of anthropogenic CO_2 into the ocean.

9.6 Compute the saturation states ΔCO_3^{2-} and Ω with respect to calcite and aragonite, respectively, for a water parcel at a depth equivalent to 6000 dbar with the following properties: $T = 0°C$, $S = 35$, $DIC = 2300\ \mu mol\ kg^{-1}$, $Alk = 2400\ \mu mol\ kg^{-1}$. Use the saturation concentrations given in table 9.3.4.

9.7 Explain the reason for the large spatial variability exhibited by the carbonate saturation horizon with respect to calcite in figure 9.3.2b, i.e., why is this horizon more than 4000 m deep in the Atlantic, and less than 1000 m deep in the North Pacific?

9.8 Compute the fraction of $CaCO_3$ exported from 100 m depth that arrives at the sediments 4000 m below, given the following characteristics for the $CaCO_3$ particles: sinking speed = 50 m day^{-1}, first-order dissolution kinetics (i.e., $n = 1$), and $k_{CaCO_3} = 0.01\ day^{-1}$. Assume two cases:

a. Constant undersaturation of $\Omega = 0.9$.

b. Saturation at 100 m, and Ω linearly decreasing with depth to reach $\Omega = 0.7$ at 4000 m.

c. How do the answers given in (a) and (b) change if you assume a sinking speed of 100 m day^{-1}, and a $k_{CaCO_3} = 0.38\ day^{-1}$?

9.9 Contrast the $CaCO_3$ content of sediments shown in figure 9.1.3 with that for opal (figure 7.1.3). Discuss what controls the burial pattern of these two minerals in the sediments. In particular, discuss the relative roles of surface production, water column dissolution, and sediment processes in determining the burial.

9.10 Schematically illustrate the expected global mean vertical profile of the carbonate ion concentration, $[CO_3^{2-}]$, for the following cases:

a. Abiological ocean with air-sea gas exchange (assume constant alkalinity).

b. Ocean with no air-sea gas exchange, but active soft-tissue pump.

c. Ocean with no air-sea gas exchange and no soft-tissue pump, but with active carbonate pump.

Assume in all cases (somewhat incorrectly) that atmospheric CO_2 is fixed at its preindustrial value of 280 ppm. Compare these cases with the observed vertical profiles shown in figure 9.1.2. Discuss also the implications of these findings for the saturation state of seawater with regard to calcite and aragonite.

9.11 What happens to the saturation horizon if the export ratio of organic carbon-to-$CaCO_3$ increases? Assume that no change occurs in the vertical remineralization and dissolution profiles.

a. Consider first a case in which there is no feedback from the sediments, i.e., the dissolution flux from the sediments remains unchanged.

b. Consider a second case where the change in saturation horizon interacts with the sediments. Assume, for simplicity, that the dissolution of $CaCO_3$ in the sediments is entirely driven by the saturation state of the waters overlying the sediments.

9.12 Observations indicate that, during the last glacial maximum, the lysocline was deeper in the Atlantic, but shallower in the Pacific relative to the present ocean.

a. What determines the depth of the lysocline in the present ocean?

b. Given your response to (a), how would you have to change these processes in order to obtain the observed lysocline changes during the last glacial maximum?

9.13 Explain the difference between the calcium carbonate compensation depth (CCD) and the lysocline. Address, in particular, what determines the difference in depth between the two (e.g., figure 9.4.1).

9.14 Discuss why a much larger fraction of the $CaCO_3$ entering the ocean realm below 1000 m gets buried in the sediments in comparison to organic matter.

9.15 Explain the process of respiration-driven $CaCO_3$ dissolution and discuss its significance for the preservation of $CaCO_3$ in the sediments.

9.16 Discuss what is meant by the terms "open system response" and "closed system response" in the context of the oceanic $CaCO_3$ cycle. In particular, explain why this difference is important in determining the response of atmospheric CO_2 to perturbations in the strength of the ocean's biological pump.

9.17 What is the timescale associated with the open system response, and which processes determine this timescale?

9.18 Consider the stacked three-box model used for problem 7.14 in chapter 7. Use this model with the same geometry and exchange rates given in problem 7.14 to investigate the impact of differential remineralization of $CaCO_3$ and organic matter on atmospheric CO_2. Assume that the $CaCO_3$-to-organic carbon mole ratio of export production is 0.06.

a. Calculate the fractions $\gamma_{C\,org}$ and γ_{CaCO_3} of the flux out of the surface box of organic carbon and $CaCO_3$ that get remineralized/dissolved in the thermocline box, assuming an exponential length-scale for organic carbon of $z^*_{C\,org} = 500$ m and for $CaCO_3$ of $z^*_{CaCO_3} = 3000$ m.

b. Calculate the steady-state surface concentrations of *DIC* and *Alk*, given a phosphate concentration in the surface box of 0 mmol m^{-3} and a global mean phosphate concentration of 2.0 mmol m^{-3}. Assume that the uptake and remineralization of phosphate is tied with a fixed stoichiometric ratio to the generation and remineralization of organic carbon. Assume also that there are no sediments, i.e., all organic matter and $CaCO_3$ entering the deep box gets remineralized/dissolved there. Use the global mean *Alk* and *DIC* given in table 8.2.4.

c. Investigate how atmospheric pCO_2 changes in response to changes in the fractions $\gamma_{C\,org}$ and γ_{CaCO_3}. How much can you draw down atmospheric CO_2 by reducing γ_{CaCO_3} to zero? Use a global mean surface temperature of 20°C and a salinity of 35 for the surface box. Hint: Use the lookup table for problem 8.14 in chapter 8 to compute pCO_2.

d. How would your answer to (c) change if you included an infinitely large $CaCO_3$ sediment pool in the deep box?

Chapter 10

Carbon Cycle, CO_2, and Climate

Throughout most of this book, our discussions have focused on the steady-state characteristics of the cycles of biogeochemically important elements in the ocean. However, as more about the past behavior of these cycles was learned, we have to come to realize that these cycles have seldom been in a "true" steady state. It appears that variability is as much a fundamental property of these biogeochemical cycles as it is a property of the climate system in general (see discussion in section 2.5). In this last chapter, we explore the variability of these cycles on a range of timescales, from a few years to several hundred thousand years. In trying to understand the mechanisms that give rise to these variations, we will apply many of the tools and concepts that we have learned in the previous nine chapters. We will focus on the carbon cycle, because of its central role in biogeochemical cycling and its interaction with climate as a result of the greenhouse gas properties of CO_2 in the atmosphere. This cycle is also of particular interest given the fact that humankind has began to perturb the global carbon cycle dramatically and likely will continue to do so.

10.1 Introduction

A very good indicator for the state of the global carbon cycle is atmospheric CO_2, since it represents the concentration of a relatively small reservoir that acts as a conveyor for the exchange of carbon between the two other important reservoirs of the global carbon cycle: the terrestrial biosphere, including the soils, and the ocean, including its sediments (figure 10.1.1). The amount of CO_2 in the atmosphere is also small compared with the gross exchange fluxes between it and the ocean and terrestrial biosphere, as reflected in the short residence time of about 6 to 8 years for CO_2 in the atmosphere with respect to these exchange fluxes. Atmospheric CO_2 is therefore expected to react sensitively to changes in the global carbon cycle.

Thanks to the preservation of tiny air bubbles in ice cores, researchers have been able to reconstruct concentrations of atmospheric CO_2 over the last 400,000 years with great precision [*Indermühle et al.*, 1999; *Petit et al.*, 1999]. A summary of such reconstructions in figure 10.1.2 shows that atmospheric CO_2 has varied by nearly 100 ppm over the last 400,000 years, indicating that substantial changes must have occured in the distribution of carbon between the different reservoirs. Several observations stand out in figure 10.1.2. First, atmospheric CO_2 has varied quasi-periodically over the last 400,000 years between approximately 180 ppm and 280 ppm. These variations occurred in close concert with the glacial-interglacial cycles, with fully glacial conditions coinciding with the lower-bound CO_2 concentration of about 180 ppm, and interglacial conditions experiencing the upper-bound concentrations of about 280 ppm. This upper bound appears to be rather stable, and also characterizes the largest part of the current interglacial, the Holocene.

The cause for these glacial-interglacial changes must be connected with the oceanic carbon cycle, since this reservoir controls atmospheric CO_2 on any timescale longer than a few hundred years. We will discuss many hypotheses and mechanisms that have been proposed over the last two decades to explain these observed changes, but as it turns out, no mechanism proposed so far has been able to explain all existing constraints. In the near future, as scientists learn more about ocean circulation, biogeochemistry, and their interaction with the climate system during ice ages from detailed analyses of climate records, we likely will be able to correctly identify the mechanisms. However, at the moment the causes of the glacial-interglacial variations in

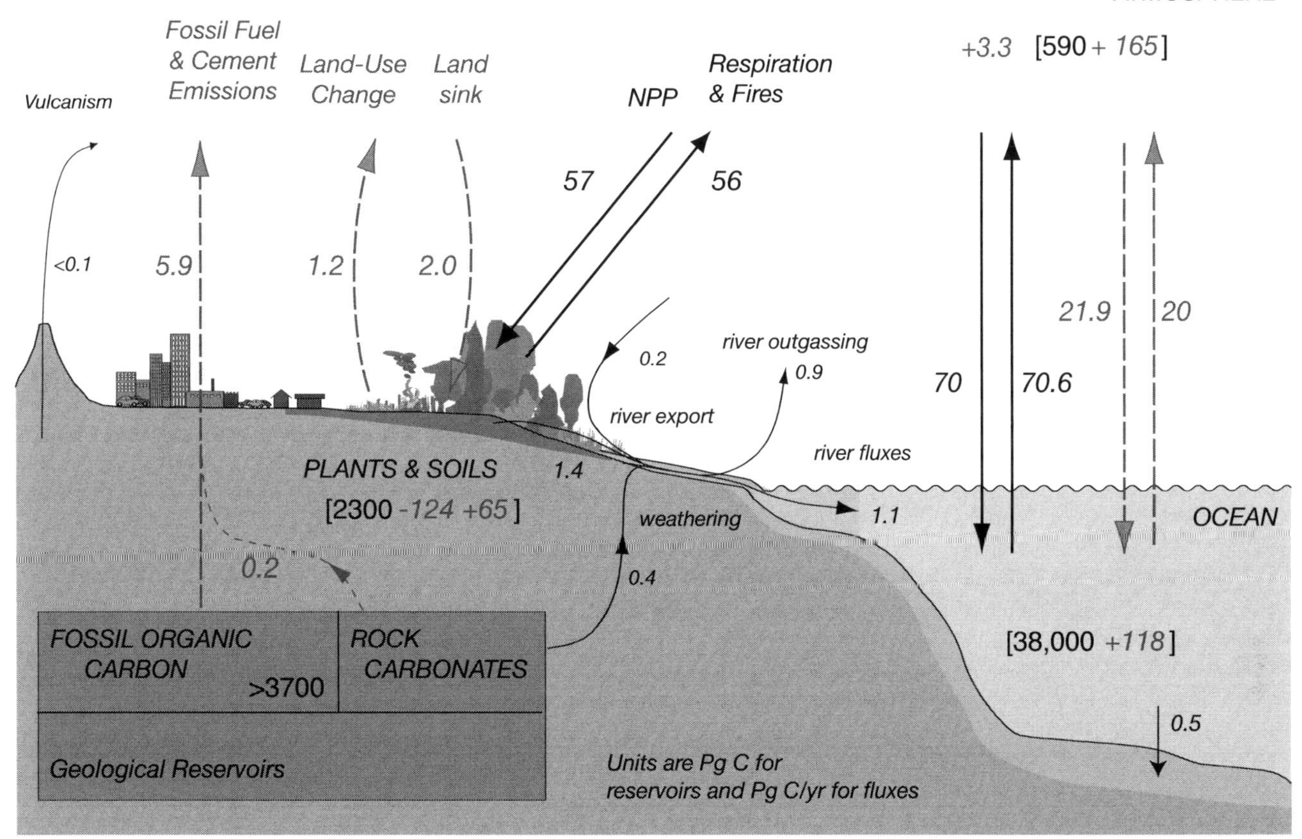

FIGURE 10.1.1: Schematic diagram of the global carbon cycle. Shown are the reservoir sizes in Pg C and the fluxes in Pg C yr^{-1} (1 Pg $= 10^{15}$ g). The solid arrows denote the preindustrial fluxes, whereas the dashed arrows denote the average anthropogenic fluxes for the period of the 1980s and 1990s (see table 10.2.1). Roman numbers in brackets refer to the preindustrial reservoir sizes, and italic numbers in brackets show the changes that occurred to these reservoirs over the Anthropocene (1800 to 1994). Based on *Sarmiento and Gruber* [2002] and *Sabine et al.* [2004].

atmospheric CO_2 remain one of the greatest puzzles of the global carbon cycle.

A second outstanding feature revealed in figure 10.1.2 is the dramatic increase of atmospheric CO_2 since the beginning of the industrial revolution in the late eighteenth century, a period we will refer to as the *Anthropocene* [*Crutzen and Stoermer*, 2000]. This increase is a consequence of anthropogenic activities, mainly the burning of fossil fuels, but also the conversion of forests and other pristine areas into lands for agricultural and other human use. About half of the emissions from these anthropogenic activities have remained in the atmosphere, leading to an increase that is now approaching the change that occurred from glacial to interglacial periods. However, this human-induced increase happened within less than 200 years, whereas it took atmospheric CO_2 several thousand years to change from glacial to interglacial levels.

CO_2 is the most important anthropogenic greenhouse gas in the atmosphere (see next subsection), and its increase is cause for concern, as it may lead to signifcant changes in climate, with major consequences for human beings. Efforts have started now at an international level to curb emissions of CO_2 and other greenhouse gases. However, since fossil fuels are the primary source of energy and heat in the industrial world, and as a result are tightly linked to our current economy, CO_2 emission reductions require major economic investments. It is therefore of great importance to be able to quantify the redistribution of the emitted CO_2 in the global carbon system, since only the fraction remaining in the atmosphere adds to the greenhouse warming.

The observation that a substantial fraction of the emitted CO_2 has remained in the atmosphere is actually surprising at first, since one would intuitively assume that the emitted CO_2 would be redistributed between the atmosphere and ocean with the same 1:65 ratio as the present distribution. Estimates of the actual oceanic uptake suggest that only about a third of the total anthropogenic CO_2 emissions has been taken up by the oceans. Why is the oceanic uptake so small? This is mainly a consequence of kinetic constraints imposed by the limited rate of ocean circulation. However, even

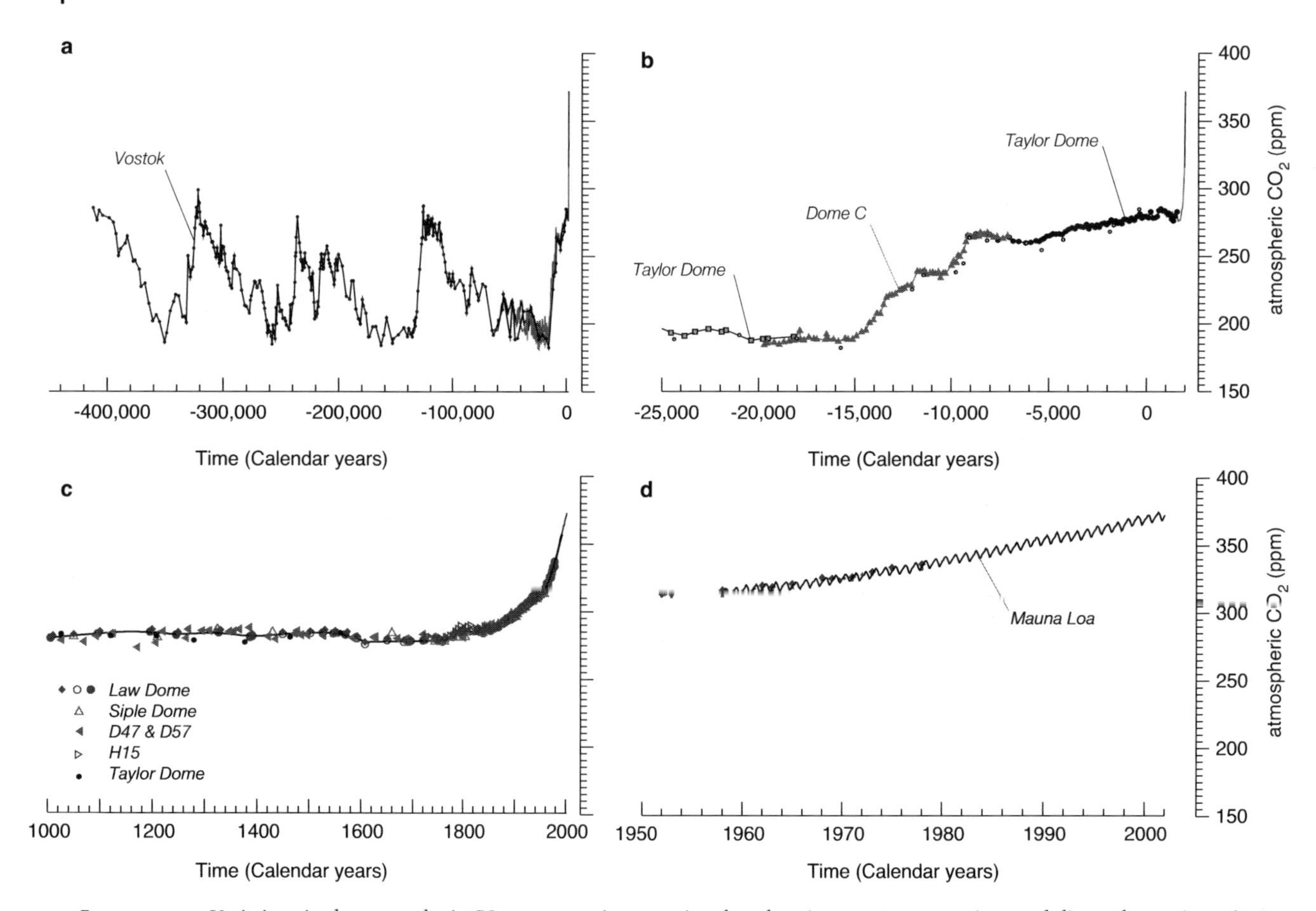

FIGURE 10.1.2: Variations in the atmospheric CO_2 concentration over time based on ice core reconstructions and direct observations since 1958. (a) CO_2 variations over the last 400 kyr (1 kyr = 1000 years). (b) CO_2 variations from 25,000 BC to present. (c) CO_2 variations during the last 1000 years as reconstructed from Antarctic ice cores. (d) CO_2 variations during the last 50 years as directly measured in the atmosphere at Mauna Loa, Hawaii. This comparison illustrates that the onset of the industrial revolution led to a dramatic increase of atmospheric CO_2 after a period of relatively stable atmospheric CO_2 during the previous 10,000 years. The present atmospheric CO_2 level is unparalleled in the last 400,000 years. More than 10,000 years ago, atmospheric CO_2 showed relatively regular variations associated with glacial-interglacial cycles. During the last four glacial maxima, atmospheric CO_2 concentrations were at their lowest concentration around 180 ppm, whereas during the interglacial periods, atmospheric CO_2 concentrations were near 280 ppm. The Mauna Loa data are from *Keeling and Whorf* [1998], the Law Dome, Antarctica, data from *Etheridge et al.* [1996], the Taylor Dome, Antarctica, data from *Indermühle et al.* [1999, 2000], the Dome C data from *Monnin et al.* [2002], and the Vostok, Antarctica, data from *Petit et al.* [1999].

if the ocean mixed extremely rapidly, the fraction taken up by the ocean without further reaction with the $CaCO_3$ in the sediments would only be around 80 to 85%, substantially smaller than the 98% that would be predicted from the present distribution. This rather surprising result is caused by one of the many peculiarities of the oceanic CO_2 system. In the following subsection, we will discuss briefly the greenhouse effect and explain why it is so important for climate.

Greenhouse Effect

Climate and the biogeochemical cycling of major elements on Earth are tightly interwoven. One of the most important coupling points exists in the biogeochemical controls on the composition of Earth's atmosphere, in particular of those gases that interact with solar and terrestrial radiation and change how the energy is redistributed within the climate system. Those gases that absorb energy at infrared wavelengths typical of terrestrial radiation are called *greenhouse gases*, with CO_2 being one of the most important ones.

Biogeochemical cycles can also influence the physical climate system through changes in other climate-relevant properties at the Earth's surface, e.g., by changing the albedo, i.e., the fraction of shortwave radiation from the sun that is directly reflected back; or by interacting with the hydrological cycle, altering the latent heat fluxes associated with condensation and evaporation (see figure 10.1.3 and, e.g., *Betts* [2000]). We will limit our discussion here to the *greenhouse effect*.

The Earth receives short-wavelength radiation emitted from the sun, and radiates this energy back to space at longer wavelengths. Figure 10.1.3 shows the flow of

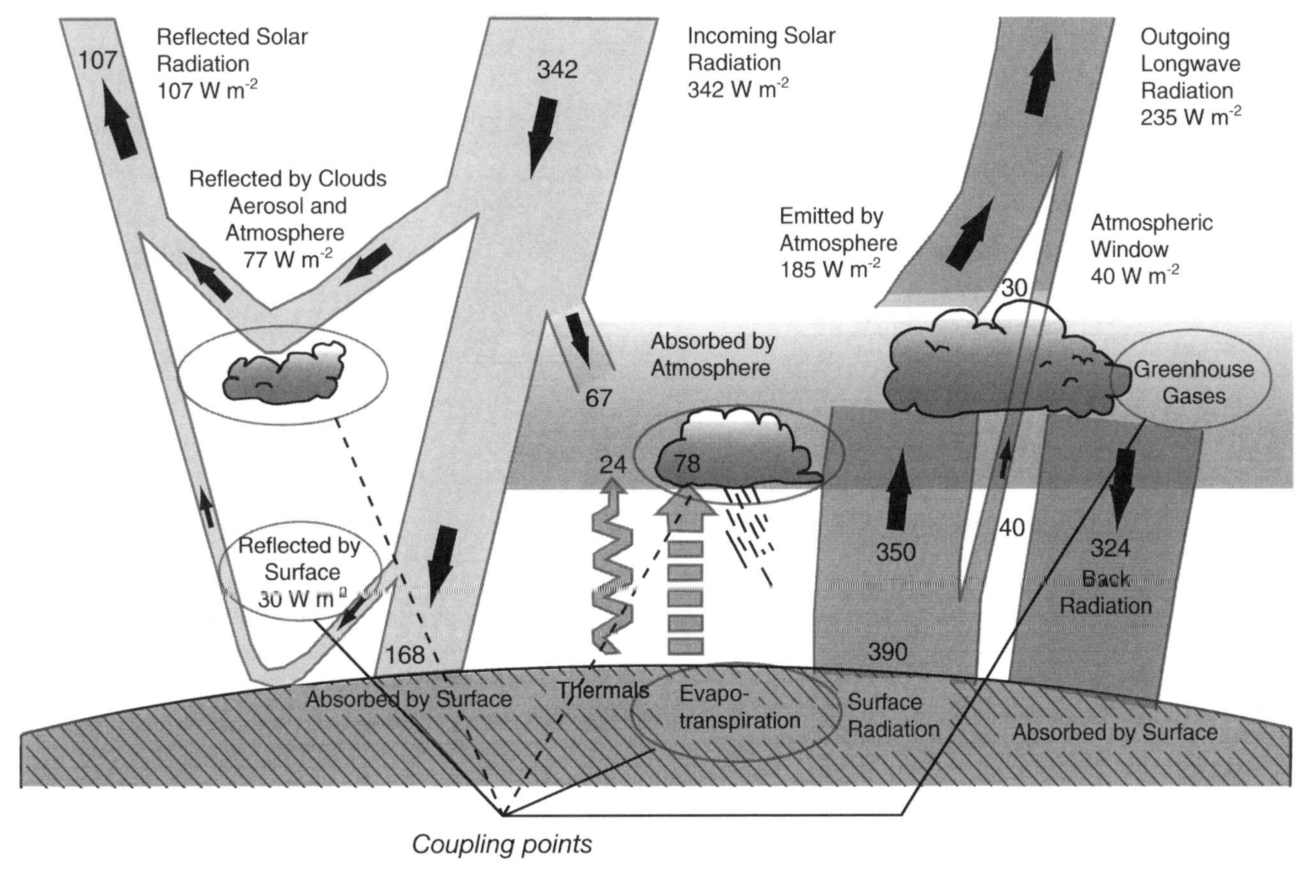

FIGURE 10.1.3: Flow of energy through the atmosphere. Numbers are in watts per square meter of Earth's surface, and some may be uncertain by as much as 20%. The greenhouse effect is associated with the absorption and reradiation of energy by atmospheric greenhouse gases and particles, resulting in a downward flux of infrared radiation from the atmosphere to the surface (back radiation) and therefore in a higher surface temperature. Also shown are the coupling points between biogeochemical cycles and the physical climate system. Adapted from *Kiehl and Trenberth* [1997].

energy through the atmosphere, starting from the short-wave solar radiation received at the top of the atmosphere proceeding down to the surface of Earth. Along the way, more than half of the radiation either gets absorbed by the atmosphere and clouds, or gets reflected back into space.

The short wavelength energy absorbed by the Earth's surface warms the planet, and the Earth itself acts as a blackbody radiator, but because its temperature is much lower than that of the sun, most of its emitted radiation occurs at much longer wavelengths (Wien's law), i.e., in the infrared band. In the absence of an atmosphere, the surface temperature of the Earth would warm until the surface radiated heat to space at the same rate as it intercepted radiation from the sun. This equilibrium temperature is given by the Stefan-Boltzmann law of blackbody radiation, $Q=\sigma T^4$, where Q is the heat flux, $\sigma=5.670\times10^{-8}\,\mathrm{W\,m^{-2}\,K^{-4}}$ is the Stefan-Boltzmann constant, and T is the temperature in degrees Kelvin. The Earth receives an average solar radiation of about $235\,\mathrm{W\,m^{-2}}$ after subtracting off the 30% of the incoming solar radiation reflected back to space (figure 10.1.3). In this case with no atmosphere, the equilibrium temperature for the Earth is thus a frigid 255 K (−18°C).

The presence of an atmosphere with gases that absorb at infrared wavelengths, i.e., greenhouse gases, turns Earth into a much more habitable planet. The impact of such gases is illustrated by figure 10.1.4, which shows the flux of energy emitted by the surface of the Earth as a function of wavelength. The dashed lines are the ideal blackbody radiation with no atmospheric interference. The solid line is the calculated clear-sky radiation flux for a surface temperature of 294 K with the effect of greenhouse gas absorption by the atmosphere included. This is what a satellite outside the atmosphere would see if there were no clouds. The deviation of the solid line from the dashed line for 294 K is due to absorption of outgoing radiation by greenhouse gases in the atmosphere. The minimum in outgoing radiation centered at 15 μm is due to absorption by CO_2, and that near 9–10 μm is due to absorption by ozone. The largest absorber of all is water vapor (table 10.1.1), which absorbs across a broad range of

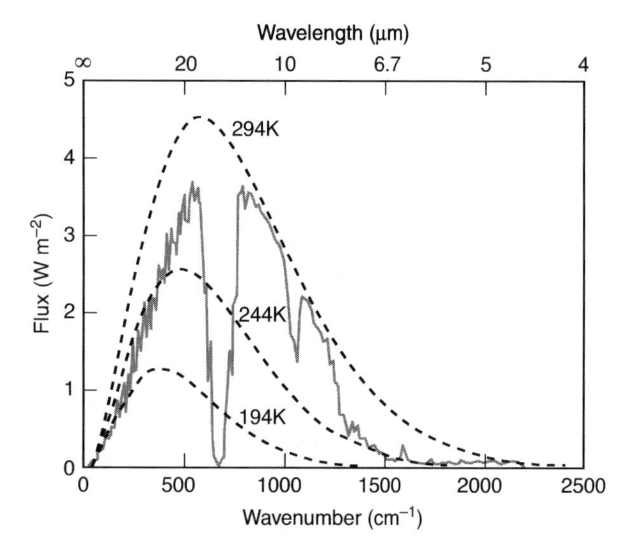

FIGURE 10.1.4 Outgoing clear sky radiation calculated by K. P. Shine for a region where the surface temperature is 294 K [*Houghton et al.*, 1990]. The dashed lines show the hypothetical energy flux, which would be observed if none of the outgoing radiation were intercepted in the atmosphere.

wavelengths. The energy absorbed by the radiatively active gases in the atmosphere is re-emitted from within the atmosphere (see figure 10.1.3). This re-emission occurs in all directions, therefore to a substantial degree also in a downward direction, reducing the efficiency with which the surface of Earth is able to cool itself. One way to think about the resulting trapping of radiation is that if you were in outer space looking at the Earth at a wavelength of 15 μm, most of what you would see would be radiation emitted from the sky.

Because radiation is trapped in the atmosphere and is re-emitted at some height above the surface, the average equilibrium radiative blackbody temperature of 255 K occurs at some elevation that is well above the surface of the Earth. Because air is compressible, the atmosphere warms as one descends from this elevation to the surface. Temperature at the surface of the Earth is thus warmer than the radiative equilibrium temperature, giving the observed average surface temperature of 288 K. The warming of the surface of the Earth due to the presence of the greenhouse effect is thus a remarkable 33 K.

GLOBAL WARMING

Human activities have increased the atmospheric burdens of many radiatively active gases, including CO_2, methane (CH_4), nitrous oxide (N_2O), and chlorofluorocarbons (CFCs). Methane, nitrous oxide, and CFCs tend to absorb terrestrial radiation in a wave-length window where the atmosphere is currently relatively transparent, i.e., where little absorption by other atmospheric constituents takes place. As a result, the release to the atmosphere of these trace gases has an important effect on the radiative balance even at very low levels of concentration. Atmospheric CO_2, on the other hand, is already absorbing most of the radiation in its absorption band, and any additional CO_2 acts by increasing the absorption toward the edges of the band. As a consequence, a proportionally larger increase in the atmospheric CO_2 burden is necessary to affect significantly the radiative balance of the atmosphere. Table 10.1.2 gives the relative efficiencies of the major greenhouse gases, expressed as forcing per molecule normalized to CO_2.

Table 10.1.1 summarizes the contributions of each of the greenhouse gases to the radiative balance of the atmosphere due to their concentration increases between 1765 and 1990. Despite its high background concentration, the addition of CO_2 to the atmosphere has been so large that it accounts for about 60% of the enhanced capacity of the atmosphere to absorb long-wavelength radiation emitted from the surface. CH_4 and

TABLE 10.1.1
Major greenhouse gases
Mixing ratios, present rate of increase, and increase in radiative forcing are from Ramaswamy et al. *[2001]. Preindustrial radiative forcing is from* Dickinson and Cicerone *[1986].*

	Mixing Ratio in Dry Air (ppm)			*Radiative Forcing ($W m^{-2}$)*	
Gas	*1765*	*1992*	*Current Rate of Increase (% yr^{-1})*	*Preindustrial (<1765)*	*Anthropogenic (1765–1990)*
H_2O	—	—	—	94	—
CO_2	278	356	0.4	50	1.46
CH_4	0.7	1.71	0.6	1.1	0.48
CFC-11	0.0	0.000268	0.0	0.0	0.07
CFC-12	0.0	0.000503	1.4	0.0	0.17
Other CFCs	—	—	—	—	0.10
N_2O	0.275	0.310	0.25	1.25	0.15
Total				146	2.43

TABLE 10.1.2
Radiative forcing per additional molecule relative to the radiative forcing due to one additional molecule of CO_2
Radiative forcing for changes in CO_2 from an initial mixing ratio C_o can be estimated from $\Delta Q = 6.3 \ln (C/C_o)$ $W m^{-2}$ for $CO_2 < 1000$ ppm. From Houghton et al. [1990].

Gas	*Relative Forcing*
CO_2	1
CH_4	21
N_2O	206
CFC-11	12,400
CFC-12	15,800

N_2O together account for about 27% of the total anthropogenic radiative forcing, and chlorofluorocarbons for the remaining 12%.

An estimate of the surface warming that will result from the increased greenhouse trapping can be obtained by use of a simple zero-dimensional heat balance model:

$$H\frac{d\Delta T}{dt} = \Delta Q - \lambda \Delta T \tag{10.1.1}$$

where H ($W m^{-2} K^{-1} s$) is the heat capacity of the combined ocean and atmosphere system, ΔT is the change in surface temperature, ΔQ ($W m^{-2}$) is the radiative forcing given in table 10.1.1, and λ ($W m^{-2} K^{-1}$), is referred to as the climate sensitivity parameter. In equilibrium, (10.1.1) gives

$$\Delta T = \frac{\Delta Q}{\lambda} \tag{10.1.2}$$

The value of λ is usually determined from the response of a variety of climate models to perturbations in radiative forcing [*Cubasch et al.*, 2001]. A value of 3.3 $W m^{-2} K^{-1}$ is obtained by assuming that absorption of infrared radiation by greenhouse gases occurs without any feedbacks. However, global warming increases the water vapor content, which further increases the absorption. It also modifies the vertical gradient of temperature (lapse rate) in the atmosphere. Further feedbacks include changes in snow and ice cover, which affect albedo; cloud cover, and its interaction with short- and long-wave radiation; and several more. Consideration of these feedbacks significantly reduces λ to about $1.8 \pm 0.7\, W m^{-2} K^{-1}$, with the uncertainty estimate reflecting the range of results from a variety of models. An analysis of the effect of individual feedbacks shows that the cloud feedback is by far the largest contributor to the overall uncertainty. Using this climate sensitivity, we obtain an equilibrium warming of 1.5 to 3.4 K for a CO_2 doubling ($\Delta Q = 3.7\, W m^{-2}$ [*Ramaswamy et al.*, 2001]).

TABLE 10.1.3
Radiative forcing changes between 1765 and early 1990s, based on *Ramaswamy et al.* [2001]

	Midpoint ($W m^{-1}$)	*Range* ($W m^{-1}$)
From table 10.1.1	2.43	±15%
H_2O[a]	0.02	
Ozone		
troposphere	0.35	(0.2 to 0.5)
stratosphere	−0.15	(−0.05 to −0.25)
Aerosols		
biomass burning	−0.2	(−0.1 to −0.6)
sulfate	−0.4	(−0.2 to −0.8)
soot (black carbon)	0.2	(0.1 to 0.4)
organic carbon	−0.1	(0.03 to 0.3)
mineral dust	—	(−0.6 to 0.4)
indirect effect on cloud albedo	−1.0	(0 to −2.0)
Change in solar forcing since 1850	0.3	(0.1 to 0.5)
total	1.5	

[a] The increase in radiative forcing for water vapor is only that portion due to the effect of increased methane on water vapor in the stratosphere.

The equilibrium climate warming for a CO_2 doubling reported by *Houghton et al.* [2001] is 1.5 to 4.5 K, which requires $\lambda = 1.7 \pm 0.8\, W m^{-2} K^{-1}$.

The total increase of $2.43\, W m^{-2}$ in infrared trapping over the last 200 years shown in table 10.1.1 should thus give an equilibrium warming of 1.0 to 3.0 K. Additional changes in radiative forcing shown in table 10.1.3 include the effect of increased stratospheric water vapor due to production from methane, changes in tropospheric and stratospheric ozone, direct solar forcing, and the cooling effect of aerosols. Aerosol cooling explains some unusual observations about the pattern of warming. One of these is that the southern hemisphere has warmed more than the northern. This is due to the larger production of aerosols in the northern hemisphere and their limited atmospheric lifetime, which prevents them from being spread into the southern hemisphere. The other is that northern hemisphere warming has been greatest at night. Aerosols increase reflection of solar radiation from the Earth, so their cooling effect should be felt more during the day. The combined impact of the additional radiative forcing terms reduces the total current radiative forcing to $1.5\, W m^{-2}$. This gives an equilibrium warming of 0.6 to 1.8 K. The observed warming over the last century is smaller, 0.3 to 0.6 K (figure 10.1.5). However, our estimate assumes that the temperature of the whole climate system has equilibrated with the increase in radiative forcing, whereas, in reality, the large heat capacity of the ocean and the slow rate of mixing has led to a substantial slowdown of the warm-

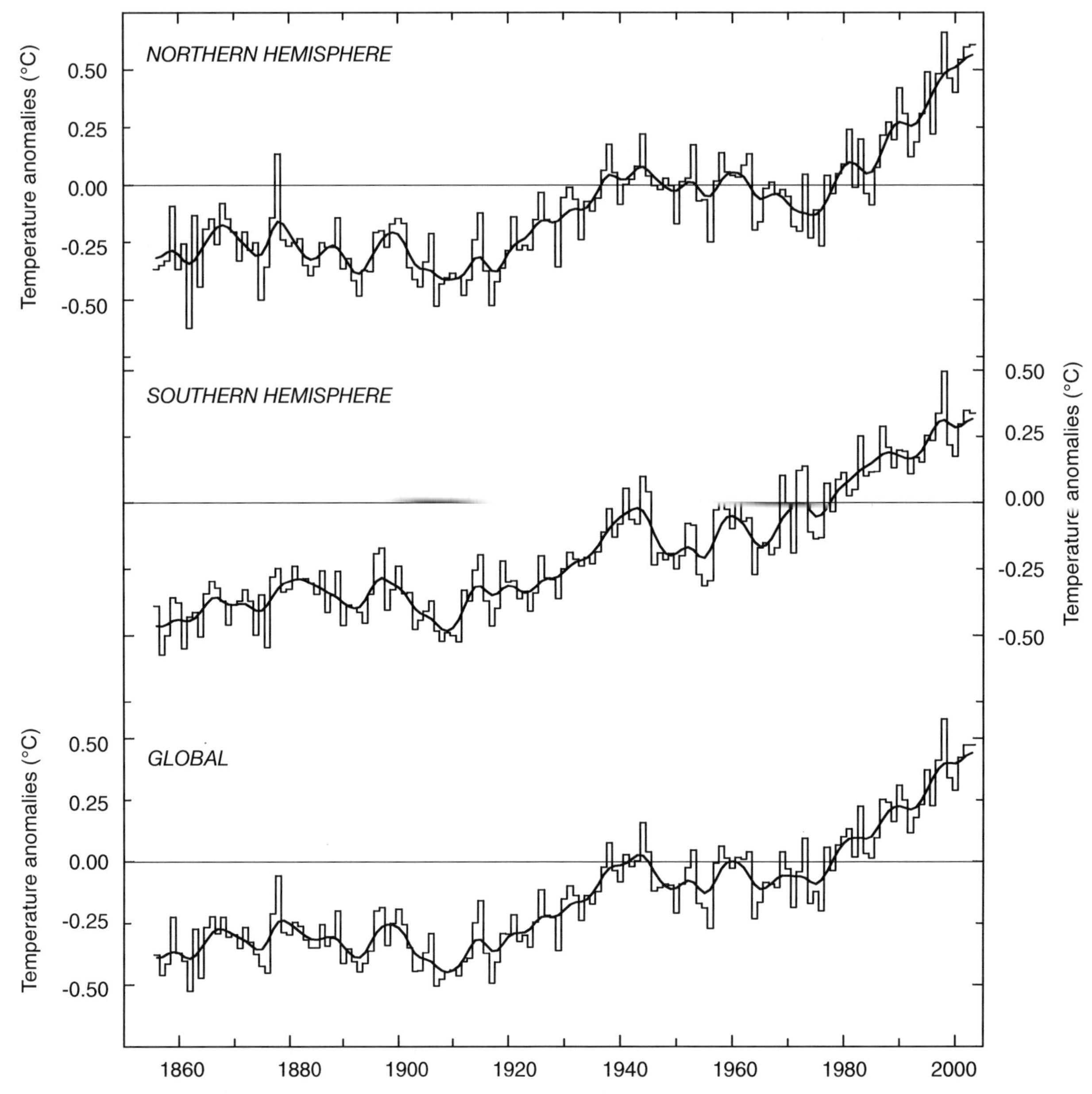

FIGURE 10.1.5: Observed warming from 1850 to 2002. Shown are the anomalies relative to the period 1961–1990. The global mean remained essentially constant from 1856 to the 1910s. A first warming of 0.3°C occurred through the 1940s, followed by a slight cooling. From the early 1970s to the present, the warming exceeded 0.5°C. Data from *Jones and Moberg* [2003].

ing rate. Including this effect in fully coupled ocean-atmosphere-land surface climate model simulations [e.g., *Mitchell et al.*, 2001] gives a present warming estimate that is very close to the observed one (figure 10.1.5). The flip side of this effect is that if the growth in radiative forcing were stopped at current levels, the Earth would continue to warm up until the equilibrium warming was reached.

While the remainder of this chapter is devoted to CO_2, it is important to recall that of the anthropogenic greenhouse gases shown in table 10.1.1, N_2O also has important oceanic sources and sinks (see also chapters 3 and 5). Its oceanic source is estimated to be 3.7 to 4.3 Tg N yr^{-1} [*Nevison et al.*, 2004], compared with total sources of 4 to 12 Tg N yr^{-1} in preindustrial times, and 13 to 20 Tg N yr^{-1} today [*Prather et al.*, 2001]. The chlorofluorocarbons have found important use in oceanography as tracers of water motion (see chapter 2), but their net uptake by the oceans is negligible because of their very low solubility and chemical inertness (see chapter 3).

Outline

This chapter is organized as follows. We first address the anthropogenic CO_2 perturbation over the last two hundred years. We will discuss in particular the reasons for the slow oceanic uptake of anthropogenic CO_2, and learn about a number of techniques that have been used to identify and quantify anthropogenic CO_2 in the ocean. We will also investigate the role of ocean biology in this process and will see that its influence becomes very important once the climate system moves significantly away from steady state as a consequence of global climate change. The response of the ocean carbon cycle to such changes is poorly understood and thus very difficult to predict. Observations of interannual to decadal timescale variability in the climate and global carbon systems provide one of the few means to study these interactions directly. Direct observations of atmospheric CO_2 since 1958 show that the growth rate of atmospheric CO_2 experienced large changes from year to year, while the growth rate of fossil fuel emissions changed only minimally. This indicates that substantial amounts of carbon are transferred between the ocean, atmosphere, and terrestrial biosphere reservoirs on interannual to decadal timescales. However, there is presently little consensus about the relative role of the oceans and of the land biosphere in controlling these changes. Section 10.3 will discuss these issues in more detail. An important avenue for improving our ability to make prediction for the future is to be able to understand the causes for the changes seen in the past. The last section of this chapter will investigate the glacial-interglacial CO_2 problem, one of the most puzzling questions of global carbon cycle research.

10.2 The Anthropogenic Perturbation

Over the Anthropocene, i.e., from about 1800 to 1994, humankind has released about 340 to 420 Pg carbon to the atmosphere in the form of CO_2 ($1\ Pg\,C = 10^{15}\ g\,C$) (see table 10.2.1) [*Sabine et al.*, 2004]. Most of this carbon stems from the combustion of fossil fuels (about 240 Pg C), but a significant albeit very uncertain fraction is associated with emissions arising from changes in land use, such as deforestation. About 165 Pg C of these emissions have stayed in the atmosphere, leading to an increase in atmospheric CO_2 from a preindustrial level of about 280 ppm to about 360 ppm in 1994 (figure 10.1.2). The remainder has been redistributed within the land and ocean (figure 10.1.1). Anthropogenic CO_2 emissions are expected to continue to rise well into the twenty-first century, despite the initiation of efforts on the international level to curb the emissions. Given the importance of CO_2 as a greenhouse gas, the scientific community needs to be able to predict how much CO_2 the ocean and land biosphere will be taking up from the atmosphere, thereby helping to mitigate the anthropogenic CO_2 burden in the atmosphere and the associated increase in radiative forcing.

Before any prediction can be attempted, it is necessary to correctly quantify the redistribution of the anthropogenic carbon that has been emitted since the beginning of the industrial revolution. The budget for the entire Anthropocene as well as that for the last two decades have been assessed in detail [*Prentice et al.*, 2001; *Le Quéré et al.*, 2003; *Sabine et al.*, 2004]. According to the budgets recently established by *Sabine et al.* [2004] and shown in table 10.2.1, the ocean represents the largest sink of anthropogenic CO_2 over the Anthropocene, having taken up about 118 ± 19 Pg C. It is rivaled only by the storage in the atmospheric reservoir. Over the same period, the terrestrial biosphere acted as a net source of 39 ± 28 Pg C, the sum of an ill-constrained source coming from land-use change (about 100 to 180 Pg C), indicating a sink of somewhere

TABLE 10.2.1
Anthropogenic CO_2 budget for the Anthropocene (1800–1994) and for the period of the 1980s and 1990s.
From Sabine et al. [2004].

	1800–1994 *Pg C*	*1980–1999* *Pg C*	*1980–1999 (Avg)* *Pg C yr^{-1}*
Constrained sources and sinks			
(1) Emissions from fossil fuel and cement production	244 ± 20	117 ± 5	5.9 ± 0.3
(2) Storage in the atmosphere	-165 ± 4	-65 ± 1	-3.3 ± 0.1
(3) Ocean uptake	-118 ± 19	-37 ± 7	-1.9 ± 0.4
Inferred net terrestrial balance			
(4) Net terrestrial balance $= [-(1)-(2)-(3)]$	39 ± 28	-15 ± 9	-0.8 ± 0.5
Terrestrial balance			
(5) Emissions from changes in land use	$100-180$	24 ± 12	1.2 ± 0.6
(6) Terrestrial biosphere sink $= [-(1)-(2)-(3)-(5)]$	-61 to -141	-39 ± 18	-2.0 ± 1.0

between 61 to 141 Pg C. Over the last two decades, i.e., from 1980 to 1999, the ocean also has acted as a strong sink for anthropogenic CO_2, taking up, on average, about 1.9 Pg C yr^{-1}, i.e., about 27% of the total anthropogenic emissions of about 7.1 Pg C yr^{-1} (table 10.2.1).

How were these oceanic uptake estimates obtained, and what controls the rate of uptake? How might the oceanic sink for anthropogenic CO_2 change in the future when climate is predicted to be significantly altered? These questions are of fundamental importance, as the current radiative forcing associated with increased atmospheric CO_2 would be about 30% higher in the absence of the oceanic sink. Furthermore, positive feedbacks could develop if the ocean became less efficient in taking up anthropogenic CO_2 in a warming world, leading to an acceleration of the growth of CO_2 in the atmosphere and therefore an acceleration of global warming.

We address these questions next. We start with an assessment of the oceanic uptake capacity after equilibrium has been reached, but will soon show that there are strong kinetic constraints that prevent the equilibrium from being reached within the next millennium. We then evaluate briefly different methods that have been employed to estimate the present uptake of anthropogenic CO_2 by the oceans. We conclude by discussing models that have been used to calculate the past uptake as well as to make predictions for the future.

Capacity Constraints

If CO_2 dissolved in seawater without undergoing chemical reactions like most other gases, about 70% of the CO_2 emitted by anthropogenic activities would remain in the atmosphere. However, the uptake capacity of the oceans is greatly enhanced by the reaction of CO_2 with the carbonate ion to form bicarbonate, making the carbonate ion concentration the primary determinant for the oceanic uptake capacity for anthropogenic CO_2. As seen above (see 8.3.8), the buffering reaction can be written as

$$H_2CO_3^* + CO_3^{2-} \rightleftharpoons 2HCO_3^- \qquad (10.2.1)$$

The carbonate ion content of the ocean is not a conservative property, however. It is tied to the inorganic carbon system in seawater, ($[CO_3^{2-}] \approx Alk - DIC$), as well as to the large reservoir of $CaCO_3$ in the sediments. The latter can absorb CO_2 directly by

$$H_2CO_3^* + CaCO_3 \rightleftharpoons Ca^{2+} + 2\,HCO_3^- \qquad (10.2.2)$$

As we will discuss in more detail below, these two buffering reactions operate on two very different timescales. The characteristic time for reaching equilibrium between the atmosphere and the ocean alone is of the order of several hundred years. The characteristic time required for reaching equilibrium between the atmosphere, the ocean, and the sea-floor sediments is on the order of several thousand years [*Broecker and Takahashi*, 1977a, *Archer et al.*, 1997]. We therefore separately consider the case where the anthropogenic CO_2 entering the ocean is neutralized by the carbonate concentration in seawater alone, and the case where it is neutralized by the carbonate ion in seawater plus the calcium carbonate in the sediments. We will also briefly discuss the very long-term sinks of anthropogenic CO_2 stemming from the weathering reaction with $CaCO_3$ and siliceous rocks on land.

Buffering by Dissolved Carbonate

Let us first consider the situation for an ocean with uniform chemistry and no sediments. On the basis of what we have learned about the CO_2 chemistry so far, we can predict how much of an incremental addition of CO_2 to the atmosphere would dissolve in the ocean under the assumption that equilibrium is maintained, i.e., that air-sea gas exchange and ocean mixing are fast enough to keep the oceanic perturbation in equilibrium with that in the atmosphere.

Let the atmospheric increment be

$$\delta N^{atm} = \delta\chi^{CO_2^{atm}} N^{atm}_{tot} \qquad (10.2.3)$$

where $\chi^{CO_2^{atm}}$ is the atmospheric mixing ratio, and where N^{atm}_{tot} is the number of moles of air in the atmosphere ($N^{atm}_{tot} = 1.8\times10^{20}$ mol). The increment in the ocean is given by

$$\delta N^{oc} = \delta DIC\, m^{oc} \qquad (10.2.4)$$

where m^{oc} is the mass of the ocean volume under consideration.

Since we assumed that the perturbation in the ocean tracks the perturbation in the atmosphere, the incremental change in oceanic pCO_2 has to be equal to the incremental change in atmospheric pCO_2, which itself is equal to the change in the mixing ratio times the total atmospheric pressure, P^{atm} (see panel 3.2.1, and where we neglect the contribution of water vapor pressure).

$$\delta pCO_2^{oc} = \delta pCO_2^{atm} = \delta\chi^{CO_2^{atm}}\, p^{atm} \qquad (10.2.5)$$

The change in the atmospheric inventory relative to the change in the oceanic inventory is thus given by

$$\frac{\delta N^{atm}}{\delta N^{oc}} = \frac{\delta\chi^{CO_2^{atm}} N^{atm}_{tot}}{\delta DIC\, m^{oc}} = \frac{\delta pCO_2^{oc}\, N^{atm}_{tot}}{p^{atm}\delta DIC\, m^{oc}} \qquad (10.2.6)$$

$$= \gamma_{DIC} \frac{1}{\alpha} \frac{N^{atm}_{tot}}{p^{atm}\, m^{oc}} \qquad (10.2.7)$$

where γ_{DIC} is the buffer factor defined by (8.3.14), and α is the ratio of mean DIC and mean surface ocean pCO_2.

TABLE 10.2.2
Distribution of carbon in a simplified ocean-atmosphere carbon system as a function of the size of a perturbation

	Preindustrial	Size of Perturbation (Pg C) 150	600	1800
Atmosphere	1.5	16.4	17.4	20.8
Ocean	98.5	83.6	82.6	79.2

Note: Totals in all cases are normalized to 100.

We have assumed that the increments δ are small enough so that they can be approximated by ∂. However, we are more interested in calculating the fraction f of the emissions taken up by the oceans, thus

$$f = \frac{\delta N^{oc}}{\delta N^{oc} + \delta N^{atm}} = \left(\frac{\delta N^{atm}}{\delta N^{oc}} + 1 \right)^{-1} = \left(\frac{\gamma_{DIC}}{\alpha} \frac{N^{atm}_{tot}}{p^{atm} m^{oc}} + 1 \right)^{-1} \quad (10.2.8)$$

If we assume that only the top 75 m of the ocean ($m^{oc} = 2.7 \times 10^{19}$ kg) is in equilibrium with the atmosphere, and insert global surface average values for DIC, pCO_2, and the buffer factor, we find a fraction f of about 0.08. If the whole ocean were available ($m^{oc} = 1.35 \times 10^{21}$ kg), then f would be 0.81, thus about 81% of an increment added to the atmosphere would be absorbed by the ocean. Table 10.2.1 informs us that the present fraction f is of the order of 30%; hence the equivalent of only about 8% of the total volume, i.e., about the top 350 m of the ocean, is currently equilibrated with the atmospheric CO_2 perturbation.

This estimate of f is for a small addition only. In order to estimate the equilibrium air-sea distribution after a pulse has been added to the atmosphere, the above equation needs to be integrated over time. In doing so, it is important to note that the buffer factor γ_{DIC} increases, i.e., the buffer capacity decreases, when DIC increases due to the addition of CO_2 from the atmosphere (see equation (8.3.16)). This effect can be understood by considering that the uptake of CO_2 from the atmosphere depletes the CO_3^{2-} ion content of the ocean, therefore decreasing its capacity for further uptake.

As a consequence of the reduction in the uptake capacity, the fraction f decreases as the perturbation grows. Table 10.2.2 illustrates this behavior for a simplified ocean-atmosphere carbon system. In the preindustrial steady state, only about 1.5% of the total carbon is found in the atmosphere, and 98.5% is in the ocean. If we add a perturbation of 1800 Pg C to this system and let it equilibrate, the fraction of the perturbation that goes into the ocean is about 79.2%, whereas the fraction now residing in the atmosphere is more than 10 times larger (20.8%) than the preindustrial steady state. *Sarmiento et al.* [1995] demonstrated that the effect of a decreasing buffer capacity on the future oceanic uptake of anthropogenic CO_2 is large, and neglect of this effect would lead to a substantial overestimation of the oceanic uptake of anthropogenic CO_2 by the ocean (see figure 10.2.1).

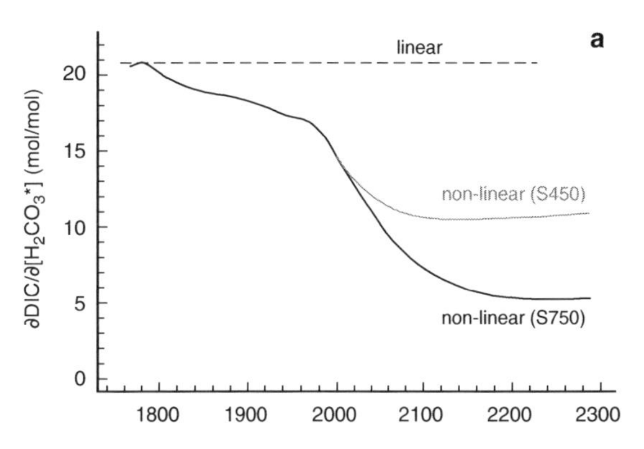

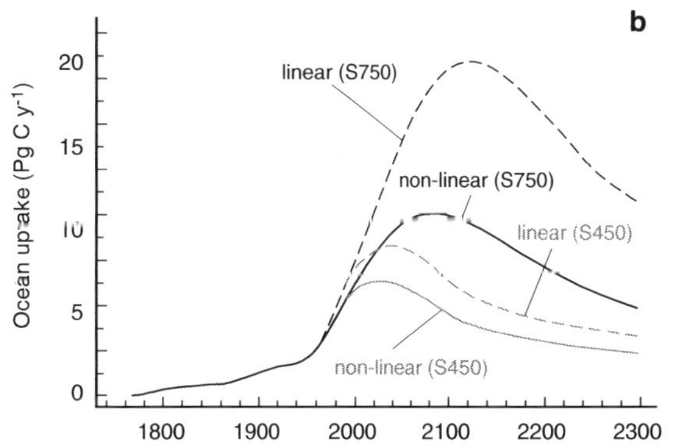

FIGURE 10.2.1: (a) The global mean instantaneous change in surface ocean DIC resulting from a given change in $H_2CO_3^*$ concentration for the S450 and S750 scenarios in which CO_2 is stabilized at 450 ppm and 750 ppm, respectively. The large reduction in this ratio through time is a measure of the reduction in the oceanic buffer capacity of surface waters. (b) The annual oceanic uptake for S450 and S750 scenarios, using either full (nonlinear) chemistry (solid lines) or simplified linear chemistry (dashed lines). The nonlinear CO_2 chemistry of seawater leads to a dramatic reduction in the future oceanic uptake of CO_2, with the effect becoming larger as the atmospheric perturbation increases. From *Sarmiento et al.* [1995].

Figure 10.2.1a actually shows the ratio $\partial DIC/\partial[H_2CO_3^*]$, which is related to the buffer factor, γ_{DIC}, by

$$\frac{\partial DIC}{\partial[H_2CO_3^*]} = \frac{\partial DIC}{\partial(K_0 \cdot pCO_2^{oc})} = \frac{DIC}{K_0 \cdot pCO_2^{oc}} \frac{1}{\gamma_{DIC}} = \frac{\alpha}{K_0} \frac{1}{\gamma_{DIC}} \quad (10.2.9)$$

where K_o is the solubility of CO_2, and where we use α as a shorthand for the ratio of ocean mean DIC and pCO_2 (see equation (10.2.7) above). The partial derivative, $\partial DIC/\partial[H_2CO_3^*]$, is therefore proportional to the inverse of γ_{DIC}, and the simulated decrease in $\partial DIC/\partial[H_2CO_3^*]$, corresponds to an increase in the buffer factor γ_{DIC}.

BUFFERING BY SEDIMENT $CaCO_3$

In all of the above calculations we have neglected so far the potential reaction of anthropogenic CO_2 with the $CaCO_3$ in the sediments at the sea floor (reaction

(10.2.2)). If all CO_2 were absorbed by this reaction, the net result of the invasion of anthropogenic CO_2 into the ocean would be to increase the total *DIC* and *Alk* content of the ocean by two moles for each mole of CO_2 taken up from the atmosphere. In this case, the CO_3^{2-} ion content, which we can approximate by *Alk* − *DIC*, would remain constant. Therefore, this buffering reaction compensates nearly completely for the decrease in the buffer capacity stemming from the CO_3^{2-} ion consumption (see panel 8.3.1), leading to almost no change in γ_{DIC} as anthropogenic CO_2 builds up in the ocean. This can be understood by considering the approximation for the buffer factor given in (8.3.16) and rearranging it to express it in terms of the *Alk*/*DIC* ratio, i.e.,

$$\gamma_{DIC} \approx \frac{3 \cdot Alk \cdot DIC - 2 \cdot DIC^2}{(2 \cdot DIC - Alk)(Alk - DIC)} \qquad (10.2.10)$$

$$\approx \frac{2 \cdot DIC - 3 \cdot Alk}{\left(\frac{Alk}{DIC} - 2\right)(Alk - DIC)} \qquad (10.2.11)$$

As *DIC* and *Alk* change equally as a result of the buffering reaction of anthropogenic CO_2 with $CaCO_3$ (10.2.2), neither the numerator nor the denominator of (10.2.11) change much, thus explaining why γ_{DIC} changes little in this case. As a consequence of this buffering with $CaCO_3$, the fraction of CO_2 ultimately ending up in the ocean is significantly larger if the dissolution of calcium carbonate at the sea floor is considered. *Archer et al.* [1997] find that the buffering by the calcium carbonates in the oceanic sediments increases ocean uptake fraction by about 10 to 15% depending on the size of the anthropogenic perturbation (see figure 10.2.2).

Buffering by Weathering

In addition to the neutralization of the anthropogenic CO_2 by $CaCO_3$ on the sea floor, an additional fraction will be neutralized by weathering reactions of atmospheric CO_2 with mineral carbonates on land. After complete neutralization with the ocean and $CaCO_3$ rocks, about 8% of a CO_2 release will remain in the atmosphere [*Archer et al.*, 1997, 1998] (see also figure 10.2.2). The remaining CO_2 will react very slowly with igneous rocks on land by the reaction

$$2CO_2 + CaSiO_3 + 3H_2O \rightleftharpoons 2HCO_3^- + Ca^{2+} + Si(OH)_4 \qquad (10.2.12)$$

Since the reservoir of these siliceous rocks is much larger than the amount of fossil fuel CO_2, all emitted CO_2 will eventually disappear from the atmosphere. This latter reaction is of fundamental importance in controlling atmospheric CO_2 on geological timescales [*Walker et al.*, 1981], since it balances the CO_2 that is degassed from volcanoes.

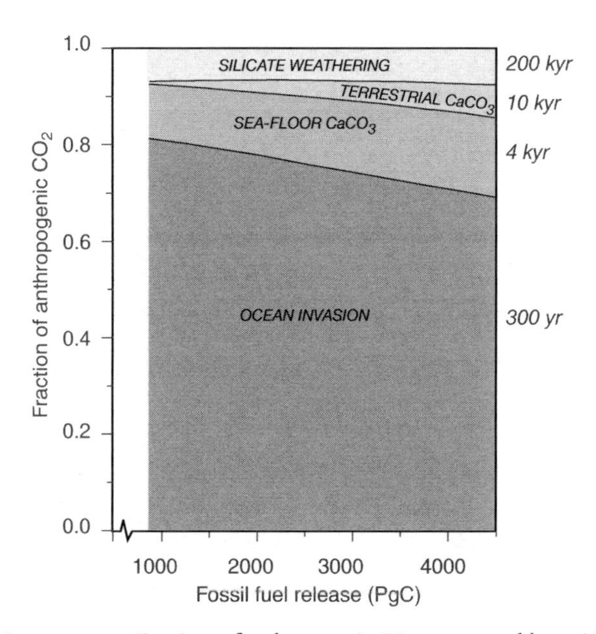

Figure 10.2.2: Fractions of anthropogenic CO_2 sequestered by various abiological processes plotted as a function of anthropogenic CO_2 release. The approximate e-folding timescales for each process are given at the right. The fraction remaining in the atmosphere for a given timescale is the difference between 1 and the level of the cumulative curve. For example, on timescales of 4000 years, the fraction remaining in the atmosphere after equilibration is determined by the magnitude sequestered by the ocean by reaction with carbonate and sea-floor $CaCO_3$. For an emission of 4000 Pg C, these two processes remove about 87% of the emission, leaving 13% in the atmosphere. From *Archer et al.* [1997].

Kinetic Constraints

The above considerations provide us with a simple tool to estimate the amount of anthropogenic CO_2 that the ocean will have taken up for a given CO_2 emission after equilibrium has been reached between the atmosphere and ocean. Our finding that only a small fraction of the ocean has equilibrated with the anthropogenic CO_2 emissions that occurred over the last 200 years indicates, however, that the timescale required for equilibration must be longer than this period. Therefore, one or several of the processes that control the oceanic uptake of anthropogenic CO_2 must be strongly rate-limiting, providing strong kinetic constraints on the oceanic uptake of anthropogenic CO_2.

Atmospheric Pulse Response

The timescales associated with the removal of anthropogenic CO_2 from the atmosphere can be assessed by investigating how quickly a pulse of CO_2 emitted into the atmosphere disappears from it [*Maier-Reimer and Hasselmann*, 1987; *Sarmiento et al.*, 1992]. Figure 10.2.3 shows the result of a simulation that followed this approach using a three-dimensional ocean general circulation model, to which a full ocean carbon cycle model was coupled. This model also includes a representation of ocean sediments, as well as a simple parameteriza-

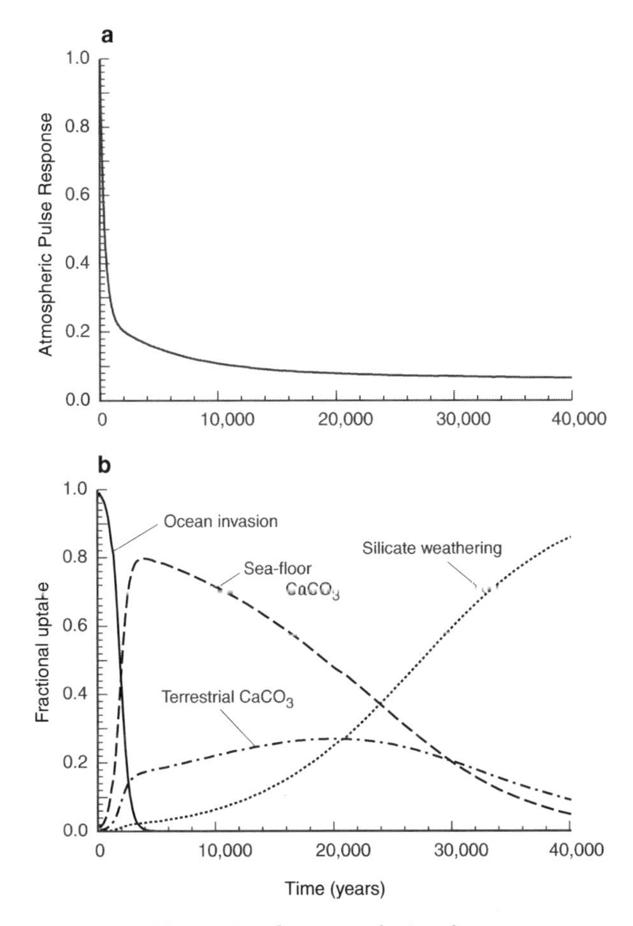

FIGURE 10.2.3: Time series of (a) atmospheric pulse response function, ($r_s = \delta N^{ML}/\delta N^{ML}_{Initial}$) and of (b) instantaneous fractional contributions of four different processes to the sequestration. The four processes are ocean invasion, reaction of the anthropogenic CO_2 with mineral calcium carbonates at the sea floor, reaction with calcium carbonates on land, and sequestration by silicate weathering. Note that the exact shape of the atmospheric pulse response depends on the size of the pulse because of the nonlinearity of the oceanic buffer factor. This pulse here has been calculated for a pulse size of 3000 Pg of carbon. Based on results by *Archer et al.* [1997].

tion of CO_2-consuming weathering processes on land [*Archer et al.*, 1997].

This simulation reveals that it takes about 400 years for the first half of the initial pulse to disappear, and that it takes another 1000 years for an additional 25% of the initial pulse to be removed from the atmosphere. This simulation also shows that after the first 1000 years, the oceanic uptake that is driven by the solution of atmospheric CO_2 in seawater and reaction with CO_3^{2-} becomes negligible, and that neutralization of the CO_2 by reaction with $CaCO_3$ on the sea floor becomes dominant. At the same time, reaction of atmospheric CO_2 with $CaCO_3$ on land begins to be a relevant removal process. After 30,000 years, the contribution of these two processes becomes small, and most of the removal of the initial pulse is governed by the consumption of atmospheric CO_2 by weathering of silicate rocks on land.

In summary, for an atmospheric CO_2 pulse of 1000 Pgc, chemical buffering by the inorganic carbon chemistry of the seawater sequesters about 80% on timescales of several hundred years, with the exact fraction depending on the size of the pulse (see figure 10.2.2) [*Archer et al.*, 1997]. Neutralization of CO_2 with the calcium carbonate on the sea floor accounts for an additional 10% decrease in the atmospheric CO_2 pulse on a timescale of about 4000 years. Reaction with $CaCO_3$ on land contributes 2%, with a timescale of about 10,000 years. After several tens of thousands of years, about 8% of the initial pulse remains in the atmosphere. On timescales of several hundred thousands of years, this remaining CO_2 will react with igneous rocks on land, so that the entire initial pulse will disappear from the atmosphere on timescales approaching a million years. What determines the dynamics of these removal processes, particularly those associated with the buffering by the ocean?

OCEAN UPTAKE AND BUFFERING WITH DISSOLVED CARBONATE

Let us first consider ocean uptake only, i.e., the removal of an atmospheric CO_2 perturbation driven solely by the dissolution of CO_2 into seawater and subsequent reactions with the CO_3^{2-} ions. There are two possible rate-limiting steps that determine the kinetics of the ocean uptake: air-sea gas exchange and transport of CO_2 from the surface layers into the interior of the ocean. We have seen in chapter 8 that the characteristic time for a 40 m deep surface layer to reach equilibrium with the atmosphere by air-sea gas exchange is about 6 months. This is much smaller than the timescale of the anthropogenic perturbation, which has an e-folding time of several decades, the exact value depending on the time period considered. One can therefore expect that air-sea exchange is generally not the rate-limiting step for the oceanic uptake of anthropogenic CO_2 from the atmosphere. This is confirmed by model simulations of oceanic uptake, which demonstrate a modest response of only approximately 10% to a doubling or halving of the gas exchange coefficient (see table 3.3.4).

It therefore must be the downward transport of the anthropogenic CO_2 burden from the surface into the interior of the ocean that is the rate-limiting step for the CO_2 uptake by the oceans. In chapter 2, we were able to establish with the help of several tracers that the timescale for the ventilation of the thermocline is between a couple of years and decades, while the ventilation timescale for the deep ocean is several hundred years. Given this long timescale for deep water ventilation, it becomes obvious that the time required for an atmospheric CO_2 pulse to reach equilibrium in an ocean-atmosphere system must be on the order of a thousand years or more.

We can investigate the transport of anthropogenic CO_2 out of the surface layer into the interior of the ocean by inspecting how quickly a pulse of anthropo-

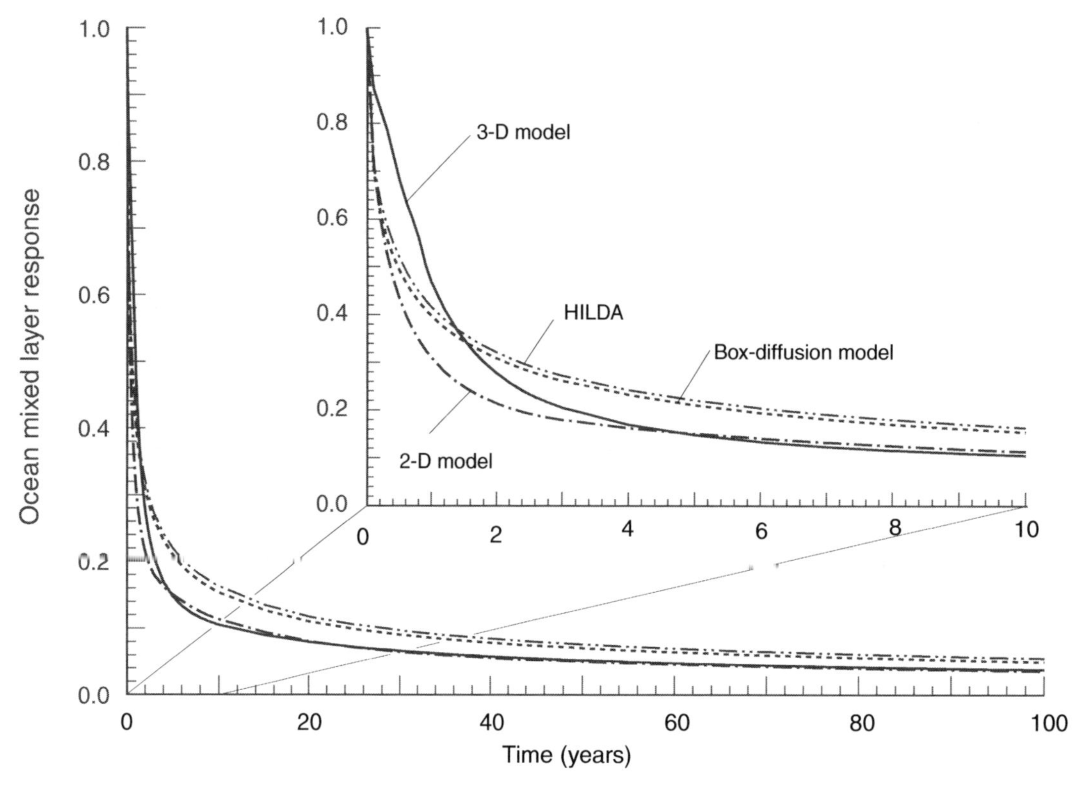

FIGURE 10.2.4: Effective mixed layer pulse-response functions, $r_s = \delta N^{ML} / \delta N^{ML}_{Initial}$, for the box-diffusion model, the HILDA model, the 2-D model, and the Princeton general circulation model. Inset shows the pulse-response functions for the first ten years. Modified from *Joos et al.* [1996].

genic CO_2 (or any other passive and conservative tracer) that is injected into the mixed layer decreases [*Joos et al.*, 1996]. In addition to its value as a tracer of oceanic transport processes, the mixed layer pulse has proved valuable in developing simple models of the ocean carbon sink. By contrast with the use of an atmospheric pulse, in which the response function depends on the amount of carbon added because of how this affects the buffer capacity, use of a mixed layer pulse response with direct calculation of the air-sea equilibration gives a model that does not depend on the pulse size (see *Joos et al.* [1996]). Figure 10.2.4 shows such mixed layer pulse-response functions for different ocean carbon models for the first 100 years. All models show a relatively rapid removal of the pulse in the first few years, followed by a strongly decelerated decrease with time.

Joos et al. [1996] demonstrated that these pulse-response functions can be very closely approximated by fitting to them a series of exponential functions. In the case of the 3-D circulation model, their fits show that about 64% of the initial pulse disappears with an e-folding time of less than a year. An additional 23% decays with an e-folding timescale of 2 years, 6% with an e-folding timescale of 15 years, 3% with an e-folding timescale of 65 years, and 2% with an e-folding timescale of 348 years. The response function asymptotes to a value of about 2% of the initial pulse, since this constitutes the ocean mean concentration after the whole ocean has equilibrated with the initial pulse. The various timescales can be interpreted as reflecting how quickly waters are transported into different depth realms of the ocean. Because upper ocean ventilation is relatively vigorous, the anthropogenic CO_2 is, at first, quickly removed from the upper ocean mixed layer. However, ocean ventilation tends to decrease with increasing depth, so it takes longer for anthropogenic CO_2 to be transported into these deeper realms.

These pulse-response functions not only provide a measure of timescales, but also quantitatively describe the entire dynamics of the transport of anthropogenic CO_2 and any other passive and conservative tracers from the surface mixed layer into the ocean's interior. This is because oceanic transport and mixing is linear, and up to now, we can assume that ocean circulation has not changed dramatically over the last 200 years. We can therefore use these pulse-response functions (also referred to as Green's functions) as a basis for a very simple model that permits us to accurately estimate the uptake of any passive and conservative tracer by the ocean without resorting to simulations with much more complex two- and three-dimensional ocean biogeochemical models (see *Joos et al.* [1996] for a complete description of the method).

Let us describe the pulse response function as $r_s(t)$. The surface concentration of anthropogenic *DIC*, $C_{ant}(t)$ (mmol m^{-3}), where t is any time after the beginning of the Anthropocene, t_o, reflects the accumulated air-sea flux of anthropogenic CO_2 from the atmosphere, Φ_{as} (mmol m^{-2} yr^{-1}), weighted by how much of this uptake has been transported away into the interior over time. This weighting is described by the pulse response function, r_s. This gives:

$$C_{ant}(t) = \frac{1}{h}\int_{t_o}^{t} \Phi_{ant}(t') r_s(t - t') dt' \qquad (10.2.13)$$

where h is the depth of the surface mixed layer. As is the case for any gas (see chapter 3, equation (3.3.6)), the air-sea flux of anthropogenic CO_2 is directly proportional to the air-sea partial pressure difference of anthropogenic CO_2 and a gas exchange coefficient, k_g, (mmol m^{-2} yr^{-1} μatm^{-1}),

$$\Phi_{ant}(t) = k_g(\delta pCO_2^{atm}(t) - \delta pCO_2^{oc}(t)) \qquad (10.2.14)$$

where δpCO_2 denotes the anthropogenic perturbation to atmospheric and oceanic pCO_2 in μatm. The anthropogenic perturbation of oceanic pCO_2 is linked to $C_{ant}(t)$ by the buffer factor γ_{DIC}:

$$\delta pCO_2^{oc}(t) = \frac{\gamma_{DIC}}{\alpha} C_{ant}(t) \qquad (10.2.15)$$

where α is the DIC/pCO_2 ratio (see eq. (10.2.9)).

By combining (10.2.13) with (10.2.14) and (10.2.15), and using a pulse-response function $r_s(t)$ determined from one of the models, the oceanic uptake of anthropogenic CO_2 for any given time trajectory of the atmospheric CO_2 perturbation can be computed with great accuracy. The only limiting assumption is that the oceanic circulation has to remain steady. *Joos et al.* [1996] provide accurate formulations for the buffer factor γ_{DIC} over a wide range of atmospheric CO_2 perturbations and also functional forms for the pulse response functions $r_s(t)$ for a suite of ocean biogeochemical models.

Buffering by Sediment $CaCO_3$

The dynamics of the buffering of anthropogenic CO_2 by sea-floor $CaCO_3$ is very slow and takes thousands of years (figure 10.2.2). This is much slower than the timescale of hundreds of years involved in taking up anthropogenic CO_2 from the atmosphere and transporting it from the surface into the interior of the ocean. An important reason for the slow dynamics of sea-floor $CaCO_3$ buffering is that most of the $CaCO_3$ in the sediments is found at mid-depths of the water column (see figure 9.1.3). Anthropogenic CO_2 must first reach these depths before it can start to dissolve the $CaCO_3$ in the sediments. In addition, the rate of dissolution induced by anthropogenic CO_2 seems to be relatively small [*Archer et al.*, 1998], a result of slow diffusion of the bottom water signal into the sediments as well as the slow kinetics of dissolution (see chapter 9). This increases the timescale for the sea-floor neutralization of anthropogenic CO_2 further.

The exact timescales for the neutralization of the anthropogenic CO_2 with calcium carbonate are associated with quite large uncertainties. Fortunately, these timescales are sufficiently long so that the contribution of calcium carbonate buffering can be neglected when considering anthropogenic CO_2 uptake over the last two centuries and well into the next century. This is supported by the results of *Archer et al.* [1998], who found that the difference between the oceanic uptake of a model with and without calcium carbonate was much less than 1%. In contrast to the anthropogenic CO_2 perturbation, these buffering reactions are of great importance when considering millennial timescale changes in the global carbon cycle, as they give rise to substantial feedbacks, as discussed in association with $CaCO_3$ compensation in chapter 9.

Anthropogenic CO_2 Uptake

So far, we have found that it will take several hundred, if not thousands of years until anthropogenic CO_2 is fully equilibrated between the atmosphere, the ocean, and the sediments on the sea floor. For human societies, however, it is of much more concern to know the current redistribution of anthropogenic carbon in the global carbon cycle and how this may change over the next few decades. This is a topic of intense research, and a number of breakthroughs in the past decade have permitted the scientific community to put relatively strong constraints on the oceanic uptake of anthropogenic CO_2 (see discussion in *Prentice et al.* [2001], *Sarmiento and Gruber* [2002], and *Sabine et al.* [2004]).

In the global budget of anthropogenic CO_2 (table 10.2.1), there are only two terms that have been measured directly. One is the emission of CO_2 by burning of fossil fuel and cement manufacture. The other is the increase in the atmospheric CO_2 concentration, which has been measured directly since 1958 at Mauna Loa, Hawaii and the South Pole [*Keeling et al.*, 1989]. These measurements have been extended back in time by measuring CO_2 in ice cores and firn [*Neftel et al.*, 1985; *Etheridge et al.*, 1996]. The remaining two budget terms (ocean uptake, net exchange with the terrestrial biosphere) have to be estimated by less direct means. We focus here on the oceanic uptake only. A good understanding of the magnitude and variability of the oceanic sink for anthropogenic CO_2 also provides important constraints for the net exchange of anthropogenic CO_2 with the terrestrial biosphere, which is very difficult to measure directly.

This section gives a brief overview of the different approaches proposed so far for estimating the current CO_2 uptake by the oceans. We fill first investigate what it takes to estimate this uptake by direct means, i.e., measuring the flux across the air-sea interface, or by measuring the long-term increase of *DIC*. Given our currently limited ability to follow these direct approaches, inventory methods that attempt to separate the small anthropogenic CO_2 signal from the large natural background have played a very important role in our deciphering of the anthropogenic CO_2 uptake by the ocean. After the discussion of these inventory methods, we turn to the atmospheric oxygen method, which has given us a lot of insight into the oceanic uptake since the early 1990s. The reader interested in more detail is referred to the IPCC summaries [*Schimel et al.*, 1995, 1996; *Prentice et al.*, 2001] and to the overviews by *Siegenthaler and Sarmiento* [1993] and by *Wallace* [1995].

Direct Estimation

The most direct means to estimate the oceanic uptake of anthropogenic CO_2 is to measure the increase of *DIC* in the oceans over time, as done for the atmosphere. How large do we expect the signal to be? We can estimate the rate of increase of *DIC* in the surface ocean using the buffer factor, γ_{DIC}, and making the assumption that the inorganic carbon system in the surface ocean stays in equilibrium with the atmospheric perturbation:

$$\frac{\partial DIC}{\partial t} = \frac{\partial DIC}{\partial pCO_2}\frac{\partial pCO_2^{oc}}{\partial dt} = \frac{1}{\gamma_{DIC}}\frac{DIC}{pCO_2^{oc}}\frac{\partial pCO_2^{atm}}{\partial t} = \frac{\alpha}{\gamma_{DIC}}\frac{\partial pCO_2^{atm}}{\partial t} \qquad (10.2.16)$$

where α is the DIC/pCO_2^{oc} ratio.

The average rate of change of atmospheric pCO_2 from 1975 to 1995 was about 1.5 μatm yr^{-1} [*Keeling and Whorf*, 1998]. Inserting the surface ocean mean *DIC* (2026 μmol kg^{-1}, see table 8.2.4), a buffer factor of 10, and a pCO_2 of 350 μatm results in a predicted *DIC* increase of about 0.9 μmol kg^{-1} yr^{-1}, i.e. a change of 9 μmol kg^{-1} over a ten-year period. The change is larger (1.1 μmol kg^{-1} yr^{-1}) in the low latitudes and smaller in the high latitudes (0.8 μmol kg^{-1} yr^{-1}) because of the spatial variation in the buffer factor. Note that this is counterintuitive, since one would expect the high latitudes, because of their colder temperature, to take up more anthropogenic CO_2 for a given atmospheric pCO_2 perturbation than the warmer low latitudes. However, in this case, the buffer capacity is the determining factor, and not the CO_2 solubility. We will discuss this in detail in the following section.

The present measurement precision of *DIC* is about 1 μmol kg^{-1}, good enough for potentially detecting such a signal over a few years. However, temporal and spatial variability of *DIC* is large, making such a detection difficult. We have seen, for example, that seasonal variations exceeding 30 μmol kg^{-1} occur at many places. Nevertheless, existing long-term observations of *DIC* in the surface ocean are in relatively good agreement with the expected increase computed from (10.2.16) (see figure 10.2.5).

Some of the limitations imposed by spatio-temporal variability can be overcome by fitting the *DIC* content in seawater at one time with a multiple linear regression model against a number of tracers unaffected by the anthropogenic transient, and then determining the difference between the predicted and the measured *DIC* at a later time [*Wallace*, 1995]. However, this approach requires two data sets of high quality separated sufficiently in time. The GEOSECS survey in the 1970s and the WOCE/JGOFS effort of the 1990s provide in principle two such data sets [*Wallace*, 2001], but the high inaccuracy and imprecision of the GEOSECS data (about 10 μmol kg^{-1}) make this approach so far only partially successful [*Sabine et al.*, 1999; *Peng et al.*, 1998, 2003; *McNeil et al.*, 2001]. An ongoing program to resample many of the cruise tracks that were sampled during the WOCE/JGOFS era will vastly expand the opportunities in the future to detect the increase of anthropgenic CO_2 in the ocean over time.

An alternative approach to assess the uptake of anthropogenic CO_2 by the ocean is estimate the exchange of CO_2 across the air-sea interface (see figure 8.1.1). Although the flux at any given location represents the superposition of a preindustrial (i.e., natural) flux and the anthropogenic perturbation flux, the globally integrated air-sea flux should be equal to the total oceanic uptake of anthropogenic CO_2 plus a global net air-sea flux of natural CO_2 associated with the balance between riverine input of organic and inorganic carbon and seafloor burial [*Sarmiento and Sundquist*, 1992] (see figure 10.1.1).

We can estimate the level of the observational challenge associated with this approach by computing the partial pressure of CO_2 difference across the air-sea interface associated with the anthropogenic CO_2 uptake. Current estimates of the globally integrated oceanic uptake of anthropogenic CO_2 uptake are about 2 Pg C yr^{-1}, or $1.6 \cdot 10^{14}$ mol C yr^{-1} [*Prentice et al.*, 2001]. Dividing this estimate by the ocean area of about 360×10^{12} m^{-2} gives a mean anthropogenic CO_2 flux, Φ^{ant}, of about 0.5 mol m^{-2} yr^{-1}. Using the air-sea flux equation (10.2.14) and solving it for the partial pressure difference gives

$$\Delta pCO_2^{ant} = \frac{\Phi^{ant}}{k_g} = \frac{0.5\ \text{mol m}^{-2}\ \text{yr}^{-1}}{0.065\ \text{mol m}^{-2}\ \text{yr}^{-1}\ \mu\text{atm}^{-1}} \approx 8\ \mu\text{atm} \qquad (10.2.17)$$

where we used the globally averaged k_g of 0.065 mol m^{-2} yr^{-1} μatm^{-1} given in table 3.3.3.

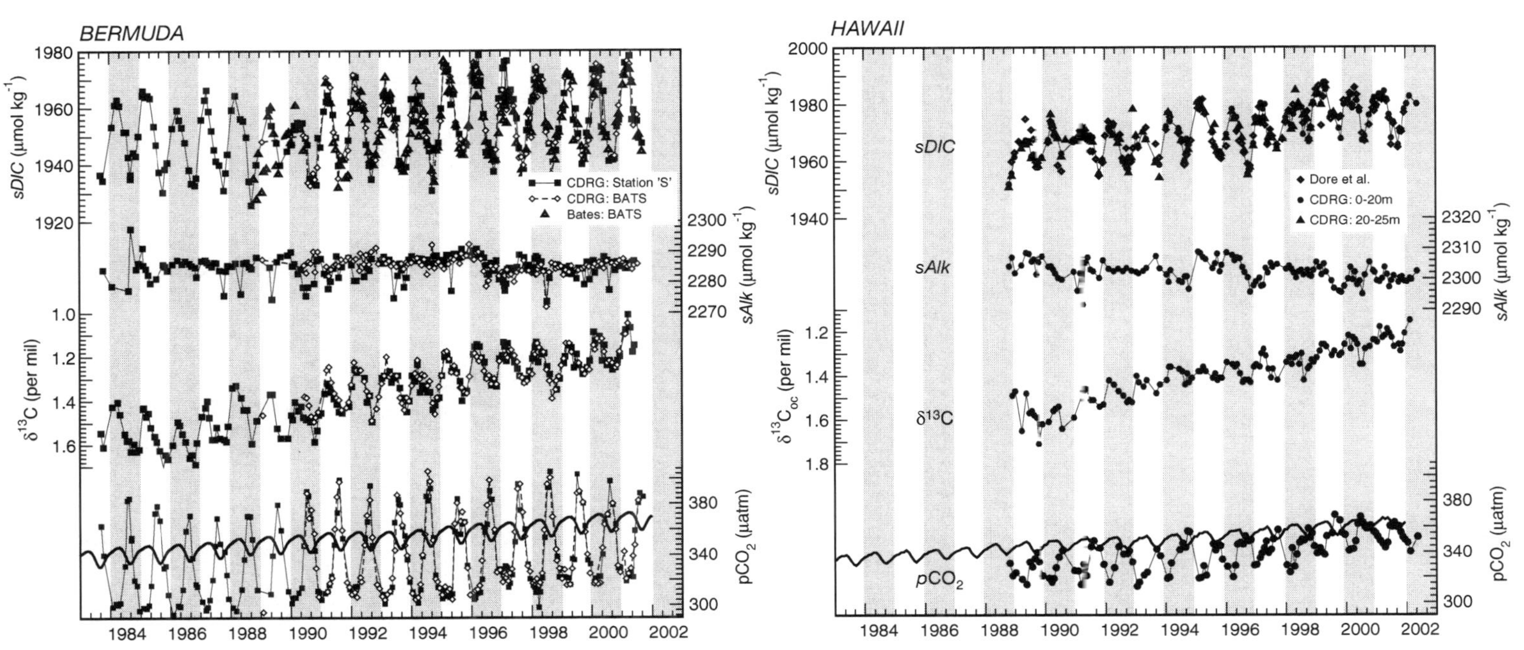

FIGURE 10.2.5 Upper ocean time series of *DIC*, *Alk*, the $^{13}C/^{12}C$ ratio of *DIC*, and calculated oceanic pCO_2 from BATS (Bermuda) and HOT (Hawaii). The *DIC* and *Alk* concentrations have been normalized to a constant salinity in order to remove the effect of freshwater fluxes. Distinct long-term trends are visible in *sDIC*, oceanic pCO_2, and the $^{13}C/^{12}C$ ratio of *sDIC*, owing to the uptake of anthropogenic CO_2 from the atmosphere. Also shown are the atmospheric pCO_2, indicating that the oceanic pCO_2 is roughly following the atmospheric trends. The trend of oceanic pCO_2 (non-salinity nomalized) is slightly elevated at HOT, likely the result of a combination of climate variability–induced changes in rainfall patterns and water mass changes that occured there around 1996/1997 (see *Dore et al.* [2003] and *Keeling et al.* [2004] for discussion). Data are from *Gruber et al.* [2002] and *Keeling et al.* [2004].

TABLE 10.2.3
Summary of Methods to Estimate the Oceanic Uptake of Anthropogenic CO_2

Method	*Present Uncertainty*†	*Spatial Coverage*	*References (Selected)*
	Direct Observational Methods		
Air-sea fluxes of CO_2	L	regional	*Tans et al.* [1990]; *Takahashi et al.* [2002]
Repeated transects	M	regional	*Peng et al.* [1998]
DIC multi-parameter analysis	M	regional	*Wallace* [1995]
Total inventory methods	M–L	regional	*Brewer* [1978]; *Chen and Millero* [1979]; *Gruber et al.* [1996]
	Indirect Observation Methods		
Atmospheric O_2 and CO_2	M–L	global	*Keeling and Shertz* [1992]
Inversion studies of atm. CO_2	L	regional	*Bolin and Keeling* [1963]; *Gurney et al.* [2002]
Oceanic change in ^{13}C inventory	L	global	*Quay et al.* [1992]
Air-sea disequilibrium of $\partial^{13}C$	M–L	global	*Tans et al* [1993]; *Gruber and Keeling* [2001]
Dynamic constraints of $\partial^{13}C$	M–L	global	*Heimann and Maier-Reimer* [1996]
Tracer analogues (^{14}C, ^{3}H, CCl_4, CFCs)	M–L	regional	*McNeil et al.* [2003]
	Model-Based Estimates		
Tracer-calibrated box models	M	global	*Oeschger et al.* [1975]; *Siegenthaler and Joos* [1992]
Dynamic 2-D/3-D models	M	regional	*Maier-Reimer and Hasselman* [1987]; *Sarmiento et al.* [1992]; *Orr et al.* [2001]

†subjectively estimated (M: medium; L: large)

The precision and accuracy of current measurements of oceanic pCO_2 is a few μatm, so this anthropogenic air-sea pCO_2 difference of about 8 ppm is fundamentally amenable to direct observation. However, we have also seen in chapters 3 and 8 (e.g., figures 3.1.3, 8.1.1, and 8.3.7) that oceanic pCO_2 has spatial and temporal variations that exceed 100 ppm in many places, therefore making the sampling requirements for adequately taking into account the observed spatiotemporal variability extremely high. At the moment, despite the most recent synthesis of more than 1 million ΔpCO_2 observations by *Takahashi et al.* [2002], many key regions of the surface ocean, such as the Southern Ocean, remain undersampled. A further problem of this approach is the large uncertainty associated with estimates of the gas exchange coefficient k_g (see discussion in chapter 3). A final source of uncertainty is the poorly known natural efflux of CO_2 from the ocean as a result of the riverine input of inorganic and organic carbon. Future improvements in the observational coverage as well as our understanding of the factors that control k_g will, however, make this approach substantially more attractive.

Because of the current difficulties of direct detection, the amount of anthropogenic CO_2 taken up by the ocean is often estimated by less direct means. The approach most commonly employed in the past has been the use of ocean models designed to take into account not only the thermodynamic capacity of seawater but also air-sea gas exchange and vertical transport. Despite large differences in complexity, most models predict fairly similar uptake rates of about 2.0 ± 0.4 Pg C yr^{-1} [*Prentice et al.*, 2001; *Matsumoto et al.*, 2004]. However, as outlined in table 10.2.3, several additional methods exist to estimate the oceanic uptake of anthropogenic CO_2. We will limit our discussion here to the inventory approach and the atmospheric oxygen method.

RECONSTRUCTION OF ANTHROPOGENIC CO_2 INVENTORY

Brewer [1978] and *Chen and Millero* [1979] were the first to point out that the anthropogenic CO_2 can be estimated by correcting the measured *DIC* in a water sample for the changes that have occurred due to the remineralization of organic matter and the dissolution of carbonates since it last lost contact with the atmosphere, ΔDIC_{bio}, and by subtracting preformed preindustrial *DIC*, DIC^0_{pi}, i.e.,

$$C_{ant} = DIC^{obs} - \Delta DIC_{bio} - DIC^0_{pi} \tag{10.2.18}$$

where the biological correction is estimated from *AOU* and the difference between *in situ* and preformed *Alk*, Alk^0:

$$\Delta DIC_{bio} = -r_{C:O_2} AOU - \frac{1}{2}(Alk - Alk^0 - r_{N:O_2} AOU) \tag{10.2.19}$$

where $r_{C:O_2}$ and $r_{N:O_2}$ are the stoichiometric C:O_2 and N:O_2 ratios during remineralization of organic matter. The first term on the right-hand side of (10.2.19) constitutes the change in *DIC* due to the remineralization of organic matter since the water parcel was last in contact with the atmosphere, and the second term represents the corresponding change in *DIC* due to the dissolution of $CaCO_3$. The factor 1/2 arises from the fact that the dissolution of $CaCO_3$ changes *Alk* and *DIC* in a 2:1 ratio. The *AOU* term in the carbonate correction term is a result of the influence of the soft-tissue pump on *Alk* (see chapter 8).

The easiest way of interpreting (10.2.18) is to view it in direct analogy to the computing of preformed properties discussed in chapter 5. The expression $DIC^{obs} - \Delta DIC_{bio}$ is directly equivalent to the computation of a preformed concentration, i.e., the biological correction takes a water sample at any given location in the ocean interior and moves it back to the surface, where it was last in contact with the atmosphere. The anthropogenic CO_2 concentration is then the difference between this preformed *DIC* concentration and the preformed concentration that existed in preindustrial times, DIC^0_{pi}.

However, the Brewer, Chen and Millero approach never found general acceptance, since the uncertainties were regarded as too large [*Shiller*, 1981, 1982; *Broecker et al.*, 1985b], particularly because of very large uncertainties associated with the estimation of DIC^0_{pi}. However, *Gruber et al.* [1996] showed that this approach can be significantly improved by taking advantage of the fact that we can estimate water mass ages on the basis of a number of transient tracers (see discussion in chapter 2).

The most relevant change to the original technique by Brewer and by Chen and Millero was to separate the DIC^0_{pi} term into a preindustrial equilibrium term, i.e., the *DIC* concentration a given water parcel would have if it had been in equilibrium with preindustrial atmospheric CO_2 for a given salinity *S*, temperature *T*, and preformed alkalinity Alk^0, i.e., $DIC_{eq\ pi}$ and a disequilibrium term ΔDIC_{diseq}, i.e.,

$$DIC^0_{pi} = DIC_{eq\ pi} + \Delta DIC_{diseq} \qquad (10.2.20)$$

Inserting (10.2.20) into (10.2.18) gives

$$C_{ant} = DIC^{obs} - \Delta DIC_{bio} - DIC_{eq\ pi} - \Delta DIC_{diseq} \qquad (10.2.21)$$

$$= \Delta C^* - \Delta DIC_{diseq} \qquad (10.2.22)$$

where we combined the first three terms on the right-hand side of (10.2.21) into ΔC^*. The tracer ΔC^* is nearly conservative in the ocean interior and can be calculated on the basis of concurrent observations of *DIC, Alk*, temperature, salinity, oxygen, and nutrients. This tracer is also directly related to the C^* tracer used in chapter 8 for separating the different pump components from the observed variations of *DIC*. The reader interested in details is referred to *Gruber et al.* [1996] and *Gruber and Sarmiento* [2002].

Therefore, in order to estimate C_{ant}, we are left with the need to estimate the disequilibrium term ΔDIC_{diseq}. Given an equilibration timescale of nearly a year for the air-sea exchange of CO_2, this disequilibrium term, which corresponds directly to a ΔpCO_2 at the time a water parcel was last in contact with the atmosphere, can be quite large.

Gruber et al. [1996] made three important assumptions to estimate ΔDIC_{diseq}. First, they argued that the expected changes of this disequilibrium over time are small, and therefore assumed it to be constant. Their argument was based on the expectation that the long-term change in this disequilibrium driven by the anthropogenic CO_2 perturbation is on the order of a few $\mu mol\ kg^{-1}$ only. Although this argument is correct, the fact that the disequilibrium is actually changing over time makes this assumption one of the largest sources of error in this method (cf. *Matsumoto and Gruber* [2005]). The second assumption is that ocean circulation and mixing occurs predominantly along isopycnal surfaces, permitting us to do the analyses on them. This requires also the consideration of the different end-members present on these isopycnals. Finally, the method assumes that the natural carbon cycle has remained in an approximate steady-state over the Anthropocene, i.e. the last 250 years.

With these assumptions, the estimation of ΔDIC_{diseq} is straightforward when examining deep isopycnal surfaces that have slowly ventilated regions where it is safe to assume that there is no anthropogenic CO_2 (see figure 10.2.6a). In those regions, C_{ant} in (10.2.22) is zero, and we can estimate the disequilibrium for the water mass *i* that dominates this portion of the isopycnal surface simply by averaging ΔC^*:

$$\Delta DIC^i_{diseq}(\sigma) = \overline{\Delta C^*}|_{\sigma = const} \qquad (10.2.23)$$

This approach fails, however, on isopycnal surfaces that are shallower and rapidly ventilated. Here anthropogenic CO_2 has already affected the entire density surface, and therefore no uncontaminated region can be found. For these density surfaces, *Gruber et al.* [1996] take advantage of the availability of age tracers that permit them to estimate the ventilation age of a water parcel. One can then compute the *DIC* equilibrium concentration at the time, *t*, a water parcel was last in contact with the atmosphere, $DIC_{eq}(t)$, which given the assumption of a constant air-sea disequilibrium differs from $DIC_{eq\ pi}$ by C_{ant}., i.e. $DIC_{eq}(t) = DIC_{eq\ pi} + C_{ant}$. One could use this "shortcut" method to estimate C_{ant} (see *McNeil et al.* [2003] for an application), but the neglect of the change in the air-sea disequilibrium makes this shorcut method biased high. Instead, *Gruber et al.* [1996] inserted $DIC_{eq}(t)$ into (10.2.21) and solved for $\Delta DIC_{diseq} = DIC_{obs} - DIC_{bio} - DIC_{eq}(t)$, where the right hand side was equated with the modified ΔC^* tracer, ΔC^*_t. Thus analyzing the variations of ΔC^*_t on

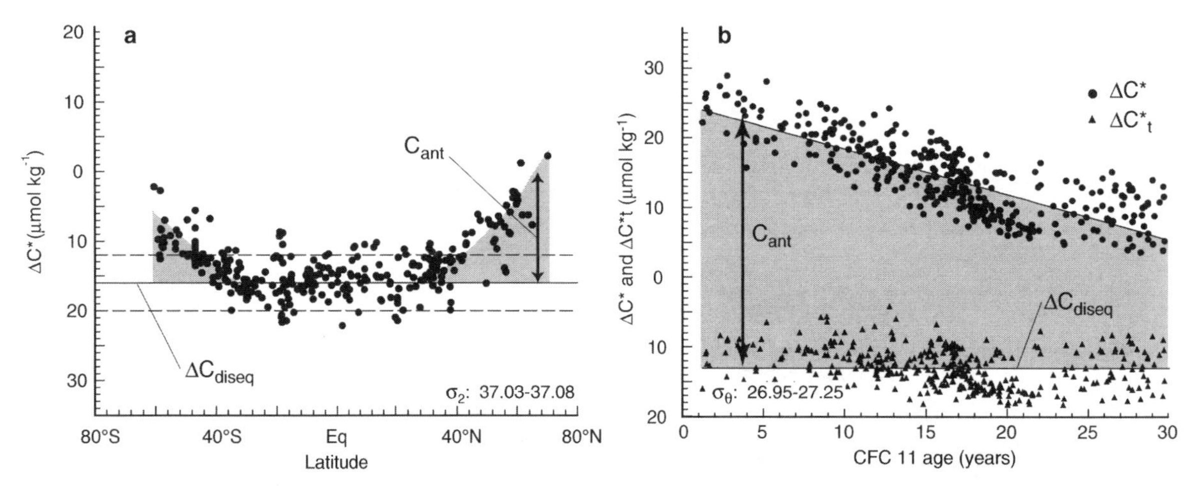

FIGURE 10.2.6: Illustration of the C_{ant} reconstruction method on the basis of ΔC^* and ΔC_t^*. (a) Estimation of C_{ant} on deep isopycnal surfaces, where one can find regions that contain undetectable amounts of anthropogenic CO_2. Plotted is ΔC^* as a function of latitude in the North Atlantic along the σ_2 interval 37.03–37.08, which represents the Denmark Strait and Iceland-Scotland Overflow Waters. The solid line depicts the average ΔC^* concentration in the regions from 20°S to the equator and from the equator to 20°N which were used to determine ΔDIC_{diseq} for the southern and northern end-member, respectively. In this case, no difference was found between these two end-members. The dashed lines are drawn at ±5 μmol kg^{-1} from the estimated ΔDIC_{diseq}. The upward trend near the outcrops in the south and the north is due to the presence of anthropogenic CO_2, which is then determined simply by differencing ΔC^* from ΔDIC_{diseq} (shaded area). (b) Estimation of C_{ant} on shallower isopycnal surfaces, where anthropogenic CO_2 has invaded the entire surface. Plotted are ΔC_t^* (triangles) and ΔC^* (circles) versus CFC-11 age along the isopycnal surface interval σ_θ 26.95–27.25 in the South Atlantic, representing Subantarctic Mode Water. The line for ΔC^* is drawn to emphasize the trend and was obtained by linear regression. The line for ΔC_t^* represents the average. In this case, C_{ant} is estimated from the difference between ΔC^* and ΔC_t^* (shaded area). Adapted from *Gruber* [1998].

these shallow isopycnal surfaces permits us to determine the ΔDIC_{diseq} terms of the different end members, *i*:

$$\Delta DIC^i_{diseq}(\sigma) = \overline{\Delta C^*_t}|_{\sigma=const} \qquad (10.2.24)$$

Once the disequilibria have been determined for all surfaces and *n* end-members, the amount of anthropogenic CO_2 can finally be calculated from (10.2.22):

$$C_{ant} = \Delta C^* - \sum_{i=1}^{n} f^i \Delta DIC^i_{diseq}(\sigma) \qquad (10.2.25)$$

where f^i are relative contributions from each end-member.

A number of investigators have applied this technique to various ocean basins [*Gruber*, 1998; *Sabine et al.*, 1999, 2002a; *Lee et al.*, 2003], with *Sabine et al.* [2004] providing a global synthesis. Figure 10.2.7 shows the resulting vertical distribution of C_{ant} along a global section from the Atlantic to the Pacific.

Near-surface concentrations of C_{ant} are highest in subtropical regions and tend to decrease toward the high latitudes. They also tend to be higher in the Atlantic than in the Pacific. This is surprising, since given the higher solubility of CO_2 in colder waters, one might at first think that the high latitudes should contain more anthropogenic CO_2. However, integrating (10.2.16) over the Anthropocene and assuming that the buffer factor, γ_{DIC}, remained constant, shows that the amount of C_{ant} in the surface ocean expected from equilibrium uptake is inversely proportional to the buffer factor:

$$C_{ant}(t) = \int_{t'=pi}^{t'=t} \frac{\partial [DIC]}{\partial t} dt' = \frac{\alpha}{\gamma_{DIC}} \int_{t'=pi}^{t'=t} \frac{\partial pCO_2^{atm}}{\partial t} dt'$$
$$= \frac{\alpha}{\gamma_{DIC}} (pCO_2^{atm}(t) - pCO_2^{atm}(t=pi)) \qquad (10.2.26)$$

As the buffer factor is substantially larger in the high latitudes, and lower in the low latitudes (mainly a consequence of variations in the *Alk*/*DIC* ratio, see (10.2.11)), the low latitudes can store substantially more anthropogenic CO_2 than the high latitudes for a given atmospheric CO_2 perturbation. The North Atlantic has a particularly low buffer factor, and therefore tends to contain the highest surface concentrations of anthropogenic CO_2. The reconstructed surface concentrations are in quite good agreement with those computed by inserting typical values in (10.2.26) for the early 1990s, when atmospheric pCO_2 was about 360 μatm (see figure 10.1.2). The indication of somewhat lower anthropogenic CO_2 content in the surface waters of the high southern latitudes compared to the expectations from equilibrium indicates that these waters are not able to completely track the atmosphere. This is presumably the result of three factors: the short residence time of these surface waters, substantial dilution by mixing with subsurface waters containing little C_{ant}, and effective inhi-

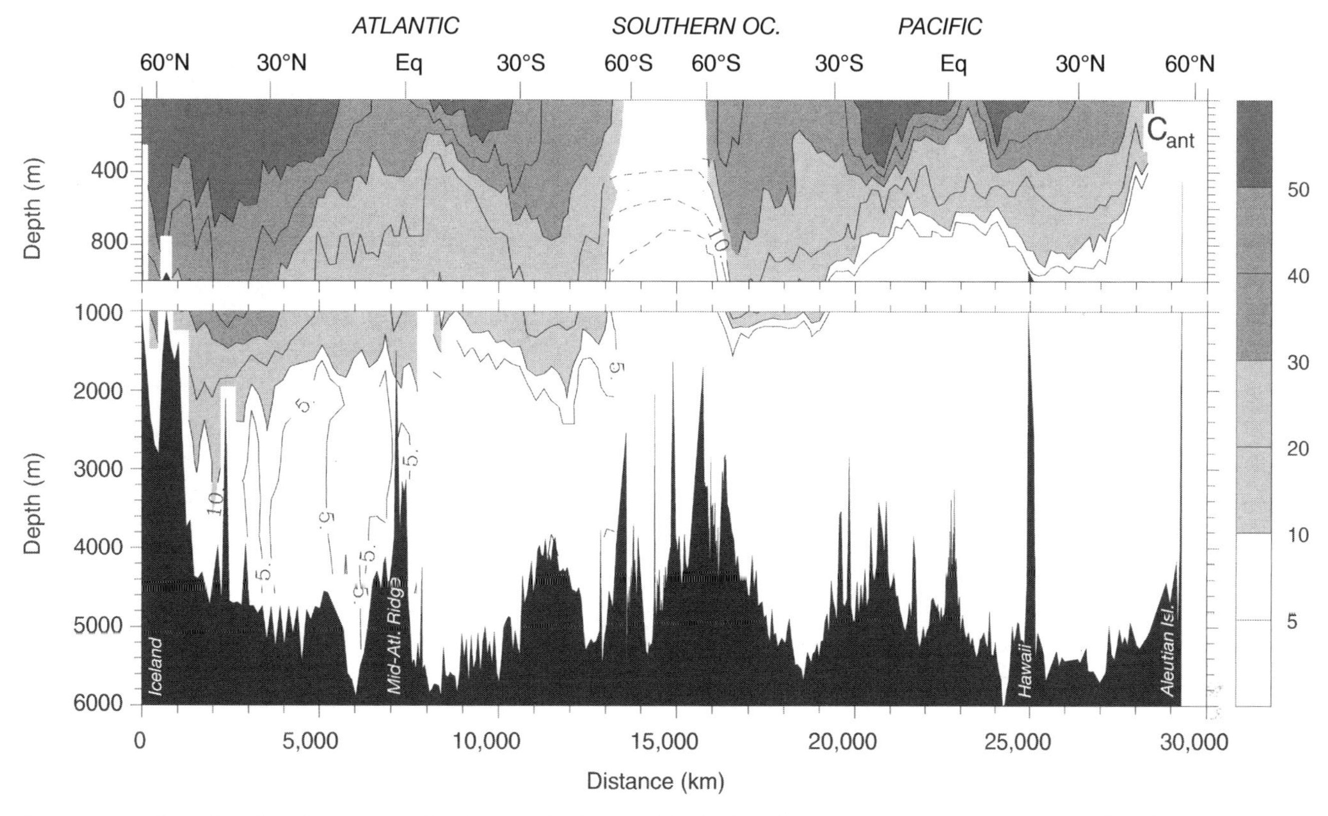

FIGURE 10.2.7: Vertical section of reconstructed anthropogenic CO_2 (μmol kg^{-1}) along the track shown in figure 2.3.3a. See also color plate 8. Anthropogenic CO_2 was estimated from the observed *DIC* distribution shown in figure 8.1.3a using the method described in the text [cf. *Gruber et al.*, 1996].

bition of air-sea gas exchange by sea ice, especially during winter.

Figure 10.2.7 also reveals substantial meridional differences in vertical penetration of the anthropogenic CO_2 signal. In the tropics, C_{ant} rapidly decreases with depth, whereas the vertical penetration of the signal is much deeper in the subtropical and temperate latitudes. The symmetry breaks down further toward the poles. In the North Atlantic north of 60°N, deep vertical penetration of C_{ant} is found, with concentrations above 10 μmol kg^{-1} down to the bottom. In sharp contrast to this, only a shallow vertical penetration of anthropogenic CO_2 is found in the Southern Ocean south of 50°S and in the North Pacific north of 50°N. Since anthropogenic CO_2 can be regarded as a conservative tracer, i.e., it is not affected by biological processes, its distribution in the interior is determined by ocean transport only. Can we understand the distribution exhibited by the reconstructed C_{ant} based on what we have learned about ocean circulation in chapter 2?

The shallow vertical penetration in the tropics and the deeper penetration in the subtropical and temperate zones are primarily reflecting the structure and ventilation of the thermocline. The sharp contrast between the North Atlantic, the Southern Ocean, and the North Pacific requires more discussion. The deep penetration of C_{ant} into the deeper layers of the North Atlantic is clearly a consequence of the formation of Labrador Sea Water and North Atlantic Deep Water and the subsequent transport of these water masses deep into the interior of the North Atlantic. In contrast, the absence of any deep water formation in the North Pacific prevents deep penetration of C_{ant} in this region.

But why is the vertical penetration of C_{ant} so low in the Southern Ocean, despite the fact that deep water formation is well known to occur around Antarctica? We have discussed already that surface waters in the Southern Ocean are lagging behind the atmospheric CO_2 increase. This deficit in anthropogenic CO_2 is then aggravated by the fact that the sinking water entrains large amount of old waters containing little or no anthropogenic CO_2, leading to only a small C_{ant} signal in Antarctic Bottom Water.

Figure 10.2.8 shows dramatically the resulting differences in column inventories of C_{ant}, i.e., vertically integrated C_{ant}. The highest column inventories are found in the mid-latitudes of the North Atlantic, and at mid-latitudes in the southern hemisphere. By contrast, the column inventories are smaller in the equatorial regions, and in the high-latitude Southern Ocean. Note that this distribution only reflects where anthropogenic CO_2 is currently stored, and not where it actually entered the ocean.

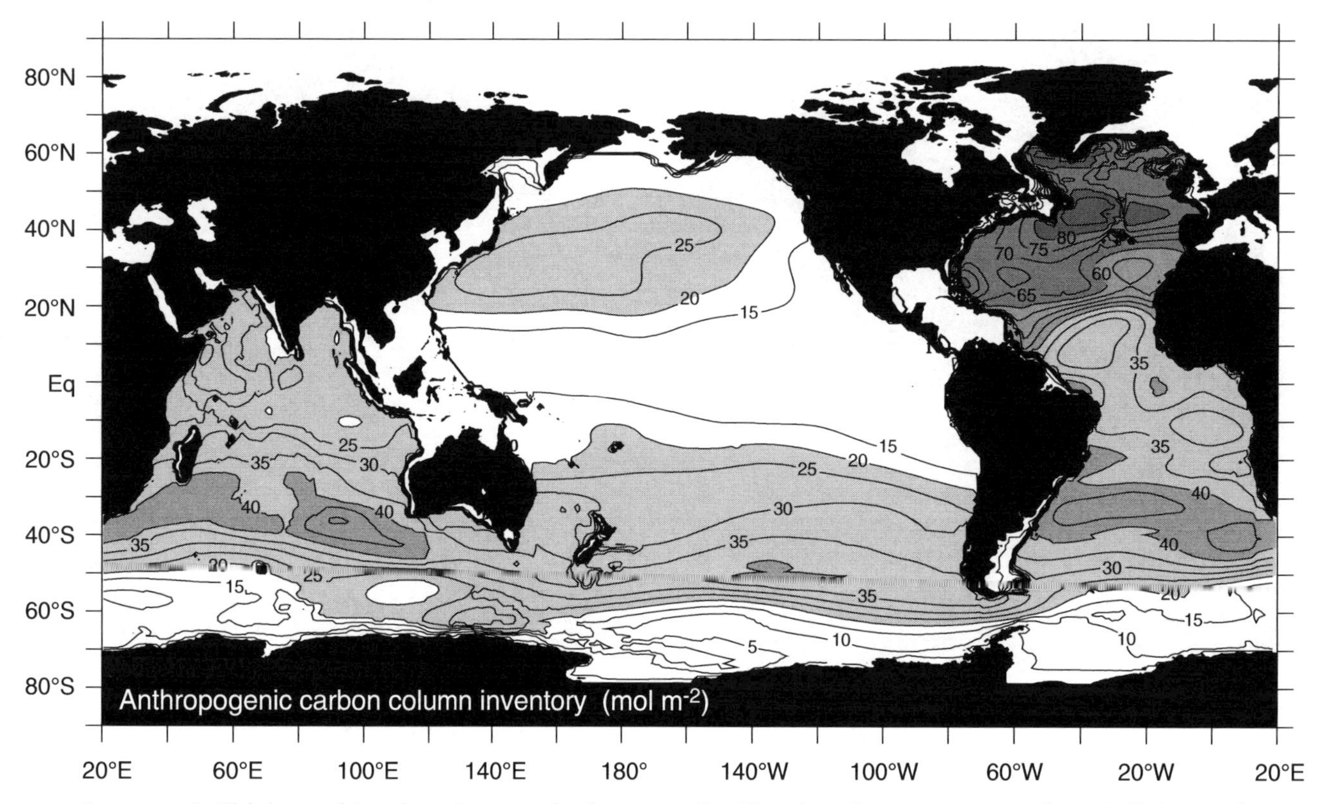

FIGURE 10.2.8: Global map of the column inventory of anthropogenic CO_2. The column inventory represents the vertically integrated concentration. Data are from *Sabine et al.* [2004].

One method to address the question of what the distribution of surface fluxes of anthropogenic CO_2 has to be in order to generate the observed pattern of C_{ant} in the ocean interior is a Green's function inversion method [*Enting*, 2002]. In such an inversion, prescribed fluxes of a dye are released at the surface in a number of predetermined regions in an Ocean General Circulation Model (OGCM) and the model is then integrated forward in time. In the case of anthropogenic CO_2, the fluxes need to be scaled in proportion to the expected increase in uptake over time, i.e., with the magnitude of the atmospheric perturbation, and the model is then integrated forward in time only for the duration of the Anthropocene, i.e., about 250 years [*Gloor et al.*, 2003]. The modeled dye concentrations are then sampled at the locations of the observations, providing a link between concentrations at these locations and fluxes in the various regions. The inversion then attempts to scale the magnitude of the fluxes in the various regions in order to obtain an optimal fit between the estimated resulting concentrations and the actually observed ones.

The results of such an inversion of the reconstructed anthropogenic CO_2 distribution are shown in figure 10.2.9 [*Gloor et al.*, 2003]. This inversion suggests, for example, that the high storage of C_{ant} in the mid-latitudes of the southern hemisphere is primarily driven by high uptake of anthropogenic CO_2 at more southerly latitudes in the subantarctic zone and subsequent transport of this C_{ant} into the ocean interior following the paths of Subantarctic Mode Water and Antarctic Intermediate Water [*Gloor et al.*, 2003]. There exists, therefore, a substantial lateral transport of anthropogenic CO_2 in the ocean, which connects uptake at places where deeper waters, generally deficient in anthropogenic CO_2, come to the surface and take up anthropogenic CO_2 from the atmosphere, with storage in the regions of deep thermoclines and net convergence in the mid-latitudes.

The column inventories shown in figure 10.2.8 sum to a global inventory of anthropogenic CO_2 of 106 ± 17 Pg C for 1994. *Sabine et al.* [2004] showed that an additional 12 Pg C is stored in marginal seas, such as the Arctic Ocean, bringing the total amount of anthropogenic CO_2 taken up by the ocean to 118 ± 19 Pg C. Table 10.2.1 shows that this oceanic sink over the Anthropocene (1800–1994) corresponds to nearly half of the total CO_2 emissions from fossil fuel burning and cement production over the 1800–1994 period (about 244 ± 20 Pg C). Together with the well-known atmospheric CO_2 increase, which corresponds to 165 ± 4 Pg C, this number permits us to constrain the net terrestrial exchange to be a net source to the atmosphere of about 39 ± 28 Pg C. This net flux is the sum of poorly known emissions from land-use change (about 100 to 180 Pg C), and an inferred

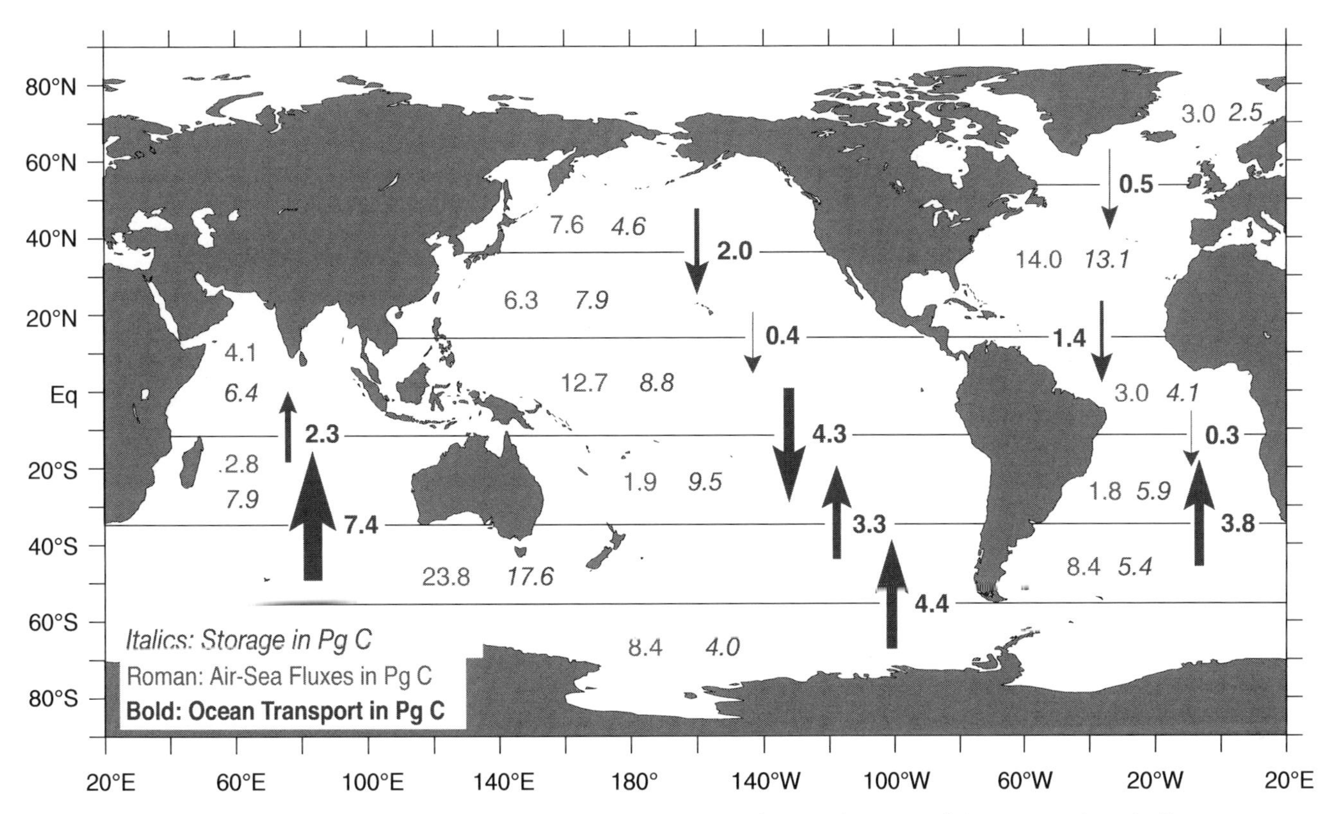

FIGURE 10.2.9: Map of anthropogenic CO_2 inversion results. Shown as roman numbers are the estimated time-integrated uptake fluxes (1750–1994). The italic numbers refer to the reconstructed inventories. The arrows depict the time-integrated lateral fluxes, computed by difference. Based on *Gloor et al.* [2003].

net sink of the terrestrial biosphere of 61 to 141 Pg C. Our oceanic sink estimate thus suggests that the ocean has taken up about between 27% and 34% of the total anthropogenic CO_2 emissions, making it the most important sink for anthropogenic CO_2 next to the atmosphere.

The Atmospheric Oxygen Method

The anthropogenic CO_2 inventory method permitted us to develop an anthropogenic CO_2 budget for the Anthropocene, but we are equally interested in constraining the budget for the last few decades. In the absence of direct measurements of the global oceanic increase in *DIC*, a very elegant method to constrain the oceanic uptake of anthropogenic CO_2 since the early 1990s is based on simultaneous measurements of atmospheric oxygen and CO_2 pioneered by *Keeling and Shertz* [1992].

The method builds upon the fact that O_2 and CO_2 are tightly coupled during the formation and oxidation of organic matter (including the burning of fossil fuels), but that these two gases are exchanged independently between the ocean and atmosphere. If one then assumes that the long-term exchange of O_2 with the ocean is zero or is small and can be estimated from independent sources, one ends up with two balance equations and two unknowns (the exchange of CO_2 with the ocean and land biosphere, respectively), which can be solved.

We begin with the mass balance equations for atmospheric O_2 and CO_2:

$$\frac{\partial N_{O_2}}{\partial t} = \Phi_{foss}(O_2) + \Phi_{as}(O_2) + \Phi_{ab}(O_2) \qquad (10.2.27)$$

$$\frac{\partial N_{CO_2}}{\partial t} = \Phi_{foss}(CO_2) + \Phi_{as}(CO_2) + \Phi_{ab}(CO_2) \qquad (10.2.28)$$

where N_{O_2} and N_{CO_2} are the number of moles of atmospheric O_2 and CO_2, respectively, and where Φ_{foss} is the fossil fuel emission of CO_2 and corresponding consumption of O_2, Φ_{as} is the exchange flux between the atmosphere and ocean, and Φ_{ab} is the exchange flux between the atmosphere and land biosphere (all fluxes have units of mol yr^{-1}). In the case of the fossil fuel emissions and the exchange with the land biosphere, the O_2 and CO_2 fluxes are linked stoichiometrically:

$$\Phi_{foss}(O_2) = r^{foss}_{O_2:C}\Phi_{foss}(CO_2) \qquad (10.2.29)$$

$$\Phi_{ab}(O_2) = r^{terr}_{O_2:C}\Phi_{ab}(CO_2) \qquad (10.2.30)$$

where $r^{foss}_{O_2:C}$ is about -1.4, and $r^{terr}_{O_2:C}$ about -1.1 [*Keeling and Shertz*, 1992; *Keeling et al.*, 1996; *Severinghaus*, 1995].

The concurrent measurements of atmospheric CO_2 and O_2 constrain $\partial N_{O_2}/\partial t$ and $\partial N_{CO_2}/\partial t$, and fossil fuel emissions of CO_2 are relatively well known [e.g., *Marland et al.*, 1998]. If we assume that the natural carbon and oxygen cycles in the ocean have remained in steady

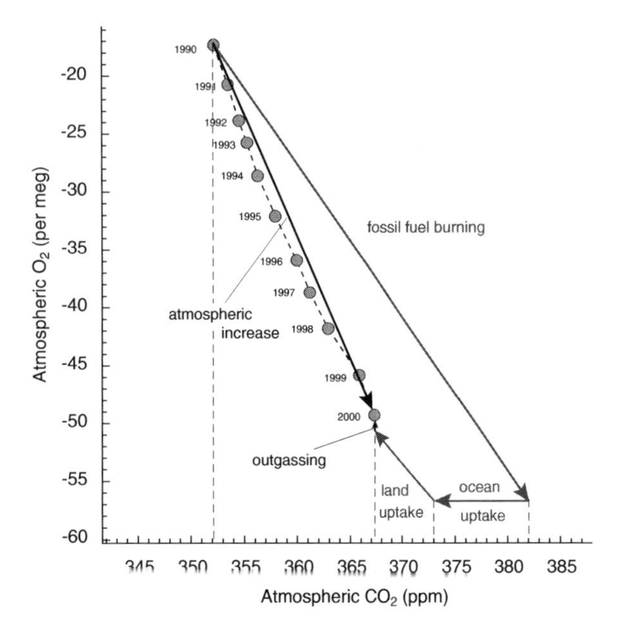

FIGURE 10.2.10: Illustration of the atmospheric oxygen method. Plotted are the observed annual mean concentrations of atmospheric O_2 and CO_2 from 1990 to 2000. If fossil fuel burning were the only process affecting these two gases, one would expect that the atmosphere would have followed the trajectory labeled "fossil fuel burning," bringing the atmosphere from the initial observation in 1990 to the point on the lower-right corner. Ocean uptake, land uptake, and ocean outgassing were responsible for the atmosphere following a different trend. Since these three processes have distinctly different slopes, a graphical solution can be obtained by vector addition once one of these fluxes is known by other means (usually ocean outgassing is assumed to be known and a solution is obtained for the uptake by land and ocean). Modified from *Prentice et al.* [2001].

state and have not been perturbed in any substantial manner, the long-term air-sea exchange flux of O_2, $\Phi_{as}(O_2)$, is very small and can be neglected. This is because O_2 has a much lower solubility than CO_2, making the perturbation air-sea flux of O_2 much smaller than the perturbation flux for CO_2 for the same magnitude of an atmospheric perturbation. If we insert (10.2.30) and (10.2.29) into (10.2.27) and (10.2.28), we end up with two equations and two unknowns, $\Phi_{as}(CO_2)$ and $\Phi_{ab}(CO_2)$. Figure 10.2.10 shows a graphical solution to these equations, where the points indicate the observed evolution of atmospheric CO_2 and O_2, and where the direction of the arrows of the different processes are given by their stoichiometric ratios.

Figure 10.2.10 actually includes a small air-sea O_2 flux contribution, $\Phi_{as}(O_2)$. This is because the steady-state assumption of the natural oxygen cycle is not entirely tenable. Long-term observations of oceanic temperatures show that the ocean has warmed significantly over the last 50 years [*Levitus et al.*, 2000]. As the solubility of oxygen is very temperature-dependent (see figure 3.1.2), this must have led to an outgassing of O_2. The warming-induced outgassing flux can be directly estimated from (3.4.4) in chapter 3. However, more recent studies indicate that the oxygen outgassing driven by ocean warming is likely amplified by an additional outgassing stemming from warming-induced changes in ocean stratification. This additional outgassing might increase the purely thermally driven flux by a factor of three or more, requiring an increase in the inferred oceanic uptake of anthropogenic CO_2 of up to 0.6 Pg C yr^{-1} relative to the original budgets [*Keeling and Garcia*, 2002; *Bopp et al.*, 2002; *Plattner et al.*, 2002].

The anthropogenic budget for the two decades from 1980 to 1999 as inferred by the oxygen method but adjusted for the ocean warming outgassing are given in table 10.2.1, both as integrated fluxes and as average fluxes per year. The most remarkable observation from inspection of this table is how strongly the global carbon cycle has been forced over the last 20 years. Almost 50% of the fossil fuel emissions over the Anthropocene occurred during the period from 1980 to 1999! Out of the total anthropogenic CO_2 emissions, the ocean has taken up about 27% over the last two decades. Interestingly, this fraction is quite similar to that over the entire Anthropocene. By contrast, the inferred net terrestrial balance suggests a sign change between the earlier part of the Anthropocene and the period from 1980 to 1999. While the terrestrial biosphere was a net source of CO_2 to the atmosphere during the Anthropocene of 39 ± 28 Pg C, it turned into a net sink of 15 ± 9 Pg C over the period from 1980 to 1999 (table 10.2.1). The reasons for this phenomenon are still controversial, but it appears as if much of the land-driven uptake is associated with regrowth of areas that have been clear-cut in the past, and have been abandoned or shifted to less intensive cultivation [*Pacala et al.*, 2001; *Schimel et al.*, 2001; *Goodale and Davidson*, 2002].

The Role of Biology

What is the role of biology in the oceanic uptake of anthropogenic CO_2? This topic has been the source of many misunderstandings and false claims [*Broecker*, 1991b; *Sarmiento*, 1991; *Longhurst*, 1991].

So far in our discussion on anthropogenic CO_2 we have neglected the marine biota. This is somewhat counterintuitive, because we know that the marine biota is responsible for a large fraction of the surface-to-deep gradient in *DIC* (biological pumps). Then why do the biological pumps not also transport anthropogenic CO_2 from the surface into the interior of the ocean? The first point to make here is that this downward transport of carbon in organic material is compensated in a steady state by an equally large upward transport of *DIC*, resulting in a zero net transport (except for the small fraction that goes into the sediments). Our neglect of the biological pumps is based on the assumption that they had been operating in steady state before the start of the anthropogenic CO_2 perturbation of the carbon cycle, and

TABLE 10.2.4
Summary of Marine Carbon Cycle Feedbacks

Process	*Sign*	*Magnitude*†	*Uncertainty/ Understanding*
	Anthropogenic CO_2 Uptake Feedbacks		
Chemical feedback	Positive	H	Low/high
Circulation feedback (anthropogenic CO_2)	Positive	H	Medium/medium
	Natural Carbon Cycle Feedbacks		
Temperature/salinity feedback	Positive	M	Low/medium to high
Circulation feedback (constant biota)	Negative	M-H	Medium/medium
Ocean biota feedback (circulation constant)	Positive/negative	M	High/low

†Subjectively estimated (M: medium; H: high)

that they have continued to do so ever since. There are two lines of evidence that support this hypothesis. First, atmospheric CO_2 remained remarkably stable for several thousand years before the eighteenth century (see figure 10.1.2), indicating that the fluxes controlling the surface ocean total carbon content were in balance. Second, according to present knowledge, marine phytoplankton are controlled by nutrients (including micronutrients), light, and vertical stability of the water column, but not by the CO_2 concentration. Therefore the biological pumps do not sequester anthropogenic carbon, but rather act as a natural background process continuing to work as in preindustrial times.

The lack of a role of marine biota in the uptake of anthropogenic CO_2 over the last two centuries is demonstrated by the model study of *Murnane et al.* [1999], who carried out a simulation of oceanic uptake in two models, one that contained just the solubility pump and a second combined simulation that included the biological pump. The combined simulation took up 5% less anthropogenic CO_2 than the solubility model, indicating that adding biology had only a minimal impact on the anthropogenic CO_2 uptake. The small reduction in uptake occurs because the biological pump reduces surface alkalinity and hence storage capacity for anthropogenic CO_2. If the solubility simulation used the same surface alkalinities as the combined model, then both would take up the same amount of anthropogenic CO_2.

One cannot assume, however, that the steady-state operation of the biological pumps will continue forever into the future. Moreover, the above discussion about the possible impact of the observed ocean warming on the oceanic oxygen cycle even suggests that the steady-state assumption for the oceanic carbon cycle over the twentieth century may need to be reevaluated.

Future climate changes associated with the buildup of greenhouse gases in the atmosphere will quite likely modify the biological pump–driven fluxes, which could then feed back on the atmospheric composition. Alterations of the biological pumps, for instance, have been proposed as a possible cause for the dramatic changes in atmospheric CO_2 between the glacial and interglacial periods. It is difficult at present to assess how the biological pump will respond to future climate changes and how it will interact with changes in ocean circulation and mixing.

FUTURE CO_2 UPTAKE

The primary factor controlling the future uptake of anthropogenic CO_2 by the ocean is the atmospheric CO_2 burden in excess over the atmospheric CO_2 concentration in equilibrium with the total emissions [*Sarmiento et al.*, 1995], as this represents the thermodynamic driving force for the uptake of CO_2 from the atmosphere. In the absence of major feedbacks, paradoxically, the higher the future anthropogenic CO_2 concentration in the atmosphere, the higher the oceanic uptake will be.

Changes in ocean temperature, salinity, circulation, and the ocean's biological pump have, however, the potential to substantially alter this uptake, both directly, by impacting the uptake of anthropogenic CO_2, and indirectly, by altering the air-sea balance of the natural CO_2 cycle. Any climate change–induced alteration in the future uptake of anthropogenic CO_2 by the ocean will lead to feedbacks in the climate system, as the altered atmospheric CO_2 concentration will either accelerate (positive feedback) or decelerate (negative feedback) the initial climate change.

It is instructive to first discuss the impact of climate change on the oceanic uptake of anthropogenic CO_2, and only thereafter investigate the impact on the natural carbon cycle. A summary of all feedbacks is given in table 10.2.4. For more detailed discussions, the reader is referred to *Plattner et al.* [2001] and *Gruber et al.* [2004].

The first feedback to consider is the chemical feedback discussed above, which arises from the decrease in the oceanic buffer capacity as a result of the uptake of anthropogenic CO_2. Figure 10.2.1 shows that this feedback is very substantial and positive in sign. It is also the best understood feedback. A second feedback arises from changes in ocean circulation, as the transport of anthropogenic CO_2 from the surface down into the ocean interior is the primary factor controlling the uptake. Many ocean-atmosphere coupled climate model simulations show a substantial increase in upper ocean stratification, primarily as a result of ocean warming in the low latitudes, and as a result of freshening in the high latitudes [*Cubasch et al.*, 2001]. This leads to a reduction of the vertical exchange and hence to a direct reduction of the uptake of anthropogenic CO_2 [*Sarmiento and LeQuéré*, 1996; *Sarmiento et al.*, 1998; *Joos et al.*, 1999b; *Matear and Hirst*, 1999; *Plattner et al.*, 2001]. This feedback thus tends to be positive as well.

These temperature, salinity, and circulation changes not only influence the uptake of anthropogenic CO_2, but also affect the natural carbon cycle within the ocean. Ocean warming will reduce the solubility of CO_2 in seawater and lead to a permanent loss of CO_2 from the oceanic reservoir of inorganic carbon. We can estimate the expected outgassing from warming alone on the basis of the temperature sensitivity of ocean *DIC*. This sensitivity can be calculated by combining the temperature sensitivity of pCO_2 with the buffer factor, γ_{DIC}:

$$\frac{\partial DIC}{\partial T} = \frac{\partial DIC}{\partial pCO_2}\frac{\partial pCO_2}{\partial T} = \frac{DIC}{pCO_2^{oc}}\frac{1}{\gamma_{DIC}} \cdot 0.04 \cdot pCO_2^{oc} = \frac{0.04 \cdot DIC}{\gamma_{DIC}} \quad (10.2.31)$$

Inserting surface ocean mean values for *DIC* and γ_{DIC} gives a temperature sensitivity of *DIC* of about $8\,\mu\text{mol kg}^{-1}\,°\text{C}^{-1}$. Dividing this number by the average heat capacity of seawater, $c_p = 4000\,\text{J kg}^{-1}\,°\text{C}^{-1}$, gives an estimate of the loss of *DIC* from the ocean for each joule of energy taken up from the atmosphere,

$$\frac{\partial DIC}{\partial Q} \approx 2\times10^{-9} mol\, J^{-1} \quad (10.2.32)$$

Levitus et al. [2000] estimated that the ocean heat content increased by about 2×10^{23} J between the mid-1950s and the mid-1990s, leading to an expected loss of *DIC* from the ocean of 4×10^{14} mol, or about 5 Pg C. This is small relative to the estimated oceanic uptake over the Anthropocene of about 118 Pg C (table 10.2.1), but future ocean warming could strongly accelerate this loss and strengthen this positive feedback.

The increase in vertical stratification and the other circulation changes not only reduce the direct uptake of anthropogenic CO_2 from the atmosphere, but also affect the natural carbon cycling substantially, even in the absence of any changes in ocean productivity. An important distinction needs to be made between the circulation feedback interacting with the natural carbon cycle and that associated with the anthropogenic CO_2 uptake. The feedback associated with the anthropogenic CO_2 uptake affects only the rate at which the atmospheric CO_2 perturbation is equilibrated with the ocean and does not change the long-term equilibrium, whereas the feedback associated with the natural carbon cycle changes the equilibrium distribution of carbon between the ocean and atmosphere.

As it turns out, in the presence of a constant biological pump, these circulation changes appear to lead to a negative feedback—that is, an increase in the uptake of atmospheric CO_2 from the atmosphere. The main reason for this somewhat surprising result is that a slowdown of the surface-to-deep mixing also reduces the upward transport of remineralized *DIC* (ΔC_{soft}) from the thermocline into the upper ocean, while the downward transport of biologically produced organic carbon remains nearly unchanged. The magnitude of this feedback tends to be linked to the magnitude of the ocean circulation feedback on the uptake of anthropogenic CO_2, as both are driven by the same changes in circulation.

We finally turn to the impact of future changes in ocean biota on the magnitude of the oceanic sink. There are many possible reasons for changes in global marine export production, including changes in surface ocean physical properties, changes in the delivery of nutrients from land by rivers and atmosphere, and internal dynamics of the ocean biota, such as complex predator-prey interactions and fisheries-induced pressures on marine predators (see *Boyd and Doney* [2003] for a comprehensive review). In general, one might expect that an increase in ocean stratification will negatively impact biological productivity in areas that are nutrient-limited, such as is the case throughout most of the low and temperate latitudes. By contrast, an increase in vertical stratification actually might lead to an increase in biological productivity in light-limited areas, as phytoplankton would tend to spend more time, on average, within the euphotic zone. *Bopp et al.* [2001] investigated these responses in a coupled physical-ecological model and indeed found such a response, with a global change in export production of only a few percent for a doubling of atmospheric CO_2, but with regional changes of much larger amplitude.

While this particular simulation has to be viewed as an illustration, since very little is known about the sign and magnitude of the changes in marine productivity and export, there is increasing evidence that the calcification rate by coccolithophorids might be significantly reduced in response to a lowering of the surface ocean pH [*Riebesell et al.*, 2000]. As a reduction of calcification leads to an increase in the ocean uptake

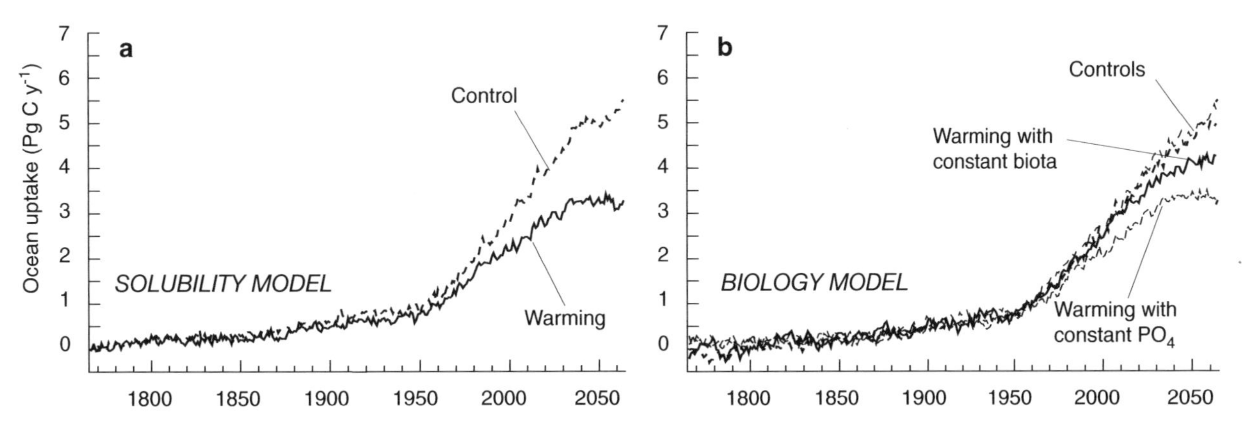

FIGURE 10.2.11: Simulated oceanic uptake of atmospheric CO_2 by the GFDL/Princeton coupled climate model. (a) Control run and ocean uptake for the solubility model only (no biology). (b) Control runs (thin dashed lines) and two scenarios for the biological pump. The thick solid line refers to a simulation where the soft-tissue pump remained constant. The thick dashed line corresponds to a scenario where the soft-tissue pump was assumed to change in a manner to keep the surface nutrient concentration constant. In the control runs, the coupled climate model was run in a preindustrial climate and only atmospheric CO_2 was changed. In the warming scenarios, the climate model was externally forced with a time series of effective atmospheric CO_2, constructed from a business-as-usual scenario (IS92a) including the negative feedback from aerosols. From *Sarmiento et al.* [1998].

capacity for atmospheric CO_2, this represents a negative feedback for climate change. A lower pH also affects the calcification rates of corals [*Gattuso et al.*, 1996; *Kleypas et al.*, 1999]. This effect, together with the enhanced sea-surface temperatures, poses a significant threat to coral reefs, with substantial implications for their ecosystem services, although their impact on atmospheric CO_2 likely will be relatively small. Another impact of the invasion on the oceanic $CaCO_3$ cycle is through its lowering of the water column saturation state, possibly affecting the rates of $CaCO_3$ dissolution [cf. *Feely et al.*, 2004a]. Given the importance of $CaCO_3$ in transporting organic matter to the deep ocean, a faster dissolution of $CaCO_3$ would reduce the strength of the biological carbon pump, and therefore lead to a positive feedback.

Figure 10.2.11 illustrates the magnitude of a number of these feedbacks and their interaction on the basis of simulations carried out by forcing a coupled ocean-atmosphere model with an interactive ocean carbon cycle with a business-as-usual emission scenario for CO_2 [*Sarmiento et al.*, 1998]. In the solubility-only model (no biology), ocean warming and the change in ocean stratification led to a very substantial reduction in the ocean uptake, highlighting the positive signs of these two feedbacks. *Sarmiento et al.* [1998] also found that the biological pump has the potential to largely offset this decreased uptake, mainly as a result of a more efficient biological pump. It has to be noted, however, that the representation of the biological pump in this model is very simplistic, and that therefore these simulations have to be regarded as case studies, rather than accurate predictions.

These simulations demonstrate that while the biological pump does not play a significant role for assessing the oceanic uptake of anthropogenic CO_2 in the past and present, these pumps might play an important role for the future uptake. However, very little is known about the possible response of the oceanic biota to climate change. In order to improve these predictions, one has to gain a clearer understanding of the processes controlling all important steps of the oceanic biological pumps, e.g., surface production, respiration, export, and remineralization. As we have seen during our grand tour through these components of the biological pumps in chapters 4 and 5, there are still considerable gaps in our understanding. Nevertheless, while we cannot foresee the future, nature has in the past performed several experiments, from which we might learn something about the future. The first set of experiments is the response of the ocean carbon cycle to climate variability on interannual to decadal timescales, to which we turn next. The second set of experiments are the large and regular atmospheric CO_2 fluctuations associated with the glacial-interglacial cycles.

10.3 Interannual to Decadal Timescale Variability

Over the last 50 years, atmospheric CO_2 has been steadily increasing by about 1.5 ppm yr^{-1} (figure 10.1.2), a direct result of the anthropogenic CO_2 emissions from the burning of fossil fuels and land use changes. However, the atmospheric CO_2 growth rate has not been constant, but has fluctuated strongly from one year to another (figure 10.3.1). These variations are not caused by changes in the amount of CO_2 emitted by

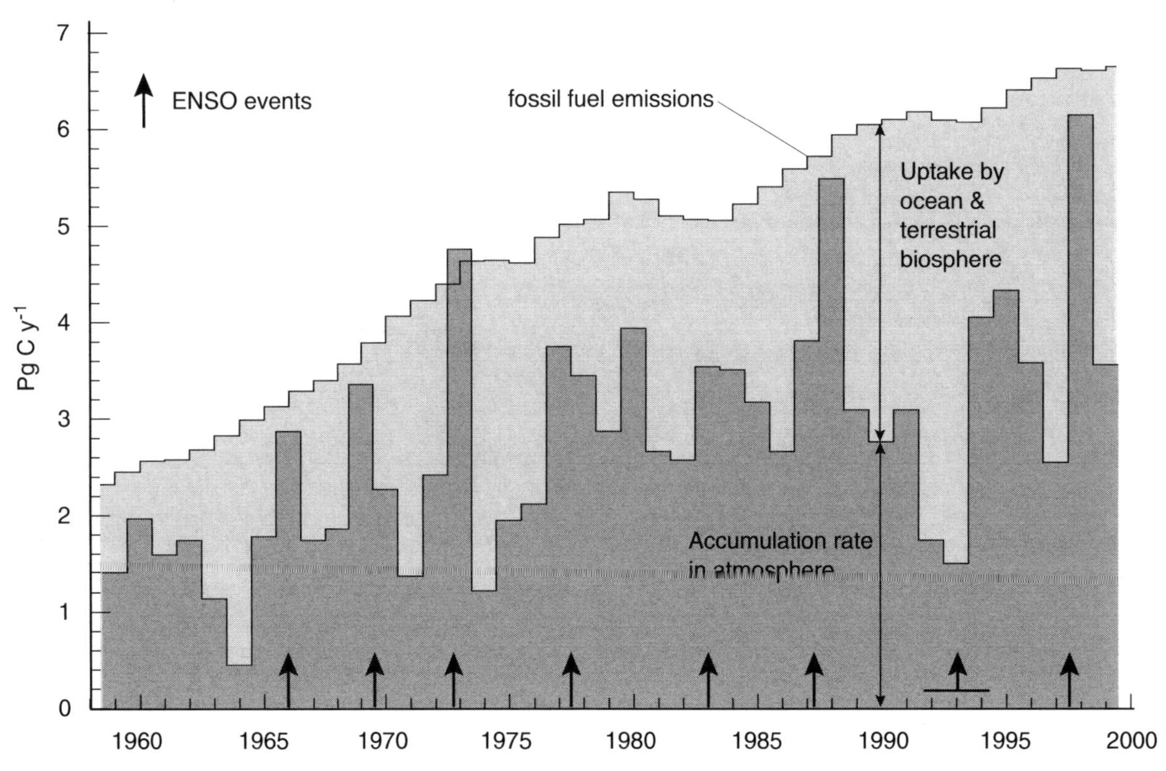

FIGURE 10.3.1: Plot of variations in growth rates of atmospheric CO_2, fossil fuel emissions, and inferred net carbon sinks. Arrows indicate the occurrence of El Niño events in the equatorial Pacific. The annual growth rates have been calculated from the annual mean atmospheric CO_2 concentrations. Data are from the Mauna Loa Observatory, Hawaii (MLO) and from the South Pole Station, Antarctica (SPO). Adapted from *Sarmiento and Gruber* [2002].

fossil fuel burning, because they are too small. Rather, these atmospheric CO_2 variations must be caused by variations in the net fluxes between the ocean and atmosphere on one hand and between the terrestrial biosphere and the atmosphere on the other hand. These variations in fluxes are presumably a consequence of interannual variations in weather and climate. Nature has thus provided us with an experiment that permits us to study how the global carbon cycle is responding to changes in the physical climate system. The study of this response provides one of the few means to learn how feedbacks between the physical climate system and the global carbon cycle operate. Only if the longer-tem feedbacks are well understood can one improve assessments of what might happen to the global carbon cycle in a future changing climate (see discussion in section 10.2 above).

Bacastow [1976] was the first to point out that these atmospheric CO_2 variations are associated with El Niño/Southern Oscillation (ENSO) events. This result has been confirmed in many subsequent studies using longer records [*Bacastow et al.*, 1980; *Keeling and Revelle*, 1985; *Keeling et al.*, 1989, *Keeling et al.*, 1995]. During an El Niño event (positive phase of ENSO), sea surface temperatures in the equatorial Pacific increase and the supply of nutrient- and carbon-rich water by upwelling ceases, thereby reducing the normally prevailing strong outgassing of CO_2 from this region (see figures 3.1.3 and 8.1.1). If net air-sea fluxes elsewhere remained constant, this reduction in outgassing would lead to a reduced growth rate in atmospheric CO_2. Figure 10.3.1 shows that the atmospheric CO_2 growth rate is, on average, indeed reduced during the early phases of an El Niño, making the link between ENSO events and atmospheric CO_2 plausible. However, figure 10.3.1 also shows that the late phases of El Niños are usually associated with very high growth rates, which are difficult to explain on the basis of the equatorial Pacific alone.

By measuring the $^{13}C/^{12}C$ ratio of atmospheric CO_2 concurrently with its concentration and employing a simple atmospheric mass balance approach, *Keeling et al.* [1989] and *Keeling et al.* [1995] found that biospheric, as opposed to oceanic, processes are the primary cause of interannual variations in atmospheric CO_2. Nevertheless, C. D. Keeling and coworkers also predicted variations in the net air-sea flux of CO_2 of up to ± 2 Pg C yr^{-1} (figure 10.3.2b). However, the timing of these variations is the opposite of what is expected from an oceanic response that is dominated by ENSO-related variations in outgassing in the tropical Pacific. *Francey et al.* [1995], *Rayner et al.* [1999], and *Joos et al.* [1999a], using similar techniques, found rather differ-

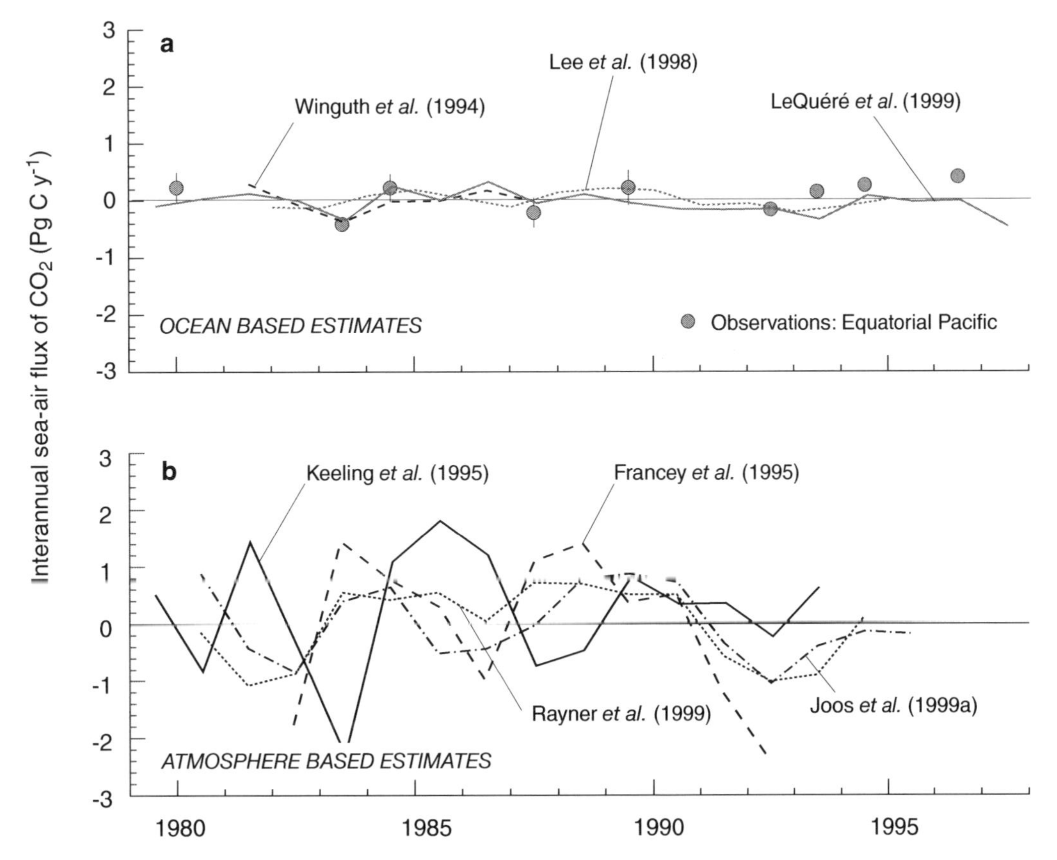

FIGURE 10.3.2: Time series of interannual sea-to-air flux of CO_2 as estimated by different studies. Estimates in the top panel are based on ocean observations and models, whereas those in the bottom panel are from atmospheric observations and models. The oceanic observations in the top panel are based on pCO_2 measurements in the equatorial Pacific only [*Feely et al.*, 1997] and are represented as deviations from a mean of 0.5 Pg C yr^{-1}. Shown are estimates from *Le Quéré et al.* [2000], *Winguth et al.* [1994], *Lee et al.* [1998], *Keeling et al.* [1995], *Francey et al.* [1995], *Joos et al.* [1999a], and *Rayner et al.* [1999]. Adapted from *Le Quéré et al.* [2000].

ent results. They predicted variations in anomalous air-sea fluxes of similar magnitude, albeit slightly smaller, but with a phasing that is nearly opposite to that of *Keeling et al.* [1995] (figure 10.3.2b). More recent estimates of global air-sea CO_2 flux variability by *Bousquet et al.* [2000] and *Roedenbeck et al.* [2003] on the basis of inversions of atmospheric CO_2 concentrations alone tend to confirm the latter estimates. Similar conclusions were found by *Battle et al.* [2000] using concurrent measurements of atmospheric O_2 and CO_2 as a constraint.

Although most recent estimates based on atmospheric constraints point toward a sizeable interannual variability in air-sea CO_2 fluxes, i.e., an amplitude of order ± 1 Pg C yr^{-1}, oceanic observations or oceanic modeling studies tend to suggest a much smaller oceanic contribution to the observed atmospheric CO_2 variations (figure 10.3.2a). Estimates that are based on measurements of the variability of pCO_2 in the tropical Pacific indicate a maximal interannual amplitude of only about ± 0.5 Pg C yr^{-1} (see summary in *Feely et al.* [1999] and *Chavez et al.* [1999]; see also figure 10.3.6). In order to compare these tropical Pacific observations with the atmospheric contraints, which are global, one needs to consider also the contribution of the extratropical oceanic regions. *Winguth et al.* [1994], *LeQuéré et al.* [2000], and *McKinley* [2002] found on the basis of global 3-D ocean biogeochemistry models that interannual variability in air-sea CO_2 fluxes is small in the extratropical regions, and that the global flux is dominated by the equatorial Pacific.

We cannot settle the controversy here, but rather would like to review the evidence for interannual variations in ocean biogeochemistry and learn the mechanisms that are in operation. We will focus on the ENSO-related variations in the equatorial Pacific because they are large, and they have been well documented. We will also discuss some recent research documenting interannual variability in the extratropical ocean carbon cycle.

Tropical Variability

ENSO is the most important mode of SST variability on interannual to decadal timescales (see section 2.5). ENSO-related SST anomalies are caused by a combi-

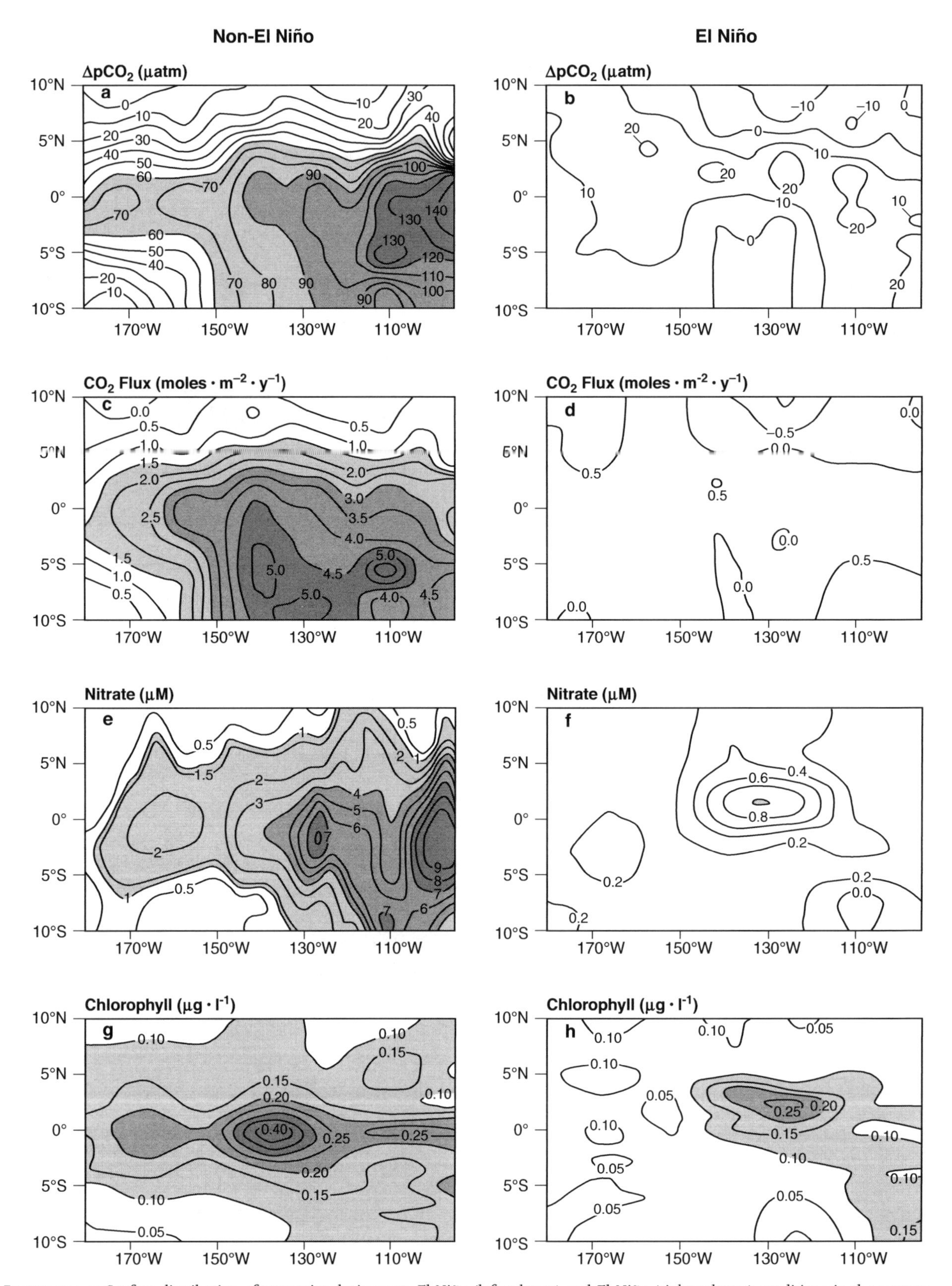

FIGURE 10.3.3: Surface distribution of properties during non–El Niño (left column) and El Niño (right column) conditions in the eastern equatorial Pacific. (a) and (b) show pCO_2 difference between the surface ocean and the atmosphere; (c) and (d) estimated sea-to-air CO_2 flux; (e) and (f) surface nitrate concentrations; and (g) and (h) surface chlorophyll. From *Chavez et al.* [1999].

nation of changes in ocean circulation (mainly a change in the strength and source of equatorial upwelling) and anomalous local air-sea heat exchanges. How does the ocean carbon cycle respond to these changes? The large positive SST anomalies that are characteristic of a mature El Niño (figure 2.5.6) should lead to an increase of oceanic pCO_2. However, observations show a dramatic decrease of pCO_2, linked to a drop in *DIC* that more than compensates for the warming effect (figure 10.3.3). Is this drop in *DIC* caused by changes in circulation, in air-sea gas exchange, or in ocean biology?

Before we can address this question we need to understand the mean state of the carbon system in the equatorial Pacific. The tropical Pacific is characterized by large positive ΔpCO_2 (figure 8.1.1). In fact, it is the region of the most intense outgassing of CO_2 from the ocean [*Takahashi et al.*, 2002]. During normal and negative phases of ENSO, the loss of CO_2 from the ocean to the atmosphere in the eastern tropical Pacific (10°S–10°N; 135°W–80°W) was estimated to be around 0.9 ± 0.1 Pg C yr^{-1} [*Feely et al.*, 1995]. What makes this region such a strong source region for atmospheric CO_2?

Divergent Ekman transport forced by the trade winds leads to strong upwelling of cold subsurface waters. The main source of the upwelling water is thought to be the Equatorial Undercurrent, located at depths of about 100 to 200 m (see figure 2.3.14). The potential for outgassing can be diagnosed by looking at a quantity referred to as potential pCO_2, i.e., the pCO_2 that would be achieved if the water at temperature $T^{in\,situ}$ were brought to the surface and isochemically warmed to ambient equatorial surface ocean temperature T^{sfc}. Potential pCO_2 can be easily computed by integrating the expression for the temperature sensitivity of pCO_2 (equation (8.3.4)):

$$pCO_2^{potential} = pCO_2^{in\,situ}\,(1 + 0.0423°C^{-1}\,(T^{sfc} - T^{in\,situ})) \quad (10.3.1)$$

Figure 10.3.4 shows that the waters of the Equatorial Undercurrent have extremely high potential pCO_2, leading to a very high potential outgassing of CO_2 once these waters come to the surface. The high potential pCO_2 in these waters is due to a combination of their cold temperatures and large amounts of remineralized carbon. However these high potential pCO_2's are, greatly reduced at the surface because after the waters arrive in the mixed layer, biological uptake fueled by the high nutrient concentrations removes inorganic carbon from the surface ocean. Additional carbon is lost to the atmosphere by air-sea gas exchange.

It is possible to separate the role of biological CO_2 uptake from the CO_2 evasion in reducing the supersaturation of the upwelling water through plots of salinity-normalized *DIC* (*sDIC*) versus salinity-normalized nitrate [*Broecker and Peng*, 1982]. As can be seen in figure

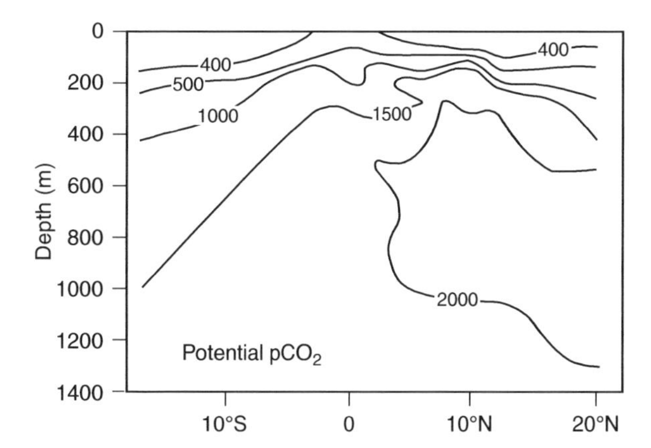

FIGURE 10.3.4: Vertical section of potential pCO_2 in the equatorial Pacific. Potential pCO_2 is the CO_2 pressure a sample would achieve if it were isochemically depressurized (to 1 atm) and warmed (to 25°C) (see text). From *Broecker and Peng* [1982].

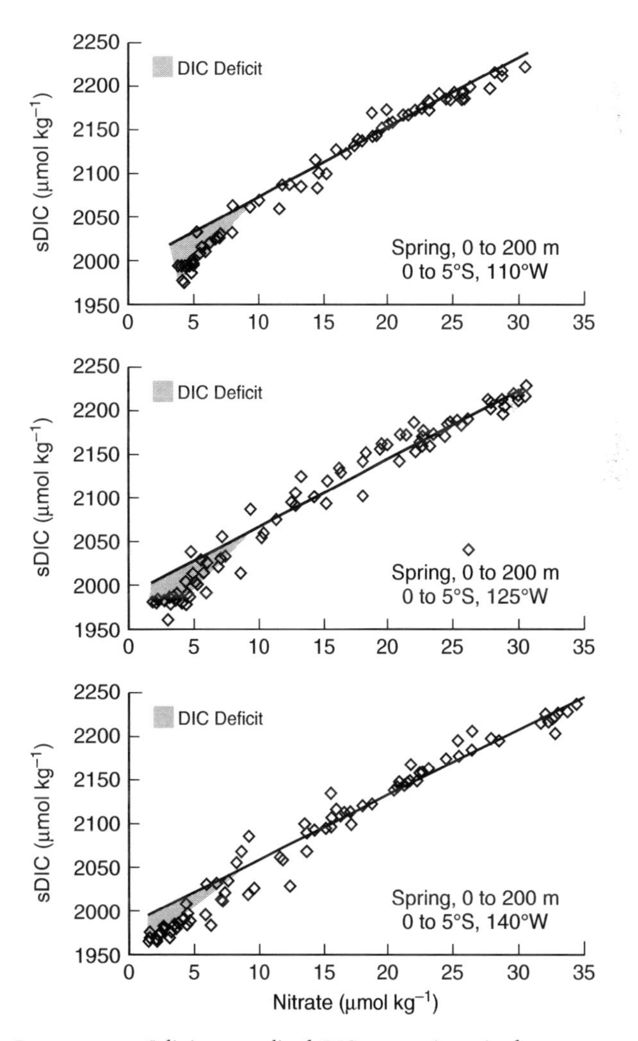

FIGURE 10.3.5: Salinity-normalized *DIC* versus nitrate in the upper 200 m between the equator and 5°S along 110°W (top), 125°W (middle), and 140°W (bottom), during boreal spring. The deviation of the surface values from the trend in the thermocline (solid line), as highlighed by the grey triangle, is attributed to the loss of *DIC* by outgassing. From *Wanninkhof et al.* [1995].

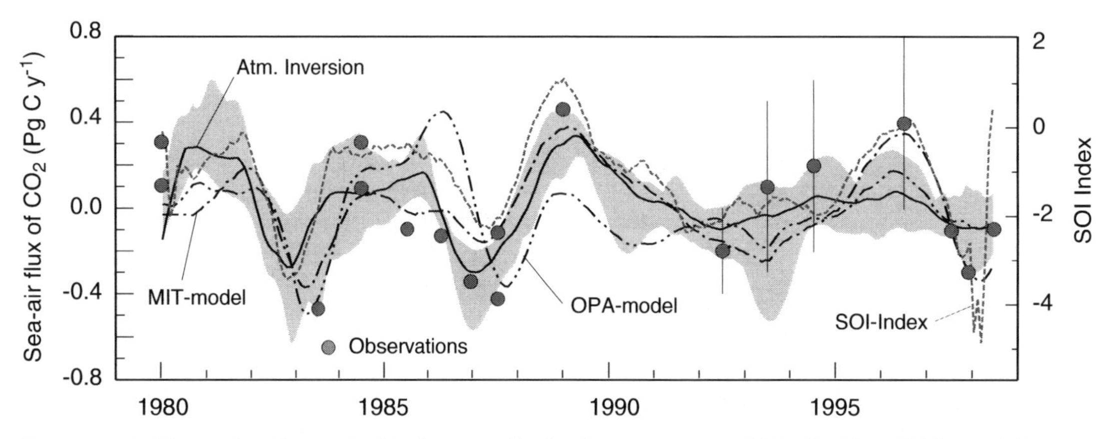

FIGURE 10.3.6: Time series of sea-to-air CO_2 flux anomalies for the eastern equatorial Pacific. The solid line and the grey band are estimates based on an inversion of atmospheric CO_2 [*Bousquet et al.*, 2000]. The dot-dashed curves represent the model-based estimates of *McKinley* [2002] and *LeQuéré et al.* [2000], respectively. The circles represent estimates based on a compilation of oceanic data [*Feely et al.*, 1999]. The Southern Oscillation Index (SOI) index is also shown as a reference (right axis). From *Peylin et al.* [2005].

10.3.5, samples taken at intermediate depths follow a trend that corresponds roughly to the stoichiometry of biological uptake and remineralization [*Anderson and Sarmiento*, 1994]. Samples taken from the surface or near the surface show a marked deviation from this trend by having significantly lower *sDIC* concentrations than would be expected from their nitrate concentration. Note that this deficiency is equivalent to the low $\Delta C_{gas\ ex}$ concentrations found in the equatorial regions (see figure 8.4.3c). This missing *DIC* presumably has been lost to the atmosphere. We can obtain a rough estimate of this CO_2 loss by multiplying the upwelling rate (about 0.7 m day^{-1} [*Wanninkhof et al.*, 1995]) with the *DIC* deficit of about 30 μmol kg^{-1}. This yields an outgassing rate of about 8 mol m^{-2} yr^{-1} or 1.3 Pg C yr^{-1} if integrated over the eastern equatorial Pacific (10°S–10°N; 135°W–80°W). This is roughly consistent with estimates based on measurements of the air-sea difference in pCO_2. A more sophisticated approach on the basis of the same concept, but using global-scale data and circulation constraints, yields an outgassing estimate of about 0.9 Pg C yr^{-1} [*Gloor et al.*, 2003].

In summary, the strong outgassing of CO_2 in the equatorial Pacific is a result of a combination of strong warming of the cold waters that upwell to the surface and an inefficient biologial pump. The latter means that surface ocean *DIC* is being replaced faster by upwelling than surface ocean biology can remove it by forming organic matter. This inefficient biological pump is manifested also by the existence of residual nutrients (see figure 10.3.3 and discussion in section 8.5).

During an El Niño, the deepening of the thermocline caused by the eastward propagating Kelvin waves push down the cold, *DIC*-rich waters that typically feed the upwelling. Instead, the upwelling is being fed by much warmer, *DIC*-poor waters. In addition, the weakening of the trade winds leads to a reduction in upwelling rates. These two changes together cause a dramatic reduction in the amount of *DIC* transported to the surface ocean, resulting in an almost complete removal of the equatorial supersaturation (see figure 10.3.3). *Feely et al.* [1999] and *Chavez et al.* [1999] estimated that the net ocean-to-atmosphere CO_2 flux during a strong El Niño event is only about 0.1 to 0.3 Pg C yr^{-1}, hence about 0.5 to 0.9 Pg C yr^{-1} lower than normal (see figure 10.3.6). Although this decrease is mainly caused by changes in ocean circulation and mixing, surface biology tends to reinforce these changes. This is, at first, counterintuitive, since the reduced upward supply of nutrients leads to a dramatic decrease in biological productivity in the area. However, it turns out that despite this reduction, the biological pump tends to become more efficient, i.e., a larger fraction of the upward supply of biologically derived *DIC* is fixed into organic matter and exported. This is most clearly expressed in the observed decrease of surface nutrients (figure 10.3.3). As a result, the anomalous fluxes associated with the changes in the biological pump during El Niños are actually directed into the ocean, i.e., contributing to the observed decrease in ocean outgassing.

Siegenthaler and Wenk [1989] developed a simple four-box model that highlights the relative roles of changes in biological productivity and upwelling (figure 10.3.7). This model consists of three oceanic boxes (equatorial surface, mid-latitude, and thermocline) overlain by a well-mixed atmosphere. Upwelling from the thermocline occurs in the equatorial surface box, from which the water is transported laterally into the mid-latitude box. The large-scale circulation is closed by downwelling in the mid-latitudes. *Siegenthaler and Wenk* [1989] included a simple parameterization of biological

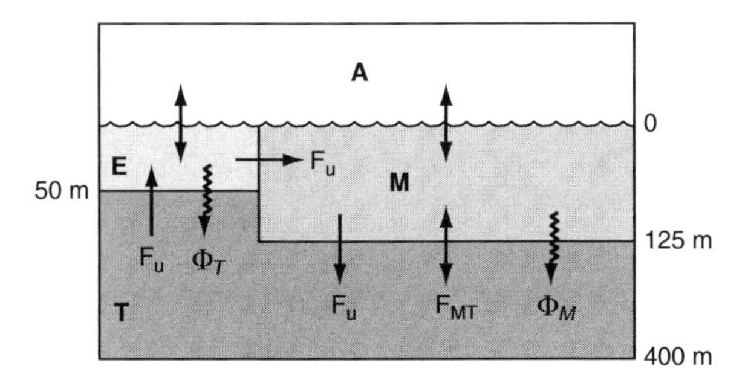

FIGURE 10.3.7: Schematic structure of the 4-box model used by *Siegenthaler and Wenk* [1989] for simulating ENSO events. *A*: atmosphere, *E*: equatorial Pacific surface water, *M*: mid-latitude Pacific surface water, *T*: thermocline. F_u and F_{mt}: water fluxes; Φ_T and Φ_M: biological particle fluxes. From *Siegenthaler and Wenk* [1989].

productivity depending on nutrient concentration. Figure 10.3.8 shows the response of this four-box model to an idealized El Niño event of reduced upwelling assumed to last for one year.

Atmospheric CO_2 starts to decrease nearly instantaneously, and the atmospheric CO_2 growth rate remains negative until upwelling resumes at the end of the event. This decrease in atmospheric CO_2 is almost exclusively a result of a dramatic decrease in the pCO_2 of the equatorial surface box, which leads to near cessation of the CO_2 evasion from this region (figure 10.3.8d and f). To understand the dynamics underlying this response, it is useful to investigate the carbon fluxes in and out of the equatorial box (figure 10.3.8f). In normal years, the large-scale transport of water constitutes a source of *DIC* for the equatorial box, since the upwelling waters have higher *DIC* concentrations than the waters leaving the box laterally. This net gain is compensated by losses due to gas exchange and by settling particles. When upwelling stops, gas exchange and biological productivity continue for a while, decreasing *DIC* and nutrient concentrations. This incrementally reduces biological export and the net outgassing of CO_2, so that at the end of the El Niño event, all carbon fluxes into or out of the equatorial box are near zero.

It is instructive to compare the model-simulated atmospheric growth rates (figure 10.3.8c) with the observed growth rates over an El Niño event (figure 10.3.1). The model predicts a minimum in the growth rate toward the end of an El Niño event, whereas atmospheric observations indicate that the minimum growth rates tend to occur early during an El Niño event if at all, and are usually followed by a very strong positive anomaly in the growth rates. It therefore appears plausible that the observed initial decline of atmospheric CO_2 growth rates may be a consequence of the reduced outgassing in the tropical Pacific. Obviously, a different mechanism must explain the unusually strong atmospheric CO_2 growth rates observed afterwards. It has been suggested that changes in extratropical waters might be involved. However, model simulations suggest that the response of the extratropical carbon cycle to ENSO variations is relatively small [*Winguth et al.*, 1994; *Le Quéré et al.*, 2000]. If this is correct, then the dominant cause for the ENSO-related atmospheric CO_2 variations are variations in the net exchange of the terrestrial biosphere.

This is a plausible hypothesis, since ENSO events in the tropical Pacific are coupled to large changes in global atmospheric circulation and precipitation patterns. The 1982/83 as well as the 1997/98 El Niños led to severe drought and forest fires in large regions, which resulted in significant release of CO_2 from soil respiration and burning of biomass [*der Werf et al.*, 2004]. The global-scale positive temperature anomalies associated with warm ENSO events tend to enhance soil respiration more strongly than photosynthesis. This hypothesis is supported by results from several simulations with prognostic terrestrial carbon cycle models forced with observed variations in temperature and precipitation [*Kindermann et al.*, 1996; *Peylin et al.*, 2005]. These models are able to reproduce successfully the observed atmospheric CO_2 growth rate variations primarily on the basis of these processes.

In summary, about two-thirds of the observed interannual variations in the atmospheric CO_2 growth rates are caused by the terrestrial biosphere, with the remainder being driven by the ocean [*Peylin et al.*, 2005]. There is currently strong consensus between observations, oceanic modeling studies, and atmospheric constraints that the equatorial Pacific contributes substantially to global air-sea CO_2 variations on interannual timescales (figure 10.3.6).

EXTRATROPICAL VARIABILITY

There is little consensus about the magnitude of interannual variations in air-sea CO_2 fluxes in the extratropics. Estimates based on atmospheric CO_2 inversions suggest substantial variations in air-sea CO_2 fluxes in the extratropical regions of both hemispheres, while currently existing ocean models suggest a very small contribution from these regions [*Peylin et al.*, 2005]. One possibility to resolve this discrepancy is to analyze observations from as many oceanic regions as possible. However, very few sites exist where the inorganic carbon system has been observed regularly enough to establish interannual variability. The two longest such records come from the subtropical gyres (see also figure 10.2.5) and will be discussed below.

Before we start investigating the subtropical records, it is instructive to first look at the processes that could lead to interannual variations in air-sea CO_2 fluxes. A good starting point is the seasonal cycle, since interannual variations in the extratropics can be viewed as perturbations of the seasonal cycle. We have seen in chapter 8 that the seasonal cycle of oceanic pCO_2 is

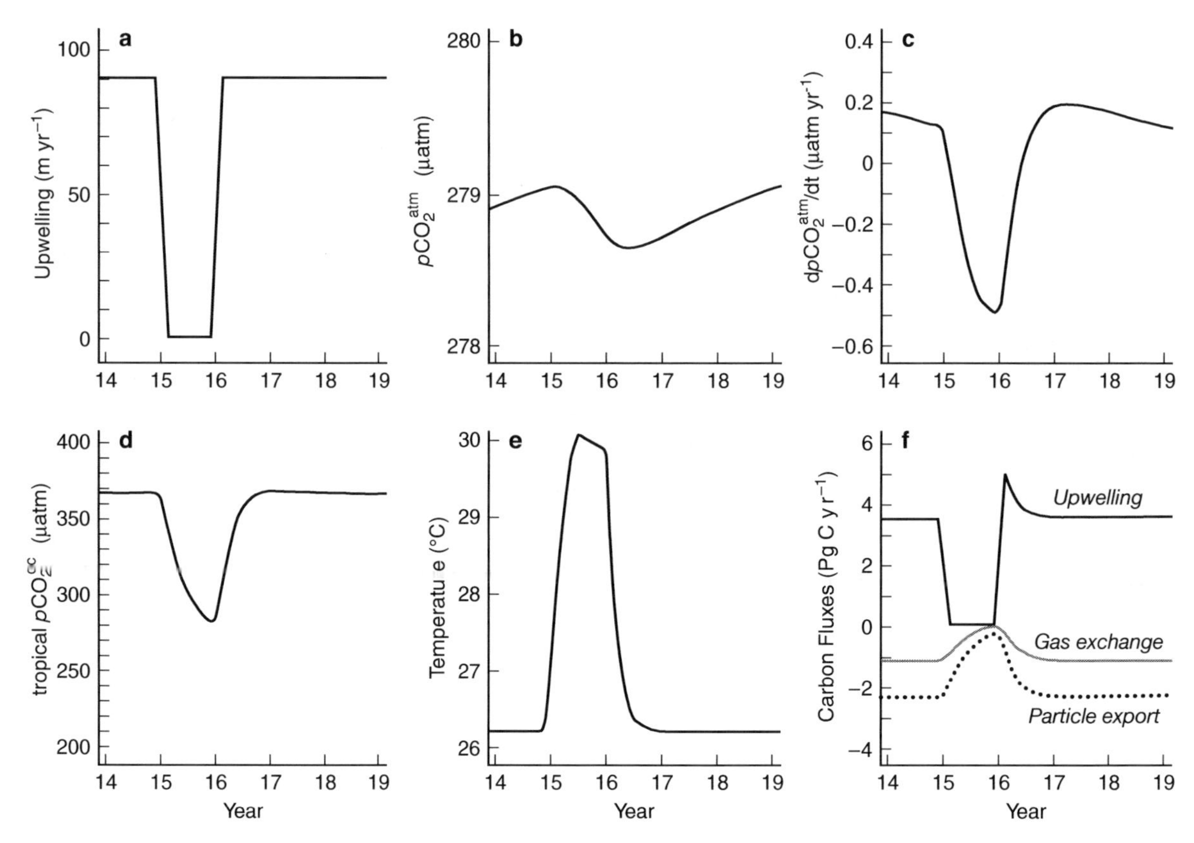

FIGURE 10.3.8: Response of the 4-box model to a simulated El Niño event starting at the beginning of model year 15 and lasting one year. (a) prescribed time evolution of the upwelling flux F_u; (b) predicted atmospheric pCO_2; (c) simulated atmospheric CO_2 growth rate variations; (d) simulated variations in pCO_2 of equatorial surface waters; (e) simulated variations in temperature of equatorial surface waters; and (f) carbon fluxes into and out of the equatorial surface water box. From *Siegenthaler and Wenk* [1989].

primarily controlled by the interplay of antagonistic tendencies between SST and surface *DIC* variations, where surface *DIC* is controlled by net community production, air-sea gas exchange, vertical entrainment (i.e., the input of thermocline *DIC* into the surface mixed layer whenever it deepens), and lateral exchange.

In the subtropical gyres, we found that the influence on pCO_2 of the seasonal warming and cooling dominates over the seasonal *DIC* changes, leading to a positive correlation between SST and *DIC* (figure 8.3.7). In contrast, the influence of the seasonal *DIC* drawdown on oceanic pCO_2 outweighs the seasonal warming in higher latitudes, leading to a negative correlation between *DIC* and SST. Therefore, if the analogy with the seasonal cycle is correct, we might expect that interannual anomalies in SST would tend to dominate interannual variations in pCO_2 in the subtropical gyres, while interannual anomalies in *DIC* would be expected to control pCO_2 anomalies at higher latitudes.

In order to get a sense of where interannual variability in oceanic pCO_2, and consequently air-sea CO_2 fluxes, might be expected to be large, we show in figure 10.3.9 maps of the variance of three factors that might drive such variability: SST, mixed layer depth, and export production. These maps were produced from satellite observations by first deasonalizing the data and then calculating the variance of the residuals (for details see *LeQuéré and Gruber* [2005]). While the SST anomalies are clearly the largest in the equatorial Pacific, the maps also suggest that there are large variations in these three driving forces in the extratropical regions, potentially giving rise to substantial interannual variations in oceanic pCO_2 there. Whether or not these extratropical variations lead to pCO_2 changes depends in part on possible correlations between the different driving factors.

For example, interannual variations in SST and *DIC* may correlate, as colder than normal SSTs are usually associated with anomalously deep surface mixed layers (both a forcing and a result of the entrainment of cold thermocline waters), which enhances the entrainment of *DIC*. This would tend to lead to a negative correlation between SST and *DIC* anomalies, suppressing the variability of oceanic pCO_2, as the reduced SST tends to decrease pCO_2, while the positive *DIC* anomalies tend to increase pCO_2. Anomalously strong entrainment also enhances the input of nutrients into the surface layer, which, in nutrient-limited ecosystems, can increase biological productivity and hence lower *DIC*.

This would tend to lead to a positive correlation between *DIC* and SST, as colder than normal conditions would be associated with anomalously low *DIC*. In such a case, these two driving factors would enhance the variability of oceanic pCO_2. The resulting relationship between SST and *DIC* and, therefore, pCO_2 critically depends, then, on the magnitude of the biological response. In contrast, in light-limited ecosystems, deeper mixed layers would tend to reduce biological production, therefore enhancing the negative correlation between SST and *DIC*. In these regions, oceanic pCO_2 variability is determined by the relative magnitude of the SST and *DIC* changes. Given these sometimes synergistic and sometimes antagonistic driving factors, it is very difficult to predict the overall response of the oceanic pCO_2 to changes, but it is useful to keep these general principles in mind as we next look at the observations from the subtropical gyres.

Figure 10.3.10 shows seasonally and long-term mean detrended time series of temperature, salinity-normalized *DIC* (*sDIC*), the reduced $^{13}C/^{12}C$ isotopic ratio of *DIC*, and calculated pCO_2 from Station "S" / BATS in the Sargasso Sea near Bermuda, and the respective records from the HOT program near Hawaii. These anomalies were computed by removing the mean seasonal cycles shown in figures 8.3.8 and 8.3.9 from the time-series records shown in figure 10.2.5. The long-term trend, primarily the result of the uptake of isotopically light anthropogenic CO_2 from the atmosphere, was taken into account by fitting a linear function of time to the data (see *Gruber et al.* [2002] and *Brix et al.* [2004] for details).

At both sites, the resulting interannual anomaly time series show substantial variations, with considerable correlations between the different variables. In particular, anomalously cold years tend to coincide with increased *sDIC* concentrations and a slight tendency for negative anomalies in the $^{13}C/^{12}C$ ratio. As a result of the anticorrelated SST and *sDIC* anomalies, variations in oceanic pCO_2 tend to be suppressed, i.e., whenever warmer than normal conditions increase oceanic pCO_2, the concurrent negative *sDIC* anomalies tend to cause a decrease of pCO_2. Overall, SST anomalies tend to prevail, causing a positive correlation between oceanic pCO_2 and SST, consistent with our hypothesis put forward on the basis of the analogy with the seasonal cycle.

What are the processes causing these covariations between SST, *sDIC*, and the $^{13}C/^{12}C$ ratio? *Gruber et al.* [2002] and *Brix et al.* [2004] investigated the processes causing the observed variations at the two sites in detail, taking advantage of the concurrently observed $^{13}C/^{12}C$ ratio variations. These data, together with a large number of other observations, such as mixed layer depth and wind speed, permitted them to estimate the interannual variability of the contribution from each of the five processes that are controlling the surface ocean *sDIC* balance: air-sea gas exchange, net community production, vertical entrainment, vertical diffusion, and lateral transport (see also section 8.3). Figure 10.3.11 shows that at both sites, air-sea gas exchange and net community production represent two of the three important contributors to the observed interannual variability. The two sites differ, however, with regard to which is the third important process. Near Bermuda, it is vertical entrainment, while it is lateral tranport near Hawaii. It thus appears that variations in convection are the dominant driving factor near Bermuda, while these variations tend to be unimportant near Hawaii. This is not entirely surprising, since mixed layer variations near Bermuda are large and play an important role for the seasonal cycle, whereas near Hawaii, mixed layer variations are small and therefore unimportant on seasonal timescales.

It is instructive to stratify the different years into two modes depending on the state of the most important external factor believed to drive interannual variations in the upper ocean carbon cycle. Variations in wintertime convection clearly emerge as the main factor near Bermuda, while SST variations appear to be the leading external forcing at the Hawaiian time-series station. Figure 10.3.12 depicts idealized seasonal cycles of the upper ocean carbon cycle at the two sites in the respective modes.

Near Bermuda, years with intense wintertime convection start their seasonal cycle with positive *sDIC* anomalies because of the anomalously large entrainment of *sDIC* from the thermocline. SSTs tend to be colder during these years. These colder temperatures outweigh the pCO_2-increasing effect of the positive *sDIC* anomalies, leading to negative oceanic pCO_2 anomalies. Coupled with the higher than normal winds, this leads to a substantially larger uptake of CO_2 from the atmosphere during these years. Biological productivity is also inferred to be higher, presumably a consequence of the higher nutrient input from below. These two factors lead to an enhanced seasonal cycle of *sDIC*, but with an overall positive *sDIC* anomaly. Years with anomalously shallow wintertime convection tend to follow the reverse pattern. The differences between the two modes are quite substantial and amount to up to 50% variations of the annual mean fluxes.

Near Hawaii, differences between the two modes are somewhat less marked, but still significant. In warmer than normal years, net community production tends to be reduced, which should lead to a positive anomaly in *sDIC*. This is outweighed by weaker uptake of CO_2 from the atmosphere caused by SST-induced positive pCO_2 anomalies and weaker winds, and a weaker than normal addition of *sDIC* by lateral transport. These two factors lead to the observed negative *sDIC* anomalies in warm years.

These observations thus indicate a substantial sensitivity of the upper ocean carbon cycle to variations in

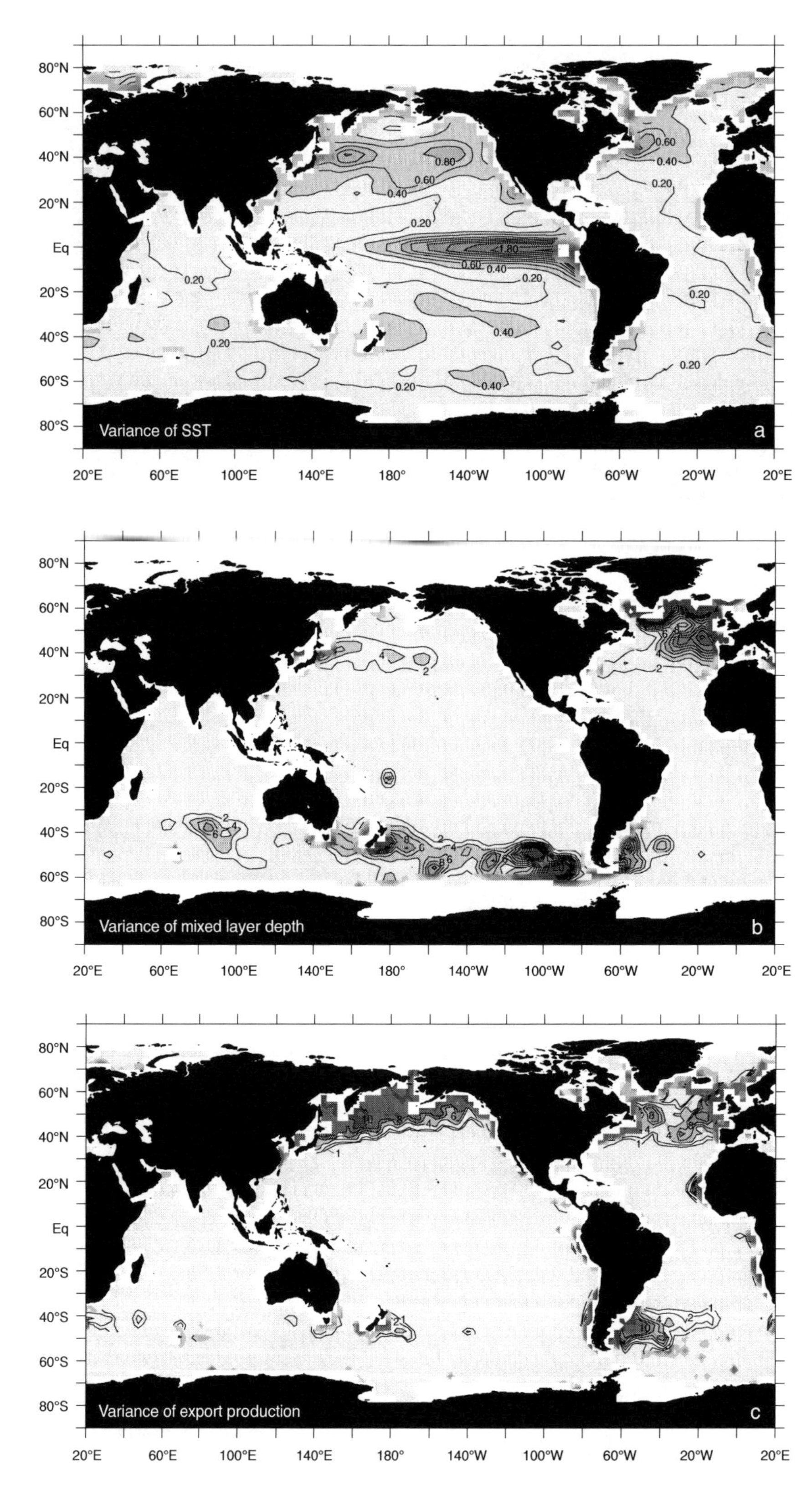

FIGURE 10.3.9: Maps of satellite based estimates of interannual variance of key drivers for the ocean carbon cycle: (a) sea-surface temperature, (b) mixed layer depth, and (c) export production. Sea-surface temperature is based on AVHRR, mixed layer depth has been deduced from Topex/Poseidon's measurements of sea surface height and AVHRR, and export production has been estimated based on SeaWiFS. See *Le Quéré and Gruber* [2005] for details.

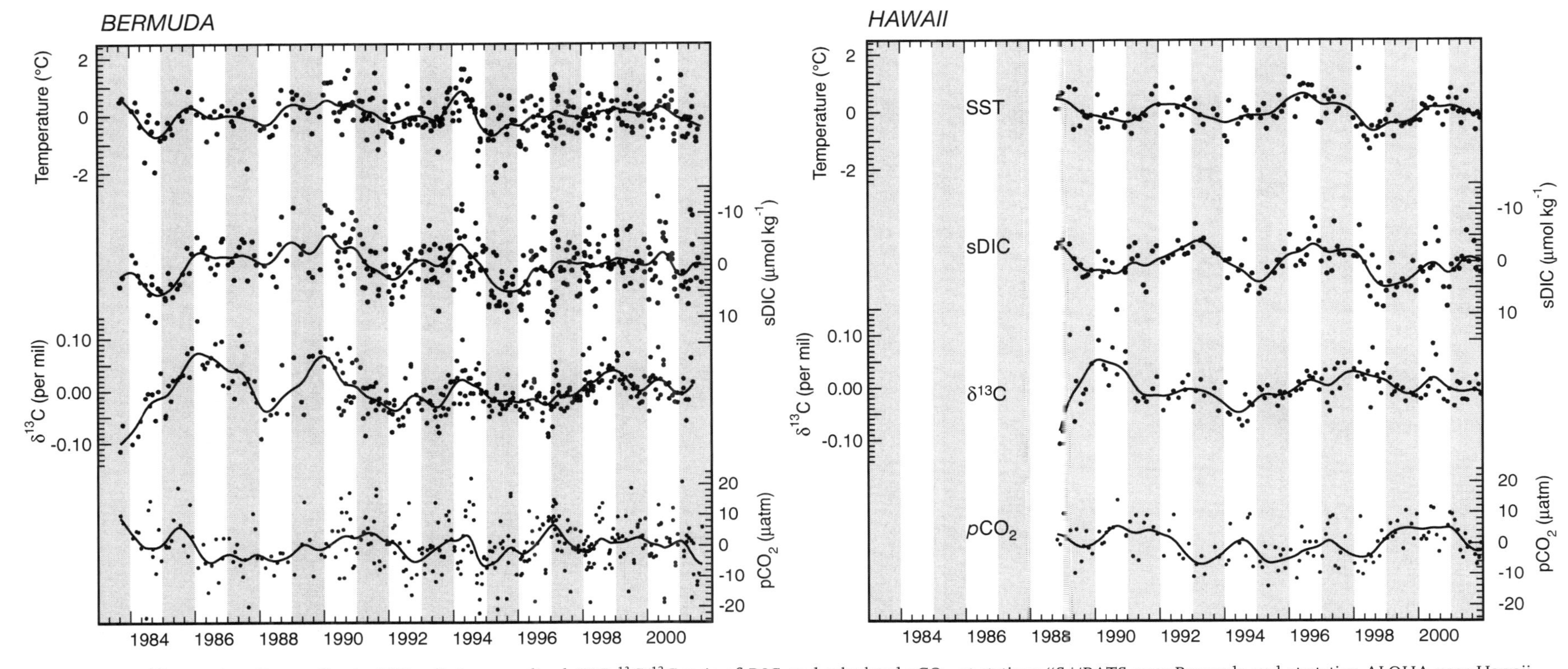

FIGURE 10.3.10: Time series of anomalies in SST, salinity-normalized *DIC*, $^{13}C/^{12}C$ ratio of *DIC*, and calculated pCO_2, at stations "S"/BATS near Bermuda and at station ALOHA near Hawaii. Anomalies were computed by removing the average seasonal cycle and a linear long-term trend. Data are from *Gruber et al.* [2002] and *Brix et al.* [2004].

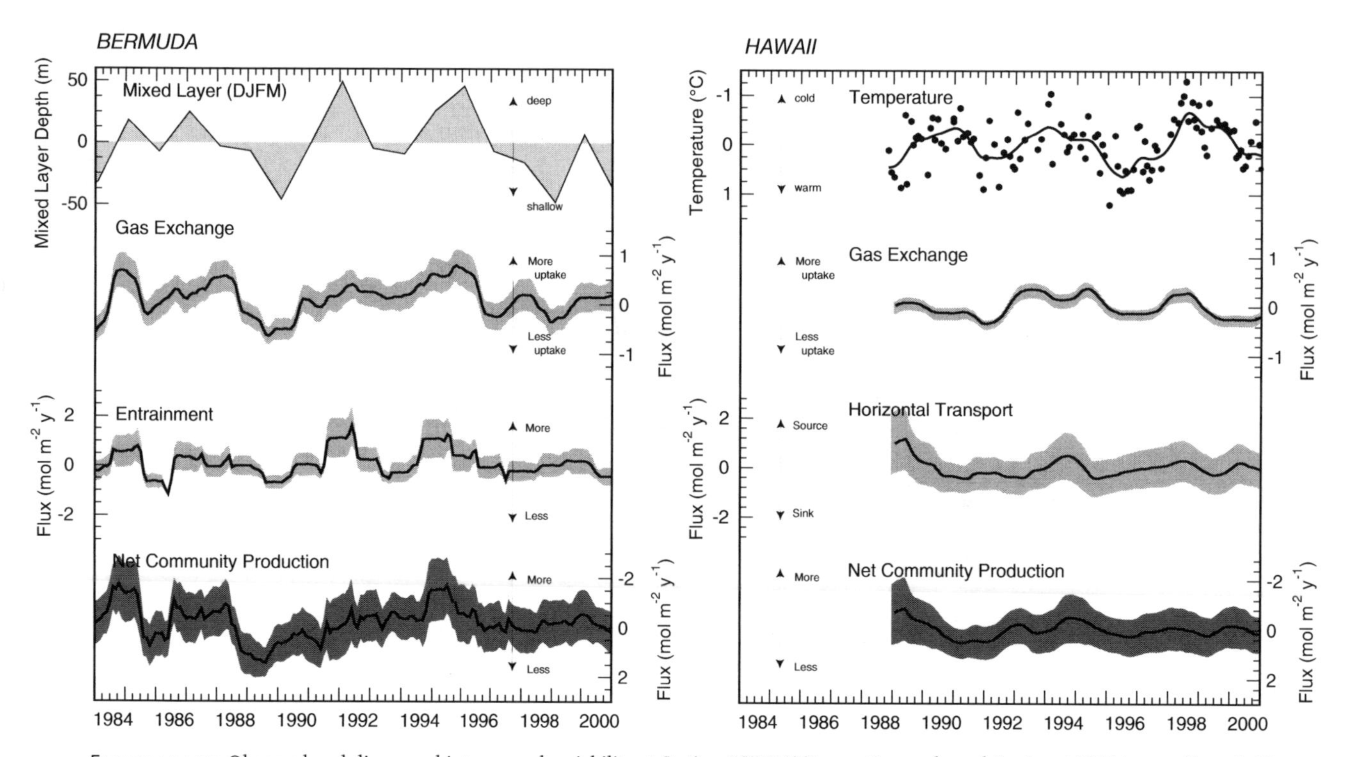

FIGURE 10.3.11: Observed and diagnosed interannual variability at Station "S"/BATS near Bermuda and Station ALOHA near Hawaii. Shown are the anomalies of all properties after the mean seasonal cycle was removed: in Bermuda, winter-time mixed layer, air-sea gas exchange, entrainment, and net community production; in Hawaii, SST anomalies, air-sea gas exchange, horizontal transport, and net community production. The contribution of diffusion and horizontal transport to interannual variability at BATS, and entrainment and diffusion at HOT are small and thus not shown. The shaded region around each line represents ± one standard deviation uncertainty estimates from Monte Carlo experiments. From *Gruber et al.* [2002] and *Brix et al.* [2004].

meteorological forcing, in particular wind and heat fluxes, but their impact on the upper ocean carbon cycle takes different routes. Near Bermuda, the influence of these two forces on the upper ocean carbon cycle is primarily expressed through changes in the amount of *sDIC* and nutrients that are entrained in the surface mixed layer. Near Hawaii, the impact is more direct, in that the heat fluxes and wind alter air-sea gas exchange, and to a lesser degree change lateral transport.

Since interannual variations in these meterological forcings tend to vary in concert over large spatial scales, it is conceivable that the observed local variations are representative over larger regions. A tentative extrapolation of the results from Bermuda and Hawii to their respective subtropical gyres suggests air-sea flux anomalies of the order of $0.3\,\mathrm{Pg\,C\,yr^{-1}}$ [*Gruber et al.*, 2002; *Brix et al.*, 2004]. It thus appears as if the extratropical ocean may be more variable than suggested by current oceanic modeling studies. However, direct comparisons suggest that the models are relatively successful in capturing the observed air-sea CO_2 flux variations near Bermuda and Hawaii, but that they tend to simulate variations at higher latitudes that tend to be opposite in phase to those at lower latitudes, thus suppressing basin-scale variability.

Too many things are still unknown or ill constrained in order to make a final assessment of what the role of the ocean is in generating the observed interannual variability in atmospheric CO_2. One can expect, however, rapid progress over the next years, as the number of sites where interannual variability can be investigated will increase dramatically. But, irrespective of the uncertainties associated with the current assessment, it is clear that the oceanic carbon cycle responds sensitively to changes in meteorological forcing. On short timescales the impact of these changes on atmospheric CO_2 will be suppressed because of the long equilibration timescale associated with the exchange of CO_2 across the air-sea interface. On timescales of decades and longer, this constraint is substantially less important. Furthermore, the reservoir of *DIC* in the ocean that can exchange with the atmosphere increases as the timescale of the variations of interest increases. On timescales of several hundreds of years, the entire oceanic reservoir of *DIC* is available for exchange, which, given the fact that this reservoir is 60 times larger than that of the atmosphere, makes the ocean the primary factor controlling atmospheric CO_2 on centennial and longer timescales. In the next section, we will investigate the most prominent example of such variations, the glacial-interglacial change in atmospheric CO_2.

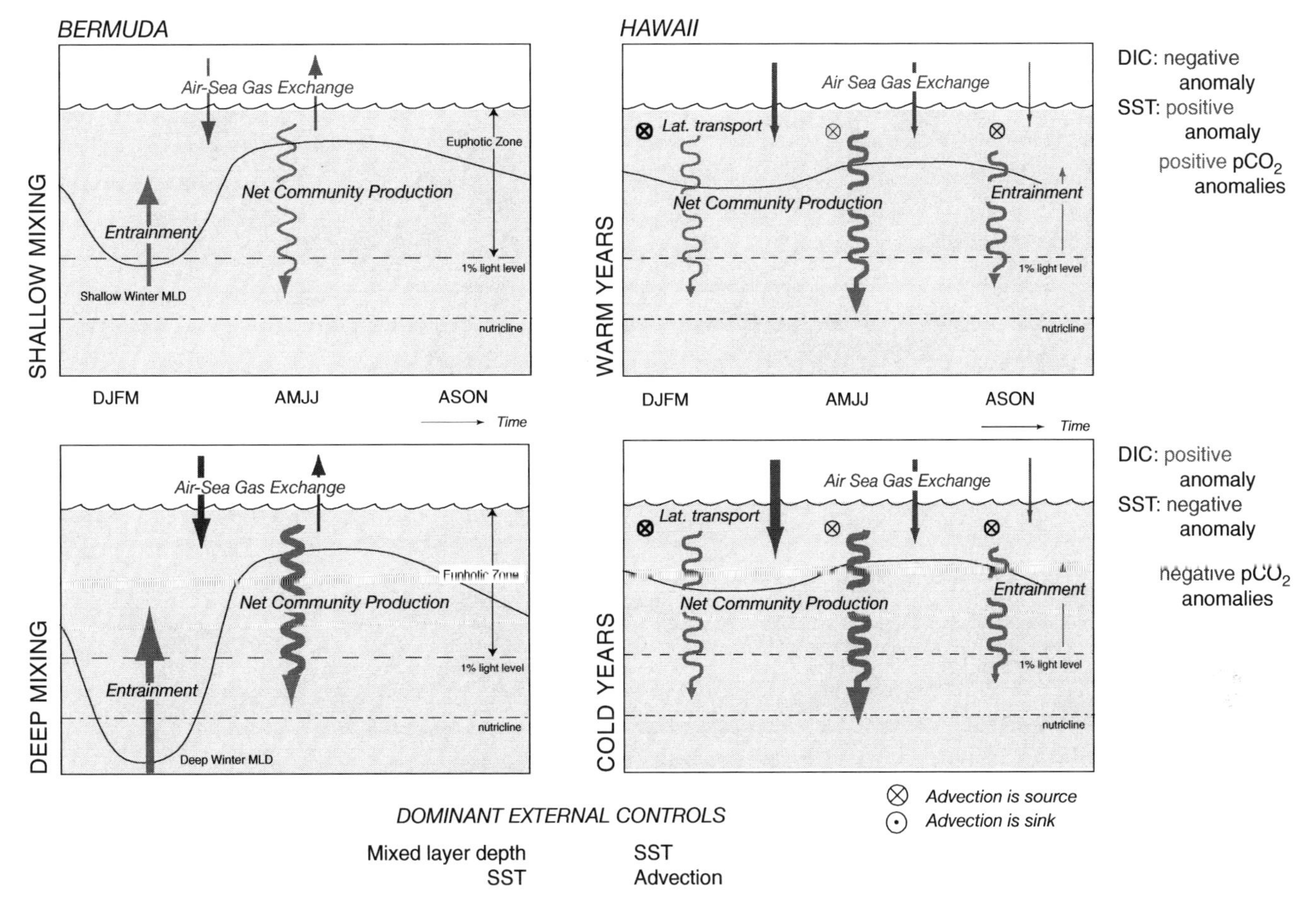

FIGURE 10.3.12: Schematic illustration of the seasonal carbon cycle and its interannual modulations at Station "S"/BATS near Bermuda and at Station ALOHA near Hawaii. The main physical forcing mechanism near Bermuda is winter mixed layer depth variations, while near Hawaii it is SST variations. Modified from *Gruber et al.* [2002] and *Brix et al.* [2004].

10.4 Glacial-Interglacial Atmospheric CO_2 Changes

The largest variations in atmospheric CO_2 in the past million years occurred in connection with the glacial-interglacial cycles that characterized Earth's climate over this period (figure 10.1.2). During full glacial conditions, such as prevailed on the Earth about 20,000 years ago, atmospheric CO_2 was about 80 ppm lower than the pre-anthropogenic interglacial value of about 280 ppm. Backward extensions of atmospheric CO_2 records from ice cores showed that the 80 ppm rise that occurred at the close of the last glaciation during a period of rapid warming also characterized the close of the preceeding three glaciations [*Fischer et al.*, 1999; *Petit et al.*, 1999]. The connection between variations in atmospheric CO_2 and temperature becomes even more evident in figure 10.4.1, where the atmospheric CO_2 record over the last 420,000 years from an Antarctic ice core drilled at Vostok is plotted together with a temperature reconstruction on the basis of isotopic variations contained in the ice. Also shown is a proxy record of ice volume, which demonstrates that not only did temperatures in Antarctica get colder during the ice ages, but also likely the whole Earth system, leading to the formation of massive ice sheets on land. Model simulations suggest that the 80 ppm change in atmospheric CO_2 accounts for more than half of the necessary change in radiative forcing in order to drive Earth's climate from interglacial to glacial conditions [*Weaver et al.*, 1998; *Webb et al.*, 1997]. Less clear is whether CO_2 is a primary driver or a secondary amplifier of the glacial cycles.

Despite the clear importance of atmospheric CO_2 as an amplifier or even as a driver of glacial cycles, the mechanisms responsible for the glacial-interglacial CO_2 changes have remained unresolved ever since these changes were first observed about 25 years ago [*Berner et al.*, 1979; *Delmas et al.*, 1980; *Neftel et al.*, 1982]. Our inability to clearly decipher the mechanisms that led to these large changes in atmospheric CO_2 in the past clearly indicates that there are important aspects of the carbon cycle we do not understand, and that may hinder our ability to predict how the global carbon cycle

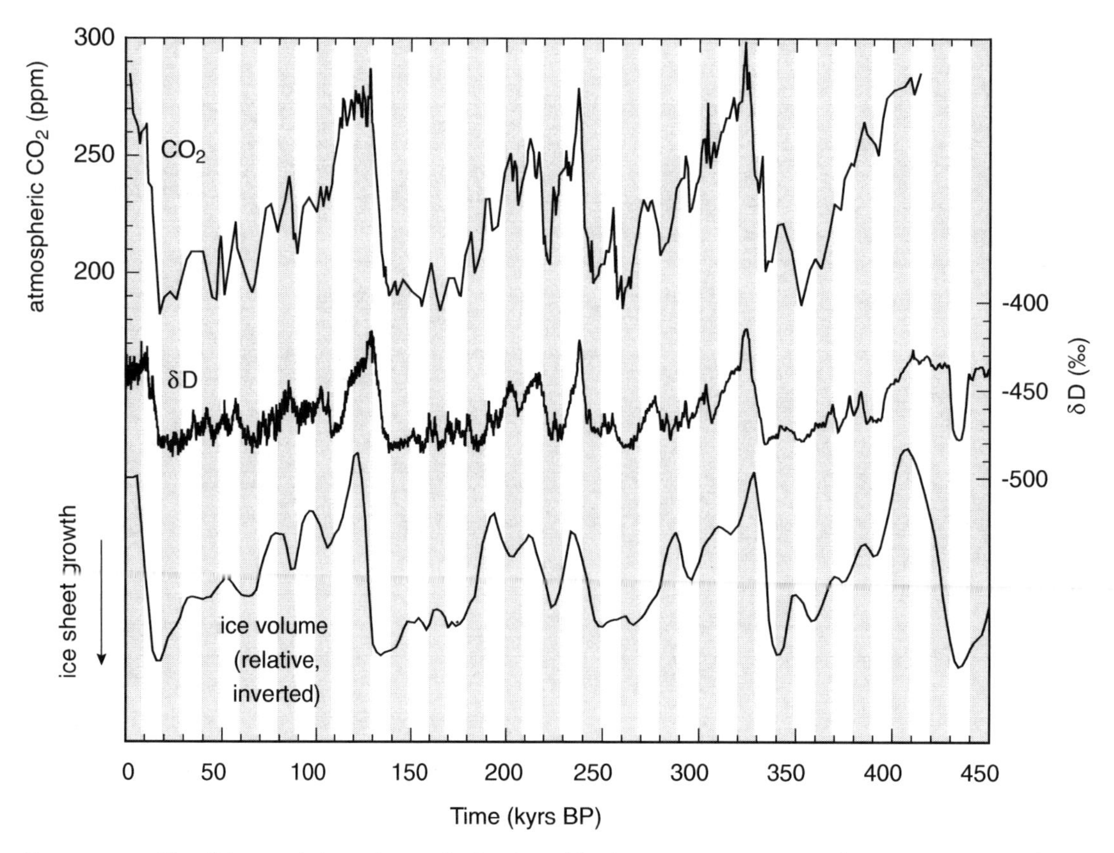

FIGURE 10.4.1: Plot of the covariations of atmospheric CO_2 and local air temperature at Vostok, Antarctica over the last 450 kyr. Local air temperature was estimated on the basis of the deuterium excess. Also shown is an estimate of ice volume, derived from benthic foraminiferal oxygen isotope data from deep-sea sediment cores. During peak glacial periods, atmospheric CO_2 is 80 to 100 ppm. lower than during peak interglacial periods, with upper and lower limits that are reproduced in each of the 100 kyr cycles. Adapted from *Sigman and Boyle* [2000].

may respond to future climate change. One of the few things that are well established is that the cause for these changes must lie in the ocean [*Broecker*, 1982]. The $CaCO_3$ rocks on land, which contain by far the largest fraction of inorganic carbon, are interacting too slowly with the rest of the global carbon cycle to account for changes over glacial cycles [*Berner et al.*, 1983]. The terrestrial biosphere could absorb additional CO_2 from the atmosphere on these timescales, but any such change would be quickly dampened by the interaction with the ocean's huge carbon reservoir (see figure 10.1.1). Furthermore, carbon-13 isotope data from the deep ocean indicate that the terrestrial biosphere actually released CO_2 during glacial times [*Shackelton*, 1977], making the need for an oceanic sink even larger.

Therefore the search for an explanation of the glacial-interglacial atmospheric CO_2 variations must focus on the factors controlling the pCO_2 in the ocean surface water. Since surface ocean pCO_2 is controlled by temperature, salinity, *DIC*, and *Alk*, changes in any of these properties alone or in conjunction could do the job. Over the last two decades many hypotheses have been proposed to explain the glacial-interglacial CO_2 variations in the atmosphere (see summary in table 10.4.1). But so far, each of them in turn has failed to conform to the increasing number of constraints placed on the glacial carbon cycle from paleoceanographic and paleoclimate records.

We begin with a discussion of the processes whose impact is relatively well known, such as temperature and salinity changes, but also the loss of carbon from the terrestrial biosphere, which needs to be compensated by the ocean. We will see that these three processes lead to nearly no change in atmospheric CO_2, because the cooling effect is nearly offset by the atmospheric CO_2 increase stemming from the loss of carbon from the terrestrial biosphere and the increase in oceanic pCO_2 resulting from the increase in salinity. This means that changes in the surface ocean *DIC* and *Alk* must be the primary means of reducing atmospheric CO_2. These concentrations are primarily controlled by the soft-tissue and carbonate pumps, interacting with ocean circulation. We will guide our search for the causes of these changes in these two pumps by first reviewing the evidence that this search should focus on the high latitudes, particularly the Southern Ocean. This Southern Ocean focus is a result of the fact that this is the primary region that connects the surface

TABLE 10.4.1
Summary of scenarios that attempt to explain the lower CO_2 content of the atmosphere during glacial periods

Hypothesis	*Problems*	*References (selected)*
	Hypotheses calling for a closure of the Southern Ocean window	
Sea-ice coverage	extent	[*Stephens and Keeling*, 2000]
Deep-ocean stratification	driver for stratification	[*Toggweiler*, 1999]
NADW-driven stratification and sea-ice coverage		[*Gildor and Tziperman*, 2001a]
Polar dominance	S. Ocean productivity	[*Sarmiento and Toggweiler*, 1984; *Siegenthaler and Wenk*, 1984]
Iron fertilization	S. Ocean productivity	[*Martin*, 1990];
S. Ocean synthesis scenario		[*Sigman and Boyle*, 2000]
	Hypotheses outside the Southern Ocean driven by physical changes	
Cooler North Atlantic	timing	[*Keir*, 1993]
Cooler subtropics	magnitude	[*Bacastow*, 1996]
Vertical redistribution	lysocline, $\delta^{13}C$	[*Boyle*, 1988]
Ocean circulation changes	magnitude	[*Archer et al.*, 2000b)
	Hypotheses outside the Southern Ocean driven by changes in the soft-tissue pump	
Shelf hypothesis	sea level	[*Broecker*, 1982]
Iron deposition and N_2 fixation		[*Falkowski*, 1997; *Broecker and Henderson*, 1998]
Denitrification		[*McElroy*, 1983; *Altabet et al.*, 1995; *Ganeshram et al.*, 1995]
	Hypotheses outside the Southern Ocean driven by changes in the carbonate pump	
Shallow $CaCO_3$	sea level	[*Berger*, 1982; *Opdyke and Walker*, 1992]
Rain ratio	lysocline	[*Archer and Maier-Reimer*, 1994]
Polar alkalinity		[*Broecker and Peng*, 1989]
Iron deposition and diatoms	S. Ocean productivity	[*Brzezinski et al.*, 2002; *Matsumoto et al.*, 2002]

ocean with the deep ocean reservoir that contains most of the carbon in the combined atmosphere-ocean system. This focus is also justified by the observation that changes in the low latitudes appear to have substantially less influence on atmospheric CO_2. If this is correct, it would greatly narrow the number of candidate processes that need to be investigated, but we will see that the high-latitude dominance is likely somewhat less strong than previously thought. Nevertheless, Southern Ocean–based hypotheses are still the primary candidates for explaining the lower atmospheric CO_2 concentrations, and we will therefore discuss these hypotheses in detail, including their evaluation with paleoceanographic constraints. Low-latitude–based hypotheses have mostly been rejected, but scenarios that invoke a change in the export of $CaCO_3$ relative to that of organic matter on a global scale can lead to large changes in atmospheric CO_2 through a process called *$CaCO_3$ compensation* (see chapter 9) and therefore need to be considered as well. We end with a possible synthesis scenario that combines several elements and appears to be consistent with most constraints. The reader interested in more detail is referred to the excellent review by *Sigman and Boyle* [2000], and the papers by *Broecker and Peng* [1998] and *Archer et al.*, [2000b]. *Broecker* [1995] provides a more general overview of the climate changes associated with glacial-interglacial cycles.

SETTING THE SCENE

TERRESTRIAL BIOSPHERE CARBON LOSS

During the last glacial maximum (LGM), about 20,000 years ago, ice sheets extended in the northern hemisphere as far south as 40°N, covering most of Alaska and Canada on the North American continent, and covering all of Scandinavia and northern Russia and Siberia on the Eurasian continent [*Peltier*, 1994]. This resulted in a substantial loss of land available for vegetation to grow, which was only partially compensated by the increase in land due to the lower sea-level stand. In addition, cooler temperatures prevailed over much of Earth, with most estimates indicating a cooling of several °C, even in the tropics [e.g., *Stute et al.* 1995], associated with a significant slowdown of the hydrological cycle [e.g., *Yung et al.* 1996]. This led to cooler and drier conditions over much of the land surface, except in a few areas such as the southwestern United States, where the presence of large inland lakes suggests wetter conditions during the LGM. These changes over land impacted global biogeochemical cycles in two major ways. First, the substantial reduction of areas suitable for ecosystems that store substantial amounts of carbon led to the loss of large amounts of carbon from the terrestrial biosphere. Second, the generally drier conditions and the existence of extensive desert-like areas caused an increase in the mobilization of dust, which presumably increased the transport and deposition of it over the ocean [*Mahowald et al.*, 1999].

Shackelton [1977] estimated the magnitude of the carbon release by the terrestrial biosphere by looking at oceanic changes in the $^{13}C/^{12}C$ ratio of *DIC*. Since terrestrial CO_2 has a much lower $^{13}C/^{12}C$ ratio than ocean *DIC*, the oceanic uptake of terrestrially derived CO_2 would lead to a whole ocean change in the $^{13}C/^{12}C$ ratio of *DIC*, whose magnitude is proportional to the $^{13}C/^{12}C$ ratio of terrestrial biosphere and the magnitude of the release. Using calcite shells of foraminifera that dwelled either in near-surface waters or in the surface sediments, he reconstructed the oceanic $^{13}C/^{12}C$ ratios of *DIC* in both near-surface and near-bottom waters, thereby eliminating changes that are due to ocean internal redistributions of carbon. He arrived at a terrestrial biosphere loss of about 500 Pg C. More recent reconstructions put the loss in the range of 300 to 700 Pg C [*Curry et al.*, 1988; *Bird et al.*, 1994, 1996; *Beerling*, 1999]. This represents up to a 30% lower carbon storage in the terrestrial biosphere relative to the amount of carbon stored in the terrestrial biosphere today, which is about 2300 Pg C (see figure 10.1.1).

An even smaller terrestrial biosphere is estimated for the LGM by biome reconstructions on the basis of terrestrial pollen data (e.g., *Prentice et al.* [2000]). These data suggest a terrestrial biospheric carbon loss of 700–1400 Pg C [*Adams and Faure*, 1998], but these estimates need to be viewed with greater caution than those based on the ^{13}C mass balance, due to large gaps in the data considered, and assumptions about the average carbon density of different biomes.

In order to evaluate the impact of this reduction in the carbon storage of the terrestrial biosphere, we need to consider how much of this biospheric loss actually remained in the atmosphere. This problem is directly analogous to the problem of the ocean's uptake of the anthropogenic CO_2 previously discussed, except that for the glacial-interglacial changes considered here, the relevant timescales are several thousand years. On such timescales, we can assume that the equilibration of the atmospheric perturbation with the oceanic reservoir of *DIC* is complete, and that the buffering by the dissolution of oceanic $CaCO_3$ is nearly complete as well. However, buffering with $CaCO_3$ and siliceous rocks on land is too slow to represent important sinks for atmospheric CO_2 on these timescales. We infer from figure 10.2.2 that the fraction remaining in the atmosphere after equilibration with the ocean and sea-floor $CaCO_3$ is about 8% for a release on the order of 1000 Pg C or less. If we adopt a terrestrial release estimate of 500 Pg C, about 40 Pg C will remain in the atmosphere even after reaction with sea-floor $CaCO_3$, leading to an atmospheric CO_2 increase of about 18 ppm. Therefore, rather than explaining a 80 ppm glacial-interglacial difference by an oceanic mechanism, we now need to explain a nearly 100 ppm difference!

It is instructive to consider exactly how the oceanic carbon cycle buffered this terrestrial release. As is the case for anthropogenic CO_2, the primary buffering by the ocean occurs by the chemical reaction of the excess CO_2 with the CO_3^{2-} ion, leading to a substantial reduction in the oceanic mean CO_3^{2-} concentration. As a result, the mean depth of the $CaCO_3$ saturation horizon shifted upward, exposing a larger fraction of the sea-floor sediments to waters undersaturated with regard to $CaCO_3$ (see figure 10.4.2). Since the $CaCO_3$ saturation horizon is the primary factor controlling $CaCO_3$ dissolution in marine sediments, this upward movement led to increasing $CaCO_3$ dissolution. This process continued until the oceanic mean *Alk* concentration increased to the level that the oceanic mean CO_3^{2-} ion concentration was the same as before the perturbation. At this point, the $CaCO_3$ saturation horizon returned to its original position, ensuring again a balance between the *Alk* input by rivers and *Alk* loss by burial (figure 10.4.2c). The oceanic $CaCO_3$ system responds therefore as a "homeostat," buffering the atmospheric CO_2 perturbation. Its response requires a temporary upward excursion of the $CaCO_3$ saturation horizon and presumably the ocean's lysocline. The opposite occurs during the buildup of the terrestrial biosphere carbon reservoir [*Broecker et al.*, 2001], leading to a preservation event in the ocean's sediments. The slow timescale associated with this "reverse" $CaCO_3$ compensation has been proposed as a mechanism to explain postglacial variations in atmospheric CO_2 during the present warm period, i.e., the Holocene [*Broecker et al.*, 2001], although this effect seems to be too small [*Joos et al.*, 2004].

Salinity Changes

During the last glacial maximum, the ocean surface was both cooler and saltier than today's ocean surface. The salinification of the ocean was a consequence of the buildup of huge continental ice sheets that locked large

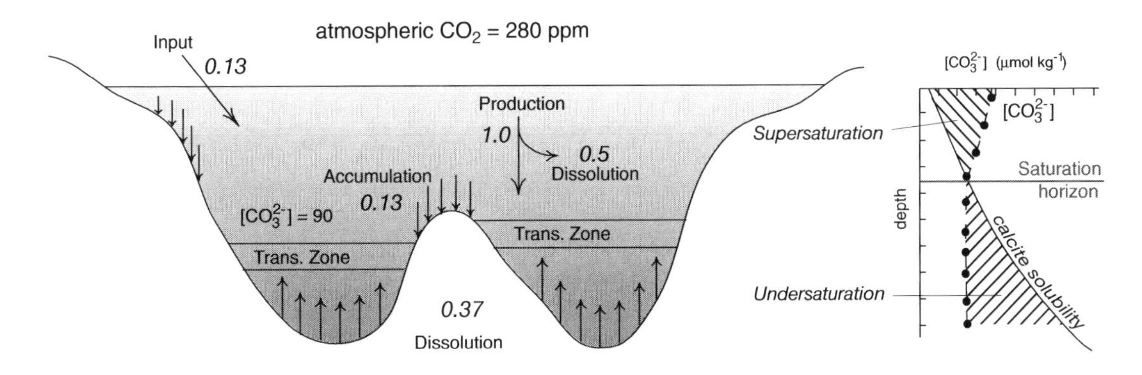

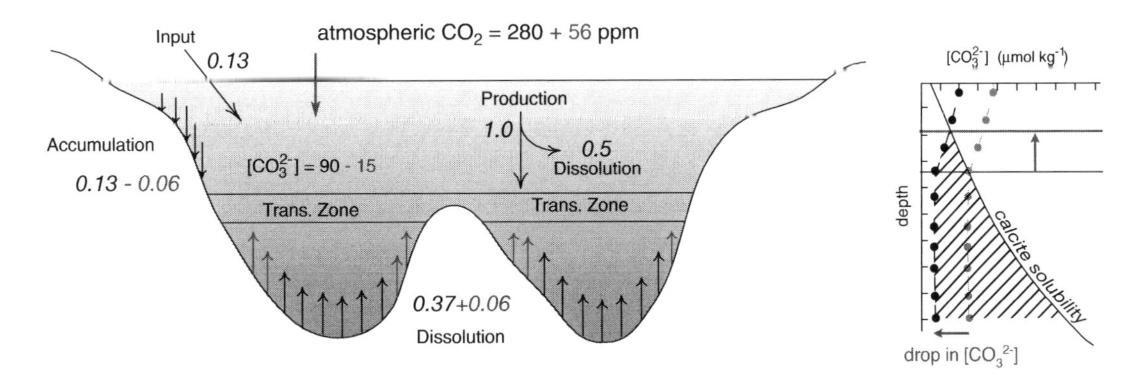

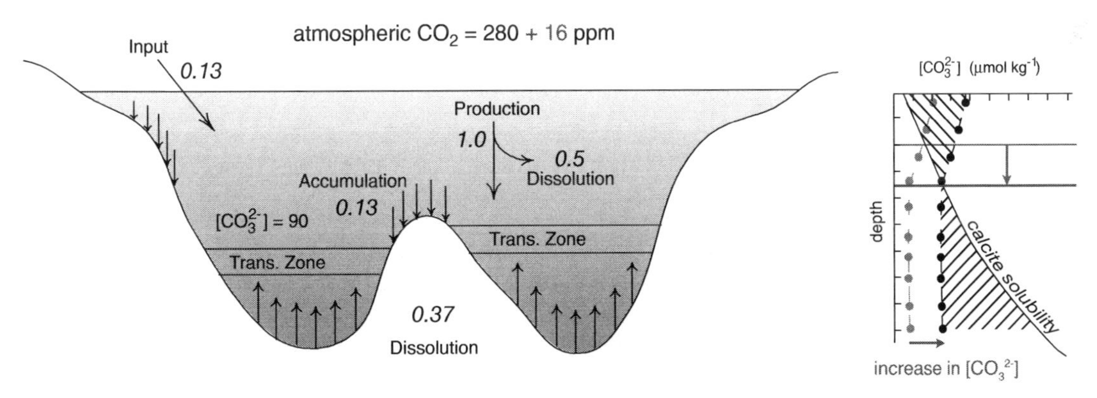

FIGURE 10.4.2: Mechanism of ocean $CaCO_3$ compensation of the loss of carbon from the terrestrial biosphere. (a) Initial steady state for interglacial conditions. (see figure 9.5.1) (b) Perturbed state after the addition of 500 Pg C to the atmosphere and equilibration of this perturbation with the ocean (but not with the sediments). (c) New steady state after $CaCO_3$ compensation. The addition of CO_2 to the ocean leads to a consumption of the carbonate ion concentration, causing a shoaling of the saturation horizon (see (b)). As a result, a positive imbalance is generated between the riverine input of *Alk* and the net removal of *Alk* by burial in the sediments. This causes the carbonate ion concentration to increase again (see (c)), and consequently the saturation horizon deepens to achieve a new steady state. The resulting net increase in oceanic alkalinity then permits the ocean to absorb additional CO_2 from the atmosphere. Adapted from *Broecker and Peng* [1998].

amounts of freshwater on land away from the ocean. The cooling and salinification of the surface ocean have opposing effects on its pCO_2, and hence on atmospheric CO_2. We know from (8.3.4) that surface ocean pCO_2 drops by about 10–12 μatm per °C of cooling. However, we cannot simply take the salinity sensitivity of pCO_2 as given in (8.3.5), because this sensitivity only takes the impact of salinity on the dissociation into account (the so-called salting-out effect), but disregards the fact that the total concentrations of *DIC* and *Alk* in the ocean were higher as a result of removing freshwater with zero *Alk* and near-zero *DIC* from the ocean. We can estimate the total change in oceanic pCO_2 that results from salinification by considering the total derivative of pCO_2 with regard to changes in salinity, *DIC*, and *Alk* while keeping temperature constant:

$$\Delta pCO_2 = \Delta S \frac{\partial pCO_2}{\partial S}\bigg|_{DIC,Alk=const} + \Delta DIC \frac{\partial pCO_2}{\partial DIC}\bigg|_{Alk,S=const} + \Delta Alk \frac{\partial pCO_2}{\partial Alk}\bigg|_{DIC,S=const} \quad (10.4.1)$$

where ΔDIC and ΔAlk are the salinity-induced changes in *DIC* and *Alk*. A simple mass balance shows that these latter two changes are proportional to the salinity changes, ΔS, with the proportionality given by the respective mean concentrations, i.e., $\Delta DIC = (DIC/S)\Delta S$, and $\Delta Alk = (Alk/S)\Delta S$. Inserting these proportionalities into (10.4.1) and using the definitions of the buffer factor, γ_{DIC} (8.3.14), the alkalinity factor γ_{Alk} (8.3.15) and the salinity dependence, $\gamma_S = (S/pCO_2)(\delta pCO_2/\delta S)$ (8.3.5), we obtain

$$\begin{aligned} \Delta pCO_2 &= \Delta S \frac{pCO_2}{S}(\gamma_S + \gamma_{DIC} + \gamma_{Alk}) \\ &\approx \Delta S \frac{350\,\mu\text{atm}}{35}(1 + 10 - 9.4) \\ &\approx 16\,\mu\text{atm} \cdot \Delta S \end{aligned} \quad (10.4.2)$$

where we used global mean values for the different γ's (see chapter 8). Therefore, the salinity-induced changes in *DIC* and *Alk* increase the salinity sensitivity to approximately 16 μatm change in pCO_2 per unit change of salinity, instead of just a 10 μatm change if *DIC* and *Alk* were held constant.

We next need to know the magnitude of the salinity change. During peak glacial times, sea level is believed to have been about 120 to 140 m lower than it is today [*Bard et al.*, 1996, *Adkins et al.*, 2002]. This is roughly equivalent to the removal of about 3% of today's volume of the ocean. Since the amount of salt remained the same, mean ocean salinity must have changed by about $0.03 \cdot 35 = 1.1$ salinity units. As this change occured uniformly over the entire ocean, we can directly estimate from (10.4.2) that this change led to an atmospheric CO_2 increase of about 18 ppm.

Temperature Changes

Sea-surface temperatures during the LGM are thought to be on average about 2 to 4°C colder than today. The exact magnitude and distribution of these changes is still intensively debated [*Broecker*, 1996]. The original findings of the CLIMAP program [*CLIMAP Project Members*, 1981] on the basis of a detailed analysis of foraminiferal assemblages suggested 2–4°C colder temperatures in the mid- to high latitudes of the southern hemisphere and in the North Pacific, 4–8°C cooling in the North Atlantic, and virtually no change in the tropics and temperate regions. Reconstructions based on oxygen isotopes preserved in pelagic foraminifera [*Broecker*, 1986; *Birchfield*, 1987] generally support such little cooling in the tropics. However, these findings, particularly the small cooling in the tropics, have come under serious challenge over the last few years. The first indication of substantially colder temperatures in the tropics came from measurements of the strontium-to-calcium ratios in marine corals [*Guilderson et al.*, 1994; *Beck et al.*, 1997] that suggest tropical cooling of up to 6°C. Additionally, *Schrag et al.* [1996] showed that the temperature change based on the oxygen isotopes might have been substantially underestimated by assuming a too small change in oxygen-18 in seawater from the growth of ice sheets on land.

Cooler tropical SSTs are also in good agreement with terrestrial temperature proxies, including snow-line elevations, noble gases in groundwater, and pollen records [e.g., *Stute et al.* 1995]. On the other hand, temperature proxy records on the basis of marine alkenones, chain-like organic molecules manufactured by marine plants [*Bard et al.*, 1997], show that the tropical cooling was likely not more than 2°C. It is presently unclear whether the disagreement between the different proxies reflects regional differences in the extent of cooling, or whether neglected or underestimated effects are affecting one proxy group or the other [*Broecker*, 1996]. We need to worry about this unresolved puzzle in our quest to understand the causes of the glacial-interglacial CO_2 variations. Temperature would appear to explain a significant fraction of the atmospheric CO_2 change if the ocean indeed cooled by 6°C over a large fraction of its surface, including the high latitudes. This is because under isochemical conditions, a 6°C change causes a pCO_2 change of about 60–70 μatm.

However, two considerations greatly diminish the importance of temperature as a possible driver for glacial-interglacial changes. The first one is that temperature changes in the high latitudes were quite certainly smaller than 6°C. Since most of these waters are already close to the freezing point, it is more likely that temperature changes in the high-latitude regions were only around 1 to 2°C. The second consideration is that changes in the low latitudes tend to have a smaller effect on atmospheric CO_2 than those in the high latitudes, despite the fact that the low latitudes cover approximately 80% of the total surface ocean area. The primary reason for the greater importance of the high latitudes is the fact that they are the primary conduit for the exchange of water and carbon between the surface ocean and the large reservoir of the deep ocean. We will discuss this issue in detail below. Current estimates of the impact of cooling suggest an atmospheric CO_2 decrease on the order of 30 ppm [*Sigman and Boyle*, 2000].

In summary, it appears that the three relatively well constrained processes, i.e., loss of carbon from the terrestrial biosphere, increase in ocean salinity, and ocean cooling, are unable to explain the 80 ppm drawdown in atmospheric CO_2 that occurred during glacial maxima. More specifically, if we take the 18 ppm increase from terrestrial biosphere carbon release, the 18 ppm increase from salinity change, and the 30 ppm decrease from cooling together, we arrive at a near-zero change in atmospheric CO_2. Clearly other processes must be at work.

Fundamental Mechanisms

Since changes in the solubility of CO_2 forced by changes in salinity and temperature can be ruled out as a cause for the glacial drawdown of atmospheric CO_2, the drawdown must be caused by processes that affect the oceanic distribution of *DIC* and *Alk*. A useful framework to start the discussion is equation (8.3.3), which links changes in *DIC* and *Alk* to pCO_2 changes:

$$pCO_2 \approx \frac{K_2}{K_0 \cdot K_1} \frac{(2DIC - Alk)^2}{Alk - DIC} \qquad (10.4.3)$$

This approximation shows that in order to decrease surface ocean pCO_2 and hence atmospheric pCO_2, the processes must cause either a decrease of surface ocean *DIC*, an increase in surface ocean *Alk*, or a combination of these two effects.

Fundamentally, changes in any of the three carbon pumps discussed in chapter 8, i.e., the soft-tissue pump, carbonate pump, or gas exchange pump, can contribute to the decrease in surface *DIC*. In thinking about the action of these pumps, it is also important to recall that the surface-to-deep gradient, $DIC_s - DIC_d$, generated by the biological pumps in a simplified one-dimensional framework depends not only on the magnitude of the downward transport of organic matter $\Phi^{C_{org}}$ and $CaCO_3$, Φ^{CaCO_3}, but also on the magnitude of vertical transport and mixing, v (see equation (1.2.3)), thus

$$DIC_s = DIC_d - \frac{\Phi^{C_{org}} + \Phi^{CaCO_3}}{v} \qquad (10.4.4)$$

Therefore, the surface ocean concentration, DIC_s, can be reduced not only by an increase in the export of carbon, but also by a reduction in vertical transport and mixing, v. Undeniably, both cases need to be considered. A reduction in DIC_d, reflecting essentially a reduction in whole-ocean *DIC*, could also be invoked, but we are unaware of a mechanism that could have caused such a decrease. In fact, the whole-ocean *DIC* must have been higher during the LGM than it is today, because of the decrease in ocean volume, because of the uptake of the carbon lost from the terrestrial biosphere, and because of the additional uptake of CO_2 from the atmosphere required to lower atmospheric CO_2.

Analogous to *DIC*, the surface ocean *Alk* concentration, Alk_s, is governed by the deep concentration, Alk_d, as well as the balance between the export of $CaCO_3$ and organic carbon, and the resupply of *Alk* by mixing and transport:

$$Alk_s = Alk_d - \frac{2\Phi^{CaCO_3} - r_{N:C}\Phi^{C_{org}}}{v} \qquad (10.4.5)$$

where $r_{N:C}$ is the nitrogen-to-carbon ratio of the organic matter exported from the surface ocean. We can therefore increase the surface concentration of *Alk* by decreasing the strength of the carbonate pump, by increasing the strength of the soft-tissue pump, or by increasing the deep concentration Alk_d. The latter change can be accomplished by dissolving $CaCO_3$ sediments, or increasing the *Alk* input into the ocean by rivers.

We thus have to consider changes in ocean circulation, in surface ocean production of organic matter and $CaCO_3$ and their subsequent export and remineralization, in river input, and in the interaction with the $CaCO_3$ in the sediments, as possible candidate processes driving the glacial-interglacial changes in atmospheric CO_2. We also have to consider changes in air-sea gas exchange, as changes in the kinetics of the exchange of CO_2 across the air-sea interface can lead to changes in surface *DIC* as well (see discussion about the gas exchange pump in chapter 8). Given these myriad possible processes, our search for a cause of the glacial CO_2 drawdown would be greatly aided if we had some guidance as to which region and which processes we need to focus on.

Southern Ocean Dominance

It was recognized early on that not all regions of the world's ocean exert an equal influence on atmospheric CO_2. A series of seminal papers in the early 1980s pointed out that the high-latitude oceans, despite their limited area, dictate the atmospheric CO_2 content to a large extent [*Knox and McElroy*, 1984; *Sarmiento and Toggweiler*, 1984; *Siegenthaler and Wenk*, 1984]. Since these independent studies were done at Harvard, Princeton, and Bern (Switzerland) respectively, *Broecker and Peng* [1998] coined the expression Harvardton-Bears to refer to them as a group. This concept of high-latitude dominance has been extremely influential in shaping the discussion about the causes for the glacial-interglacial changes in atmospheric CO_2.

The dominance of the high-latitude regions is a consequence of the direct connection that exists between the high-latitude ocean and the deep ocean through deep water formation and isopycnal ventilation, pinning the chemistry of the high-latitude surface ocean very close to that of the deep ocean. By contrast, the link between the cold and dense waters of the deep ocean (cold water sphere) and the warm and light waters of the low latitudes (warm water sphere) is much weaker, a direct consequence of ocean interior flow occuring along isopycnal surfaces. An additional reason for the high-latitude dominance stems from the fact that the transport of CO_2 via the atmosphere between the low-latitude and high-latitude regions is sufficiently rapid to bring the pCO_2 of the low-latitude surface ocean close to that of the high latitudes. As a result, any anomalous CO_2 created in the low latitudes will escape into the atmosphere, and tend to be taken up by the high-latitude oceans and ultimately buffered in the deep ocean, leading to only a small net change in atmospheric pCO_2. In order to illustrate this constraint, we use the three-box model introduced in chapter 1 (figure 10.4.3) and perturb either the high latitudes or low latitudes with a warming of 2°C. Figure 10.4.4 shows that in the new steady state, the 2°C warming in the high latitudes causes an atmospheric CO_2 change of 15 ppm, whereas the same warming in the low latitudes causes an atmospheric CO_2 change that is 5 times smaller, i.e., only 3 ppm. In the latter case, most of the CO_2 that is expelled from the low latitudes is taken up by the high latitudes and buffered by the deep ocean reservoir. We term this the *equilibration* effect.

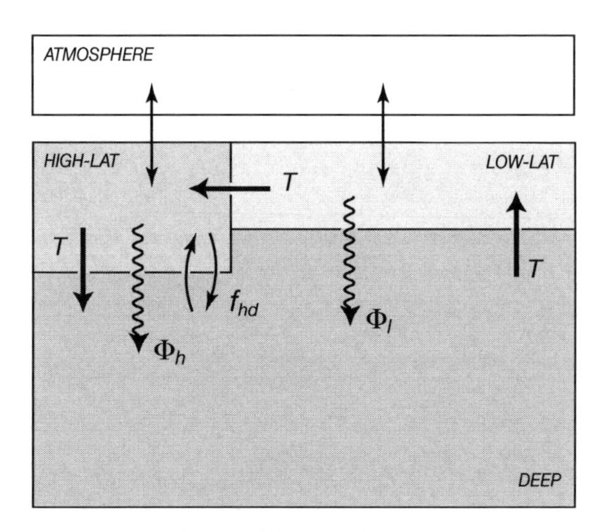

FIGURE 10.4.3: Schematics of the high-latitude outcrop three-box model [c.f. *Sarmiento and Toggweiler*, 1984; *Knox and McElroy*, 1984; and *Siegenthaler and Wenk*, 1984] (c.f. figure 1.2.5).

Equilibration of Low-Latitude Changes

What determines the magnitude of this low-latitude equilibration effect, i.e., the substantial reduction in the impact of low-latitude changes on atmospheric CO_2? It turns out that the degree of equilibration depends on the relative magnitude of air-sea gas exchange versus the magnitude of the exchange between the surface and deep ocean. If air-sea gas exchange is efficient in equilibrating the atmosphere and surface ocean, atmospheric pCO_2 will be almost entirely determined by the high latitudes and will not respond to low-latitude changes. Conversely, if air-sea gas exchange is slow relative to the exchange between the cold and warm water spheres, changes in low-latitude chemistry and temperature can lead to substantial changes in atmospheric CO_2.

Given the importance of this equilibration constraint, it behooves us to investigate it further. We return to the three-box model (figure 10.4.3), since it captures the relevant processes, while being simple enough to understand straightforwardly. The defining characteristics of this model are: (i) separation of ocean surface into a low-latitude box and a high-latitude box, (ii) a deep box, whose volume comprises nearly 97% of the total volume of the ocean, rendering its chemistry nearly invulnerable to internal redistributions of carbon, but not to external perturbations, such as associated with the dissolution of $CaCO_3$, (iii) an exchange term, f_{hd} that couples the high-latitude box very closely to the deep ocean, and (iv) an overturning term, T, which represents sinking in high latitudes and upwelling in low latitudes. While various shortcomings of this model have been noted, such as the fact that there is in reality little upwelling of deep waters directly in the warm low latitudes (see discussion in chapter 7 and *Toggweiler* [1999]), this model is nevertheless very instructive. Temperatures of this model are set to 21.5°C for the low-latitude box and 2°C for the high-latitude box. By design, the deep box will have the same temperature as the high-latitude box, i.e., 2°C.

We investigate the equilibration effect in this model by initializing it with homogeneous *DIC* and *Alk* and then determining how the carbon is redistributed within the system depending on the relative rates of air-sea exchange and ocean mixing (see figure 10.4.5). As the equilibration effect is primarily determined by physical processes, it is sufficient to neglect biology and consider only an abiotic ocean. Alkalinity is set to 2340 μmol kg^{-1} in all boxes and *DIC* is initialized to 2150 μmol kg^{-1}, yielding an initial pCO_2 of 280 μatm at the surface of the high-latitude box and an initial pCO_2 of 630 μatm at the surface of the low-latitude box. The model is then run forward in time until it reaches a steady state.

We first investigate a case where the transport of *DIC* between the surface and deep ocean is far more rapid than the transport of CO_2 from the warm surface reservoir via the atmosphere to the cold surface reservoir (figure 10.4.5c). This is accomplished by reducing the gas exchange coefficient to a very small number. No equilibration will occur in this case, as the *DIC* concentration in the low-latitude surface box will remain relatively close to the mean concentration of the deep box. As some carbon will be lost to the atmosphere, the

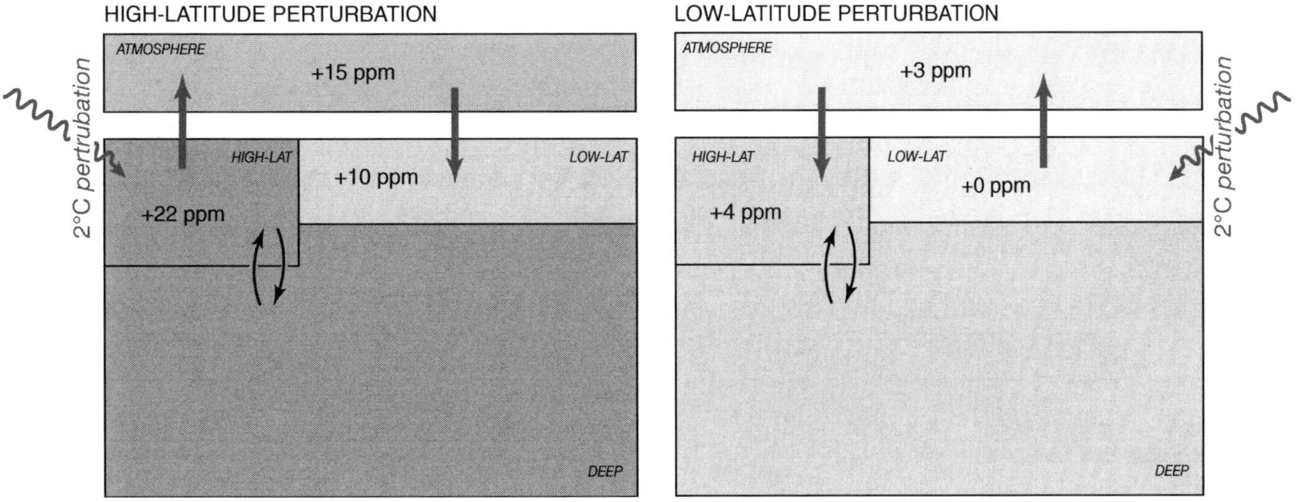

FIGURE 10.4.4: Comparison of the impact of a 2°C warming in the high versus the low latitudes on atmospheric CO_2. The changes were computed using the standard configuration of the three-box model shown in figure 10.4.3

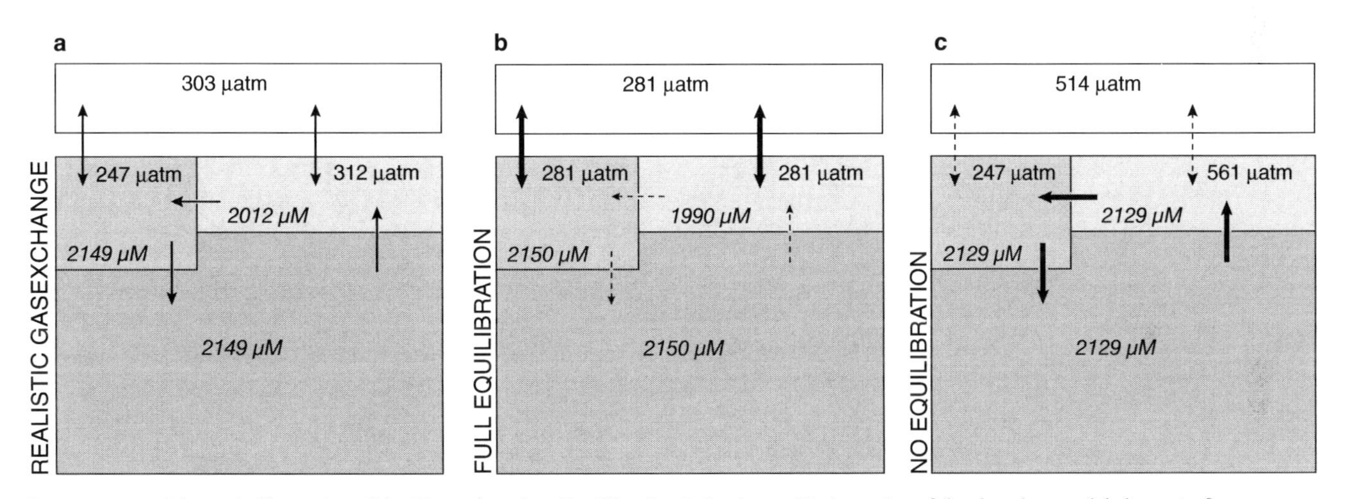

FIGURE 10.4.5: Schematic illustration of the Harvardton-Bear Equilibration Index in an abiotic version of the three-box model shown in figure 10.4.3 where differences in CO_2 chemistry are driven entirely by the solubility pump. (a) Case with realistic gas-exchange; (b) Full equilibrium case, where gas exchange is infinitely rapid; and (c) case with strongly inhibited gas exchange. If the air-sea gas exchange is much more rapid than the transport of *DIC* around the thermohaline circuit, pCO_2 in the atmosphere will be the same as the pCO_2 of the high- and low-latitude boxes (case b). At the other extreme, if the transport of CO_2 through the atmosphere from the low latitudes to the high latitudes is much slower than the transport of *DIC* between the reservoirs by the thermohaline circulation, the *DIC* concentrations in all reservoirs will be equal and the pCO_2 variations in the two surface boxes will be determined by temperature only (case c). In this case, atmospheric CO_2 equilibrates at the weighted average of 514 μatm. Hence in this simple model, atmospheric CO_2 depends on the ratio of the rate of air-sea gas exchange to the rate of thermohaline circulation. Using "realistic" parameters, the model gives an atmospheric CO_2 of 303 ppm (case a). The Harvardton-Bear Equilibration Index is a measure of the extent to which the transport of CO_2 through the atmosphere is able to remove a signal in the low latitudes and buffer it in the deep ocean. It can be calculated from the ratio of the difference between the "realistic" and the "full equilibration" cases, divided by the difference between the "no equilibration" and the the "full equilibration" cases, hence $(303-281)/(514-281)=0.09$. Adapted from *Broecker et al.* [1999a].

mean oceanic *DIC* concentration drops somewhat to about 2129 μmol kg^{-1}, yielding a low-latitude pCO_2 of 561 μatm, and a high-latitude pCO_2 of 247 μatm. Given the slow exchange of CO_2 between the ocean and atmosphere, the atmospheric pCO_2 is in this case simply equal to the area weighted mean ($0.20 \cdot 247$ μatm $+ 0.80 \cdot 561$ μatm $= 514$ μatm). The other extreme is a case where the rate of CO_2 transfer between atmosphere and ocean far outstrips the rate of water circulation and mixing (figure 10.4.5b). In this case, full equilibration will occur, and the pCO_2 in all surface boxes will be equal to that of the atmosphere. Integration of the model gives an atmospheric pCO_2 of 281 μatm. If the model parameters are set to "realistic" values of 20 Sv for the transport of water

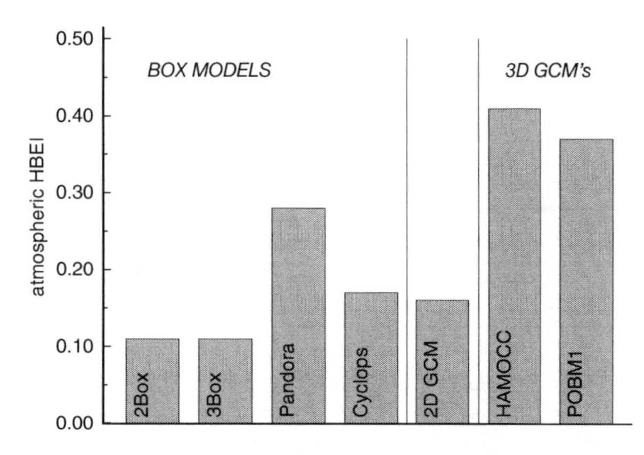

FIGURE 10.4.6: Comparison of estimates of HBEI from a range of models. From *Broecker et al.* [1999a].

around the thermohaline circuit, to 60 Sv for the high latitude-to-deep-ocean mixing term, and to 13 cm h^{-1} for the piston velocity of CO_2, atmospheric pCO_2 stabilizes at 303 μatm (figure 10.4.5a).

Broecker et al. [1999a] attempted to capture the degree of this deep ocean equilibration effect of low-latitude changes by the Harvardton-Bear Equilibration Index (HBEI). They defined this index in such a way that it describes how close the realistic case is to the full equilibration case weighted by the difference between the two extreme cases of full and no equilibration:

$$\mathrm{HBEI} = \frac{pCO_2^{atm}|_{realistic} - pCO_2^{atm}|_{full\ equil}}{pCO_2^{atm}|_{no\ equil} - pCO_2^{atm}|_{full\ equil}} \tag{10.4.6}$$

$$= \frac{303 - 281}{514 - 281} = 0.09 \tag{10.4.7}$$

The value of the HBEI can be interpreted as an approximate measure of how effective a given change in the low latitudes is in changing atmospheric CO_2. An HBEI value of 1 means that no equilibration with the deep ocean occurs and therefore atmospheric CO_2 will respond proportionally to any change in low-latitude pCO_2. A value of 0 means that full equilibration occurs, i.e., changes in low-latitude pCO_2 would have no impact on atmospheric CO_2, since atmospheric CO_2 is controlled by the high latitudes in connection with the deep ocean reservoir. Note that the HBEI is only an approximation of the degree of equilibration and atmospheric effectiveness, as the exact values depends on the magnitude of the perturbation because of the associated changes in the buffer factor.

The low HBEI value of 0.09 found by this simple three-box model implies that for a given change in low-latitude pCO_2, only approximately 9% of this change multiplied with the fractional area of the low latitudes, i.e., $0.85 \cdot 0.9 = 0.08$, would show up as a change in atmospheric CO_2. If this value of HBEI is correct, we can immediately rule out any low-latitude–based hypothesis. For example, the discussion about the exact magnitude of low-latitude cooling would be largely irrelevant, since even a very large cooling of 6°C, leading to a low-latitude pCO_2 reduction of 70 μatm, would cause only a 6 μatm reduction in atmospheric CO_2. However, before we can have any confidence in ruling out low-latitude–based hypotheses, we need to ensure that this extremely low sensitivity of low-latitude changes is real and not just a peculiarity of this simple three-box model.

Broecker et al. [1999a] computed the HBEI of a wide spectrum of models, ranging from simple box models to three-dimensional ocean biogeochemistry models (figure 10.4.6). They found that three-dimensional coupled ocean circulation/biogeochemistry models show a substantially larger influence of the low latitudes (HBEI indices up to 0.4) than the box models. It is presently not clear how large the HBEI of the real ocean is, but it is instructive to consider which processes determine its magnitude [see also *Archer et al.*, 2000a; *Ito and Follows*, 2003; *Follows et al.*, 2002]. Since the HBEI is a measure of the degree of equilibration between CO_2 anomalies in the low latitudes and the deep ocean via exchange with the high latitudes, any process that affects this equilibration changes the HBEI. Possible candidates are variations in the rate of CO_2 exchange across the air-sea interface, the magnitude and location of the mass exchange between the warm and cold water spheres, and the reservoir sizes of these two spheres.

For example, as pointed out by *Toggweiler et al.* [2003], the exchange of CO_2 between the atmosphere and the high-latitude surface ocean is kinetically unconstrained in this three-box model, while this exchange experiences strong kinetic limitation in three-dimensional ocean models. This is primarily a result of the fact that the high latitudes in the three-box model were assigned a fractional area of 15%, while a census of the surface area of the ocean that has surface densities equal to or denser than the deep ocean in a typical three-dimensional ocean model gives fractional areas of only a few percent. Restricting the high-latitude area so that the slowness of air-sea gas exchange becomes limiting for the exchange between the atmosphere and high-latitude box will work against equilibration, and therefore increase the HBEI. Figure 10.4.7b shows that if the high-latitude surface area in the three-box model is set to 3%, similar to what it is in three-dimensional ocean circulation models, the HBEI index goes up to above 0.3. The effective surface area of the high latitudes can also be decreased by increasing the fraction that is covered by sea ice, leading to a similar increase in the HBEI index (figure 10.4.7a).

The magnitude of the exchange between the warm and cold water spheres and their relation to their reservoir sizes represents a second set of factors controlling the HBEI. Any increase in the exchange between

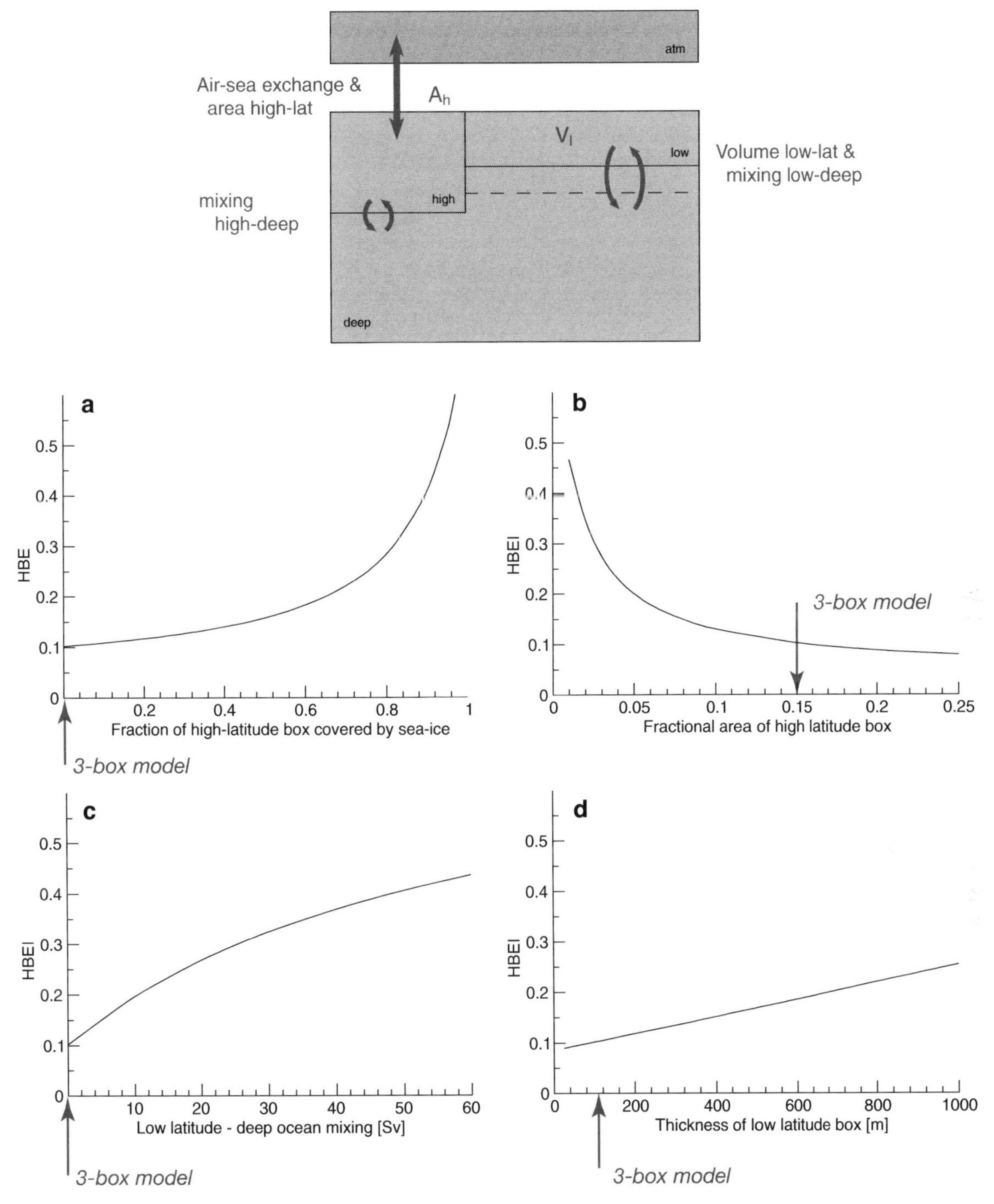

FIGURE 10.4.7: Sensitivity of HBEI in the Harvardton-Bear three-box model to changes in (a) fractional ice cover of high-latitude box, (b) fractional area of high latitudes, (c) low-latitude–deep ocean mixing, and (d) thickness of the low-latitude box. All simulations were done following the protocols of *Broecker et al.* [1999a].

the two reservoirs, e.g., by adding a low-latitude deep ocean exchange (figure 10.4.7c), would diminish the atmospheric equilibration pathway, and therefore increase HBEI. *Archer et al.* [2000a] proposed that this mechanism is the primary factor explaining the large difference between the low HBEI of the original Harvardton-Bear box models, and the high HBEI of the three-dimensional ocean circulation models that participated in the comparison study of *Broecker et al.* [1999a]. *Archer et al.* [2000a] argued that because of their relatively coarse vertical resolution, these three-dimensional models tend to be overly diffusive vertically, therefore overestimating the HBEI index. On the other hand, *Follows et al.* [2002] pointed out that these three-dimensional models represent the warm-water sphere much more realistically than the Harvardton-

Bear box models, as the warm-water sphere in the real ocean includes much of the upper and mid-thermocline, whereas the Harvardton-Bear box models restrict it to a very thin surface layer. As a consequence, the warm-water sphere has almost no capacity in the Harvardton-Bear box models, while it may comprise more than 20% of the total ocean volume in the real ocean. Figure 10.4.7d shows that the HBEI increases as the volume of the warm-water sphere increases. In fact, expanding the volume of the low-latitude box and increasing the exchange between the low-latitude and deep boxes are structurally analogous, as both tend to increase the effective carbon reservoir size associated with the low-latitude surface, making equilibration with the deep ocean more difficult.

A last measure to consider is the magnitude of the meridional overturning circulation, which is not captured in the Harvardton-Bear box models, but is included in some of the four- to seven-box models (e.g., *Toggweiler* [1999] *Lane et al.* [2005]). For example, *Lane et al.* [2005] pointed out recently that the inclusion of a dynamic feedback of the changes in ocean temperature and salinity on the meridional overturning circulation during the last glacial led to a substantial increase in the sensitivity of atmospheric CO_2 to changes in the low latitudes. This suggests a note of caution when interpreting HBEIs computed for current interglacial conditions, as these values might be considerably different during glacial conditions.

In summary, it appears that the Harvardton-Bear box models likely underestimate the HBEI, and therefore underestimate the possible role of low-latitude changes. However, given the fact that the reservoir associated with the low latitudes is much smaller than the high-latitude surface and deep ocean reservoir, the equilibration of a change in the low-latitude surface ocean with the deep ocean is always going to be larger than the equilibration of a change in the high-latitude surface ocean with the low-latitude reservoir. Therefore, the low latitudes are always going to be less effective in changing atmospheric CO_2 than the high latitudes. In addition, substantial leverage exists in the high latitudes to change surface ocean pCO_2 there, making the high latitudes the primary region of focus for explaining the large atmospheric CO_2 changes associated with G/IG transitions. We therefore focus our discussion on this region next.

Closing the Southern Ocean Window

We have seen in chapter 8 that the Southern Ocean represents an area of strong biologically driven outgassing of CO_2 (figure 8.5.1). This outgassing is primarily a consequence of upwelling that brings waters with high concentrations of respired *DIC* (ΔC_{soft}) to the surface near the Southern Ocean Polar Front. Since biological productivity in this region is too slow to fix all this excess CO_2, a substantial fraction of it escapes to the atmosphere. We also highlighted in chapter 8 that the existence of residual nutrients at the surface is a good indicator of this biologically induced CO_2 leak into the atmosphere. In steady state, the CO_2 lost to the atmosphere in the Southern Ocean is taken up elsewhere, particularly in those regions where the residual nutrients are taken up and used to fuel photosynthesis. The strength of the Southern Ocean leak is directly controlled by the balance between the upward supply of ΔC_{soft} and the downward export of organic matter. A reduction in the upward supply of ΔC_{soft}, while keeping organic matter export the same, will reduce surface ocean *DIC*, close the window, and pull down atmospheric CO_2. Another option to close the window is to strengthen export production, while keeping the upward supply of ΔC_{soft} and nutrients the same. As the Southern Ocean represents the main window of the surface ocean to the deep ocean *DIC* reservoir, the closure of this window has a large impact on atmospheric CO_2.

We investigate this Southern Ocean window and its impact on atmospheric CO_2 with the same three-box model that we used above for the discussion of the HBEI. In this case, we have to consider biological production and export, which we tie to phosphate. It is assumed that all phosphate is consumed in the low-latitude box, linking its biological export directly to the rate of supply of phosphate. In contrast, only a fraction of the phosphate is used in the high latitudes. This leaves the biological production in the high latitudes a free parameter and makes the high-latitude phosphate concentration, $[PO_4]_h$, a key diagnostic of the model. The full set of equations and a solution method is given by *Toggweiler and Sarmiento* [1985]. A steady-state solution for the high-latitude ocean *DIC*, DIC_h, can be obtained in a manner analogous to that we have used for O_2 (see equation (1.2.11)):

$$DIC_h = DIC_d - r_{C:PO_4}([PO_4]_d - [PO_4]_h) \qquad (10.4.8)$$

The high-latitude PO_4^{3-} concentration is controlled by the balance between the rate of supply from the deep ocean and the rate of biological removal by high-latitude export of organic phosphorus, Φ_h^P:

$$[PO_4]_h = \frac{[PO_4]_d\, f_{hd} - \Phi_h^P}{f_{hd} + T} \qquad (10.4.9)$$

where f_{hd} is the high-latitude–deep ocean mixing term, and T the overturning term (see also figure 10.4.3). Inserting (10.4.9) into (10.4.8) gives:

$$DIC_h = DIC_d - r_{C:PO_4}\frac{[PO_4]_d\, T + \Phi_h^P}{T + f_{hd}} \qquad (10.4.10)$$

Thus any reduction in the mixing term, f_{hd}, or increase in biological export, Φ_h^P, or overturning circulation, T,

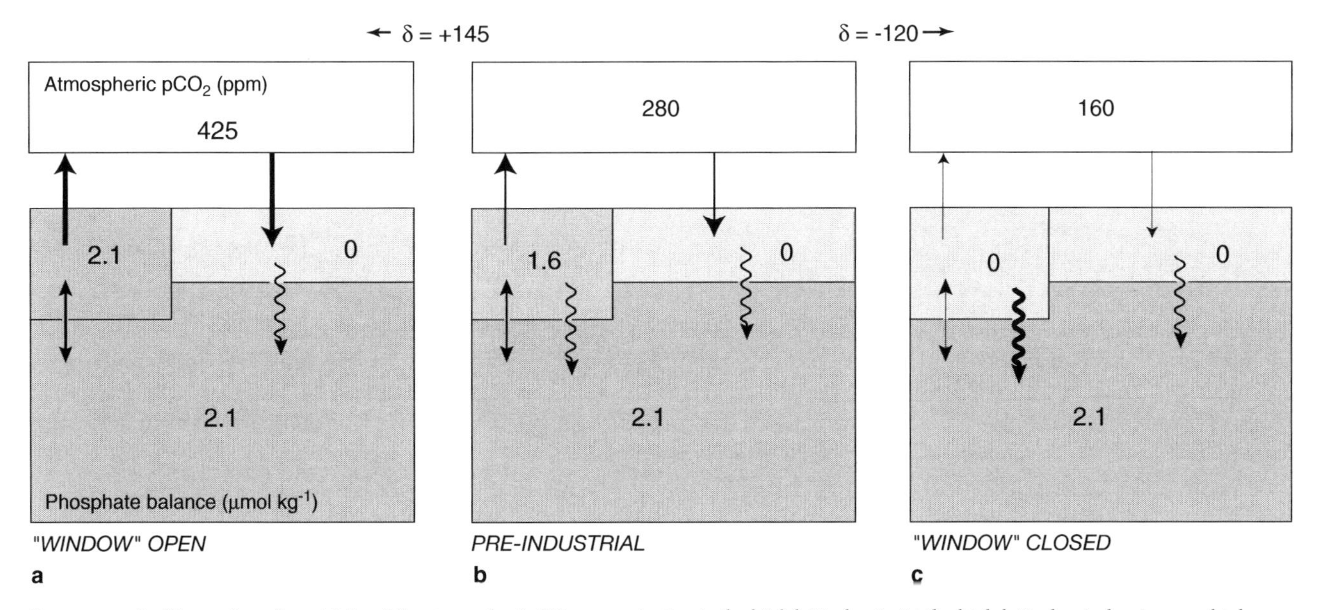

FIGURE 10.4.8: Illustration of sensitivity of the atmospheric CO_2 concentration to the high latitudes. In (a) the high-latitude window is open, high-latitude PO_4^{3-} high, and atmospheric pCO_2 attains concentrations of 425 ppm. (b) represents the preindustrial ocean, where the window is half-open, PO_4^{3-} in the high latitudes is at intermediate levels, and atmospheric CO_2 is 280 ppm. In (c) the high-latitude window is closed, high-latitude PO_4^{3-} is low, and atmospheric pCO_2 decreases to about 160 ppm.

will lead to a decrease in high-latitude DIC. Note that in the case of $T = 0$, this equation is directly analogous to equation (10.4.4), except that the mixing term here is called f_{hd} instead of v. In order to determine atmospheric CO_2, one also would need to specify the changes in the low-latitude box. However, we have seen above that changes in the low latitudes are much less effective in changing atmospheric CO_2, since changes in this reservoir tend to be compensated by equilibration with the deep ocean. This is not the case for high-latitude changes, so that any change in high-latitude pCO_2 leads to nearly corresponding changes in atmospheric CO_2. We thus have a very effective way of changing atmospheric CO_2, with the high-latitude surface concentration of PO_4^{3-} being a good indicator of the Southern Ocean window.

This "high-latitude dominance" and the role of the Southern Ocean window is illustrated in figure 10.4.8. Today, the Southern Ocean window is semi-open, as diagnosed by the fact that high-latitude surface ocean PO_4^{3-} is below that of the deep ocean. If the window were completely opened, e.g., by setting Φ_h^P, to zero and making f_{hd} very large, atmospheric CO_2 would shoot up by nearly 150 ppm. Closing the Southern Ocean window by setting $[PO_4]_d \cdot f_{hd}$ equal to Φ_h^P, i.e., setting high-latitude PO_4^{3-} to zero, lowers atmospheric CO_2 by about 120 ppm. We thus have a mechanism with a very high leverage on atmospheric CO_2.

Figure 10.4.9 shows that glacial levels in atmospheric CO_2 can be attained in the three-box model by reducing the mixing between the deep ocean and the high-latitude ocean from about 60 Sv to about 19 Sv without

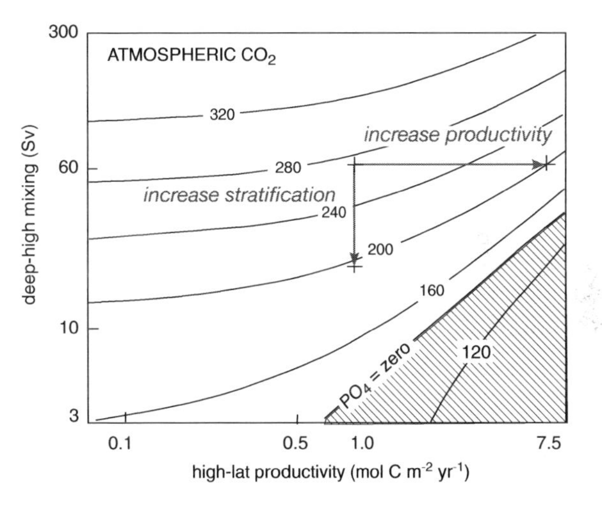

FIGURE 10.4.9: Contour plot of atmospheric CO_2 in ppm as computed by the three-box model depicted in figure 10.4.3. Atmospheric CO_2 is contoured over a parameter space defined by variations in the high-latitude–deep box exchange term, f_{hd}, and the high-latitude productivity, Φ_h. Note that the axes are plotted logarithmically. From *Toggweiler* [1999].

any change in biological productivity, or by increasing biological productivity about sixfold from about 1 mol·C m^{-2} yr^{-1} to about 6.3 mol C m^{-2} yr^{-1} without changing mixing. However, these estimates are based only on the "closed system" responses, and therefore do not include the "open system" effects associated with $CaCO_3$ compensation (see chapter 9), which tends to amplify initial changes. We therefore need to review $CaCO_3$ compensation in the context of the G/IG problem before moving on to a discussion of how the Southern Ocean window can be closed.

We have seen in chapter 9 that any change in the ocean's biological pumps, either by changing the downward transport of organic matter or $CaCO_3$, or by changing ocean circulation, will lead to imbalances between the input of *Alk* by rivers and the removal of *Alk* by carbonate burial on the sea floor. On the timescale of several thousand years as considered for the glacial-interglacial CO_2 problem, the oceanic $CaCO_3$ cycle will respond to this perturbation and restore the balance by altering the depth of the lysocline either temporarily or permanently and by changing ocean mean *Alk*. These "open system" changes impact the ocean-atmosphere balance of CO_2 and are therefore of great importance for thinking about glacial-interglacial variations in atmospheric CO_2.

The two possibilities mentioned so far to close the Southern Ocean window, i.e., increasing Southern Ocean stratification and/or increasing biological productivity, are equivalent in their impact on the vertical distribution of *DIC*, as both tend to increase the vertical gradient. If we assume that the strength of the $CaCO_3$ pump remains the same, the vertical gradient of *Alk* will change only minimally in the changing productivity case, and increase in the stratification case, but these changes are much smaller than those of *DIC*. As a result, the concentration of the CO_3^{2-} ion, approximately given by $[CO_3^{2-}] = Alk - DIC$, will decrease in the ocean's interior. This will induce an upward shift of the $CaCO_3$ saturation horizon, causing anomalous $CaCO_3$ dissolution. The resulting excess of *Alk* input by rivers over *Alk* loss by $CaCO_3$ burial will increase oceanic *Alk* until the CO_3^{2-} ion concentration is restored to its original concentration, and the $CaCO_3$ saturation horizon restored to its original position. As a result of a net dissolution of $CaCO_3$ on the sea floor, this open system response increases the mean ocean *Alk* twice as strongly as it increases mean ocean *DIC*, therefore enhancing the initial perturbation as shown in figure 9.5.2. In the case of the Southern Ocean window closure and using the three-box model, the enhancement turns out to be relatively small, only of the order of a few percent. This is primarily because the CO_3^{2-} ion changes in the deep reservoir of the three-box model are very small, simply a result of the large volume of this reservoir. We will see below, however, that in the case of changes of the $CaCO_3$ pump, the open system response can be quite large. Another way to enhance the $CaCO_3$ compensation effect is to reduce the volume of the deep reservoir, so that its chemistry changes are larger (an effect used by *Toggweiler* [1999], see below).

Physical Mechanisms

We have learned so far that there are two fundamentally different mechanisms to close the Southern Ocean window and reduce atmospheric CO_2 (figure 10.4.9). One is physically driven by reducing the exchange between the high-latitude surface box and the deep box, while the other is biologically driven and requires a substantial increase in biological fixation of CO_2 in the surface ocean and subsequent export. A third possibility exists in the form of increased sea-ice coverage, which would close the Southern Ocean window by simply impeding the physical exchange of CO_2 across the air-sea interface [*Stephens and Keeling*, 2000]. The three-box model allows us to investigate the ramifications of these changes, but it does not give us insight into the possible processes leading to these changes. We focus first on the physical mechanisms.

The exchange between the high-latitude surface box and the deep ocean can be reduced by decreasing the mixing in the Southern Ocean in response to, for example, an increase in vertical stratification there. An alternative mechanism is to separate the deep ocean vertically into several reservoirs, and add a finite amount of mixing between them. In this latter case, vertical stratification is added to the entire deep ocean, but the net effect is essentially the same, in that it reduces the amount of mixing between the surface ocean of the high latitudes and the deep ocean reservoir of *DIC*. The real-world analog is the separation between the cool, fresh Antarctic Bottom Water (AABW) and the warm, salty North Atlantic Deep Water (NADW).

Toggweiler [1999] made use of the latter strategy and proposed that deep ocean stratification between NADW and AABW was substantially larger during the LGM, resulting in reduced upwelling of ΔC_{soft}-rich waters in the Southern Ocean that prevented CO_2 escape into the atmosphere. Supporting evidence of increased AABW density comes from chlorinity and $\delta^{18}O$ measurements of pore fluids, which suggest that AABW was close to the freezing point and much more salty relative to the rest of the ocean during the LGM in comparison to today [*Adkins et al.*, 2002]. At the same time, although NADW appears to have been somewhat cooler than today, its increase in density was likely smaller than that of AABW, leading to a stronger density contrast between Southern Ocean source waters occupying the bottom ocean, and the deep waters lying above.

Toggweiler [1999] attempted to capture this physical separation between waters of southern and waters of northern high-latitude origin by extending the classical three-box models to seven boxes. Of particular relevance for this model is not only the physical separation, but also the fact that the chemical composition of northern and southern source waters is very different. We have seen in chapter 8 that North Atlantic Deep Water has a very low ΔC_{soft} concentration for its temperature (figure 8.4.6), while the ΔC_{soft} concentration of Southern Ocean source waters is very high. By increasing vertical stratification in the deep ocean, one also separates waters with a high potential for inducing outgassing from those with a much smaller potential. *Toggweiler* [1999] demonstrated with this model that a decrease in the ventilation of the bottom waters together with $CaCO_3$ compensation can cause a nearly 60 ppm drawdown in atmospheric CO_2 (21 ppm from ventilation and 36 ppm from $CaCO_3$ compensation). This model has a much larger enhancing effect from $CaCO_3$ compensation than the Harvardton-Bear models, because most of the chemical changes occur in the intermediate-size bottom water box in this seven-box model, leading to substantial changes in *DIC* and *Alk*. This requires a substantially larger amount of $CaCO_3$ dissolution to restore the ocean's CO_3^{2-} ion concentration, thereby enhancing the CO_2 drawdown in the atmosphere to a much larger extent. Although successful in creating a large drawdown in atmospheric CO_2, *Toggweiler* [1999] did not provide a physical mechanism to explain what might have caused the reduction in bottom ocean water ventilation.

Stephens and Keeling [2000] suggested instead that extended sea-ice coverage closed the Southern Ocean window. Their hypothesis is based on the arguments that the glacial Southern Ocean was much colder than today, and that the warm layer derived from NADW sitting underneath the surface, which tends to prevent sea ice from growing too rapidly in the present ocean [*Gordon*, 1981], was likely absent. Both factors would have led to a substantial increase in sea-ice coverage in the Southern Ocean. Their argument is hampered by the fact that in order for this mechanism to be effective, sea ice needs to cover nearly the entire Southern Ocean. In their model, a substantial reduction was only achieved if sea ice covered 99% of the entire Southern Ocean, for which there is only limited evidence [*Crosta et al.*, 1998].

The sea-ice extent and deep ocean stratification mechanisms were combined and put into a unified framework by *Gildor and Tziperman* [2001a] and *Gildor et al.* [2002]. They used a meridional box model of the ocean, atmosphere, land-ice, and sea-ice climate system, and combined it with a relatively simple representation of the oceanic carbon cycle. With present-day seasonal forcing from the sun, the physical part of the climate model develops 100,000 year timescale variations that mimic the glacial-interglacial cycles observed in ice cores (see figure 10.4.1). As explained by *Gildor and Tziperman* [2001b], these oscillation emerge from a combination of feedbacks between temperature and precipitation, and temperature and albedo [*Ghil et al.*, 1987; *Ghil*, 1994], but are modified by a sea-ice feedback in the northern hemisphere, which influences the growth and retreat of the continental ice sheets. When run with Milankovitch forcing, i.e., the small variations in solar forcing that result from slight but regular changes in Earth's orbit around the sun (eccentricity, timescale about 100,000 years), in the tilt of Earth's axis of rotation relative to the plane of Earth's orbit around the sun (obliquity, 41,000 years), and in the absolute direction of this tilt (precession, 23,000 years), the model's main periodicity locks in with eccentricity, as observed.

Gildor and Tziperman [2001a] propose that the physical changes in the Southern Ocean needed to draw down atmospheric CO_2 during glacial periods emerge naturally from the physical model and are driven by changes in the northern hemisphere. They suggest that NADW becomes cooler during the buildup stage of northern hemisphere ice sheets, and that it maintains this property as it moves along its journey to the Southern Ocean. This results in a cooling of the Circumpolar Deep Waters, making the stratification in the Southern Ocean more stable during glacial maxima. *Gildor and Tziperman* [2001a] then use simple formalisms linking the stratification to vertical mixing, thereby generating the mechanism needed to draw down atmospheric CO_2 according to the hypothesis of *Toggweiler* [1999]. The reduced vertical mixing and the lower deep water temperature result in a larger sea-ice extension during cold periods. This, in turn, can again reduce atmospheric CO_2 according to the mechanism of *Stephens and Keeling* [2000].

Although compelling and successful in explaining the atmospheric CO_2 drawdown, it is not possible yet to evaluate whether the mechanism proposed by *Gildor and Tziperman* [2001a] and *Gildor et al.* [2002] is correct. Their results need to be scrutinized more carefully with various paleoclimatological constraints and investigated in greater detail with models of higher complexity. We will discuss paleoclimatological constraints below, but note that the success of the hypotheses by *Toggweiler* [1999], *Stephens and Keeling* [2000], and *Gildor and Tziperman* [2001a] in explaining the glacial drawdown in atmospheric CO_2 depends critically on the existence of a strong high-latitude dominance. *Archer et al.* [2003] examined this high-latitude dominance with a particular emphasis on the sea-ice mechanism, and found that atmospheric CO_2 shows virtually no response to an increase in sea-ice coverage in the Southern Ocean in models with more spatial resolution, such as two- and three-dimensional ocean circulation models. This finding is consistent with our discussion above of these models having a higher low-latitude sensitivity, i.e., higher HBEI. Given that we do not know the high versus low latitude sensitivity of the real ocean, *Archer et al.* [2003] concluded that until these sensitivities can be resolved, glacial CO_2 hypotheses based on Southern Ocean barrier mechanisms are "walking on thin ice."

Biological Mechanisms

An alternative set of mechanisms to close the Southern Ocean window is to increase biological productivity and carbon export by taking advantage of the large unused

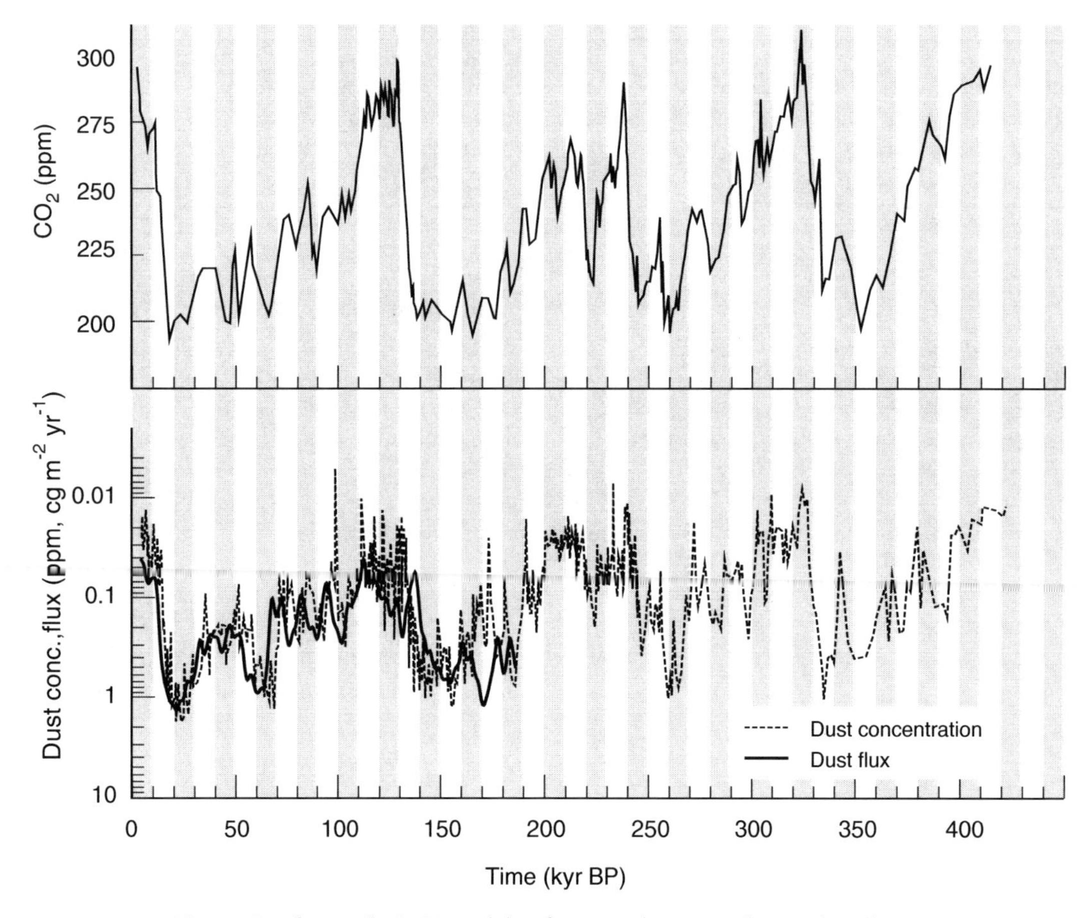

FIGURE 10.4.10: Time series of atmospheric CO_2 and dust from Vostok. Data are from *Petit et al.* [1999].

nutrient pool in this HNLC region as first suggested by the Harvardton-Bears [*Knox and McElroy*, 1984; *Sarmiento and Toggweiler*, 1984; *Siegenthaler and Wenk*, 1984]. We have seen already that in order to decrease atmospheric CO_2 to glacial levels in the Harvardton-Bear three-box model, high-latitude productivity has to increase about sixfold (figure 10.4.9). While the Harvardton-Bears did not know how such a large increase in biological productivity could have happened, the discovery that iron might be the factor limiting phytoplankton in this region (see discussion in chapter 4) provided *Martin* [1990] with the basis for proposing that the higher dust levels during the glacial periods (see figure 10.4.10) fertilized the Southern Ocean with the necessary iron to increase biological export. *Watson et al.* [2000] investigated this hypothesis in a multi-box model of the ocean carbon cycle, and found that iron fertilization on the basis of the observed variations of dust flux in the Vostok ice core could explain a drawdown of about 40 ppm in atmospheric CO_2. By contrast, *Bopp et al.* [2002], using a three-dimensional coupled ocean circulation–ecosystem model and using the LGM dust flux reconstruction of *Mahowald et al.* [1999], found a maximum effect of only 15 ppm. These results need to be viewed tentatively, since uncertainties with regard to iron availibility and the magnitude of the iron fluxes during the LGM are very large. However, some of the differences can also be explained by differences in the high-latitude sensitivity of these two models. *Watson et al.* [2000] used a model with a low HBEI, whereas *Bopp et al.* [2002] employed a model with a high HBEI, thereby having, by design, a lower effectiveness of high-latitude changes in export production.

Observational Constraints

Each hypothesis attempting to explain the observed glacial-interglacial variations in atmospheric CO_2 needs to be carefully evaluated against available paleoceanographic constraints. However, these evaluations are seldom very straightforward, since very few constraints come from direct measurements of the property of interest, e.g., temperature, but rather are derived from another, related property, e.g., the $\delta^{18}O$ of $CaCO_3$. The relationship of such proxies with the property of interest is sometimes tight and well understood, but often based only on empirical correlations. In addition, a given proxy is seldom uniquely determined by the property of interest, but often influenced by other interfering factors. In the case of temperature, for example, the $\delta^{18}O$ of CO_3^{2-} in water and hence in $CaCO_3$

is also influenced by the magnitude of the continental ice sheets, requiring this contribution to be removed before the $\delta^{18}O$ of $CaCO_3$ can be used to infer water temperature. Keeping these limitations in mind, we investigate next how the hypotheses centered around the closure of the Southern Ocean compare to these constraints. A critical prediction of the Southern Ocean vertical stratification and productivity hypotheses is that the surface-to-deep gradient of PO_4^{3-} has to increase substantially. This prediction can be tested with proxy observations.

An important proxy is the $^{13}C/^{12}C$ ratio of seawater *DIC* recorded by planktonic and benthic foraminifera. During the photosynthetic uptake of CO_2, marine plants discriminate against the heavy isotope ^{13}C by about 20‰, whereas respiration and remineralization is associated with only a very small fractionation. Because phosphate is controlled by photosynthesis and respiration/remineralization as well, a relatively good correlation exists between the $^{13}C/^{12}C$ isotopic ratio of surface water *DIC* and phosphate. Deviations in the correlation are a consequence of fractionations occuring during the air-sea gas exchange of CO_2 and due to variations in the biological fractionation. Nevertheless, to first approximation, the $^{13}C/^{12}C$ ratio of *DIC* can be regarded as a good proxy for phosphate. Any increase in the phosphate gradient between the surface waters and the deep ocean should therefore lead to an increase in the surface-to-deep gradient in the $^{13}C/^{12}C$ ratio of *DIC*. However, the existing data show little change in the surface-to-deep difference in this region [*Charles and Fairbanks*, 1990].

The conclusion based on the marine $^{13}C/^{12}C$ ratio are also supported by measurements of the cadmium content recorded in foraminifera. No significant change is found in pelagic and benthic foraminifera from the Southern Ocean [*Boyle*, 1992]. As there exists a very tight correlation between cadmium and phosphate in the modern ocean, this again suggests that nutrient concentrations in the Southern Ocean did not drop significantly during glacial time.

A third constraint, albeit less strong, is the $^{13}C/^{12}C$ ratio change in atmospheric CO_2 [*Leuenberger et al.*, 1992; *Marino et al.*, 1992]. The closing of the high-latitude window increases the $^{13}C/^{12}C$ ratio of *DIC* in the surface waters and as a consequence increases the $^{13}C/^{12}C$ ratio of atmospheric CO_2. For example, the three-box model predicts an increase of about 1‰ in the $^{13}C/^{12}C$ of atmospheric CO_2. In sharp contrast, the reconstructions suggest that the $^{13}C/^{12}C$ ratio of atmospheric CO_2 was lower by about 0.3–0.7‰ during the last glacial [*Leuenberger et al.*, 1992, *Marino et al.*, 1992]. However, there are several additional processes to consider that ease the constraint [*Broecker and Henderson*, 1998]. First, we have to take into account the whole ocean isotopic shift of about 0.4‰ that is usually interpreted as resulting from the transfer of isotopically light carbon from the terrestrial biosphere into the ocean during glacial periods [*Shackelton*, 1977; *Curry et al.*, 1988]. In addition, a recently discovered pH artifact in foraminifera [*Spero et al.*, 1997] and variations in the biological fractionation conspire together with the decrease from the isotopically light carbon from the terrestrial biosphere to reduce the $^{13}C/^{12}C$ ratio of atmospheric CO_2 by about 0.9‰ [*Broecker and Henderson*, 1998]. Hence, the corrected model-predicted increase in the $^{13}C/^{12}C$ ratio of atmospheric CO_2 is only 0.1‰, making the discrepancy with the observations relatively small.

Toggweiler [1999] demonstrated that these constraints may not be as strong as they appear at first. In particular, he pointed out that the representation of the entire deep ocean as one large box in the Harvardton-Bear three-box models neglects the possible existence of vertical stratification in the ocean. In his model, most of the changes needed to drive down atmospheric CO_2 come from the Southern Ocean bottom water box, which is much smaller in volume than the large deep box in the Harvardton-Bear three-box models. It therefore depends critically on where certain constraints are coming from in order to establish their strength.

The need for consideration of the spatial nature of the constraints becomes even more evident when considering how different export production was relative to today. Figure 10.4.11 shows a map of changes in export production inferred from a large number of paleo-productivity proxies. With regard to the Southern Ocean, a relatively clear pattern emerges, with the region south of the Polar Front (the Antarctic zone) generally showing a decrease in export production, and the region between the Polar Front and the Subantarctic Front (the Subantarctic Zone) showing an increase (see chapter 7 for a discussion of these fronts and figure 7.3.3 for illustration of the zonal mean meridional circulation in the Southern Ocean). How do these observations compare to the Southern Ocean window closure hypotheses? The consideration of changes in export production alone is insufficient to answer this question, as the closing of the window requires an increase in the efficiency of the biological pump, i.e., an increase in the downward transport of organic matter relative to the upward supply. We have seen above that a good proxy for this balance, i.e., the degree of nutrient utilization, is the surface concentration of macronutrients.

It turns out that the ratio of ^{15}N to ^{14}N measured in organic matter deposited on the sea floor is a good recorder for the degree of nutrient utilization in the overlying surface ocean [*Francois et al.*, 1997]. Due to an isotopic fractionation during the photosynthetic uptake of nitrate, the ^{15}N content of near-surface organic nitrogen exhibits in the modern ocean a very strong correlation with the concentration of nitrate [*Altabet and Francois*, 1994; *Sigman et al.*, 1999b]. As this signal gets transmitted and incorporated into the sediments, bulk-

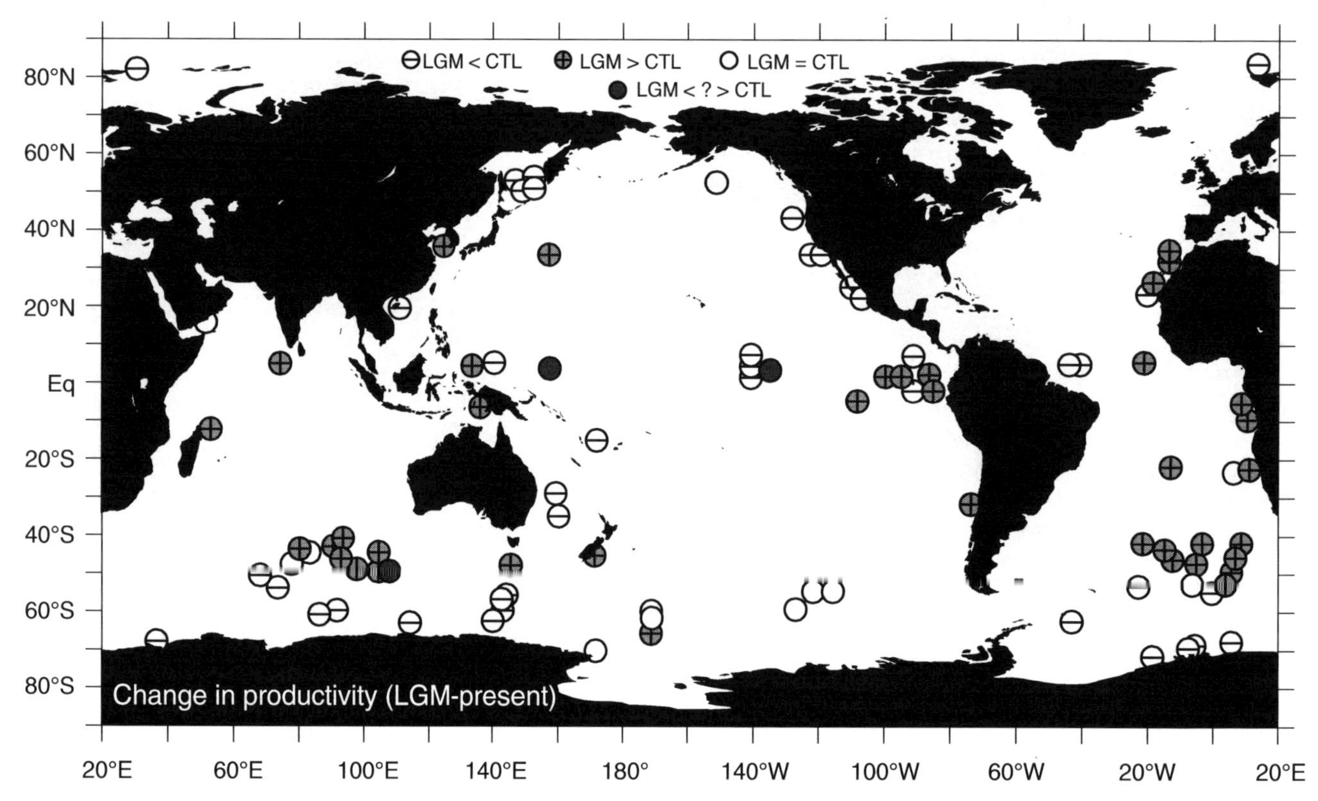

FIGURE 10.4.11: Map of differences of marine biological productivity between the present interglacial and the LGM. Modified from *Kohfeld et al.* [2005].

sediment $^{15}N/^{14}N$ provides a good proxy of the degree of nitrate utilization in the overlying surface waters, i.e., the balance between upward physical supply of nitrogen and downward export of organic nitrogen [*Sigman et al.*, 1999, 2000].

Observations of the $^{15}N/^{14}N$ ratio in the Southern Ocean reveal a substantial increase in the nitrate utilization south of the modern Polar Front, i.e., in the Antarctic [*Francois et al.*, 1997; *Sigman et al.*, 1999a], whereas little change in nitrate utilization was found north of the modern Polar Front. The former change is very consistent with the Southern Ocean window closure hypotheses, as an increase in nutrient utilization reduces the outgassing of biologically derived CO_2. Combining the increased nutrient utilization in the Antarctic with the lower biological export in this area (figure 10.4.11) suggests that the primary mechanism responsible for the closure is a reduced supply of nutrients, probably due to an increase in vertical stratification, in agreement with the hypothesis by *Toggweiler* [1999]. The increase in biological export in the Subantarctic Zone coupled with a nearly unchanged nutrient utilization suggests that the supply of nutrients to this zone increased but had little influence on the evasion of CO_2, since the increase in supply was compensated by an increase in the downward export of organic matter.

In summary, the available paleoceanographic constraints indicate a spatially inhomogeneous picture, one that paleoceanographers are still grappling to fully understand. Most recent constraints suggest that the Southern Ocean experienced substantial changes during the LGM, with a distinct difference between the Antarctic and Subantarctic zones. While the available constraints tend to indicate changes that are consistent with the Southern Ocean window being more closed during the LGM than today, the changes they permit seem insufficient to explain the entire drawdown of atmospheric CO_2. As we will see below, however, an important argument in favor of the Southern Ocean playing an important role is the strong connection between southern hemisphere climate changes and atmospheric CO_2.

A Role for the Regions outside the Southern Ocean?

With paleoceanographic constraints limiting the contribution of the Southern Ocean, many researchers have given increased attention to mechanisms outside the Southern Ocean. We review the most important hypotheses in the next subsections, organized into whether the hypotheses primarily invoke changes in ocean circulation, the soft-tissue pump, or the carbonate pump. We will discuss their strengths and weaknesses at the same time. A summary of these hypotheses is given in table 10.4.1. We will see that few of these hypotheses are consistent with existing constraints, while still large enough to explain a substantial drawdown of atmospheric CO_2, except those that invoke a modest change in the $CaCO_3$-to-organic carbon rain ratio. In the last section, we will develop a synthesis scenario that combines elements from the Southern Ocean hypotheses with some of those discussed next.

Circulation Scenarios

We have so far neglected changes in circulation outside the Southern Ocean, in particular the possibility of changes in the meridional overturning circulation associated with NADW. Although there is abundant evidence that the meridional overturning circulation in the Atlantic was rather different during full glacial and interglacial conditions, with substantial variability in between [*Alley et al.*, 2003], it turns out that these changes appear to have a relatively small impact on atmospheric CO_2. *Archer et al.* [2000b] found a reduction of only 6 ppm for the LGM, based on a three-dimensional ocean biogeochemistry model whose circulation was optimized by an adjoint method to reflect paleoceanographic constraints for the glacial ocean [*Winguth et al.*, 1999]. However, given the relatively small number of paleoceanographic constraints, it is possible that this adjoint model underestimates the changes that occurred between glacial and interglacial periods. This is indeed suggested by the recent six-box modeling study of *Lane et al.* [2005], who obtained a very strong drawdown of atmospheric CO_2 when they permitted their meridional overturning circulation to adjust dynamically to the changed boundary conditions. In particular, they found that the data constraint for the glacial ocean asked for a moderate reduction in the overturning circulation associated with NADW from about 16 Sv to 12 Sv, and a very strong reduction from about 17 Sv to 9 Sv in the formation of abyssal waters in the Southern Ocean. They did not provide a separation into how much of the atmospheric CO_2 drawdown was driven by the Southern Ocean pathway versus the North Atlantic pathway, but if it turns out that the majority of the drawdown is driven by the Southern Ocean, their explanation would belong to the category of physical mechanisms closing the Southern Ocean window (as do the explanations of *Toggweiler* [1999] and *Gildor and Tziperman* [2001a]). As mentioned before, an important point noted by *Lane et al.* [2005] was their finding that the reduction in ocean circulation reduced atmospheric CO_2 not only by sequestering it in the interior ocean through a more efficient biological pump, but also because these ocean circulation changes made atmospheric CO_2 respond more sensitively to the low-latitude cooling.

A substantial change in the oceanic circulation would indeed be needed if the nutrient redistribution hypothesis of *Boyle* [1988] was correct. His argument is based on the observation that the reorganization of the Atlantic conveyor circulation during glacial periods shifted nutrients and metabolic CO_2 from intermediate waters to deeper waters. This causes a substantial $CaCO_3$ compensation response, which would be mainly responsible for the atmospheric CO_2 drawdown. *Boyle* [1988] argued that this mechanism can explain at least half of the 80 ppm difference between glacial and interglacial periods. However, there are two arguments that speak against this hypothesis. *Sigman et al.* [1998] showed that such a nutrient redistribution associated with $CaCO_3$ compensation would lead to a significant steady-state deepening of the lysocline (see chapter 9). Such a deepening was observed in the Pacific and Indian Oceans, whereas the opposite was the case in the Atlantic Ocean, resulting in an overall deepening of the lysocline during the LGM of only a few hundred meters [*Catubig et al.*, 1998]. A second argument is the observation that during the deglaciation, i.e., between the LGM and about 10,000 years ago, temperature in the high latitudes of the northern hemisphere and especially Greenland evolved rather differently than atmospheric CO_2 (figure 10.4.12). In particular, atmospheric CO_2 started to increase several thousand years before the Greenland temperature, and by the time Greenland began to warm markedly about 14,500 years ago, atmospheric CO_2 had already changed by nearly 50% of its total glacial-interglacial difference. Since the northern hemisphere high-latitude temperatures from Greenland are a good indicator of the strength of the Atlantic overturning circulation, this lack of similarity between the evolution of Greenland temperature and atmospheric CO_2 speaks against the hypothesis of *Boyle* [1998]. In contrast, temperatures in Antarctica and atmospheric CO_2 co-evolved remarkably synchronously during the deglaciation, providing support for Southern Ocean–driven hypotheses.

Soft-Tissue Pump Scenarios

If circulation changes are unable to explain the glacial CO_2 drawdown, maybe changes in the strength of biological export outside the Southern Ocean could represent the key mechanism. In fact, initial attempts to solve the mystery of reduced atmospheric CO_2 focused from the very beginning on the role of the soft-tissue pump. This is not surprising, given the importance of this pump in generating the surface-to-deep gradients in *DIC* (see figure 8.4.2) and the many ways this pump could respond to changes in environmental factors. As the level of nutrients in the surface waters constitute one of the primary factors controlling the biological productivity, many scenarios have been developed that are coupled to changes in nutrients (see table 10.4.1). The primary idea behind most of these hypotheses is to increase ocean productivity throughout the ocean by increasing the total ocean nutrient inventory. This can be accomplished by a reduction of the fraction of nutrients that are

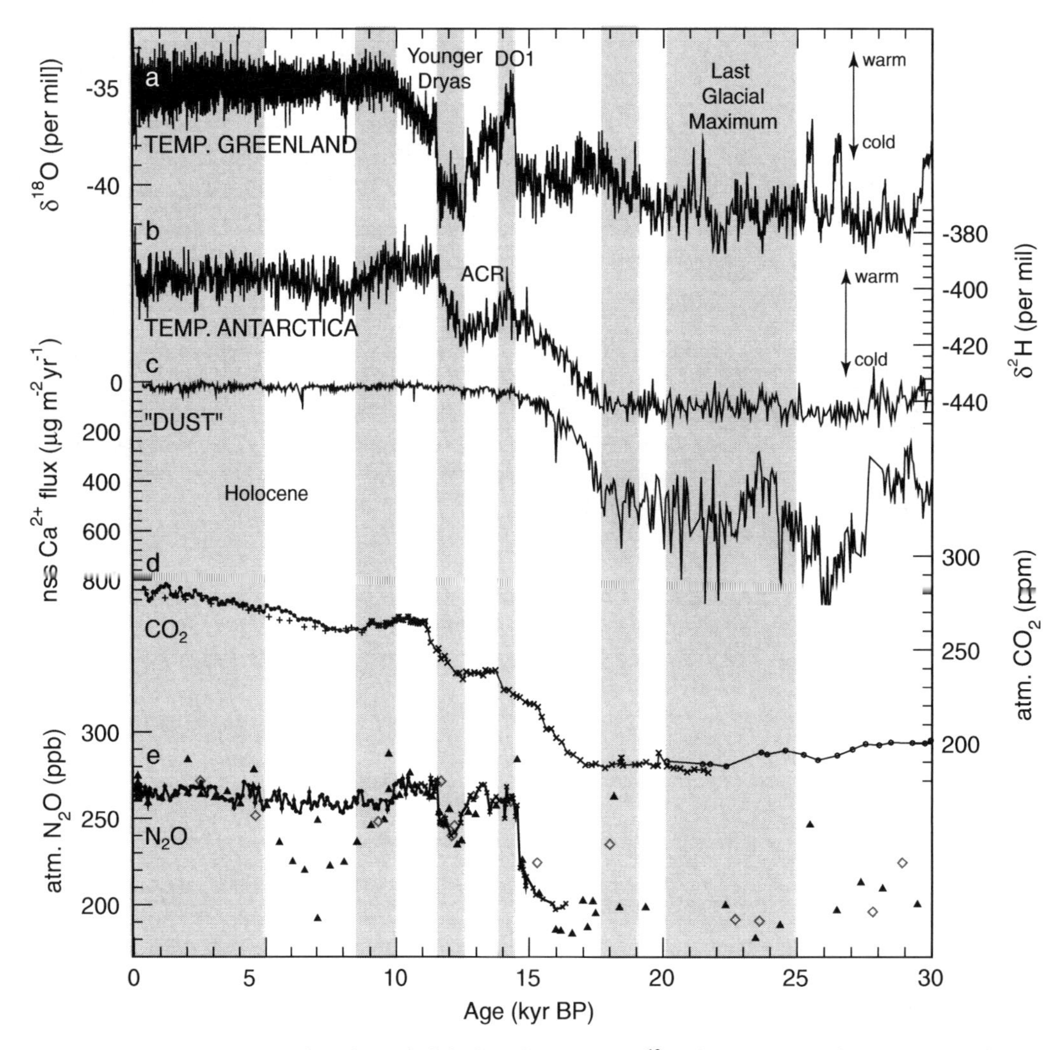

FIGURE 10.4.12: Events surrounding the end of the last glaciation. (a) $\delta^{18}O$ of ice as a proxy for temperature, from GRIP (Greenland) [*Johnsen et al.*, 1997]. (b) δ^2H of ice as a proxy for temperature, from Dome C (Antarctica) [*Jouzel et al.*, 2001]. (c) Dust flux (inverted scale) at Dome C site, as reconstructed from the sea-salt–corrected Ca^{2+} and the accumulation rate [*Roethlisberger et al.*, 2002]. (d) Atmospheric CO_2, as recorded in air bubbles from various Antarctic ice cores [*Indermühle et al.*, 1999, 2000; *Flückiger et al.*, 2002; *Monnin et al.*, 2002]. (e) Atmospheric N_2O, as recorded in air bubbles from various Antarctic and Greenland ice cores. The filled triangles denote data from the GISP ice core drilled in Greenland [*Sowers et al.*, 2003], the open circles are from the Taylor Dome [*Sowers et al.*, 2003], the solid line with the crosses denotes data from the GRIP ice core [*Flückiger et al.*, 1999], and the solid line with the circles is from Dome C [*Flückiger et al.*, 2002]. The data are plotted versus age before the year 1950, using age-scales of the respective records. ACR is a period of cooling in Antarctica, DO1 stands for Dansgaard-Oeschger event 1, a period of rapid warming in Greenland with little change in Antarctica.

buried in the shelf sediments [*Broecker*, 1982] or in the case of nitrogen, by changing the balance between nitrogen fixation and denitrification [*McElroy*, 1983; *Broecker and Henderson*, 1998] or changing stoichiometry.

We can immediately exclude the shelf-nutrient hypothesis of [*Broecker*, 1982] because it is tied to sea-level change, which clearly occurred after the rise in atmospheric CO_2 [*Broecker and Henderson*, 1998]. By contrast, the recent proposal of *Broecker and Henderson* [1998] that iron-controlled imbalances between nitrogen fixation and denitrification can lead to drastic changes in the fixed-nitrogen inventory in the ocean is in agreement with the sequence of events as the Earth's climate went from glacial conditions into interglacial ones. During glacial periods, high dust fluxes bearing large amounts of iron are thought to fuel intense nitrogen fixation in the subtropical gyres [*Falkowski*, 1997], leading to a gradual buildup of the ocean inventory of fixed nitrogen. Such a nitrogen buildup could alternatively be caused by lower denitrification in the glacial ocean, as proposed by *Ganeshram et al.* [1995]. If we neglect the limiting role of phosphate and other (micro)nutrients for the moment, this increase in fixed nitrogen would support higher biological production, leading to higher export production and hence lower pCO_2. After the abrupt cessation of the dust flux, nitrogen fixation would sharply drop as well, and the fixed nitrogen inventory would start to decrease, with a timescale of approximately 3000 years [*Gruber and Sarmiento*, 1997], consistent with the observed slow increase of atmospheric CO_2 over about 8000 years (figure 10.4.12). Though this scenario is in agreement with the sequence of events during deglaciation, its feasibility hinges also on the magnitude of the changes in nutrient inventory necessary to cause a significant drop of atmospheric CO_2.

In order to reduce atmospheric CO_2 by 80 ppm in the three-box model, the limiting nutrient in this model has to increase by more than 70%. As it turns out, this low sensitivity is again a consequence of the low HBEI exhibited by the three-box model. We mentioned above that three-dimensional models have a substantially larger HBEI, opening the possibility that smaller changes in nutrient inventories could suffice to change atmospheric CO_2 dramatically. However, we still have to consider the role of other limiting nutrients, especially phosphate (see review by *Gruber* [2004]).

In strong contrast to the fixed-nitrogen inventory, the phosphate inventory of the ocean cannot be changed directly by biological processes. This inventory is controlled by river input and burial, which set the residence time at about 50,000 years [*Delaney*, 1998]. Thus, for the timescale of interest here, the phosphate inventory can be viewed as being constant. Therefore, unless we accept that the stoichiometric ratios between phosphate and nitrate uptake during photosynthesis have changed from the standard 1:16 relationship, ocean productivity can only change within rather tight limits as a function of the fixed-nitrogen inventory before phosphate becomes limiting. There exists some leeway, because the phosphate inventory is about 10% larger than expected from the nitrate inventory. We have already encountered this phosphate excess as the nonzero intercept of the nitrate-versus-phosphate plot in figure 5.1.4. This excess also shows up as a positive constant in the definition of N^* in equation (5.3.8b). It is therefore possible to increase the ocean fixed-nitrogen inventory by about 10% without running into a phosphate limitation. Changing the nutrient inventory in the three-box model by this amount decreases atmospheric CO_2 by only 10 ppm. However, if the "true" HBEI was four times bigger than that of the three-box model, atmospheric CO_2 would decrease by about 40 ppm!

We used observations of the $^{13}C/^{12}C$ ratio in the ocean and atmosphere before as an argument against the proposal that changes in high-latitude ventilation have reduced atmospheric CO_2. Any increase in surface productivity enriches the *DIC* remaining in the surface waters in ^{13}C and hence also enriches atmospheric CO_2, quite contrary to the observations which suggest a reduction in the $^{13}C/^{12}C$ ratio of atmospheric CO_2, and little change in the $^{13}C/^{12}C$ ratio of planktonic foraminifera [*Broecker and Peng*, 1998]. However, we have also seen, above, that the input of isotopically light carbon from the terrestrial biosphere, a pH related artifact, and a decrease in the photosynthetic fractionation work together to more than offset the increase in the $^{13}C/^{12}C$ ratio of atmospheric CO_2 caused by the higher productivity. Thus, the proposal that a significant increase in the nutrient inventory was at the core of the processes that reduced atmospheric CO_2 to glacial levels is not inconsistent with the $^{13}C/^{12}C$ records of atmospheric CO_2 and surface ocean *DIC*.

There exists, however, at least one serious caveat. *Sigman et al.* [1998] and *Archer et al.* [2000b] show that $CaCO_3$ compensation could strongly offset any atmospheric CO_2 drawdown caused by an increase in the nutrient inventory. This compensation has to do with the fact that an increase in the surface ocean productivity tends also to enhance the formation of $CaCO_3$ shells, thus increasing the flux of $CaCO_3$ to the sea floor. This is equivalent to saying that the $CaCO_3$-to-organic matter rain ratio remains constant. In addition to an increase in $CaCO_3$ rain to the sea floor, the increased oxygen demand in the sediments shifts much of the sedimentary respiration from oxic chemistry, which promotes $CaCO_3$ dissolution, to anoxic reactions, which have a much lesser influence on pore-water pH. The sum of these effects is to sharply increase the rate of $CaCO_3$ burial in the ocean, throwing off the balance with the *Alk* input. In order to restablish the equilibrium between input and burial, the ocean has to dissolve more $CaCO_3$, which is accomplished by a reduction of the oceanic CO_3^{2-} ion content, permitting a much more shallow lysocline. A lower CO_3^{2-} ion content increases the mean ocean buffer factor, which reduces the ocean's ability to hold CO_2 and therefore increases atmospheric CO_2. This strongly offsets the initial response of the ocean carbon cycle to the increase in the nutrient inventory. *Sigman et al.* [1998] find that a 30% increase of the nutrient inventory initially draws atmospheric CO_2 down by 34 ppm, but that the compensation response increases pCO_2 of the atmosphere back up, for a total of only 18 ppm. Similar results are found by *Archer et al.* [2000b]. If the negative feedback by $CaCO_3$ compensation is indeed as strong as suggested by these studies, the nutrient inventory scenario can be rejected as a possible driver for the low glacial atmospheric CO_2. However, if the formation and export of $CaCO_3$ is uncoupled with general biological productivity, nutrient inventory changes can play a role.

Alkalinity and Carbonate Pump Scenarios

We have seen in the last example how changes in the ocean's carbonate pump can have substantial influence on atmospheric CO_2, particularly in the presence of $CaCO_3$ compensation feedbacks. Alternative mechanisms affecting the ocean's alkalinity balance include changes in river input and burial, which are independent of carbonate pump changes. Given the high sensitivity of atmospheric CO_2 to changes in the ocean's alkalinity balance, alkalinity-based mechanisms were among the first to be proposed (e.g., *Berger* [1982]; *Opdyke and Walker* [1992]; *Milliman* [1993]; see table 10.4.1). All of these intial hypotheses are linked to sea-level change, permitting us to reject them immediately, since atmospheric CO_2 started to rise several thousand years before sea level started to rise significantly during the last two deglaciations [*Broecker and Henderson*, 1998].

The sea-level constraint does not affect any scenario that calls for a reduction in the carbonate pump. Everything else being the same, a reduction in the carbonate pump would at first drastically reduce the burial of $CaCO_3$ at the sea floor. In order to compensate, the lysocline has to deepen to suppress dissolution until a new steady state between burial and alkalinity input has been achieved (see figure 9.5.2). As we noted previously, such a change in the lyscocline depth has not been observed. However, the situation may be more complex, since factors other than the deep ocean CO_3^{2-} content can control deep ocean dissolution of $CaCO_3$. We discussed in chapter 9 that the lysocline is only poorly related to the magnitude of the dissolution flux, particularly in cases where the flux of organic matter reaching the sediments is larger than the flux of $CaCO_3$. Respiration-enhanced dissolution of $CaCO_3$ in the sediments provides a means by which the dissolution of $CaCO_3$ in the sediments can be decoupled from the CO_3^{2-} content of the overlying waters.

Archer and Maier-Reimer [1994] argued that if the rain rate of organic matter to the sediment was higher during glacial times, the CO_3^{2-} concentration in the deep ocean could increase without a large change in the lysocline and the depth of the transition zone. Their proposed sequence is a follows: The increased flux of organic matter leads to intensified respiration-driven dissolution, which if not complete, reduces the CO_3^{2-} content of the overlying waters lifting the lysocline temporarily to shallower depths. Both this increased dissolution of $CaCO_3$ and the smaller downward flux of $CaCO_3$ lead to dramatically smaller deposition of $CaCO_3$ on the sea floor, thus resulting in a large alkalinity imbalance with the input. In order to compensate, the deep ocean CO_3^{2-} concentration has to increase until the bottom water CO_3^{2-} content balances the respiration-induced deficit of CO_3^{2-} in the upper sediments. The lysocline would then move back to its original position as observed. This would lead to a situation where the depth of the saturation horizon is strongly separated from the depth of the lysocline.

Support for the rain ratio scenario is provided by paleo-reconstructions of ocean pH with the help of boron isotopes [*Sanyal et al.*, 1995]. Boron pH values from the glacial deep Pacific and Atlantic seem to show an increase of about 0.1 to 0.3 pH units, which would correspond to an increase in the CO_3^{2-} concentration of about 40–100 μmol kg^{-1}, consistent with the rain ratio hypothesis. However, according to *Sigman et al.* [1998], the predicted strong separation between the saturation horizon and the lysocline cannot be sustained under such circumstances. When *Sigman et al.* [1998] halved the $CaCO_3$ rain in the low latitudes, the lysocline in their box model deepened by 1000–1500 m, thus drastically increasing the area of high-$CaCO_3$ sediments. This has not been observed in the sediment cores, thus putting the rain ratio scenario in doubt. However, *Archer et al.* [2000b] argue on the basis of their 3-D biogeochemistry model that the deepening of the lysocline is much smaller than modeled by *Sigman et al.* [1998] and not at odds with the observations. If the lysocline constraint by *Sigman et al.* [1998] is indeed not as strong as they suggested, then the rain ratio hypothesis emerges as one of the likely candidates explaining the low glacial atmospheric CO_2. However, many questions are still unanswered, and it is too early to either accept or reject this scenario. For instance, what are the reasons for the shift in the $CaCO_3$-to-organic carbon rain ratio? *Archer et al.* [2000b] show that such a change could occur as a consequence of a drastic increase in the input of $Si(OH)_4$ from weathering, which would promote the growth of diatoms at the expense of coccolithophorids, hence reducing the $CaCO_3$ formation rate in the surface ocean.

Brzezinski et al. [2002] recently proposed another plausible scenario for how a shift in the $CaCO_3$-to-organic carbon export ratio could have occurred, building on the premise that given extra $Si(OH)_4$, diatoms would tend to outcompete coccolithophorids. In this case, however, the extra $Si(OH)_4$ comes from the Southern Ocean as a result of a leakage of the normally existing Southern Ocean trap for $Si(OH)_4$ (see chapter 7). Under present-day conditions, $Si(OH)_4$ in the Southern Ocean is tighly confined to the Antarctic continent (see figure 7.1.1a), since it gets drawn down very quickly once it is upwelled south of the Polar Front and pushed equatorward by Ekman drift (see figure 7.3.3). In contrast, the region of high NO_3^- in the Southern Ocean extends much further north (figure 7.1.1b). When these waters reach the latitudes where Antarctic Intermediate Water (AAIW) and Subantarctic Mode Water (SAMW) are formed, $Si(OH)_4$ concentrations are essentially zero, while NO_3^- concentrations are still elevated. We have seen in chapter 7 that $Si^* = Si(OH)_4 - NO_3^-$ is a useful indicator of this strong deficiency of $Si(OH)_4$ relative to NO_3^- (figure 7.3.4). We also found that the resulting low-Si* waters, indicative of high concentrations of preformed NO_3^-, but containing no preformed $Si(OH)_4$, can be traced throughout the southern hemisphere thermocline, and even into the thermocline of the North Atlantic (figure 7.3.5). This extraordinary reach, coupled with the fact that diapycnal mixing in the ocean

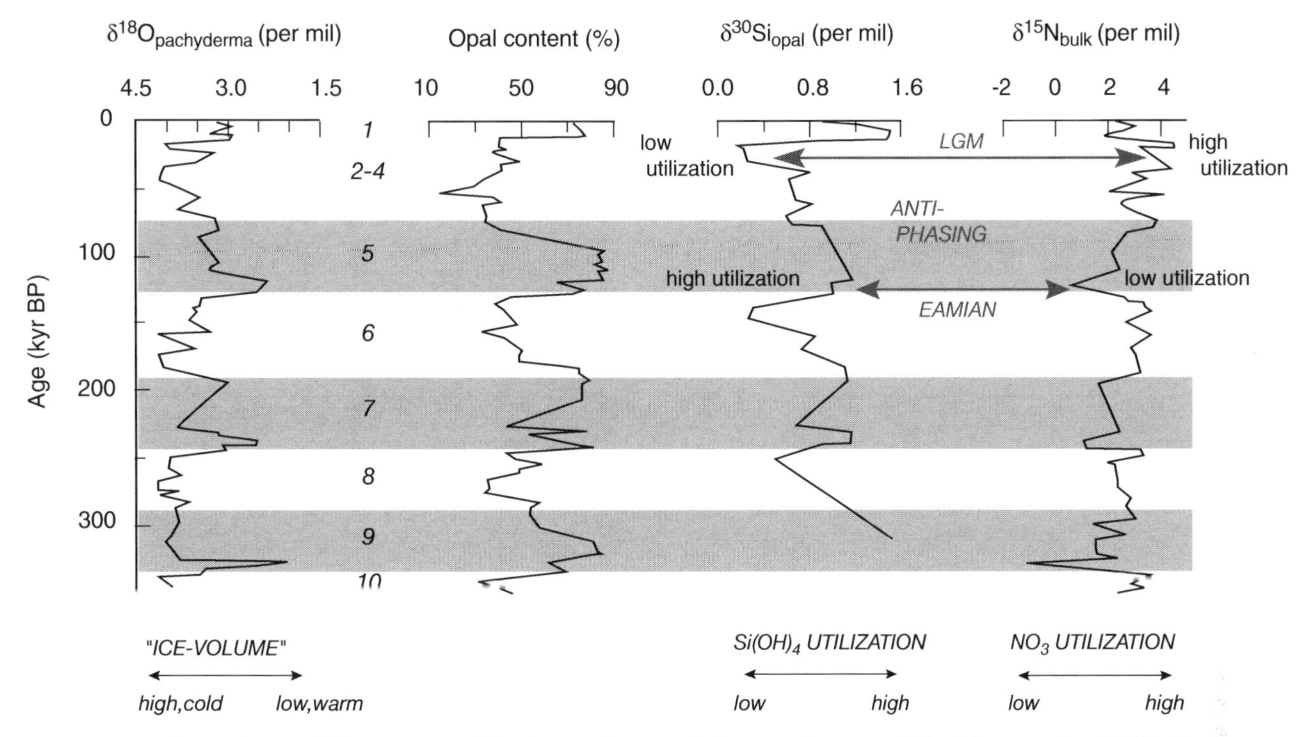

FIGURE 10.4.13: Time evolution of $\delta^{18}O$ in $CaCO_3$ of the foraminifera *N. pachyderma*, opal content, $\delta^{30}Si$ of diatom opal and $\delta^{15}N$ of bulk sediment, from a sediment core taken in the Antarctic zone of the South Atlantic (core RC13-259 [53°53'S, 4°56'W, 2677 m]). Numbers indicate oxygen isotope stages 1–10. The precision of the isotope analyses is 0.13 and 0.20‰ (±1 s.d.) for $\delta^{30}Si$ and $\delta^{15}N$ measurements, respectively. From *Brzezinski et al.* [2002].

tends to be small, has provided the foundation for the proposal that changes in the regions where AAIW and SAMW form could lead to substantial alterations of the nutrient distribution in the thermocline of the world's oceans, ultimately affecting biological productivity throughout the low latitudes [*Sarmiento et al.*, 2004a]. To illustrate the importance of this pathway for providing nutrients to the low latitudes, *Sarmiento et al.* [2004a] set all nutrients entering this pathway to zero, and found that productivity north of 30°S dropped by about 75%.

We discussed in chapter 7 that the trapping of $Si(OH)_4$ in the Southern Ocean, while substantial amounts of NO_3^- get exported into the low latitudes, is likely the result of iron limitation of diatom growth in the Southern Ocean, as diatoms are known to silicify much more heavily under such conditions. This opens the possibility that an increase in the availability of iron in the Southern Ocean through higher aeolian deposition during the LGM could have relieved the diatoms from their iron stress. Since diatoms tend to take up NO_3^- and $Si(OH)_4$ in a 1:1 ratio under iron-replete conditions, and since the $Si(OH)_4$ concentration of the water upwelled into the Southern Ocean is much higher than this ratio (see figure 7.2.4), this could have led to a situation where the surface waters feeding AAIW and SAMW were replete in both $Si(OH)_4$ and NO_3^-. Thus, the Southern Ocean trap for $Si(OH)_4$ would have turned into a region of $Si(OH)_4$ export, enriching the thermocline of the low latitudes with $Si(OH)_4$. When these waters make it to the surface in the low latitudes, they tend to stimulate the growth of diatoms at the cost of coccolithophorids, leading to a reduction in the export of $CaCO_3$ from the surface ocean. Through $CaCO_3$ compensation, this would lead to a substantial drawdown of atmospheric CO_2 [*Brzezinski et al.*, 2002]. Therefore, without changing biological productivity, an iron-induced change in diatom physiology in the Southern Ocean could lead to a change in phytoplankton community composition "downstream" in the low latitudes, with global carbon-cycle implications.

Matsumoto et al. [2002] investigated the magnitude of this effect in a multi-box model and concluded that it could explain an atmospheric CO_2 drawdown of about 30 to 40 ppm. They noted, however, that given the much higher low-latitude sensitivity or higher HBEI, three-dimensional models could generate a much larger drawdown, perhaps as large as 100 ppm. Evidence in support of this hypothesis comes from the observation that sediment records in the Antarctic region of the Southern Ocean show an antiphasing of $\delta^{30}Si$ and $\delta^{15}N$ (see figure 10.4.13). Since variations of $\delta^{30}Si$ are primarily reflecting the degree of Si utilization, i.e., directly analogous to $\delta^{15}N$ for N, this antiphasing has puzzled researchers for a long time, during the LGM, $\delta^{30}Si$ showed a lower degree of nutrient utilization, while $\delta^{15}N$ showed a higher degree of nutrient utilization [De La *Rocha et al.*, 1998]. The hypothesis of *Brzezinski et al.* [2002] now provides a good explanation for this puzzle, because it exactly predicts this antiphasing in the utilization of these two nutrients in the Antarctic zone.

The present debate about the role of the $CaCO_3$ pump as a driver for low glacial atmospheric CO_2 concentration is still associated with many uncertainties even at relatively fundamental levels. This is reflected in the fact that small changes in the formulation of how $CaCO_3$ compensation operates can make drastic changes in the model results [*Sigman et al.*, 1998]. Much progress can be expected in the next years, as more quantitative information about $CaCO_3$ cycling in the sediments becomes available, thus drastically reducing the number of uncertainties in the model parameterizations.

A Synthesis Scenario

The causes for the 80 ppm variations in atmospheric CO_2 associated with glacial-interglacial cycles remain elusive and represent one of the largest unsolved puzzles of the ocean carbon cycle. We have reviewed many hypotheses, rejected many of them because they are not in accordance with paleoceanographic evidence, and identified a few that look promising. It's important to note that we focused largely on single mechanisms to explain the entire glacial-interglacial change or at least a good fraction of it. However, the current evidence suggests that it is more likely that the atmospheric CO_2 change is caused by the interaction of several mechanisms rather than a single one. The remarkable degree of coupling between Earth's temperature and atmospheric CO_2 and the regularity of the global carbon cycle response to physical climate changes suggest that these mechanisms must be linked to each other in a predictible manner. In addition, mechanisms centered on or connected with Southern Ocean changes remain at the forefront. Not only do they tend to have much more impact on atmospheric CO_2 changes, but they would also explain the near-synchronicity between Antarctic temperature and atmospheric CO_2 during the LGM-Holocene transition, which does not exist for northern hemisphere high-latitude temperature (figure 10.4.12).

Following these arguments, *Sigman and Boyle* [2000] suggested a synthesis scenario centered around the Southern Ocean that combines several proposed scenarios into a single framework. We extend this synthesis scenario with the Si leakage mechanism of *Brzezinski et al.* [2002]. Figure 10.4.14a and b show the condition during interglacial (today's) conditions, while figure 10.4.14c and d depict the proposed conditions during the LGM. *Sigman and Boyle* [2000] suggest that the cooler conditions during the LGM relative to today caused a northward shift of the westerly wind belt, resulting in a decrease in upwelling of deep water in the Antarctic. In response, this region was more likely to develop a cold but fresh stable surface layer with frequent substantial ice coverage. This further reduced the ventilation of deep waters. In response, biological productivity decreased in the Antarctic, but the utilization of nitrate went up, indicating that the drop in the upward supply was larger than the reduction in biological export. The higher dust input from the atmosphere was not large enough to overcome the physically induced reduction in phytoplankton reduction in the Antarctic, but likely was large enough to alleviate diatoms from iron stress. This led to a lower Si utilization in this region. The glacial subantarctic was more productive, perhaps because of an increased supply of iron from dust. Subantarctic utilization of nitrate and phosphate did not change much, possibly because of increased nutrient supply from the thermocline below the subantarctic surface, offsetting the effect of the observed increase in productivity. These changes suggest a combined physically and biologically induced closure of the Southern Ocean window, thereby reducing atmospheric CO_2. In this scenario, the effect is dominated by the Antarctic.

These changes in the Southern Ocean were linked with changes throughout the rest of the ocean. Some of these changes outside the Southern Ocean are a consequence of the fact that changes in the nutrient concentrations of the Subantarctic have a disproportionally large impact on low-latitude productivity. In particular, we propose that the alleviation of iron stress of diatoms led to a leakage of the Southern Ocean Si trap, providing Si to the low latitudes. This is suggested to have altered the balance from coccolithophorids to diatoms, thereby reducing the export of $CaCO_3$ from the surface, and affecting atmospheric CO_2 through $CaCO_3$ compensation. Figure 10.4.14c also suggests that a stronger vertical separation of *DIC* and nutrients occurred during the LGM, with the mid-depth ocean losing *DIC* and nutrients, while the bottom ocean waters gained them. This would have brought changes of opposite sign to ocean interior O_2. This vertical separation would have caused a $CaCO_3$ dissolution event in the abyssal ocean at the onset of ice ages, further reducing atmospheric CO_2.

This synthesis hypothesis is consistent with many paleoproxies, but is far from being validated. Clearly, one of the major obstacles in the evaluation of this and other hypotheses is the uncertainty and ambiguities associated with the interpretation of proxy data. However, as more proxies are developed and the existing ones are expanded, these uncertainties will become smaller, permitting us to give more scrutiny to this hypothesis and the many others. For example, the information contained in the sequence of events during the initiation of glaciations or during the deglaciations has only begun to be used. Without doubt, the quest for the causes of the low glacial atmospheric CO_2 concentrations will remain an exciting research topic for the future.

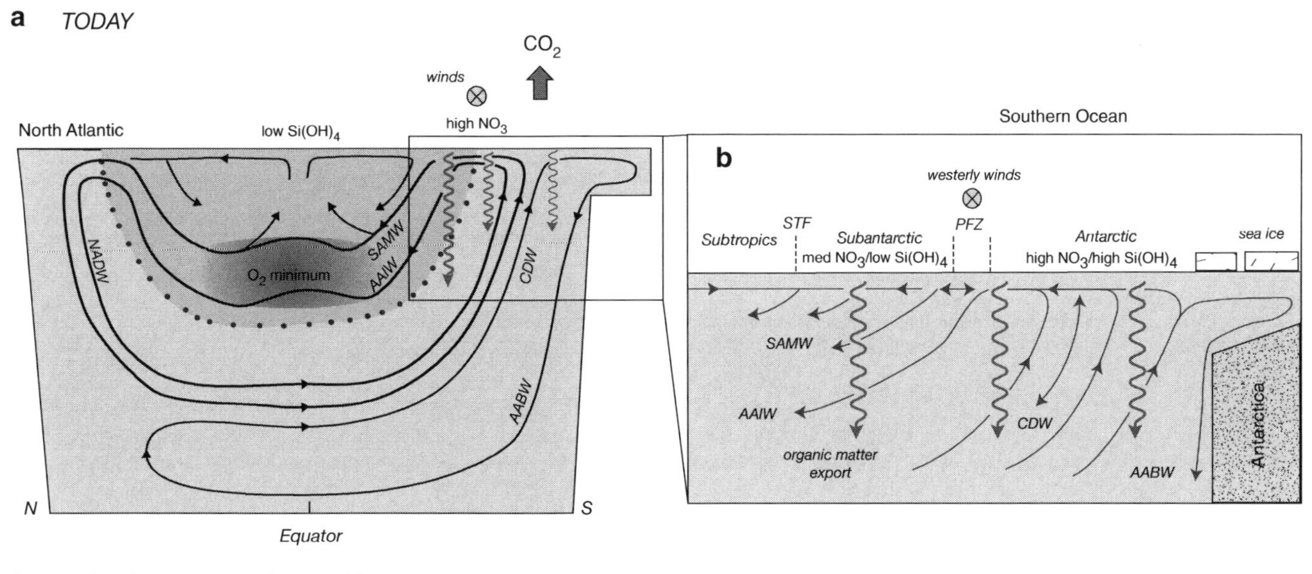

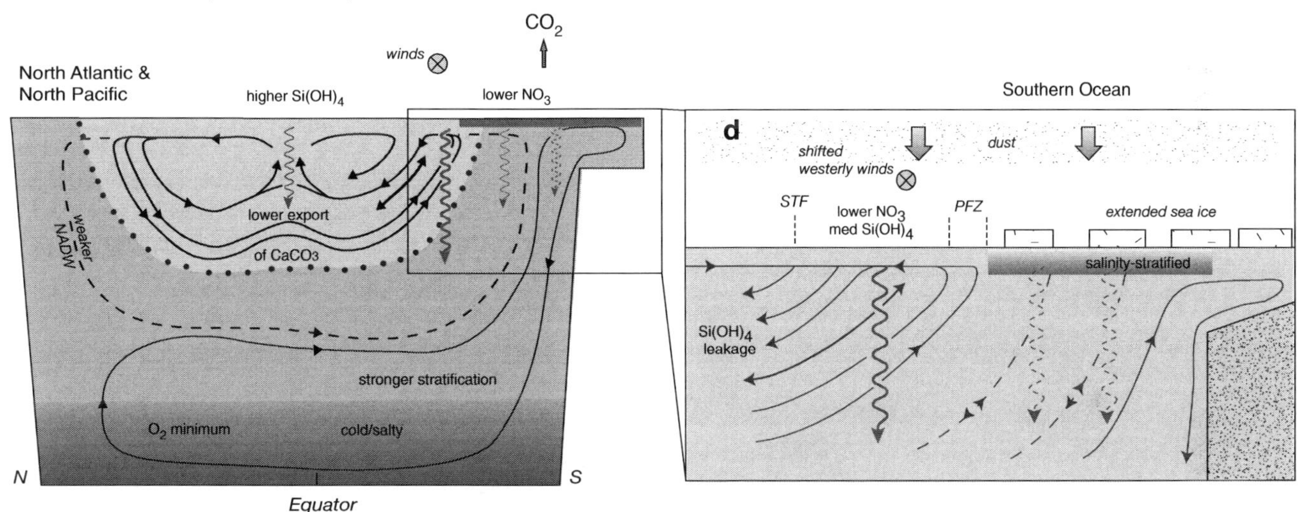

FIGURE 10.4.14: The modern ocean (a, b) and a Southern Ocean–based synthesis hypothesis for reduced levels of atmospheric CO_2 during glacial times (c, d). The figure shows a generalized depth section running from north to south (a, c), with an expanded view of the Southern Ocean (b, d). The meridional components of ocean circulation are shown as lines, while the intensity of circulation is depicted by the shading, with the lighter shade representing more vigorous circulation. This illustrates the hypothesis that ice-age circulation was less vigorous in the cold, dense, deep ocean but more vigorous in the warmer, less dense, upper ocean. In the modern ocean (a, b), circulation of the interior can be characterized as follows. New deep water forms in the high-latitude North Atlantic (North Atlantic Deep Water, NADW). NADW flows south and mixes with other deep waters (including Antarctic Bottom Water, AABW) to generate Circumpolar Deep Water (CDW). Eastward winds (a circled cross) over the Polar Frontal Zone (PFZ) drive upwelling of CDW into the Antarctic surface, releasing deep-sequestered CO_2 to the atmosphere (upward arrow in (a)). Antarctic surface water flows northward to the PFZ, where it sinks into the subsurface as Antarctic Intermediate Water (AAIW) or mixes with subantarctic surface water, some of which forms Subantarctic Mode Water (SAMW). The biological production and export of organic matter (arrows) extracts nutrients and inorganic carbon from surface waters. AAIW and SAMW ventilate the mid-depth ocean and supply nutrients to the lower-latitude surface. This nutrient supply fuels biological export production, the degradation of which causes a subsurface deficit in dissolved oxygen, which is most intense at intermediate depths (the O_2 minimum in (a)). The hypothetical ice-age Southern Ocean described in the text includes the following changes (c, d). Upwelling into the Antarctic surface decreased, possibly due to an equatorward shift in the belt of eastward winds, and/or a freshening of the surface ("salinity stratification") related to an increase in sea ice. In the subantarctic, the winds may have driven the upwelling of intermediate-depth waters, which formed in the North Atlantic and/or North Pacific. The supply of iron from dust was high (downward arrows and stippling). Organic carbon export was higher in the subantarctic but lower in the Antarctic (arrows). Possibly due to the greater supply of dust-borne iron, the degree of nitrate utilization was higher in both regions, leading to lower surface nitrate and less nutrient transport to the lower latitudes. This pattern of nutrient supply and export production caused the O_2 minimum to migrate into the abyssal ocean and reduced the release of CO_2 from the Southern Ocean. It is also proposed that the higher dust flux to the glacial Southern Ocean lowered the Si-to-NO_3^- uptake ratio of diatoms there, leading to a lower $Si(OH)_4$ utilization in the Subantarctic. As a result, some of the currently trapped $Si(OH)_4$ could have leaked into the low latitudes, causing a phytoplankton community shift from coccolithophorids to diatoms. The resulting lowered export of $CaCO_3$ relative to organic matter then would have aided in reducing atmospheric CO_2 through a rain ratio mechanism involving $CaCO_3$ compensation. Modified from *Sigman and Boyle* [2002].

Problems

10.1 Define and explain these terms:

a. Greenhouse effect

b. Climate sensitivity

c. Climate feedback

10.2 Calculate the expected equilibrium increase in global mean tropospheric temperature since preindustrial times for the following cases:

a. Radiative forcing from the increase in atmospheric CO_2 alone for the year 2000 (atmospheric CO_2 mixing ratio of 369 ppm)

b. Radiative forcing from the increase in atmospheric methane alone for the year 2000 (northern hemisphere atmospheric CH_4 mixing ratio of 1.8 ppm)

c. Radiative forcing from the increase in atmospheric CO_2 alone for the year 2100 under the IS92a scenario (CO_2 mixing ratio of 788 ppm)

Assume a climate sensitivity of $1.7\,W\,m^{-2}\,K^{-1}$. Hint: Use the caption in table 10.1.2 to obtain the change in radiative forcing for the change in atmospheric CO_2, and then use the information in this table and table 10.1.1 to obtain the change in forcing for CH_4.

10.3 Discuss why the addition of anthropogenic CO_2 to the ocean decreases the capacity of the ocean to take up further anthropogenic CO_2.

a. Show the relevant reactions and equations.

b. Compute the oceanic uptake fraction f for the preindustrial mean ocean (see table 8.2.4).

c. Compute how much the uptake fraction is reduced when 100 $\mu mol\,kg^{-1}$ DIC is added uniformly to the preindustrial ocean.

10.4 Explain why the surface ocean concentration of anthropogenic CO_2 is higher in low latitudes than it is in high latitudes. Why is it higher in the Atlantic than in the Pacific?

10.5 How long will it take for a pulse of CO_2 emitted into the atmosphere to be reduced to 50%, 20%, 10%, and 1% of its original value? For each answer list which process is the primary one responsible for the removal of CO_2 from the atmosphere at the point in time the threshold is crossed.

10.6 Direct injection into the ocean of liquid CO_2 captured from flue gases such as from fossil fuel–based power plants has been proposed as a means for reducing the buildup of atmospheric CO_2, while still permitting the use of fossil fuel to generate electrical power. Discuss this management option considering the following issues:

a. Compute the increase of the partial pressure of CO_2 if you add 10 ml of liquid CO_2 to 1 m^3 of seawater at a temperature of 20°C. Assume that the initial *DIC* is 2000 mmol m^{-3}, while *Alk* remains constant at 2300 mmol m^{-3}. The density of liquid CO_2 is about 760 kg m^{-3} at this temperature and 1 atm pressure. The molecular weight of CO_2 is 44. Hint: Use the lookup table for problem 8.14 in chapter 8 for computing the oceanic pCO_2 for a given *DIC* and *Alk*.

b. Assume that we want to inject 10% of the current global CO_2 emissions, which are about 7 Pg C yr^{-1}. How much seawater do we need per year to dilute these injections with the same dilution factor given above (which is about 1:10^5)? Convert your answer into Sverdrups, i.e., 10^6 m^3 s^{-1}. Contrast this number with the flow of the Amazon, which is about 0.2 Sv.

c. Given what you know about oceanic circulation, where would you inject the liquid CO_2 in order to maximize the oceanic retention?

d. Where would most of the CO_2 that the ocean cannot retain be emitted back into the atmosphere?

10.7 Explain the atmospheric oxygen method to estimate the oceanic uptake of anthropogenic CO_2. Discuss its major assumptions.

10.8 Explain the apparent paradox that the tropical Pacific is viewed as being a large sink for anthropogenic CO_2, despite the fact that it is a region of net outgassing of CO_2.

10.9 The role of ocean biology in taking up additional CO_2 from the atmosphere in response to rising atmospheric CO_2 has been an issue of intense debate. To address this question, it is helpful to clearly distinguish a number of different cases:

a. Constant ocean circulation, constant temperature, and constant oceanic biota

b. As (a), but reduced biological export production

c. As (a), but a decrease in upper-ocean thermocline ventilation

d. As (a), but reduced rate of $CaCO_3$ formation and export

For each case discuss the role of ocean biology for the net exchange of CO_2 between the ocean and the atmosphere.

10.10 Climate models suggest that the ocean will take up more than 1 petajoule (10^{15} J) of heat over the next few decades. Estimate the net loss of CO_2 from the ocean in response to this heat uptake. Hint: See equation (10.2.32).

10.11 Explain why, in the climatological mean state, the partial pressure of CO_2 in the tropical Pacific is inversely proportional to SST, i.e., why high pCO_2

occurs in regions of low SST and vice versa. Discuss the implications of this finding for the response of the surface ocean pCO_2 to variations associated with ENSO.

10.12 Discuss the impact of ENSO on the air-sea exchange of natural and anthropogenic CO_2 in the tropical Pacific.

10.13 Use the three-box model of the equatorial Pacific (figure 10.3.7) for investigating the impact of ENSO on surface ocean properties and atmospheric CO_2. (Note that this model is structurally very similar to the three-box model of the global ocean, except that the overturning circulation is reversed.) The area of the equatorial surface box is 22×10^{12} m^2 and that of the mid-latitude box is 100×10^{12} m^2. The depths are given in figure 10.3.7.

a. Write down the phosphate balance equations for the three boxes given the assumption that biological uptake and export depend linearly on the phosphate concentration, i.e., $\Phi = \lambda \cdot [PO_4^{3-}]$, and that all organic phosphorus exported from the surface boxes remineralizes within the thermocline box.

b. Plot the phosphate concentration in the equatorial and mid-latitude box as a function of the upwelling mass flux, F_u. Set the exchange flux F_{MT} to 25 Sv. Assume a mean phosphate concentration of 1 mmol m^{-3}. Assume for λ a value of 1/30 day^{-1}.

c. Discuss the expected biologically induced air-sea exchange fluxes, considering what you learned in chapter 8 about the relationship between residual nutrients and this biologically induced flux. Assume a mean *DIC* concentration of 2100 mmol m^{-3} and a constant carbon-to-phosphorus stoichiometry, $r_{C:PO4}$ of 117:1.

d. Add *DIC* and *Alk* to the model, but do not yet consider air-sea exchange. Compute the *DIC* concentration, and the pCO_2 in the two surface boxes. Assume a constant *Alk* of 2320 μmol kg^{-1} and use the lookup table for problem 8.14 in chapter 8. The temperature of the equatorial upwelling box is 22°C and that of the mid-latitude box is 26°C.

e. Compute the exchange flux of CO_2 between the two surface boxes and the atmosphere for a fixed atmospheric CO_2 of 360 μatm. Assume that this exchange is a small fraction of the total carbon balance, permitting you to neglect it in the computation of the surface *DIC*. Use the global mean gas exchange coefficient for CO_2 given in chapter 3.

f. Investigate how the air-sea gas exchange varies as you change the upwelling flux from the thermocline, F_u, from 55 Sv (for La Niña condition) to 5 Sv (for El Niño condition). (To keep the problem manageable, keep atmospheric CO_2 constant as before.) If you keep the anomalous air-sea flux constant for a year, how much would you change atmospheric CO_2, actually?

g. Finally, consider the fully coupled system, i.e., make the carbon in the atmosphere an integral part of the carbon balance in the system. This problem requires a numerical solution.

10.14 List and discuss the various lines of evidence which suggest that the amount of carbon stored in the terrestrial biosphere was much lower during the last glacial maximum than it is today.

10.15 Why can we assume that only a small fraction of the carbon lost from the terrestrial biosphere during glacial times remained in the atmosphere?

Explain in detail the role of the oceanic $CaCO_3$ cycle, and what impact the loss of terrestrial CO_2 has on the location of the $CaCO_3$ saturation horizon.

10.16 Explain why hypotheses to explain the lower atmospheric CO_2 during the last glacial maximum have focused so strongly on the high latitudes, in particular the Southern Ocean.

10.17 Refer to the global three-box model depicted in figure 10.4.3.

a. Derive equations (10.4.8) through (10.4.10), i.e., derive an expression that relates the high-latitude surface ocean *DIC* concentration to the deep ocean *DIC* and phosphate concentrations as well as to a number of mixing terms.

b. Assuming that deep ocean *DIC* and phosphate concentrations are constant, draw a contour plot of the high-latitude surface *DIC* and phosphate concentration as a function of high-latitude productivity, Φ_h^P, and the high-latitude–deep ocean mixing term, f_{hd}.

c. Given the idea of high-latitude dominance in determining atmospheric CO_2, what do your results in (b) suggest with regard to how changes in the high-latitude productivity and the high-latitude–deep ocean mixing term affect atmospheric CO_2?

10.18 Explain how $CaCO_3$ compensation works, and why this process is believed to be important when considering glacial-interglacial atmospheric CO_2 variations.

10.19 Investigate the impact of a more efficient biological carbon pump on atmospheric CO_2 with the two-box model depicted in figure 1.2.3. In contrast to how we parameterized biological export production in chapter 1, i.e., by setting the surface ocean concentration of phosphate to zero, assume that biological export depends linearly on the phosphate concentration, i.e., $\Phi = \lambda \cdot [PO_4^{3-}]$. Compute how atmospheric CO_2 changes as you vary λ. For simplicity, assume that the atmospheric pCO_2 is equal to the surface ocean pCO_2. Assume a constant *Alk* of 2320 μmol kg^{-1} and a surface ocean temperature of 20°C.

Appendix

Table A.1
Earth-System Quantities

	Description	Value#	Units
M_{oc}	Mass of ocean	1.35×10^{21}	kg
V_{oc}	Volume of ocean	1.34×10^{18}	m^3
$V_{oc\ sfc}$	Volume of surface ocean (0–50 m)	1.81×10^{16}	m^3
$V_{oc\ deep}$	Volume of deep ocean (> 1200 m)	9.44×10^{17}	m^3
A_{oc}	Area of ocean	358×10^{12}	m^2
$A_{oc\ icefree}$	Annual mean ice-free area of ocean	$\sim 332 \times 10^{12}$	m^2
A_{ATL}	Area of Atlantic (>45°S)	75×10^{12}	m^2
A_{PAC}	Area of Pacific (>45°S)	151×10^{12}	m^2
A_{IND}	Area of Indian Ocean (>45°S)	57×10^{12}	m^2
A_{SO}	Area of Southern Ocean (<45°S)	60×10^{12}	m^2
A_{ARCTIC}	Area of Arctic	9.6×10^{12}	m^2
A_{MED}	Area of enclosed seas (Mediterranean, etc)	4.5×10^{12}	m^2
H_{oc}	Mean depth of ocean	3690	m
h_{ml}	Global mean mixed layer depth	67	m
M_{atm}	Dry mass of atmosphere	5.132×10^{18}	kg
N_{atm}	Moles in atmosphere	1.773×10^{20}	moles
V_{river}	River flow	3.7×10^{13}	$m^3\ yr^{-1}$

Depth and volumetric data of the ocean were computed from ETOPO2 [*National Geophysical Data Center*, 2001]. Sea-ice coverage is from *Ropelewski* [1995]. Mixed layer depth is from the climatology of *Kara et al.* [2003].

Table A.2
Biogeochemical/Physical Properties

	Description	Value	Units
C_D	Typical drag coefficient	10^{-3}	
g	Gravitational acceleration of Earth	9.81	$m\ s^{-1}$
Ω	Angular velocity of Earth	7.3×10^{-5}	radians s^{-1}
R	Gas constant	0.082053	liter atm $K^{-1}\ mol^{-1}$
V_{ideal}	Molar volume of an ideal gas	22.4136	liter mol^{-1}
k_w	Global mean gas transfer velocity at $Sc = 660$ (cf. table 3.3.2)	21	cm hr^{-1}
S_{CO_2}	Solubility of CO_2 at 15°C (cf. Table 3.2.3)	38,300	mmol $m^{-3}\ atm^{-1}$
$\lambda_{^{14}C}$	^{14}C decay constant	1.21×10^{-4}	yr^{-1}
$\tau_{^{14}C}$	Half-life of ^{14}C	5730	yr
$\tau_{^3He}$	Half-life of ^3H (Tritium)	12.4	yr
$\tau_{^{222}Rn}$	Half-life of ^{222}Rn	3.825	day

TABLE A.3
Average Ocean Properties#

Tracer	Global Mean	Global Mean Surface (0–50 m)	Low-latitude Surface Mean (45° S–45° N)	NH High-Latitude Surface Mean (>45° N)	SH High-Latitude Surface Mean (<45° S)	Global Mean Thermocline (50–1200m)	Global Mean Deep Ocean (>1200 m)
Temperature (°C)	3.53	17.88	22.85	4.21	3.22	7.42	1.92
Salinity	34.72	34.77	35.13	33.12	34.06	34.70	34.72
Density (kg m^{-3})	1027.51	1024.67	1023.96	1026.12	1027.02	1026.97	1027.75
NO_3^- (mmol m^{-3})	31.0	5.7	1.8	7.8	21.4	26.8	32.8
PO_4^{3-} (mmol m^{-3})	2.17	0.57	0.31	0.89	1.54	1.93	2.28
$Si(OH)_4$ (mmol m^{-3})	92.0	7.7	2.9	12.3	26.0	43.8	110.3
O_2 (mmol m^{-3})	178	249	222	329	323	163	182
AOU (mmol m^{-3})	154	3	0	3	12	141	161
DIC (μmol kg^{-1}) (~1990s)	2255	2026	2003	2049	2119	2203	2280
Alk (μmol kg^{-1})	2364	2308	2315	2257	2291	2322	2381
^{14}C (per mil) (~1990s)	−140	61	82	47	−21	−63	−173
DIC_{pi} (μmol kg^{-1}) (preindustrial, DIC-C_{ant})	2249	1983	1958	2006	2082	2184	2279
C_{ant} (μmol kg^{-1}) (anthropogenic perturbation, estimate)	6	43	45	44	37	19	0
$^{14}C_{pre\text{-}bomb}$ (per mil) (ca 1950, estimate)	−152	−70	−62	−87	−98	−114	−170
$^{14}C_{bomb}$ (per mil) (bomb perturbation, estimate)	12	131	144	134	77	51	0

Temperature, salinity, nutrient, and oxygen data are from the World Ocean Atlas 2001 [*Conkright et al.*, 2002]. *DIC*, *Alk*, ^{14}C and the derived values are from GLODAP [*Key et al.*, 2004]. As GLODAP is based primarily on WOCE data, the values listed for *DIC*, *Alk*, and ^{14}C are therefore representative for the early to mid-1990s. Note that GLODAP does not include the Arctic and some of the enclosed basins (e.g. Mediterranean). Therefore the estimates are biased slightly toward the lower latitudes, particularly the estimates for the Northern hemisphere high-latitude region. Anthropogenic CO_2, C_{ant} has been estimated using the ΔC^* technique of *Gruber et al.* [1996], and the separation of the observed ^{14}C into a bomb and pre-bomb component was done using the potential alkalinity method of *Rubin and Key* [2002] (see *Key et al.* [2004] for details).

References

Abraham, E. R., C. S. Law, P. W. Boyd, S. J. Lavender, M. T. Maldonado, and A. R. Bowie (2000), Importance of stirring in the development of an iron-fertilized phytoplankton bloom, *Nature, 407,* 727–730.

Adams, J. M., and H. Faure (1998), A new estimate of changing carbon storage on land since the Last Glacial Maximum, based on global land ecosystem reconstruction, *Global Planet. Change, 16–17,* 3–24.

Adams, R.L.P., J. T. Knowler, and D. P. Leader (1986), *The Biochemistry of the Nucleic Acids,* Chapman and Hall, New York.

Adkins, J. F., K. McIntyre, and D. P. Schrag (2002), The salinity, temperature, and $\delta^{18}O$ of the glacial deep ocean, *Science, 298,* 1769–1773.

Agawin, N.S.R., C. M. Duarte, and S. Agustí (2000a), Nutrient and temperature control of the contribution of picoplankton to phytoplankton biomass and production, *Limnol. Oceanogr., 45,* 591–600.

Agawin, N.S.R., C. M. Duarte, and S. Agustí (2000b), Nutrient and temperature control of picoplankton biomass and production, *Limnol. Oceanogr., 45,* 1891.

Alldredge, A. L., and Y. Cohen (1987), Can microscale chemical patches persist in the sea? Microelectrode study of marine snow, fecal pellets, *Science, 235,* 689–691.

Alldredge, A. L., and C. C. Gotschalk (1989), Direct observations of the mass flocculation of diatom blooms: Characteristics, settling velocities and formation of diatom aggregates, *Deep-Sea Res., 36,* 173.

Alldredge, A. L., and M. W. Silver (1988), Characteristics dynamics and significance of marine snow, *Prog. Oceanogr., 20,* 41–82.

Aller, R. C. (1994), Bioturbation and remineralization of sedimentary organic matter: Effects of redox oscillation, *Chem. Geol., 114,* 331–345.

Aller, R. C. (1998), Mobile deltaic and continental shelf muds as suboxic, fluidized bed reactors, *Mar. Chem., 61,* 143–155.

Aller, R. C., N. E. Blair, Q. Xia, and P. D. Rude (1996), Remineralization rates, recycling, and storage of carbon in Amazon shelf sediments, *Continental Shelf Research, 16,* 753–786.

Alley, R. B., et al. (2003), Abrupt climate change, *Science, 299,* 2005–2010.

Alperin, M. J., D. B. Albert, and C. S. Martens (1994), Seasonal variations in production and consumption rates of dissolved organic carbon in an organic-rich coastal sediment, *Geochim. Cosmochim. Acta, 58,* 4909–4929.

Alperin, M. J., C. S. Martens, D. B. Albert, I. B. Suayah, L. K. Benninger, N. E. Blair, and R. A. Jahnke (1999), Benthic fluxes and porewater concentration profiles of dissolved organic carbon in sediments from the North Carolina continental slope, *Geochim. Cosmochim. Acta, 63,* 427–448.

Altabet, M. A., and R. Francois (1994), Sedimentary nitrogen isotopic ratio as a recorder for surface ocean nitrate utilization, *Global Biogeochem. Cycles, 8,* 103–116.

Altabet, M. A., R. Francois, D. W. Murray, and W. L. Prell (1995), Climate-related variations in denitrification in the Arabian Sea from sediment $^{15}N/^{14}N$ ratios, *Nature, 373,* 506–509.

Amon, R.M.W., and R. Benner (1994), Rapid cycling of high-molecular-weight dissolved organic matter in the ocean, *Nature, 369,* 549–552.

Amon, R.M.W., and R. Benner (1996), Bacterial utilization of different size classes of dissolved organic matter, *Limnol. Oceanogr., 41,* 41–51.

Anderson, L. A. (1995), On the hydrogen and oxygen content of marine phytoplankton, *Deep-Sea Res. I, 42,* 1675–1680.

Anderson, L. A., and J. L. Sarmiento (1994), Redfield ratios of remineralization determined by nutrient data analysis, *Global Biogeochem. Cycles, 8,* 65–80.

Anderson, L. A., and J. L. Sarmiento (1995), Global ocean phosphate and oxygen simulations, *Global Biogeochem. Cycles, 9,* 621–636.

Andersson, J. H., J.W.M. Wijsman, P.M.J. Herman, J. J. Middelburg, K. Soetaert, and C. Heip (2004), Respiration patterns in the deep ocean, *Geophys. Res. Lett., 31.*

Archer, D. [E.] (1991), Modeling the calcite lysocline, *J. Geophys. Res., 96,* 17037–17050.

Archer, D. E. (1996a), An atlas of the distribution of calcium carbonate in sediments of the deep sea, *Global Biogeochem. Cycles, 10,* 159–174.

Archer, D. [E.] (1996b), A data-driven model of the global calcite lysocline, *Global Biogeochem. Cycles, 10,* 511–526.

Archer, D. [E.], and A. H. Devol (1992), Benthic oxygen fluxes on the Washington Shelf and Slope: A comparison of *in situ* microelectrode and chamber flux measurements, *Limnol. Oceanogr., 37,* 614–629.

Archer, D. E., and K. Johnson (2000), A model of the iron cycle in the ocean, *Global Biogeochem. Cycles, 14,* 269–279.

References

Archer, D. [E.], and E. Maier-Reimer (1994), Effect of deep-sea sedimentary calcite preservation on atmospheric CO_2 concentration, *Nature, 367,* 260–263.

Archer, D. [E.], S. Emerson, and C. Reimers (1989), Dissolution of calcite in deep-sea sediments: pH and O_2 microelectrode results, *Geochim. Cosmochim. Acta, 53,* 2831–2845.

Archer, D. [E.], M. Lyle, K. Rodgers, and P. Froelich (1993), What controls opal preservation in tropical deep-sea sediments? *Paleoceanogr., 8,* 7–21.

Archer, D. [E.], H. Kheshgi, and E. Maier-Reimer (1997), Multiple timescales for the neutralization of fossil fuel CO_2, *Geophys. Res. Lett., 24,* 405–408.

Archer, D. [E.], H. Kheshgi, and E. Maier-Reimer (1998), Dynamics of fossil fuel CO_2 neutralization by marine $CaCO_3$, *Global Biogeochem. Cycles, 12,* 259–276.

Archer, D. E., G. Eshel, A. Winguth, W. Broecker, R. Pierreumbert, M. Tobis, and R. Jacob (2000a), Atmospheric *p*CO_2 sensitivity to the biological pump in the ocean, *Global Biogeochem. Cycles, 14,* 1219–1230.

Archer, D. [E.], D. Lea, and N. Mahowald (2000b), What caused the glacial/interglacial atmospheric pCO_2 cycles? *Reviews of Geophysics, 38,* 159–189.

Archer, D. E., P. A. Martin, J. Milovich, V. Brovkin, G.-K. Plattner, and C. Ashendel (2003), Model sensitivity in the effect of Antarctic sea ice and stratification on atmospheric pCO_2, *Paleoceanogr., 18,* 1012, doi:10.1029/2002PA000760.

Armstrong, R. A. (1994), Grazing limitation and nutrient limitation in marine ecosystems: Steady state solutions of an ecosystem model with multiple food chains, *Limnol. Oceanogr., 39,* 597–608.

Armstrong, R. [A.], S. Bollens, B. Frost, M. Landry, M. Landsteinerr, and J. Moisan (1994), Food webs in biological/physical modeling of upper ocean processes, Technical Report, 25–35 pp., Woods Hole Oceanographic Institute, WHOI-94-32, Woods Hole.

Armstrong, R. A., C. Lee, J. I. Hedges, S. Honjo, and S. K. Wakeham (2002), A new, mechanistic model for organic carbon fluxes in the ocean based on the quantitative association of POC with ballast mineral, *Deep-Sea Research II, 49,* 219–236.

Arnarson, T. S., and R. G. Keil (2001), Organic-mineral interactions in marine sediments studied using density fractionation and X-ray photoelectron spectroscopy, *Org. Geochem., 32,* 1401–1415.

Arnosti, C. (2004), Speed bumps and barricades in the carbon cycle: Substrate structural effects on carbon cycling, *Mar. Chem., 92,* 263–273.

Arrigo, K. R., D. H. Robinson, D. L. Worthen, R. B. Dunbar, G. R. DiTullio, M. VanWoert, and M. P. Lizotte (1999), Phytoplankton community structure and the drawdown of nutrients and CO_2 in the Southern Ocean, *Science, 283,* 365–367.

Arrigo, K. R., G. DiTullio, R. Dunbar, D. Robinson, M. VanWoert, D. L. Worthen, and M. Lizotte (2000), Phytoplankton taxonomic variability in nutrient utilization and primary production in the Ross Sea, *J. Geophys. Res., 105,* 8827–8846.

Asher, W., and R. Wanninkhof (1998), Transient tracers and air-sea gas transfer, *J. Geophys. Res., 103,* 15,939–15,958.

Aufdenkampe, A. K., J. J. McCarthy, M. Rodier, C. Navarette, J. Dunne, and J. W. Murray (2001), Estimation of new production in the tropical Pacific, *Global Biogeochem. Cycles, 59,* 101–112.

Aumont, O., J. C. Orr, P. Monfray, G. Madec, and E. Maier-Reimer (1999), Nutrient trapping in the equatorial Pacific: The ocean circulation solution, *Global Biogeochem. Cycles, 13,* 351–369.

Azam, F. (1998), Microbial control of oceanic carbon flux: The plot thickens, *Science, 280,* 694–696.

Azam, F., and R. A. Long (2001), Sea snow microcosms, *Nature, 414,* 495–498.

Azam, F., T. Fenchel, J. G. Field, J. S. Gray, L. A. Meyer-Reil, and F. Thingstad (1983), The ecological role of water-column microbes in the sea, *Marine Ecology Progress Series, 10,* 257–263.

Bacastow, R. B. (1976), Modulation of atmospheric carbon dioxide by the Southern Oscillation, *Nature, 261,* 116–118.

Bacastow, R. B. (1996), The effect of temperature change of warm surface waters of the oceans on atmospheric CO_2, *Global Biogeochem. Cycles, 10,* 319–333.

Bacastow, R. B., and E. Maier-Reimer (1990), Ocean circulation model of the carbon cycle, *Climate Dynamics, 4,* 95–125.

Bacastow, R. B., J. A. Adams, C. D. Keeling, D. J. Moss, T. P. Whorf, and C. S. Wong (1980), Atmospheric carbon dioxide, the Southern Oscillation, and the weak 1975 El Niño, *Science, 210,* 66–68.

Bacon, M. P., and R. F. Anderson (1982), Distribution of thorium isotopes between dissolved and particulate forms in the deep sea, *J. Geophys. Res., 87,* 2045.

Bacon, M. P., C. A. Huh, and R. M. Moore (1989), Vertical profiles of some natural rationuclides over the Alpha Ridge Arctic Ocean, *Earth Planet. Sci. Lett., 95,* 15–22.

Baes, C. F. (1982), Effects of ocean chemistry and biology on atmospheric carbon dioxide, in *1982 Carbon Dioxide Review,* edited by W. C. Clark, pp. 189–211, Oxford University Press, Oxford.

Balch, W. M., and K. Kilpatrick (1996), Calcification rates in the equatorial Pacific along 140°W, *Deep-Sea Res. II, 43,* 971–993.

Baldauf, S. L. (2003), The deep roots of eukaryotes, *Science, 300,* 1703–1706.

Bange, H., Rapsomanikis, S., and Andreae, M. O. (2001), Nitrous oxide cycling in the Arabian Sea, *J. Geophys. Res. 106,* 1053–1066.

Barber, R. T., and F. P. Chavez (1983), Biological consequences of El Niño, *Science, 222,* 1203–1210.

Bard, E., B. Hamelin, M. Arnold, L. Motaggioni, G. Cabioch, G. Faure, and F. Rougerie (1996), Deglacial sea-level record from Tahiti corals and the timing of global meltwater discharge, *Nature, 382,* 241–244.

Bard, E., R. Rostek, and C. Sonzogni (1997), Interhemispheric synchrony of the last deglaciation inferred from alkenone paleothermometry, *Nature, 385,* 707–710.

Barnes, R. O., and E. D. Goldberg (1976), Methane production and consumption in anoxic marine sediments, *Geology, 4,* 297–300.

Baskaran, M., P. H. Santschi, G. Benoit, and B. D. Honeyman (1992), Scavenging of thorium isotopes by colloids in seawater of the Gulf of Mexico, *Geochim. Cosmochim. Acta, 56,* 3375–3388.

Bates, N. R., A. F. Michaels, and A. H. Knap (1996a), Alkalinity changes in the Sargasso Sea: geochemical evidence of calcification? *Mar. Chem.*, 51 (4): 347–358.

Bates, N. R., A. F. Michaels, and A. H. Knap (1996b), Seasonal and interannual variability of the oceanic carbon dioxide system at the U. S. JGOFS Bermuda Atlantic time-series study site, *Deep-Sea Res. II*, *43*, 347–383.

Battle, M., M. L. Bender, P. P. Tans, J.W.C. White, J. T. Ellis, T. Conway, and R. J. Francey (2000), Global carbon sinks and their variability inferred from atmospheric O_2 and $\delta^{13}C$, *Science*, *287*, 2467–2470.

Bauer, J. E., and E.R.M. Druffel (1998), Ocean margins as a significant source of organic matter to the deep open ocean, *Nature*, *392*, 482–485.

Bauer, J. E., P. M. Williams, and E.R.M. Druffel (1992), ^{14}C activity of dissolved organic carbon fractions in the north-central Pacific and Sargasso Sea, *Nature*, *357*, 667–670.

Bauer, J. E., E.R.M. Druffel, P. M. Williams, D. M. Wolgast, and S. Griffin (1998), Temporal variability in dissolved organic carbon and radiocarbon in the eastern North Pacific Ocean, *J. Geophys. Res.*, *103*, 2867–2881.

Beck, J. W., J. Récy, F. Taylor, R. L. Edwards, and G. Cabioch (1997), Abrupt changers in early Holocene tropical sea surface temperature derived from coral records, *Nature*, *385*, 705–707.

Beerling, D. J. (1999), New estimates of carbon transfer to terrestrial ecosystems between the last glacial maximum and the Holocene, *Terra Nova*, *11*, 162–167.

Behrenfeld, M. J., and P. G. Falkowski (1997), A consumers guide to phytoplankton primary production models, *Limnol. Oceanogr.*, *42*, 1479–1491.

Bender, M. L., and D. T. Heggie (1984), Fate of organic carbon reaching the deep sea floor: A status report, *Geochim. Cosmochim. Acta*, *51*, 1345–1364.

Bender, M. [L.], R. Jahnke, R. Weiss, W. Martin, D. T. Heggie, J. Orchardo, and T. Sowers (1989), Organic carbon oxidation and benthic nitrogen and silica dynamics in San Clemente Basin, a continental borderland site, *Geochim. Cosmochim. Acta*, *53*, 685–697.

Benner, R. (2002), Chemical composition and reactivity, in *Biogeochemistry of Marine Dissolved Organic Matter*, edited by D. A. Hansell and C. A. Carlson, pp. 59–90, Academic Press, New York.

Benner, R., J. D. Pakulski, M. McCarthy, J. I. Hedges, and P. G. Hatcher (1992), Bulk chemical characteristics of dissolved organic matter in the ocean, *Science*, *255*, 1561–1564.

Berelson, W. M. (2001), The flux of particulate organic carbon into the ocean interior: A comparison of four U.S. JGOFS regional studies, *Oceanography*, *14*, 59–67.

Berelson, W. M. (2002), Particle settling rates increase with depth in the ocean, *Deep-Sea Res. II*, *49*, 237–251.

Berelson, W. M., D. E. Hammond, and K. S. Johnson (1987), Benthic fluxes and the cycling of biogenic silica and carbon in two southern California borderland basins, *Geochim. Cosmochim. Acta*, *51*, 1345–1363.

Berelson, W. M., D. E. Hammond, D. O'Neil, X.-M. Xu, C. Chin, and J. Zukin (1990), Benthic fluxes and pore water studies from sediments of the central equatorial north Pacific: Nutrient diagenesis, *Geochim. Cosmochim. Acta*, *54*, 3001–3012.

Berelson, W. M., D. E. Hammond, J. McManus, and T. E. Kilgore (1994), Dissolution kinetics of calcium carbonate in the equatorial Pacific sediments, *Global Biogeochem. Cycles*, *8*, 219–235.

Berger, W. H. (1970), Planktonic foraminifera: Selective solution and the lysocline, *Marine Geology*, *8*, 111–138.

Berger, W. H. (1982), Increase of carbon dioxide in the atmosphere during deglaciation: The coral reef hypothesis, *Naturwissenschaften*, *69*, 87–88.

Berger, W. H., and G. Wefer (1990), Export production: Seasonality and intermittency, and paleoceanographic implications, *Paleoceanogr.*, *89*, 245–254.

Berner, E. K., and R. A. Berner (1987), *The Global Water Cycle*, Prentice-Hall, Englewood Cliffs, NJ.

Berner, R. A. (1964), An idealized model of dissolved sulfate distribution in recent sediments, *Geochim. Cosmochim. Acta*, *28*, 1497–1503.

Berner, R. A. (1980), *Early Diagenesis: A Theoretical Approach*, 241 pp., Princeton University Press, Princeton, NJ.

Berner, R. A. (1982), Burial of organic carbon and pyrite sulfur in the modern ocean: Its geochemical and environmental significance, *Am. J. Sci.*, *282*, 451–473.

Berner, R. A. (1995), Sedimentary organic matter preservation: An assessment and speculative synthesis—a comment, *Mar. Chem.*, *49*, 121–122.

Berner, R. A. (2001), Modeling atmospheric O_2 over Phanerozoic time, *Geochim. Cosmochim. Acta*, *65*, 685–694.

Berner, R. A., and D. E. Canfield (1989), A new model of atmospheric oxygen over Phanerozoic time, *Am. J. Sci.*, *289*, 333–361.

Berner, R. A., and S. Honjo (1981), Pelagic sedimentation of aragonite: Its geochemical significance, *Science*, *211*, 940–942.

Berner, R. A., A. C. Lasaga, and R. M. Garrels (1983), The carbonate-silicate geochemical cycle and its effect on atmospheric carbon dioxide over the past 100 million years, *Am. J. Sci.*, *283*, 641–683.

Berner, R. A., et al. (2000), Isotope fractionation and atmospheric oxygen: Implications for Phanerozoic O_2 evolution, *Science*, *287*, 1630–1633.

Berner, W., B. Stauffer, and H. Oeschger (1979), Past atmospheric composition and climate, gas parameters measured on ice cores, *Nature*, *275*, 53–55.

Betts, J. N., and H. D. Holland (1991), The oxygen content of ocean bottom waters, the burial efficiency of organic carbon, and the regulation of atmospheric oxygen, *Palaeogeogr., Paleoclimatol., Palaeoecol.*, *97*, 5–18.

Betts, R. A. (2000), Offset of the potential carbon sink from boreal forestation by decreases in surface albedo, *Nature*, *408*, 187–190.

Betzer, P. R., W. J. Showers, E. A. Laws, C. D. Winn, G. R. DiTullio, and P. M. Kroopnick (1984), Primary productivity and particle fluxes on a transect of the equator at 153°W in the Pacific Ocean, *Deep-Sea Res., Part A*, *31*, 1–11.

Biddanda, B., S. Opsahl, and R. Benner (1994), Plankton respiration and carbon flux through bacterioplankton on the Louisiana shelf, *Limnol. Oceanogr.*, *39*, 1259–1275.

Bidle, K. D., and F. Azam (1999), Accelerated dissolution of diatom silica by marine bacterial assemblages, *Nature*, *297*, 508–512.

Bidle, K. D., M. Manganelli, and F. Azam (2002), Regulation of oceanic silicon and carbon preservation by temperature control on bacteria, *Science, 298,* 1980–1984.

Birchfield, G. E. (1987), Changes in deep-ocean water $\delta^{18}O$ and temperature from the last glacial maximum to the present, *Paleoceanogr., 2,* 431–442.

Bird, M. I., J. Lloyd, and G. D. Farquhar (1994), Terrestrial carbon storage at the last glacial maximum, *Nature, 371.*

Bird, M. I., J. Lloyd, and G. D. Farquhar (1996), Terrestrial carbon storage from the last glacial maximum to the present, *Chemosphere, 33,* 1675–1685.

Bishop, J.K.B., J. C. Stepien, and P. H. Wiebe (1987), Particulate matter distributions, chemistry, and flux in the Panama Basin: Response to environmental forcing., *Prog. Oceanogr., 17,* 1–59.

Blackman, F. F. (1905), Optima and limiting factors, *Annals Bot., 19,* 281–295.

Bolin, B., and C. D. Keeling (1963), Large-scale atmospheric mixing as deduced from the seasonal and meridional variations of carbon dioxide, *J. Geophys. Res., 68,* 3899–3920.

Bopp, L., P. Monfray, O. Aumont, J.-L. Dufresne, H. Le Treut, G. Madec, L. Terray, and J. C. Orr (2001), Potential impact of climate change on marine export production, *Global Biogeochem. Cycles, 15,* 81–100.

Bopp, L., C. Le Quéré, M. Heimann, A. Manning, and P. Monfray (2002), Climate-induced oceanic oxygen fluxes: Implications for the contemporary carbon budget, *Global Biogeochem. Cycles, 16,* doi:10.1029/2001GB001445.

Bopp, L., K. E. Kohfeld, and C. L. Quéré (2003), Dust impact on marine biota and atmospheric CO_2 during glacial periods, *Paleoceanogr., 18,* 1046, doi:10.1029/2002PA000810.

Boudreau, B. P. (1987), A steady-state diagenetic model for dissolved carbonate species and pH in the porewaters of oxic and suboxic sediments, *Geochim. Cosmochim. Acta, 51,* 1985–1996.

Boudreau, B. P. (1991), Modeling the sulfide-oxygen reaction and associated pH gradients in porewaters, *Geochim. Cosmochim. Acta, 55,* 145–159.

Boudreau, B. P. (1996), Is burial velocity a master parameter for bioturbation? *Geochim. Cosmochim. Acta, 58,* 1243–1249.

Boudreau, B. P. (1997), *Diagenetic Models and Their Implementation,* 414 pp., Springer-Verlag, New York.

Boudreau, B. P., and B. R. Ruddick (1991), On a reactive continuum representation of organic matter diagenesis, *Am. J. Sci., 291,* 507–538.

Boulahdid, M., and J. F. Minster (1989), Oxygen consumption and nutrient regeneration ratios along isopoycnal horizons in the Pacific Ocean, *Mar. Chem., 26,* 133–153.

Bousquet, P., P. Peylin, P. Ciais, C. Le Quéré, P. Friedlingstein, and P. P. Tans (2000), Regional changes in carbon dioxide fluxes of land and oceans since 1980, *Science, 290,* 1342–1346.

Boutin, J., and J. Etcheto (1995), Estimating the chemical enchancement effect on the air-sea CO_2 exchange using the ERSI scatterometer wind speeds, in *Air-Water Gas Transfer,* edited by B. Jahne and E. C. Manahan, pp. 827–841, AEON Verlag & Studio, Hanau, Germany.

Bown, P. R., J. A. Lees, and J. R. Young (2004), Calcareous nannoplankton evolution and diversity through time, in *Coccolithophores: From Molecular Processes to Global Impact,* edited by H. Thierstein and J. R. Young, pp. 481–501, Springer-Verlag, Berlin.

Boyd, P. W., and S. C. Doney (2002), Modelling regional responses by marine pelagic ecosystems to global climate change, *Geophys. Res. Lett., 29,* doi:10.1029/2001GL014130.

Boyd, P. W., and S. C. Doney (2003), The impact of climate change and feedback processes on the ocean carbon cycle, in *Ocean Biogeochemistry: The role of the Ocean Carbon Cycle in Global Change,* edited by M.J.R. Fasham, pp. 157–187. Springer-Verlag, Berlin.

Boyd, P. W., et al. (2000), A mesoscale phytoplankton bloom in the polar Southern Ocean stimulated by iron fertilization, *Nature, 407,* 695–702.

Boyd, P. W., et al. (2004), The decline and fate of an iron-induced subarctic phytoplankton bloom, *Nature, 428,* 549–553.

Boyle, E. A. (1988), Vertical oceanic nutrient fractionation and glacial/interglacial CO_2 cycles, *Nature, 331,* 55–56.

Boyle, E. A. (1992), Cadmium and $\delta^{13}C$ paleochemical ocean distributions during the stage 2 glacial maximum, *Ann. Rev. Earth Planet. Sci., 20,* 245–287.

Brandes, J. A., and A. H. Devol (2002), A global marine fixed nitrogen isotopic budget: Implications for Holocene nitrogen cycling, *Global Biogeochem. Cycles, 16,* 67.61–67.14, doi: 10.1029/2001GB001856.

Brewer, P. G. (1978), Direct observation of the oceanic CO_2 increase, *Geophys. Res. Lett., 5,* 997–1000.

Brewer, P. G., G.T.F. Wong, M. P. Bacon, and D. W. Spencer (1975), An oceanic calcium problem? *Earth Planet. Sci. Lett., 26,* 81–87.

Brewer, P. G., A. L. Bradshaw, and R. T. Williams (1986), Measurements of total carbon dioxide and alkalinity in the North Atlantic Ocean in 1981, in *The Changing Carbon Cycle: A Global Analysis,* edited by J. R. Trabalka and D. E. Reichel, pp. 348–370, Springer-Verlag, Berlin.

Brink, K. H., F.F.G. Abrantes, P. A. Bernal, R. C. Dugdale, M. Estrada, L. Hutchings, R. A. Jahnke, P. J. Müller, and R. L. Smith (1995), Group report: How do coastal upwelling systems operate as integrated physical, chemical, and biological systems and influence the geological record? The role of physical processes in defining the spatial structures of biological and chemical variables, in *Upwelling in the Ocean: Modern Processes and Ancient Records,* edited by C. P. Summerhayes et al., pp. 103–124, John Wiley & Sons Ltd., New York.

Brix, H., N. Gruber, and C. D. Keeling (2004), Interannual variability of the upper ocean carbon cycle at station ALOHA near Hawaii, *Global Biogeochem. Cycles, 18,* GB4019, doi:10.1029/2004GB002245.

Broecker, W. S. (1971), A kinetic model for the chemical composition of sea water, *Quarternary Research, 1,* 188–207.

Broecker, W. S. (1974), 'NO,' A conservative water mass tracer, *Earth Planet. Sci. Lett., 23,* 100–107.

Broecker, W. S. (1979), A revised estimate for the radiocarbon age of North Atlantic deep water, *J. Geophys. Res., 84,* 3218–3226.

Broecker, W. S. (1982), Glacial to interglacial changes in ocean chemistry, *Progr. Oceanogr, 11,* 151–197.

Broecker, W. S. (1986), Oxygen isotope constraints on surface ocean temperatures, *Quaternary Research, 26,* 121–134.

Broecker, W. S. (1987), The role of $CaCO_3$ compensation in the glacial to interglacial atmospheric CO_2 change, *Global Biogeochem. Cycles, 1,* 15–29.

Broecker, W. S. (1991a), The great ocean conveyor, *Oceanography, 4,* 79–90.

Broecker, W. S. (1991b), Keeping global change honest, *Global Biogeochem. Cycles, 5,* 191–192.

Broecker, W. S. (1995), *The Glacial World According to Wally,* Lamont-Doherty Earth Observatory of Columbia University, Palisades, NY.

Broecker, W. S. (1996), Glacial climate in the tropics, *Science, 272,* 1902–1904.

Broecker, W. S., and E. Clark (2003), Pseudo dissolution of marine calcite, *Earth and Planetary Sciences, 26,* 291–206.

Broecker, W. S., and G. M. Henderson (1998), The sequence of events surrounding Termination II and their implications for the cause of Glacial-Interglacial CO_2 changes, *Paleoceanogr., 13,* 352–364.

Broecker, W. S., and T.-H. Peng (1982), *Tracers in the Sea,* 690 pp., Lamont-Doherty Geological Observatory, Palisades, NY.

Broecker, W. S., and T.-H. Peng (1987), The role of $CaCO_3$ compensation in the glacial to interglacial atmospheric CO_2 change, *Global Biogeochem. Cycles, 1,* 15–29.

Broecker, W. S., and T.-H. Peng (1989), The cause of the glacial to interglacial atmospheric CO_2 change: A polar alkalinity hypothesis, *Global Bigeochem. Cycles, 3,* 215–239.

Broecker, W. S., and T.-H. Peng (1992), Interhemispheric transport of carbon dioxide by ocean circulation, *Nature, 356,* 587–589.

Broecker, W. S., and T.-H. Peng (1998), *Greenhouse Puzzles,* 2nd ed., Eldigio Press, Lamont-Doherty Earth Observatory of Columbia University, Palisades, NY.

Broecker, W. S., and A. Sanyal (1998), Does atmospheric CO_2 police the rate of chemical weathering? *Global Biogeochem. Cycles, 12,* 403–408.

Broecker, W. S., and T. Takahashi (1977a), Neutralization of fossil fuel CO_2 by marine calcium carbonate, in *The Fate of Fossil Fuel CO_2 in the Oceans,* edited by N. R. Anderson and A. Malahoff, pp. 213–239, Plenum, New York.

Broecker, W. S., and T. Takahashi (1977b), The solubility of calcite in seawater, in *Thermodynamics in Geology,* edited by D. G. Graser, pp. 365–379, Reidel, Dordrecht, Netherlands.

Broecker, W. S., T.-H. Peng, G. Ostlund, and M. Stuiver (1985a), The distribution of bomb radiocarbon in the ocean, *J. Geophys. Res., 90,* 6953–6970.

Broecker, W. S., T. Takahashi, and T.-H. Peng (1985b), Reconstruction of past atmospheric CO_2 contents from the chemistry of the contemporary ocean, an evaluation, DOE Tech. Rep., U. S. Dep. of Energy, Washington, DC, DOE/OR-857, 79 pp.

Broecker, W. S., T. Takahashi, and T. Takahashi (1985c), Source and flow patterns of deep-ocean waters as deduced from potential temperature, salinity, and initial phosphate concentration, *J. Geophys. Res., 90,* 6925–6939.

Broecker, W. S., J. R. Ledwell, T. Takahashi, R. Weiss, L. Merlivat, L. Memery, T.-H. Peng, B. Jahne, and K. O. Munnich (1986), Isotopic versus micrometeorologic ocean CO_2 fluxes: A serious conflict. *J. Geophys. Res., 91,* 10517–10528.

Broecker, W. S., S. Blanton, W. M. Smethie, and G. Ostlund (1991), Radiocarbon decay and oxygen utilization in the deep Atlantic Ocean, *Global Biogeochem. Cycles, 5,* 87–117.

Broecker, W. S., S. Sutherland, W. Smethie, T.-H. Peng, and G. Ostlund (1995), Oceanic radiocarbon: Separation of the natural and bomb components, *Global Biogeochem. Cycles, 9,* 263–288.

Broecker, W. S., et al. (1998), How much deep water is formed in the Southern Ocean? *J. Geophys. Res., 103,* 15833–15843.

Broecker, W. [S.], J. Lynch-Steiglitz, D. Archer, M. Hofmann, E. Maier-Reimer, O. Marchal, T. Stocker, and N. Gruber (1999a), How strong is the Harvardton-Bear constraint? *Global Biogeochem. Cycles, 13,* 817–820.

Broecker, W. S., S. Sutherland, and T.-H. Peng (1999b), A possible slowdown of deep water formation in the Southern Ocean during the present century, *Science, 286,* 1132–1135.

Broecker, W. S., J. Lynch-Steiglitz, E. Clark, I. Hajdas, and G. Bonani (2001), What caused the atmosphere's CO_2 content to rise during the last 8000 years? *Geochem. Geophys. Geosys.,* doi:10.1029/2001GC00177.

Bronk, D. A. (2002), Dynamics of DON, in *Biogeochemistry of Marine Dissolved Organic Matter,* edited by D. A. Hansell and C. A. Carlson, pp. 153–247, Academic Press, New York.

Bronk, D. A., P. M. Gilbert, and B. B. Ward (1994), Nitrogen uptake, dissolved organic nitrogen release, and new production, *Science, 265,* 1843–1846.

Brown, C. W., and J. A. Yoder (1994), Coccolithophorid blooms in the global ocean, *J. Geophys. Res., 99,* 7467–7482.

Bruland, K. W., and E. L. Rue (2001), Analytical methods for the determination of concentrations and speciation of iron, in *The Biogeochemistry of Iron in Seawater,* edited by D. R. Turner and K. A. Hunter, pp. 255–289, John Wiley & Sons, Ltd., Chichester.

Bruland, K. W., P. K. Bienfang, J. K. B. Bishop, G. Eglinton, V.A.W. Ittekkot, R. Lampitt, M. Sarnthein, J. Thiede, J. J. Walsh, and G. Wefer (1989), Flux to the seafloor, in *Productivity of the Ocean: Present and Past,* edited by W. H. Berger et al., pp. 193–215, John Wiley & Sons, Ltd., New York.

Bruland, K. W., J. R. Donat, and D. A. Hutchins (1991), Interactive influences of bioactive trace metals on biological production in oceanic waters, *Limnol. Oceanogr., 36,* 1555–1577.

Bruland, K. W., K. J. Orians, and J. P. Cowen (1994), Reactive trace metals in the stratified central North Pacific, *Geochim. Cosmochim. Acta, 58,* 3171–3182.

Bryan, K. (1969), A numerical method for the study of the circulation of the world ocean, *J. Comput. Phys., 4,* 347–376.

Brzezinski, M. A. (1985), The Si:C:N ratio of marine diatoms: Interspecific variability and the effect of some environmental variables, *J. Phycol., 21,* 347–357.

Brzezinski, M. A., and D. M. Nelson (1989), Seasonal changes in the silicon cycle within a Gulf Stream warm-core ring, *Deep-Sea Res., 36,* 1009–1030.

Brzezinski, M. A., T. A. Villareal, and F. Lipschultz (1998), Silica production and the contribution of diatoms to new and primary production in the central North Pacific, *Marine Ecology-Progress Series, 167,* 89–104.

Brzezinski, M. A., D. M. Nelson, V. M. Franck, and D. E. Sigmon (2001), Silicon dynamics within an intense open-ocean diatom bloom in the Pacific sector of the Southern Ocean, *Deep-Sea Res. II, 48,* 3997–4018.

Brzezinski, M. A., C. J. Pride, V. M. Franck, D. M. Sigman, J. L. Sarmiento, K. Matsumoto, N. Gruber, G. H. Rau, and K. H. Coale (2002), A switch from $Si(OH)_4$ to NO_3^- depletion in the glacial Southern Ocean, *Geophys. Res. Lett., 29,* doi:10.1029/2001GL014349.

Brzezinski, M. A., M.-L. Dickson, D. M. Nelson, and R. Sambrotto (2003a), Ratios of Si, C and N uptake by microplankton in the Southern Ocean, *Deep-Sea Res. II, 50,* 619–633.

Brzezinski, M. A., J. J. Jones, J. Bidle, and F. Azam (2003b), The balance between silica production and silica dissolution in the sea: Insights from Monterey Bay, California, applied to the global data set, *Limnol. Oceanogr., 48,* 1846–1854.

Buesseler, K. O. (1991), Do upper-ocean sediment traps provides an accurate record of particle flux? *Nature, 353,* 420–423.

Buesseler, K. O. (1998), The decoupling of production and particulate export in the surface ocean, *Global Biogeochem. Cycles, 12,* 297–310.

Buesseler, K. O., M. P. Bacon, J. K. Cochran, and H. D. Livingston (1992), Carbon and nitrogen export during the JGOFS North Atlantic Bloom Experiment estimated from ^{234}Th: ^{238}U disequilibria, *Deep-Sea Res., 39,* 1115–1138.

Buesseler, K. O., D. K. Steinberg, A. F. Michaels, R. J. Johnson, J. E. Andrews, J. R. Valdes, and J. F. Price (2000), A comparison of the quantity and composition of material caught in a neutrally buoyant versus surface-tethered sediment trap, *Deep-Sea Res. I, 47,* 277–294.

Buitenhuis, E. T., H.J.W. de Baar, and M.J.W. Veldhuis (1999), Photosynthesis and calcification by *Emiliani huxleyi* (Prymnesiophyceae) as a function of inorganic carbon species, *J. Phycol., 35,* 949–959.

Bullister, J. L., and D. P. Wisegarver (1998), The solubility of carbon tetrachloride in water and seawater, *Deep-Sea Res. I, 45,* 1285–1302.

Bullister, J. L., D. P. Wisegarver, and F. A. Menzia (2002), The solubility of sulfur hexafluoride in water and seawater, *Deep-Sea Res. I, 49,* 175–188.

Burdige, D. J. (2001), Dissolved organic matter in Chesapeake Bay sediment pore waters, *Org. Geochem., 32,* 487–505.

Burdige, D. J. (2002), Sediment pore waters, in *Biogeochemistry of Marine Dissolved Organic Matter,* edited by D. A. Hansell and C. A. Carlson, pp. 611–663, Academic Press, New York.

Burdige, D. J., W. M. Berelson, K. H. Coale, J. McManus, and K. S. Johnson (1999), Fluxes of dissolved organic carbon from California continental margin sediments, *Geochim. Cosmochim. Acta, 63,* 1507–1515.

Caldeira, K., and P. B. Duffy (2000), The role of the Southern Ocean in uptake and storage of anthropogenic carbon dioxide, *Science, 287,* 620–622.

Cane, M. A. (1992), Tropical Pacific ENSO models: ENSO as a mode of the coupled system, in *Climate System Modeling,* edited by K. E. Trenberth, pp. 583–614, Cambridge University Press, Cambridge.

Canfield, D. E. (1993), Organic matter oxidation in marine sediments, in *Interactions of C, N, P and S Biogeochemical Cycles and Global Change,* edited by R. Wollast et al., pp. 333–363, Springer-Verlag, Berlin.

Canfield, D. E. (1994), Factors influencing organic carbon preservation in marine sediments, *Chem. Geol., 114,* 315–329.

Carlson, C. A. (2002), Production and removal processes, in *Biogeochemistry of Marine Dissolved Organic Matter,* edited by D. A. Hansell and C. A. Carlson, pp. 91–151, Academic Press, New York.

Carlson, C. A., and H. W. Ducklow (1995), Dissolved organic carbon in the upper ocean of the central equatorial Pacific Ocean, 1992: Daily and finescale vertical variations, *Deep-Sea Res. II, 42,* 639–656.

Carlson, C. A., and H. W. Ducklow (1996), Growth of bacterioplankton and consumption of dissolved organic carbon in the Sargasso Sea, *Deep-Sea Res. II, 10,* 69–85.

Carlson, C. A., H. W. Ducklow, and A. F. Michaels (1994), Annual flux of dissolved organic carbon from the euphotic zones in the northwestern Sargasso Sea, *Nature, 371,* 405–408.

Carlson, C. A., N. R. Bates, H. W. Ducklow, and D. A. Hansell (1999), Estimation of bacterial respiration and growth efficiency in the Ross Sea, *Antarctic, Aquatic Microbial Ecology, 19,* 229–244.

Carlson, C. A., D. A. Hansell, E. T. Peltzer, and W. O. Smith, Jr. (2000), Stocks and dynamics of dissolved and particulate organic matter in the southern Ross Sea, Antarctica, *Deep-Sea Res. II, 47,* 3201–3225.

Caron, D. A. (1994), Inorganic nutrients, bacteria, and the microbial loop, *Microbial Ecology, 28,* 295–298.

Carpenter, E. J., and K. Romans (1991), Major role of the cyanobacterium trichodesmium in nutrient cycling in the north Atlantic Ocean, *Science, 254,* 1356–1358.

Carr, M.-E. (2002), Estimation of potential productivity in Eastern Boundary Currents using remote sensing, *Deep-Sea Res. II, 49,* 59–80.

Catubig, N. R., D. E. Archer, R. Francois, P. deMenocal, W. Howard, and E.-F. Yu (1998), Global deep-sea burial rate of calcium carbonate during the last glacial maximum, *Paleoceanogr., 13,* 298–310.

Cember, R. (1989), Bomb radiocarbon in the Red Sea: A medium-scale gas exchange experiment, *J. Geophys. Res., 94,* 2111–2123.

Chai, F., R. C. Dugdale, T.-H. Peng, F. P. Wilkerson, and R. T. Barber (2002), One-dimensional ecosystem model of the equatorial Pacific upwelling system. Part I: Model development and silicon and nitrogen cycle, *Deep-Sea Res. II, 49,* 2713–2745.

Chai, F., M. Jiang, R. T. Barber, R. C. Dugdale, and Y. Chao (2003), Interdecadal variation of the transition zone chlorophyll front: A physical-biological model simulation beween 1960 and 1990, *Journal of Oceanography, 59,* 461–475.

Chang, P., L. Ji, and H. Li (1997), A decadal climate variation in the tropical Atlantic Ocean from thermodynamic air-sea interactions, *Nature, 385,* 516–518.

Charles, C. D., and R. G. Fairbanks (1990), Glacial to interglacial changes in the isotopic gradients of southern ocean surface water, in *Geological History of the Polar*

Oceans, edited by U. Bleil and J. Thiede, pp. 519–538, Kluwer Academic Publ., Dordrecht, Netherlands.

Charlson, R. J., J. E. Lovelock, M. O. Andreae, and S. G. Warren (1987), Oceanic phytoplankton, atmospheric sulphur, cloud albedo and climate, *Nature, 326*, 655–661.

Chavez, F. P., P. G. Strutton, G. E. Friederich, R. A. Feely, G. C. Feldman, D. G. Foley, and M. J. McPhaden (1999), Biological and chemical response of the Equatorial Pacific Ocean to the 1997–98 El Niño, *Science, 286*, 2126–2131.

Chavez, F. P., J. Ryan, S. E. Lluch-Cota, and M. Ñiquen C. (2003), From anchovies to sardines and back: Multidecadal change in the Pacific Ocean, *Science, 299*, 217–221.

Chelton, D. B., R. A. DeSzoeke, M. G. Schlax, K. ElNaggar, and N. Siwertz (1998), Geographical variability of the first baroclinic Rossby radius of deformation, *J. Phys. Oceanogr., 28*, 433–460.

Chen, C.-T.A. (2004), Exchanges of carbon in the coastal seas, in *The Global Carbon Cycle*, edited by C. B. Field and M. R. Raupach, pp. 341–351, Island Press, Washington, DC.

Chen, G. T., and F. J. Millero (1979), Gradual increase of oceanic CO_2, *Nature, 277*, 205–206.

Chisholm, S. W. (1992), Phytoplankton size, in *Primary Productivity and Biogeochemical Cycles in the Sea*, edited by P. G. Falkowski and A. D. Woodhead, pp. 213–237, Plenum Press, New York.

Chisholm, S. W., and F.M.M. Morel (1991), What controls phytoplankton production in nutrient-rich areas of the open sea? *Limnol. Oceanogr., 36*, 1507–1965.

Chisholm, S. W., R. J. Olson, E. R. Zettler, R. Goericke, J. B. Waterbury, and N. A. Welschmeyer (1988), A novel free-living prochlorophyte abundant in the oceanic euphotic zone, *Nature, 334*, 340–343.

Cho, C. C., and F. Azam (1988), Major role of bacteria in biogeochemical fluxes in the ocean's interior, *Nature, 332*, 441–443.

Christian, J. R., M. A. Verschell, R. Murtugudde, A. J. Busalacchi, and C. R. McClain (2002), Biogeochemical modelling of the tropical Pacific Ocean. II: Iron biogeochemistry, *Deep-Sea Res. II, 49*, 545–565.

Chung, S.-N., K. Lee, R. A. Feely, C. L. Sabine, F. J. Millero, R. Wanninkhof, J. L. Bullister, R. M. Key, and T.-H. Peng (2003), Calcium carbonate budget in the Atlantic Ocean based on water column inorganic carbon chemistry, *Global Biogeochem. Cycles, 17*, 4.1–4.16, 1093, doi:10.1029/2002 GB002001.

Ciais, P., P. P. Tans, M. Trolier, J.W.C. White, and R. J. Francey (1995), A large northern hemisphere terrestrial CO_2 sink indicated by the $^{13}C/^{12}C$ ratio of atmospheric CO_2, *Science, 269*, 1098–1102.

Claquin, P., V. Martin-Jézéquel, J. C. Kronmkamp, M.J.W. Veldhuis, and G. W. Kraay (2002), Uncoupling of silicon compared with carbon and nitrogen metabolisms and the role of the cell cycle in continuous cultures of *Thalassiosira pseudonana* (Bacillariophyceae) under light, nitrogen, and phosphorus control, *J. Phycol., 38*, 922–930.

Clayton, T. D., and R. H. Byrne (1993), Spectrophotometric seawater pH measurements: Total hydrogen ion concentration scale calibration of m-creosol purple and at-sea results, *Deep-Sea Res., 40*, 2115–2129.

Clegg, S. L., and M. Whitfield (1990), A generalized model for the scavenging of trace metals in the open ocean. I: Particle cycling, *Deep-Sea Res., 37*, 809–832.

Clegg, S. L., M. P. Bacon, and M. Whitfield (1991), Application of a generalized scavenging model to thorium isotope and particle data at equatorial and high-latitude sites in the Pacific Ocean, *J. Geophys. Res., 96*, 20655–20670.

CLIMAP Project Members (1981), Seasonal reconstruction of the earth's surface at the last glacial maximum, Geol. Soc. Am., Tech. Re. Map and Chart Serv., MC 36, Boulder, CO.

Cloern, J. E., C. Grenz, and L. Vidergar-Lucas (1995), An empirical model of the phytoplankton chlorophyll: carbon ratio: The conversion factor between productivity and growth rate, *Limnol. Oceanogr., 40*, 1313–1321.

Coale, K. H., et al. (1996), A massive phytoplankton bloom induced by an ecosystem-scale iron fertilization experiment in the equatorial Pacific Ocean, *Nature, 383*, 495–501.

Coale, K. H., et al. (2004), Southern ocean iron enrichment experiment: carbon cycling in high- and low-Si water, *Science, 304*, 408–414.

Cochran, J. K. (1985), Particle mixing rates in sediments of the eastern equatorial Pacific: Evidence from 10Pb, 239, 240Pu and 137Cs distributions at MANOP sites, *Geochim. Cosmochim. Acta, 49*, 1195–1210.

Codispoti, L. A., and J. P. Christensen (1985), Nitrification, denitrification and nitrous oxide cycling in the Eastern Tropical South Pacific Ocean., *Mar. Chem., 16*, 277–300.

Codispoti, L. A., J. A. Brandes, J. P. Christensen, A. H. Devol, S.W.A. Naqvi, H. W. Paerl, and T. Yoshinari (2001), The oceanic fixed nitrogen and nitrous oxide budgets: Moving targets as we enter the anthropocene? *Scientia Marina, 65*, 85–105.

Cohen, Y., and L. I. Gordon (1978), Nitrous oxide in the oxygen minima of the eastern tropical North Pacific: Evidence for its consumption during nitrification and possible mechanisms for its production., *Deep-Sea Res., 25*, 509–524.

Colman, A. S., F. T. Mackenzie, and H. D. Holland (1997), Redox stabilization of the atmosphere and oceans and marine productivity, *Science, 275*, 406–407.

Conkright, M. E., R. A. Locarnini, H. E. Garcia, T. D. O'Brien, T. P. Boyer, C. Stephens, J. I. Antonov (2002), *World Ocean Atlas 2001: Objective Analyses, Data Statistics, and Figures, CD-ROM Documentation*. National Oceanographic Data Center, Silver Spring, MD, 17 pp.

Conte, M. (1998). The Oceanic Flux Program: Twenty years of particle flux measurements in the deep Sargasso Sea, *Oceanus, 40*, 15–19.

Craig, H. (1971), The deep metabolism: Oxygen consumption in abyssal ocean water, *J. Geophys. Res., 74*, 5491–5506.

Crosta, X., J. Pichon, and L. Burckle (1998), Application of modern analog technique to marine Antarctic diatoms: Reconstruction of maximum sea ice extent at the last glacial maximum, *Paleoceanogr., 13*, 284–297.

Crutzen, P. J., and E. F. Stoermer (2000), The Anthropocene, *IGBP Newsletter, 41*, 12–13.

Cubasch, U., G. A. Meehl, G. J. Boer, R. J. Stouffer, M. Dix, A. Noda, C. A. Senior, S. Raper, and K. S. Yap (2001), Projections of future climate change, in *Climate Change 2001: The Scientific Basis*, edited by J. T. Houghton et al., pp. 525–582, Cambridge University Press, New York.

Cullen, J. J. (1982), The deep chlorophyll maximum layer: Comparing vertical profiles of chlorophyll *a*, *Can. J. Fish. Aquat. Sci.*, *39*, 791–803.

Cullen, J. J. (1996), Status of the iron hypothesis after the open-ocean enrichment experiment, *Limnol. Oceanogr.*, *40*, 1336–1343.

Curry, W. B., J. C. Duplessy, L. D. Labeyrie, and N. J. Shackelton (1988), Changes in the distribution of $\delta^{13}C$ of deep water ΣCO_2 between the last glaciation and the Holocene, *Paleoceanogr.*, *3*, 317–341.

D'Hondt, S., S. Rutherford, and A. J. Spivak (2002), Metabolic activity of subsurface life in deep-sea sediments, *Science*, *295*, 2067–2070.

Danckwerts, P. V. (1951), Significance of liquid-film coefficients in gas absorption, *Ind. Engng. Chem.*, *43*, 1460–1467.

Dauwe, B., J. J. Middelburg, P.M.J. Herman, and C.H.R. Heip (1999), Linking diagenetic alteration of amino acids and bulk organic matter reactivity, *Limnol. Oceanogr.*, *44*, 1809–1814.

Davis, C. S., and J. H. Steele (1994), Biological/physical modeling of upper ocean processes, Tech. Rept., Woods Hole Oceanog. Inst., WHOI-94-32.

de Baar, H.J.W. (1994), Von Liebig's Law of the Minimum and plankton ecology (1899–1991), *Prog. Oceanogr.*, *33*, 347–386.

de Baar, H.J.W., and J.T.M. de Jong (2001), Distributions, sources and sinks of iron in seawater, in *The Biogeochemistry of Iron in Seawater*, edited by D. R. Turner and K. A. Hunter, pp. 123–253, John Wiley & Sons, Ltd., Chichester.

de Baar, H.J.W., P. M. Saager, R. F. Nolting, and J. van der Meer (1994), Cadmium versus phosphate in the world ocean, *Mar. Chem.*, *46*, 261–281.

de Baar, H.J.W., J.T.M. de Jong, D.C.E. Bakker, B. M. Löscher, C. Veth, U. Bathmann, and V. Smetacek (1995), Importance of iron for plankton blooms and carbon dioxide drawdown in the Southern Ocean, *Nature*, *373*, 412–415.

de Haas, H., T.C.E. van Weering, and H. de Stitger (2002), Organic carbon in shelf seas: Sinks or sources, processes and products, *Continental Shelf Research*, *22*, 691–717.

Deacon, E. L. (1977), Gas transfer to and across an air-water interface, *Tellus*, *29*, 363–374.

Del Amo, Y., and M. A. Brzezinksi (1999), The chemical form of dissolved Si taken up by marine diatoms, *J. Phycol.*, *35*, 1162–1170.

Delaney, M. L. (1998), Phosphorus accumulation in marine sediments and the oceanic phosphorus cycle, *Global Biogeochem. Cycles*, *12*, 563–572.

De La Rocha, C. L., M. A. Brzezinski, M. J. DeNiro, and A. Shemesh (1998), Silicon-isotope composition of diatoms as an indicator of past oceanic change, *Nature*, *395*, 680–683.

De La Rocha, C. L., C. L. Hutchins, and M. A. Brzezinski (2000), Effects of iron and zinc deficiency on elemental composition and silica production by diatoms, *Mar. Ecol. Prog. Ser.*, *195*, 71–79.

Delmas, R. J., J.-M. Ascencio, and M. Legrand (1980), Polar ice evidence that atmospheric CO_2 29,000 BP was 50% of present, *Nature*, *284*.

DeMaster, D. J. (2002), The accumulation and cycling of biogenic silica in the Southern Ocean: Revisiting the marine silica budget, *Deep-Sea Res. II*, *49*, 3155–3167.

Denman, K. L. (1976), Covariability of chlorophyll and temperature in the sea, *Deep-Sea Res.*, *23*, 539–550.

Denman, K. L., and J. F. Dower (2001), Patch dynamics, in *Encyclopedia of Ocean Sciences*, edited by J. H. Steele, et al., pp. 2107–2114, Academic Press, New York.

der Werf, G.R.V., J. Randerson, G. J. Collatz, L. Giglio, P. Kasibhatla, A. Arellano, S. Olsen, and E. Kassichke (2004), Continental-scale partitioning of fire emissions during the 1997 to 2001 El Niño/La Niña period, *Science*, *303*, 73–76.

Deser, C., M. A. Alexander, and M. S. Timlin (1996), Upper-ocean thermal variations in the North Pacific during 1970–1991, *J. Clim.*, *9*, 1840–1855.

Deuser, W. G. (1986), Seasonal and interannual variations in deep-water particle fluxes in the Sargasso Sea and their relation to surface hydrography, *Deep-Sea Res.*, *33*, 225–246.

Deutsch, C., N. Gruber, R. M. Key, and J. L. Sarmiento (2001), Denitrification and N_2 fixation in the Pacific Ocean, *Global Biogeochem. Cycles*, *15*, 483–506.

Deutsch, C., D. M. Sigman, R. C. Thunell, A. N. Meckler, and G. H. Haug (2004), Isotopic constraints on glacial/interglacial changes in the oceanic nitrogen budget, *Global Biogeochem. Cycles*, *18*, GB4012, doi:4010.1029/2003GB002189.

Dhakar, S. P., and D. J. Burdige (1996), A coupled, non-linear, steady state model for early diagenetic processes in pelagic sediments, *Am. J. Sci.*, *296*, 296–330.

Dickinson, R. E., and R. J. Cicerone (1986), Future global warming from atmospheric trace gases, *Nature*, *319*, 109–115.

Dickson, A. G. (1981), An exact definition of total alkalinity and a procedure for the estimation of alkalinity and total inorganic carbon from titration data, *Deep-Sea Res.*, *28A*, 609–623.

Dickson, A. G. (1990), Thermodynamics of the dissociation of boric acid in synthetic seawater from 273.15 to 318.15K, *Deep-Sea Res.*, *37*, 755–766.

Dickson, A. G. (1993), The measurement of sea water pH, *Mar. Chem.*, *44*, 131–142.

Dickson, A. G., and C. Goyet (1994), *Handbook of Methods for the Analysis of the Various Parameters of the Carbon Dioxide System in Sea Water*, ORNL/CDIAC-74, Oak Ridge Natl. Lab., Oak Ridge, TN.

Dickson, A. G., and F. J. Millero (1987), A comparison of the equilibrium constants for the dissociation of carbonic acid in seawater media, *Deep-Sea Res.*, *34*, 1733–1743.

Dickson, R., J. Lazier, J. Meincke, P. Rhines, and J. Swift (1996), Long-term coordinated changes in the convective activity of the North Atlantic, *Prog. Oceanogr.*, *28*, 241–295.

Dixit, S., and P. Van Cappellen (2002), Surface chemistry and reactivity of biogenic silica, *Geochim. Cosmochim. Acta*, *66*, 2559–2568.

Dixit, S., and P. Van Cappellen (2003), Predicting benthic fluxes of silicic acid from deep-sea sediments, *J. Geophys. Res.—Oceans*, *108*, doi:10.1029/2002JC001309.

Dixit, S., P. Van Cappellen, and A. J. van Bennekom (2001), Processes controlling solubility of biogenic silica and pore water build-up of silicic acid in marine sediments, *Mar. Chem.*, *73*, 333–352.

Doney, S. C. (1999), Major challenges confronting marine biogeochemical modeling, *Global Biogeochem. Cycles, 13,* 705–714, doi:10.1029/1999GB900039.

Doney, S. C., and W. J. Jenkins (1988), The effect of boundary conditions on tracer estimates of thermocline ventilation rates, *J. Mar. Re., 46,* 947–965.

Doney, S. C., D. M. Glover, and W. J. Jenkins (1992), A model function of the global bomb tritium distribution in precipitation, *J. Geophys. Res., 97,* 5481–5492.

Doney, S. C., D. M. Glover, and R. G. Najjar (1996), A new coupled, one-dimensional biological-physical model for the upper ocean: Applications to the JGOFS Bermuda Atlantic Time Series (BATS) Site, *Deep-Sea Res., 43,* 591–624.

Doney, S. C., I. Lima, K. Lindsay, J. K. Moore, S. Dutkiewicz, M.A.M. Friedrichs, and R. J. Matear (2001), Marine biogeochemical modeling, *Oceanography, 14–4,* 93–107.

Doney, S. C., D. M. Glover, S. J. McCue, and M. Fuentes (2003), Mesoscale variability of Sea-viewing Wide Field-of-view Sensor (SeaWiFS) satellite ocean color: Global patterns and spatial scales, *J. Geophys. Res., 108,* 3024, doi:1029/2001JC000843.

Döös, K., and D. J. Webb (1994), The Deacon cell and the other meridional cells of the Southern Ocean, *J. Phys. Oceanogr., 24,* 429–442.

Dore, J. E., R. Lukas, D. W. Sadler, and D. M. Karl (2003), Climate-driven changes to the atmospheric CO_2 sink in the subtropical North Pacific Ocean, *Nature, 424,* 754–757.

Doval, M. D., and D. A. Hansell (2000), Organic carbon and apparent oxygen utilization in the western South Pacific and the central Indian Oceans, *Mar. Chem., 68,* 249–264.

Dreisigacker, E., and W. Roether (1978), Tritium and ^{90}Sr in North Atlantic surface water, *Earth Planet. Sci. Lett., 38,* 301–312.

Dreybrodt, W., J. Lauckner, U. Svensson, and D. Buhmann (1996), The kinetics of the reaction $CO_2 + H_2O \rightarrow H^+ + HCO_3^-$ as one of the rate limiting steps for the dissolution of calcite in the system H_2O-CO_2-$CaCO_3$, *Geochim. Cosmochim. Acta, 60,* 3375–3381.

Dring, M. J. (1982), *The Biology of Marine Plants,* 199 pp., Edward Arnold (Publishers) Ltd., London.

Druffel, E.R.M., P. M. Williams, J. E. Bauer, and J. R. Ertel (1992), Cycling of dissolved and particulate organic matter in the open ocean, *J. Geophys. Res., 97,* 15,639–15,659.

Duce, R. A., and N. W. Tindale (1991), Atmospheric transport of iron and its deposition in the ocean, *Limnol. Oceanogr., 36,* 1715–1726.

Duce, R. A., et al. (1991), The atmospheric input of trace species to the World Ocean, *Global Biogeochem. Cycles, 5,* 193–260.

Ducklow, H. W. (2001), Bacterioplankton, in *Encyclopedia of Ocean Sciences,* edited by J. Steele et al., pp. 217–224, Academic Press, New York.

Ducklow, H. W. (2003), Biogeochemical provinces: Towards a JGOFS synthesis, in *Ocean Biogeochemistry,* edited by M.J.R. Fasham, pp. 3–17, Springer, New York.

Ducklow, H. W., and C. A. Carlson (1992), Oceanic bacterial production, in *Advances in Microbial Ecology,* edited by K. C. Marshall, pp. 113–181, Plenum Press, New York.

Dugdale, R. C., and J. J. Goering (1967), Uptake of new and regenerated forms of nitrogen in primary productivity, *Limnol. Oceanogr., 12,* 196–206.

Dugdale, R. C., and F. P. Wilkerson (1998), Silicate regulation of new production in the equatorial Pacific upwelling, *Nature, 391,* 270–273.

Dugdale, R. C., F. P. Wilkerson, and H. J. Minas (1995), The role of a silicate pump in driving new production, *Deep-Sea Res., 42,* 697–720.

Dunne, J. [P.], J. W. Murray, M. Rodier, and D. A. Hansell (2000), Export flux in the western and central equatorial Pacific: Zonal and temporal variability, *Deep-Sea Res. I, 47,* 901–936.

Dunne, J. P, R. A. Armstrong, C. A. Deutsch, A. Gnanadesikan, N. Gruber, J. L. Sarmiento, and P. S. Swathi (2005a), Diagnosing production, export and size structure from nutrient and alkalinity distributions in a global ocean model, *in preparation.*

Dunne, J. P., R. A. Armstrong, A. Gnanadesikan, and J. Sarmiento (2005b), Empirical and mechanistic models for the particle export ratio, *Global Biogeochem. Cycles, in press.*

Dunne, J. [P.], J. L. Sarmiento, A. Gnanadesikan, R. Armstrong, and N. Gruber (2005c), Global particle export of the major elements and their cycling in the ocean interior, *in preparation.*

Dutay, J.-C., et al. (2001), Evaluation of ocean model ventilation with CFC-11: Comparison of 13 global ocean models, *Ocean Modelling, 4,* 89–120.

Dutay, J. C., et al. (2004), Evaluation of OCMIP-2 ocean models' deep circulation with mantle helium-3, *Journal of Marine Systems, 418,* 15–36.

Edmond, J. M., S. S. Jacobs, A. L. Gordon, A. W. Mantyla, and R. F. Weiss (1979), Water column anomalies in dissolved silica over opaline pelagic sediments and the origin of the deep silica maximum, *J. Geophys. Res., 84,* 7809–7826.

Elrod, V. A., W. M. Berelson, K. H. Coale, and K. S. Johnson (2004), The flux of iron from continental shelf sediments: A missing source for global budgets, *Geophys. Res. Lett., 31,* doi:10.1029/2004GL020216.

Emerson, S. (1985), Organic carbon preservation in marine sediments, in *The Carbon Cycle and Atmospheric CO_2: Natural Variations Archean to Present,* edited by E. T. Sundquist and W. S. Broecker, pp. 78–88, American Geophysical Union, Washington, DC.

Emerson, S. (1987), Seasonal oxygen cycles and biological new production in surface waters of the subarctic Pacific Ocean, *J. Geophys. Res., 92,* 6535–6544.

Emerson, S., and M. Bender (1981), Carbon fluxes at the sediment-water interface of the deep-sea: Calcium carbonate preservation, *J. Mar. Re., 39,* 139–162.

Emerson, S., and J. I. Hedges (1988), Processes controlling the organic carbon content of open ocean sediments, *Paleoceanogr., 3,* 621–634.

Emerson, S., and J. Hedges (2003), Sediment diagenesis and benthic flux, in *Treatise on Geochemistry,* vol. 6, edited by H. Elderfield, pp. 293–320, Elsevier, Amsterdam.

Emerson, S., P. Quay, C. Stump, D. Wilbur, and M. Knox (1991), O_2, Ar, and ^{222}Rn in surface waters of the subarctic ocean: Net biological O_2 production, *Global Biogeochem. Cycles, 5,* 49–69.

Emerson, S., P. D. Quay, C. Stump, D. Wilbur, and R. Schudlich (1995), Chemical tracers of productivity and respiration in the subtropical Pacific Ocean, *J. Geophys. Res., 100,* 15,873–15,888.

Emerson, S., P. Quay, D. Karl, C. Winn, L. Tupas, and M. Landry (1997), Experimental determination of the organic carbon flux from open-ocean surface waters, *Nature, 389,* 951–954.

Emerson, S., S. Mecking, and J. Abell (2001), The biological pump in the subtropical North Pacific Ocean: Nutrient sources, Redfield ratios, and recent changes, *Global Biogeochem. Cycles, 15,* 535–554.

Enfield, D. B. (1989), El Niño past and present, *Reviews of Geophysics, 27,* 159–187.

Enfield, D. B., and D. A. Mayer (1997), Tropical Atlantic sea surface temperature variability and its relation to El Niño-Southern oscillation, *J. Geophys. Res., 102,* 929–945.

England, M. H. (1993), Representing the global-scale water masses in ocean general circulation models, *J. Phys.Oceanogr., 23,* 1523–1552.

Enting, I. G. (2002), *Inverse Problems in Atmospheric Constituent Transport,* Cambridge University Press, Cambridge.

Eppley, R. W. (1972), Temperature and phytoplankton growth in the sea, *Fish. Bull., 70,* 1063–1085.

Eppley, R. W., and B. J. Peterson (1979), Particulate organic matter flux and planktonic new production in the deep ocean, *Nature, 282,* 677–680.

Eppley, R. W., J. N. Rogers, and J. J. McCarthy (1969), Half-saturation constants for uptake of nitrate and ammonium by marine phytoplankton, *Limnol. Oceanogr., 14,* 912–920.

Esbensen, S. K., and Y. Kushnir (1981), The heat budget of the global ocean: An atlas based on estimates from surface marine observations, Clim. Res. Inst. Tech. Rep. 29, Oreg. State Univ., Corvallis.

Etheridge, D. M., L. P. Steele, R. L. Langenfelds, and R. J. Francey (1996), Natural and anthropogenic changes in atmospheric CO_2 over the last 1000 years from air in Antarctic ice and firn, *J. Geophys. Res., 101,* 4115–4128.

Evans, G. T., and J. S. Parslow (1985), A model of annual plankton cycles, *Biological Oceanography, 3,* 327–347.

Fabry, V. J. (1990), Shell growth rates of pteropod andheteropod molluscs and aragonite production in the open ocean—implications for the marine carbonate system, *Journal of Marine Research, 48,* 209–222.

Falkowski, P. (1997), Evolution of the nitrogen cycle and its influence on the biological sequestration of CO_2 in the ocean, *Nature, 387,* 272–275.

Falkowski, P. G., and J. A. Raven (1997), *Aquatic Photosynthesis,* 375 pp., Blackwell Science, Malden, MA.

Falkowski, P. G., D. Ziemann, Z. Kolber, and P. B. Bienfang (1991), Role of eddy pumping in enhancing primary production in the ocean, *Nature, 352,* 55–58.

Falkowski, P. G., E. A. Laws, R. T. Barber, and J. W. Murray (2003), Phytoplankton and their role in primary, new and export production, in *Ocean Biogeochemistry,* edited by M.J.R. Fasham, pp. 99–121, Springer, New York.

Fan, S.-M., M. Gloor, J. Mahlman, S. Pacala, J. L. Sarmiento, T. Takahashi, and P. Tans (1998), A large terrestrial carbon sink in North America implied by atmospheric and oceanic CO_2 data and models, *Science, 282,* 442–446.

Fanning, K. A. (1992), Nutrient provinces in the sea:Concentration ratios, reaction rate ratios, and ideal covariation, *J. Geophys. Res., 97,* 5693–5712.

Farmer, D. M., C. L. McNeil, and B. D. Johnson (1993), Evidence for the importance of bubbles in increasing air-sea gas flux, *Nature, 361,* 620–623.

Fasham, M.J.R. (1992), Modelling the marine biota, in *The Global Carbon Cycle,* edited by M. Heimann, Springer-Verlag, New York.

Fasham, M.J.R., H. W. Ducklow, and S. M. McKelvie (1990), A nitrogen-based model of plankton dynamics in the oceanic mixed layer, *J. Mar. Re., 48,* 591–639.

Fasham, M.J.R., J. L. Sarmiento, R. D. Slater, H. W. Ducklow, and R. Williams (1993), A seasonal three-dimensional ecosystem model of nitrogen cycling in the North Atlantic euphotic zone: A comparison of the model results with observation from Bermuda Station "S" and OWS "India," *Global Biogeochem. Cycles, 7,* 379–415.

Feely, R. A., R. Wanninkhof, C. E. Cosca, P. M. Murphy, M. F. Lamb, and M. D. Steckley (1995), CO_2 distributions in the equatorial Pacific during the 1991–1992 ENSO event, *Deep-Sea Res., 42,* 365–386.

Feely, R. A., R. Wanninkhof, C. Goyet, D. E. Archer, and T. Takahashi (1997), Variability of CO_2 distributions and sea-air fluxes in the central and eastern Equatorial Pacific during the 1991–1994 El Niño, *Deep-Sea Res. II, 44,* 1851–1867.

Feely, R. A., R. Wanninkhof, and T. Takahashi (1999), Influence of El Niño on the equatorial Pacific contribution to atmospheric CO_2 accumulation, *Nature, 398,* 597–601.

Feely, R. A., et al. (2002), In-situ calcium carbonate dissolution in the Pacific Ocean, *Global Biogeochem. Cycles, 16,* doi:10.1029/2002GB001866.

Feely, R. A., C. L. Sabine, K. Lee, W. Berelson, J. Kleypas, V. J. Fabry, and F. J. Millero (2004a), Impact of anthropogenic CO_2 on the $CaCO_3$ system in the oceans, *Science, 305,* 362–366.

Feely, R. A., C. L. Sabine, R. Schlitzer, J. L. Bullister, S. Mecking, and D. Greeley (2004b), Oxygen utilization and organic carbon remineralization in the upper water column of the Pacific Ocean, *Journal of Oceanography, 60,* 45–52.

Fenchel, T., and T. H. Blackburn (1979), *Bacteria and Mineral Cycling,* 225 pp., Academic Press, New York.

Fiadeiro, M. [E.] (1980), The alkalinity of the deep Pacific, *Earth and Planetary Sciences Letters, 49,* 499–505.

Fiadeiro, M. E., and H. Craig (1978), Three-dimensional modeling of tracers in the deep Pacific Ocean. I: Salinity and oxygen, *J. Mar. Re., 36,* 323–355.

Fine, R. A., W. H. Peterson, and H. G. Östlund (1987), The penetration of tritium into the tropical Pacific, *J. Phys. Oceanogr., 17,* 553–564.

Fischer, H., M. Wahlen, J. Smith, D. Matroianni, and B. Deck (1999), Ice core records of atmospheric CO_2 around the last three glacial terminations, *Science,* 283: 1712–1714.

Fisher, N. S., J.-L. Teyssie, S. Krishnaswami, and M. Baskaran (1987), Accumulation of Th, Pb, U, and Ra in marine phytoplankton and its geochemical significance, *Limnol. Oceanogr., 32,* 131–142.

Fitzwater, S. E., G. A. Knauer, and J. H. Martin (1982), Metal contamination and its effect on primary production measurements, *Limnol. Oceanogr., 27,* 544–551.

Flückiger, J., A. Dällenbach, T. Blunier, B. Stauffer, T. F. Stocker, D. Raynaud, and J.-M. Barnola (1999), Variations in atmospheric N_2O concentration during abrupt climate changes, *Science, 285,* 227–230.

Flückiger, J., E. Monnin, B. Stauffer, J. Schwander, T. F. Stocker, J. C. D. Raynaud, and J.-M. Barnola (2002), High-resolution Holocene N_2O ice core record and its relationship with CH_4 and CO_2, *Global Biogeochem. Cycles, 16,* doi:10.1029/2001GB001417.

Flückiger, J., T. Blunier, B. Stauffer, J. Chappellaz, R. Spahni, K. Kawamura, J. Schwander, T. F. Stocker, and D. Dahl-Jensen (2004), N_2O and CH_4 variations during the last glacial epoch: Insight into global processes, *Global Biogeochem. Cycles, 18,* doi:10.209/2003GB002122.

Fofonoff, N. P. (1977), Computation of potential temperature of seawater for an arbitrary reference pressure, *Deep-Sea Res., 24,* 489–491.

Fofonoff, N. P. (1978), Erratum, *Deep-Sea Res., 25,* 335.

Fofonoff, N. P. (1985), Physical properties of seawater: A new salinity scale and equation of state for seawater, *J. Geophys. Res., 90,* 3332–3342.

Follows, M. J., T. Ito, and J. Marotzke (2001), The wind-driven, subtropical gyres and atmospheric pCO_2, *Global Biogeochem. Cycles, 16,* 1113, doi:10.1029/2001GB001786.

Francey, R. J., P. P. Tans, C. E. Allison, I. G. Enting, J. W. C. White, and M. Troller (1995), Changes in oceanic and terrestrial carbon uptake since 1982, *Nature, 373,* 326–330.

Franck, V. M., M. A. Brzezinski, K. H. Coale, and D. M. Nelson (2000), Iron and silicic acid concentrations regulate Si uptake north and south of the Polar Frontal Zone in the Pacific Sector of the Southern Ocean, *Deep-Sea Res. II, 47,* 3315–3338.

Francois, R., M. A. Altabet, E.-F. Yu, D. M. Sigman, M. P. Bacon, M. Frank, G. Bohrmann, G. Bareille, and L. D. Labeyrie (1997), Contribution of Southern Ocean surface-water stratification to low atmospheric CO_2 concentrations during the last glacial period, *Nature, 389,* 929–933.

Francois, R., S. Honjo, R. Krishfield, and S. Manganini (2002), Factors controlling the flux of organic carbon to the bathypelagic zone of the ocean, *Global Biogeochem. Cycles, 16.*

Frew, N. M. (1997), The role of organic films in air-sea gas exchange, in *The Sea Surface and Global Change,* edited by P. S. Liss and R. A. Duce, pp. 121–171, Cambridge University Press, Cambridge.

Froelich, P. N., G. P. Klinkhammer, M. L. Bender, N. A. Luedtke, G. R. Heath, D. Cullen, P. Dauphin, D. Hammond, B. Hartman, and V. Maynard (1979), Early oxidation of organic matter in pelagic sediments of the eastern equatorial Atlantic: Suboxic diagenesis, *Geochim. Cosmochim. Acta, 43,* 1075–1090.

Froelich, P. N., V. Blanc, R. A. Mortlock, S. N. Chillrud, W. Dunstan, A. Udomkit, and T.-H. Peng (1992), River fluxes of dissolved silica to the ocean were higher during glacials: Ge/Si in diatoms, rivers, and oceans, *Paleoceanogr., 7,* 739–767.

Frost, B. W. (1987), Grazing control of phytoplankton stock in the open subarctic Pacific Ocean: A model assessing the role of mesozooplankton, particularly the large calanoid copepods Neocalanus spp., *Marine Ecology Progress Series, 39,* 49–68.

Frost, B. W., and N. C. Franzen (1992), Grazing and iron limitation in the control of phytoplankton stock and nutrient concentration: A chemostat analogue of the Pacific equatorial upwelling zone, *Marine Ecology Progress Series, 83,* 291–303.

Fuchs, F., W. Roether, and P. Schlosser (1987), Excess ^{3}He in the ocean surface layer, *J. Geophys. Res., 92,* 6559–6568.

Fuhrman, J. [A.] (1992), Bacterioplankton roles in cycling of organic matter: The microbial food web, in *Primary Productivity and Biogeochemical Cycles in the Sea,* edited by P. G. Falkowski and A. D. Woodhead, pp. 361–383, Plenum Press, New York.

Fuhrman, J. A., S. G. Horrigan, and D. G. Capone (1988), Use of ^{13}N as a tracer for bacterial and algal uptake of ammonium from seawater, *Marine Ecology Progress Series, 45,* 271–278.

Fung, I., S. Meyn, I. Tegen, S. Doney, J. John, and J. Bishop (2000), Iron supply and demand in the upper ocean, *Global Biogeochem. Cycles, 14,* 281–295.

Gallinari, M., O. Ragueneau, L. Corrin, D. J. DeMaster, and P. Tréguer (2002), The importance of water column processes on the dissolution properties of biogenic silica in deep-sea sediments I. Solubility, *Geochim. Cosmochim. Acta, 66,* 2701–2717.

Ganachaud, A. (2003), Large-scale mass transports, water mass formation, and diffusivities estimated from World Ocean Circulation Experiment (WOCE) hydrographic data, *J. Geophys. Res., 108,* doi:10.1029/2002JC001565.

Ganeshram, R. S., T. F. Pedersen, S. E. Calvert, and J. W. Murray (1995), Large changes in oceanic nutrient inventories from glacial to interglacial periods, *Nature, 376,* 755–757.

Gao, Y., Y. J. Kaufman, D. Tanre, D. Kolber, and P. G. Falkowski (2001), Seasonal distributions of aeolian iron fluxes to the global ocean, *Geophys. Res. Lett., 28,* 29–32.

Gao, Y., S.-M. Fan, and J. L. Sarmiento (2003), Aeolian iron input to the ocean through precipitation scavenging: A modeling perspective and its implication for natural iron fertilization in the ocean, *J. Geophys. Res., 108,* 4221, doi:10.1029/2002JD002420.

Garabato, A. C. N., K. L. Polzin, B. A. King, K. J. Heywood, and M. Visbeck (2004), Widespread intense turbulent mixing in the Southern Ocean, *Science, 303,* 210–213.

Garcia, H., and L. I. Gordon (1992), Oxygen solubility in seawater: Better fitting equations, *Limnol. Oceanogr., 376,* 1307–1312.

Gargett, A., and J. Marra (2002), Effects of upper ocean physical processes (turbulence, advection and air-sea interaction) on oceanic primary production, in *Biological-Physical Interactions in the Sea,* edited by A. R. Robinson et al., pp. 19–49, John Wiley & Sons, Inc., New York.

Garrels, R. M., and A. Lerman (1984), Coupling of the sedimentary sulfur and oxygen cycles–an improved model, *Am. J. Sci., 284,* 989–1007.

Gattuso, J. P., M. Frankignaille, S. V. Smith, J. R. Ware, R. Wollast, R. W. Buddemier, and H. Kayanne (1996), Coral reefs and carbon dioxide, *Science, 271,* 1298–1300.

Gehlen, M., L. Beck, L. Calas, A.-M. Flank, A. J. Van Bennekom, and J.E.E. Van Beusekom (2002), Unraveling the atomic structure of biogenic silica: Evidence of the structural association of Al and Si in diatom frustules, *Geochim. Cosmochim. Acta, 66,* 1601–1609.

Geider, R. J., H. L. MacIntyre, and T. M. Kana (1996), A dynamic model of photoadaptation in phytoplankton, *Limnol. Oceanogr., 41,* 1–15.

Gent, P., and J. C. McWilliams (1990), Isopycnal mixing in ocean circulation models, *J. Phys. Oceanogr., 20,* 150–155.

Gervais, F., U. Riebesell, and M. Y. Gorbunov (2002), Changes in primary productivity and chlorophyll *a* in response to iron fertilization in the Southern Polar Frontal Zone, *Limnol. Oceanogr., 47.*

Ghil, M. (1994), Cryothermodynamics: The chaotic dynamics of paleoclimate, *Physica D, 77,* 130–159.

Ghil, M., A. Mullhaupt, and P. Pestiaux (1987), Deep water formation and quaternary glaciations, *Clim. Dyn., 2,* 1–10.

Gildor, H., and E. Tziperman (2001a), Physical mechanisms behind biogeochemical glacial-interglacial CO_2 variations, *Geophys. Res. Lett., 106,* 2421–2424.

Gildor, H., and E. Tziperman (2001b), A sea ice climate switch mechanism for the 100-kyr glacial cycles, *J. Geophys. Res., 106,* 9117–9133.

Gildor, H., E. Tziperman, and J. R. Toggweiler (2002), Sea ice switch mechanism and glacial-interglacial CO_2 variations, *Global Biogeochem. Cycles, 16,* doi:10.1029/2001GB001446.

Gille, S. T. (1997), The Southern Ocean momentum balance: Evidence for topographic effects from numerical model output and altimeter data, *J. Phys. Oceanogr., 27,* 2219–2232.

Gloor, M., N. Gruber, J. L. Sarmiento, C. S. Sabine, R. Feely, and C. Rödenbeck (2003), A first estimate of present and pre-industrial air-sea CO_2 flux patterns based on ocean carbon measurements, *Geophys. Res. Lett., 30,* doi:10.1029/2002GL015594.

Glover, D. M., and W. S. Reeburgh (1987), Radon-222 and radium-226 in southeastern Bering Sea shelf waters and sediment, *Cont. Shelf Res., 5.*

Glover, D. M., N. M. Frew, S. J. McCue, and E. J. Bock (2002), A multi-year time series of global gas transfer velocity from the TOPEX dual frequency, normalized radar backscatter algorithm, in *Gas Transfer at Water Surfaces,* edited by M. A. Donelan et al., pp. 325–331, American Geophysical Union, Washington, DC.

Gnanadesikan, A. (1999a), A global model of silicon cycling: Sensitivity to eddy parameterization and dissolution, *Global Biogeochem. Cycles, 13,* 199–220.

Gnanadesikan, A. (1999b), A simple predictive model for the structure of the oceanic pycnocline, *Science, 283,* 2077–2079.

Gnanadesikan, A., and R. Hallberg (2002), Physical oceanography, thermal structure and general circulation, in *Encyclopedia of Physical Science and Technology,* edited by R. A. Meyers, Academic Press, San Diego.

Gnanadesikan, A., R. D. Slater, N. Gruber, and J. L. Sarmiento (2002), Oceanic vertical exchange and new production: A comparison between models and observations, *Deep-Sea Res. II, 49,* 363–401.

Gnanadesikan, A., J. P. Dunne, R. M. Key, K. Matsumoto, J. L. Sarmiento, R. D. Slater, and P. S. Swathi (2004), Oceanic ventilation and biogeochemical cycling: Understanding the physical mechanisms that produce realistic distributions of tracers and productivity, *Global Biogeochem. Cycles, 18,* GB4010, doi:4010.1029/2003GB002097.

Goldman, J. C., and M. R. Dennett (1983), Carbon dioxide exchange between air and seawater: No evidence for rate catalysis, *Science, 220,* 199–201.

Goldman, J. C., J. J. McCarthy, and D. G. Peavey (1979), Growth rate influence on the chemical composition of phytoplankton in oceanic waters, *Nature, 279,* 210–215.

Goldman, J. C., D. A. Hansell, and M. R. Dennett (1992), Chemical characterization of three large oceanic diatoms: Potential impact on water column chemistry, *Mar. Ecol. Prog. Ser., 88,* 257–270.

Goodale, C. L., and E. A. Davidson (2002), Carbon cycle: Uncertain sinks in the shrubs, *Nature, 418,* 593–594.

Gordon, A. L. (1981), Seasonality of Southern Ocean sea ice, *J. Geophys. Res., 86,* 4193–4197.

Goreau, T. J., W. A. Kaplan, S. C. Wofsey, M. B. McElroy, F. W. Valois, and S. W. Watson (1980), Production of NO_2 and N_2O by nitrifying bacteria at reduced concentrations of oxygen, *Appl. Environ. Microbiol.,* 526–532.

Gran, H. H., and T. Braarud (1935), A quantitative study of the phytoplankton in the Bay of Fundy and the Gulf of Maine, *Journ. Biol. Board Canada, 1,* 279–467.

Griffin, J. J., H. Windom, and E. D. Goldberg (1968), The distribution of clay minerals in the world ocean, *Deep-Sea Res., 15,* 433–459.

Gruber, N. (1998), Anthropogenic CO_2 in the Atlantic Ocean, *Global Biogeochem. Cycles, 12,* 165–192.

Gruber, N. (2004), The dynamics of the marine nitrogen cycle and its influence on atmospheric CO_2, in *The Ocean Carbon Cycle and Climate,* edited by M. Follows and T. Oguz, pp. 97–148, NATO ASI Series, Kluwer Academic, Dordrecht, Netherlands.

Gruber, N., and C. D. Keeling (2001), An improved estimate of the isotopic air-sea disequilibrium of CO_2: Implications for the oceanic uptake of anthropogenic CO_2, *Geophys. Res. Lett., 28,* 555–558.

Gruber, N., and J. L. Sarmiento (1997), Global patterns of marine nitrogen fixation and denitrification, *Global Biogeochem. Cycles, 11,* 235–266.

Gruber, N., and J. L. Sarmiento (2002), Large-scale biogeochemical-physical interactions in elemental cycles, in *The Sea,* edited by A. R. Robinson et al., pp. 337–399, John Wiley & Sons, Inc., New York.

Gruber, N., J. L. Sarmiento, and T. F. Stocker (1996), An improved method for detecting anthropogenic CO_2 in the oceans, *Global Biogeochem. Cycles, 10,* 809–837.

Gruber, N., C. D. Keeling, and T. F. Stocker (1998), Carbon-13 constraints on the seasonal inorganic carbon budget at the BATS site in the northwestern Sargasso Sea, *Deep-Sea Res. I, 45,* 673–717.

Gruber, N., E. Gloor, S.-M. Fan, and J. L. Sarmiento (2001), Air-sea fluxes of oxygen estimated from bulk data: Implications for the marine and atmospheric oxygen cycles, *Global Biogeochem. Cycles, 15,* 783–804.

Gruber, N., C. D. Keeling, and N. R. Bates (2002), Interannual variability in the North Atlantic ocean carbon sink, *Science, 298,* 2374–2378.

Gruber, N., P. Friedlingstein, C. B. Field, R. Valentini, M. Heimann, J. E. Richey, P. Romero-Lankao, D. Schulze, and C. Chen (2004), The vulnerability of the carbon cycle in the 21st century: An assessment of carbon-climate-human interactions, in *Toward CO_2 Stabilization: Issues, Strategies, and Consequences*, edited by C. B. Field and M. R. Raupach, pp. 45–76, Island Press, Washington, DC.

Gruber, N., H. Frenzel, S. C. Doney, P. Marchesiello, J. C. McWilliams, J. R. Moisan, J. Oram, G.-K. Plattner, and K. D. Stolzenbach (2005), Simulation of plankton ecosystem dynamics and upper ocean biogeochemistry in the California Current System. Part I: Model description, evaluation, and ecosystem structure, *Deep-Sea Res., in press.*

Guider, R. J., and J. L. Roche (1994), The role of iron in phytoplankton photosynthesis, and the potential for iron-limitation of primary productivity in the sea, *Photosynthesis Research, 39,* 275–301.

Guilderson, T. P., R. G. Fairbanks, and J. L. Rubenstone (1994), Tropical temperature variations since 20,000 years ago: Modulating interhemispheric climate change, *Science, 263,* 663–665.

Gundersen, J. K., and B. B. Jorgensen (1990), Microstructure of diffusive boundary layers and the oxygen uptake of the sea floor, *Nature, 345,* 604–607.

Gurney, K. R., et al. (2002), Towards robust regional estimates of CO_2 sources and sinks using atmospheric transport models, *Nature, 415,* 626–630.

Hackbusch, J. (1979), Eine Methode zur Bestimmung der Diffusions, Loslichkeits und Permeabilitats Konstanten von Radon-222 in Wasser und Meerwasser, Ph.D. thesis, University of Heidelberg, Heidelberg, Germany.

Haeckel, M., I. Konig, V. Reich, M. E. Weber, and E. Suess (2001), Pore water profiles and numerical modelling of biogeochemical processes in Peru Basin deep-sea sediments, *Deep-Sea Res. II–Topical Studies in Oceanography, 48,* 3713–3736.

Hairston, N. G., F. E. Smith, and L. B. Slobodkin (1960), Community structure, population control, and competition, *American Naturalist, 94,* 421–425.

Hale, M. S., and J. G. Mitchell (2001), Functional morphology of diatom frustule microstructures: Hydrodynamic control of Brownian particle diffusion and advection, *Aquatic Microbial Ecology, 24,* 287–295.

Hales, B. (2003), Respiration, dissolution, and the lysocline, *Paleoceanogr., 18,* 23.21–23.14, 1099, doi:10.1029/2003 PA000915.

Hales, B., and S. Emerson (1996), Calcite dissolution in sediments of the Ontong-Java Plateau: In situ measurements of pore water O_2 and pH, *Global Biogeochem. Cycles, 10,* 527–541.

Hales, B., and S. Emerson (1997), Evidence in support of first-order dissolution kinetics of calcite in seawater, *Earth and Planetary Sciences Letters, 148,* 317–327.

Hales, B., S. Emerson, and D. Archer (1994), Respiration and dissolution in the sediments of the western North Atlantic: Estimates from models of in situ microelectrode measurements of porewater oxygen and pH, *Deep-Sea Res., 41,* 695–720.

Hall, A., and M. Visbeck (2002), Synchronous variability in the southern hemisphere atmosphere, sea ice, and ocean resulting from the annular mode, *J. Clim., 15,* 3043–3057.

Hall, T. M., and T.W.H. Haine (2002) On ocean transport diagnostics: The idealized age tracer and the age spectrum, *J. Phys. Oceanogr., 32,* 1987–1991.

Hall, T. M., T.W.N. Haine, and D. W. Waugh (2002), Inferring the concentration of anthropogenic carbon in the ocean from tracers, *Global Biogeochem. Cycles, 16,* 1131, doi:1110.1029/2001GB001835.

Hallberg, R., and A. Gnanadesikan (2001), An exploration of the role of transient eddies in determining the transport of a zonally reentrant current, *Physical Oceanography, 31,* 3312–3330.

Hamm, C. E., R. Merkel, O. Springer, P. Jurkojc, C. Maier, K. Prechtel, and V. Smetacek (2003), Architecture and material properties of diatom shells provide effective mechanical protection, *Nature, 421,* 841–843.

Hammond, D. E., J. McManus, W. M. Berelson, T. E. Kilgore, and R. H. Pope (1996), Early diagenesis of organic material in equatorial Pacific sediments: Stoichometry and kinetics, *Deep-Sea Res. II, 43,* 1365–1412.

Hanawa, K., and L. D. Talley (2001), Mode waters, in *Ocean Circulation and Climate,* edited by G. Siedler and J. Church, pp. 373–386, Academic Press, San Diego.

Hansell, D. A. (2002), DOC in the global ocean carbon cycle, in *Biogeochemistry of Marine Dissolved Organic Matter,* edited by D. A. Hansell and C. A. Carlson, pp. 685–715, Academic Press, New York.

Hansell, D. A., and C. A. Carlson (1998), Deep-ocean gradients in the concentration of dissolved organic carbon, *Nature, 395,* 263–266.

Hansell, D. A., and C. A. Carlson (2001), Biogeochemistry of total organic carbon and nitrogen in the Sargasso Sea: Control by convective overturn, *Deep-Sea Res. II, 48,* 1649–1667.

Hansell, D. A. and C. A. Carlson (2002), *Biogeochemistry of Marine Dissolved Organic Matter,* Academic Press, New York.

Hansell, D. A., and H. W. Ducklow (2003), Bacterioplankton distribution and production in the bathypelagic ocean: Directly coupled to particulate organic carbon export? *Limnol. Oceanogr., 48,* 150–156.

Hansell, D. A., and E. T. Peltzer (1998), Spatial and temporal variation of total organic carbon in the Arabian Sea, *Deep-Sea Res. II, 45,* 2171–2193.

Hansell, D. A., and T. Y. Waterhouse (1997), Controls on the distribution of organic carbon and nitrogen in the eastern Pacific Ocean, *Deep-Sea Res. I, 44,* 843–857.

Hansell, D. A., K. Gundersen, and N. R. Bates (1995), Mineralization of dissolved organic carbon in the Sargasso Sea, *Mar. Chem., 51,* 201–212.

Hansell, D. A., N. R. Bates, and C. A. Carlson (1997), Predominance of vertical loss of carbon from surface waters of the equatorial Pacific Ocean, *Nature, 386,* 59–61.

Hansell, D. A., C. A. Carlson, and Y. Suzuki (2002), Dissolved organic carbon export with North Pacific Intermediate Water formation, *Global Biogeochem. Cycles, 16,* doi:10.1029/2000GB001361.

Hansell, D. A., N. R. Bates, and D. B. Olson (2004), Excess nitrate and nitrogen fixation in the North Atlantic Ocean, *Mar. Chem., 84,* 243–265.

Hardy, A. C. (1924), The herring in relation to its animate environment. Part I: The food and feeding habits of the herring with special reference to the east coast of England, *Fish. Invest. Lond. (Ser. 2), 7,* 1–53.

Harrison, K. G. (2000), Role of increased marine silica input on paleo-pCO_2 levels, *Paleoceanogr., 15,* 292–298.

Harrison, W. G., F. Azam, E. H. Renger, and R. W. Eppley (1977), Some experiments on phosphate assimilation by coastal marine phytoplankton, *Mar. Biol.*, *40*, 9–18.

Harrison, W. G., T. Platt, and M. R. Lewis (1987), f-ratio and its relationship to ambient nitrate concentration in coastal waters, *Journal of Plankton Research*, *9*, 235–248.

Hartmann, D. L. (1994), *Global Physical Climatology*, 411 pp., Academic Press, San Diego.

Hartnett, H. E., and A. H. Devol (2003), Role of a strong oxygen-deficient zone in the preservation and degradation of organic matter: A carbon budget for the continental margins of northwest Mexico and Washington State, *Geochim. Cosmochim. Acta*, *67*, 247–264.

Hartnett, H. E., R. G. Keil, J. I. Hedges, and A. Devol (1998), Influence of oxygen exposure time on organic carbon preservation in continental margin sediments, *Nature*, *391*, 572–575.

Harwood, D. M., and R. Gersonde (1990), Lower Cretaceous diatoms from ODP Leg 113 Site 693 (Weddell Sea). Part 2: Resting spores, Chrysophycean cysts, an endoskeletal dinoflagellate, and notes on the origin of diatoms, in *Proceedings of the Ocean Drilling Program, Scientific Results*, edited by P. F. Barker et al., pp. 403–425, Ocean Drilling Program, College Station, TX.

Hasse, L. (1971), The sea surface temperature deviation and the heat flow at the sea-air interface, *Boundary-Layer Meteorology*, *1*, 368–379.

Hedges, J. I. (2002), Why dissolved organics matter? in *Biogeochemistry of Marine Dissolved Organic Matter*, edited by D. A. Hansell and C. A. Carlson, pp. 1–33, Academic Press, New York.

Hedges, J. I., and R. G. Keil (1995), Sedimentary organic matter preservation: An assessment and speculative synthesis, *Mar. Chem.*, *49*, 137–139.

Hedges, J. I., F. S. Hu, A. H. Devol, H. E. Hartnett, and R. G. Keil (1999a), Sedimentary organic matter preservation: A test for selective oxic degradation, *Am. J. Sci.*, *299*, 529–555.

Hedges, J. I., R. G. Keil, C. Lee, and S. G. Wakeham (1999b), Atmospheric O_2 control by a "mineral conveyor belt" linking the continents and ocean, in *Geochemistry of the Earth's Surface*, edited by H. Armannsson, pp. 241–244, A. A. Balkema, Rotterdam.

Hedges, J. I., J. A. Baldock, Y. Gelinas, C. Lee, M. L. Peterson, and S. G. Wakeham (2002), The biochemical and elemental compositions of marine plankton: A NMR perspective, *Mar. Chem.*, *78*, 47–63.

Heggie, D., C. Maris, A. Hudson, J. Dymond, R. Beach, and J. Cullen (1987), Organic carbon oxidation and preservation in NW Atlantic continental margin sediments, in *Geology and Geochemistry of Abyssal Plains*, edited by P.P.E. Weaver and J. Thomson, p. 215, Geological Society, Washington, DC.

Heimann, M., and E. Maier-Reimer (1996), On the relations between the oceanic uptake of CO_2 and its carbon isotopes, *Global Biogeochem. Cycles*, *10*, 89–110.

Heinze, C., A. Hupe, E. Maier-Reimer, N. Dittert, and O. Ragueneau (2003), Sensitivity of the marine biospheric Si cycle for biogeochemical parameter variations, *Global Biogeochem. Cycles*, *17*, 1086, doi:10.1029/2002BG001943.

Henrichs, S. M. (1992), Early diagenesis of organic matter in marine sediments: Progress and perplexity, *Mar. Chem.*, *39*, 119–149.

Hesshaimer, V., M. Heimann, and I. Levin (1994), Radiocarbon evidence for a smaller oceanic carbon dioxide sink than previously believed, *Nature*, *370*, 201–203.

Higbie, R. (1935), The rate of absorption of a pure gas into a still liquid during short periods of exposure, *Trans. Am. Inst. Chem. Engr.*, *35*, 365–373.

Hildebrand, M., K. Dahlin, and B. E. Volcani (1998), Characterization of a silicon transporter gene family in *Cylindorotheca fusiformis*: Sequences, expression analysis, and identification of homologs in other diatoms, *Mol. Gen. Genet.*, *260*, 480–486.

Hinga, K. R. (1985), Evidence for a higher primary productivity in the Pacific than in the Atlantic Ocean, *Deep-Sea Res.*, *32*, 117–126.

Hinrichs, K.-U., and A. Boetius (2002), The anaerobic oxidation of methane: New insights in microbial ecology and biogeochemistry, in *Ocean Margin Systems*, edited by G. Wefer et al., Springer, Heidelberg.

Hirst, A. C., and T. J. McDougall (1998), Meridional overturning and dianeutral transport in a z-coordinate ocean model including eddy-induced advection, *J. Phys. Oceanogr.*, *28*, 1205–1223.

Hirst, A. C., D. R. Jackett, and T. J. McDougall (1996), The meridional overturning cells of a world ocean model in neutral density coordinates, *J. Phys. Oceanogr.*, 775–791.

Holland, H. D. (1984), *The Chemical Evolution of the Atmosphere and Oceans*, Princeton University Press, Princeton.

Honjo, S. (1978), Biogenic carbonate particles in the ocean: Do they dissolve in the water column? in *The Fate of Fossil Fuel CO_2 in the Oceans*, edited by N. Andersen and A. Malahoff, pp. 295–321, Plenum Press, New York.

Hoover, T. E., and D. C. Berkshire (1969), Effects of hydration in carbon dioxide exchange across an air-water interface, *J. Geophys. Res.*, *74*, 456–464.

Houghton, J. T., G. J. Jenkins, and J. J. Ephraums (1990), *Climate Change, The IPCC Scientific Assessment*, 365 pp., Cambridge University Press, New York.

Houghton, J. T., Y. Ding, D. J. Grigg, M. Noguer, P. J. van der Linden, X. Dai, K. Maskell, and C. A. Johnson (Eds.) (2001), *Climate Change 2001: The Scientific Basis*, 881 pp., Cambridge University Press, Cambridge.

Howell, E. A., S. C. Doney, R. A. Fine, and D. B. Olson (1997), Geochemical estimates of denitrification in the Arabian Sea and the Bay of Bengal during the WOCE, *Geophys. Res. Lett.*, *24*, 2549–2552.

Huang, R. X. (1990), On the three-dimensional structure of the wind-driven circulation in the North Atlantic, *Dyn. Atmos. Oceans*, *15*, 117–159.

Huettel, M., W. Ziebis, and S. Forster (1996), Flow-induced uptake of particulate matter in permeable sediments, *Limnol. Oceanogr.*, *41*, 309–322.

Hulth, S., R. C. Aller, and F. Gilbert (1999), Coupled anoxic nitrification/manganese reduction in marine sediments, *Geochim. Cosmochim. Acta*, *63*, 49–66.

Huntley, M. E., and M.D.G. Lopez (1992), Temperature-dependent production of marine copepods: A global synthesis, *American Naturalist*, *140*, 201–242.

Hupe, A., and J. Karstensen (2000), Redfield stoichiometry in Arabian Sea subsurface waters, *Global Biogeochem. Cycles, 14,* 357–372.

Hurd, D. C., and S. Birdwhistell (1983), On producing a more general model for biogenic silica dissolution, *Am. J. Sci., 283,* 1–28.

Hurtt, G. C., and R. A. Armstrong (1996), A pelagic ecosystem model calibrated with BATS data, *Deep-Sea Res. II, 43,* 653–683.

Hutchins, D. A., and K. W. Bruland (1998), Iron-limited diatom growth and Si:N uptake ratios in a coastal upwelling regime, *Nature, 393,* 561–564.

Hutchinson, G. E. (1961), The paradox of the plankton, *Amer. Nat., 95,* 137–145.

Iglesias-Rodríguez, M. D., R. Armstrong, R. A. Feely, R. Hood, J. Kleypas, J. D. Milliman, C. L. Sabine, and J. L. Sarmiento (2002a), Progress made in study of ocean's carbonate budget, *EOS Trans. AGU, 83,* 365, 374–375.

Iglesias-Rodríguez, M. D., C. Brown, S. C. Doney, J. Kleypas, D. Kolber, Z. Kolber, P. K. Hayes, and P. G. Falkowski (2002b), Representing key phytoplankton functional groups in ocean carbon cycle models: Coccolithophorids, *Global Biogeochem. Cycles, 16,* 1100, doi:10.1029/2001GB001454.

Iler, R. K. (1973), Effect of adsorbed alumina on the solubility of amorphous silica in water, *J. Colloid Interface Sci., 43,* 399–408.

Imbrie, J., and J. Z. Imbrie (1980), Model showing how Milankovitch cycles might produce the observed glacial cycles, *Science, 207,* 943–953.

Indermühle, A., et al. (1999), Holocene carbon-cycle dynamics based on CO_2 trapped in ice at Taylor Dome, Antarctica, *Nature, 398,* 121–126.

Indermühle, A., E. Monnin, B. Stauffer, T. F. Stocker, and M. Wahlen (2000), Atmospheric CO_2 concentration from 60 to 20 kyr BP from the Taylor Dome ice core, Antarctica, *Geophys. Res. Lett., 27,* 735–738.

Irigoien, X., et al. (2002), Copepod hatching success in marine ecosystems with high diatom concentrations, *Nature, 419,* 387–389.

Ito, T., and M. J. Follows (2003), Upper ocean control on the solubility pump of CO_2, *J. Mar. Re., 61,* 465–489.

Ivlev, V. (1945), Biologicheskaya produktivnost' vodoemov, *Usp. Sovr. Biol, 19,* 98–120.

Jackson, G. A. (2001), Effect of coagulation on a model planktonic food web, *Deep-Sea Res. I, 48,* 95–123.

Jackson, G. A., and A. B. Burd (2002), A model for the distribution of particle flux in the mid-water column controlled by subsurface biotic interactions, *Deep-Sea Res. II, 49,* 193–217.

Jähne, B. (1980), Zur Parameterisierung des Gasaustausches mit Hilfe von Laborexperimenten, Ph.D. thesis, University of Heidelberg, Heidelberg, Germany.

Jähne, B., G. Heinz, and W. Dietrich (1987a), Measurements of the diffusion coefficients of sparingly soluble gases in water with a modified Barrer method, *J. Geophys. Res., 92,* 10,767–10,776.

Jähne, B., K. O. Münnich, R. Bösinger, A. Dutzi, W. Huber, and P. Libner (1987b), On parameters influencing air-water gas exchange, *J. Geophys. Res., 92,* 1937–1949.

Jahnke, R. A. (1990), Ocean flux studies: A status report, *Reviews of Geophysics, 28,* 381–398.

Jahnke, R. A. (1996), The global ocean flux of particulate organic carbon: Areal distribution and magnitude, *Global Biogeochem. Cycles, 10,* 71–88.

Jahnke, R. A., and G. A. Jackson (1987), Role of sea floor organisms in oxygen consumption in the deep North Pacific Ocean, *Nature, 329,* 621–623.

Jahnke, R. A., and D. B. Jahnke (2004), Calcium carbonate dissolution in deep-sea sediments: Reconciling microelectrode, pore water, and benthic flux chamber results, *Geochim. Cosmochim. Acta, 58,* 47–59.

Jahnke, R. A., and G. B. Shimmield (1995), Particle flux and its conversion to the sediment record: Coastal ocean upwelling systems, in *Upwelling in the Ocean: Modern Processes and Ancient Records,* edited by C. P. Summerhayes et al., pp. 83–100, John Wiley & Sons Ltd, New York.

Jahnke, R. [A.], D. Heggie, S. Emerson, and V. Grudmanis (1982a), Pore waters of the central Pacific Ocean: Nutrient results, *Earth Planet. Sci. Lett., 61,* 233–356.

Jahnke, R. A., S. R. Emerson, and J. W. Murray (1982b), A model of oxygen reduction, denitrification, and organic matter remineralization in marine sediments, *Limnol. Oceanogr., 27,* 610–623.

Jahnke, R. A., S. R. Emerson, C. E. Reimers, J. Schuffert, K. Ruttenberg, and D. Archer (1989), Benthic recycling of biogenic debris in the eastern tropical Atlantic Ocean, *Geochim. Cosmochim. Acta, 53,* 2947–2960.

Jahnke, R. A., D. B. Craven, and J.-F. Gaillard (1994), The influence of organic matter diagenesis on $CaCO_3$ dissolution at the deep-sea floor, *Geochim. Cosmochim. Acta, 58,* 2799–2809.

Jenkins, W. J. (1980), Tritium and ^3He in the Sargasso Sea, *J. Mar. Re., 38,* 533–569.

Jenkins, W. J. (1987), ^3H and ^3He in the Beta Triangle: Observations of gyre ventilation and oxygen utilization rates, *J. Phys. Oceanogr., 17,* 763–783.

Jenkins, W. J. (1988), Nitrate flux into the euphotic zone near Bermuda, *Nature, 331,* 521–523.

Jenkins, W. J. (1998), Studying subtropical thermocline ventilation and circulation using tritium and ^3He, *J. Geophys. Res., 103,* 15,817–15,831.

Jenkins, W. J., and J. C. Goldman (1985), Seasonal oxygen cycling and primary production in the Sargasso Sea., *J. Mar. Re., 43,* 465–491.

Jenkins, W. J., and P. B. Rhines (1980), Tritium in the deep North Atlantic Ocean, *Nature, 286,* 877–880.

Jenkins, W. J., and D.W.R. Wallace (1992), Tracer based inferences of new primary production in the sea, in *Primary Productivity and Biogeochemical Cycles in the Sea,* edited by P. G. Falkowski and A. D. Woodhead, pp. 299–316, Plenum, New York.

Jickells, T., and L. Spokes (2001), Atmospheric iron inputs to the oceans, in *The Biogeochemistry of Iron in Seawater,* edited by D. R. Turner and K. A. Hunter, pp. 85–121, John Wiley, New York.

Jin, X., and N. Gruber (2003), Offsetting the radiative benefit of ocean iron fertilization by enhancing N_2O emissions, *Geophys. Res. Lett., 30,* 2249.

Jin, X., C. D. Nevison, and N. Gruber (2005a), Constraining oceanic N_2O formation mechanisms by inverting oceanic N_2O data. *In preparation.*

Jin, X., N. Gruber, J. Dunne, R. A. Armstrong, and J. L. Sarmiento (2005b), Diagnosing $CaCO_3$ and opal export and phytoplankton functional groups from global nutrient and alkalinity distributions, *Global Biogeochem. Cycles, in press.*

Johnsen, S. J., et al. (1997), The $\delta^{18}O$ record along the Greenland Ice Core Project deep ice core and the problem of possible Eemian climatic instability, *J. Geophys. Res., 102,* 26397–26410.

Johnson, K. S., and H. W. Jannasch (1994), Analytical chemistry under the sea surface: Monitoring ocean chemistry *in situ, Naval Research Reviews, 46,* 4–12.

Johnson, K. S., R. M. Gordon, and K. H. Coale (1997), What controls dissolved iron concentrations in the world ocean? *Mar. Chem., 57,* 137–161.

Jones, P. D., and A. Moberg (2003), Hemispheric and large-scale surface air temperature variations: An extensive revision and an update to 2001, *J. Clim., 16,* 206–223.

Joos, F., M. Bruno, R. Fink, U. Siegenthaler, T. F. Stocker, C. Le Quéré, and J. L. Sarmiento (1996), An efficient and accurate representation of complex oceanic and biospheric models of anthropogenic carbon uptake, *Tellus, 48B,* 397–417.

Joos, F., R. Meyer, M. Bruno, and M. Leuenberger (1999a), The variability in the carbon sinks as reconstructed for the last 1000 years, *Geophys. Res. Lett., 26,* 1437–1440.

Joos, F., G.-K. Plattner, T. F. Stocker, O. Marchal, and A. Schmittner (1999b), Global warming and marine carbon cycle feedbacks on future atmospheric CO_2, *Science, 284,* 464–467.

Joos, F., S. Gerber, I. C. Prentice, B. L. Otto-Bliesner, and P. J. Valdes (2004), Transient simulations of Holocene atmospheric carbon dioxide and terrestrial carbon since the last glacial maximum, *Global Biogeochem. Cycles, 18,* doi:10.1029/2003GB002156.

Jouzel, J., et al. (2001), A new 27 ky high resolution East Antarctic climate record, *Geophys. Res. Lett., 28,* 3199–3202.

Kamatani, A., J. P. Riley, and G. Skirrow (1980), The dissolution of opaline silica of diatom tests in sea water, *J. Oceanogr. Soc. Jpn., 36,* 201–280.

Kamykowski, D., and S.-J. Zentara (1985), Nitrate and silicic acid in the world ocean: Patterns and processes, *Mar. Ecol. Prog. Ser., 26,* 47–59.

Kara, A. B., P. A. Rochford, and H. E. HurlbMurt (2003), Mixed layer depth variability over the global ocean, *J. Geophys. Res., 108,* doi:10.1029/2000C000736.

Karl, D. M. (1999), A sea of change: Biogeochemical variability in the North Pacific Subtropical Gyre, *Ecosystems, 2,* 181–214.

Karl, D. M. (2002), Nutrient dynamics in the deep blue sea, *Trends in Microbiology, 10,* 410–418.

Karl, D. M., and K. M. Bjorkman (2002), Dynamics of DOP, in *Biogeochemistry of Marine Dissolved Organic Matter,* edited by D. A. Hansell and C. A. Carlson, pp. 250–366, Academic Press, New York.

Karl, D. M., R. Letelier, D. V. Hebel, D. F. Bird, and C. D. Winn (1992), *Trichodesmium* blooms and new nitrogen in the North Pacific gyre, in *Marine Pelagic Cyanobacteria: Trichodesmium and Other Diazotrophs,* edited by E. J. Carpenter, pp. 219–237, Kluwer Academic Publishers, Dordrecht, Netherlands.

Karl, D. M., R. Letelier, D. [V.] Hebel, L. Tupas, J. Dore, J. Christian, and C. Winn (1995), Ecosystem changes in the North Pacific subtropical gyre attributed to the 1991–92 El Niño, *Nature, 373,* 230–234.

Karl, D. M., J. R. Christian, J. E. Dore, D. V. Hebel, R. M. Letelier, L. M. Tupas, and C. D. Winn (1996), Seasonal and interannual variability in primary production and particle flux at Station ALOHA, *Deep-Sea Res. II, 43,* 539–568.

Karl, D. [M.], R. Letelier, L. Tupas, J. Dore, J. Christian, and D. [V.] Hebel (1997), The role of nitrogen fixation in biogeochemical cycling in the subtropical North Pacific Ocean, *Nature, 388,* 533–588.

Karl, D. M., R. R. Bidigare, and R. M. Letelier (2001), Long-term changes in plankton community structure and productivity in the North Pacific Subtropical Gyre: The domain shift hypothesis, *Deep-Sea Res. II, 48,* 1449–1470.

Karl, D., A. Michaels, B. Bergman, D. Capone, E. Carpenter, R. Letelier, F. Lipschultz, H. Paerl, D. Sigman, and L. Stal (2002), Dinitrogen fixation in the world's oceans, *Biogeochemistry, 57/58,* 47–98.

Karl, D. M., et al. (2003), Temporal studies of biogeochemical processes determined from ocean time-series observations during the JGOFS era, in *Ocean Biogeochemistry,* edited by M.J.R. Fasham, pp. 239–267, Academic Press, New York.

Kawamura, R. (1994), A rotated EOF analysis of global sea surface temperature variability with interannual and interdecadal scale, *J. Phys. Oceanogr., 24,* 707–715.

Keeling, C. D., and R. Revelle (1985), Effects of El Niño/Southern Oscillation on the atmospheric content of carbon dioxide, *Meteoritics, 20,* 437–450.

Keeling, C. D., and T. P. Whorf (1998), *Atmospheric CO_2 Concentration—Mauna Loa Observatory, Hawaii, 1958–1997* (revised August 1998), ORNL NDP-001, Oak Ridge Natl. Lab., Oak Ridge, TN.

Keeling, C. D., S. C. Piper, and M. Heimann (1989), A three-dimensional model of atmospheric CO_2 transport based on observed winds: 4. Mean annual gradients and interannual variations, in *Aspects of Climate Variability in the Pacific and the Western Americas,* edited by D. H. Peterson, Geophys. Monogr., *55,* pp. 305–363, AGU, Washington, DC.

Keeling, C. D., T. P. Whorf, M. Wahlen, and J.V.d. Plicht (1995), Interannual extremes in the rate of rise of atmospheric carbon dioxide since 1980, *Nature, 375,* 666–670.

Keeling, C. D., H. Brix, and N. Gruber (2004), Seasonal and long-term dynamics of the upper ocean carbon cycle at Station ALOHA near Hawaii, *Global Biogeochem. Cycles, 18,* GB4006, doi:10.1029/2004GB002227.

Keeling, R. F. (1993a), Heavy carbon dioxide, *Nature, 363,* 399–400.

Keeling, R. F. (1993b), On the role of large bubbles in air-sea gas exchange and supersaturation in the ocean, *J. Mar. Re., 51,* 237–271.

Keeling, R. F., and H. E. Garcia (2002), The change in oceanic O_2 inventory associated with recent global warming, *Proceedings of the National Academy of Sciences, 99,* 7848–7853.

Keeling, R. F., and T.-H. Peng (1995), Transport of heat, CO_2 and O_2 by the Atlantic's thermohaline circulation, *Philosophical Transactions of the Royal Society, 348,* 133–142.

Keeling, R. F., and S. R. Shertz (1992), Seasonal and interannual variations in atmospheric oxygen and implications for the global carbon cycle, *Nature,* 358, 723–727.

Keeling, R. F., S. C. Piper, and M. Heimann (1996), Global and hemispheric CO_2 sinks deduced from changes in atmospheric O_2 concentration, *Nature, 381,* 218–221.

Keeling, R. F., B. B. Stephens, R. G. Najjar, S. C. Doney, D. Archer, and M. Heimann (1998), Seasonal variations in the atmospheric O_2/N_2 ratio in relation to the kinetics of air-sea gas exchange, *Global Biogeochem. Cycles, 12,* 141–164.

Keil, R. G., D. B. Montlucon, F. G. Prahl, and J. I. Hedges (1994), Sorptive preservation of labile organic matter in marine sediments, *Nature, 370,* 549–551.

Keir, R. S. (1980), The dissolution kinetics of biogenic calcium carbonates in seawater, *Geochim. Cosmochim. Acta, 44,* 241–252.

Keir, R. S. (1993), Cold surface ocean ventilation and its effect on atmospheric CO_2, *J. Geophys. Res., 98,* 849–856.

Keir, R. S. (1995), Is there a component of Pleistocene CO_2 change associated with carbonate dissolution cycles? *Paleoceanogr., 10,* 871–880.

Keller, M. D., W. K. Bellows, and R.R.L. Guillard (1989), Dimethyl sulfide production in marine phytoplankton, in *Biogenic Sulfur in the Environment,* edited by E. S. Saltzman and W. J. Cooper, pp. 167–182, American Chemical Society, Washingon, DC.

Kelts, K., and K. J. Hsü (1978), Freshwater carbonate sedimentation, in *Geochemical Processes, Water and Sediment Environments,* edited by A. Lerman, pp. 219–312, Springer, Berlin.

Kemp, A.E.S., R. B. Pearce, I. Koizumi, J. Pike, and J. Rance (1999), The role of mat-forming diatoms in the formation of Mediterranean sapropels, *Nature, 398,* 57–61.

Kemp, A.E.S., J. Pike, R. B. Pearce, and C. B. Lange (2000), The "fall dump"—a new perspective on the role of a "shade flora" in the annual cycle of diatom production and export flux, *Deep-Sea Res. II, 47,* 2129–2154.

Kennedy, M. J., D. R. Pevear, and R. J. Hill (2002), Mineral surface control of organic carbon in black shale, *Science, 295,* 657–660.

Kennish, M. J. (Ed.) (1989), *CRC Practical Handbook of Marine Science,* 710 pp., CRC Press, Inc., Boca Raton, FL.

Kent, E. C., T. N. Forrester, and P. K. Taylor (1996), A comparison of oceanic skin effect parameterizations using shipborne radiometer data, *J. Geophys. Res.—Oceans, 101,* 16649–16666.

Key, R. M., A. Kozyr, C. L. Sabine, K. Lee, R. Wanninkhof, J. Bullister, R. A. Feely, F. Millero, C. W. Mordy, and T.-H. Peng (2004), A global ocean carbon climatology: Results from Global Data Analysis Project (GLODAP), *Global Biogeochem. Cycles, 18,* GB4031, doi:4010.1029/2004GB002247.

Kiefer, D. A., and C. A. Atkinson (1984), Cycling of nitrogen by plankton: A hypothetical description based upon efficiency of energy conversion, *J. Mar. Re., 42,* 655–675.

Kiehl, J. T., and K. E. Trenberth (1997), Earth's annual global mean energy budget, *Bull. Am. Meteor. Soc., 78,* 197–208.

Kindermann, J., G. Würth, G. H. Kohlmaier, and F.-W. Badeck (1996), Interannual variation of carbon exchange fluxes in terrestrial ecosystems, *Global Biogeochem. Cycles, 10,* 737–755.

Kirchman, D. L., C. Lancelot, M.J.R. Fasham, L. Legendre, G. Radach, and M. Scott (1993), Dissolved organic matter in biogeochemical models of the ocean, in *Towards a Model of Ocean Biogeochemical Processes,* edited by G. T. Evans and M.J.R. Fasham, pp. 209–226, Springer-Verlag, New York.

Klaas, C., and D. E. Archer (2002), Association of sinking organic matter with various types of mineral ballast in the deep sea: Implications for the rain ratio, *Global Biogeochem. Cycles, 16,* 1116, doi:1110.1029/2001GB001765.

Kleypas, J. A., and S. C. Doney (2001), Nutrients, Chlorophyll, Primary Production and related biogoechemical properties in the ocean mixed layer, 55 pp, NCAR, TN-447+STR, Boulder, CO.

Kleypas, J. A., R. W. Buddemeier, D. Archer, J. P. Gattuso, C. Langdon, and B. N. Opdyke (1999), Geochemical consequences of increased atmospheric carbon dioxide on coral reefs, *Science, 284,* 118–120.

Knauss, J. A. (1997), *Introduction to Physical Oceanography,* 2nd ed., Prentice-Hall, Upper Saddle River, NJ.

Knox, F., and M. McElroy (1984), Changes in atmospheric CO_2, influence of marine biota at high latitudes, *J. Geophys. Res., 89,* 4629–4637.

Kohfeld, K.E., C. Le Quéré, S.P. Harrison, R. F. Anderson, (2005). Role of marine biology in glacial-interglacial CO_2 cycles. *Science, 308,* 74–78.

Kolber, Z. S., R. T. Barber, K. H. Coale, S. E. Fitzwater, R. M. Greene, K. S. Johnson, S. Lindley, and P. C. Falkowski (1993), Iron limitation of phytoplankton photosynthesis in the equatorial Pacific Ocean, *Nature, 371,* 145–149.

Kortzinger, A., W. Koeve, P. Kahler, and L. Minitrop (2001), C:N ratios in the mixed layer during the productive season in the northwest Atlantic Ocean, *Deep-Sea Res., 48,* 661–687.

Kristensen, E., and M. Holmer (2001), Decomposition of plant materials in marine sediment exposed to different electron acceptors (O_2, NO_3^-, and SO_4^{2-}), with emphasis on substrate origin, degradation kinetics, and the role of bioturbation, *Geochim. Cosmochim. Acta, 65,* 419–433.

Kristensen, E., S. I. Ahmed, and A. H. Devol (1995), Aerobic and anaerobic decomposition of organic matter in marine sediment: Which is fastest? *Limnol. Oceanogr., 40,* 1430–1437.

Kromer, B., and W. Roether (1983), Field measurements of air-sea gas exchange by the radon deficit method during JASIN (1978) and FGGE (1979), *"Meteor" Forsch. Ergebnisse, A/B, 24,* 55–75.

Kröger, N., and R. Wetherbee (2000), Pleuralins are involved in theca differentiation in the diatom Cylindrotheca fusiformis, *Protist, 151,* 263–273.

Kump, L. R. (1988), Terrestrial feedback in atmospheric oxygen regulation by fire and phosphorus, *Nature, 335,* 152–154.

Kump, L. R., and R. M. Garrels (1986), Modeling atmospheric O_2 in the global sedimentary reddox cycle, *Am. J. Sci., 286,* 337–360.

Kurz, K. D. (1993), Zur saisonalen Variation des ozeanischen Kohlendioxidpartialdrucks, Ph.D. thesis, 107 pp., Examarbeit Nr. 18, Max-Planck-Institute für Meteorologie, Hamburg, Germany.

Kustka, A. B., S. Sanudo-Wilhelmy, E. J. Carpenter, D. Capone, J. Burns, and W. G. Sunda (2003), Iron requirements for dinitrogen- and ammonium-supported growth in cultures of trichodesmium (IMS 101): Comparison with nitrogen fixation rates and iron:carbon ratios of field population, *Limnol. Oceanogr., 48,* 1869–1884.

Kuypers, M. M. M., *et al.* (2003), Anaerobic ammonium oxidation by anammox bacteria in the Black Sea, *Nature, 422,* 608–611.

Kuypers, M.M.M., G. Lavik, D. Woebken, M. Schmid, B. M. Fuchs, R. Amann, B. B. Jorgensen, and M.S. Jetten (2005), Massive nitrogen loss from the Benguela upwelling system through anaerobic ammonium oxidation, *Proc. Natl. Acad. Sci., 102,* 6478–6483.

Lalli, C. M., and T. R. Parsons (1993), *Biological oceanography: an introduction,* 301 pp., Pergamon Press, New York.

Lampitt, R. S. (1992), The contribution of deep-sea macroplankton to organic remineralization: Results from sediment trap and zooplankton studies over the Madeira Abyssal Plain, *Deep-Sea Res., 39,* 221–234.

Lancelot, C. (1984), Extracellular release of small and large molecules by phytoplankton in the Southern Bight of the North Sea, *East Coast Shelf Science, 18,* 65–77.

Lancelot, C., E. Hannon, S. Becquevort, C. Veth, and H.J.W. de Baar (2000), Modeling phytoplankton blooms and carbon export production in the Southern Ocean: Dominant control by light and iron in the Atlantic sector in Austral spring 1992, *Deep-Sea Res., 47,* 1621–1662.

Landry, M. R. (2002), Integrating classical and microbial food web concepts: Evolving views from the open-ocean tropical Pacific, *Hydrobiologia, 480,* 29–39.

Landry, M. R., et al. (1997), Iron and grazing constraints on primary production in the central equatorial Pacific: An EqPac synthesis, *Limnol. Oceanogr., 42,* 405–418.

Lane, E., S. Peacock, and J. M. Restrepo (2005), A dynamic box-flow model and high-latitude sensitivity, *Tellus, Series B., submitted.*

Lasaga, A. C. (1989), A new approach to isotopic modeling of the variation of atmospheric oxygen through the Phanerozoic, *Am. J. Sci., 289,* 411–435.

Lasaga, A. C., and H. Ohmoto (2002), The oxygen geochemical cycle: Dynamics and stability, *Geochim. Cosmochim. Acta, 66,* 361–381.

Law, B. A. (1980), Transport and utilization of proteins by bacteria, in *Microorganisms and Nitrogen Sources,* edited by J. W. Payne, pp. 381–409, Wiley, New York.

Laws, E. A. (1991), Photosynthetic quotients, new production and net community production in the open ocean, *Deep-Sea Res., 38,* 143–167.

Laws, E. A., P. G. Falkowski, W. O. Smith, Jr., H. Ducklow, and J. J. McCarthy. (2000), Temperature effects on export production in the open ocean, *Global Biogeochem. Cycles, 14,* 1231–1246.

Lawson, D. S., D. C. Hurd, and H. S. Pankratz (1978), Silica decomposition rates of decomposing phytoplankton assemblages at various temperatures, *Am. J. Sci., 278,* 1373–1393.

Le Borgne, R., and M. Rodier (1997), Net zooplankton and the biological pump: A comparison between the oligotrophic and mesotrophic equatorial Pacific, *Deep-Sea Res. II, 44,* 2003–2023.

Ledwell, J. R., A. J. Watson, and C. S. Law (1993), Evidence for slow mixing across the pycnocline from an open-ocean tracer-release experiment, *Nature, 364,* 701–703.

Ledwell, J. R., A. J. Watson, and C. S. Law (1998), Mixing of a tracer in the pycnocline, *J. Geophys. Res., 103,* 21449–21529.

Ledwell, J. R., E. T. Montgomery, K. L. Polzin, L. C. St. Laurent, R. W. Schmitt, and J. M. Toole (2000), Evidence for enhanced mixing over rough topography in the abyssal ocean, *Nature, 403,* 179–182.

Lee, K. (2001), Global net community production estimated from the annual cycle of surface water total dissolved inorganic carbon, *Limnol. Oceanogr., 46,* 1287–1297.

Lee, K., and F. J. Millero (1995), Thermodynamic studies of the carbonate system in seawater, *Deep-Sea Res. I, 42,* 2035–2062.

Lee, K., F. J. Millero, and R. Wanninkhof (1997), The carbon dioxide system in the Atlantic Ocean, *J. Geophys. Res., 102,* 15693–15707.

Lee, K., R. Wanninkhof, T. Takahashi, S. C. Doney, and R. A. Feely (1998), Low interannual variability in recent oceanic uptake of atmospheric carbon dioxide, *Nature, 396,* 155–158.

Lee, K., D. M. Karl, R. Wanninkhof, and J.-Z. Zhang (2002), Global estimates of net carbon production in the nitrate-depleted tropical and subtropical oceans, *Geophys. Res. Lett., 29,* doi:10.1029/2001GL014198.

Lee, K., et al. (2003), An updated anthropogenic CO_2 inventory in the Atlantic Ocean, *Global Biogeochem. Cycles, 17,* 1116, doi:10.1029/2003GB002067.

Lee, M. M., and A. C. Coward (2003), Eddy mass transport for the Southern Ocean in an eddy-permitting ocean model, *Ocean Modelling, 5,* 249–266.

Lee, S., and J. A. Fuhrman (1987), Relationships between biovolume and biomass of naturally derived marine bacterioplankton, *Appl. Environ. Microbiol., 53,* 1298–1303.

Lee, Z. P., K. L. Carder, J. Marra, R. G. Steward, and M. J. Perry (1996), Estimating primary production at depth from remote sensing, *Appl. Opt., 35,* 463–474.

Lefevre, N., and A. J. Watson (1999), Modeling the geochemical cycle of iron in the oceans and its impact on atmospheric CO_2 concentrations, *Global Biogeochem. Cycles, 13,* 727–736.

Lehodey, P., M. Bertignac, J. Hampton, A. Lewis, and J. Picaut (1997), Southern oscillation and tuna in the western Pacific, *Nature, 389,* 715–718.

Lenton, T. M., and A. J. Watson (2000a), Redfied revisited. 1: Regulation of nitrate, phosphate, and oxygen in the ocean, *Global Biogeochem. Cycles, 14,* 225–248.

Lenton, T. M., and A. J. Watson (2000b), Redfield revisited. 2: What regulates the oxygen content of the atmosphere? *Global Biogeochem. Cycles, 14,* 249–268.

Le Quéré, C., and N. Gruber (2005), Satellite observation constraints on air-sea CO_2 flux variability, *in preparation.*

Le Quéré, C., J. Orr, P. Monfray, and O. Aumont (2000), Interannual variability of the oceanic sink of CO_2 from 1979 through 1997, *Global Biogeochem. Cycles, 14,* 1247–1265.

Le Quéré, C., et al. (2003), Two decades of ocean CO_2 sink and variability, *Tellus, 55B,* 649–656.

Lerman, A., and D. Lal (1977), Regeneration rates in the ocean, *Am. J. Sci., 277,* 238–258.

Leuenberger, M., U. Siegenthaler, and C. C. Langway (1992), Carbon isotope composition of atmospheric CO_2 during the last ice age from an Antarctic ice core, *Nature, 357.*

Levitus, S., M. E. Conkright, J. L. Reid, R. G. Najjar, and A. Mantyla (1993), Distribution of nitrate, phosphate and silicate in the worlds oceans, *Prog. Oceanogr., 31,* 245–273.

Levitus, S., R. Burgett, and T. Boyer (1994a), *World Ocean Atlas 1994*, vol. 3: *Salinity*, 99 pp., U.S. Government Printing Office, Washington, DC.

Levitus, S., R. Burgett, and T. Boyer (1994b), *World Ocean Atlas 1994*, vol. 4: *Temperature*, 117 pp., U.S. Government Printing Office, Washington, DC.

Levitus, S., T. P. Boyer, M. E. Conkright, T. O'Brien, J. Antonov, C. Stephens, L. Stathoplos, D. Johnson, and R. Gelfeld (1998), *World Ocean Database 1998*, vol. 1: *Introduction*, 346 pp., NOAA NESDIS, Washington, DC.

Levitus, S., J. I. Antonov, T. P. Boyer, and C. Stephens (2000), Warming of the world ocean, *Science*, *287*, 2225–2229.

Lewin, J. C. (1961), The dissolution of silica from diatom walls, *Geochim. Cosmochim. Acta*, *21*, 182–195.

Leynaert, A., E. Bucciarelli, P. Claquin, R. C. Dugdale, V. Martin-Jézéquel, P. Pondaven, and O. Ragueneau (2004), Effect of iron deficiency on diatom cell size and silicic acid uptake kinetics, *Limnol. Oceanogr. 49*, 1134–43.

Li, W.K.W., D. V. Subba-Rao, W. G. Harrison, J. C. Smith, J. J. Cullen, B. Irwin, and T. Platt (1983), Autotrophic picoplankton in the tropical ocean, *Science*, *219*, 292–295.

Li, Y.-H., and T.-H. Peng (2002), Latitudinal change or remineralization ratios in the oceans and its implication for nutrient cycles, *Global Biogeochem. Cycles*, *16*, doi:10.1029/2001GB001828.

Lipschultz, F. (2001), A time-series assessment of the nitrogen cycle at BATS, *Deep-Sea Res. II*, *48*, 1897–1924.

Lipschultz, F., N. Bates, C. A. Carlson, and D. A. Hansell (2002), New production in the Sargasso Sea: History and current status, *Global Biogeochem. Cycles*, *16*.

Liss, P. S., and L. Merlivat (1986), Air-sea gas exchange rates: Introduction and synthesis, in *The Role of Air-Sea Exchange in Geochemical Cycling*, edited by P. Buat-Menard, pp. 113–127, D. Reidel, Dordrecht, Netherlands.

Liss, P. S., and P. G. Slater (1974), Flux of gases across the air-sea interface, *Nature*, *247*, 181–184.

Liss, P. S., and W.G.N. Slinn (1983), *Air-Sea Exchange of Gases and Particles*, 561 pp., D. Reidel, Dordrecht, Netherlands.

Liu, X. W., and F. J. Millero (2002), The solubility of iron in seawater, *Mar. Chem.*, *77*, 43–54.

Logan, B. E., and J. R. Hunt (1987), Advantages to microbes of growth in permeable aggregates in marine systems, *Limnol. Oceanogr.*, *32*, 1034–1048.

Lohrenz, S. E., G. A. Knauer, V. L. Asper, M. Tuel, A. F. Michaels, and A. H. Knap (1992), Seasonal variability in primary production and particle flux in the northwestern Sargasso Sea: U.S. JGOFS Bermuda Atlantic Time-Series Study, *Deep-Sea Res.*, *39*, 1373–1392.

Longhurst, A. R. (1991), A reply to Broecker's charges, *Global Biogeochem. Cycles*, *5*, 315–316.

Louanchi, F., and R. G. Najjar (2001), Annual cycles of nutrients and oxygen in the upper layers of the North Atlantic Ocean, *Deep-Sea Res. II—Topical Studies in Oceanography*, *48*, 2155–2171.

Lovelock, J. E. (1995), *The Ages of Gaia: A Biography of Our Living Earth*, Oxford University Press, Oxford.

Lueker, T. J. (1998), The ratio of the first and second dissociation constants of carbonic acid determined from the concentration of carbon dioxide in gas and seawater at equilibrium, Ph.D. thesis, University of California at San Diego, La Jolla.

Lueker, T. J., A. G. Dickson, and C. D. Keeling (2000), Ocean pCO_2 calculated from dissolved inorganic carbon, alkalinity, and equations for K_1 and K_2: Validation based on laboratory measurements of CO_2 in gas and seawater at equilibrium, *Mar. Chem.*, *70*, 105–119.

Lupton, J. (1998), Hydrothermal helium plumes in the Pacific Ocean, *J. Geophys. Res.—Oceans*, *103*, 15853–15868.

Lutz, M., R. Dunbar, and K. Caldeira (2002), Regional variability in the vertical flux of particulate organic carbon in the ocean interior, *Global Biogeochem. Cycles*, *16*, 1037, doi:10.1029/2000GB001383.

Luyten, J. L., J. Pedlosky, and H. Stommel (1983), The ventilated thermocline, *J. Phys. Oceanogr.*, *13*, 292–309.

MacIsaac, J. J. (1978), Diel cycles of inorganic nitrogen uptake in a natural phytoplankton population dominated by *Gonyaulax polyedra*, *Limnol. Oceanogr.*, *23*, 1–9.

MacIsaac, J. J., and R. C. Dugdale (1969), The kinetics of nitrate and ammonia uptake by natural populations of marine phytoplankton, *Deep-Sea Res.*, *16*, 45–57.

MacIsaac, J. J., and R. C. Dugdale (1972), Interactions of light and inorganic nitrogen in controlling nitrogen uptake in the sea, *Deep-Sea Res.*, *19*, 209–232.

Mackin, J. E. (1987), Boron and silica behavior in salt-marsh sediments: Implications for paleo-boron distributions and the early diagenesis of silica, *Am. J. Sci.*, *287*, 197–241.

Mahowald, N., K. Kohfeld, M. Hansson, Y. Balkanski, S. P. Harrison, I. C. Prentice, M. Schulz, and H. Rodhe (1999), Dust sources and deposition during the last glacial maximum and current climate: A comparison of model results with paleodata from ice cores and marine sediments, *J. Geophys. Res.*, *104*, 15,895–15,916.

Maier-Reimer, E., and K. Hasselmann (1987), Transport and storage of CO_2 in the ocean—an inorganic ocean-circulation carbon cycle model, *Climate Dynamics*, *2*, 63–90.

Maier-Reimer, E., U. Mikolajewicz, and K. Hasselmann (1993), Mean circulation of the Hamburg LSG OGCM and its sensitivity to the thermohaline surface forcing, *J. Phys. Oceanogr.*, *23*, 731–757.

Mantua, N. J., and S. R. Hare (2002), The Pacific decadal oscillation, *Journal of Oceanography*, *58*, 35–44.

Marchal, O., P. Monfray, and N. R. Bates (1996), The spring-summer imbalance of dissolved inorganic carbon in the mixed layer of the northwestern Sargasso Sea, *Tellus*, *48*, 115–134.

Marino, B. D., M. B. McElroy, R. J. Salawitch, and W. G. Spaulding (1992), Glacial-to-interglacial variations in the carbon isotopic composition of atmospheric CO_2, *Nature*, *357*, 461–466.

Marra, J., and R. T. Barber (2004), Phytoplankton and heterotrophic respiration in the surface layer of the ocean, *Geophys. Res. Lett.*, *31*, L09314, doi:10.1029/2004GL019664.

Marra, J., C. Ho, and C. C. Trees (2003), *An Algorithm for the Calculation of Primary Productivity from Remote Sensing Data*, LDEO Technical Report, 27 pp., Lamont-Doherty Earth Observatory, LDEO-2003-1, Palisades, NY.

Marshall, J. C., A. J. G. Nurser, and R. G. Williams (1993), Inferring the subduction rate and period over the North Atlantic, *J. Phys. Oceanogr.*, *23*, 1315–1329.

Martel, F., and C. Wunsch (1993), The North Atlantic circulation in the early 1980s—an estimate from inversion of a finite-difference model, *J. Phys. Oceanogr.*, *23*, 898–924.

Martens, C. S., and R. A. Berner (1977), Interstitial water chemistry of Long Island Sound sediments: I. Dissolved gases, *Limnol. Oceanogr.*, *22*, 10–25.

Martin, J. H. (1990), Glacial-interglacial CO_2 change: The iron hypothesis, *Paleoceanogr.*, *5*, 1–13.

Martin, J. H. (1991), Iron, Liebig's Law and the greenhouse, *Oceanography*, *4*, 52–55.

Martin, J. H., and S. E. Fitzwater (1988), Iron deficiency limits phytoplankton growth in the north-east Pacific Subarctic, *Nature*, *331*, 341–343.

Martin, J. H., G. A. Knauer, D. M. Karl, and W. W. Broenkow (1987), VERTEX: Carbon cycling in the northeast Pacific, *Deep-Sea Res.*, *34*, 267–285.

Martin, J. H., R. M. Gordon, S. E. Fitzwater, and W. W. Broenkow (1989), VERTEX: Phytoplankton/iron studies in the Gulf of Alaska, *Deep-Sea Res.*, *36*, 649–680.

Martin, J. H., S. E. Fitzwater, and R. M. Gordon (1990a), Iron deficiency limits phytoplankton growth in Antarctic waters, *Global Biogeochem. Cycles*, *4*, 5–12.

Martin, J. H., R. M. Gordon, and S. E. Fitzwater (1990b), Iron in Antarctic waters, *Nature*, *345*, 156–158.

Martin, J. H., S. E. Fitzwater, and R. M. Gordon (1991), Iron deficiency limits phytoplankton growth in Antarctic waters, *Global Biogeochem. Cycles*, *4* , 5–12.

Martin, J. H., S. E. Fitzwater, R. M. Gordon, C. N. Hunter, and S. J. Tanner (1993), Iron, primary production, and carbon-nitrogen flux studies during the JGOFS North Atlantic bloom experiment, *Deep-Sea Res. II*, *40*, 115–134.

Martin, J. H., et al. (1994), Testing the iron hypothesis in ecosystems of the equatorial Pacific Ocean, *Nature*, *371*, 123–129.

Martin, W. R., and F. L. Sayles (1996), $CaCO_3$ dissolution in sediments of the Ceara Rise, western equatorial Atlantic, *Geochim. Cosmochim. Acta*, *60*, 243–263.

Martin-Jézéquel, V., M. Hildebrand, and M. A. Brzezinski (2000), Silicon metabolism in diatoms: Implications for growth, *Journal of Phycology*, *36*, 821–840.

Matear, R. J., and A. C. Hirst (1999), Climate change feedback on the future oceanic CO_2 uptake, *Tellus*, *51B*, 722–733.

Matear, R. J., and G. Holloway (1995), Modeling the inorganic phosphorus cycle of the North Pacific using an adjoint data assimilation model to assess the role of dissolved organic phosphorus, *Global Biogeochem. Cycles*, *9*, 101–119.

Matsumoto, K. and N. Gruber (2005). How accurate is the estimation of anthropogenic carbon in the ocean? An evaluation of the ΔC^* method, *Global Biogeochemical Cycles*, *19*, GB3014, doi:10.1029/2004GB002397.

Matsumoto, K., and R. M. Key (2004), Natural radiocarbon distribution in the deep ocean, in *Global Environmental Change in the Ocean and on Land*, edited by M. Shiyomi et al., TERRAPUB, Tokyo.

Matsumoto, K., J. L. Sarmiento, and M. A. Brzezinski (2002), Silicic acid leakage from the Southern Ocean as a possible mechanism for explaining glacial atmospheric pCO_2, *Global Biogeochem. Cycles*, *16*, 1031, doi:10.1029/2001 GB001442.

Matsumoto, K., et al. (2004), Evaluation of ocean carbon models with data-based metrics, *Geophys. Res. Lett.*, *31*, L07303, doi:07310.1029/2003GL018970.

Mauritzen, C., K. L. Polzin, M. S. McCartney, R. C. Millard, and D. E. West-Mack (2002), Evidence in hydrography and density fine structure for enhanced vertical mixing over the Mid-Atlantic Ridge in the western Atlantic, *J. Geophys. Res.—Oceans*, *107*, doi:10.1029/2001JC001114.

Mayer, L. M. (1994), Surface area control of organic carbon accumulation in continental shelf sediments, *Geochim. Cosmochim. Acta*, *58*, 1271–1284.

Mayer, L. M. (1999), Extent of coverage of mineral surfaces by organic matter in marine sediments, *Geochim. Cosmochim. Acta*, *63*, 207–215.

McCarthy, J. J. (2002), Biological responses to nutrients, in *Biological-Physical Interactions in the Sea*, edited by A. R. Robinson et al., pp. 219–244, John Wiley & Sons, Inc., New York.

McCartney, M. S. (1977), Subantarctic mode water, in *A Voyage of Discovery*, supplement to *Deep-Sea Res.*, *George Deacon 70th Anniversary Volume*, edited by M. V. Angel, pp. 103–119, Pergamon, Oxford.

McCartney, M. S. (1982), The subtropical recirculation of Mode Waters, *J. Mar. Res.*, *40*, *Supplement*, 127–161.

McDougall, T. J. (1987), Neutral surfaces, *J. Phys. Oceanogr.*, *17*, 1950–1964.

McElroy, M. F. (1983), Marine biological controls on atmospheric CO_2 climate, *Nature*, *302*, 328.

McGillicuddy, D. J., A. R. Robinson, D. A. Siegel, H. W. Jannasch, R. Johnson, T. D. Dickey, J. McNeil, A. F. Michaels, and A. H. Knap (1998), Influence of mesoscale eddies on new production in the Sargasso Sea, *Nature*, *394*, 263–266.

McGillis, W. R., and R. Wanninkhof (2005), Boundary layer CO_2 gradients expressed in terms of aqueous CO_2 concentration, *Mar. Chem.*, *in press*.

McGillis, W. R., J. B. Edson, J. E. Hare, and C. W. Fairall (2001), Direct covariance air-sea CO_2 fluxes, *J. Geophys. Res.—Oceans*, *106*, 16729–16745.

McGowan, J. A. (2004), Sverdrup's biology, *Oceanography*, *17*, 106–112.

McKinley, G. A. (2002), Interannual variability of the air-sea fluxes of carbon dioxide and oxygen, Ph.D. thesis, Massachusetts Institute of Technology, Cambridge.

McManus, J., D. E. Hammond, W. M. Berelson, T. E. Kilgore, D. J. DeMaster, O. G. Ragueneau, and R. W. Collier (1995), Early diagenesis of biogenic opal: Dissolution rates, kinetics, and paleoceanographic implications, *Deep-Sea Res. II*, *42*, 871–903.

McNeil, B. I., B. Tilbrook, and R. J. Matear (2001), Accumulation and uptake of anthropogenic CO_2 in the Southern Ocean, south of Australia between 1968 and 1996, *J. Geophys. Res.–Oceans*, *106*, 31431–31445.

McNeil, B. I., R. J. Matear, R. M. Key, J. L. Bullister, and J. L. Sarmiento (2003), Anthropogenic CO_2 uptake by the ocean based on the global chlorofluorocarbon dataset, *Science*, *299*, 235–239.

McPhaden, M. J., and D. X. Zhang (2002), Slowdown of the meridional overturning circulation in the upper Pacific Ocean, *Nature*, *415*, 603–608.

McWilliams, J. C., and G. Danabasoglu (2002), Eulerian and eddy-induced meridional overturning circulation in the tropics, *J. Phys. Oceanogr.*, *32*, 2054–2071.

Measures, C. I., and S. Vink (2001), Dissolved Fe in the upper waters of the Pacific sector of the Southern Ocean, *Deep-Sea Res. II*, *48*, 3913–3941.

Mehrbach, C., C. H. Culberson, J. E. Hawley, and R. M. Pytkowicz (1973), Measurement of the apparent dissociation constants of carbonic acid in seawater at atmospheric pressure, *Limnol. Oceanogr.*, *18*, 897–907.

Michaels, A. F., N. R. Bates, K. O. Buesseler, C. A. Carlson, and A. H. Knap (1994), Carbon-cycle imbalances in the Sargasso Sea, *Nature*, *372*, 537–540.

Michaels, A. F., D. Olsen, J. Sarmiento, J. Ammerman, K. Fanning, R. Jahnke, A. H. Knap, F. Lipschultz, and J. Prospero (1996a), Inputs, losses and transformations of nitrogen and phosphorus in the pelagic North Atlantic Ocean, *Biogeochemistry*, *35*, 181–226.

Michaels, A. F., J. L. Sarmiento, and J. Prospero (1996b), Excess nitrate and the rate of nitrogen fixation in the Sargasso Sea, *EOS*, *76*, OS84.

Michaels, A. F., D. M. Karl, and A. H. Knap (1999), Temporal studies of biogeochemical dynamics in oligotrophic oceans, in *The Changing Ocean Carbon Cycle*, edited by R. B. Hanson et al., pp. 392–413, Cambridge University Press, Cambridge.

Middelburg, J. J. (1989), A simple rate model for organic matter decomposition in marine sediments, *Geochim. Cosmochim. Acta*, *53*, 1577–1581.

Middelburg, J. J., T. Vlug, F. Jaco, and J.W.A. van der Nat (1993), Organic matter mineralization in marine sediments, *Global and Planetary Change*, *8*, 47–58.

Middelburg, J. J., K. Soetaert, P.M.J. Herman, and C.H.R. Heip (1996), Denitrification in marine sediments: A model study, *Global Biogeochem. Cycles*, *10*, 661–673.

Middelburg, J. J., K. Soetaert, and P. M. J. Herman (1997), Empirical relationships for use in global diagenetic models, *Deep-Sea Res.*, *44*, 327–344.

Millero, F. J. (1982), The thermodynamics of seawater at one atmosphere, *Ocean Sci. Eng.*, *7*, 403–460.

Millero, F. J. (1983), Influence of pressure on chemical processes in the sea, in *Chemical Oceanography*, edited by J. P. Riley and R. Chester, pp. 1–86, Academic Press, London.

Millero, F. J. (1995), Thermodynamics of the carbon dioxide system in the oceans, *Geochim. Cosmochim. Acta*, *59*, 661–677.

Millero, F. J. (1998), Solubility of Fe(III) in seawater, *Earth Planet. Sci. Lett.*, *154*, 323–329.

Millero, F. J., R. H. Byrne, R. Wanninkhof, R. A. Feely, T. Clayton, P. Murphy, and M. F. Lamb (1993), The internal consistency of CO_2 measurements in the equatorial Pacific, *Mar. Chem.*, *44*, 269–280.

Millero, F. J., K. Lee, and M. P. Roche (1998), The distribution of total alkalinity in the surface waters, *Mar. Chem.*, *60*, 111–130.

Milligan, A. J., and F.M.M. Morel (2002), A proton buffering role for silica in diatoms, *Science*, *297*, 1848–1850.

Milligan, A. J., D. E. Varela, M. A. Brzezinksi, and F.M.M. Morel (2004), Dynamics of silicon metabolism and silicon isotopic discrimination in a marine diatom as a function of pCO_2, *Limnol. Oceanogr.* *49*, 322–329.

Milliman, J. D. (1993), Production and accumulation of calcium carbonate in the ocean: Budget of a nonsteady state, *Global Biogeochem. Cycles*, *7*, 927–957.

Milliman, J. D., and A. W. Droxler (1996), Neritic and pelagic carbonate sedimentation in the marine environment: Ignorance is not bliss, *Geol. Rundschau*, *85*, 496–504.

Milliman, J. D., P. J. Troy, W. M. Balch, A. K. Adams, Y.-H. Li, and F. T. Mackenzie (1999), Biologically mediated dissolution of calcium carbonate above the chemical lysocline? *Deep-Sea Res.*, *46*, 1653–1670.

Mills, M. M., C. Ridame, M. Davey, J. La Roche, and R. J. Geider (2004), Iron and phosphorus co-limit nitrogen fixation in the eastern tropical North Atlantic, *Nature*, *429*, 292–294.

Minster, J.-F., and M. Boulahdid (1987), Redfield ratios along isopycnal surfaces—a complementary study, *Deep-Sea Res.*, *34*, 1981–2003.

Miralto, A., et al. (1999), The insidious effect of diatoms on copeod reproduction, *Nature*, *402*, 173–176.

Mitchell, B. G., E. A. Brody, O. Hom-Hansen, C. McClain, and J. Bishop (1991), Light limitation of phytoplankton biomass and macronutrient limitation in the Southern Ocean, *Limnol. Oceanogr.*, *36*, 1662–1677.

Mitchell, J.F.B., D. J. Karoly, G. C. Hegerl, F. W. Zwiers, M. R. Allen, and J. Marengo (2001), Detection of climate change and attribution of causes, in *Climate Change 2001: The Scientific Basis*, edited by J. T. Houghton et al., Cambridge University Press, Cambridge.

Moffett, J. W. (2001), Transformations between different forms of iron, in *The Biogeochemistry of Iron in Seawater*, edited by D. R. Turner and K. A. Hunter, pp. 343–372, John Wiley & Sons, Ltd., Chichester.

Moloney, C. L., and J. G. Field (1989), General allometric equations for rates of nutrient uptake, ingestion, and respiration in plankton organisms, *Limnol. Oceanogr.*, *34*, 1290–1299.

Moloney, C. L., and J. G. Field (1991), The size-based dynamics of plankton food webs. I: A simulation model of carbon and nitrogen flows, *Journal of Plankton Research*, *13*, 1003–1038.

Monahan, E. C. (2002), The physical and practical implications of a gas transfer coefficient that varies as the cube of the wind speed, in *Gas Transfer at Water Surfaces*, edited by M. A. Donelan et al., pp. 193–197, American Geophysical Union, Washington, DC.

Monahan, E. C., and M. Spillane (1984), The role of oceanic whitecaps in air-sea gas exchange, in *Gas Transfer at Water Surfaces*, edited by W. Brutsaert and G. H. Jirka, pp. 495–503, Reidel, Boston.

Monnin, E., A. Indermühle, A. Dällenbach, J. Flückiger, B. Stauffer, T. F. Stocker, D. Raynaud, and J.-M. Barnola (2002), Atmospheric CO_2 concentrations over the last glacial termination, *Science*, *291*, 112–114.

Monod, J. (1949), The growth of bacterial cultures, *Ann. Rev. Microbiol.*, *3*, 371–394.

Moore, J. K., S. C. Doney, D. M. Glover, and I. Y. Fung (2002a), Iron cycling and nutrient-limitation patterns in surface waters of the World Ocean, *Deep-Sea Research II*, *49*, 463–507.

Moore, J. K., S. C. Doney, J. A. Kleypas, D. M. Glover, and I. Y. Fung (2002b), An intermediate complexity marine ecosystem model for the global domain, *Deep-Sea Research Part II—Topical Studies in Oceanography*, *49*, 403–462.

Moore, J. K., Doney, S. C., and Lindsay, K. (2004), Upper ocean ecosystem dynamics and iron cycling in a global 3D model, *Global Biogeochem. Cycles*, *18*, GB4028, doi:10.1029/2004GB002220.

Moore, L. R., A. F. Post, G. Rocap, and S. W. Chisholm (2002), Utilization of different nitrogen sources by the marine cyanobacteria *Prochlorococcus* and *Synechoccocus*, *Limnol. Oceanogr.*, *47*, 989–996.

Morel, A. (1988), Optical modeling of the upper ocean in relation to its biogenous matter content (Case I Waters), *J. Geophys. Res.*, *93*, 10,749–10,768.

Morel, A., and R. C. Smith (1974), Relation between total quanta and total energy for aquatic photosynthesis, *Limnol. Oceanogr.*, *19*, 591–600.

Morel, F.M.M. (1987), Kinetics of nutrient uptake and growth in phytoplankton, *Journal of Phycology*, *23*, 150–156.

Morel, F.M.M., and J. G. Hering (1993), *Principles and Applications of Aquatic Chemistry*, 588 pp., John Wiley & Sons, Inc., New York.

Morel, F.M.M., R.J.M. Hudson, and N. M. Price (1991a), Limitation of productivity by trace metals in the sea, *Limnol. Oceanogr.*, *36*, 1742–1755.

Morel, F.M.M., J. G. Rueter, and N. M. Price (1991b), Iron nutrition of phytoplankton and its possible importance in the ecology of ocean regions with high nutrient and low biomass, *Oceanography*, *4*, 56–61.

Morse, J. W., and R. S. Arvidson (2002), The dissolution kinetics of major sedimentary carbonate minerals, *Earth-Science Reviews*, *58*, 51–84.

Morse, J. W., and R. A. Berner (1972), Dissolution kinetics of calcium carbonate in seawater. II: A kinetic origin for the lysocline, *Am. J. Sci*, *272*.

Mucci, A. (1983), The solubility of calcite and aragonite in sea water at various salinities, temperatures, and one atmosphere total pressure, *Am. J. Sci.*, *283*, 780–799.

Munk, W. H., and G. G. Carrier (1950), The wind-driven circulation in ocean basins of various shapes, *Tellus*, *2*, 158–167.

Munk, W. [H.], and C. Wunsch (1998), Abyssal recipes. II: Energetics of tidal and wind mixing, *Deep-Sea Res. I*, *45*, 1977–2010.

Murnane, R. J., J. L. Sarmiento, and M. P. Bacon (1990), Thorium isotopes, particle cycling models, and inverse calculations of model rate constants, *J. Geophys. Res.*, *95*, 16,195–16,206.

Murnane, R. J., J. K. Cockran, and J. L. Sarmiento (1994), Estimates of particle- and thorium-cycling rates in the northwest Atlantic Ocean, *J. Geophys. Res.*, *99*, 3373–3392.

Murnane, R., J. L. Sarmiento, and C. L. Quéré (1999), Spatial distribution of air-sea CO_2 fluxes and the interhemispheric transport of carbon by the oceans, *Global Biogeochem. Cycles*, *13*, 287–305.

Musgrave, D. L., J. Chou, and W. J. Jenkins (1988), Application of a model of upper-ocean physics for studying seasonal cycles of oxygen., *J. Geophys. Res.*, *93*, 15679–15700.

Nagasaki, K., Y. Tomaru, N. Katanozaka, Y. Shirai, K. Nishida, S. Itakura, and M. Yamaguchi (2004), Isolation and characterization of a novel single-stranded RNA virus infecting the bloom-forming diatom *Rhizosolenia setigera*, *Appl. Environ. Microbiol.*, *70*, 704–711.

Najjar, R. G. (1990), Simulations of the phosphorus and oxygen cycles in the world ocean using a general circulation model, Ph.D. thesis, Princeton University, Princeton, NJ.

Najjar, R. G., and R. F. Keeling (1997), Analysis of the mean annual cycle of the dissolved oxygen anomaly in the world ocean, *J. Mar. Re.*, *55*, 117–151.

Najjar, R. G., and R. F. Keeling (2000), Mean annual cycle of the air-sea oxygen flux: A global view, *Global Biogeochem. Cycles*, *14*, 573–584.

Najjar, R. G., and J. C. Orr (1998), Design of OCMIP-2 simulations of chlorofluorocarbons, the solubility pump and common biogeochemistry, http://www.ipsl.jussieu.fr/OCMIP/.

Najjar, R. G., J. L. Sarmiento, and J. R. Toggweiler (1992), Downward transport and fate of organic matter in the oceans: Simulations with a general circulation model, *Global Biogeochem. Cycles*, *6*, 45–76.

Najjar, R. G., N. Gruber, and J. Orr (2001), Predicting the ocean's response to rising CO_2: The Ocean Carbon Cycle Model Intercomparison Project, *U.S. JGOFS Newsletter*, *11*, 1–4.

Nakabayashi, S., K. Kuma, K. Sasaoka, S. Saitoh, M. Mochizuki, N. Shiga, and M. Kusakabe (2002), Variation in iron(III) solubility and iron concentration in the northwestern North Pacific Ocean, *Limnol. Oceanogr.*, *47*, 885–892.

Nakamura, T., T. Awaji, T. Hatayama, K. Akitomo, T. Takizawa, T. Kono, Y. Kawasaki, and M. Fukasawa (2000), The generation of large-amplitude unsteady lee waves by subinertial K_1 tidal flow: A possible vertical mixing mechanism in the Kuril Straits, *J. Phys. Oceanogr.*, *30*, 1601–1621.

National Geophysical Data Center, U.S. Department of Commerce, National Oceanic and Atmospheric Administration, 2001. *2-minute Gridded Global Relief Data (ETOPO2)* http://www.ngdc.noaa.gov/mgg/fliers/01mgg04.html.

Neftel, A., H. Oeschger, J. Schwander, B. Stauffer, and R. Zumbrunn (1982), Ice core measurements give atmospheric pCO_2 content during the past 40,000 years, *Nature*, *295*, 220–223.

Neftel, A., E. Moor, H. Oeschager, and B. Stauffer (1985), Evidence from polar ice cores for the increase in atmospheric CO_2 in the past two centuries, *Nature*, *315*, 45–47.

Nelson, D. M., and L. I. Gordon (1982), Production and pelagic dissolution of biogenic silica in the Southern Ocean, *Geochim. Cosmochim. Acta*, *46*, 491–505.

Nelson, D. M., J. A. Ahren, and L. J. Herlihy (1991), Cycling of biogenic silica within the upper water column of the Ross Sea, *Mar. Chem.*, *35*, 461–476.

Nelson, D. M., P. Treguer, M. A. Brzezinski, A. Leynaert, and B. Queguiner (1995), Production and dissolution of biogenic silica in the ocean: Revised global estimates, comparison with regional data and relationship to biogenic sedimentation, *Global Biogeochem. Cycles*, *9*, 359–372.

Nevison, C. [D.], J. H. Butler, and J. W. Elkins (2003), Global distribution of N_2O and the DN_2O-AOU yield in the subsurface ocean, *Global Biogeochem. Cycles*, *17* (4), 1119, doi:10.1029/2003GB002068.

Nevison, C. D., T. J. Lueker, and R. F. Weiss (2004), Quantifying the nitrous oxide source from coastal upwelling, *Global Biogeochem. Cycles*, *18*, GB1018, doi:1029/2003 GB002110.

Nightingale, P. D., P. S. Liss, and P. Schlosser (2000a), Measurements of air-sea gas transfer during an open ocean algal bloom, *Geophys. Res. Lett.*, *27*, 2117–2120.

Nightingale, P. D., G. Malin, C. S. Law, A. J. Watson, P. S. Liss, M. I. Liddicoat, J. Boutin, and R. C. Upstill-Goddard (2000b), In situ evaluation of air-sea gas exchange parameterizations using novel conservative and volatile tracers, *Global Biogeochem. Cycles*, *14*, 373–387.

Nobre, P., and J. Shukla (1996), Variations of sea surface temperature, wind stress, and rainfall over the tropical Atlantic and South Atlantic, *J. Clim.*, *9*, 2464–2479.

Nozaki, Y. (1997), A fresh look at element distribution in the North Pacific Ocean, *Eos, Trans. AGU*, *78*, 221.

Nozaki, Y., Y. Horibe, and H. Tsubota (1981), The water column distributions of thorium isotopes in the Western North Pacific Atlantic Ocean, *J. Mar. Re.*, *39*, 119–138.

Nozaki, Y., J. S. Yang, and M. Yanada (1987), Scavenging of thorium in the ocean, *J. Geophys. Res.*, *92*, 772–778.

Oeschger, H., U. Siegenthaler, U. Schotterer, and A. Gugelmann (1975), A box diffusion model to study the carbon dioxide exchange in nature, *Tellus*, *27*, 168–192.

Olsen, D. B., H. G. Ostlund, and J. Sarmiento (1986), The western boundary undercurrent off the Bahamas, *J. Phys. Oceanogr.*, *16*, 233–240.

O'Neil, R. V., D. L. DeAngelis, J. J. Pastor, B. J. Jackson, and W. M. Post (1989), Multiple nutrient limitations in ecological models, *Ecological Modelling*, *46*, 147–163.

Opdyke, B. N., and J. C. G. Walker (1992), Return of the coral reef hypothesis: Basin to shelf partitioning of $CaCO_3$ and its effect on atmospheric CO_2, *Geology*, *20*, 733–736.

Orcutt, K. M., F. Lipschultz, K. Gundersen, R. Arimoto, A. F. Michaels, A. H. Knap, and J. R. Gallon (2001), A seasonal study of the significance of N_2 fixation by *Trichodesmium* spp. at the Bermuda Atlantic Time-series Study (BATS) site, *Deep-Sea Res. II—Topical Studies in Oceanography*, *48*, 1583–1608.

Orr, J., et al. (2001), Estimates of anthropogenic carbon uptake from four three-dimensional global ocean models, *Global Biogeochem. Cycles*, *15*, 43–60.

Oschlies, A. (2001), Model-driven estimates of new production: New results point towards lower values, *Deep-Sea Res. II*, *48*, 2173–2197.

Oschlies, A., and V. Garcon (1998), Eddy-induced enhancement of primary production in a model of the North Atlantic Ocean, *Nature*, *394*, 266–269.

Paasche, E. (1962), Coccolith formation, *Nature*, *193*, 1094–1095.

Paasche, E. (2002), A review of the coccolithophorid *Emiliana huxleyi* (Prymnesiophyceae), with particular reference to growth, coccolith formation and calcification-photosynthesis interaction, *Phycologia*, *40*, 503–529.

Pacala, S. W., et al. (2001), Consistent land- and atmosphere-based US carbon sink estimates, *Science*, *292*, 2316–2320.

Pacanowski, R. C., and S. M. Griffies (1999), The MOM 3 Manual, Alpha Version, 580 pp., NOAA/Geophysical Fluid Dynamics Laboratory.

Pace, M. L., G. A. Knauer, D. M. Karl, and J. M. Martin (1986), Vertical flux of particulate organic matter in the Northeast Pacific: Relationship with depth, primary production, and new production, *EOS Trans. AGU*, *67*, 1036.

Papadimitriou, S., H. Kennedy, I. Bentaleb, and D. N. Thomas (2002), Dissolved organic carbon in sediments from the eastern North Atlantic, *Mar. Chem.*, *79*, 37–47.

Parekh, P., M. J. Follows, and E. Boyle (2004), Modeling the global ocean iron cycle, *Global Biogeochem. Cycles*, *18*, doi: 10.1029/2003GB002061.

Parsons, T. R., and C. M. Lalli (1988), Comparative oceanic ecology of the plankton communities of the sub-Arctic Atlantic and Pacific Oceans, *Oceanogr. Mar. Biol. Ann. Rev.*, *26*, 317–359.

Parsons, T. R., M. Takahashi, and B. Hargrave (1984), in *Biological Oceanographic Processes*, 3rd ed., 330 pp., Pergamon Press, New York.

Pedlosky, J. (1987), *Geophysical Fluid Dynamics*, 710 pp., Springer, New York.

Pedlosky, J. (1996), *Ocean Circulation Theory*, 453 pp., Springer, New York.

Peixoto, J. P., and A. H. Oort (1992), *Physics of Climate*, American Institute of Physics, New York.

Peltier, W. R. (1994), Ice Age Paleotopography, *Science*, *265*, 195–201.

Peng, T.-H., and W. S. Broecker (1987), C/f ratios in marine detritus, *Global Biogeochem. Cycles*, *1*, 155–161.

Peng, T.-H., T. Takahashi, and W. S. Broecker (1974), Surface radon measurements in the North Pacific at Ocean Station Papa, *J. Geophys. Res.*, *79*, 1777–1780.

Peng, T.-H., W. S. Broecker, G. Kipphut, and N. Shackleton (1977), Benthic mixing in deep sea cores as determined by ^{14}C dating and its implications regarding climate statigraphy and the fate of fossil fuel CO_2, in *The Fate of Fossil Fuel CO_2 in the Oceans*, edited by N. R. Anderson and A. Malahoff, pp. 355–373, Plenum Publ., New York.

Peng, T.-H., W. S. Broecker, G. G. Mathieu, Y. H. Li, and A. E. Bainbridge (1979), Radon evasion rates in the Atlantic and Pacific Oceans, *J. Geophys. Res.*, *84*, 2471–2486.

Peng, T.-H., T. Takahashi, and W. S. Broecker (1987), Seasonal variability of carbon dioxide, nutrients and oxygen in the northern North Atlantic surface water: Observations and a model, *Tellus*, *39B*, 439–458.

Peng, T.-H., R. Wanninkhof, J. L. Bullister, R. A. Feely, and T. Takahashi (1998), Quantification of decadal anthropogenic CO_2 uptake in the ocean based on dissolved inorganic carbon measurements, *Nature*, *396*.

Peng, T.-H., R. Wanninkhof, and R. A. Feely (2003), Increase of anthropogenic CO_2 in the Pacific Ocean over the last two decades, *Deep-Sea Res. II*, *50*, 3065–3082.

Petersen, E. E. (1965), *Chemical Reaction Analysis*, 276 pp., Prentice-Hall, Englewood Cliffs, NJ.

Petit, J. R., et al. (1999), Climate and atmospheric history of the past 420,000 years from the Vostok ice core, Antarctica, *Nature*, *399*, 429–436.

Peylin, P., P. Bousquet, C. Le Quéré, S. Sitch, P. Friedlingstein, G. A. McKinley, N. Gruber, P. Rayner, and P. Ciais (2005), Multiple constraints on regional CO_2 flux variations over land and oceans, *Global Biogeochem. Cycles*, *19*, GB1011, doi:10.1029/2003GB002214.

Philander, S. G. (1990), *El Niño, La Niña, and the Southern Oscillation*, 293 pp., Academic Press, San Diego, CA.

Pickard, G. L., and W. J. Emery (1990), *Descriptive Physical Oceanography*, 5th ed., Pergamon Press, New York.

Pickart, R. S., N. G. Hogg, and J.W.M. Smethie (1989), Determining the strength of the Deep Western Boundary Current using the chlorofluoromethane ratio, *J. Phys. Oceanogr.*, *19*, 940–951.

Pickett-Heaps, J., A.M.M. Schmid, and L. A. Edgar (1990), The cell biology of diatom valve formation, *Prog. Phycol. Res.*, *7*, 1–186.

Pitchford, J. W., and J. Brindley (1999), Iron limitation, grazing pressure and oceanic high nutrient–low chlorophyll (HNLC) regions, *Journal of Plankton Research*, *21*, 525–547.

Platt, T., and W. G. Harrison (1985), Biogenic fluxes of carbon and oxygen in the ocean, *Nature*, *318*, 55–58.

Platt, T., and A. D. Jassby (1976), The relationship between photosynthesis and light for natural assemblages of coastal marine phytoplankton, *J. Phycol., 12,* 421–430.

Platt, T., D. V. Subba-Rao, and B. Irwin (1983), Photosynthesis of picoplankton in the oligotrophic ocean, *Nature, 301,* 702–704.

Plattner, G. K., F. Joos, T. F. Stocker, and O. Marchal (2001), Feedback mechanisms and sensitivities of ocean carbon uptake under global warming, *Tellus, 53B,* 564–592.

Plattner, G. K., F. Joos, and T. F. Stocker (2002), Revision of the global carbon budget due to changing air-sea oxygen fluxes, *Global Biogeochem. Cycles, 16,* 1096, doi:10.1029/2001GB001746.

Polzin, K. L., J. M. Toole, J. R. Ledwell, and R. W. Schmitt (1997), Spatial variability of turbulent mixing in the abyssal ocean, *Science, 276,* 93–96.

Pomeroy, L. R. (1974), The ocean's food web: A changing paradigm, *BioScience, 24,* 499–504.

Pond, S., and G. L. Pickard (1983), *Introductory Dynamical Oceanography,* 2nd ed., 329 pp., Pergamon, New York.

Pondaven, P., D. Ruiz-Pino, J. N. Druon, C. Fravalo, and P. Treguer (1999), Factors controlling silicon and nitrogen biogeochemical cycles in high nutrient, low chlorophyll systems (the Southern Ocean and the North Pacific): Comparisions with a mesotrophic system (the North Atlantic), *Deep-Sea Res., 46,* 1923–1968.

Pondaven, P., O. Raqgueneau, P. Tréguer, L. Dezileau, and J. L. Reyss (2000a), Resolving the 'opal paradox' in the Southern Ocean, *Nature, 405,* 168–172.

Pondaven, P., D. Ruiz-Pino, C. Frabalo, P. Treguer, and C. Jeandel (2000b), Interannual variability of Si and N cycles at the time-series station KERFIX between 1990 and 1995—a 1-D modelling study, *Deep-Sea Res., 1,* 223–257.

Pope, R. H., D. J. DeMaster, C. R. Smith, and H. Seltmann, Jr. (1996), Rapid bioturbation in equatorial Pacific sediments: Evidence from excess ^{234}Th measurements, *Deep-Sea Res. II, 43,* 1339–1364.

Prather, M., et al. (2001), Atmospheric chemistry and greenhouse gases, in *Climate Change 2001: The Scientific Basis,* edited by J. T. Houghton et al., pp. 239–288, Cambridge University Press, Cambridge.

Precht, E., and M. Huettel (2003), Advective pore-water exchange driven by surface gravity waves and its ecological implications, *Limnol. Oceanogr., 48,* 1674–1684.

Prentice, I. C., D. Jolly, and BIOME 6000 Participants (2000), Mid-Holocene and glacial-maximum vegetation geography of the northern continents and Africa, *J. Biogeogr.*, 27, 507–519.

Prentice, I. C., G. D. Farquhar, M.J.R. Fasham, M. L. Goulden, M. Heimann, V. J. Jaramillo, H. S. Keshgi, C. Le Quéré, R. J. Scholes, and D.W.R. Wallace (2001), The carbon cycle and atmospheric carbon dioxide, in *Climate Change 2001: The Scientific Basis,* edited by J. T. Houghton et al., pp. 183–237, Cambridge University Press, Cambridge.

Quay, P. [D.], and J. Stutsman (2003), Surface layer carbon budget for the subtropical N. Pacific: δ^{13}C constraints at the station ALOHA, *Deep-Sea Res., 50,* 1045–1061.

Quay, P. D., B. Tilbrook, and C. S. Wong (1992), Oceanic uptake of fossil fuel CO_2: Carbon-13 evidence, *Science, 256,* 74–79.

Quigg, A., Z. V. Finkel, A. J. Irwin, Y. Rosenthal, T.-H. Ho, J. R. Reinfelder, O. Schofield, F.M.M. Morel, and P. G. Falkowski (2003), The evolutionary inheritance of elemental stoichiometry in marine phytoplankton, *Nature, 425,* 291–294.

Quinby-Hunt, M. S., and K. K. Turekian (1983), Distribution of elements in sea water, *EOS,* 130.

Rabouille, C. [R.], and J.-F. Gaillard (1991a), A coupled model representing the deep-sea organic carbon mineralization and oxygen consumption in surficial sediments, *J. Geophys. Res., 96,* 2761–2776.

Rabouille, C. R., and J. F. Gaillard (1991b), Towards the EDGE: Early diagenetic global explanation. A model depicting the early diagenesis of organic matter O_2, NO_3, Mn, and PO_4, *Geochim. Cosmochim. Acta, 55,* 2511–2525.

Rabouille, C., J.-F. Gaillard, P. Tréguer, and M.-A. Vincendeau (1997), Biogenic silica recycling in surficial sediments across the Polar Front of the Southern Ocean (Indian sector), *Deep-Sea Res. II, 44,* 1151–1176.

Rabouille, C., J. F. Gaillard, J. C. Relexans, P. Treguer, and M. A. Vincedeau (1998), Recycling of organic matter in Antarctic sediments: A transect through the polar front of the Southern Ocean (Indian sector), *Limnol. Oceanogr., 43,* 420–432.

Rabouille, C., F. T. Mackenzie, and L. M. Ver (2001), Influence of the human perturbation on carbon, nitrogen, and oxygen biogeochemical cycles in the global coastal ocean, *Geochim. et Cosmochim. Acta, 65,* 3615–3641.

Ragueneau, O., et al. (2000), A review of the Si cycle in the modern ocean: Recent progress and missing gaps in the application of biogenic opal as a paleoproductivity proxy, *Global and Planetary Change, 26,* 317–365.

Ragueneau, O., M. Gallinari, L. Corrin, S. Grandel, P. Hall, A. Hauvespre, R. S. Lampitt, D. Rickert, H. Stahl, A. Tengberg, and R. Witbaard (2001), The benthic silica cycle in the Northeast Atlantic: Annual mass balance, seasonality, and importance of non-steady-state processes for the early diagenesis of biogenic opal in deep-sea sediments, *Prog. Oceanogr., 50,* 171–200.

Ragueneau, O., N. Dittert, P. Pondaven, P. Treguer, and L. Corrin (2002), Si/C decoupling in the world ocean: Is the Southern Ocean different? *Deep-Sea Res. II, 49,* 3127–3154.

Raimbault, P., M. Rodier, and I. Taupier-Letage (1988), Size fraction of phytoplankton in the Ligurian Sea and the Algerian Basin (Mediterranean Sea): Size distribution versus total concentration, *Mar. Microb. Food Webs, 3,* 1–7.

Ramaswamy, V., O. Boucher, J. Haigh, D. Hauglustaine, J. M. Haywood, G. Myhre, T. Nakajima, G. Y. Shi, and S. Solomon (2001), Radiative forcing of climate change, in *Climate Change 2001: The Scientific Basis,* edited by J. T. Houghton et al., pp. 349–416, Cambridge University Press, Cambridge.

Ransom, B., D. Ki, M. Kastner, and S. Wainright (1998), Organic matter preservation on continental slopes: Importance of mineralogy and surface area, *Geochim. Cosmochim. Acta, 62,* 1329–1345.

Rasmusson, E. M., and T. H. Carpenter (1982), Variations in tropical sea surface temperature and surface wind fields associated with the Southern Oscillation/El Niño, *Monthly Weather Review, 110,* 354–384.

Raven, J. A. (1983), The transport and function of silicon in plants, *Biol. Rev.*, *58*, 179–207.

Raymo, M. E., and W. F. Ruddiman (1992), Tectonic forcing of late Cenozoic climate, *Nature*, *359*, 117–122.

Rayner, P. J., I. G. Enting, R. J. Francey, and R. Langerfelds (1999), Reconstructing the recent carbon cycle from atmospheric CO_2, $\delta^{13}C$ an O_2/N_2 observations, *Tellus*, *51B*, 213–232.

Redfield, A. C., B. H. Ketchum, and F. A. Richards (1963), The influence of organisms on the composition of seawater, in *The Sea*, vol. 2, edited by M. N. Hill, pp. 26–77, Wiley Interscience, New York.

Reeburgh, W. S. (1976), Methane consumption in Cariaco Trench waters and sediments, *Earth Planet. Sci. Lett.*, *28*, 337–344.

Reeburgh, W. S. (2003), Global methane biogeochemistry, in *The Atmosphere*, edited by R. F. Keeling, vol. 4: *Treatise on Geochemistry*, edited by H. D. Holland and K. K. Turekian, pp. 65–89, Elsevier-Pergamon, Oxford.

Reeburgh, W. S., S. C. Whalen, and M. J. Alperin (1993), The role of methylotrophy in the global methane budget, in *Microbial Growth on C-1 Compounds*, edited by J. C. Murrell and D. P. Kelly, Intercept Press, Andover, UK.

Reid, J. L. (1997), On the total geostrophic circulation of the Pacific Ocean: Flow patterns, tracers, and transport, *Prog. Oceanog.*, *29*, 263–352.

Reid, J. L., and R. J. Lynn (1971), On the influence of the Norwegian-Greenland and Weddell seas upon the bottom waters of the Indian and Pacific oceans, *Deep-Sea Res.*, *18*, 1063–1088.

Reimers, C. E. (1998), Feedbacks from the sea floor, *Nature*, *391*, 536–537.

Reimers, C. E., and K. L. Smith, Jr. (1986), Reconciling measured and predicted fluxes of oxygen across the deep sea sediment-water interface, *Limnol. Oceanogr.*, *31*, 305–318.

Reimers, C. E., R. A. Jahnke, and D. C. McCorkle (1992), Carbon fluxes and burial rates over the continental slope and rise off central California with implications for the global carbon cycle, *Global Biogeochem. Cycles*, 6, 199–224.

Rhein, M., O. Plän, R. Bayer, L. Stramma, and M. Arnold (1998), Temporal evolution of the tracer signal in the Deep Western Boundary Current, tropical Atlantic, *J. Geophys. Res.*, *103*, 15869–15883.

Rhines, P. B., and W. R. Young (1982), A theory of the wind-driven circulation. I: Mid-ocean gyres, *J. Mar. Re.*, *40* (*Suppl.*), 559–596.

Ribbe, J. (1999), On wind-driven mid-latitude convection in ocean general circulation models, *Tellus Series A—Dynamic Meteorology and Oceanography*, *51*, 517–525.

Richardson, T. L., A. M. Ciotti, and J. J. Cullen (1996), Physiological and optical properties of *Rhizosolenia formosa* (Bacillariophyceae) in the context of open-ocean vertical migration, *J. Phycol.*, *32*, 741–757.

Riebesell, U., I. Zondervan, B. Rost, P. D. Tortell, R. E. Zeebe, and F.M.M. Morel (2000), Reduced calcification of marine plankton in response to increased atmospheric pCO_2, *Nature*, *407*, 364–367.

Riley, G. A. (1951), Oxygen, phosphate, and nitrate in the Atlantic Ocean, *Bulletin of the Bingham Oceanographic Collection*, *1*, 1–126.

Rintoul, S. R., and C. Wunsch (1991), Mass, heat, oxygen and nutrient fluxes and budgets in the North Atlantic Ocean, *Deep-Sea Res.*, *38*, S355–S377.

Robbins, P. E., J. F. Price, W. B. Owens, and W. J. Jenkins (2000), On the importance of lateral diffusion for the ventilation of the lower thermocline in the subtropical North Atlantic, *J. Phys. Oceanogr.*, 30, 67–89.

Robertson, J. E., and A. J. Watson (1992), Thermal skin effect of the surface ocean and its implications for CO_2 uptake, *Nature*, *358*, 738–740.

Robinson, A. R., and H. Stommel (1959), The oceanic thermocline and the associated thermohaline circulation, *Tellus*, *3*, 295–308.

Roedenbeck, C., S. Houwelling, M. Gloor, and M. Heimann (2003), Time-dependent atmospheric CO_2 inversions based on interannually varying tracer transport, *Tellus*, *55B*, 488–497.

Roemmich, D. (1980), Estimation of meridional heat flux in the North Atlantic by inverse methods, *J. Phys. Oceanogr.*, *10*, 1972–1983.

Roethlisberger, R., R. Mulvaney, E. W. Wolff, M. A. Hutterli, M. Bigler, S. Sommer, and J. Jouzel (2002), Dust and sea salt variability in central East Antarctica (Dome C) over the last 45 kyrs and its implications for southern high-latitude climate, *Geophys. Res. Lett.*, *29*, doi:10.209/2002GL015186.

Roman, M. R., H. A. Adolf, M. R. Landry, L. P. Madin, D. K. Steinberg, and X. Zhang (2002), Estimates of oceanic mesozooplankton production: A comparison using the Bermuda and Hawaii time-series data, *Deep-Sea Res. II*, *49*, 175–192.

Ropelewski, C.F. 1995. *NOAA/NMC/CAC Arctic and Antarctic monthly sea ice extent*. Boulder, CO: National Snow and Ice Data Center. Digital media.

Ropelewski, C. F., and M. S. Halpert (1987), Global and regional scale precipitation patterns associated with the El Niño/Southern Oscillation, *Monthly Weather Review*, *115*, 1606–1626.

Round, F. E., R. M. Crawford, and D. G. Mann (1990), *The Diatoms: Biology and Morphology of the Genera*, Cambridge University Press, Cambridge.

Rubin, S., and R. M. Key (2002), Separating natural and bomb-produced radiocarbon in the ocean: The potential alkalinity method, *Global Biogeochem. Cycles*, *16*, 1105, doi: 10.1029/2001GB001432.

Rudnicki, M. D., H. Elderfield, and M. J. Mottl (2001), Pore fluid advection and reaction in sediments of the eastern flank, Juan de Fuca Ridge, 48 degrees N, *Earth Planet. Sci. Lett.*, *187*, 173–189.

Rue, E. L., and K. W. Bruland (1997), The role of organic complexation on ambient iron chemistry in the equatorial Pacific Ocean and the response of a mesoscale iron addition experiment, *Limnol. Oceanogr.*, *42*, 901–910.

Ryther, J. H. (1969), Photosynthesis and fish production in the sea, *Science*, *166*, 72–77.

Sabine, C. L., R. M. Key, K. M. Johnson, F. J. Millero, A. Poisson, J. L. Sarmiento, D. W. R. Wallace, and C. D. Winn (1999), Anthropogenic CO_2 inventory of the Indian Ocean, *Global Biogeochem. Cycles*, *13*, 179–198.

Sabine, C. L., H. W. Feely, R. M. Key, J. L. Bullister, F. J. Millero, K. Lee, T.-H. Peng, B. Tilbrook, T. Ono, and C. S. Wong (2002a), Distribution of anthropogenic CO_2 in the

Pacific Ocean, *Global Biogeochem. Cycles, 16,* 1083, doi:10.1029/2001GB001639.
Sabine, C. L., R. M. Key, R. A. Feely, and D. Greeley (2002b), Inorganic carbon in the Indian Ocean: Distribution and dissolution processes, *Global Biogeochem. Cycles, 16,* 1067, doi:10.1029/2002GB001869.
Sabine, C. L., et al. (2004), The oceanic sink for anthropogenic CO_2, *Science, 305,* 367–371.
Sakamoto, C. M., D. M. Kalr, H. W. Jannasch, R. R. Bidigare, R. M. Letelier, P. M. Walz, J. P. Ryan, P. S. Polito, and K. S. Johnson (2004), Influence of Rossby waves on nutrient dynamics and the plankton community structure in the North Pacific subtropical gyre, *J. Geophys. Res., 109,* doi: 10.1029/2003JC001976.
Sambrotto, R. N., G. Savidge, C. Robinson, P. Boyd, T. Takahashi, D. M. Karl, C. Langdon, D. Chipman, J. Marra, and L. Codispoti (1993), Elevated consumption of carbon relative to nitrogen in the surface ocean, *Nature, 363,* 248–250.
Samelson, R., and G. K. Vallis (1997), Large-scale circulation with small diapycnal diffusion: The two-thermocline limit, *J. Mar. Re., 55.*
Santschi, P. H. (1991), Measurements of diffusive sublayer thickness in the ocean by alabaster dissolution and their implications for the measurements of benthic fluxes, *J. Geophys. Res., 96,* 10,641–10,657.
Sanyal, A., N. G. Hemming, G. Hanson, and W. S. Broecker (1995), Evidence for a higher pH in the glacial ocean from boron isotopes in forminifera, *Nature, 373,* 234–236.
Sarmiento, J. L. (1983), A tritium box model of the North Atlantic thermocline, *J. Phys. Oceanogr., 13,* 1269–1274.
Sarmiento, J. L. (1991), Oceanic uptake of anthropogenic CO_2: The major uncertainties, *Global Biogeochem. Cycles, 5,* 309–313.
Sarmiento, J. L., and M. Bender (1994), Carbon biogeochemistry and climate change, *Photosynthesis Research, 39,* 209–234.
Sarmiento, J. L., and N. Gruber (2002), Sinks for anthropogenic carbon, *Physics Today, 55,* 30–36.
Sarmiento, J. L., and C. Le Quéré (1996), Oceanic carbon dioxide uptake in a model of century-scale global warming, *Science, 274,* 1346–1350.
Sarmiento, J. L., and J. C. Orr (1991), Three-dimensional simulations of the impact of Southern Ocean nutrient depletion on atmospheric CO_2 and ocean chemistry, *Limnol. Oceanogr., 36,* 1928–1950.
Sarmiento, J. L., and E. T. Sundquist (1992), Revised budget for the oceanic uptake of anthropogenic carbon dioxide, *Nature, 356,* 589–593.
Sarmiento, J. L., and J. R. Toggweiler (1984), A new model for the role of the oceans in determining atmospheric pCO_2, *Nature, 308,* 621–624.
Sarmiento, J. L., C.G.H. Rooth, and W. Roether (1982), The North Atlantic tritium distribution in 1972, *J. Geophys. Res., 87,* 8047–8056.
Sarmiento, J. L., T. D. Herbert, and J. R. Toggweiler (1988a), Causes of anoxia in the world ocean, *Global Biogeochem. Cycles, 2,* 115–128.
Sarmiento, J. L., J. R. Toggweiler, and R. Najjar (1988b), Ocean carbon cycle dynamics and atmospheric pCO_2, *Phil. Trans. R. Soc. Lond., 325,* 3–21.
Sarmiento, J. L., G. Thiele, R. M. Key, and W. S. Moore (1990), Oxygen and nitrate new production and remineralization in the North Atlantic subtropical gyre, *J. Geophys. Res., 95,* 18,303–18,315.
Sarmiento, J. L., J. C. Orr, and U. Siegenthaler (1992), A perturbation simulation of CO_2 uptake in an ocean general circulation model, *J. Geophys. Res., 97,* 3621–3646.
Sarmiento, J. L., R. D. Slater, M.J.R. Fasham, H. W. Ducklow, J. R. Toggweiler, and G. T. Evans (1993), A seasonal three-dimensional ecosystem model of nitrogen cycling in the North Atlantic euphotic zone., *Global Biogeochem. Cycles, 7,* 417–450.
Sarmiento, J. L., C. Le Quéré, and S. W. Pacala (1995), Limiting future atmospheric carbon dioxide, *Global Biogeochem. Cycles, 9,* 121–138.
Sarmiento, J. L., T. M. C. Hughes, R. J. Stouffer, and S. Manabe (1998), Simulated response of the ocean carbon cycle to anthropogenic climate warming, *Nature, 393,* 245–249.
Sarmiento, J. L., J. Dunne, A. Gnanadesikan, R. M. Key, K. Matsumoto, and R. Slater (2002), A new estimate of the $CaCO_3$ to organic carbon export ratio, *Global Biogeochem. Cycles, 16,* 1107, doi:1029/2002GB00191.
Sarmiento, J. L., N. Gruber, M. A. Brzezinski, and J. P. Dunne (2004a), High latitude controls of the global nutricline and low latitude biological productivity, *Nature, 427,* 56–60.
Sarmiento, J. L., et al. (2004b), Response of ocean ecosystems to climate warming, *Global Biogeochem. Cycles, 18,* GB3003, doi:1029/2003GB002134.
Sathyendranath, S., and T. Platt (1989), Computation of aquatic primary production: Extended formalism to include effect of angular and spectral distribution of light, *Limnol. Oceanogr., 34,* 188–198.
Sayles, F. L., and W. R. Martin (1995), In situ tracer studies of solute transport across the sediment-water interface at the Bermuda Time Series site, *Deep-Sea Res., 42,* 31–52.
Sayles, F. L., W. R. Martin, Z. Chase, and R. F. Anderson (2001), Benthic remineralization and burial of biogenic SiO_2, $CaCO_3$, organic carbon, and detrital material in the Southern Ocean along a transect at 170W, *Deep-Sea Res. II, 48,* 4323–4383.
Scheffer, M., S. Rinaldi, J. Huisman, and F. J. Weissing (2003), Why plankton communities have no equilibrium: Solutions to the paradox, *Hydrobiologia, 49,* 9–18.
Schimel, D., D. Alves, I. Enting, M. Heimann, F. Joos, D. Raynaud, and T. Wigley (1996), CO_2 and the carbon cycle, in *Climate Change 1995,* edited by J. T. Houghton et al., pp. 76–86, Cambridge University Press, Cambridge.
Schimel, D., I. G. Enting, M. Heimann, T.M.L. Wigley, D. Raynaud, D. Alves, and U. Siegenthaler (1995), CO_2 and the carbon cycle, in *Climate Change 1994,* edited by J. T. Houghton et al., pp. 35–71, Cambridge University Press, Cambridge.
Schimel, D. S., et al. (2001), Recent patterns and mechanisms of carbon exchange by terrestrial ecosystems, *Nature, 414,* 169–172.
Schink, D. R., and N. L. Guinasso, Jr. (1977), Effects of bioturbation on sediment seawater interaction, *Marine Geology, 23,* 133–154.
Schink, D. R., N. L. Guinasso, Jr., and K. A. Fanning (1975), Processes affecting the concentration of silica at the

sediment-water interface of the Atlantic Ocean, *J. Geophys. Res.*, *80*, 3013–3031.

Schlitzer, R. (2002), Carbon export fluxes in the Southern Ocean: Results from inverse modeling and comparison with satellite-based estimates, *Deep-Sea Res. II*, *49*, 1623–1644.

Schmitz, W. J. (1995), On the interbasin-scale thermohaline circulation, *Reviews of Geophysics*, *33*, 151–173.

Schmitz, W. J. (1996), On the world ocean circulation, vol. II, Woods Hole Oceanographic Institution, Technical Report, WHOI-96-08, Woods Hole, MA.

Schneider, B., R. Schlitzer, G. Fischer, and E. Nolthig (2003), Depth-dependent elemental compositions of particulate organic matter (POM) in the ocean, *Global Biogeochem. Cycles*, *17*, 1032, 10.1029/2002GB001871.

Scholten, J. C., J. Fietzke, S. Vogler, M.M.R. van der Loeff, A. Mangini, W. Koeve, J. Waniek, P. Sotffers, A. Antia, and J. Kuss (2001), Trapping efficiencies of sediment traps from the deep Eastern North Atlantic: The Th-230 calibration, *Deep-Sea Res. II*, *48*, 2383–2408.

Schrag, D. P., G. Haupt, and D. W. Murray (1996), Pore fluid constraints on the temperature and oxygen isotopic composition of the glacial ocean, *Science*, *272*, 1930–1932.

Schudlich, R., and S. Emerson (1996), Gas supersaturation in the surface ocean: The roles of heat flux, gas exchange, and bubbles, *Deep-Sea Res. II*, *43*, 569–589.

Scott, J. R., J. Marotzke, and A. Adcroft (2001), Geothermal heating and its influence on the meridional overturning circulation, *J. Geophys. Res.*, *106*, 31141–31154.

Seiter, K., C. Hensen, J. Schröter, and M. Zabel (2004), Organic carbon content in surface sediments: Defining regional provinces, *Deep-Sea Res. I*, *51*, 2001–2026.

Severinghaus, J. P. (1995), Studies of the terrestrial O_2 and carbon cycles in sand dune gases and in Biosphere 2, Ph.D. thesis, Columbia University, New York.

Shackleton, N. J. (1977), Carbon-13 in Uvigerina: Tropical rainforest history and the equatorial Pacific carbonate dissolution cycles, reprinted from *The Fate of Fossil Fuel CO_2 in the Oceans*, edited by N. R. Anderson and A. Malahoff, pp. 401–427, Plenum Publ., New York.

Shaffer, G. (1996), Biogeochemical cycling in the global ocean. 2: New production, Redfield ratios, and remineralization in the organic pump, *J. Geophys. Res.*, *101*, 3723–3745.

Sharp, J. H. (2002), Analytical methods for total DOM pools, in *Biogeochemistry of Marine Dissolved Organic Matter*, edited by D. A. Hansell and C. A. Carlson, pp. 35–58, Academic Press, New York.

Sharp, J. H., R. Benner, L. Bennett, C. A. Carlson, R. Dow, and S. E. Fitzwater (1993), Re-evaluation of high temperature combustion and chemical oxidation measurements of dissolved organic carbon in seawater, *Limnol. Oceanogr.*, *38*, 1774–1782.

Shaw, D. A., and T. J. Hanratty (1977), Turbulent mass transfer rates to a wall for large Schmidt numbers, *AIChE J.*, *23*, 28–37.

Shiller, A. M. (1981), Calculating the oceanic CO_2 increase: A need for caution, *J. Geophys. Res.*, *86*, 11083–11088.

Shiller, A. M. (1982), Reply to comment by Chen et al. on "Calculating the oceanic CO_2 increase: A need for caution" by A. M. Shiller, *J. Geophys. Res.*, *87*, 2086.

Siedler, G., J. Church, and J. Gould (Eds.) (2001), *Ocean Circulation and Climate*, 715 pp., Academic Press, San Diego.

Siegel, D. A., and T. D. Dickey (1987a), Observations of the vertical structure of the diffuse attenuation coefficient spectrum, *Deep-Sea Res.*, *34*, 547–563.

Siegel, D. A., and T. D. Dickey (1987b), On the parameterization of irradiance for open ocean photoprocesses, *J. Geophys. Res.*, *92*, 14,648–14,662.

Siegel, D. A., S. C. Doney, and J. A. Yoder (2002), The North Atlantic spring phytoplankton bloom and Sverdrup's critical depth hypothesis, *Science*, *296*, 730–733.

Siegenthaler, U. (1986), Carbon dioxide: Its natural cycle and anthropogenic perturbation, in *The Role of Air-Sea Exchange in Geochemical Cycling*, edited by P. Buat-Menard, pp. 209–248, D. Reidel, Dordrecht, Netherlands.

Siegenthaler, U., and F. Joos (1992), Use of a simple model for studying oceanic tracer distributions and the global carbon cycle, *Tellus*, *44B*, 186–207.

Siegenthaler, U., and J. L. Sarmiento (1993), Atmospheric carbon dioxide and the ocean, *Nature*, *365*, 119–125.

Siegenthaler, U., and T. Wenk (1984), Rapid atmospheric CO_2 variations and ocean circulation, *Nature*, *308*, 624–626.

Siegenthaler, U., and T. Wenk (1989), Modeling the ocean's role for El Niño related CO_2 variations, in *Third International Conference on Analysis and Evaluation of Atmospheric CO_2 Data Present and Past*, pp. 189–194, WMO Environmental Pollution Monitoring and Res. Prog. Report No. 59, Geneva.

Siever, R. (1991), Silica in the oceans: Biological-geochemical interplay, in *Scientists on Gaia*, edited by S. H. Schneider and P. J. Boston, pp. 287–295, MIT Press, Cambridge, MA.

Sigman, D. M., and E. A. Boyle (2000), Glacial/interglacial variations in atmospheric carbon dioxide, *Nature*, *407*, 859–869.

Sigman, D. M., D. C. McCorkle, and W. R. Martin (1998), The calcite lysocline as a constraint on glacial/interglacial low-latitude production changes, *Global Biogeochem. Cycles*, *12*, 409–427.

Sigman, D. M., M. A. Altabet, R. Francois, D. C. McCorkle, and J.-F. Gaillard (1999a), The isotopic composition of diatom-bound nitrogen in Southern Ocean sediments, *Paleoceanogr.*, *14*, 118–134.

Sigman, D. M., M. A. Altabet, D. C. McCorkle, R. Francois, and G. Fischer (1999b), The $\delta^{15}N$ of nitrate in the Southern Ocean: Consumption of nitrate in surface waters, *Global Biogeochem. Cycles*, *13*, 1149–1166.

Sigman, D. M., M. A. Altabet, D. C. McCorkle, R. Francois, and H. Fischer (2000), The $\delta^{15}N$ of nitrate in the Southern Ocean: Nitrogen cycling and circulation in the ocean interior, *J. Geophys. Res.*, *105*, 19599–19614.

Sigman, D. M., S. L. Jaccard, and G. H. Haug (2004), Polar ocean stratification in a cold climate, *Nature*, *428*, 59–63.

Sikes, C. S., R. D. Roer, and K. M. Wilbur (1980), Photosynthesis and coccolith formation: Inorganic carbon source and net inorganic reaction of disposition, *Limnol. Oceanogr.*, *25*, 248–261.

Sillén, L. G. (1961), The physical chemistry in sea water, in *Oceanography*, edited by M. Sears, pp. 549–581, AAAS Publ, Washington, DC.

Silver, M. W., and M. M. Gowing (1991), The "particle" flux: Origins and biological components, *Prog. Oceanog.*, *26*, 75–113.

Sloyan, B. M., and S. R. Rintoul (2001), Circulation, renewal, and modification of Antarctic Mode and Intermediate Water, *J. Phys. Oceanogr.*, *31*, 1005–1030.

Smetacek, V. S. (1985), Role of sinking in diatom life-history cycles: Ecological, evolutionary and geological significance, *Marine Biology*, *84*, 239–251.

Smetacek, V. [S.] (1998), Diatoms and the silicate factor, *Nature*, *391*, 224–225.

Smetacek, V. [S.] (1999), Diatoms and the ocean carbon cycle, *Protist*, *150*, 25–32.

Smethie, W. M., T. Takahashi, D. W. Chipman, and J. R. Ledwell (1985), Gas exchange and CO_2 flux in the tropical Atlantic determined from Rn-222 and pCO_2 measurements, *J. Geophys. Res.*, *90*, 7005–7021.

Smethie, W. M., R. A. Fine, A. Putzka, and E. P. Jones (2000), Tracing the flow of North Atlantic Deep Water using chlorofluorocarbons, *J. Geophys. Res.—Oceans*, *105*, 14297–14323.

Smith, C. R., R. H. Pope, D. J. DeMaster, and L. Magaard (1993), Age-dependent mixing of deep-sea sediments, *Geochim. Cosmochim. Acta*, *57*, 1473–1488.

Smith, D. C., M. Simon, A. L. Alldredge, and F. Azam (1992), Intense hydrolytic enzyme activity on marine aggregates and implications for rapid particle dissolution, *Nature*, *359*, 139–142.

Smith, J. N., and C. T. Shafer (1984), Bioturbation processes in continental slope and rise sediments delineated by Pb-210, microfossil, and textural indicators, *J. Mar. Re.*, *42*, 25–36.

Smith, S., and F. MacKenzie (1987), The ocean as a net heterotrophic system: Implications for the carbon biogeochemical cycle, *Global Biogeochem. Cycles*, *1*, 187–198.

Smith, W. O., R. F. Anderson, J. K. Moore, L. A. Codispoti, and J. M. Morrison (2000), The U.S. Southern Ocean Joint Global Ocean Flux Study: An introduction to AESOPS, *Deep-Sea Res. II*, *47*, 3073–3093.

Soetaert, K., P. M. J. Herman, J. J. Middelburg, and C. Heip (1998), Assessing organic matter mineralization, degradability and mixing rate in an ocean margin sediment (Northeast Atlantic) by diagenetic modeling, *J. Mar. Re.*, *56*, 519–534.

South, G. R., and A. Whittick (1987), *Introduction to Phycology*, 341 pp., Blackwell Scientific Publications, Boston.

Sowers, T., R. B. Alley, and J. Jubenville (2003), Ice core records of atmospheric N_2O covering the last 106,000 years, *Science*, *301*, 945–948.

Speer, K. G. (1989), The Stommel and Arons model and geothermal heating in the South-Pacific, *Earth Planet. Sci. Lett.*, *95*, 359–366.

Speer, K. [G.], and E. Tziperman (1992), Rates of water mass formation in the North Atlantic Ocean, *J. Phys. Oceanogr.*, *22*, 93–104.

Speer, K. [G.], S. R. Rintoul, and B. Sloyan (2000), The diabatic Deacon Cell, *J. Phys. Oceanogr.*, *30*, 3212–3222.

Spero, H. J., J. Bijma, D. W. Lea, and B. E. Bemis (1997), Effect of seawater carbonate concentration on foraminiferal carbon and oxygen isotopes, *Nature*, *390*, 497–499.

Spitzer, W. S., and W. J. Jenkins (1989), Rates of vertical mixing, gas exchange and new production: Estimates from seasonal gas cycles in the upper ocean near Bermuda, *J. Mar. Re.*, *47*, 169–196.

Staal, M., F.J.R. Meysman, and L. Stal (2003), Temperature excludes N_2-fixing heterocystous cyanobacteria in tropical oceans, *Nature*, *425*, 504–507.

Stammer, D. (1998), On eddy characteristics, eddy transports, and mean flow properties, *J. Phys. Oceanogr.*, *28*, 727–739.

Stammer, D., C. Wunsch, R. Giering, C. Eckert, P. Heimbach, J. Marotzke, A. Adcroft, C. N. Hill, and J. Marshall (2002), The global ocean circulation during 1992–1997, estimated from ocean observations and a general circulation model, *J. Geophys. Res.*, *107*, 3118, doi:3110.1029/2001JC000888.

Steele, J. H. (1998), Incorporating the microbial loop in a simple plankton model, *Proceedings of the Royal Society of London Series B—Biological Sciences*, *265*, 1771–1777.

Steele, J. H., and E. W. Henderson (1992), The role of predation in plankton models, *Journal of Plankton Research*, *14*, 157–172.

Steemann Nielsen, E. (1952), The use of radio-active carbon (C^{14}) for measuring organic production in the sea, *J. Cons.*, *18*, 117–140.

Stephens, B. B., and R. F. Keeling (2000), The influence of Antarctic sea ice on glacial-interglacial CO_2 variations, *Nature*, *404*, 171–174.

Stephens, B. B., R. F. Keeling, M. Heimann, K. D. Six, R. Murnane, and K. Caldeira (1998), Testing global ocean carbon cycle models using measurements of atmospheric O_2 and CO_2 concentrations, *Global Biogeochem. Cycles*, *12*, 213–230.

Stocker, T. F. (1996), An overview of century time-scale variability in the climate system: Observations and models, in *Decadal Climate Variability: Dynamics and Variability*, edited by D.L.T. Anderson and J. Willebrand, pp. 379–406, Springer-Verlag, Berlin.

Stoll, H. M., P. Ziveri, M. Geisen, I. Probert, and J. R. Young (2002), Potential and limitations of Sr/Ca ratios in coccolith carbonate: New perspectives from cultures and monospecific samples from sediments, *Phil. Trans. R. Soc. Lond. A*, 360, 719–747.

Stommel, H. (1948), The westward intensification of wind-driven ocean currents, *Transactions of the American Geophysical Union*, *29*, 202–206.

Stommel, H. (1958), The abyssal circulation, *Deep-Sea Res.*, *5*, 80–82.

Stommel, H. (1965), *The Gulf Stream: A Physical and Dynamical Description*, 2nd ed., 248 pp., University of California Press, Berkeley.

Stommel, H. (1982), Is the South-Pacific He-3 plume dynamically active? *Earth Planet. Sci. Lett.*, *61*, 63–67.

Strickland, J.D.H. (1965), Production of organic matter in the primary stages of the marine food chain, in *Chemical Oceanography*, edited by J. P. Riley and G. Skirrow, pp. 447–610, Academic Press, London.

Stuiver, M., and H. A. Polach (1977), Discussion: Reporting of ^{14}C data, *Radiocarbon*, *19*, 355–363.

Stuiver, M., and P. D. Quay (1981), Atmospheric ^{14}C changes resulting from fossil fuel CO_2 release and cosmic ray flux variability, *Earth Planet. Sci. Lett.*, *53*, 349–362.

Stuiver, M., P. D. Quay, and H. G. Ostlund (1983), Abyssal water carbon-14 distribution and the age of the world oceans, *Science, 219,* 849–851.

Stumm, W., and J. J. Morgan (Eds.) (1981), *Aquatic Chemistry,* 2nd ed., 523–598 pp., Wiley and Sons, New York.

Stute, M., M. Forster, H. Frischkorn, A. Serejo, J. Clark, P. Schlosser, W. S. Broecker, and G. Bonani (1995), Cooling of tropical Brazil (5°C) during the last glacial maximum, *Science, 269,* 379–383.

Suess, E.(1980), Particulate organic carbon flux in the ocean: Surface productivity and oxygen utilization, *Nature, 288,* 260–283.

Suess, H. E. (1955), Radiocarbon concentration in modern wood, *Science, 122,* 415–417.

Sugimura, Y., and Y. Suzuki (1988), A high temperature catalytic oxidation method for the determination of non-volatile dissolved organic carbon in seawater by direct injection of a liquid sample, *Mar. Chem., 24,* 105–131.

Sun, M.-Y., R. C. Aller, C. Lee, and S. G. Wakeham (2002), Effects of oxygen and redox oscillation on degradation of cell-associated lipids in surficial marine sediments, *Geochim. Cosmochim. Acta,* 66, 2003–2012.

Sunda, W. G. (2001), Bioavailability and bioaccumulation of iron in the sea, in *The Biogeochemistry of Iron in Seawater,* edited by D. R. Turner and K. A. Hunter, pp. 41–84, John Wiley & Sons, Ltd., Chichester.

Sundquist, E. T. (1991), Steady- and non-steady-state carbonate-silicate controls on atmosphere CO_2, *Quarternary Science Review, 10,* 283–296.

Suntharalingam, P., and J. L. Sarmiento (2000), Factors governing the oceanic nitrous oxide distribution: Simulations with an ocean general circulation model, *Global Biogeochem. Cycles, 14,* 429–454.

Suntharalingam, P., J. L. Sarmiento, and J. R. Toggweiler (2000), Global significance of nitrous-oxide production and transport from oceanic low-oxygen zones: A modeling study, *Global Biogeochem. Cycles, 14,* 1353–1370.

Suttle, C. A., J. A. Fuhrman, and D. G. Capone (1990), Rapid ammonium flux and concentration-dependent partitioning of ammonium and phosphate: Implications for carbon transfer in planktonic communities, *Limnol. Oceanogr., 35,* 424–433.

Suzuki, Y. (1993), On the measurement of DOC and DON in seawater, *Mar. Chem., 16,* 287–288.

Sverdrup, H. U. (1953), On conditions for the vernal blooming of phytoplankton, *Journal du Conseil Permanent International pour l'Exploration de la Mer, 18,* 287–295.

Sverdrup, H. U. (1955), The place of physical oceanography in oceanographic research, *J. Mar. Re., 14,* 287–294.

Sverdrup, H. U., N. W. Johnson, and R. H. Fleming (1941), *The Oceans,* Prentice Hall, Englewood Cliffs, NJ.

Sweeney, C., et al. (2000), Nutrient and carbon removal ratios and fluxes in the Ross Sea, Antartica, *Deep-Sea Research II, 47,* 3395–3421.

Takahashi, T., W. S. Broecker, S. R. Werner, and A. E. Bainbridge (1980), Carbonate chemistry of the surface waters of the world oceans, in *Isotope Marine Chemistry,* edited by E. D. Golberg, Y. Horibe, and K. Saruhashi, pp. 291–326, Uchinda Rokakuho Publ. Co. Ltd., Tokyo.

Takahashi, T., W. S. Broecker, and A. E. Bainbridge (1981), The alkalinity and total carbon dioxide concentration in the world oceans, in *Scope 16: Carbon Cycle Modellling,* edited by B. Bolin, pp. 271–286, John Wiley, New York.

Takahashi, T., W. S. Broecker, and S. Langer (1985), Redfield ratio based on chemical data from isopycnal surfaces, *J. Geophys. Res., 90,* 6907–6924.

Takahashi, T., J.Olafsson, J.G. Goodard, D.W. Chipman, and S.C. Sutherland (1993), Seasonal variation of CO_2 and nutrients in the high-latitude surface oceans: A comparative study, *Global Biogeochem. Cycles, 7,* 843–878.

Takahashi, T., R. H. Wanninkhof, R. A. Feely, R. F. Weiss, D. W. Chipman, N. Bates, J. Olafsson, C. Sabine, and S. C. Sutherland (1999), Net sea-air CO_2 flux over the global oceans: An improved estimate based on the sea-air pCO_2 difference, in *Proceedings of the 2nd International Symposium, CO_2 in Oceans,* edited by Y. Nojiri, pp. 9–14, Center for Global Environmental Research, National Institute for Environmental Studies, Tsukuba.

Takahashi, T., et al. (2002), Global sea-air CO_2 flux based on climatological surface ocean pCO_2, and seasonal biological and temperature effects, *Deep-Sea Res. II, 49,* 1601–1622.

Takeda, S. (1998), Influence of iron availability on nutrient consumption ratio of diatoms in oceanic waters, *Nature, 393,* 774–777.

Talley, L. D. (1991), An Okhotsk Sea water anomaly: Implications for ventilation in the North Pacific, *Deep-Sea Res., 38 Suppl. 1,* S171–S190.

Talley, L. D. (1993), Distribution and formation of North Pacific Intermediate Water, *J. Phys. Oceanogr., 23,* 517–537.

Talley, L. D. (1997), North Pacific Intermediate Water transports in the mixed water region, *J. Phys. Oceanogr., 27,* 1795–1803.

Tani, H., J. Nishioka, K. Kuma, H. Takata, Y. Yamashita, E. Tanoue, and T. Midorikawa (2003), Iron(III) hydroxide solubility and humic-type fluorescent organic matter in the deep water column of the Okhotsk Sea and the northwestern North Pacific Ocean, *Deep-Sea Res. I, 50,* 1063–1078.

Tans, P. P., I. Y. Fung, and T. Takahashi (1990), Observational constraints on the global atmospheric CO_2 budget, *Science, 247,* 1431–1438.

Tans, P. P., J. A. Berry, and R. F. Keeling (1993), Oceanic $^{13}C/^{12}C$ Observations: A new window on ocean CO_2 uptake, *Global Biogeochem. Cycles, 7,* 353–368.

Taylor, A. H., D. S. Harbour, R. P. Harris, P. H. Burkill, and E. S. Edwards (1993), Seasonal succession in the pelagic ecosystem of the North Atlantic and the utilization of nitrogen, *Journal of Plankton Research, 15,* 875–891.

Taylor, G. T. (1991), Vertical fluxes of biogenic particles and associated biota in the eastern north Pacific: Implications for biogeochemical cycling and productivity, *Global Biogeochem. Cycles, 5,* 289–303.

Tegen, I., and I. Fung (1994), Modeling of mineral dust in the atmosphere: Sources, transport and optical thickness, *J. Geophys. Res., 99,* 22897–22914.

Thiele, G., and J. L. Sarmiento (1990), Tracer dating and ocean ventilation., *J. Geophys. Res., 95,* 9377–9391.

Thomas, F., V. Garcon, and J.-F. Minster (1990), Modeling the seasonal cycle of dissolved O_2 in the upper ocean at Ocean Weather Station P, *Deep-Sea Res., 37,* 463–491.

Thompson, D.W.J., and J. M. Wallace (2000), Annular modes in the extratropical circulation. Part I: Month-to-month variability, *J. Clim., 13,* 1000–1016.

Thomson, J., S. Colley, and P.P.E. Weaver (1988), Bioturbation into a recently emplaced deep-sea turbidite as revealed by ^{210}Pb excess, ^{230}Th excess and planktonic foraminifera distributions, *Earth Planet. Sci. Lett., 90,* 157–173.

Thurman, H. (1990), *Essentials of Oceanography,* 3rd ed., Merrill Publishing Company, Columbus, OH.

Titman, D., and P. Kilham (1976), Sinking in freshwater phytoplankton: Some ecological implications of cell nutrient status and physical mixing processes, *Limnol. Oceanogr., 21,* 409–417.

Toggweiler, J. R. (1999), Variation of atmospheric CO_2 by ventilation of the ocean's deepest water, *Paleoceanogr., 14(5),* 571–588.

Toggweiler, J. R., and B. Samuels (1993a), Is the magnitude of the deep outflow from the Atlantic Ocean actually governed by southern hemisphere winds? in *The Global Carbon Cycle,* edited by M. Heimann, pp. 303–331, Springer-Verlag, Berlin.

Toggweiler, J. R., and B. Samuels (1993b), New radiocarbon constraints on the upwelling of abyssal water to the ocean's surface, in *The Global Carbon Cycle,* edited by M. Heimann, pp. 333–366, Springer-Verlag, Berlin.

Toggweiler, J. R., and B. Samuels (1995), Effect of Drake Passage on the global thermohaline circulation, *Deep-Sea Res. 1, 42,* 477–500.

Toggweiler, J. R., and J. L. Sarmiento (1985), Glacial to interglacial changes in atmospheric carbon dioxide: The critical role of ocean surface water in high latitudes, in *The Carbon Cycle and Atmospheric CO_2: Natural Variations Archean to Present,* edited by E. Sundquist and W. S. Broecker, pp. 163–184, American Geophysical Union, Washington, DC.

Toggweiler, J. R., K. Dixon, and K. Bryan (1989), Simulation of radiocarbon in a coarse-resolution world ocean model. 1: Steady state prebomb distributions, *J. Geophys. Res., 94,* 8217–8242.

Toggweiler, J. R., K. Dixon, and W. S. Broecker (1991), The Peru upwelling and the ventilation of the South Pacific thermocline, *J. Geophys. Res., 96,* 20467–20497.

Toggweiler, J. R., R. Murnane, S. Carson, A. Gnanadesikan, and J. l. Sarmiento (2003a), Representation of the carbon cycle in box models and GCMs. 1: Solubility pump, *Global Biogeochem. Cycles, 17,* 1026, doi:10.1029/2001GB001401.

Toggweiler, J. R., R. Murnane, S. Carson, A. Gnanadesikan, and J. L. Sarmiento (2003b), Representation of the carbon cycle in box models and GCMs, 2. Organic pump, *Global Biogeochem. Cycles, 17,* 1027, doi:10.1029/2001GB001841.

Treguer, P., and P. Pondaven (2000), Silica control of carbon dioxide, *Nature, 406,* 358–359.

Treguer, P., D. M. Nelson, A.J.V. Bennekom, D. J. DeMaster, A. Leynaert, and B. Queguiner (1995), The silica balance in the world ocean: A reestimate, *Science, 268,* 375–379.

Tremblay, J.-E., B. Klein, L. Legendre, R. B. Rivkin, and J.-C. Therriault (1997), Estimation of f-ratios in oceans based on phytoplankton size structure, *Limnol. Oceanogr., 42,* 595–601.

Trenberth, K. E., and T. J. Hoar (1997), El Niño and climate change, *Geophys. Res. Lett., 24,* 3057–3060.

Tromp, T. K., P. V. Cappellen, and R. M. Key (1995), A global model for the early diagenesis of organic carbon and organic phosphorus in marine sediments, *Geochim. Cosmochim. Acta, 59,* 1259–1284.

Tsuda, A., et al. (2003), A mesoscale iron enrichment in the western Subarctic Pacific induces a large centric diatom bloom, *Science, 300,* 958–961.

Tsunogai, S. (2002), The Western North Pacific playing a key role in global biogeochemical cycles, *Journal of Oceanography, 58,* 245–257.

Turner, D. R., and K. A. Hunter (Eds.) (2001), *The Biogeochemistry of Iron in Seawater,* 396 pp., John Wiley & Sons, Ltd., Chichester.

Turner, J. (2004), The El Nino-southern oscillation and Antarctica, *International Journal of Climatology, 24,* 1–31.

Turpin, D. H. (1991), Effects of inorganic N availability on algal photosynthesis and carbon metabolism, *J. Phycol., 27,* 14–20.

Tyrrell, T. (1999), The relative influences of nitrogen and phosphorus on oceanic primary production, *Nature, 400,* 525–531.

UNESCO (1985), The international system of units (SI) in oceanography, Technical Papers, UNESCO, Paris France, 45.

Uppström, L. (1974), The boron/chlorinity ratio of deep-sea water from the Pacific Ocean, *Deep-Sea Res., 21,* 161–162.

Usbeck, R. (1999), Modeling of marine biogeochemical cycles with an emphasis on vertical particle fluxes, Ph.D. thesis, University of Bremen, Bremen, Germany.

U.S. Department of Commerce, National Oceanic and Atmospheric Administration, National Geophysical Data Center (2001), *2-minute Gridded Global Relief Data (ETOPO2), http://www.ngdc.noaa.gov/mgg/fliers/01mgg04.html.*

Valiela, I. (1984), *Marine Ecological Processes,* 546 pp., Springer-Verlag, New York.

van Bennekom, A. J., J. H. F. Jansen, S. J. van der Gaast, J. M. van Iperen, and J. Pieters (1989), Aluminum-rich opal: An intermediate in the preservation of biogenic silica in the Zaire (Congo) deep-sea fan, *Deep-Sea Res., 36,* 173–190.

van Bennekom, A. J., A. G. Buma, and R. F. Nolting (1991), Dissolved aluminum in the Weddell-Scotia confluence and effect of Al on the dissolution kinetics of biogenic silica, *Mar. Chem., 35,* 423–434.

van Brakel, J., and P. M. Heertjes (1974), Analysis of diffusion in macroporous media in terms of a porosity, a tortuosity and a constrictivity factor, *Int. J. Heat Mass Transfer, 17,* 1093–1103.

Van Cappellen, P., and E. D. Ingall (1994), Benthic phosphorus regeneration, net primary production, and ocean anoxia: A model of the coupled marine biogeochemical cycles of carbon and phosphorus, *Paleoceanogr., 9,* 677–692.

Van Cappellen, P., and E. D. Ingall (1996), Redox stabilization of the atmosphere and oceans by phosphorus-limited marine productivity, *Science, 271,* 493–496.

Van Cappellen, P., and L. Qiu (1997a), Biogenic silica dissolution in sediments of the Southern Ocean. I. Solubility, *Deep-Sea Res. II, 44,* 1109–1128.

Van Cappellen, P., and L. Qiu (1997b), Biogenic silica dissolution in sediments of the Southern Ocean. II: Kinetics, *Deep-Sea Res. II, 44,* 1129–1149.

Van Cappellen, P., S. Dixit, and J. van Beusekom (2002), Biogenic silica dissolution in the oceans: Reconciling experimental and field-based dissolution rates, *Global Biogeochem. Cycles, 16,* doi:10.1029/2001GB001431.

Van Scoy, K. A., K. P. Morris, J. E. Robertson, and A. J. Watson (1995), Thermal skin effect and the air-sea flux of carbon dioxide: A seasonal high resolution estimate, *Global Biogeochem. Cycles, 9,* 253–262.

Venegas, S. A. (2003), The Antarctic Circumpolar Wave: A combination of two signals? *J. Clim., 16,* 2509–2525.

Villareal, T. A. (1991), Nitrogen-fixation by the cyanobacterial symbiont of the diatom genus *Hemialus, Marine Ecology Progress Series, 76,* 201–204.

Villareal, T. A., M. A. Altabet, and K. Culver-Rymaza (1993), Nitrogen transport by vertically migrating diatom mats in the North Pacific Ocean, *Nature, 363,* 709–712.

Volk, T., and M. I. Hoffert (1985), Ocean carbon pumps: Analysis of relative strengths and efficiencies in ocean-driven atmospheric CO_2 changes, in *The Carbon Cycle and Atmospheric CO_2: Natural Variations Archean to Present,* edited by E. T. Sundquist and W. S. Broecker, pp. 99–110, American Geophysical Union, Washington, DC.

von Liebig, J. (1840), *Die organische Chemie in ihrer Anwendung auf Agricultur und Physiologie,* F. Vieweg und Sohn, Braunschweig.

Waite, T. D. (2001), The biogeochemistry of iron in seawater, in *The Biogeochemistry of Iron in Seawater,* edited by D. R. Turner and K. A. Hunter, pp. 291–342, John Wiley & Sons, Ltd., Chichester.

Walker, G. T. (1924), Correlation in seasonal variations of weather. IX: A further study of world weather, *Mem. Indian Meteorological Dept., 24 (9),* 275–333.

Walker, G. T., and E. W. Bliss (1939), World Weather V, *Mem. Royal Soc., IV,* 53–84.

Walker, J.C.G., P. B. Hays, and J. F. Kasting (1981), A negative feedback mechanism for the long term stabilization of earth's surface-temperature, *J. Geophys. Res., 86,* 9776–9782.

Walker, S. J., R. F. Weiss, and P. K. Salameh (2000), Reconstructed histories of the annual mean atmospheric mole fractions for the halocarbons CFC-11, CFC-12, CFC-11, and carbon tetrachloride, *J. Geophys. Res., 105.*

Wallace, D.W.R. (1995), Monitoring global ocean inventories, OOSDP Background Rep. 5, 54 pp., Ocean Observ. Syst. Dev. Panel, Texas A&M Univ., College Station, TX.

Wallace, D.W.R. (2001), Storage and transport of excess CO_2 in the ocean: The JGOFS/WOCE Global CO_2 survey, in *Ocean Circulation and Climate,* edited by G. Siedler, J. Church, and J. Gould, pp. 489–521, Academic Press, San Diego.

Wallace, D.W.R., and C. D. Wirick (1992), Large air-sea gas fluxes associated with breaking waves, *Nature, 356,* 694–696.

Wallace, J. M., and D. S. Gutzler (1981), Teleconnections in the geopotential height field during the northern hemisphere winter, *Monthly Weather Review, 109,* 784–812.

Wang, Y., and P. V. Cappellen (1996), A multicomponent reactive transport model of early diagensis: Application to redox cycling in coastal marine sediments, *Geochim. Cosmochim. Acta, 60,* 2993–3014.

Wanninkhof, R. (1992), Relationship between wind speed and gas exchange over the ocean, *J. Geophys. Res., 97,* 7373–7383.

Wanninkhof, R., and W. McGillis (1999), A cubic relationship between air-sea CO_2 exchange and wind speed, *Geophys. Res. Lett., 26,* 1889–1892.

Wanninkhof, R., J. R. Ledwell, and W. S. Broecker (1985), Gas exchange-wind speed relation measured with sulfur hexafluoride on a lake, *Science, 227,* 1224–1226.

Wanninkhof, R., W. Asher, R. Weppernig, H. Chen, P. Schlosser, C. Langdon, and R. Sambrotto (1993), Gas transfer experiment on Georges Bank using two volatile deliberate tracers, *J. Geophys. Res., 98,* 20,237–20,248.

Wanninkhof, R., R. A. Feely, D. K. Atwood, G. Berberian, D. Wilson, P. P. Murphy, and M. F. Lamb (1995), Seasonal and lateral variations in carbon chemistry of surface water in the eastern equatorial Pacific during 1992, *Deep-Sea Res. II, 42,* 387–409.

Wanninkhof, R., et al. (1997), Gas exchange, dispersion, and biological productivity on the west Florida shelf: Results from a Lagrangian tracer study, *Geophys. Res. Lett., 24,* 1767–1770.

Wanninkhof, R., S. C. Doney, T.-H. Peng, J. L. Bullister, K. Lee, and R. A. Feely (1999a), Comparison of methods to determine the anthropogenic CO_2 invasion into the Atlantic Ocean, *Tellus, 51B,* 511–530.

Wanninkhof, R., E. Lewis, R. A. Feely, and F. J. Millero (1999b), The optimal carbonate dissociation constants for determining surface water pCO_2 from alkalinity and total inorganic carbon, *Mar. Chem., 65,* 291–301.

Wanninkhof, R., S. C. Doney, T. Takahashi, and W. McGillis (2002), The effect of using time-averaged winds on regional air-sea CO_2 fluxes, in *Gas Transfer at Water Surfaces,* edited by M. A. Donelan et al., pp. 351–356, American Geophysical Union, Washington, DC.

Wanninkhof, R., K. F. Sullivan, and Z. Top (2004), Air-sea gas transfer in the Southern Ocean, *J. Geophys. Res., 109,* CO8S19, doi:10.1029/2003JC001767.

Warner, M. J., and R. F. Weiss (1985), Solubilities of chlorofluorocarbons 11 and 12 in water and seawater, *Deep-Sea Res., 32,* 1485–1497.

Warren, B. A., and C. Wunsch (Eds.) (1981), *Evolution of Physical Oceanography,* 623 pp., MIT Press, Cambridge, MA.

Watson, A. J. (2001), Iron limitation in the oceans, in *The Biogeochemistry of Iron in Seawater,* edited by D. R. Turner and K. A. Hunter, pp. 9–39, John Wiley & Sons, Ltd., Chichester.

Watson, A. J., R. C. Upstill-Goddard, and P. S. Liss (1991), Air-sea exchange in rough and stormy seas measured by a dual-tracer technique, *Nature, 349,* 145–147.

Watson, A. J., D. C. E. Bakker, A. J. Ridgwell, P. W. Boyd, and C. S. Law (2000), Effect of iron supply on Southern Ocean CO_2 uptake and implications for glacial atmospheric CO_2, *Nature, 407,* 730–733.

Waugh, D. W., T. M. Hall, and T.W.N. Haine (2003), Relationship among tracer ages, *J. Geophys. Res.,* 108(C5), 3138, doi:10.1029/2002JC001325.

Weast, R. C., and M. J. Astle (Eds.) (1982), *CRC Handbook of Chemistry and Physics,* 63rd ed., CRC Press, Inc., Boca Raton, FL.

Weaver, A. J., M. Eby, A. F. Fanning, and E. C. Wiebe (1998), Simulated influence of carbon dioxide, orbital forcing and ice sheets on the climate of the last glacial maximum, *Nature, 394,* 847–853.

Webb, R. S., D. H. Rind, S. J. Lehman, R. H. Healy, and D. Sigman (1997), Influence of ocean heat transport on the climate of the last glacial maximum, *Nature, 385,* 695–699.

Webster, P. J., and R. G. Peterson (1997), The past and the future of El Niño, *Nature, 390,* 562–564.

Weiss, R. F. (1970), The solubility of nitrogen, oxygen, and argon in water and seawater, *Deep-Sea Res., 17,* 721–735.

Weiss, R. F. (1971), Solubility of helium and neon in water and seawater, *J. Chem. Eng. Data, 16,* 235–241.

Weiss, R. F. (1974), Carbon dioxide in water and seawater: The solubility of a non-ideal gas, *Mar. Chem., 2,* 203–215.

Weiss, R. F., and T. K. Kyser (1978), Solubility of krypton in water and seawater, *J. Chem. Eng. Data, 23,* 69–72.

Weiss, R. F., and B. A. Price (1980), Nitrous oxide solubility in water and seawater, *Mar. Chem., 8,* 347–359.

Weiss, R. F., H. G. Ostlund, and H. Craig (1979), Geochemical studies of the Weddell Sea, *Deep-Sea Res., 26A,* 1093–1120.

Weiss, R. F., J. L. Bullister, R. H. Gammon, and M. J. Warner (1985), Atmospheric chlorofluoromethanes in the deep equatorial Atlantic, *Nature, 314,* 608–610.

Welschmeyer, N. A., S. Strom, F. Goericke, G. DiTullio, M. Belvin, and W. Petersen (1993), Primary production in the subarctic Pacific Ocean: Project SUPER, *Prog. Oceanogr., 32,* 101–135.

Westbroek, P., E. W. De Jong, P. van der Wal, A. H. Borman, J.P.M. Devrind, D. Kod, W. C. Debruijn, and S. B. Parker (1984), Mechanism of calcification in the marine alga Emiliani-Huxleyi, *Phil. Trans. R. Soc. Lond. B, B304,* 435–444.

Wheeler, P. A., and D. L. Kirchman (1986), Utilization of inorganic and organic nitrogen by bacteria in marine systems, *Limnol. Oceanogr.,* 998–1009.

Whitaker, S. (1967), Diffusion and dispersion in porous media, *Am. Inst. Chem. Eng. J., 13,* 420–427.

White, W. B., and R. G. Peterson (1996), An Antarctic circumpolar wave in surface pressure, wind, temperature and sea-ice extent, *Nature, 380,* 699–702.

Whitman, W. G. (1923), The two-film theory of gas absorption, *Chem. Met. Eng., 29,* 146–148.

Wiebe, H. C., F. Kooistra, and L. K. Medlin (1996), Evolution of the diatoms (Bacillariophyta). IV: A reconstruction of their age from small subunit rRNA coding regions and the fossil record, *Molecular Phylogenetics and Evolution, 6,* 391–407.

Wiesenburg, D. A., and J.N.L. Guinasso (1979), Equilibrium solubilities of methane, carbon monoxide, and hydrogen in water and sea water, *J. Chem. Eng. Data, 24,* 356–360.

Wilke, C. R., and P. Chang (1955), Correlation of diffusion coefficients in dilute solutions, *AIChE J., 1,* 264–270.

Willebrand, J., and D. B. Haidvogel (2001), Numerical ocean circulation modelling: Present status and future directions, in *Ocean Circulation and Climate,* edited by G. Siedler et al., pp. 547–556, Academic Press, San Diego.

Williams, P.J.le B. (1990), The importance of losses during microbial growth: Commentary on the physiology, measurement and ecology of the release of dissolved organic material, *Marine Microbial Food Webs, 4,* 175–206.

Williams, R.J.P. (1981), The Bakerian Lecture, 1981: Natural-selection of the chemical-elements, *Proceedings of the Royal Society of London Series B—Biological Sciences, 213,* 361–397.

Wimbush, M. (1976), The physics of the benthic boundary layer, in *The Benthic Boundary Layer,* edited by I. N. McCave, pp. 3–10, Plenum Publ., New York.

Winguth, A.M.E., M. Heimann, K. D. Kurz, E. Maier-Reimer, U. Mikolajewicz, and J. Segschneider (1994), El Niño-Southern Oscillation related fluctuations of the marine carbon cycle, *Global Biogeochem. Cycles, 8,* 39–64.

Winguth, A.M.E., D. E. Archer, E. Maier-Reimer, U. Mikolajewicz, and J. C. Duplessy (1999), Sensitivity of paleonutrient tracer distribution and deep sea circulation to glacial boundary conditions, *Paleoceanogr., 14,* 304–323.

Wischmeyer, A. G., Y. Del Amo, M. Brzezinksi, and D. A. Wolf-Gladrow (2003), Theoretical constraints on the uptake of silicic acid species by marine diatoms, *Mar. Chem., 82,* 13–29.

Wollast, R. (1993), Interaction of carbon and nitrogen cycles in the coastal zone, in *Interactions of C, N, P and S Biogeochemical Cycles and Global Change,* edited by R. Wollast et al., pp. 195–210, Springer, Berlin.

Wollast, R., and R. M. Garrels (1971), Diffusion coefficient of silica in seawater, *Nature, 229,* 94.

Woolf, D. K. (1997), Bubbles and their role in gas exchange, in *The Sea Surface and Global Change,* edited by P. S. Liss and R. A. Duce, pp. 173–205, Cambridge University Press, Cambridge.

Woolf, D. K., and S. A. Thorpe (1991), Bubbles and the air-sea exchange of gases in near-saturation conditions, *J. Mar. Re., 49,* 435–466.

Wroblewski, J. S. (1977), A model of phytoplankton plume formation during variable Oregon upwelling, *J. Mar. Re., 35.*

Wroblewski, J. S., J. L. Sarmiento, and G. R. Flierl (1988), An ocean basin scale model of plankton in the North Atlantic. 1: Solutions for the climatological oceanographic conditions in May, *Global Biogeochem. Cycles, 2,* 199–218.

Wu, J., W. Sunda, E. A. Boyle, and D. M. Karl (2000), Phosphate depletion in the Western North Atlantic Ocean, *Science, 289,* 759–762.

Wunsch, C. (1984), An eclectic Atlantic Ocean circulation model. I: The meridional flux of heat, *J. Phys. Oceanogr., 14,* 1712–1733.

Wunsch, C. (1994), Dynamically consistent hydrography and absolute velocity in the eastern North Atlantic Ocean, *J. Geophys. Res., 99,* 14,071–14,090.

Wunsch, C. (2002), What is the thermohaline circulation? *Science, 298,* 1179–1181.

Wunsch, C., D. Hu, and B. Grant (1983), Mass, salt, heat and nutrient fluxes in the South Pacific Ocean, *J. Phys. Oceanogr., 13,* 725–753.

Yamanaka, Y., and E. Tajika (1996), The role of the vertical fluxes of particulate organic matter and calcite in the ocean carbon cycle: Studies using an ocean biogeochemical general circulation model, *Global Biogeochem. Cycles, 10,* 361–382.

Yamanaka, Y., and E. Tajika (1997), Role of dissolved organic matter in the marine biogeochemical cycle: Studies using an ocean biogeochemical general circulation model, *Global Biogeochem. Cycles, 11,* 599–612.

Yasuda, I., Y. Hiroe, K. Komatsu, K. Kawaski, T. Joyce, F. Bahr, and Y. Kawasaki (2001), Hydrographic structure and transport of the Oyashio south of Hokkaido and the formation of North Pacific Intermediate Water, *J. Geophys. Res., 106,* 6931–6942.

Yasuda, I., S. Kouketsu, K. Katsumata, M. Ohiwa, Y. Kawasaki, and A. Kusaka (2002), Influence of Okhotsk Sea Intermediate Water on the Oyashio and North Pacific Intermediate Water, *J. Geophys. Res., 107,* 3237, doi:10.1029/2001 JC001037.

Yool, A., and T. Tyrrell (2003), Role of diatoms in regulating the ocean's silicon cycle, *Global Biogeochem. Cycles, 17,* 1103,doi:10.1029/2002GB002018.

Yu, E. F., R. Francois, M. P. Bacon, S. Honjo, A. P. Fleer, S. J. Manganini, M.M.R. van der Loeff, and V. Ittekot (2001), Trapping efficiency of bottom-tethered sediment traps estimated from the intercepted fluxes of Th-230 and Pa-231, *Deep-Sea Res. II, 48,* 865–889.

Yung, Y. L., T. Lee, C. H. Wang, and T. Z. Shieh (1996), Dust: A diagnostic of the hydrologic cycle during the last glacial maximum, *Science, 271,* 962–963.

Zebiak, S. E. (1993), Air-sea interaction in the equatorial Atlantic region, *J. Clim.,* 6, 1567–1586.

Zehr, J. P., and B. B. Ward (2002), Nitrogen cycling in the ocean: New perspectives on processes and paradigms, *Appl. Env. Microbiol., 68,* 1015–1024.

Zheng, M., W. J. De Bruyn, and E. S. Saltzman (1998), Measurements of the diffusion coefficients of CFC-11 and CFC-12 in pure water and seawater, *J. Geophys. Res., 103,* 1375–1379.

Zondervan, I., R. E. Zeebe, B. Rost, and U. Riebesell (2001), Decreasing marine biogenic calcification: A negative feedback on rising atmospheric pCO_2, *Global Biogeochem. Cycles, 15,* 507–515.

Index